LIST OF SECTIONS

Numerical

THIS VOLUME IS ONE OF AN EXTENSIVE GROUP
OF REFERENCE HANDBOOKS PUBLISHED BY

THE RONALD PRESS
A DIVISION OF JOHN WILEY & SONS, INC.

PRODUCTION HANDBOOK

THIRD EDITION

GORDON B. CARSON
ALBION COLLEGE

HAROLD A. BOLZ
THE OHIO STATE UNIVERSITY

HEWITT H. YOUNG
ARIZONA STATE UNIVERSITY

Editorial Consultants

A RONALD PRESS PUBLICATION

JOHN WILEY & SONS, New York • Chichester • Brisbane • Toronto

TS
155
. P747
1972

ISBN 0 471 06651-6
Library of Congress Catalog Card Number: 71–137775
PRINTED IN THE UNITED STATES OF AMERICA
10 9 8 7 6 5 4 3 2

CONTRIBUTING AND CONSULTING EDITORS

B. H. AMSTEAD, PH.D., P.E.
  President, The University of Texas of the Permian Basin

JAMES M. APPLE, P.E.
  Professor of Industrial and Systems Engineering, Georgia Institute of Technology

JOHN R. BANGS, M.E.
  Visiting Professor, Division of Continuing Education, University of Florida

JAMES W. BARANY, PH.D.
  Professor and Associate Head, Industrial Engineering, Purdue University

DAVID D. BEDWORTH, PH.D., P.E.
  Professor of Engineering, Arizona State University

KENNETH L. BLOCK, P.E., C.P.A., C.M.C.
  President, A. T. Kearney & Co., Inc.

JOHN S. CROUT
  Vice President (retired); Consultant, Battelle Memorial Institute

MARSHALL G. EDWARDS
  Director, Olivetti North America Purchasing Coordinating Office

JOHN W. ENELL, M.E., ENG. SC.D.
  Vice President for Research, American Management Association

C. B. GAMBRELL, PH.D., P.E.
  Vice President for Academic Affairs, Florida Technological University

J. W. GAVETT, PH.D.
  Associate Professor of Management, University of Rochester

HENRY P. GOODE
  Professor of Industrial Engineering and Operations Research, Cornell University

HERBERT F. GOODWIN
  Senior Lecturer at the Sloan School of Management, Massachusetts Institute of Technology

JOHN V. GRIMALDI, PH.D.
  Director, The Center for Safety, New York University

WILLIAM W. HINES, PH.D., P.E.
  Professor of Industrial and Systems Engineering, Georgia Institute of Technology

THOMAS T. HOLME, D.ENG., I.E.
  Professor of Industrial Engineering, Yale University

RICHARD A. JACOBS
  Principal, A. T. Kearney & Co., Inc.

CHARLES F. JAMES, JR., PH.D.
  Professor and Chairman of Industrial Engineering, University of Rhode Island

GEORGE E. KANE, P.E.
  Professor of Industrial Engineering, Lehigh University

iii

PREFACE

Since the emergence at the turn of the century of the pioneering concepts known as "scientific management," the *Production Handbook* through its immediate precursors has played an historic role in the development of modern industrial management.

Because of the wide interest in his many articles dealing with these new concepts in the early volumes of *American Machinist*, the late Leon Pratt Alford encouraged a group of his peers to join forces in producing under his editorship a reference book designed to apply them to a broad range of problems in industry. This was published in 1924 as *Management's Handbook*. The immediate and enthusiastic acceptance of this first handbook of its kind led to publication in 1934 of the *Cost and Production Handbook,* edited by Alford, followed ten years later by the First Edition of the *Production Handbook,* co-edited by Alford and John R. Bangs. Thus, for approximately fifty years the present volume and its predecessors have been directly associated with the betterment of manufacturing planning, operations, and control. Adapting to the new and retaining the time-tested, the Handbook records the cumulative insight of generations of leading authorities, all adhering to the original plan—the compact, systematic presentation of essential information drawn from successful practice.

At whatever stage or with whatever problem industrial production is approached, the *Production Handbook* continues to provide sound orientation, reliable policies, standard practices, a diversity of implementation and control procedures, and tested criteria for evaluating results. The Handbook offers an encyclopedic distillation of the best experience in manufacturing systems and procedures, principles of organization and manpower utilization, and specialized techniques for achieving greater productivity and reduced costs.

Completely up to date, the Handbook aids in the quantification of many practical problems through clearly demonstrated applications of modern statistical analysis and operations research techniques. The man-machine interrelationship is thoroughly discussed wherever relevant, with special stress on the implications for scheduling, quality control, and cost control. The contributions to manufacturing efficiency of properly designed physical plant, improved material handling methods, utilization of new machines and materials, adequate safety requirements, and sound maintenance and housekeeping practices are described and illustrated.

Today, as in previous editions, the Handbook provides a synthesis of the vast literature of the field. For each production function analyzed, the basic concepts are formulated, the rules and principles explained, planning and designing described, and systems, methods, and operating procedures concretely detailed with cases, forms, tables, and charts. The Handbook recognizes that the ultimate purpose of the production phase of industry is to satisfy a competitive market efficiently and economically. Thus, the inseparable roles of management and engineering in the direction and operation of an industrial enterprise are analyzed and integrated.

v

The Handbook fully reflects the upsurge in recent years of science and technology, with new approaches, new tools and methods, and new insights into human behavior. Most significant is the growing application of management science in all branches of production. Operations research and the other methods of the quantitative analyst are brought to bear to solve problems, produce data, forecast, and control. Full recognition is given to the ever expanding role of the computer, but without ignoring the need for and application of manually maintained records and charts.

The automated factory is no longer a matter of conjecture; it exists and can produce anything from transistors to automobile bodies. But rather than diminish need for the Handbook's guidance, automation has expanded the scope of the production man's involvement and increased his need for authoritative information. Effective decision making in industry increasingly depends upon persons whose knowledge and resources extend well beyond their own immediate experience. To serve the demanding requirements of modern decision makers is a major function of the Handbook.

The Publishers' debt of gratitude to the Editorial Consultants—Gordon B. Carson, Harold A. Bolz, and Hewitt H. Young—and to the many individuals who have given unsparingly of their time and talents in preparing this Third Edition, is freely acknowledged and is detailed in the following pages.

Due recognition is accorded to the members of the Ronald staff for the many hours of planning, editing, proof reading, and supervisory effort that were required to bring the Third Edition to completion.

Scores of manufacturing companies in many industries have generously provided illustrative and procedural material and factual data. Manufacturing and trade associations, management consultants, and professional engineering societies have also made available various types of information. Finally, hundreds of books, periodicals, and other publications have been drawn upon in the preparation of this Handbook. Specific credit is given these companies, organizations, and publications wherever possible within the text, and in the Acknowledgments preceding the Index.

THE RONALD PRESS COMPANY
Publishers

ABOUT THE EDITORS

Issuance of this Third Edition has been accomplished with the cooperation of many persons. Of special importance among these are the Editorial Consultants whose advice on technical questions and critical review of individual sections provided invaluable help in closing out the manuscript.

Gordon B. Carson, who ably served as Editor of the prior edition, is Executive Vice President of Albion College and former Vice President, Business and Finance, of The Ohio State University. At Ohio State Dr. Carson was Dean of the College of Engineering and Director of the Engineering Experiment Station for a number of years. A Professional Engineer and past President and Fellow of the American Institute of Industrial Engineers, Dr. Carson is a Director of Industrial Nucleonics Corporation, and is widely known for his continuing contributions to industrial engineering.

Harold A. Bolz, Editor of the *Materials Handling Handbook*, is Dean of the College of Engineering and Director of the Engineering Experiment Station at The Ohio State University. President (1971–72) of the American Society for Engineering Education, Dr. Bolz is a Professional Engineer and a Fellow of the American Society of Mechanical Engineers. Dr. Bolz headed the General Engineering Department at Purdue University prior to joining the faculty at Ohio State, and has served in industry as a development engineer.

Hewitt H. Young is Professor of Engineering and Chairman of the Faculty of Industrial Engineering at Arizona State University. Formerly Professor of Industrial Engineering at Purdue University, Dr. Young has a diversified background of industrial and consulting experience. His research in production systems design, industrial mechanization and control, and related areas has been published widely. A Professional Engineer, Dr. Young has long been prominent in the activities of the American Institute of Industrial Engineers.

The Contributing and Consulting Editors were selected on the basis of direct experience and outstanding expertise in the various interrelated phases of modern production planning, operations, and management. While the contributions of individuals are frequently reflected in many parts of the book, special credit is due for the preparation of all or substantial portions of specific sections as follows:

B. H. Amstead .... Industrial Organization and Systems; Production Planning
James M. Apple .... Material Handling
James W. Barany .... Work Measurement and Time Study
David D. Bedworth .... Inventory Control
Marshall G. Edwards .... Purchasing
C. B. Gambrell .... Work Measurement and Time Study
Herbert F. Goodwin .... Work Simplification
John V. Grimaldi .... Industrial Safety
William W. Hines .... Quality Control and Reliability
Richard A. Jacobs .... Production Control

vii

Charles F. James, Jr. .... Tools, Jigs, and Fixtures
Peter J. Kolesar .... Statistical Methods
William G. Lesso ... Industrial Organization and Systems; Production Planning
Lawrence Mann, Jr. .... Plant Maintenance
Gilbert A. Marshall .... Inspection
Harlan C. Meal .... Operations Research
Ravinder Nanda .... Motion and Methods Study
John A. Nattress .... Process Charts
George Post .... Plant Layout and Facilities Planning
Marc H. Richman .... Manufacturing Processes and Materials
Raymond C. Salling .... Capital Investment Analysis
Joseph Stanislao .... Manufacturing Processes and Materials; Tools,
                                                        Jigs, and Fixtures
John M. Stewart .... Industrial Research and Development
Roland E. Stumpff .... Materials Management
Dwight L. Totten .... Plant Layout and Facilities Planning
Roger C. Vergin .... Production Planning
Dan Voich, Jr. .... Industrial Organization and Systems

CONTENTS

INDUSTRIAL ORGANIZATION AND SYSTEMS

CONTENTS

CONTENTS *(Continued)*

INDUSTRIAL ORGANIZATION AND SYSTEMS

Nature of Organization

SCOPE. Organizing is the process of determining and establishing a structure or system of authority and activity relationships among the resources being used to meet the company's objectives. The end result is an organization, a system of limited human and non-human resources capable of producing goods and services. The **subject of organization** in its broadest definition includes (1) the human resources who utilize the physical resources of the company to achieve its objectives, (2) the respective positions they occupy and the functions they perform, (3) the range of authority and responsibility they individually exercise, (4) the framework of formal and informal relationships through which they communicate and deal with one another, and (5) the mechanisms through which they operate and coordinate their activities in the enterprise. It is upon the basis of persons, physical resources, positions, authority, contacts, operations, communications, and coordination that successful work is carried on.

PURPOSE OF ORGANIZATION. There are two general viewpoints concerning the nature and purpose of organization. First, an organization is a conscious plan of subtasks which defines the relationships between these subtasks so as to coordinate the efforts of people in the accomplishment of the overall task effectively and efficiently. Second, an organization is the pattern of ways in which numbers of people relate themselves to each other in the planned systematic accomplishment of a complexity of tasks. The first viewpoint places more emphasis on job content, definition, analysis, and relationships while the second viewpoint tends to emphasize the personal relationships and human element to a greater extent.

Implicit in both viewpoints is the fact that the **purpose** of establishing an organization is to enable the human resource to work more effectively as a unit. Employee morale and performance tend to be better in situations characterized by order and certainty. Voich and Wren (Principles of Management) state that: "To contribute his best, every employee needs to know what he is expected to do, who his superior is, and what his relationship is to other employees and other parts of the organization. An assortment of machines, raw materials, and men does not make an organization. A proper combination of resource inputs and a soundly conceived organizational structure are required to achieve increased productivity and synergy." A sound organization tends to alleviate certain types of personnel problems such as (1) duplication of effort, (2) nonperformance by anyone, (3) unbalanced workloads, (4) defining training needs, (5) dormant programs, and (6) lack of coordination.

THE IMPORTANCE OF OBJECTIVES. Whenever two or more persons unite to attain a common purpose an organization is formed. This common purpose or objective is the primary integrating element of the organization and it establishes the framework within which individuals cooperate. Defining the objectives of the organizational unit is **the initial step** in the organizing process, and performance and evaluation of the members of the organization must be related to the objectives of the organization. Since objectives change, most organizations are short lived in their original form.

FRAMEWORK OF ORGANIZATION. The structure of an industrial organization under the usual conditions of operation is the result of growth. Concerns start in a small way and gradually expand. One or two men have authority and all responsibilities at first and, by force of circumstances, operation centers in and around them. As growth continues, this method of operation and control becomes ineffective and a planned or formal framework of organization becomes imperative. Determining the organizational structure to satisfy this demand is not a one-time decision but rather a continuing consideration.

With the passage of time, any fixed organization will degenerate under mechanical and routinized operation and hence the organizational structure must be continuously studied and periodically revamped to prevent stagnation or stunted growth.

Dale (Planning and Developing the Company Organization Structure) identifies seven major stages of a company's growth and relates these to organizational changes and problems. A summary of these stages is as follows:

Stage of Growth	No. of Employees	Organizational Problem and Its Possible Consequences
I	3–7 (any size)	Formulation of objectives: Division of work
II	25 (10)	Delegation of responsibility: The accommodation of personalities
III	125 (50–100)	Delegation of more management functions: Span of control
IV	500 (50–500)	Reducing the executive's burden: The staff assistant
V	1,500 (100–400)	Establishing a new function (functionalism): The staff specialist
VI	5,000 (100–500)	Coordination of management functions: Group decision-making
VII	465,000 (over 500)	Determining the degree of delegation: Decentralization

The figures in parentheses show the size of the company when problems may arise. Other figures indicate the actual size of the company studied.

ASPECTS OF INDUSTRIAL ORGANIZATION. The industrial organization like the proverbial elephant studied and described by the three blind men, takes on an apparent form depending upon the angle from which it is viewed. In order that a true picture may be obtained it is necessary to study an industrial organization from at least **three different aspects.**

Factors or Components. Certainly the initial viewpoint of the organization should be through its factors and components, such as men, money, machines, materials, methods, and ideas. If one is to understand the whole he must have knowledge of the parts.

Functions and Activities. In contrast to the factors of an industrial organization are its functions and activities, including processes, work flow, authority, reward and penalty, evaluation, and perpetuation. Coordination, communication, and control can, when integrated, present the most valuable viewpoint of all. Of course, implied here under functions is that undesirable feature of malfunctions.

Structural Relationships. The most common aspect of organization is the structural relationships of the factors involved in the enterprise. In reality it is a view of the **delegation of responsibility,** along the lines of product, sub-products, territory, function, or some other rational basis. Related to this are the problems of **span of management** and **chain of command.**

Other Characteristics. Timing is a most important aspect in the study of an industrial organization. Since an organization is dynamic in character, it is only possible to picture or to chart it as of some moment. Such a record is useful but certainly not essential. It is the organization, its objectives, factors, functions, and relationships, which comes first, not the picture. The picture serves as a baseline for change and a communicating device.

Another aspect of organization unseen and intangible but none the less real is the so-called **informal organization.** It recognizes the endowments, character, and personality of the individuals who form the organization personnel. If the informal organization is good, it will function with little friction. Many of the activities of industrial managers are concerned with this informal organization, the interaction of persons, and the endless chain of effects that arise from human associations.

DEFINITIONS. Litterer (The Analysis of Organizations) broadly summarizes the differing concepts of organization as (1) classical or planned and (2) behavioral or naturalistic. The **classical approach** to organization places emphasis on job content and specifies what the behavior of individuals ought to be. Compensation, adjustments, and controls are necessary incentives. Under the classical approach an organization is a set of logically planned tasks or activities and interrelationships based on specific objectives and a formal authority structure. The **behavioral approach** views an organization as a thing that arises from the needs of its members. People behave as they do to satisfy certain fundamental needs. An effective manager is one who understands and provides for these needs and is aware of the face-to-face relationships and interactions that occur.

Fayol (General and Industrial Management) states: "To manage is to forecast and plan, to organize, to command, to coordinate, and to control." The Management Division of the American Society of Mechanical Engineers has defined management as the "art and science of preparing, organizing, and directing human effort applied to control the forces and to utilize the materials of nature for the benefit of man." Voich and Wren (Principles of Management) define management as "the activity that allocates and utilizes resources to achieve organizational goals, and it is the organ of our society specifically charged with making resources productive."

Implicit in these (and other) definitions of management are **functions, methods,** and **resources** applied and utilized in an **organizational setting** to achieve objectives. The organization serves as a means to an end.

Policies and Organization

RELATION OF POLICIES AND ORGANIZATION. A necessary preliminary to all activity in an industrial enterprise is a clear, complete statement of the aims of the activity, formulated as a policy or set of instructions. In this connection an **industrial policy** can be thought of as a general rule or guideline which states the established procedure to follow in a recurring situation. Policies considered here are those set up by production executives as part of the managerial procedure of operating the plant to produce the product. A policy of this kind defines objectives and formulates plans for achieving them. Thus all rules, regulations, and systems should be explicit expressions of formulated policies.

Determination of policies is an important step; if policies are wrong, the subsequent results may bring confusion and losses. Policies must be based on (1) thorough investigation and analysis and (2) a full appreciation of consequences, both good and bad. **Adoption of a policy** requires on the part of managers both ability and courage, for fundamentally a policy at the time of its adoption is intended to initiate a change, to improve conditions, to correct ineffectiveness, or to eliminate inefficiencies.

THE POLICY MANUAL. The policy manual or "policy book" is a valuable management device. It can be employed for any one or all of the several kinds of policies found in a production organization—managerial or supervisory, long range, general, or departmental. Its advantages are: (1) it avoids misunderstandings, friction, lost motion, and expense, for the policies are written down; (2) it facilitates a check on compliance; (3) it aids in indoctrinating or inculcating throughout the executive and supervisory personnel the principles and procedures necessary to put the policies into effect.

Mooney (The Principles of Organization) emphasizes the importance of **indoctrination** if freedom of action is to be secured in a production organization:

Industrial indoctrination simply means thorough definition of the principles governing the industrial policy. It includes the application of the principles through line delegation of authority, the staff function, the duties and responsibilities of each in relation to the others, the place and purpose of rules and procedures, and the comprehension of this doctrine throughout the organization.

POLICY ENFORCEMENT. Inasmuch as a policy is a guideline statement of what is to be done, it has within itself no force to bring results. **Executive action** is needed to make any policy effective. The stronger and more effective the leadership exerted, the greater the probability that policies will be adequately enforced and predetermined results realized. In other words, achievements in manufacturing enterprises come from decisions put into effect by the will of the one who is responsible for policy enforcement and who exercises leadership.

Whatever an organization accomplishes by way of achievement or economic results depends upon the **quality of leadership** that organizes, directs, and manages the efforts of every individual, from president to laborer. A plant cannot develop sound objectives or achieve adequate results without effective leaders.

Development of a Plan of Organization

INDUSTRIAL ORGANIZATION DESIGN. Industrial organization design recognizes: (1) levels of authority and (2) degrees of responsibility. The **line of authority** (line of command, or line of instruction) goes down from a higher to a lower level of authority. The **line of response** (line of performance, or line of accountability) comes up from a lower to a higher level of authority. These lines are also called **lines of communication,** and may be used to ask questions directed up or down the lines. In addition to these principal lines there is another function in the functional, or expert, staff.

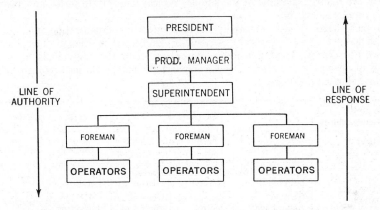

Fig. 1. Lines of authority and response in an industrial organization.

A typical line of industrial authority is shown in Fig. 1, with the titular positions of president, production manager, superintendent, foreman, and operator giving five **levels of authority.** This number of levels is seldom exceeded except in very large organizations. The arrow at the left of the diagram indicates the flow of authority and issuing of orders from the president down, level by level, until the operators are reached. The arrow at the right indicates the line of response whereby reports and returns flow upward from operators to whatever level these reports are directed. This line of response should be open for operators to send suggestions, make complaints, and ask for adjustment of grievances. A way must be provided for these communications to reach the level of authority and responsibility at which action can be taken. These lines of communication hold the organization together and make a coordinated operating unit. They make possible the enforcement of policies and execution of orders with economy and dispatch.

Methods, procedures, and techniques of communication are important factors in carrying on an industrial organization, and their excellence and effectiveness of use mean much to the efficiency of organizational performance. In this connection Barnard (The Functions of the Executive) says:

In an exhaustive theory of organization, communication would occupy a central place, because the structure, extensiveness, and scope of organization are almost entirely determined by communication techniques.

Certain **controlling factors** are essential to frictionless operation of organizational lines of communication. Barnard recognizes seven subfactors:

1. The channels of communication should be definitely known.
2. A definite, formal channel of communication is required to every member of an organization.
3. The line of communication must be as direct or short as possible.
4. The complete line of communication should usually be used.
5. The competence of the persons serving as communication centers, that is, officers, supervisory heads, must be adequate.

6. The line of communication should not be interrupted during the time when the organization is to function.

7. Every communication transmitted along the line of communication should be authenticated.

There are many variations of the simple, typical lines of authority and response diagrammed in Fig. 1.

Typical degrees of responsibility of the operating organization, and general duties assigned to the several positions, are indicated in Fig. 2. This diagram shows the degrees of responsibility devolving upon the individuals at several levels of authority and also, in its lower part, the general duties attached to each of the indicated positions.

Degree of Responsibility	First	Second	Third	Fourth	Fifth
Level of Position	PRESIDENT	PROD. MANAGER	SUPER-INTENDENT	FOREMAN	OPERATORS
Range of Duties	Administrative policies General management	Managerial policies Managerial control	Operating control General control of production	Detail control of production	Performance of an assigned job

Fig. 2. Degrees of responsibility and corresponding duties in an industrial organization.

WORK DIVISION IN INDUSTRIAL ORGANIZATION. The relation of duties to be performed and the selection of individuals to whom duties are to be assigned in a production organization are based on the principle of work division. Assignment of separate duties is necessary because of: (1) the volume or amount of work to be done in an industrial plant; (2) the differences in nature, capability, and skill of men; and (3) the wide range of knowledge required in an organization.

Important advantages of work division are:

1. Efficiency tends to increase as the number and variety of mental or manual operations per worker decreases.
2. Training of a worker is easier and less material is wasted in the process.
3. Less skilled labor may be utilized.

Although work division is the foundation of organization, there are **limitations** beyond which it should not be carried: (1) no advantage is gained by subdividing work so minutely that the resulting task is less than that which a man can perform when working continuously; (2) technology and custom make it impractical to subdivide certain kinds of work, although the influence of these factors is subject to change; and (3) subdivision must not be carried to the point of organic subdivision.

The answer to overdivision of work has been the concept of **job enlargement** which attempts to expand a worker's responsibility for his job and hence his interest in and identification with it. McGregor's (The Human Side of Enterprise) **Theory Y** concept is an example of increasing motivation and productivity through job enlargement.

Voich and Wren (Principles of Management) state that job enlargement may be accomplished in basically two ways (1) increasing **job breadth** or (2) increasing **job depth.** As the breadth of the job is enlarged the worker can better see the significance of each step of job performance, and he has some relief from short work cycles. Expanding job depth allows the worker to plan some of his own work, gather materials and tools, and execute the task. The manager must consider the technical and human aspects of organizing and strive for a balance between the advantages of division of work with the human consequences of job enlargement.

COORDINATION IN ORGANIZATION. When duties and activities are subdivided and allotted throughout a production organization, means must be provided to have all of them performed and completed according to a predetermined schedule. This process of timing activities and reuniting subdivided work in a factory is called **coordination.** Mooney (Principles of Organization) felt that coordination was the first principle of organization and defined it as "the orderly arrangement of group effort, to provide unity of action in the pursuit of a common purpose." The mechanism of normal routine through which coordination is achieved is the **system** of authority, policies, organization, methods, procedures, and communications developed in line with objectives.

A basic question to be answered in attempting to achieve coordination is how to get individuals to perform the right action at the right time. This may be attained by voluntary arrangements or a directed method. The authority structure relates to the **directed method.** Mooney felt that coordination has its foundation in authority. Directive coordination can be of two types (1) **hierarchical coordination,** where tasks are linked together by a central authority, and (2) **administrative system,** which is used when there is a horizontal flow of work of a routine nature. Formal policies and procedures facilitate the routine coordinative work. Hierarchical coordination tends to become cumbersome as the number of organization levels increase.

In **voluntary arrangements,** the individual sees the need to coordinate and does whatever is necessary to attain the desired goal. Barnard (The Function of the Executive) lays emphasis on the willingness of persons to contribute their efforts as an indispensable factor in the successful operation of organizations. All of the components and forces of the organization must work in harmony and unison. Activities must be kept in balance and properly timed at each level of authority.

The necessity for scheduling work, or coordinating efforts, is apparent by observing a gang of men hauling on a rope, or moving a heavy object. Members of the group must pull together, or heave together, if the work is to be done. Otherwise their efforts are wasted. Coordination means to combine activity into a consistent and harmonious action.

INFORMAL ORGANIZATION. The development of a plan of organization to achieve effective organization performance must take into account the aspect of informal organization. The need patterns of individuals vary. Maslow (Motivation and Personality) classifies human needs into five categories:

1. **Physiological needs.** These needs are basic for survival—food, water, air.
2. **Safety needs.** These needs include physical and psychological, or real or imagined, desire to be safe or secure.
3. **Belonging and love needs.** These needs include the desire for friendly interpersonal relations, attention, and acceptance.
4. **Esteem needs.** The need for prestige, respect, or esteem from other people; reflected in the individual's needs for achievement, self-respect, adequacy, and independence and freedom.
5. **Need for self-actualization (self-realization).** Everyone desires to become that which he can become.

Maslow's classification of needs reflects a hierarchy of needs, moving from the most basic physiological to self-actualization. The effectiveness of the ideal organization design from the standpoint of work division, authority relationships, and standard operating procedures is influenced by these needs or properties of informal group relationships. Several of the more evident and significant effects of these properties on organization structure and performance are summarized by Voich and Wren (Principles of Management) as follows:

1. **Informal Group Norms.** People within organizations generally tend to develop their own standards about how much work should be done. The basis for informal group standards is found in the group's desire to protect itself and to enhance job security.
2. **Informal Group Discipline.** Procedures to enforce group norms, such as sarcasm, ridicule, and ostracism are generally found in most large organizations.
3. **Reluctance to Change.** The threat or unknown factor presented by organization changes creates insecurity in the minds of workers.
4. **Informal Leaders.** The informal organization generally selects a member of the group, usually not a formal leader in the organization, to express their needs and wants. The informal leader may be a **task leader** (one who influences what will be done), or a **social leader** (one who influences the type of social interaction that occurs).
5. **Informal Communications Channels.** Informal groups communicate among themselves because they desire to be kept informed of the operations and developments within the organization. Often, the informal channel of communication is faster than the formal chain of command and response.

The foregoing properties of informal organization and their effects on organization performance occur in most large organizations. The manager must recognize the fact that they cannot be eliminated and (1) identify the informal group leaders and communications channels, and (2) provide for more open communications between management and workers to reduce the distortions of the informal communication channel. In this way, two primary objectives of organization performance are achieved: (1) human resources are guided toward the accomplishment of organization goals, and (2) the employees' lower and higher level needs are satisfied, work becomes more meaningful, and individual development is improved.

DETERMINATION OF FUNDAMENTALS. In setting up a plan of organization, the first step is the determination of fundamentals such as objectives, policies, authorities, responsibilities, and duties or activities, and their relationships. Objectives and policies have already been discussed. Authorities, responsibilities, and duties are discussed here.

Authority. The term authority may be broadly defined as the right of a person to utilize resources. In an organizational sense, authority is the right of one person to require another to perform certain duties.

Authority is the right to act, decide, and command. In a corporation it emanates from the stockholders, flows to the elected board of directors, whence it is delegated to designated persons who issue orders and instructions to subordinates. Authority may be classified as either direct or delegated. **Direct** authority exists where the line between the issuer and acceptor is unbroken; **delegated** authority exists where an intermediate agency is between the issuer and acceptor.

Authority may have one or more of these aspects:

1. It may be **formal,** that is, conferred by law or delegated within an organization.

2. It may be **functional** or intrinsic, because it is based on special knowledge or skill.
3. It may be **personal**, that is, accorded because of seniority, popularity, or outstanding qualities of leadership.

While the **classical approach** to organization treats authority as the right to require performance of duties by another, the **behavioral approach** states that authority rests upon the acceptance of the orders or instructions by the person to whom they are addressed. Barnard (The Functions of the Executive) has this to say:

The necessity of the assent of the individual to establish authority for him is inescapable. A person can and will accept a communication as authoritative only when four conditions simultaneously obtain: (1) he can and does understand the communication; (2) at the time of his decision he believes it is not inconsistent with the purpose of the organization; (3) at the time of his decision he believes it to be compatible with his personal interest as a whole; and (4) he is able mentally and physically to comply with it.

Responsibility. A clear conception of the significance of the term responsibility is presented in the following definition: In an organizational sense, responsibility is accountability for the performance of assigned duties.

Responsibility is a moral attribute. It implies fulfillment of a task, duty, or obligation according to orders given or promises made. Authority is commonly delegated only to persons of proven responsibility.

Organizational design calls for setting up **limits of responsibility** for each activity and effort; otherwise shortcoming or failure cannot be traced to its source and cause. Executives of weak responsibility cannot carry the burden of many simultaneous obligations or make the multitude of decisions necessary in the operation of an industrial concern. Barnard (The Functions of the Executive) supports this statement:

Executive positions (1) imply a complex morality, and (2) require a high capacity of responsibility, (3) under conditions of activity, necessitating (4) commensurate general and specific technical abilities as a moral factor—in addition there is required (5) the faculty of creating morals for others.

Duties. The activities assigned to a person in an organization are best specified in the form of duties, a term which may be thus defined: In an organizational sense, the duties allotted to an individual are the activities he is required to perform because of the place he occupies in the organization.

A duty is that which a person is bound by obligation to do. In a factory it is often called "a piece of work," a job, a task, or a **work assignment.** In an organizational sense it is a contribution to the goal or objective, and an organization can be thought of as a system of coordinated contributions, or a system of coordinated activities.

PRINCIPLES OF ORGANIZATION. The express purpose of organization and organizational principles is to achieve objectives by developing coordination and high morale. Organizational principles can be broadly grouped into those which provide for (1) cooperation (related to what has to be done), (2) coordination (related to communications, policies, and procedures) and (3) action (related to authority and incentives). The following principles have been developed and applied in a variety of organization types (industry, government, military, etc.) and serve as general guidelines for improving organizational design and performance:

1. The **Principle of the Objective:** Every organization must be an expression of the purpose of the undertaking concerned or it is meaningless and there-

fore redundant. You cannot organize in a vacuum; you must organize for something.

2. The **Principle of Specialization:** The activities of every member of any organized group should be confined, as far as possible, to the performance of a single function. Implicit in this principle is the concept that the more a task is limited to a few manual or mental operations, the higher will be the efficiency of task performance. However, McGregor (Human Side of Enterprise) feels that continued division of labor will reach a point of diminishing returns in efficiency. This occurs at the point where the task is so small that the worker loses interest and may also lose sight of the overall objective. Therefore, job enlargement can be an effective incentive for achieving higher job motivation and performance.

3. The **Staff Principle:** The advisory or service staff is a further development of the specialization concept. It does the things the line managers would do if they had the time and specialized knowledge. The staff should not become an end in itself, however.

4. The **Principle of Coordination:** The purpose of organizing per se, as distinguished from the purpose of the undertaking, is to facilitate coordination; unity of effort. The following are related to coordination:

   a. **Principle of functional similarity.** Duties should be grouped and assigned in accordance with their similarities in how they contribute to the unit's or organization's objectives. This departmentation should facilitate coordination and control.

   b. **Principle of organization relationships.** The number of organizational relationships increases at a much greater rate than do the number of new personnel added to the organization.

   c. **Exception principle.** Each manager at each level should make all decisions within the scope of his authority, and only those matters which he is not competent to decide (because of authority limitations) should be referred to his superior for decision.

5. The **Principle of Authority:** In every organized group the supreme authority must rest somewhere. There should be a clear line of authority from the supreme authority to every individual in the group. This principle has also been called **The Scalar Principle** (Mooney and Reiley, Onward Industry), **The Hierarchical Principle** (Fayol, General and Industrial Management) and **The Chain of Command** (military writers). The following are related to this principle.

   a. **Principle of organization levels.** The number of levels of authority should be kept to a minimum.

   b. **Principle of functions.** Functions (tasks) are the main entities around which a manager builds an effective organization structure. Only functions which are absolutely necessary, related in a clear and concise manner, and accompanied by clear statements of authority and responsibility of persons charged with performing them, should be incorporated in the organization.

   c. **Principle of complete assignment.** Responsibility and authority for every activity necessary for effective operation of an organization should be definitely assigned to someone.

   d. **Principle of delegation.** Authority should be delegated to the greatest possible extent, consistent with necessary control, so that coordinating and decision-making can take place as close as possible to the point of action.

   e. **Principle of unity of command.** Each subordinate should have but one immediate, direct supervisor.

6. The **Principle of Responsibility:** The responsibility of the superior for the acts of his subordinate is absolute. The process of delegation does not relieve the responsible executive from his responsibility, authority, or accountability.

7. The **Principle of Definition:** The content of each position—the duties involved, the authority and responsibility contemplated, and the relationships

with other positions should be clearly defined in writing and published to all concerned.

8. The **Principle of Correspondence**: In every position the responsibility and the authority should correspond. A person should not be given a responsibility unless he is given enough authority to use resources to discharge it.

9. The **Span of Control**: No person should supervise more than five, or at the most six direct subordinates whose work interlocks. This is also referred to as **span of management** or **management visibility**. The demand for communications by subordinates tends to limit the span of management. The choice of a narrow span means that most of the coordination and control comes from the superior.

10. The **Principle of Balance**: It is essential that the various units of an organization should be kept in balance. The development of the various functions and organizational units must conform as closely as practicable to their contribution to overall objectives.

11. The **Principle of Continuity**: Reorganization is a continuous process; in every undertaking specific provisions should be made for it. Organizing should take into account changes in objectives, scope, content, and personnel.

Types of Organization

LINE ORGANIZATION. Line of authority, or command, in its simple form is often referred to as the **military type of organization** or **chain of command.** Its prototype is the organization of an army in its line or operational activities, apart from the present-day staff or planning and strategy functions, and probably it is as old as the combining of individuals for a joint activity, such as hunting or war. In its simple, typical form it is not extensively used in industry, except in small shops. In larger companies it is usually combined with the functional or expert staff.

Requirements for Command. Because of the ancient origin and continued use of line organization, its relations are well defined. Henri Fayol, the French industrialist, laid down the requirements of command for the functioning of a line organization as follows:

1. There must be a thorough knowledge of the working force.
2. Incompetence must be eliminated.
3. There must be a sound knowledge of the agreements between the management and its employees.
4. Those in authority must set a good example to the working force.
5. The organization must be periodically examined with the help of charts.

Features of Line Organization. A business controlled under the line form of organization may act more quickly and effectively in changing its direction and policy than any other form of organization. Authority is passed down from the owners through a board of directors to a general manager, to whom report the heads of the various departments. Each department in most instances is a complete self-sustaining unit, its head being responsible for the performance of its particular process, product, or function. This means that the foreman must (1) direct its techniques, (2) formulate the necessary work specifications, (3) sometimes purchase materials, (4) plan and schedule the work, (5) oversee the necessary material handling, and (6) keep the necessary shop cost and production records. This same procedure would be repeated in all other departments with complete control centered in each head, subject only to the will of the general manager. What little research, planning, or central record-keeping is absolutely required falls upon the general manager. Fig. 1 is a partial presentation of such a plan in diagrammatic form.

The line organization is very stable and ideas and orders travel strictly according to the line of authority. There is never any question as to who is boss. Each division, department, or section is under a supervisor or foreman who is completely responsible for the work of his unit, except for those particular items which the general manager reserves for his own attention. The only interrelationship between the various departments is such as the general manager may establish. In short, he must be in constant touch with all the details of the business, and make decisions constantly, based upon, and involving, these details. It is obvious that this plan of organization grows more unwieldy and inefficient the larger the company becomes.

Advantages and Disadvantages of Line Control. The advantages and disadvantages of line control may be summarized briefly as follows:

Advantages:
1. It is simple.
2. There is a clear-cut division of authority and responsibility.
3. It is extremely stable.
4. It makes for quick action.
5. Discipline is easily maintained.

Disadvantages:
1. The organization is rigid and inflexible.
2. Being an autocratic system, it may be operated on an arbitrary, opinionated, and dictatorial basis.
3. Department heads carry out orders independently and often in accordance with their own whims and desires.
4. As division of labor is only incidental, crude methods may prevail because of lack of expert advice.
5. There is undue reliance upon the skill and personal knowledge of operators.
6. Foremen may offer resistance to much-needed changes.
7. Key men are loaded to the breaking point.
8. The loss of one or two capable men may cripple the entire organization.
9. Difficulty of operation occurs in large or complex enterprises.

TAYLOR SYSTEM OF FUNCTIONAL FOREMANSHIP. In the process of his investigations, Frederick W. Taylor (Shop Management), who developed what became known as "scientific management," made an analysis of the duties of a first-class foreman as found in the organization of his day. In his writings he states that such a foreman must:

1. Be a good machinist.
2. Be able to read drawings readily.
3. Plan the work of his department and see that it is properly prepared.
4. See that each man keeps his machine clean and in good order.
5. See that each man turns out work of the proper quality.
6. See that the men work steadily and fast.
7. See that the work flows through the work centers in the proper sequence.
8. In a general way, supervise timekeeping and rate setting.
9. Maintain discipline and adjust wages.

Since men possessing three, four, or five of these qualities were readily obtainable, he assigned them to specialized duties in harmony with their characteristics and training, such that they would act as **functional** rather than **all-round** foremen. He gave this explanation of his idea: "Functional management consists in so dividing the work of management that each man from the assistant superintendent down shall have as few functions as possible to perform."

The plan really means loading each man to capacity. Taylor discovered, moreover, that the typical foreman of his day was loaded with much clerical duty as well as operating responsibility. He found it necessary to remove the **plan-**

ning activities from the shop, where they were performed at low efficiency and hampered production, into the hands of men who could specialize in such work. Thus production could be speeded up and costs radically reduced.

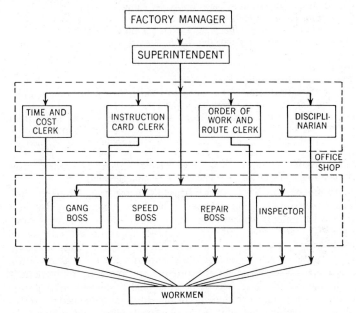

Fig. 3. Taylor plan of functional foremanship.

The **separation of the functions** was made as shown in Fig. 3. The time and cost clerk, instruction card clerk, and order-of-work and route clerk attended to the mental and clerical functions of production, while the gang boss, speed boss, repair boss, and inspector looked after the actual production in the shop. A disciplinarian was appointed to look after the disciplinary functions of the whole plant.

The duties of the gang boss were to see that machines were set up for jobs and that work was moved efficiently from machine to machine. The speed boss served as an instructor to the workmen and saw that they maintained the specified rates of production. The repair boss and the inspector performed the duties that their titles indicate.

Evaluation of Taylor's Functional Organization. The marked advantage of this type of organization was that each function was administered by specialists. Expert advice was available to each individual worker, and division of labor was carefully planned. In fact, functional organization was highly instructional. The specialists in the various fields served as instructors, with suitable authority.

The marked disadvantage of Taylor's functional plan of organization was that it gave eight foremen or supervisors, in turn as occasion arose, direct authority or temporary supervision of some form over the individual workmen, whereas it is now clearly recognized that no man can work satisfactorily for, or obey, the instructions of more than one foreman or executive. The plan, however, called for so many interrelationships and such integrated coordination that it became cumbersome and topheavy.

An analysis of the advantages and disadvantages of this system of organization follows:

Advantages:
1. Functional organization is based on expert knowledge.
2. Division of labor is planned, not incidental.
3. The highest functional efficiency of each person is maintained.
4. The manual work is separated from the mental—a separation initiated by Taylor.

Disadvantages:
1. A relative lack of stability is manifest.
2. The coordinating influences needed to insure a smoothly functioning organization may involve heavy overhead expenses.
3. The inability to locate and fix responsibility may seriously affect the discipline and morale of the workers through apparent or actual contradiction of orders.
4. Overlapping authority may give rise to friction between foremen and supervisors.
5. The initiative of supervisors may become stifled. Men may become mere automatons and routine may become very complicated.

Importance of the Functional Idea. The functional foremanship plan or organization was applied to plants mainly by Taylor, the group of men who were associated with him as consultants, and certain executives who saw the importance of Taylor's work and early adopted his principles and methods. Experience, however, showed the seriousness of its disadvantages as a physical arrangement, and as such it had gradually disappeared by about 1920. It did provide for specialists to do all the preliminary planning of work in the shop offices, leaving the foremen free to become highly efficient in their four major responsibilities—getting work done according to plan and schedule, setting standards for work, training workers for higher proficiency and gradually increased earnings, and directly handling grievances and all other immediate personnel problems with their workers. This was one of the most progressive steps in the entire field of industrial management.

Some concept of the influence of functional foremanship is gained by examining closely any modern organization chart. Therein, to one familiar with the Taylor plan, it is clearly evident that the functions are all provided for in the following manner:

1. The gang boss is now usually two men—the setup man, and the move man, trucker, or craneman.
2. The speed boss is the assistant foreman.
3. The repair boss has become a maintenance engineer.
4. The inspector heads the inspection organization. There is now often a supervisor of quality control.
5. The time and cost clerk is replaced by two men—the payroll clerk and the cost accountant.
6. The instruction card man has become the time-and-motion-study engineer.
7. The order-of-work and route clerk has developed into two men—one scheduling work and doing machine loading, and the other the methods engineer or individual who plans procedures and prepares operation lists for the parts and assemblies.

Taylor's developments along the above lines became the basis of the present-day **line and staff plan of organization,** in which the recognition of functions, and the corresponding assignment of specialized advisory and facilitation duties to staff individuals who give their attention solely to such work, has tremendously aided and increased the efficiency of line performance in the industrial plant.

Without the staff function modern industry and business could not operate except at tremendous disadvantages. Every industrial organization gets its real work done by action according to a preplanned program set up and operated by the line organization. The aim is **accomplishment**—quickly, efficiently, and at reasonable cost. The basis of the most successful action, however, is **preplanning** to set up goals which inspire men to accomplishment and then devise the ways for attaining the objectives. Those who work out and recommend these ways are the **specialists** in the organization. They are appointed by the operating executives to concentrate their thinking and to develop their funds of information and their analytical skills so as to become experts or authorities upon the respective special subjects to which they have been assigned. Executives who actively operate the organization to get the work actually done then take the suggestions and recommendations of the various specialists, modify them where necessary, and convert them from possibilities into actualities by translating the ideas into practical forms of action to attain the goals. Ideas and information from several or all the specialists are pooled and **translated from sound theory into practical workable plans.** The line executives know how to build up such plans from the information provided them and know what means and which individuals to use for getting action under the plans and achieving objectives.

LINE AND STAFF ORGANIZATION. In the line and staff organization, the line serves to maintain discipline and stability; the staff serves to bring in expert information. The staff function is strictly advisory and carries no power or authority to put its knowledge into operation.

The duties of the **staff organization** are as follows:

1. Research into technical, operating, or managerial problems.
2. Determination and recommendation of the various standards of performance.
3. Keeping of records and statistics on the above activities as a measuring stick of performance.
4. Advice and aid in carrying out plans and programs.

Separation of Operating Authority and Advisory Service. Line and staff control makes a clear distinction between doing and thinking—between getting work done in the line departments, and analyzing, testing, researching, investigating, and recording activities of the staff departments. It permits specialization by desirable functions but at the same time maintains the integrity of the principle of undivided responsibility and authority throughout the line organization. A simple diagrammatic illustration of the idea is given in Fig. 4,

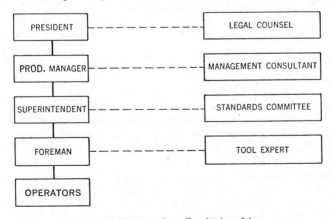

Fig. 4. Line and staff relationship.

where the line executives are listed at the left and the staff experts who may assist them with advice are at the right.

Here at the level of president is shown his legal counsel; at the level of production manager, a management consultant, who may be either an outside engineer or a resident industrial engineer; at the level of superintendent, a standards committee; and at the level of foreman, a tool expert. The relation of each one of these staff experts is shown at only one level. However, it is evident that advice from the legal counsel may be given not only to the president but also to any other individual in the line of authority. Similarly, the management consultant may advise not only the production manager but also the superintendent or foreman. Again, the shop standards committee may advise the superintendent and also the foreman, while the tool expert, whose primary responsibility is to advise the foreman, may also assist the operators. That is, Fig. 4, which puts in relationship the line of authority and expert staff, indicates that there are **cross relationships in the line and staff organization** for the purpose of more immediate contact without breaking down the line authority or scattering control.

Advantages and Disadvantages. The advantages and disadvantages of the line and staff organization may be summarized as follows:

Advantages:

1. It is based upon planned specialization.
2. It brings expert knowledge to bear upon management and operating problems.
3. It provides more opportunity for advancement for able operators, in that a greater variety of responsible jobs is available.
4. It makes possible the principle of undivided responsibility and authority, and at the same time permits staff specialization.
5. It repays its added costs many times over through the savings resulting from increased efficiency of operations.

Disadvantages:

1. Unless the duties and responsibilities of the staff members are clearly indicated by charts and manuals, there may be considerable confusion throughout the organization as to the functions and positions of staff members with relation to the line supervisors.
2. The staff may be ineffective for lack of authority to carry out its functions or intelligent backing in the application of its recommendations.
3. The inability to see each other's viewpoint may cause difficulty and friction between the line supervisors and staff members.
4. Although expert information and advice is available, it reaches the operators through line officers and thus runs the risk of misunderstanding and misinterpretation.
5. Line supervisors sometimes may resent the activities of staff members, feeling that the prestige and influence of line men suffers from the presence of the specialists.

STAFF PLAN OF COORDINATION. Thus far only one form of the staff organization has been mentioned. Actually four different types of staff can generally be found in most large industrial organizations. Litterer (The Analysis of Organizations) summarizes these types as assistant-to, general staff, specialized staff, and operating service staff. Each of these types contributes support to line activities.

Assistant-To. The assistant-to or **personal** staff exists primarily at the upper levels of an organization. The concept is derived from the military idea of aide-de-camp to the commander. His work is varied and general, depending on his superior. The **assistant-to** carries no line authority compared to an **assistant manager** (such as the Assistant Plant Manager, Assistant Controller, etc.). An assistant manager, such as the Assistant Plant Manager, has line authority since he is between the Plant Superintendent and the Foremen in the scalar chain. The assistant-to provides a valuable service by handling much of the administrative burden of his superior, by projecting his superior's ideas and wishes to others in the organization, and by sensing what is going on in the organization.

General Staff. Although not widely used in business (compared to the military and government), there is a growing tendency to add general staff positions concerned with overall planning at the top of large industrial corporations. While the assistant-to is intended to aid his superior in a personal way, the **general staff** is more concerned with preparing plans. In large organizations, divisions and departments are often so extensive as to demand the concentrated attention of the division or department heads to the exclusion of their cooperative participation in achieving total organizational goals. The general staff concept is a medium through which diversified work of large segments of the organization may be coordinated in the planning process.

Specialized Staff. The specialized staff (discussed previously under Line and Staff Organization) serves in an advisory capacity for line activities, based on its specialized knowledge. The functions of this staff are many, and it may do all those things that the line managers would do if they had unlimited time and specialized knowledge. In aiding planning, the specialized staff may gather facts, study alternatives and develop plans and programs for the line managers. In aiding control, it may advise on work standards, gather information on performance, and call deviations to the attention of management. Note, however, that the specialized staff does not take corrective action since this is a line function. The specialized staffs are often excluded from a staff designation, but merely referred to as **service units.**

Operating Service Staff. Operating service staffs and specialized staffs have the common feature of serving the organization as a whole, or at least large sectors of it. The operating service staff, however, facilitates the operation of the organization in a physical sense (they are custodial in nature).

COMMITTEE ORGANIZATION. Supplementary to both line and staff organization is the committee form of organization. A committee may be defined as a group of people who collectively make decisions or present viewpoints and whose conduct is governed by a set of rules. The committee represents an additional degree and **type of delegation** of line or staff authority. The use of committees may effectively increase the span of management and maintain or decrease the number of communications levels. A committee may be considered interchangeable with an individual, for organizing purposes, although they do not necessarily have the same effectiveness.

The committee is used for the most part in addition to, rather than in lieu of, line and staff organization. The primary functions of committees are to define and solve complex problems (generally interdepartmental), to facilitate execution of decisions, and to improve coordination, cooperation, and control throughout the entire organization.

Classification of Types. Committee types may be classified in numerous ways, such as by authority or power, by responsibility or function, by objective or purpose, by organizational levels represented, or by degree of permanence.

Principles of Operation. The effectiveness of a committee can be increased if certain **basic principles** are followed:

1. Members of the committee should be selected from those persons possessing an open mind and are willing to consider all of the facts.

2. Members of the committee should possess a wide range of knowledge concerning the operations of the business.
3. The number of members should generally be small. The effectiveness of committees depends on a high degree of interaction among members. Too large a committee may tend to limit participation, or make for lengthy and cumbersome committee proceedings.
4. Exploration of differences and alternatives may be more desirable than reaching a decision without sufficient facts.
5. An agenda and related materials should be prepared and distributed to members in sufficient time for study prior to a meeting.
6. Committee meetings should follow the agenda and adhere to a time schedule. The chairman is responsible for the behavior of the members and the adherence to the agenda and schedule.

Committee Functions. The real functions of a committee are to:

1. Interchange ideas.
2. Secure a meeting of minds.
3. Supply important information to the members or through them to the departments and the organization.
4. Receive and act on reports from committee members or departments which have been asked for data or information.
5. Assemble facts from many sources and put them together into a combined plan.
6. Assay the results of operations, arrive at conclusions, and formulate reports or suggestions.
7. Make intelligent and expert studies of important factors, activities, or problems.
8. Develop and recommend procedures of operation.
9. Coordinate, or set up, time relationships between the operations of different departments.
10. Correlate or associate together the activities of different departments.
11. Provide for cooperation—or special efforts in performance—between the different departments.
12. Formulate and set up standards of various kinds.
13. Act as a clearing house for matters for which no other channel has been provided.

Advantages and Disadvantages. The major advantages and disadvantages of committee activities may be summarized as follows:

Advantages:

1. Under a strong executive chairman, a committtee may quickly marshal many valuable points of view.
2. In conducting investigations, the several phases of the various questions may be quickly assigned to responsible members, with a reasonable assurance of speedy action if a time schedule and proper follow-up are instituted.
3. There is a stimulus toward cooperative action.
4. The members of the committee know better what is going on in the plant so that they can spread the information and team up with other individuals or departments.

Disadvantages:

1. Committees may be too large for constructive action.
2. Committees are expensive in time and may have to be prodded to prevent delays.
3. Important executives may be called so frequently from their work for meetings that the operations of the enterprise slow down.
4. The members of the committee may be unfamiliar with important details of questions at issue and therefore may make wrong or ineffective decisions.
5. Action may often be superficial because of the lack of time or the lack of. interest of committee members.

6. Committees may weaken individual responsibility and make for compromise instead of clear-cut decisions.
7. The decisions may conform with what it is assumed some executive wants, or enable the members to avoid direct responsibility for any bad results.
8. Aggressive and outspoken members may dominate committee meetings and unduly influence the action, often adversely.

EXAMPLES OF COMMITTEES. General Executive Committee. An excellent example of the need for a committee is shown at the level of authority which includes the sales manager, production manager, controller, chief engineer, and director of industrial relations. These men, with the general manager, represent the various **functions of the executive power.** No one of them alone can intelligently decide difficult major questions of policy. But when the five men are assembled, each as an executive thoroughly familiar with his own field, the discussions are authoritative, and the decisions have every chance of being sound.

The general manager is logically the chairman of such a committee, and the matters that usually come before the committee pertain to the **general policy** of operating the factory. Thus, the committee might decide the character and size of the articles to be manufactured. It might approve all manufacturing orders for either stock or special products. It would decide all questions of extraordinary expenditures and would consider all economic problems.

Joint Labor–Management Committee. In an effort to provide for better communications and working relations between labor and management a labor-management committee may be used. Matters concerning working conditions, compensation, grievances, and management prerogatives are generally discussed in this type of committee.

Shop Conference Committee. Under some conditions, a committee composed of a few of the shop foremen and a representative from the order department, with the superintendent of manufacturing as chairman, is most effective in solving **production problems.** Such a committee would discuss all matters pertaining to the operation of the factory and the status of production orders, and the discussion would bring to light any portion of a given production order that was behind schedule, together with the reasons for the delays. The findings of such a committee, in fact, may constitute a **progress report** of all work in process, and thus put into the hands of the superintendent first-hand information as to what should be done to speed up the work. Committees of this kind, composed of men who are actually in touch with the work, are of great value if they are properly conducted.

Equipment or Standards Committees. The equipment committee (also called a standards committee) is made up of men drawn from different ranks and may consist of a representative of the shop superintendent's office, the foreman of the tool design and tool-making department, the methods engineer, and any other men from the production control, production engineering, or shop departments who may be of service. The chief engineer or his representative may also be included with advantage in this committee. If the plant is large enough to employ an equipment manager, he would be logically the chairman of the committee.

Such a committee would discuss all problems concerning **new tools** or **improvement of existing equipment.** When ways and means of reducing the cost of manufacture of any particular line of goods are under discussion, the engineer who is familiar with the line should always sit with the committee. An engineer, a good foreman toolmaker, and a good manufacturing foreman together can often reduce costs to an extent which would be beyond the power of any one of them singly. A committee of this kind—when no other specific committees exist—is also very valuable in **establishing standards** and in advising the executive committee regarding the standardization of products.

Plant Safety Committee. The plant safety committee, with the production manager as chairman, is the form that the safety organization usually takes in the small concern. This committee is charged with making the work force "safety conscious," but the actual safety work is the responsibility of the operating organization (see section on Industrial Safety).

CRITICISM OF COMMITTEES. Committee organization has been the target for much abuse and the subject of many caustic remarks. Committees have been a particularly useful tool for those who wish to avoid decision: "Let's refer it back to committee."

The advantages and disadvantages of committees have already been summarized along with some basic principles which tend to improve committee performance. The important point to remember is that the committee represents a delegation of authority that permits management to bring special talent to bear on problems and frees management from some of the initial demands for communications. Many of the criticisms of the committee can be directed at other organizational arrangements such as the staff concept. As an organizational structural device it is only as effective as the members assigned to it.

MULTIPLE MANAGEMENT. In any company with more than approximately 100 employees, the problem arises of maintaining close relationships between management and employee. Important questions are: How can employees be given a feeling of participating in the company? How can management keep employees informed? How can management be extended down to the intermediate levels of authority as the company grows?

The concept of multiple management may work toward providing answers to these questions. Charles P. McCormick (The Power of the People) originated this concept and put it into operation in McCormick & Company. He felt that a major advantage of the system was its ability to train employees to take over responsibility as rapidly as they were able. Under the McCormick plan, management is extended to a large number of executive and supervisory men in the company through a system of four boards.

1. The **senior board,** comparable to the board of directors and elected by stockholders, sets up and controls company policy; and acts through its members as a clearing house for final decisions on all recommendations from other boards.

2. The **junior board,** a group of younger executives representing the office and executive group, meets regularly once a week to consider any matters dealing with company affairs it cares to investigate. Its recommendations must be unanimous, and become final only upon approval of the senior board member concerned with the suggestion.

3. The **factory board,** composed of active members and associate members, represents the factory, warehouse, and shipping department unit of the business. Problems concerning production schedules, stock control and shipping, machinery, and maintenance are discussed at regular weekly meetings. Recommendations must be unanimous and approved by the senior board member concerned.

4. The **sales board,** composed of active and associate members chosen from the outside sales force actually calling on the trade, plus inside sales executives, meets less frequently but for longer periods. Cross-section opinions of the sales force are secured. Resolutions passed unanimously for senior board

approval at sales board meetings represent the opinion of the salesmen on merchandising, advertising, sales training, and similar subjects.

Election to the Boards. Provision is made to give eligible men in the company an opportunity to be elected to the boards by a merit-rating system whereby each board member rates on a rating chart the abilities of every other member. The three members who are lowest on the semi-annual rating chart on each board are dropped off and replaced.

The employees do not elect the members of the board, because the tendency would be to select on popularity rather than on merit. However, the employees have a feeling of participation in that men they know and work with in their departments every day are actually directors in management.

Benefits to Employees and Company. Many claims are made concerning the benefits to be derived from the multiple management system. McCormick (The Power of the People) includes the following appraisal from a worker's viewpoint:

Through this channel you receive valuable information regarding the so-called inner facts of the organization that would take you years to get otherwise. Because of this added knowledge, you take a greater interest in the business and try to grasp more threads which will help you to become of greater service.

A similar, but less formal, system of multiple management is found in the **regular management meetings** of some companies. At these meetings, usually attended by all managers and held during off-duty hours, problems are discussed according to a formal agenda. The off-duty atmosphere tends to generate freer communications between managers and supervisors. The advantage of this type of system over McCormick's multiple management system is that all managers participate rather than a select few.

Developing an Organization

ASSIGNMENT OF DUTIES IN AN ORGANIZATION. Duties or activities in an industrial organization are assigned to individuals by several different methods, of which the following are the most important:

1. **By persons.** An executive or supervisor is given authority over, and made responsible for, certain subordinates.
2. Within **physical boundaries.** An executive or supervisor is given supervision over a room, a department, or a production center.
3. **By production.** An executive or supervisor is given supervision over the manufacture of a particular item of product, or a certain line of product.
4. **By process.** An executive or supervisor is given supervision over a particular manufacturing process, or over a series of such processes.
5. **By equipment.** An executive is given supervision over a particular group of machines or class of equipment.

The above assignments are in the nature of **vertical subdivisions** where each executive's or supervisor's authority is exercised within his determined sphere of control but subject to higher line authority and to functional or expert staff advice.

THE LADDER OR BRIDGE OF FAYOL. In addition to vertical, there are **horizontal relations** which have received but little study and managerial attention. In this connection there is a nearly unexplored field, where the vertical lines of authority cross and conflict with the horizontal lines of relations. The situation is made plain from a consideration of Fig. 5.

This diagram represents two lines of authority apexing in the president and in each case following through a production manager, superintendent, depart-

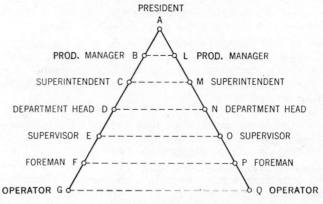

Fig. 5. Cross contacts or relationships illustrated by the "ladder" or "bridge" of Fayol.

ment head, supervisor, and foreman to the operator. It is evident that, to perform duties assigned to each line—assuming that they are in the same industrial organization—there must be some contact, communication, and relationship between individuals at each level of authority in the two converging lines. Such **cross relations** may be concerned with: (1) jurisdiction, that is, a determination of which line is to do certain work; (2) coordination of policies and operation methods needed to secure uniform operating results; (3) review and criticism of work, which may occur where work is transferred from one line of authority to another in order to complete succeeding operations; and (4) division of overlapping duties. The points where vertical authority and cross relations meet are illustrated by the small circles in the diagram. In practice these relations are real and continuing, and give rise to **points of possible friction and conflict.**

Essentials for Maintaining Intersecting Relationships. Fayol, whose "requirements for command" have been previously mentioned, set up the essentials for establishing and maintaining these intersecting relationships on a frictionless and properly managed basis. Essentially his solution is a **process of self-adjustment.** He pointed out that executives at any level in line of authority may contact one another, reach decisions, and initiate action, provided these requirements are satisfied:

1. Contact or relationship should be initiated only with the consent of the immediate line superiors.
2. Before any action is taken, it must be approved by the immediate line superiors.

SPAN OF MANAGEMENT. The term **span of management, or span of control,** refers to the number of subordinates who can be successfully directed by a supervisor or superior. In the direction and control of industrial enterprise it has long been apparent that delay, friction, and confusion can be traced to the fact that too many subordinates are assigned to one superior.

The theory of Graicunas (Relationships in Organization) states that in an organization there are **three kinds of relationships:** direct single, direct group, and cross.

In almost every instance a supervisor measures his responsibility by the number of **direct single relationships** between himself and his subordinates. He thinks of a group of 12 employees as requiring twice as much work of supervision as a group of 6. However, there are **direct group and cross relationships** to be considered as well as the simple direct. To illustrate:

Designate the supervisor as A, and assume that he has only two subordinates, B and C. It is evident that A can deal individually with B and with C, or he can deal with them as a pair. The behavior of B in the presence of C will differ from his behavior if he alone is with his superior, A. Furthermore, the attitude of B toward C, and C toward B, constitutes cross relationships which A must keep in mind in arranging the work of B and C. It is evident that B and C might have widely different racial, political, or trade union affiliations, which would have no influence upon either as a good workman working alone, but which might prevent them from working together harmoniously. Then again, some individuals, the so-called "lone wolf type," cannot work well with others, no matter who they may be.

Graicunas presented a number of formulas for finding the maximum number of relationships concluding with the following formula which covers all possible relationships:

$$N = n(2^{n-1} + n - 1)$$
$$\text{where} \quad N = \text{total relationships}$$
$$n = \text{number of subordinates}$$

Obviously the purely mathematical approach does not recognize such factors as the type of contact made, the extent or duration of the contact, the frequency of contact and the like. But the magnitude of the answers derived from the application of the above formula does indicate the wisdom of maintaining as small a span of control as possible.

In real situations there are no fixed quantitative values to assign to the appropriate span of management. The manager must consider each general situation separately. Voich and Wren (Principles of Management) summarize the factors influencing the span of management as:

1. **Level of the organization.** The span tends to become more narrow as we move to higher organizational levels because of growing complexity and increasing responsibilities. At lower levels the span tends to be wider because of less responsibility and decisions have a smaller impact on total operations.
2. **Ability and training of labor and management.** Better trained or more capable subordinates will permit a manager to delegate more authority, thus reducing the amount of communications needed. A more capable manager is able to plan better and supervise more people.
3. **Managerial attitude toward delegation of responsibility.** Managers who tend to limit delegation will create a narrow span because of the increasing need for communications.
4. **Degree to which work is interdependent.** If work of one unit is highly interdependent on other units, the need for coordination and communications increases, resulting in a narrower span of management.

EXECUTIVE OR MANAGEMENT LEVELS. A study of the organization from the standpoint of the respective levels of executive rank is important. The basic aim is to have the fewest number of levels and yet not create problems of too wide a span of management. The use of staff authority tends to alleviate the problems of too long a chain of command and too wide a span of management. Executive levels are usually indicated by the **descent of authority** down the organization line. Some divisions and departments have few levels between the executive who reports to the president, and the supervisors ranking just above the operator. Other departments have several such levels.

Determining the Levels. This involves more than a determination of respective duties and responsibilities; it covers the general sphere of influence which the individual governs and the relative rating of his position as compared with other jobs throughout the entire company. If a good job evaluation plan is in effect in the organization, the basis for rating positions on their relative levels is well established. In fact, this is one of the elements in job evaluation. Fine shades of distinction in rank are unnecessary as well as impossible.

Probably the best approach to the problem, with or without a job evaluation plan in effect, is to set up three or four general ranges of rank or levels under which the respective positions will be grouped.

THE FIVE IMPORTANT LEVELS. There are five important levels into which executives and supervisors may be classed. These levels are:

1. Top executives.
2. Senior executives.
3. Intermediate executives (often divided into two or three ranks according to the nature and size of company).
4. Junior executives.
5. Supervisors and foremen.

Top Executives. In the ranks of top executives are the officers of the company. Typical titles are: President, Vice-President in Charge of Engineering, Vice-President in Charge of Manufacturing, Vice-President in Charge of Sales, Vice-President in Charge of Industrial Relations, Secretary, Treasurer, General Manager.

Holden, Fish, and Smith (Top Management Organization and Control), differentiate between three zones of top management as follows:

1. The **trusteeship function** (Board of Directors) performs a **judicial** and **intermittent** function in appraising and approving major **company-wide** proposals and results.
2. The **general management or administrative function** exercises **active** and **continuous** initiation, formulation, coordination and development of **company-wide** proposals and results.
3. The **divisional or departmental management function** is responsible for active direction and management of the **respective divisions or parts of the company** within the policies and delegated authority of general management.

These men discharge **major responsibilities** and exercise a **wide range of authority.** They are directly concerned with the application of the basic policies of the company and the direction of its respective major lines of activity. Likewise they establish the central coordination between the principal activities and should have delegated to them practically full authority—subject to company policies and the responsibilities of, and relationships with, other activities —to carry on their work according to plans which they themselves largely develop.

Since there is no definite standard for executive titles among industrial organizations, the titles employed here are descriptive and indicative rather than specific and exact. Thus, a general manager, could exercise immediate direction over the activities of engineering, manufacturing, sales, and accounting, and perhaps over industrial relations. The vice-presidents, treasurer, and secretary, might cover corporate duties of a special nature, rather than operating responsibilities, although a vice-president might also be the general manager.

Senior Executives. The class of senior executives might include the chief engineer, plant manager, sales manager, purchasing agent, controller, personnel director, and two or three others with related titles. While these men do not have the official rank entitling them to the designation of top executives, they carry heavy responsibilities and exercise full authority in their respective areas. They are rated as executive heads of the divisions or departments of which they have charge. Their tasks are to break down the company's **basic policies,** particularly the ones governing their lines of work, into directive regulations and to develop the **fundamental procedures** for their respective divisions or departments.

Intermediate Executives. In moderate- and large-sized companies the senior executives have immediate assistants who are qualified in some cases to take over the work of the division or department in the absence of their chiefs, or to aid them by performing certain of their duties or handling special assignments. Such men usually are chosen to succeed their chiefs if the latter leave or are promoted to higher responsibilities. Reporting to the chief engineer, for example, would be an assistant chief engineer or perhaps an executive engineer; to the plant manager such men as a manufacturing manager, production control superintendent or manager, chief inspector, works engineer, and perhaps others with corresponding duties. Under the sales manager would be an assistant sales manager; under the purchasing agent, an assistant purchasing agent; under the controller, a chief accountant, cost accountant, statistician, or budget director and office manager; and under the personnel director, an employment manager, training director, employee-service manager, a specialist in industrial relations and collective-bargaining, and perhaps a safety director.

First-rank intermediate executives are often assigned **particular areas of work** in the division or department. They become, in effect, specialists in such lines and usually assume practically full direction of the activities which they supervise. They are responsible for the direct application of the immediate policies of the company, the development of specific procedures for the performance of the work, and direct supervision over assistants carrying more detailed assignments.

The second rank of intermediate executives are important subexecutives who work under the direction of those in the first rank. The assistant chief engineer would have reporting to him perhaps a chief draftsman, in some cases project engineers, and engineering specialists. The manufacturing manager may have superintendents who head departments. The planning superintendent would have a methods engineer, a chief time-study engineer, a chief of planning, and related heads of units under his jurisdiction. Under the assistant sales manager might be an advertising manager, a field sales manager, etc. Under the assistant purchasing agent would be buyers and others in responsible positions connected with purchasing; and under the employment manager, a director of employee tests, and a chief of shop training, and heads of other personnel sections. The chief accountant would have assistants specializing in specific phases of accounting, such as accounts receivable and accounts payable.

In all cases these subexecutives would be in charge of **detailed sections of work** which, in large companies, would be still further subdivided into units to be handled by junior executives. Certain decisions are required of the intermediate executives, and they are responsible for the development and improvement of various techniques and procedures. They have direct control over their assistants and are held responsible for the proper performance of the work.

Junior Executives. The level of junior executives includes those who have begun their progress through the organization because of experience or training which qualifies them to head a **smaller unit of the enterprise** and direct the work of a few assistants or supervisors. Under the chief draftsman may be a chief checker who may direct the work of drawing inspection and checking; under the superintendents would be assistant superintendents; under chief of planning there may be a schedule man who lays out the program for factory work. The field sales manager often has branch managers. Buyers sometimes have assistant buyers. The chief of shop training may have a head instructor in machine-tool work, and a head safety-training instructor.

Junior executives in some cases would direct the work of supervisors or foremen, especially in the factory departments. Assistant superintendents, for example, may have general shop foremen covering particular units of the plant engaged in distinctive kinds of work such as foundry, forge-shop, machine-shop, etc. In other cases the junior executive may head a unit of workers, sometimes operating under group leaders, who are performing some definite kind of work, especially in the office, engineering department, or laboratory. His responsibilities include the making of decisions, giving of advice, and the development of procedures. He is therefore doing an executive class of work rather than conducting an operation. Consequently he should be rated in the executive category rather than as a supervisor or foreman.

The junior rank involves activities in many of which the individual is on his own, but in which he has the ready aid of some chief close enough to his work to guide him in his decisions and procedures. At the junior level, therefore, the executive applies on his own responsibility many of the things he learned as a worker and supervisor, but at the same time he is on the "proving ground," being tested in his ability to use what he knows, to train others, and to be a leader in getting work done and in commanding not only the obedience but also the respect, high regard, and loyalty of his subordinates. No man who demonstrates mere "drive" for accomplishment but fails to continue his own training and lacks the power of building his assistants into a loyal and efficient team should be promoted into the intermediate or higher ranks.

Supervisors and Foremen. In immediate charge of employees are supervisors and foremen. Many companies distinguish between these two terms, implying by the title of "supervisor" that such men have **semi-advisory or assisting duties** included in their range of activity. The term "foreman" is then used to indicate that the holder of this title actually **directs employees** in their work. The designation "foreman" is likewise used with different implications in different plants. Sometimes a foreman heads only a small group of workers. In other cases an individual with this title may head a department of one or two hundred, or more, workers, in which case he would be aided by assistant foremen, under whom subforemen or group or squad leaders would work. Each person immediately in charge of the performance of the work, therefore, would have perhaps from 5 to 20 workers under him.

Persons in supervisory or foreman ranks are responsible mainly for four important functions:

1. Getting work done in the time, and of the quality, set by careful and coordinated planning.
2. Setting standards for the work in conjunction with the special units (methods, time study, inspection, etc.) particularly concerned with such problems.
3. Training workers to perform their tasks better so that spoilage will be cut down, quality maintained, and output increased, thus enabling employees to add to their earnings.
4. Handling grievances in the department, thus eliminating causes of dissatisfaction as soon as they arise and thereby maintaining good labor relations.

INTERCOMMUNICATION BETWEEN LEVELS. A good organization functions with freedom of communication among its executives regardless of their respective positions. In well-run companies a junior executive may ask for advice or aid from some senior executive in another department by direct approach, when his problem concerns, or is affected by, some element under the senior's control. Senior executives may likewise secure the aid of junior executives. There is no violation of line control in such contacts, and they improve the efficiency and speed with which work is done.

QUALIFICATIONS OF EXECUTIVES. Considerable discussion has been devoted to the question as to what are the important qualifications of a competent executive or manager. Character considerations, such as honesty, integrity, loyalty, fairness, experience and knowledge of work, adaptability, ability to motivate, and self-control are some of the frequently cited characteristics of a good manager. These characteristics apply generally to all levels of management. Although all managers perform basically the same functions and work with similar resources, the scope of their responsibility increases from lower levels to the upper levels.

Katz (Harvard Business Review, Vol. 33) summarizes the capabilities, foundation skills, or abilities needed by the manager or executive as conceptual, human, and technical. Katz uses **"technical"** skill to apply to knowledge of the functional areas of business; **"human"** skill to apply to the "ability to work effectively as a group member and to build cooperative effort with the team he (the executive) leads"; and **"conceptual"** skill as the "ability to see the enterprise as a whole" and the various relationships in the business environment.

Voich and Wren (Principles of Management), using this basic framework, define **"technical"** capability as "knowledge and ability to work with functional activities and resources used in those activities"; **"managerial"** capability as the "knowledge and ability to perform the basic management process of planning, organizing, controlling, and administering in the utilization of all resources"; and **"conceptual"** capability as the "ability to take various parts of a situation or problem and place them in their proper relationship so as to see and create a better whole."

Conceptual capability, compared with technical and managerial capabilities, is more important at upper levels or for top and senior executives. They must be able to see parts of the total firm and how they are related internally and externally. Technical capability is more important for lower levels or for supervisors, junior executives, and to some extent intermediate executives. Managerial capability pervades every organizational level, top executives to supervisors, since it is related to the basic or universal functions of any manager. Included in this managerial capability is the factor of human- or people-oriented skills.

Organization Charts and Manuals

FUNDAMENTAL CONSIDERATIONS. The package of aids consisting of the organization chart, organization manual, and standard operating procedures serves as an important tool for communication and coordination. The **organization chart** shows where in the organization a function is performed and by whom; the **manual** explains what the nature of the function is and the duties and responsibilities involved; and the **standard operating procedures** describe how and when it is to be performed. In planning an organization and developing the charts, manual, and standard operating procedures, it is necessary to keep in mind the following basic requirements:

1. The **organization is a means to an end,** and should reflect the critical functions and subfunctions required to achieve the objectives of the company. The organization must be built around functions, not individuals. The activities or tasks which must be carried on should be blocked out following a typical or representative chart which carries the major functions normally necessary in most manufacturing enterprises. Modifications may be introduced at any point

to adapt the chart to the particular company. There is no such thing as a standard chart of organization and no organization chart can be copied in whole from any other company, although it may be in the same industry or even in another plant of the same company. There are chart patterns but none can be applied directly without alteration. The **main functions** in a manufacturing concern are those of engineering, manufacturing, selling, purchasing, industrial relations, accounting, and financing.

2. **Functions which are closely related** must come under the same head.

3. When the major and related functions are thus distinguished and identified, they should be arranged so that **each individual will be definitely provided for** and all necessary duties may be properly performed.

4. In smaller companies—or in cases where the extent or the amount of work in some duties is limited—similar **duties may be combined** and handled by one person.

5. Only after the duties have been defined and arranged should the question of personnel enter into the picture. Since it is obviously impossible to secure in all cases individuals who are ideally qualified to handle the various functions or tasks, those who have most of the necessary qualifications are selected or engaged to take over the individual assignments. At this point it may be necessary to **readjust the setup** so that the best results may be secured in performance. While some realignments may be necessary and certain unassociated activities delegated to the same individual, two points should be rigidly adhered to:

   a. The functions and subfunctions must be clearly identified and retained in the framework.

   b. No assignments should be made which allow any two individuals to cross lines of authority and come into conflict. The authority and responsibility accompanying each function and subfunction, therefore, should be definitely delimited in the organization manual.

6. No more than 5 or 6 persons, in general, should report to an individual on executive or subexecutive level, where these subordinates, in turn, direct the work of others. Departments should be broken down into sections and the sections into units or groups, each headed by capable assistants. Where immediate direction of workers enters, the limit to the number of subordinates reporting to the supervisor or foreman should be 10 to 15 in most cases. If work is highly routine or a large number of workers are engaged in repetitive work, the limit may be raised to about 20. It is better to break up any larger units into two or three groups, each headed by a leader who reports to the supervisor or foreman.

7. While it is desirable to limit the number of executives, supervisors, and foremen in a company for reasons of efficiency and economy, no executive should be unduly overloaded with contacts and work. Especially in a growing organization, each executive should have some time for **constructive planning** and for growing in his job.

8. The organization plan should be developed with the idea of **future expansion** in mind, so that major reorganizations at a later period may be avoided.

9. All persons in executive capacity should have copies of the charts and manual and everyone in the organization should have access to these items at all times. There is nothing secret about the information. It is advisable to change all copies of charts and manuals, or at least issue change notifications, as soon as such shifts occur. At least the master copies should be altered promptly to keep them up to date.

10. **Freedom of contact** among all members of the organization for quick

interchange of information and arriving at prompt decisions (**Fayol's Principle**) should be provided for in developing the charts and writing the organization manual.

11. Since no organization stays in a permanent form, it should be realized that **frequent changes** in the charts and manual will become necessary, beginning shortly after the original setup is made.

ORGANIZATION CHARTS. It is advisable, even in small companies, to draw up and post organization charts so that all persons may know how the various duties and activities are assigned and can find out where they fit into the setup. While the worker levels are not indicated on such charts, workers know their own supervisor or foreman and therefore can see the relationship of their units or groups to the remainder of the company. In large companies it is necessary to draw a **series of charts**, starting with one for the top organization, which shows the relationships among top and senior executives and the breakdown into divisions. Other charts may show the organization of the respective departments and sections, indicating intermediate and junior executives. A further breakdown may detail the individual sections and units to indicate how authority and responsibility are carried down to supervisors, foremen, and group leaders.

The **breakdown charts** should include a skeleton outline of the top organization to indicate the line of authority from the president down through the divisions, and then through the particular departments, sections, units, and groups down to the individuals to whom the workers report. The workman in a company needs to know how the industrial team of which he is a member, and in which he earns his living, is organized and the positions which he and his boss play in the enterprise.

Constructing Organization Charts. The drawing and revising of organization charts requires considerable work and expense, but these are many times repaid by the benefits derived from keeping employees informed. Constructing an organization chart is not difficult, once the general structure of the company has been determined. George (Management in Industry) presents the following guidelines which are useful for refining the organization chart:

1. The **vertical arrangement of functions** should be according to grade or position in the company hierarchy, at least for major segments.
2. A rectangle should be used to indicate a unit or person on a chart.
3. Straight vertical and horizontal lines should be used to show **flow of authority.** Lines of authority should not run through the rectangles.
4. Heavy and solid lines may be used to show line authority relationships, and light and solid lines for functional supervision. Broken lines are generally used to show staff (advisory) and service relationships.
5. Rectangles on the same level should be made the same size, and placed so that the same authority levels are horizontally aligned. Rectangles may be reduced in size for each succeeding lower level.
6. On any individual chart, show at least one level above and two levels below the primary organization or person being charted.

Since the organization chart is an **aid to communication,** it is essential that it can be understood. Guidelines, such as the above, seem very basic, yet they can quickly clarify the organizational relationships that exist within a company.

Typical Organization Charts. There are **five typical ways** in which organization charts may be drawn, as shown in Fig. 6. Fig. 6a represents the most common arrangement while Fig. 6c is a variant sometimes used for saving space. The other types are used less frequently. Besides the **title of the function,** organization charts may also contain information about the person

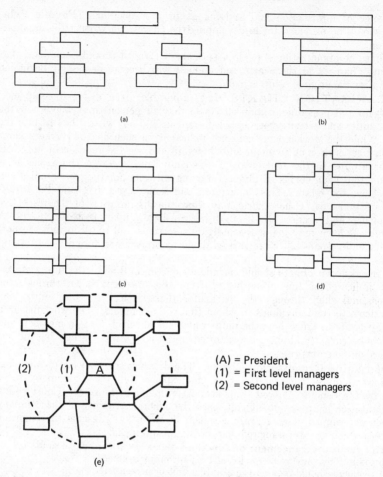

Fig. 6. Five typical ways of drawing organization charts.

occupying the position, such as **age, years with company,** and **promotable status** of incumbents (personnel department use).

Figure 7 (pages 32 and 33) is a typical organization chart for manufacturing plants in that it shows the principal functions and subfunctions usually provided for, and the range of work covered by the various departments and sections. As stated before, these are no standard organization charts, but merely representative charts which serve as guides in individual cases. This chart, however, merits careful study because of its comprehensive coverage of functions and the descriptive nature of its titles and the kinds of work indicated for the respective departments. It is seen that the names of individuals are not important in the development of a chart, because obviously there will be changes occurring constantly in the personnel of the respective companies. In some companies, the names are added to make the charts clearer to the members of the entire organization.

THE APPROACH TO GENERAL PRODUCTION ORGANIZATIONS. Although the specifics relating to particular manufacturing companies may vary, the general nature of the overall production function is similar for most companies. Resources are acquired, stored, used, and disposed of by a series of conversion processes. In a large manufacturing firm these conversion activities may be divided in several ways. Voich and Wren (Principles of Management) summarize these as follows:

1. **By function**—grouping of input efforts into departments according to similar occupational skills (see Fig. 8, page 34). Advantages include simplicity and relatively easy coordination. The major limitation is achieving coordination between functions as an organization grows.
2. **By activity** (sometimes called **process** departmentation)—grouping of input activities essential to the flow of work (see Fig. 9). Advantages and limitations are similar to functional departmentation.
3. **By product**—grouping by output or service performed (see Fig. 10). Advantages of product departmentation result from centralizing all aspects of production under one manager thus facilitating coordination. A major limitation is that some duplication of functions (especially overhead) occurs.
4. **By territory**—grouping by output area served (see Fig. 11). The advantage of territory departmentation is that greater attention to particular markets can be achieved. Limitations arise chiefly out of economies of scale since the area may not be large enough for a production plant or the cost of shipping or raw materials and finished goods is too great.
5. **By customer served**—grouping of efforts according to specific users of products (see Fig. 12). Advantages and limitations are similar to product and territory methods.
6. **Conglomerate structures**—as the lower levels of the organization are taken into consideration, generally some combination of the five foregoing methods of departmentation are used (see Fig. 13). The functional method appears in almost every company to some degree.

A Complex Organization. Despite its size, General Motors Corporation has an enviable record of managerial excellence and profitable operations and service. The organization of General Motors provides an example of decentralization of authority and utilization of the line and staff concept. The top management organization of General Motors is shown in Fig. 14 (pages 36 and 37). Three general levels of operation reflected in the organization are as follows:

1. The **Major Control Level** is the authority down to, but not including, the President. This level includes the stockholders, Board of Directors (including the Audit and Salary and Bonus Committees), the Finance Committee and Executive Committee. The **Finance Committee** is responsible for formulating financial policy for the whole corporation. The **Executive Committee** (headed by the President) deals with the operating affairs of the business. All major issues in the field of policy and administration are handled by the Finance and Executive Committees.
2. The **Executive Control Level**, headed by the President, has the entire responsibility of interpreting the policies formulated by the Major Control Level and distributing them throughout the organization. The Operations Staff, Financial Staff, and Legal Staff support the President in his responsibility by **performing** those functional activities that can be accomplished more effectively by one activity, rather than in every operating segment, and **coordinating** similar functional activities of different operating segments. Certain segments of these staffs cooperate directly with those departments of the operating units whose activities are in the same functional areas.
3. The **Operating Level,** composed of major product groups (and divisions, departments, and functions within each group), is responsible for the acquisition and utilization of resources in the production and distribution of products and services. Within each of the operating divisions (similar to separate companies in many respects) is found an organization quite complex in itself.

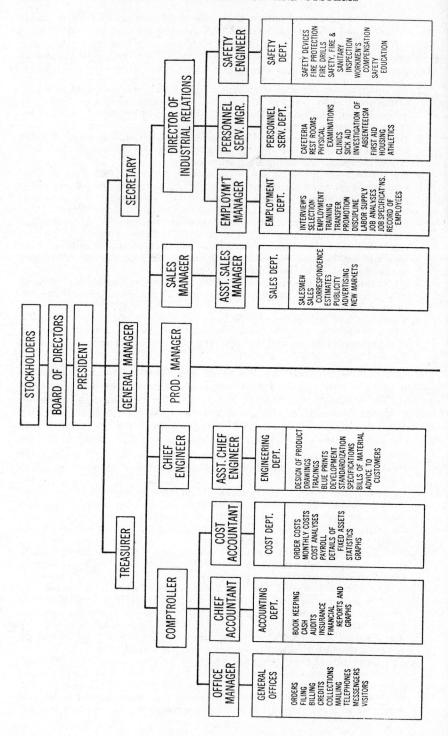

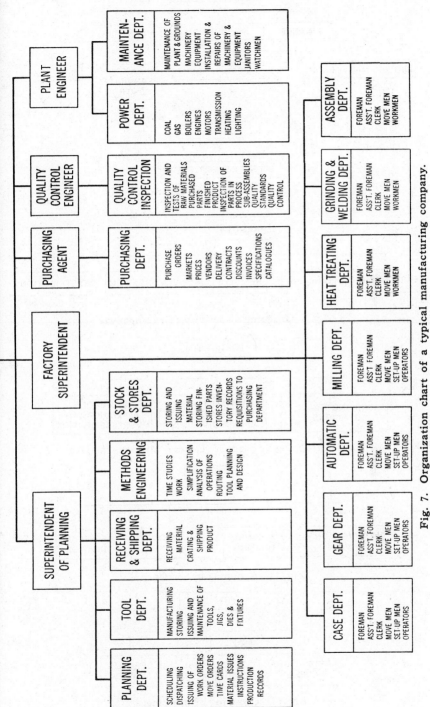

Fig. 7. Organization chart of a typical manufacturing company.

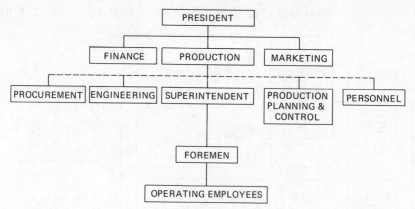

Fig. 8. Grouping by function.

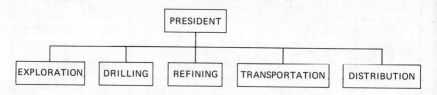

Fig. 9. Grouping by activity (oil company).

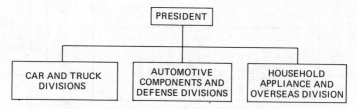

Fig. 10. Grouping by product (General Motors example).

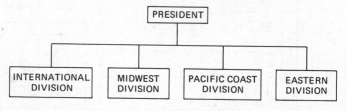

Fig. 11. Grouping by territory.

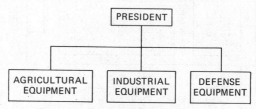

Fig. 12. Grouping by customer.

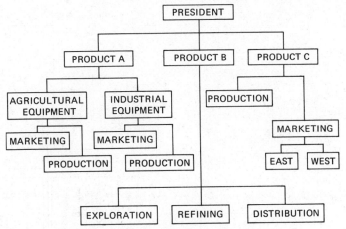

Fig. 13. Conglomerate structure.

A proper balance between the benefits of centralization and decentralization is the General Motors approach to organization. Management is established on centralized policy making and staff assistance and decentralized administration of operations. Centralization makes possible long-range control, coordination, and specialization through the staff concept. Decentralization develops initiative and responsibility, distributes the decision-making process to all levels of management, and provides for flexibility and cooperative effort.

THE APPROACH TO SPECIFIC PRODUCTION ORGANIZATIONS. The following discussions describe in more detail the activities of the production manager, or vice-president of production, found in most large manufacturing companies. The organization shown in Fig. 8 is used as the point of departure for analysis.

Production Manager. The production manager is generally a vice-president in a manufacturing company whose major responsibilities are related to the physical processing and conversion of materials into finished products.

The **production line organization** would generally be as shown in Fig. 15. The line organization is responsible for directing the work of the operating employees in the operation of the plant facilities and production equipment. Because of size, the production manager is assisted in this responsibility by intermediate executives (superintendent and foremen), the number of which will vary depending on the number of operating employees involved. The line organization is conversion oriented, that is, it deals with the physical conversion of materials into finished products.

The **production staff organization** serves the production line organization in the physical conversion process. A number of activities dealing with planning and controlling of production work, acquiring labor and materials, and general design and development are included in the production staff organization. Fig. 16 shows the production staff organization.

The **staff segments** of the production organization assist the line managers by providing advice; development of better production processes; evaluation of operations, facilities and personnel; and furnishing the required materials and personnel when needed for the production activity. A representative list of the

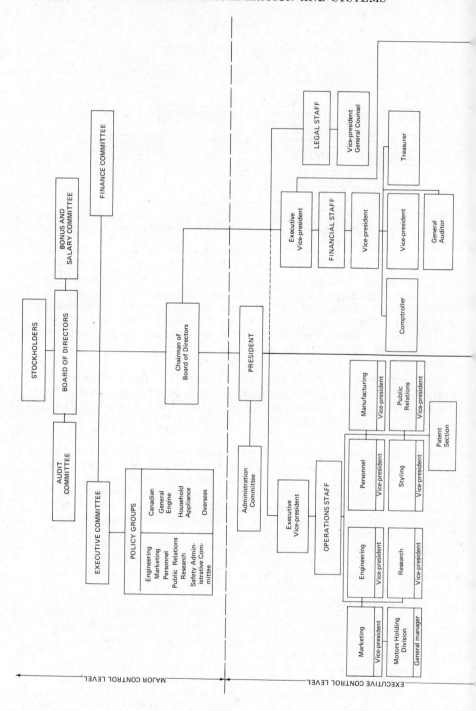

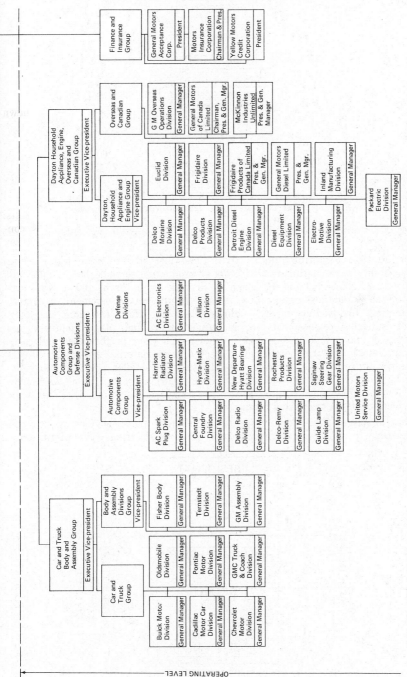

Fig. 14. Top management organization chart of General Motors (summary form).

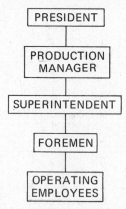

Fig. 15. Partial line organization for production.

types of activities performed by each of the production staff segments are shown in Fig. 17.

Production Planning and Control. This is a staff function whose major responsibility is to coordinate all factors of production so that the products will be produced efficiently and on schedule. Serving as the brain and nervous system of the production organization, it helps coordinate procurement, personnel, production, and engineering by planning when and where the work should be done; initiating requirements for materials, labor, and plant facilities; and developing and operating a system for controlling work performance and costs. In larger manufacturing organizations, scheduling, materials control, cost control and budgeting are separate sub-activities. In smaller organizations, they may be grouped under one individual. However, the important point to remember is that they must be performed by someone, therefore they must be provided for in the organizing process.

Engineering. This function generally includes plant engineering, process engineering, product engineering, and industrial engineering. The **plant engi-**

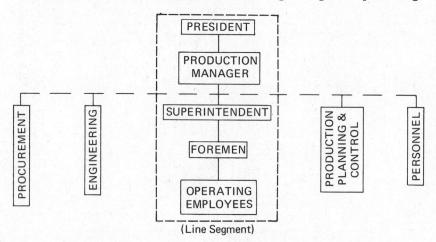

Fig. 16. Partial line and staff organization for production.

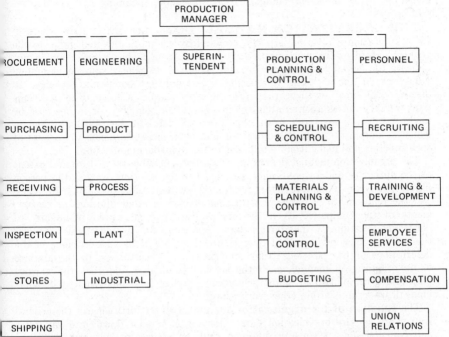

Fig. 17. Activities of production staff organization.

neer is generally responsible for designing, selecting, installing, evaluating, and monitoring the physical facilities of the production organization. His work is influenced by the production plans and breakdowns in machines, equipment, or facilities. The **process engineer** focuses his attention on maintaining effective and economical physical production conditions and methods. The **industrial engineer** is primarily concerned with maintaining effective and economical human conditions and work methods. Plant layout, motion and time studies, and work procedures development are several activities related to industrial engineering. The **product engineer** is responsible for the product design and development. These four aspects of engineering are highly interrelated and interdependent, and require close coordination. Therefore, they generally are placed under a **chief engineer**. **Quality specifications** are developed in the engineering area and influence production and procurement activities, as well as production planning and control and personnel.

Procurement. The procurement function is responsible for acquiring from vendors all materials, supplies, parts, and tools needed by the production organization. Procurement activities, generally placed under a procurement manager, include **purchasing** (locating sources, negotiation, bidding, buying); **receiving; inspection** (quantity and quality); **stores** (inventory management); and **shipping** (finished products). These activities are influenced by production planning and control and engineering operations since the procurement requirements originate in these areas.

Personnel. This function is responsible for securing, training, and maintaining an adequate production work force. As with the procurement activity, personnel requirements are influenced by production planning and control requirements.

THE ORGANIZATION MANUAL. Organization charts show where in the organization a function is performed and by whom. Titles may indicate the general nature of the work carried on by each person, but often they do not indicate the actual functions performed. There is no general agreement among industries on the meaning of specific titles and few companies attempt to develop definitions for the various titles in their plants. A foreman in one company would rank as a superintendent in some other company, or in one case he might be over 5 men and in another over 200 but still be called a foreman. Thus the organization chart requires an additional means to define the nature of each function and make it more relevant to the dynamic organization.

The organization manual defines as precisely as possible the nature and extent of the authority and responsibility assigned to each position, so that the person filling it may know his duties, his relationship with those above and below him, and the relationship of his work with that of men in other divisions or departments of the company. Even the organization manual cannot eliminate all borderline questions on the above points. The continual shifting of the organization to meet new conditions of operation call for additions and revisions in the descriptions of the positions from time to time. Nevertheless, in the absence of an organization manual, even the smaller companies find that the various positions overlap, two or more persons are trying to handle some particular kinds of work, and certain other work is neglected.

The **function of the organization manual** is to set forth clearly the general framework of the organization; state how it is broken down into divisions, departments, etc.; state the different ranks or levels of authority, officers, division heads, department heads, etc.; indicate the difference between line authority and the staff function; define the terms used in the above explanations; and give, for each division down through its successive breakdowns, the titles of the different positions, their relationships to the immediate superior and subordinate positions, and the range of responsibility, authority, and duties carried under the title. Though it defines the lines of authority and responsibility it cannot directly show the many crisscrossing channels of contact between individuals in different parts of the organization. These are provided for according to Fayol's principle, in a general statement authorizing all individuals to secure aid or information from any proper source within the company.

Organization manuals are generally developed as a part of the company's Policies and Procedures Guide or Manual, generally the initial section. Standard Oil of California's "Management Guide" is considered a leader in this area. The Policy and Procedures Manual generally contains an organization section, a policies section, and a standard operating procedures section. Included are references to (or actual copies of) formal company regulations and technical manuals.

Administration of a Company Manual. The company-wide manuals are generally administered by the Policies and Procedures Department in the office of the Executive Vice-President. The manual is revised constantly to keep it up to date. Changes and additions are announced and authorized by the appropriate division head as soon as an appointment or transfer has been made. These changes are followed as quickly as possible by the revised pages of the manual.

In order to facilitate the mechanics of issuing and maintaining organization pages, except for the top, overall company chart, the manual itself may be set up on the basis of **typewritten organization charts.** Levels of organization are indicated by indentations. Alongside each position is shown the department number, the name of the incumbent, and an **approval code.** This code denotes the general range and degree of authorizations which the incumbent of the particular position may approve. Such authorizations may be in such areas as budgets and expenses, personnel hiring and transfer, timecards, and capital equipment. Revisions and additions to the entire manual may be handled in a typewriter thus eliminating time-consuming and costly drafting. It also results in a considerable saving in space. A typical page in such a manual is shown in Fig. 18.

Position Descriptions. These descriptions specify the title, the objectives, the specific duties and responsibilities, and special relationships, if any, of each position. In order to stress the nature of managerial responsibilities inherent in all supervisory jobs, and in order to economize on space for individual descriptions, a listing of common duties and responsibilities **applicable to all management positions** precedes all descriptions.

Organization charts, position descriptions, schedules, budgets, and other management tools are to be relied upon as guides for decisions; they do not relieve any supervisor of the obligation to exercise common sense and good judgment in the accomplishment of his mission. The functions listed below apply only to the degree required in a given position. The scope of their applicability is stated in the "specific" section of the position description of which this "common" section constitutes a part.

A. PLANNING:

1. **To Formulate Policy**
   a. To formulate policy relative to his position.
   b. To establish short and long term objectives and to develop programs to implement these objectives.
   c. To be guided by the applicable Divisional and General SOP (Standard Operating Procedures) in regard to objectives, policies, procedures, forms, organization, etc.

2. **To Forecast Requirements**
   a. To forecast personnel, machine, material, space, financial and other requirements pertinent to his function.
   b. To prepare estimates, as required, on such matters as cost of work, delivery date, etc.

3. **To Keep Abreast**
   a. To keep abreast of developments in the field of management, of the technical areas under his supervision, and of developments within the Company.

4. **To Establish an Organization**
   a. To insure that each subordinate has a clean-cut responsibility to a single superior only.
   b. To clarify and define the objectives, functions, responsibilities, and relationships inherent in the positions under his jurisdiction; to make no change in responsibilities without notifying all parties concerned; to insure that his subordinates understand how they fit into the overall picture of the Division.
   c. To delegate responsibility (with commensurate authority) to the greatest possible extent in order to get decisions made as close to the point of operation as possible; in order to utilize effectively the talents and capabilities under his jurisdiction; and in order to develop them to assume the fullest possible responsibilities for the Division's as well as their own benefit.

	Dept. No.	Names of Incumbents	Approval Code	Night Shift
(MANAGER, AERONAUTICAL EQUIPMENT DIVISION)	(3100	———	D)	
(MANUFACTURING MANAGER)	(3400	———	C)	
GENERAL FOREMAN, FLIGHT INSTRUMENT ASSEMBLY–MECHANICAL	3401	———	C	
FOREMAN, S-3 GYROSYN ASSEMBLY	3475	———	B	
SECTION FOREMAN, GYRO ASSEMBLY	3473	———	A	
ASST. FOREMAN, GYRO ASSEMBLY	3473	———	A	
ASST. FOREMAN, GYRO ASSEMBLY	3473			——— A
ASST. FOREMAN, GYRO ASSEMBLY	3473			
SECTION FOREMAN, S-3 GYROSYN ASSEMBLY	3475	———	A	
ASST. FOREMAN, GYRO HORIZON ASSEMBLY	3482	———	A	
ASST. FOREMAN, FLIGHT INSTRUMENT PREPARATION	3439	———	A	
FOREMAN, MECHANICAL ASSEMBLY FOR AUTOMATIC PILOTS	3465	———	B	
ASST. FOREMAN, MOTOR & SYNCHRO ASSEMBLY	3425	———	A	
SECTION FOREMAN, C-5-C ASSEMBLY	3451	———	A	
ASST. FOREMAN, FLIGHT INSTRUMENT ASSEMBLY	3472	———	A	
SECTION FOREMAN, MECHANICAL ASSEMBLY FOR AUTOMATIC PILOTS	3465	———	A	
ASST. FOREMAN, MECHANICAL ASSEMBLY FOR AUTOMATIC PILOTS	3465	———	A	
ASST. FOREMAN, FLIGHT INSTRUMENT REPAIR	3483	———	A	
ASST. FOREMAN, MECHANICAL ASSY. FOR AUTOMATIC PILOTS.	3465	———	A	
GENERAL FOREMAN, FLIGHT INSTRUMENT ASSEMBLY–ELECTRICAL	3402	———	C	
FOREMAN, ELECTRONIC EQUIPMENT ASSEMBLY	3434	———	B	
ASST. FOREMAN, A-12 MODIFICATION	3484	———	A	
SECTION FOREMAN, ELECTRONIC EQUIPMENT ASSEMBLY	3434	———	A	
ASST. FOREMAN, ELECTRONIC EQUIPMENT ASSEMBLY	3434	———	A	
SECTION FOREMAN, ELECTRONIC EQUIPMENT ASSEMBLY	3434	———	A	
ASST. FOREMAN, ELECTRONIC EQUIPMENT ASSEMBLY	3434	———	A	
ASST. FOREMAN, ELECTRONIC EQUIPMENT ASSEMBLY	3434	———	A	
SECTION FOREMAN, ELECTRONIC EQUIPMENT ASSEMBLY	3434			——— A
ASST. FOREMAN, ELECTRONIC EQUIPMENT ASSEMBLY	3434			——— A

AERONAUTICAL EQUIPMENT DIVISION CHART
Issue Date: June 7, 19—
Page: 7

Signed ______________
Manager, Aeronautical Equipment Division

Fig. 18. Typewritten organization chart from organization manual

   d. To designate one subordinate to act in his absence or to meet this requirement by rotation among his subordinates.
   e. To develop at least one potential successor.
   f. To appoint all personnel under his immediate jurisdiction with the approval
   of his superior.

5. **To Establish Internal Controls**
   a. To insure optimum utilization of men, machines, and materials.
   b. To develop procedures, standards, methods, internal controls, budgets, and
   schedules pertinent to his function.
   c. To insure that management controls are governed by a philosophy of "prevention" and good planning, rather than merely by "after-the-fact control."

B. EXECUTING:

1. **To Integrate Activities Under His Jurisdiction**
   a. To operate in accordance with the above objectives and policies.
   b. To disseminate to his subordinates all possible general information, plans,
   and policies for their guidance.
   c. To encourage suggestions and complaints.
   d. To direct, motivate, integrate, and coordinate all personnel under his
   jurisdiction. To assign work, to promote the closest possible teamwork, and
   to insure appropriate quality and quantity.
   e. To assure the safety of his subordinates.
   f. To insure compliance with Company and Division policies and regulations,
   as well as labor and other laws, etc.
   g. To approve all authorizations as prescribed for his position.
   h. To initiate, maintain, and dispose of appropriate records.
   i. To maintain and protect Company assets under his jurisdiction.

2. **To Achieve Cost and Delivery Objectives**
   a. To operate within the approved budget.
   b. To insure that individuals under his jurisdiction make responsible commitments and promises regarding costs, schedules, etc., and that they live up to
   these commitments.
   c. To promote cost-consciousness to the utmost, but not to promote short-
   term economies at the expense of long-term growth.
   d. To maintain and promote a questioning and experimental attitude toward
   his operations.

3. **To Coordinate His Activities with Those of Other Supervisors**
   a. To work out problems relevant to his position directly with other supervisors affected, but to keep his principal informed in regard to matters:
      (1) For which his superior may be properly held accountable by others.
      (2) Which are likely to cause disagreements or controversy.
      (3) Which require the superior's advice or his coordination with other com-
      ponents of the organization.
      (4) Which involve recommendations for change of, or in variance from,
      established policies.
      To present (in the event that such problems cannot be resolved without a
      higher level decision) complete recommendations to his principal, rather
      than merely appeal for a decision.
   b. To promote management "by exception" wherever possible. (This entails
   that each supervisor use a maximum of discretion within the authority
   delegated to him, and within the established plans, policies, budgets, schedules, etc., and bring to his principal's attention only those matters which
   deviate from and are exceptions to these established plans.)
   c. To coordinate his activity with those of other departments not under his
   control.
   d. To insure that operations and services not under his control are brought to
   bear effectively on his operation.

4. To Promote Sound Human Relations

a. To build positive, productive human relations. To consider the dignity and well-being of the individual employee as a principal factor governing management decisions. To promote meaningful job contents and constructive relationships down the line. To treat subordinates as individuals. To promote a team spirit and feeling of loyalty and "belonging" on the part of employees. To maintain effective two-way communication with the subordinates and superior.

b. To direct the selection, hiring, placing, evaluating, transferring, promoting, disciplining, training, developing, and discharging of personnel, with emphasis on proper placement and development of employees.

C. REVIEW:

1. To Review Performance

a. To review and evaluate continuously the performance of his function, and to make recommendations for any action whose need arises from this review.

2. To Improve Performance

a. To develop a habit for continuous self-criticism; to improve his job; to seek, accept, and transmit suggestions; and to make suitable recommendations; and to coach and counsel his subordinates constructively in the performance of their work.

3. To Report Performance

a. To render a periodic account of progress to his superior; to insure that the benefits of the above reviews are "fed back" into the planning of the next phase of operations.

In general, position descriptions are intended to **define the mission** of the incumbent of a particular position. They specify what he is expected to accomplish and the functions he is to perform. They do not include the "how" of the job. The manner in which it is to be accomplished is left for procedural write-ups contained in the Standard Operating Procedures. Thus, by preceding the position descriptions with a common section and by limiting them to the mission of the job, the individual descriptions can be held relatively brief. An example of the specific position description of a **division manager** follows:

TITLE:

        Division Manager

OBJECTIVE:

        To operate the division within the objectives and policies specified by the Vice President for Operations; to attain an optimum profit at the highest possible quality level, and at competitive cost and delivery, consistent with good customer service and good human relations practice.

DUTIES AND RESPONSIBILITIES:

The following are in addition to the statement of "Duties and Responsibilities Common to All Management Positions" which constitutes a part of this position description:

1. To be fully responsible for the operations and, inasmuch as possible, for the profitability of his division within the limits of policies established by the Vice President for Operations.

    Note: All authority not retained specifically by top management (as specified in the Policy Manual, or in other written or verbal directives) shall be vested in Division Managers. Policies, standards, and procedures which are not stated specifically by the Policy Manual as mandatory, may be taken exception to by division. However, such policies, standards, and procedures, as developed by central groups or even by other divisions, shall be used to the fullest possible extent consistent with the objectives of the division.

2. To develop long- and short-term sales, engineering, manufacturing, financial, and industrial relations goals and policies for the division in consultation with the members of the Management Committee.
3. To develop, productize, manufacture, price, and sell the product line assigned to his division.
4. To conduct public, employee, customer, and vendor relations pertinent to the division; to represent the Company locally (if the division is removed geographically from the parent plant) and to participate in local affairs for the purpose of promoting the welfare of the division and its environment.
5. To review continuously the division's performance, with particular emphasis on its profit and loss position.

SPECIAL RELATIONSHIPS:
  1. **Other Division and Plant Managers**
     To cooperate with the managers of other divisions and plants of the Company on all matters of mutual interest and concern.
  2. **The Management Committee**
     To consult with the members of the Management Committee on all matters in which they might be of assistance by virtue of their specialized experience, and to utilize to the extent necessary the centralized, specialized services available. To bring to the attention of the Vice President for Operations only those problems which cannot be resolved satisfactorily at the above level, and to present to him, in such exceptional cases, complete recommendations for his decision.

The description of the duties, responsibilities, and membership of the **management committee** cited above, follows:

TITLE:
  Management Committee
OBJECTIVE:
  A. As a Committee: To assist the Executive Vice President in planning and appraising Corporate Group performance.
  B. As Individuals: To assist in preserving corporate know-how and experience and in bringing the best possible information relative to specialized skills to bear throughout all corporate groups and divisions. To make the experience of the whole corporation available to all groups and divisions.

DUTIES AND RESPONSIBILITIES:
A. **As a Committee:**
  1. To develop long-range goals and objectives for the Corporate Group and to recommend policies and programs for their implementation.
  2. To keep abreast of the Corporate Group's general competitive and profit positions.
  3. To appraise the performance of the Corporate Group and to make recommendations for improvement.
  4. To maintain balance among, and recommend growth rates of, the divisions of the Corporate Group.
  5. To recommend the location, size, nature, and product areas of all plants and facilities.
  6. To resolve policies relating to compensation and status.
  7. To resolve problems referred to it by Divisional Management.

B. **As Individuals:**
  1. To generate long-range objectives and programs within their individual fields of specialization and to coordinate programs of mutual interest among corporate groups and divisions.
  2. To monitor and audit divisional performance relative to their individual fields of specialization and to make recommendations for improvement.
  3. To offer consultation within their specialty.

4. To issue technical manuals within their field of specialization, if necessary and desirable.
5. To keep abreast of current developments within their field, both within the Corporation and outside, and to act as a clearinghouse in disseminating this information.
6. To operate specialized central services for the benefit of the Corporate Group and its divisions.
7. To undertake special studies and projects within their field of specialty.
8. To relieve the President, the Executive Vice-President, and the Management Committee, through consultation with Division Managers, of all decisions which can be resolved by virtue of their specialized individual know-how and experience and to bring to the attention of the Corporate Group's top management only those problems which cannot be resolved satisfactorily at this level; to present in such exceptional cases complete recommendations for decision.

MEMBERSHIP:
1. Executive Vice President (Chairman)
2. Vice President for Operations
3. Vice President for Sales
4. Vice President for Industrial Relations
5. Vice President and Treasurer
6. Vice President for Manufacturing
7. Vice President for Research and Development
8. Assistant to the Executive Vice President (Secretary)
9. President (Ex officio)

SPECIAL RELATIONSHIPS:
Close liaison with Patent Committee.

STANDARD OPERATING PROCEDURES. An organization is further implemented by what are known as standard operating procedures, which constitute write-ups of established procedures for the carrying out of various activities or the performance of certain kinds of work. Such write-ups constitute the "system" of the enterprise, but should not be allowed to degenerate into the class of unnecessarily complicated routines commonly called "red tape." It is readily seen that the organization chart tells "where" in the company a function is placed, the organization manual tells "what" the detailed nature of the function or position is and indicates its relationships to other functions and positions, while the standard operating procedures tell "how" the functions are to be carried on and the duties and responsibilities are to be discharged.

It is necessary to have such procedures written up not only for the guidance of employees performing the work—so that it is all done according to instructions—and the training of new employees, but also because many of the important procedures cover several sections and departments some of which are feeders-in of information that others must then compile and use in carrying on their work. Thus the procedure of handling time tickets covers not only getting the workers' time on jobs but also the checking off of completed work from production schedules, the making up of the payroll to pay workers, the charging of time to jobs or kinds of product to get current costs and record data for future estimating, and the posting of entries in the general accounts. In addition these time tickets form the basis for social security payroll deductions and taxation and for reports to workers and to the government on earnings and withholdings. The time tickets are only one of hundreds of items for which standard procedures are not only helpful but also imperative.

The various divisions and departments may have **handbooks** in which these standard procedures are kept on file for frequent reference, copies being placed in the hands of executives and supervisors who are concerned with phases of such standard practices. Like charts and manuals, these write-ups must also be frequently checked and revised to keep them up-to-date.

Analysis of Organization and Procedures

REVIEW OF POLICY AND ORGANIZATION. The production organization of a plant employing a large number of workers is complex and made up of many individuals, each with numerous duties. Because of this complexity, continual changes in projects, methods, and procedures, and the developments and evaluations which come about with passage of time, it is easy for an organization to depart from the polices, authorities, responsibilities, and duties as they have been established.

The need for organization change generally arises from two major sources (1) dynamic nature of industrial organizations and (2) inefficiencies of existing structures.

Dynamic Nature of Industrial Organizations. An industrial organization may find it necessary, because of **competitive and environmental** reasons, to change the emphasis given certain products, activities, or territories. **Personnel** changes, particularly at the top management level, may lead to organization changes since personal abilities, attitudes toward organization, and strengths and weaknesses vary among executives, therefore leading to modifications in organization. **Technological** changes also may affect the production processes and production planning and control techniques, thereby influencing organization make-up.

Inefficiencies in Existing Organization. Some of the common deficiencies or weaknesses in existing organizations are as follows:

1. Lack of clear-cut **objectives** and inadequate long-range **planning** and research for new ideas.
2. **Slowness** in decision-making and implementation of decisions due to too many channels, too much red tape, and poor communications in general.
3. **Unqualified executives** resulting in turnover, excessive absences, under utilization, and general dissatisfaction of qualified executives.
4. Interdepartmental and **personality** problems, especially between line and staff.
5. **Excessive span of management** leading to lack of knowledge of results, poor planning and control, and ineffective coordination.
6. **Inefficient committees** which tie up valuable time of executives.
7. **Production breakdowns** and failure to meet deliveries and schedules due to shortages of materials and labor, poor production planning and control, and ineffective engineering.

ORGANIZATION ANALYSIS. An organization, and particularly its procedures, whether existing or in process of being designed, can be analyzed in detail. The aim is to simplify and standardize duties, activities, methods, and procedures. As a result conflict and friction, wasted time and energy, work performance times, and costs can be reduced. The general framework for organization analysis and change are as follows:

1. **Assigning responsibility** and providing authority for conducting the analysis, and informing the affected areas, in order to generate cooperation.
2. **Analyzing the existing organization** using defects, breakdowns, or change in purpose or objectives as the points of entry for the study.
3. **Designing the new organization** from the standpoint of what would be ideal. The ideal structure serves as a type of model or organization objective.
4. **Modifying the ideal structure,** recognizing the limitations of the available

human resources. Modification of the ideal structure should generally be considered as a temporary or short-run solution. The ideal structure should continue to be the objective as changes in the human resource capability occurs.

5. **Implementing the new organization,** with special attention to timing, acceptance, and need.

The analytical procedures are similar to those used in methods analysis or work simplification (see sections on Motion and Methods Study and Work Simplification). These are summarized as follows:

1. **Selecting the activity** to be improved.
2. **Securing and recording facts.** Recording all details of the activity as presently practiced with process charts (see section on Process Charts).
3. **Questioning** each and every detail. Using the "five prompters":
   a. What is done? Why?
   b. Where is it done? Why?
   c. When is it done? Why?
   d. Who does it? Why?
   e. How is it done? Why?
4. **Developing new procedure.** This requires using the four principal tools of methods improvement: elimination, combination, change in sequence, and simplification.
5. **Applying new procedure to the activity.** It is human nature to resent criticism and to resist change, therefore "selling" the new method to those who must work with it becomes of paramount importance in the improvement of an activity. Before any new procedure can be made to work, it must be accepted and understood by those whom it affects. Participation is a means to cooperation and therefore it is most desirable that those who must live with the new procedure be consulted in its birth. Those who understood the reason and need for a change and have been able able to participate in the development of a new procedure will have a paternalistic feeling toward its success. A change in procedure by fiat has little chance of survival.

Human Factors in Organization Change. Since the organization structure is a reflection of the formal interrelationships of its members, organization analysis and change involves **values** for both the organization and its members. The organization, to be effective, must provide for the achievement of (1) organizational purpose, (2) individual self-maintenance and growth, and (3) social satisfactions. An awareness of these different values and goals permits a more careful analysis and design of organizations.

Management Audit. The review of policy and organization, in its broadest context, is the function of the management audit. Someone in the organization (generally a staff member) should be assigned the duty of continually watching and observing the objectives, policies, organization, procedures, and working relationships that exist in an organization, to determine and report deficiencies to top management. This **function** of the management audit is no different in principle than the inspection of product. In the latter case the inspector determines whether or not the product as produced conforms to the specifications; in the former case, the reviewer determines the degree of conformance of operation with established policies, delegation of authority and responsibility, and allotment of duties as originally laid out.

The management audit concept is a **diagnostic technique,** widely used in industrial firms; however, the details of an audit program vary. Typically the audit is a top management device for reviewing managerial and organizational performance. Voich and Wren (Principles of Management) present the following conceptual framework of the management audit activity:

1. **Economic considerations** relating to the criteria of profitability, efficiency, productivity, and market position.

2. **Organizational considerations** relating to how to best coordinate the members of the organization in order to utilize the non-human resources and to supervise the other human resources.
3. **Behavioral considerations** relating to techniques of motivating the human resource.
4. **Procedural considerations** relating to the mix of policies, procedures, and the general area of data management. The emphasis is on the organization's system of communications and the methods of evaluating performance.

An example of a **framework for the management audit concept** applied to the production operator is as follows:

1. What has been the growth in productivity per production employee?
2. Are the firm's production costs competitive with the industry?
3. Do production executives operate on the floor or from behind desks?
4. What employee grievance procedures exist?
5. What labor disputes have occurred recently?
6. How important is engineering to overall production?
7. What procedures are followed in job performance evaluation?

The benefits of the management audit concept are summarized by Voich and Wren (Principles of Management) as follows:

1. It permits a more objective, complete, and regular evaluation of the total organization and operations.
2. It permits management to reconsider the basic problems of what, how, and for whom to produce, in terms of short-run and long-run changes in external and internal conditions.
3. It identifies potential major problem areas before serious malfunctions occur.
4. It provides a mechanism for continually updating the total structure and operations.

EFFECT OF ORGANIZATION ON OPERATING RESULTS. Fixing definite tasks and responsibilities in an organization is a requirement for satisfactory production control. Such determination likewise is a prerequisite to the personal efficiency of every industrial executive, supervisor, and foreman. Only by removing all uncertainties and conflicts of responsibility and authority can a smooth-running organization be set up. Once established and maintained it must constantly be renewed to operate with a minimum of executive effort and to yield the lowest obtainable manufacturing costs. In this way the most can be made of both human and physical resources. The former include the personalities, strong points, training, and experience of every member. The latter comprehend the practical work of bringing together and directing the use of materials, tools, machines, equipment, working space, power, and all other physical agencies of production. Thus organization affects every activity in industrial operation, and is a powerful aid in obtaining economic results.

The Systems Approach in Industry

DEFINING THE SYSTEM. A system can be defined as a set of objectives or entities that possess an essential interrelationship and, due to their arrangement or assembly, achieve a unique purpose or combination of purposes.

Nadler (Work Design) uses a definition that he feels is more prescriptive, universally applicable to every conceivable situation, and understandable by all: A system of any size or type is the specified and organized conditions for the elements of function, inputs, outputs, sequence, environment, physical catalysts, and human agents detailed for each element in physical, rate, control, and state dimensions.

A system, in Nadler's view, is not totally independent, even though there is an implication of wholeness in the concept of the system.

Thus the idea of a system is closely related to that of the organizations which may be comprised of many systems or actually comprise a system itself. For example, a clock is an assembly of parts (an organization) designed to keep time. The assemblage constitutes a system. The General Motors Corporation is a large organization that uses many systems to take raw materials and turn out automobiles (among many other things). One of the characteristics of an effective industrial organization is its ability to function as an interlocked entity, a system.

Total Production Function. When the general definition of a system is applied to production, a **production system** becomes the combination of men, materials, machines, money, and time required to manufacture products.

Within this definition, the exact nature of the interrelationships is not explicit. It is necessary to define a series of **subsystems** to illustrate the mechanism that makes the production system work at high efficiency. Figure 19 illustrates the

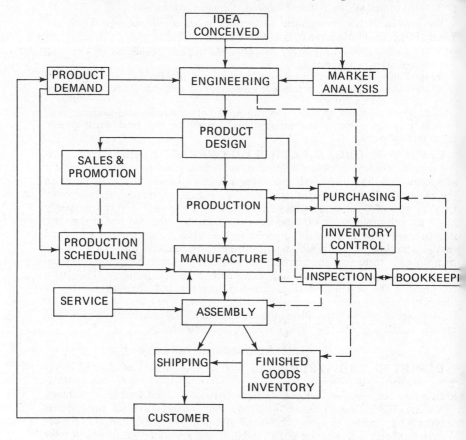

Fig. 19.   The systems approach.

basic systems approach. The various subsystems which represent the different areas of responsibility are shown with their interactions. In studying the production operation, note that it is also necessary to include the effects from marketing, sales engineering, and the customer both from the demand and the service standpoint.

SYSTEMS VERSUS TRADITIONAL APPROACH. The systems engineering approach differs from the traditional engineering approach to the extent that a global look at the system and its goals is objectively made before subsystems are studied in more detail. A traditional approach is more of a "building block" design from components to assembled system; interactions are recognized but subsystem optimization often precedes system optimization and goals are therefore suboptimized.

In a systems engineering approach, the first step is to identify the nature, extent, and purpose of the system being studied and the outputs sought. After this evaluation is completed, modifications of individual components and subsystems can be considered, each interaction being evaluated to determine its effect on all components of the system and the ultimate output of the system. Problems to which the system engineering approach seem to be well suited include those where large volumes of facts or data must be collected, organized, and utilized, and where there are obvious interactions between subsystems.

Systems engineering includes **cybernetics**, the science of information flow. In fact, Norbert Wiener (Cybernetics) states that all systems are information systems and that systems can best be understood by focusing attention on the information inputs and outputs. The system is thus conceived of as an information processor, with the energy and material conversions playing subsidiary roles.

Systems are also affected by environment. This includes more than the natural environment (temperature, pressure, heat, etc.); it includes also logistical and management policies under which a system must operate. A system which may function well within one environment may not function well in another. The well-designed production system will self-adjust to a variety of environmental changes —natural, political, economical, and managerial.

SYSTEMS DESIGN. The development of a system necessitates a logical, step-by-step design procedure if the resulting system is to be efficient and productive. The design process is commonly termed **systems engineering,** or alternatively systems analysis or systems design. In its application to production or man-productive systems, systems engineering begins with the development and specification of selected management objectives.

Management Objectives. One objective of systems engineering is to provide management with a better "feel" for what is going on when a system is in an operational mode. A second objective is to provide selected types of information to management, relating past decisions to the results obtained, which will aid management in making better decisions for the future. Finally, an important management objective of systems engineering is to make optimal time-use of resources. Decisions must be made as to when and where an industry's time and dollars are to be invested, decisions which may be improved upon if management has better organized information on the operational systems of the enterprise.

SYSTEMS MODELING. A logical and often quantitative approach is taken in modeling the real-world system, so that the system's response to external and internal changes can be better understood. Such information is invaluable to management when faced with decisions affecting resource allocation or program direction. The model may be a physical, mathematical, graphical, or analog representation of the real-world system. Modeling of the system and its decision processes requires a thorough understanding of the **transfer processes** which will exist in the real-world system when it is made operational. This understanding can be achieved through a critical analysis of the types of data needed, the data sources, and the possible methods for data transmission and reduction.

Information Requirements. Information is needed to determine what possible future states should be considered within the decision model and the probability of each of these future states occurring, what alternatives relative to systems configuration or subsystem design are most likely to achieve criteria established for the system, and what utility or value to the organization is likely to accrue for each alternative given when any one of the possible futures happens. An effective information system can also spark new areas for inquiry and research and can guide program management in the selection of logical sequences for program development.

Computer Simulation. The digital computer offers a valuable tool for simulating all or part of the system relative to the flow transmission processes inherent in the system. Simulation techniques provide the advantages of **time condensation** in evaluating response to input changes and in the establishment of unknown systems characteristics through input–output analysis. The simulation may also permit a comprehensive evaluation of various alternatives for combining or for improving on system components.

In a simulation, the model of a real-world system is allowed to perform over a period of time with selected input and control parameters. Outputs are measured, and input–output comparisons are made. The systems analyst must then decide whether the model is performing as the real-world system might be expected to perform. If he is not satisfied with the simulation results, the model may be modified and another simulation run. Thus an iterative process between modeling and simulation allows the analyst to develop a model which, in his opinion, best represents the real-world system.

THE PRODUCTION SYSTEM. The productive system has three basic ingredients: **utility** (the output is useful); **interactions** among components (several different components working together, functioning together); and **control.**

Utility. The usefulness of the output of a productive system is sometimes incapable of being defined objectively. To environmentalists the automobile's utility may be subject to debate. When the output of one system is used to destroy another system, utility becomes a matter of interpretation depending on the side on which one finds himself.

Interactions Among Components. One of the most difficult parts of systems design is to make the various components in the system interact in the most efficient way and to take advantage of these interactions to build the strongest possible output system. A term often used to describe this is **synogism:** if two components are put together so that they contribute to each other, the productive efficiency of the two as a system may be far greater than either one operating by itself. An important objective of good systems design is the synogism of two or more subsystems into a combination which enhances the usefulness of the individual components beyond that realizable if they were acting independently.

Most productive systems in existence are quite inefficient and provide ample opportunity for improvement. Consider the average American-made automobile which utilizes 200 horsepower, and weighs one and a half tons, to move a 165-pound man from one location to another. Much of the weight and horsepower are a result of competitive design rather than of transportation requirements. In his book, *Unsafe at Any Speed*, Ralph Nader correctly notes that one of the problems in modern-day automotive design is that the goals for transportation have been modified by engineers and managers over the years to a very different set of goals than were initially conceived by automobile pioneers.

Control. An important ingredient of a well-designed production system is control. Consider the control elements present in a typical management information system. The **environment** includes customers, suppliers, creditors, stockholders, workmen, and government. The **inputs** include orders received, materials supplied, labor, power, and types of machinery, and tools that are available, buildings, space, etc. The **outputs** are parts, products, service, scrap, and data or information. The **process** includes the functions of the system which convert form—taking in inputs and producing outputs. Processing, assembling, packaging, maintenance, marketing, design, sales, and the general office functions are all subsets of the process by which form is changed.

In general, the control actions available to management include advertising, sales promotion, redesign, modification, inspection, acquisition, dispersal, and/or liquidation. Optimum control action is selected and the system is designed to automatically adapt to an optimizing procedure through the application of operations research methodology in combination with control technology.

OPERATIONS RESEARCH AND SYSTEMS ENGINEERING. Operations research provides procedures for maximizing or minimizing a set of objective functions under environmental constraints within which a system must operate. Operations research methodologies permit the representation of real-world systems by mathematical or analog models, thus permitting decisions to be made from an interpretation of the model which will be useful in the real world. The future conditions under which a system must operate are often defined as future states. If it is assumed that there is only one future state of interest, usually because it is most likely to occur, then the decision problem may be interpreted as one under **certainty.** It is assumed that this state is going to occur with probability one, and the choice of an alternative can then be expressed by a simple mathematical equation. For instance, in determining the number of machines to put into a production line, if the production standards are known, if the available hours in which the machine has to operate over a period are known, and if the quantity to be produced is known and fixed, then the exact number of machines needed can be calculated. Of course, such ideal conditions seldom, if ever, exist in real life. Treating a decision as one under certainty, using such devices as averages, other possible future states are discounted when making the decision (selecting an alternative) on the basis of highest value. The averages are considered representative of long-term results.

Decision models may also be developed for conditions of risk or **uncertainty.** In both cases, several future states are recognized. If a reasonable probability of occurrence can be ascertained for each future state, the model is considered a risk model. If no reasonable probabilities can be associated with the future states, the decision must be made under uncertainty. The development of probabilities for decision modeling under risk usually suggests some type of sampling procedure as a basis for trend prediction. It is a difficult part of the systems engineering design function, for it involves an interpretation of future events based on past events. A good understanding of statistical estimation and prediction is essential.

TYPES OF PRODUCTION SYSTEMS. Production systems can be classified by the **flow characteristics** of the materials and products going through the system. In **serial production,** the product passes along an assembly line from one operation to another so that at the end of the line it is a completed unit. **Parallel systems,** usually used for smaller production runs, are less structured, less automatic, less dependent upon past and future operations on the part. A parallel system can be pictured as a number of craftsmen seated side by side, each producing the same product. Naturally this is an oversimplification since many parallel systems are much more sophisticated. There are numerous situations involving both serial and parallel systems.

Choice of a System. In considering a serial system or assembly line versus parallel production, a number of factors must be included: size of order, complexity of job, demand rate, production rate, relative costs, and whether the work is principally labor or machine intensive. Ultimately, it is a cost comparison that dictates which system is chosen, and the key is **relevant costs.** They are defined as those costs which will be incurred or avoided depending on which decision is made. For example, if the basic assembly line structure exists, then the cost associated with it is not relevant. However, if there is no existing assembly line, the cost of setting up the line is relevant as it would not be incurred if a parallel system were chosen. Likewise, if a piece of equipment would be used by both, then the cost would not be relevant to the decision.

The size of the production run affects the choice since small production runs do not justify setting up a production line. The more complex the job, the more likely a production line should be used. If the demand rate is about equal to the production rate, an assembly line is suggested, particularly if the rates are high. If the demand rate is considerably lower than the production rate, the size of production run to justify the setup cost may be completely overshadowed by the inventory holding cost. Here the better choice may be to set up one or more workers making the product. In series systems involving a product composed of many parts, it is almost inevitable that some parallel as well as series systems will feed the assembly line.

Effects of Learning Rate on Systems. The dominant factor that gave rise to the use of assembly lines was the **learning rate effect.** While implicitly recognized by Henry Ford, it was not until about thirty years later, in the aircraft industry, that this concept became well defined and the **learning curve** was born. It was observed that the labor effort required to produce an aircraft decreased with each subsequent unit. This effect was attributed to learning on the part of workers. While this is a major cause, it is by no means the only one. As a product enters production, the production facilities are usually adequate, although later improvements may be made that will increase production rates. Also, design changes are introduced to make the product more reliable and easier to produce. While these are all in some sense part of the process, they are distinct from the learning associated with repetitive operations.

Perhaps the most useful aspect of the learning rate is that it can be qualified. Based on observations, the time required to produce a product is found to decrease by a given percentage each time the total production doubles. For example, if the process is assumed to have an 80 percent learning rate and the first unit required 1,000 man hours, then the second unit (double production) would require 800 hours and the fourth unit (doubled again) would require 640 hours. By the time production reached the 1,000th unit, the production time would be approximately 110 man hours. The usefulness of this concept is that it is possible to determine the average time to complete a production run. The critical factor is the learning rate. Eighty percent is a very favorable rate more indicative of simple tasks. As the job becomes more complex, the rate of decrease in time decreases. A 100 percent rate would imply no learning and represents an extreme bound.

As may be expected, the learning rate for serial and parallel lines will be different, with the serial rate almost always lower. It is this fact that makes assembly lines so attractive. When the process for making a product consists of many operations, the learning factor for a single worker performing all operations will be high. If the process is broken into sets of tasks and each set is performed by a single worker, learning is more rapid and the combined learning factor will be lower for the line.

The number of workers required for an assembly line is determined from the design of the line; the number of workers required in a parallel system is governed by production and the demand rate. In each case the number of production hours for a run is predicted from a learning curve.

ESTABLISHING INITIAL CONTROL SYSTEMS. After the type of production system has been selected and designed, it is not sufficient just to start operation. Any system, especially a new one, must have effective controls.

The master budget represents a plan for operation since the production manager needs to know how well he is adhering to the plan or whether the basic plan itself is changing outside of his range of control (see section on Production Planning).

Feedback Systems. As with any mechanical or electrical system, control begins with feedback information. Since production levels are set from a sales forecast, it is necessary to determine whether the actual orders are as predicted. Inventory levels are often reliable indicators and so represent one of the best sources of feedback to management. Changes in inventory should, ideally, prompt changes in production levels. For optimum production, feedback information should indicate what has been produced, the status of jobs in process, and present or potential problem areas. There are several means of obtaining feedback. Among the more traditional are:

1. Inspection reports
2. Control charts
3. Variance reports
4. Finished product inventory
5. In-process inventory

Rebalancing Operations To Minimize Delays. After a new facility has been in operation for a time, it is usually necessary to "retune" the system. Initial estimates of production and learning rates are often inaccurate. Extra operations may have been added to overcome unforeseen difficulties. The net effect can be an uneven flow of material and finished products. The first step in rebalancing the system is to update and prepare new **flow diagrams** for all subsystems. Then, from actual production data, the flow rates through each "station" in the diagram are determined along with some measure of its variability (either the statistical variance or, at least, the possible range). The flow through the subsystem will be governed by the slowest process rate. If the difference between the slowest and fastest process stations is not too great, then there is little need for rebalancing. Where it is necessary, parallel stations can be added for slow operations, tasks can be reassigned to reduce the work load at slow stations and to increase the load on fast ones, and if necessary the equipment or product can be redesigned.

Systems Applications

LARGE HIGH-VOLUME PRODUCERS. This type of operation is characterized by long production runs and well-established manufacturing processes. A **sequential process flow** can be established. This may take the form of a continuous assembly line or a sequence of work stations without continuous flow. The latter may involve moving partially finished products in batches from one work station to the next. The objective in such systems should be to establish procedures that require little control.

For an assembly line to operate efficiently, the work level must be reasonably balanced along the line. The **basic steps in line balancing** can be summarized as follows:

1. **Determine standard times** for all operations including all of the tasks at each work station and not just the totality of tasks done at the station. This is necessary in order to allow for possible transfer of tasks from one operation to another.
2. **Establish any precedence relationships** that exist to determine if certain tasks must be performed prior to others. The network methods of CPM are useful for this purpose.
3. Starting with the finished product, **re-establish the operations** to even out the time or production rate for an operation. There are a number of heuristic methods to accomplish the line balance, mostly in the realm of trial and error.

Line balancing requires that consideration be given to the following factors.

Inventory Control and Scheduling. For a production line to operate efficiently, it is necessary to supply parts to each work station at the proper rate. A balance is necessary between the amount of inventory on hand at each station and the probability of having to shut down an operation due to lack of parts.

In-Process Parts. Since standard times are used to establish the balanced line, parts should be supplied to match the calculated production rate. For example, if an operation normally requires five minutes per subassembly, and uses 5 units of part X and 7 units of part Y in the assembly, 12 subassemblies are made per hour, using 60 units of part X and 84 units of part Y. Unfortunately, the standard time of an operation is only an average and the actual variation may be from four to ten minutes; hence, a larger bank of parts must be provided. Since the output of one station becomes the input for the next, the variation in production rate may alternately flood or starve the following station.

Some industries provide large banks of materials at each station. When the items are small and of low value, this may be an appropriate solution but there is a cost associated with large banks. They require floor space and money is tied up in in-process inventory. The cost of maintaining an adequate bank must be weighed against the cost of lost production if the system is shut down due to a lack of parts.

SMALL-VOLUME MANUFACTURING. Where small production runs are the rule, it may be advantageous to set up production lines if conversion from product to product is not too difficult. It may be desirable to schedule larger production runs and thus warehouse finished products against future orders if, as has been discussed, the warehousing costs are not overly burdensome to the price structure.

Job Shop Operations. The ability of a plant to produce miscellaneous quantities of various products on the same equipment, and at irregular intervals, presents scheduling difficulties and often makes economic production difficult. There may be long delays in processing and equipment may be idle for long periods.

A number of studies (Eilon, Elements of Production Planning and Control, and Holt et al., Planning Production, Inventories, and Work Force) have been made of scheduling rules for job shop operations. Attempts to obtain optimum solutions are impractical except for problems involving a few machines, operations, or jobs. Machine loading or scheduling is used for more complex applications and the methods employed generally consist of one of the following rules:

1. Select the job with the shortest operation time (SS rule).
2. Select the job with the longest operation time (SL rule).
3. Select the job in some random order (SIRO rule).
4. Select the job by first-in, first-out (FIFO rule).
5. SS rule with priority given to jobs that have waited a certain time.
6. SL rule with priority times.
7. SIRO rule with priority times.

Eilon (Elements of Production Planning and Control) has evaluated these rules by using computer simulation. The measures of performance were (1) through-put time or elapsed time in the shop, (2) delay factor or the ratio of through-put time to standard machine time, (3) total idle time per machine, and (4) number of jobs completed per unit time. An important parameter in such a problem is the **load factor,** which is defined as the ratio of total machine time to complete a batch of operations to the total machine time available. As this factor approaches 1.0, the system becomes saturated and excessive delays build up. As the factor exceeds 0.75, there is a sharp increase in average through-put and in the delay factor. In a comparison study, the results obtained by Eilon indicate that the rules can be ranked in the following order.

The SIRO rule, FIFO rule, and SS rule with priority given to jobs that have waited beyond a certain time give about equally good evaluations; the SL rule is next best, and surprisingly the SS rule consistently gives the poorest results. This is contrary to intuition and there is no obvious explanation for this result, except that a job may receive good service on one machine where its operating time is low, and then may be delayed a long time at the next sequence.

The important aspect from the viewpoint of production schedules is an awareness of the load factor and the need to find extra machine capacity should this factor exceed 0.75.

HIGH-VOLUME, SMALL PARTS PRODUCTION. The Aerosol Research Company provides a good **case study** of this type of production. This company manufactures the small valves used on aerosol spray cans. Aside from minor machining operations, manufacturing involves only the assembly of purchased components. Since the valves must be compatible with the material they are supposed to spray, a single standard valve is not possible. Aerosol Research in fact produces thousands of different valves and ships almost 2.5 million valves a day to producers or fillers of hundreds of different aerosol products.

The company's **customer service** and **research and development** departments help determine the proper valve needed for a particular product. If new equipment or modifications to existing machines are needed to produce a new type of valve, the company's **automation department,** implemented by a maintenance machine shop and a group of machine builders, does the work. All production equipment is designed by Aerosol Research engineers. Although all components are made by outside vendors, Aerosol Research owns and provides all the dies and molds required to produce the components.

All lots of incoming parts must pass a rigorous sampling inspection. Since sorting of the small parts used is too expensive, an entire lot that fails to pass inspection is returned to the supplier. When a lot passes inspection it is labeled as approved and sent to parts storage. Only parts with approval labels can be released to production. Parts are withdrawn from storage on a first-in, first-out basis. Parts are identified with color coded tags to indicate the month they were received; thus one month's supply of parts must be depleted before the following month's parts can be withdrawn. The **quality control department** rates vendors on a month-by-month system. Further quality control is maintained by withdrawing an entire lot from production if production problems appear to be developing due to defective parts. The troublesome lot is then reinspected.

Manufacturing Process. Valves are assembled automatically on production lines which consist of a sequence of carefully integrated and joined units that perform all operations, including testing. An operator services each line to observe the performance of the units and watch a control board for any signs of a malfunction on the line. The operator also restocks the various input bins to the line and removes cartons of finished products. Assemblies are inspected automatically without interrupting production. Faulty assemblies are ejected from the assembly machines automatically. Machines are designed to err on the safe side, that is, they will tend to reject good assemblies before accepting bad ones. The control panels continuously indicate the status of each part of the line. If a defect is found, a yellow light goes on; if four consecutive defects of the same type occur, the line shuts itself off and a red light goes on. The lights also pinpoint the area of the line that is having trouble. Automatic counters on each line record the number of good valves as well as the number of defective valves by types of defect. The machines are programmed to shut off when the required number of good valves in a particular order is reached. The machine will also shut off automatically if any of the component parts runs short.

For some very small orders semi-automatic machines are used requiring some hand operations.

Order Records and Control. The status of each order is easily determined because of the highly automated nature of production. **Inventory control** is able to determine the status of parts and finished product instantly.

Since a given type of valve may be produced during a particular run for more than one order, a simple method is used to control order quantities. Boxes and box labels are preprinted for the number of boxes required to fill the order. When a box is finished, it is labeled; when the operator runs out of labels he knows that the order quantity has been reached and begins on another order following the same labeling procedure. Together with the automatic counters on the machines themselves this provides for a foolproof method of preventing shortages or overages.

CONTINUOUS PRODUCTION. A manufacturer of electrical wire provides an example of production on a continuous basis resembling, in some ways, a process system. The wire plant contains more than 150,000 sq. ft. of space and employs about 300 people.

Each step in the production of wire is directly related to the steps that precede it. Since each run of wire remains an unbroken continuous link throughout the processing stages, unique "line balancing" problems must be solved before a successful run can be made. The only steps in the manufacturing process that do not involve the wire itself are those involving the various materials used as the outer insulation for the wire. The basic raw material used in the production of wire is electrolytic tough-pitch copper rod, a nominal 5⁄16 in. diameter, purchased in 250 lb. coils.

The prospective vendor of the copper must submit samples for examination by the company's **quality control** and **engineering** departments before the material is purchased. Purchased material is given a receiving inspection and subjected to other in-process inspections throughout the various manufacturing stages. Finished wire is grouped into lots varying in length from 10,000 ft. to several million feet. All lots are given a final continuity test. The test, performed during a respooling operation, checks the number of strands, diameter, color, etc. Other special quality control tests are made at this point for certain types of wire.

Manufacturing Process. Most of the operations involved in wire making are automatic and continuous. Material handling is necessary when transporting spools of wire from one process to another. The principal metalworking process involved in wire manufacture is drawing. From its initial 5⁄16 in. diameter, the copper rod is drawn down to 8 gauge (slightly more than ⅛ in.) through cemented tungsten carbide dies. Further draws can bring the wire down to as small as 38 gauge (about 0.004 in.). Additional processing involves tinning, stranding, insulating, coating, extruding plastic insulation, taping, and braiding. Most finished wire is wound on disposable wooden spools while some wire is shipped coiled in drums. Packaging specifications are given to the shipping department by the engineering department.

Although the wire undergoes numerous changes in diameter as it proceeds through the manufacturing process the various winding and takeup speeds are adjusted to prevent abnormal snagging and pulling each step of the way. **Production** and **inventory control** are key elements in the production systems since machine loading constitutes a major scheduling problem. The wide range of wire sizes, stranding, and insulation types requires careful attention to in-process inventories.

The final product appears to be rather simple, yet it can present complex production problems requiring elaborate and highly sophisticated techniques and machines.

MULTI-PRODUCT, MULTI-PLANT OPERATIONS. The Delco-Remy division of General Motors has 11 manufacturing plants, 145 manufacturing departments and thousands of products. Some items are produced at the rate of a few a month while others are produced at the rate of hundreds of thousands per day. With so vast an organization and so varied an output, one of the key elements in the production system is **inventory control.** At Delco-Remy **production control, material control,** and **purchasing** are grouped under a single Director of Purchasing and Production Control as shown in Fig. 20. A central data processing section uses computers to develop production schedules as well as to assist the production control, material control, and purchasing functions of the company. These operations center about three data processing systems: material control system, order status system, and a daily departmental scheduling system. The material control system keeps track of orders and inventories and includes information about the status of the manufacturing processes. The order status system keeps track of customer orders, as well as future requirements. Schedules and shipments of orders are recorded in this system. The purpose of the daily departmental scheduling system is to keep track of daily inventories and requirements and develop detailed production information for the manufacturing departments. A **daily report** is issued to

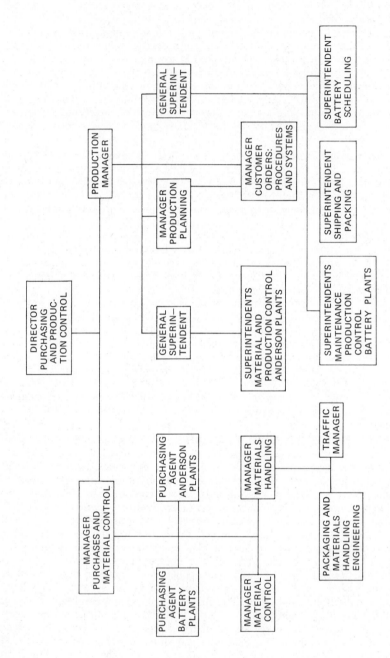

Fig. 20.   Organization chart for the purchasing and production control functions of a large multi-plant company.

each production foreman giving such information as: part number and name; total inventory of parts with a breakdown of parts assigned to the production schedule and parts available but not assigned, scrap totals, completed units, daily and weekly schedules for a 40-working-day period, etc.

Computer Applications. The computer spends more than half of its time servicing these production functions. The other part of its time is devoted to other data processing tasks such as accounting and payroll preparation.

Computer processing of production information relies heavily on accurate up-to-date reporting by all personnel involved. The input information to the computer includes production figures, scrap, inventory status, inspection reports, requisitions, move orders, stock and toolroom records, etc.

Production control begins with the issuance of a **bill of materials** developed from engineering drawings and routings prepared by the work standards group. The computer uses this information as well as data on lead times, inventory control, line balancing requirements, manpower, economical production lots, etc., to prepare the daily and weekly schedules. Material control and purchasing functions use the information to determine the timing and quantities of material purchases.

PRODUCTION PLANNING

CONTENTS

PRODUCTION PLANNING

Nature of Planning

DEFINITION. Production planning involves the organization of an **overall manufacturing system** to produce a product. Specifically, it consists of designing the output, determining equipment and capacity requirements, designing the physical layout and material handling system, determining the sequence of operations together with the proper methods of performance and time requirements, and specifying certain production quantity and quality levels. The **objective** of production planning is to provide a physical system together with a set of operating guidelines to efficiently convert raw materials, human skills, and other inputs into finished products. Production planning does not include the scheduling of specific orders and products to particular work centers and times; such decisions fall within the scope of production control and will be covered in the next section of the Handbook.

FACTORS DETERMINING PLANNING PROCEDURES. The particular procedures used in production planning vary from company to company and even within each individual firm. At one extreme production planning may begin with a **product idea** and plan the design of the product and the entire manufacturing system to produce it. At the other extreme the task may be to merely plan for the manufacturing of a slightly modified version of an existing product on existing facilities. The apparently wide difference between planning procedures in one company and another arises primarily from differences in the economic and technological conditions under which the firms operate, not from attaching different degrees of importance to the planning function.

Volume. The amount and intensity of production planning is determined primarily by the volume and character of operations and the nature of the manufacturing process. One may view production planning as an investment. Planning is undertaken in the expectation that it will reduce manufacturing costs. If only a few units of a product are to be produced as in a custom order job shop, it would be uneconomical to invest in an elaborate production planning effort. Planning may be limited to acquiring the parts and raw materials and determining which work centers have the capability of manufacturing the product. However, in a high volume operation, such as automobile manufacturing, thousands of hours of planning go into the design of both the product and the manufacturing process since even small improvements may result in large cost reductions when hundreds of thousands of units are produced.

Nature of Production Process. In the job shop the production planning may be all **informal and unrecorded.** The development of work methods and even design details are left to the individual workman who is usually highly skilled. In the high volume operations, teams of highly trained specialists may concentrate on even the minutest details. In the automobile industry, dozens of propulsion engineers, mechanical engineers, metallurgists, industrial designers, etc., spend full time on designing the product. Other process engineers, methods engineers, and equipment designers work continuously on planning the manufacturing process. Special purpose equipment is designed and built to perform particular specialized tasks.

Type of Operations. Repetitive operations are more conducive to detailed production planning than varied operations. In practice, there are many variants between these two extremes. These may be represented by continuous production of a single standardized product on the one hand and the custom-order business on the other. Some of the principal variants are:

1. **Manufacturing to order,** which may or may not be repeated at regular intervals. Examples: jobbing foundries, printing plants, bleaching and dyeing, jobbing and repair machine shop, manufacturers of locomotives, conveying machinery, large special machines·in general, and machine shops that contract for batches of products for other plants.
2. **Manufacturing for stock, under repetitive or mass production, with some choice between the type of process and assembly.** Examples: automobiles, watches, clocks, and typewriters. Custom orders may be intermingled with repetitive work, but this is not considered good practice if it can be avoided.
3. **Manufacturing for stock, where the product is made up of parts but the processes are not optional.** Examples: shoe manufacture, clothing, and other non-machine shop industries. Custom orders may be intermingled.
4. **Manufacturing for stock, under continuous process manufacturing.** Examples: chemical and food products, glass, soap, paper, and synthetic yarn.

The degree to which production planning is carried out varies with the nature of the process. It is the greatest in a highly automated system where a single homogeneous product is treated by a fixed sequence of operations in a continuous flow. Modern examples on a vast scale are in paper, oil, and chemical industries. Flow sheets in these industries exhibit a continuous stream of production in which many operations are performed, materials added, and by-products and wastes eliminated, but without a break in flow or exceptions in work or processes. Close planning is necessary to assure that the various component parts meet at the proper time and in the proper quantities to assure an uninterrupted flow. New specialized equipment may be required and developed for seemingly minor changes in product specifications. In contrast to the close planning of the product and process, very little production scheduling is required, since the **schedule is embodied in the equipment itself.** It is not even necessary to send route sheets and operations to the shop since the route and operations are the same day in and day out.

In contrast to the continuous process industries are the repetitive operations in plants making automobiles, typewriters, sewing machines, and similar complex mechanisms. Here a great variety of material is used in many ways and for many purposes. There are hundreds and even thousands of parts, on each of which one or many operations take place on a diversity of machines. Production planning requirements are less than in the continuous process case but are still substantial. It is necessary to design the products and plan the sequence and methods of operations. However, equipment is of a more general purpose nature and capacity levels are not as critical since the various stages of the production process are often disconnected in time through the use of inventories of semi-finished goods. When many parts or products follow similar sequences of operations, equipment location and material handling decisions are important. In **repetitive operations** the production scheduling decisions are complex since it is necessary to bring together in proper sequence and at the right time and place the results of numerous complex subactivities.

In custom manufacturing to order, less accurate production planning is possible than when manufacturing for stock. In many cases, however, it is possible to forecast probable business rather closely, based on past experience and known trade conditions. In some custom manufacturing, the magnitude of the differences of the products is much smaller than the magnitude of their similarities. In such cases substantial production planning occurs with flexibility designed into the manufacturing process to accommodate the differences from order to order. In other custom manufacturing situations, little similarity in products occurs from one order to the next. In a small foundry formal production planning might be limited to acquiring inputs of materials, equipment and manpower, and limited product design with the remainder of the planning left to the workman.

Benefits of Planning. In evaluating the costs and benefits of production planning it is necessary to recognize that, in every production organization, someone is performing the planning function. Whether it is done by a group of specialists or by the superintendents, foremen, and workmen is a matter for each organization to decide, after a consideration of the costs and benefits of each method. The volume and nature of the production process are predominant factors in influencing these costs and benefits.

The **benefits of effective production planning** are seen in the reduction of manufacturing costs. The costs of poor production planning can be summed up as poorly designed products, low rates of production, high costs, poor morale, and disappointed customers.

PRODUCTION PLANNING SYSTEM. The production planning system may be divided into two interrelated subsystems:

1. Product planning system.
2. Process planning system.

H. L. Timms (The Production Function in Business) illustrates the two subsystems as shown in Fig. 1. This diagram presents an idealized flow of decisions and information such as might occur for a completely new system which is created solely as a result of detecting some unfulfilled need in the marketplace. In most situations the work stations and other facilities already exist, and it is necessary to design products with characteristics which will allow them to be produced by existing equipment, perhaps with minor additions or modifications.

The product planning and process planning subsystems are closely interrelated. The indicated lines of **information feedback** illustrate the need to coordinate the systems. In considering available processes and technology in the process planning system it often becomes evident that it will be difficult or even impossible to produce according to original product specifications. This information is then passed back to the product planning function where adjustments are made in the product design.

Within each subsystem, a wide variety of decisions must be made. These too are interrelated. The work methods depend on the content of operations which in turn depend on flow of products and the capabilities of individual work stations. There exists no overall concise theory or science of designing optimal production systems. At the present time the task is simply too complex to permit solution by following a set of rules. There are, however, rules and tools for dealing with each of the individual segments of the production planning process. Thus a scientific approach may be taken in dealing with the segments. But, finding the most effective product and process design remains an art requiring the production manager's best judgment and intuition.

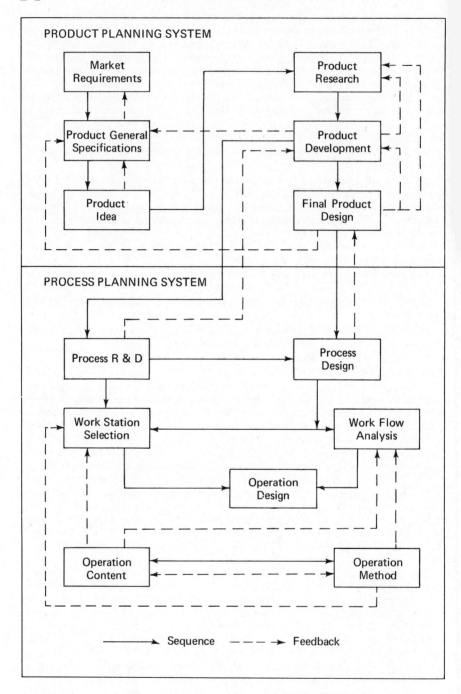

Fig. 1. Production planning system.

Master Planning Budget

IMPORTANCE OF BUDGET. Small production systems can operate on an informal "fill-orders-as-received" basis. For larger organizations it is necessary to use more sophisticated methods. The fundamental tool of the planner is the master planning budget. It is the basis for all planning, production scheduling, and control. Although the term **budget** is normally applied to financial matters, in its broader sense it applies to all types of planning. A budget is defined in master planning as an expression of a plan in quantitative terms. These quantitative terms may be items produced, number of workers required, amount of raw material, and other factors. Ultimately the quantities are expressed in dollar equivalents.

Systems Input—The Sales Budget. The sales budget or forecast is the first item presented in the master budget since it is a key factor in determining the production rate and the size of a business. Firms which produce a single product must estimate sales on a yearly, quarterly, or monthly basis. In general, an estimate of the per-unit sales price is necessary to prepare the total sales budget.

The sales budget represents a loose commitment to deliver the product in stated quantities within the planning periods. The expected sales volume and price can be used to estimate the revenue.

When considering more than one product, the master sales budget must consist of a set of single product budgets. Separate identity of each budget is necessary since they are used as a basis for material, labor, equipment, and cash budgets.

Manpower Budget by Skills. The **direct labor** required can be determined from a detailed knowledge of the process to be employed with the quantity to be produced. The direct labor usually is expressed in terms of number of workers and man-hours. A further breakdown separates workers as machinists, assemblers, process control operators, etc. In addition to direct labor there is support or **indirect labor,** the amount of which is usually established as a ratio to direct labor. For instance, there may be one group leader for every eight production workers, a ratio of 1:8. Other indirect labor includes machine setup, test and evaluation, inspection and quality control, material handling, and custodial services. The cost of some support activities remains fixed and does not change appreciably with increases or decreases in production output.

The manpower budget provides the scheduled demand for production skills according to customer needs. By using standard labor rates and overhead factors, the budget will provide an estimate of labor costs and a basis of cost variance analysis.

Facilities Budget. In most cases, the amount of labor can be varied. However, one of the purposes of the manpower budget is to smooth out the use of this resource. On the other hand, facilities, such as machinery, process equipment, and utilities, have a finite limit or capacity. Often, different products produced in the same plant compete for these facilities. Some machines or facilities are overloaded during certain periods and are unused during others. The facilities budget can be improved by using machine utilization tables and a machine scheduling plan, often referred to as a production control board (see section on Production Control).

Raw Material Budget. Once the production budget and schedule have been determined, the raw material budget should be prepared. The term raw material is extended to include every item or substance used in the production process that is not a product made in that process. What is considered a raw material in the production system of one plant may be the finished product of another plant. Thus in one plant, nuts, bolts, and transistors may be the product but in another they are the raw material. Such items may be produced by an outside source or by another division or department of the same organization. The usual distinction in the latter case is that if an item is procured by a purchase order or requisition, or if stock control records are maintained, or the item is placed in inventory, then it is considered a raw material.

In determining the raw material budget the following factors must be considered:

1. **Lead time on material.** By using the production budget, it is possible to determine when an item must be introduced into the production system. Not all raw material is "dumped into the hopper" at the beginning of the process. Normally, there is a continual flow of material. In the procurement of the material, sufficient time must be allowed for: (a) finding sources of supply, (b) determining prices and terms of delivery, (c) placing an order, (d) the supplier's production process (he may be "making to an order" rather than supplying "off the shelf"), (e) the transportation or delivery time, and (f) the receiving function (i.e., keeping of records, inspection, stock control). Some of these steps require little time in a continuous one-product operation but there is an inherent time necessary to procure material. Though problems can be alleviated by carrying a sufficient stock of raw material at all times, the determination of what is sufficient and the added cost of such a policy will usually affect profits. This factor is discussed in the section on Inventory Control.

2. **Proper inventory levels.** Raw material inventory must be adequate since a shortage can be costly. Excessive inventory ties up capital and space and may become obsolete if a product undergoes redesign or if the shelf life is short. If it were possible to obtain instant delivery, then orders would be placed only when an item was needed. However, there is a lead time on materials that may vary from hours to months. Some items ordered and procured on an "as needed" basis require considerable time, effort, and money due to the process of ordering. There are economies to be obtained by "batch" ordering policies. The timing of the order is another factor in the determination of proper inventory levels. If the items are small or inexpensive, then the cost of holding extra minimum stock would not be excessive and would be inexpensive insurance against a stock-out. If the items are large or expensive, a more refined policy would be in order. The general strategy should be to increase the minimum stock level to the point where the cost of holding an additional unit just balances the expected cost of a stock-out unit.

Other Items in the Master Budget. There are other sub-budgets included in the master budget which, while not directly concerned with the production process, do have an impact. They include **overhead sales expense** and **cash flow budgets.** The first of these includes indirect expenses which are incurred in the course of carrying out the operations. The sales budget reflects the cost of maintaining a sales office, salesmen's salaries and commissions, and advertising and other promotional activities. The cash flow budget can be considered an inventory budget of dollars and measures the timing and amounts of inflow and outflow of cash.

COORDINATING AND SCHEDULING. Even for the relatively simple situation of single-product production, by the time the manpower and facilities budgets are made, there is a good possibility of conflicts and constraints on the availability of resources. The master budget approach will expose such problems and, by using some of the techniques to be covered, the production schedule can be modified to improve performance. The manpower and facilities budgets also provide a means to coordinate the uses of those resources and to communicate the requirements to personnel and purchasing.

Product Design

ORGANIZATION OF THE DESIGN FUNCTION. Fundamentally the manufacturing activity results from the need and desire for a product in the marketplace. The flow of a design project starts from an initial stage of nebulous ideas and generalities and ends with a product spelled out in detail in the form of models, blueprints, charts, etc.

This product design process is executed within an organizational framework divided into a sequence of stages that subdivide the total design activity in time and responsibility. There is a basic sequence to the stages and the design project evolves from one stage to the next in progressive degrees of detail.

The organizational location of responsibility for each stage varies among firms. In some instances, **product design is a one-man project** in which the basic problem is simple but nonetheless implicitly flows through the stages, even though these stages may be recognizable only to the individual. In other cases the project may involve hundreds of people of various skills and specialties and responsibility for the stages may be rigidly assigned along strict organizational boundaries.

Within each stage the product designers execute a methodology involving various phases including the statement of problems, the analysis of problems, and the synthesis of alternative designs and their evaluation leading to design decisions. In the early stages, basic economics and marketing are important and the people involved are marketing managers and accountants, as well as engineers. In the later stages of design, responsibility will be centered in engineers, technicians, and line managers—the people who will have to implement the design and make it work.

Origination of Product Idea. The first stage in product design is the recognition of some unfulfilled want or need. Products are produced in order to satisfy human needs and wants either directly, as in consumer goods, or indirectly, as in capital goods or materials to be made into consumer goods. Man's needs for physical well-being are few. His wants are many and insatiable. While the wants are largely emotional and based on psychological drives, they are very real and provide the market for a majority of the products produced today.

A famous motto for success in business is "Find a need and fill it." Successful new products are those which effectively provide new methods of satisfying needs or newly recognized wants. Most successful products are the result of looking at man's wants and needs from a new viewpoint. The viewpoint may be the result of a technical invention, a new material or method, a new cultural level, a new awareness, or a new proportioning.

New product ideas are generated from a variety of sources. Figure 2 from a Booz, Allen & Hamilton report (Management of New Products) illustrates some common sources. Some come from salesmen in the field. Customers describe to them new uses for existing products and suggest improvements which will make the products more effective. Others come from the factory. As excess capacity appears in departments, management seeks out potential product ideas which will be compatible with existing facilities. Still other new products result from copying or improving on a competitor's new development. Most of these "new" products are simply revisions or improvements on existing products. These re-

Functional Area	Number of Ideas (%)
Marketing	32.4
Research and Development	26.5
Top Management	13.1
New Product Department	6.7
Management	3.7
Others	17.6

Fig. 2. Sources of new product ideas.

vised and new products are essential if a firm is to survive. Few existing products are immune to obsolescence.

Determining Market Potential. After a product idea has been generated, the next stage in the design process is to determine the potential market. Actually there will probably be a series of market oriented tests in a complete design process. At this point in time, however, the decision is whether or not to commit resources to the design process. This in turn depends on the need for the product. The validity and strength of the need is established by **market research and analysis.** They provide information about potential demand. In addition, they begin to provide the basic product characteristics which will be desired by customers.

Often new products are developed because of some technological breakthrough and discovery of new materials, devices, or methods. In many cases the products are designed to take advantage of the technology without the economic need being investigated. The new discovery might act as a trigger for an economic study, but it cannot be a substitute for it.

In some cases wants can be **stimulated through advertising.** This has been effectively done in the cosmetic industry. Thus, even if no market currently exists, the product itself and the promotion of it may create its own market.

RESEARCH AND DEVELOPMENT. With a potential market identified and some general product characteristics determined, the product design now proceeds to the research and development stage. This stage may be divided into three substages of (1) product research, (2) product development, and (3) final design. While precise dividing lines between the substages do not exist in practice, it is convenient to separate them for discussion purposes.

Product Research. Research undertaken by industrial firms is either pure or applied. **Pure research** or fundamental research is aimed towards finding knowledge for its own sake. It is not undertaken to solve a problem associated with the design of a particular product. It provides the basis or foundation for applied research by unlocking the secrets of nature and developing new knowledge. Most companies do not do sufficient business to afford any substantial amount of pure research since the costs are high, the risks substantial, and the payoff uncertain.

Many firms can economically afford applied research. **Applied research** is research designed to solve a particular set of problems which will allow production of a specific product. The payoff on applied research is quicker and more certain than on pure research. While the results of pure research may or may not eventually lead to a profitable product idea, applied research is directed to solving particular problems in designing specific products. Hence

:here are fewer blind alleys and costly false starts than in pure research. Still, applied research can be quite costly and a large volume of research work may be necessary in order to justify maintaining a staff of scientists and technicians and specialized research equipment. Some firms find it economical to contract out their applied research to specialized research organizations. Other small volume firms find they are unable to afford any applied research and are content to let larger competitors develop new products and play the role of a follower rather than an industry leader.

Since research is designed to discover new knowledge, the best starting point for new laboratory efforts is from the limit of existing knowledge. A good deal of **library research** should be undertaken before the physical research is even designed. This can often produce the desired information without requiring any physical research.

The scientific and technical journals are filled with detailed reports of other research projects. The pace of growth of technical publications has been such that it is almost impossible for researchers to keep abreast of even very specialized fields. Thus, one of the primary research needs is an effective method of bringing to the attention of researchers all the published reports that relate to their work.

This need is met in a number of ways. Some research staffs employ a research librarian whose task is to circulate relevant articles from current publications to the appropriate members of the research staffs. Large-scale computerized **information retrieval systems** are also used. In such systems all new articles from some list of sources are categorized by taking certain key words from the article titles. These categorizations are then inserted into the computer system. A researcher desiring all the relevant articles pertaining to his current research project describes the project by a set of key words. These words are then matched with the categorization in the computer file and a list of relevant articles printed out. The cost of developing and maintaining such a system is high but the cost of using it is low. Thus few individual organizations can afford their own information retrieval systems. Several large-scale systems can be used on a subscription or fee basis. Examples are those at Case Western Reserve University supported by the American Society for Metals, Chemical Abstracts at Columbus, Ohio, specializing in chemistry literature, and one at Indiana University which includes the scientific reports of the National Aeronautics and Space Administration.

Product Development. Product development is the function of translating the general specifications desired in the product into a prototype or working model and a set of preliminary technical specifications describing the product. The preliminary technical specifications described are those which are critical from the standpoint of use by the customer. Non-critical technical specifications are left to the next stage of design.

There are few hard and fast rules to suggest how this stage should be performed. H. L. Timms (The Production Function in Business) characterizes the process as "essentially one of directed trial and error through successive designs of the product until one model finally meets product general specifications as closely as is economically feasible." Since it is difficult to anticipate the kinds of structural, mechanical, electrical, and other kinds of problems and bottlenecks that will occur in developing the product, it is impossible to closely plan the successive stages of development. However, the term **directed trial and error** implies that the trials are not haphazard. Rather they are guided by the technical knowledge of the engineers and technicians. Through the use of **cre-**

ative experimentation the developers apply their knowledge of science and technology to create new configurations that will meet the required general specifications dictated by the market needs. Successive trials are taken and models developed, each progressively introducing improvements in the product.

As each model is developed, it must be tested to determine how closely it meets the general specifications. These **performance tests** may be quite extensive with the testing engineers making detailed measurements of all the operating characteristics.

The process of developing and testing models could continue indefinitely with each successive round bringing an improved product. However, somewhere along the line a point of diminishing returns is reached. Subsequent improvements achieve less and less while the cost of each improvement remains approximately the same. A good deal of intuitive judgment must be applied in establishing a cut-off point since the amount of improvement which will occur on another round is impossible to predict and the value of the improvement difficult to measure. Sales forecasts have an important effect on the length of the product development stage. The cost of each round of model development and testing will be approximately the same for both the high and low volume product, but the value of resulting improvements from each round will be proportional to the volume of expected sales so that the high volume products will undergo a more lengthy product development process.

Product Final Design. The last subfunction of product planning involves the development of the technical specifications that go to production. The **starting point** for this last stage is the final model from the product development stage, its critical specifications and test history. The specifications which are non-critical from a functional standpoint must now be developed.

The final specifications are critical to the production process since they affect the cost of manufacturing as well as the marketability of the product.

The final specifications include **working drawings** for each part and sub-assembly as well as for the final product itself. They also include bills of material for each component and assembly.

Some of the major decisions in final design are concerned with quality, reliability, alternative inputs and conversion processes, and standardization.

QUALITY AND RELIABILITY. Most products can be produced at different quality and reliability levels. In order to choose the appropriate level a company needs information concerning the market demand at the various levels. For example, in the case of transistors, market research might reveal that the market is composed of three segments: 4,000,000 transistors of high quality and reliability per year for use primarily in the aerospace industry, 10,000,000 medium-quality transistors used for computers, color television sets, etc., and 28,000,000 low-quality transistors for radios and similar inexpensive products. From this information management must decide which of the markets it wishes to serve.

The three different quality levels represent three different products to the production man. While they may all look similar, each has its own raw materials and production methods; each its own set of technical specifications. The high-quality transistors will require higher quality raw materials, more highly skilled workers and more accurate methods than the lower quality products.

Design quality should not be confused with production quality. **Production quality** is concerned with how well the production process is working, i.e., to what degree the technical specifications of the product are being met. **Design quality** is reflected in the technical specifications. Production quality may be high or low no matter what the design quality level. In establishing the design quality, the product designers must consider the company's existing capacity and technology and its production capabilities. Specifying a high quality and reliability level is a waste of time unless the production process is capable of producing at that level. Similarly, designating low-quality technical specifications in a firm that has the man and machine capability for high quality production may be uneconomical since such a production process is more accurate, and more expensive, than the product demands.

Clearly, process design must become a consideration during, rather than after, the product design stage. This interaction is shown in Fig. 1. The product designers must be familiar with process capabilities in the factory in order to assure that the technical specifications can be met at a reasonable cost. Although the production process is not designed until the product design is complete, information concerning process capabilities must be utilized in designing the product.

All too frequently, products are "overdesigned"; the technical specifications are set too rigidly without concern for process capabilities or even for the anticipated product usage. Tolerances should be determined by balancing the need for close fits against the cost of making products to such exact specifications. The tolerances set should be determined by the usage of the product, modified by the ability of the available processes to do the work at a reasonable cost. Not all tolerances deal with dimensions but the same need-vs.-cost balancing problem occurs whether one is dealing with the strength of material, exactness of chemical composition, degree of smoothness, or some other technical specification. Engineers must determine how square is square enough, how smooth is smooth enough, etc. In practice such matters are often decided by the draftsman and not the engineer and often without much appreciation of the cost of meeting the tolerances set. Furthermore, because the draftsman sets the tolerances and sets them closely, he determines the production processes. His tight tolerances rule out all but the most exact and costly methods of producing the product. A part of the normal process of reviewing final design specifications should be an **evaluation of all tolerances** to determine if they are necessary and if they can be achieved at a reasonable cost. Such reviews often lead to an easing of the tolerances and a resulting lowering of production costs.

Product **reliability** is the probability that a product will operate successfully for some specified period of time under specified operating conditions. It depends on both design and production quality. Product reliability is extremely important in such complex and expensive systems as missiles and electronic equipment. The failure of a single inexpensive component in a multimillion dollar space vehicle can cause a tremendous loss of not only money but even human life. As a result such systems are designed with automatic testing equipment that will reveal a likely failure before it occurs. Another design feature is the use of redundant or spare components that can automatically take the place of those that fail during operation (see section on Quality Control and Reliability).

Product reliability considerations do not end when the product leaves the factory. Procedures must be designed to assure that the product is properly installed and maintained. With some products, the producing firm will employ and train its own installers and servicemen to assure that the specified installation and maintenance procedures are followed. In other cases it is sufficient to provide written instructions to the product user. The installers and servicemen should not be overlooked as information sources on product quality and reliability. Their firsthand observations on the durability and operations of products and the causes of failure can provide the information required for redesigning and improving existing products and perhaps indicate what types of new products are desired.

STANDARDIZATION AND SIMPLIFICATION. Standardization is the process of establishing basic specifications for a set of commonly used characteristics of size, shape, and performance for products. For example, any garden hose purchased will fit on any house faucet since the threads of each are of a standard size. This standard size is accepted for all houses in the United States. Similarly, one encounters no difficulty in screwing a light bulb into a socket because light bulb bases are standardized. A closely related process, **simplification,** refers to the reduction in the number of different sizes and shapes produced or stocked.

Standardization (including simplification) is an important economic concept in product design. The use of standard components and materials is more economical than the use of non-standard items. Standardization may occur in the product line itself as well as in the choice of components and materials used in the production process. When standardization of the product line occurs, non-standard items will not be produced unless the customer specifically orders them and pays for the extra cost of producing them.

Many standards are established outside of the individual firm. Some standards have been enacted into law for safety or health reasons. Automobile seat belts and windshield safety glass are examples. **Codes** regulating electrical wiring, plumbing, and construction also fall into this category. Many standards are established through cooperation of industry members. Such efforts are generally sponsored by industry-wide associations, professional associations, or governmental agencies. The American National Standards Institute is a national organization supported by over 100 trade associations, professional societies, and consumer organizations. The ANSI provides facilities for testing and aids in the development of nationwide standards. It publishes and disseminates the various standards established by the groups using its facilities. The American Society for Testing and Materials, the Society of Automotive Engineers, and the National Bureau of Standards are other important standard-setting organizations.

Preferred Numbers. When the end products are standardized, a system of **preferred numbers** is often used to determine sizes. Products such as light bulbs, tin cans, packaged food, etc., can be made in a continuous series of sizes from very small to very large. Since this would be impractical from both a production and sales standpoint, manufacturers of such products must concentrate on certain particular sizes in order to permit economical manufacture and sale. The sizes chosen usually follow a preferred number progression. Preferred numbers are geometric progressions in size from small to large with each number in the series being proportionally, rather than absolutely, larger than the preceding number. A geometric series of size differences (4, 16, 64, 256, etc.) generally meets the requirements of customers better than an arithmetic progression (4, 8, 12, 16, etc.). The geometric series provides several small sizes but holds the number of large sizes to a reasonable number. To determine the proper proportion to use, the product designer must specify the largest and the smallest size and the number of sizes in between. The **proportionality factor** is

$$\sqrt[n-1]{\frac{\text{Largest size}}{\text{Smallest size}}}$$

here n is the total number of sizes to be made. The answer obtained is the ratio between sizes. Thus, 1.50 would mean that each size is 50 percent larger than the next smaller size. If a sugar refiner wanted to produce packaged sugar in five different sizes with the smallest package at 1 pound and the largest at 100 pounds, the formula would yield a ratio of 3.16 and the sizes used would be 1, 3, 10, 30, and 100 pounds. By comparison, an arithmetic progression would produce sizes of 1, 25, 50, 75, and 100 pounds. The geometric preferred number series would obviously better satisfy customer needs.

Component Standardization. The standardization and simplification program should extend beyond the end product to include component parts and raw materials. In designing end products, **standard components** should be used whenever possible. Even if the standard part does not fit or work quite as well as a specially designed part would work, it will be far less costly to produce or acquire and therefore its use may be justified. Standardization throughout the system reduces the kinds, types, and sizes of raw materials which have to be purchased. Thus, larger quantities of the required sizes can be purchased at lower prices. Standardization also reduces manufacturing costs by allowing longer production runs on the smaller number of products and components. Setup costs are reduced and more specialized machinery can be utilized. Also, fewer blueprints, manufacturing instructions, patterns, and tools are required.

Standardization is not without its disadvantages. In many cases, the "perfect part" may be more effective in operation and easier to install so that it is less expensive in use even though it may cost substantially more to acquire. When production volume is large, the specially designed non-standard parts may eventually be used in such quantities that they can be mass produced and cost no more than standard parts.

Industry-wide standardization programs are often only partially successful. Standardization tends to favor the large companies which realize the greatest benefits because of the large quantities they produce. The smaller firms may be unable to match the costs of their larger competitors. They may succeed by **offering non-standard sizes and models** to customers at slightly higher costs. Though this undercuts the industry program it may be the most successful program for the smaller firms.

Another disadvantage of standards is that they may **retard progress.** New products may be standardized too soon, before effective designs have been developed. The development of color television was probably retarded by the Federal Communications Commission in trying to force standardization on the industry while this new process was still in the evolutionary stage. Established standards present an obstacle to change. Yet improvements must be made if progress is to continue. Standards should not be maintained for their own sake; they are of value only as long as they allow progress in design and lower production costs.

Organizing Standardization Program. An effective standardization and simplification program in a firm requires that responsibility be clearly designated. While the particular organizational arrangement may vary among firms, it is vital that a continual effort to implement standardization and simplification be made somewhere in the organization. All too often firms institute a standardization program on a **crash basis** only after it has become apparent that production quantities are too small and manufacturing is inefficient. Although these crash programs are able to produce impressive statistics on the number of components and products eliminated and on an expected large annual saving, the figures actually point out how expensive the lack of consistent standardizatic can be. Moreover, the reduction of products will require extensive revampin of production plans and sales promotional literature with its attendant cost And, once the crash program has been completed the firm is likely to graduall; slip back into its former position as special orders are received.

A consistent, continual **standards program** requires cooperation among ε number of organizational entities. The sales department must regularly analyze the sales volume of products and periodically drop items that fail to generate sufficient sales. It can also work along with the engineering department to help develop products that may not be standardized in appearance because of differences in trim, coloring, and accessories, but which can be made from standardized components. Inventory control should supply engineering design with information on slow moving materials and components which are likely candidates for elimination. In the end, the final responsibility lies with engineering design. It uses the information supplied by other sources, but it is here that the standardization program must be implemented. As each new product is designed a continual effort to use standardized parts must be maintained. And, a continual review of existing designs must be made not only to see that sizes and models do not proliferate but also to assure that existing products are continually improved.

VALUE ENGINEERING AND VALUE ANALYSIS. Value analysis is a formal concerted effort to reduce costs through analysis of existing products or products in the design stage. Focus is placed on the **value of a product**— what function is to be performed by the product—and how that value can be achieved at the lowest cost. Although value analysis is applied to all phases of the production process, primary attention is devoted to the materials and components going into the product. If this were not the case, it would be difficult to distinguish value analysis from traditional methods improvement which concentrates on the process. Both deal with the same types of problems, have the same objectives, and even use similar tools in resolving the problems. Part of the impetus behind value analysis is the rapid and dramatic growth of materials technology. New materials are constantly being made available through product research.

Value analysis typically follows a rather close structured pattern of analysis. A **value analysis team** or **committee** takes a product which has been designed or produced and attempts, first of all, to define what function the product is to fulfill. While this sounds simple, it may prove to be a formidable task. It is usually comparatively easy to define the purpose of the product but often rather difficult to state precisely how effectively it must perform this function. There are physical evaluations and consumer requirements. While one can state how many springs are in a mattress and the sharpness of a carving knife, how can one state how comfortable the mattress should be or how sharp the knife should be? Presumably there is some relationship between measurable, physical factors and consumers' evaluation of the functional utility but it may be difficult to pinpoint. In addition, quality, reliability, ease of maintenance, appearance, and a host of similar requirements must be defined.

Once performance requirements of the product have been established by the value analysis committee, the product is examined to determine the cheapest way to achieve those requirements, often through a structured series of **leading questions** such as:

1. Is there a less expensive product which will perform the required function?
2. What does each component contribute?

3. Can a less expensive material be used?
4. Can parts be combined?
5. Can it be designed for easier assembly?
6. Can it be fabricated by other methods?

The effectiveness of value analysis will be dependent upon the knowledge and creative insight that the individuals who are doing the study can bring to bear. The value analysis approach has been designed to release such insights by providing a structural framework to encourage the development of alternative strategies. Value alternatives are derived by analogy. For example, comparisons might be made between various joining methods such as welding, adhesion, brazing and mechanical fasteners such as nuts and bolts, screws, pins, etc. As can be seen, there is really nothing new about the decision processes involved. Value analysis is essentially a new organizational form that is useful in order to assure that products are periodically analyzed to take advantage of changes in technology.

Applications of value analysis frequently produce **large savings.** General Electric, which began both the term value analysis and the formalized approach (Miles, Techniques of Value Analysis and Engineering), cites the following examples. A screw of special design cost 15 cents. Value analysis found a way to make it for $1\frac{1}{2}$ cents with an annual savings of $20,000. A handmade gasket costing $4.15 each when made at G.E. was found to cost 15 cents from an outside supplier. A die-cast cover cost 60 cents. Changed to a stamping it proved better and cost only 20 cents producing a $39,000 annual savings.

Similar savings can also be obtained by relaxing unnecessarily tight tolerances. In some cases parts are found to be made too reliable. It would make little sense, for example, to produce an automobile generator that would last for a million miles when the rest of the automobile would not last one-fourth that long.

The savings of value analysis are not obtained without expenditures. It is possible to spend dollars analyzing areas where only pennies can be saved. However, by intelligent choice of products to analyze, value analysis should produce savings several times its cost.

COMPUTERS IN DESIGN. Computers are used in product design work in two important areas: (1) providing information to assist in design, and (2) formulation and evaluation of alternative designs.

In a large company, the individual designer cannot be expected to learn or remember all the different components and materials used in the firm. When designing a new product he would obviously like to use standard components and materials that the firm regularly keeps in stock. The information about all standard materials can be stored in the computer. When a designer is considering a new material, the computer can quickly tell him what materials are available with specifications close to those he requires. One of the major automobile companies has developed a system to store the configuration and dimensions of the thousands of different forgings it produces in a computer data system. When a designer needs a forging in a new design, he consults the computer which searches its data and prints out the blueprint numbers of the forgings similar to the one required.

In addition, computers are actually developing drawings and designing new products. General Motors uses computers in automobile design. General Electric uses them to design electric motors and transformers. Airplane companies use them to compute contours of wings and lay out electronic circuits. In most computer-aided design processes, experienced engineers lay out the general rules to be followed in designing the product. The operating characteristics of the product and the input requirements are specified. The computer then follows the set of **design rules** and determines the best combination of inputs to achieve the desired output specifications. When there are alternative combinations of inputs, the computer can evaluate hundreds of combinations in the time it would require a human designer to evaluate one.

Beyond formulating the design, the computer can be programmed to issue instructions to automatic drafting machines to make up shop drawings and even provide all the instructions for manufacturing the product that the factory requires. The computer is even capable of producing perspective drawings of the product viewed from any desired angle.

UTILIZATION OF PLANT FACILITIES. At the time a product is being designed, preliminary consideration is given to the capacity of the firm to manufacture the product. The technology required and the particular machines and processes needed are determined and compared to what is available. The decision may be to continue designing the product, if its production will be compatible with existing manufacturing capacity or if the needed new capacity can be acquired. If it appears that the new product will not be compatible with existing operational equipment, the decision may be to terminate the design effort.

After the final design has been completed, the manufacturing process must be planned. Design of the manufacturing process again gives rise to capacity questions. As the process plans are firmly established, **processing time requirements** of specific machinery and equipment are generated. These time requirements must then be evaluated against the available capacity and against the cost of acquiring the new machines and equipment required before a final decision is made to produce the product.

PROCESS PLANNING. Process planning is thus concerned with planning the work that directly results in conversion of materials into products. A production **process** is a series of **operations** performed at work stations to achieve the design specifications of the planned output. In producing a complex product there may be hundreds of different operations and dozens of different kinds of equipment and machines. Part of the operations may be performed at one geographical location with the semifinished parts shipped to another part of the country for the remainder of the production process. Simpler products may require only a few operations.

Process planning consists of two parts: (1) process design and (2) operations design. Both stages provide information which is required to effectively utilize existing equipment and to determine what new equipment is required.

The tools and techniques of process planning are covered in detail in other sections of the Handbook. Process planning will be considered here to the extent that it affects the use of plant facilities.

Process Design. Process design is concerned with the overall sequence of operations to realize the product specifications. It specifies the type of work stations that are to be used, the machines and equipment necessary, and the quantities of each required.

The **sequence of operations** in the manufacturing process is determined by the nature of the product, the materials used, the quantities being produced, and very often, the existing physical layout of the plant. In certain industries equipment arrangements are changed frequently, even for short-run products. The sequence of operations in these cases is often determined by the physical nature of the components that make up the final product. The furniture industry is an example of manufacturing requiring frequent relayouts in equipment to facilitate processing of components and assemblies. In some heavy industries such relayout is impractical because of the size and structural requirements of the machines involved. Heavy forges, presses, casting equipment and furnaces, and rolling mills are examples of processing elements that are rarely moved to fit the sequence of the manufacturing process. Rather, processing is designed to follow as efficiently as possible the existing layout. In most cases, automation has resulted in more rigidity in the equipment layout. Numerical control machining centers, transfer machines, automatic assemblers, and the like are arranged according to general material flow into and out of the plant rather than the processing of a particular product. Only in cases of high-volume, continuous processing are there likely to be major realignments in the physical positions of these work stations.

Operations Design. Operations design is concerned with the makeup of the individual manufacturing operation and not the interrelationships between operations. This does not imply that a systems approach is not applicable in the design of operations. It does mean that ultimately the final step in the system, the man or machine that is actually changing the raw material into the finished or semi-finished product, must be examined and decisions made based on this step. Operations design must specify how much **man or machine time** is required for each unit of production. This subject is discussed in further detail in the sections on Motion and Methods Study and Work Measurement and Time Study.

Work Station Selection. Most products can be produced under a wide variety of alternative conversion processes ranging from an individual craftsman system where each individual produces a complete unit of product from start to finish, to a production line system where each worker performs only a very small portion of the work on each unit. While particular machines and equipment may be required to perform some operations because of their unique technological characteristics, most operations can be done at several different kinds of work stations. The **selection of work stations** to make up the process then is primarily a choice based on economics among several man-machine systems, each of which has the required technological capabilities.

The least-cost process will be selected from among the available alternatives. Expected sales volume strongly influences the choice. Suppose that a process designer is faced with a choice of three alternative methods of machining a part, all of which can perform the required operation. The three work stations have costs as shown in Fig. 3. The two elements of cost are the fixed setup cost

Machine	Setup Cost (Labor and Tooling)	Production Cost (Labor, Material, Power, etc., per Unit)
Engine Lathe	$ 20.00	$1.30
Turret Lathe	80.00	1.05
Numerically Controlled Lathe	180.00	0.85

Fig. 3. Fixed and variable costs for machining a part.

which will occur once for each production run and the variable cost for labor, material, power, etc., which will vary with the number of units produced. These costs are illustrated and compared in Fig. 4.

The appropriate choice can now be made by looking at Fig. 4. If the quantity to be produced is less than 240 units, the engine lathe is the least costly; between 240 and 500, the turret lathe is least expensive; and in lots of more than 500, the numerically controlled machine yields the minimum cost.

The analysis above assumes that processing time is available on all three machines. If one of the machines had to be purchased, an allocation of a portion of the capital costs would obviously have to be included in the analysis. The decision might also be influenced by other factors. If one machine was more accurate in holding dimensional tolerances it might be selected even at a slightly higher direct cost.

If a process is being selected for a new plant the number of alternatives may be quite large. In an existing plant, the set of alternatives will be limited to the work stations now in place plus new equipment which the firm is willing to acquire. Among the alternatives now in place, choice must be limited to those that will have extra time available. Thus in selecting the viable alternatives it may be necessary to examine current production schedules and to project future sales patterns of existing products and those currently being designed to determine the availability of processing time.

Material Handling. The alternative work stations now in place are located at specific points in the plant. Some type of material handling system transports material between the stations. Thus, there is a pattern of **work flow** associated with each set of choices of alternative stations and the material handling system in use. These flow patterns will each have different costs because of differing traveling distances. The process designer must include these transport costs in his analysis and seek the process that achieves the product design specifications at the lowest cost combination of operations performed at particular work stations and movement of material between stations.

One goal in process planning is to reduce handling and movement of goods as much as possible since movement adds cost but does not directly contribute to the value of the product. When a process is being designed for a new product with large expected sales volume, it may be economical to acquire new equipment which will be used solely in producing the new item. In such a case, the machines can be arranged in the same sequence as required in the process thus reducing material handling. In the typical case, the new product is just one of many that will be produced intermittently on existing facilities with perhaps one or two new pieces of equipment added. While the possibility of moving equipment and revising the layout in order to reduce movement of goods in process exists, it is usually not advisable to do so. Any reduction in travel for the new product may well be offset in increase in travel for other products. In addition the cost of moving heavy equipment and the disruption of production during the movement may overweigh the resulting economies in material handling.

ADAPTING TO MAJOR PRODUCT CHANGES. It is inevitable that changes to a product will be made. The impetus may be from customer use: performance experience, reliability, aesthetic design, or maintainability. It may come from manufacturing: to ease a production problem or to reduce costs. The changes may be proposed by engineering: to improve characteristics, to prevent a potential problem, to improve the functioning of the product, or to make it more competitive. Whatever the cause, the effects of what appears to be an insignificant change may be quite pronounced.

In the early stages of product development, changes in design can be made relatively easily. However, once a design has been released, there are factors that must be considered before changes are made. First, a change may affect the tooling to be used, the material from which the product is made, and the method of

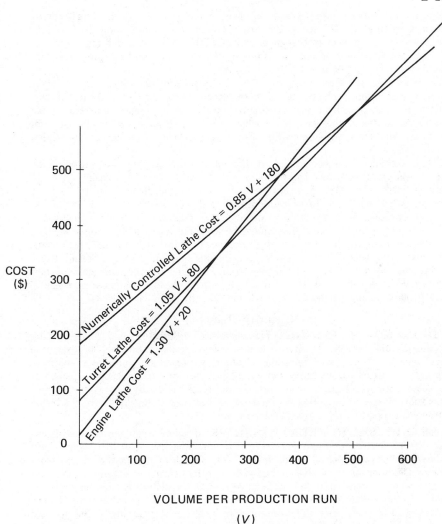

Fig. 4. Breakeven chart for work station selection.

production. In the case of a previously manufactured product, interchangeability may be adversely affected and it may be necessary to **retrofit** units already produced.

Introducing Changes. The Department of Defense has established a rather elaborate and effective system for controlling design changes. Many organizations producing items for the government are contractually bound to follow the structure of the system though at times the system appears to be overburdened with red tape and paper work. However it does provide a basis for examining the proper controls for introducing changes.

The system considers changes on two levels. First, if the product is in the

design and development stage, changes can be made by an **engineering change notice, ECN.** After the product goes into production, changes then are made by a **change in design notice, CIDN.** The basic difference between the two is the amount of control exercised. This control can be expressed by the number of persons who must approve a change.

ECN changes must have the approval of **design engineering.** Their concern is with dimensions, tolerances, aesthetic appearance, and mechanical fits that affect interchangeability. Engineering is concerned with such matters as the materials used and the stresses to which the part will be subjected. In some organizations, materials engineering and manufacturing engineering also review ECN changes.

Once a change has been approved, the product drawing or the computer design cards are altered and the point of introduction of the change into the system is fixed. The part identification may need to be changed and, at times, the parts list may need to be re-identified to avoid confusion.

If the product is in production and a CIDN is necessary, the change process includes all of the foregoing plus added review and approval by groups such as production planning (disposition of present parts: continued use, rework, scrap), and marketing (disposition of parts in customers' hands). A major consideration is the cost involved: cost of existing parts, cost of retrofit, and cost of new tooling. The degree of control and, consequently, the elaborateness of the system depend on the products involved and the effects of changes.

Quantity Planning

NEED FOR PLANNING. The relative complexity of modern production is caused chiefly by the division of labor, or specialization, both at the executive level and in the factory. Division of labor, within limits, promotes economy through increased productivity per man-hour, but it usually leads to problems of coordination. Efficient production is rarely achieved by the voluntary cooperation of specialists acting on their own initiative. It must be brought about through the operation of a preconceived plan.

ESTABLISHING RELATIONSHIPS. Production planning must establish the basic relationships among production capacity, inventory levels, and sales rates for some period in the future. The length of this period ranges from one or two months to twelve or eighteen months. Its exact extent depends on the type of business and the scale of operation involved. In any case, plans which extend much more than a year and a half ahead are usually better classified as general operations planning or investment planning. This type of planning is somewhat different from production planning and is not the area of planning with which we are principally concerned here. **Short-range scheduling** of one or two months ahead falls into the category of production control and is covered in the next section of the Handbook.

Within an area properly classified as production planning, the outward appearance of the problems dealt with can vary widely. However, there is no basic difference between the problem of balancing the raw material output of one division with the supply requirements of a second and the problem of balancing the output of three fabricating departments with the parts requirements of an assembly department.

Although the physical facilities available for production are usually fixed for the interval of the planning cycle, output can often be varied through a considerable range. Changes in the number of shifts or the length of the work week, for example, can have a marked effect. Inventory levels also can fluctuate from practically nothing to the upper limit of permanent storage capacity and, in some cases, beyond. The relationships between production rates and inventory levels can be established quite exactly as soon as sales rates are known. In most cases, however, the level and composition of future demand are uncertain. Despite this uncertainty, the objective of production planning is to strike the most economic balance possible between production rates and inventory levels. This involves:

1. Evaluation of the **uncertainty** involved in the sales estimate; determination of expected levels of demand and the probable range of error on each side of these levels.
2. Establishment of **production rates and inventory levels** which offer the best probability of meeting estimated sales requirements with the most economic combination of labor, facilities, materials, and working capital.
3. Determination of the **changes needed** in these relationships if actual demand begins to vary by predetermined amounts from expected demand.

ANALYZING TRENDS. Continually changing demand makes the task of planning difficult. The factor of change manifests itself in a variety of ways. Sometimes it occurs with sufficient regularity and frequency to show an underlying rhythm. Sometimes long-term trends can be discovered and therefore predicted with a reasonable degree of certainty. Often qualitative as well as quantitative changes in demand are important.

Secular changes are upward or downward trends over long periods of time, often traceable to some profound technological or competitive disturbance. They may affect the interests of the enterprise adversely or favorably. They play a large part in the long-run investment plans of industry, but their impact on shorter-range production plans is generally limited. Where **cyclical changes** are concerned, the production planning problem is one of foreseeing significant turns in business activity with sufficient accuracy to avoid serious maladjustments.

Seasonal fluctuations may often be predicted with reasonable accuracy since the causes lie in the regularly changing seasons. Figure 5 shows the way in which current production may be adjusted to provide for seasonal demands. The production line is held uniform at three hundred units per month. Sales vary from two hundred at the beginning and end of the period to a maximum of five hundred units toward the middle of the period. There are in stock two hundred units at the beginning of the period, and cumulative lines show the relation of production and stocks to products sold. Production begins and ends with two hundred in stock and keeps ahead of sales, although at the end of the peak period, only fifty units are in stock. This margin might or might not be considered safe. If not, overtime would probably be indicated as the peak period was being passed.

The dependence of steady production on reliable forecasts of sales is shown by Fig. 5. When the nature of the business prevents reasonably accurate forecasts of sales, the risks involved in a steady rate of production for twelve months may be too great for a manufacturer to assume. Firms producing highly stylized products or products whose sales depend on weather conditions at some season fall into this category. Here, too, are many contract manufacturers and companies with perishable products. Considerably greater fluctuations in production levels than shown in Fig. 5 are usually found in these firms.

Weekly and daily fluctuations are important chiefly in industries producing very perishable products or services which must be supplied at the instant demand develops. This situation presents a highly specialized problem of planning in public utilities such as electric and gas companies. The problem of plan-

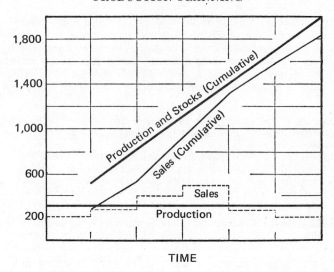

TIME

Fig. 5. Adjusting production to seasonal sales.

ning and control in such instances consists chiefly of **maintaining suitable standby equipment** which may be cut in when required and cut out when peak-demand periods are passed.

SALES FORECASTS. Several different forecasts are needed in most companies. Plossl and Wight (Production and Inventory Control) classify the forecasts according to the time span covered as follows:

Long-Range Forecasts: for plant expansion and equipment acquisition, in order to plan capital investment five years or more in advance.

Intermediate-Range Forecasts: for procurement of long lead time materials or planning of operating rates; taking into account seasonal or cyclical products one or two years in advance.

Short-Range Forecasts: to determine the proper order quantities and order timing for purchasing or manufacturing components, and to plan the proper manufacturing capacity, taking into consideration the desirability of leveling man-load three to six months in advance.

Immediate-Future Demands: for assembly schedules and finished goods inventory distribution on a weekly or daily basis.

The intermediate-range and short-range forecasts are needed for the production planning discussed in this section. The long-range forecasts are used primarily for investment planning. They require an understanding of economic factors, competitive and technological influences, and capital expansion plans of top management. The immediate future demands are needed in production control. They come from a study of orders on hand and current inventory positions.

Both intermediate-range and short-range forecasts are usually developed by extrapolating past demand data into the future. Complete reliance on past statistics assumes that the pattern of factors influencing demand will remain the same in the future. Economic factors, competitive and technological influences do change, however. Therefore, sometimes the statistical forecasts should be adjusted to reflect these changes. At a minimum, procedures should be established to continually check the accuracy of the extrapolated forecasts to determine the relevancy of the existing forecast techniques.

Past Performance Data. Adequate records of past performance are essential to production planning. These will certainly include data on production, shipments, and inventory for each item manufactured. They may also cover quality or other factors pertinent to a particular business. Weekly records as well as summary reports for specified periods should be available.

Three points are important for **effective handling of data** on past performance:

1. Records must be reasonably accurate and available as soon as possible after the period covered.
2. Data recording, correlation, and extraction procedures should be simple, rapid, and as automatic as the economics of a given situation will permit.
3. Information must be presented in forms which highlight significant factors and promote decisions.

One of the commonest failings of systems for the collection and presentation of production planning data is that planning personnel are forced to work with information that is so voluminous and so unimaginatively presented as to confuse planning decision processes rather than simplify them. The exact method of **presentation** must of course be selected to fit the requirements of a given situation. The best way will vary from one firm to another.

The past data is then used to develop a sales forecast or budget which forms the basis for production plans. Such a forecast is an attempt to predict what changes in demand will occur during a coming period.

Statistical Forecasting Techniques. Averages provide the simplest statistical method of forecasting for many individual products. The average may be simple or weighted. For example, if demand had been 100 units in January and 120 in February, an unweighted average forecasting method would produce a March forecast of 0.5 (100) + 0.5 (120) = 110 units. If it was decided to weight the more current information more heavily than the older information, the most recent month might be given a weight of 0.8. In that case, the March forecast would be 0.2 (100) + 0.8 (120) = 116 units. If a trend exists, any kind of average, no matter how heavily weighted, will lag behind the actual demand. Thus a trend correction effect is often added to the forecasting process. A simple trend adjustment is to measure the change in average from one period to the next (which is the trend) and estimate demand for the next period equal to the average plus the trend correction.

Exponential smoothing is a forecasting technique based on the work of R. G. Brown (Smoothing, Forecasting and Prediction of Discrete Time Series). It provides a routine method for updating forecasts regularly. A computer programmed for this technique can keep track of thousands of products and provide information for production planning and inventory control far better than it could be done otherwise. Consider a first-order smoothing method. Suppose that our forecast for February in the previous discussion had been 110 units. Recall that actual demand turned out to be 120 units. The new estimate of average demand for March is found by taking the old February average and adjusting it by some proportion of the amount of the error in the old estimate as revealed by comparing it to the actual demand as follows:

$$\text{New average} = \text{Old average} + \alpha \, (\text{Demand} - \text{Old average})$$

where α is amount of weight given to the most recent demand data or the **smoothing constant.** This equation can be rewritten as follows:

$$\text{New average} = (1 - \alpha) \, \text{Old average} + \alpha \, \text{Demand}$$

Using $\alpha = 0.5$, and the data from the above example,

$$\text{March forecast} = (1 - 0.5)\ 110 + (0.5)\ 120 + 115 \text{ units}$$

It can be seen that the exponential smoothing method is actually a weighted average. The most recent period is given a weight of α. The old average was originally developed from demand data from previous months.

This may be expressed algebraically using

$$d_{n-1} = \text{actual demand during last time period}$$
$$d_{n-2} = \text{actual demand during period before last}$$
$$d_{n-3} = \text{actual demand during second period before last}$$

Successively substituting into the previous smoothing formula the new forecast becomes

$$\text{New average} = (1 - \alpha)d_{n-1} + (1 - \alpha)\alpha d_{n-2} + (1 - \alpha)\alpha^2 d_{n-3} + \ldots + (1 - \alpha)\alpha^n d_0$$

Notice that the terms further removed from the present time carry less and less weight since the term $(1 - \alpha)\alpha^n$ becomes smaller as n increases.

The above form of exponential smoothing works well in predicting demand for fairly stable items. It will detect a trend quite rapidly but it will produce forecasts that lag behind the trend. However, it can be adjusted easily by adding a trend factor. The current trend is the difference between the average demand figures for the latest two periods.

$$\text{Current trend} = \text{New average} - \text{Old average}$$

The average trend may then be determined by the same smoothing method used to compute the average:

$$\text{New average trend} = \alpha\ (\text{Current trend}) + (1 - \alpha)\ (\text{Old trend})$$

The forecast is then based on the new average estimate plus the trend correction.

$$\text{New forecast} = \text{New average} + \left(\frac{1 - \alpha}{\alpha}\right) \text{New average trend}$$

The term $(1 - \alpha)/\alpha$ is necessary to properly project the trend adjustment into the next time period.

An illustration of the exponential smoothing method with trend correction is shown in Fig. 6 using $\alpha = 0.2$.

Month	Demand	New Average	Current Trend	New Average Trend	Next Period Forecast
January	200	200	0	0	200
February	240	208	8	1.6	214
March	300	226	18	4.9	246
April	340	249	23	8.5	283
May	450	289	40	14.8	348
June	520	335	46	21.0	419
July	610	390	55	27.8	501

$$\alpha = 0.2$$

Fig. 6. Exponential smoothing with trend adjustment.

The results obtained by the exponential smoothing method depend to a large extent on the value chosen for α, the smoothing constant. Fluctuations in demand from period to period result from two causes: (1) changes in the economic, competitive, and technological conditions affecting demand, and (2) random, unpredictable factors. An ideal forecasting method would adjust immediately and fully to causes of the first type since their effect will persist into the future but would ignore causes of the second type because their effect will be dissipated immediately. Unfortunately, any statistical forecasting technique that closely follows changes of type 1 will also follow changes of type 2. And, any that ignores changes of type 2 will also ignore type 1 changes. Thus, forecasting represents a compromise between the two objectives. The choice of an α value depends on which objective is to be emphasized. A relatively small α, say 0.1, means that recent demand will exert comparatively little influence and therefore the forecast will react rather slowly to demand changes. If a high α, such as 0.4, is used, the forecast will react sharply to demand changes and will be highly erratic if there are sizable random fluctuations in demand. In practice, the choice of α is often determined by computer simulation with several α values being tested against actual sales data.

Controlling the Forecast. A frequent question is, "How often should a forecast be revised?" The answer is that a forecast should be revised only when it has to be. The method for determining when it should be revised involves the calculation of control limits for forecast errors and a plot of the individual forecast values on a control chart (see sections on Quality Control and Reliability and Statistical Methods).

One company has used an interesting adaptation of statistical control techniques to meet its needs. A trend line is established on cumulative monthly product shipments and extended into the future along with $2S$ control limits as shown in Fig. 7. In practice, this trend would be based on more data than the four months used in this example.

Cumulative sales for January, February, March, and April are available. These months are used as the X coordinates and are numbered 1, 2, 3, and 4. The Y coordinates are cumulative monthly sales. The standard error of estimate, S_Y, is calculated from the following formulas:

$$\Sigma(Y) = Na + b\Sigma(X)$$

$$\Sigma(XY) = a\Sigma(X) + b\Sigma(X^2)$$

$$S_Y^2 = \frac{\Sigma(Y^2) - a\Sigma(Y) - b\Sigma(XY)}{N}$$

where N is the number of items (months in this case). Substitution of the values from Fig. 7 gives:

$$13 = 4a + 10b$$
$$39 = 10a + 30b$$

from which

$$a = 0; \qquad b = 1.3$$

$$S^2 = \frac{51 - 0 - 50.7}{4} = 0.075$$

$$S = 0.274$$

$$2S = \pm 0.55$$

The trend line, $Y = 1.3X$, and the $2S$ control limits around that line ± 0.55, are plotted as shown in Fig. 7. In May and June further plots are made and these points stay within established limits. By the end of July, however, it is evident that the old trend line no longer applies. The probability is quite small that the July plot (11 million pounds), is simply a chance variation from the January–June trend. Thus, if any monthly plot falls outside a control limit, or if six successive plots approach the same limit, a new trend line is established together with its control limits. The number six is an arbitrary selection, chosen simply because it worked well in this company's application. Other products and markets might well require other choices. The number of points selected in calculating the second trend line is that number between 4 and 12 which gives the smallest total difference between values of Y computed on the old trend line and actual values of Y. Again this rule is empirical and based on this company's situation.

Calculations for the second set of lines in Fig. 7 are:

$$\underset{x=7}{\overset{x=3}{\Sigma d}} = -1.5 \qquad \underset{x=7}{\overset{x=2}{\Sigma d}} = -1.9 \qquad \underset{x=7}{\overset{x=1}{\Sigma d}} = -1.6$$

$\therefore$ 2d Trend based on March–July ($X = 7$ to $X = 3$) plots

$$34 = 5a + 25b$$
$$187 = 25a + 135b$$

from which

$$a = -1.7; b = +1.7$$

$$S^2 = \frac{262 + 57.8 - 318}{5} = 0.36$$

$$S = \pm 0.6$$
$$2S = \pm 1.2$$

The trend line equation is

$$y = -1.7 + 1.7x.$$

August and September sales remain within the new limits. The company in question found that a trend line usually lasted from 5 to 14 months before out-of-limit plots required a new one.

In this particular case, the charts were confined to a relatively few major products. Analyzing its shipments the company determined that about 10 percent of the items accounted for 80 percent of shipments. Charts were applied only to these items, since manual calculations and charting of more items at monthly intervals would quickly have grown too cumbersome. Since the arithmetic involved is quite simple, the system can be adapted to punched cards with elimination of manual calculating and the need for charts. This will allow additional coverage. Computers open an even wider range.

It should be emphasized that historical data are not the sole bases for an estimate of future shipments. They provide a check on the forecasts supplied by sales, and they direct planning attention to the items which need attention most. Properly presented, they minimize the risk of overcontrolling production operations.

PRODUCTION PLANS. To the production department, the sales forecast is a statement of desired results. If it is to serve as the basis for production, the figures in the forecast must be translated into specific production requirements at specific periods of time. The resulting production plan will establish relationships between end products, required quantities of parts and materials,

Month	Sales	X	Y	XY	X^2	Y^2	Y_c	$d*$	XY	X^2	Y^2
Jan.	1	1	1	1	1	1	1.3	+0.3			
Feb.	2	2	3	6	4	9	2.6	−0.4			
Mar.	1	3	4	12	9	16	3.9	−0.1	12	9	16
Apr.	1	4	5	20	16	25	5.2	+0.2	20	16	25
		10	13	39	30	51					
May	1	5	6				6.5	+0.5	30	25	36
June	2	6	8				7.8	−0.2	48	36	64
July	3	7	11				9.1	−1.9	77	49	121
		25	34						187	135	262
Aug.	1	8	12								
Sept.	2	9	14			$*d = Y_c - Y$					

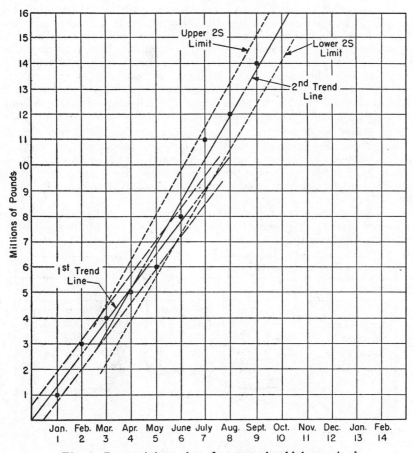

Fig. 7. Determining when forecast should be revised.

necessary lead times for manufacture or procurement of each item, the load on the plant's facilities, and the capacity available to meet this load. It will also determine the extent and timing of the contributions which all other groups in the company must make to implement the plan successfully. These considerations are readily developed once the amount of each product to be produced and the times when these quantities will be required have been determined.

The **most useful production plans** are those which are reviewed and changed at frequent intervals. No one can guarantee that a given forecast is completely accurate. The occasional forecast which does check exactly with actual developments is more the result of chance than of technique. The intervals between revisions of forecasts and plans vary among different companies. Plans may be reviewed monthly, quarterly, semi-annually, or merely once a year. Annual review would be adequate only in a company whose market was very stable; monthly changes of significant scope would be too frequent for most businesses. The majority of firms, whose markets permit detailed planning and forecasting of production, review and alter established requirements either quarterly or semi-annually.

Those companies, for example, that prepare a new forecast every quarter always have at hand a year's forecast which is never more than three months old. The general accuracy of each forecast is constantly checked by the actual rate of sales during the first quarter of its life. Succeeding forecasts may be carbon copies of the first, but it is more likely that they will vary from it as they make use of actual sales data during the early quarters and of evidences of new trends in the economy. Production plans, of course, are also changed to reflect revisions in the sales forecasts. The companies which follow this pattern of planning always have a general plan of operations extending a year in the future and a fairly accurate idea of operations during the next few months.

Plan Based on Forecasts. A manufacturer of tools and shop equipment developed planning procedures based on quarterly sales forecasts. Figure 8 shows a portion of the form in which each quarter's forecast appears. At the left are listed all the products in the company's line. The columns at the right provide space for monthly sales estimates by product classes and individual items. These sheets are prepared in the sales department and are passed on to the production group and top management about four weeks before the start of the new quarter. The second quarter forecast shown in Fig. 8 would be released around March 1.

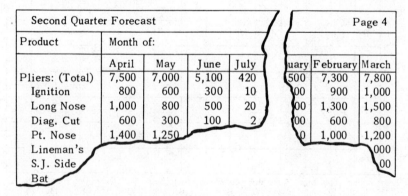

Fig. 8. Sales forecast of tool manufacturer.

After discussion of the new forecast, and any necessary revisions, the production planning group translates it into the **production and inventory schedule** illustrated in Fig. 9. Based on the sales forecast for each item for each month,

Production and Inventory Schedule								
Second Quarter Forecast								
	April		May		June		July	Au
	End. Inv.	Sched. Prod.	End. Inv.	Sched. Prod.	End. Inv.	Sched. Prod.	End. Inv.	Sched. Prod.	End. Inv.
Pliers: (Total)	4,000	6,000	2,800	5,800	3,300	5,600	4,000	4,900	
Ignition	500	700	400	500	600	500	900	400	
Long Nose	700	900	600	700	800	700	1,100	500	
Diag. Cut								

Fig. 9. Production and inventory schedule based on sales forecast.

this plan takes account of such factors as equipment loading, availability of materials, vacation schedules, economic lots, desired average inventory levels, and the like. The plan is an attempt to minimize the pertinent costs of carrying inventory, working overtime and hiring and laying off employees. The planner sets down production totals and estimates of closing inventories for each month for every item and product class.

The final results of the production and inventory schedule are charted as shown in Fig. 10 for control purposes. Note that variations in production range roughly between 5,000 and 6,000, while sales and inventories vary much more sharply. In addition to plots of current estimates for sales, inventories, and production, the chart includes the sales estimate from preceding forecasts of the current year. Since only one other forecast has been prepared so far, only one appears here. The old estimates are indicated by the triangles (▲) and were generally lower than those of the second quarter forecast. Although these charts are prepared only for product classes, the inventory control group, of course, follows the figures for each individual item in each class and gives warning if changes in product mix have left total inventories of a class about as predicted but have made individual item inventories out of balance.

The process described here is repeated every quarter. Actual results are also plotted (Fig. 10) during the intervening months, and changes can be made if sales depart sharply from estimated levels.

ECONOMICAL MANUFACTURING QUANTITY. Many production planning and control problems lend themselves to mathematical analysis. Subsequent Handbook sections deal with the details of many of these techniques. The discussion which follows indicates how this type of analysis can promote more effective planning in determining manufacturing quantities.

The economical manufacturing quantity is the quantity of any one item which should be manufactured at any one time to realize the lowest unit cost. There is a balance of cost in **determining the economical manufacturing quantity** just as there is in the computation of economical purchasing quantities. (See sections on Purchasing and Inventory Control.) Some of the factors involved are as follows:

1. Costs which tend to decrease as the size of the manufacturing order is increased.

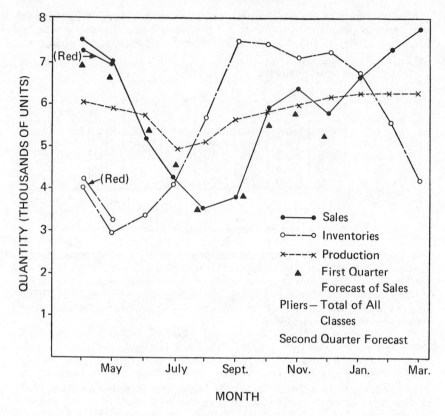

Fig. 10. Control chart of forecast sales, inventories, and production.

a. Setup and make-ready costs in the shop.
b. Costs associated with the issuance and control of shop orders.
2. Costs which tend to increase as the size of the order is increased.
a. Inventory storage charges.
b. Inventory carrying charges.

The same cost considerations hold true for storing and carrying manufactured items as for purchased items.

Setup Costs. Setup costs may range from less than $1.00 to hundreds of dollars each time a manufacturing order is processed in the shop. Hoehing (in AMA Special Report No. 4) suggests that "a reasonable assumption for the job order and job lot type of mechanical production is that about 5 percent of productive time in the factory is consumed in preparing and setting up for the job." In many factories the percentage is much higher. It is not sufficient to use an average cost of setup since the value may vary from product to product by considerable amounts. In one firm, the setup cost ranged from $20 on one product to $3,500 on another. While setups cannot be eliminated entirely, they can usually be reduced by better planning and by producing in economic lot quantities. The industrial engineering department often sets a standard setup time for each type of job. Multiplying this by an average setup rate per hour gives the setup cost for that part. If no time standards are available, the foreman can usually estimate the setup time for each job rather closely.

Cost of Shop Orders. The costs of planning and initiating shop orders together with the handling and make-ready costs involved in issuing and processing an order through the shop are hard to estimate because of the many variables involved. However, it is certain that not only production control, but supervision, cost accounting, inspection, timekeeping, and other functions are affected adversely by a larger number of orders. Some companies use the rule of thumb of about $5.00 to $10.00 per order in computing savings. Other companies neglect this factor in computing savings because they say that a decrease in the number of shop orders will not result in any reduction in work force.

Calculating the Economic Manufacturing Quantity. The formula for the least-cost quantity will be derived and illustrated by solving a specific problem. Assume a product which starts at zero inventory at time t_0 as shown in Fig. 11. The product is produced for a length of time $t_1 - t_0$, during which inventory builds to its maximum level at t_1. After production stops, inventory is sold off until it is depleted at t_2 at which time another production run is begun.

The following symbols will be used:

d = annual rate of demand for the product = 6,400
p = annual rate of production = 9,600
s = cost of setting up or preparing for a production run = $100
c = cost per unit of producing the product = $40
i = cost of carrying inventory per year = 20%
Q_p = production quantity
Q^*_p = economic manufacturing quantity
TC = total cost of policy
TC^* = minimum total cost policy when $Q_p = Q^*_p$

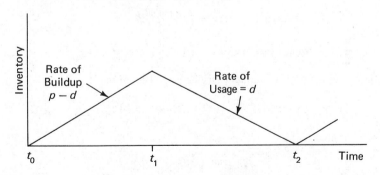

Fig. 11. Inventory position.

There are two opposing sets of costs: **carrying cost** which increases linearly as Q_p increases and the **setup cost** which is inversely proportional to Q_p.

$$\text{Annual setup cost} = \frac{d}{Q_p} s$$

In this particular problem, annual setup cost is equal to $6,400(100)/Q_p$ which results in the curve shown in Fig. 12. In determining the carrying cost, notice that a triangular pattern exists with inventory going from zero to a maximum and back to zero. Thus, by geometry the average inventory is equal to maximum inventory divided by 2. To determine maximum inventory, note that since production is at the rate p and sales the rate of d, inventory increased at

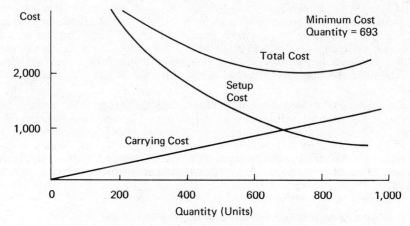

Fig. 12. Cost of manufacturing various quantities.

the rate of $p - d$ during production and is depleted at the rate d during non-production times. Thus the maximum inventory is $(p - d)(t_1 - t_0)$. Note that $Q_p = (t_1 - t_0)p$ and substituting,

$$\text{Maximum inventory} = (p - d)\frac{Q_p}{p} = \left(1 - \frac{d}{p}\right)Q_p$$

$$\text{Average inventory} = \left(1 - \frac{d}{p}\right)\frac{Q_p}{2}$$

$$\text{Annual carrying cost} = \left(1 - \frac{d}{p}\right)\frac{Q_p}{2}\,ic$$

For the example problem, annual carrying cost is equal to

$$\left(1 - \frac{6,400}{9,600}\right)\left(\frac{Q_p}{2}\right)(0.2)40 = \left(\frac{4}{3}\right)Q_p.$$

The total annual cost is

$$TC = \frac{ds}{Q_p} + \frac{icQ_p(1 - d/p)}{2}$$

The minimum point on the total cost curve is found by taking the first derivative with respect to Q_p and setting it equal to zero. Solving for $Q^*{}_p = Q_p$

$$Q^*{}_p = \sqrt{\frac{2ds}{(1 - d/p)ic}}$$

Substituting $Q^*{}_p = Q_p$ into the **cost equation** where $TC = TC^*$

$$TC^* = \sqrt{2ds(1 - d/p)ic}$$

Using the values of the example

$$Q^*_p = \sqrt{\frac{2(6,400)(100)}{(1 - 6,400/9,600)(0.20)(40)}} = 693 \text{ units}$$

$$TC^* = \sqrt{2(6,400)(100)(1 - 6,400/9,600)(0.20)(40)} = \$1,848$$

The **number of production runs** for the year is

$$N = \frac{d}{Q^*_p} = \frac{6,400}{693} = 9.24$$

The **length of each run** is

$$t_1 - t_0 = \frac{Q^*_p}{p} = \frac{693}{9,600} = 0.072 \text{ year}$$

The **maximum inventory level** is

$$(1 - d/p)Q^*_p = \left(1 - \frac{6,400}{9,600}\right) 693 = 231 \text{ units}$$

Use of Nomographs and Slide Rules. Many companies have developed nomographs and special slide rules to speed up the computation of economic manufacturing quantities or economic purchasing quantities. These nomographs show the economic lot quantities to be produced if the setup cost and the rate of consumption are known. Each nomograph is based on a particular inventory carrying charge, and if the carrying charge changes, a new nomograph must be made up.

Slide rules are also manufactured commercially for computing both economic manufacturing quantities and purchasing quantities. Scales are designed for calculating manufacturing lot size so that by knowing the setup cost, the unit cost, and the monthly usage, the economic lot size is read directly from the slide rule. Purchasing lot size calculators are based on the principle of equating the unit cost saving with the added order cost to show a minimum monthly usage for which the discount is justified. If the company's expected monthly usage is higher than this minimum point, it is economical to buy at the higher discount level. A new calculator must be made up whenever the carrying charges change.

Modifications in Economic Production Quantity Theory. There are a number of other considerations which may modify or outweigh the objective of economic lot cost for the company. Some of these considerations include the following:

1. Possibility of change in future demand compared to present average usage.
2. Limitation of available storage facilities; reluctance or inability to finance additional ones.
3. Possibility of sizable changes in unit price.
4. Maintenance of continuity of supply on a scheduled basis so as to help vendors.
5. Correlation with operating schedules and seasonal conditions.
6. Adjustment of ideal quantity to average tooling productivity. (Example: A die may have to be ground at a certain number of pieces. Hence it is not feasible to specify a larger number of pieces.)
7. Limitation of working capital with consequent necessity of operating with a higher unit cost of goods than optimum because of lack of capital.

Capacity Planning

FACTORS IN PLANNING. Whether they are based on sales forecasts or specific customer's orders, production plans must be related to the actual productive capacity of the plant and to technical requirements of sequence and timing. The former may be accomplished through analysis of capacity, the latter through preliminary scheduling and knowledge of process limitations.

Although the objectives are the same, production planning problems and procedures in this phase vary considerably from industry to industry and from one type plant to another. In many chemical plants, for example, questions of timing relate to raw material supplies, inventory tank capacities, bulk transport schedules, and geographic availability of finished inventories in future periods. Planning problems emphasize **inventory planning and transportation economics** to a much greater degree than factors inherent in the production process itself. In a custom machine shop, on the other hand, machine capacity, alternative routing, and process balance occupy the planner's attention. Between the extremes lies a whole range of industrial situations which mix these elements in varying proportions.

ANALYSIS OF CAPACITY. When the necessary information on materials and methods of manufacture has been developed, the next step is to acquire equally accurate data on the capacity of the machine which will operate on the material. In considering a machine, the first question likely to arise is:

> How long will this machine take to perform its operation on a unit quantity of material?

This question can be answered either (a) by actual experiment and trial, or (b) by reference to records of past performance. Though this appears a simple enough procedure, it is not infrequently complicated by the fact that speed of operation varies according to the nature of material and according to the degree of finish or accuracy required.

In machining operations, for instance, the cutting speed varies according to the material worked on. The condition of castings may cause wide variations in possible operating time. Greater speed may be expected in coarse work than in finely finished work. In working near the limit of capacity of a machine, insufficient power may be available, or the awkward character of the job may prevent power being applied in the right amount. In textile manufacture operation speeds vary with the nature of yarn or fabric. In many paper plants the particular material being worked on is a controlling factor in operating speed.

Preparation for the Job. Actual time consumed on a job is made up of two factors in all machine tool work and in many other cases. It takes time, called setup or make-ready time, to prepare the equipment for the job. It is frequently difficult to determine this time accurately, since it varies with the nature of the job.

In the "make-ready" of a large press for printing a three-color job, long and tedious work is sometimes necessary. The amount of such work may be wholly unexpected and may be due to some slight peculiarity in plates, speed, pressure, ink, or paper. Once these are mutually adjusted the operation proceeds at an expected speed; but, if the run is not a long one, the cost of the job may be affected unfavorably. In setting up machine tools, **proper work holding** is frequently a problem that can be solved only by the design of a special fixture or jig. Cutting speeds and feeds have to be determined and much preliminary work done before the job is actually started. In other industries, no such preparation is necessary; material is placed in a hopper or fed into grips to be drawn into the machine without further attention. Where preparation time is considerable, the question of the economic size of lots to be worked on at one setting becomes important.

In machine tool and similar work, the following questions can be asked:

How long does it take to set up this machine for a new job?

How long does this machine take to perform the actual operation on a unit quantity of material, once it has been set up?

From what has already been said it is apparent that the first question can be answered only in an average way. It may be assumed that an average job takes n minutes for setup. But, when it comes to the consideration of a specific job, this factor requires careful scrutiny to make sure that there are no exceptional difficulties inherent in it. In many cases only experiment will disclose the proper time allowance. This phase of the problem may require **operation analysis**, inasmuch as it is a characteristic of specific jobs.

The answer to the second question can be determined quite accurately, either by observation or from records of past performance.

Total Capacity in Terms of Time. The second question in machine analysis is:

How many units of each variety of material can be processed on this machine per unit of time?

Summation of the number of units which can be processed by similar machines gives the total plant capacity in units of product for one process. When all processes have been analyzed and tabulated, a third question can be answered:

What is the maximum plant capacity for each process on each variety of material per unit of time?

Units of Capacity. Units of product in which capacity may be expressed vary necessarily according to industry. In textile and other fabric industries, poundage of yarn and yardage of fabric are usual. In foundries, a tonnage basis may be expected. In machine shops, the problem is, generally, more difficult. Machine tool analysis implies determination of the interaction of several factors: spindle speed, torque, maximum chip area, length of piece that can be operated on, swing-over, and available feeds. All these variables must be taken into account before actual performance can be determined. The capacity of machine tools, therefore, cannot be stated accurately in general terms, except in cases where long runs of identical pieces allow output to be stated as dozens or hundreds per hour. It can, however, be reduced to tables which aid in the solution of any particular problem.

Machine analysis, therefore, has **two important objectives**:

1. To determine approximate maximum capacity of each process, and hence of all processes, and the plant as a whole.
2. In machine shops and industries where several factors enter into machine capacity, to provide a basis for calculating the time of operation on specific jobs.

This calculation is usually effected by embodying the results of machine analysis in tabular or in graphic form. When both setup and operating time requirements are determined for a job, the period during which the machines will be occupied by such jobs will be proportional to the size of the lots going through. Machine loading for all planned jobs can then be determined easily. The amount of work ahead of each machine, expressed in working hours, becomes a known quantity.

Bottlenecks are a serious obstacle to economical production. When the output of each process for each variety of material has been charted or tabulated, flow sheets can be drawn which will show at once if excess capacity, or insufficient capacity, exists at any point.

SIMULTANEOUS PERFORMANCE AND OPERATING EFFICIENCY. When several operations are performed at the same time, it does not necessarily follow that each one is performed with maximum efficiency. The secondary or tertiary operations, performed simultaneously with the primary operation by which the rate of performance is set, may be permitted to lag in certain points of efficiency, provided that the work is accomplished within the time limits set by the primary operation. In other words, the problem is to obtain a net advantage on the whole series even though particular operations might be more efficient if conducted singly. An **example of simultaneous operation** is afforded by a baking oven through which the product moves slowly on a conveyor. Such an operation may be adjusted to any rate of flow within reasonable limits and, as long as the material is in the oven just long enough to be completely baked, dried, or ripened, the required results are accomplished. At any one time, individual pieces may be at all stages of drying, but the net cost is far below that of small batches handled in individual ovens.

Simultaneous operations in assembly work are common practice. Assembly operations are divided into units capable of being handled by single operators. A belt or rotary conveyor carries the product forward, each operator performing his unit task simultaneously with all the others. Thus, in drug and similar industries, bottles are loaded on a conveyor, filled, labeled, sealed, and then removed from the table. The five successive operations overlap. A further stage is reached when some or all of the steps are performed automatically.

Simultaneous production can also be effected in the case of lots, containing a large number of pieces, without special tools or devices. As soon as one day's product is free from machine A, it is passed to machine B for the next process. A and B are thus working on the same job at the same time, though on different individual pieces of it. This procedure does not economize labor or machine time, but it does speed the passage of the job through the plant. It is better described as **overlapping or telescoping** rather than as simultaneous production. From the control viewpoint, simultaneous operations on the same unit of product simply shorten the time during which the machine is occupied by the job. When the process is applied to different units, control boards and route sheets must be so arranged as to permit grouping of these units for simultaneous processing.

PROCESS INDUSTRIES. Analysis of capacity in terms of machine analysis and preparation times is of slight interest in many process industries. The physical relationships and problems of timing which concern the planners here are apt to revolve around raw material supplies, physical limitations on finished inventories, shipping schedules, and transportation economics.

In a typical process plant, for example, a petrochemical operation, these points should be understood:

1. The process is entirely enclosed within a system of pipes, towers, and tanks. Acetic acid, formaldehyde, acetone, and other alcohols and ketones are the end products derived from mixed feed gases.
2. Within certain limits, the mix of end products can be altered by changes in the proportions of gases in the feedstock.
3. Feedstock of a specified composition must be contracted for two to three months in advance of processing in order to assure supply and to buy economically. Spot purchases can usually be made only in limited quantities to achieve small variations in feedstock composition.

4. With the plant built, the process is established. Information on capacity is immediately available. The primary products must be made; their volume may be altered somewhat by production of secondary products, but they will always be present in large amounts.
5. Storage capacity is expensive and limited. Some tanks can be used for different products, some cannot.
6. Bulk movements to the terminal must fit the tanker schedule. Inventories of products shipped in bulk must be high at the plant when the tanker is to be loaded; similarly, tanks at the terminal must be almost empty when the tanker arrives there to unload.
7. Freight economics determine whether a customer is served from the plant or the terminal. Inventories of various products at each point must therefore meet the demands of specific groups of customers.

In a typical case, production planning might involve:

1. Estimates, by periods, of inventory requirements for each product at each storage location so that forecast customer requirements can be met most economically.
2. Establishment of plant operating rates at a level for each period which will:
   a. Maintain inventories at planned levels.
   b. Promote stable plant operations and minimize fluctuations from period to period.
   c. Take into account tanker schedules for bulk movements to the terminal.
3. Indication of the quantity and composition of feedstock required for each period.
4. Establishment of operating periods for secondary operations which produce several products alternately from the same facilities. This must be accomplished in such a way that operating and cleaning cycles are economic but physical limitations on product storage are not exceeded.
5. Preservation of maximum flexibility to meet sudden surges or drops in demand for specific products.

ALTERNATIVES FOR INCREASING CAPACITY. When it is determined that insufficient capacity exists to accommodate the production of new products or increased volume of existing products, a number of alternatives exist for accomplishing the extra work, some of which call for the acquisition of additional equipment and others which provide additional processing capacity without additional equipment.

Machinery and Equipment Alternatives. A number of things can be done to get extra output out of existing equipment. Improved scheduling of orders and material flow can reduce downtime. Longer production runs will increase the proportion of running time to idle or setup time. Increased preventive maintenance during non-operating hours may reduce breakdowns.

New machinery may be acquired. If the extra capacity is needed only temporarily, it may be possible to rent equipment rather than purchase it. Some types of equipment can readily be obtained through the normal equipment rental channels. Even if the equipment is not normally handled by the rental agency, they can often be induced to acquire it if they are assured that your initial rental will cover some portion of the purchase price. Occasionally, it will be possible to arrange rental of machine hours in another firm's factory. Firms with temporary excess capacity on expensive equipment are able to use the rental fees to absorb the overhead fees assigned to the idle equipment. Such rental agreements have even occasionally extended to the temporary loan of machine operators from the firm with excess capacity to the firm needing more capacity. The same effect can be achieved under a different pricing arrangement by subcontracting work.

Use of Inventory. Inventories can be utilized as a temporary substitute for capacity and also in place of capacity in seasonal industries. If demand increases beyond forecast levels and beyond the production capacity of the firm, the increase can be handled by allowing inventories to become depleted. The inventories exist because they are performing a function (see section on Inventory Control) and any reduction below normal levels will soon create costly problems in the firm. Such measures should, therefore, be only temporary and some other permanent solution must be sought to the capacity problem.

If firms producing products like lawn mowers, boats, anti-freeze, etc., which sell only during a short portion of the year, attempted to match the production schedule to the demand schedule, they would require large amounts of capacity which would be utilized only a small portion of the time. Instead they can substitute inventory for capacity by producing at a more level rate and building inventories during normally slack sales periods. These seasonal inventories can then be used up during the peak sales periods while production and capacity is maintained at a lower rate.

Make-or-Buy Alternative. The capacity of a firm to produce finished goods depends on at what stage they begin the manufacturing process. If it starts with the basic raw materials, the total output will be much less than if many of the parts and components are purchased. The firm is faced with a decision of whether to make or buy each part.

Make-or-buy decisions are basically questions of specialization and vertical integration. Some parts differ so markedly from the remainder of parts in the product that there is no question that a producer specializing in the foreign part can produce it more economically. A small manufacturer of fiberglass boats, for example, would probably purchase his glass windshields from a glass manufacturer rather than make them. If the firm reached the size level where it was purchasing $1,000,000 worth of windshields a year, it might then begin manufacturing them. The make-or-buy question must be evaluated periodically because of changing conditions.

Control Functions in Production Planning

RELATIONSHIPS OF CONTROL FUNCTION. Production planning is basically concerned with designing products and the manufacturing systems to produce them. When the manufacturing system goes into operation, several control systems act to monitor the system and to make the adjustments which are required to assure smooth operation. Foremost among these are production control, inventory control, quality control, and cost control. Although these control systems will be considered individually in detail in other sections of the Handbook, their interrelationships will be briefly evaluated here.

A **control system** consists of four parts: (1) a goal or objective, (2) operations to achieve the goal, (3) measurement and comparison of the operations to the goal, and (4) corrective action to adjust the operations (or the goal). The establishment of goals for the control systems is part of the production planning task. In designing the product, the design quality is specified. This is used to determine the goal for the quality control system. The establishment of economic manufacturing quantities helps define goals for production control and inventory control. Capacity analysis and the design of the product provide the essential information for establishing goals for the cost control system. Thus, the control systems must be designed to fit into the framework established by production planning in order to have a cohesive production system.

The control systems are similarly interrelated to each other. If the quality control system detects a large number of rejects or defective items, this will soon be felt in production control since production will fall behind schedule; in inventory control since extra inventory of materials will be required; and in cost control since the result of the extra defective items will be excess production cost.

Thus, it can be seen that although the manufacturing system may be viewed as a group of semi-independent functions on an organization chart, in reality it consists of a group of very closely intertwined functions all working together to achieve the overall production goal—the production of needed goods as economically as possible. Because the interrelationship of the various functions are so complex and so dependent on economic, technological, and competitive factors, the best manufacturing systems cannot be simply and arbitrarily determined. Rather the production manager must combine his knowledge of the operations of the various control systems with his fundamental knowledge of production planning to develop and maintain an effective manufacturing system.

PRODUCTION CONTROL

CONTENTS

CONTENTS (*Continued*)

CONTENTS *(Continued)*

PRODUCTION CONTROL

———

Nature and Scope of Control

IMPORTANCE OF CONTROL FUNCTION. Production control serves as the "eyes and ears" of manufacturing. Its role is to relieve the manufacturing supervisor of non-operating and recording duties so he can concentrate on supervision, production, and quality.

The production control responsibility extends to every element of the manufacturing process. Its function is to:

1. Provide for the production of the proper quality and quantity of materials or product at the required time.
2. Coordinate, monitor, and feed back to manufacturing management the results of the production activity, analyzing and interpreting their significance, and causing action to be taken as indicated.
3. Provide for optimum utilization of facilities and people.
4. Achieve the broad objectives of reliable customer service and low-cost operation.

Costs and Benefits of Production Control. The cost of production control varies with the complexity of the manufacturing operation. Factors affecting the complexity include:

1. Varied rather than repetitive processes.
2. Intermittent rather than continuous processes.
3. Size and dispersion of operations.

All companies provide production control services either through specialists or by incorporating the activities into the responsibilities of line supervisors. As a result, it is often difficult to isolate all costs of production control.

The ratio of persons engaged in specific production control activities to the total number of productive personnel payroll varies from about 1 in 30 to 1 in 70. The amount of work required for production control is more a factor of complexity of the manufacturing process than size of operation. Timms (The Production Function in Business) supports this position, stating that the control of output is usually undertaken entirely by manufacturing supervision in small firms, and in companies where one type of product prevails. He adds that in those firms with mixed types of products and varied processes, the volume and complexity of scheduling, dispatching, and follow-up work will usually be great enough to warrant the use of a specialized group to perform production control.

There are many and important benefits to be obtained from an effective program of production control. Ramlow and Wall (Production Planning and Control) categorize these benefits as follows:

1. **Improvements in profits**
   a. Overall understanding and teamwork.
   b. Maintenance of a balanced inventory of materials, parts, work-in-process, and finished goods.
   c. Balanced and stabilized production.

    d. Maximum utilization of manpower, equipment, tooling, and manufacturing and storage space.
    e. Minimum inventory investment.
    f. Reduction of indirect costs.
    g. Reduction in setup costs.
    h. Reduction in scrap and rework costs.
    i. Reduction in inventory losses.
    j. Reduction of expediting costs.

2. **Competitive advantages**
    a. Reliable delivery promises to customers.
    b. Shortened delivery schedules to customers.
    c. Lower production costs and greater pricing flexibility.
    d. Orderly planning and marketing of new or improved products.

ELEMENTS OF PRODUCTION CONTROL. Effective production control can provide data for analysis and prompt managerial action and warn of incipient trouble or delay in the production process. To do this it relies on procedures that quickly relate actual progress to previously established plans and schedules. These procedures involve the following:

1. **Control of planning.** Assure receipt of latest forecast data from sales and production planning, bill of material data from product engineering, and routing information from process engineering.
2. **Control of materials.** Determine availability of materials in organizations that include inventory management as part of production control, and provide for issues to and movement of materials within the shop.
3. **Control of tooling.** Check on the availability of tooling and provide for issuance of tools to shop departments from the tool crib.
4. **Control of manufacturing capacity.** Determine the availability of men and machines in order to prepare and issue meaningful schedules for the entire system, and provide a means for recording completed work.
5. **Control of activities.** Release orders and information at assigned times.
6. **Control of quantity.** Follow orders to see that correct quantities are processed at each production step and to assure that action is initiated where work fails to pass each stage of inspection.
7. **Control of material handling.** Release orders for movement of work to assure availability of material as required at each stage of the operation.
8. **Control of due dates.** Check on the relation of actual and planned schedules, and determine the cause of delays or stoppages that interfere with weekly schedules of work assigned to each machine.
9. **Control of information.** Distribute timely information and reports showing deviations from plan so that corrective action can be taken, and provide data on production performance measurement for future planning.

PRODUCTION CONTROL SYSTEM. The production control system is a grouping of procedural elements that operate as a whole to fulfill the four functions listed previously under Importance of Control Function.

A well-structured system operating within carefully defined boundaries consists of the following elements:

1. Means for setting the system in motion, such as manufacturing orders.
2. Methods to determine lead time.
3. Methods to control and monitor manufacturing operations, including means to:
    a. Determine what and where work is to be done.
    b. Determine when work is to be done.
    c. Provide orders to the shop and assure that work is completed.
4. Techniques for measuring and recording data on machine utilization, scrap, and indirect labor that can serve as a basis for manufacturing action leading to optimum utilization of facilities and low cost operation.

5. An information system providing for display, filing, and retrieval of data, as well as processing and flow of data.

Prerequisites for Production Control

SYSTEMS AND PROCEDURES. The increasing complexity of manufacturing operations and the importance of having information upon which to base decisions indicates the need for written statements of goals, objectives, and policies developed by top management relative to production control.

The **systems approach** is a means for achieving the production control goals and objectives. This approach uses a set of integrated procedures performed in sequence to achieve written objectives.

Flow charts, decision tables, and written procedures are among the techniques used for analysis, design, implementation, and maintenance of the production control system. Flow charts illustrate in graphic form, decision tables illustrate in tabular form, and procedures relate in narrative form, the series of steps needed to operate the system.

Development of Systems. The Systems Workshop, a series of seminars developed by A. T. Kearney & Company, Inc., management consultants, suggests use of a logical sequence of steps in development of a production control system. These include:

1. **Problem definition.** The problem is clearly defined and objectives of the system to be developed are written.
2. **Preliminary survey.** All facts pertaining to present system and procedures are gathered and analyzed. Facts are evaluated to determine whether:
   a. All activities are necessary and productive.
   b. Flow of information is logical, direct, and controlled.
   c. All procedures are simple, easily followed, and accomplish desired objectives.
3. **Design.** The new or revised system and procedures are developed to accomplish specified objectives.
4. **Implementation.** This includes:
   a. Developing and writing the detailed standard instructions required to accomplish the procedures.
   b. Testing procedures.
   c. Training operating personnel in the system and procedures.
   d. Installing new or revised methods and debugging as necessary.
5. **Audit.** The new system is checked to determine that objectives are achieved.

Components of System Design. Procedural flow charts are useful supplements to written descriptions of control procedures. Their most important contributions are in describing and evaluating existing systems and designing new systems. Flow charts provide a graphic presentation of the system, the control it provides, and the effectiveness of its operation.

Figure 1 illustrates a **flow process chart** frequently used for manual analysis and clerical work distribution studies. An illustration of a segment of a **computerized system** is shown in Fig. 2. Each of the small circled numbers on the **flow chart** refers to a written portion of the procedure. Taken together, the flow chart and procedure provide a concise packet of information for personnel training or for evaluating a system.

For analysis and solution of a problem, it is very useful to reduce the problem statement to a table of condition/action statements. Such a table is called a **decision table.**

The condition statement is always a question requiring a yes (Y) or no (N) answer. Based on the answer, the action is read on the checked line among the action statements. A decision table follows, on page 5.

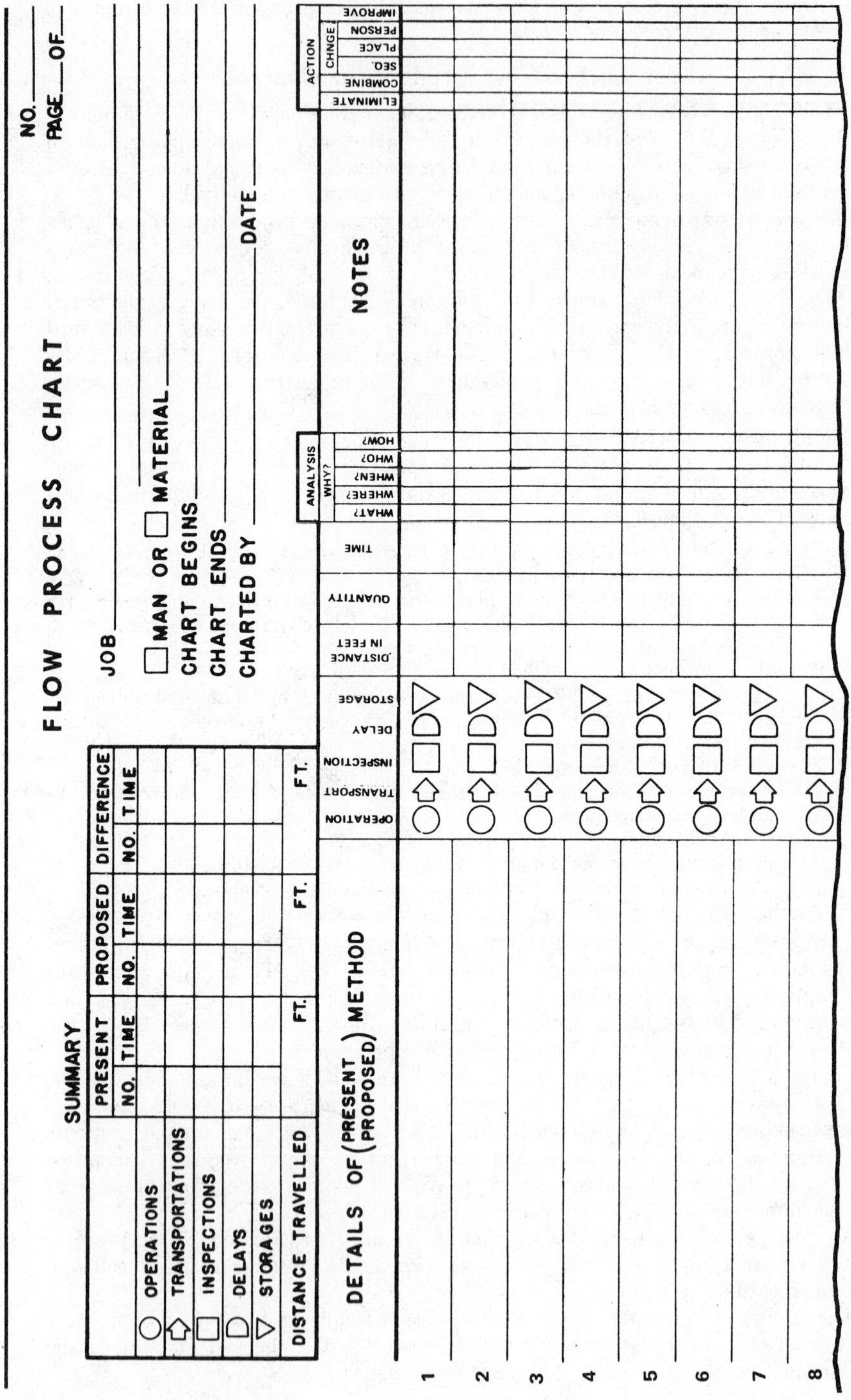

FLOW PROCESS CHART

NO.
PAGE OF

JOB

☐ MAN OR ☐ MATERIAL

CHART BEGINS
CHART ENDS
CHARTED BY   DATE

SUMMARY

	PRESENT		PROPOSED		DIFFERENCE	
	NO.	TIME	NO.	TIME	NO.	TIME
◯ OPERATIONS						
⇧ TRANSPORTATIONS						
☐ INSPECTIONS						
D DELAYS						
▽ STORAGES						
DISTANCE TRAVELLED	FT.		FT.		FT.	

DETAILS OF (PRESENT / PROPOSED) METHOD

OPERATION	TRANSPORT	INSPECTION	DELAY	STORAGE	DISTANCE IN FEET	QUANTITY	TIME	WHAT?	WHERE?	WHEN?	WHO?	HOW?	NOTES	ELIMINATE	COMBINE	SEQ.	PLACE	PERSON	IMPROVE

ANALYSIS — WHY?

ACTION — CHNGE

1
2
3
4
5
6
7
8

Fig. 1. Flow process chart.

Condition statement: Credit limit is OK?   Y   N

Action statement: Approve order   X

Return order   X

Preparation of Procedures and Standard Instructions. Written procedures should include the following sections:

1. Purpose of the procedure.
2. Basis of the procedure, including a statement of the fundamental principles upon which the procedure is based.
3. Scope of the procedure, including a brief outline of contents.
4. Responsibility for administration and performance of the procedure.
5. Reporting relationships.
6. Detailed instructions arranged in outline form for ease of reference and maintenance.
7. Exhibits, flow charts, decision tables, and forms used with the procedure.
8. Audit checklists providing a basis for maintaining the system.
9. Approval of the procedure by the affected department managers.

A **standard instruction** is used to prescribe in detail the operations required to perform each step in a procedure. Figure 3 shows a sample page from a set of standard instructions and the form to which this part of the procedure applies. Written procedures and standard instructions are best developed from the viewpoint of the user. Care is exercised to assure clarity, proper utilization of skills, and on-time performance in the manner specified.

It is best to print procedures and instructions on loose-leaf paper so manuals can be easily updated, as necessary. Each page carries a reference heading and the date of the latest revision of the page. A central file shows the distribution of each copy of the manual. This file is checked each time a revision is made to assure that new pages are distributed and old pages recalled.

The production control manager participates in development of and approves all procedures and standard instructions affecting the production control department.

FACTORS DETERMINING CONTROL PROCEDURES. Systems and procedures for production control are influenced by several factors. These factors include the nature of the manufacturing process, the degree of complexity of the production operation, and the magnitude of operations. Each of these factors has a number of variations tending to cause production control procedures to become more sophisticated, complex, and formal.

Nature of Manufacturing. Manufacturing companies are often categorized according to the length of time machines are run without setup changes. This distinguishing characteristic is used to identify a manufacturing operation as either intermittent, continuous, or composite production, although few companies exhibit characteristics totally of one type.

Sophistication of production control procedures is generally at a minimum in the **continuous flow process operation.** Examples of this type of operation are found in the petrochemical, soap, and synthetic fiber industries. Flow sheets in these industries show a continuous stream of production in which many operations are performed without break in flow. In this type of production, quality control is highly developed and planning for raw materials, finished inventory levels, and markets is extremely important. The production control function in such industries is generally embodied in the process equipment itself.

Intermittent, multi-operation production contrasts sharply with continu-

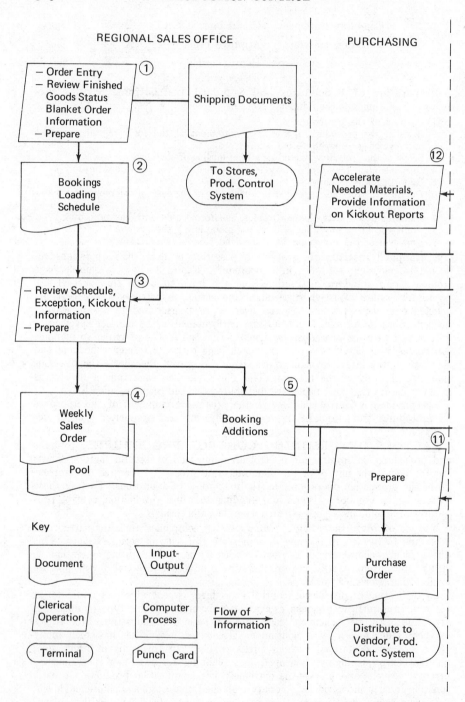

Fig. 2. Computer system

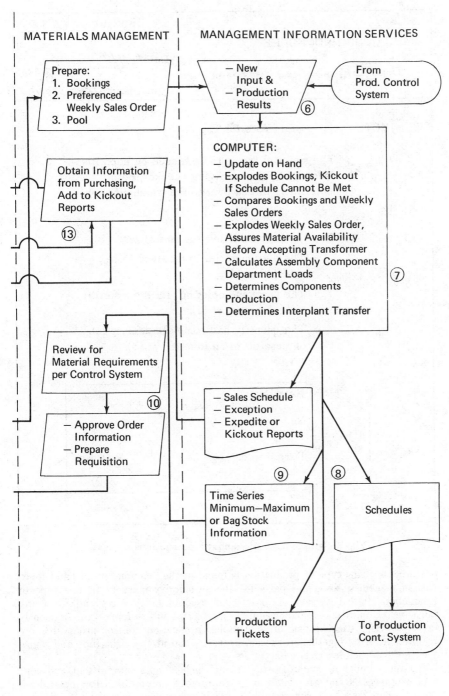

MATERIALS MANAGEMENT

MANAGEMENT INFORMATION SERVICES

Prepare:
1. Bookings
2. Preferenced Weekly Sales Order
3. Pool

— New Input &
— Production Results ⑥

From Prod. Control System

COMPUTER:
— Update on Hand
— Explodes Bookings, Kickout If Schedule Cannot Be Met
— Compares Bookings and Weekly Sales Orders
— Explodes Weekly Sales Order, Assures Material Availability Before Accepting Transformer
— Calculates Assembly Component Department Loads
— Determines Components Production
— Determines Interplant Transfer ⑦

Obtain Information from Purchasing, Add to Kickout Reports ⑬

Review for Material Requirements per Control System ⑩

— Approve Order Information
— Prepare Requisition

— Sales Schedule
— Exception
— Expedite or Kickout Reports ⑨ ⑧

Time Series Minimum—Maximum or Bag Stock Information

Schedules

Production Tickets

To Production Cont. System

flow chart.

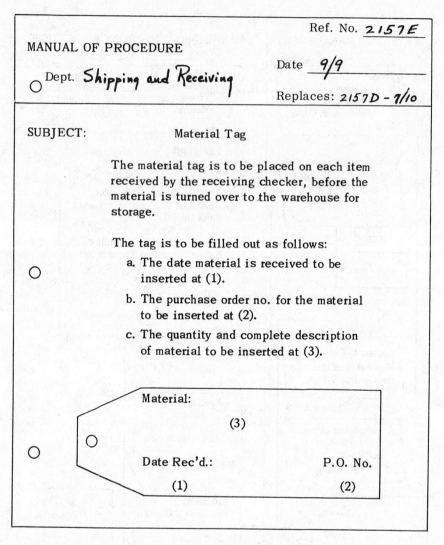

MANUAL OF PROCEDURE

Ref. No. *2157E*

Dept. *Shipping and Receiving*

Date *9/9*

Replaces: *2157D - 7/10*

SUBJECT: Material Tag

The material tag is to be placed on each item received by the receiving checker, before the material is turned over to the warehouse for storage.

The tag is to be filled out as follows:

a. The date material is received to be inserted at (1).

b. The purchase order no. for the material to be inserted at (2).

c. The quantity and complete description of material to be inserted at (3).

Material:

(3)

Date Rec'd.:   P.O. No.

(1)   (2)

Fig. 3. Sample page of standard instructions manual.

ous process. This type of production is found in the manufacture of hand tools, custom machines, toys, and parts for the automotive market. In this type of production a great variety of material is used in many ways and for many purposes. Hundreds, even thousands of parts and subassemblies are frequently involved. Sophisticated and often complex production control procedures are required to bring together in proper sequence, and at the right time and place, the results of such interrelated activities.

A large number of manufacturing plants include both intermittent and continuous processes and are classified as **composite** or **combination** operations. Such a plant may have subassembly departments making parts in a continuous operation while the final assembly department works on an intermittent basis (as in the furniture and custom packaging industries). Production control procedures often become sophisticated and complex in situations where a smooth flow in the continuous process operation must be maintained while simultaneously minimizing inventories ahead of assembly.

Complexity of the Operation. Generally, the greater the variety of operations, the more complex is the problem of effective production planning and control. The extremes of complexity range from continuous production of a single standardized product to production of special orders.

Factors affecting the complexity of production control procedures are:

1. Number of ultimate parts in the end product.
2. Number of different operations on each part.
3. Extent to which processes are dependent, that is, processes that cannot be performed until previous operations have been completed.
4. Variations in production rates of machines used in the process.
5. Number of discrete parts, subassemblies, and assemblies that cannot be stocked.
6. Degree to which customers' orders with specific delivery dates occur.
7. Receipt of many small lot orders.

Production control procedures become more simplified as these factors decrease.

Magnitude of Operations. The size of the operation and the distance parts must travel from operation to operation are important in establishing production control procedures. In the small, single-plant enterprise, procedures are typically informal. Procedures tend to become formal where assembly plants are large and depend on parts from subassembly plants located hundreds of miles away. Generally, greater need for centralized production control organization and formal procedures is found as the size of the operation increases and dependent operations become more physically separated.

LONG-RANGE PLANNING. Long-range planning is an essential prerequisite for design and maintenance of the production control system. It specifies the unified goals of the company, providing production control with the direction required to establish departmental objectives, prepare an organization plan, and update operating procedures.

Robert H. Schaffer (Harvard Business Review, vol. 45) stresses the importance of planning, and states that:

The essential purpose of corporate planning should be to enable managers to act today with increased skill, speed, and confidence to produce desired results tomorrow. . . . It (planning) must become an overall strategic framework for moving forward on many fronts toward the goal of increased managerial control. Of especial importance, it must help management to gain practical, successful experiences in setting a goal, creating a plan to reach it, and making the plan succeed.

To be successful long-range plans must be developed and supported by top management. Each year's portion of the long-range plan is called the **annual target.** This includes the coordinated individual plans of marketing, engineering, manufacturing, and finance. Together they help establish boundaries within which the production control system will operate.

The **marketing** plan is a forecast of the type, quantity, and timing of products to be sold. **Engineering** plans provide for orderly development and introduction of new products and improvement of existing products. **Production** plans specify the methods, processes, materials, and capacity used or developed to support the marketing plan. Each functional department prepares salaried and hourly organization plans, including overhead expenses needed to support the annual target. The **financial** plans summarize the capital investment, inventory investment, operating and burden expenses. These financial plans are consolidated into a forecast of earnings and cash flow.

SHORT-RANGE PLANNING: SALES, INVENTORY, AND OPERATIONS PLANNING. In the course of the year each functional department adjusts its plans in an attempt to keep pace with operations in other departments. Effectiveness of communications, delay in reacting to off-standard conditions, and results achieved by each department in meeting their modified plans, are factors in a company's ability to keep the activities of all functional departments in balance.

The need to adjust plans quickly to effectively correct off-standard conditions leads some companies to create a **short-range planning committee.** The committee's purpose is to provide regular and systematic integration of sales needs, operating capabilities, and inventory requirements in a unified short-term corporate plan. Meetings are held each two to four weeks depending on corporate needs. The committee is composed of key men representing the vice-presidents of sales-marketing, manufacturing, and materials management. The small size of the committee contributes to its effectiveness.

Each member of the committee independently develops a detailed short-term plan within the objectives and parameters of the company's long-range plan.

The **sales plan** is developed by marketing and consists of a forecast of shipments based on information developed from customer interviews, industry projections, current backlog and sales history. The **operating plan** is developed by production control. It consists of a production plan of end units and parts to be produced in each work center in each time period. The operating plan provides data for development of the manpower and machine loads required in each work center. The **inventory plan** is developed by inventory control and is based on information on raw materials, purchased parts, components, and end units required over the time span of the plan. The inventory data are developed by "exploding" the bills of material of each end unit for the quantity of required parts and components.

The short-range planning committee resolves department conflicts or imbalances and produces an approved, coordinated plan or **master plan.** The committee reports directly to the chief executive of the company or division who judges the committee on its ability to develop **realistic plans.**

Communications improve because committee members are given the authority to act and make decisions for their respective vice-presidents, and are charged with responsibility to take the coordinated plans back to their departments for implementation.

Among **benefits** claimed by users of this planning technique are:

1. Improved sales because of greater delivery service reliability and shorter lead times.
2. Reduced inventories.
3. Improved utilization of facilities and manpower.
4. More stable work schedules.

PRODUCTION CONTROL ORGANIZATION. The strategic plans of a successful company are usually dynamic and changing. The organization of the production control activity, consequently, must be analyzed and updated at least annually to assure attainment of corporate objectives.

Department Framework. The size and character of a production control organization depends on the duties specifically delegated by top management, the degree of control specified by corporate objectives, and the size and complexity of the manufacturing operation.

The manager of a production control department often reports directly to the factory or division manager in operations in which the production control activity is vital to success. A popular variation of this organization structure involves use of the **materials management** concept (see section on Materials Management). In this concept, production planning, purchasing, inventory control, production control, stores, and traffic activities are organized under a single individual who is a part of or reports directly to top management. The functions included will, of course, vary among companies. Generally, the relative importance of the production control activities compared to other activities in the materials management concept is highest in a large, intermittent, custom operation where the cost of direct labor is higher than the cost of material in the end product.

Typical examples of the many ways companies organize production control activities are shown in Figs. 4, 5, and 6. In the combination organization shown in Fig. 6, each assembly department assembles a different product and supplies many required parts from its own stockroom. Production is controlled by planners in the individual departments.

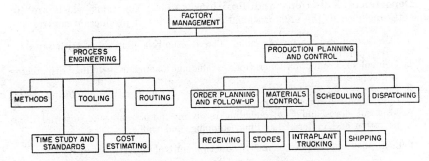

Fig. 4. Production control organized along functional lines.

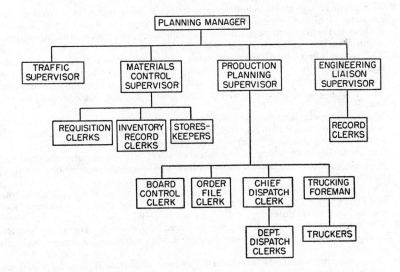

Fig. 5. Centralized production control organization.

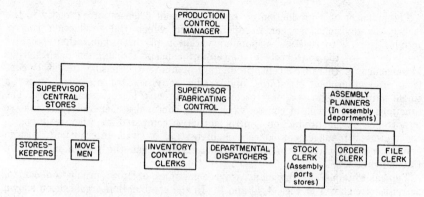

Fig. 6. Combination centralized–decentralized production control organization.

Department Functions and Positions. Among functions which are frequently included in the organizational responsibilities of production control are:

1. Forecasting and master planning. This is usually a joint effort with sales and production management.
2. Control of production operations, including loading, scheduling, dispatching, and expediting.
3. Preparation of shop paper such as copies of routings, manufacturing orders, production schedules, and requisitions.
4. Material stores and flow, including receiving and warehousing, internal transportation and material handling, shipping, and traffic.

Typical positions and duties in a production control department are:

1. The **manager** of the production control department is responsible for administration and supervision.
2. The **planner** ascertains schedule quantities on each order, loads and records progress of orders, issues additional work or approves overtime as required, authorizes replacement of spoiled work, and supervises preparation of purchase requisitions for materials or parts bought outside.
3. The **materials control clerk** receives and records data from stores, makes reservations of materials for orders being planned for production, and prepares purchase requisitions for materials to be bought for replenishment of stores. This position exists where material control is combined with production control.
4. The **routing clerk** prepares copies of route sheets for each part or component order, checks due dates for each stage of the work (after scheduling), and supervises preparation of work orders and time tickets.
5. The **order-writing clerk** writes work or operation orders, material and tool issue slips, inspection orders, move orders, and purchase requisitions for items bought outside.
6. The **scheduler** maintains machine load charts, calculates and posts loads from route sheets, establishes dates for each operation, and maintains progress data on production schedules.
7. The **dispatcher** maintains progress and control records for the shop; receives and issues shop paper; and reports delays, idle machines, shortages and substitutions.
8. The **traffic man** controls intraplant movement of materials, schedules pick-ups and deliveries, and may control and follow up interplant shipments.

Many of the above positions may be combined or eliminated in a computer-based production control activity. Computers, for example, are used for preparing route sheets, loading and scheduling, issuing shop orders and requisitions as required, and maintaining production progress data.

Two Approaches to Production Control. There are two basic approaches to production control. Its activities can be incorporated into the jobs of line foremen, or they can be conducted by specialists in a production control department. These approaches have the following advantages and disadvantages:

1. Foreman control
   Advantages:
   a. Working knowledge of current capabilities and limitations of men and machines in his department.
   b. Knowledge of the sequencing of jobs for the most effective production efficiency within his department.
   Disadvantages:
   a. Best sequence of orders in one department may be poor in another.
   b. No means for developing an accurate picture of shop load.
   c. Order follow-up is difficult.
   d. Little opportunity for overall production planning and coordination.
   e. Foremen are diverted from their primary function of supervision.
2. Central office control
   Advantages:
   a. Can provide an accurate picture of present and future loads.
   b. Effect of each new order on the shop can be determined more easily.
   c. Accurate delivery commitments possible.
   d. Responsibility may be assigned to this group for evaluation of delays.
   e. Provides means for centralized collection of data on which corrective action may be based.
   f. Provides for central coordination of production and supporting activities.
   Disadvantages:
   a. Tendency for central office personnel to lose contact with actual shop conditions.
   b. High installation and operating costs.
   c. Risk of inflexibility.
   d. Danger of encroaching on areas of line authority.

In actual practice, production control procedures attempt to capitalize on the advantages of both approaches and minimize the disadvantages of each.

Manufacturing Orders

THE PRODUCTION ORDER. The production order is a document initiated by production control based on information provided by data from individual customer orders, sales demand for stock, or inventory replenishment rules. It **authorizes the start of production.** The function of the production order is to:

1. Convey information about the customer, his requirements, specifications, and promised time and place of delivery.
2. Provide a nucleus for cost collection, either for the order as a whole, or for individual components and processes on components.
3. Form a starting point for the production control mechanism.

There are many variations on the means employed to develop production orders. For example, in the steel industry it is customary to **reserve a block of capacity** in a specific time period by means of a **virtual** or forecasted order. The sales department will then convert the virtual order to an **actual** order as firm data are received from the customer.

B82

PRODUCTION ORDER

CHARGED TO ~~ORDERED BY~~ John Doe

SHIP PREPAID ~~COLLECT~~ TO John Doe

New York

C

SHIP. ORDER NO. 38700

WEIGHT ______ HOW PACKED ______

VIA Express

F. O. B. Providence   SERIAL NO. ______

DATE ORDERED BY CUSTOMER			SHIPPING ORDER ISSUED			SHIPMENT DUE		
MONTH	DAY	YEAR	MONTH	DAY	YEAR	MONTH	DAY	YEAR
4	15		4	15		4	17	

TERMS: Net 30 Days   CUSTOMER'S ORDER NO. 40   REQ. 750

Item	Quantity	Description	Drawing No.	Check	Pattern No.	Stock M'f'g. Purch.	Symbol	On Hand		Ship-ped
1	100	#2 Change Gears 72T. 16P. 7/16" F.	11701-D-72			WM	Rav. 11701-72 G2R	✓	✓	
						PM		–	–	
						P		–	–	

CUSTOMER'S ORDER REC'D		SHEET WRITTEN AND APPROVED		ORDER WRITTEN		DRAWINGS AND DATA CHECKED		ISSUES WRITTEN		MATERIAL APPORT'ND		PASSED	SHIPMENT COMPLETE		BILLED		ENTERED ON RECORD	
GL		CSB		MM		CSB		RT		JS		FG	RH		DH		JH	
DAY	HOUR	DAY	HOUR	DAY	HOUR	DAY	HOUR	DAY	HOUR	DAY	HOUR	INSPECTOR	DAY	HOUR	DAY	HOUR	DAY	HOUR
4/15	1	4/15	2	4/15	3	4/15	2	4/15	3	4/15	3	4/15 – 3	4/15	4	4/15	5	4/15	5

WM = worked material. *PM* = plant foundry material. *P* = purchased material. The two checks in the "On Hand" column indicate that material is on hand and ready to ship.

Fig. 7. Production order and progress record.

One type of production order is illustrated in Fig. 7. The format of such orders will vary depending on the needs of the production operation.

Figure 8 illustrates a production order for stock. The authorization for production is for a quantity of each size and kind of fabric during one week. The production order is based on a consolidation of orders from customers, or on anticipated demand.

KNITTING ORDER
for

Serial No. 46

Mill No. 6   Week of 10/10/ to 10/17/   Date 10/11/

FABRIC	7½	8	8½	9	9½	10	10½	11	12	12½	13	14	15	16	17	18	19	20	21	22	23	24	25	26	GRAND TOTAL
4639 Machine Numbers													2-16	3-5	7-10	8-11	9-12	14-17	16	17					
4639 Cloth Order													600	700	700	1700	240	300	300	360					4900
4940 Machine Numbers												18-1	24-26	28-30	30-40	42-41	46-47	49-60	68-66	67-65	63-61	79-78			
4940 Cloth Order												400	500	600	700	700	700	800	400	400	400	400			6000

Fig. 8. Production order authorizing weekly outputs.

OTHER MANUFACTURING ORDERS. In many companies a production order initiates a series of other manufacturing orders for parts and components. These may include orders for patterns and castings; tool, fixture, and jig orders; material requisitions; store-issue orders; move orders; inspection orders and replacement orders.

Computer programs, utilizing on-line bill of material, routing, and tool files, rapidly produce complete sets of orders with the single entry of the finished part number of the production order. Figure 9 illustrates a computer-prepared set of orders including: summary shop order, material requisition, and multi-use job-move-location ticket, all produced by the computer from an input of a part number and production week.

COMBINATION MANUFACTURING ORDERS. In small plants and in industries with fixed processes, manufacturing orders are sometimes **consolidated** with the routing document and even the scheduling procedures. Figure 10 illustrates a manually-prepared set of manufacturing orders. This form serves as production order, job order, and routing. The top stub represents the production order; the lower stubs, other types of manufacturing orders. Figure 11, on page 18, illustrates a production order combined with a schedule.

THE JOB TICKET. The job ticket or job order is a manufacturing order authorizing the operator to start production. The job ticket may include information to describe and plan the job. Job tickets can also be designed to record work progress by the operator and become the source document for status control, cost accounting, and payroll.

Data typically recorded on the job ticket includes, but is not limited to:

1. Order number and part number.
2. Operation number and description.
3. Machine number, operator's name and clock number. (This may be scheduled or assigned at time of dispatching.)

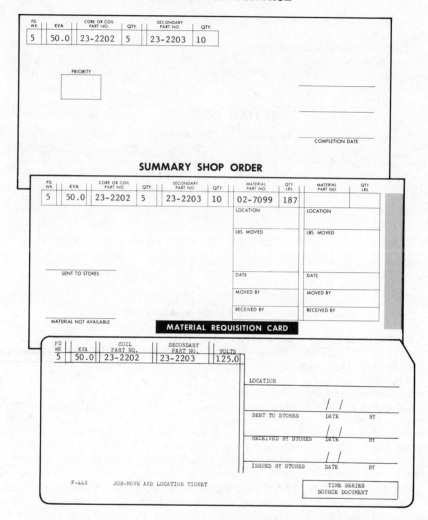

Fig. 9. Manufacturing orders produced by a computer program.

4. Number of pieces to be processed, actual good production, and scrap.
5. Date jobs should start.
6. Standard or estimated time allowed for setup (if applicable) and operation.
7. Start, finish, and elapsed production time on the job.

Job tickets are frequently combined with other forms. **Combination form design** permits multiple use of the same data. A place to record inspection results is often included on the ticket. Move coupons are frequently attached in piece-work industries. In some cases, the job ticket is used as the time record.

Figure 12 illustrates a job ticket in the form of a tabulating card, which, when punched, is used for costing, payroll, and other purposes. Figure 13 shows a job

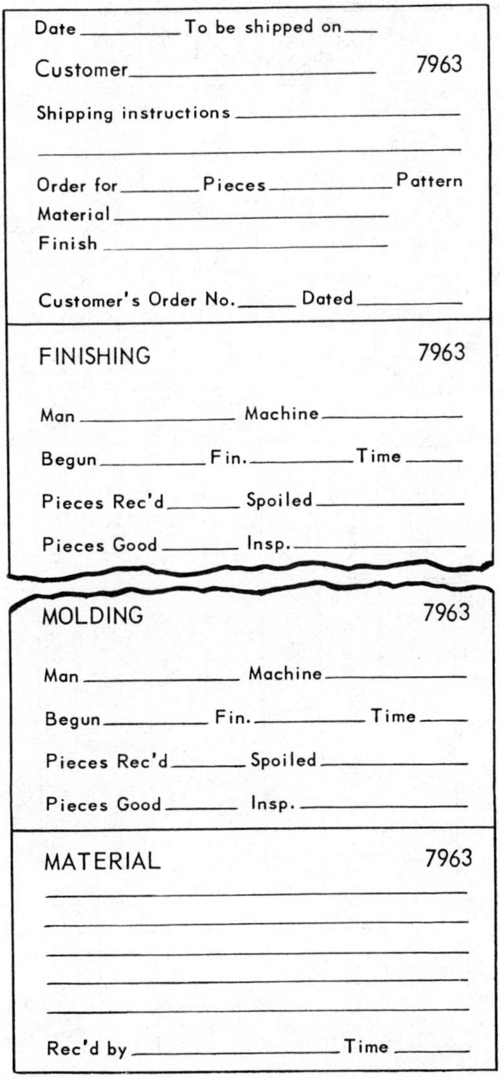

Date _____ To be shipped on ___

Customer_____   7963

Shipping instructions _____
_____

Order for _____ Pieces _____ Pattern

Material _____

Finish _____

Customer's Order No. _____ Dated _____

FINISHING   7963

Man _____ Machine _____

Begun _____ Fin. _____ Time _____

Pieces Rec'd _____ Spoiled _____

Pieces Good _____ Insp. _____

MOLDING   7963

Man _____ Machine _____

Begun _____ Fin. _____ Time _____

Pieces Rec'd _____ Spoiled _____

Pieces Good _____ Insp. _____

MATERIAL   7963

_____
_____
_____
_____
_____

Rec'd by _____ Time _____

Fig. 10. Combined production and job order where routing is fixed.

ticket, called a production card, used in a large plant manufacturing automobile bodies. The lower portion of the form is an inspector's report where provision is made for half-hourly quality checks.

An **operation card,** similar to the document shown in Fig. 14, is another type of combination job ticket. Operation cards are sometimes made as complete order copies. Individual order numbers in such cases are circled to indicate to which operation the card is applicable.

DISTRIBUTION AND CONTROL OF INFORMATION. Manufacturing orders are usually distributed to all departments working on the order

CHARGE TO: *John Widdecomb & Co.* LOT: *5* ORDER NO. *6000*

SHIP TO: *Grand Rapids, Mich.* SOLD BY: *J. L.* CUST. NO. *4001*

VIA:

RECEIVED	MO	DAY	YR	~~WANTED~~ PROMISED	MO	DAY	YR
	Oct.	1	19–		Oct.	31	19–

SIZE		BEV.	SILV.	PATTERN	QUAL.	PLATES	REMARKS	DATE CUT	SCHEDULING: ACID	POL.	BEV.	PE	MIT.		SILV.
18"	x 30"	—	√	plain	I	50		3		A-50					A-50
12"	x 20"	—	√	"	I	200		4,5 & 6		A-50 B-50					B-100
22"	x 40"	—	√	"	I	150		8 & 9		B-150					B-150
15"	x 28"	—	√	"	III	100		10 & 11		B-100					C-100
36"	x 60"	1"	—	342-a	I	10	See previous order #3562	10			C-10	C-10			

Fig. 11. Production order combined with schedule.

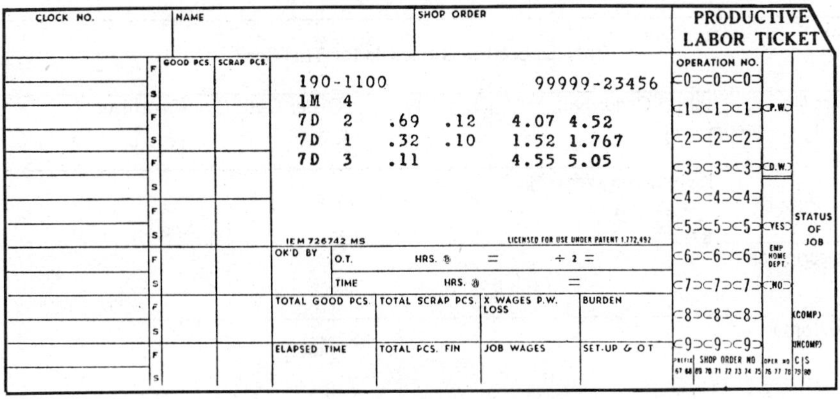

CLOCK NO.		NAME			SHOP ORDER			PRODUCTIVE LABOR TICKET
		GOOD PCS	SCRAP PCS					OPERATION NO.
	F			190-1100		99999-23456		⊂0⊃⊂0⊃⊂0⊃
	S			1M 4				⊂1⊃⊂1⊃⊂1⊃ P.W.
	F			7 D 2	.69 .12	4.07 4.52		⊂2⊃⊂2⊃⊂2⊃
	S			7 D 1	.32 .10	1.52 1.767		⊂3⊃⊂3⊃⊂3⊃ D.W.
	F			7 D 3	.11	4.55 5.05		⊂4⊃⊂4⊃⊂4⊃
	S							⊂5⊃⊂5⊃⊂5⊃ YES. STATUS OF JOB
	F			IEM 726742 MS		LICENSED FOR USE UNDER PATENT 1,772,492		⊂6⊃⊂6⊃⊂6⊃ EMP HOME DEPT
	S			OK'D BY	O.T. HRS. @ = + 2 =			⊂7⊃⊂7⊃⊂7⊃ NO
	F				TIME HRS. @ =			⊂8⊃⊂8⊃⊂8⊃ (COMP)
	S			TOTAL GOOD PCS.	TOTAL SCRAP PCS.	X WAGES P.W. LOSS	BURDEN	⊂9⊃⊂9⊃⊂9⊃ (NCOMP)
	F			ELAPSED TIME	TOTAL PCS. FIN	JOB WAGES	SET-UP & O T	PREFIX SHOP ORDER NO OPEN NO C S 67 68 69 70 71 72 73 74 75 76 77 78 79 80
	S							

Fig. 12. Job order tabulating card.

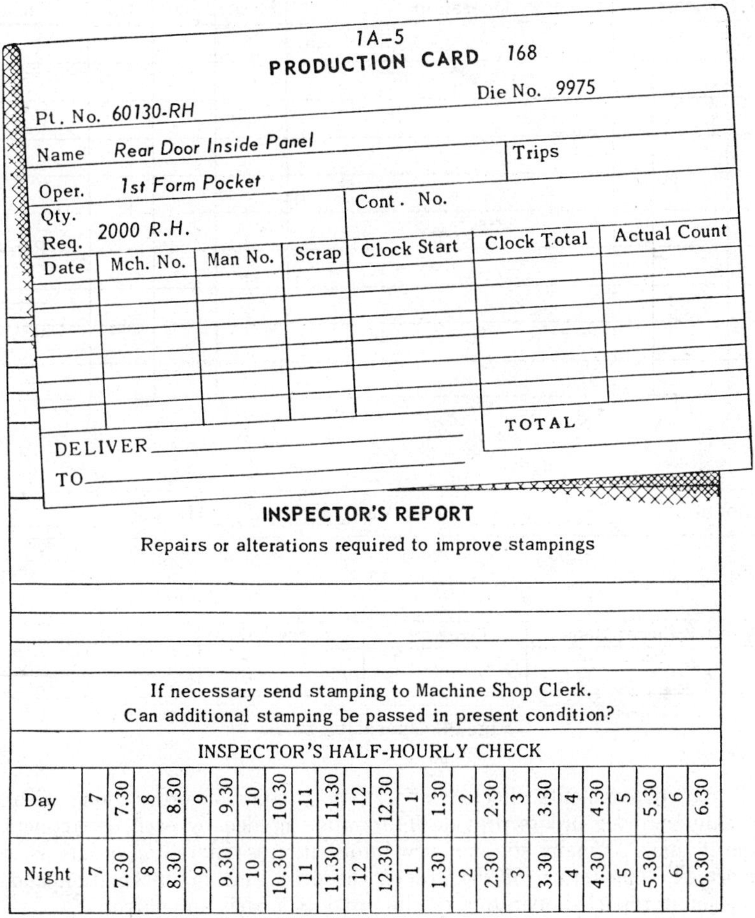

1A-5
PRODUCTION CARD 168

Die No. 9975

Pt. No. 60130-RH

Name Rear Door Inside Panel

Trips

Oper. 1st Form Pocket

Cont. No.

Qty. Req. 2000 R.H.

Date	Mch. No.	Man No.	Scrap	Clock Start	Clock Total	Actual Count

TOTAL

DELIVER

TO

INSPECTOR'S REPORT

Repairs or alterations required to improve stampings

If necessary send stamping to Machine Shop Clerk.
Can additional stamping be passed in present condition?

INSPECTOR'S HALF-HOURLY CHECK

Day	7	7.30	8	8.30	9	9.30	10	10.30	11	11.30	12	12.30	1	1.30	2	2.30	3	3.30	4	4.30	5	5.30	6	6.30
Night	7	7.30	8	8.30	9	9.30	10	10.30	11	11.30	12	12.30	1	1.30	2	2.30	3	3.30	4	4.30	5	5.30	6	6.30

Fig. 13. Job order combined with inspection report.

Only Operation Circled Is for This Card				
Part No.	Shop Order No.	Quantity	Order Date	Promised
Page of	Scheduled Week: Forge	Fitting		Assem.
Description				
Customer				

Dept.	Op. Seq.	M e n	Operation	Mach.	Fce.	Set Up	Std. Hrs. Per. C.

Dir. Labor at Std. _____   Rate Var. _____ Use Var. _____

Date	Pieces Made	Position	Clock No.	Punch	
				F	
				S	

Fig. 14. Operation card.

and all staff departments affected by it. The production order along with the other manufacturing orders provide information needed by each department to perform its job. At many stations, new information is entered on orders as work is performed. The return of the orders, the processing of information and interpretation of reported operating results provides a basis for control.

Five elements need to be considered in establishing a workable system for **control of shop orders,** including:

1. **Timing** of document flow control activity. Schedules and transmittals, similar to Fig. 15, are provided to indicate when, by whom, and what information is to be forwarded. Logs are used to make sure that all required information is received prior to beginning an activity.
2. **Responsibility** for handling information. Each individual performing an activity in the document flow has a defined responsibility according to established procedures.
3. **Auditing** of system. It is essential that each document have a predefined flow in order to fix responsibility for errors and initiate training or other remedial action.
4. **Balancing procedures.** These procedures are used to check that all documents are processed accurately. Two types of control techniques are used:
    a. **Document counts** are appropriate when no new posting is required. This involves a simple count of documents received.
    b. **Control totals** are appropriate when significant numbers are to be transcribed. A precise formal balancing of accumulated numbers is required in this situation.
5. **Accuracy of data.** The degree of accuracy of counts and the proper time, place, and method of measuring varies with individual situations. The financial impact of accuracy is considered before selecting the method of control. The cost of obtaining control should not exceed the potential loss if no control existed.

Routing

THE ROUTING FUNCTION. Routing defines the work to be done and where it is to be performed. This information is usually provided to production control by a product or process engineering function. It is used by production control to load and schedule, and may be sent to manufacturing to be used during production.

Typical of the information and data received from engineering are:

1. Detail drawings of parts and assemblies.
2. Bills of material and quality specifications.
3. Sequence of operations to be followed at each work center.
4. Equipment to be used.
5. Tools, jigs, and fixtures required.
6. Feeds and speeds for machine tools.
7. Machine setup and operating standards.
8. Direct and indirect labor time standards.

PREPARATION OF ROUTE SHEETS. A route sheet is a document providing information and working data for translating materials into finished parts or products. It defines each step in the production operation.

Route sheets for distribution to the shop may be prepared manually or mechanically. Generally, in custom manufacturing, engineering develops route sheets manually with each new order. In continuous-process production and in production for stock, route sheets are often fixed and developed by mechanical means. In some **computer-based systems** it is possible to enter only specifications for the end product and the computer develops and prints the route sheet based on stored information.

The format of a route sheet is designed to suit the needs of the production operation. A detailed route sheet is shown in Fig. 16. Data typically included on route sheets are:

1. Number and order identification.
2. Identification of the part and drawing reference numbers, including the date of the last engineering revision.

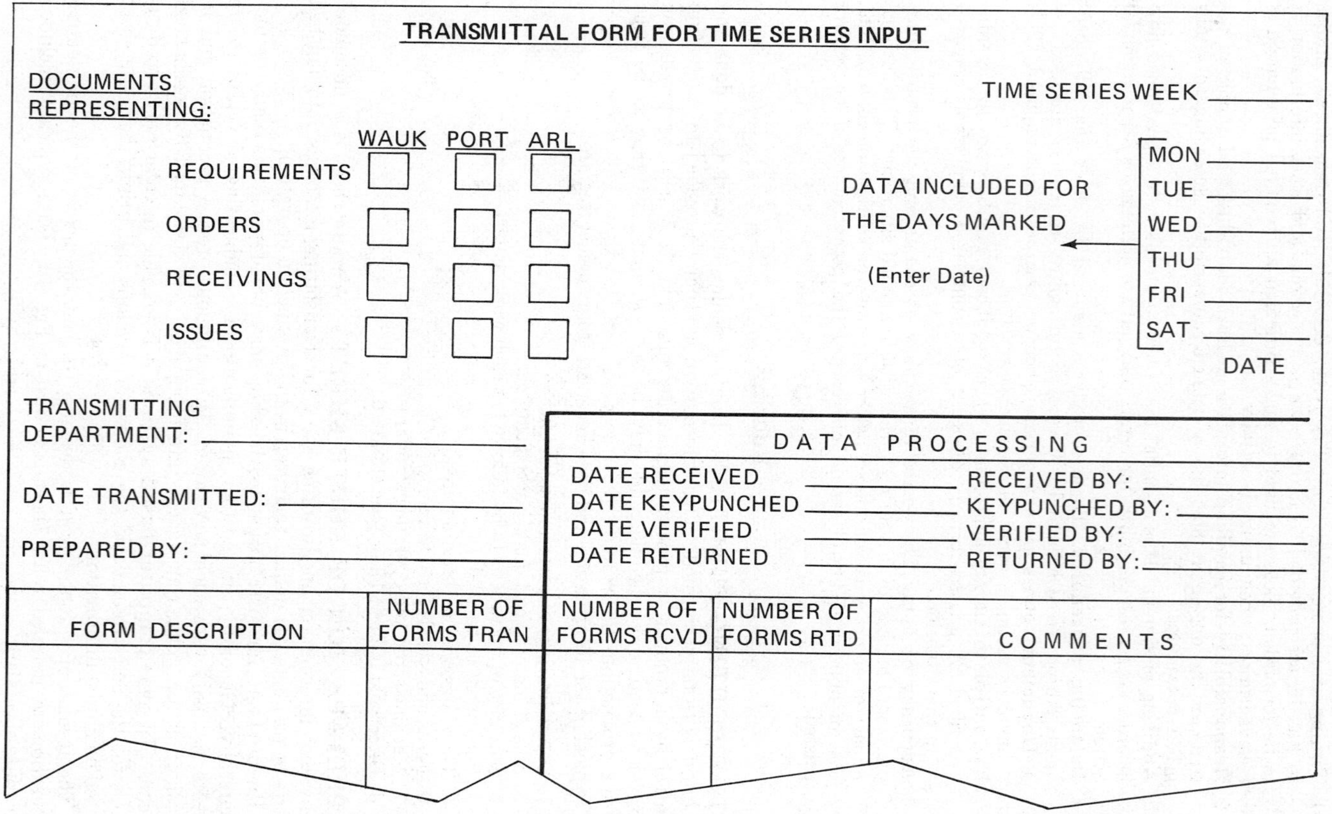

TRANSMITTAL FORM FOR TIME SERIES INPUT

TIME SERIES WEEK ——————

DOCUMENTS
REPRESENTING:

	WAUK	PORT	ARL
REQUIREMENTS	☐	☐	☐
ORDERS	☐	☐	☐
RECEIVINGS	☐	☐	☐
ISSUES	☐	☐	☐

DATA INCLUDED FOR
THE DAYS MARKED

(Enter Date)

MON ——————
TUE ——————
WED ——————
THU ——————
FRI ——————
SAT ——————

DATE ——————

TRANSMITTING
DEPARTMENT: ——————

DATE TRANSMITTED: ——————

PREPARED BY: ——————

DATA PROCESSING

DATE RECEIVED ——————   RECEIVED BY: ——————
DATE KEYPUNCHED ——————   KEYPUNCHED BY: ——————
DATE VERIFIED ——————   VERIFIED BY: ——————
DATE RETURNED ——————   RETURNED BY: ——————

FORM DESCRIPTION	NUMBER OF FORMS TRAN	NUMBER OF FORMS RCVD	NUMBER OF FORMS RTD	COMMENTS

Fig. 15. Transmittal form.

PRODUCT ORDER NO.	PART NUMBER	CUSTOMER		QUAN.	DATE	PROD. WK.	DUE DATE	DRAWING NUMBER	LATEST ENG. REV.		OF
									NO.	DATE	PAGES
BASIC MATERIAL	MATERIAL FORM	SPEC. NO.	OUT. DIAM.	INSIDE DIAM.	LENGTH	ISSUE QUAN.	UNIT	WASTE ALLOWANCE	SCRAP ALLOW.		
			MTL.	FIN.	MTL.	FIN.	MTL.	FIN.			
SEQ. NO.	WORK CTR.	OPERATION DESCRIPTION	IN-PROCESS DIMEN.	TOOL NAME	TOOL NO.	MACH. NO.	MAN-NING	FIXED SEQ.	SET-UP TIME	CYCLE TIME	TAKE-DOWN TIME	IND. STD. HRS.	TOTAL HRS. THIS SEQ.
			REC. MIN	MAX									

Fig. 16. Sample of a route sheet.

PRODUCTION ORDER NO. _____   PART NUMBER _____

PRODUCTION HISTORY

SEQ. NO.	WORK CTR.	QTY. START.	QTY. COMP.	DATE-TIME STARTED	DATE-TIME COMPLETED	TOTAL ELAP. TIME	TOTAL STD. HRS. THIS SEQ.	OPERATOR NO.

INSPECTION HISTORY

QTY. OK	QTY. SCRAP	QTY. HELD	DATE-TIME INSPECTED	INSPECTOR NO.

Fig. 17. Reverse side of route sheet shown in Fig. 16. The recorded information serves as a history of the job through the process.

3. Number of pieces to make, when routing is combined with the production order.
4. Material to be used.
5. Operation data including:
    a. List of operations on the part.
    b. Work center or department doing the work.
    c. Description of operation.
    d. Machines and tools to be used.
    e. Feeds and speeds of machine tools.
6. Standard crew size.
7. Time for setup, unit cycle or batch process, and takedown or clean-up.
8. Provision for recording data on production and inspection results.

Standards and Allowances. Development and maintenance of meaningful standards are fundamental to the production process. Production schedules are developed based on the standard time information provided on the routings. Subsequently, reporting is used to monitor production rates against these standard times and feedback information on off-standard conditions.

A waste allowance for converting raw materials and a spoilage allowance for in-process fabrication is important in planning. Depending on the operation, these allowances may be combined into a single percentage. These allowances are shown on the route sheet to assist production supervisors in recognizing abnormal conditions.

Techniques for standards setting are covered in the section on Work Measurement and Time Study. The importance of reliable standards, and a program for updating standards with methods change, is vital to the routing of a process.

Sequence of Operations. Routing begins with determination of the **sequence of operations.** Methods of work and machine analysis are preliminary to this determination. Routing can be performed properly only by a person who is thoroughly familiar with the character of the work to be done and with the processes available for doing it. The sequence of operations adopted for any class of work can have a noticeable effect on the time and cost of production. **Changing the sequence** of a single operation in production often affects the performance of all subsequent operations on the part.

Consideration of plant layout and material-handling procedures is needed to secure the most efficient routing for production of the item. Proper routing is also an aid in the inspection program. It may be possible to minimize inspection by selection of the operation sequence. Inspection is listed at the proper point on the routing sheet to ensure that added production operations are not performed on defective parts.

USE OF ROUTE SHEETS. In some production control systems, route sheets serve as the basis for recording progress of a part through a cycle of operations. For example, they can register progress of the part from stores to completion, and back to stores.

The route sheet, illustrated in Fig. 16, moves with the material from work center to work center. The reverse side of the sheet, Fig. 17, is used to record production progress and inspection results. Note that in these fully developed forms, **time** and **quantity** are included. In some cases, a copy of the route sheet is maintained in the production control office and data provided by dispatchers are used to record the job history.

Route sheets are also used to load and schedule, determine labor efficiency, calculate payrolls, secure tooling, and arrange material moves. Actually the same objectives can be served by any well-designed route sheet, regardless of

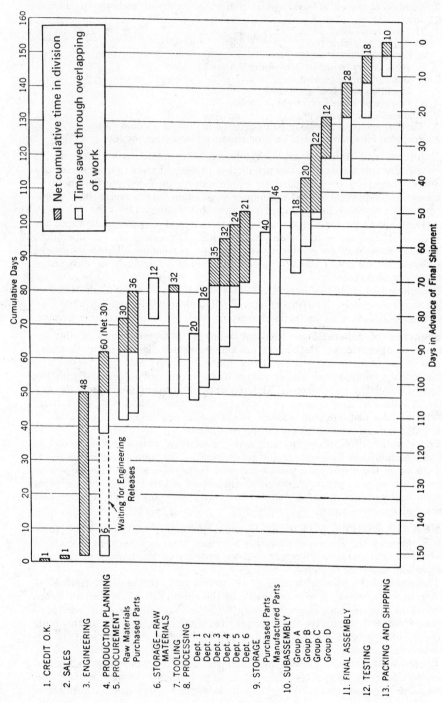

Fig. 18. Time cycle schedule of a production order in Gantt chart form.

its format. Identification of different operations on a single component, with some provision for checking them off as completed, is the minimum requirement.

Lead Time

DETERMINATION OF LEAD TIME. Total lead time for processing a production order includes the sum of all elapsed time from receipt of order through the time the last item processed on the order is shipped to the customer.

Typically, production control is charged with responsibility for providing sales with lead time information. Determination of lead time is accomplished by a detailed analysis of all elements of the manufacturing process. The analysis includes various clerical activities, each work element in the production process, possible overlapping of work, move and queue times, and reasonable allowances for unexpected delay for each operation. Reliability in determining overall manufacturing lead time is essential for good customer service. Potential customers may place orders elsewhere if the conservative estimate of the lead time is too long and non-competitive. Overly optimistic dating is equally damaging to good customer service. The objective from the standpoint of production control is to set up a working timetable of shop operations so that:

1. Every order is completed in the shortest possible time.
2. Promised delivery dates are based on definite information. The sales department may then be furnished with information to provide meaningful delivery promise dates.
3. A constant supply of work is kept ahead of each process or machine.

Determination of lead time may be routine and performed only periodically in some manufacturing operations. In custom manufacturing it must be determined with each production order. When the custom order is similar to a project, PERT or CPM may be employed to determine lead time. In nearly all situations it is a basic manufacturing objective to comply with established lead time.

ELEMENTS INFLUENCING LEAD TIME. Each element of the manufacturing process must be carefully controlled in order to assure that a production order will be produced within a desired time period. Elements of the manufacturing process include all activities of both line and staff groups from the time of preliminaries to manufacturing to final shipment. In custom manufacturing involving a number of separate processes, analysis of lead time may become complex, as illustrated by Fig. 18.

Order Acceptance. Credit checking is performed on all new accounts as a first step of order entry. Where a production order is prepared for stock replenishment or for established customers, the credit check is not required. The sales department's handling of an order in custom production consists of rewriting it in the company's format and providing production control with data required to prepare a production order. In situations involving production to forecast or replenish stock, production control initiates the production orders stating the desired due date and authorizes the manufacturing department to proceed with operations.

The engineering department in custom manufacturing receives its copy of the production order concurrently with manufacturing. Engineering proceeds (as required) with design, preparation of specifications, bills of material, drawing lists, route sheets, and all other matters concerned with the fundamental structure and method of manufacture of the product.

Computer systems, in plants where finished goods are made for stock, are designed to permit customer data links or computer-to-computer ordering. In such cases order acceptance, planning, procurement, and issuance of manufacturing orders are fully automated.

Production Planning Time. The initial task in the production control department is **analyzing the production order** in light of existing commitments and machine loads to determine a probable completion date. This analysis is often complex and is discussed in depth under Loading and Scheduling in this section.

In estimating production planning time, it is necessary to determine how long the order is expected to be in the planning department. If portions of the order can be released early, considerable overall time may be saved by telescoping or overlapping activities. It is useful to estimate, for planning purposes, the net time in each department for which there is no overlapping work in other departments that will advance the actual completion of the order.

Procurement. In custom manufacturing, procurement generally extends the overall lead time due to the need for special purchased parts, long vendor lead times, and non-stock materials requiring initial or lengthy processing. Conversely, if raw materials already in storage can be withdrawn to start manufacture of the parts with the longest production cycle, the fact that other materials arrive later may have little or no effect on extending the total lead time.

Tooling. In certain situations tooling can often be a cause of delay unless the need is recognized in the planning cycle. Once the need for new types of tools, jigs, or fixtures is determined, design and tool production follows promptly. Methods and tooling parallel procurement and sometimes processing, depending on where and when tooling is required. The total time in which tooling extends the total production cycle is added to the lead time. In many cases the necessary tools are available. The condition of these tools is ascertained promptly and any required adjustments and repairs made at the time the order is being prepared for production.

Factory Processing Cycle. Factory processing is planned and scheduled to give the shortest overall cycle time commensurate with existing load and the most economic utilization of equipment capacity. If the processing of new production orders is delayed because existing orders must be processed, the delay time is added as part of the factory cycle. The total delay is determined by an analysis of:

1. Available machine capacity.
2. Existing or desired queues.
3. Availability of material when needed.
4. Setup times.
5. Sequence in which parts are needed.
6. Telescoping permissible with the subassembly and assembly program.

Handling and Storage Time. The time consumed in **moving work** from station to station in the department or between departments can be an important lead time factor. It is futile to set close time values on machining operations only to have parts sit idle or be in transit an unduly long time. A well-designed and implemented production control system provides a means for minimizing this non-productive time.

Component parts storage exists only to the extent that parts are conveniently finished or delivered ahead of actual need by subassembly or assembly. In these cases the manufacturing cycle time is unchanged. Components being built far in advance of actual need may delay subsequent orders requiring similar processing or lead to additional setup costs. The cost of carrying inventories tends to discourage long storage times for work in process.

Assembly. **Parts assembly** and **subassembly** frequently overlap both factory processing time and final assembly time. The process time for the longest subassembly group is usually the element of the process that will add to the lead time. Overlapping is reduced when long subassembly operations must await several parts assemblies, that in turn follow on deliveries from some of the longest parts manufacturing cycles.

On an order for a number of identical or very similar items, **final assembly** may overlap subassembly to a considerable degree. One typical example is in the assembly plants of the radio and television industry. The building of special individual machine tools, by contrast, permits little overlapping time. In this latter case, it is best to take any overlapping from subassembly and obtain a net assembly time.

Testing and Inspection. Various forms of inspection are employed throughout the process cycle. Sampling of in-process batches increases the production cycle to the extent that the whole lot is delayed from move until inspection is passed. Certain products require instrument testing. Scheduling of these instruments may be a significant factor adding to the overall cycle. Careful planning and preferencing for the instrument testing of critical parts could permit overlapping the testing of other parts or assemblies. The net time for final assembly testing is added to the cycle.

An important consideration in estimating the cycle for custom work deals with inspection rejects. It is common that job lots are shipped to the customer complete. Thus, a reject of one part may delay shipment of the full lot. In these cases emphasis is given to minimizing inspection time, expediting decisions on held parts, and assuring preference to rejected parts in reprocessing so the delayed lot can be shipped as soon as possible. Planning for a certain percentage of reject delay time must be built into the overall cycle. Estimates are often based on historic data.

Packing and Shipping. Time for shipping and packing of single unit products is added to the overall cycle. Disassembly time and time for protecting against damage, rust, the elements, overseas packaging and other special provisions, as required, is included in the cycle plan. Where a number of items are shipped to a single customer location, sequential packing and shipping operations may overlap final assembly so that a net time is determined.

Loading and Scheduling

CAPACITY AND LOADING. Loading is defined as that portion of the production control function that allocates required production time against available shop capacity. Loading is often accomplished in two steps. **Rough loading,** involving allocation of a block of time against a department's capacity, is done as a first step. Rough loading typically spans a period of one to six months in the future. **Fine loading** follows, and involves allocation of machine time in daily or hourly increments.

Machine load control has the main objectives of:

1. Keeping machines working continuously.
2. Assigning work to the machine that will result in an optimum balance between meeting schedule dates and minimizing manufacturing costs.

A capacity analysis and machine loading procedure are required to accomplish the objectives.

Capacity Analysis. Capacity analysis is the procedure for determining the productive capacity of equipment. **Productive capacity** is generally classified into the broad categories of **machine-paced** or **man-paced.** Equipment whose output is expressed in terms of pounds, yards, or units per hour are illustrative of machine-paced capacity. Machine load in hours is determined by dividing desired output by fixed capacity.

General purpose machine tools, hand tools, and assembly operations illustrate man-paced capacity. Man-paced work centers require an **operation study** to determine the time for making each product. Capacity is expressed in hours of work. Total machine load on a piece of equipment is determined by accounting for total machine time of the different units to be produced on that piece of equipment.

Total capacity available in a work center in a given period is defined by the following formula:

$$C = D \times T \times M \times E[8 - (S + U + F + R)]$$

where C = Machine hours available for loading in that work center in a given time period.

D = Number of working days in the period being determined.

T = Number of shifts the work center is expected to run.

M = Number of machines in work center.

E = Average production efficiency at all machines in the work center expressed as a decimal percent.

8 = Number of hours in a normal shift.

S = Average setup time in hours per shift normally experienced in that work center.

U = Average lost time in hours per shift normally experienced in that work center. This includes time for maintenance, delays, and other downtime.

F = An operations loss factor in hours per shift for that work center. This factor considers such elements as non-scheduled time because of mix work, loss time for major maintenance, inability to obtain manning for the machine and other factors.

R = A reserve factor in hours per shift for that work center. This factor is usually assigned by management and is used as a safeguard against such occurrences as major unpredictable downtime that might result from acts of God, or receipt of unforeseen emergency orders from key customers.

Machine Loading. Machine loading is the procedure for assigning hours of work to each machine. Two general techniques used for loading are called finite loading and infinite-capacity loading.

Finite loading uses a procedure in which the hours of operation for each job in a specific work center are scheduled backward from the date required by final assembly. Hours for each period are accumulated until a predetermined capacity is reached. If further work is required, jobs are moved to another time period, or provisions are made for overtime or sub-contracting. Finite loading usually presents the shop with a more level load and cuts down the time for continual capacity adjustments. The problem with finite loading is that manufacturing cycle times are constantly changing and delivery reliability may decrease.

In loading to **infinite capacity** the load hours are scheduled forward from the first operation and allowed to fall where they may. Initially, no consideration is given to overloads that may develop. The production planners attempt to adjust for overload after preliminary loading by means of adjusting plant capacity. Infinite capacity procedures are definitely easier to program and have the advantage of focusing attention on bottlenecks. The plant must have good procedures for adjusting overloads if manufacturing cycle variance is to remain low.

An effective **combination** technique is sometimes used in situations where components may be stocked. In this technique, **critical machines** are finite loaded. Critical machines are those machines that are frequently fully loaded and often bottleneck in the operation. High-value components are loaded by means of the finite loading technique against these machines. These components are loaded in discrete quantities sufficient to fulfill sales demand for the next month. Little or no safety stocks are carried on the high-value components. Low-value components are loaded against remaining available capacity of all machines using an infinite capacity technique. These low-value components carry sufficient safety stocks such that problems are not created if these components are not ordered immediately when the stock reaches the order point. If, as a result of infinitely loading the low-value components, abnormal overloads develop, order quantities are reduced or spread to another time period.

Machine loads are recorded through the use of ledgers and by graphic means. Figure 19 illustrates a simple ledger form for recording machine loads. Ledgers are easy to maintain and are recommended where balancing work loads on a number of machines is not a problem.

Graphic techniques, such as the Gantt-type graph in Fig. 20, provide a convenient means for analyzing over- and underload conditions, and for determining out-of-balance conditions.

Analysis of **overload** and **underload** conditions is made to keep machines working continuously. Short-term overload relief is obtained by scheduling overtime or shifting work to other machines, where possible. Persistent overload problems are indicative of the need for additional capacity.

Underload problems may arise from insufficient work, a change in product mix, or as a result of delay from a feeder department. Remedies include: improving planning techniques and tightening control against out-of-balance conditions. The processing of stock parts, when permitted, provides another short-term solution to underload problems.

LOADING PROCEDURES FOR VARIOUS TYPES OF MANUFACTURING. The selection of the proper loading procedure in the plant is determined by the nature or type of manufacturing, its complexity, and the degree of control required. Manufacturing types include: intermittent, continuous, and composite manufacturing.

Intermittent Manufacturing. In this type of manufacturing, products are routed through the plant in batches or lots. Machines run for a relatively short period of time between setups and the variety of items is large. An order control system is employed. Individual orders may vary in quantity, form, and material in order control systems. Typically, machines are loaded based on the requirements of each order.

The plant is arranged on a work center basis. Similar-purpose machines are often grouped in a single work area. Parts accompanied by detailed route sheets move in lots from operation to operation. Assembly is dependent on all parts being ready simultaneously. Effort is directed toward achieving short manufacturing cycles and balancing work center production to avoid excess accumulation of in-process inventories.

Scheduling and loading procedures are significantly different for the intermittent production of custom orders as against intermittent manufacturing for stock. The need for scheduling flexibility is emphasized in custom production. Production is a continuous compromise between new customer demands and existing commitments. Companies that manufacture to specific customer order generally cannot carry end product or component inventories. This absence of a stock of finished or semi-finished products means that the fulfilling of production requirements is not undertaken until each order is received. The interval involved in filling the order corresponds to the total planning and manufacturing lead time.

MACHINE NO. _____ DEPT. _____
PRODUCTION WEEK _____
STANDARD LOAD HOURS _____
SCHEDULED HOURS ADJUSTED FOR EFFICIENCY _____

PART NO.	NO. OF PIECES	STANDARD HOURS	AVAILABLE HOURS

Fig. 19. Ledger method for recording machine loads.

The loading procedure in this situation must emphasize the importance of meeting delivery commitments. Accurate load data and rapid information feedback are required to permit detailed scheduling and quick response to off-schedule conditions. These factors tend to make loading and scheduling complex and increase the need for maximum control.

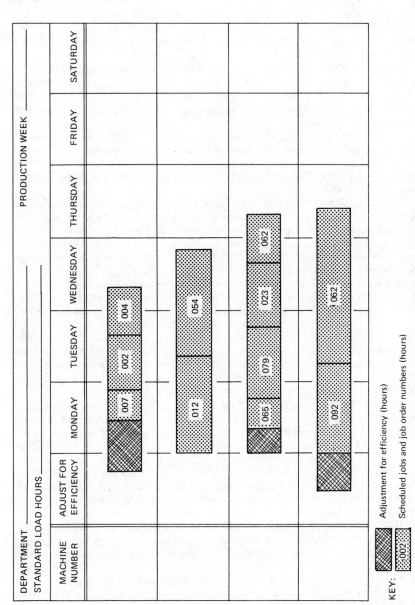

Fig. 20. Graphic method for recording machine loads.

Production control procedures are simplified when **stocking of finished goods and components** are permitted. Production loading procedures for this type of manufacturing emphasize replenishment of inventories and avoidance of out-of-stock conditions. Although these factors make loading problems less complex, effective control must be exercised to balance cost of inventory investment with costs of production delays and resulting loss in utilization of men and machines.

Continuous Manufacturing. Characteristics of continuous manufacture include large volumes of similar products produced with few if any interruptions in the process. Loading procedures are simplified and a system called **flow control** is often used. The problem of loading is one of balancing demand and capacity for each process.

Specialized equipment and machines are arranged according to the fixed sequence of operations. Parts and materials are frequently conveyed or moved by mechanical means from operation to operation. Work-in-process inventories, consequently, are minimized. Production control procedures emphasize careful pre-production planning.

As with intermittent manufacturing, there are two general types of continuous production. The first type is the **mass production of a single product** requiring no assembly. Production rates are established well in advance by providing the required equipment to fulfill the demand for capacity. Variations in demand are met by activating or shutting down certain pieces of equipment, thereby adjusting existing capacity.

Assembly line production is the second type of continuous manufacturing. It may involve either a number of separate independent product lines, or mass-produced components feeding a common assembly line. Each line, in the former case, is treated as a unit with no attention to individual work stations within the line. Monthly schedules for assembly operations are merely lists of daily production rates for each item. The latter situation is more complex and involves balancing banks of parts ahead of assembly. Coordination and control of raw material inputs and balanced outputs feeding the main assembly line usually involves sophisticated loading procedures.

Composite Manufacturing. A large number of manufacturing plants have characteristics of both continuous and intermittent manufacturing. An example of the combination or composite situation is a fabricating department on an intermittent or job order basis and a subassembly department on a continuous flow basis, both feeding a common assembly line. The assembly line has intermittent characteristics, but within a period or "block," operates on a continuous basis on each product. The loading procedure uses a block and load system by allotting specific amounts of machine time in a particular operating period for the production of a specific quantity of each product.

SCHEDULING. Scheduling is defined most simply as the portion of the production control function that determines when each operation is to start and finish. Optimum start and finish times are based on satisfying these principal objectives:

1. Meeting customer and finished stock inventory delivery commitments.
2. Balancing production loads in order to obtain maximum utilization of available machines and manpower.
3. Sequencing production in a manner that will produce minimum production and inventory carrying costs.

Scheduling evolves as a by-product of machine loading. The three elements included in the loading and scheduling procedure are:

1. **Machine loading,** which is the determination of when each part can start on the required machine based on available capacity and existing machine load.
2. **Parts scheduling,** which is the determination of relative dates for each part to start and finish at each work center in order to provide a smooth flow of work at subsequent, dependent operations.
3. **Order scheduling,** which is the determination of priorities for each manufacturing order in each work center.

When all processes on each part have been assigned to a machine in a manner permitting optimization of scheduling objectives, the scheduling procedure is complete.

Establishing priorities for orders involves a series of scheduling decisions. The planner must possess accurate load data for each manufacturing order, a knowledge of machines available for processing and the remaining equipment capacity. Constraints on possible decisions include: **desired completion** dates; **costs of operating** machines; **existing loads** on machines available for processing the order; and **most economical sequence** that will minimize setup costs.

The schedule results from analysis of the interrelationship of facilities, orders, and time. The decision is reached by the best balance of available capacity against the above capacity constraints.

Rules for Scheduling. Scheduling decisions, to be effective, must be based on rules designed to meet company objectives. These rules are established and supported by top management.

Scheduling rules and methods found in an intermittent type of manufacturing company provide for:

1. Standard queue times for parts manufacture in each work center.
2. Methods and responsibility for adjusting standard work center capacity to actual capacity.
3. Methods for loading a work center in a time period, as, for example, the use of finite loading versus infinite capacity loading.
4. Priorities for handling emergency, overdue, and reprocessed orders from rejects.
5. Priorities for handling regular orders, such as the use of FIFO, earliest due date, minimum slack, or shortest process time.
6. Methods for controlling rate of output from each subassembly department into final assembly.

Load Data. The effectiveness of scheduling decisions depends on the accuracy of standard load data. Methods employed for collecting and maintaining load data depend on the level of control required. The most accurate type of loading stems from calculations of the time required for a machine or operator to perform each operation. These calculations are based on time studies or built up from reliable tables of synthetic elemental times. Operating times based on a summation of estimates are likely to be inaccurate unless the estimator has had many years of experience on the same type of order in a particular shop.

As the size of the scheduled production unit increases from one machine to a battery of machines or to a department as a whole, the shortest practical schedule interval also increases. When this interval reaches a day or week, finer gradations of load measurement are no longer necessary. If, for example, schedules consist of weekly production quotas for a whole department, it is unnecessary to calculate loads from the elemental times required to do each operation on every order in that department. An overall estimate of processing time for each order will probably be sufficient.

USE OF COMPUTERS IN LOADING AND SCHEDULING. Analysis of interrelationships between facilities, orders, and time frequently involves extensive mathematical calculation. Use of computers as an aid to analysis is practical and economical in situations involving a complex manufacturing process, multi-level bills of materials, many parts, and many transactions. This involves programming into the computer the rules used for decision making, route sheets, and elemental times. Techniques used for computer analysis often depend on the nature of the manufacturing process.

Typically, simulation techniques are used for solutions of **production-for-stock schedules.** In this analysis the best configuration of machines to meet required production is determined based on an input of the sales forecast and the degree of uncertainty in both the probable rate of receiving orders and the probable quantities required for each order.

Linear programming techniques are often used for solutions of **custom order schedules.** In this situation, the solution is stated in terms of the most economic assignment and sequence of products to available machine hours.

AUTOMATED PLANNING AND NUMERICAL CONTROL. Automated manufacturing planning is a computer technique designed to enable conversion of engineering designs to precise manufacturing instructions. These instructions include directions as to operations performed, raw materials used, exact tools and fixtures required, in-process dimensions and temperatures, if applicable, time standards, and cost estimates (A.M.A. Management Bulletin No. 107).

At the same time as the loading and routing instructions are being prepared, the computer can calculate and cut a numerical control tape. Thus if the operation involves a tracer lathe, the template configuration and dimension information are generated. Following the preparation of route sheets, the computer may use the data for machine loading.

There are two prime factors in determining application for automated manufacturing planning and numerical control. First, standardization is required with a logical flow from input to output. Second, the nature of the product must require sufficient planning and be of a volume to support automation.

In describing automated manufacturing planning and numerical control, Charles Porten (Business Management, vol. 29) foresees two broad categories of users. First are those firms manufacturing a group of parts or assemblies having common characteristics. The second includes those firms in which the products are made by similar methods and have common qualities.

Dispatching

DEFINITION. Dispatching may be defined as setting production activities in motion through the release of orders and instructions in accordance with previously planned time schedules and process routings. Dispatching also provides a means for comparing actual progress with planned production progress and communicating this information to the production functions concerned.

The dispatching section is the **representative of production control** in the shop. Production control can adjust plans or cause action to be taken to improve results, based on information feedback from the dispatching section.

PRINCIPAL FUNCTIONS. Principal dispatching functions include:

1. Providing for movement of material from stores to the first operation and from operation to operation.

2. Issuing orders instructing the tool department to collect, make ready, and furnish tools, jigs, and fixtures to the using department before the operation starts.
3. Issuing job orders authorizing operations in accordance with dates and times previously planned and entered on machine loading charts, route sheets, and progress and control sheets or boards.
4. Sequencing manufacturing orders with the production foreman, in order to achieve the most effective utilization of men and machines within schedule parameters.
5. Issuing time tickets, instruction cards, drawings, and other necessary items to workers who will perform various production operations.
6. Providing for inspection in production control systems in which inspection is a dispatching function.
7. Collecting drawings and bills of material after completing the production operation and returning them to the production control department.
8. Monitoring work progress by recording start, finish, and status of jobs and acting as liaison between manufacturing and scheduling on off-schedule conditions.
9. Collecting job and time tickets and forwarding them to the data center or time-checker. If applicable, returning job records to the production control department.
10. Recording and reporting idle time on machines and of operators, and requesting action on delays.

VARIATIONS BASED ON TYPE OF MANUFACTURING. The scope of dispatching, as with most production control procedures, varies with the nature of manufacturing, degree of complexity of process, and magnitude of operations. Although some degree of dispatching is needed for every type of production, it is most needed for scheduling and control of intermittent production.

In **intermittent manufacturing** the dispatcher is typically responsible for issuing manufacturing orders and prints for each part and assembly. The sequence of dispatching operations is illustrated in Fig. 21. The number of work centers and points monitored by the dispatcher depends on the allowable time delay for information feedback and degree of control desired. Progress and status information on each work center may be monitored by dispatchers as frequently as every hour in some custom manufacturing operations.

Dispatching is less complex in **continuous manufacturing.** Routings and operations instructions are routine and repetitive. Work is assigned according to the **general monthly schedule,** and emphasis is placed on assuring smooth material flow. Since there is little opportunity to deviate from standard work flow, the scope of dispatching is at a minimum. In continuous production it is not unusual for the foreman to perform the dispatching role.

FOREMAN–DISPATCHER RELATIONSHIP. The foreman's relationship with the dispatcher in any type of production process must be clearly defined with emphasis on cooperative efforts. Foreman–dispatcher conflict is minimized when production control provides the foreman with a **written procedure** of the dispatcher's responsibility, method of operation, and bases for scheduling decisions. The foreman is responsible for the personnel of the manufacturing department and has specific objectives for the quantity and quality of work performed. Typically, the foreman and dispatcher jointly **sequence** the scheduled manufacturing orders in order to obtain the best utilization of men and machines. In addition, the dispatcher assists the production foreman by assuring that tools and materials needed by the production department are available as required.

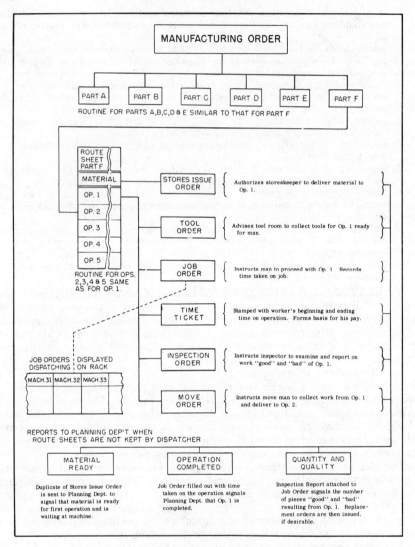

Fig. 21. Sequence of dispatching operations.

If for any reason the foreman desires to change the agreed-upon sequence of work, he communicates with the dispatcher to have the necessary changes made. Sequence changes, then, are mutually agreed to or resolved at a higher level.

DISPATCHING SYSTEM ROUTINES. The procedures for receiving, filing, and issuing shop paper are vital to an effective dispatching system. These procedures are discussed at length in the paragraphs that follow.

Receipt and Filing of Shop Paper. Production control usually gives the dispatcher the production order and all manufacturing orders, routing, engineering drawings, and other required shop paper in advance of the scheduled production date.

There are two basic methods for filing shop paper: by production order number or by machine number or matching group to which the order is assigned. Either file method is further subdivided by required date. It is considered good practice to employ a **production calendar** for dating. A production calendar, essential in computer-based systems, is one in which each day of the fiscal year is sequentially numbered. These calendars provide a convenient means for determining lead times and generally minimize dating errors.

In continuous process work, or production primarily for stock, the assigned-machine file method is often convenient. Where a schedule accompanies the manufacturing orders, the order number file method is generally the safest and quickest method for filing and retrieving information.

Another filing method **combines the two systems.** The basic filing is by order number, but when advice is received from stores that the material is ready for shop delivery, corresponding shop paper is withdrawn from the number file and placed in the machine file. Regardless of choice, the method adopted should be consistent, convenient, and permit papers next in sequence to be found easily.

Machine Assignments. The central feature of many dispatch offices is the **dispatch board** or rack, providing a visual means for control. A designated space on the board is assigned to each machine in the department. Typically, the space is divided into **three compartments** for:

1. Jobs in process. A duplicate of the **job order** actually being worked on occupies this compartment.
2. Next job to dispatch. This space is occupied by a job order and its duplicate to be transferred to compartment 1 as the current order is completed.
3. Orders next in sequence. Job orders for which all material is checked are filed in the sequence that they will eventually be transferred first to 2 and finally to 1.

A glance at the board shows: (1) whether all machines are actually engaged in jobs; (2) whether any machine lacks a next job; (3) whether each machine has a fair supply of work in reserve after the next job is run. Indications of approaching shortages of work for any machine are relayed to the planning department for confirmation. Opportunity is provided, in this manner, to reroute jobs to idle machines and relieve pressure on other equipment.

Various methods may be used to **display** job orders. The principal consideration is to preserve visibility so that a single glance takes in the entire series of open orders and detects any machine where work shortage is threatening.

Shop paper that accompanies the job order is given to the operator at the time the job order is placed in the 1 position on the dispatch board. Shop paper typically dispatched to the operator includes the **route sheet, job-labor** ticket, operation card, and engineering drawings and instructions.

Requisitioning and Issuing Material. At the time the shop paper is forwarded to the dispatcher, a **material requisition** and **stores-issue** ticket is forwarded to stores. These tickets may be separate or combined as shown in Fig. 9. **One of the following methods** generally is used for the issuance of material:

1. A date of delivery is marked on the requisition, and the stores department automatically moves the material to the department on the date required.
2. A delivery date is marked on the requisition, and the stores department collects and kits the material, but awaits authority from the dispatcher for delivery time.

3. The dispatcher advises stores, generally a day or two ahead, of the next sequenced jobs. Stores gather the material, and either delivers it when ready or awaits delivery authority.

The last method for issuance of material is the most flexible, but has the disadvantage of providing little forewarning if any material is discovered short in stores. The first method gives maximum warning time, but is least flexible. The second method is a popular compromise, but may be difficult if there is a shortage of make-ready area in the stores department. The choice should be that which best serves the operation. The essential objective is to deliver the material where needed, neither too early nor too late for the operation.

Tooling Orders. It is customary to specify cutting tools as well as to list all jigs, fixtures, and gauges required for the job. The information usually is included on the route sheet from which a **tool order** is then prepared. The tool order may be prepared manually by the dispatcher, or, in computer-based systems, is printed with other manufacturing orders as part of machine output. An illustration of a manually-prepared tool order is shown in Fig. 22.

TOOL ORDER									
							Date___3–3___		
No. Req'd	Tools and Gages	Sect.	Bin	No. Req'd	Tools and Gages	Sect.	Bin		
1	Set of 18-38 Vise Jaws	10	28	1	x-467	6A	28		
	VF160			1	7/16" Plug Gage				
1	3" dia. x 7/16" face x 1" hole Side Milling Cutter								
1	#50 Vise	1E	54						
1	1" Dia. Arbor	Rack							
				1	Drwg. E-97				

Descr. Rocker Arm for Operating Distributing Roll			Oper. Mill inner surface in vise jaws.		
Part No. 18-33	Oper. No. 6	Dept. 4	Mach Hand Miller.	Equip No. 981	

Fig. 22. Tool order form.

Typically, dispatchers issue tool orders in advance of the production operation requiring the tools. The tool room collects the tools and holds them until requested by the assigned operator. Authority for tool release generally is given by presentation of the job order or job ticket for the operation in question.

Inspection Orders. Often, the dispatcher is responsible for arranging for inspection, unless it is being carried on continuously in the process. Inspection forms for this purpose are frequently combined as illustrated in Figs. 13 and 17. Inspection orders are sometimes arranged to specify: (1) order and part, (2) operation and sequence, (3) operator's number, (4) number of good pieces, (5) number of bad pieces, (6) number capable of repair, and (7) inspector's number. Inspection orders are usually returned to the dispatcher and forwarded to production control for use in providing data for recording order progress.

Material is often delayed because of rejection at the point of inspection. It is important for the dispatcher and scheduling to be informed so that subsequent operations can be properly dispatched and necessary corrective action on the rejected material initiated. A **material reject ticket** is sometimes used to tag-identify the material against unauthorized use and a copy of the ticket sent to the dispatcher. A basic form of this ticket is shown in Fig. 23.

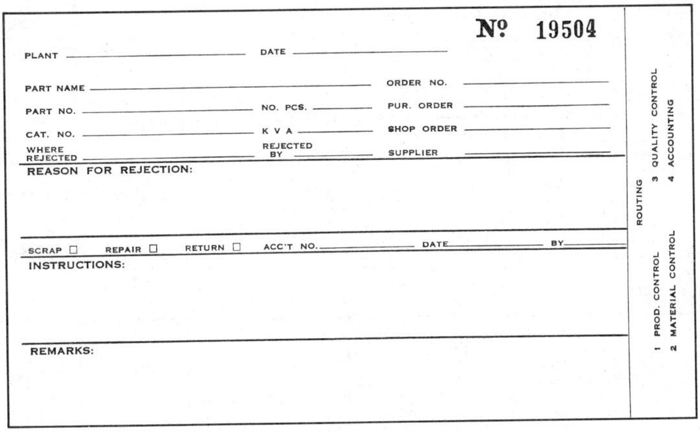
№ 19504
PLANT ______ DATE ______
PART NAME ______ ORDER NO. ______
PART NO. ______ NO. PCS. ______ PUR. ORDER ______
CAT. NO. ______ K V A ______ SHOP ORDER ______
WHERE REJECTED ______ REJECTED BY ______ SUPPLIER ______
REASON FOR REJECTION:
SCRAP ☐ REPAIR ☐ RETURN ☐ ACC'T NO. ______ DATE ______ BY ______
INSTRUCTIONS:
REMARKS:
ROUTING 1 PROD. CONTROL 2 MATERIAL CONTROL 3 QUALITY CONTROL 4 ACCOUNTING

Fig. 23. Material reject ticket.

The dispatcher has the responsibility of checking on the disposition of inspection-delayed material. As a result of reject, a replacement order may be required. These orders usually take the form of a job order, but are generally distinguished by color and marked **"Replacement"** in order to emphasize priority.

Move Orders. The final task of the dispatcher, before a job is disposed of, is to issue a move order. The move order is the authority for material handling to deliver the parts to the next work station. Move orders are frequently combined with other tickets. In Fig. 13 the words on the upper part of the form are moving instructions. In Fig. 9 the Job-Move-Location ticket illustrates another form of a combined move ticket.

PROGRESS REPORTS AND FOLLOW-UP. The dispatcher plays a key role in providing a means for control by performing the **monitoring function.** The dispatcher advises the responsible foremen and production control of any off-standard conditions and initiates corrective action as required.

To be effective, a monitoring system depends on:

1. Attainable plans established by scheduling.
2. Means to detect non-standard conditions.
3. Means to feed back information to production and central control in sufficient time to initiate corrective action.

Monitoring can be provided by means of a **Progress and Status Control Report** (Fig. 24). Data are recorded in the columns as follows.

DISPATCHER'S PROGRESS AND STATUS CONTROL REPORT

MACHINE NO. ____ DEPARTMENT ____ PRODUCTION WEEK ____

DAY	SHIFT	DAILY REQUEST	MADE AVAILABLE	RELEASED	AVAILABLE BALANCE	PRODUCTION	% COMPL.	CONTROL LIMITS		REMARKS
								UNDER	OVER	
MONDAY	1									
	2									
TUESDAY	1									
	2									
WEDNESDAY	1									
	2									
THURSDAY	1									
	2									
FRIDAY	1									
	2									
SATURDAY	1									
	2									

Fig. 24. Progress and status control report.

Daily request is the number of hours per day the machine is scheduled. The space under the diagonal is for recording accumulated Daily request.

Made available is the total standard hours of work in compartments **3, 2,** and remaining standard hours in **1** on the dispatch board at the start of the week (see Machine Assignments, above). When additional work in the form of manufacturing orders is delivered to the dispatcher on subsequent days, the hours are added to Made available.

Released time is the sum of the hours moved from compartment **2** to compartment **1** during the shift.

Available balance is the work ahead of the machines and is equal to the hours in Made available less Released.

Production is the number of standard hours completed each shift. The space under the diagonal is used to record the accumulated hours of Production.

% Complete is determined by dividing the accumulated hours of Production by the accumulated Daily request.

Control limits are given in terms of **Underproduction** and **Overproduction** in percent. These figures are to be compared with % Complete. If production falls below the **Under** figure, corrective action is initiated. A figure exceeding the **Over** percent triggers the need for additional orders.

The dispatcher prepares a separate sheet for each machine in the department.

The techniques used for information feedback vary depending on the magnitude of operations and the extent that the data is required by control. At one end of the spectrum are manual techniques, used where the time span is not critical. At the other end are the electronic, visual, and data processing links that are used where quick response is vital to successful operations. These communication techniques, as well as EDP dispatching are discussed under those topics elsewhere in this section.

Additional follow-up means are also used in some operations. **Expediters** are used to locate a delayed order, follow its course through the shop, and see that it is handled with all possible speed. In a plant manufacturing helicopter components, for example, a special expediter was assigned the role of follow-up on government orders containing penalty clauses. Modern directions in production control system design, however, tend to minimize the role of the expediter by incorporating his activities into scheduling and dispatching functions.

Loss Control

OBJECTIVES. A prime objective of the production function is maximum utilization of facilities with minimum cost of operation. This can be accomplished, in part, by reducing waste and scrap and decreasing idle operator time. Production control can make a valuable contribution to the production objective by collecting, recording, analyzing and reporting the data upon which decisions for programs of corrective action may be based. Some of the techniques used in analyzing data on machine utilization, scrap, and indirect labor are discussed below.

IDLE EQUIPMENT. Studies indicate that idleness of 20% to 60% in intermittent production and 20% to 30% in continuous production are not unusual. However, regardless of the type of cost system employed, effective utilization of equipment is essential to profitability.

Production control can assist manufacturing in making a profit contribution by providing data and suggesting improvements in machine utilization.

An analysis of existing and projected machine utilization can often lead to increased utilization. The analysis provides data on out-of-balance conditions

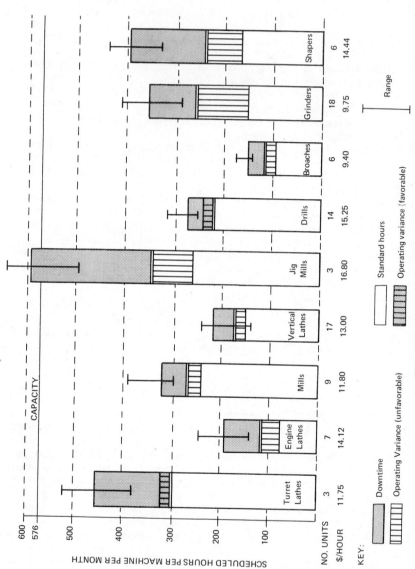

Fig. 25. Machine analysis chart.

due to product mix and loss because of downtime or operating inefficiencies. A **machine analysis chart,** similar to Fig. 25, provides a graphic representation of the data from the sales forecast. The chart in the illustration was developed by a job order machine shop from a history of order mix and use of equipment. In cases of production for stock, analysis of this type is likely to be even more accurate. Among the points illustrated by the chart are:

1. Only existing general production equipment is included in the analysis. Equipment maintained for special jobs and not available for the general mix of work is not included.
2. Each machine group is averaged on a per-machine basis.
3. Each bar graph shows average historical hours of production at standard, a double line for standard operating variance adjustment, and the average historical maintenance, delay, and operating downtime. Downtime when the machine is not scheduled is not included.
4. The range of hours per month is shown by the center line through each bar.
5. The numbers of machines in the group and the average cost per hour per machine are shown under the bar.
6. The 576 hour line, is the "ideal" capacity for a machine group.

An optimum condition is reached when the top range of demand is the same for all machine groups, no operating variance exists, and downtime is a practical minimum.

In a typical situation the chart provides information on:

1. Number of pieces of additional equipment needed in any group.
2. Number of pieces of equipment that could be stored, classified obsolete, or sold.
3. Areas for concentration to improve operating variance, thereby providing additional capacity.
4. Areas for concentration to minimize downtime.

Programs for minimizing downtime are typically pursued by manufacturing and maintenance. They involve preventive maintenance and priorities in maintenance and schedule delay for both critical capacity and high-cost equipment. Additional industrial engineering or process engineering programs may be established as a result of the analysis. These programs are typically designed to concentrate on training, methods, and jigs and fixtures in order to reduce unfavorable, direct operating variances.

Generally, production control will collect data during the year so that production-performance-to-standard reports may be prepared periodically. These reports provide manufacturing with data on results of machine utilization programs. The reports serve as a measure of effectiveness of current programs and suggest areas of need for further emphasis.

WASTE, SCRAP, AND YIELD CONTROL. The degree of control exerted to improve material utilization depends on the cost of material compared to the total manufacturing cost, the value of scrap, and the extent to which loss is inevitable due to the nature of the process.

Collection of data for waste, scrap, and yield control are typically assigned to production control in companies in which inventory control is a part of production control. Production control may also make a contribution to material control programs by providing piece-count data.

The following are selected methods for improving materials utilization:

1. Establish and maintain standards for waste, scrap, and yield.
2. Provide the minimum quantity of material necessary for the operation. For example, if 18 inches of round bar stock is required for machining, issue only an 18-inch length.

3. Establish a scrap classification and control program to assure that saleable scrap is not contaminated.
4. Determine yields on the basis of the value of good material output divided by total material value of input. Make sure that good material output plus the sum of scrap and waste equals the total material input.
5. Analyze scrap loss relative to operator, machine, and operation to detect unusual circumstances that may require remedial action.
6. Determine lost machine hours as a result of scrap and off-standard yield. Relate these results to machine analysis data to provide a basis for engineering effort.

OPERATOR CONTROL. Operator productivity utilization is determined by dividing the standard hours of good production by the total hours the operator works. Total hours worked include time spent on both production operations and indirect activities.

Production control exerts a substantial influence on productivity utilization by providing manufacturing with data that can be used in programs for minimizing indirect labor.

Indirect Labor Controls. Indirect labor is comprised of lost production time due to machine downtime, unavoidable non-machine reasons, and waiting for services provided by other functional activities.

Specific machine-related elements of indirect labor time may be influenced by the operator, including:

1. Machine setups.
2. Material setups.
3. Material changes.
4. Fluid changes.
5. Machine maintenance (when performed by or with the assistance of the operator, such as jams, welding breaks, etc.).
6. Tool, die, and fixture maintenance (when performed by the operator, such as tool marks, burrs, nicks, dirt).
7. Operator inspection.
8. Samples.
9. Machine adjustments, greasing, and oiling.
10. Machine changes.
11. Takedown.
12. Cleanup.

A standard is set on each operational delay element for each type of job and operators are measured against these standards. Standards are influenced by production control, based on the effectiveness of loading, scheduling, and sequencing. For example, jobs utilizing the same die, but with stripper or punch adjustments should be run sequentially, where possible. In examples of plastic extrusion and in printing processes, color sequencing if not properly scheduled, is a significant element leading to higher indirect labor times.

Other indirect labor times include lost time over which the operator has no influence, and lost time for unavoidable reasons. The latter elements include accidents, power failures, meetings, and training.

Production control can provide manufacturing management with data upon which to base programs for indirect labor reduction in other functional departments. In many cases, significant improvements are possible where indirect labor is due to:

1. **Scheduling.** For example, no scheduled work on one machine when standards call for the operator to run two or more pieces of equipment.
2. **Delay.** For example, waiting for maintenance; material; tools, dies, or fixtures; information; or inspection.

In many such situations, the dispatcher is in a position to cause action to be taken to minimize delay length.

Manual Collection of Data. Clerical records, such as operator time tickets, are frequently used to obtain the data required in manual recording systems. A code is generally established to facilitate identifying causes of indirect labor. The foreman and the dispatcher can often assist operators in improving the quality of reported data. As with any manual recording system, the validity of results is limited by operator accuracy in recording counts, part numbers, and cause of downtime. In some cases, companies use time checkers, dispatchers, and foremen to record results. Automatic recorders, where justified, are often used as a means for recording data.

Periodic spot audits of the correctness of counts and other recorded data are often desirable to assure continued accurate information.

Automatic Recorders. There are many instruments and automatic recorders available for recording downtime. These instruments are used for counting events, lag time, and frequency; recording mass and weight, linear and angular displacement, area, fluid volume, pressure, liquid level, temperature, fluid flow rate, power consumption, and electrical, physical, and chemical properties.

A list of **desirable features** for an instrument used to control downtime would include:

1. A continuous, graphic, and automatic record, throughout the shift, of the productive and idle periods of each machine.
2. An indication on this record of the reasons for each period of idleness shown.
3. The reasons for idleness recorded semi-automatically in such a way that the wrong reason (or no reason) could result only from inattention or indifference on the part of the operator.
4. A sensing device on the machine that could be used to differentiate positively between those times when the machine was actually producing and when it was merely turning over.
5. Some form of cycling or timing arrangement that could be set at a predetermined standard and record as idleness, thus necessitating an explanation, any productive time in excess of the standard.
6. A record of the number of productive cycles and a piece count.
7. Safeguards to prevent tampering or overriding by operator.
8. All records easily tabulated and correlated for control purposes.
9. Simplicity of design and operation.

An example of such an instrument is the Productolog produced by the Meylan Stopwatch Corporation. This system comes equipped with a stylus which draws a continuous line on a disc graph. Production time and downtime are differentiated by the position of the line on the graph. A totalizing register keeps a cumulative count of production time and permits a direct digital readout of total production time at any point. The total downtime can be calculated easily by subtracting the recorded production time from the total elapsed time as indicated by the disc graph.

The Telecontrol System produced by Hancock Telecontrol Corporation provides automatic counting at the operation and transmits data either electrically or electromechanically to the central dispatch office. Counters on a central control board provide a cumulative piece count during elapsed running time, elapsed downtime, and the quantity of units remaining on the job. Downtime is checked by the foreman, causes noted and relayed to the control center. By integrating these controls with other automatic data collection systems, a means for automatic exception reporting, downtime cause reports, individual man and machine reports, and automatic scheduling are obtained.

Users of such systems claim advantages including provisions for improved machine efficiencies, reduced indirect labor loss, and avoidance of unauthorized production. Recorder selection depends on which features of control are most vital, cost, available alternatives, and the potential loss without control.

Reports. Data on direct and indirect labor, whether collected by manual or automatic means, must be in a form that can be tabulated quickly, correlated in various ways, and presented to management as an exception report for initiating corrective action.

Format of these reports should highlight productivity and utilization in each department, and permit management to evaluate:

1. Direct operating efficiencies of each department.
2. Efficiencies of certain service functions, such as inspection, and maintenance.
3. Amount of idle time in comparison to operating time.
4. Seriousness and frequency of each type of delay.
5. Effectiveness of the foreman in holding idleness to a minimum.
6. The ability, skill, and effort of the operator.
7. Necessity for corrective action and points where it may best be applied.

Manual and Schematic Techniques for Scheduling and Control

SELECTION OF TECHNIQUES. Bowman and Fetter (Analysis for Production and Operations Management), in discussing alternative scheduling and control techniques, comment:

The economic problems facing the production planner is one of allocating various resources (men, machines, material) to the attainment of some objective. This problem most often calls for the programming of several interdependent activities in order to obtain some given or alternative outputs. In some cases the number of choices is few, and various manual or schematic techniques may be used. In other cases, the number of distinct solutions might run into the many thousands. In a case of this latter type, mathematical programming has been helpful.

There are a vast number of techniques for scheduling and control available to the planner. The selection of the scheduling and control techniques to be employed in any company depends on objectives, degree of accuracy required, and time and means available to the company. Among manual and schematic techniques commonly used are graphs, assembly charts, network techniques, Gantt charts, line-of-balance charts, and critical-ratio computations.

GRAPHS. Simple graphs are often used to show the general relationship between two or more factors and to record variations in the relationship. Two popular variations of the schematic technique are line and bar graphs. Scheduled versus actual progress can be portrayed by either method.

Figure 26 illustrates a **line graph** of progress in fulfilling a major contract for machinery. The vertical axis represents the percentage complete on the contract and the horizontal axis, the number of weeks remaining before scheduled completion. The solid line plot illustrates planned progress and the dotted line, actual status. Although the chart does not show results in detail, it does convey a general picture.

Figure 27 is a bar graph showing scheduled and actual production of components. It can thus pinpoint the source of the recent delay in fulfilling the contract illustrated in Fig. 26. Subassembly department C, which was to have produced 450 units by August 16, has produced only 340 units.

ASSEMBLY CHARTS. Assembly charts are often used as aids in planning and monitoring assembly operations. A type of assembly chart that has proved useful in practice is illustrated in Fig. 28. This chart shows the sequence of operations on a tractor and the production intervals of the various

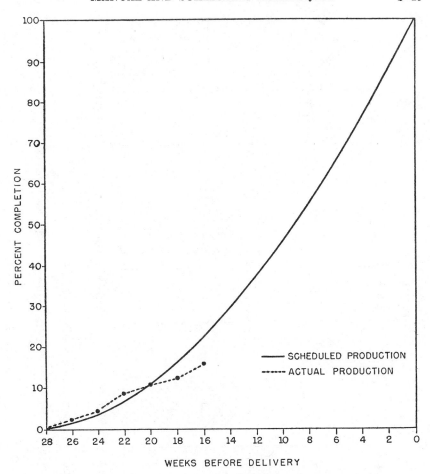

**Fig. 26. Scheduled versus actual progress in fulfilling
a major contract for machinery.**

operations. Only a few of the tractor components are shown. The total production interval needed for a complete operations cycle is 10 weeks.

In addition to its use as a guide to required sequence of operations and lead times involved, the chart can be posted to follow progress on each group of tractors released for production. On the basis of results reported to it, the production control group fills in the space between the double horizontal lines to indicate whether progress is behind, ahead of, or on schedule.

NETWORK TECHNIQUES. Network techniques are applicable to custom operations in which the production order is processed like a major project. The techniques are also used in planning programs or determining lead times involving numerous events (see section on Operations Research).

A number of approaches, primarily developed to solve particular problems, are available. Each has its distinctive acronym: PERT (Program Evaluation and Review Technique—though this originally stood for Program Evaluation Research Task); CPM (Critical Path Method); IMPACT (Integrated Management Plan-

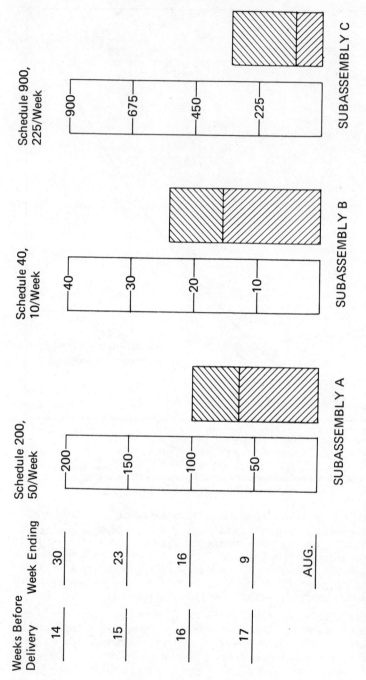

KEY: Shaded bar equals actual production for the week ending August 16

Fig. 27. Weekly subassembly schedules and production for three departments.

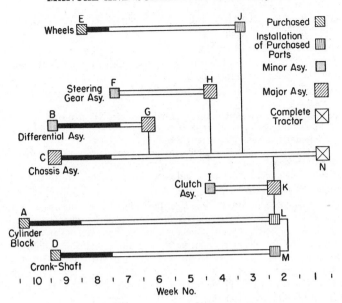

Fig. 28. Assembly chart: tractor production.

ning and Control Technique); SCANS (Scheduling and Control by Automated Network Systems); and others. In 1966, Pritsker and Happ (Journal of Industrial Engineering, Vol. 17) described a generalized network technique which they called GERT (Graphical Evaluation and Review Technique) (see section on Operations Research).

Horowitz (Critical Path Scheduling) sets forth the following prerequisites for the use of network techniques:

. . . the project must have a definite beginning and a definite end. It must also be made up of a series of smaller jobs or operations and must be done in an orderly sequence to complete the project. . . . The method does not lend itself to flow-type work, such as an assembly line, although it would be useful for setting up a new assembly line.

Organizing for control is essential for network planned projects, with responsibility specified for each activity. Project review meetings are held at regular intervals to determine the status of network completion. For best control, individual activities are defined with **milestones**, especially when the activity is lengthy. The milestone might be used, for example, in an activity involving the fabrication of ten subsections. Each subsection may require ten days to complete. By noting a milestone after fabrication of each subsection, control is maintained without waiting for the total one-hundred days to elapse.

Computerized Network Control. Programs for network control are available for most computers. Computerized control is practical when:

1. Networks are large and complex.
2. Frequent status updatings are required.
3. Need for advanced computational work, such as cost calculation, is required.
4. Costs of computer solutions are less than costs of manual control over the duration of the project.

Tables, such as a **criticality table,** are printed as part of these programs and provide a basis for control.

Advantages and Disadvantages of Network Techniques. Moder and Phillips (Project Management with CPM and PERT) list the benefits of network and critical path computations as a means to:

1. Encourage a logical discipline in the planning scheduling, and control of projects.
2. Encourage more long-range and detailed planning or projects.
3. Provide a standard method of documenting and communicating project plans, schedules, and time and cost performance.
4. Identify the most critical elements in the plan, focusing management attention on the 10 to 20 percent of the project that is most constraining on the schedule.
5. Illustrate the effects of technical and procedural changes on the overall schedule.

Among the disadvantages of network control is the inability of management to grasp and understand its value. Complex networks appear as a maze and may be discarded as a tool by management. Improperly designed computational tables may fail to clearly specify the area of needed attention. The failure to properly **organize** for network control is a frequent cause of failure. As with most techniques, PERT and CPM only provide a means of control, they do not in themselves control.

Edward B. Roberts (The Dynamics of Research and Development) describes how PERT and CPM are oversold:

Project completion is far more seriously influenced by such factors as the company's risk failing propensity and general quality, the customer's decision process, and the ability of the firm to expand its organizations rapidly while maintaining effectiveness. Since these aspects, rather than the schedule itself, strongly influence project performance, it is unfortunate that so much government and industry attention has been misdirected to the scheduling problems.

GANTT CHARTS. The charts devised by Henry L. Gantt, on the basis of many years experience in a wide range of industries, are perhaps the most widely used device for planning and monitoring the use of a firm's facilities. The Gantt chart makes use of two fundamental elements: a left vertical axis on which the **kinds of capacity or orders** are listed and a horizontal axis representing the available **time** for the parameters on the vertical axis.

The value of the Gantt chart stems from its ability to show clearly and quickly the relationships among several variables and, properly applied, focus on those situations which need attention.

There are a number of variations of Gantt charts, the more popular adaptions of which are for reserved time planning, machine loading, and progress monitoring. Figure 29 illustrates the notations and symbols commonly used in constructing Gantt charts. These symbols are by no means fixed, and are often modified to suit the needs of the company or the planner.

Reserved-Time Planning Charts. A reserved-time planning chart is illustrated in Fig. 30. This chart, developed for a milling department, shows work scheduled and completed as of the end of work on Wednesday. The scheduled work is shown by a light-lined rectangle, the completed work by a shaded rectangle. The numbers above the line refer to job order numbers to be processed. This Gantt chart illustrates the following:

1. Machine No. 1 is slightly behind schedule.
2. Machine No. 2 is idle, having completed order 940. No work is scheduled until Saturday morning, so it is possible for a new order to be loaded on this machine.
3. Machine No. 3 is ahead of schedule, so order 933 has been added.

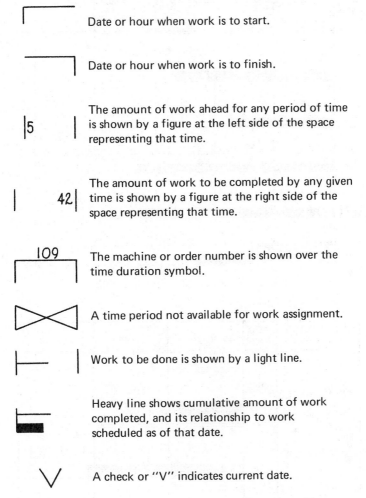

Date or hour when work is to start.

Date or hour when work is to finish.

The amount of work ahead for any period of time is shown by a figure at the left side of the space representing that time.

The amount of work to be completed by any given time is shown by a figure at the right side of the space representing that time.

The machine or order number is shown over the time duration symbol.

A time period not available for work assignment.

Work to be done is shown by a light line.

Heavy line shows cumulative amount of work completed, and its relationship to work scheduled as of that date.

A check or "V" indicates current date.

Fig. 29. Conventional symbols and notations used for Gantt charting.

4. Order 941 on Machine No. 4 was delayed in completion, but 942 is ahead of schedule.
5. Work delivered late from a preceding routing delayed the start of order 926 on Machine No. 5.

Load Charts. A load chart is often used to illustrate the **balance of work** in a firm or department. Figure 31 portrays such a chart for a packaging materials division.

The chart, updated as of mid-May, indicates a number of potential operating problems. This division normally attempts to maintain a two- to three-week backlog of work. The wax laminators and the foil laminators enjoy a five- and seven-week backlog, respectively, as indicated by the heavy, shaded bars of cumulative work. The light lines above the bars represent individual orders. The dashed bars indicate work not completed, and therefore carried over from

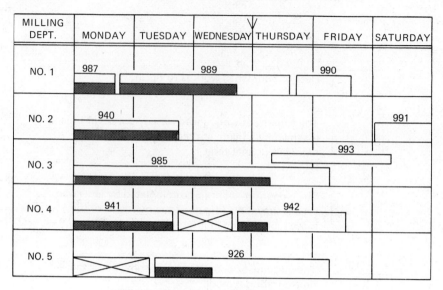

Fig. 30. Reserved-time planning chart.

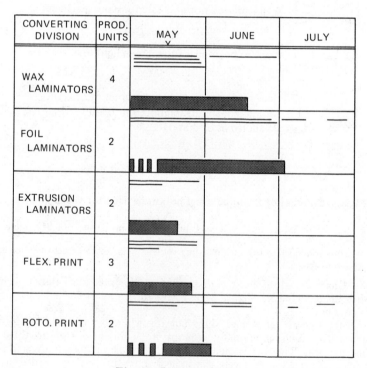

Fig. 31. Gantt load chart.

previous periods. The extrusion department, by contrast, has less than a week's work backlog.

Detail load charts, similar to the illustration previously shown in Fig. 20, are used for department loading. The method for assigning work to common machines is one of trial and error. Orders are moved among machines until the best work balance in the department is achieved.

Progress Charts. Progress charts, as with other Gantt-type charts, take a variety of forms. Figure 32 illustrates two forms of a chart used by a job order company manufacturing explosive fuses. Each manufacturing order for components or subassemblies is indented in the left-hand column to indicate manufacturing operations preceding final assembly. In the upper progress chart, the small hand-inserted numerals represent individual steps on each operation. There are four steps in detonator stock manufacture and no detonator assembly work may be performed until the fourth and last operation on the stock is complete. Detonator stock production for this reason, must lead detonator assembly to assure smooth work flow in the assembly department.

When the operation begins, a heavy line is drawn part way across the first step and when any portion of the subsequent steps begin, a heavy line is drawn part way across that step. When all operations have started, there is a series of broken bars whose length simply indicates that work has commenced on these operations. After the full order of pieces has passed through a step, the bar is drawn through that space to connect with the next portion. Ultimately, there is a continuous bar, when all steps are completed on the job order. Marginal indentations and step progress line shadings are helpful in illustrating the necessity for coordinating various elements of manufacturing.

The lower chart of Fig. 32 depicts a more abbreviated method of showing progress. In this case, individual operations are omitted and the initial brackets are placed to show when deliveries from the final steps of each operation are to begin.

The time difference between the start of each part, subassembly, and final assembly are designed to include proper reserve stocks for uninterrupted production. All preceding operations, in this way, are keyed to assembly operation. Consequently, in reading the chart it is necessary only to compare relative positions of any component bar with the assembly bar to determine the relative number of days an operation is ahead or behind schedule. All bars should be abreast of the current date, October 12, indicated by the V, for production to be at the scheduled rate.

Modified Gantt Charts. A wide variety of charts embodying Gantt-chart principles has been developed for special applications. A modified chart (Fig. 33) developed by E. A. Boyan for monitoring development work and prototype production at the M.I.T. Radiation Laboratory is still widely used. This technique utilizes a **target schedule** (▼) to provide information on the relative priority of each operation. Where schedule completion dates (⌐) and target schedules coincide, these operations are limiting. **Limiting operations** are those in which a subsequent operation depends on completion of a limiting operation. Delay of a limiting operation delays the entire project. A **safety margin** appears between scheduled date and target schedule in non-critical operations. Delays of these items minimize the need for redoing the entire chart.

The purpose of this chart is to give early warning of delay. The chart is designed to show each engineer how his work affects, and is affected by, the work of others.

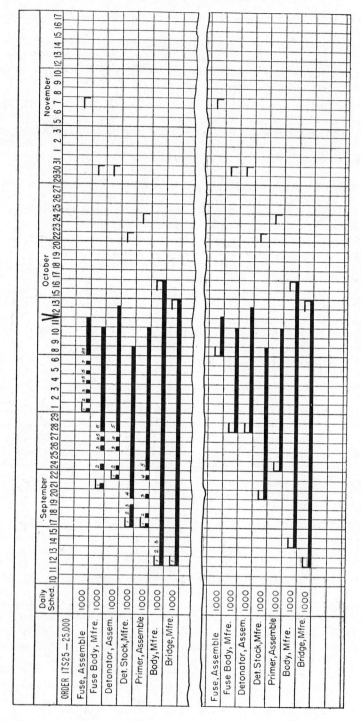

Fig. 32. Gantt progress chart for production of a fuse.

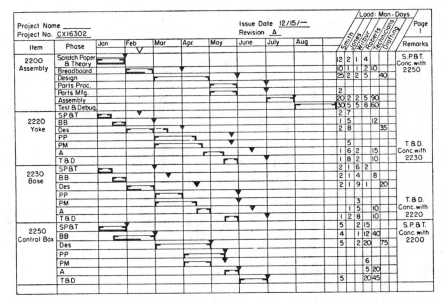

Fig. 33. Modified Gantt chart used as a target and commitment schedule.

LINE-OF-BALANCE CHARTS. A line-of-balance technique, similar to PERT, was pioneered by the military as a means for balancing inventory acquisition with the production process and delivery requirements.

The line-of-balance technique is often effective in those types of intermittent manufacture where a final assembly line is fed by a number of job or stock component lines, and delivery of end units is required at predetermined specified intervals.

Christian (Proceedings of the APICS Annual Conference) divides line-of-balance studies into four elements:

1. Objectives of project.
2. Program plan and a schedule for achieving it.
3. Current status.
4. A comparison between where you are and where you are supposed to be.

Objectives of manufacture are defined as the desired rate of final assembly. A line graph, similar to Fig. 26, illustrates the objective in a simple form.

Figure 34 illustrates an **assembly or program plan** for production of tractors. The underlying assumptions in preparing the plan include:

1. Components of the end item must be produced **at the same rate** as the end item.
2. Components must be available sooner than the end item.

In other words, if the components going to the end item are plotted on an objective graph, the lines are parallel and precede it by a lead time.

Christian emphasizes "The effectiveness of the entire line-of-balance technique hinges on the design of the program plan."

The plan is prepared in "assembly tree" form working from **right to left.** This method correlates final assembly to the objective graph, and by working backwards permits a calculation of start time for each part and component. In this way, resulting production time will be as short as possible.

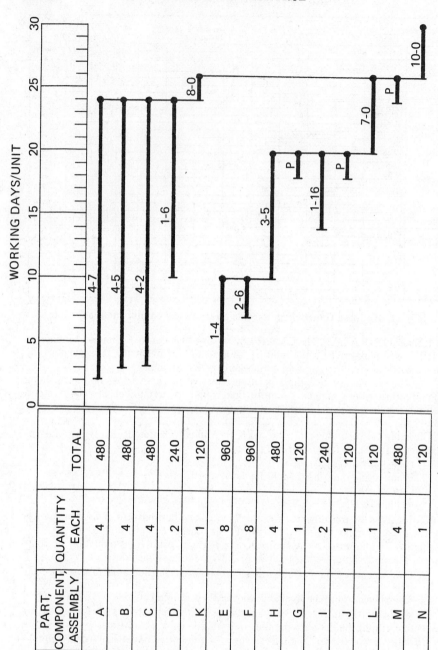

Fig. 34. Assembly plan for the production of tractors.

The working days for each part or component represent the number of days required to complete the quantity required for each unit. The numerals over the time span for each component indicate the department and the machine scheduled to produce the part or component. Departments 7 and 8 are sub-assembly departments, and department 10 is final assembly. A "P" represents a purchased component. Setup time, process time, move time and reserve time are included for each component. The **assembly plan** shows clearly when each part or subassembly will have to be put into operation in order to complete production in the minimum time period.

The third step of the line-of-balance technique is to monitor and record **actual progress.** A bar graph, appropriately scaled, serves this purpose.

The final step is to "strike the line of balance" on the bar chart. Figure 35 illustrates the bar chart and **line of balance** for production of tractors and components through December.

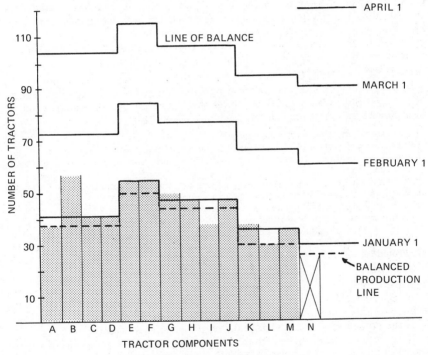

Fig. 35. Line-of-balance chart.

The line balance chart reveals that tractors are two units behind schedule. By using the line of balance chart and assembly plan, the source of the lost production is traced to subassembly L, and in turn to part I made on machine 16 in department 1. In a similar manner, component B, G, and K are observed ahead of schedule.

A fundamental principle of this technique is that all line-of-balance profile lines are parallel. By applying this principle, a parallel line, similar to the dashed line illustrated in Fig. 35 may be drawn to monitor balanced production. This line is called the **balanced production line.** Since all department outputs are controlled to the rate of assembly, the balanced production line indicates which department's output must be adjusted in the coming time period relative to assembly. In Fig. 35, only component I is below the balanced production line. The planners' task is to rebalance production. Alternative choices available to him include: scheduling overtime for component I and, subsequently, on each required component until the line of balance is regained; or in extreme cases, as with component B, reducing the production effort, thereby reducing output until the line of balance is regained.

Line-of-balance charts are often excellent tools in providing a basis for improved control. The technique shows status and relative imbalance at a glance and provides forewarning for possible schedule deviations. Its primary disadvantage is the inflexibility of the charts, once prepared. A change made early in the production process usually necessitates a major overhaul of all subsequent operations, and preparation of a new chart.

An electrical transformer manufacturer has achieved excellent results by means of a line-of-balance report. This report is computer-generated and provides loading information and weekly schedules for each component and subassembly department, geared to output of final assembly.

Line-of-balance effectiveness, as with most techniques, depends on how imaginatively it is used and how well it is understood and applied.

CRITICAL RATIO. Critical ratio is a production scheduling technique for establishing and maintaining relative priorities on jobs. Putnam (Proceedings of the APICS Annual Conference) states that the critical-ratio technique can be incorporated into most production scheduling systems to:

1. Determine the status of a specific job.
2. Establish relative priorities among jobs on a common basis.
3. Provide the ability to relate both stock and made-to-order jobs on a common basis.
4. Provide the capability of adjusting priorities (and revising schedules) automatically for changes in both demand and job progress.
5. Permit dynamic tracking of job progress and location.
6. Eliminate the expediting functions of job progress look-up, special hand-carrying, etc., by providing foremen and dispatchers with proper job sequence based on most current information.
7. Provide basic data for overall queue control and manning decisions.

The critical ratio is defined and determined by dividing **working days to required completion date** by **days of work required for completion.**

If the critical ratio is **greater than 1.0,** the job is ahead of schedule and takes a lesser priority since it will be completed earlier than needed.

If the critical ratio is **equal to 1.0,** the job is on schedule.

If the critical ratio is **less than 1.0,** the job is behind schedule and takes sequence priority or it will be finished later than the required date.

The decision policy is founded on the proposition that the **lower** the critical ratio the more critical is the job. Critical jobs in this manner are given preference while jobs ahead of schedule lag. The net effect is to bring production more nearly in balance and improve the reliability for on-time completion.

The dispatcher, in many cases, calculates the critical ratio and makes the sequencing decision. The numerator of the formula is calculated by subtracting the current date and nonwork days from the due date. The denominator, often called **total lead time remaining,** is furnished by central scheduling. The total lead time is calculated in hours and converted to days by dividing by the number of hours normally worked in a day.

The total lead time remaining is calculated as follows:

$$L = O + \Sigma \left[Q + \frac{(U + R)}{NP} \right] + \Sigma T$$

where
L = Total lead time in hours.
O = Order preparation time in hours. This includes the time to get all the information, tools, fixtures, and special materials to each work center to the extent it adds to lead time.
Q = Standard queue in hours. A department is expected to have a normal backlog of work to avoid interrupted production and to provide a pool for efficient sequence.
U = Setup time in hours.
R = Running time in hours.
N = Hours normally worked in a day.
P = Percent effectiveness in meeting standards in the particular work center.
T = Transit times between work centers.

For example, if six work days remain before an order is due and total lead time (calculated from the above formula) is eighty hours, the critical ratio can be calculated as follows:

$$\frac{80 \text{ (total lead time)}}{8 \text{ (hours normally worked in a day)}} = 10 \text{ days}$$

$$\frac{6}{10} = 0.6 \text{ critical ratio}$$

A **queue analysis report** can be developed by the production control department using the critical ratio technique. This report (Fig. 36) often proves to be an excellent management tool.

Queue analysis reports compare an actual queue with a standard queue. The actual queue is equal to the total lead time calculated according to the above formula. Figure 36 could be converted to a critical ratio report by dividing

Work Center	Days			Date: 6/15	
	Minimum Queue	Standard Queue	Maximum Queue	Actual Queue	Criticality
101	3	5	7	5.3	
102	3	5	7	5.9	
103	2	4	6	2.1	Lo
201	5	7	9	9.3	2-Hi
202	5	3	7	7.4	1-Hi
301	4	6	8	6.2	
401	3	5	7	6.9	Hi
501	4	6	8	6.0	

Fig. 36. Queue analysis report.

standard queue by actual queue. For example, the critical ratio of work center number 202 is:

$$\frac{3}{7.4} = 0.41 \text{ critical ratio}$$

When requirements exceed capacity for any period of time, critical ratios shrink and queues increase. Critical ratios become high and queues decrease when capacity exceeds requirements for any period of time. Management attention is directed to the exception work centers. In Fig. 36 both work centers 201 and 202 are high queues and require immediate attention to avert late deliveries. Work center 103 is approaching low criticality and is in danger of either running out of work or losing efficiency for lack of sufficient orders for optimum sequencing.

Analytical Techniques for Scheduling and Control

DEVELOPMENT OF ANALYTICAL METHODS. Complexity of many production systems poses a major problem for production control. Historical solutions have frequently depended upon experience, judgment, intuition, or expensive experimentation on the production line.

The use of modern analytical methods for the solution of production control problems has evolved only since the end of the Second World War. Forrester (Industrial Dynamics) discusses four specific points that have made modern approaches possible:

1. **The understanding of information feedback.** This includes the action taken as a result of information contained in production progress reports; the accuracy of reported information; delays in compiling information and initiating action; and the interaction of information, orders, materials, facilities, personnel, and money.

2. **Knowledge of the decision-making process.** This point recognizes the type and quality of information that managers must gather to establish policy and make decisions. The vast majority of decisions made by a production planner, for example, depend on the apparent condition of the operation at any point in time. Based on each apparent condition, the planner typically makes a specific decision.

3. **Experimental model approach to complex systems.** Understanding of business systems has evolved with the introduction of the systems analyst whose job includes investigation of the behavior of production operations. The systems analyst typically constructs a model of the system to help understand its behavior, then designs new or improved business systems to improve that behavior. The **model** itself is a symbolic (flow chart), mathematical (equation), or descriptive portrait of the business system under study.

4. **Use of digital computer to simulate production systems.** The computer, perhaps more than any other tool, has made possible the solution of problems with great speed and minimal cost and has expanded the applications of mathematical models to complex production systems.

The following discussions are intended to provide information on the general purpose of a number of the more common techniques as related to production control.

Management Responsibility for Use. Limited application of analytical methods in production control systems is due in part to misunderstanding its objectives. Magee (Production Planning and Inventory Control) discusses the reluctance of management people to accept and be convinced of the value of analytical methods. The complexities of mathematics and the management impression that the techniques are theoretical rather than practical for real-life situations are still prevalent. Nevertheless, applications are in use. Perry (Automation, vol. 14) states that linear programming, where properly used, provides the more progressive company with advantages over their less sophisticated competitors.

Jacobs at an American Production and Inventory Control Society Spring Seminar discussed the use of management science techniques in decision making and stressed: "The eventual success of the small company, as well as the large company, may rest with the ability to recognize the need for advanced technical analysis, pursue solutions, and implement the results."

INDUSTRIAL DYNAMICS. Industrial dynamics evolved out of the M.I.T. classrooms shortly after 1956. This technique, among other purposes, is used to understand policy-making decisions (rates) in a system, how they affect the present apparent condition of operations (levels), and how, in turn, rates are affected by new levels at a later time.

Using **simulation** techniques, improved planning decisions are developed aimed at reducing delays or utilizing different or improved information for decisions. Among the more practical applications that may be made by the production control planner using the industrial dynamics technique are analyses of when and how many people to hire, when to use overtime, and when to add facilities or equipment to support production operations.

Industrial dynamics stresses use of **macro analysis** (the broad overview) as a means for **improved behavior.** Operations research and linear programming techniques differ from industrial dynamics by using **micro analysis** (the examination of details) in finding an **optimum solution** to a specific problem.

LINEAR PROGRAMMING. Linear programming is a mathematical method for determining the optimum combination of a limited set of resources to maximize some objective, such as profit. There are a number of variations of linear programming techniques, including: the simplex method, transportation method, distribution method, stepping-stone method, north-west corner rule method, and MODI (modified distribution).

Figure 37 illustrates the general method for problem solutions by use of linear programming. In the initial solution matrix, for example, various orders are listed in the top row with each respective order processing time in the bottom row. The orders may be processed entirely on any one machine or split between machines. The machines and hours available on each machine are listed in the left and right hand rows, respectively. The subsquares contain the cost per hour of running each order on each of the various machines. The objective is to determine the least cost of assigning orders to machines.

The "dummy" column represents excess machine capacity over demand.

The best solution is determined by rearranging the hours required to process each order to each of the various machines until the least cost is reached.

The model used in solving this type of problem by linear programming is a set of mathematical equations. These equations specify costs, values of parameters, and constraints of the system. The model is then programmed into a computer that processes these equations until the best solution is reached. Computer processing, at great speed, simulates the trial and error efforts of moving portions of the orders around the matrix and computing the cost after each move.

Orders / Machines	#101		#102		#103		Dummy		Standard Hrs. Available
#1	10	10.00	10	4.00		8.00		0	20
#2		12.00	30	17.00		8.00		0	30
#3		15.00		8.00	30	16.00	10	0	40
Standard Hrs. of Demand	10		40		30		10		90

INITIAL SOLUTION

Orders / Machines	#101		#102		#103		Dummy		Standard Hrs. Available
#1	10	10.00	10	4.00		8.00		0	20
#2		12.00		17.00	30	8.00		0	30
#3		15.00	30	8.00		16.00	10	0	40
Standard Hrs. of Demand	10		40		30		10		90

FINAL SOLUTION

Fig. 37. Matrix used in a linear programming solution of the least cost, optimum assignment of orders to machines.

INITIAL SOLUTION

ORDER	UNITS × MACHINE COST		COST
#101	10 × 10.00	=	$ 100.00
#102	10 × 4.00	=	40.00
	30 × 17.00	=	510.00
#103	30 × 16.00	=	480.00
			$1,130.00

FINAL SOLUTION

ORDER	UNITS × MACHINE COST		COST
#101	10 × 10.00	=	$100.00
#102	10 × 4.00	=	40.00
	30 × 8.00	=	240.00
#103	30 × 8.00	=	240.00
			$620.00

The application of this technique does not require writing a new model for each scheduling problem. The basic cost and loading information is stored in the computer. Each week only the available machine capacity and standard hours for each new order are required in order for the computer to solve the scheduling problem. Processing of this type is of particular value in complex **intermittent, custom** manufacturing situations.

Perry (Automation, vol. 14) discusses in detail the means for computer-aided decision-making in **custom work.** He suggests programming into the computer a matrix of the hours to run each required product plotted against available capacity of each machine capable of producing the product. The setup and operating costs of each machine is added, and the minimum cost solution of products to machines is determined using a **distribution** method of linear programming.

QUADRATIC PROGRAMMING. Constructing a model of first-order equations (linear programming) does not always produce high validity with real-life situations in more complex problems. Quadratic programming on the other hand contains second-order equations making possible the simulation of curves other than straight lines and resulting in a more realistic model. Unfortunately, the mathematical complexities of the higher order equations are often beyond the technical capabilities of even trained mathematicians thus limiting the applications of this technique in production control.

QUEUEING THEORY. Queueing, or waiting-line theory, as applied to production control, involves a mathematical model representing the different order processing times and the numbers of orders waiting to be processed on a machine.

Queueing theory is used to solve problems that involve a random rate of orders being delivered to a work center (from an earlier process station), and each order having a random processing time. Solutions are stated in terms of the number of machines (or people) to place in the centers to minimize costs (cost of the added facility or the cost of carrying work-in-process inventory, whichever is less).

MONTE CARLO METHOD. Monte Carlo analysis is used for solving problems of **stock production** involving equipment decisions in the face of an uncertain flow of incoming orders. The Monte Carlo technique is a process which uses data generated by a set of random numbers. The random numbers are tested against a curve of historical data to simulate, for example, the expected orders in a given day.

Alternative decisions on equipment or people could be tested for least cost against simulated receipt of orders.

GAME THEORY. Game theory attempts to determine which strategy has the most chance of achieving the highest payoff. Most typical games are "zero-sum" involving a situation where the losses of one party are equal to the gains of the other party.

Situations of the "zero-sum" type are not often encountered in production control situations. Nevertheless, game theory provides the analyst with an insight into the quantification of production control processes and possible analytical solutions rather than arbitrary decisions.

SYMBOLIC LOGIC. Symbolic logic is a method of logical reasoning in formal, mathematical terms. The technique proves valuable in analyzing a set of assumptions since it checks on inconsistencies or incompleteness in logic.

Symbolic logic might be used in determining the best sequence of orders based on pre-established rules stored in a computer.

Control Boards

USE OF CONTROL BOARDS. Control boards are used to provide a graphic display of information. They are designed to present a comprehensive picture of operations that serves as an aid for effective planning, scheduling, and control.

Among the **advantages** claimed by users of control boards, are:

1. Information easy to read and interpret quickly.
2. Effective planning and scheduling decisions facilitated by visual display of information.
3. Provides an effective guide to dispatching.
4. Provides a flexible means visualizing and recording rescheduling.
5. Simplifies paperwork connected with scheduling and loading.

The **problems** of using control boards include:

1. Maintaining information on a current basis.
2. Lack of permanent information or duplicate records.
3. Difficulty in maintaining accurate information where manufacturing cycles are short (one or two hours), or long (greater than four weeks).
4. Tendency to become unwieldy in large operations.

TYPES OF CONTROL BOARDS. The various forms of control boards are closely related to charts in purpose and concept. The different types and applications of control boards in common use are discussed in the following paragraphs.

Pocket Racks; Hook Boards. Pocket racks and hook boards, although dating back to Taylor's time, are still used in many applications. Both types can serve as useful guides to dispatching.

Figure 38 shows a typical hook-type dispatch board. Each machine or workspace has three sets of hooks, one for the job in process, one for the next job to be run, and one for jobs assigned but not yet ready for work. The work-in-process hook carries a duplicate of the work order, the original being given to the operator. Other hooks carry either the duplicate work orders, with the original filed, or both duplicate and original, depending on the system of filing used. Pocket racks are used in exactly the same fashion except the three sets of hooks for each machine are replaced by wooden racks having three pockets.

Pocket or Grooved Strip Boards. Remington Rand's Sched-U-Graph system employs the principles of the Gantt chart for machine loading. The chart shows each work center or machine vertically and uses the horizontal scale for time.

The Sched-U-Graph utilizes either an 18-inch or 40-inch-long by 5-inch-high pocket for each work center or machine to be loaded. A 5-inch by 3-inch operation record card, as illustrated in Fig. 39, is inserted in the left side of each pocket. An opaque signal is inserted to show how many hours per day the machine works. A time scale insert, extending the length of the pocket, shows each day of the week divided into ten periods.

To load a machine, the planner determines the running time for each assigned job by referencing the table on the operation card. Time periods on the table are calculated by dividing **hours required for the job** by **number of machine hours available per day** and then multiplying the result by ten. Use of the scheduled "period" technique minimizes time scaling problems when machines

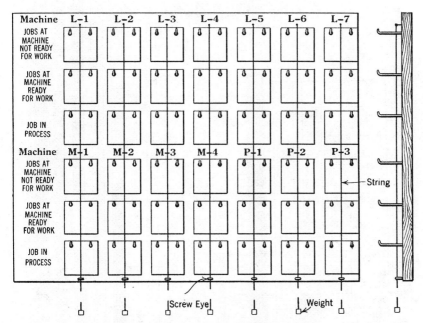

Fig. 38. Typical hook-type dispatch rack.

are loaded different hours per day. The operation cards then are cut to length, according to the period scale at the bottom, and inserted under the top of the pocket. The exposed tip of the pocket highlights machine loads, since the bottom of the operation card is red, while at the same time the white portion of the strip (where no card is inserted) indicates available capacity. If rescheduling becomes necessary, the cut-to-length operation cards are moved easily to the new period.

The Sched-U-Graph uses a vertical progress line to identify current time, and a green progress card to note actual status. Initially, the progress card is covered by a shield card. As production accumulates, the green progress card is slid from behind the shield card, and over the red span card.

The Sched-U-Graph system proves extremely useful in operations where flexibility in rescheduling is important.

Another type of pocket strip board is featured by Acme Visible Records in their Visual Control Panel System. Figure 40 shows a job loading panel as used by a large rug and carpet manufacturer. The cards at the left, indexed "Machine #" are used to record description, capacity, and all other pertinent data. A time index, printed to show an hourly scale, runs the length of the board. Individual job cards containing order specifications have their lower edge printed in various colors to distinguish between different product types. The cards are cut to show the time required for each job, and are easily moved or changed to different machines when rescheduling is required. The visible edge of the job cards slide into and are held in place by a tube strip located above the time index. Colored plastic signals, using either inserts or carrying notes written on the signal, are snapped onto the tube and slide along its length. Different colored signals are used for showing actual production and forecast.

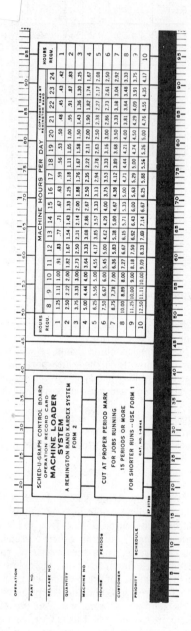

Fig. 39. The Sched-U-Graph operation record card.

HOURS REQU.	MACHINE HOURS PER DAY																	HOURS REQU.
	8	9	10	11	12	13	14	15	16	17	18	19	20	21	22	23	24	
1	1.25	1.11	1.00	.91	.83	.77	.71	.67	.63	.59	.56	.53	.50	.48	.45	.43	.42	1
2	2.50	2.22	2.00	1.82	1.67	1.54	1.43	1.33	1.25	1.18	1.11	1.05	1.00	.95	.91	.87	.83	2
3	3.75	3.33	3.00	2.73	2.50	2.31	2.14	2.00	1.88	1.76	1.67	1.58	1.50	1.43	1.36	1.30	1.25	3
4	5.00	4.44	4.00	3.64	3.33	3.08	2.86	2.67	2.50	2.35	2.22	2.11	2.00	1.90	1.82	1.74	1.67	4
5	6.25	5.56	5.00	4.55	4.17	3.85	3.57	3.33	3.13	2.94	2.78	2.63	2.50	2.38	2.27	2.17	2.08	5
6	7.50	6.67	6.00	5.45	5.00	4.62	4.29	4.00	3.75	3.53	3.33	3.16	3.00	2.86	2.73	2.61	2.50	6
7	8.75	7.78	7.00	6.36	5.83	5.38	5.00	4.67	4.38	4.12	3.89	3.68	3.50	3.33	3.18	3.04	2.92	7
8	10.00	8.89	8.00	7.27	6.67	6.15	5.71	5.33	5.00	4.71	4.44	4.21	4.00	3.81	3.64	3.48	3.33	8
9	11.25	10.00	9.00	8.18	7.50	6.92	6.43	6.00	5.63	5.29	5.00	4.74	4.50	4.29	4.09	3.91	3.75	9
10	12.50	11.11	10.00	9.09	8.33	7.69	7.14	6.67	6.25	5.88	5.56	5.26	5.00	4.76	4.55	4.35	4.17	10

COPYRIGHT 1948 BY REMINGTON RAND

SCHED-U-GRAPH CONTROL BOARD
OPERATION RECORD CARD
MACHINE LOADER
SYSTEM
A REMINGTON RAND KARDEX SYSTEM
FORM 2

CUT AT PROPER PERIOD MARK
FOR JOBS RUNNING
15 PERIODS OR MORE
FOR SHORTER RUNS -- USE FORM 1

OPERATION
PART NO
RELEASE NO
QUANTITY
MACHINE NO
HOURS
CUSTOMER
PRIORITY
PERIODS
SCHEDULE

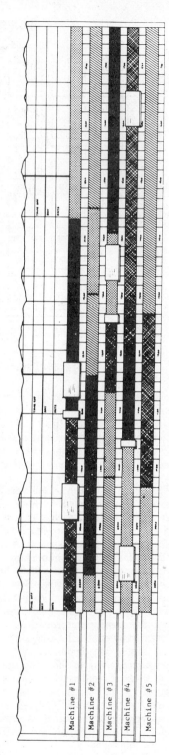

Fig. 40. Visual control panel used as a machine loading chart.

Machine #1
Machine #2
Machine #3
Machine #4
Machine #5

Spring Clip Boards. Spring clip control boards, such as the Acme-Mc-Caskey control board shown in Fig. 41, are used as combined dispatch and schedule boards.

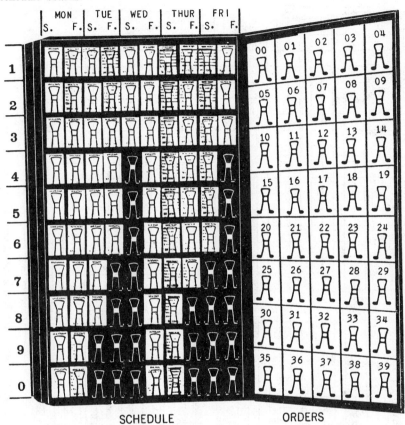

Fig. 41. Board used for filing orders and machine scheduling.

The control board is used with a multi-part job order form. Orders released to the dispatcher are filed, according to the last two digits of the order number, behind the clips on the right side of the board. The left side of the board contains separate rows for the various machines in the department. The dispatcher places an original and a colored copy of the order behind a start **(S)** clip on the day to start, and a different colored copy behind a finish **(F)** clip on the expected finish day. The original copy is filed behind the **S** clip and then given to the machine operator for processing when the job is ready to start.

The dispatcher, in this manner, has a convenient technique for filing orders and loading machines. Control is provided by observing any orders not started at the end of the start day, or not completed by the end of the finish day.

Tape or Peg Boards. Another variety of control board employs moveable tapes or strings that run horizontally across it to record progress of work. The Produc-Trol boards shown in Fig. 42 (Wassell Organization) are used extensively for production control purposes. On the left-hand side of the board is a

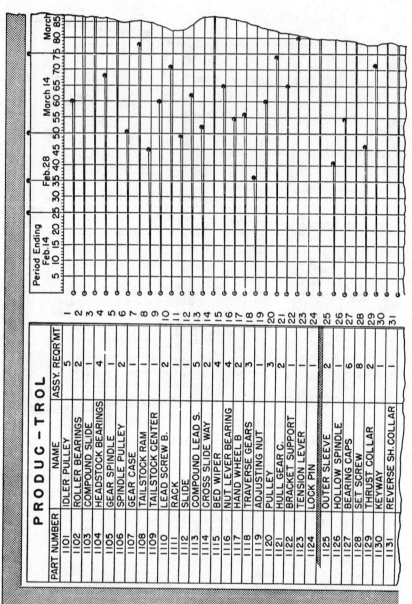

Fig. 42. Order schedule control board.

pocket panel in which record cards are inserted so their bottom margin is visible. This margin carries whatever identification is required. The right-hand side of the board is divided into a series of peg holes for visible charting. A heading strip is available for listing across the top of the board section divisions of time or quantity as required. The heading strip may also be used to divide the board into departmental sections.

Each card in the visible index panel is aligned with an "item line" on the peg board section. The "item line" consists of two lines of peg holes; the upper of these two lines is used to indicate, by means of signal pegs, the scheduled operations to be performed. The lower of the two peg hole rows in each "item line" is used to record progress against schedule.

The tape peg is attached to a tape or cord, which is always in tension no matter where the peg is inserted, from the first peg hole to the 200th hole. Each tape peg is numbered to correspond with the item line it covers.

By use of a colored vertical "today" line cords (representing the present day) a means for control may be introduced. The relationship between these vertical today line cords and the signal and tape pegs indicate, at a glance, whether work is behind or ahead of schedule and by how much in terms of divisions set on the heading strip.

Pictorial Boards. Pictorial boards are also used to show work progress. These boards use a background of a layout drawing or other schematic. Signals, such as colored pins or tacks, indicate progress. One manufacturer of large machine tools uses this principle to track assemblies through the production process. Units often require a number of weeks to be fully fabricated. The board employed is a visual picture of the plant layout. As a machine tool is started, a colored tack showing the order number is placed in the first assembly department. The tack is moved along the assembly line board as information on progress is received. A standard queue, stated in terms of optimum numbers of orders in any assembly department, is recorded on the board. The board shows, at a glance, department underloads and overloads and serves as a basis for required management effort.

Magnetic Boards. Metal boards are often used in a manner similar to pictorial boards with magnets used to record progress. The Magna-Chart system (Magna-Visual, Inc.) uses magnetic blocks of various size, coated on one side with a colored plastic material. Order specifics are recorded on the plastic face with grease pencils, then erased and reused after job completion.

Many companies employing this system use a time-scaled chalkboard as a background. The board is usually lined or segregated by department. Chalk lines are drawn using Gantt symbols and the magnets moved to show progress. Magna-Visual blocks are also available in PERT symbols and provide a convenient means to plan a project on a chalkboard using the network technique. After the planner is satisfied with his network, permanent lines may be drawn and other magnetic blocks used as a means to monitor progress.

Other Types of Boards. A variety of other boards are available for planning and control.

The **Rol-A-Chart** system (Wm. A. Steward Co.) employs a mylar plastic sheet that rolls around the board. Under the sheet are guide lines, and loads are written on the mylar with colored pencils using Gantt notations. A "today" line hangs vertically on the face of the board. A stationary index, lined to correspond to the board lines, is on the left for recording data such as order numbers, machines, or men. As the sheet is rolled from right to left, status of projects is easily read and action initiated as required.

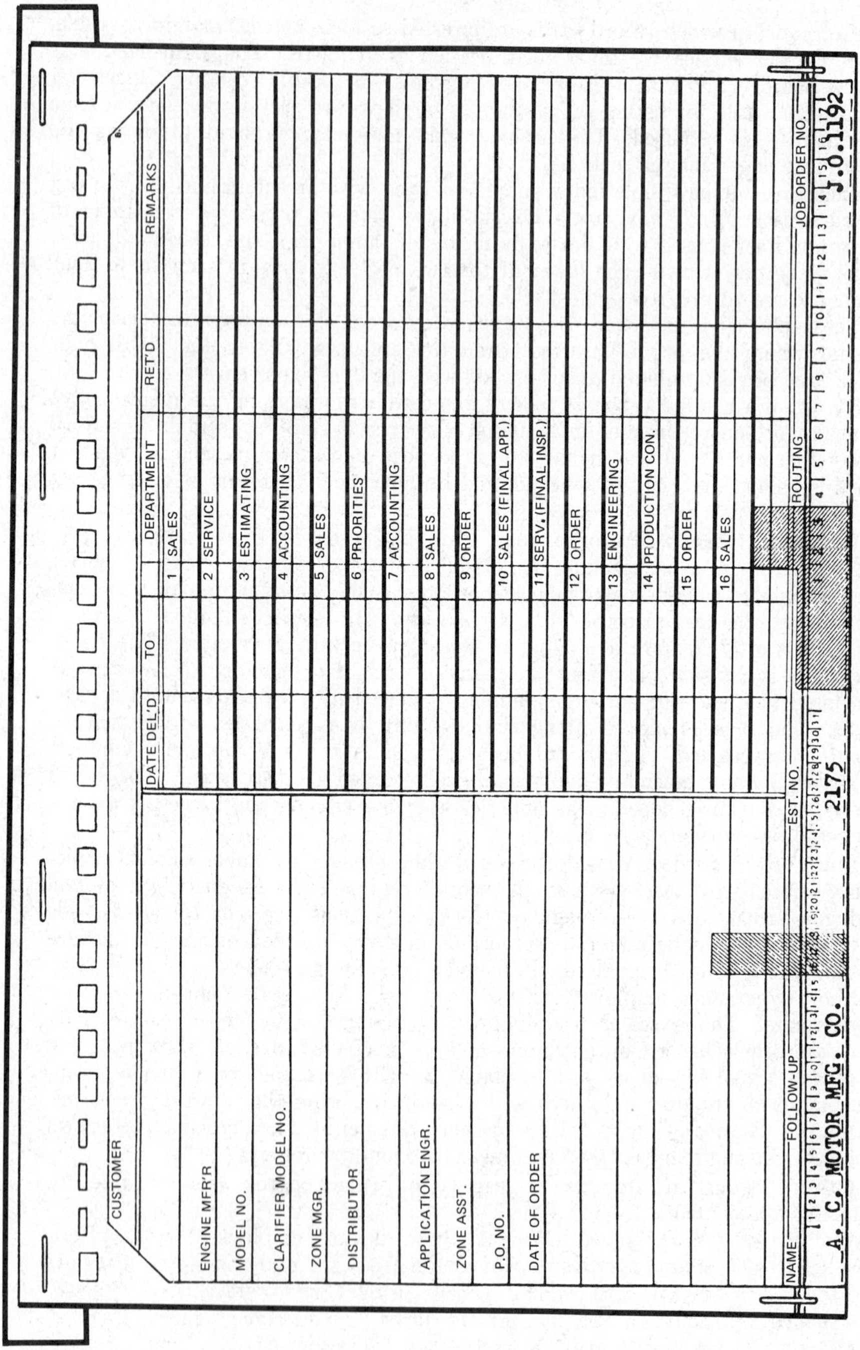

Fig. 43. Card for controlling progress on an order.

Index-Visible boards (Remington Rand) use cards that fit into a special binder. The pertinent index information is shown and progress is recorded on the exposed index line. The cards may be raised or lowered in the binder, without removing them, to view additional information. Cards stay in the binder until the job is completed, and are then removed and filed.

VCA boards (Visual Control Associates) employ a means for line graphing performance comparisons similar to Fig. 27. Colored elastic cords are anchored in place with clips. Both forecast and actual progress are charted by moving the clips.

Data File and Retrieval Systems

USE OF CARD FILES. Cards are used extensively in production control systems to record data. They may be filed and selected data retrieved as required. Cards occupy much less space than control boards, are accessible, can bring pertinent information together in condensed form, offer greater facility for checking and comparisons, and reduce personnel required for record keeping. Card file data on the other hand are generally more difficult to interpret and inflexible as a means for loading.

TYPES OF CARD FILE SYSTEMS. Each type of file system offers specific user advantages. Selection of a system for production control depends on department needs and the cost it is willing to incur for filing and retrieving information.

Vertical Card Files. Vertical card files are most often used for indexing and recording such information as orders, route sheets, and standard machine data. The machine load ledger shown in Fig. 19 is maintained in a vertical card file.

Vertical card file systems are usually the least costly to establish. **Disadvantages** of these systems include: extra time required to locate, remove, post, and refile, and the danger of losing or misfiling cards. In many production control installations these disadvantages have led to abandonment of this type of card file in favor of the visible index form of record.

Efforts to improve standard vertical files have also led to such developments as the Simplafind system (Wheeldex Mfg. Co.), a motorized, automatic card file designed to rapidly bring the desired card to the operator.

Visible Index Files, Vertical Overlapping. This type of system features cards or forms held in pockets that overlap each other in vertical rows. File systems offered by Kardex (Remington Rand), Acme, and a number of others are in wide use because of economy of time and space, ease of reference and posting, and safety from loss or misfile.

The vertically overlapping cards carry identifying information printed along the exposed index for ease of reference. The cards are carried in holders provided with flexible hinges. When necessary to consult a card, the proper card is located and those covering it are thrown back. Posting is done by hand, typewriter, or tabulating machine.

Visible index cards are frequently provided with transparent coverings over the index section so cards are kept clean, and do not become dog-eared or damaged. Colored, transparent signals can be used along the index. A follow-up signal is typically located on the left edge of the index to indicate, for example, a scheduled date for releasing a job order for run. A progress signal is on the right and is moved along the edge of the index to denote cumulative progress.

Figure 43 illustrates a visible index, vertical overlapping card for recording order progress. The left signal is a follow-up date for return of the production order. The right progress signal traces order flow through each department that must act on the order.

Visible Index Files, Horizontal Overlapping. Horizontal overlapping systems, such as Diebold's VRE system feature many of the advantages of vertical systems, and, in addition, are particularly well suited for records that are removed and used with automatic tabulating and posting equipment.

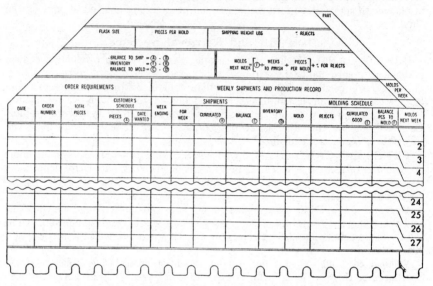

Fig. 44. Card with visible triple margins.

The card illustrated in Fig. 44 has three visible margins. They are cut on either or both upper corners for indexing. Cards are offset in their holding tray by means of a series of notches on the bottom of the card that fit over spacer rods attached to the tray. Detailed operations data, such as daily production quantities, are posted to the columns normally covered on the card, leaving the visible side margin for totals. A complete history of the job status in successive operations may be viewed down the side margin. This progressive, visible margin is a unique feature of these cards. Various kinds of signals may be applied to the card, including progressive signals along the diagonal or vertical margins.

Diebold, among others, features a series of mechanized, horizontal overlapping files. Where an operator must maintain many thousands of records, **power files** may prove economical. The operator, sitting in the normal posting position, is able to retrieve any record by the push of a button in less than three seconds.

Rotary Files. Rotary file systems provide another means of reducing retrieval and storage time of the operator. Cards are mounted by snapping slot-punched holes over specially formed bars or rings located around the circumference of a drum. The desired card is brought to the reference position by revolving the drum. A special stabilizing device permits hand posting without holding the wheel or removing the card.

Rotary files range in size from small, desk-top hand models that revolve either vertically or horizontally, to large self-contained power units.

MICROFILM AND MICROFICHE SYSTEMS. Microfilm systems use a reduced photographic negative to record, file, and retrieve data. They are most often used in production control for storage and retrieval of engineering information, routings, and drawings. The advantage of microfilm storage is economy of file space. Images on microfilm are enlarged and viewed on special screens. When hard copy is required, machines are available to enlarge and print the negative. More **common methods for filing data** on microfilm are: roll film, aperture cards, and microfiche.

A spool of microfilm **roll film** consists of a continuous series of negatives which are the photographed records. Roll film methods are generally useful in situations where fixed sequences of records are required or when retrieval of a specific record is not essential to the file system.

Another type of microfilm system uses an **aperture card.** Aperture cards are punched cards containing a window in which the microfilm is held in place. Aperture cards provide for retrieving specific records and serve as a document on which to record additional data pertinent to the record.

Aperture cards are retrieved by means of ADP sorting equipment. They are used with equipment that can enlarge the retrieved card for viewing or with equipment used for duplicating or printing the enlarged negative.

Microfiche systems use cards containing, typically, a series of 60, 70, or 98 standard page formats of microfilm negatives on a single card. This system further condenses storage. The applications for viewing, enlarging and printing are similar to other microfilm applications. Figure 45 shows a microfiche viewer with provision for reading two adjacent documents. Kish and Morris (Microfilm in Business) sum up the advantages of microfiche over other types of microfilm as follows:

1. Provides a quicker and more economical means of preparing and distributing multipage reports.
2. Filing and retrieval of bulky material are easier and filing space requirements are much less. Reduction of all material to one standard size eliminates the need for special filing equipment.
3. Adaptable to most microfilm readers and reader/printers. Existing microfilm systems easily converted to microfiche applications by means of cutting and stripping operation.

AUTOMATIC MICROIMAGE RETRIEVAL SYSTEMS. Advanced technology in file systems is best described as automatic microimage retrieval systems. The SD-500 (Sanders Associates Inc./Diebold, Inc.) system claims the following features:

1. Stores intermixed sizes and types of format, including microfilm, microfiche, filmstrips and frames of 35 mm or other size.
2. Utilizes any number of remote cathode-ray tube (television type) viewing terminals. Each terminal contains a control panel for calling a record and incorporates a means for close-up viewing of fine print lines.
3. Provides for dry hard copy printers at the central repository or remote terminals.
4. Permits accessing of any one of 5,000,000 records in an average of 8 seconds. Prints the image, if desired, in less than 10 seconds.
5. Stores records in a central repository that occupies space equivalent to 6 standard file drawers. The equivalent hard copy storage space would require 140 standard file drawers.
6. Interfaces with real-time computers.
7. Permits random accessing of a single microfilm or a module of forms for editing, leaving the rest of the system operative.

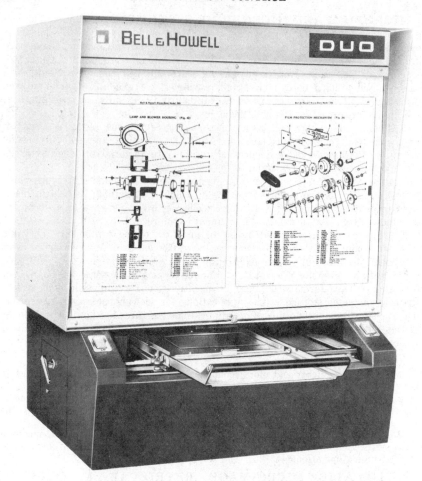

Fig. 45. Microfiche viewer with dual-image feature.

8. Provides for a security lock-out system that makes selected information inaccessible to unauthorized viewers.

Automatic Data Processing

INFORMATION SYSTEMS. Information is considered by many as a prime resource of any business, and essential to profitable operations. Production control needs information to plan and control. In particular it requires feedback information to properly assess the results of prior planning and provide a basis for new planning. The formalized routine for accomplishing this is the **information system** which deals with the regular and timely processing and flow of data.

AUTOMATIC DATA PROCESSING SYSTEMS. Automatic data processing (ADP) is data processing done by any type of machine and includes all non-manual systems. Integrated equipment systems, unit record systems, and electronic computer systems all are automatic data processing.

Investments in ADP are substantial. Dean (Harvard Business Review, vol. 46) reports the average company spent $5,600 per million dollars of sales in 1967. Companies performing fabrication and assembly, where production control tends to be more complex, spent an average of $9,700 per million dollars of sales. Dean's opinion is: "The day may not be far distant when those who analyze annual business failures can add another category to their list of causes—failure to exploit the computer."

Withington (The Use of Computers in Business Organizations) lists seven factors which, in the long run, are essential to the installation of a successful computer application. These include:

1. Tackling the critical problems of the business in preference to the easy jobs.
2. Adapting the paperwork procedures to the machine in order to take advantage of the capabilities of the equipment.
3. Providing services for all significant functional units of the business.
4. Assuring that systems, procedures, and decision rules are sound before being converted to the computer.
5. Utilizing the best people available in the organization for the development of systems.
6. Assuring that the implemenation of any new system includes careful planning and detailing, training of people, and testing before final conversion.
7. Involving the firm's top management and obtaining their support of the project from beginning to end.

ELEMENTS OF THE SYSTEM. ADP systems are used in manufacturing to produce the reports and documents essential to planning and controlling production, inventories, costs, and other elements of plant operation. The common elements of ADP systems are:

1. An **input medium.** This is a physical substance on which data is recorded in machine-readable form. Examples include punched cards, punched paper tape, magnetic tape, magnetic discs, or documents that can be electronically or optically sensed.
2. An **input device.** The most common type of device is a key punch unit operated by a keyboard similar to that of a typewriter or ten-key adding machine. The key punch unit records data by punching holes in a card. Other input devices process data directly on tape, or obtain data by use of an optical scanner or telecommunication equipment.
3. **Processing machines,** guided by instructions on the input medium and in the machine itself, automatically perform an assortment of functions. These include: storing data, moving data within storage, comparing data for magnitude, and performing arithmetic operations (addition, subtraction, multiplication, and division).
4. An **output medium.** Examples of output media include punched cards, punched tape, magnetic tape, magnetic discs, and printed forms.

MARGINALLY PUNCHED CARD SYSTEMS. The McBee **Keysort** system for production control (Litton Automated Business Systems) provides for processing data on machines, materials, and tools needed in production of parts, assemblies, and end products.

The system is based on pre-printed forms and marginally punched cards, similar to Fig. 46. The forms assure uniformity of information and accuracy of reporting final cost and production information. As many forms as are required in the penetration are created from an original "master copy" form. Among **applications** of marginally punched cards are: programs to classify and tabulate machine loads, production figures, material requirements, material usage, and direct and indirect labor cost summaries. Processing of cards

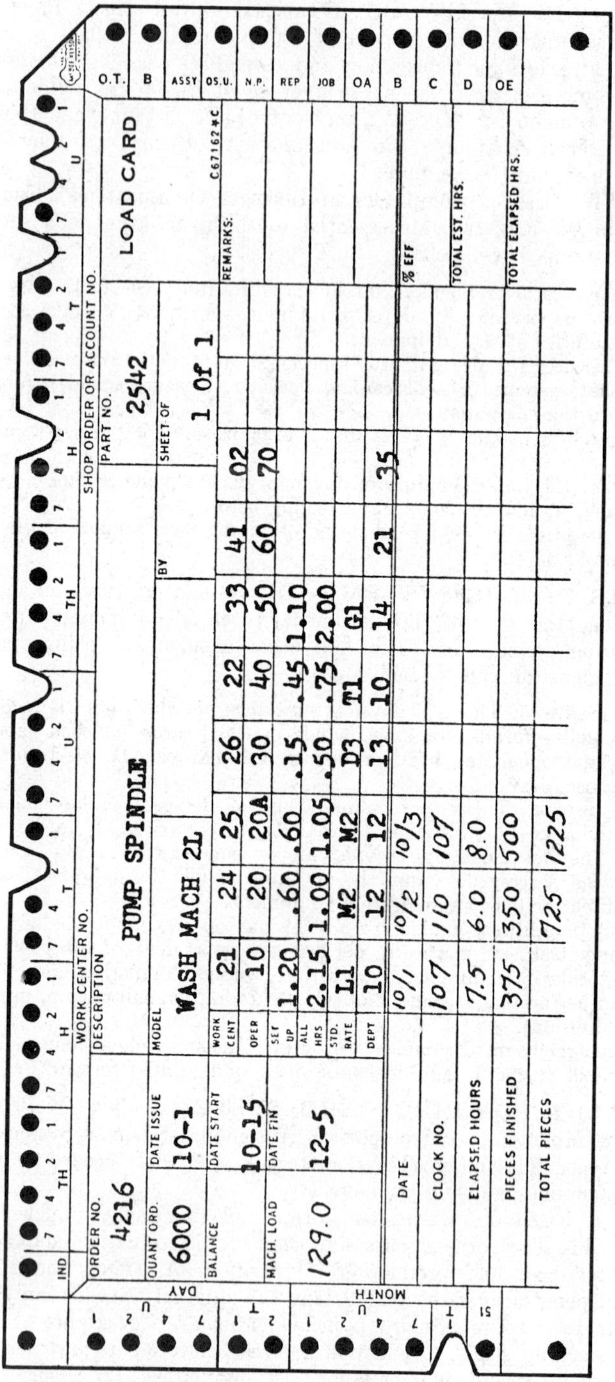
LOAD CARD
O.T. B ASSY OS.U. N.P. REP JOB OA B C D OE
IND 7 4 2 1 7 4 2 1 7 4 2 1 7 4 2 1 7 4 2 1 7 4 2 1 7 4 2 1
TH H T U TH H T U
DAY 1 2 4 7 1 2 T
MONTH 1 2 4 7 1 T 51
ORDER NO. 4216
QUANT ORD. 6000
BALANCE
MACH. LOAD 129.0
DATE ISSUE 10-1
DATE START 10-15
DATE FIN 12-5
WORK CENTER NO.
DESCRIPTION PUMP SPINDLE
MODEL WASH MACH 2L
BY
SHOP ORDER OR ACCOUNT NO.
PART NO. 2542
SHEET OF 1 Of 1
REMARKS:
C67162*C
WORK CENT 21 24 25 26 22 33 41 02
OPER 10 20 20A 30 40 50 60 70
SET UP 1.20 .60 .60 .15 .45 1.10
ALL HRS 2.15 1.00 1.05 .50 .75 2.00
STD. RATE L1 M2 M2 D3 T1 G1
DEPT 10 12 12 13 10 14 21 35
% EFF
TOTAL EST. HRS.
TOTAL ELAPSED HRS.
DATE 10/1 10/2 10/3
CLOCK NO. 107 110 107
ELAPSED HOURS 7.5 6.0 8.0
PIECES FINISHED 375 350 500
TOTAL PIECES 725 1225

Fig. 46. Marginally punched card used as a load card.

is handled by a needle-sort selection system, or as the operation grows, converted to a unit record system.

UNIT RECORD SYSTEMS. A unit record system is one based on punched cards. The holes in the cards enable the equipment to perform functions such as calculating, sorting, classifying, comparing, and summarizing. All processed information, such as the results of calculations, are stored on the punched cards.

Unit record systems are used extensively in providing data for planning and controlling production. **Typical programs** utilizing punched cards in an assembly operation are:

Demand Forecast Reports. These are based on information punched on cards from sales orders, shipments, and inventory transactions.

Material and Inventory Planning Reports. In order to obtain this information a permanent **bill of material** deck of punched cards is established. One card is punched for each component item, and contains the part number of the component used. Other clarifying data may be included as needed.

The quantity of finished product desired is punched in an order card, along with its part number and scheduled production date. The exploded bill of material is used to generate a deck of requirements cards. These requirements decks are accumulated for all scheduled products, sorted by part number, and then used to print quantities required of each part in each time period.

Facility and Manpower Planning Reports. A deck of labor specification cards containing information on labor times and descriptions, tools required, and area performed for each step of the manufacturing process is established for these reports. Information is generated based on the quantity of finished products desired, similar to the explosion process used for material planning.

By-Product Information. One of the advantages of automatic data processing systems is the ability to generate new information and additional reports from a common data base. For example, once the data is stored in punched card form, the cards may be collated, sorted, or manipulated in any way to produce the additional information. This can be accomplished at great speed in an error-free manner. Among by-product reports or documents generated are: production orders, similar to Fig. 47; a series of subsidiary orders, including a summary shop order, material requisition cards and a job-move-locator ticket, shown in Fig. 9; and a machine tool load summary, illustrated in Fig. 48.

ADVANTAGES AND DISADVANTAGES OF UNIT RECORD SYSTEMS. Once the information is correctly placed on punched cards, a unit record system is capable of providing fast, accurate information designed to serve as a basis for improved decision-making. Applications described above are typical of the capabilities of these systems.

There are several disadvantages of unit record equipment, however. Care must be taken in planning for the equipment. Capabilities of the equipment are often limited by the ingenuity and abilities of the planner. These planners are specialists, and because of their training are costly. Finally, because people handle cards, wire the panels, and operate equipment needed for a unit record system, there are possibilities for human error.

INTEGRATED EQUIPMENT SYSTEMS. An integrated equipment system involves automatic generation of an input media for an ADP system by means of a by-product machine as the primary operation is performed.

PART NO.	ORD.NO.	QUANTITY	DATE ISSUE		SHEET	MANUFACTURING ORDER
27002	125	325	5	12	1	

CHASSIS	PWR UNIT

OPER. NO.	DEPT. NO.	GRP. NO.	DESCRIPTION OF OPERATION		TOOL NUMBER	S C E	C O N D	START DATE	
1	500		RAW STORES	2110064 LB		5		115	
3	500	122	SHEAR	TO FIT DIE			6	117	
5	170	112	BLANK	PIERCE HOLES	1087	6	6	118	
10	170	20	FORM		87603	6	6	119	
20	190	810	SPOT WELD	CORNERS				120	
25	030	201	CSK & BURR					122	
35	010	142	FINISH	BRIGHT CADMIUM PLATE				124	
40	830		STOCK					125	
			ENG.CHG.	#4263					

Fig. 47. A manufacturing order.

The attached by-product machine, often called a "slave" unit punches a card or paper tape as the main unit, such as a bookkeeping machine, automatic writing machine, or typewriter-calculator, is operated.

The by-product punched cards or paper tape are frequently sent to a service bureau. Typically, programs, similar to those described for unit records, are processed by the service bureau and returned to the company.

The main **advantage** of the integrated equipment system is capturing data at the earliest opportunity. With data entered only once, chances for error are greatly reduced. Once data is captured, it can be used repeatedly for any type of by-product report or document, thereby reducing redundant clerical effort.

ELECTRONIC DATA PROCESSING. An electronic data processing system utilizes a computer, has "memory" capacity, is capable of split-second speed and can store information. The electronic computer combines many of the capabilities of electro-mechanical machines of a unit record system.

Computers have developed through "three generations." The first generation used vacuum tubes as the heart of the system. This marked the transition from punched card, or electro-mechanical system, to all-electronic, automated data processing. First generation computers were capable of performing operations in thousandths or ten-thousandths of a second. Unfortunately, vacuum tubes required a good deal of space and large amounts of power, generated vast amounts of heat, and often failed. Normally, these machines were operated by batching information to be processed, as is done with unit record equipment.

Second generation computers overcame many of the problems of first generation equipment by substituting solid state devices such as transistors for vacuum tubes. Solid state devices are smaller, require much less power, are more reliable and generate considerably less heat than vacuum tubes. In addition, solid state devices permit even faster operating speeds than vacuum tube ma-

MACHINE TOOL LOAD SUMMARY

MACHINE SHOP A

DEPT. NO.	GRP. NO.	DESCRIPTION	NO. OF MACHS.	EFFIC'Y	WK.	CAPACITY	LOAD	AVAILABLE CAPACITY	OVER-LOAD
1	1	BENCH MILLS	5	85%	1	136.0	130.0	6.0	
					2	170.0	160.0	10.0	
					3	170.0	165.5	4.5	
					4	170.0	179.0		9.0
					5	170.0	162.3	7.7	
					6	170.0	185.1		15.1
					7	170.0	150.0	20.0	
					8	170.0	162.8	7.2	
						1326.0*	1294.7*	55.4*	24.1*
1	3	SMALL HORZ MILLS	8	80%	1	204.8	198.0	6.8	
					2	256.0	250.0	6.0	
					3	256.0	251.9	4.1	
					4	256.0	269.5		13.5
					5	256.0	256.0		
					6	256.0	240.0	16.0	
					7	256.0	263.0		7.0
					8	256.0	248.0	8.0	
						1996.8*	1976.4*	40.9*	20.5*
1	5	MED HORZ MILLS	7	80%	1	179.2	178.1	1.1	
					2	224.0	221.0	3.0	
					3	224.0	222.0	2.0	
					4	224.0	225.6		1.6
					5	224.0	218.4	5.6	
					6	224.0	221.0	3.0	
					7	224.0	226.8		2.8
					8	224.0	223.0	1.0	
						1747.2*	1735.9*	15.7*	4.4*
1	7	LGE HORZ MILLS	6	80%	1	153.6	149.2	4.4	
					2	192.0	193.2		1.2
					3	192.0	194.1		2.1
					4	192.0	191.5	.5	
					5	192.0	187.2	4.8	
					6	192.0	191.0	1.0	
					7	192.0	193.2		1.2
					8	192.0	190.0	2.0	
						1497.6*	1489.4*	12.7*	4.5*

Fig. 48. A machine tool load summary.

chines. Second generation computers also introduced the ability to process information in a random manner as opposed to a sequential one.

Third generation computers, with greater internal speeds and memory, removed equipment limitations on applications. The new equipment is smaller in size for comparable capacity, has greater efficiency and improved reliability. These gains were achieved through integrated circuitry combined with the solid state devices. Third generation equipment further developed the important technique of directly or randomly accessing information. Thus the computer is able to rapidly obtain any information in storage independently of other data. Direct access capability is vital to on-line computer systems.

Some third generation machines can also receive and process information from a remote location, and process more than one set of data simultaneously.

It is said that third generation equipment or **hardware** is capable of doing anything man instructs it to do. If anything, computers are limited by the abilities of specialists to develop instructions, called programs or **software,** to take advantage of computer capabilities.

On-Line, Real-Time Systems. On-line, real-time systems have developed in response to management's need for more information, and for greater accessibility to the data stored in the computer.

On-line means all elements of the computer system, including any remote terminals, are connected directly with the computer. **Real-time** is a term describing the ability of an information system to collect data on events as they occur, process data immediately, and use information to influence succeeding events.

The two terms become interrelated, for example, in an airline reservation system. The clerk at the counter calls the computer by means of a teletypewriter linked to the computer by a telephone line to place a reservation for a seat on a specific flight. The ability to call the computer illustrates on-line capability. The computer directly accesses the inventory of seats on the desired flight, and if available, adjusts the inventory and tells the clerk (by typing out the acknowledgment on the remote typewriter) the reservation is confirmed. This illustrates real-time principle. It is possible to have on-line capabilities without real-time.

Most information processing is done by means of **sequential** or "batch" processing. In this technique, data is collected, transmitted to the data center, and processed based on a schedule determined by the data center manager. A call to the computer by an operator at an on-line terminal would provide information on the status as of the last batch processing.

According to Withington (The Use of Computers in Business Organizations) real-time systems are divided into three broad functional classes, including:

1. **Inquiry systems.** These are the simplest systems having only a limited amount of real-time information, the balance being sequentially processed. Finished goods inventory status might be linked on real-time so that the order entry department is capable of making a shipping decision while handling a telephone inquiry from a customer. All other systems in the operation could be batch processed.

2. **Dispatching systems.** These are more complex and involve automatic computer action based on a stored decision table as part of the process. Continuing the previously described example, the order entry clerk places the release for the available finished goods requested by the customer. The computer then generates shipping papers and invoice. It adjusts the inventory for the new level, determines if the order point is reached, and if reached, generates the production order and subsidiary orders for replenishment. A further process involves loading each machine required in processing the new order, and reserving the components to process the assembly. As reservations for components are processed, the computer checks the balance of raw stock and, based on stored decision tables, generates replenishment purchase orders as required.

A more sophisticated extension of the above example by-passes manual order entry. The customer's computer scans its inventory status records, and notes that a reorder point is reached. As a result, the customer's computer calls the manufacturing plant's computer and generates all of the above described processes.

3. **Decision-making systems.** These are the most complex real-time systems. The system operates similar to a dispatching system, allocating resources to meet demand. The difference lies in the method employed. The dispatching system applies pre-established decision rules; the decision-making system finds an optimum answer to every demand. To do this it may employ a linear programming model.

In the previous example, the customer's computer is linked to all of the manufacturers' computers supplying the desired part. As the customer's computer notes a reorder is required, it determines the least-cost situation of fulfilling all of its requirements from each of its suppliers.

Right-Time (Semi-Real-Time) Processing. There are a number of disadvantages of real-time computing, including:

1. Great expense.
2. Any error in transmitting data is processed. To minimize error processing, strict disciplines are required at all operating terminals. Enforcing these disciplines adds additional costs.
3. Accuracy of "real-time" data is limited by the longest lead time of getting all influencing data into the machine. In the finished goods example described previously, great expense is incurred by the order entry department in order to determine the instantaneous level of inventory. While the order entry department is making inquiries and changing levels in real-time, the data center may be processing the finished goods produced by the factory only once each shift. Obviously reliability of inventory level is accurate only at shift changes.

The concept of **right-time** or semi-real-time processing may circumvent these disadvantages. Right-time systems work on the principle of determining the maximum aging of data permitted in order to make a valid probabilistic decision. For example, if finished inventories are brought up to date daily, decisions made in order entry would be valid 98% of the time. This level of accuracy or assuredness may satisfy management requirements. The level may be close to the accuracy of real-time because of normal transmission of errors. If this policy is adopted, batch processing becomes possible, transmission controls are applied for error minimization before processing, and data processing costs are greatly reduced.

This position is supported in a report by the American Telephone and Telegraph Company (Data Communications in Business, Edgar C. Gentile, Jr., ed.). They state: "These systems provide many of the advantages of on-line systems without all the costly complexities of the real-time operation. One of the prime advantages of the semi-real time systems is that it is possible to maintain some control over the input data at a modest cost."

Time-sharing System. A time-sharing system permits a number of users simultaneous access to the same computer from remote terminals. This system reduces the cost of individual use, while providing each user with large computer capability.

Each user in the time-sharing system has a teletypewriter as an input-output terminal. The system concept works on the principle that any individual user requires only a portion of the computer's total capability at any time, and that processing time for most problems may be a fraction of a second for a powerful computer.

This principle applies to more than intra-company use. A number of small companies enter into time-shared systems as joint ventures thus gaining the capability of a large computer at a fraction of the cost.

Typically, companies using the joint, time-shared venture process the bulk of the input data on a sequential process basis. Similarly, lengthy reports are run off-line. Consoles residing in the production control office can then be used to provide data for improved decisions, such as in the following situation:

1. An urgent order from a key account is received for immediate processing. The computer is queried as to what existing orders may be late if the new order is inserted in the schedule.

2. Five orders are likely to be late as a result of the inserted order. The production control manager considers two of these critical and queries the computer on the amount of overtime required and in which departments it is needed to avoid delay on the two critical orders. The computer provides this information and appropriate dispatch orders are prepared.

ELECTRONIC COMPUTERS IN PRODUCTION CONTROL. Trends in industry are toward the concept of the integrated management information system. The principle of this system is that all essential management information is accumulated in one information center. Development of the electronic computer with direct access capability, coupled with on-line data availability for the decision-making manager, makes such a system possible.

Required information of the system is stored in a **data base** in the computer. Production control both receives information from and contributes information to the data base. Production control information, as well as information from other functional parts of the business, comprise the data base. An integrated information system is illustrated in Fig. 49. The data base interre-

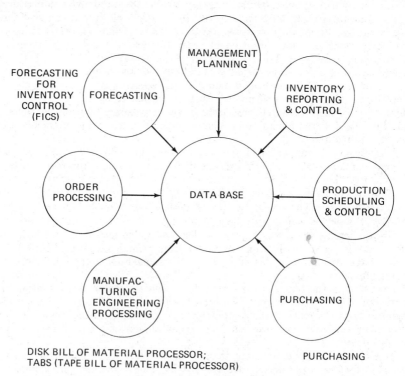

Fig. 49. Data base concept as an integral part of a management information system.

lationship illustrated in this figure is based upon a design by Honeywell Inc., used to illustrate their approach to an integrated management information system. Most computer manufacturers provide the purchaser with literature and technical assistance designed to aid the company in developing the programs required for a management information system.

The production control interrelationship in the management information system is shown in Fig. 50. This flow chart is significant in that it also portrays the integration of activities in all facets from long-range planning to floor control and back. The following describes the flow chart process:

1. The executive committee evaluates the company's potential and establishes strategic and long-range plans. These are converted into annual targets of desired results.

2. The inventory, sales, and operations committee considers the annual targets and the new and modified sales forecasts and prepares a unified, short-term master plan of expected operating results. The master plan includes the **capacity plan** for rough loading.

3. The capacity plan and the production orders are used by the production control department to prepare detailed work center **schedules and machine loads**. The computer also generates subsidiary orders from the input of the production order.

4. The central production control office dispatches the schedules, loads, and orders to the floor by use of on-line terminals.

5. Dispatchers report the results of **production progress** back to the central office through on-line terminals. This information is used to update and revise loads and schedules, as necessary.

6. The production control office prepares a **production performance** report, based on production progress information. This information is used to up-date the data base and provide the inventory, sales, and operations committee with information on the company's ability to perform according to the master plan.

7. The ability to perform is considered in new master plans and provides a base for the master plan performance report. The master plan performance report is used to update the data base and provide the executive committee with information on the company's ability to perform according to the annual target.

8. The executive committee modifies targets, as required, based on the company's ability to perform. The modified targets are then forwarded to the inventory, sales, and operations committee, and the cycle continues.

Capacity Plans. The fundamental principle of the capacity plan is that at any point in time, work centers should be loaded to a desired level. The master plan establishes the future level of activity. Orders from customers are received and booked. They are rough loaded into a work center by use of the routing file, then compared with the capacity plan to detect problems of under- or overload, based on actual receipt of work.

Figure 51 illustrates a computer-issued capacity plan that provides data for control decisions. The report was issued during week 36 and illustrates the expected load position for week 47.

The **Capacity** columns show the **Maximum** capacity for the work center and the **Desired** load as of the current week. In the **Load** portion of the report, the **Forecast** column represents the difference between Desired and Maximum. **Booked** orders are firm orders received from customers as of that date. The **Total** is the sum of Booked and Forecast and is compared to Maximum Capacity. This comparison is calculated as a **Utilization Factor**.

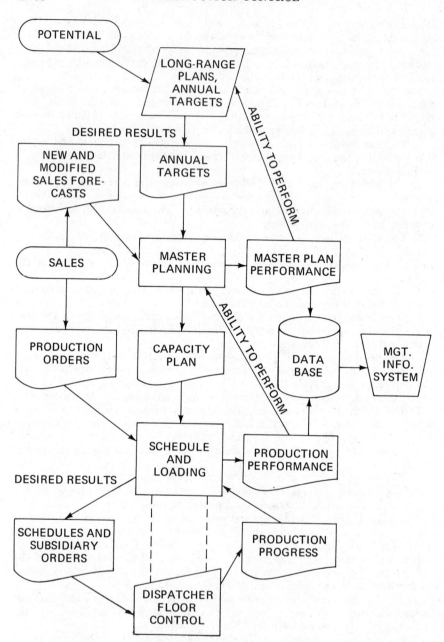

Fig. 50. Production control system interrelationship with the
management information system.

Current Booking Week ___36___

Capacity Plan for Booking Week ___47___

| Dept. | Work Center | Capacity | | Load | | | Utilization Factor |
		Desired	Maximum	Booked	Forecast	Total	
14	14-1	135	225	57	90	147	0.65
	14-2	150	250	140	100	240	0.96
15	15-1	180	300	234	120	354	1.18

Fig. 51. Computer generated capacity plan.

This report reveals a significant underload situation in work center 14-1 and an overload condition in work center 15-1. In actual practice, this report is designed as an exception report. For the periods ten to fifteen weeks in the future, actual-to-forecast results are expected to vary by 20 percent. If these parameters were established, only work centers with a utilization outside the range 0.80 to 1.20 are reported for management attention.

Schedule and Load. Computers are used to prepare detailed schedules and loads. Figure 52 illustrates a computer report issued from central production control to the dispatchers in order to provide information for machine loading.

The report lists each order in the work center's queue or those due to arrive within the next two days. The critical ratio provides information on priority. Orders with critical ratios over 1.0 can be run in a sequence permitting maximum production efficiency.

Arrival and due time are described by the day and the hour to the nearest tenth. **Due Time** dates and hours remain fixed for the order. **Arrival Time,** however, is updated based on on-line production progress reported by the dispatcher.

Production Progress Reports. Production progress reports are essential to production control as an aid for minimizing costs and providing answers to customer inquiries. The basic information serves the needs, not only of production control, but of sales and cost accounting as well. By means of the computerized integrated management information system, data are used for multiple purposes, such as: order and work center control, payroll, estimating, cost variance analysis reports, on-time delivery performance, and late-order analysis. Typical of reports designed to provide data for improved decisions in production, and answer customer inquiries are the following:

1. **Queue Analysis Report,** Fig. 36, is an action report. It assists management in recognizing current work center over- or underload conditions.

Work Center 14—2			SCHEDULE AND LOAD REPORT						Date 257—8.0	
Order Number	Part Number	Oper. Number	Critical Ratio	Set-up Hours	Run Hours	Lot Size	From WC	Arrival Time	To WC	Due Time
1453	328765	020	0.47	2.4	5.0	40	14-1	AVAIL	14-3	256-1.3
1457	324875	020	0.86	1.7	2.8	30	14-1	AVAIL	14-3	256-18.4
1611	324111	030	1.23	1.3	6.6	110	12-7	AVAIL	14-3	259-0.0
1612	324115	030	1.25	1.2	1.3	20	12-7	AVAIL	14-3	259-2.6
1613	324109	020	1.34	2.5	2.0	75	12-2	AVAIL	16-1	260-8.0
1725	331117	016	0.56	2.4	2.4	80	14-1	257-16.0	14-3	257-15.0
1814	324106	020	1.24	2.0	8.4	140	14-1	258-3.0	16-1	262-3.4

Fig. 52. Machine load report.

PRODUCTION PERFORMANCE REPORT						Through Production Week 6			
Department	WC	Scheduled Standard Hours	Actual Hours Produced	Actual/Standard %	Hours in Queue Overdue	Desired Load	Current Available Load	Smoothed Perform. %	Year to Date %
14	01	114.0	97.6	79.5	14.3	72.0	84.2	95.2	93.4
	02	126.2	123.4	97.8	0	72.0	68.4	105.0	105.1
15	01	121.0	126.0	103.7	0	72.0	74.3	100.1	102.1

Fig. 53. Production performance report.

2. **Production Performance Report,** Fig. 53, provides data on the work center's ability to perform according to schedule. In turn, these results are used to update master plans. Performance reports become control and action reports on a delayed-time basis. These reports are an index to the need for engineering effort, methods analysis, and supervisory appraisal.

The report includes a number of significant figures. The **Scheduled Standard Hours** are hours loaded in the past week. The normal load is 120.0 hours, but this was varied based on the **Smoothed Performance percent.** Smoothed performance differs from average performance, recorded in the last column, by accounting for the work center production trend. The trend figure proves more reliable as a loading guide.

3. **Order Status Report,** upper part of Fig. 54, is an exception report, printed daily, of any order expected to be delivered to stores more than two days late.

4. **Delivery Performance Report,** lower part of Fig. 54, is a summary report measuring on-time delivery-to-stores reliability. The orders delivered late are recorded in three classifications to provide sales with a profile of late orders.

Automatic Data Collection and Dispatching. A number of computer manufacturers have addressed themselves to the problem of automatic data collection for improved shop control. Typical is IBM's START system approach (Status and Reporting Technique for Shop Floor Control) which provides for a company's initial entry into automatic data collection. The START approach can be enlarged and integrated into an on-line system as the company grows.

The system utilizes a number of input terminals located throughout the shop, a badge reader used by the production operator at a terminal for transmitting employee information, a keyboard at the terminal for recording quantity produced, and a central output station. The system uses a data collection system and punched input cards, including: a control center card, a job traveller card, and a labor control card. Advantages claimed for this type of system are:

1. Quick job locating and expediting.
2. Uncomplicated schedule changes.
3. Easy routing changes.
4. Unauthorized work accountability.
5. Operation-by-operation count verification.
6. Joint work assignment reviews.
7. Automatic data collection.
8. Job consolidation.
9. Quick determination of open orders.
10. Better planning.

Combining ideas of the START system with on-line scheduling and computer-issued and collected shop paper, provides a means for loading and unloading and automatic dispatching and data collection.

Reproduction and Communications Systems

FORMS DESIGN, USE, AND CONTROL. Production control needs production progress information to plan and monitor the manufacturing process. The company's information systems effort, which deals with processing and data flow assists production control in meeting these needs. Time and money are wasted and chance of error is increased if data are repeatedly copied. Effective production control frequently calls for forms design and control, use of standard printed forms, and computer-generated forms.

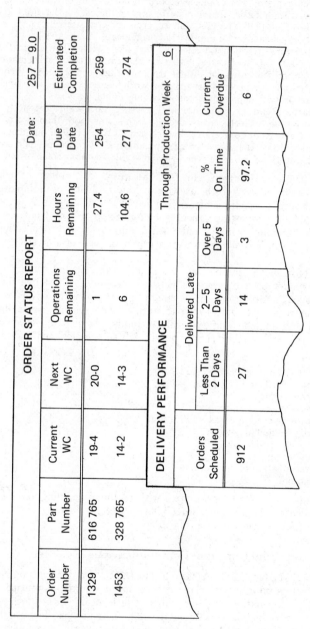

ORDER STATUS REPORT

Date: 257 – 9.0

Order Number	Part Number	Current WC	Next WC	Operations Remaining	Hours Remaining	Due Date	Estimated Completion
1329	616 765	19·4	20·0	1	27.4	254	259
1453	328 765	14·2	14·3	6	104.6	271	274

DELIVERY PERFORMANCE

Through Production Week 6

Orders Scheduled	Delivered Late			% On Time	Current Overdue
	Less Than 2 Days	2–5 Days	Over 5 Days		
912	27	14	3	97.2	6

Fig. 54. Order status and delivery performance reports.

All companies, at some time, require special forms for various elements of the company's operations. Marien (Forms Design, Ideas for Management) suggests these factors to consider in **forms design**:

1. Layout of data.
2. Arrangement and positioning of data.
3. Size of form.
4. Paper to be used.
5. Filing considerations.
6. Standardization of data from form to form.
7. Type styles and sizes.

Pre-design considerations of these factors lead to more effective use and faster processing of forms.

A **forms control** procedure, including assignment of an individual responsible for its administration, is an essential part of the production control system. As a first step the existing forms should be flow charted, duplication eliminated, and delays minimized. New proposed forms are then assessed for their need, use, and purpose before approval. A **forms data sheet** is often used by companies for forms control. The information on the data sheet includes:

1. Purpose of form.
2. Benefits derived from use.
3. Reference to supporting procedures.
4. Source of information to be entered on the form.
5. Numbers of copies of each form.
6. Distribution, frequency of handling, and method of file or disposition.
7. Quantity used per time period.
8. Approval by affected department managers and individual responsible for forms control.

Typically, a procedural system flow chart of the processing the form will undergo is attached to the forms data sheet. The flow chart illustrates the movement and delays in **handling** information. A final consideration, before approval of the form, assures that all essential data reaches the individual requiring the information by the fastest practical means.

Forms design and special printing is expensive. **Standard printed forms** should be considered before any original design. A number of pre-printed forms for production control are available from the Standard Register Company, Moore Business Forms, Inc., Acme Visible Records, Inc., and other well-known forms producers. These companies also assist customers in analyzing control procedures, writing methods, designing forms, and effecting improvements and simplification, wherever possible.

In EDP systems a source of basic data may be contained in the computer. Computers, as a result, are often used to generate basic data and forms. The advantage of this type of reproduction method is speed and elimination of clerical copy errors. Figures 9 and 46 contain samples of **computer-generated forms**. The job-move-locator ticket in Fig. 9 is a multiple-use form. It is produced as a by-product output with the schedule, and provides computer input data later in the process.

Care must be exercised in using computer-generated forms. First, the fundamentals of forms control are as important in computer-produced forms as in manually-produced forms. Second, design is a consideration. The need for clarity often dictates a special design be developed for the output document from the computer. As with all special design forms, they may be expensive.

Finally, computer printers are sometimes used as "multiple-copy printing presses" in some companies. This is rarely economical.

REPRODUCTION METHODS. Manufacturers of reproducing and duplicating equipment are valuable sources of assistance in design of reproduction systems. They often employ systems analysts to assist customers in analyzing procedures, designing systems, writing methods and effecting simplification and improvement practices.

Diazo Reproduction. In a typical application of the diazo method an operations sheet is printed on translucent paper. This sheet is maintained by process engineering which provides production control with a matte film copy of the master.

When a production order is released, the matte film copy is retrieved from file and variable information entered. The matte film copy is then used to produce required copies of the final assembly order. This packet generally includes an assembly copy, requisition copy, finished copy, finished stores copy, cost accounting copy for cost accumulation of the assembly order, and an engineering copy. The engineering copy advises process engineering to prepare a new film copy for the production control file.

Spirit Duplicator. In this system a master order copy is prepared by handwriting or typing and then duplicate copies are run using a spirit duplicator. Copies typically include a production order original, file and cost accounting copy, a machine load copy, material requisition set, and parts identification and move copy that follows the order through the process.

The master is used to prepare the operations cards. In order to duplicate the operations cards to show the order heading, the single operation to be performed, and the next operation, a multiple line selector blockout device is used on the duplicating machine. The adjustable blockout consists of a holder and moveable tapes. It permits selective blockout of any one or combination of lines on the master.

A master strip containing variable information facilitates reuse of the basic operation card master.

Xerography. Systems for shop order reproduction using a xerography principle are in broad use. In this type of reproduction system, a basic master of the shop order is maintained in the production control office. A card stock copy of the material is run when an order is received from sales. The variable information of data and quantity are entered on the card stock, which becomes the new master. The original master is returned to the basic file. The card stock copy is used to run the other manufacturing orders which are often reproduced in different colors for easy identification, and typically include a dispatch copy, material requisition, stores receipt-accounting copy, parts identification and move copy, and an inspection copy.

The card stock copy is maintained for machine loading and to record order progress as the order is routed through the process. The card stock copy is filed and, at a later time, provides input data for standards review.

Duplicating Plates. The manual preparation of bills of materials, listing component parts of a given assembly, is a common requirement in many kinds of manufacturing. These listings become complex when many parts have a common usage in similar end products.

In a process developed by Addressograph-Multigraph master plates are used in combination with offset duplicating masters to permit such listings to be reproduced mechanically. This in turn provides the masters that are used to produce the required number of copies. The master plates are classified by means of tabs attached to the plates so that an automatic selector device on the writing machine may be used to select and prepare a listing of common parts plates comprising a given assembly.

Automatic writing machines distribute this information across the bill of materials form when processed and thereby eliminate manual typing and checking of parts data.

COMMUNICATIONS SYSTEMS. Communications systems are used in production control as a means of transmitting data required to plan and monitor the production operation. Typically, a number of communication techniques are employed in production control systems, including: mechanical, voice, and printed communication, and pictorial systems.

Mechanical Communication. The use of pneumatic tubes provides a fast, constantly available messenger service between departments and operations. It requires a network of tubes between production control and the various reporting locations in the shop. Information is placed in capsules, inserted in the tubes, and propelled by air pressure to their destination.

Although initial investment for installing tube systems is high, it can be very beneficial in production situations that are spread over a wide area and where speed in written communications is highly important.

Voice Communication. Two-way radio communications (transceiver systems) provide excellent means for rapid transmission of instructions to remote areas of the plant. Typically, applications of two-way radio communications in production control systems are used between dispatchers and the central office, and between dispatchers and lift truck drivers. The advantage of this system is that it permits complete mobility of both the sender and the receiver.

Telephone systems, more than all others, are used for remote communications. Interplant communications systems may use regular telephone sets or a private system.

Plants placing a large volume of long distance telephone calls may justify **WATS** (Wide Area Telephone Service) systems. This permits unlimited long distance phone service to a designated zone of the country at a fixed cost. Companies with WATS service find this capability very beneficial in situations requiring the central production control department to be in contact with remote plants many times a day.

Instantaneous Printed Communications. Various printed communications systems are used in production control departments to exchange information and instructions between points. Among such communications systems in common use are Teletype, Tel Autograph Telescribes, and computer input/output terminals.

Teletypewriters and associated equipment manufactured by companies such as the Teletype Corporation are used by a central production control office to issue orders to different plants, modify schedules at widely separated points in the plant, and similar applications. These plants, or remote points, can in turn report back information on production, inventories, equipment loading, orders to be processed, and other data on the manufacturing situation.

The basic teletypwriter resembles and is operated like an ordinary keyboard. The message is received by a teletypewriter that automatically types the message as sent. Information is also transmitted and received in the form of **perforated tape.** Production progress information may be accumulated on perforated tape and transmitted later at a rapid rate. The punched paper tape is often used in connection with integrated data processing equipment systems. Similar applications of communications systems accumulate and transmit data on **magnetic tape.** The advantages of magnetic tape include higher transmission speed than paper tape and computer compatibility.

Other electrical devices permit instantaneous transmission of handwritten messages from production control to one or more remote terminals in the plant or vice versa. Operation is simple and requires no special training. The writer uses a special stylus. As the stylus is moved, small pens incorporated in both the sending and receiving units record the exact stylus movement on paper, providing a permanent written record at both locations.

These communications systems, beside providing speed in transmission, are valuable in situations where oral communications would be difficult. As with all hand-written records, legible writing is necessary in order to avoid misunderstandings. Use of a transceiver permits sending and receiving of such hand-written messages and sketches via telephone communications lines.

Use of **Data-phone** with a computer data transmission terminal illustrates another application of automated transmission equipment for production control. In these systems the terminal reads punched cards containing standard data, and accepts a keyboard entry of variable information. Telephone systems also permit use of the push-button, **touch tone dialing system** as the variable keyboard entry. An application of this system permits a remote plant to send variable order quantities of standard parts into the central office production control output terminal. The received card is in machine-readable form for direct entry into a computerized production control system.

Automatic data transmission by **computer and/or input-output** terminals is in popular use in production situations. Systems for automatic data collection and dispatching described under Electronic Computers in Production Control is one such illustration.

Pictorial Systems. Many situations in production control systems require pictorial as well as voice communication. A number of modern methods have been developed to accomplish this.

Closed circuit **television** is used in production control systems as part of an automatic monitoring and dispatching procedure. Cameras, for example, are located at critical assembly line points to view production progress. A manufacturer of large machine tools is using it to monitor special machine tools as they progress through assembly. Automatic switching in the central office permits the chief dispatcher to visually determine which tools are ahead or behind a pre-determined point in the assembly line by a certain time.

Facsimile terminals are another means for transmitting documents and pictures from one location to another. Facsimile terminals are basically duplicating devices. The transmitting facsimile converts black and white signals on a document to electric signals. These signals are transmitted to the receiving terminal where the picture is reproduced. Speeds of transmission vary from one page in six minutes to sixteen pages per minute.

The Xerox Tele-copier is an example of a system for transmitting pictures over regular telephone channels as shown in Fig. 55.

In other applications it is possible to obtain **visual displays** of information stored in a computer system. The terminal consists of a keyboard for mak-

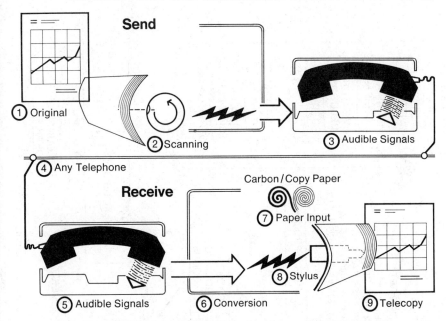

Fig. 55. Method for sending documents over telephone lines.

ing the inquiry, a signal generator-interpreter, and a visual display screen similar to a television screen. The user queries the computer by pressing the keys and receives his answer on the screen.

Other methods of automatic transmission for production control include **microfilm transmission,** which will permit automatic microfilm-microfiche retrieval systems to transmit and display the file over telephone lines; **optical scanning systems,** which will permit original documents to be interpreted at a remote terminal, transmitted over telephone lines, accumulated in a buffer, and automatically fed into a computer system; and **picture-phone** service, which will permit telephone transmission of detailed documents.

Comprehensive Communications Systems. Comprehensive systems involving **real** or **semi-real-time,** on-line data processing with various communication systems are in wide industrial use. Figure 56 (page 96) represents an abbreviated view of a system employed by the Westinghouse Electric Corporation. It demonstrates the various communication techniques used by designers in developing a materials management system. It is used for automatic message switching, order processing, production control, inventory control, corporate accounting applications, and automatic transmission of data.

The system permits any of more than 300 outlying stations to make inquiries, direct messages to other stations, and enter orders. The computer processes answers, notices of shipments, and backlog reports. Computer-issued bills of lading, and packing lists are printed at the appropriate field warehouse by means of the remote terminals. Real-time order action is directed to the factory, as required, based on review of inventory records. Batch processing is used to generate stock bulletins, analysis reports, invoices and accounting information.

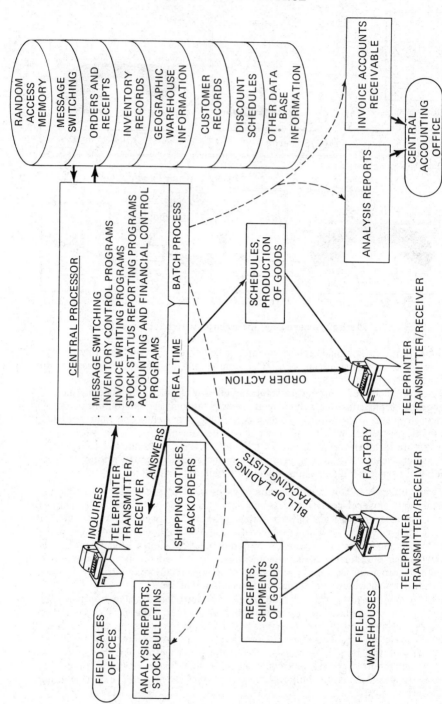

Fig. 56. Principal parts of a comprehensive communications system.

MATERIALS MANAGEMENT

CONTENTS

CONTENTS (*Continued*)

MATERIALS MANAGEMENT

The Materials Management Concept

SCOPE OF MATERIALS MANAGEMENT. The American Production and Inventory Control Society (APICS Dictionary of Production and Inventory Control Terms) defines **materials management** as: "The grouping of management functions related to the complete cycle of **material flow,** from the purchase and internal control of production materials to the planning and control of work-in-process to the warehousing, shipping, and distribution of the finished product. Differs from **materials** control in that the latter term, traditionally is limited to the internal control of production materials." The key words here are **materials flow.** In order for raw materials to be converted into a finished product, materials must flow through a production or manufacturing process.

The grouping of all functions concerned with the **flow** of materials from planning for its procurement to ultimate delivery to customer, on a horizontal basis, is shown in Fig. 1. The traditionally separate organizational functions integrated under this system include purchasing, production planning and control, scheduling, material handling, warehousing, traffic and distribution.

Fundamental Activities. Basically the materials management span of activities are:

1. **Planning:** what is needed, how much, where and when to fulfill the objectives of the organization. Scheduling is considered a detailed, time-phased plan in this context.

2. **Obtaining:** the raw materials, subassemblies or parts to meet the schedules on time from whatever sources required.

3. **Control:** the follow-up, or tracking and warning to insure that the plans and schedules are adhered to. Expediting, record keeping, data collection, and information feedback loops are included here.

4. **Storing:** receiving, incoming inspection, and the storing and issuing of the raw materials, component parts, subassemblies and work in process required by the manufacturing function. Also includes storage and handling of the finished goods and office supplies or nonproductive stores items and records.

5. **Handling:** movement, packaging, transfer, and delivery of the materials involved, from incoming receipt to final distribution.

6. **Distribution:** of the finished goods, warehousing, packing, assembly, palletizing, and shipping to the customer.

7. **Transportation and traffic control:** routing standards, carrier selections, commodity rates, consolidation, and dispatching shipments.

Sometimes other nonrelated areas, such as operation of the company store (which sells finished products to employees at special prices), order entry activities, and perhaps communication activities are grouped under the jurisdiction of the materials management group.

Principal Phases. There are four principal phases of the materials management function.

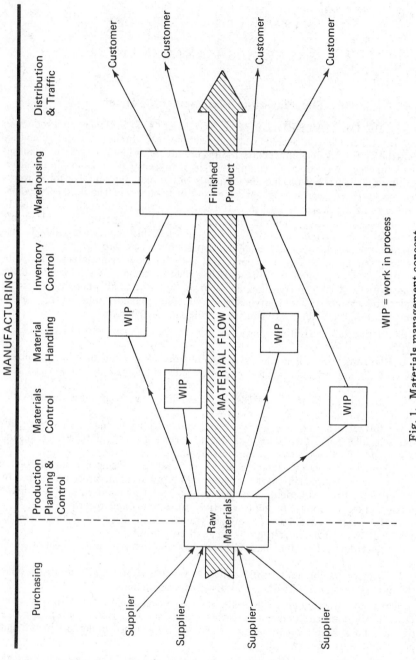

Fig. 1. Materials management concept.

1. **Planning**: The basic planning elements are plans for capacity or production levels (or rates) and the required inventory levels. Integral to these plans is the forecast. The detailed, time-sequenced plan called the schedule results to meet the sales forecast and inventory requirements.

2. **Materials utilization**: The efficiency of the flow of materials through the process is a phase of materials management. The amount of material leaving the plant as finished goods divided by the amount of material started through the production process determines the materials utilization.

3. **Physical**: This involves the physical storing, receiving, and issuing of materials and the related function of physically counting the inventories of raw materials, work in process, and finished goods and record keeping.

4. **Control or follow-up**: This is one of the most important of the functions of materials management. It includes the information feedback loops and the corrective action generated by this information monitoring the production rates, schedules, plant loads, dispatching, expediting and resultant follow-up. Once a plan or schedule is made, control is required to insure that they are met or a valid reason exists for a deviation.

ADVANTAGES. A study in Factory (Vol. 125) shows that companies with true materials management organizations realized such satisfactory results as reduced lead times, fewer shortages, better delivery performance, reductions in inventories, and increased inventory turnover. One of the first benefits obtained, according to a survey in Production (Vol. 64) is a reduction in raw material on hand. Each of these benefits derives from five broad factors directly influenced by this concept.

Reduction in Materials Cost. Where the purchasing, production control, and materials control functions are separated there is often a conflict in overall goals. The purchasing function wants to optimize its procedures to make purchasing look good. The same holds true for materials control. This tends to increase inventories while raising the cost of materials. Under a unified system, the conflicts in authority are minimized by the department goals which are for the **total least cost**. Quantity discounts can be taken advantage of when they contribute to the overall least cost and will not create surplus inventory. The **materials utilization program** also will produce lower material costs through a controlled program to achieve better efficiency in the use of materials. The prevention of physical loss of materials through the physical inventory program and the inventory control system also will drive the material cost factor down. Reduction of waste due to theft, breakage, and deterioration from improper storage is also accomplished. A **value analysis** program reduces material costs, not only of the final product but also of the component parts, when coupled with a materials substitution program.

Efficient, Effective Inventories. The purpose of inventories are to provide a desired service level to the customer at a minimum of investment or at the best dollar turnover. To accomplish this requires proper inventories at each stage of the production process not only at the finished goods level. Having the entire flow of materials including the inventories at each stage under one head provides a much greater efficiency of materials and makes for much more effective control. As stated above, left to their own devices purchasing, materials control, and sales all would tend toward higher and higher inventories. Each would be able to pass responsibility for this to another group. Materials management eliminates this by having common goals.

Increase in Labor Utilization. Lack of proper materials, in sufficient quantity and on time, is the frequent cause of interruption of the manufacturing process. Having the proper materials when and where needed is the function of the materials management group. Elimination of interruptions due to materials or parts shortages leads to an increase in labor utilization.

Organizational Efficiency. The materials management organization itself is discussed later in the section. Elimination of overlapping jobs between purchasing, production planning, inventory control, distribution, and traffic can produce tangible reduction in the number of people involved. The approach also results in better coordination and communication within all the organizations concerned. The conflict of interests are eliminated. As Ammer points out (Harvard Business Review, Vol. 47), the materials manager will pay much more attention to storage, material handling, receiving, and shipping than will the manufacturing manager who is more concerned with the manufacturing process.

Improved Control and Flexibility. Improvement in control is achieved through integration of all material flow functions. By having one group responsible for all points in the system conflicting objectives are eliminated and better control results. Flexibility also follows due to fewer barriers to communication across false departmental lines. Communication within a department is usually more informal (and less costly) and faster. The materials manager has the flexibility to implement improvements within his own department without having to sell the idea to several other department heads.

DISADVANTAGES. The Production (Vol. 64) survey reveals that companies introducing the materials management concept in an on-going system experience some disruption in their organization due to shifts in levels and responsibility. The biggest loser in the realignment necessary to implement the changeover is often the purchasing department which drops down a full level and must report to the materials manager.

Materials management was pioneered and used successfully by electronic and aerospace companies. In general, the concept works best with large-size companies having a more formal and structured organization. A small company does not usually have the communications barriers and divergent departmental objectives mentioned above. As Plossl and Wight (Production and Inventory Control Principles and Techniques) caution, this approach is not a cure-all. An organizational change will not solve the problems in the design of a system, or in the quality of information, or in the production process itself. Most failures of materials management can be attributed to misapplication and poor implementation rather than to failure of the concept itself.

The Materials Management Program

ESTABLISHING POLICY. The materials management program must be based in some degree on **flexible policies** because, as pointed out by the American Management Association (AMA Research Study 76), they:

1. Integrate functions and activities and encourage teamwork.
2. Promote consistency of management decisions.
3. Permit managers to handle problems with freedom and speed.
4. Define the constraints within which the team must operate.

Well-defined **specific management policies** must be established in order to meet the goals of:

1. Customer service levels.
2. Inventory turnover.
3. Efficient manufacturing operation.

Although the customer service level referred to above usually is defined in allowable stockouts over some period of time or "finished goods in stock" some percentage of time, it should be remembered that this reflects the stock situation at all inventory stages, raw material and work in process, too. This must be considered in the policy writing.

Similarly inventory turnover refers to the aggregate of all inventories although in most companies finished goods comprise greatest percentage of the total.

GENERAL POLICIES. The responsibility for material inventory policies must rest with top management. Unless the broad and basic policies are laid down by top management, the detailed operating policies and procedures necessary for effective provision and control of materials cannot be developed properly by the manufacturing organization.

The establishment of inventory policies provides a guide for all operating personnel who have to make the day-to-day decisions regarding material and inventory. Such policies, therefore, should provide full flexibility to permit the operating people to exercise judgment in meeting new conditions as they arise.

Factors Affecting Policies. Policies set by top management will vary with the type of industry, type of company, characteristics of the product line, competitive practices, and the current state of business. The current financial condition of the particular company and the amount of capital available also may cause variations in policy.

For example, in some of the **process industries,** such as steel and chemicals, where the price of raw materials procured has a decided effect on the profit and loss statement, the company may engage in a certain amount of speculative or forward buying. Since this is a vital factor in the profit of the business, the decision as to when and how much raw material to buy is usually made by top management.

In **job shop operations,** on the other hand, the usual procedure is to do as little speculative or forward buying as possible. Although the policies of these companies often call for the maintenance of certain inventories of raw materials and parts in order to achieve purchasing and manufacturing economies and provide for availability of material, little attempt is made to try to outguess the market on price or to hedge against possible advances in prices.

Economic conditions or the general state of business often cause companies to change their inventory policy. During periods of high demand the steel industry traditionally produces to a large order backlog, and steel is started in the open hearths only after specific customer orders have been received for that "heat" of steel. During periods of low demand, however, the mills have found it necessary and advisable to produce certain standard specifications of steel for stock before receipt of customer orders. Thus, on many standard sizes, gauges, and specifications they are in a position to offer almost immediate shipment on customer orders.

Seasonal businesses must accumulate material and inventory in expectation of high sales periods. Policies must be laid down concerning the production level or levels and the planned accumulation of inventory during periods of low sales activity. When new product lines or models are introduced, it may be desirable to build up certain quantities of inventories so as to be prepared for a **"surge" demand** due to promotional efforts and stocking by dealers.

In some industries, such as **automobiles and appliances,** where the policy is to keep stocks or raw materials and parts in the assembly plants as low as possible, schedules of authorized shipments from suppliers sometimes are changed as often as several times a week in order to reduce to an absolute minimum the amount of inventory which is carried on hand in the assembly plants.

OPERATING POLICIES. Policies set by top management cover a wide variety of subjects relative to materials control. The following are some of the areas in which the top management policies are most essential:

1. The delivery time quoted to customer for each product line.
2. The maximum amount of capital to be tied up in inventory.
3. The desired inventory turnover.
4. The degree of protection desired for normal manufacturing operations. Also the protection desired in case of threatened shortages due to strikes by supplier, possible increases in prices of raw materials, etc.
5. The time lag on shipments of service parts to customers.

In a divisional organization structure, a **separate inventory policy** must be set for each of the divisions since policy may well differ in regard to the products and problems involved in each segment of the business.

MATERIALS MANAGEMENT ORGANIZATION. As shown in Fig. 1 the materials management concept is basically one of organizational realignment. The shepherding of the flow of materials from selection of suppliers to shipment to customer is quite vital to the company's profitability. The organization in a conventional (non-materials management oriented) company is shown in Fig. 2.

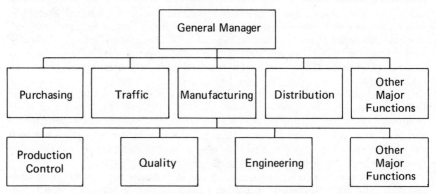

Fig. 2. Organization without materials management.

A company using the materials management approach might have the organization structure shown in Fig. 3.

The organizational efficiency gained by this approach should be very apparent. The span of control of the general manager is greatly improved and the attention to the vital areas of materials efficiency, handling, receiving, storage, value analysis, and effective inventories are assured by grouping the functions shown under one man. The use of electronic data processing and computers plus the systems approach to management add strength to this approach. An integrated materials system treats the flow of materials and therefore must be able to ignore the artificial boundaries to that flow caused by departmental lines.

Most of the functional breakdowns of materials management such as purchasing, inventory control, material handling, and production control are discussed as separate sections in this Handbook.

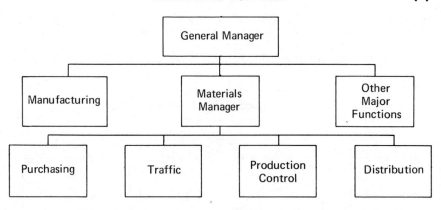

Fig. 3. Typical organization with materials management.

Materials Control

DEFINITION. The American Production and Inventory Control Society (APICS Dictionary of Production and Inventory Terms) defines **materials control** as: "The function of maintaining a constantly available supply of raw materials, purchased parts and supplies that are required for the manufacture of products. Functional responsibilities include the requisitioning of materials for purchase in economic quantities at the proper time, and their receipts, storage and protection; the issuing of materials to production upon authorized request; and the maintenance and verification of inventory records."

ORGANIZATION FOR MATERIALS CONTROL. Providing raw materials and parts in proper quantities and when they are needed is one of the major concerns of any manufacturing business. Unless this is done properly people will be idle and delivery promises will not be met. Therefore, it is vitally important to provide proper delineation of authority and responsibility for this function and so to place it in the company's organization structure that it will function most effectively.

In **process industries** the procurement of raw materials and the follow-up to see that these materials arrive at the plant when they are needed is generally a responsibility of a vice-president of materials or director of purchases who reports to the president of the company. In these industries large quantities of material must be scheduled into the plant in ample time for production needs but without excessive accumulation in stock piles.

In most **manufacturing operations** the provision of materials is a function of the production control manager, who usually reports directly to the chief manufacturing executive or the materials manager. The **production planning and control manager** is responsible for planning the availability of manpower, materials, and production facilities in such a manner that predetermined product goals or objectives can be attained.

The responsibility of the production control manager in regard to materials is usually considerably greater than that of the **materials control manager** who is concerned only with raw materials and parts. The production control manager is responsible for seeing that all parts are produced in the plant according to schedule and that the finished product is brought through the plant to completion and shipped to the customer on time. Thus, placing materials control under the materials manager maintains the flow of materials into the plant through production and into finished products storage or into the shipping department. It is a completely centralized and integrated cycle.

A typical organization of the production control department is shown in Fig. 4.

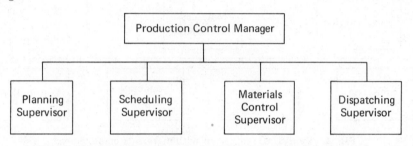

Fig. 4. **Organization of typical production control department.**

In some cases the production planning and scheduling functions might be combined under one supervisor and frequently the inventory control function is combined with the materials control supervisor. The entire production planning and inventory control function is becoming systematized and in the more sophisticated companies, computerized. Figure 5 (Stumpff, Proceedings of the APICS National Technical Conference) shows the inventory control module of the production planning and inventory control system used by a major electronic company. In this organization orders are entered or received and shipment follows from the available field stock. The company has several field warehouses. The orders are received centrally and through the communication system the proper field warehouse is advised to ship the order. Data are kept on the following:

1. Orders shipped complete.
2. Orders shipped on time.
3. Items not shipped or out of stock.
4. How long items were out of stock.

This information is screened against the "Management Customer Service Inventory Policy."

History is accumulated and entered into the short-range forecast module. This is an adaptive (self-correcting) statistical forecast module. The statistical forecast is then filtered through several screens to make it as effective as possible. The forecast is checked against the order board which is the file of open orders for shipment at a future date. The field salesmen's forecast is fed in as field intelligence. The usual extrinsic forecast factor checks and inputs are employed and then the results of this are fed into the system. Again the system is adaptive so the new sales inputs automatically adjust the reorder points, reorder quantities, and safety stocks.

The control parameters which are a basic function of management's materials policies are reviewed and kept updated as an offshoot of the Customer Service Inventory Policy subsystem discussed above. The various inventory stages are replenished as required by the tripping of reorder quantities as material is shipped from the field stock. The inventory stages are chained together. Field stocks are replenished in economical reorder quantities from the Central Finished Stock. It likewise is replenished in economical reorder quantities from bulk stock.

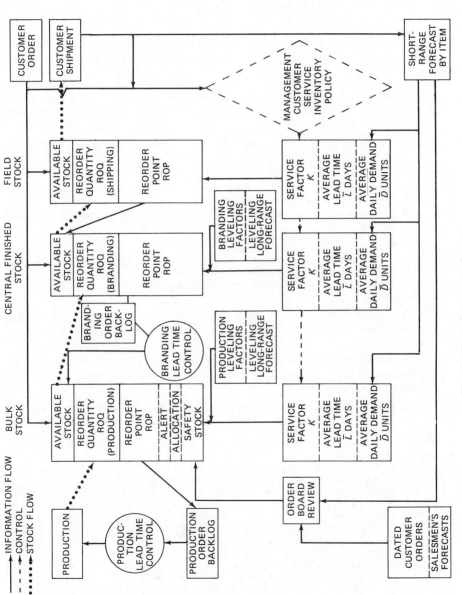

Fig. 5. Materials control system.

This basic bulk stock calls for economic or production order quantities from production. As EOQ's are called for they go into a queue of production backlog which has a control subsystem governing the size and length of time that an order may remain in the queue.

Controls are also maintained on the subproduction functions, assembly and branding or decorating.

A production leveling model is superimposed at the production levels or stages. The long-range or strategic plan is involved as the device to determine production levels and leveling factors.

Systems of the size and scope of this one have to be custom designed for each application and are almost always computerized. Modules of systems or subsystems can be manually operated and human intervention, monitoring, and control points must be provided.

ESTABLISHING THE BASES FOR MATERIALS CONTROL. There are several basic decisions which have to be made before a materials control system can be properly set up. The most important of these are as follows:

1. Group material into classes and decide which department or section is responsible for each class of material.
2. Determine inventory status with reference to policy of producing finished products:
   a. To individual customer order only.
   b. To a grouping of customer orders.
   c. To a grouping of customer orders plus stock orders for finished items.
   d. To stock by means of a sales forecast or manufacturing schedule.
3. Modify control methods and stock status according to the value of the items involved.

Once these steps are taken, the materials control system can be set up on a proper basis. The considerations involved in each of these decisions are explained below.

Grouping Materials into Classes. Materials may be grouped or classified according to their general nature, use, or condition. Five such classes cover practically all cases:

1. **Raw materials** comprise items which must be purchased and processed to convert them into component parts or to prepare them to go into (or to be converted into) a finished product.
2. **Component parts** are used in finished assemblies or for repair parts and are of two kinds:
   a. Parts purchased from vendors in completed form.
   b. Parts produced in the plant from raw materials.
3. **Supplies, expense, nonproductive,** or **"indirect" items** are used in the manufacturing process but do not form a part of the finished product. Such supplies are usually purchased. They include cutting and lubricating oils, cleaning or pickling solutions, waste and wiping rags, janitors' supplies, office supplies, construction materials, repair parts for machines and equipment, other maintenance items, etc.
4. **Work in process** includes all materials, parts, subassemblies, and assemblies which are being processed or assembled into finished products. These items are those actually undergoing productive operations or in temporary storage between processes or operations in manufacturing departments.
5. **Finished products** are units or assemblies carried in stock in completed form ready for delivery to customers. They are usually items which have been manufactured or processed by the company, but they may be items purchased in finished condition for purpose of resale.

The above classes, except for supplies (item 3), are **direct materials** because they go into or are the product which is delivered to the customer. Supplies do not form a part of the finished product and are therefore indirect materials.

Materials control is generally concerned with the procuring of raw materials and purchased parts and the supplying of these to production. Usually when parts manufactured in the shop are completed, they are returned to stores, where they are under the control of the materials control department until they are issued to the assembly floor.

The materials control section sometimes stores and controls supplies and indirect items, reordering these when necessary. In other cases, supplies are ordered and stored by the maintenance department or by factory supervision.

Work in process is the primary concern of the manufacturing department since this comprises all material being worked. Finished units are often shipped directly to customers by the shipping department, or they may be put into a stockroom or warehouse under the control of the sales or production control department. These two classifications of inventory are usually not a concern of the materials control section.

Materials Utilization. In the flow of material through the manufacturing process, certain losses occur along the way. Errors, shrinkage, scrap, and the like use up valuable, paid-for material. In the design process expected losses, scrap, and shrinkage are specified and taken into consideration in the costing process. Material utilization provides a measure of the effectiveness of the estimates of required materials. It can be quantified in the following formula:

$$\text{Material utilization} = \frac{\text{Output} + \text{Expected scrap}}{\text{Input}}$$

As an example, suppose it takes 16 tons of glass to make a 12-ton mirror, and the utilization of material is 100%. If 18 tons were now used to produce the mirror, the utilization would be

$$\frac{16}{18} = 0.88 \quad \text{or} \quad 88\%$$

In addition to spotlighting the gaps in materials efficiency and highlighting high-loss areas, a good material utilization program tends to sharpen up estimates of expected or designed-in losses.

Speeding Materials Procurement. Since there is generally a reluctance on the part of the customer to wait any longer than necessary, some companies have developed ways to speed up the ordering of material. The engineering department itself often places the necessary requisitions for the special parts or material as soon as they design the specific part, or they may give an **advance notice** to the production department so that they can order the parts. This can save the time normally required to complete the final design and engineering paperwork and to process the order through the usual channels.

Optimum Stock Requirements. The most important determinant of which material to stock, therefore, is the lead time between receipt of a customer order and the time when the finished product must be shipped. This sets up a **minimum condition for inventory** which the company must carry by classes of material. The shorter the quoted delivery time, the more stock must be carried, and it must be carried in a more advanced stage of completion, as shown in Fig. 6.

Delivery Time Quoted to Customers	Items Maintained in Stock			
	Finished Products	Standard Sub-assemblies	Standard Purchased and Manufactured Parts	Raw Material
1. As much time as the manufacturer wants............				
2. Relatively long time......				X
3. Relatively short time.....		X	X	X
4. Immediate shipment upon receipt of customer order..	X	X	X	X

Fig. 6.   Relationship of stock policy to delivery time.

There are, however, several other **reasons for stocking standard parts and raw material** even though there would be sufficient lead time to purchase them after receipt of a customer order:

1. **Price advantage** due to purchasing or producing parts in larger quantities than presently needed.
2. **Flexibility advantage** for producing on short notice additional quantities of finished goods over and above those forecast or scheduled. Some companies insist that a certain percentage of reserve raw material and purchased parts be carried on hand at all times to allow for sudden increases in schedules which could not be accommodated by maintaining normal inventories.
3. **Protection against market change.** In some cases, manufacturers carry base stocks of raw material and purchased parts which they can use to produce customer orders if suppliers should take longer than the normal lead time provided or if shortages of these materials should develop. Material is taken from these base stocks and replaced when it is received.

The maintenance of a stock position is also a worthwhile idea where an assembly line or process flow is involved. The base stock acts as protection against the line or process being shut down in case material is not received according to schedule.

MATERIALS CONTROL CYCLE. The materials control cycle comprises all of those procedures which are necessary for the provision of materials for the manufacturing process with a minimum of investment and at lowest cost possible. A knowledge of this cycle is fundamental to an understanding of the principles and practices of materials control. This **materials control cycle** is as follows:

1. Determining material needs.
2. Preparing requisitions for purchased items and requests for work orders for parts made in the shop.
3. Receiving purchased materials and finished parts into the plant.
4. Inspecting purchased material and parts, and inspection of finished shop parts. Delivering all material and parts to the storeroom for storage.
5. Entering receipts in store records or perpetual inventory records; apportioning material in the records to current orders; authorizing requisitions of materials from stores for production of shop parts and requisitions of parts from stores for assembly into finished items.
6. Issuing of parts and material to the shop for production and assembly.
7. Recording the issue in store records or perpetual inventory records.

8. Entry of receiving and issuing transactions into cost and accounting records.
9. Determination of necessity for replenishment of stores, which leads to step No. 1 above.

There are three **auxiliary materials control activities** which parallel the materials control cycle. They are as follows:

1. Determining the proper quantity to requisition for each item of material and for each purchased and manufactured part.
2. Physically checking the quantity on hand in the storeroom of each part and each item of raw material in order to verify the balances shown in stock on store records or perpetual inventory cards.
3. Standardizing materials and parts for lowest cost manufacturing.

Records and Procedures for Materials Control

BILL OF MATERIALS. Most fabrication and assembly manufacturing organizations are required to keep large volumes of records that describe the structure or make up of the products to be manufactured. These are vital records in the planning and implementation of the manufacturing process. Product structure records are usually referenced by such names as bills of materials, parts lists, or where-used lists. Within each different organization the basic product structure documents are re-sorted, re-formatted or abstracted to suit the individual use by the department involved.

Often engineering, financial, purchasing and manufacturing maintain separate product structure files for their own uses. This introduces inaccuracies and differences, not to speak of the duplication and increased costs. A **central information file** which can serve all the different functional areas in the company should be strongly considered. Whether or not this central bill of material file should be computerized would depend on the complexity and size of the file.

A centralized bill of material file would provide the following **advantages**:

1. Reduce overall costs to company.
2. Assure timely information to all using groups by having a one-time recording of data at the earliest time possible.
3. Eliminate duplicate entries and minimize possibilities of errors in data transfers.

Normally the product structure records would be kept or organized in two sequences:

1. Assembly or bill-of-material sequence.
2. Where used.

Product structure data originate in the engineering department from the design process. The data are usually released as a series of lists of materials or assemblies showing a detailed breakdown of the following:

1. Top level assemblies into subassemblies.
2. Subassemblies into fabricated detailed parts and purchased part numbers.
3. Fabricated parts into raw materials used.

This information is usually contained on the assembly drawing or blueprint. The bill of material lists the components of the assembly along with the quantity used per assembly. Also demonstrated is the next assembly where-used list, showing the higher level assemblies on which this particular assembly is used and the quantity required; and end item use, showing the end items or end products on which this assembly is used, and the total quantity used.

Multilevel Bills of Materials. The product structure can be displayed as a "tree" or a series of levels or tiers as shown in Fig. 7. This product structure tree shows that Product X breaks down into major components A, B, and C.

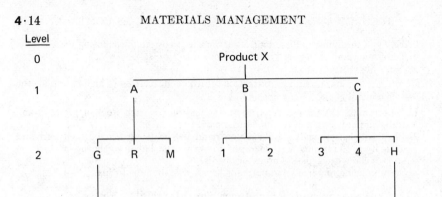

Fig. 7. Product structure tree.

These in turn break down into multiple levels or tiers of subassemblies. A complete description of Product X requires that all of the branches be included. As shown there are three component 1's, two component 2's, and three component 4's. Such level or tier breakdowns of the materials components and subassemblies lend themselves very well to electronic data processing.

The Bill of Materials Explosion. This is the name given to a retrieval process that divides an assembly or product into its immediate components or lower level components through their subassemblies. The outcome of the explosion usually results in an assembly order depicting the quantity per assembly of the various components required. A sample bill of material is shown in Fig. 8, a typical assembly order is shown in Fig. 9. The assembly order can be divided into pick tickets to be used in issuing component parts needed for an assembly.

BILL OF MATERIALS

Drawing No. _____   Assembly No. _____

Charge No. _____   Description _____

Drawing Source_____   _____

Part No.	Description	Unit Spec.	Quantity Per Assembly	Source Dept.	

Fig. 8. Sample of a bill of materials.

The Bill of Materials Implosion. This retrieval technique traces the direct or indirect uses of a part on higher level assemblies. As can be surmised, this is the explosion process worked backwards. It is useful when a change is to be

```
ASSEMBLY ORDER

Order No. _____        Assembly No. _____

Order Quantity _____        Description    _____
                                                   _____
```

Part No.	Description	Unit of Measure	Quantity Per Assembly	Total Quantity

Fig. 9.   Typical assembly order.

made to a basic component. The implosion process shows where the part is used throughout the assembly. It is also a useful technique to show which assemblies will be affected and when by late delivery of a basic component.

The key to effective use of the bill of material and the product structure file is coding. The **coding** of the levels or tiers in the product structure tree is of great importance. Properly coded, maintenance of the central file can be most efficient.

DETERMINING MATERIAL NEEDS. Material control reorder systems for standard items, parts, components are described below. As is the case with any system, the method of execution or processing should be determined by a feasibility study. Integrated data processing has become practically a necessity in the information and control systems area, due to the sheer volume of data the system must digest to affect its control and reporting function.

The system described on the following pages might be either manual, semi-automated, or computerized.

Magee (Harvard Business Review, vol. 34) distinguishes two different methods for ordering standard stock items. While the two are basically similar in concept, they have a somewhat different effect on reserve or safety stocks. The objective of each is the same, that is, to determine "scientifically" when to reorder so that production levels may be maintained and inventory held to proper levels.

The first is the **fixed order system,** by which "the same quantity of material is always ordered, but the time at which an order is placed is allowed to vary with fluctuations in usage. The objective is to place an order whenever the amount on hand is just sufficient to meet a reasonable maximum demand over the course of the lead time which must be allowed between placement of the replenishment order and receipt of the material."

The second method, the **periodic reordering system,** is based on reviewing stocks at fixed time intervals and varying the order quantity according to the usage since the last review, or according to predicted usage. This is often used in monthly ordering for assembly schedules, such as in appliance manufacturing.

Magee says that the fixed order system is advantageous:

1. Where some type of continuous monitoring of the inventory is possible, such as with perpetual inventory records or bin-tag methods.

2. Where the inventory consists of items of low unit value purchased infrequently in large quantities compared with usage rate, or where otherwise there is less need for tight control.

3. Where the stock is purchased from an outside supplier and represents a minor part of the supplier's total output, or is otherwise obtained from a source whose schedule is not tightly linked to the particular item or inventory in question, and where irregular orders for the item from the supplier will not cause production difficulties.

He considers the periodic reordering system useful under the following conditions:

1. Where tighter and more frequent control is needed because of the value of the items.

2. Where a large number of items are reordered jointly, as in the case of a warehouse ordering many items from one factory for freight advantage.

3. Where items representing an important portion of the supplying plant's output are regularly reordered.

One of the modifications of this scheme is the **base stock system,** by which inventory stocks are reviewed on a periodic basis but replenishment is made only when the stocks on hand and on order have fallen to or below some specified level. When this happens, an order is placed to bring the amount on hand and on order up to a specified maximum level.

FIXED ORDER SYSTEMS. On the fixed order basis, material is reordered whenever the quantity on hand reaches a certain predetermined point, established by usage and reserve policy. Orders are placed whenever the balance declines to this reorder point figure. At that time a new order is placed for an amount equal to or greater than an **economical order quantity.** In some companies the reorder at this point is called the **"review" point,** since each time it is reached the current and projected usage is reviewed and the point is changed if usage has changed or policy in regard to reserves has been changed. The balance in stock continues to decline while the replacement order is in process, and at the time when it is received, the balance in stock is at a minimum figure. The addition of the quantity received raises the balance in stock to a maximum figure, and the balance again starts to decline as additional quantities are issued.

There are several methods of control, ranging from very simple to fairly complex, which use the basic principles of the fixed order system. The simplest methods merely involve keeping track of quantities on hand and on order, and reordering whenever the balance of available material is "getting low," as judged by the man watching the quantity on hand. Other methods attempt to allocate stock to customer orders or manufacturing schedules, and the most sophisticated methods attempt to determine "scientifically" reorder points, order quantities, and reserve stocks. Due to gradations and adaptations it is not easy to make an exact classification, but four general types, ranging from simple to complex, are discussed below:

1. Bin-tag methods.
2. Stores record card methods.
3. Allocation methods.
4. Demand and supply methods.

Bin-Tag Methods. These methods are based on reordering when the quantity of parts remaining in the bin reaches a low level, as shown on the bin tag. A representative bin tag is shown in Fig. 10. These tags are hung on hooks at the bins or shelves where the items are kept. They are usually posted when the material is withdrawn from the bin and when material is received and placed in

BIN TAG

SYMBOL _____ S F B S F ½ x 1½ _____

NAME _____ Bolts, F.F. Hd _____

_____ Steel   U.S. Std. _____

SIZE OR DESCRIPTION _____ ½ x 1½ _____

LOCATION _____ A B 2 G 4 _____

MIN. STOCK _____ 1,500 _____

| DATE | | ORDER NO. | | REC'D | ISSUED | BALANCE |
MO.	DAY	NO.	LOT			
10	5			Brought	forward	3,968
10	7	S.O 9874			415	3,553
10	14	S.O.9940			700	2,653
10	18	Count (Weight) loss			33	2,620
10	21	S.O.9980			1000	1,620
10	25	P O 11,352		10,000		11,620
10	29	S.C.9980		25		11,645
			(Cont over)			

Fig. 10. Bin tag.

the bin or on the shelf. Tags are used front and back. The S.O. numbers in the illustration represent shop orders on which items are issued from stores or are made in the plant and placed in stores. The P.O. numbers represent purchase orders on which shipments have arrived. The S.C. numbers represent stores credits for materials returned from jobs of the same number.

The lot column on the tag may be used when incoming shipments on a purchase order or stock replenishment order come in by lots, or when a job order is put through in successive lots which it is desirable to distinguish. Otherwise it may be omitted.

Fig. 11 shows a bin tag for entering receipts or withdrawals and making additions or subtractions on the tag. In the company where this bin tag is used it is the practice for the storeskeeper to enter the balance from the bin tag onto the receiving slip or stores requisition each time a transaction takes place so that the stores record clerk receiving these slips can check his balance on stores record cards which are also maintained. Thus, any discrepancies can be caught immediately by calling for a recount and then making the necessary changes in both the stores records and the bin tags so that the balances correspond with existing material or parts on hand. The bin tag in Fig. 11 is designed to fit in a metal slot or frame so that it can be posted without removal, an advantage not always obtained with the bin tag on a hook.

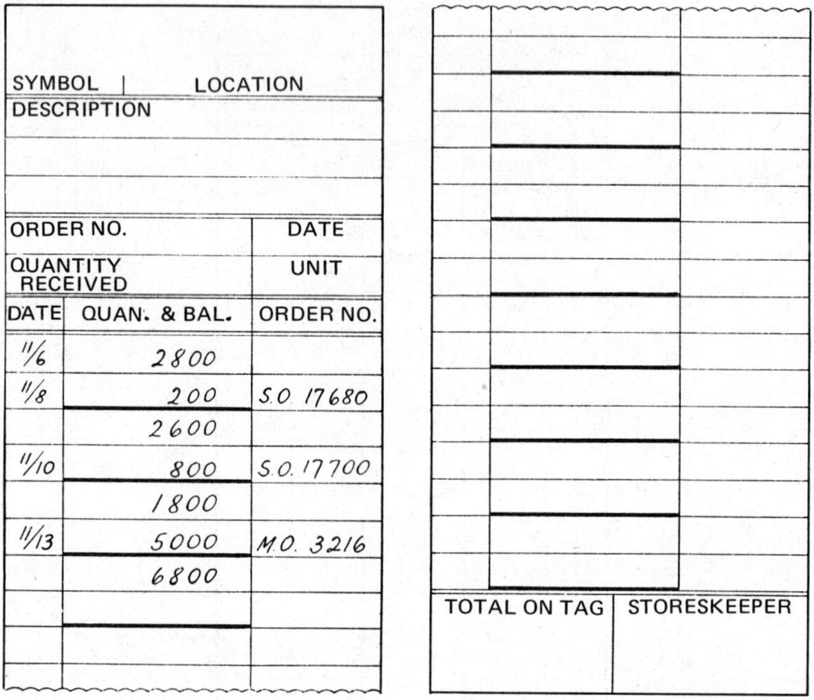

SYMBOL | LOCATION
DESCRIPTION
ORDER NO. | DATE
QUANTITY RECEIVED | UNIT

DATE	QUAN. & BAL.	ORDER NO.
11/6	2800	
11/8	200	S.O. 17680
	2600	
11/10	800	S.O. 17700
	1800	
11/13	5000	M.O. 3216
	6800	

TOTAL ON TAG | STORESKEEPER

Fig. 11. Bin tag with sections for addition or subtraction to get new balance.

Where the first-in first-out method is followed, physically (or in the records) using up all the units received on a purchase or replenishment order before using any stock from later orders or lots, a separate bin tag may be used for each such lot. When the last units are issued, which sometimes means taking some from a new lot to make up the required quantity, the old bin tag is canceled, the tag for the new lot comes into use, and the same procedures are then followed. In the case of materials subject to deterioration, the actual lots are properly tagged and are issued in order of arrival. If changes in design have made component parts subject to withdrawal from use, the corresponding bin tags will be so marked.

Stores Record Card Methods. The simplest form of a stores record or **perpetual inventory card** merely shows the receipts, issues, and balance on hand (see Fig. 12).

Receipts are posted from receiving reports or from finished goods tickets and initialed by the stock room, with any discrepancies noted thereon. Issues are posted from material requisitions which are turned in at the store room for mate-

DESCRIPTION				ITEM NO.	
RECEIVED		ISSUED		BALANCE IN STOCK	
DATE	QUANTITY	DATE	QUANTITY	DATE	QUANTITY

Fig. 12.   Stores record card.

rial or parts. The storeskeeper initials these forms and sends them to the stores record section where they are posted to stores records cards. The difference between the quantity received and the quantity issued is the **balance in stock.**

The posting of receipts and issues and the balance in stock may be done manually or by posting machines, punched-card machines, and electronic data processing equipment.

The stores record clerk is responsible for watching the quantity in stock for each item and for reordering when the balance gets below "a safe level." This is usually a quantity the stores record clerk has in mind at which to place the order so that the stock will not be depleted before new stocks are received. The clerk usually makes some notation of this quantity on the stock record card and increases it if he runs out of stock before the new order is received.

Where requirements are computed for several months ahead or where requirements are likely to be recomputed before outstanding orders are received in the plant, information regarding orders for replacements which have not yet been received are also placed on the card (see Fig. 13).

								STORES SYM. SVMS	UNIT Ea	MIN. 5,000	
Washer - 3/4"								LOCATION IN STORES AD - 4 - F - 5			
DATE	ORDER NO.	QUAN. ORD'RD	BAL. ON ORDER	QUAN. RECD.	QUAN. ISSUED	ISSUED TO	BAL. ON HAND	DATE	PRICE	TOTAL COST	
5-11	7428	30,000					5,000	6-1	.005	$50.00	
6-1	7428		20,000	10,000			15,000				
6-8					5,000	50-1875	10,000				
								DATE	APPOR'TD	ORDER NO.	
								5-11	10,000	50-1875	

Fig. 13.   Component parts stores record.

The determination of whether it is necessary to order material based on the balance on hand or the available balance is determined by the **replenishment cycle.** If the replenishment cycle is long compared to the amount which will be ordered, some purchase orders are always outstanding and must be taken into account in placing other orders. For example, if the replenishment cycle is three months but only a one-month quantity is purchased at a time, it is obvious that there will be several orders outstanding at all times.

If, however, the replenishment cycle is fairly short compared to the amount on order and, in general, the replenishment will be received before another order is placed, there will very seldom be any orders outstanding. In this case the available column would substantially be the balance on hand, since there would be no quantities on order.

In some plants the stores record keeping function is decentralized to the storeroom location. This is perhaps permissible for certain types of inexpensive and expendable material, but records are normally centralized.

Allocation Methods. Earlier in this section, emphasis was placed upon the importance of preplanning to fit the materials control program definitely into the schedules of production, which in turn are planned in advance to conform to future sales demands. Under this plan, an allocating or apportioning of materials is carried on when the production programs are originated. When the kinds and quantities of products to be made in the forthcoming period are scheduled, materials are ordered in kinds and amounts to carry on such production and to deliver finished products at the times specified. In other words, the materials flow in at a rate to keep production going and are thus actually allocated when ordered under purchased contracts.

Where conditions such as the nature of the business, inefficiencies in making sales forecasts or controlling production, or lack of a capable staff to administer the procedure do not permit a coordinated master plan to be put into effect, a more detailed and localized control is needed, but the clerical cost of the work is relatively higher. A greater burden is likewise placed on the stores record system itself. Materials should be allocated to production in advance so that material shortages may be eliminated.

The plan followed is to set up the stores records so that, by inclusion of **allocation** and **available** columns, materials may be allotted to current orders in advance of production. This plan avoids running short, which may happen when the records show only the balance on hand of the needed items and not how much of each will be withdrawn for work still to go into manufacturing.

The allocation procedure is to send the materials requisition slips written during the planning of the order for production, or the bill of materials where this form can be more readily applied, to the stores record clerk so that he may enter in the allocated column on the card for each item the quantity to be reserved and the order number to which it will be applied. He then subtracts this same quantity from the balance in the available column to show how much remains for future orders. When asked subsequently to check a bill of materials for a new order coming through to find out the materials status, he can indicate by checks against the items on the list the ones for which there are sufficient materials available, knowing that these materials will not be required for any work previously planned to go into production. Stores are replenished when the available balances reach stated reduced levels and there is thus maintained a cushion stock sufficient for the regular course of production. Any unusual demands can be met by putting through purchase requisitions while such large new orders are being planned in time to get deliveries made before the materials must actually go into production.

Figure 14 is a typical **allocation record form,** accompanied by an explanation of entries, which may be printed on cards or loose-leaf ledger sheets in any convenient standard size. The headings are representative and may be changed as desired to fit any stores system. The columns are the minimum number needed and may be increased in number if additional data are required, or if it is desired to have order numbers and dates appear immediately adjacent to the respective postings instead of all at the left. One line across the entire sheet is used for each transaction even though this plan leaves spaces blank in several of the columns. The records can be inspected to obtain data much more quickly under this plan.

Figure 14 is the general type of form, a result of the practice of Taylor, Gantt, Barth, and others of this group. Where considered helpful, the sequence of columns shown in the form can be changed, a good arrangement being: 1—Ordered, 2—Apportioned, 3—Available, 4—On Hand.

The **stores record principle** behind this form is expressed in the following equation:

$$\text{Amount on order (total)} + \text{Amount on hand (balance)}$$
$$= \text{Amount allocated (total)} + \text{Amount available}$$

All postings may be checked against this equation to test the correctness of the record, the last entry in each column being the one used in the equation.

In the form illustrated, the unit cost or price of the item and the total value of the amount (inventory) on hand is included. In the case of received materials the stores record clerk posts such data from information supplied by the purchasing department through copies of the purchase order which indicates prices and amounts ordered. Receiving reports show the quantities actually received. The amount of any cash discount is deducted and the cost of freight added to get the cost delivered. For issues, he uses the unit costs thus determined. In the illustration the **average method of costs** is used as shown in one of the calculations. Other methods—**standard cost, first-in first-out,** etc.—are also used, according to the accounting procedures of individual companies. The stores record clerk performs these calculations and entries so that he can price the materials requisition slips as he posts issues of material, thus saving duplication of certain handlings and operations by the cost accounting department. **Inventory value summaries** also can be made up rapidly for accounting reports and financial statements from the source which has the latest assembled data on this point at any one time. When a **standard cost system** is in use in the company, the current cost of the item is entered at the top of the sheet or card and the stores record clerk does not have any calculations to perform. When the standard is changed, he is notified to change his records, and accordingly, he makes his entry on the card.

When a company prefers to omit the inventory value data from the stores records, even though the dollar value of the inventory is desired monthly, the quantity balance can be extended to a price balance when the need arises. A good way to secure total inventory value, usually by material classifications, is to carry a separate summary record for each stores ledger or group of stores cards, with each classification as a minimum for a group. Under this method, a **daily dollar value** is secured by (1) costing requisitions as they clear through the stores records, (2) sorting costed requisitions by material classification, (3) totaling them, and (4) posting the totals to the disbursement columns of the summary cards. Receipts are accumulated in a similar manner. While more than one item will usually appear on the posting media (invoices), the fact that most vendors' materials fall into only one classification simplifies the distribution. Transportation costs and discounts, of course, must be prorated.

STORES RECORD

Name Bushing, Bronze
Description 1" i.d. x 1¼" o.d. x 2" long
Specification No. 2,240
Minimum economic order quantity 6,000   Requisition when available shows 1,500

Symbol SBBB 1 x 1¼ x 2
Unit Each
Dwg. No. C-14750   Part No. W-16842
Location in stores AC3F4

Date	Order No. (Shop or Purchase Order)	1—Ordered (But not delivered)			Received	2—On Hand (In the storeroom)				3—Allocated (Not yet issued)		4—Available (For new orders)
		Req. No.	Quantity	Total		Issued	Balance	Unit Price	Total Value	Quantity	Total	Quantity
Jan. 3					Inv'y		3,000	$ .09	$270.00			3,000
4	S.O. 115					300	2,700	.09	243.00			2,700
5	S.O. 131									+200	200	2,500
7	S.O. 156									+700	900	1,800
10	S.O. 131					200	2,500	.09	225.00	-200	700	
11	S.O. 210									+400	1,100	1,400
11		Req. 230	6,000									
14	S.O. 115				(credit)	+50	2,550	.09	229.50			1,450
15	S.O. 156					700	1,850	.09	166.50	-700	400	
18	P.O. 471		+6,000	6,000								7,450
19	S.O. 275									+300	700	7,150
21	S.O. 210					400	1,450	.09	130.50	-400	300	
27	P.O. 471		-6,000	0,000	6,000		6,000	.10	600.00			
							7,450	.0980	730.50			
28	S.O. 131				(Repl.)	20	7,430	.0980	731.96			
31	S.O. 275					300	7,130	.0980	702.56	-300	000	7,130

EXPLANATION

Entry Date	Kind of Entry	Procedure in Making Entry	Entry Date	Kind of Entry	Procedure in Making Entry
Jan. 3	Inventory	Assume that sheet is started from physical inventory (Equation: 0 + 3,000 = 0 + 3,000).	Jan. 18	Purchase	Standard order placed. Ordered (= 6,000) is considered to be bought to arrive before stock falls below cushion or reserve, therefore to be available for planning future manufacturing (= 7,450). Some authorities do not count purchases as available until actually received.
4	Issue	System just started so this order was not preapportioned.			
5	Apportion	Balance on hand (2,700) unchanged but quantity available reduced (= 2,500).	19	Apportion	Add to apportioned (= 700), subtract from available (= 7,150).
7	Apportion	Add to total apportioned (= 900), subtract from available (= 1,800).	21	Issue	Subtract from balance (= 1,450) and from apportioned (= 300).
10	Issue	Subtract from balance (= 2,500) and from apportioned (= 700).	27	Receipt	Subtract from ordered (= zero), post to received, insert in balance column, put unit price and total value of new order in their respective columns.
11	Apportion	Add to total apportioned (= 1,100), subtract from available (= 1,400).			
11	Requisition	Available (= 1,400) is now below 1,500, the reorder point. Standard purchase amount is requisitioned for ordering by purchasing department (if a purchased item).	28	Calculation	Add old balance and new receipt (= 7,450), add old and new values (= $730.50). Divide: $730.50 ÷ 7,450 = .0980, new unit price under average method.
14	Credit	Received back from previously issued shop order. Add to balance (= 2,550) and to available (= 1,450).	28	Replacement	Spoilage or shortage on shop order. Subtract from balance (= 7,430), price valuation of balance at new unit rate, subtract from available (= 7,130).
15	Issue	Subtract from balance (= 1,850) and from apportioned (= 400).	31	Issue	Subtract from balance (= 7,130) and from apportioned (= zero).

The total value is recalculated whenever the balance is increased by receipts or reduced by issues, and at the unit price of that date. The equation, Ordered (total) + On hand (balance) = Apportioned (total) + Available (quantity), can be checked at any point by taking the last entry in each of these columns and inserting it in the equation. If the equation does not balance, an error has been made in posting.

Fig. 14. Stores record for allocation method.

The weakness of all of these allocation methods is that, although information is supplied relative to the status of each item controlled, the reorder points are usually not set very scientifically, and hence more stock may be on hand than needed, thus raising operating costs and inventory levels; or too little material may be on hand, thus resulting in costly expediting and permitting possible shutdowns of production lines or work stations.

Demand and Supply Methods. Demand and supply methods of materials control utilize principles of planning the demand for and supply of each item at the lowest cost possible and the lowest possible inventory consistent with operating requirements.

The basic difference between these and less "scientific" types of recordkeeping and ordering is that much more time and effort is devoted to developing for each part or item an **order point** based on objective analysis of past and projected usage. **Economic purchasing and manufacturing lot sizes** are also developed to minimize the total cost of procuring, storing, and utilizing each specific item of material and parts. (See sections on Purchasing, Production Planning, and Production Control.)

This method is usually used to keep track of quantities only, and dollar or financial controls are maintained in the accounting department.

In order to use this method of materials control, certain information must be known or estimated for each part and raw material item controlled:

1. **Average Usage.** The expected usage of the particular part or raw material in the future. Since this is difficult to determine, past usage is usually taken as the best indication of the future and this quantity per time period is usually used. If usage is expected to increase or decrease substantially in the future, the usage figure should reflect this trend.
2. **Lead Time.** The average elapsed time between the initiation of the order paperwork and the receipt of the material from the shop or from vendors.
3. **Reorder Point.** The quantity expected to be consumed during the replenishment lead time plus a reserve. It is computed by multiplying the lead time in periods by the average usage per period and adding the reserve quantity.
4. **Reserve Stock.** An extra amount which is kept on hand to take care of greater than normal usage during the replenishment lead time or an average usage during a greater than normal lead time, or a combination of the two. It is usually expressed as usage during some additional time period, although it may also be a certain fixed quantity determined in some empirical way.
5. **Economical Order Quantity.** The quantity which is most economical to order and to stock, considering all factors bearing on the situation. It is peculiar to each part, as is explained in detail in the section on Inventory Control.

The demand and supply control card, shown in Fig. 15, provides a detailed and comprehensive history of all transactions and planning related to each particular part. The types of information kept on the card are:

1. Actual balance in stock.
2. Quantity available to cover future usage.
3. Detailed record of open orders.
4. Reorder information.
5. Usage information.

Information on the heading of the card includes part number and name, material to be furnished, approximate cost to make or to buy, economic order quantity, average usage per month, order point, reserve quantity, lead time requirement, inventory account and classification, unit measure and weight, and stockroom location.

The following information is posted to the card as transactions are reported to materials control:

1. **Cumulative Demand.** The cumulative requirements for this particular part as exploded from bills of material or drawings to cover manufacturing schedules or as posted from customer orders. This is an allocation of all material presently available in stock or soon to become available through replenishment orders.
2. **Cumulative Supply.** The total quantity of the item which has been ordered plus any physical inventory on hand when the card was opened. This is sometimes called the "ordered" column.
3. **Demand Coverage.** The quantity of material which is available to cover future demands for the part. It is the excess of the supply over the demand or a quantity ordered and in stock over the amount apportioned and used. This figure is compared with the order point, and whenever the demand coverage declines to or below the order point a new order is placed for the economical order quantity.
4. **Cumulative Receipts.** A posting of the quantity as received in the storerooms.
5. **Cumulative Issues.** The total quantity of material issued from the storeroom to the shop.
6. **Balance in Stock.** The actual physical balance in the storeroom. It is the difference between receipts and issues.
7. **Open Purchase or Shop Orders.** The details of all open orders which have not been received.

After posting, the following information is available on the card:

1. **Balance on Order.** Equals cumulative supply minus cumulative receipts. (This figure should check with the totals shown in the "balance open" columns of open purchase or shop orders.)
2. **Actual Balance in the Storeroom.** Equals cumulative receipts minus cumulative isues.
3. **Unfilled Demand.** Equals cumulative demand minus cumulative issues. This is the quantity allocated to orders which has not been requisitioned or issued to the shop.
4. **Demand Coverage.** Equals cumulative supply minus cumulative demand. This is the quantity which can be allocated to future orders or manufacturing schedules.

In this example the **calculation of the order point** is as follows:

$$\text{Order point} = [\text{lead time (in months)} \times \text{average usage per month}] + \text{reserve}$$
$$= \tfrac{1}{2} \times 2{,}000 + 500$$
$$= 1{,}500$$

Rather than automatically reordering whenever the order point is reached, the expected usage should be reviewed in light of current conditions and the order point revised if necessary. Whenever the lead time is changed, the purchasing department and the shop should advise the materials control section, which will revise the affected order point accordingly.

In **mechanized** or **computerized systems** the recalculation of both reorder quantities and reorder points, safety stock, and reserve stocks can be made automatically as part of the system. The system is then referred to as being an adaptive system.

Last Bag Method. Often the posting of receipts, issues, and balance on stores record can be eliminated completely by the use of the so-called **last bag, sealed quantity,** or **double-bin method.** A quantity sufficient to last during the lead time necessary to get a replacement order and allow a generous reserve is bundled into a separate bag or carton, kept in a separate bin, or in some way identified clearly as being the reorder stock.

DEMAND AND SUPPLY CONTROL CARD

PART NAME	INVENTORY		UNIT		PART NO.
	ACCOUNT	CLASS	MEASURE	WEIGHT	

MATERIAL TO BE FURNISHED		
APPROXIMATE COST		ORDER POINT 1,500
ECONOMICAL ORDER QUANTITY 6,000		RESERVE 500
AVERAGE USAGE PER MONTH 2,000		LEAD TIME 1/2 mo.
APPROVED BY		STOCKROOM LOCATION 2 G 12

SUMMARY

SUPPLY REFERENCE (1)	DEMAND REFERENCE (1)	DETAIL ENTRY (2)	CUMULATIVE SUPPLY (3)	CUMULATIVE DEMAND (4)	DEMAND COVERAGE (5)	CUMULATIVE RECEIPTS (6)	CUMULATIVE ISSUES (7)	ACTUAL BALANCE IN STOREROOM (8)
1/3 Inventory			3,000		3,000	3,000		3,000
4	S.O. 115	300		300	2,700			2,700
5	S.O. 131	200		500	2,500		300	
7	S.O. 156	700		1,200	1,800			
10	S.O. 131	200						2,500
11	S.O. 210	400		1,600	1,400		500	
11 P.O. 471		6,000	9,000		7,400			
14 S.O. 115		50	9,050		7,450	3,050		2,550
15	S.O. 156	700		1,900	7,150		1,200	1,850
19	S.O. 275	300						
21	S.O. 210	400					1,600	1,450
27 P.O. 471		6,000				9,050		7,450
28	S.O. 131R	20		1,920	7,130		1,620	7,430
31	S.O. 275	300					1,920	7,130

OPEN PURCHASE OR SHOP ORDERS (9)

	BALANCE ON ORDER		
	DATE OF ENTRY	RECEIPTS	BALANCE OPEN
ORDER NO. P.O. 471	1/11	–	6,000
DATE DUE 1/25	1/27	6,000	0
DATE RELEASED 1/11			
DATE STARTED 1/18			
ORDER NO.			
DATE DUE			
DATE RELEASED			
DATE STARTED			
ORDER NO.			
DATE DUE			
DATE RELEASED			
DATE STARTED			

EXPLANATION

Entry Date	Kind of Entry	Procedure in Making Entry	Entry Date	Kind of Entry	Procedure in Making Entry
Jan. 3	Inventory	Physical Inventory of 3,000 recorded as shown.	Jan. 14	Credit	Received 50 back from previously issued shop order. Add to cumulative receipts (=3,050), to balance (=2,550), to cumulative supply and to demand coverage (=7,450).
4	Issue	System just started so this order for 300 was allocated at time of issue.	15	Issue	Issue 700 and subtract from balance (=1,850).
5	Allocation	Balance in storeroom (2,700) unchanged but demand coverage reduced (=2,500) by order for 200.	19	Allocation	Add 300 to cumulative demand (=1,900), subtract from demand coverage (=7,150).
7	Allocation	Add 700 to cumulative demand (=1,200), subtract from demand coverage (=1,800).	21	Issue	Issue 400 and subtract from balance (=1,450).
10	Issue	200 shown as issue and subtract from balance in storeroom (=2,500).	27	Receipt	Post 6,000 to cumulative receipts, increase balance in storeroom to 7,450. Record receipt in open purchase order column.
11	Allocation	Add 400 to cumulative demand (=1,600), subtract from demand coverage (=1,400).	28	Replacement	Spoilage or shortage on shop order. Add to cumulative demand, subtract from demand coverage (=7,130), add to issued (=1,620). Subtract from balance (=7,430).
11	Requisition	Demand coverage (=1,400) is now below 1,500, the reorder point. Standard purchase amount 6,000 is requisitioned for ordering by purchasing department (if a purchased item).	31	Issue	Add to cumulative issues (=1,920) and subtract from balance (=7,130).

Fig. 15. Demand and supply control card.

When the stores department uses the first piece of this quantity, they forward the identification and reorder tag attached to the quantity to the purchasing or production control department, which places a replacement order. The reorder stock is large enough to supply production and leave some parts on hand when the new order is received.

The usage of these parts should be reviewed periodically and new reorder quantities set. This can be done by reviewing the quantity ordered by the purchasing or production control departments during the previous year or six months. This gives an average usage per month which can be used in determination of the reorder quantity.

ADJUSTMENT OF REORDER POINT. It is, of course, advisable to keep the reserve quantity as low as possible in case of model or engineering changes. But this always involves the risk of running short, since the shop or vendors may be late in supplying the parts ordered or actual usage may exceed predicted usage. A method for allowing for these factors is described in Factory (vol. 113), using a chart which takes three factors into account:

1. Excess use over estimated production requirements.
2. Excess use plus a slight delivery delay by vendors or the plant.
3. Maximum delivery delay only.

Excess usage is the difference between maximum usage and normal usage for the maximum delivery time. Slight delivery delay is expressed as normal or excess usage times the amount of the delivery delay. Thus, by selecting the combination of factors which is most likely to happen, the proper quantity of material can be ordered. The steps in using this chart (Fig. 16) are as follows:

1. Determine normal delivery time and maximum delivery time.
2. Determine maximum estimated usage.
3. Decide the degree of protection needed for the particular order.
4. Follow a vertical line down the column (indicated by arrow) for the appropriate delivery time, maximum delivery time, and estimated maximum usage.
5. Follow a horizontal line across the row (indicated by arrow) for the degree of protection required.
6. Read the order point factor at the intersection of the vertical and horizontal lines you are following.

This factor is used to find the quantity to order as shown in Fig. 17. The estimated usage is multiplied by the order point factor to get the order point quantity. From this are deducted any quantities on hand and on order to determine the quantity to order.

PERIODIC REORDER SYSTEM. By the periodic reorder system, manufacturing or production schedules are reviewed on a periodic basis and the quantity to reorder of each part and raw material is determined. The basic feature of this system is that each part and each item of raw material necessary to produce the schedule is ordered each time a schedule is issued.

Parts Required and In Stock. In order to determine the quantity of each part required, the manufacturing schedule is exploded by multiplying the total quantity of each finished item by the quantity required per unit of each part and each item of raw material. These requirements are sorted and totaled by part number to get the **total requirement** for each part and for each item of raw material for the coming period or for several periods to come.

Under this system it is desirable to know the quantity of each item remaining in stock (if any) so that this quantity may be deducted from the requirements of the schedule and a lesser quantity ordered. This information may be obtained by **physically counting** the parts of material in stock each time a schedule is

Degree of protection needed								
Normal delivery time (weeks)	3	3	3	4	4	4	4	4
Maximum delivery time (weeks)	4	4	4	5	5	5	6	6
Estimated maximum usage (normal usage=1.0)	1.1	1.2	1.3	1.1	1.2	1.3	1.1	1.2
A. Excess usage only $= \left(\begin{array}{c}\text{Normal delivery}\\ \text{time}\end{array}\right) \times \left(\begin{array}{c}\text{Estimated}\\ \text{maximum usage}\end{array}\right) =$	3.3	3.6	3.9	4.4	4.8	5.2	4.4	4.8
B. Excess usage plus slight delivery delay $= \left(\begin{array}{c}\text{Normal}\\ \text{delivery time}\end{array}\right) + \left[\left(\begin{array}{c}\text{Maximum}\\ \text{delivery time}\end{array}\right) \times \left(\begin{array}{c}\text{Estimated}\\ \text{maximum usage}\end{array} - 1.0\right)\right] =$	3.4	3.8	4.2	4.5	5.0	5.5	4.6	5.2
C. Excess usage and maximum delivery delay $= \left(\begin{array}{c}\text{Maximum}\\ \text{delivery time}\end{array}\right) \times \left(\begin{array}{c}\text{Estimated}\\ \text{maximum usage}\end{array}\right) =$	4.4	4.8	5.2	5.5	6.0	6.5	6.6	7.2
D. Maximum delivery delay $= \left(\begin{array}{c}\text{Maximum}\\ \text{delivery time}\end{array}\right) =$	4.0	4.0	4.0	5.0	5.0	5.0	6.0	6.0

Formulas for finding order-point factors beyond chart range

Fig. 16. Determination of order point factor.

Item Description	A Est. current usage (weekly)	B Ordering code. (See order-point factor chart)	C Order-point factor (from order-point factor chart)	D Order-point quantity (A×C)	E Quantity on hand	F Quantity on order	G Total (E+F)	H *Quantity to order (D−G)
0.008 in. × 25.25 in. sheet aluminum coils	5,000 lb.	4-5-1.2-B	5.0	25,000	8,500	10,500	19,000	6,000 lb.
4-in. strap hinge	5	4-5-1.2-C	6.0	30	6	10	16	14
⅝ × 2¼ hex. head screw caps	10 gross	3-4-1.2-A	3.6	36	6	15	21	15 gross
200-watt, 120-v. med. base incandescent lamp bulb	75	3-4-1.2-B	3.8	285	150	75	225	60
0.016 in. × 28.25 in. sheet aluminum coils	5,000 lb.	4-5-1.2-C	6.0	30,000	6,000	28,000	34,000	0

*Or economical minimum-ordering quantity, whichever is greater

Fig. 17. Use of order point factors to order material.

issued, or some record may be kept of receipts, issues, and balances in stock. Using the information from these records, the schedule requirements are then compared with the total in stock and on order by placing these quantities on a periodic reordering worksheet, Fig. 18. Since this worksheet serves as a record

Part No.	Description	In Stock	On Order	Total Available	Required for Schedule	Place Orders for
136	Slide rod	1,000	0	1,000	3,000	2,000
179	Gasket	500	3,000	3,500	6,000	2,500
9966	Tube plug	1,000	1,000	2,000	3,000	1,000
710	Clip	0	0	0	3,000	3,000

REORDER SCHEDULE   NO. D63

Fig. 18. Period reordering schedule.

of the requirements, it is usually not necessary to post the requirements to inventory control cards. Thus, either three- or four-column cards (including an on-order column) can be used, and no allocated and available columns are required.

If the schedule is projected for several months in advance, it is important to indicate the dates on which specific quantities of material are required. Thus in Fig. 13 it would be necessary to have several extra columns headed place orders for and to show the quantity required for each schedule and the date when that material was due. It is also advisable to show the date when the on-order material is due so as to be able to verify that the material will be in the plant in time to complete the monthly or weekly schedule.

One appliance manufacturer using this method schedules his production of castings, sheet metal stampings, and spot-welded subassemblies 30 days in advance. He schedules the enameling of the stampings and spot-welded assemblies and the production of electrical and other subassemblies three days in advance of his assembly line schedule so that these items will be available on the line when the finished units are assembled.

PUNCHED-CARD MATERIALS CONTROL. Because of the volume of transactions and the number of items involved, mechanized materials control systems are frequently used. The methods employed by various companies vary widely but the results are the same.

The **three methods of punched-card materials control** most frequently used are:

1. The balance-forward method.
2. Unit-record method.
3. Batch method.

The Balance-Forward Method. A card is created for each classification of stock. If there are 6,000 items in the line, 6,000 cards would be required. Information carried in the card would be as follows.

1. Item identification information
   a. Material code and description
   b. Item class
   c. Units of measure
   d. Cost per unit
2. Date
3. Opening balance
4. Minimum inventory
5. Reorder point
6. Receipts
7. Issues
8. On hand
9. On order
10. Available

The figures in the card are changed daily according to the transactions that occur. The opening balance remains constant, the on-hand figure is the balance after the receipts have been added to the opening balance and the issues subtracted.

$$\text{Opening balance} + \text{Receipts} - \text{Issues} = \text{Balance on hand}$$

The available figure is the result of adding the total on hand to the total on order and subtracting the minimum inventory figure (or comparing to the reorder point).

After the transactions are received each day, they are punched in cards. For example, receipts cards would be punched from receiving reports, issue cards from issue slips or shipping orders, and on-order cards would be punched from purchase orders. Often these cards are produced for other purposes and can be used in the materials control system as a by-product. These transaction cards are accumulated for each day, sorted by item identification and merged with the balance-forward cards. The deck is processed through the tabulating equipment and new balance-forward cards are summary punched. These new cards are used for the day's transactions. Where the reorder point is reached, the particular item card can be identified with an order punch and used to start the processing to replenish the item. Also with the unit cost contained in each item card, the inventory can be priced out by multiplying the on hand by the price. Aggregate totals can also be carried and, in fact, are required as a control check using the same formula.

$$\text{Opening balance} + \text{Receipts} - \text{Issues} = \text{Balance on hand}$$

If the totals check out, it can be assumed that the individual transactions are in order.

Prepunched cards can be employed to reduce the amount of punching required. The system must be maintained. Stock adjustments due to rotating counts or physical inventory must be made as they are identified. Also discrepancies in counts on issues, receipts, etc., must be corrected as soon as they are discovered. Periodic reports such as shown in Fig. 19 can be generated by the balance-forward cards.

The Unit Method. This method requires that a card be punched for each item in the stock. If there were 18,000 units in the inventory of all items, there would be 18,000 cards in the tub file.

A card is added for each new item received and as units are issued the equivalent number of cards are withdrawn. The remaining cards in the tub file represents the physical inventory in stock.

SUMMARY STOCK STATUS REPORT

ɔ: _____   Date _____

⸱m ⸱de	Description	Opening Balance	Actual Transactions			Future Transactions		
			Rec'd	Issued	On Hand	On Order	Req'd	Not Committed

Fig. 19. Summary stock status report.

The tub files are usually broken down with divider cards for each item class or description. The individual cards for each unit of inventory are end printed with the item description or code number and can also be serially numbered to maintain the proper sequence for issuance.

Where a shipping unit (dozen, gross, 100, etc.) is prepackaged in cases or standard packages, the number of cards involved can be reduced. In such cases a card is prepared for each package or standard unit rather than for a single item.

Using the unit method a visual reorder scheme can be used by having a distinctive card placed at the predetermined reorder point. When all the cards up to that card are pulled, the order card is started through the order process and a new order card placed in the proper place for future reorder.

Beside the visual advantages, inventory status reports can be generated by machine processing the cards in the file.

The Batch Method. Of the three methods listed above, the batch method permits the most complete machine processing. As issues or sales are made a copy of the issue slip or sales order is processed through keypunch; if a billing or accounting machine is used, the tape or card is entered into the system. The following information is contained relative to each issue:

1. Customer's code or receiving department code.
2. Quantity.
3. Unit selling or transfer price.

Periodically these transactions are sorted by item code number and represent the withdrawals of that period. The same is done with the other transactions receipts, transfers, and the like. A balance-forward card is created by batch rather than individually as in the balance-forward method described earlier.

Summary reports can be created more frequently when the batch method is used.

COMPUTER APPLICATIONS. According to an American Management Association study (AMA Research Study No. 77) production and inventory control uses lead all other applications of EDP in production. Material control and inventory control are often used interchangeably. Seventy-four percent of the companies in the AMA study reported using EDP to maintain stock status reports. As shown in the discussion of the punched-card systems described above, the effective control of materials requires records, huge amounts of detailed records and transactions. The computer fits into the massive data-handling picture quite well. For example, the materials control techniques discussed in the previous few paragraphs can be transferred directly onto a computer emulating tabulating equipment. Cards can be replaced by tapes or discs and the processing time significantly decreased.

Where a feasibility study indicates that it is warranted (using the least cost approach), a custom designed material control system executed on a computer can be very effective. Usually the system goes beyond just the material control area and includes billing, order entry, shipping, scheduling, purchasing, forecasting, and the other relevant phases of the business cycle.

REQUISITIONS FOR NEW ORDERS. Requisitions for the purchase of production material or for making a part in the shop are usually initiated by the materials control section. There are several ways to **initiate requisitions for procuring materials,** depending upon the method of control used:

1. On the fixed order basis, purchase orders and shop requisitions are initiated whenever the reorder point is reached for these stock items, as explained above. Special items are purchased or produced as needed.
2. On the periodic order basis, requisitions for purchased raw materials and parts are initiated whenever manufacturing schedules are reviewed and the need to place orders determined, as explained above.
3. In commodity purchasing, purchase orders may be initiated for planned programs, whenever it is decided to buy material on a speculative basis, or when there is a threatened shortage due to a strike or excessive demand.

Purchase Requisitions. The **sources of requisitions** for specific purchases are mainly (1) the materials control section, for items carried regularly in stores and controlled through stores records; (2) the planning or scheduling section, for production items which are not regularly stored or which are controlled by a production schedule; and (3) individuals authorized to issue purchase requisitions as needs arise. The nonstocked items are usually ordered by the planning section from data on bills of materials or parts lists, with the exception of those requisitioned by the superintendent or other operating authority for materials or parts specially required for experimental jobs or other unusual uses.

With a good system of materials control, the operating departments are not responsible for the kinds or quantities of production items stored and do not do any ordering or requisitioning of materials for production, this work being the function of the planning and stores records section of production control.

Purchase requisitions for stock items are usually issued by the materials control section whenever a reorder point is reached or a new schedule is worked. Sometimes requisitions are originated by the storeskeeper if he keeps the stores record. The purchase requisition is often made out in two copies, with one copy being retained by materials control until they receive a copy of the purchase order. An example of a purchase requisition is shown in Fig. 20 (also see section on Purchasing).

Repeating Purchase Requisitions. Some companies have developed a repeating purchase requisition for stock material and parts which are ordered on a repetitive basis. This saves making out an individual requisition every time material is required. The heading is typed only when the card is originally made up and lists all information regarding that particular part. Alternate vendors, prices, and price breaks are also shown (Fig. 21).

The repeating purchase requisition is normally filed with the materials control card. When a replenishment order is needed, the card is pulled from the file and all of the information shown under "To Be Filled In Only by Production Planning Office" is inserted by the materials control man. The card is then sent to the purchasing department which places an order and records the information "To Be Filled In Only by Purchasing Department." They then return the card to the materials control section.

JONES MANUFACTURING COMPANY

Purchase Requisition

TO _____   REQ. NO. _____

Please purchase the following items: Date _____

Quantity	Description		

Ship to _____   Date Required _____

Via _____

Signed _____   Approved _____

Fig. 20. Purchase requisition.

Other **advantages** of the repeating requisition are that it insures greater accuracy in the writing of requests by eliminating errors in copying descriptions, and it facilitates review of the request by the materials control supervisor and by the purchasing department, since previous order quantities, dates, and prices are shown. It also saves time of the buyer in looking up vendors and prices. Repeating requisition cards can be used for all items which are likely to be reordered.

Blanket Orders. Some industries place contracts covering future requirements for a certain time period such as a year. They are given the privilege of specifying the quantity which is to be released and shipped to them at any one date. They frequently control their materials on a "float" basis. The **factory float** consists of materials for a certain number of days of production, which are stored on the manufacturing floors. The **stores float** is the number of days' production of materials kept in the storeroom, and the **purchase float** is the number of days' production still on order but not yet received. The production program usually is built up by months, and the quantity to be produced per month is stated for each product. An analysis is made of the materials needed

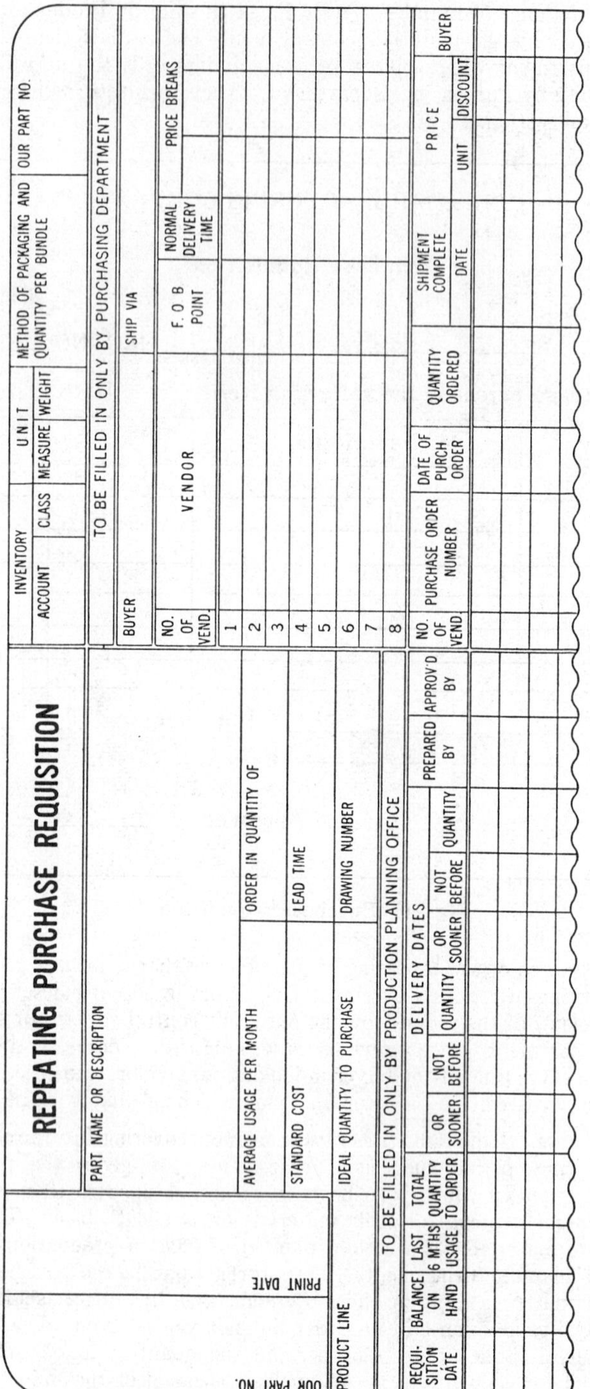

Fig. 21. Repeating purchase requisition.

to produce on schedule. If purchase contracts covering future commitments have been made for such materials, the requisitioning process consists of issuing authorizations from the purchasing departments to the vendors to release certain quantities against the contract. New contracts are placed to cover future periods whenever the quantities in stores, plus undelivered quantities on current contracts, fall to stated levels. These practices apply to process industries and those engaged in repetitive manufacture of assembled products, notably automotive plants.

The **objective** in issuing blanket orders is to assure a good price, or to protect manufacturing against market shortages, as well as to meet current needs. Pure speculative purchasing, that is, buying larger quantities than will be needed for production with the intention of selling surpluses to other users when markets rise, is not a manufacturing venture and ordinarily is not engaged in by manufacturing companies.

Requests for Shop Order. Many companies manufacture some or many of their own parts. These are produced from raw materials withdrawn from stores in the usual way. Production of these parts is authorized by the materials control section of the production control department when the quantity on hand reaches a reorder point or, where the storeroom carries on the records function, by the chief storeskeeper. Material requisitions for the withdrawal of the necessary raw materials from stores are prepared by the planning section, the materials control section, the dispatcher, or the foreman.

A **repeating request for shop orders,** which follows the same principle and has the same advantages as the repeating request for purchase orders, is shown as Fig. 22. All heading information remains constant and the person requesting a

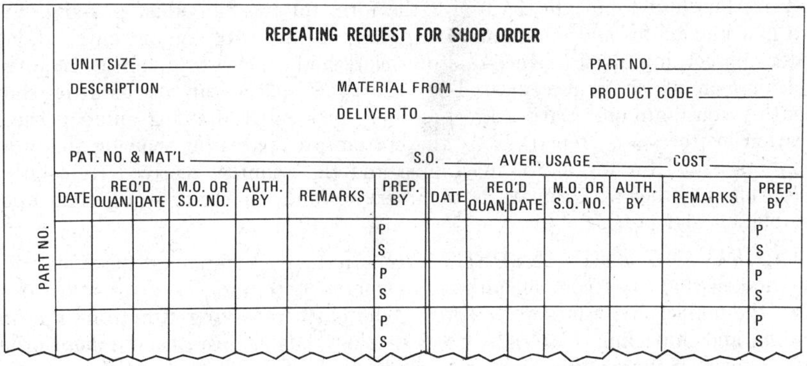
REPEATING REQUEST FOR SHOP ORDER

UNIT SIZE ______ PART NO. ______

DESCRIPTION MATERIAL FROM ______ PRODUCT CODE ______

DELIVER TO ______

PAT. NO. & MAT'L ______ S.O. ______ AVER. USAGE ______ COST ______

DATE	REQ'D QUAN.	REQ'D DATE	M.O. OR S.O. NO.	AUTH. BY	REMARKS	PREP. BY	DATE	REQ'D QUAN.	REQ'D DATE	M.O. OR S.O. NO.	AUTH. BY	REMARKS	PREP. BY
						P S							P S
						P S							P S
						P S							P S

Fig. 22. Repeating request for shop order.

shop order has only to date the card, indicate the quantity required, indicate the date required, assign a shop order number, and initial the card. The **shop order preparation section** can then pull the appropriate repeating master and run off the required shop order and material requisition.

A combined purchase or production order requisition and low stock notice used to request the replenishment of either stores or manufactured component parts carried in stock is shown in Fig. 23. It is issued to the purchasing depart-

ORDER REQUISITION											
AND											
LOW STOCK NOTICE							**N<u>o</u>**				

ISSUED TO   DATE   CHARGE SYMBOL

NAME OF PART

                 DATE WANTED

FORGE	CANCEL	BASIS									
MACHINE	MANUF'URE										
ASSEMBLE	PURCHASE										
OLD MAX.	OLD MIN.	AMOUNT ON HAND	LAST ORDER NUMBER		BAL. CLERK	SUPERVISOR		QUANTITY		UNIT	
NEW MAX.	NEW MIN.										
1	2	3	4	5	6	7	8	9	10	11	12

SYMBOL   DRAWING NO.   SHEET   LOCATION IN STORES

REMARKS

FORGING ORDER NO.	QUANTITY		
MACHINE ORDER NO.	QUANTITY		
		BALANCE CLERK	SUPERVISOR

Fig. 23. Order requisition and low stock notice.

ment or to the particular shop in the plant which makes the part in question. Checks beside the notations "Forge, Machine, Assemble" and "Cancel, Manufacture, Purchase" indicate the use of the form for the individual purpose. Old and new maximums and minimums are given, together with amount on hand, last order number, initials of balance-of-stores clerk and supervisor, quantity required, and unit in which item comes. The numbered spaces can be used to show monthly consumption. Entries are provided for symbol, drawing number, sheet, location in stores, and remarks. At the bottom are spaces for showing the order numbers issued for forging or machining and the quantity received from either operation, together with signatures of supervisor and balance-of-stores clerk upon completion and posting.

RECEIVING AND INSPECTION. In many companies receiving is a function of materials control, although sometimes materials control takes over after the materials are in and accepted. The **basic receiving functions** are unloading and unpacking materials, checking materials against the purchase order and invoice, having any necessary inspection done, making out receiving and inspection reports, putting materials into standard containers (often in standard lots), and delivering the materials to the storeroom. One copy of the receiving and inspection report is kept in the receiving department and others go to the purchasing department and the storeroom, the latter copy finally to the stores record section. Fig. 24 is a typical receiving report.

In some cases, copies of purchase orders are used as receiving reports, particularly if the number of copies necessary for the internal procedure is limited. If the purchase order is for one item, only one or two copies of the purchase order are furnished to the receiving department, which files and holds them until the material is received. If there are several items on the purchase order, or if the quantities are large enough to be sent in several shipments, several additional copies of the receiving reports are sent to receiving, and one is used for each receipt. In cases where a ditto master is ordered to prepare the purchase order, the master itself may be sent to the receiving department and copies run off as required. Usually the quantity column is blocked out in the reproduction process so that the receiving personnel have to count the exact quantity and record their count in the receiving report.

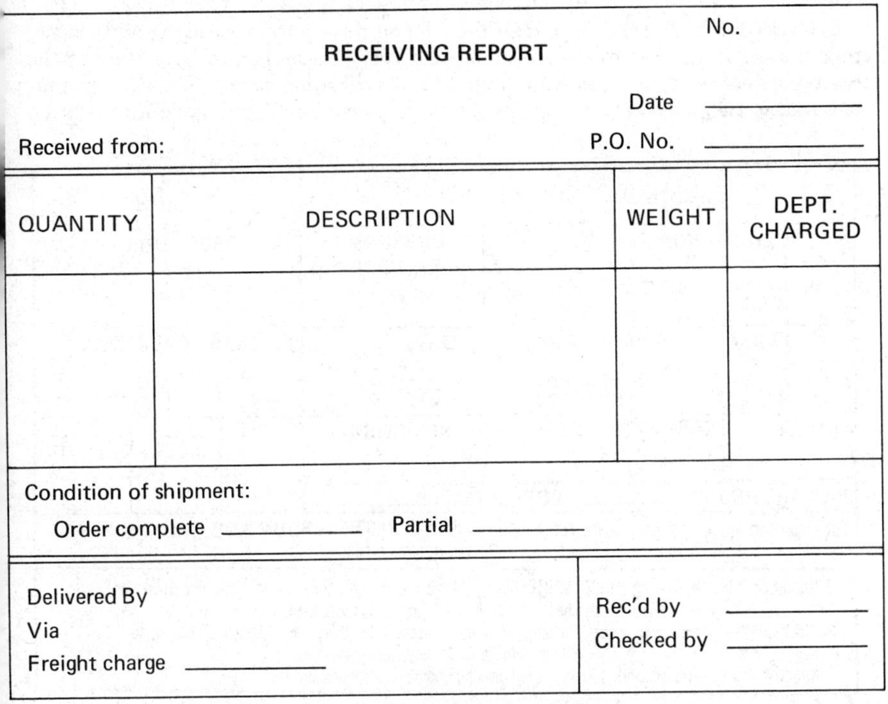

	RECEIVING REPORT		No.	
			Date _____	
Received from:			P.O. No. _____	
QUANTITY	DESCRIPTION		WEIGHT	DEPT. CHARGED
Condition of shipment: Order complete _____ Partial _____				
Delivered By Via Freight charge _____			Rec'd by _____ Checked by _____	

Fig. 24. Receiving report.

The number of copies of each receiving report to be made up and distributed depends on the organization, procedures, and internal control of the particular companies. Copies may be sent to the following departments:

1. **Purchasing Department.** To indicate receipt so that expediting is informed and to clear the files (see section on Purchasing).
2. **Materials Control Department.** For entry into stores ledgers.
3. **Accounting Department.** To confirm receipt so invoices can be paid.
4. **Receiving Department.** File copy.
5. **Storeroom.** To identify and accompany material.

In some cases one copy may be routed through several departments.

Adjustments. If the material received is rejected and is to be returned or if there is a shortage or an overshipment on the order, correspondence with the supplier about the discrepancy is usually conducted by the purchasing group. In certain industries a variation on the count of shipped versus received is fixed at a certain percentage of the total. Typically 5 percent but as high as 10 percent is sometimes allowed. The purchaser pays the amount indicated. Theoretically the shortages and overages average out over time. On material orders where spoilage may occur, it is customary to order more than is required to cover the shrinkage.

Defective material to be returned should be accompanied by a copy of the inspection report showing in detail the nature of the defect. A replacement order (if required) should also be instituted.

Returns for defects and requests for adjustment should be made only with authorization. A history of discrepancies and deviations from regular procedure should be noted and kept in the vendor rating file.

DELIVERY TO STOREROOM. Upon release from receiving and inspection, materials are sent to the storeroom (or user), accompanied by a copy of the receiving report or sometimes by a special identification form. For parts manufactured in the plant and sent to the storeroom, a **finished goods report** is often

FINISHED GOODS REPORT				No. 18231		
T O	STOREROOM		STOREROOM AREA CODE	SHOP ORDER NO.		
F R O M	AREA	MADE BY	DATE	INVENTORY ACCT. NO.		
		REC'D & CHKD. BY	DATE			
QUANTITY	PART NO. OR SIZE		DESCRIPTION	UNIT	UNIT COST	TOTAL COST
PARTIAL ORDER _____ COMPLETE _____						
PREPARED IN SHOP BY	DATE		REC'D IN STOREROOM AND COUNTED BY	DATE		
INSTRUCTIONS ON USE OF FORM: WHITE COPY - Mailed to Central Dispatch to close out Shop Order. Matched with Pink, then mailed to Cost Dept. PINK COPY - Travels with pieces. Mailed to Central Dispatch by Storeroom. Matched with white Posted to D. & S. Card, then filed. YELLOW COPY - Travels with pieces. Used in Storeroom as Identification Tag.						

Fig. 25. Finished goods report.

made out by the dispatcher or foreman. A typical form is shown in Fig. 25. This is often made out in several copies which may be used as follows:

1. Travels with the pieces to identify them and is used as an identification tag in the storeroom.
2. Travels with the pieces and is then sent to the materials control section for posting receipt of materials after they are checked in and counted by the storeskeeper.
3. Sent directly to the production control section to inform them to close out shop orders. Then sent to the cost department for closing out their records.

When items to be stored are received from the receiving department or from the manufacturing departments, the kind, condition, and count should be checked with the accompanying receiving report, finished goods ticket, or materials credit ticket. Discrepancies should be noted and reported at once for adjustment. The storeskeeper is responsible for what he receives and, in fact, must usually sign or initial such forms to acknowledge his acceptance and responsibility. When he has accounted for the delivery and posted his records, he will send the form, so initialed, to the stores record section for entry in the materials control records.

ISSUING FROM STOREROOM. Materials should be issued from stores only upon the presentation of duly authorized requisitions. These **requisitions for production materials** may originate and the materials may be delivered in the following ways:

1. Where no central production control system exists, foremen or department heads make out requisitions as materials are needed for jobs.

2. Under regular production control, the materials requisitions are written in the materials control or planning section, usually sent to the stores record section for apportionment or reservation on the records against the job order, are delivered with other papers to central dispatching, and are sent from there to the manufacturing department where the materials will be needed. Here the local dispatcher, or the foreman, if there is no dispatcher, will send them to the storeroom in advance of setting up the job and will thus have them delivered in time to start the work.

3. For continuous production (chemicals, cement, etc.), a production schedule often is made out and materials are forwarded to the using departments according to daily needs in accordance with this schedule. Changes in amounts forwarded will be authorized by changed production schedules. These schedules may be issued monthly or as authorizations to run as specified until further notice.

4. For continuous production, also, materials may be stored in the production areas and placed under control of the manufacturing departments, which make periodic checks on quantities and originate requests or requisitions for replenishment at predetermined reorder points.

5. For mass production of repetitive lots, as in automobile assembly plants, materials sufficient for each day's run may be sent in from parts plants or outside vendors, under planned control from the central order control division. They are delivered directly to subassembly stations or to the main assembly lines. Some materials, usually smaller items which are not economically handled in small lots, may be stored in the assembly plants in quantities for a month's (or more) regular production and withdrawn to supply the assembly lines as required.

6. For special or irregular use, foremen, department heads, or executives may authorize the issuing of materials or parts which may be needed to replace those damaged in production or found unsuitable, or to compensate for shortages on a previous requisition.

Supplies and materials for any nonproductive purposes in the plant, office, construction, or service departments, and parts and supplies used in maintenance work, are withdrawn with **special materials requisitions** authorized by the heads of any of these units. Often the maintenance department will have its own storeroom and control its own materials because of the recurrent nature of its work and the need for parts and supplies at hours when the regular storerooms are closed.

In all well-run plants, it is a fixed rule that storerooms must be kept locked, outsiders must not be admitted, and no materials must be given out except upon duly authorized requisitions.

Stores Requisitions. The important information on any materials or stores requisition includes the following:

1. Description of item wanted.
2. Quantity wanted, and unit of issue (piece, foot, pound, etc.).
3. Point to which material should be delivered (department, location, etc.).
4. Date wanted.
5. Charge or order number.
6. Signature of person making out or authorizing requisition.
7. Date of signing.

8. Initials or signature of person receiving material.
9. Initials or signature of storeskeeper.

Simpler forms carry the above data, but often additional entries are needed. Figure 26, for example, has space also for stores location, new balance on the bin

MATERIAL ISSUE					CHARGE OR ORDER NO.	
CREDIT OR TRANSFER					CREDIT	
DELIVER TO	SHOP SYMBOL	NAME SHOP	LOCATION		DATE WRITTEN	
DESCRIPTION	(Specify UNIT WANTED as lbs., pieces, coils, barrels, each, etc.)				DATE WANTED	
					UNIT WANTED	
					QUANTITY WANTED	
					APPROVED BY	

NEW BAL. BIN TAG	ISSUED			PRICE	STORES VALUE	HANDLING	TOTAL COST	EXCESS
	QUANTITY	UNIT	BOOK					
			SPEC.					
APPORTIONED	STOREKEEPER			BALANCE CLERK		COST CLERK	MATERIAL REC'D BY	

Fig. 26. Material issue, credit, or transfer slip.

tag, amount issued, cost data, initials of stores record clerk who may apportion material in advance of issue, initials of balance or stores record clerk who posts the actual issue, and initials of the cost clerk who enters the issue on the cost records for the job. It is the usual practice to put the cost data on these slips as they are posted to the stores record (if they are shown on these records), to save unnecessarily repeated work for the cost clerk in looking up materials costs. The space for apportionment provided on the stores issue form eliminates the need for a special reservation slip to assign and hold quantities sufficient for the orders covered.

Forms like these for issuing single kinds of material are used most frequently when the items come under different cost-charge classifications, when the items are in different storerooms, when they are issued at different times, and when job order methods of costing are used.

Group Stores Issues. When a number of items for the same production order are to go to the same manufacturing department at the same time and from the same storeroom, considerable clerical work and stores and transportation labor are saved by the use of a **group requisition or materials issue form** on which all such items are listed, the remainder of the data being the same as with single issues. Stores record and cost record work on these forms are handled in the same way as for single issues.

When the bill of materials required for a complete order can be conveniently arranged to serve as a group stores issue, as in the case of an assembly order, a form similar to Fig. 27 may be employed. This form is prepared as a master from which duplicates can be made, the shaded areas showing up white on the print. Ink or pencil data can then be inserted to cover any particular order, giving quantity, date, lot number, a check column to show delivery, balance in stock of each part, amount required and delivered, cost, unit on which cost is based, and total cost amount for each item. Total cost may be added up and brought forward from sheet to sheet and finally summarized. Special notations may be made under remarks. Authorization and filling, with dates, are indicated by names or initials of the requisitioner and storeskeeper.

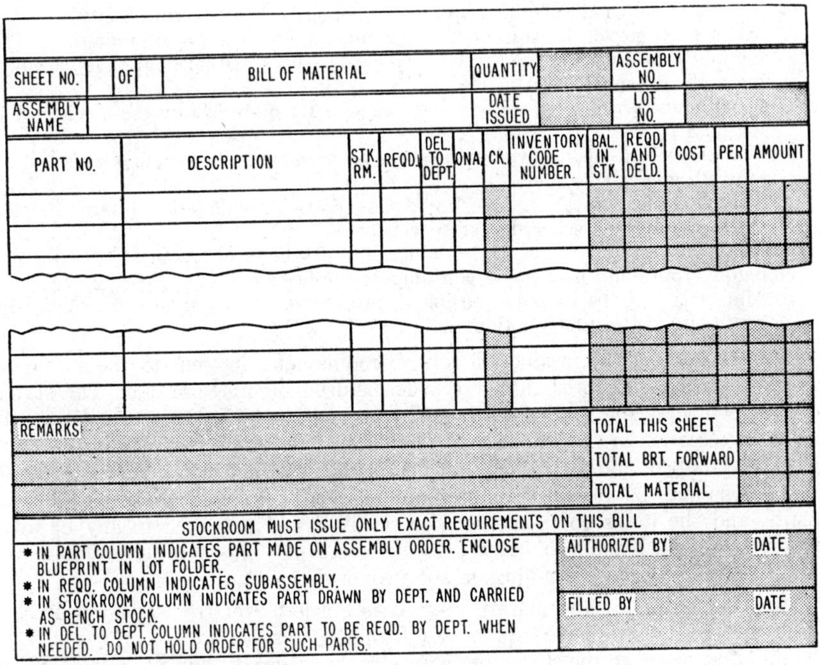

SHEET NO.	OF	BILL OF MATERIAL	QUANTITY		ASSEMBLY NO.	
ASSEMBLY NAME			DATE ISSUED		LOT NO.	
PART NO.	DESCRIPTION	STK. RM.	REQD.	DEL. TO DEPT.	ONA	CK.	INVENTORY CODE NUMBER	BAL. IN STK.	REQD. AND DELD.	COST	PER	AMOUNT

REMARKS	TOTAL THIS SHEET	
	TOTAL BRT. FORWARD	
	TOTAL MATERIAL	

STOCKROOM MUST ISSUE ONLY EXACT REQUIREMENTS ON THIS BILL

* IN PART COLUMN INDICATES PART MADE ON ASSEMBLY ORDER. ENCLOSE BLUEPRINT IN LOT FOLDER.
* IN REQD. COLUMN INDICATES SUBASSEMBLY.
* IN STOCKROOM COLUMN INDICATES PART DRAWN BY DEPT. AND CARRIED AS BENCH STOCK.
* IN DEL. TO DEPT. COLUMN INDICATES PART TO BE REQD. BY DEPT. WHEN NEEDED. DO NOT HOLD ORDER FOR SUCH PARTS.

AUTHORIZED BY   DATE

FILLED BY   DATE

Fig. 27. Assembly bill of materials, master for group stores issue requisitions.

Some companies **bundle** parts, that is, they tie together a fixed number of parts to utilize space better, to facilitate counting of, and to eliminate the posting of detail requirements and issues for individual releases. **Bundling quantities** are determined from the best available information to represent roughly a month's usage, considering weight, size, or other factors. The bundling quantity should usually be in multiples of 10 or more. Bundling quantities are reported in whole bundles only and the remaining pieces are disregarded in the posting to quantity records. The parts are considered issued from the quantity records at the time that the tag is broken and forwarded to the material control for posting as an issue. Individual issues are still made by the storeroom, based on the quantity shown on the bills of material, parts list, or individual requisitions.

Material Transfer. In cases where materials issued for one order are to be switched to another order, a material transfer (Fig. 26) is used to make the charge to the new order and credit the old. Since a new stores requisition slip will be needed to get other materials for the old order if it is still to be put through, the issuing and cost accounting for the new transactions will proceed automatically under regular procedures. The material transfer serves merely to switch the charges on the material already issued so that the cost accounting records may be correct.

Material Deliveries. The required materials asked for on a requisition will be taken from the bins or shelves for delivery to the requisitioner. There are various ways for **handling stores or materials requisitions** and moving stores to the required point:

1. Workmen in small plants or where no organized production control system exists may go to the storeroom with the requisitions to get materials. This method is also usually used in emergencies, to get a special item or replacement materials for spoiled work.
2. Move men carrying requisitions may be sent for materials by the dispatcher or foreman.
3. The storeroom may have its own move men who deliver materials to the requisitioning department.
4. The plant transportation department may make the deliveries through its shop express system on scheduled or special trips.
5. Conveyors may connect the storeroom and the user departments where a regular route and need for a continuous supply exist.
6. Materials may be stored in using departments, as stated above, when large quantities are regularly required.

In cases 3 and 4 the production control requisitions are sent to the storeroom by the dispatcher or foreman a day or so before deliveries are due. The storeskeeper then sorts and groups the requisitions and has his assistants make up the orders in a section of the storeroom. In this way all orders from one department may be grouped and deliveries can be made economically. For convenience, requisitions may be filed in the storeroom by due dates and according to departments, and the items on requisitions to be filled and delivered in any one day may be assembled and held until time for removal. This **requisition date file** usually insures prompt handling on the proper dates.

Stores Credits. If any materials or parts are left over from any job because the order was cut down or quantities supplied were in excess of needs, they should be returned to the storeroom with a **material credit slip** (Fig. 26). When it is not desired to place the items in stores, they may be sent to the salvage planner, who determines their disposition and enters the notation and his signature on the form.

Entry in Cost and Accounting Records. Materials issue and stores credit slips are sorted in the stores record section by classes or groups in the manner in which the accounts are kept, and posted and priced if the price file is maintained in the stores records. The slips are then sent to the cost accounting department for financial control purposes. Receiving reports are similarly classified and charged to the accounts. Discrepancies between stores records and stores counts must likewise be posted as adjustments to obtain agreement.

The stores requisitions are sorted by classes of products or by job numbers and overhead expense accounts, in accordance with the way costs are set up. They are then entered on the cost records, together with direct labor costs from time tickets. With the addition of proper overhead costs, those totals show the **manufacturing costs of products or specific jobs.** They are used for checking actual costs against estimates or previous records and for studying new work so that plans for holding costs within limits, or reducing them, can be applied.

Under a standard cost plan, of course, procedures are simplified by the use of the predetermined cost of each item.

Miscellaneous Stores. Under the designation of miscellaneous stores are the following kinds of items:

1. Parts or materials temporarily stored for current work.
2. Parts ordered for special jobs and not used because they are surplus or are replaced by other items.
3. Materials ordered for special jobs but not used.
4. Repair or replacement parts on old models or obsolete models of the company's products for which orders are only infrequently received.
5. Items or materials no longer used on current products and for which there is no demand.
6. Parts from miscellaneous dismantled products or equipment which may be considered usable for certain purposes around the plant.
7. Maintenance parts and items little used or superseded.

Such items are usually kept in separate bins in the regular storage locations to prevent them from becoming mixed with or issued in place of the items regularly carried, to make room for regular items, and to simplify stores issuing. Because they may be used from time to time, however, they are kept in the storeroom and entered on the stores records so that they are known to exist and will be available if wanted. Since they are not regularly issued and stocks are not replenished, it is necessary only to sort them into groups, put them in a suitable place, and record them on simple records of standard size but specially printed and filed in a separate section of the regular stores record system. Bin tags or identification tickets should be placed with them in the storeroom.

Physical Inventories

NEED FOR PHYSICAL COUNT. As explained above, the basic record in a materials control system is the stores record or perpetual inventory record of raw materials and parts. Receipts and disbursements are entered on these records which show the quantity of each part or item of raw material in the storeroom or in the plant ready for use. It is very important that these records show the exact physical quantity of raw materials and parts which are available for use, or production may be delayed with consequent loss of labor efficiency and customer good will.

Occasionally, however, clerical errors, erroneous descriptions, wrong counts or weights, and perhaps omission of materials issues result in inaccurate records. It is therefore necessary to verify the perpetual inventory and accounting records by a physical inventory.

Accuracy of Count. Taking physical inventory is not a casual operation. The need for accuracy is as important here as in other production operations. The results of the count are compared with stock records and ultimately translated into dollar figures. Discrepancies in the two figures—actual count and stock record—have far-reaching effects.

Emma (Journal of Industrial Engineering, vol. 17) stresses the need for an accuracy level consistent with the uses to which the inventory information is to be put:

The requirement for accuracy in the correlation between the stock records and physical stock is derived from the special needs of three separate levels in the supply system: requisitioner, purchaser, and administrator. The need for accuracy at each of the organization levels is not equal. A requisitioner may experience nearly equal amounts of frustration for each requisition not honored because of an error of even a few units in a stock record. At the buying level, where strategic actions are determined, an error of a few units could be tolerated in a certain proportion of the records without serious overall operating problems, and without committing an unacceptable percentage of valuable stock funds inefficiently in respect to actual need. On the administrative level, where the total asset investment must be evaluated, an even greater percentage of error in the stock records could be tolerated, since the purposes to which such information is put would not normally be sensitive to small discrepancies.

The physical inventory not only provides the basic information for making this correlation, but also serves as a corrective measure. It is therefore important that the seriousness of accurate counts be emphasized to those personnel involved in the operation.

METHODS OF TAKING INVENTORY. There are several methods of taking a physical inventory:

  1. A complete or year-end count.
  2. Rotating counts of material.
  3. Out-of-stock counts of material.

Complete or Once-a-Year Count. The most common methods of checking inventory records and accounting records is to make a complete or once-a-year count of all materials, parts, work in process, and finished products in the plant. By this method each item in stock, and usually those on the production floor as well, are counted during the same short period of time. If the inventory is taken in this manner, it should occur at or near the end of the fiscal year for accounting purposes.

Rotating Physical Counts. Regular stores personnel may make physical counts of the items in stock during lulls in storeroom activity or whenever specific requests for a check are forwarded from the stores record section. This is often a planned periodic program in which each item in the storeroom is physically counted at least once a year, and the more important items are often counted semiannually, quarterly, monthly, or sometimes oftener.

This rotating count is sometimes made by counters who are attached to the staff of the controller and who do nothing else except take counts on a planned basis.

Where this method is used, a **stores count report** is sent by the storeskeeper or counter to the stores record section where any corrections necessary are made in the stores record or perpetual inventory cards, after which the reports with the notation of discrepancies are sent to the accounting department for periodic adjustment in the materials accounts.

Where this method is used, the plan of physical count should be related to the A–B–C value classification of inventory so that the more expensive and important items under closer control are counted more often than those of low value.

Out-of-Stock Counts. The quantity is reported under this method whenever the supply of a particular part is exhausted, when only a small quantity remains in the bin, or when the "last bag" is broken in the case of the "sealed minimum" or "last bag" method out-of-stock counts are made and reported. The storeskeeper notifies the stores record or perpetual inventory department of the count, and corrections are made in the records.

The advantages of rotating and out-of-stock count methods are as follows:

  1. The plant does not have to be shut down.
  2. The count is not made under pressure; hence it may be more accurate.
  3. Records are kept more nearly up to date when subject to a continuous check.
  4. Errors and irregularities are discovered and adjusted more quickly.

One of the difficulties of this type of physical check is that both the stores record section and the accounting department must have an accurate cut-off date for each transaction in order to coordinate the count with the exact status of receipts of material and issues of material. Thus, the **time of the count** in relation to receipts and issues immediately preceding and following it must be accurately specified.

PROGRAM FOR PHYSICAL INVENTORY. The taking of a physical inventory requires that certain arrangements and instructions be prepared in advance. Cashin and Owens (Auditing) point out that advance planning will greatly aid in producing an orderly and accurate inventory. One of the most important jobs in this regard is the preparation of inventory instructions. If the company takes inventories at frequent intervals, there may be **permanent instructions** covering matters of count, arrangement of inventory, and other points that do not change, and **temporary instructions** to include variable points such as the date, time, etc. Some of the points to be included in the **inventory instructions** are:

1. **Shutdown of Operations.** As far as practicable the operations should facilitate the count. Conversely the count should not hinder operations. For example, work in process should be completed as far as possible.
2. **Arrangement of Materials.** Materials should be arranged to facilitate the count.
3. **Obsolete, Damaged, Scrap and Waste Materials.** These should be disposed of, if possible, before the inventory.
4. **Personnel.** The person writing descriptions should have a knowledge of the stock and the descriptions used by the purchasing and accounting departments. The count should be made by one employee and, if possible, independently verified by another employee.
5. **Assuring Count.** A means of assuring that all goods are counted should be used. Any goods received or shipped during the inventory should be carefully verified.

The date of a complete physical inventory should be as close to the end of a month as possible to facilitate comparison with the financial records. It should also be as close to the end of the fiscal year as possible so as to serve for verification of the inventory on the financial statement. Physical inventory should be made when there is the least possible interference with operations, but sufficient time must be taken to complete the inventory accurately or the whole purpose is lost. It is, of course, desirable to take inventory at a time when stocks in the plant are at a low level if this is possible.

Physical inventories in many plants take longer than one day but should be completed as rapidly as possible to shorten the shut-down of the plant and to minimize the movement of materials if some of the count has to be made when the plant is not idle.

Organization and Personnel. Since the year-end physical inventory is primarily a financial check, it is usually under the general supervision of a financial executive, often the controller. The actual physical direction of the counting is usually a responsibility of the factory or production manager, and detailed supervision of the counters is usually by the various factory department heads. A good technique to insure responsible coverage of all areas is to make up a map of every area to be inventoried showing the supervisors who are responsible for the inventory in each location. Counters should be familiar with the shop and the material and very accurate in their counts. It is wise to have people available who can identify material and accurately describe it on the inventory tickets.

Programs, Instructions, and Forms. The need of adequate preplanning and preparation for taking the annual physical inventory cannot be overemphasized. The larger the inventory the more important the preparation. Complete written instructions should be prepared in ample time for all individuals to become familiar with the plan.

Some large concerns have complete **inventory manuals** containing very explicit instructions on every phase of inventory-taking. Such manuals are advantageous in that procedures become standardized and are consistent from year to year except for modifications to suit current conditions. Thus, the inventory personnel become more expert in carrying out the procedure, fewer mistakes are made, and the inventory is completed more quickly.

Before the inventory starts, meetings should be held with all of the personnel concerned to go over the instructions in detail and to answer any questions. An **inventory schedule** should be prepared detailing areas to be inventoried and including a time-table for each of the departments to be covered. One of the primary functions of the inventory supervisory force during the actual physical inventory is to check continually to see that this schedule is being maintained. If any department is lagging other crews may be transferred in to help.

Preparation of Stock for Counts. It is very important that a number of preparations be made before the date on which the counting is to begin in order that the counting may be accurate and proceed according to plan. In addition to the points previously noted the following measures should be taken:

1. Restacking and sorting of material into piles which are easy to count.
2. Returning material to storerooms from machines and production lines.
3. Proper identification of all material.
4. A general cleanup of the plant so that accurate counts are possible.
5. Listing by the person responsible for subcontracting of all materials at the subcontractors.
6. All paperwork cleared and stores and accounting records brought up to date, including latest receipts and issues.

Forms To Be Used. The principal forms to be used during the physical inventory should be described in writing and detailed written instructions issued as to how to handle and fill them out. The most important forms used are **inventory tags** and **inventory sheets.**

The tag method is considered the best way of inventorying materials since it provides a visible check of material which has been inventoried and is a convenient way to work to summary sheets and company records. The form shown as Fig. 28 is suitable for all classes of materials. Spaces are provided for name, part number, descriptions, operations performed (if the item is work in process), unit quantity, location, and counter's initials. The tag is divided into two sections by perforations. The reverse side provides space for recording movement of materials after count is made, which, in the case of slow-moving parts, is frequently done several days before actual date of inventory.

One person should be responsible for assigning **blocks of tags** to the supervisors of the physical inventory in the individual areas. He should keep an accurate record on a **summary sheet**, as shown in Fig. 29, of who has received the tags, where they were issued, and how they are being returned. When the tags are returned they should be checked in on the summary sheet and any missing tags posted to a sheet such as Fig. 30 so that a search can be made for the missing tags.

Those actually counting should work in teams of two or more, one person to write tags and place them upon the materials, and the others to identify and to

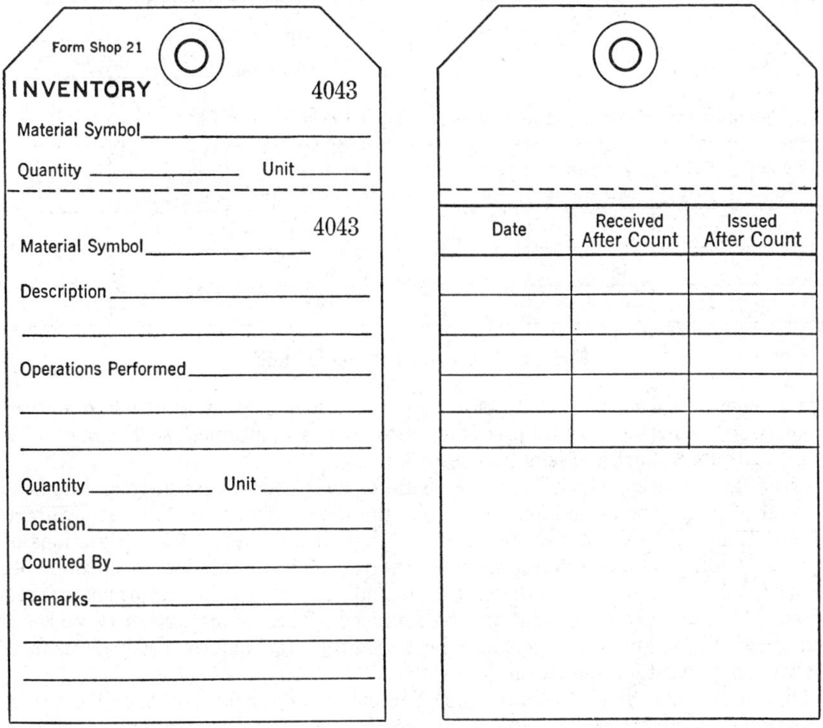

Form Shop 21 ○	
INVENTORY 4043	
Material Symbol_____	
Quantity _____ Unit_____	

4043	
Material Symbol_____	
Description_____	
Operations Performed_____	
Quantity_____ Unit_____	
Location_____	
Counted By_____	
Remarks_____	

| ○ |
| Date | Received After Count | Issued After Count |

Fig. 28. Inventory tag.

Date _____

Supervisor _____

Issued	Group & Location	Issued			Ret'd Used			Ret'd Unused (Red)		
		From (Incl)	By	To (Incl)	From (Incl)	By	To (Incl)	From (Incl)	By	To (Incl)

Fig. 29. Issues and receipts of inventory tags.

count or measure. When a team has completed its work, each departmental supervisor should check each kind or lot of material in his department or storeroom to see that it is covered by an inventory tag completely filled out. Tags for work-in-process materials should clearly indicate the stage of completion or progression of that material.

In inventories of precious metals, or when incomplete stores records are maintained, a tag containing three sections is sometimes used, and a double count is made of all items. The two lower sections of the tags are filled in by separate

Date_____

Supervisor _____

When corrected, cross out and initial in Red and post to Form "A."

Ticket No.	Entry By	Storeroom and Group	Issued To	Remarks (To aid in checking)	Block of Tickets Returned	Checked

Fig. 30. Record of missing tags.

sets of counters and reconciled before the departments are permitted to resume production. Where perpetual inventory records are maintained, such a method is a duplication of effort and is not recommended.

After these checks, which should be made as nearly simultaneously and as near the end of inventory period as practicable, the lower sections of the tags are torn off and returned to the person responsible for their issuance. The top sections of the tags remain with the materials to facilitate check for lost tags and to enable recounts to be made when necessary. Spoiled tags should be returned marked "Void." All unused tags should also be returned. The issuing authority makes a numerical check and all tags must be accounted for before a department is authorized to resume production.

The process described above can be done using prepunched, interpreted cards. A section of the card can be used to record the actual count in pencil. The data is later keypunched directly into the same card. Mark-sense cards can also be used; the mark-sensed quantity is punched into the card or "read" by special equipment. Using prepunched or mark-sense cards can speed up the physical inventory considerably by having the arithmetic done automatically by the data processing equipment. Other **machine readable** tags are available where most if not all of the data can be read by the data processing equipment. The summarization and cross checking of the punched cards can be done by machines, speeding up that process also.

Establishment of Controls. After the tags have all been checked in and missing tags accounted for they are sorted in different ways for various purposes and in general are used as follows:

1. To check stores location records so that locations of all items in the storeroom are known. If two or more locations are shown in the storeroom area the stores-keeper should investigate and try to consolidate these amounts unless the storeroom is run by having the reserve stock segregated from an active location.
2. To check in the stores records or perpetual inventory records the quantities shown as balance in stock. The book records should be adjusted to the actual count in such a way that the nature and date of the adjustments are clearly shown. It may be desirable to set up limits to this automatic adjustment of the book records.

A record should be kept to tabulate all adjustment to the perpetual records and to arrive at a total difference or adjustment from the book record to the physical. A form for this purpose is shown in Fig. 31.

In cases where unit prices are not kept on the inventory records, the price and extension are filled in by the accounting department.

Department: Assembly   Date: December 31, 19xx

Stock No.	Item	Unit	Quantity-Units		Variance (AC-CR)	Unit Price	Amount	
			Control Records	Actual Count			Inc.	Decr.
11641	Manifold	Each	62	42	− 20	$3.50		$70.00
11694	Hinges	Pair	210	225	+15	1.00	$15.00	
11752	Handles	Pair	18	2	−16	2.00		32.00
11840	Knives	Doz.	61	55	− 6	3.00		18.00
11841	Knives	Doz.	48	54	+ 6	2.00	12.00	

Fig. 31. Listing of inventory adjustments.

The inventory tags are then listed on **inventory sheets** which may be compiled according to classes of material so that a group of sheets will correspond with the items included in one accounting control account. These sheets are sent to the accounting department for pricing and extension and for adjustment of material control accounts. Work-in-process tags are reconciled to work-in-process accounts and records.

Storeskeeping

OBJECTIVES. Storeskeeping is primarily a service function in which the storeskeeper acts as a custodian of all items carried in the store. It should aim at providing this service as efficiently and as effectively as possible with a minimum of cost. Directly related to these costs are the physical characteristics of the storage space and the type and quantity of material to be stored.

Failure to locate storerooms properly and the inefficient use of available storeroom space can be most costly. These failures may require excess handling of materials, greater handling personnel and facilities, difficulty in locating materials, and overstocking and its attendant costs. Problems in materials control and time involved in supplying materials when needed are frequently the results of improper storeroom location and use.

STORESKEEPING FUNCTIONS. The detailed functions of storeskeeping are:

1. Receipt of materials into storage.
2. Record keeping of materials in storage.
3. Storage of materials.
4. Maintaining stores.
5. Issuing stores.
6. Coordinating storeskeeping with related production functions.

Raw materials, supplies, and purchased parts are usually received at a designated **receiving dock,** unloaded, inspected, and then moved into storeskeeping. The storeskeeper identifies the material in accordance with the stores classification, dispatches it to the appropriate storage area and position, and records the receipt of materials into storage.

In general, semiprocessed parts in transit between operations are the responsibility of manufacturing or material handling. However, semiprocessed materials may be placed in storeskeeping when the parts are very valuable and subject to pilferage, very fragile, or in need of special protection or aging.

The maintenance of **storeskeeping records** in an efficient and orderly manner is of the utmost importance in order to locate incoming and outgoing materials quickly and accurately, to provide the necessary information concerning the exact whereabouts of the materials, and to supply the materials control and cost-keeping departments with correct and timely data.

The proper storage of materials involves holding the materials for safekeeping and protection until they are required elsewhere by **authorized requisitions.** Therefore, it is necessary to have a sufficient area for the materials and suitable storage devices such as boxes, bins, and racks. The storage facilities should be selected so as to provide the maximum of protection and accessibility and to utilize a minimum amount of space.

Materials in storage may require not only protection but maintenance as well. To keep the stores in the desired condition over a period of time may require very simple or very elaborate measures, depending on the nature of the material, the length of time in storage, and the rates of deterioration. Proper **stores maintenance measures** may range from special covering or periodic lubrication to controlled atmospheric conditions.

Another primary function of storeskeeping is issuing of stores. When this function is performed efficiently, the materials will be delivered quickly with a minimum of delay, and records will be correctly adjusted immediately. The physical and clerical facilities for storeskeeping must be carefully preplanned if the issuance of stores is to be prompt, positive, and accurate.

Coordination of storeskeeping with materials control is an essential aspect of a properly functioning storeskeeping department. Coordination must be planned or designed into a good storeskeeping system. Duly authorized storage requisitions must originate at the proper source and records of storage changes must be maintained for use by the materials and cost control centers. The responsibility for such coordination may rest only in part with the storeskeeping center, depending on the type of production in the plant, the size of the company, and the organizational makeup of the company.

PERSONNEL. The number of persons required to operate a storeskeeping system varies widely, depending on the size of the company and the plan of the overall company organization. Typical storeroom personnel would consist of a **chief storeskeeper, assistant storeskeeper,** and one or more **stores clerks** if stores records are kept in the storeroom. A material handling crew might be included if the storeroom operates its own delivery system to manufacturing departments, moves incoming materials from the receiving areas to the storeroom, or contains such integrated systems as stacker cranes and automatic order pickers. However, the **stores records** are often kept by a clerk in the materials control section of the production control department, in which case the storeroom clerk would merely enter receipts, issues, and balances on the bin tags. **Transportation** also is often handled by a shop express system under a dispatcher or material handling manager, or by truckers in manufacturing departments who call at the storeroom for materials.

Planning for Storeskeeping

SPACE PLANNING. The storeskeeping department does not determine such factors as quantities of materials, order periods, rates of use and similar information but these data must be acquired and reviewed in order to plan for storeskeeping. In the case of an existing storeskeeping department with fixed storage capacity, the present limitations of space and facilities will influence the planning of the materials control department. When new factories are planned, however, much careful preplanning will have to be done for storeskeeping.

Planning for storeskeeping involves gathering all the available data relative to the items to be stored, the space requirements, the storeroom equipment and facilities, and the probable flow paths into, through, and out of, stores. Where exact information is not available, careful estimates should be made. Allowances for variability should be included. Finally, alternative **storeroom layouts** should be planned and cost analyses made before deciding on the final storeskeeping plan.

The first stage of planning, as suggested by Sitler (Materials Handling Manual No. 1, Laughner, ed.), is to determine the overall **cubical space requirements for storeskeeping.** These requirements should be estimated on both a short-term and long-term basis in order to anticipate future as well as present requirements. It is necessary to determine the quantities of materials to be stored and the specifications which may affect storage, such as size, weight, fragility, susceptibility to pilferage, and other factors which may require special handling or storage treatment. Then data must be gathered covering the purchasing lead time, and the rates of delivery and rates of use in the factory of the materials to be serviced by storeskeeping. This information must be translated into space requirements. Changes in anticipated quantities of certain materials may or may not affect the total space required since there may be compensating changes in other material requirements. Thus, it is necessary to project the space requirements over a long period of time, such as five to twenty years, depending on the classification of industry for which a given type of factory is being designed.

Space Forecasts. Several analytical techniques are employed in the area of space forecasting for stores. Volpel (Industrial Engineering, vol. 1) discusses calculating the **bulk factor** (the space required to warehouse the parts that go into a product) pallets per set of parts for 100 machines. This is then converted into cubic feet of storage space required from the marketing department's sales forecast.

Another technique of forecasting space requirements is through **simulation.** A model of the process, in this case the storage or warehouse function, must be designed and programmed for simulation, usually on a computer. If the model and process are not very involved, it could be simulated manually or on a desk calculator. Usually, however, the number of calculations involved require the use of a computer. There are special programming languages for the purpose of simulating this type of situation. IBM's GPSS (General Purpose Systems Simulator) and RCA's Simscript are two examples. As an example, a highly complex simulation of a Sperry & Hutchinson Company warehouse operation in Hillside, Illinois, was developed prior to its being built (Transportation and Distribution Management, Vol. 9).

A third approach to forecasting discussed by Volpel is the use of network analysis, or network diagrams for planning future space and materials handling systems. Using the PERT approach, probability can be introduced into the analysis. The ease of comparing alternative systems and the ability to pinpoint potential bottlenecks are the main advantages of this approach.

Moore (Plant Layout and Design) recommends the use of the following techniques for determining both long-term and short-term space requirements:

1. Graphical methods
2. Conditional probability
3. Queuing theory

Detailed discussions on the principles of these approaches are given in the sections on Statistical Methods and Operations Research. Accurate historical records must be analyzed in conjunction with data from the latest forecasts to determine the limits within which space must be provided for storeskeeping. Fig. 32 is a typical example of such a forecast. The upper and lower limits may have a **non-uniform variation**. It is obvious that the further into the future such a projection is made, the smaller the chance of complete accuracy.

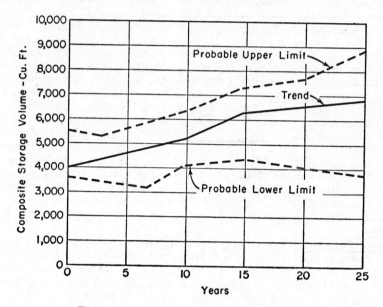

Fig. 32. Forecast of storage requirements.

Cost of Space. The cost of **excess storage space** derives from building and land costs, the interest on capital invested, and the possible additional handling costs due to extra travel distance. The costs due to **inadequate space** may be the cost of ordering more frequently, lack of large lot discounts, cost of extra warehouse rental, possible temporary shutdown or slowdown on operations, and customer dissatisfaction due to inability to supply product or spare parts as rapidly as desired.

The cubage required will then have to be increased to allow for storing different items independently and to allow for space which cannot be utilized because of shelving, aisle, and handling equipment considerations.

The proper layout for storage results from the determination of the most effective **space utilization**. Consequently, planning for storage and layout of storage space are interdependent activities and should either be done concurrently or separately, but with full understanding of considerations involved in each activity. Material handling equipment selection is also directly related to the planning and layout of storage and therefore should be considered (see sections on Material Handling and Plant Layout and Facilities Planning).

DETERMINATION OF AREA REQUIREMENTS. If an existing building is to be used for the manufacturing enterprise, the first phase of planning for storeskeeping may be largely predetermined. Both available space within the present building structure and its limitations are known.

If a new building is to be constructed, there is usually much more latitude in the available space for storeskeeping. The problem becomes an economic determination of the amount of space required within the framework of manufacturing needs, available capital, and potential site limitations.

A rough estimate of storeskeeping cubage requirements would first be made as described previously, followed by a more detailed breakdown of cubage space requirements, still in rough form.

It is necessary to classify the materials to be stored in a manner that allows for a judicious **grouping** or **combining** of different materials. This involves considerations such as separation of dissimilar types of materials, separate storage locations due to different points of usage, and separation of storage areas for different materials because of special handling, safety requirements, etc. Thus, grouping of materials in a rough manner at this phase of the layout usually results in a cubage requirement larger than the first rough estimate. The next percentage of cubage allowance may be estimated by means of rough calculations of an **allowance for unused space** caused by the type of shelving to be used. For example, the shelving itself takes up space, and different quantities of the same parts can be stored in one large container as against several shelves occupying the same area. Another allowance should be added for aisles and material handling equipment. It is emphasized that the allowances must be crude at this phase. The interdependence of these allowances can be seen from the fact that usually a different depth of shelving will result in a different percentage of space required for aisles. In general, the deeper the shelving, within practical limits, the less aisle space will be required. Volpel (Industrial Engineering, vol. 1) indicates that less than 50 percent of the available floor area is actually used for storage because of other warehouse functions.

LOCATION OF STORESKEEPING DEPARTMENT. The determination of the final location of the storeskeeping areas will depend on the needs of the plant and will vary greatly from plant to plant. Certain basic **storage location factors,** however, should be taken into account in order to arrive at the best decision:

1. **Materials classification.** It may be desirable to locate materials of the same classification together, regardless of differences in size, weight, and destination. For example, one area may be designated for supplies, raw materials, and purchased parts; another area may be designated for semiprocessed parts, and another for finished inventory.

2. **Similarity of materials.** Articles of similar size, shape, and weight may be stored together even though they may be classified differently or be required at different production locations. All small parts may be kept in one place, bar stock in another, and castings and bulky items in still another place.

3. **Point of use.** It is sometimes desirable to locate all materials to be used at each production center as close as possible to that particular center. This may involve storeroom **decentralization** or merely special grouping of materials within a centralized storeroom.

4. **Material handling considerations.** Materials which require similar handling equipment may be grouped together. For example, liquids which must be specially measured or moved by special handling equipment may be placed together. Items lending themselves more efficiently to truck handling or conveyorized handling may be placed together. Also, items which are to be handled more often because of the greater frequency of requisition may be placed at more accessible locations than infrequently handled items.

5. **Special requirements.** Certain materials have to be handled or stored in a special manner because of their physical characteristics. They may be fragile, explosive, extremely valuable, or may require special **atmospheric conditions.** Such materials will have to be stored in locations affording the proper facilities even though, in many cases, the facilities may not be ideally located for point of use delivery or for best material handling utilization.

Centralized Storerooms. Usually, a combination of the preceding storage location factors must be taken into consideration in order to achieve the best internal storeroom location. Also, there may be building limitations which require a storeroom location other than the optimum location because of these factors. This is especially true in the case of adapting an existing building to the needs of a new manufacturing enterprise. Not only is there the problem of deciding the best location of adjacent storage groupings, there is also the necessity of deciding whether all stores should be located in one centralized storeroom or in two or more decentralized areas. There are advantages and disadvantages either way. The **advantages** of centralized storing are:

1. Fewer persons are required in the storeroom.
2. The entire stores personnel becomes familiar with all materials stored. There is less trouble if someone quits or is absent.
3. Better control is afforded over inventory and storeroom records all in one place.
4. Less total inventory is usually carried because of lack of duplication.
5. There is more leveling off of load on personnel.
6. Less total space is occupied.
7. Less obsolete stores are carried and there is quicker discovery of practically duplicate items and of items declining in use.
8. Periodic physical inventory checks against balance-of-stores records are easier to take.
9. Clerical costs are lessened.

The **disadvantages** of centralized storing are:

1. The storeroom may be far away from some departments so that greater transportation is required to and from the storeroom by those departments. More material handling equipment and greater time and labor are involved for distant departments.
2. Centralized stores clerks are less familiar with the needs and peculiarities of each individual area being served.
3. Longer waiting lines tend to form with only one storage area as opposed to several areas.
4. It may not be economically feasible to duplicate sophisticated systems such as stacker cranes and automatic pickers.

In order to make the best decisions both as to internal grouping of storage areas and as to centralization or decentralization, it is necessary to attempt to total the factors involved and to express the alternative advantages and disadvantages in terms of costs. This can best be done by the use of sound engineering judgment coupled with the use of analytical techniques.

ALLOCATION OF STORAGE SPACE. After deciding on the degree of centralization of storage areas, blocking out subgroupings of materials, and crudely determining the relative locations of each area, there remains the detailed layout of storeroom space. The objective of good utilization of space are as follows:

1. Intensive utilization of space.
2. Optimum accessibility of materials.
3. Greatest base of control of materials.

4. Maximum flexibility of arrangement.

5. Complete protection of materials.

Designation of Unit Loads. It is highly desirable to arrive at a proper unit load designation for all materials to be stored. Actually, it would be most convenient if it could be arranged so that materials arrive packaged in the best unit load containers, packs, or pallet loads. Thus, they could be taken from receiving through stores and on to the manufacturing department in the same package or unit so as to eliminate unpackaging and repackaging, and where possible this should be done. Unit loads should be specified and arranged in **standard containers or pallets** as soon as possible after receipt of materials, where prepackaging in unit loads cannot be done. In this manner, the standard size pallet or container, rather than the small, individual article, becomes the unit in which the item is handled. The article itself may be the standard unit, where the article is large, such as bags of cement or large castings. The exact size of the unit load will depend on the size and shape of the individual item, the handling equipment to be used, and the rate at which the item is required at the production center.

Methods of Storing. Before detailing the dimensions of the storage area, both the unit loads and the methods of storage will have to be worked out. The materials may be piled, stacked, or placed in bins, racks, or shelving.

If the material is to be piled or stacked, it may be placed directly on the floor or on pallets or skids. The practice of piling on the floor either in cubical piling or in pyramid form should be discouraged. There are few cases where some sort of palletization or stacking on skids cannot be substituted with substantial savings. The time and cost of eliminating floor piling will be greatly offset by reduced handling costs, ease of counting, and more efficient utilization of space. **Stacking height** will depend on the floor-load capacity, the crushability of the materials, the floor-to-ceiling height and the speed and lifting capacity of the equipment to be used. Use of stacker cranes often increases the height available for storage. The height permitted should not interfere with the effective operation of the sprinkler or lighting systems.

Where large quantities of materials are to be stacked, block stacking is most effective. **Block stacking** is the stacking of pallets or containers in rows such that each row contacts the adjacent row. For small quantities, row stacking is preferable. **Row stacking** is the stacking of materials in rows with sufficient space between rows so that any row of pallet stacks can be withdrawn without interference. Row stacking is more flexible than block stacking but requires greater total aisle space. Also, it is important to consider the orientation of the rows or blocks of tiered materials relative to the aisles. If the long dimension of the rows of pallets is perpendicular to the cross aisle, the ratio of occupied area to total area is greater than an arrangement whereby the long dimension is parallel to the aisle as may be seen in Fig. 33.

A large proportion of the materials serviced by the storeskeeping department are best kept in **bins, shelves,** or **racks.** This is because so many items can be handled in and out of the storage area without the aid of mechanical handling devices, even though these same items will often be transported to the production center, sometimes together with other items, by means of handling equipment. The type of bin, shelf, or rack selected is important in determining the most efficient use of the storeroom. Bins, racks, and shelving vary in size, material, and construction, and it is advisable to standardize them, once the proper specifications have been determined. The dimensions of storage areas can then be proportioned in multiples of unit dimensions. For example, if unit dimensions are

3×5 ft., and another unit which might be stored in the same area is 4×7 ft., a storage area depth of 11 ft. would always have some unused space, but a depth of 12 ft. would provide efficient accommodations for both items.

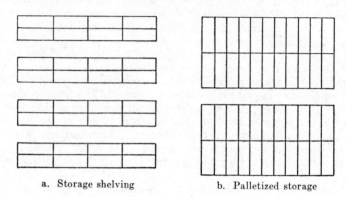

a. Storage shelving   b. Palletized storage

Fig. 33. Good layout of aisles in a storeroom.

Adjustable and portable bins and racks provide a most efficient storage medium where they can be used to advantage. Adjustable racks should combine structural rigidity and still lend themselves to rapid assembly and dismantling. The addition of **caster bases** to standard bins or tiering racks is most effective where they can be used for storage at the storerooms and then moved to the production center for direct **point-of-use storage.** Detailed examples of these common types of storage equipment will be given in the treatment of storeroom equipment which follows.

Moving Materials Into and Out of Stock. Consideration in the layout must be given to the system of moving materials into and out of stock. In many instances it is important that a first-in first-out system be used where deterioration may result from keeping materials in stock too long. There are several means of achieving **first-in first-out storage:**

1. **Coupon systems.** Two coupons are made out for each container. One coupon is attached to the container and the other coupon, with the storeroom location on it, is placed in a file. When material is requisitioned, the oldest coupon is referred to first and thus the corresponding material is removed first.

2. **Moving-division system.** Goods are removed from one end of a tiered row and added to the other end. In this manner, the pile will shift in position over a period of time. This can be allowed to go on up to a limit line and then the remaining section will have to be shifted.

3. **Gravity-feed systems.** Certain types of materials dispensed in bulk may be stored in gravity-feed storage bins. Materials in drums or circular packages can be moved by gravity first-in and first-out by constructing a sloping rack of the pitch required by the type of goods.

4. **Wheel or roller conveyor system.** A most satisfactory method of obtaining first-in first-out storage for a wide variety of materials packed individually or on platforms or pallets is obtained by storing the materials on rollers. Aisles are required at the feeding and pick-up ends of the lines.

Even where first-in first-out storage is not required, it is important to have a specified manner of adding to and withdrawing materials from rows of stored materials to achieve efficient utilization of space. If materials are removed and added at random to different rows of a section, empty spaces which cannot be utilized will occur in this section. Such **honeycombing** can be minimized by stacking in short rows and withdrawing or adding to only one row at a time.

Aisles. Aisle space is required to provide passageways for personnel and handling equipment to, from, and within the storage area. Aisle space should be planned to give adequate operating space for handling and stacking supplies and should provide the proper relationship to doors and fire protection equipment. The **turning radius** of fork trucks and other handling gear must be checked on the layout before starting work on the area.

Main aisles should run the entire length of the storeroom. These aisles should be wide enough to permit easy passage of material handling equipment going in opposite directions. Widths of 6 to 12 ft. should be adequate. Large storerooms often require **subaisles** parallel to the main aisle at distances of about 20 ft.

Cross aisles usually run perpendicular to the main aisles, connecting them and providing access to bins, racks, or tiered pallets. Cross aisles need accommodate one-way traffic only and should allow sufficient space for loading and unloading as shown in Fig. 33. Additional service aisles should be provided to break down areas for accessibility of small lots as required. However, in order to conserve aisle space, small lots should be stored in pallet racks or in box pallets at the end of large storage sections at the intersections of main and cross aisles. Aisles should be straight, clear of obstacles, and clearly marked with highly visible tape or paint stripes.

Auxiliary Storeskeeping Areas. Distributing counters are usually located adjacent to or along the stores issue windows. Additional tables or counters may be located near the issue windows or booths and in the receiving areas. Such counters are convenient for inspection, sorting, weighing, or temporary storage. An area should also be allowed for receiving, unpacking, counting, loading containers, and similar operations. Figure 34 portrays a typical storeroom layout.

Storage Area Identification. The numbering or marking system for the sections, rows, and bins, should start at a permanent corner and increase in the direction toward which there may be storeroom expansion. **Letters and numbers** may be used alternately, letters being more convenient for sections and columns or stacks, and numbers for rows and tiers. Corresponding stacks or bins should be given corresponding designations. A **standard symbol system** should also be followed for labeling subdivisions of bins or stacks throughout the system. For example, if one bin in a row should be omitted for some reason, its number should be omitted in order to preserve the regularity of notation. Figure 34 shows a typical storeroom layout with sections and rows marked. Figure 35 shows a simple method of designating bin columns and tiers.

Storeroom Equipment

SHELVES, BINS, AND RACKS. Effective storeskeeping depends in large measure on the correct selections of storeroom equipment. Such equipment consists of a variety of shelving, bins, and racks; material handling equipment; and auxiliary storeroom equipment such as scales, ladders, and record keeping equipment. Shelves and bins are usually constructed from wood or steel. Each has advantages and disadvantages. Either type may be fixed or adjustable and may have dividers for maximum flexibility.

Wood shelving has several advantages. It can be made from ordinary lumber obtainable locally and can usually be bought in specified lengths so as to minimize waste. Wood shelves can be readily adapted to unusual layouts where special steel equipment might be required. Wood also has a softer and more resilient surface

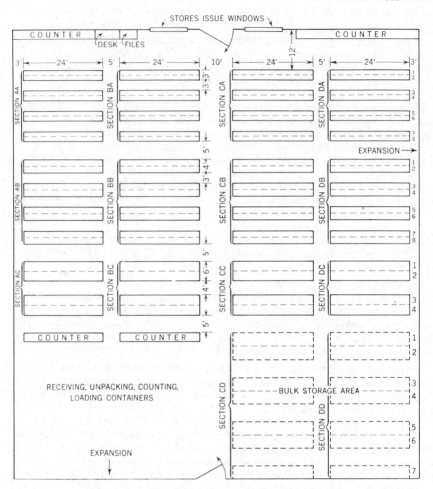

Fig. 34. Typical method of storeroom layout with area designations.

than steel and thus provides greater protection for delicate equipment. However, combinations of wood and metal shelving may be used to gain the advantages each has to offer.

Steel shelving is widely used in standard forms and sizes that can be mass produced at low cost and can be shipped in knocked down form for compactness and ease of handling. Standard steel shelves are easy to assemble at the point of use and, because they can be easily taken apart again, provide maximum flexibility, adaptability, and expandability. Steel shelving requires thinner members than wood and thus provides more space for storage. Steel is more durable and is no fire hazard. It can be readily cleaned and affords greater protection against vermin.

In general, bins and shelving are used for storage of items to be issued in small lots which cannot be economically palletized. Items taken from several

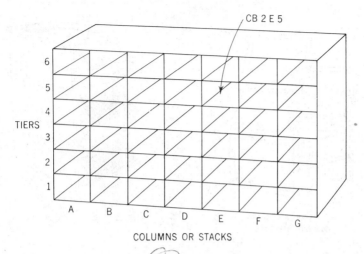

Fig. 35. Method of designating bin columns and tiers by symbols (section CB, row 2, column E, tier 5).

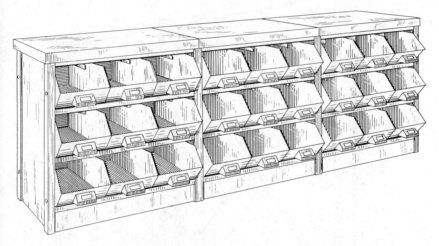

Fig. 36. Open-type removable bins in counter unit.

bins or shelves can often be placed on pallets or skids for removal to points of use throughout the factory.

Storage Equipment Types. A variety of types of bins, shelves, and racks are used in industry.

Open-type shelving may be used where the loading is light. Sway braces may be employed for greater rigidity. **Closed-type shelving** provides greater stability and is available in various heights. **Counter-height ledges** provide convenient temporary storage space. These shelf sections may consist of a variety of combinations as desired. Drawers or dividers can be added in numerous arrangements.

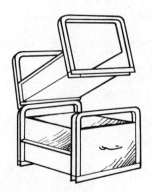

Fig. 37. Stack racks with
drawer inserts.

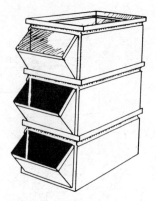

Fig. 38. Stacking boxes.

Fig. 39. Rack for lumber, bar stock, or pipe.

Bins may be provided with doors, either sliding or on hinges, to provide protection against theft, dust, etc. **Open-type removable bins** (Fig. 36) are useful for small items such as nails and bolts. These bins may be in a counter unit providing accessibility for storing and issuing unpackaged items.

Stacking racks and **boxes** are widely used to provide maximum flexibility of storage arrangements. Fig. 37 shows a typical stacking rack so constructed that it may be locked together without tools and allow for permanent or temporary storage with easy assembly and disassembly of racks. Stacking boxes are shown in Fig. 38. They are essentially tote boxes with an interlocking construction allowing for the formation of a stable, self-supporting pile. The type of stacking box illustrated has sloping bin fronts allowing ready access to the contents of all boxes without unstacking. These boxes may be used in the stockroom and are also useful in the manufacturing or assembly area. A variety of stacking boxes are available, with or without sloping bin fronts. They may be made of wood, metal, plastic, or rubber. Some are constructed in such a way that they may be stacked when in use and nested to conserve space when they are empty. Portable shelf and bin storage provides a high degree of versatility in the use of storage equipment for small parts.

Storage racks for lumber, bar stock, or pipe are also available in many varieties in both stationary and portable models. Fig. 39 illustrates one consisting of posts with horizontal arms attached. Other racks are made of structural shapes and may be more solidly built. An A-frame structure, with or without a caster base for portability, is another widely used rack form. Fig. 40 shows another type of portable **stock-selecting rack** which makes order picking easy for items that will slide or roll. This rack has three shelves that can be raised or lowered to vary the angle of slope or distance between shelves. Dividers on each shelf are adjustable for articles of different widths.

STORESKEEPING HANDLING EQUIPMENT. The selection of suitable material handling equipment for receiving, storing, and delivery of materials is of utmost importance for storeskeeping. The common types of handling equipment used are various types of trucks for skid and pallet handling, conveyors for storage and handling of stores, overhead crane systems, and tractor-trailer systems.

An economic analysis should precede the selection of material handling equipment for storeskeeping (see section on Material Handling). The analysis should take into consideration the characteristics of the items to be handled, the distances to be covered, the volume to be handled, and the relative cost of alternative combinations of handling equipment and labor. Where possible, standardization of handling equipment should be attempted both for storeskeeping and the whole plant.

There are a number of common types of **pallet, skid,** and **fork lift truck** equipment. Such trucks may be manually operated or may be gasoline or electrically power driven. They may be of a counterbalanced design where the load is carried ahead of the front axle or of a straddle arm design where the forks are located between two outriggers or straddle arms which extend alongside the pallet load. (Details concerning these trucks may be found in the Material Handling section.)

The use of these trucks in conjunction with skids and pallets is often of great value to storeskeeping. It makes possible the receipt, storage, and movement of materials in standard unit load sizes. Materials suitable for such handling should be made up into standard skid or pallet loads upon their arrival at the plant. In some cases, the vendor may be·asked to package the items in standard

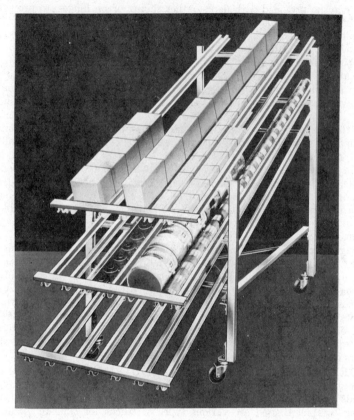

Fig. 40. Portable gravity-feed stock-selecting rack.

quantities of issue. In other cases the items may be put in standard quantities into boxes or containers or directly on to pallets or skids as the shipment is unpacked and inspected in the receiving department. The practice of having vendors actually ship the goods on skids or pallets in standard loads is increasing. This allows the unloading, checking of materials through receiving, and placing them in stores to be done by fork or lift trucks without manual handling.

The **skid** is a wood or metal platform with wood or metal runners or legs. The use of skids is generally much more limited than that of pallets, primarily because skids usually do not allow tiering.

Pallets are usually made of wood, metal, plastic, fiberboard, or combinations of these materials. The double-faced pallet is most widely used. It consists of stringers attached to an upper and lower platform. Because the bottom platform distributes the weight of the load evenly rather than at concentrated points, stacking or tiering of these pallets to heights of about 16 ft. is common. Use of stacker cranes greatly increases the height palletized loads are stored. The actual height, of course, will depend on the density of the material, the desired unit load size, the floor loading allowed, the capacity of the truck, and the floor-to-ceiling space available. These pallets may be constructed for two-way entry of the truck forks, or for four-way entry for greater flexibility. The single-faced pallet, the straddle-truck-type pallet, the disposable pallet, and the box pallet are other types which may prove useful (see section on Material Handling).

Special pallets and racks may be designed to accommodate articles which cannot be stacked in cubical piles, such as drums, pipe stock, or irregular castings.

Also in widespread use for storeskeeping order-picking are numerous types of **stock-selecting trucks.** These trucks are often designed to accommodate the types of items to be handled. They may have several shelves, be open or closed, and have many different types of wheel combinations. Fig. 41 shows a typical shelf truck. These trucks may have special coupling pins for use with dragline conveyors, or they may have trailer couplings for movement in tractor-trailer combinations.

Fig. 41. Shelf truck for storing and handling material.

Conveyors of various types lend themselves to efficient storeskeeping handling. Gravity conveyors, powered conveyors, and dragline conveyors are most widely used for storeskeeping. There are two basic types of gravity conveyors, roller and skate-wheel, which may be used for storage or for transporting **order-picking boxes.**

Standard power-driven belt conveyors or overhead conveyors may be used for moving stores over longer distances, or for raising or lowering materials. Dragline conveyors may be of the overhead type, or may be imbedded in the floor. This type of conveyor is constantly in motion at slow speeds, so that flat bed trailers, or portable bins or trucks with the necessary coupling pins, may be quickly and easily attached or removed. In this manner, stores may be transported with great flexibility, without using an operator or one or more tractors, as would be required for tractor-train operation.

Tractor-train operation for moving one or more trailers or bins offers a power-driven means of conveyance of materials directly between point of receipt, storage, and use of materials. Radio controlled operatorless tractor-trains may be employed, guided by signals relayed by a cable imbedded below the surface of the floor. The tractor is equipped with a set of hand controls for operator use when desired.

Overhead cranes may be used with hoists for lifting and placing heavy materials or used with a traveling pallet stacker. This combination of crane, pallet stacker, and pallet shelving permits a highly efficient utilization of storage space.

AUXILIARY STORESKEEPING EQUIPMENT. A number of other items, such as scales, ladders, package-labeling equipment, and record keeping equipment, may commonly be found in the storeroom.

Fig. 42. Self-supporting movable ladder.   Fig. 43. Movable ladder operating on overhead rail guide.

Weighing scales are used for measuring bulk materials issued by weight and also for measuring quantities of small items like screws and bolts, where the approximate weight is a sufficient indication of the count. Scale types used range from counter scales to in-the-floor scales for weighing trucks or portable bins of materials.

Movable ladders are required in storerooms to provide access to materials where shelving and bins are over 8 ft. high. These ladders may be the movable type mounted on platforms equipped with wheels or casters as in Fig. 42, or they may run on overhead tracks or guides with wheels at the bottom, to be moved alongside fixed rows of shelving as in Fig. 43.

Stores Records and Protection

CLASSIFICATION. In addition to receipt, storage, and delivery of materials, the storeskeeping department must be concerned with the classification and nomenclature of materials, the maintenance of up-to-date records, and the protection and safekeeping of all stores which are entrusted to it.

Attention should be given to the designation of materials by the adoption of names and symbols which will be used throughout the plant to avoid confusion when referring to materials.

A typical method of categorizing materials, which usually ties in with the accounting system, the engineering department, and the manufacturing department, is to classify the materials by type as follows:

1. **Raw materials**—such as casting, forgings, bar stock, sheet stock, and all other materials to be processed in the plant.
2. **Supplies**—such as paper, typewriter ribbons, and ink for the offices, and lubricants and packing materials for the shop.
3. **Purchased parts**—all parts not requiring manufacturing operations, which can be directly assembled or used in conjunction with other parts.
4. **Semiprocessed parts**—purchased parts or work in process which have had one or more manufacturing operations performed on them.

5. **Finished inventory**—finished parts or subassemblies, or finished products awaiting shipment.
6. **Miscellaneous**—Office storage for active and inactive records. Small jigs and fixtures, gages and special tools (see discussion of toolrooms in the section on Tools, Jigs, and Fixtures).

RECORD KEEPING. The responsibility of the storeskeeping department regarding record keeping is to check incoming materials to see that the kind, count, and conditions are as indicated; to enter their arrival on records; to record withdrawals and enter the balance on hand of each item; and to make periodic and systematic check counts to verify balances and report such counts to the stores record or materials control section of the production control department.

The responsibility for storeroom records should be centralized, usually under the **materials control section,** even where storeroom operation is quite decentralized. Thus, all records and forms relative to the issue and receipt of stores should be standardized. These forms should be as simple as possible, kept to a minimum number, and tied in with the cost and accounting records. Details as to the type and composition of storeskeeping records and forms may be found elsewhere in this section and in the section on Inventory Control.

A Simplified Stores Control System. Mattison (Industrial Engineering, vol. 1) describes an inexpensive but accurate control system used in the 36,000-pallet warehouse of Rubbermaid, Inc. There are 18,000 storage locations in the warehouse, each containing two identical pallets. The **physical components** of the control system are color-coded plastic cards, a plywood file divided into spaces representing each of the locations in the warehouse, and a rotary file.

The warehouse is divided into 6 vertical levels; each level contains 18 aisles, and each aisle has 190 positions. A three-part number code: aisle—position—level, identifies each location. The location code is indelibly printed on the plastic cards. Blank cards are stored in the plywood box in the proper position. The card also contains space for item identification.

The color of the plastic card signifies the access time to a particular warehouse location. The time it takes to reach an opening in the warehouse from a pallet location was determined by time studies. Three time zones were established: fast, medium, and slow. The color for the slow area also indicates the fifth- and sixth-level locations which can only be reached by a special high lift truck. Thus the color helps the foreman assign the proper truck for this area.

When a pallet of merchandise enters the warehouse, it is assigned an access time and an appropriately colored plastic card is chosen with a suitable storage position. The lift truck operator delivers the merchandise to its storage location, fills in the card in pencil with the item description, and returns the card for filing in a rotary file. The electrically operated rotary file has space for 20,000 cards and can bring a card into position for retrieval within four seconds. The cards are filed in the rotary file according to item type and color. The plastic cards are placed at the rear of a particular group and withdrawn from the front of the group. Thus providing a **first-in, first-out inventory system.** When an item is requisitioned from the warehouse, the first card in the rotary file for that item and color is pulled. The lift truck operator retrieves the two pallets from the position indicated by the card, erases the item information, and returns the blank card to its position in the plywood file.

According to Mattison the system's major drawback is its lack of historical records on inventories. However, the Rubbermaid data processing department does have complete inventory information based on production records and shipments.

The major advantages cited by Mattison are the system's simplicity; its low cost; the ease of handling and storing the cards, as well as their durability; the absence of paper forms; and the adaptability of the system to future electronic or mechanical data processing equipment.

PROTECTION OF STORES. An important aspect of storeskeeping is the protection of stores against hazardous conditions, weather, deterioration, dust, and theft. Each kind of material, such as textiles, rubber goods, leather goods, and metals, requires its own kind of care.

Hazardous Conditions. Extreme care must be taken with hazardous and explosive commodities. Hazardous conditions should be minimized as far as possible. Commodities such as oxidizing agents should be kept apart from combustible materials. Strong acids should be segregated or kept in a **fire resistant enclosure.** Adequate explosion venting and approved fire doors should be provided.

Adequate fire-fighting equipment, such as high-pressure water lines, modern sprinkler systems, suitable chemical extinguishers, fire pails, sand, axes, and ladders, should be provided (see section on Industrial Safety).

Storage areas should be kept orderly and clean, with clear aisles and passageways.

Weather. Materials whose bulk and nature require that they be stored out of doors, such as lumber and steel shapes and bars, can be protected by open sheds or by textile or plastic coverings.

Special rooms with temperature and humidity controls should be available for materials which are affected by changes in temperature. Some commodities require heat to prevent possible freezing, some may require freezing to prevent spoilage, and others may require a high relative humidity to prevent drying out.

Protection against weather may be provided on a temporary or emergency basis by use of **inflatable buildings.** These structures are made of fabric or other rubberized material and come complete with entrance doors and windows where required. The structure is maintained in its inflated state by a constant blower action. When no longer needed they store in a compact shape ready for immediate reuse.

Deterioration. Some materials deteriorate with age. Such materials should not be overstocked and a first-in first-out storage system should be adopted. Often, aging can be retarded by temperature and moisture control. Some materials which deteriorate due to the absorption of odors can be protected by suitable **segregation measures.** Another form of deterioration is the damage which may result from improper material handling. The adoption of the most suitable containers, handling equipment, and methods can do much to reduce this source of damage.

Theft and Vandalism. Pilferage of materials by employees and outsiders may constitute a sizable loss in money, inventory control, and delay in production. Thus, it is advisable to provide **closable storage areas.** Valuable items should be kept in locked cabinets or in a safe. Materials should not be issued without proper requisitions. Materials stored in yards and out-buildings are subject to damage or destruction by vandalism. Fenced-in areas may be needed together with security systems for their protection.

Traffic Management and Physical Distribution

OBJECTIVES AND RESPONSIBILITIES. Under the materials management concept, the movement of materials into and out of the plant—and ultimately to the customer—is the responsibility of the materials manager, although administratively a separate traffic manager may be the operational head of the traffic department or division.

The **traffic department** is assigned the day-to-day control of freight movement. Its primary objective is to provide the proper transport facilities to meet the physical distribution requirements of the company. These requirements include timing, size and form of shipment, and in the case of multi-plant operations, the determination of the point of origin of the shipment. Traffic management requires an enormous amount of up-to-date information concerning shipping and handling methods, rates, schedules, and routes.

PHYSICAL DISTRIBUTION SYSTEM. The Definitions Committee of the American Marketing Association has made a special effort to distinguish "physical distribution" from "distribution" by stating that physical distribution is the **movement and handling of goods** from the point of production to the point of consumption or use. They caution that the word "distribution" to describe this activity should be avoided since its use in economic theory, by the Bureau of the Census, and by marketing and business men actually refers to "marketing."

The **elements** of physical distribution systems common to large manufacturing companies include:

1. An organization for communicating ideas, transmitting questions and data, and implementing orders. This would include the chain of command beginning with top management, continuing through the materials manager, and terminating at the traffic manager or other operating management.
2. Warehouses or other storage facilities.
3. Receiving and shipping services. These would include all activities and operations including handling of incoming traffic, expediting, packaging for shipment, and the like.

Physical distribution activities are especially concerned with classification of shipments, choice of carriers, scheduling, and documentation.

CLASSIFICATION OF SHIPMENTS. Whether a company makes a single product or many products, the type of shipment will depend on quantity, speed of delivery, destination, and other variable factors. This in turn will affect the physical nature of the shipment in many ways.

The classification of a product takes into consideration many factors in order to arrive at a rating on which **freight rates** are based. Thus the **rating** of the shipment (based on its classification) determines the cost of shipping.

The basic motor carrier classification system is called the **National Motor Freight Classification.** It includes 23 classes of freight, though local or regional groups of carriers may have additional classifications. Classifications give weight to risk of loss or damage, waste or theft in transit, danger of damaging other freight, handling expense, packaged weight per cubic foot, handling methods, quantity, packaging, and basic movement costs. The rating takes into consideration the volume of product to be moved—truck-load (TL) or less-than-truck-load (LTL). For example, the rating for vinegar shipped in barrels LTL is 60; shipped TL, the rating is 35.

A similar classification, the **Uniform Freight Classification,** is used for rail carriers. In this case the two volume classifications are car-load (CL) and less-than-car-load (LCL).

CHOICE OF CARRIERS. A number of carrier options are open to most shippers. In practice, combinations of two or more different types of carriers are employed. If a company owns its own truck fleet or other transport, the traffic department is still responsible for selecting the proper means and for scheduling shipments.

Air Transport. The outstanding advantage of air transport is its speed over long distances, particularly to overseas points. Within continental United States this advantage is not always clear-cut, especially when one considers door-to-door delivery times.

Speed in shipment can permit fewer distribution points and a lower inventory. But this advantage may be diminished by the fact that air freight is still fairly limited with regard to local pick up and delivery. Air facilities are available in the most heavily concentrated industrial and consumer regions.

The greatest limitation on air transport is the size of the aircraft itself, which restricts the size, weight, and type of objects that can be shipped intact by air.

Air transport is by most measures the least dependable of commercial carriers. Equipment characteristics, on-ground facilities, and weather are special problems that can hamper service.

Motor Transport. Trucks dominate the short-haul transport of products of all sizes. They provide frequent service between most points in the country, particularly the smaller towns and rural communities not served by rail or air. In general, service is dependable though weather does hamper schedules to some extent. Part of this dependability is due to the fact that trucking operations depend on so few people (primarily the drivers). As with air transport, space and motor power limit the shipping capabilities of trucks. Tandem operations, permitted in some states, extend these capabilities somewhat. Perhaps the outstanding feature of motor transport is its ability to provide door-to-door service without transferring to other carriers. As a result, trucking delivery speeds are competitive with most other carriers.

Railroads. For various reasons, freight movement by rail is comparatively slow. This takes into account handling of the shipment at the rail terminals, delays in freight classification yards, and other delays unique to railroading. Rail transport is generally available on a once-a-day basis, but this does not mean daily service can be expected by the shipper. Railroad trains must use a fairly limited system of tracks to move between points and thus the possibility of "bottlenecks" exists. Modern railroad systems use electronic classification systems to increase volume over a fixed set of tracks. In addition, piggyback shipments can often by-pass these classification systems completely. Although the railroads at one time covered the country, much of the trackage to smaller, less populated areas has been abandoned. A large shipper may have a railroad siding directly at his door. However, most shipments are trucked to the rail terminal for long-distance hauling.

Other Means of Transport. Geographic conditions and special products may permit the use of other effective means of transport. Typically these means are used for the movement of raw materials rather than finished products. Oil, natural gas, and coal (in the form of a slurry) are transported long distance via pipelines. Barges are used for river and lake transport. Except for overseas freight movement, water carriers are generally restricted to coastwise transport and a few very large riverways.

PREPARATION FOR SHIPMENT. The rating classification discussed previously depends partially on the type of packaging and dunnage provided by the shipper. **Dunnage** is the protective material or devices used in packing products in the carrier or warehouse. Usually the more packaging provided, the lower will be the ratings. Very often the carrier will specify the type, quantity, and arrangement of packing in order to avoid loss or damage during shipment. In the "damage-free" railroad car, dunnage is designed into the equipment of the car itself. Cross-bracing and other means of compartmentalizing the car are used.

Containerization. Heskett, Ivie, and Glaskowsky (Business Logistics) describe containerization as the packing of goods of like or unlike characteristics in an enclosed box to eliminate rehandling of materials in their transportation from point to point. These containers are fitted with fixtures which permit easy handling, often by specialized handling equipment, from one carrier to another or to storage. The advantages of containerization are:

1. Lower handling costs and thus lower freight rates.
2. Lower in-transit insurance costs.
3. Reduction in loss or damage and elimination of pilferage while in transit.

The various types of carriers have been fitted with equipment which permits ready interchange of containers between truck and ship, railroad car and truck, and the like.

DISPATCHING TO CUSTOMER. Shipment of orders can be appreciably expedited if a number of steps are taken in conjunction with the customer. These include adjustment of customer order patterns with regard to timing and quantity of orders. **Consolidation** of smaller orders into a single large shipment may result in a significant saving in freight charges. The **material handling systems** of the customer should be coordinated with the shipping and packaging specifications at the shipping end. Containerization, palletization, and other means of grouping shipments are more effective if the customer has similar facilities for handling the shipment at the receiving end. Ordering procedures and reordering patterns if properly coordinated with the production and inventory practices of the producer can prevent shortages and stockouts.

OPERATIONS RESEARCH AND PHYSICAL DISTRIBUTION. The design of a complete and efficient physical distribution system requires application of a number of operations research techniques. The discussion that follows touches upon applications of these methods in a qualitative sense only. Detailed treatment can be found in the sections on Operations Research, Purchasing, and Material Handling.

Optimization of physical distribution systems lends itself to such techniques as linear programming and multiple regression. Mathematical models can be formulated for inventory control systems and solved for different inputs and outputs. Linear programming techniques are particularly useful in problems involving allocation of resources, numbers and location of storage or warehousing facilities, and transportation routes. The application of queueing theory has proven successful in material handling, servicing, and scheduling problems.

The use of models and computers has permitted traffic management to simulate various alternatives available, especially under multiple plant and warehouse conditions. When available, historical data can be analyzed and forecasts made using statistical methods.

Network techniques such as PERT and CPM are especially effective and have gained widespread use in the planning of coordinated physical distribution systems.

INVENTORY CONTROL

CONTENTS

CONTENTS *(Continued)*

INVENTORY CONTROL

Nature of Inventory Control

DEFINITION. Inventory control has been defined by the American Production and Inventory Control Society (APICS Dictionary of Production and Inventory Control Terms) as, "The technique of maintaining stockkeeping items at desired levels, whether they be raw materials, work-in-process, or finished products." The determination of **desired levels** and maintaining inventory at these levels is the heart of the inventory control problem. Not stipulated in the APICS definition, but mandatory as a requirement for effective inventory control, is the fact that inventory control should be considered in light of the **total production control system.** Errors in **forecasting**, with schedules being made on the basis of the forecasts, will result in fluctuations of inventories. Even though inventory control is considered as an autonomous item in this section, certain decisions regarding levels of inventory to maintain will depend on expected fluctuations in demand and upon the known fact that errors will occur in forecasting.

SCOPE AND IMPORTANCE OF INVENTORY CONTROL. Inventory control covers all aspects of the production or business operation. This section will be concerned with all facets of the inventory control problem for the **production** situation, but certain topics, such as **stocking requirements,** are equally applicable to the general inventory problem as might be found in such operations as supermarkets, warehousing, and service station stock control, among others. As stated by Magee and Boodman (Production Planning and Inventory Control), "The problems occur at almost every step in the production process, whether purchasing, production of in-process materials, finished production, distribution of finished product, or service of customers." In this context, the scope of inventory control covers **analysis** and **planning** as regards:

1. Raw stock inventory levels.
2. Batch production of component items.
3. In-process inventory requirements.
4. Final product inventories for customer satisfaction.

The importance of inventory analysis and control is apparent to anyone giving even a cursory look at business publications. As inventories rise, possibly due to overproduction or unforeseen decreased demand, **capital** is tied up and is not available for optimum return investments such as plant expansion, product research and development, and so on. An idealistic yet ideal situation would be to have sufficient inventory on hand at all times to just match the demand at any given instant in time—an obvious impossibility. Forecasts of demand are subject to error because of the inherent uncertainty of customer whims and economic conditions. Most production/demand systems are **dynamic** not static, thus necessitating a dynamic and not static inventory policy.

Plant Efficiency. Plant efficiency may be measured in terms of the **satisfaction of customer requirements** commensurate with the **minimization of**

overall production costs, which includes the cost of inventory. Also, efficient operation of any industry depends upon proper turnover of investment. Receipt of inventory must be scheduled so as to be available when needed, and in sufficient quantities to maintain production and on-time shipment for customer satisfaction. At the same time, the investment in inventory must not increase beyond that required to meet current needs and maintain a reasonable factor of safety. **In-process** inventories allow downtime to occur in successive manufacturing operations while still maintaining a continuous flow of production.

Customer Service. Inventories are required to allow delivery to customers by pre-determined due dates. However, customer service could always be satisfied by maintaining excessive inventories. From a plant efficiency point of view it was seen that this would not allow proper turnover of investment. Therefore, inventory control allows evaluation of **expected** customer demand to provide **sufficient** but not excessive stock for that demand. If demand is seasonal, employment and production can be stabilized with judicious inventory build-up in the off-season to balance production which is below demand during the peak requirement period.

Control of Raw Material and Parts. It is necessary to maintain raw materials and parts in sufficient quantities to maximize plant efficiency and provide optimum customer service. Inventory control procedures allow determination of **economic order quantities** that balance the costs of **storing** the inventory against the cost of **ordering** an adequate quantity of inventory. However, the control procedures should be sufficiently flexible to allow **quantity discounts** or to allow large quantities of items to be purchased when such conditions as possible strikes threaten to curtail future supply (see section on Purchasing).

If quantity discounts are available, inventory control analysis allows a determination to be made as to whether the added carrying charges will be offset by decreased unit purchase costs. Not all raw materials will justify tight control from an economic point of view. The **ABC** classification system, discussed elsewhere in this section, is a procedure for evaluating raw material inventory from a **value** point of view. Low value items, which constitute on the average a majority of the inventory, should be stocked in sufficient quantities to insure no runout possibility. High value items would warrant close control.

Control of Manufactured Goods. Control of raw materials and parts is mandatory if customer demand is to be satisfied. The scheduling of manufactured **standard items** should be evaluated in a similar fashion as the economic order quantity for raw materials. Annual demand for the standard items is forecast and production of the items is accomplished in **batches,** with the batch sizes being determined by balancing production preparation and setup costs against the carrying costs of holding the final products in storage. Products manufactured in this manner would not be produced for **specific** customer orders but would be produced for **expected** demand. Control of manufactured goods (as well as new materials and parts) is dependent upon the type of production. These include production for customer items, standard products, stock items, or modified products.

Effective Budgeting. The **long-range** planning inherent in inventory control allows accurate estimates of inventory costs to be made. As stated by Stockton (Basic Inventory Systems: Concepts and Analysis), "Inventories are an asset in the firm and, as such, appear in dollar form on the balance sheet. From a financial standpoint, inventories represent a capital investment and must, therefore, compete with other asset forms for the firm's limited capital funds." In manufacturing plants, 16 percent to 50 percent of the sales dollar is spent for material, and in some processing plants as much as 75 percent of the sales dollar is so spent. Therefore, it is important that material costs be closely controlled, for they may mean the difference between profit and loss on operations.

The **investment in inventories** is often equivalent to 50 percent of the current assets of a company and may be the largest item on the balance sheet, with the exception of plant and equipment. Having knowledge of the expected costs of this item through inventory control goes a long way to developing an accurate and effective budget.

Planning for Inventory Control

LEAD TIME. Competitive practices often set limits on the delivery time that a company can quote to customers. Fig. 1 shows the normal lead time, exclusive of transportation, required for delivery from factory, as quoted to customers by companies in several different industries. In some industries, such as **appliances** or **foods,** it is customary to provide immediate shipment of finished items "off the shelf." In industries such as **elevator manufacturing** or **steel tank fabrication,** where substantially longer delivery times are quoted to customers, almost all manufacturing or production may be accomplished after the customer order is received. But even in these product lines with long delivery times there is a tendency for one of the competitors to break the customary quoted delivery time and offer products on a more immediate basis. Usually other competitors are then forced to follow suit. The shorter the quoted delivery time, the greater the amount of inventory that has to be carried by the manufacturer; and this inventory must be processed closer to the finished goods stage.

PRODUCTION BASIS FOR DETERMINING INVENTORY POLICY. A decision to hold sufficient inventory to ensure delivery of a product

Type of Industry	Days Required for Delivery Exclusive of Transportation	
	Standard Items	Special Items
Gas Ranges	3–4	28–35
Highway Trailers	14–21	35–42
Steel Sheets	28–35	
Furniture	35–42	
Gymnasium Equipment	30–45	
Steel Lockers	60–90	
Steel Tanks (glass-lined).....................	14–28	70–98
Cosmetics	2	
Powder Metal Products	14–21	28–42
Pianos	7–14	
Pumps	60–90	60–90
Valves	14	
Pipe—Seamless	90	
Pipe Fittings	7–14	30

Fig. 1. Typical lead times for delivery from factory.

upon customer demand, or to order raw stock and component items only upon receipt of a firm customer order, depends, to an extent, on the type of production. Logically, the latter policy would be fallacious for a semi-conductor device manufacturer, and the former would be erroneous for a manufacturer of custom stereo equipment.

Custom-made Line. If the product is designed specifically for the customer and is not part of the manufacturer's regular product line, no material or parts can be ordered until the design is finalized and the customer has approved the design. This does not mean that some raw material common to the line of custom-items will not be held in inventory. A university research machine shop will manufacture custom items for research projects, such as a positioning table to be controlled by a digital computer. Pulse-drive motors, the table itself, and sequencing equipment would not be ordered until design approval has been obtained. Raw stock and hardware utilized in connecting the pulse-drive motors to the table will be standard material applicable to a myriad of other projects.

Standard Lines. If the company has a standard line of products but the customers are willing to wait long enough, the manufacturer does not have to order any material until the customer order is received. At that time, or upon completion of any engineering or planning necessary, each part and each item of raw material can be purchased specifically for the particular customer order. Usually, standard lines will involve continuous production with components and raw materials being ordered and/or manufactured in batches in order to minimize the overall inventory interest charges, and order and setup costs (see Economic Purchasing Quantities, later in this section).

Batch Production. In other cases, customer orders may be grouped after receipt by similar specifications and produced on a batch basis. This is especially true in steel production where several orders are grouped to make up a "heat" of steel with the exact specifications for that group of orders. Later, each order may be rolled to a different size specification called for by each customer order. In this case, raw material is usually kept on hand in sufficient quantities to produce for any combination of orders.

Production for Stock. In many cases, customers want immediate delivery, and finished products must be produced in advance of receipt of their orders. These finished products are held in stock and shipped when customer orders are received. In this case, the manufacturer usually offers a standard product line which is produced to predetermined specifications. In this type of manufacturing it is necessary to predict what customers will want, since the manufacturing process must be completed before customer orders are received. The usual procedure is for the sales department to prepare a sales forecast which gives the number of units of each type and model which are expected to be sold in each successive time period.

Modified Products. In some industries customers want standard products modified in some way so that they will better suit their particular requirements. In many cases, this is also coupled with a requirement for quick delivery.

In highway trailer manufacturing, for example, a standard trailer is often modified in some way to suit customer requirements. Standard raw materials and parts are carried in stock, but special items are ordered specifically from vendors upon receipt of customer orders. Therefore, the delivery time quoted to the customer must include the fabrication time required for the modification, plus the lead time for any special items that are purchased.

OPTIMUM STOCK REQUIREMENTS. The most important determinant of which material to stock is the lead time between receipt of a customer order and the time when the finished product must be shipped. This sets up a **minimum condition for inventory** which the company must carry by classes of material. The shorter the quoted delivery time, the more stock must be carried, and it must be carried in a more advanced stage of completion.

There are, however, several other reasons for **stocking standard parts and raw material** even though there would be sufficient lead time to purchase them after receipt of a customer order:

1. **Price advantage** due to purchasing or producing parts in larger quantities than presently needed.
2. **Flexibility advantage** for producing on short notice additional quantities of finished goods over and above those forecast or scheduled. Some companies insist that a certain percentage of reserve raw material and purchased parts be carried on hand at all times to allow for sudden increases in schedules which could not be accommodated by maintaining normal inventories.
3. **Protection against market change.** In some cases, manufacturers carry base stocks of raw materials and purchased parts which they can use to produce customer orders if suppliers should take longer than the normal lead time provided or if shortages of those materials should develop. Material is taken from these base stocks and replaced when it is received.

The maintenance of a stock position is also a worthwhile idea where an assembly line or process flow is involved. The base stock acts as protection against the line or process being shut down in case material is not received according to schedule.

MATERIALS CONTROL CYCLE. The materials control cycle comprises all of those procedures which are necessary for providing materials for manufacturing with a minimum of investment and cost. This **materials control cycle** is as follows:

1. Determining material needs.
2. Preparing requisitions for purchased items and requests for work orders for parts made in the shop.
3. Receiving purchased materials and finished parts into the plant.
4. Inspecting purchased material and parts, and inspecting finished shop parts. Delivering all material and parts to the storeroom for storage.
5. Entering receipts in store records or perpetual inventory records; apportioning material in the records to current orders; authorizing requisitions of materials from stores for production of shop parts and for assembly into finished items.
6. Issuing of parts and material to the shop for production and assembly.
7. Recording the issue in stores records or perpetual inventory records.
8. Entering receiving and issuing transactions into cost and accounting records.
9. Determining the necessity for replenishment of stores, which leads to step No. 1 above.

There are three **auxiliary materials control activities** which parallel the materials control cycle. They are as follows:

1. Determining the proper quantity to requisition for each item of material and for each purchased and manufactured part.
2. Physically checking and verifying the quantity of each part and each item of raw material in stock against stores records or perpetual inventory cards.
3. Standardizing materials and parts for lowest cost manufacturing.

Determining Purchase Quantities

FACTORS INFLUENCING ORDER QUANTITIES. Optimum order quantities can be determined analytically. Basically, the procedures involve balancing holding costs against purchase costs. However, as with all analytical techniques, the results must be evaluated in terms of current operating conditions. **Price breaks**, for example, might well present significant savings even though holding costs will be increased.

Reasons for Inventories. Inventories may be held higher than dictated by some analytical order quantity calculation because:

1. Quantity discounts can often provide substantial economies.
2. Higher costs are anticipated because of price increases for raw material and higher labor rates.
3. Difficulties are anticipated in procurement due to possible strikes or periods of high demand. In this case, inventories may be built up on a temporary basis.

Disadvantages in Large Inventories. There are a number of disadvantages in having too large a stock on hand. Most of these are concerned with the increased costs which may result from carrying large inventories and include:

1. Interest on investment in inventory.
2. Storage or space charges.
3. Taxes.
4. Insurance.
5. Physical deterioration or its prevention.
6. Housekeeping.
7. Handling and distribution.
8. Recordkeeping costs.
9. Obsolescence.
10. Repairs to product.

It is necessary that close control be exercised over the inventory so that procurement and holding costs are the lowest possible consistent with the availability of **material** according to predetermined lead times, the availability of **capital** for investment in inventory, and the availability of **space** in which to store the inventory. As Magee and Boodman point out (Production Planning and Inventory Control), ". . . costs, and the balancing of opposing costs, lie at the heart of all production and inventory control problems. The cost elements essential to a production or inventory problem are characteristically not those reported in summary accounting records."

Physical Factors. The physical factors in the analysis of when and how much to buy are shown graphically in Fig. 2.

A simple **example** will illustrate how the physical factors alone can be used to determine the size of an order. Suppose the following conditions apply to a material used in production.

Amount on hand..........	750 units
Delivery time.............	3 weeks
Reserve stock (in time) (for use in emergencies)......	4 weeks
Average rate of use per month	300 units
Maximum supply to be kept on hand (in time)........	17 weeks

From this information it is possible to calculate the proper quantity to purchase and the timing of the order. The number of months the current supply will last is

$$\frac{750 \text{ units}}{300 \text{ units/mo.}} = 2.5 \text{ months' supply}$$

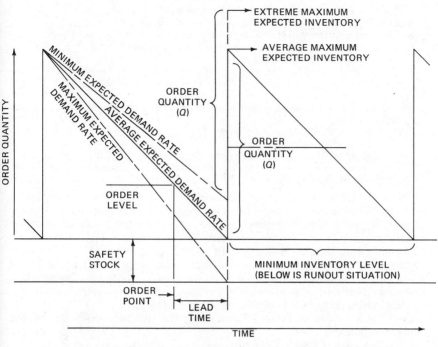

Fig. 2. Typical inventory model.

On a weekly basis this is

$$2.5 \times 4.33 \text{ (average weeks per month) } = 10.8 \text{ weeks}$$

say 11 weeks' supply on hand.

A 4-week reserve supply is required and 3 weeks are required for delivery so that the order must be placed in

$$11 - (4 + 3) = 4 \text{ weeks}$$

The number of units used in 7 weeks is

$$\frac{300 \text{ units/month}}{4.33 \text{ weeks/month}} \times 7 \text{ weeks} = 485 \text{ units}$$

The maximum number of units to be kept on hand is

$$\frac{300}{4.33} \times 17 \text{ weeks} = 1{,}177 \text{ units, including the 4-week reserve stock}$$

Thus the order quantity represents a 13 weeks' (or 3 months') supply and an order should be placed four times a year. The minimum stock point is 4 weeks or 277 units.

The above figures have not taken into account a calculation of the economic purchase lot, which besides serving production needs and providing proper cushions and order points, also introduces the analysis of discounts or reduced unit prices on quantity purchases versus the costs incurred in the owning and storing of the material. In determining the **economic purchase lot size** for an item the following factors must be taken into consideration.

1. Quantity on hand and available (not apportioned) in stores.
2. Rate of use, and whether use is steady, periodic, or irregular.
3. Time required for delivery of new lot (included in preceding discussion).
4. Amount to be kept as a cushion or reserve in case of delayed delivery (included in preceding discussion).
5. Amount to be kept for emergencies such as a jump in demand, excessive spoilage in processing, and desire of sales department to build up extra finished stocks.
6. Length of manufacturing period which it is desired to protect by having materials on hand.
7. Quantity discount obtainable in large lots.
8. Rental cost of storage area occupied.
9. Amount of working capital tied up and loss of earnings on this capital.
10. Other costs and losses in owning materials—taxes, insurance, depreciation, obsolescence, extra handling, etc.

UNIT COSTS. There are several auxiliary procedures necessary to implement the materials cycle. One of these is to determine the economic purchasing quantity for each purchased part and item of raw material and an economic manufacturing quantity for each part produced in the shop.

The **economic lot size** is that quantity which will give lowest unit cost, considering all costs involved. It is the overall unit cost of the material as delivered to the production department, rather than the specific price quotation at which material is purchased or produced which is important. Thus the cost of storing, handling, and keeping records of the part must be considered.

There are a number of **factors influencing the ultimate unit cost of the product** including:

1. The price paid for a purchased item or the cost of a manufactured item.
2. The cost of placing the order for replenishment quantities.
3. The cost of carrying the item in inventory.
4. The amount of money tied up in inventory.
5. Physical deterioration or obsolescence of the item while held in storage.

The effects of these factors may be reduced to a fairly definite **schedule of costs** which may be classified as follows:

1. Unit costs which tend to decrease as the size of the order is increased:
   a. Purchase price or cost of manufacturing.
   b. Cost of purchase or shop orders.
2. Unit costs which tend to increase as the size of the order is increased:
   a. Inventory storage costs, which include expenses for rent, heat, light, janitor service, plant protection, etc. If materials are stored in buildings owned by the company, space charges such as depreciation, repairs, maintenance, insurance, and taxes on the building replace the rental charge.
   b. Inventory carrying charges, which include interest on investment in the larger average inventory carried, personal property taxes and insurance on the materials, risk of spoilage, depreciation, and possible obsolescence, excess materials handling, record-keeping costs, etc.

The factors involved in economic purchasing quantities for raw materials and purchased parts and the computation relative to economic lot sizes for purchased items are covered first.

ECONOMIC PURCHASING QUANTITIES. The possibility of securing advantageous terms by buying in larger quantities is the most important single factor in determining an economic purchasing quantity. In many cases, purchase discounts are given for buying in larger quantities.

If larger quantities are purchased on each order, it means that fewer orders will be placed during the year. This may provide savings which should also be taken into account in determining the economic lot size. The **average cost of**

placing a purchase order may be computed by adding the costs of requisitioning, purchasing, receiving, and paying for materials during a given period, and dividing this total cost by the number of purchase orders placed during that period. Or purchase orders may be classified as to type and the total cost may be allocated to each group depending upon the estimated time taken to purchase each particular group. From these figures, in turn, average purchase order costs per group are determined.

COSTS OF INVENTORY. In determining the quantity to order, the cost of purchasing is balanced against the interest charges and storage costs associated with holding the items until used. The final decision regarding the quantity to purchase must be made in light of current conditions as presented earlier. The costs associated with inventory, namely storage costs and carrying charges, are discussed here.

Computing Storage Costs. Inventory storage costs such as rent, heat, light, janitor service, and plant protection may be computed by allocating all the costs of renting and operating storerooms to the items in **proportion** to the space occupied or on some other reasonable basis. Inventory storage costs are usually converted to a percent of the value of the items carried in inventory. The dimensions for storage costs, therefore, are in terms of (dollars/inventory unit) (time unit). Since all storage charges increase with respect to the time inventory is held in storage, the dimension of costs must contain a basic time unit, such as days, weeks, etc.

If several different products are stored in the same basic storage space, possibly in random fashion, then charges are usually calculated on the basis of the average amount of inventory held during batch consumption time. This is generally one half of the batch purchase size plus any safety stock. However, if only one product is used in a particular storage area, then all charges associated with that storage facility will be assigned as storage costs for the item. By convention, costs then will be based on the total batch purchase quantity plus safety stock. This would be equivalent to the maximum expected inventory. Graphically, storage costs are linear with respect to quantity and time.

Carrying Charges. Inventory carrying charges are concerned with such items as taxes and insurance on the material, and the interest charges resulting from the investment of capital in inventory. The carrying charges are generally based on the **average inventory** on hand during the batch-use cycle, or one half the batch purchase quantity plus the safety stock. Scheele, Westerman, and Wimmert (Principles and Design of Production Control Systems) list the items to be considered in the determination of carrying cost, other than storage charges just discussed, as:

1. The cost of upkeep of material in stock and allowances for possible deterioration.
2. The cost of keeping the stock records:
   a. Posting the records.
   b. Taking physical counts.
3. The cost of insurance.
4. The cost of obsolescence. This cost depends upon the type of items being stored—for example, it would be insignificant for hardware items, but very high for style goods.
5. The cost of the money invested in inventory (interest).
6. The cost of taxes on the items being stored. In many states there is a property tax on inventories.

Taxes and insurance can usually be allocated on a value basis directly. **Interest on investment** is a function of the dollars invested in each item and can be computed in several ways. In some cases a nominal interest charge of 5 or 6 percent is used for the capital tied up in inventory. Other companies charge a much higher rate in the belief that inventories should be expected to contribute to profit in the same manner as any other investment or asset. Thus, if the company is achieving a return of 25 percent on its other assets, it would include a charge of 25 percent as the interest cost of the inventory.

Damage, depreciation, and spoilage can sometimes be computed by classes and assigned on a percent of value basis. **Obsolescence** may be computed on an accounting or judgmental basis and may differ for various classes of material. Once the figure is determined, however, it is usually expressed as a percent of value for each class of inventory. The possibility of obsolescence increases with time, and some managements have made it a policy never to order more than a year's supply of any item on the theory that almost any item might be changed after a year and, therefore, the obsolescence factor rises to 100 percent at that time.

Total Cost of Carrying Inventory. The total cost of carrying inventory is a function of the number of times the items are purchased and the actual quantities purchased. As larger quantities are purchased at one time, the quantity comprising the average inventory also increases. For example, suppose that an item is purchased for stock and the usage is $2,000 per year. One order could be placed for $2,000, or 12 orders might be placed for $167 each. In the first case, the cost of ordering is only one-twelfth of that in the second case. However, the average inventory in the first case is $1,000 plus safety stock whereas the average inventory in the second case is $83.50 plus safety stock. In each case the average inventory on hand is computed as one-half the ordered inventory on the theory that depletion of inventory with respect to time is essentially linear. It is apparent that the carrying costs for interest, taxes, insurance, space charges, etc., will be substantially higher for the first case than for the second. Figure 3 shows this graphically.

Scheele, Westerman, and Wimmert (Principles and Design of Production Control Systems) state that the industrial average for the total carrying cost charge, including storage costs, is between 15 and 25 percent of the average yearly inventory. One large automotive manufacturer uses 25 percent per year as its figure. One commonly used estimate of the **percentage cost per year of carrying inventory** is as follows:

Storage facilities	2%
Taxes and insurance	1
Material handling and record-keeping	4
Interest on investment	5
Depreciation, obsolescence, and shrinkage	5
Total	17%

A manufacturer of highway trailers who figured the cost of carrying inventory in the light of his own experience found that his costs were as follows:

Storage	3.00%
Insurance	0.05
Taxes	1.50
Material handling and record-keeping	2.70
Depreciation	2.50
Interest	5.00
Obsolescence and shrinkage	1.00
Total	15.75%

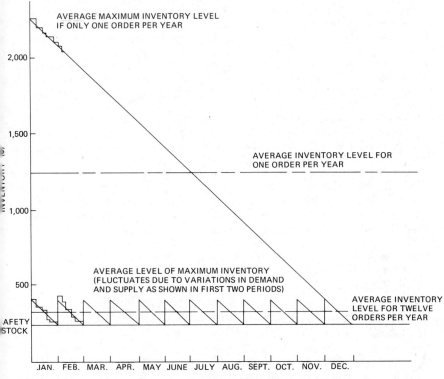

Fig. 3. Comparison of average inventories—12 orders per year versus 1 order per year for a specified demand.

This manufacturer added a **contingency factor** to the total figure. When he switched over to using economic purchasing quantities he was afraid that the lowest cost inventory might require more capital than he had available. He therefore inserted a contingency factor of 20 percent in computing his economic lot quantities, which raised the total carrying charge to 35.75 percent. When he saw that his inventories were not going out of line he reduced this factor to 10 percent. He will probably reduce it still further, since he has proved that the use of economic lot quantities did not increase inventory but when used with other techniques for better materials control, gave him a better-balanced inventory with no added investment.

INVENTORY COST COMPUTATION. Several methods are available for costing inventory. If items in inventory were used immediately upon receipt of the item then each item could be priced at its input value. In reality inventory is in storage for varying lengths of time, with the input prices of the inventory varying relative to the time certain lots were received. The question to be answered is, on what basis is the price of outgoing inventory going to be established, knowing that all items are not equally priced. Two of the more familiar methods of pricing inventory are the FIFO and LIFO techniques.

First-in First-out (FIFO). Using the FIFO method, items are priced on the basis of the **oldest lot** currently in storage, until that particular lot is exhausted. Items subsequently withdrawn are priced on the basis of the next oldest lot and so on. MacNiece (Production Forecasting, Planning, and Control) feels that this method is applicable when manufacturing costs must reflect the actual sequence of price fluctuations for raw materials, and can also be used when raw materials account for a relatively small part of the total production costs. He also points out that the FIFO approach is applicable for **perishable** items which are actually rotated in a first-in first-out sequence to prevent deterioration. Pricing then actually corresponds to the physical movement of the items.

Last-in First-out (LIFO). The logical inverse of the FIFO system is an inventory pricing method that prices items according to the **latest lots** received until an amount of inventory corresponding to that lot is released. Subsequent items will then be priced according to the next most recent lot until that is completely released, and so on. MacNiece finds that LIFO permits stock to be carried on the books at old prices, so that inventory values do not change greatly from one accounting period to the next. Current fluctuations in the price of raw material are more realistically reflected in the current cost of sales.

FIFO and LIFO Example. An example of FIFO and LIFO given by Greene (Production Control: Systems and Decisions) demonstrates the actual consequences of using either of the pricing procedures. Consider Fig. 4, which represents in tabular form pricing of items which are **increasing** in value.

	In	Out	Balance	Price/Gallon	Transaction
FIFO Method:			00 gal.		
	50 gal.		50	50 cents	+$25.00
	100		150	75 cents	+ 75.00
		60 gal.	90	{ 50 at 50 cents { 10 at 75 cents	− 32.50
LIFO Method:			00 gal.		
	50 gal.		50	50 cents	+$25.00
	100		150	75 cents	+ 75.00
		60 gal.	90	60 at 75 cents	− 45.00

Fig. 4. FIFO/LIFO example.

In commenting on his example, Greene states of the FIFO method: "Since the prices are going up, we can expect to pay at least the most recent price of 75 cents a gallon for the next purchase, or $45 to replace the 60 gallons sold. This same amount of material was priced out at $32.50, or $12.50 of any profit we might have anticipated is lost just by the act of replacing inventory." As Greene points out, this is not very profitable, so in an increasing market, it would be more appropriate to price on the basis of LIFO. Now, the remaining inventory is out of date as regards asset value. In summary, Greene points out that: "A business man is permitted to use any recognized method of inventory evaluation as long as he uses the same methods consistently."

UNCERTAINTY OF SUPPLY AND DEMAND. Analysis of order quantities is based on deterministic or stochastic assumptions. The **deterministic** case allows no leeway for demand or delivery **uncertainty**. In real life this is not the case. Demand for an item is not constant, thus nullifying the constant consumption period assumed. Similarly, suppliers have situations in their own facilities, such as overdemand in certain time periods or machine failures, which prevent delivery of orders at exactly predetermined points in time. The existence of uncertainty in supply and demand requires the use of **stochastic, or probabilistic,** techniques.

Deterministic Analysis. The determination of order sizes with deterministic assumption will be developed prior to the discussion of the uncertainty situation itself.

Fig. 5. Variation of purchase order cost/unit as order quantity varies.

Considering only the purchase cost and carrying charges, as depicted in Figs. 5 and 6, the **total unit cost** to be minimized is as shown in Fig. 7. The unit cost to be minimized in equation format, with symbols as defined in Figs. 5, 6, and 7, is:

$$UC = \frac{S}{Q} + KQ + C$$

Minimization of the total cost is obtained by taking the derivative of UC with respect to Q, setting the results equal to zero, and solving for that batch quantity which then minimizes unit costs. This leads to the **optimum** batch quantity:

$$\frac{S}{Q_0} = KQ_0;$$

$$Q_0 = \sqrt{\frac{S}{K}}$$

where Q_0 represents the optimum purchase quantity. This of course indicates that the optimum quantity occurs at the intersection of the carrying cost line and purchase cost line, as shown in Fig. 7.

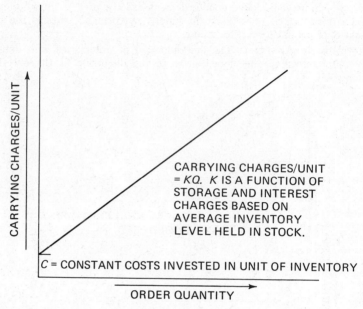

Fig. 6. Variation of carrying charges/unit as order quantity varies.

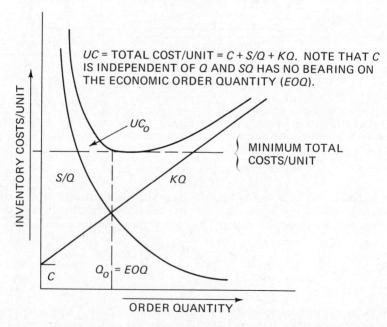

Fig. 7. Variation of total inventory cost/unit as order quantity varies.

Economical Batch Quantity. If management decides to allow a slight deviation above the minimum unit costs, say from UC_0 up to UC_u, it is often found that a much wider range of freedom can be obtained for the allowable

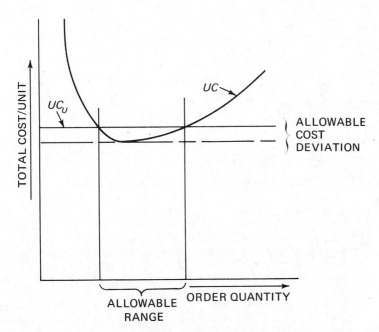

Fig. 8. **Effect on allowable purchase quantity of increase in total cost/unit.**

purchase quantity. This is shown in Fig. 8. The economical batch quantity can be found from:

$$\text{(Lower value of } Q) = \frac{(UC_u - C) - \sqrt{(C - UC_u)^2 - 4KS}}{2K}$$

$$\text{(Upper value of } Q) = \frac{(UC_u - C) + \sqrt{(C - UC_u)^2 - 4KS}}{2K}$$

Use of the economical batch quantity allows decisions to be made regarding purchasing without resorting to several batch quantity computations.

Probability and Statistical Techniques. The probabilistic characteristics inherent in the economic purchase quantity are created by two basic problems:

1. The vendor of a particular item may not always deliver the items at exactly the required delivery date due to **uncertainty** in his own operation.
2. Runout of the particular item by the user will not be exactly on a specified date, therefore creating the possibility of running out of the item for a certain length of time. Conversely, the possibility of creating inventory levels higher than desired is also a distinct possibility if delivery is made prior to a desired due date.

Hopefully, if an analysis is made to determine a logical order point in time, statistical data is available regarding delivery dates as contrasted to order dates for particular vendors. The one major problem in performing statistical analyses

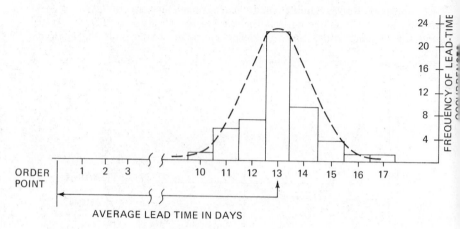

Fig. 9. Histogram of historical lead-time data.

in industry is lack of applicable data. A record of the delivery dates in relation to order dates can be maintained on order cards for each item, and updated each time an item is ordered and each time an order is delivered.

An example of a particular item's **lead time** characteristics might be as shown in Fig. 9. The lead times are plotted for 50 values obtained from historical files. The continuous dotted line plotted through the frequencies of lead times defines the shape of the **probability distribution** of lead times. Assuming the total area under the dotted curve has a value of one, then the **cumulative area** working from left to right, represents the empirical probability of a particular delivery being within a certain lead time. Techniques are available for analyzing probability distributions of many varieties (see section on Statistical Methods) but for the purposes of this section, the data will be assumed to be **normally distributed**. The normal distribution has a bell-shaped curve as depicted in Fig. 9.

The standard deviation for a set of data gives a measure of dispersion of the data around the mean, and allows the tabular probabilities to be determined for specified data values. In the case of the problem being discussed, this would be specific lead times. The standard deviation is given by the formula

$$\sigma_x = \sqrt{\dfrac{\displaystyle\sum_{i=1}^{N} f_i (X_i - \overline{X})^2}{\displaystyle\sum_{i=1}^{N} f_i - 1}}$$

where σ_x is the calculated standard deviation

       X_i represents one item of historical data (lead time)

       $\overline{X}$ represents the mean of the historical data and is calculated by

$$\dfrac{\displaystyle\sum_{i=1}^{N} f_i X_i}{\displaystyle\sum_{i=1}^{N} f_i}$$

N is the number of different lead times considered; in the problem it has a value of 8

f_i is the frequency of occurrence of the ith lead time (X_i)

$\sum\limits_{i=1}^{N} f_i$ is the number of pieces of data being analyzed, in this case 50

Lead Time X_i	Frequency (f_i) of Occurrence of X_i	f_iX_i	Dispersion of X_i About Mean $(X_i - \overline{X})$	$(X_i - \overline{X})^2$	$f_i(X_i - \overline{X})^2$
10	1	10	−3	9	9
11	6	66	−2	4	24
12	7	84	−1	1	7
13	22	286	0	0	0
14	9	126	1	1	9
15	3	45	2	4	12
16	1	16	3	9	9
17	1	17	4	16	16
Totals	50	650			86

Fig. 10. **Lead-time data calculations required for finding standard deviation.**

Fig. 10 summarizes the calculations for the lead time data graphed in Fig. 9. The standard deviation of the data is found to be:

$$\sigma_x = \sqrt{\frac{86}{49}} = 1.3$$

The mean is:

$$\overline{X} = \frac{650}{50} = 13$$

Typical probability values, Prob. (Z), for a normal distribution, for values (Z) standard deviations away from the mean distribution are as follows:

Prob. 3 standard deviations below: 0.0013, or 0.13%
Prob. 2 standard deviations below: 0.0228
Prob. 1 standard deviations below: 0.1587
Prob. 0 standard deviations (i.e. mean):   0.5000
Prob. 1 standard deviations above: 0.8413
Prob. 2 standard deviations above: 0.9772
Prob. 3 standard deviations above: 0.9987

From probability values such as these, a desired lead time can be determined to satisfy certain delivery probabilities. If for some reason, a 50-percent probability of being delivered within a particular lead time is satisfactory, then the items would be ordered that number of days equal to the mean of the historical data, before the current inventory is expected to hit a minimum level. For the present example this would be 13 days. A more realistic situation would be a probability possibly of only 2 or 3 percent for runout. From the probabilities given, it can be seen that designing a lead time for ordering on the basis of 2 standard deviations above the mean would be logical. This would give a value of approximately 13 + (2) (1.3) days, which rounded to the nearest day would suggest ordering 16 days before the present inventory depletes to the desired minimum.

Demand Variability. The problem just presented becomes more complex when it is realized that in certain manufacturing situations, the **demand characteristics** are by no means certain as regards runout time prediction. Fig. 11

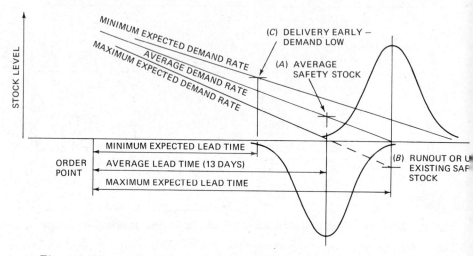

Fig. 11. **Effect of uncertainty of both supply and demand in the inventory model.**

demonstrates graphically how uncertainty for both lead time and demand can affect the decision regarding when and how much to order. If demand characteristics can be predicted exactly then the preceding analysis would require just a determination of the lead time, based on some probability requirement. The quantity to be ordered could be some fixed amount determined through a deterministic economic order quantity calculation. However, with the demand behaving stochastically, the amount to order also becomes a variable.

If the inventory replenishment cycle were designed on the basis of expected arrival and expected runout times, as defined in Fig. 11, then most frequently the safety stock of inventory would lie at point *A*. A small percentage of the time it could be much smaller, say when the demand is at the maximum and the delivery is at the latest expected time. For this case the safety stock would lie at point *B*. Sometimes, the stock will be considerably above the safety stock at point *C*, but very seldom will this be so.

Shifting the arrival distribution to the right, as shown in Fig. 12, will allow this average safety stock to be reduced, but it would also increase the probability of depletion before arrival of the next batch. This is denoted by point *D* on Fig. 12.

Greene (Production Control: Systems and Decisions) suggests a method where the reorder point quantity rather than the lead time is varied and further suggests that "The quickest and easiest way of obtaining a reorder point quantity would appear to be the product of the **maximum order lead time** (order time to latest arrival time) multiplied by the **maximum demand rate.**" This will give the largest expected maximum inventories but will minimize runout possibilities. As Greene points out, if inventories become excessive then this quantity can be adjusted down according to experience. The order lead time would be T_1 as indicated by Fig. 12.

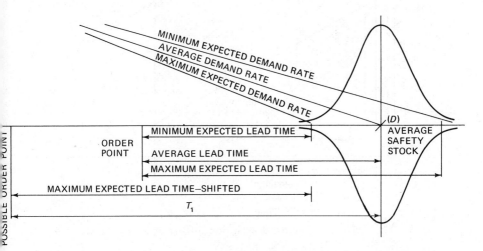

Fig. 12. Effect of shifting order point to attempt to minimize increasing safety stocks.

Inventory Policies Under Uncertainty. There are several commonly accepted inventory policies which can handle the problem of uncertainty of demand quite well. Two of these are considered here. The first is frequently called the (s,S) policy. The so-called **two-bin policy** falls into the (s,S) classification. As soon as inventory depletes to a level (s) then an order is made such that the expected level of inventory upon receipt of the order is S. Using the symbols given in Fig. 13, and realizing that S would have a value of Q added to ΔQ, the order quantity would be:

Order quantity $= S - s +$ (Expected demand in a lead time)

The average expected demand in a lead time would be $s - \Delta Q$ of course. The advantage in the (s,S) policy lies in the fact that ordering is initiated only when necessary from a demand point of view. The disadvantage is that a relative constant monitoring of inventory is needed.

One way to offset the monitoring problem is to use a (t,S) policy. Inventory with this policy is checked on a regular cyclic basis every t time periods, where t could be:

$$t = \frac{Q}{\text{Expected demand rate}}$$

An amount, determined by the order quantity equation, is ordered except that s is now replaced with the current inventory on hand at the review period, not some pre-determined value. A possible disadvantage of the (t,S) policy lies in the fact that ordering of very small quantities can ensue if the policy is followed too closely. A variant procedure is to only reorder if the cyclic review reveals that the current level of inventory is equal to or less than some pre-determined value, say s.

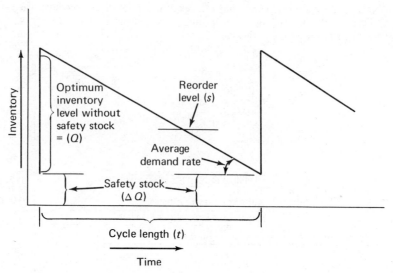

Fig. 13. Symbol description for inventory policies.

Maximum Demand Approach. One method of handling uncertainty in demand is to order a batch of items when current inventory decreases to a certain "reorder level." A specific economic order quantity is ordered at the time this occurs. This is quite similar to the simplest of the inventory control procedures, the **two-bin** system. In this approach inventory is theoretically stored in two bins. When one bin is depleted, the next batch is ordered and inventory is drawn from the second bin.

A variant of this is proposed by Brown (Smoothing, Forecasting and Prediction of Discrete Time Series). This is a very simple technique to apply and satisfies the possible variations that might occur in both demand and supply.

Records are maintained of demand occurring in past lead times. This can be accomplished by indicating on the order card current accumulated demand when the order is made. Upon receipt of the order, accumulated demand is again recorded and the calculated difference is the demand during the latest lead time. Brown shows that if records are maintained for the last N lead time demands, X_k, and these N demands are ranked such that:

$$X_1 > X_2 > \cdots > X_k > \cdots > X_{N-1} > X_N$$

then the probability that X_k will be exceeded in the next lead time period is equal to:

$$P[(X_{N+1}) > X_k] = \frac{k}{N+1}$$

For example, if the probability of demand exceeding a particular value is desired to be 0.05, then the value chosen could be the highest in the last 19 demands during a lead time. The probability would be $1/(19+1)$. Also, if more data records are available then the demand to use could be the second highest in the last 39 pieces of data such that the probability equals $2/(39+1)$. Brown shows that the more data held, the smaller the resulting variance of estimate. The next batch would be ordered when the current inventory depletes to a level equal to the demand chosen from historical data, plus the desired safety stock. Of course, this approach is really only feasible when there is no significant serial correlation between demands during a lead time.

Make-or-Buy Decision Under Uncertainty. One decision to be made in the production cycle is whether to purchase a batch from an outside vendor or whether to produce in a batch fashion within the production facility. In the deterministic case this problem is quite straightforward. It is a matter of comparing two alternatives, where one has a purchase cost and the other a setup cost. An advantage for production is the elimination of the vendor's profit. However, inventory carrying charges usually increase with the make situation due to the length of time required to manufacture the batch, whereas purchased items have an instantaneous delivery characteristic. In general, for the deterministic case, it would probably be advantageous to make if batch requirements are high. However, a vendor supplying to various customers will share his fixed costs among the customers, which if high would indicate a purchase situation might be more feasible.

Thuesen (Engineering Economy) discusses the following **hypothetical** example.

Assume that make-buy figures are as follows for a particular product:

	Purchase	Produce
Item cost	$ 6.00	$ 5.90
Purchase cost	10.00	–
Setup cost	–	50.00
Carrying cost per unit	1.32	1.30

The total costs can be plotted versus quantity required giving the plot of Fig. 14, which indicates the break-even quantity between make or buy.

When the facility is operating in the uncertainty phase of purchasing it may be advantageous to make rather than buy for several reasons:

1. In order to eliminate runout, inventories of purchased items tend toward their maximum value.

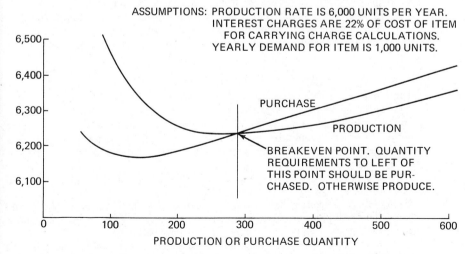

Fig. 14. Graphical evaluation of the make-buy problem.

2. If the item is batch produced within the facility then closer control can be maintained over delivery times of the items thus tightening the probability distribution for the availability of inventory.

If runout times and supply times can be predicted within narrow tolerances then the decision can be treated as a deterministic case. If not, then decisions have to be made regarding the maximum allowable probability of runout in order to determine the maximum expected inventory levels of the two alternatives.

EXAMPLE OF ECONOMIC PURCHASE QUANTITY. The probability and statistical techniques presented earlier were developed primarily to give methods for determining a lead time for ordering that minimizes the probability of running out of the current batch. An example will now be presented to cover the calculation of an economic order quantity based on a deterministic approach.

It was previously shown that the economic order quantity, Q_o, can be found from $\sqrt{\dfrac{S}{K}}$, where S represents the purchase order cost of ordering and K represents the carrying charges, both per unit of inventory. The carrying charges are a function of the average inventory held during batch consumption, and the length of time for batch consumption. In fact, K can be written as:

$$K = \frac{H_c T_b}{2Q}$$

where H_c = holding charges/unit per unit of time, and is a function of interest and storage charges based on the amount invested in one item of inventory

T_b = batch consumption time which of course is a function of batch quantity

Since T_b is in terms of Q, the above equation can be rewritten as:

$$K = \frac{H_c}{2R_c}$$

where R_c is the average demand during the same time period as is used in H_c. This might be month, day, week, year, etc., but has to be consistent.

For an example, assume that H_c, determined from finance and accounting department data, is $0.002 per unit-month. Also, monthly demand for the item has averaged 1,000 units per month, for the past 2 years. K is found to be

$$\frac{0.002}{(2)(1,000)} = 1 \times 10^{-6}$$

Assume purchase cost charges are $16.00 per order.

The optimum purchase quantity, Q_0, is then found to be 4,000 units, or 4 months' supply. Obviously, if the purchase order costs increase so will the batch quantity in order to minimize the number of times the items are ordered. Conversely, as the carrying charges increase, the batch quantity decreases to minimize inventory charges.

The problem is solved in tabular form as shown in Fig. 15 and represented graphically in Fig. 16.

Safety stock of course has to be planned as a storage function above and beyond the 4,000 units ordered. If the average safety stock is to be a 2-months' supply, then storage would have to be planned for 6,000 items.

It is interesting to see the effects of a small change in total costs per unit versus allowable purchase quantity as discussed earlier. From Fig. 15, it can be

Batch Quantity (Q)	Carrying Charges/Unit (KQ)	Purchase Charges/Unit (S/Q)	Total Costs/Unit (KQ) + (S/Q)
1,000	$0.0010	$0.0160	$0.0170
2,000	0.0020	0.0080	0.0104
3,000	0.0030	0.0053	0.0083
4,000*	0.0040	0.0040	0.0080*
5,000	0.0050	0.0032	0.0082
6,000	0.0060	0.0027	0.0087
7,000	0.0070	0.0023	0.0093
8,000	0.0080	0.0020	0.0100
9,000	0.0090	0.0018	0.0108

* Optimum batch conditions.

Fig. 15.   Tabular solution for determining economic order quantity.

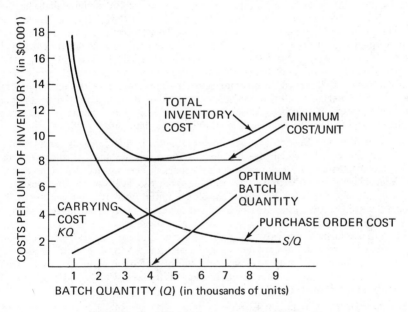

Fig. 16.   Graphical representation of Fig. 15.

seen that allowing total costs to vary from the optimum about 3 or 4 percent, the purchase quantity is allowed to vary as much as 20 percent. This might be quite a leeway in purchase planning.

Manual Records and Procedures for Inventory Control

INVENTORY LEVELS. Any facility with a large number of inventory items needs to keep a **recursive** record of all transactions for each item of inventory so that it is possible to determine current inventory levels at any period of time. Accurate knowledge of current inventory levels is mandatory as the value of inventory, according to Greene (Production Control: Systems and Decisions), is of such great importance to the manufacturer that it shows up in the most important financial statements, the **balance sheet** and the **profit and loss statement.** The inventory level records are also mandatory for obtaining data for economic order quantities, such as average demand data.

Record Keeping for Inventory Systems. The minimum information for an item inventory record can be classified into two parts: **constant information,** consisting of such data as part number and name, unit of measure, the unit price, minimum reorder point, the economic order quantity, and possible suppliers; and **variable transaction information,** consisting of the transaction date, type of transaction (receipt or issue), and current balance.

Additional information may be valuable in specific cases.

TYPES OF INVENTORY RECORDS. There are a variety of record forms and cards that could be used with inventory systems. Each has some advantage in terms of ease of working, visual characteristics, access to information, and the like. It should be emphasized however that these records are merely tools for preserving and presenting information in a convenient usable form. No control is obtained unless the records are properly analyzed and decisions made and implemented in line with the policies and objectives of the company. Although the physical form of the record (cards, charts, padded sheets, etc.) may vary, the type of information recorded often varies very little from one plant to another. Scheele et al. (Principles and Design of Production Control Systems) designed a **material record form** similar to the one shown in Fig. 17. The variable information allows orders to be compared with receipts to check partial shipments. Scheele also suggests a **usage recap sheet** that might be used to provide historical demand data, as shown in Fig. 18.

Rotary Drums. Record systems are obtainable in the form of large rotary drums on the rims of which are holders for mounting notched or slotted cards separated into groups by index guides or markers. These mechanisms (for example, Wheeldex) come in various sizes, ranging from the desk variety up to cabinet sizes holding 5,000 cards. With the simplified units, over 100,000 cards can be brought within the working area of an operator. A typical drum inventory card is shown in Fig. 19 (page 26).

For quicker distinction between kinds of items and greater ease of operation and control, cards of different colors are used for different general classes of items. One company utilizing the drum variety of equipment has buff cards headed "Raw Material," salmon cards headed "Purchased Part," and green cards headed "Manufactured Part."

Visible Index Cards. A Kardex system (Sperry Rand) using a series of **three foldover forms** can be used to indicate different aspects of the inventory system. A buff-colored form for standard commercial parts has a foldover front showing commitments (upper part of Fig. 20, on page 27) and the main inventory portion underneath carrying the visible index (lower part of Fig. 20). The raw materials card is the same (commitments on the front, inventory underneath), except for the name heading and the color, which is salmon. The third form, for consigned material, is light blue.

Another visible index card is shown in Fig. 21 (page 28). The cards are laid in a wide file with a series of pockets. In each pocket a number of cards are overlapped horizontally so that the part number or balance on order on each card is visible. The notches in the bottom of the card fit over a series of rods which hold

PART NO. _____ S 178 _____   PART NAME _Compression Spring_

BIN LOCATION _A-8-4_   UNIT OF MEASURE _Gross_ UNIT PRICE _#22_

REORDER POINT _____ 15 _____   ORDER QUANTITY (EOQ) _____ 45 _____

VENDORS _____ Acme Supply, 433 Plum St., Cincinnati _____
_____ Allison Spring Co., 1743 E. Main St., Indianapolis _____

SUBSTITUTE PART NOS. _____ S179, S230 _____

Date	Purchase/Shop Order No.	Purchases Amt. Purch.	Total Due In	Receipts	Issues	Balance	Allocations Allocated	Available
4-13	P.O. 1783			40		40		40
4-23	S.O. 1612				10	30		30
5-2	S.O. 1783				5	25		25
5-19	S.O. 1891						10	15
5-21	P.O. 1213	45	45					15(60)
6-2	S.O. 2017		45		10	15		15(50)
6-3	P.O. 1213		20	25		40		40(50)
				20		60		60(50)

Fig. 17. Material record form.

PART NO. S 178

MONTH	19____	19____	19____	19____	19____	19____
JANUARY	5	6	8	7		
FEBRUARY	9	8	11	12		
MARCH	6	7	17	15		
APRIL	8	12	13	10		
MAY	7	9	8	5		
JUNE	13	14	17	20		
JULY	10	11	16			
AUGUST	9	12	12			
SEPTEMBER	12	14	15			
OCTOBER	14	16	19			
NOVEMBER	13	16	17			
DECEMBER	10	14	15			

Fig. 18. Usage recap sheet.

the card in position. The cards may have various types of signal devices cut into the exposed margin, which is clearly visible.

Control Boards. Another system for handling records involves four types of forms: a **receiving slip,** which can also be used as a materials credit slip, a **materials requisition and stock record,** a **stock inventory,** and **permanent stock cards.** The two former are carbon-backed and are made out in triplicate. The stock inventory is in duplicate and the permanent record is a folded tab card made out singly. Various colors are used to code the type and destination of the forms.

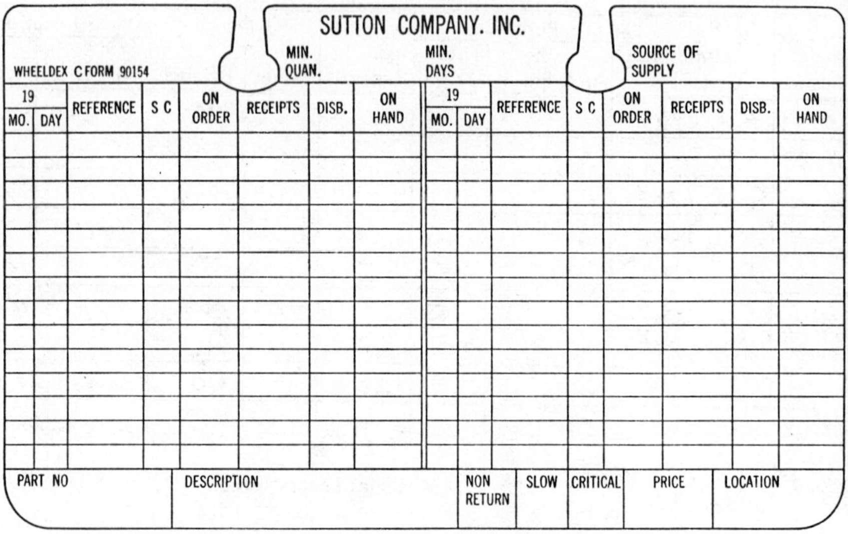

19 MO.	DAY	REFERENCE	S	C	ON ORDER	RECEIPTS	DISB.	ON HAND	19 MO.	DAY	REFERENCE	S	C	ON ORDER	RECEIPTS	DISB.	ON HAND

SUTTON COMPANY. INC.

WHEELDEX C FORM 90154   MIN. QUAN.   MIN. DAYS   SOURCE OF SUPPLY

PART NO	DESCRIPTION	NON RETURN	SLOW	CRITICAL	PRICE	LOCATION

Fig. 19. Inventory control card for use in a rotary drum.

A board is usually located in the stores area and another in the stores record department, but a single control is used in some cases. Clips are mounted on the board in rows running horizontally and vertically; the boards constitute panels which are used on both sides, and they are mounted on wall brackets so that a series can be grouped together and swung back to give access to whatever board is wanted. The clips are coded with the bin or shelf-location number where the material is stored. A visible index system shows the bin number and the corresponding clip-board number of any item in stores.

Receiving slips are made out in the receiving department (or if used for stores credits, in the department from which the material is returned) and sent to the purchasing department and to the storeroom with the material. When the material is checked and stored, the storeskeeper notes from the top (last-posted) ticket under the corresponding clip the previous balance in stock and enters it as the in-stock amount on the new receiving slip, which shows the amount just received. He adds the two amounts and obtains the new total. He then files a copy on top under the clip to show the latest condition of the material and sends the carbons to the stores records section to be clipped on the board there. As the storeskeeper receives material requisitions, he fills them, notes from the top card under the appropriate clip the previous balance, which

STANDARD COMMERCIAL PARTS COMMITMENTS

CONTRACT REQUIREMENTS   +   %   **TOTAL REQUIRED**

SPECIFICATIONS

DATE	PUR. ORDER	REC. REPORT	FOLIO	VENDOR	ON ORDER			RECEIPTS		DUE ON ORDER	
					QUANTITY	UNIT	VALUE	QUANTITY	VALUE	QUANTITY	VALUE

STANDARD COMMERCIAL PARTS INVENTORY

	RECEIPTS				ISSUED					BALANCE		
DATE	REC. REPORT	QUANTITY	UNIT	VALUE	DATE	SOURCE	QUANTITY	UNIT	VALUE	DATE	CHECKED	QUANTITY	AV. UNIT	VALUE

WORK ORDER   **PART NO.**   **DESCRIPTION**

Fig. 20. Foldover card forms.

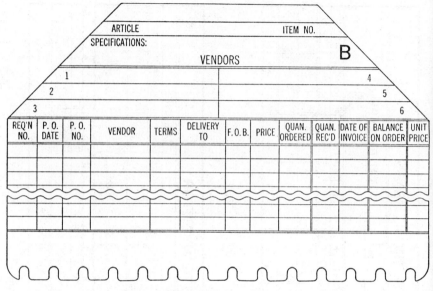

Fig. 21. Visible index record card.

he enters as in-stock amount on the new requisition slip, subtracts the new amount issued, and enters the new total. He then forwards the carbons to the stores record section. Stores credits are handled like receiving slips (Fig. 22).

The stores record clerk checks each slip as it comes in and verifies it with the previous tickets for the same item. Then he enters the necessary cost data, which are not recorded in the storeroom, posts his copy of the ticket on his board according to bin number, and sends the carbon to the cost department, accounting department, or wherever designated. Additional copies of tickets can be provided in the system if more are needed. The boards in the storeroom and stores record department are identical as to postings. **Stock inventory checks** (and periodic physical inventories, for which special report and summary forms are used) can be made in the same way as under other systems and the tickets handled as above to show the check or make corrections in quantities. The stock inventory tickets call for quantity, price, units, and amount entries. For group issues, apportioning of material, and other purposes, the system is varied slightly to conform to needs.

The **tabbed permanent stock card** (Fig. 23) is used only in the stores record section. It is folded at the vertical line beside the price tab when filed, and is placed behind all of the tickets under the clip, with the tabs showing above the tickets. The minimum (reorder) quantity shows on the middle tab so that when the balance falls to this figure a reorder requisition can be placed. A copy of this requisition may go on the board under the clip as a record. The price tab is used for quick reference in entering costs on requisitions. A permanent record of purchase or manufacturing orders to replenish stock is entered with sources, deliveries, prices, etc.

ABC INVENTORY CONTROL METHOD. There are many situations where a large percentage of cost in an enterprise is contributed by a relatively

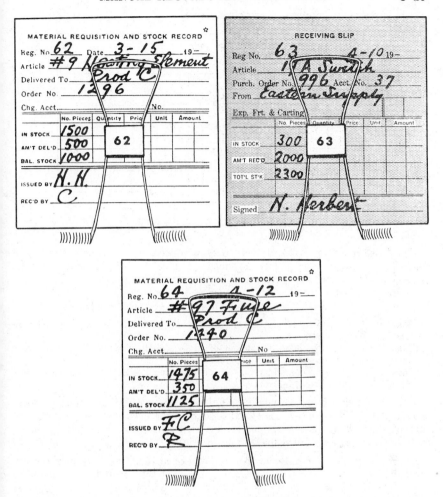

Fig. 22. Control board for inventory control (duplicate boards in storeroom and stores record section).

small percentage of components in the enterprise. As Plossl and Wight (Production and Inventory Control Principles and Techniques) state, ". . . about twenty percent of the people in the United States have eighty percent of the wealth; about twenty percent of the items in the family budget account for eighty percent of the dollar expenditures." This can be extended to a multitude of cases. This type of relationship was widely publicized by Vilfredo Pareto (1848–1923) in his empirically derived law covering the distribution of incomes.

Inventories exhibit the same kind of relationships with a **large proportion** of the cost of inventory contributed by a **small percentage** of items in inventory. Thus controlling the costs of only these few items will contribute to an effective control of a large amount of costs, clerical costs are reduced and inventory costs will be well controlled.

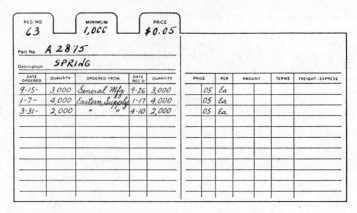

REG NO	MINIMUM	PRICE
63	1,000	$0.05

Part No. A 2875

Description SPRING

DATE ORDERED	QUANTITY	ORDERED FROM	DATE REC'D	QUANTITY	PRICE	PER	AMOUNT	TERMS	FREIGHT-EXPRESS
9-15-	3,000	General Mfg	9-26	3,000	05	Ea			
1-7-	4,000	Eastern Supply	1-17	4,000	05	Ea			
3-31-	2,000	" "	4-10	2,000	05	Ea			

Fig. 23. Permanent stores card.

The ABC inventory control method is a procedure in which inventory is classified into high-value (Class A), medium-value (Class B), and low-value (Class C). Obviously the system does not have to be limited to three classifications but this is the usual policy. Magee and Boodman (Production Planning and Inventory Control) divide inventory into the following three classes:

Class A. The top 5 to 10 percent of items, which accounts for the **highest dollar** inventory investment.

Class B. The middle 20 to 30 percent of items, which accounts for a **moderate** share of the inventory.

Class C. The large remaining group of stockkeeping items, which accounts for a **small fraction** of total investments.

The key difference in the handling of these parts is that large reserve stocks should be established for Class C items so that there is never a possibility of running short. On A items, the reserve quantities should be held to a minimum in order to decrease the investment in inventories.

Example of ABC Analysis. An example of an ABC analysis of inventory presented by Barnett (Industrial Management Society Bulletin) is described below. The "average usage" and "factory or purchase cost" of each item in inventory is extended to get a **"usage times cost"** value as shown in Fig. 24. Separate sheets should be used for raw materials, purchased parts, and shop parts. Items are then recapped and grouped by dollar classifications as shown in Fig. 25. The classifications on the recap sheet are then combined at logical break points. In this example, the final ABC classification of inventory is:

Class	No. of Items	Percent of Items	Value	Percent of Value
A	328	9	$630,000	75
B	672	20	142,000	17
C	2,421	71	67,000	8

From the above table it will be noted that 71 percent of the total quantity or number of items in stock comprise only 8 percent of the total dollars expended

Part No.	Average Usage	Factory or Purchase Cost	Usage Times Costs
F10	5	$ 2.10	$ 10.50
F11	75	.15	11.25
F14	2	30.10	60.20
F15	2,000	.05	100.00
F16	700	.80	560.00
F17	1	180.00	180.00
F19	250	1.10	275.00
F21	10,000	.005	50.00
F22	400	.30	120.00
F23	650	.25	162.50
F31	10	.08	.80
F32	25	.60	15.00
F35	90	1.10	99.00
F36	200	8.50	1700.00
F38	50	.80	40.00
F40	1,500	.40	600.00
F41	150	.10	15.00
F42	20	.50	10.00

Fig. 24. Worksheet for ABC inventory analysis.

Usage Times Cost Value	No. of Items	Cumulative Total	Cumulative Percent	Value of Items	Cumulative Value	Cumulative Percent
$700 and over	242	242	7	$580,000	$580,000	69
500–700	86	328	9	50,000	630,000	75
400–500	55	383	11	25,000	655,000	78
300–400	95	478	14	34,000	689,000	82
200–300	170	648	19	42,000	731,000	87
100–200	352	1,000	29	41,000	772,000	92
Under 100	2,421	3,421	100	67,000	839,000	100

Fig. 25. Recap of items in Fig. 24.

for material, and 9 percent of the total number of items used during the year comprise 75 percent of the total dollars spent for material.

AVOIDING STOCK OUTS. One of the best preventives against customer dissatisfaction or production line shutdown is the elimination of stock shortage in raw materials, shelf items, and component parts. In most cases, both can be prevented, though customer dissatisfaction is harder to convert to a cost value than is a physical line shutdown. A major step toward avoiding stock outs is the realization that demand and supply are probabilistic rather than deterministic. Holding buffer stock against possible shortages due to uncertainly will alleviate the situation. This buffer stock is commonly classified as safety stock.

Safety Stock. In the ideal situation everything can be forecast exactly; safety stock is therefore unnecessary. In reality it is the unexpected that creates a need for safety stock. Order quantities are predicated for the most part on avoiding runout of inventory. However, it should be understood that a safety stock should be available under this runout condition in order to protect against the **worst case** possibilities—the receipt of a batch at the latest possible time and runout of present stock at the maximum possible rate.

Magee and Boodman (Production Planning and Inventory Control) define safety stock as "the average level of minimum balances on hand. The safety stock is designed to absorb fluctuations in the minimum balances, as under a fixed order system. In a fixed order system the amount of stock on hand at the minimum points will fluctuate by the same amount as the differences between actual and expected demand over the period of a lead time plus cycle time. Thus the required safety stock can be determined from an analysis of these fluctuations with a specified risk runout."

If an inventory system under uncertainty is designed as presented previously in Fig. 12, the lead time is calculated on the basis of an allowable risk of a runout. On the average, the safety stock would equal the difference between the expected runout time of the current demand and the expected arrival time of the next batch, multiplied by the average demand rate. The largest safety stock that should be considered would be the latest expected runout time minus the earliest arrival time, multiplied by the smallest expected demand rate. With the approach given in Fig. 12, a lead time can be designed whereby the probability of never being below a particular safety stock is minimized. However, the logical approach is to design the lead time on the basis of rarely running out, as just mentioned, and then the safety stock will be a function of the lead time and average demand.

Automatic Inventory and Order Processing

INVENTORY AND REQUISITION RECORDS. Complex and varied inventory often justifies the use of more sophisticated equipment for maintaining records and control of inventory. On the other hand, many small production-inventory systems might not justify even the manual type of record system as all expenditures for these systems should be justified from an economic point of view. The use of **digital computers** for data acquisition, storage, and processing represents the most sophisticated technique.

First, consideration will be made to record media that lend themselves to automatic processing and possibly, but not necessarily, to digital computer manipulation. Again, the type of recording should be determined by an **economic** consideration of the complexity of the current inventory process. As noted by Evens (Engineering Data Processing System Design), "A roll of microfilm or a set of aperture cards for a single microfilm roll in an office system costs as much as many items which are subject to rigid inventory control in manufacturing plants. This type of system would tend to highlight the amount of scrap and wastage, figures about which are often unknown in office operations."

Punched Cards. There are various forms of punched cards, but the most familiar is the 80-column IBM card. This is $3\frac{1}{4}$ by $7\frac{3}{8}$ inches and has a thickness of 0.007 inch. Approximately 145 to 150 cards will take up a space width of an inch. Assuming that up to four cards can store the pertinent inventory information for one item, it follows that a considerable number of item records could be maintained in a relatively compact space.

It is relatively simple to enter data on a punched card using a key-punch machine with conventional typewriter symbols. They can be processed by a variety of office equipment which sort, collate, merge, update, and so on. Robichaud (Understanding Modern Business Data Processing) discusses a punched card inventory control system that requires four basic punched cards:

1. Inventory order card.
2. Master inventory card.
3. Inventory balance card.
4. Inventory issues or receipts card.

Robichaud's control procedure is discussed in some detail under Inventory Information Processing in this section.

Magnetic Tape. Awad (Business Data Processing) presents the basic characteristics of magnetic tape for data processing storage:

1. Tape is usually ½ to 1 inch wide and 1,200 to 3,000 feet per reel.
2. Data is stored and read by read/write heads. Format for the data is in conventional binary logic and therefore is directly compatible with computer systems.
3. Generally, the contents of about seven, 80-column punched cards can be stored per inch of tape—high density storage. An average reel of tape, say 2,400 feet, could store the information contained in up to 100,000 punched cards. It could be quite a bit less.

Advantages of magnetic tape are numerous, ranging from **cost** to **density of storage.** Awad indicates that the cost of a reel of tape to be about $38.00 as contrasted to $1.00 per 1,000 cards. Of course, updating the tape and printing stored information requires rather expensive equipment. Use of magnetic tape with a computer inventory control system is presented elsewhere in this section.

Other Methods. Many other methods for storing inventory data are available. **Microfilming** is a common technique utilizing 100-foot rolls of film each frame of which contains a recorded document. Searching for a particular document is relatively slow. However, equipment is available that can automatically retrieve and reproduce the document on a microfilm reader.

A cross between punched cards and microfilm is the **aperture card.** A 35-mm film frame, representing one document, is mounted in a conventional 3¼ by 7⅜-inch punched card. Particular documents are retrieved by sorting the aperture cards in a manner similar to punched cards.

Microfiche is a process favored by libraries since many Federal documents are now available in this format. Up to 60 documents can be recorded photographically on a 4 by 6-inch sheet mm. A special reader allows the documents to be sequenced through a viewer.

Commercial storage processes employing some of the above techniques usually incorporate special features designed to increase the speed of retrieval and the amount of information that can be stored. Some processes have magnetic coding of document images to allow very rapid search for particular items. Other features include built-in automatic reproduction of the stored document. The permanent aspect of these types of records make them applicable to storage of historical and physical inventory data such as vendors and their performance, material characteristics, and so on.

Computer memories also fall into the category of storage media, and have the advantage of being easily changed and so are not permanent. The three main categories of memories are **magnetic core, magnetic disk,** and **magnetic drum.** The latter two would generally be considered back-up storage, though in early computers drums formed the central processor memory. Core memory would be utilized for calculation storage with information from disks or drums being transferred in and out of core storage much like information stored in magnetic tapes.

The advantage of disk storage is its rapid retrieval time for specific data. Disk storage has an average access time up to 1,000 times faster than magnetic tape. Magnetic drums are inexpensive and allow large amounts of data to be stored. Their average access time is also much better than magnetic tape.

Core memory is used as intermediary storage while inventory information stored on disks or drum is updated. This approach to inventory information is feasible for inventory systems with large numbers of different items, such as might be encountered in a military or aerospace facility.

INVENTORY INFORMATION PROCESSING. Automatic inventory information processing can be considered in two categories—punched-card processing and digital computer processing. The latter might use punched cards for data input but it would be quite automatic with pre-programmed decisions triggering information output, such as items to reorder, quantities, and when and from what vendors. The punched-card system is **open-loop** in character, requiring human intervention between most of the operations. The punched-card system requires tabulating equipment commonly found in business offices.

Tabulating Machines. A punched-card inventory control system requires the following processing equipment:

1. **Key-punch** for punching inventory information on cards.
2. **Sorter** to sort cards according to some specified information, typical of which would be the item number.
3. Punched-card calculator **(tabulator)** or accounting machine used to develop inventory reports.
4. **Lister** for printing in summary report format specific items from punched card decks.

Many other pieces of equipment might be required for specific needs.

A punched-card control procedure developed by Robichaud (Understanding Modern Business Data Processing) is presented in outline form to indicate the type of punched card operations possible.

1. Figure 26 shows the four basic cards required for the system: the inventory order card, master inventory card, inventory balance card, and inventory issues or receipts card.
2. **Inventory File.** Data to be kept in the inventory file and sources of data: item code number; item description; inventory minimum quantity for each item, as established by inventory control; starting quantity on hand, from actual warehouse count.
   a. For each item, reference data is keypunched on a master inventory card (Fig. 26) and verified.
   b. The cards are arranged in item-number sequence by a sorter.
   c. The cards are filed in a drawer classified as the **reference-card file.**
   d. The reference-card file is maintained up to date by keypunching new cards for new items or changes in data. The new cards are filed and the old cards pulled out.
   e. The balance-on-hand quantity for each item, along with the item's code number, is keypunched on a **starting inventory-balance** card (Fig. 26).
3. **Recording Issues from Inventory.** Information on a product-sales card is input. This card is prepared by sales and contains product number and description as well as invoice number, customer, quantity in transaction, unit price, net after discount, and so on.
   a. Product sales-cards are filed for later use in preparing stock reports (see item **5).**

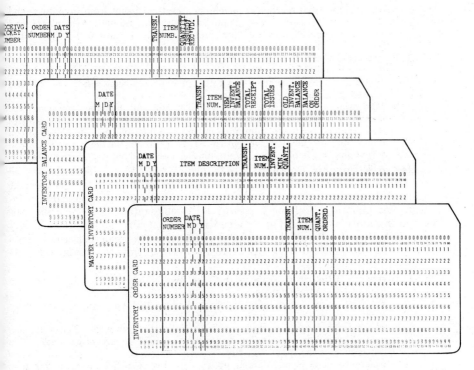

Fig. 26. Basic cards for punched-card and computer inventory control systems.

4. Recording Receipts and Orders. Receiving department forwards receiving ticket to the inventory control department. This ticket shows date of receipt, receiving ticket number, inventory item number and description, quantity received in stock, order number for receipts representing deliveries on orders. The manufacturing department forwards to inventory control a copy of each order issued to replenish inventory including such information as date of order, order number, item number and description, and quantity ordered.

    a. An inventory receipt card is punched from the receiving ticket with date; receiving ticket number; order number, if receipt fills an order; item number; and quantity received. From order documents, an order card is punched with order number, date, item number, and quantity ordered. Of course, if keypunches are available in manufacturing and sales the above two cards could be prepared at the source.

    b. The inventory-receipt and order cards are filed for future use in stock reporting (see item 5).

5. Preparing Inventory Stock Reports. Data comes from the cards prepared in steps 2, 3, and 4. Basically there are four cards: issues or product-sales cards, receipts cards, order cards, and old inventory-balance cards. Initially, the starting inventory-balance cards would be used.

    a. All four sets of cards are sorted together by item number.

    b. The cards are tabulated to produce an inventory stock report with headings such as item number, old balance, quantity issued or received, and balance on hand. At the same time, a summary punch is used to produce an inventory-balance card containing the date, the item number, the old inventory balance, the total receipts to inventory, the total issues from inventory, and the new balance on hand. Simultaneously, the tabulator calculates and summary punches into the new inventory-balance card the new balance needed for the next procedure.

6. **Preparing Inventory Control Reports.** Data required is on the master inventory card and the new inventory-balance card.
   a. The master inventory cards and the inventory-balance cards are collated by item number.
   b. The cards are placed in a punched-card accounting machine to prepare the inventory control report which would indicate item number, item description, balance on hand, minimum inventory allowable, net inventory, open orders, and net available inventory. This report would then trigger control action as regards ordering, determining causes of inventory bottlenecks, etc.

Use of Electronic Computers. The digital computer is able to handle all the functions discussed for the punched-card operation but is also able to make some of the control decisions triggered by the output reports in the card operation. For example, exception reporting is possible rather than reporting the entire inventory situation. For an indication of what a computer system might do, the system developed by Robichaud (Understanding Modern Business Data Processing) for punched cards will be repeated for a computer system (also developed by Robichaud).

1. **Inventory File.** It is assumed that the four basic cards in Fig. 26 will be utilized in the computer system.
   a. For each item, the reference data is keypunched on a master inventory card.
   b. The master inventory cards are read into the computer. The reference data is stored in random-access storage assigned to inventory data. This is classified as the **inventory file.**
   c. Reference data is maintained up to date by keypunching cards for new items or changes. The new data is read into the computer and the old data is deleted.
   d. The balance-on-hand quantity is keypunched for each item, along with the item's code number, on a starting inventory-balance card. The data is read into the computer under program control and the starting inventory-balance card is stored in the random access memory.
2. **Recording Issues from Inventory.** Same input cards as with punched-card system.
   a. Product-sales cards are read into the computer. Sales data is transferred into random storage. Quantity sold is subtracted from balance on hand and result is stored as new balance on hand.
3. **Recording Receipts and Orders.** Same data and source as for punched-card procedure.
   a. Receipt and order cards are prepared as with the punched-card system. These would probably be prepared at the source.
   b. Receipt and order cards are entered into the computer which then reads receipt data and automatically adds the quantity received to the old balance on hand stored in random storage and stores the new balance on hand. If a receipt fills an order, the computer subtracts the quantity received from the old balance on order and stores the new balance on order to form a new balance on order in memory. All functions are performed for any item having some type of transaction.
4. **Preparing Inventory Stock Reports.** All data required for this operation is now stored in a memory, such as magnetic tape.
   a. The computer is directed to print an inventory stock report (see item 5b of the punched-card procedure). No other functions are required as with the punched-card system since balances are already computed (see item 3).

5. **Preparing Inventory Control Reports.** All data required for this operation is now stored in memory.
   a. The computer is directed to print an inventory stock report for all items or only for those items that have a negative inventory position (for control report see item 6b of punched-card procedure).

In summarizing the punched-card and computer inventory control systems Robichaud (Understanding Modern Business Data Processing) makes some realistic comments: "The punched-card and computer systems for inventory control are practical only when there is a need for current and frequent stock reports and inventory control reports. These are required when there is a large volume of inventory transactions. Under the punched-card system, the stock reports and inventory control reports are prepared on a schedule—once a week, for example. The computer system has a distinct advantage over the punched-card system in that stock and inventory control quantities can be made available on an up-to-the-minute basis. . . . Again, however, it must be pointed out that a computer system requires more expensive equipment than that used for other systems. Many business organizations find it neither necessary nor practical to use the more costly equipment because up-to-the-minute data on inventory is not required for the effective operation of their businesses."

It should be pointed out that the computer could also be used for calculating economic order quantities as an integral function of the system. As companies move to on-line control digital computers in their production operations, these same computers can be utilized for inventory and production control functions in both an on-line and free-time basis.

INFORMATION HANDLING. As technology becomes more complex, the information system that services that particular technology becomes more complex. The procedures utilized in an inventory system should be related to the **complexity** of the inventory system itself, as discussed previously in terms of the punched-card or computer procedures. Speed of retrieval required for inventory information has to be a function of the number of items held in inventory and the complexity of the manufacturing process. For example, **continuous**, or assembly, production, has a fixed train of inventory requirement. Utilization of the economic lot size formulation, usually based on some hypothetical conditions, is justified in this situation, and less subject to uncertainties except for supply. However, a complex job-shop is another matter. The number of stock items required by the United States Air Force is something over 1.5 million, all subject to economic order quantity manipulation, of course by exception. This would be completely unrealistic without an automated type of inventory control procedure using computers. The manufacturer of a single product requiring a few standard components, with the product having a forecasted large and stable demand, would be able to handle his inventory with only pencil and paper.

Information Storage and Retrieval. The storage and retrieval for a punched-card or computer system was discussed in terms of case-problem systems earlier in this section. Storage for the punched-card system is on punched cards with the cards themselves being in files. Retrieval is handled somewhat manually with the cards being sorted and listed for the information required. The digital computer system stores information in an internal memory. Original information can be input by punched cards or by on-line input consoles. Retrieval can be almost instantaneous with output being printed or displayed on an output console device.

Ease of storage and speed of retrieval are based on the need for these characteristics. Microfilm and other document storage procedures were presented earlier. These items have a variety of values as regards storage and retrieval characteristics. The approach used for inventory information storage and retrieval should be geared to the particular application. Lipetz (in Scientific American Information) gives an interesting contrast for certain record approaches as shown in Fig. 27.

Characteristics	Microfiche	Aperture Cards	Recordak Miracode	Ampex Videofile
Normal storage configuration	4″ × 6″ sheet film in jacket	35-mm film frame in 3¼″ × 7⅜″ card	100-ft. reel, 16-mm film	7,200-ft. reel, 2″ wide
Size of storage container	30 jacketed sheets per inch of file	125 cards per inch of file	4″ × 4″ × 1″	16″ × 15″ × 3″
Storage capacity, 8½″ × 11″ documents (Retrieval ratio)	60 per sheet (20:1)	Up to 8 per card (24:1)	2,000 per reel (24:1)	250,000 per reel (1,280 lines per frame)
Maximum documents per cubic inch	75	42	125	350
Sequential search rate	(Manual)	Up to 33 cards per second	200 documents per second	1,050 documents per second

Fig. 27. Characteristics of various record storage systems.

In commenting on the data, Lipetz states, "Storage mediums are here compared by storage density and other characteristics. Microfiche contain no provision for machine searching. Aperture cards are usually stored in large open bins called **tub files.** A tub file 26 inches wide and 16 feet long will hold about 50,000 cards. If the file is well organized a hand search can retrieve a given card at random in less than 10 seconds. To extract the same card by machine search of a small batch of cards will often take much longer."

Evaluation of Inventory Control Programs

REVIEW OF CONTROL PROCEDURES. Any dynamic control system should be subject to review to determine its effectiveness. In an inventory control scheme, the simplest review would be to evaluate lost production or lost sales versus stability of on-hand inventory. Large inventories or accumulating inventories should be an anathema to the inventory control department. Scheele, Westerman, and Wimmert (Principles and Design of Production Control Systems) list six criteria by which a production control department can be evaluated. Five of these (taken somewhat out of context) pertinent to inventory control are:

1. **The system must furnish timely, adequate, and accurate information.** This can be reviewed by determining the worth of the economic order quantities in providing valid order points and inventory purchase quantities to satisfy demand without creating excessive inventories and storage requirements.

2. **The system must be flexible to accommodate necessary changes.** This has to be true for any dynamic process. Inventory accumulation and depletion is, by the very nature of production, dynamic. If additions and deletions of items or changes to existing item data cannot be made simply and effectively then the system is poorly designed.

3. **The system must be simple and understandable to operate.** If this is not the case then the system will simply not be applied by the inventory control people. Review of this point will be a function of the acceptance of the scheme. One that is not accepted cannot be implemented and therefore is a poorly designed inventory control scheme and should be modified.

4. **The system should be economical to operate.** If the overhead associated with the inventory control procedure is more costly than the inventory itself then obviously there is something wrong with the system. This would be the case if a computer were used where a simpler punched-card system was adequate. Or a punched-card system was designed where a manual system would suffice. Obviously, discovering that a system is too expensive after it is installed might be too late. Proper planning is essential. Examination of inventory control costs would give some indication of this problem.

5. **The system must permit "management by exception."** This should be qualified except for the most complex systems. The Air Force example involving more than 1.5 million items subject to economic order quantity review could only be accomplished by an exception procedure.

In summary, if an order cannot be met in the facility due to lack of inventory then the scheme has to be changed. If inventories are being built-up then the scheme has to be re-evaluated. If the scheme is not accepted by those who have to apply it then it is either too complex and should be simplified, or the personnel have not been educated sufficiently in the procedures required.

REPORT TO MANAGEMENT. The fundamental purpose of inventory control has to be the **minimization of overall costs** associated with inventory, commensurate with allowing for a smooth uninterrupted flow of sales or production. An inventory control policy cannot accomplish this by itself. Environmental conditions create the need for management decisions. Inventory control procedures provide certain possible alternatives, but management has to evaluate all alternatives in terms of the entire operation. The basic inventory problem is just one segment of the forecasting/scheduling/inventory interface. Each affects the other. Reporting inventory status and requirements allows management to coordinate all activities toward the best operation of the **entire** organization.

Since the management of inventories must reflect the overall operations of the company, top management is usually the focal point for all policy decisions in this area. In order to properly execute this function there must be a free interchange of information between the various groups in the company. Toan (Using Information to Manage) states that it is vital to the improvement of inventory management that all information which reflects the company's position at a given point in time be known: information showing how much is on order, how much is on hand, the current status of production, the workload ahead, old and new estimated customer delivery dates, quality problems and other facts of this nature. Much of this can be accomplished with **periodic status reports.**

Coordination with Production Control. A total logistics system is presented by Sims (Automation, Vol. 14) as shown in Fig. 28. Even though a

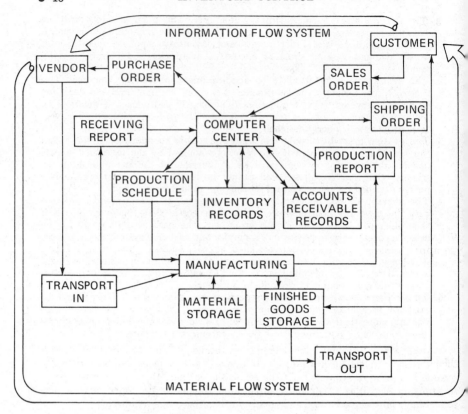

Fig. 28. Total logistics system for a production system.

computer center is given as the focal point of the information process, this should be construed as the decision center as management makes the decisions based on the information from the computer center. All functions listed in Fig. 28 have to be designed for the benefit of the entire organization. A production schedule cannot be firmed unless required inventory will be available when needed. Similarly, inventory cannot be planned until demand for items are forecast, probably through the sales organization. An effective forecasting procedure cannot be developed without feedback, in terms of inventory levels, regarding the validity of the forecasts. Data for the economic order quantities cannot be obtained except in terms of past demand from the sales or production departments. Therefore, it should be remembered that two-way communication is mandatory between all interacting departments. From an inventory control point of view this mandatory communication has to be with the production department.

PURCHASING

CONTENTS

CONTENTS (*Continued*)

SECTION 6

PURCHASING

Purchasing Functions and Policies

DEFINITION. Industrial purchasing is the procuring of materials, supplies, machines, tools, and services required for the equipment, maintenance, and operation of a plant.

The **purchasing department** is intrusted with this procurement duty. The function of the purchasing department is to procure these needed materials supplies, machines, tools, and services at an ultimate cost consistent with economic conditions surrounding the item being purchased; while safeguarding the standard of quality, continuity of service, competitive position, and the company's reputation for fairness and integrity. The purchasing department must also insure the company against the violation of all laws governing commercial transactions (Aljian, ed., Purchasing Handbook).

Procurement of goods through purchase accounts for about half the money spent by the average industrial concern, the range among different industries being from about 20 percent to 90 percent. The financial aspect of purchasing, therefore, is obviously of great importance.

MAJOR IMPORTANCE OF PURCHASING. Purchasing is of major importance because:

1. It is a primary function, directly influencing the major cost of operating a business.
2. Efficient operation of any industry depends upon proper turnover of investment. The purchasing department must insure receipt of proper materials, when needed, in sufficient quantities to maintain production; at the same time it must not increase investment beyond that required to meet current needs and maintain a reasonable factor of safety.
3. By its constant contact with many other companies and the general market, the purchasing department is in a position to:
   a. Discover new materials which may be used to advantage as substitutes for materials in use.
   b. Identify possible new lines of products to be added.
   c. See changes in trends, either in prices or other factors, that will affect the sales of the company.
   d. Build up goodwill in the business world with which it deals.
4. Its knowledge of vendors, market trends, and the manufacturing and marketing policies of other industries make it possible for the purchasing department to contribute invaluable help in framing plans for initiation of new products, scheduling of production, and determination of marketing policies.

The relative importance of the purchasing function, compared with the functions of other departments, will vary with the industry and, usually, the ratio between material cost and the value added by manufacture.

DUTIES OF THE PURCHASING DEPARTMENT. The duties of the purchasing department cover all dealings with vendors. No contacts regarding the purchase of any goods or services should be made without the knowledge of this department. Only with the consent of, and preferably in the presence of, some member of the purchasing department should other departments confer with vendors' representatives.

Its principal duties, not necessarily in order of importance, are:

1. Locating and selecting sources of supply for materials or services required.
2. Interviewing suppliers' representatives, arranging conferences and plant visits.
3. Requesting quotations and conducting negotiations.
4. Procuring materials and services when required.
5. Verifying quality and quantity received.
6. Handling rejections and adjustments.
7. Maintaining records necessary for proper operation of its function.
8. Keeping informed on business trends, assembling and analyzing pertinent data on markets, supply, demand, price trends, etc.
9. Disposing of scrap and surplus.

Objectives. Primary objectives of the purchasing department are:

1. To obtain the necessary materials, supplies, etc., of proper quality.
2. To procure them in time for plant requirements and have them delivered to the proper place.
3. To procure them at the lowest possible ultimate cost.

Briefly stated as a single **purchasing objective** it is to obtain what is wanted, when it is wanted, where it is wanted, of the right quality and at the right cost.

This objective is basically a matter of ultimate cost. Interruption of operations and poor quality of material are undesirable because they add to cost. Obviously cost is not the same as price, since the price of a material is only one element in its ultimate cost.

The impact of many factors considered together finally determines ultimate cost. Among these factors, in addition to delivered price, are: value in use, cost of carrying inventory, cost of errors and losses, cost of interruption of production, and departmental expense. As an example, the purchasing department must weigh the savings to be effected by buying in quantity at low spots in commodity markets against the expense of storing quantities in excess of immediate requirements until used.

GENERAL POLICIES. Every organization needs purchasing **guidelines** within which the daily actions of the buyers will be carried out. The existence of well-established policies reduces the number and complexity of the purchasing decisions that must be made by the buyers and insures a reasonable uniformity of action.

Purchasing policies define the basic decisions of the top organization manager as they relate to buying actions. Such policies show buyers how the purchasing manager would decide repetitive buying actions. In practice the top purchasing manager in the organization prepares the purchasing policies for the approval of the top management executive.

The number and detail of the purchasing policies will vary with the **organization philosophy** of top management. In most companies the following areas are covered by policy statements.

1. A definition of authority and responsibility.
2. Relationships with vendor companies.
3. Treatment of sales representatives.
4. Proper handling of competitive bidding.
5. Proper handling of vendor technical service and design work.
6. Reciprocity.
7. Employee purchases.
8. Ethical practices.

PURCHASING MANUALS. Purchasing manuals are designed to avoid conflicts between departments, to clarify responsibility, and to provide consistent instructions covering the regular activities of the purchasing department. There are two types of purchasing department manuals in common use: the policy manual and the procedures manual. These manuals may be issued separately or together.

Policy Manual. This manual is a written statement of the company's general purchasing policies for use by everyone both inside and outside of the company. These policies must be coordinated with general company policy and must have the approval of top management. Because of the many implications purchasing policies have toward relations with vendors it is common practice to make copies of the Purchasing Policy Manual available to vendors.

Procedures Manual. The Procedures Manual is a detailed precise statement of the **intracompany** procedural responsibilities. Such a statement, when properly communicated, insures that all repetitive actions carried out in the purchasing department will be performed in a consistently efficient manner. The procedure for purchasing discussed elsewhere in this section can serve as the basis for the development of a Procedures Manual.

Format of Manuals. Purchasing Manuals should be issued in **looseleaf or similar form** which permits easy separation and revision. Such revised manuals can only be effective if they are maintained according to current practice at all times. Neither policies nor procedures should be considered to be sacred and unchangeable but should be revised whenever careful consideration indicates that a better practice is available.

ETHICAL PRACTICES. Probably more than any other employee of a company, the buyer is in a position of ethical temptations. He is liable to ethical errors covering a wide range of situations ranging from outright dishonest personal gain to a completely unconscious failure to properly consider the company's interests.

The most common ethical problem facing the buyer is the avoidance of any action that places him in a position of obligation to a vendor. Outright commercial bribery obviously falls in this category but there are many more subtle ways that the buyer may be improperly influenced in favor of a vendor. Gifts or entertainment may constitute this kind of influence; the close personal friendship of a sales representative may lead to favoritism. Not only must the buyer avoid being improperly influenced in fact, he must also avoid the **appearance** of such influence because this in itself can hurt his company's best interests over a period of time.

Another ethical problem is that of maintaining an atmosphere of complete **competitive fairness** among vendors. In his desire to get the best "deal" for his company a buyer may be tempted to use various tricks commonly referred to as "sharp practice" in his negotiations with vendors. Among these is the practice of giving a vendor misleading information either about competitive offers or about the conditions of the contract in order to force him to improve his proposal. To exert undue pressure on a supplier is also an unethical practice which the buyer may be tempted to use.

It is apparent that in actual practice flagrant violations are less of a problem than the many subtle and often unconscious unethical situations that may occur in purchasing departments. The maintenance of completely ethical conduct in a purchasing department is essential but in some respects difficult. It can best be maintained by having strict, positive policy statements covering the area of ethics.

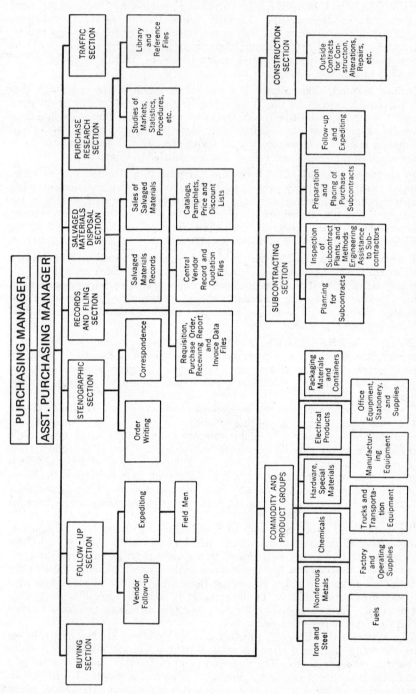

Fig. 1. Functional organization chart of a purchasing department.

Company policy and its implications should be discussed frequently with the purchasing personnel to assure complete understanding followed by strict enforcement.

Purchasing Organization

FUNCTIONAL ORGANIZATION. Fig. 1 is a **functional organization chart** for the purchasing organization of a large metalworking plant. Most of the duties usually assigned to a purchasing department are shown on the chart. Note that the Buying Section includes, in addition to the Commodity and Product Groups, sections specializing in Subcontracting and Construction. The Subcontracting Section is indicative of a trend in modern industry to buy from outside specialty vendors more of the components and assemblies while manufacturing in plant only the more critical components.

The **Purchase Research Section** is becoming common in large purchasing departments because it enables people who are specialists in research methods to conduct valuable studies leading to lower costs and better value. Although commodity buyers are usually involved in such studies they are normally under too much daily pressure to make the studies themselves.

Fig. 1 also includes a **Traffic Section.** Whether or not the traffic function is a part of purchasing will usually depend upon whether the company's major transportation problems involve incoming or outgoing materials. In some companies the Purchasing Department is given responsibility for the overall control of materials up to the time they are delivered to the production lines. Such an organization might include material control and warehousing as integrated functions and would then tend toward the **material management** concept of organization.

FORMS OF ORGANIZATION. The purchasing functions shown in Fig. 1 may be grouped differently or assigned as part of other departments but all of them must be performed in some way in any company. Responsibility for the direction of purchasing may come from a centralized point or it may be localized through the individual plants or divisions of large companies.

Centralized Purchasing. Responsibility for the purchasing function should be centralized whenever it can be done without substantial sacrifice in efficiency due to restriction of local initiative. There are great advantages to: undivided responsibility, consistent buying policies, and large buying power. With centralized purchasing, all purchase records are in one place and under one supervision. Compiling and consulting the records is facilitated. Thus more effective action can be taken to meet changing market conditions.

Centralization aids in the standardization of specifications and tends toward lower inventory investment. Vendor selling costs are reduced under centralized purchasing because only one buyer must be solicited. This, in turn, can lead to lower buying prices. Centralized purchasing should be the rule for any concern operating a single plant.

If the company has more than one plant, many factors must be considered to determine if centralization is possible: geographical location of plants, similarity of products manufactured, type of materials forming the bulk of purchases, special market conditions relating to large volume materials, location of suppliers, and other conditions peculiar to the industry or company.

It follows then that a company having several plants in the same section of the country, manufacturing similar products requiring the same general types of material, should have centralized purchasing (National Association of Purchasing Management, Guide to Purchasing).

Localized Purchasing. At the other extreme is the concern operating several plants in widely separated locations and manufacturing different products, so that each has individual material requirements quite different from those of any other. Here separate purchasing departments as part of the local plant organization may be preferable. While reporting on a line basis to the plant manager and serving directly the needs of their respective plants, these purchasing agents will usually have a functional control relationship with the corporate purchasing director. He will establish and enforce general purchasing policies and procedures and see that there is a proper interchange of ideas and information. The corporate purchasing director will execute most of the contracts applicable to more than one plant and will check or audit the work of the local purchasing agents, who will submit periodic reports.

Centralized-Localized Purchasing. Between these two extremes are many companies operating several plants whose geographical locations may not be too widely scattered and whose product and material requirements, while to a considerable degree heterogeneous, may nevertheless cover purchases of a large number of similar parts and materials used in common and in large quantities. Under such conditions it is desirable partially to centralize and partially to localize the purchasing function. A general purchasing department should be set up to establish general policies, to do actual buying when advisable, and to supervise and direct the work of local purchasing departments which will be set up at each plant.

Where a large degree of localized purchasing is practiced, consideration must be given to items used by all plants, on which preferential discounts for quantity contracts are available, but which can be obtained only by contracting for the requirements of all plants and specifying delivery to individual plants against the contract. Such situations are covered by a form of **pooled buying.** The individual plant which is the largest user of the material, especially if it is in the locality from where this material comes, buys for all plants on the basis of estimates, although such contracts may actually be written and put through by the central purchasing organization. Goods are shipped to each plant as specified and charged to the plant receiving them. Here again there must be coordination through a central purchasing department, because the plant using the major portion of the requirements may not be in the best position to buy the total amount needed.

Conditions Governing Centralized-Localized Buying. The main requirements for successful operation of centralized-localized buying are:

1. **Adequate central control** by the corporate purchasing director through established policies, procedures, basic forms, recordkeeping methods, and other fundamentals necessary for uniformity throughout purchasing. Delineation of authority and range of buying of the local purchasing agents.
2. **Effective communications** carried out so that the central office will promptly receive adequate information regarding local transactions and market conditions.
3. **Central office approval** of any deviation from the range of authority given to the local purchasing managers.
4. **Issuance of orders** by local plant purchasing managers against company wide contracts covering the needs of two or more plants. The actual negotiation and administration of such national contracts can vary greatly according to conditions but in all cases they should be approved by the corporate purchasing director.
5. **Flexibility** so that specific buying authorities can be reassigned as conditions change.

While companies with one plant naturally tend to establish centralized purchasing, there is a definite trend toward greater centralized control even in companies having more than one plant (National Association of Purchasing Management, Guide to Purchasing).

COORDINATION WITH OTHER PRODUCTION FUNCTIONS.
The purchasing department has a relationship with nearly every other production function. Basically it provides a service to those functional areas by securing the goods and services required by them. In the case of the eight functions discussed below there is a continuing relationship directly involved with production efficiency and profitability.

Production Control. Production control is concerned with scheduling the overall production of the factory and with coordinating individual departmental requirements. It is imperative that production control consult with purchasing in developing its schedules to insure that delivery requirements for materials are possible and economical. Likewise, purchasing should keep production control fully advised on delivery times of major materials so that schedules can be developed accordingly. The production control function can further help by providing projections of future requirements as far in advance as possible. This will enable purchasing to plan future supply sources and begin negotiations at the best time for the most economical purchase.

Engineering. The engineering department normally develops the technical specifications for the company's materials, components, and equipment. It should bring purchasing in at an early stage so that the established specifications can be purchased economically. Naturally this can be done best if commercial standards are used and if the production facilities of the logical suppliers are considered. The purchasing department should provide information to engineering about new processes, materials, and vendor capabilities so that these facts can be considered as designs are developed. Likewise, it is necessary that engineering understand leadtime requirements on critical items so that it can give priority to these specifications in developing bills of materials. In the area of leadtime requirements and in that of tolerance allowances purchasing and engineering will often have different viewpoints. Much can be accomplished if there is mutual understanding and respect.

Material Control. The material control function determines what items are to be stocked and what levels of inventory should be maintained. Although it is primarily concerned with direct production materials it is often responsible for supply and maintenance stores. It sends purchasing the requisitions that authorize the purchase of all such items. There must be a constant flow of information between purchasing and material control especially regarding leadtimes and economic order quantities. Failure to interchange such information effectively can be extremely costly to the company's operations.

Because of the vital relationship between purchasing and material control modern business organization will often have both functions integrated and reporting to a materials manager.

Receiving and Warehousing. Incoming materials must be promptly and accurately checked in, recorded, and stored safely until they are used. Purchasing can expedite these functions by insisting that vendors ship materials properly packaged and marked for effective handling. The receiving department must quickly report to purchasing all information on receipts or on damage so that unnecessary inquiries to vendors can be avoided or prompt claim action taken.

Quality Control. Quality control is often responsible for determining the acceptability of the materials supplied by vendors. Prompt incoming inspection and reporting is essential so that there is reasonable time for purchasing to settle claims for defective materials without severely disrupting the supplier and the department requiring the material.

Disputes sometimes develop between purchasing and quality control involving deviations from engineering specifications. Usually the dispute centers on whether or not a minor deviation is justification for rejection of a shipment of material. Although the quality control department normally has the right to reject material for any deviation from specifications this can at times be very costly for both the company and the vendor. Resolving such matters therefore requires good judgment as well as a sense of proportion on the parts of all parties.

Manufacturing. It is essential that purchasing secure delivery of production materials in time for their need on the production line. If this is not possible the manufacturing function must be notified far enough in advance so that production schedules can be altered without undue disruption of men and machines.

The manufacturing department should keep purchasing fully informed of all matters which may affect purchasing, such as schedule changes, usage rate changes, unusual wastage, or problems with the working characteristics of competitive materials.

Traffic. As mentioned earlier the traffic function is sometimes a part of the purchasing department especially if the transportation problems of incoming materials are substantial. Purchasing should work closely with traffic to insure that materials are being delivered with the lowest possible transportation cost. Buyers should always be aware of the proportion of the delivered price represented by transportation charges and seek to hold this proportion to a minimum. Since purchasing is always involved with expediting delivery of critical materials, a close working relationship with traffic is essential to insure proper routing and efficient tracing when necessary.

Accounting. Purchasing and accounting are traditionally closely related simply because most of the expenditures other than payroll are processed through purchasing. The most active relationship is with the accounts payable section of the accounting department where all vendor invoices are processed for payment. Both functions are concerned with the smooth flow of paperwork, permitting vendors to be paid promptly and cash discounts to be taken. Poor processing of payments to vendors results in much illwill and can lead to serious and costly disputes.

Purchasing normally assists accounting in establishing standard costs for all production and stores items. It also provides considerable cost information for budgeting purposes.

PURCHASING DEPARTMENT ORGANIZATION. The purchasing department of a small or medium-sized company will consist of a purchasing manager, buyers or assistants to the purchasing manager, an expediting section, and a clerical force. In a large company the buying division of an integrated purchasing department may consist of a director or manager of purchases, buyers, junior buyers, and a general service section (under the direction of a purchasing office manager). The general service section in turn may be subdivided into a correspondence unit, expediting unit, price-checking unit, file unit, and stenographic unit, each headed by a unit head responsible to the purchasing office manager. In such an organization, the purchasing office manager reports directly to the head of the purchasing department. The service section handles all matters regarding purchase orders occurring subsequent to the placing of the order, but operates in cooperation with the interested buyer and subject to his direction as far as any action affecting vendor relations or purchasing policy is concerned.

Purchasing Manager. The purchasing manager should be selected with careful regard for his ability, personality, versatility, and breadth of vision. He must be an organizer and leader for his department, a worthy representative of the company in its contacts with other concerns, a keen student of business, a man capable of consideration and prompt decision, having balanced judgment and clear foresight. He should be of the executive, not the clerical, accounting, or mechanical type, although insight into these fields is an asset.

The importance of the purchasing job was expressed by Mark Shepherd, Jr., President of Texas Instruments, in the following statement:

At the same time as our purchasing job becomes more complicated and more difficult, the delegation of the responsibility to the individual purchasing manager increases.

We look to our purchasing departments to support our engineering efforts with ideas and suggestions and liaison with outside vendors, to support our marketing effort by pointing out new business opportunities and suggesting marketing approaches (not reciprocity) which can be helpful. In general we expect our purchasing departments, because they are in the position to be so well informed, to be a source of innovations in its own operations and for the other operations within the company.

Responsibilities of the Purchasing Manager. The principal responsibilities of the purchasing manager, some delegated to assistants, are:

1. Keeping up the company's standard of quality production by his share in the choice of materials used.
2. Organizing and directing the purchasing department and acting as sole head of its personnel.
3. Spending a large portion of the company's money and being responsible for its wise expenditure.
4. Preserving the operation of the company's production schedules without interruption.
5. Representing the company in one branch of its major contacts with other firms.
6. Maintaining the company's reputation for integrity and fair dealing by his method of negotiating with vendors.
7. Acting as an executive of the company and a partner in its councils, particularly in preparing the purchase budget.
8. Keeping the company in step with progress and competition by research and openmindedness on new materials, new tools, etc.
9. Acting as final check, in the interest of economy, on all goods requisitioned, questioning need, quantity, and quality specifications.

These functions are in large part delegated by him to members of his department. In some small plants the purchasing manager retains responsibility for keeping in touch with and signing all contracts. The purchasing manager's actual **personal duties** are as follows:

1. Interviewing salesmen to obtain up-to-date information, securing and comparing quotations, and placing orders for such main commodities as he shall reserve to himself to purchase.
2. Establishing purchasing policies for his department to execute.
3. Preparing, or at least overseeing, all general reports on purchasing presented to management.

4. Conducting all major adjustment negotiations which are sufficiently vital to affect his company's goodwill.

5. Taking part in interdepartment conferences, whether for planning, formulating of company policies, or other purposes.

6. Approving material specifications on major commodities.

7. Supervising other functions of his department.

Assistant Purchasing Manager. The assistant purchasing manager is, under the purchasing manager's supervision, responsible for aiding in department operation, conduct of staff functions, and buying assigned classes of purchased goods. His duties in buying are to inform himself thoroughly about the market and manufacturing conditions of the class of goods so assigned, to interview salesmen, to secure and compare quotations, and to place orders for all materials entrusted to him. In a smaller company he usually conducts correspondence relating to shortages, defects, adjustments and, in some companies, follow-ups on orders. When general service or follow-up sections are maintained, the respective correspondence duties may be delegated to these sections.

Buyers. The buying section in a centralized purchasing department includes the purchasing manager, his immediate assistants, and a number of buyers. Items to be purchased are divided between buyers by **types of material** rather than by point of use or any other consideration. For example, a large purchasing department might have a buyer of ferrous metals, of nonferrous metals, of tools, of stationery and supplies, of textile products, and so on. Division of items to be purchased by this method allows specialization on related materials, reports on which are generally grouped and analyzed together in trade papers, forecasts, etc. It also provides for opportunity to become acquainted with the vendors' representatives and the vendors' plants by continued interest in and association with the same type of materials.

In some organizations the purchasing department may be set up on a **divisional or functional** rather than a commodity basis. In that case one buyer would handle all purchasing for a particular division or function of the business. The theory behind this plan is that the buyer thus becomes familiar with the particular needs and specifications of that division and learns to work with the plant organization responsible for that function. This plan also eliminates separation or double handling of requisitions containing several items belonging in different commodity groups. Except under unusual conditions, however, commodities rather than divisions or function should be the basis of buyers' assignments. Otherwise, duplication of effort and loss of total purchasing influence in a commodity field are inevitable.

Commodity Assignment. In assigning commodities to be purchased among buyers, it is well to have in mind groupings which have market tendencies in common and which are capable of being studied together to the best advantage. The related use of products may also be a factor in combining them for buying purposes. How far such grouping should be carried is affected by volume of purchases of various items, total volume of purchases, size of organization, and many other factors. Fig. 2 illustrates a method of dividing major procurement classes among several buyers so as to provide:

1. Ease of sales interview through relationship of items handled.

2. Full utilization of the technical ability of each buyer and easier study of new developments.

3. Close relationship with personnel of using departments.

Expediting. It is the responsibility of the expediting section to see that all goods ordered are received in time for requirements or in accordance with specified delivery dates and to keep a record of each order acknowledgment and promise of delivery. Some companies have found definite advantages in making the expediting function part of each buyer's duties. If such routines are well established, a buyer can perform more efficiently, exert more mature judgment, and feel greater responsibility.

Buyer 1	Buyer 2	Buyer 3
Subcontracts	Resale items	Machine tools
Construction	Hardware	Capital equipment
Fuel	Fasteners	Machine maintenance
Utilities	Abrasives	Lubricants
Automobiles	Electrical	Power transmission
Car rentals	Photographic	Perishable tools
Lumber		

Buyer 4	Buyer 5	Buyer 6
Metals	Chemicals	Paper
Castings	Cleaning supplies	Advertising
Forgings	Paints and lacquers	Office supplies
Molded parts	Hospital supplies	Office equipment
Plastic materials	Textiles and fibers	Packaging
Fiber	Leather	Dining room and kitchen
Precious metals	Safety equipment	supplies and equipment
Precious stones	Clothing	Containers
	Laundry	

Fig. 2. Grouping of materials for purchase by several buyers.

Clerical Force. The specific duties of the clerical force related to purchasing are to type purchase orders, record receipts, and keep all necessary departmental or interdepartmental records. Clerical work in the purchasing department should be directed and checked in exactly the same manner as any other clerical routine. In large departments the clerical force may be pooled under an office manager serving all departmental operations. In some cases clerical people work within each buying section and report to the buyers.

Classification of Purchases

RAW MATERIALS. Raw materials are basic, unfabricated materials bought in large quantity. They include such items as pig iron, copper, lead, tin, cotton, rubber, lumber, sand, leather, steel (except in fabricated forms), etc. In the main they are materials from which products are fabricated. Coal, coke, and fuel oil should be included with raw materials. Although they do not enter directly into the product, the methods and conditions of their purchase and their nature classify them as raw materials.

Raw materials as a rule are bought in large quantities, and usually contracts are made covering requirements for a considerable period. Prices are governed by quotations in commodity markets and are subject to rapid and wide change. The use of raw materials in a plant is relatively constant and information indicating price trends is readily available. Such materials, therefore, lend themselves to forward buying in anticipation of future needs. They should be bought on specification by chemical analysis or physical characteristics. It is the practice in the automotive industry to buy material to do the job rather than by physical specifications. For example, steel is purchased to make a certain fender, and so on. Raw materials are bulky, so storage space is an item to be considered. Transportation rates and handling charges form a large part of their cost.

SUPPLIES. Supplies are the many items necessary for the operation and maintenance of plant, shipping department, office, etc. In general they are items which do not enter into the product but which are necessary in daily operations. They include stationery, electrical supplies, pipe and pipe fittings, shipping containers, transmission supplies, bolts, nuts, washers, screws, packings, lubricants, hardware, abrasives, etc.

The purchase of supplies is characterized by the presence of many small items which, generally speaking, are standard items of manufacture, subject to published lists and discounts, though there may be specialties among them. There is little occasion to buy supplies for future requirements. Storage space is not a factor, but the clerical labor of checking, store keeping, and accounting is great because of the variety of items in comparison with the value involved.

FABRICATED PARTS. Under the designation of fabricated parts are included parts and small tools or accessories which are bought for resale, either as a part of the product manufactured or in connection with it. This class may include bearings, chucks, abrasive wheels, pumps, wrenches, tools, etc. It also includes special small parts which can be manufactured elsewhere more profitably than in the plant. Purchases of fabricated parts should be made on competitive quotations, which, wherever possible, should be checked against shop estimates for making the same parts in the plant. The accurate analysis of such make-or-buy decisions requires careful and rather sophisticated techniques. In-plant costs calculated by normal accounting methods can be somewhat misleading because specific allocation of burden items, such as supervision and technical personnel, is difficult to make. For standard catalog items special pricing arrangements are available to manufacturers and the buyer should be certain that he is purchasing at the best pricing.

INDUSTRIAL EQUIPMENT. Equipment items include machine tools, furnaces, boilers, automobiles, trucks, blowers, safes, and similar major items bought for the plant or office. Equipment purchase requisitions should be approved by the management before being handled by the purchasing department. Technical considerations are supreme in buying equipment, and such purchases should be covered by careful specifications, preferably by a description of purpose and result or performance demanded. Contributions made by the purchasing department in the purchase of equipment are:

1. Obtaining quotations.
2. Offering alternate sources.
3. Negotiating the contract.
4. Verifying the specification to prove conclusively that the machine will perform as required.
5. Maintaining an up-to-date library of catalog information.
6. Being alert to new developments, which might be more efficient than present processes, and seeing that interested persons are informed.

SERVICES. Every company has major expenditures for services performed by employees of outside companies. Such services can range from major construction work to cleaning service or the repair of office machines. Supervision of such work is under the direction of the department directly involved but the purchasing department should cooperate in negotiating the contract under which the service will be performed.

Purchasing Methods and Techniques

FUNDAMENTAL APPROACHES TO PURCHASING. Nearly any buying situation involves a complex set of conditions under which the purchase must be made. Many approaches to these buying situations have been developed and are used by the skillful buyer. The true test of a buyer is his thorough knowledge of all of the methods and techniques of purchasing and his ability to apply the one most appropriate to a given situation.

It is difficult to precisely define these approaches to purchasing because they tend to overlap and often several may be applicable to a given situation. However, the key factors are timing, quantity, and duration of commitments.

Morris (Analysis of Management Decisions) suggests an extension of the economic lot size calculations to problems involving economic purchasing quantities. The quantitative models can provide insight into a wide class of operating decisions concerned with the matching of inputs and outputs (see section on Operations Research).

Purchasing by Requirements. Purchasing by requirements means that no purchase is made until a need arises, and then the quantity bought covers only the existing need. This method applies principally to emergency requirements or to goods used so infrequently that they would not be stocked. It is essentially specialty buying and ordinarily makes the procurement of the goods the outstanding requirement. The task of the purchasing department is to have vendor connections that can be depended upon to fill such orders promptly and without taking advantage of the situation.

Manufacturers of specialty products produced on contract to customer specifications may use this type of purchasing routinely. Only when it obtains a customer contract can it begin to purchase the special materials needed, such as a special walnut veneer for a television cabinet.

Purchasing for a Specified Future Period. Purchasing for a specified future period is standard practice for goods regularly used, but not in great quantity, and on which price variations are negligible. Most supplies are bought by this method. The period for which the purchase is made may be fixed by a production schedule or by the record of past use or by a combination of both. Savings to be gained by the purchase of a given quantity also affect the determination of the period, as does the cost of carrying the items in inventory. It is important to note that no fixed period should be set for all purchases but that a separate and flexible period should be set for each item.

Purchasing According to Market. Market purchasing is designed to take advantage of price fluctuations. So long as market purchasing conforms to the production schedule and its possible changes, or to the demands of the plant or business, it cannot be classed as speculative purchasing. It is entirely possible to purchase wholly with reference to demand and yet take reasonable advantage of market fluctuations. This procedure is followed in the case of railroads, public utilities, and some manufacturing corporations, which have definite construction or manufacturing programs mapped out for long periods ahead. By constant study of market statistics and factors that affect prices, an efficient purchasing department will be able to forecast the trend of market prices and buy to best advantage. When its studies indicate that the price range is at a reasonably low point and that the future trend will be toward higher levels, the purchasing department will cover its requirements for a considerable period ahead. If indications are that prices are close to peak and that the trend will be downward, a hand-to-mouth purchasing program is indicated until prices have become stabilized at a lower level. There are cases where it is possible to do market purchasing and not take spot delivery or make immediate payment. This condition applies especially to fabricated or partly fabricated materials where the fabricator can "cover" at an advantageous price for his raw materials but does not want to make a commitment without being sure of an outlet for at least part of his finished goods.

This method is especially applicable to purchases of raw materials generally.

Advantages of this method are:

1. Large savings in purchase prices.
2. Greater margin of profit on the finished product, the price of which does not fluctuate as does that of the raw material.
3. Consolidation of purchases of a given material into one transaction with resulting saving in purchase expenses.

Disadvantages are:

1. Higher inventories with consequent higher carrying charges and tying up of storage space.
2. Liability to obsolescence in case radical changes are made in specifications.
3. Possible error in judging of market tendencies, which may mean large losses.

Speculative Purchasing. Strictly speaking, speculative purchasing consists in buying when the market is low, more than can possibly be used in manufacturing, with the idea of later reselling much of the material at a considerable price advance to users who may come on the market when the price is high. The term, however, is often applied to a more-than-normal purchase risk in acquiring an excess of materials on low markets in the belief that the price will advance very substantially, thus saving the company considerable money. It goes a step further than market purchasing, makes price trends in commodity markets the primary factor, and gives less regard to a fixed program of use as a basis for buying. It does not base decisions on demands of the business itself but on the possibility of market price savings. In some conversion industries the cost of a single raw material alone is more than 50 percent of the total cost of production. In some branches of the textile industry, cost of cotton outweighs all other elements of cost in producing cotton cloth. Here a saving of a few cents a pound on raw cotton offers a greater chance of profit than does any other activity of the business. In such a case successful speculation is often a primary means of earning dividends.

Speculative purchasing is not properly a function of the purchasing department. It should be authorized only by direct action of a financial executive or the directors. The purchasing manager should present the full facts, including hazards as well as possible advantages, together with his conclusions as to the advisability of taking the gamble. For ordinary manufacturing, the method is to be discouraged. Its single advantage is the possibility of huge speculative profits. Its **disadvantages** are many, including:

1. Tying up large amounts of capital.
2. Endangering the manufacturing schedules by waiting for profitable buying points.
3. Using large storage spaces.
4. Running the risk of obsolescence in case of radical change in specifications.

Contract Purchasing. Contract purchasing offers advantages comparable to those of market or speculative purchasing without some of the latter's disadvantages. By a contract calling for deferred delivery over a period, advantage can be taken of low prices in effect on materials at the time of placing the contract, while spreading delivery of the materials over a schedule consistent with estimated future requirements. Thus the price advantage is obtained without adding unduly to inventory. When contracting is possible it should be done in cases of raw materials fluctuating widely in price from time to time, although often contract prices will not be as favorable as prices for spot purchases of the same quantities. Sometimes contracts are made to purchase items at current prices with a fixed top price.

Contract purchasing may be a means of **assuring continuous supply** as well as a method of getting price advantages. It then becomes applicable to the purchases of parts, tools, etc. It is particularly helpful on those items where the production program is known, but the timing of it is not entirely certain. In effect it then becomes **scheduled purchasing**, as described below, but is embodied in contract form for the better protection of both vendor and purchaser.

Grouping Items. Group purchasing of small items is an interesting development in purchasing offering the possibility of large savings. Every purchasing manager finds that he must buy hundreds of small items so trivial in value that the cost of placing an order often exceeds the value of the goods purchased. Such purchases should be handled as quickly and as inexpensively as possible.

The purchasing department must continually watch the items included in the group, since the demand for an individual item may grow so large that it should be removed from the group and bought separately. Arrangements may be made to send orders for all group items to a single supplier who agrees to handle and bill them at a fixed percentage of profit above cost, his cost records being open to inspection by the buyer on demand. A considerable saving in clerical and purchasing expense is achieved by this plan.

The method is used chiefly in such fields as pipe fittings, general hardware, electrical supplies, and stationery. Often the practice is to secure from a reasonable number of bidders their quotations on a list of these small items with the understanding that the prices will be guaranteed on all requirements in the class of items covered for a period of three months. Then all orders go to the successful bidder without further inquiry or bid. An interesting variation is that some purchasing managers forward merely a copy of the requisitions to the supplier to eliminate the expense of orders and accept a single monthly bill to avoid the checking of multiple invoices. The small order is one of the most costly elements in buying. When the costs of the extra multicopy purchase order forms, the typing, the addressing, and the handling and filing operations are added up, the excessive expense is readily understood.

This concept has been expanded and combined with contract buying into a technique sometimes called **Systems Contracting,** a title registered by the Carborundum Company. Under this approach the supplier virtually takes over the complete inventory responsibility for a given category of materials for an agreed renewable period of time. Thus, with a minimum of paperwork the supplier keeps the inventory required to service all small orders for a repetitive category of maintenance or supply type materials. The exact terms of such an agreement can vary considerably to fit a specific situation.

Scheduled Purchasing. The schedule plan for purchasing materials used regularly in large quantities can be a source of important savings. It was devised to reduce investments in stocks. Essentially it consists in giving suppliers approximate estimates of purchase requirements over a period of time, thus placing them in a position to be able to anticipate orders and be prepared to fill them when received.

Although minimum inventory is probably the most important objective in this plan, other objectives are good quality, timely deliveries, and low cost. Good quality can be obtained by giving suppliers enough time to produce. Timely arrival of materials can be assured by laying down in advance a definite material requirement. Low cost results from giving suppliers advance information on requirements, thus permitting them to produce materials in the most economical manner.

The danger inherent in this plan is that requirements or specifications may be changed, and goods made up or allocated but not now required may become a matter of dispute between vendor and purchaser. Scheduled purchases should be restricted, therefore, to items definitely known to be required within a closely limited period and should be established as to specifications. Also, the purchaser should make very clear in his correspondence the responsibility he assumes and the risk the supplier assumes.

Blanket orders are purchase orders placed and accepted for large quantities of materials to be delivered as later specified. By the agreement the vendor agrees to furnish and the purchaser agrees to accept a stated number of units, usually within a given period. Orders based on customers' requirements are characteristic of the automotive industry. The vendor is then in a position to manufacture or procure the full amount in the assurance that the purchaser will authorize shipment in due time. The blanket order saves some of the formality of the contract method and achieves many of the desirable features of scheduled purchasing, but with a greater certainty as to the legal rights involved. Blanket orders often cover semifabricated parts or tools and contain a binding price based on the vendor's best estimate of his costs and the desirability of having a certain outlet for his product.

VALUE ANALYSIS. Value analysis is a systematic study which concentrates on the functional value of a material, part, or component with the objective of reducing costs. From the buyer's viewpoint value analysis can start when he clearly identifies the precise function of any material or component he has to purchase. With his knowledge of available materials and suppliers he then begins to ask what he can find that would perform the function as well or better at a lower or equal cost.

A typical industrial value analysis study involves many people both inside and outside the company at some stage in the project and the use of some complex and sophisticated techniques. The ten points of analysis developed by L. D. Miles, of General Electric Company, are typical of the customary approach:

1. Does use of the item contribute value?
2. Is the cost of the item proportionate to its usefulness?
3. Does the item need all of its features?
4. Is there anything better for the intended use?
5. Can a usable part be made by a lower-cost method?
6. Can a standard product be found which will be usable?
7. Considering the quantities used, is the item made on proper tooling?
8. Do materials, reasonable labor, overhead, and profit total its cost?
9. Will another dependable supplier provide the item for less?
10. Is anyone buying it for less?

In larger organizations value analysis activity is great enough to warrant full-time specialists who coordinate their activities with the buyers in the purchasing department. Nevertheless, smaller companies are training their buyers in the technique of value analysis so that they can apply these principles in their day-to-day purchasing.

NEGOTIATION. At some point in each of the preceding methods and techniques the process of negotiation must take place before final agreement is reached. Skillful negotiation of a contract will result in clearer understanding of all of its terms and conditions and usually in greater satisfaction on both sides during the period of its operation. The only time that negotiation does not occur is when a seller offers a product for sale and the buyer immediately accepts at the price and terms offered. It is unlikely that there will be such complete agreement on any significant transaction.

Negotiation is then the communication process by which the buyer and the vendor's representatives, in a face-to-face discussion, arrive at the exact terms of the contract. It is natural for the vendor to submit, as his first sales proposal, terms which are somewhat more favorable to himself than the minimum he might be willing to accept. The degree of "padding" in the proposal will vary with the competitive situation at the time. It is apparent that the buyer's negotiating skill will enable him to remove this padding from the proposal and arrive at an agreement which is more favorable to the buyer and still fair and acceptable to the supplier.

Skillful negotiation becomes most important in reaching final agreement in situations where there is little or no opportunity for direct competitive bidding. In these cases it is important for the buyer and his consultants to very carefully analyze every detail of the proposal to insure that it offers full value. He must have available complete detailed information about every aspect of the service to be rendered to be able to judge its value with reasonable accuracy. During the negotiation sessions the seller can be asked to justify the items which seem unreasonable. Concessions or changes in specifications that reduce costs on one side can result in more favorable terms on the other. Often the buyer or his team can show the seller how he can reduce his costs by eliminating unnecessary operations.

The principles and techniques of negotiation are extensive and require considerable study and practice under the direction of a skilled negotiator. Because the advantages of effective negotiation are so great all buyers should be required to undertake this study.

Establishing Purchase Prices

METHODS OF PRICING. Different commodities tend to be priced in different ways primarily because of the conditions under which they are bought and sold. The purchasing department must understand these conditions and should develop a system for obtaining and recording price information on all of the commodities which it buys. Prices are normally established by:

1. Market movement, often related to supply and demand.
2. Catalog price lists.
3. Competitive bidding.

Price Established by Market. In this group are commodities on which prices are established by market movements, information about which may be drawn from market reports, published quotations, etc. These commodities include the majority of raw materials, such as pig iron, steel in commercial grades, nonferrous metals, hides, cotton, and rubber. The purchasing department follows market reports, trade papers, and other sources of published information and maintains charts or graphs on prices and conditions which govern prices, such as production and demand. Daily reports are available in trade papers of metal industries and in general business dailies. In specific instances it is necessary even to have cabled or wired reports from the markets. A purchasing department library of **market information** on such items should be built in accordance with the necessity for prompt and accurate data.

Even though items are covered by catalogs and price lists, the buyer, in the case of important orders, should determine the factors that go to make up the price and if conditions warrant, try for a reduction, not waiting for conditions to force a revised price. As an example, in New York City there is a fairly well-established price list on paper bags. The major item of cost is kraft paper. When the market on kraft makes a steady decline the buyer should not wait for a revised price list but should find a vendor who, because of small stocks, can give him a reduction on the price of bags warranted by the drop in the price of kraft paper.

There are many such commodities in which the price of the finished goods is dependent very directly on **raw material prices.** Some typical examples are steel drums (which fluctuate in price with the market on steel sheets), cardboard containers (which follow the market on chipboard or kraft), leather belting (which varies with the market on hides), and many secondary chemical products (which follow very closely the price of their chief constituent primary chemical). Other commodities depend on the primary constituents.

Price Data Given in Catalogs. Prices of commodities in this group are governed by issuance of a catalog and periodical discount or net price sheets. This group includes most standard mill supply and hardware items such as regular drills, taps and dies, wrenches, and many commodities of this general type. The purchasing department must have a **catalog file** and a **file of current prices on catalog items,** carefully indexed and kept strictly up to date. The clerk or assistant responsible for catalogs and net prices or discounts should check his files with the vendor's at regular periods. All trade papers should be followed for news items about new or revised publications affecting items bought.

The exact method of arriving at the appropriate net price from the list price shown in the catalog or price list will vary with the commodity and industry practice. The usual practice is to offer a discount from the list price to a customer based either on his end use of the product or upon the volume of his purchases. It is the buyer's responsibility to thoroughly understand the pricing policy of the suppliers so that he will be sure he is obtaining the lowest possible pricing schedule to which he is legally entitled.

Competitive Quotations. On many items such as special tools, forgings, castings, stampings and subassemblies there are no market quotations or standard catalog prices or discounts available. Such items are nonstandard, being made to the customer's specifications. On the large majority of such items the only safe way to buy is by inquiry to a number of sources and comparison of their competitive quotations. To be effective in getting the best price, quotations should be:

1. Considered on a truly competitive basis. Award should be made to lowest bidder, except under unusual circumstances, such as doubts arising about the lowest bidder's financial ability or ability to complete on time or within specifications, or cases where the money value of bids does not represent a proper comparison in the light of final cost.

2. Considered as final at the first figure submitted. Allowing continual revision of bids or quotations after submission may bring a better price on that one transaction, but it militates against truly competitive quotations with the best obtainable prices in future transactions.

3. Considered as confidential by buyer. The policy of hinting to competitors of a bidder as to the nature of quotations destroys bona fide competition and in the long run is expensive.

Quotations, after receipt and action, become part of the purchasing department information file in the hands of the price clerk. Losing as well as winning quotations should be recorded to give a proper price perspective on future purchases.

COST-PRICE ANALYSIS. Frequently situations arise in buying special items when it is difficult to know whether the price which has been quoted is a proper or fair price to pay. One of the tools which has been developed to help make such evaluations is called cost-price analysis. To learn to use this tool effectively requires intensive study but it should be understood by buyers who work with such products because its results can be highly productive.

Using this technique the buyer or his cost specialist would study the item to be purchased thoroughly enough to develop a cost breakdown very similar to the one the seller should develop in establishing his price. The basic elements of a cost breakdown as reported by Purchasing Magazine (vol. 62) are rather standard and well known in any given industry. Generally they include: material, direct labor, overhead or burden, and profit. A detailed analysis will subdivide these elements much further but the individual situation will determine how precise the cost analysis study can be.

It is usually not too difficult for the purchaser to estimate rather accurately the amount of material required for the manufacture of a part, and therefore its cost. Direct labor is somewhat more difficult because it will depend upon the process of manufacture and the degree of automation being used but an estimator who knows the process can come reasonably close in making such estimates. The allocation of overhead cost is usually a percentage factor frequently applied to the direct labor cost. In some processes this is not logical and can be argued but it is usually possible to learn the industry practice and to accept it if it is reasonable. Likewise the profit factor is a percentage which can be checked against the results shown on the supplier's annual report.

By using this technique the buyer can develop an estimated price for the item he has out for quotations. If his estimate varies greatly from the quotations received, he can ask questions in an attempt to explain the difference. It is quite in order to ask the supplier to explain where his costing differs from the buyer's estimate and to justify the difference. In the process it is often possible to negotiate lower prices for several reasons: errors in the supplier's costing process may be uncovered, unnecessary costly operations may be identified and eliminated, or it may reveal that the vendor's costing system is too rough, requiring that he refine it.

Some vendors may refuse to discuss cost breakdowns claiming that this is private information but many vendors who are seeking long term relations will cooperate. In order to secure cooperation for this technique the buyer must convince the supplier that he is not simply trying to beat down prices but rather is trying to eliminate unnecessary cost elements. If the negotiation is carried on in this spirit the vendor will see that both companies can gain by increased efficiency and greater production.

Bargain or Cut Prices. Many "bargains" and "cut prices" are offered by houses which are not dealt with regularly. It is a safe rule to regard such offers with suspicion and demand of the supplier proof of the quality of goods and his credit standing. There are exceptional cases in which a purchasing agent has the opportunity to buy below the normal price through some special concession. But as a rule, all that the purchaser can expect to do is to buy at the right price consistent with the volume of the purchase, the credit standing of the vendor, and market conditions.

Buying at the Best Price. The best assurance that a purchasing department can have that it is buying at the best price is mutual confidence between purchaser and vendors. Sales representatives are excellent sources of information on price trends, contemplated revisions, and vendors' price policies. While price checks may be made periodically, it is, nevertheless, highly desirable that the purchasing department should choose and stay with vendors who have proven that they intend to give the buyer good prices. There are many items to be purchased on which there is no available reliable price information and the volume of purchases is insufficient to justify an exhaustive study of costs. Sometimes competition is impractical because of lack of time to secure competitive bids, because only one source of supply is considered capable, or because open competition would disturb the market and bring about higher prices. In such cases dependence on sources of supply is the only course open to the purchasing department. A record of previous purchases of an item over a period, together with knowledge of changes in general conditions over the same period, offers a valuable means of price checking. "Price alone is not a sufficient reason for us to change a source of supply" are the words of one purchasing manager.

Purchasing Agreements

TYPES OF PURCHASING AGREEMENTS. A buyer should be thoroughly familiar with **contract law** because all of his buying transactions ultimately are governed by this law. Legally an enforceable contract exists whenever a buyer and seller come to a meeting of the minds and one of the parties acts in good faith in accordance with the agreement. It is apparent from this that a purchasing agreement could take various forms including: a verbal exchange, correspondence, a purchase order form and acceptance, or a formal signed contract. There are obvious pitfalls to carrying on business by verbal agreement since the final result is dependent upon the memory and the good faith of both parties. It is general practice for purchasing agreements to be reduced to writing using purchase order forms specifically designed to meet the needs of the buying company. Verbal modifications of such agreements should also be avoided but instead written confirmation using the proper forms should be made. Such purchase order forms and their specific provisions will be discussed below under Purchasing Procedure.

CONTRACT PROVISIONS. The contract should set forth clearly in definite terms: quantity, specifications, price, terms, time, special conditions, inspection, guaranties, penalties.

Quantity may be expressed as a definite figure in pieces, pounds, or other unit or it may be expressed as "purchaser's requirements, estimated at _____." It may be expressed as a rate of manufacture to be maintained by the vendor or in terms of the stock he binds himself to keep on hand for the purchaser. It is common to have quantity expressed in terms of tolerances, by which the vendor agrees to furnish not more than a certain maximum or less than a stated minimum. A usual expression is "purchaser's requirements not to exceed _____." From the purchasing standpoint the greater the margin allowed on quantity, to allow for possible error or change in estimate, the better the contract.

Specifications. Specifications should follow the same rules laid down for buying specifications. When specifications are vital and rigid, provision should be made in the contract for a clear and definite agreement on the basis of inspection and rejection, including where and how inspection is to be made. Unusual inspection procedures must be known to the buyer and specifically covered by the contract, as they may be an item of cost overlooked by the supplier. In some cases inspection by an outside agency or company may be required. In such cases the outside company should be named and clear provision made as to who should bear the expense of the inspection.

Price. Price may be fixed at a predetermined figure by contract. In the case of goods whose price fluctuates beyond the control of the vendor, definite agreement on a fixed price cannot be had. In such instances it is often desirable to make the provision that price shall be based on market price at date of shipment, stating accurately the means of determining the market price, such as the standard price reports in a reputable trade paper. When possible to do so, **sliding-scale agreements** should be safeguarded by inserting a maximum price, above which a vendor shall not charge, regardless of market changes. A most favorable type of price agreement is that by which a vendor fixes current prices as a maximum price and agrees to give the buyer advantage of all declines in market prices. Price terms in contracts sometimes include so-called **escalator clauses,** which provide for advances in the price under certain conditions which may concern the vendor's cost of material or of labor. It is essential that such escalator clauses be carefully worded, so as to avoid any misunderstanding between the parties.

Terms. Terms such as **f.o.b. point** and **cash discount** should be plainly stated. The f.o.b. point is preferably on cars at the purchaser's freight station or siding, since this provision leaves title to goods in the vendor's hands and at the vendor's risk during transport. In instances where the contract is for delivery of goods at more than one plant, delivery f.o.b. point of origin is preferable, title to the goods then transferring to the purchaser. In some instances vendors refuse to accept hazards of ownership during transport but are willing to allow the purchaser the transportation charges, and terms being f.o.b. shipping point with freight allowed.

A cash discount is offered by the seller as a financial incentive to the buyer to pay the invoice promptly so that the seller can reduce the cash requirements of his business by not having large accounts receivable outstanding. The financial incentive is considerable because the terms offered usually are 1 or 2 percent discount if the payment is made within ten days. Since the normal payment term without discount is thirty days the cash discount is the equivalent of either 18 percent or 36 percent per annum for earlier payment.

This favorable financial incentive should encourage buyers to negotiate cash discounts. It is important to keep such discounts separate from any pricing consideration and think of them strictly on the conditions for which they are given.

Time Clauses. Time clauses are important in purchase contracts. The contractors may agree upon delivery within a certain period, on a certain date, or in accordance with purchaser's instructions. The most favorable kind of contract in this respect is one that calls for delivery "within period from _____ to _____ as called for by purchaser." The statement of the delivery agreement should be clear.

Time should be of the essence in purchase contracts. A clear statement of the right to cancel and refuse deliveries if they are not made on time should be included in the contract. Every contract should contain a definite statement as to its termination, either by lapse of time or by action of the parties. Many contracts are so loose in their conditions covering time of delivery that they have little effect, except that of a price agreement during the period for which the contract runs. Many purchasers favor this kind of contract, since it binds the vendor to responsibility, while placing no responsibility upon the purchaser. It is a mistake, however, to think of such limited price agreements as contracts of purchase.

Special conditions are sometimes necessary in purchase contracts for the protection of the purchaser's or vendor's interest. Such conditions should be stated specifically and not left to mutual understanding. Examples of such special conditions are given in the following paragraphs.

Statements of Ownership. Statement of ownership of special patterns, tools, fixtures, etc., required for the manufacture of certain products is an extremely important point to cover definitely, whether in a contract or an order form. Custom varies widely as to ownership of such items. In certain instances the purchaser of the goods pays for these facilities, and they are his property, subject to removal by him at any time that he desires to place his orders elsewhere. Often their cost is included in the price of the first order, and subsequent orders are filled at a lower price. A purchaser, in such a case, may regard himself as owner with the right of taking them away whenever he desires. It has been held, however, that special devices are not necessarily the property of the purchaser unless specific arrangements to this effect have been made at the time of the initial order. Where their number and cost are large, ownership by the vendor practically dictates placing all orders with him. Even were the purchaser willing to go to the expense of providing a second set, the time involved in their production might preclude this step.

In some cases special tools and fixtures may not be adapted to the equipment of another manufacturer. In other instances, while patterns, tools, etc., are the purchaser's property, he has no right to remove them from the vendor's possession. Such a contract may be of advantage to the purchaser, however, if it prevents the vendor from using them to make products for other customers.

Purchase contracts should cover, in considerable detail, the following points in regard to special equipment:

1. Ownership.
2. If title rests with the vendor, whether or not the purchaser has a right of purchase of this equipment.
3. Limiting use of special equipment to the purchaser's product and forbidding their use on material manufactured for competitors.
4. The party who should bear expenses of maintenance and repairs of special patterns, tools, and fixtures.

Inspection at Vendor's Plant. Inspection at the vendor's plant should be included in the purchase contract if such inspection is contemplated. The vendor should agree, and the extent of inspection should be stated as specifically as is practical. Such equivocal expressions as "satisfactory quality," "reasonable inspection," etc., should be avoided. Inspection standards and privileges must be clearly stated with a clean-cut agreement as to whose decision shall be final. The most satisfactory form of agreement covering inspection is a definite printed specification relating to such inspection in the contract.

Guaranties. Guaranties are contract obligations of the vendor to furnish quality, quantity, or service. A guarantee adds nothing to a purchase contract unless it is a definite guarantee of a specific item not covered by the general contract. A guarantee of service for a fixed period after delivery, or against defect discovered within a fixed period, is a valuable protection to the purchaser in some cases.

Penalty Clauses. Penalty clauses in contracts obligate the vendor to make good any specified losses suffered because of his failure to meet the time obligations in the contract. The vendor may file a bond to guarantee performance of the contract, or the contract may contain a statement of the amount of payment which shall. be accepted as liquidated damages in full settlement for defaulting on the contract. Such a clause is seldom used in buying equipment, materials, or supplies.

BLANKET CONTRACTS. Blanket contracts are usually developed to serve as the overall agreement against which specific purchases of a particular commodity or group of commodities are made during the period of time covered by the contract. By means of a blanket contract the total volume of a commodity used in several plants is combined in an agreement to gain a price advantage and other favorable terms. In such cases the blanket contract may be negotiated by the central purchasing department and releases are ordered against it by the plants (Aljian, ed., Purchasing Handbook).

As in all contracts the terms and conditions should be as specific as possible and spelled out in detail. However, to allow the flexibility required by the different locations making releases against the contract some conditions must be stated in somewhat general terms. It is usually possible to word the agreement so that it can still be workable under these circumstances.

STANDARD PURCHASE CONTRACT FORMS. Standard purchase contract forms are adopted by some large manufacturers for the sake of assurance that their interests will be safeguarded without the necessity of legal advice on each purchase agreement. It is difficult, however, to make up a standard purchase contract which will cover all circumstances and be satisfactory to all vendors. Special clauses or insertions are required, therefore, when some unusual condition is included in a contract.

LEGAL ASPECTS OF CONTRACTS. Legal aspects of purchase contracts demand close attention. In drawing up contracts which involve liquidated damages or any clauses of a complicated legal nature, legal advice should be sought. Important factors in the writing of contracts follow.

1. Only the matter which appears above the signature is binding.
2. Any additions or revisions in the original text of a contract should be initialed or signed by the parties to avoid any claim of alteration after signature.
3. All necessary provisions should be placed in the contract. It is not safe to depend on formal clauses printed at the bottom of the page or on its back. If they are used, a clause such as the following should be printed above the signature on the contract: "The clauses printed on the back (at the bottom) of this order are hereby made a part of this contract."
4. Printed matter in the body of the contract and above the signature is binding. Such printed matter, no matter how small the type, should be read carefully before signing.
5. The entire agreement between the parties should be expressed in the contract.
6. Every clause of the contract should be made definite to the point of making misunderstanding impossible.

Purchasing Procedure

STEPS IN THE PROCEDURE. For most purchases, issuing the purchase order is only one step in the overall process. The general plan of operation, sources of supply, period to be covered, and other significant factors must all be determined in advance. The principal steps in procuring goods for industrial production are as follows:

1. Determining **what to purchase** and in **what quantity.**
2. Initiating the purchase request. A **purchase requisition** is issued by the department needing the goods to the purchasing department.
3. Studying **market conditions** to determine when conditions are favorable to buy.
4. Determining the **source of supply.** Reference is made to catalogs, indexes, quotation and price records, and vendor records in the purchasing department.
5. Selecting a **favorable price** and a **specific vendor.** Prices obtained by inquiry, quotation, or bid.
6. Entering into a **contractual relationship** with vendor through the issuance of a purchase order.
7. **Following up vendor performance** especially with respect to delivery date.
8. **Receiving, inspecting,** and **delivering** the goods to the storeroom or the department requiring it.
9. **Checking** and **completing** the transaction.

PURCHASE SPECIFICATIONS. Buying proper quality depends on: (1) having proper specifications from which to work, (2) placing the order with a reliable vendor, and (3) checking material bought against specifications. A specification is no more than an accurate description of material to be purchased. Definiteness tends to minimize costs.

There are many forms of specifications indicated by:

1. Brand or trade name.
2. Blueprint or dimension sheet.
3. Chemical analysis or physical characteristics.
4. Detail of material and method of manufacture.
5. Description of purpose or use.
6. Identification with standard specification known to the trade generally and to the vendor.
7. Vendor's sample.
8. Buyer's sample.

Specification by **brand or trade name** reduces competition and makes the buyer entirely dependent upon the vendor's reputation for quality. It should be used only in cases where the branded product has been found to be superior to all others for the purpose intended, or where it is deemed to be satisfactory but its composition is secret or unknown, or the desired characteristics of quality can only be tested by destroying the material. It also may be used in cases of standard products used in unimportant processes or in small quantities, when the extent of use does not justify expense of investigation and detailed specifications. The purchasing department should attempt to have at least two, and preferably more, approved brands. There are comparatively few brands that do not now have a competitive or equal grade.

Specification by **blueprint or dimension sheet** is advisable in the purchase of tools and fixtures or of components of an item to be manufactured, to meet special requirements worked out by the engineering department of the purchasing company. Blueprints provide a safe and easy method of checking against specifications when items are received and inspected.

Specifications by **chemical analysis or physical characteristics,** or both, are ideal for raw materials in the metallic class. Such specifications can be checked accurately by laboratory tests.

Specifications by **detail of material and method of manufacture** are usually confined to subcontract purchases. Because this type of specification is often used in conjuncture with inspection at the vendor's plant, it is considered too expensive for extensive use.

Specification by **description of purpose or use** is highly effective. If the vendor is dependable and accepts such a specification, the responsibility is entirely his. This form of specification is the least difficult to prepare and is recommended especially in the purchase of machines or tools about which the purchaser has no particular technical knowledge.

Specification by identification with some **standard specification** already published and accepted is a most satisfactory form of specification, provided such a standard specification can be accepted without undue and unnecessary expense.

Specification by **vendor's sample** is used where particular characteristics of quality are not readily measurable or easily described in words. Natural products, such as basic ores, chemicals, leather, and products involving color, are examples. Buyers should be careful that all characteristics of desired quality present in the sample are known to the seller at the time of purchase.

Specification by **buyer's sample** is recommended only where no other type of specification is possible, as in buying printed forms and containers, where copy is necessary or where color is involved. Samples are too often subject to change, loss, or misinterpretation, and no adequate record of the purchase is available for reference.

Requirements for an Industrial Specification. Some companies, usually large ones, develop their own **standard industrial specifications.** Although difficult to prepare and expensive to develop and maintain, they effectively increase competition and reduce rejections and disputes. An industrial specification should:

1. Be as simple as is consistent with exactness. Unnecessary detail in the specification is expensive.
2. Incorporate nationally recognized commercial standards wherever possible.
3. Contain reasonable tolerances. Unnecessary precision is expensive.
4. Be capable of being met by several vendors for the sake of competition.
5. Be capable of being checked and contain suitable testing methods as well as characteristics to be checked.
6. Include packaging specifications and units of purchase.
7. Include, wherever possible, the use to which the material or equipment will be put by the buyer.

Where commodities purchased are covered by **commercial standards,** the purchaser's task is greatly lessened, since it is necessary only to specify the desired grade according to the standard, and the seller may guarantee conformance. Frequently provision is made for a **certifying label** as an assurance to open-market buyers. These labels normally state the name of the guarantor, the commodity and grade covered, the name and number of the commercial standard, and a definite, concise guarantee of conformance to all requirements.

Federal specifications can be secured from the National Bureau of Standards of the U.S. Department of Commerce, which publishes a directory of such specifications. **Military specifications** are available through the Department of Defense. Many agencies prepare specifications which are largely accepted by manufacturers, among them, the American Society of Testing Materials, the Underwriters' Laboratories, and the American Society of Mechanical Engineers. The amount of standard specification material available is great and always growing. No purchasing department library is complete without references to the various directories of specifications and copies of individual specifications applying to the particular industry.

PURCHASE REQUISITIONS. The purchase requisition may be originated by the stores records supervisor, production control department, head of some operating department, the chief storekeeper, plant or maintenance engineer office manager, or other responsible person authorized to request items which he needs from an outside vendor. Limitation of power to issue purchase requisitions is necessary, and a general procedure order should be set up by the management, stating definitely which persons in the company are permitted to issue or sign purchase requisitions for specific kinds of items. Requisitions from all other persons must be turned back to be countersigned by the proper authorized person or canceled. It is likewise standard practice to require that all purchase requisitions be "edited," that is, reviewed to see whether the items are carried in the storeroom, whether the request for some unusual items could not be satisfactorily filled by something carried in the storeroom, or whether it could not be replaced by a standard item more readily obtainable and costing less.

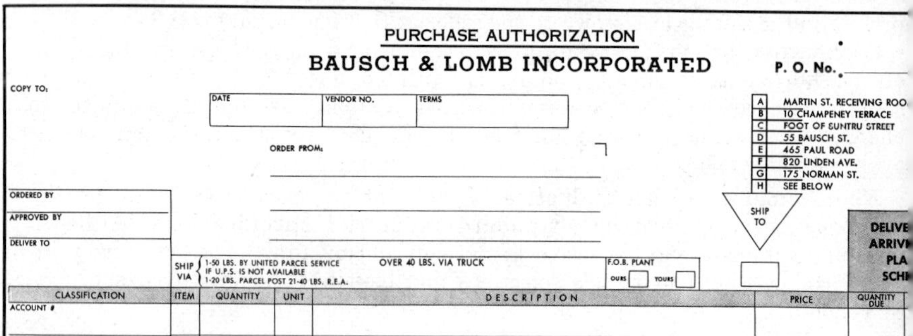

PURCHASE AUTHORIZATION
BAUSCH & LOMB INCORPORATED
P. O. No.
COPY TO:
DATE | VENDOR NO. | TERMS
ORDER FROM:
A MARTIN ST. RECEIVING ROO
B 10 CHAMPENEY TERRACE
C FOOT OF SUNTRU STREET
D 55 BAUSCH ST.
E 465 PAUL ROAD
F 820 LINDEN AVE.
G 175 NORMAN ST.
H SEE BELOW
SHIP TO
ORDERED BY
APPROVED BY
DELIVER TO
DELIVE ARRIV PLA SCH
SHIP VIA 1-50 LBS. BY UNITED PARCEL SERVICE IF U.P.S. IS NOT AVAILABLE 1-20 LBS. PARCEL POST 21-40 LBS. R.E.A.
OVER 40 LBS. VIA TRUCK
F.O.B. PLANT OURS YOURS
CLASSIFICATION | ITEM | QUANTITY | UNIT | DESCRIPTION | PRICE | QUANTITY DUE
ACCOUNT #
DEPT. CHARGED
FACTORY ORDER #
REF. #
S.E. #
INVENTORY #
CONTRACT #
SUBMIT IN DUPLICATE TO PURCHASING DEPT.

Fig. 3. Purchase requisition.

Form of Requisition. The purchase requisition (Fig. 3) will vary somewhat in form according to its point of origin and the accounting system in use. The form should contain the following essential information:

1. Name and accurate description of goods wanted.
2. Quantity needed and shipping units to save repacking upon receipt.

3. Date on which goods will be needed.
4. Account to which goods are to be charged.
5. Amount on hand and past or current rate or record of use.
6. Point of delivery within the plant to which the shipment should go either after passing through receiving and inspection or directly, as in the case of such items as machinery which should be unloaded as close to point of use as possible.
7. Vendor's name (in certain special cases, as in repair parts for equipment). Selection of the vendor is ordinarily a purchase duty.
8. Authority for purchase.
9. A space or form for use of the purchasing department in adding vendor's name, price, terms, routing, estimated weight, number of purchase order issued, and all other information necessary for typist in making out purchase order. Later, entry of delivery date promised by vendor.

In the case of requisitions from the control department, production reference to the order or lot number of a production job is sufficient justification, and this information is sometimes conveyed by the account number to which goods are to be charged. In requisitions from the plant engineer, office manager, etc., enough information should be included to enable the purchasing department to pass intelligently on questions of need and quantity.

The form of purchase requisition will vary with many factors. Requisition forms should be of a size to fit standard files and preferably should be distinctive in color from other forms in the purchasing system.

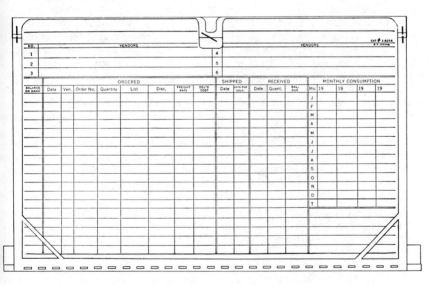

Fig. 4. Traveling requisition form.

The traveling or permanent requisition (Fig. 4) has been found to be an economical and practical short cut on many regularly used and readily available items carried in stock. Although there are many types in use, all have certain characteristics in common:

1. The complete and accurate specification for ordering is typed on the form and so reduces the possibility of error from faulty copying by the stores record clerk.

2. The accepted sources are shown, relieving the buyer of the need to refer to other records to select the source.
3. Previous purchase records of delivery, price, and costs are shown in order to aid in source selection.
4. Monthly consumption provides a means of determining consumption trends as an aid in determining proper order quantities.

Whatever system of stores record is adopted, the traveling requisition procedure is the same. The form is stored with the record card and pulled when the order point is reached. The stores clerk enters the balance on hand, brings the consumption record up to date, and forwards the card to the buyer. The buyer determines the vendor, order quantity, price, and delivery schedule. The requisition is then ready for the order typist. If further approval is required, the traveling requisition again saves time by providing all information necessary to permit qualified decision. After the replenishment order is written, the traveling requisition, with order number and date filled in, is returned to the stock record clerk, who enters receiving data as the order is filled and refiles the requisition for subsequent use.

Route of the Requisition. The purchase requisition may go first to the purchase or quotation record desk on entering the purchasing department, for comparison with records of previous transactions for the same commodity. If the goods asked for by some department are carried in stores, the requisition can be returned to the requisitioning department with instructions to secure the items from stock. If the quantity wanted is in excess of previous requirements as indicated by the record, the reason for such unusual requirements can be checked. If the quantity varies from a standard package or economical ordering quantity, the quantity can be adjusted. The purchase record or quotation card for the particular item may be attached to the purchase requisition, and both may be submitted for the buyer's consideration, since the purchase record will also show possible sources of supply, past experience with these sources, price variations over a period, terms on previous orders for the goods, and all essential facts to make a complete picture of the transaction contemplated.

An alternative routing of purchase requisitions takes them first to the buyer's desk. Only when the buyer feels that comparison with the record of previous transactions is desirable are they sent to the record desk. Fig. 5 is a **flow chart** showing the course of the purchase authorization (requisition) and other forms and procedures in the purchasing department.

REQUEST FOR QUOTATION. The buyer must decide—once a purchase requisition has reached him and received his approval as to quantity, form, and propriety—with what vendor the order shall be placed. His primary interest is to secure, from a list of vendors in whom he has confidence as to quality of product and ability to make proper delivery, one who will give him the best or, at least, the right price. He may have price information available either in the form of catalogs and price lists submitted by vendors or in the general quotations already received.

If he is not satisfied with his knowledge of the right price to be paid, or if there is any doubt as to which vendor will give the best price, the buyer should issue a request for a quotation on the items covered by the purchase requisition (Fig. 6). This request should be specific in describing the material and in identifying it with the vendor's product. When possible, the catalog designation should be given.

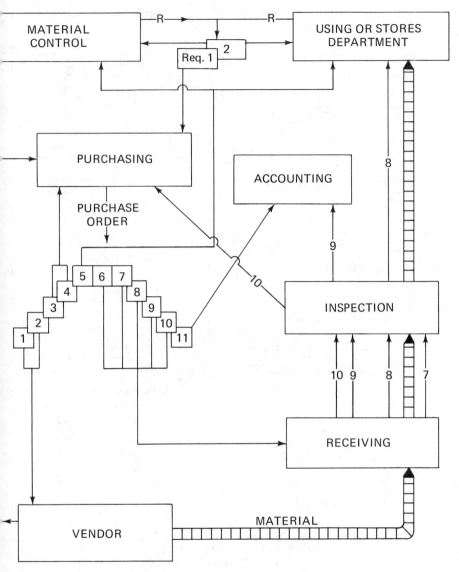

Fig. 5. Purchase order procedure flow chart.

Information on the Request. The request for quotation should give the following information, essentially in the form in which it will appear later on a purchase order.

1. Quantity to be ordered.
2. Name and full specifications of material, including inspection procedure.
3. Point to which goods are to be delivered.

REQUEST FOR QUOTATION   REQUEST NO. _____
                                   ISSUANCE DATE _____

TO	SUMMARY OF QUOTATIONS			
	QUANTITY	1	2	3
TO				
TO				
	DELIVERY			
	F.O.B.			
	TERMS			

D E S C R I P T I O N	QUANTITY	UNIT PRICE	TOTAL PRICE

INITIAL DELIVERY REQUIRED _____
Tooling should be provided to maintain
    delivery of _____

The above quotation is based on vendor supplying
all materials unless noted

Any order placed as a result of this quotation
will be subject to "Conditions of Purchase" on order form

TOOL CHARGES   Please quote tool charges separately.

The following information must be filled in over authorized
signature.

WE CAN FURNISH THE ABOVE ITEMS

F. O. B. _____

For Shipment From _____

On _____

Terms _____

Company Name _____

Signature _____

Dated _____

Fig. 6.   Request for quotation.

4. Delivery time which is to be allowed.
5. Date on which quotations will be considered.
6. Purchaser's conditions.

It should ask for the following information:

1. Price.
2. F.o.b. point.
3. Terms of payment.
4. Delivery time.

5. Any special condition or terms the vendor wishes to make. This point is often covered by a clause on the request for quotation form which reads: "Except as stated specifically in your reply, your quotation will be considered as subject to the following conditions." All conditions of purchase desirable from the buyer's point of view are then listed.
6. Approximate shipping weight, provided this information is necessary. Often this question applies to purchases for export.

To Whom Quotation Requests Are Sent. The buyer's experience and records will usually determine to whom requests for quotations are sent. The purchase or quotation record card should indicate any sources of supply that previous experience has shown to be unsatisfactory. It also should reveal **past experience with vendors** regarding price, quality, delivery, and service. In the case of a new item of purchase, dependence must be placed on the buyer's knowledge of the vendor field or on manufacturers' or dealers' registers, files of advertising matter, catalogs, etc.

In most cases requests for quotations should be sent to not less than three and not more than five possible sources of supply, the companies selected being those giving the **broadest possible price picture.**

As indicated by Fig. 6 it is common practice to send requests for quotation **in duplicate,** one copy to be filled in and returned as a quotation by the vendor. The advantages of reply forms for quotations are:

1. Uniformity of quotations for tabulation and filing.
2. Assurance that all points will be covered in quotation.
3. Certainty that buyer's terms of purchase will be either accepted or specifically rejected.
4. Economy in quotation contact.

One method of determining when and to what extent quotations should be requested in advance of placing an order is to establish **a dollar amount as the dividing line.** If the order exceeds this amount, quotations must be received and the price shown on the purchase order sent to the vendor. Below this amount no quotations need be secured. Such a procedure allows small orders to go out unpriced.

QUOTATION ANALYSIS. The quotations received should be carefully tabulated and compared before making the buying decision. The prices, deliveries, and terms should be reduced to a uniform basis and shown in columnar form. In the process all deviations from specifications or essential items should be noted and evaluated.

It is usually convenient to use the file copy of the Request for Quotation form as the analysis sheet especially when it is designed with this in mind as in the case of Fig. 6.

SELECTING THE VENDOR. From the data thus assembled and analyzed the buyer selects the vendor with whom he wishes to place the purchase order. Frequently the purchase requisition form is used as the basis of writing the order. This form, after the buyer adds to it the necessary information on source of supply, price, etc., passes to the typist who prepares the purchase order. All information should be in such clear form that the typist need have no questions or misunderstanding regarding the order. The requisition is given to the buyer with the purchase order for comparison and checking, after which the latter is signed and mailed. The purchase requisition is then marked with the purchase order number for reference purposes and is filed under department of origin. If in duplicate, the copy is returned to the department which originated it, with purchase order number and other needed information added.

Often, however, a copy of the purchase order, perhaps with certain information blanked out, is sent to the requisitioner to indicate that his request is being filled.

THE PURCHASE ORDER. The purchase order is the vendor's authority to ship and charge for the goods specified and is the buyer's commitment to the vendor for the value of the goods ordered. It is an integral part of the sales agreement and establishes a **contractual relationship** immediately upon issuance, when it is an acceptance of a previous quotation or offer. Otherwise it is an offer to negotiate such a contractual relationship which is completed by its acknowledgment or acceptance by the vendor. It is the most important of purchasing forms, and its provisions should be planned with care and continually reviewed for possible revision. A typical purchase order form is shown in Fig. 7.

On the reverse side of the original and acceptance copies of the Purchase Order form appear the terms and conditions which the buyer expects to be a part of the purchase contract. These terms and conditions often are extensive and detailed aimed at covering most buying situations in the industry. In developing a new purchase order form these terms and conditions should be carefully worded as a joint effort of the purchasing manager and the legal officer of the company. Typical terms and conditions are discussed in the following paragraphs.

The purchase order should cover definitely and precisely the essential elements of purchase to be made, in a manner which will render future misunderstandings impossible and minimize the necessity of correspondence. It should be worded as far as possible exactly like the request for quotation, if one was sent out, and should include the following data:

Number. The purchase order number will identify the transaction in future correspondence, help the vendor in shipping and billing the goods, and enable the contract to be recorded and filed for future reference. The vendor should be instructed to put the purchase order number on all packages and on all his invoices, for comparison and checking purposes. Since the receiving department will have a copy or a record of the order, including its number, all receipts of goods against it will be identified.

Quantity Ordered. The quantity of goods ordered is expressed in the units proper for the commodity. A clause should appear in the order form, limiting the vendor's right to overship or undership the order.

Description. Description of goods ordered should be specific and wherever possible should state the basis and means by which quality is to be checked. Existing standards of quality and standard specifications are the most satisfactory for this purpose. Any unusual test, inspection, etc., to which goods are to be subjected should be stated. Material specifications of quality should be fortified by inserting in the purchase order form a clause stating the **rights of rejection and return** which the purchaser reserves. A specimen clause of this kind is:

All material furnished must be as specified and will be subject to inspection and approval of buyer after delivery. The right is reserved to reject and return at the risk and expense of the supplier such portion of any shipment as may be defective or fail to comply with specifications, without invalidating the remainder of the order. If rejected it will be held for disposition at expense of and risk of the seller.

Delivery and Shipping Instructions. Delivery specification may be stated either as a date on which goods are to be shipped or as a date on which goods are to arrive at the buyer's plant. The latter date is preferable because it can be taken directly from the purchase requisition without computation of time

REQ. NO.	PURCHASING APPVL	ACCOUNT NO.	APPROPRIATION	DEPT.		REQUISITIONER		DESTINATION

olivetti Corporation of America   **purchase order**

IMPORTANT: This number must appear on all shipping containers, shipping documents, correspondence and invoices.

P.O. No.   69960

DATE OF ORD.	☐ NOT CONFIRMING	☐ CONFIRMING

VENDOR NO.	F.O.B.	SHIP VIA	PPD	COL	TERMS	* SUBJ. TO TAX	BUYER NAME
						YES ☐ NO ☐	

Address any questions or correspondence pertaining to this order to buyer at issuing location indicated by code below.
SUBMIT INVOICES IN DUPLICATE to insure prompt payment. Submit to OLIVETTI CORP. OF AMERICA, attention ACCOUNTS PAYABLE at address indicated by "INVOICE TO" box.

TO:

☐ ISSUED AT	☐ SHIP TO	☐ INVOICE TO
A. 500 Park Ave., New York, New York 10022		Tel. 212-371-5500
B. 85 Nutmeg Rd., S. Windsor, Conn. 06074		Tel. 203-528-9531
C. 2800 Valley Rd., Harrisburg, Pa. 17112		Tel. 717-652-2363
D. 720 Old Willets Path, Hauppauge, N.Y. 11787		Tel. 516-234-1771
E. 415 St. Mary's St., Burlington, N.J. 08016		Tel. 609-386-3200
F. 1 Park Ave., New York, New York 10016		Tel. 212-679-3400
G. 155 White Plains Rd., Tarrytown, N.Y. 10591		Tel. 914-631-8100
H.		

REF. NO.	QUANTITY	UNIT	OUR ITEM/PART NO.	OUR ITEM/PART NO. TO APPEAR ON PACKING SLIPS, INVOICE & LABELS. DESCRIPTION	UNIT PRICE	PER	SHIP DATE

* SALES AND/OR TAX PERMIT NOS.
CONNECTICUT 5-159-675   NEW JERSEY H 135-591 901
NEW YORK 13-559 190 1   PENNSYLVANIA 99-01917

OLIVETTI CORPORATION OF AMERICA

AUTHORIZED REPRESENTATIVE _____

TEAR OFF THIS STUB AND RETURN PER INSTRUCTIONS BELOW★

On the reverse side hereof are terms and conditions to which Vendor agrees by acceptance of this order. This order including the terms and conditions on the face and the reverse side hereof constitute the complete and firm agreement between Purchaser and Vendor. No other agreement in any way modifying said terms and conditions will be binding upon Purchaser unless made in writing and signed by Purchaser's authorized representative.

MAY BE RETURNED IN *
STANDARD WINDOW
ENVELOPE #8¼, #8-5/8, #9

THIS ORDER IS NOT BINDING UNTIL
ACCEPTANCE IS EXECUTED AND
RETURNED TO PURCHASER

**PLEASE SIGN AND RETURN
THIS ACKNOWLEDGMENT**

PLEASE TYPE IN ADDRESS BELOW
FROM "ISSUED AT" BOX ABOVE ●

VENDOR ORDER NO.

APPROXIMATE DEL. DATE

ACCEPTANCE

olivetti corporation of america

●
●

SUPPLIER

FIRM NAME _____
BY _____   DATE _____

P.O. 69960

Fig. 7. Purchase order form.

allowed for transit and also because it shifts to the vendor the responsibility for transportation delays. A specific date should be stated, and expressions such as "Urgent," "Rush," "Prompt Shipment," "At Once," etc., should be avoided.

Delivery specifications may be buttressed by a clause in the purchase order stating the buyer's rights of **cancellation for default in time of delivery.**

Shipping instructions include a statement of the point at which goods are to be delivered, routings, instructions concerning packing material, and designation of **method of shipping**—parcel post, express, freight, etc. Many firms state in their orders: "Ship cheapest way unless otherwise specified" and reserve the right to charge back transportation charges in excess of the cheapest rate for shipment. A clause in the purchase order form may protect the buyer from charges which he is unwilling to bear in connection with packing and shipping.

Billing and Terms. Billing instructions cover the number of copies of invoice required, information desired on invoices, how invoices should be marked, to what point they should be mailed, etc. A typical provision is: "Invoice in duplicate and mail with shipping papers the day shipment is made." Billing instructions are usually uniform for all orders and may be included in the printed part of the purchase order form.

A definite statement of the f.o.b. point avoids misunderstanding. It is generally stated in connection with the price if the price is placed in the purchase order.

Terms of payment, including cash discount terms, are usually at the option of the vendor. They are stated in connection with quoted prices mentioned in the purchase order, but otherwise omitted. Understanding of cash discounts should be established by correspondence with vendor firms and recorded in a file for checking invoices.

Prices. Prices should be stated in the purchase order if the order is based on a quotation or an agreement as to price, together with a reference to quotation or agreement. Some vendors price all orders as received. This practice has the advantage of certainty of cost prior to billing. It has the disadvantages of added clerical expense for pricing and the tendency of the vendor's billing clerks to follow the price in the order and not give the advantage to declines which might not be known to the buyer.

Stock clauses in the purchase order form are used to afford the buyer **price protection.** A typical clause is:

If price is not stated on this order, material must not be billed at a price higher than last paid, without notice to buyer and its acceptance thereof.

Miscellaneous Clauses. Miscellaneous clauses and conditions will vary with trade conditions or customs, nature of goods, and many other factors. **Patent infringement** is often the subject of a clause, such as follows:

Seller shall indemnify and save harmless the buyer and/or its vendees from and against all cost, expenses, and damages arising out of any infringement or claim of infringement of any patent or patents in the use of articles or equipment furnished hereunder.

Other typical clauses on various conditions are:

The material on this order must be furnished only by the person or firm to whom the order is addressed unless otherwise authorized by the purchasing agent.

Seller agrees that no part of this order shall be sublet without purchaser's approval.

When the cost of tools involved in the manufacture of parts covered by this order is included in the price per unit, tools become the property of this company upon the completion of our orders.

Any controversy or claim arising out of or relating to his contract or the breach thereof, shall be settled by arbitration in accordance with the rules then obtaining of the American Arbitration Association, and judgment upon the award rendered may be entered in the highest court of the forum, state or federal, having jurisdiction.

It is agreed that the waiver or acceptance by us of any breach on seller's part of any of the terms of this order shall not operate to relieve seller of responsibility hereunder for any prior or subsequent breach.

It is impracticable to list or illustrate all special conditions which may require mention in a purchase order. Child labor laws, federal food and drug acts and revenue acts, and other statutes may require provisions in particular cases. In addition, there may be clauses covering responsibility for patterns, special causes for cancellation, penalties for default, etc., and other requirements. Clauses which are general and apply to all purchase orders of the company will be made part of the printed purchase order form. Special conditions peculiar only to the order in question must be typed in.

Many clauses have recently been developed as a result of state sales and use taxes, federal social security tax provisions, federal and state wage and hour laws, civil rights legislation, etc. Ordinarily it is sufficient to include a general provision that "nothing in the purchase order shall vary the legal responsibilities of either vendor or purchaser under federal or state laws." It is generally true that any attempt to insert clauses which shift such liability is ineffective.

Acceptance by the Vendor. Acceptance or acknowledgment of the order must be made by the vendor to bind the agreement unless the purchase order is itself acceptance of a quotation or offer. It is desirable to insert as a stock clause in the purchase order form, a provision requiring acknowledgment or acceptance by the vendor. A typical provision is (Treadway and Affleck in National Association of Purchasing Management, Guide to Purchasing):

Please sign and return acceptance immediately. On the reverse side hereof are terms and conditions to which Vendor agrees by acceptance of this order. This order including the terms and conditions on the face and the reverse side hereof constitute the complete and firm agreement between Purchaser and Vendor. No other agreement in any way modifying said terms and conditions will be binding upon Purchaser unless made in writing and signed by Purchaser's authorized representative. This order is not binding until acceptance copy is executed and returned to Purchaser.

Other Data. Accounting information given on the purchase requisition tells the purchasing department the account or job number to which the purchase is to be charged. This information need not go to the vendor. It should be typed into the purchase order, however, so that it will appear on duplicate, triplicate, etc., copies which are used internally for receiving and charging material.

Date, signature, vendor's name, etc., obviously are essential features. The signature should show the company name and the name and position of the purchasing officer authorizing the purchase. Each order should be signed either by him personally or with his name and "By _____," with the name of the person actually signing.

Form and Number of Copies. A specific form of purchase order is advisable as long as it is readily understood, economically handled by the vendor, and suited to the requirements of the purchaser's accounting system.

A typical setup calls for eight copies:

1. Vendor's copy.
2. Acknowledgment form, to be filled in by vendor and returned.

3. File copy for purchasing department, used for record purposes and for checking of invoice.
4. Follow-up copy.
5. Accounting department copy.
6. Receiving department copy.
7. Inspection department copy.
8. Requisitioner's copy.

There is a growing need for a greater number of purchase order copies to simplify receiving department procedures, satisfy requirements of government contracts, etc. For example, it is common practice to include three additional copies which go to the receiving department and are then used in lieu of the preparation of separate receiving reports (see Fig. 5). This method has the advantage of reducing clerical work in the receiving department as well as minimizing errors and confusion in describing the material received.

FOLLOW-UP OF PURCHASE ORDER. The importance of follow-up is obvious, since maintenance of uninterrupted production schedules depends primarily on receipt of material on time. Its importance is greatly increased in a seller's market. Every order should bear a delivery date, and the follow-up section should see that that date is kept or learn of unavoidable delays in time to prevent crippling the plant for lack of materials. The follow-up section should:

1. Obtain a copy of the acknowledgment or acceptance of order.
2. Obtain a promise of delivery consistent with requirements.
3. Check with vendor on progress of filling the order—several times if the order is highly important.
4. See that the delivery promise is kept.

Some companies send the vendor two copies of the order, one of which is an acknowledgment form for signature and return. In general, the full size 8½ × 11 in. order copy is easier to handle, record, and file. It is good practice to demand acknowledgment by return mail as a condition of the order. If acknowledgment is not received after a reasonable lapse of time for exchange of mail, the follow-up section should write requesting acknowledgment. Form letters and postcards may be used for this purpose. However, the personally directed, individually written letter will usually get better attention. Routine action anticipates only clerical attention.

After acknowledgment is secured with a statement of delivery date, the follow-up section, if the order is important or special, will make later checks on the vendor by letter, postcard, or in urgent cases by telephone or telegraph. Reply forms may be included with requests from the follow-up section to get a specific reply. The clerk, preparing the form to cover a specific purchase, enters the order number information and specifications and checks the box or boxes covering the specific information requested. The supplier is requested to reply on the same form in the simplest manner possible.

When a broken delivery promise endangers shop schedules, emergency action should be taken through the purchasing manager or a buyer. Follow-up is usually based on a carbon copy of the order, with proper headings for notation of delivery information, etc. Follow-up copies should be filed in a tickler file under dated headings, which will bring each order up for review as to delivery conditions in ample time to allow for action. The review period for follow-up is important to insure that shipment was actually made when promised and then that no delay occurs during shipment. In critical cases special expediting action includes long distance telephone calls and personal field trips.

Many companies follow the practice of requiring all major expediting action to be carried out by the buyer who placed the order. This practice is based on the theory that the buyer will have greater influence on the supplier for prompt emergency action. It also confines the contact with supplier to one person and enables the buyer to have first-hand knowledge of the supplier's delivery performance for his future buying decisions.

RECEIVING. Duties of the receiving department include:

1. Checking incoming shipments to see whether items and quantities conform to order.
2. Recording receipt.
3. Taking necessary steps to insure inspection or testing when required.
4. Notifying department or storeroom of receipt of shipment and its amount and condition.
5. Informing purchasing manager or buyer of all facts which require an adjustment with vendor or carrier, whether for overshipment, shortage, or defective material.
6. Delivering material to proper point in plant for storage or use.

The receiving department should receive a copy of the purchase order with which to check incoming shipments for proper goods, quantity, quality, and all other essentials. Any variation from shipping instructions on the order should be noted. Although the packing slip may be a good guide it should not be considered proof of contents and verification of actual counts and specifications should be made.

The record made of goods received should be in a form suitable for checking against the vendor's invoice and against bills for transportation charges. It is usually given in the form of a **receiving report.** Such a record should show:

1. Purchase order number.
2. Date of receipt.
3. Vendor's name and address.
4. Kind of goods.
5. Quantity received.
6. Condition when received (whether damaged, etc.).
7. Units in which the goods arrived (pieces, quantity per box or package, etc.).
8. Type of container.
9. Any necessary identifying marks on shipment.
10. Medium of transportation.
11. Transportation charges.
12. By whom received.
13. To whom delivered.

Four copies of the receiving report may be made, one copy remaining as a receiving record, one going to the purchasing department for comparison with the order, and one to the department to which the shipment is sent (usually the storeroom), and perhaps one to the accounting department as a voucher for invoices. It is a good practice to send two copies of the receiving report with the goods, having one copy signed by the receiver and returned to the receiving department.

As noted earlier the receiving clerical load is reduced considerably when the receiving report is made a part of the Purchase Order. In this manner much of the required information is typed as a part of the Purchase Order typing and only the additional information verifying the actual receipt must be added by the receiving department.

In the case of multiple receipts against a single purchase order additional copies of the receiving report section can be easily made by any convenient reproduction method.

INSPECTION. If inspection or test of the items after receipt is required by the specifications, the receiving department must see that this inspection takes place. Often it is practical to hold goods in the receiving department pending reports on tests or inspection. Often it is more practical to transfer the goods to an inspecting or test room where the necessary apparatus is present.

Inspection prior to receipt concerns the receiving department only in that it should have on file an approval of goods before accepting them. The extent and rigidity of inspection cannot be dictated except in individual cases. Practice varies with requirements and circumstances. Sometimes every unit must be inspected. In other cases, **representative samples** can be selected, and lot quality determined by statistical methods (sequential analysis or Dodge-Romig tables). Reports of inspection results should be available to the purchasing department as guides to placing future orders.

If goods are rejected and are to be returned, or if there is a shortage or overshipment on the order, all correspondence relating to an adjustment by the vendor should be conducted by the purchasing manager or the buyer. What constitutes an adjustable shortage or overshipment depends on the value of goods and trade customs. In certain trades there is a custom that a variation of 10 percent either way in quantity must be accepted. Regardless of trade custom if such variations in quantity represent a serious problem this factor should be negotiated to an acceptable agreement.

Defective material to be returned should be accompanied by the **inspection report** showing in detail the nature of the defect, and an order for replacement, if desired, should be issued. It is usually good practice to pay for goods shipped, pending inspection, and obtain credit memoranda for any defective goods returned. Some companies obtain authorization and shipping instructions from vendors before shipment and bill back to vendor at the same time. In the case of vendors whose credit rating is doubtful, it is safer to hold invoices until inspection is completed and deduct for defective goods.

Returns for defects and all forms of requests for adjustments should be made only on authorization by the purchasing manager or buyer. In a large purchasing department there may be a claims clerk whose sole duty relates to defects, shortages, etc.

If goods are returned for credit, there should be a **record of the goods returned and the claims made,** this record to be closed only by receipt of a credit memorandum or replacement on a no-charge basis. This claims record preferably should be in the accounting department as a check on the purchasing department's closing of claims. It is also desirable that the purchase record contain a notation of all claims and their disposal as part of the history of the transaction, to guide buyers in the placing of future business.

ACCOUNTS PAYABLE. The purchase transaction is completed when payment is made for the merchandise. The responsibility for the payment of invoices is normally with the accounting department. The **objective** of the accounts payable function is very basic: to determine that the invoice is correct in every respect and then to issue payment to the supplier within the time limit agreement of the purchase contract.

It is common business procedure for the accounts payable department to receive a copy of the purchase order when it is issued. Likewise it receives copies of receiving reports and the supplier's invoices. The three documents are matched and if the invoice is in agreement with the quantity received and with the terms of the purchase order, the invoice can be processed for payment. If there is any point of disagreement among the documents the invoice is referred to the purchasing department to either approve the discrepancy or work out an adjustment with the vendor which would normally result in a corrected invoice.

Ideally the material covered by the invoice should be received and inspected by quality control before payment is authorized. Sometimes delays in transit or in inspection over which the vendor has no control make this procedure impractical, because it would delay payment past the cash discount period or even beyond the normal period of payment. Some companies process payment by checking the invoice against the purchase order and assume that any claim for shortage or defective material can still be resolved at a later date with a reputable vendor.

To assist the accounts payable department the purchasing department should insist that vendors provide invoices in the proper form, hold discrepancies to a minimum, and when discrepancies occur adjust them promptly. Purchasing is interested in the processing of invoices so that advantage is taken of the cash discounts that have been negotiated. It is also concerned that all payments are made on time so that the company's reputation with its suppliers is maintained.

TRANSFER TO STORAGE. After the material has been properly checked in and approved for quality it is physically transferred to storage in accordance with the procedures of the company and the nature of the material. Usually the storage area is separate from production and individual lots of material are drawn from storage to meet schedule requirements. Sometimes material which is difficult to move, such as steel, is stored adjacent to its first point of processing.

Purchasing is interested in the transfer to storage since this signifies that the material no longer requires its attention and is completely ready for use. Up to this point purchasing must be concerned with expediting the material to insure its availability for its intended use.

Purchasing Department Records

TYPES OF RECORDS. The most important records to be kept by the purchasing department are:

1. Purchase record.
2. Contract record.
3. Vendor record.
4. Summary of purchase work.
5. Miscellaneous records.

PURCHASE RECORD. There should be a separate purchase record card for each commodity and usually for each size or variety of each commodity. On the **purchase record card** are entered all orders placed and sometimes all deliveries or shipments against them. Copies of purchase orders are used as sources of information, such as date, quantity, price, requisition number which authorizes purchase, account number to which goods are to be charged, vendor's name, and all other facts pertinent to a complete record of the orders.

The record should be arranged to allow quick and accurate comparison between different purchases of the same commodity. This record is in many respects the heart of the purchasing routine, as it shows the buyers the facts regarding each order as to choice of vendors, previous experience as to volume of purchases, prices, etc. Fig. 8 shows a good example of a purchase record card. The purpose of the record card is served in another way by the traveling requisition (Fig. 4). Generally the traveling requisition contains most of the information that would be shown on the purchase record. The only disadvantage of the traveling requisition is that it is not readily available at times other than

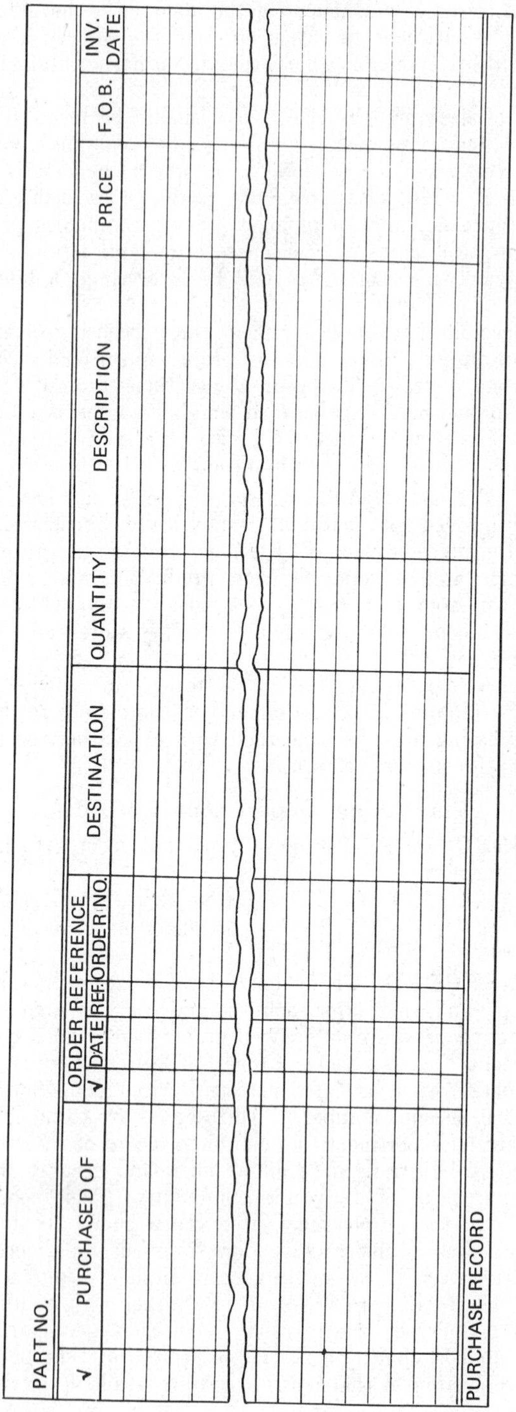

Fig. 8. Purchase record card.

when an actual article is to be purchased. This is not a major problem since it can usually be obtained from the material control section which is often located near the purchasing department. When the traveling requisition is used as a source of purchasing information there will be a purchasing department reference file showing additional information about the entire category of materials.

CONTRACT RECORD. It is essential to have available at all times a complete record of purchase contract commitments. The contract record should show commodity, vendor, order number, total quantity contracted for, time limits of contract, price and unit, and all other necessary information for filling the contract properly and recontracting upon expiration. There should also be spaces for posting quantities ordered or received against contract, together with a perpetual balance column.

VENDOR RECORD. Records of vendors on the purchase record card may be sufficient when supplemented by directories, registers, etc., which are available. Many purchasing managers prefer, however, to keep a separate card file of vendors, arranged and filed by commodity headings. This file constitutes a list of potential sources of supplies on various commodities and gives an opportunity for notations of confidential facts affecting each vendor's desirability as a source of supply.

Such cards become more valuable with time. Folders rather than cards, covering vendors with whom business is being done or where special investigations have been made, are sometimes used.

Vendor Rating. Many companies have built up a file in which they evaluate their suppliers. The vendor characteristics evaluated for this file depend on the particular industry, but Westing and Fine (Industrial Purchasing) suggest the following as a general guide:

1. Plant facilities.	6. Performance.
2. Personnel.	7. Geographical location.
3. Housekeeping.	8. Reserve facilities.
4. Procedures (efficiency).	9. Quality standards.
5. Production specialization.	10. Service.

It is obvious that the first five items require a visit to the vendor's plant. In fact some companies schedule a visit to each of their vendors annually, during which a technical expert representing the purchasing department evaluates these characteristics.

Some form of systematic vendor rating system is highly desirable at least for major suppliers of critical components and materials. Such systems have been developed and used by many companies for their major suppliers. The problem which faces most purchasing departments is to determine how complex a vendor rating system can be justified under existing circumstances.

Smith (National Association of Purchasing Management, Guide to Purchasing) suggests that the small purchasing department begin with a basic plan by which major vendors are reviewed periodically. During the period all comments—good or bad—about the vendor are carefully recorded. The review evaluates these comments and establishes a rating for the period. Although this plan is informal it does provide for systematic review and evaluation which should be the primary objective of vendor rating.

Vendor rating during the period of performance should cover the three basic factors of quality, service, and price. In making comparisons between vendors or in deciding when a vendor should be replaced it is often difficult to evaluate the quality and service factors. When two vendors supply the same item **statistical quality control** methods may permit fairly accurate comparisons but many characteristics are still difficult to evaluate. A rating based on service is even more difficult because the factors to be evaluated are more intangible. The most common service factor, delivery, can be measured with some degree of accuracy but even here it is necessary to evaluate how difficult the delivery requirement was before a true comparison can be made. Other service factors, such as technical support, are even more difficult to measure in precise terms.

Once a satisfactory method for measuring these factors has been established the final problem in vendor rating is that of properly weighing each of the three basic factors and then deciding what is a minimum acceptable level for maintaining a vendor as active.

SUMMARY OF PURCHASING WORK. Certain statistical records should be maintained in the purchasing department to summarize the activity going on in the department. Generally the figures themselves are not very meaningful but the month-by-month trends can be used effectively by the purchasing manager. Typical of the records of this type are: purchase orders and order changes issued, requisitions processed, dollar value purchased, quotation requests issued, letters and telegrams sent, salesmen interviewed, etc. Such records may also be used to check the department workload.

MISCELLANEOUS RECORDS. Various other records may be essential to certain purchasing departments depending on the nature of the industry and other factors. Records of pattern location, possession of tools needed for manufacture of purchased goods, and special arrangements of various kinds are often necessary.

ELECTRONIC DATA PROCESSING. There are many ways in which the computer and its accessory equipment can be extremely useful to the purchasing department. The primary objective in programming the computer should be to assemble the data required by the buyer in a meaningful manner so that he can make the best possible buying decision.

All of the records which have been discussed in this section can be provided most efficiently through a computerized data collection system. The data processing system is also most effective when it is integrated into the material and production control system so that requirements, usage rates, and leadtimes are provided to purchasing with the requisition.

Performance Evaluation

PURPOSE OF REPORTS. Reports to management serve to keep management informed of the key factors that affect the efficiency of the operation. This is particularly true of reports from the purchasing function. However it is difficult to report on the purchasing function because some of the most vital factors are intangible and therefore very difficult to evaluate and report upon objectively. As examples: some of the purchasing department's greatest contributions to the company will come from how well it negotiates major contracts, how well it selects the best long-term suppliers, and how well it communicates with and maintains supplier good will. Yet none of these factors can be easily measured.

Reporting to management has a secondary benefit because it requires the purchasing manager and those involved in preparing the reporting to do a self evaluation. In a good reporting plan the purchasing manager will be constantly analyzing and measuring his performance in order to anticipate management's questions. Thus the reporting process to a good manager will almost automatically result in performance improvement.

SCOPE OF PURCHASING DEPARTMENT REPORTS. The extent of the purchasing reports in any organization will vary with the size of the operation, the particular requirements of the executive to whom the report is addressed, and to the purchasing manager himself. He should report all significant problems faced by purchasing and should describe the approach used by purchasing in handling its major projects.

As a basic rule the purchasing manager should report, to some extent, on every area that reflects on the efficiency of the purchasing operation. Among the factors that determine **purchasing efficiency** are:

1. Proved savings. These will include:
   a. Proved savings from purchases under market prices for goods bought.
   b. Proved savings achieved by initiation of improved methods or substitution of better or cheaper materials.
2. Intangible savings.
3. Expense of operating purchasing department.
4. Expense caused by purchasing department failures—which include:
   a. Purchasing loss and error account.
   b. Cost of failure to have material on hand when needed.
5. Inventory expense.

Proved Savings on Price. Proved savings on price should be established by charts or tabular reports on the principal commodities purchased, showing comparatively the average market prices of the commodities and the actual prices paid over the period covered by the report. It is impractical to keep such charts or reports on all items bought. Since published market prices are a necessary part of such a study, it is usual to maintain them on basic commodities only and assume that all other items are bought at market prices. A summary of proved savings on price may then be prepared from these data.

Proved Savings by Substitution. Proved savings by substitution cannot be established merely on the claim of the purchasing department. Other departments may have an equal claim to credit for substitutions made. It is necessary, therefore, to initiate claims for such savings in the purchasing department, then have them substantiated by the proper plant authority, and finally have each claim approved by the management before it is credited to the purchasing department.

Another way of proving savings is to show the differential between the price paid for a fabricated article and the market price of the component materials. For example, for solder, grade 50–50, the metal content is established, and there are published prices on lead and tin. The purchase records will show the price paid over the price of metals. If by developing a new source of supply or by developing a new method of purchasing, the differential is lowered, the purchasing department can claim the credit for the difference between the new and old differential.

Departmental Expenses. The departmental expense of the purchasing department should be as inclusive as possible and include salaries and wages, stationery and supplies, light, heat, charges for premises occupied, telephone, telegraph, and postage expenses, and all other expenses incurred by the company in operating the department. The report showing this expense will originate in the accounting or finance department.

Loss and Error Account. The purchasing department loss and error account covers all expenses caused by errors in specifying material, loss in receiving, shortages not adjusted by vendors, defective goods not replaced or adjusted by vendors, etc. The accounting department should keep this record and support a report which summarizes it. Goods returned for defects should be charged to this account, and credit memoranda received, or replacements should be credited to it. All transportation charges on such transactions as well as handling charges should be debited to the extent that they are not borne by the vendor.

Failures To Receive Material on Time. There should be a definite report on failures of the purchasing department to receive material on time. So far as the accounting system provides a means for charging actual expenses caused to the company by such failures, actual expenses should be used. When the actual expense is not ascertainable, as is generally true, some method for making a fixed charge against the purchasing department for delivery failures should be adopted. For the purpose of determining its efficiency in this respect, the purchasing department should maintain a record of orders received on time and orders received late, with the ratio of each to total orders. Fig. 9 shows a purchasing department report on delivery of purchase orders.

	Jan.	Feb.	Mar.	Apr.	May	etc.	Total
Open orders in file 1st of month.........	645	580	613	575	560	.....	.....
Orders issued during month...............	3,600	3,155	3,325	3,267	3,159		29,697
Orders received during month.............	3,665	3,122	3,363	3,282	3,224		29,950
Orders open at end of month.............	580	613	575	560	495		
Orders received on time.................	3,209	2,850	3,113	3,169	2,998		27,134
Orders received 1 day to 1 week late......	318	176	150	76	145		1,977
Orders received 1 week to 2 weeks late....	90	82	69	33	60		599
Orders received later than 2 weeks........	48	14	31	4	21		240
% received on time.......................	87.6%	91.3%	92.6%	96.6%	92.9%		90.6%
% received 1 day to 1 week late...........	8.7	5.6	4.5	2.3	4.5		6.6
% received 1 week to 2 weeks late........	2.4	2.6	2.0	1.0	1.9		2.0
% received later than 2 weeks.............	1.3	.5	.9	.1	.7		.8

Fig. 9. Summary of delivery of orders.

Purchasing Department Activity. Management should also receive a detailed report showing the activities of purchasing. Such a report should include the number of orders placed, average money value per order, and all other pertinent facts indicating the volume and scope of the purchasing department's activities (Fig. 10). This report should be initiated in the purchasing department.

	January	February	March	etc.	Total for Year
No. in Dept.	9	9	9	....	9
Orders issued	3,600	3,158	3,325		29,697
Total purchases	$213.117.00	$175,765.48	$164,859.41		$1,234,567.89
Value per order	$59.20	$55.71	$49.58		$41.57
Invoices rec'd	4,487	3,996	4,416		38,062
Letters written	1,160	912	1,125		9,306
Telegrams sent	143	100	103		843
Carloads rec'd	11	8	15		131
L. C. L. rec'd	114 tons	149 tons	115 tons		1,025 tons
Parcel post	2,340	2,134	2,419		23,813
REA	524	447	429		3,868
Truck	1,160	490	690		6,352
Returns	206	175	227		1,848

Fig. 10. Purchasing department activities report.

EVALUATION OF DEPARTMENT EFFICIENCY. The statistical and accounting reports described here play an important role in the evaluation of the purchasing department. However, it is extremely important that company management in making its evaluation not look at these figures outside of their proper perspective. For example, it is common for purchasing departments to keep records of the cost of placing an order. Usually this figure is a simple average determined by dividing the total cost of operating the department by the number of purchase orders placed.

Such a figure has some significance if it is examined as a trend from period to period with the proper explanation of the reasons for changes. However, it cannot be significant if viewed as a raw figure and compared with other companies and other industries. Here comparison becomes invalid because seldom are the same cost figures used and the scope of the purchasing function varies considerably from company to company.

A similar discussion would show that nearly all of these statistical measures can give a distorted picture if viewed by themselves. Therefore, in evaluation, statistics should be used to compare trends and to identify problems which should then be analyzed and explained.

To supplement the statistical reports of efficiency it is extremely important that management look closely at all indications of the proficiency of the purchasing department. As indicated earlier many of these are intangible but this is the common problem of the business executive as he evaluates his management functions. Much can be learned about the purchasing function by having its members prepare reports describing their approach to unusual or difficult buying situations. When these reports show careful consideration of all of the factors and a high degree of creative ingenuity in seeking the best purchase contract, it can reasonably be assumed that the department is performing well.

SECTION 7

INSPECTION

CONTENTS

CONTENTS (*Continued*)

INSPECTION

Nature of Inspection

DEFINITION. Inspection has been defined by Doyle (Manufacturing Processes and Materials for Engineers) as a function whose purposes it is to interpret specifications, verify conformance to these specifications, and communicate the information obtained to those responsible for making necessary corrections in the manufacturing process. This definition in a broad sense implies all specifications: physical dimensions; electrical, mechanical, chemical and other properties; and performance. However, the term inspection is often used in a narrower sense and is often interpreted to mean only the inspection of physical **dimensions and appearance.** The term **testing** is then used to describe the process of comparing performance and physical, chemical or other properties of the part or material. The choice of terminology is primarily dependent on the major use or need for inspection or testing, and upon the organization of the company. If the major concern is dimensions and appearance then all comparison processes are usually referred to as inspection. If performance or detection of properties of the product are a major concern then a clear distinction between inspection and testing is usually made, especially so if the company is large enough to justify separate departments for these functions.

Metrology is usually defined as the science of measurement. In common practice metrology is concerned with the calibration of measuring devices and gages. Thus a metrology laboratory or department is concerned with the calibration of the instruments used to inspect the product. Metrology does not involve inspection of parts and materials themselves.

INSPECTION AND QUALITY CONTROL. A very clear distinction must be made between inspection and quality control. Inspection is primarily a comparison with established standards, whereas quality control is concerned with any function which contributes to the quality of the goods produced. Amrine, Ritchey, and Hulley (Manufacturing Organization and Management) state that "Quality control refers to all those functions or activities that must be performed to fill the company's quality objectives." This complete involvement requires detailed knowledge of the established standards and a voice in establishing such standards. A major function of quality control is the establishment of criteria and interpretation of various specifications.

Inspection is **one phase of quality control.** Inspection depends upon other phases of quality control to establish definitions, interpretations, and procedures. In turn inspection provides information required by the other phases of quality control. In addition to the interpretation of specifications and inspection of these specifications, quality control may also include establishment of criteria for the selection of production equipment, tooling, and personnel. Quality control often involves statistical analysis and procedures for controlling the many functions carried on in industry.

OBJECTIVES. S. L. Nisbett states (Manufacturing Engineering and Man agement, vol. 60), "The purpose of inspection is to assure that components are o the desired quality . . . Ultimately the role of inspection is to check on the reli ability of manufacturing processes." Traditionally inspection has been aimed a discovering faults and defects, separating the good from the bad, and keeping records of the inspected parts. These continue to be primary objectives o inspection, but there are others. In many industrial plants the major objectiv of inspection is the **prevention of defects.** The need for this objective come about in two ways. Competition forces management to scrutinize closely thei manufacturing costs, and high reject rates are costly. Furthermore, tight manu facturing tolerances require close control of the manufacturing processes. On method of achieving close control is repeated inspection during the processing Thus, inspection becomes a part of the manufacturing process, and not an after thought. Inspection in itself never adds value to a product but it can be used effectively to reduce the production of defective parts.

The type of product or part, the nature of the processes, and the level o quality to be maintained all have a bearing on the goals and objectives of in spection. Therefore, objectives will not be the same for all plants or even fo all inspection areas within one plant. The following are the most common objectives:

1. Detect defects as they occur in processing.
2. Detect trends in the process which might lead to defects.
3. Validate the ability of machine and/or operator to perform an operation satisfactorily.
4. Monitor a process.
5. Remove defective parts from production to stop further handling and proc essing costs.
6. Remove defective parts to prevent poor performance of finished product
7. Inform all levels of management on the performance of manufacturing de partments or units.
8. Provide records and information to enable management to study and correc poor performance.
9. Provide records and information necessary for establishing inventory contro and product control scheduling.
10. Provide records for evaluation of individual machine or worker performance

Any one of the above objectives may justify some form of inspection. Mos inspection activity will satisfy or contribute to several of these objects.

QUALITY STANDARDS. The word "quality" does not imply the highes standards obtainable. Similarly quality control does not imply maintaining highest standards obtainable. Quality control implies the maintenance of tha level of quality which has been designated by management. Establishing the level of quality is a management function.

High standards of quality require precision equipment, highly skilled workers and much time and care in the manufacture of the product. Obviously this entails high cost. On the other hand, low standards of quality will result in poor performance of the product and dissatisfied customers. Most companies then, will adopt a standard somewhere between these two extremes. It ther becomes the function of the design engineer and the quality control engineer to relate this standard of quality in terms of dimensions and specifications which can be produced most efficiently.

Scope of Inspection

TESTING STAGES AND OBJECTIVES. There are several stages or levels at which inspection may be performed. These may be referred to in broad categories as **receiving inspection, pre-production inspection, production inspection,** and **product tests.** These several activities may be completely separate entities or they may be parts of one function. The point of distinction made here is not their physical or organizational relationships but rather their purpose and objectives.

Receiving Inspection. Receiving inspection is the process of examining purchased goods and materials as they are received into the plant. Its primary purpose is to establish whether or not these goods and materials are acceptable.

Such an inspection may be only a cursory examination or it may be very thorough. It should be an inspection for quantity as well as quality. Receiving inspection reports are valuable documents and copies should be sent to the purchasing department, to production engineering, to the stockroom and often to several other functions throughout the company (see section on Purchasing). The value of this function has often been underestimated. Inadequate inspection or poor receiving inspection records may result in the acceptance of defective materials. This may result in the expenditure of time and money in the handling and processing of materials already defective. The purchasing department often finds this function very valuable in establishing purchasing policies.

The receiving inspection function should be located near or even in the receiving area. The ideal layout of this area would have material passing from the receiving dock to receiving inspection and then on to its next destination. The layout should provide no opportunity for material to by-pass receiving inspection. At the same time receiving inspection should not interfere with activities in the dock or storage areas.

The place of receiving inspection in the company organization has been the subject of great controversy in many plants. Where receiving inspection is an outgrowth of a need by the purchasing department for such a function, it usually reports its results directly to the purchasing department. As the value of receiving inspection becomes apparent to other groups this organization often proves inefficient. Most companies seem to favor an organization in which receiving inspection is combined with the other inspection functions. Another concept brings the receiving inspection function under a group known as **materials distribution.** Such a group includes all of the functions concerned with receiving, inspecting, storing, distributing and shipping all materials. The subject is discussed further in this section under Organizing for Inspection.

Pre-Production Inspection. Pre-production inspection does not denote a particular inspection group but rather inspection of materials prior to their actual fabrication or production. This includes first-piece inspection, inspection of equipment and tooling, and pilot lot or pilot plant operation. The primary purpose of pre-production inspection is to check on the ability of a machine or process to produce parts to specifications. It must be recognized that this is only a check and is not an assurance that this ability will continue or be fully utilized. The ability to produce parts to specifications depends upon many factors, such as wear of the tools and equipment, looseness and vibration, temperature variations, and the skill and attitude of the operator. An initial inspection should provide assurance that the potential for this ability exists, but will not guarantee its continued existence.

First-piece inspection refers to the practice in which an operator of a machining process brings the first part he produces on a new setup to an inspector to verify its correctness before he proceeds with the rest of his work order. Thus, such an inspection is of value primarily to the operator since it assures him that he has properly set up the tools and equipment. It remains his responsibility, obviously, to maintain this condition. In many plants, particularly when expensive processes are involved, first-piece inspection is required. In other cases where less costly processing is involved, first piece inspection is not required but is available as a service to the operator.

Inspection of equipment and tooling is not often thought of as a function of the inspection group. It involves a periodic examination of all tooling and equipment used in certain processes to see that they are still capable of producing parts to specification. When a piece of equipment or tooling becomes defective for any reason it is removed from production and not used again until the defect has been repaired. Again such an inspection does not insure that equipment will be used properly or that it will maintain its ability to produce to specifications. It is merely a check to see that the potential exists.

A **pilot lot** or a **pilot plant** makes use of regular production equipment and personnel. It is used to check out a new product design, a new process or series of new processes before it is used on a production basis. New designs and new processes usually develop faults and difficulties which are not anticipated. It is the purpose of a pilot lot or pilot plant to discover these weaknesses or defects and correct them before the process or part is put on a regular production schedule. Some types of industries, particularly the process industries such as textile, paper and chemical industries, maintain permanent pilot lot or pilot plant facilities. In the metal-working industries pilot lots usually exist only when a new series of processes or a new line is being established. Satisfactory operation of such a line depends upon the balance and successful operation of each process within the line. Once again such an inspection process does not assure the continued production of parts to specifications. But it provides assurance of the potential for this ability. This method of pre-production inspection is often considered too expensive unless a major change is involved.

Production Inspection. There are several forms of inspection that take place during manufacturing. Such inspection functions may be scheduled as regular operations or they may be random checks. As regular operations they may occur at the end of a fabrication process or at critical points along the line. It is not uncommon to find some form of inspection at the end of each operation although it may not be complete or even very thorough. Where critical dimensions must be maintained during certain high-speed manufacturing processes, continuous 100 percent inspection may be necessary. For example, in the manufacture of paper, thickness is continuously being monitored automatically. When a defective run is detected, the machine may be automatically stopped or the defective section of paper marked for future identification. See also Automatic Inspection Methods elsewhere in this section.

Two of the more common forms of production inspection are referred to as **patrol or floor inspection** and **centralized inspection.** These are discussed in detail under Systems of Inspection.

In addition to inspections for dimensions, appearance, and quantity there are often many inspections which can be labeled **functional tests** of individual parts and components. These include tests for various physical and chemical properties; tests for performance, endurance, and efficiency of operation; and the like. Functional tests are usually performed after parts and components have been fabricated and often after they are assembled into a finished unit. Although this operation is frequently referred to as product test it is actually a form of inspection.

Occasionally certain properties and conditions of a part cannot be examined without destroying the part in the process of testing it. This process, called **destructive testing**, was commonly used for examining castings and welded fabrications. Several nondestructive techniques can now be used to ascertain the same information without destroying parts. These techniques involve the use of x-rays, ultrasonics, eddy currents, dyes, and the like.

Sampling vs. 100 Percent Inspection. One-hundred percent inspection obviously gives more information than partial inspection but it is also much more costly. The choice between 100 percent inspection and sampling rests primarily on the magnitude of the risk that a defect will occur that management is willing to take. Obviously this is not a decision for an inspector, but rather one for management to make. Such a decision may well result in 100 percent inspection of a few specifications and sampling inspection for others. Sampling is often justified in receiving inspection, production inspection, and functional tests, and a necessity in destructive tests. The decision to use sampling raises another question: how large should the sample be? A sampling program must be based upon sound statistical analysis for its results to be meaningful. Establishing sample sizes and sampling procedures is a quality control function and is therefore discussed in detail in the sections on Quality Control and Reliability, and Statistical Methods.

INSPECTION RECORDS. An extremely important part of inspection is the keeping of accurate records. These records may involve a variety of forms depending upon the type of inspection and the objective. Regardless of the form used, however, it should be emphasized that the value of inspection may be destroyed or seriously undermined by inaccurate or poorly prepared records and reports.

Inspection Tickets. An inspection ticket is the form used by the inspector to record his findings. Such a form may be a separate piece of paper or card, as shown in Fig. 1, or part of a routing sheet as in Fig. 2. The inspection ticket need not accompany the work order. An inspection record kept on the routing sheet itself is common if inspection is listed as a regular operation. This arrangement has several advantages since it usually requires less paper work, has less chance of getting lost, and is readily available for examination. Its chief disadvantage is that it does not allow duplicate copies to be retained by the inspector, or distributed to other people in management. Duplicate copies discourage alterations of an inspection record that can be made more easily on a routing sheet. A separate form usually permits a more complete record with notes and comments. Inspection records may also be combined with other forms such as move tickets or production control forms. The latter may be advantageous in a small company where inspection is more closely associated with production control.

An inspection ticket should contain the following information:

1. Part number and name.
2. Work order number and the date.
3. Operation number, if any, or point in the process.
4. Total number of parts in the order.
5. Total number of pieces inspected.
6. The number of parts accepted and the number rejected.
7. Signature of the inspector.

Other information may be added that is appropriate for the particular part or process. For example, the inspector may wish to note the reason for some of the rejects. He may also wish to note the disposition of the rejected material.

```
 _ _ _ _ _ _ _ _ _ _ _ _ _ _ _ _ _ _ _ _ _ _ _ _

   Part No. _____ Work Order _____ Date _____

   Part Name _____

                  I N S P E C T I O N   T I C K E T

   Last oper. completed _____ Inspector No. _____
   Quantity required in order _____
   Quantity received _____
   Quantity accepted _____
   Quantity rejected _____
   Disposition of rejected parts _____

   Comments _____

   _____

   _____

                         _____

                                          Inspector

   Copy for Production
```

Fig. 1. Typical triplicate form inspection ticket.

Since many defective parts may be salvaged it is important to note what has been done with rejects.

Inspection Reports. An inspection report can be distinguished from an inspection ticket by the fact that it is sent through the mail or messenger service and does not accompany the work. An inspection report need not contain the same information as that on an inspection ticket. Very often its information is more detailed than that on an inspection ticket. The form and information contained in the report depends upon the purpose of the inspection itself and to whom the report will be sent (Fig. 2).

One common type of inspection report is known as the **receiving inspection report.** This is primarily a confirmation of the receipt of a purchased order, the results of the inspection of the materials received, and the disposition of the material. It is essential that the purchasing department receive this type of inspection report. Copies may also be sent to the one who initiated the purchase request, and the major user of the material, such as the manager of production.

Another type of inspection report is the **performance and endurance test report.** Such reports provide a thorough description of the tests performed,

Operation No.	Quantity Received	Quantity Accepted	Quantity Rejected	Date	Inspector	Notes
1						
2						
3						
4						
5						
6						
7						
8						
9						
10						
11						
12						
13						
14						
15						
16						

INSPECTION RECORD

Part No. _____ Part Name _____ Work Order _____

Fig. 2. Typical inspection record form printed on the back of a production routing sheet.

the results obtained, and an analysis and interpretation of the results. They are of great value to the engineering, production, and sales departments.

There are several other quality control functions which require what might be considered inspection reports. For example, reports on the periodic examination of tools, fixtures, machines, and other equipment may be regarded as inspection reports.

Reports such as these serve as a basis for the entire quality control program and are especially valuable in the design of the product, selection of equipment, and assignment of personnel.

RELATION TO PRODUCTION METHODS. Inspection for the sake of inspection is generally of little value. If it were only a matter of separating defective parts from acceptable parts few inspection operations could be justified. Almost all inspection operations yield information which can and should be used constructively to produce a better product at a lower cost.

Reporting Defects. Although inspection cannot correct defects already made, it can help in the prevention of further defects. A complete, accurate inspection ticket or inspection report can provide valuable clues to the reasons for defects. It may point to a poor design specification, difficult-to-produce shapes, poor materials, inadequate process control, defective equipment, or human errors. It should be made clear that it is not the responsibility of the inspector to determine the cause of defect, but merely to provide a report or record of these defects.

The supervisor or manager of the production function must be continually aware of defects discovered in the many inspection processes. It is then his responsibility to initiate action which will correct the cause of these defects.

Design Modification. With manufacturing processes and machines continually undergoing change, it is difficult for design personnel, especially in medium and large-size plants, to keep completely abreast of their company's manufacturing capabilities. It is no wonder then that parts are designed that are difficult, if not impossible, to manufacture or that are unnecessarily complex because they are based on outmoded manufacturing methods. Since difficult or obsolescent operations often lead to increases in the reject and scrap rate, a careful study and analysis of inspection records should be made to reveal if a particular design characteristic is at fault. Quality control can then confer with product engineering to determine whether a design modification is in order. The new design should still satisfy the required functional characteristics while at the same time making the most effective use of the plant's manufacturing facilities. Inspection can best assist in this procedure by keeping accurate records of the reasons for each reject without trying to interpret or imply the underlying causes of each reject.

Process Modification. In many industries a part may be manufactured in a variety of ways. The planner has a choice of processes. Since each process may have certain capabilities not attainable in the other processes the choice is often a compromise. Such a choice is usually made from the experience of the planner. However actual production and inspection records may indicate that the chosen process will not achieve some of the specifications that could be attained with another process.

This may be illustrated with an example. A certain steel casting required that a slot be cut across one side. The tolerance on the slot was +0.0005 inch. Since other milling operations were done on the part, milling was also used to cut this slot. Inspection records showed that 40% of one lot was scrapped because the tolerance on this slot was not attained. On the basis of this analysis it was determined that broaching the slot would cost less than the cost of this 40% scrap. Broaching resulted in a 2% scrap loss.

Equipment Modification. The planner not only has a choice of processes, but often a choice of equipment as well. Several machines capable of doing the same type of process may not have equal rigidity or speed capabilities. Tooling also provides varying capabilities for doing essentially the same type of work. Faulty tooling, whether due to poor design or from wear, is a common cause of scrap.

Certain military ordnance contracts require periodic inspection of machines and tooling to assure proper maintenance of this equipment. This type of inspection will not assure good design nor proper use of the equipment, however.

Through records of workpiece inspection and tooling inspection one manufacturer was able to pinpoint the cause of several related defects on a certain part. The milling fixture used on this operation was found to be well maintained, but the design of the clamping arrangement was poor. A modification of the clamps resulted in less vibration and much greater machining accuracy and practically eliminated the cause of the former rejects.

HANDLING DEFECTIVE PARTS. The term "defective part" is used here in a broad sense meaning a part or product which does not meet all of the specifications established for it. Inspection will usually reveal some defective parts. When a defective part is discovered the facts of its rejection are recorded and it is normally segregated from the acceptable parts and disposed of in some manner. It is not economically sound to assume that all defective parts are worthless. A defective part may be disposed of in one of several ways.

Repairs and Reprocessing. It was mentioned earlier that many parts which are rejected by inspection may be salvaged. Most companies find it beneficial to maintain a **Materials Review Board** (MRB). It is the function of such an MRB to review and analyze inspection tickets and reports along with part specifications. They decide whether or not rejected parts can be salvaged, whether salvage is feasible, and how it can best be accomplished.

The salvaging of rejected parts may involve one or more courses of action:

1. Tolerances and specifications were tighter than necessary and therefore the part may be used as is.
2. Parts with dimensions out of tolerance may be fitted to mating parts by selective assembly and therefore used as is.
3. The defect may be of such a nature that the part can be reworked.

If the reworking of the defective part involves considerable machining or other expensive processing it may not be economically feasible to salvage the part. On the other hand if simple processing will satisfy the requirements, the value of the part may be well worth the extra effort. Without the benefit of such a materials review many acceptable parts might otherwise be lost to scrap.

Scrap. When it has been determined that the cost of salvaging a rejected part is equal to or exceeds the value of the part it should be disposed of in some suitable way. At this point, of course, inspection serves no further purpose. However, it may be the responsibility of a quality control group of the Materials Review Board to dispose of this material.

Although it might be possible to machine a rejected part to make it into a different usable part this practice is seldom feasible. The extra burden of paper work and precautions to avoid confusion make the cost of the newly formed part excessive.

The simplest, most common way, and probably least expensive in the end, is to sell the rejected parts as scrap material along with chips and other unusable materials.

Organizing for Inspection

PLACE IN PLANT ORGANIZATION. The place of inspection in the organization depends upon such factors as the size of the company, its principal products, and the type of manufacturing operations it conducts. These factors also determine the extent of the quality control function and the nature of the inspection process.

Some form of inspection and quality control is needed in every industry to achieve both precision manufacturing and economical operation. In small companies, quality control functions, if they are identifiable at all, are performed by a few individuals involved in other functions, such as design, production control, manufacturing engineering, and the like. However medium and large-size plants need a separate quality control group in order to function effectively. In such organizations inspection is part of the quality control group, and not a separate entity in itself. Figures 3 and 4 reflect the relationships that exist between quality control and inspection in typical plant organizations.

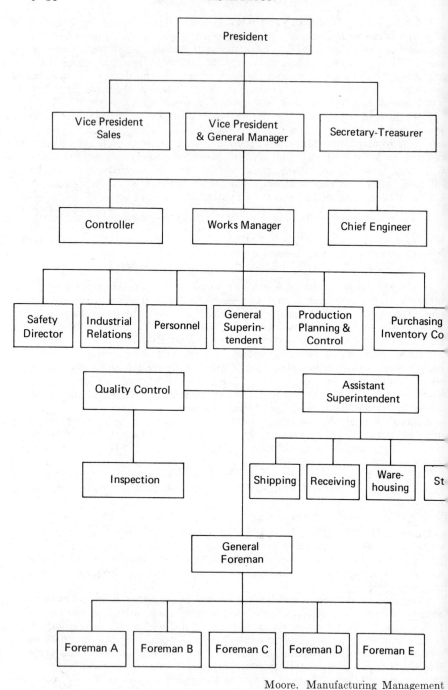

Moore. Manufacturing Management

Fig. 3. Typical manufacturing organization of a medium-size company.

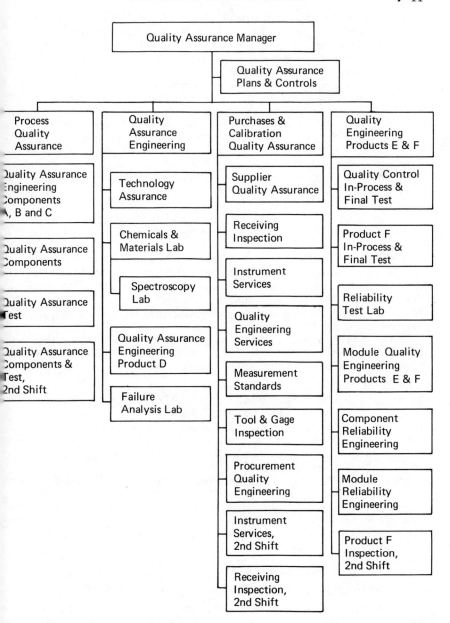

Fig. 4. Quality assurance organization of a large corporation.

Defining Inspection Responsibility. As mentioned earlier, the purpose of inspection will vary greatly from one situation to another. Similarly, responsibilities will vary according to the purposes. Amrine (Amrine, Ritchey, and Hulley, Manufacturing Organization and Management) states that "The inspection department possesses the authority to reject those items that fail to meet standards, . . . it must pass on purchased materials, the processing of materials, and the finished product. Also it may be responsible for the inspection of all jigs, fixtures, dies, gages, and measuring devices used by both the inspectors and production workers. Supervision of the procedure for the disposal of rework or rejected items may be another responsibility assigned to the inspection department."

He also states that "To carry out its responsibilities, the inspection department will need to establish a number of policies and procedures. It must decide what to inspect, where to inspect, how to inspect, and many other pertinent matters."

It is important that inspection not be responsible for any part of the actual manufacture of a part. There must be no conflict of interests within the inspection function.

If a quality control department or group exists within the organization it is usually their responsibility to define the functions and responsibilities of inspection. In the absence of a separate quality control department the responsibilities of inspection should be outlined at a high level in management. The details of specific functions and responsibilities should be planned by the works manager and/or the chief inspector.

In any case, the overall objectives of the company must be recognized when planning the broad scope of inspection. Specific responsibilities must reflect the need and purpose of each of the several inspection functions.

Level of Authority of Inspection. Amrine points out that inspection fulfils four primary functions or responsibilities: "(1) it **checks** the quality of incoming materials, (2) it **checks** on all finished goods to insure that only acceptable goods reach the customer, (3) it **aids** in maintaining process control and attempts to locate the flaws in manufacturing that would cause subsequent difficulties, and (4) it **serves** in an advisory capacity in attempting to correct or prevent quality control problems."

Some of these functions and responsibilities are staff functions and as such carry no authority. Therefore, the level of authority of inspection must refer to those functions which check or control production. It is important that the manager or supervisor of inspection be on at least an equal level with the manager or supervisor of manufacturing. He should never be subservient to manufacturing.

When a quality control department or group exists its manager should be on a level equal to that of Works Manager. Inspection then, being a part of quality control, should be on a level equal to that of manufacturing.

Organization of Inspection Departments. A major factor affecting the organization of the inspection department is whether or not the inspection department stands as an entity in itself. Figures 5 and 6 show the functions and the organizational plan of the inspection department of a typical medium-size company that does not have a quality control department. The three functions, examination, administration, and services may conceivably be performed under the supervision of one person in a very small plant or they could be divided into three or even more groups in a larger plant.

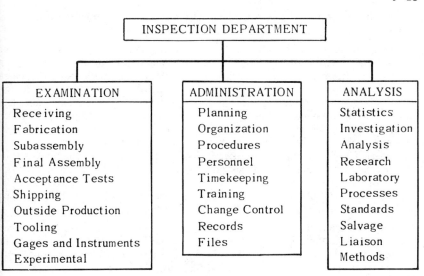

Fig. 5. Inspection department functions.

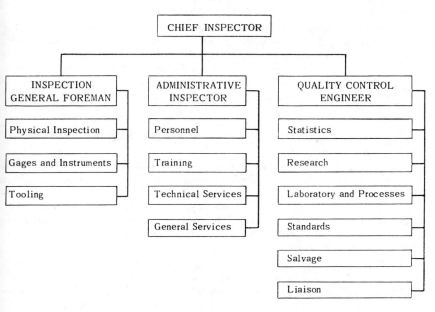

Fig. 6. Typical organization chart for inspection department.

If a quality control department exists, several of the functions shown performed by the inspection department would be performed by branches of quality control other than inspection.

PERSONNEL. The personnel involved in an inspection department will vary considerably depending upon the functions of the department and the size of the department. The personnel titles discussed here are typical of those found in manufacturing plants.

Chief Inspector. The head of the inspection department is usually given the title of chief inspector. Other titles in common use for this position are manager of inspection and supervisor of inspection. In most cases the chief inspector's duties are supervisory and administrative and he is responsible for all of the activities carried on within his department. All communications between the inspection department and other departments or groups within the company are carried on through the chief inspector. He may, however, delegate this responsibility and authority to a subordinate in the department.

The chief inspector must be well acquainted with most other functions within the company including product design, manufacturing capabilities, and the cost accounting procedures of his company. It is vital that he be technically trained and familiar with all inspection procedures and techniques so that he is able to talk the language of the people under him.

Foreman Inspectors. A foreman inspector is the immediate supervisor over a group of inspectors. In a small plant he may do inspection work himself. He is looked upon by other inspectors as the authority when problems and questions arise. Accordingly, he is expected to be the most highly skilled and experienced member of the inspection department. In some small companies the title "foreman inspector" does not exist and the chief inspector assumes the responsibilities otherwise assigned to such a person.

Floor Inspectors. A floor inspector is one who moves about the production area inspecting parts at the work stations as they are being made. He carries with him only the most basic inspection tools and often uses the workers' tools and gages for his inspection. His primary responsibility is to inform the production worker whether or not he is meeting specifications.

The floor inspector should also be an experienced person since he normally works alone and is not in immediate contact with his foreman or other inspectors.

Other titles such as **roving inspector** or **toll gate inspector** are sometimes used in place of floor inspector.

Bench Inspectors. A bench inspector is so named because he is assigned to a specific location. It is the bench inspector who does the routine inspection work separating the bad from the good. He is usually closely associated with other inspectors and with the foreman inspector. Therefore, this position does not require the high degree of skill and experience needed by the floor inspector. However, integrity and industriousness are required of every inspector. A beginner is normally assigned to the position of a bench inspector.

Special Inspecting Personnel. A number of other titles are used for a wide variety of inspection tasks. A **first-piece inspector** is often made available to production workers to check the first piece produced by a new machine or process setup. He performs his duties at a fixed work station usually consisting of a large surface plate or table similar to that used by the floor inspector. His primary responsibility is to inform the production worker whether or not the setup is producing to specifications.

Personnel working in the area of product test are commonly referred to as **technicians** rather than inspectors. These men often work closely with product engineers and other quality control personnel. A product test inspector needs a more highly technical background than does a floor or bench inspector. He must be capable of setting up, using, and interpreting a wide range of instrumentation.

Many companies find a need for inspectors in special categories such as receiving inspection, metallurgy, tooling and gaging, and packaging inspection.

In the absence of a quality control group the inspection department may require personnel for statistical analysis, research, and the development and maintenance of standards.

Women as Inspectors. Women are generally more patient, more dexterous and better suited to repetitive types of inspection then are men. They are not as well adapted as men to inspection tasks requiring a wide variety of responsibilities nor are they as well adapted to highly technical inspection tasks. They are most commonly employed as bench inspectors and especially when many of the production workers are women.

Handicapped Persons. Many companies find that handicapped persons make excellent inspectors. They possess many qualities which are highly desirable in an inspector. They are generally very conscientious workers, patient, and thorough. Most handicapped persons having lost one or more physical capability have developed others far above that of the average person. Blind persons normally have a very keen sense of touch. A person with one hand or arm will normally have unusual strength and dexterity in the other hand. Handicapped persons have been found to be very cooperative and dependable. There are many associations which are very helpful in selecting and even training handicapped persons for industrial jobs. Inspection is one of the more common tasks for which handicapped persons are well qualified.

TRAINING INSPECTION PERSONNEL. Many authorities have pointed out that proper instruction of inspectors is one of the most essential steps in the establishment of an efficient inspection department. Of course, such instruction must include the proper use and care of measuring instruments and other inspection equipment. But inspection operations should also be stressed. The keeping of complete, accurate records is extremely important. Perhaps the most essential things that an inspector must learn about his job is the philosophy of inspection: Why is inspection done? What are we looking for? How will the results be used? By whom will they be used? In spite of all the elaborate equipment available for the job, inspection still involves a great deal of human thinking and decision making. Unless the purpose of inspection is understood the human effort is largely lost. Inspection then becomes an inefficient, costly, and almost worthless endeavor. The importance of integrity and proper attitude of the inspector cannot be overemphasized.

Training of the inspector should begin as soon as a new inspector is employed. Unskilled personnel should be taught not only by formal teaching methods but by example through his fellow workers and especially from the foreman inspector. He should become thoroughly acquainted with company procedures and policies as well as the records and equipment that will be made available to him.

The education of inspection personnel is a continuing process. Experienced inspectors as well as new inspectors must be kept abreast of new product developments, current manufacturing capabilities, and the latest inspection practices and devices. Furthermore, there should be a constant effort to make inspection a more efficient and profitable process. This is primarily the responsibility of the chief inspector. Each member of the inspection department can contribute to this through his own efforts. The management and supervision of an inspection department should be such that it will foster team work among all personnel in the department. This attitude of team work and cooperation is perhaps more important in inspection than it is in most other phases of an industrial enterprise.

Training Course for Inspectors. Juran (Factory, vol. 110) suggests as a minimum the following items for coverage in a training course:

1. How to use the measuring instruments, how to read them, how to care for them.
2. How to identify what is good and what is defective for qualities that cannot be measured.
3. The why of the defects, that is, what will happen if the defects get through unnoticed.
4. How to decide on the sample size, how to select the sample, how many defects to allow in a sample.
5. What records to keep and why. Show him how the records are used (the temptation to fake inspection data is greatest where subsequent use is not known).
6. How to identify the product so it is not mixed up or confused with other products.
7. How to do the incidentals of the inspection job, that is, materials handling, cleaning, and counting.
8. How to deal with his industrial neighbors on the factory floor: machine operators, setup men, foremen, engineers, and many others.
9. How to earn the respect that is the just reward of the ethical and factual inspector.

Some of the points mentioned above may not be the responsibility of each and every inspector, nevertheless he should be aware of these points and understand them. The determination of sample size, the number of defects to allow in a sample, and many other statistical procedures associated with sampling are usually the responsibility of other quality control personnel. It should not be the responsibility of an inspector to determine these sample sizes and other statistics.

COST OF INSPECTION. O. W. Ehrhardt (Manufacturing Engineering and Management, vol. 57) points out that the company will never know the real cost of inspection unless it also knows the cost of lack of inspection. In analyzing the total cost aspect of quality control and inspection he points out that there are two major cost categories involved in such an analysis, "quality failure" costs, and "appraisal and prevention" costs. Quality failure costs include in-plant failure costs (scrap), and exterior failure costs (the cost of replacing or repairing defective parts in service). What we normally think of as inspection functions make up the **appraisal** costs, and other quality control functions including some inspection activities contribute to the **prevention** costs. Ehrhardt makes it clear that complete and accurate records are necessary in order to analyze these costs of inspection.

Compensating the Inspector. The inspector's wages are often only a small part of the total cost of inspection. Juran (Quality Control Handbook) points out that by the very nature of his task the inspector's status is above that of the operator whose work he judges. Yet the inspector's pay scale may be less than the operator's, especially if a high level of skill is required of the operator.

To compensate for any inequity in pay Juran suggests three alternatives:

1. Provide the inspection job with opportunity to earn incentive pay.
2. Make the inspection job a steppingstone toward promotion.
3. Provide nonfinancial incentives. These are in part inherent in the inspection job.

Wage incentives have proven satisfactory for simple routine inspection work, but as the inspection task becomes more dependent upon mental decisions and skill direct incentives become troublesome. Very often incentives for inspectors are not piecework or direct incentives, but rather are applied to several weeks or months of work. Occasionally measured daywork is used as a basis for incentives.

Optimizing Quality Costs. Ehrhardt furnishes a graph, Fig. 7, which shows that in order to reduce the number of defective parts increased costs for

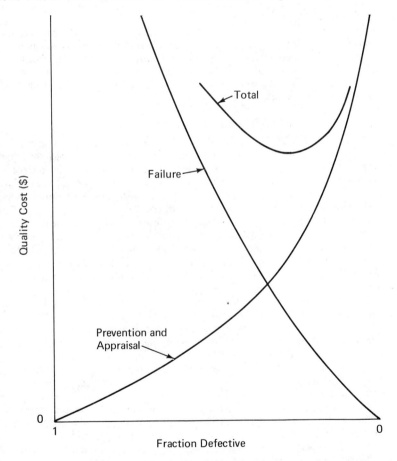

O. W. Ehrhardt. Manufacturing Engineering and Management. Vol. 57

Fig. 7. Increased expenditures for quality appraisal are offset by reduced total quality cost.

prevention and appraisal are necessary. This graph also shows that as the number of defective parts is reduced the cost of failures is also reduced. The total cost is the sum of prevention and appraisal costs and failure costs. The optimum cost is that of lowest total cost. Beyond that point increased expenditures for prevention and appraisal are not economically justified. It would be cheaper to risk accepting a certain percentage of defective parts than to increase the inspection necessary to find that number of defective parts. If the failure of a part jeopardizes human safety then failure costs become exhorbitant and 100 percent inspection is feasible. It should be understood, however, that even with 100 percent inspection a defective part could be passed as acceptable through human or mechanical error.

The cost of precision inspection equipment often increases at a greater rate than the corresponding increase in precision. At the same time the cost of failures for high precision parts becomes disproportionately higher than parts with less precision.

Statistical studies have pointed out that failures are never distributed uniformly over the quality characteristics. That is, a small percentage of causes account for a large percentage of the total failure cost. The significance of this maldistribution is that it permits concentration on the few characteristics having the greatest risk of defects. It is better to concentrate efforts on those areas where the greatest number of defects may occur.

Systems of Inspection

PATROL OR FLOOR INSPECTION. Although there are many approaches to the inspection process most may be classified into two or three general systems of inspection. One of these systems is referred to as **patrol or floor inspection.** In some plants this may also be referred to as **roving inspection.** This system requires the inspector to move about the manufacturing area inspecting parts at the work station where they are being made. The inspector normally carries with him the most basic measuring devices such as micrometers, calipers, and simple gages. In some cases he may have a cart enabling him to carry more elaborate equipment, although he would seldom have all of the facilities that are available at a centralized inspection area. In many cases he may depend upon gages and other special inspection equipment used by the worker himself.

The floor inspector is expected to keep the same complete records required of inspectors in other systems. Most of his records will be written at the work station. However he may return periodically to a central location to prepare more complete records and reports. Generally these are not as detailed or as complete as reports and records in other systems; on the other hand records of the floor inspector can reveal situations which would not be observed or recorded in other systems.

The floor inspector is generally assigned one area in the manufacturing plant. Thus he becomes well acquainted with that particular line of work or process. He becomes keenly aware of the capabilities and limitations of the machines and the workers in that area. As a result of this awareness he may quickly spot errors and defects which would not be noticed by an inspector in a centralized area. This will enable him to concentrate his time on the more common and more serious problems and spend less time on problems of less consequence. In this sense he is a much more efficient inspector than is the inspector in a centralized area.

At the same time the floor inspector is subjected to problems which centralized inspectors do not have. Every worker wants to have his work accepted. The foreman in that manufacturing area will also want a good inspection record. Thus there is a considerable amount of pressure upon the inspector to pass the work that he inspects. Though this pressure may be brought upon him in a very subtle manner it is nevertheless very real.

Advantages and Disadvantages of Floor Inspection. To summarize, the advantages and disadvantages of floor inspection may be listed as follows:

Advantages:
1. Errors are detected sooner. This provides an opportunity to correct errors early in the process and therefore avoid further rejects.
2. The inspector develops the ability to detect errors peculiar to a particular machine or process. He can recognize deviations from the norm more quickly.
3. As a result of the inspector's familiarity with a particular type of work he will spend less time on specifications which seldom are in error.
4. Handling materials to and from a centralized inspection station may be eliminated. Furthermore the space required for an extra inspection station may be eliminated also.

Disadvantages:
1. The lack of sophisticated equipment or even a good inspection environment may hamper the work of the inspector.
2. The inspector is subjected to pressure from both the worker and his foreman to pass work. The inspector is isolated from his supervisor and other inspectors and may lack the moral support he needs to do a good inspection job.
3. Much time may be lost in moving from one work station to another. This of course will depend upon the layout of the work stations, the type of work stations involved, and the amount of equipment the inspector must carry with him.
4. It may be difficult for the inspector to dispose of rejected materials. In many cases he may only be able to mark the rejected workpiece. The rejected workpiece may subsequently be placed among the good workpieces and continue on to other processes, incurring further expense.

CENTRALIZED INSPECTION. The term "centralized" is not used here in the geographical sense but rather to indicate that several inspectors and their equipment are grouped together in a particular area where the inspecting takes place. This is in contrast with the patrol or floor system discussed above where inspection takes place at work stations throughout the manufacturing area. The centralized system is generally used where inspection is required at a specific point in the manufacturing sequence of operations. When the part reaches the point it is sent to the centralized inspection area where it is inspected and, if acceptable, returned to production for the next operation.

Centralized inspection is often used in job lot production. In this case an entire lot is sent to inspection before moving on to the next manufacturing operation. This does not necessarily mean that 100 percent inspection takes place. In fact, the use of sampling techniques is common in a centralized inspection system. In general, however, a larger sampling is used in centralized inspection than in floor inspection.

Since the inspector in centralized inspection does not move about the manufacturing area, he is not in as close contact with the worker as is the floor inspector. As mentioned above, this will produce some advantages and some disadvantages. This system of inspection provides an opportunity for better communications among inspectors and better supervision. It also permits and helps justify the use of more sophisticated equipment. All of these facts make it possible to use less skilled inspectors than is possible with floor inspection.

Advantages and Disadvantages. In general the disadvantages of floor inspection become the advantages of centralized inspection, and vice versa. The advantages and disadvantages of centralized inspection may be summarized as follows.

Advantages:
1. Sophisticated inspection equipment may be used.
2. Good communication between inspectors and close supervision possible.
3. Recordkeeping facilitated.
4. Pressure from workers and production foremen minimized.
5. Less skilled inspectors may be used.

Disadvantages:
1. Errors will not be detected until the entire lot is processed. Often several operations are performed before the parts reach this inspection operation.
2. Extra handling of parts to and from the inspection area is required.
3. Space is required for the centralized inspection area and workpiece storage.

COMBINED SYSTEMS. Since floor inspection provides only limited inspection information and centralized inspection does not produce inspection information quickly enough in many cases, a combination of the two systems is ordinarily used. The majority of all parts manufactured have some critical operation which must be immediately followed by inspection. This type of inspection operation is usually performed in a centralized inspection area. However, if defects produced during the critical operation become serious a floor inspection may be justified. Such defects may render the part unsalvageable or salvageable only at high cost. If the material in the part is expensive or if the part has become expensive because of many previous operations then a defect at a critical operation becomes serious. Under these circumstances floor inspection is definitely justified.

It must be remembered that floor inspection involves sampling. To achieve maximum effectiveness sampling should be based upon a specific statistical analysis. In many cases, however, floor inspection can be of value even without statistical analysis. In a great many plants a conscientious floor inspector moving about the manufacturing area can accomplish the desired results. Relatively few manufacturing plants make use of a thorough statistical analysis for floor inspection.

There are several other types of inspection which might be considered a combination of floor inspection and centralized inspection. **First-piece** inspection is usually done at a centralized inspection area; yet it retains some of the advantages of floor inspection. The first piece produced on a new tool or machine setup is inspected to determine if the worker and setup are capable of meeting the specifications for that part. The purpose of a first-piece inspection differs from that of a regular floor inspection or centralized inspection since it is primarily for the benefit of the worker. Sometimes no permanent record of this inspection is kept.

Another form of inspection called **line inspection** might be considered a combination of floor inspection and centralized inspection. Line inspection occurs at the end of a production line or following a critical operation in a production line. The inspector is located at a fixed work station and materials are brought to him. Thus it resembles centralized inspection. Furthermore, it is specified as one of the regular operations in the manufacturing process. Generally it involves 100 percent inspection, although, sampling is sometimes used. In line production, parts do not move in lots, but rather one at a time from one operation to the next. Therefore, even if the line inspection is located at the end of the line, only a few defective parts will be produced before the first one is caught.

Line inspection however lacks some of the advantages of centralized inspection inasmuch as each line inspector is removed from the others. Communication between inspectors and supervision of these inspectors becomes difficult. Also, sophisticated inspection equipment is usually not available since duplication of this expensive equipment at each inspection station is seldom justified.

With the development of highly automated processes in line production manual inspection of any form or system has proven to be a bottleneck. Therefore, several **automatic** inspection machines have been developed. As with line inspection an automatic inspection machine may be placed at the end of a production line or it may be placed after any critical operation in the production line. Automatic inspection methods are discussed more thoroughly elsewhere in this section.

The automatic inspection machines mentioned above merely replace the human inspector in the production line. Another type of automated inspection, called **adaptive control**, interfaces directly with the production machine with the output of the machine undergoing continuous 100 percent inspection. Feedback from the inspection process is used to regulate the machine in such a way as to compensate for any errors detected.

Physical Requirements

PHYSICAL LAYOUT FOR INSPECTION. The location and layout of an inspection area depends largely on the manufacturing methods used, the inspection system and equipment required, and the nature of the workpiece to be inspected. Careful consideration must be given to special problems presented by bulky or delicate instrumentation, oversize or microscopic workpieces, and critical environmental conditions. The physical arrangement of the inspection area must be conducive to effective inspecting and designed so as to facilitate the movement of workpieces into and out of the area. Space requirements for inspection may vary from as little as 50 or 60 square feet to as much as 200 or 300 square feet per person. Adequate provision must be made for workpiece storage during the inspection process.

Determining Locations. In general inspection should be centrally located with respect to the area in the plant that it serves. Since there are several types or systems of inspection, the requirements of each should be considered.

Although a floor inspector does not have a fixed location in the manufacturing area he must have a location where he can care for and store his equipment and prepare and maintain records and reports. This base of operations may be part of a centralized inspection area or a room near the manufacturing area. The inspector will need a desk or bench and cabinets in which to store equipment. He may also need a filing cabinet or he may share a common filing system with the other inspectors.

A centralized inspection area should be assigned in conjunction with the other production facilities giving consideration to any special requirements that may exist. The cost of material handling plays an important role in this determination. For this reason, a centralized inspection area is usually located on a main aisle. Two types of layouts are common for centralized inspection areas. In one case each inspector has his own bench and work storage area. Some equipment, such as optical comparators and other sophisticated or expensive equipment are shared in common with other inspectors. Even though each work station is separate they are all located in one general area under one supervisor. In the other case, each inspector is provided with his own work bench but a common storage area for work materials is shared by all inspectors as are major pieces of inspection equipment. The latter case provides somewhat greater flexibility in that each inspector has access to any work brought to the area. On the other hand, this may cause some confusion. It is not as neat an area, and it may require more supervision (Fig. 8).

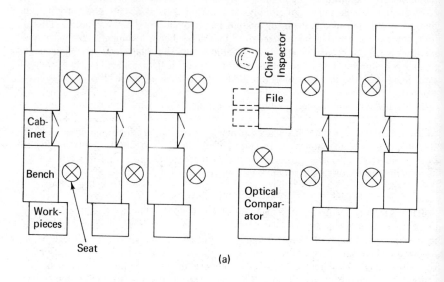

(a)

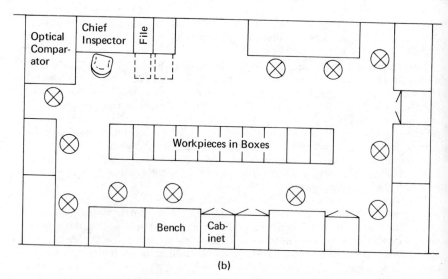

(b)

Fig. 8. Two typical layouts for a centralized inspection area.

First-piece inspection should be located near the major manufacturing operations it serves and readily accessible to all workers in the area. This does not mean that it has to be located on a main aisle; sometimes location on a side or cross aisle may be more convenient. The nature of first-piece inspection usually results in workers standing idle while their work is inspected. While this idleness may be unavoidable, it should be kept to a minimum. Furthermore, such idleness can have bad side effects. Management often objects to seeing workers standing idle. Workers sometimes become intrigued with the inspection equipment and procedures and they may actually interfere with the inspection process. Thus the location and layout of a first-piece inspection area should be such as to discourage excessive idleness. Locating this area near a production foreman may help. If the worker's station is nearby he may be encouraged to return to it while his work is being inspected. It may also be helpful to provide a counter or other barrier to separate the workers from the inspection area. Such a counter should be far enough from the aisle to provide standing room for the workers. In many respects the inspection area resembles that of a tool crib in both location and layout.

A large surface plate or table of steel or granite is commonly the center of a first-piece inspection area especially where machining operations are involved in the manufacturing area served. Other inspection machinery, equipment, and furniture are arranged around this table. It is very desirable, however, to leave at least three sides of the surface table accessible. Other inspection equipment and cabinets such as optical comparators, dividing heads, test stands, etc., form the remaining boundaries of the inspection area. Fig. 9 illustrates a typical layout.

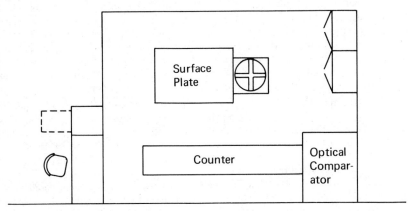

Fig. 9. Typical layout of a first-piece inspection area.

Line inspection, of course, is located in the proper sequence in the production line. The layout of the inspection station must be coordinated closely with the rest of the line to insure a smooth flow of workpieces to and from inspection. The location of storage space for workpieces and the arrangement of inspection equipment require careful study and analysis by both production and inspection management. Operations research techniques such as queuing theory and motion and methods study are helpful in determining the proper flow rates for workpieces and the optimum physical layout for processing workpieces.

Most often a line inspection station is involved with the inspection of only one part or product. In many cases however, two or more lines are channeled through one inspection area. In this case the inspection area may be similar to a centralized inspection area except for the manner of handling materials.

Enclosed Inspection Area. There are many advantages in providing an enclosed area for inspection. Some areas such as clean rooms, are enclosed to provide special environmental conditions. But even when these special conditions are not required an enclosed area offers a cleaner atmosphere than an open space in the manufacturing area as well as freedom from disturbances. The inspection may be totally enclosed with partitions running to the ceiling or by a false ceiling built under the main ceiling. It may be only partially enclosed, i.e. with partitions seven feet high or less, which cut out most of the noise and it presents a barrier to intruders. Even fencing may present a sufficient barrier to help in some cases. A partial wall or fencing also permits the use of the general plant lighting and ventilation systems. Supplemental lighting systems such as local lighting at the inspection station will almost always be required whether the inspection area is enclosed or not.

Enclosures always hamper the movement of materials. If plant disturbances are minimal, the atmosphere adequate for the inspection process involved, and security is not a problem, an enclosure is of little value.

SPECIAL PHYSICAL REQUIREMENTS. L. O. Heinold (Manufacturing Engineering and Management, vol. 59) discusses the effects of tolerances specified in millionths. He states ". . . it's a formidable challenge. Not only does it mean new machines but also new gages and much more critical techniques in using them. More than a few manufacturers suddenly find themselves unprepared to cope with the accuracies implicit in such a step up of the tolerance scale." Heinold points out that many special physical requirements are necessary in order to maintain control of production under these circumstances. Most commonly it means controlling atmospheric conditions such as temperature, humidity, and air cleanliness. It may also include the control of noise, light, and vibration. Not only must the environment of such a room or area be controlled but materials brought into this area and the people who work there (as well as their clothing) must be rigidly monitored.

Clean Rooms. One of the most common ways of controlling conditions for inspection is with the use of clean rooms. Several governmental agencies including the National Aeronautics and Space Administration, the U.S. Air Force, and the General Services Administration have established clean room standards. However, Federal Standard 209 issued by General Services Administration has gained the widest acceptance.

Federal Standards 209a defines a clean room as follows: "A clean room is an enclosed area employing control over the particulate matter in air with temperature, humidity and pressure control as required." This clean room standard is not limited to inspection use, but is applicable wherever control of these elements is required.

Three classes of clean rooms are established by this standard: Class 100; Class 10,000; and Class 100,000. The class designation pertains to the particle count per cubic foot. For example, a class 100 clean room must contain a particle count not to exceed a total of 100 particles per cubic foot of a size 0.5 microns and larger. Explicit specifications are established for the measurement and monitoring of particle count. This standard also allows for the specification of other criteria such as air pressure, air change rate, temperature range, humidity range, audio noise level, vibration, microbial contamination, and other environmental factors.

Most clean rooms are maintained at a temperature of 72°F. However, clean rooms for inspection should be maintained at a temperature of 68°F. This is necessary since the standard for precision measurement has been established at this temperature. Normally a controlled temperature should vary at a rate not to exceed 2°F per hour, with a maximum change not exceeding ±1°F. Since rusting is a serious problem with gages and other measuring equipment, the relative humidity should not exceed 45%.

An important characteristic of clean rooms is the pressure differential that must be maintained. Most standards specify that a clean room must maintain a pressure above that of its surrounding area to assure that all air leakage is directed outward. Doors and other openings must be kept closed as much as possible. Entrance to clean rooms is made through air-locks. The movement of materials in and out of the clean room must be done with care to prevent contamination. Since conveyors, hand trucks, and other types of material handling equipment are common sources of contamination manual handling is often preferred over mechanical handling. If manual handling is not feasible tubular type trolley conveyors may be used or simple hand trucks. The greater the exposed mechanisms on the handling equipment the greater the possibility for contamination.

Clean rooms were once thought of as necessary only for metrology, i.e. for the inspection and calibration of measuring equipment. However, modern manufacturing uses clean rooms for both inspection and production.

Screen Rooms and Quiet Rooms. Many electronic instruments and devices used for testing and inspection must be calibrated to extremely close tolerances making it necessary to completely isolate the instrument and the calibration equipment from stray electrical current. Screen rooms are so named because they involve the use of wire mesh screens or barriers to conduct stray electrical currents to ground. A screen room may be located in an otherwise open area. However, in most circumstances if a screen room is necessary, a clean room is also necessary. Thus the screen or electrical barrier is often found within a clean room.

Tests of noise levels are often conducted in quiet rooms or anechoic chambers. As the name implies, a quiet room is one in which sounds and reverberations are carefully controlled usually at a very low level. Sound control is achieved through the use of acoustical materials in walls, floors, and ceiling or by locating sound baffles at specific points throughout an area. Quiet rooms or chambers are often segregated physically from the main building structure by literally constructing a room within a room. The inner, or quiet, room is supported and connected to the outer room by means of vibration and sound damping materials.

Special Environmental Conditions. Most of the environmental conditions required for inspection or testing can be maintained in a clean room, a screen room, or a quiet room. It was pointed out that clean room standards take into account such things as vibration, noise level, and lighting conditions. In some cases these requirements may be quite apart from the need for a clean room. **Vibration** may be a serious problem in precision measurement, especially when heavy machinery is operating in the area. Some types of building construction transmit vibrations a great distance. When it is practical to do so, vibrations should be absorbed at their source, but this is not always feasible. Therefore it may be necessary to provide the isolation at the inspection area. Surface tables may be supported on vibration absorbing pads. This remedy is often satisfactory for minor disturbances and at certain frequencies. When the problem is more severe the whole inspection area may have to be isolated from this vibration as discussed under Screen Rooms and Quiet Rooms. This may not be practical or even possible, particularly if several such areas are involved. Many manufacturers of precision equipment have had to construct an entirely separate building to house their metrology laboratory or test areas. If the product is a large piece of equipment it may be desirable to test this product in a separate building.

Lighting is an important factor in any inspection operation. In some situations, however, lighting takes on special significance. It is sometimes difficult to get sufficient quantity of lighting without getting glare and shadows. Modern lighting sources and lighting fixtures make this problem an easier one to solve. Enclosed recessed troffer lighting fixtures with plastic or glass diffusers are usually recommended for clean rooms and other enclosed areas. Such lighting fixtures greatly reduce glare and shadows and are relatively easy to maintain. Many inspection operations also require localized lighting, which again presents a problem with shadows, glare, and heat. Low-voltage high-intensity work lamps and fixtures can provide a flexible, economical solution. Where necessary, color correction may be achieved through the proper selection of lamps.

Fundamentals of Inspection

RELATING METHODS TO OBJECTIVES. One of the common criticisms directed toward inspectors is that they too often fail to understand or show concern for the purpose of a particular inspection operation. One reason behind this situation is the selection of personnel. Poor supervision and the lack of motivation from the supervisor are also factors. But more fundamental than these is a lack of understanding of the basic objectives of inspection. This problem seems to exist at all levels from top management right down to the inspector himself in a great many industrial companies.

The industrial newspaper Metalworking News has on several occasions quoted Ted Busch who, among others, is concerned about this lack of understanding on the part of management. One reason given for this lack of understanding is the inability to communicate. Mr. Busch states (Metalworking News, vol. 9) "The machine operator cannot communicate with the inspector, the inspector cannot communicate with the quality control statistician, etc. Worst of it is that they all think that they can. But even the most basic terms as accuracy and precision have different meanings at different levels." On another occasion Busch expressed his concern that many companies treat measurement merely as overhead with little knowledge of what it can do to cut costs.

S. L. Nisbett (Manufacturing Engineering and Management, vol. 60) also expresses his concern for a lack of adequate communication. He states: "The design engineer often produces inadequate drawings. These have to be interpreted first by manufacturing and then by quality control. We now have two interpretations not necessarily identical and with no assurance that either is correct. The search for the design intent can be frustrating. It can also cost a lot of money."

Nisbett also points out that relatively few design engineers, or quality control engineers for that matter, realize that there must be a close relationship between method of processing and methods of inspection. He points out that quality control engineers, the people who establish what inspection methods should be used, so often lack a professional approach to the subject. "Too often he is a user of techniques rather than a professional who understands the reasons behind the techniques."

Responsibility for Establishing Methods. In any company in which a quality control group exists the quality control engineer must establish what methods of inspection are to be used. In the absence of a quality control engineer the chief inspector is expected to establish these inspection methods. Regardless of who establishes the methods it is important that he recognize and properly interpret this "design intent."

Every production drawing carries many specifications for the part to be manufactured. Such things as length, diameter, and other linear dimensions are usually specified very clearly. Many other characteristics of the part are usually not expressed clearly. Most companies follow one or more standards for a variety of physical characteristics of manufactured parts. Companies working on government contracts are usually required to follow the so-called Military Standards (MIL-STD).

After years of use and numerous revisions these sets of standards still lack clear-cut interpretation. Often the reason can be traced to the imprecision of the language where words mean different things to different persons.

Thus selecting the proper methods for inspection first requires a complete understanding of what is to be inspected and why. The next step is to fully understand the capabilities and limitations of the equipment and techniques available. Only then can the proper equipment and techniques be selected to satisfy the objectives of inspection. Selection of the best inspection methods becomes more complex as manufacturing specifications become more precise. New processing methods and new inspection methods and techniques are continually being developed. New inspection methods make possible a more effective and efficient inspection operation. But **selection** of the best method may become more complicated and difficult.

MEASUREMENT STANDARDS. The instruments and techniques of inspection are of little value unless they are calibrated and performed according to accepted standards. These standards may be related to physical dimensions, chemical and physical behavior, specifications, procedures, and the like.

While no single agency exists for establishing and maintaining standards in the field of production, professional societies, governmental bureaus, industrial groups, special commissions, and others have issued many that are widely accepted in industry.

International Standards. It took nearly 70 years, from 1890 to 1959, for the United States, Canada, and England to resolve their differences in the exact delineation of the inch. Until world trade became an important factor in the economy of the United States there seemed to be little reason for interest in international standards. The same of course was true with other countries. However dependency on world trade has brought about a great need for developing standards which can and will be accepted worldwide.

In 1968, the International Organization for Standardization had over 100 technical committees but few standards which were internationally accepted. However, much work has been done in establishing conversions from one to another of the several sets of standards used throughout the world. With very few exceptions manufacturing in the United States depends solely upon standards which are strictly American.

Metric System. The two systems of measurement currently in use throughout the world are the **metric system** and the **English system**; practically all of the leading industrial nations use the metric system. Numerous American companies have voluntarily set up **metrication programs** to facilitate the international exchange of engineering data. Technical societies and trade associations are using metric units in all articles appearing in their journals and reports. In 1971 a government study recommended that the United States officially begin conversion to the metric system.

Metric units, or more accurately **SI units** (Système International d'Unités) include the **meter**, defined as 1,650,763.73 wavelengths of the radiation emitted from krypton-86 gas; the **kilogram**, equal to an international platinum–iridium standard mass; the **second**; the **ampere** (the electric current unit); the **candela** (the unit of luminous intensity); and the **Kelvin** temperature scale. Other units, such as the **liter** and the **Celsius** temperature scale, are derived from these basic units. An international temperature standard for testing and other purposes was set at 20°C (68°F). In 1959, agreement was reached on a universal conversion factor which established the inch equal to 25.400000 millimeters.

American National and Military Standards. In the United States standards have been adopted for a great many manufacturing situations. These standards cover such areas as linear and angular dimensions; surface conditions; dimensional characteristics, such as eccentricity, squareness, and parallelness; and many others. Product design standards cover preferred basic sizes, systems of tolerances, systems of fits, surface finishes, and even product performance. Although it is impractical to list the many standards and their sources here a few deserve special attention because they are of vital concern to the inspection process.

The American National Standards Institute is generally considered the clearing house for the numerous recommendations and standards issued by professional and industrial groups. It is also the United States representative on many international committees concerned with the problems of standardization.

American National Standard Y14.5–1966 "Dimensioning and Tolerancing for Engineering Drawings," based on recommendations of the International Organization for Standardization (ISO), identifies such characteristics as surface flatness, straightness, roundness, cylindricity, and the like, and provides a system for specifying them. Fig. 10 illustrates some of the symbols used for this purpose

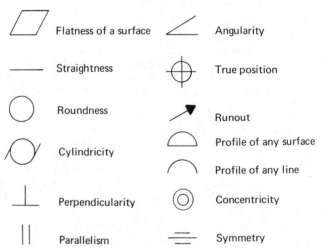

Flatness of a surface   Angularity

Straightness   True position

Roundness   Runout

Cylindricity   Profile of any surface

Profile of any line

Perpendicularity   Concentricity

Parallelism   Symmetry

Fig. 10. Symbols used in the dimensioning and tolerancing of engineering drawings, ANSI Y14.5–1966.

on engineering drawings. This system also provides for "size modifiers" referred to as "maximum material condition" (MMC) and "regardless of feature size" (RFS) and the designation of datum surfaces, datum features, and basic dimensions.

Another standard that is of particular interest to inspection is known as **gagemakers tolerances.** Figure 11 gives the gagemakers tolerances used for plug, ring, and snap gages. In practice these four classes are used as follows:

Class XX. Precision lapped to very close tolerances and used principally for master gages and very close toleranced product.

Class X. Precision lapped to very close tolerances and used principally as master and reference gages. They are not used as working gages except when especially close tolerances are required.

Class Y. Precision lapped and used in working gages and inspection gages of accurately ground parts.

Class Z. Used when extreme accuracy is not essential and when the working tolerances are fairly liberal. These gages have a ground and lapped finish.

Size (inches)		Gagemakers' Tolerance Classes			
Above	To and Including	XX	X	Y	Z
0.029	0.825	0.00002	0.00004	0.00007	0.00010
0.825	1.510	0.00003	0.00006	0.00009	0.00012
1.510	2.510	0.00004	0.00008	0.00012	0.00016
2.510	4.510	0.00005	0.00010	0.00015	0.00020
4.510	6.510	0.000065	0.00013	0.00019	0.00025
6.510	9.010	0.00008	0.00016	0.00024	0.00032
9.010	12.010	0.00010	0.00020	0.00030	0.00040

Fig. 11. Standard tolerances for gage blanks.

Several military standards (MIL STD 110, 111, 112, 113, 115, 116, 133, and 134) cover essentially the same classes of gages and tolerances as the gagemakers tolerances mentioned above with one major difference. The military standards specify a manufacturing tolerance and a **wear tolerance** for the gage whereas the commercial gagemakers tolerances specify only manufacturing tolerance.

Any plug, ring or snap gage must slide over a portion of the work in order to be used. This of course produces wear. The purpose of the wear allowance, then, is to assure a reasonable length of life to the gage. Many people feel, however, that specifying a separate wear allowance is not feasible. Since that extra wear allowance must either increase the total tolerance of the gage or reduce the manufacturing tolerance of the gage, they feel that neither alternative is justified.

Industry Standards. There are many industries for which well-established standards such as ANSI Y14.5–1966 and the gagemakers tolerances are not applicable or practical. Therefore, separate standards have been developed for their own use. Many such standards are voluntarily adopted by all companies in the industry. In other cases individual companies have developed their own set of standards. The most effective standards are those with the widest possible application. While standards with limited use and value are generally discouraged, much effort has been given to the development of standards which can be accepted and used universally.

PRECISION AND ACCURACY. Repetitive operations cannot be expected to produce exactly the same end results. There are always bound to be some deviations from a prescribed objective. The reasons for the deviations can usually be traced back to either the equipment used in the operation or the human operator. For example, a metal-cutting tool will wear very gradually and cause inaccuracies in the work it produces. A gage can lose its adjustment from constant use and pass defective parts. Or an inspector may tighten the thimble of a micrometer excessively resulting in an inaccurate reading. Theoretically these errors can be prevented; practically it may be very difficult or costly to attempt to eliminate them completely. However, considerable improvement is possible through better equipment design, proper personnel training, and increased interest on the part of the operator or inspector.

Deviations from a norm involve the concepts of accuracy and precision. Both terms have very specific meanings and great significance in inspection. **Accuracy** is defined as the closeness to a specified goal while **precision** is defined as the ability to repeat a specified goal.

Figure 12 will help to relate the two terms. A floor inspector has measured twelve parts being produced in a certain operation. The tolerance specified is plus or minus 0.001 inches. The operation is sufficiently accurate if 90 percent of the parts fall within the tolerance. It is noted in Fig. 12a that all 12 parts fall well within the limits specified. It must be concluded, then, that the operation is sufficiently accurate. If the tolerance specified were ±0.0005 inches the results would indicate that the operation is not sufficiently accurate. In addition the twelve measurements indicate that the deviation from the norm is increasing with each successive operation. It is conceivable that the next part and all those thereafter might well fall outside the prescribed limits. This indicates that it is not a precise operation. Figure 12b shows the results of inspection on another twelve parts. In this case all twelve measurements fall within a very close range, but none of them are within the prescribed limits. This operation, then, is precise but not accurate.

In many cases it is of little concern to the inspector whether or not an operation is precise or accurate. However, if his record or report is to be most effective it should supply sufficient information so that these factors may be seen. Since it is rarely feasible to manufacture a part to an exact dimension some variation from this basic value must be anticipated, specified, and accepted. This specified variation of the basic value is the **tolerance.** The tolerance for any particular dimension may be specified on the part drawing or it may be specified in a reference file as a company standard. Tolerances for all dimensions should be specified in some manner. The lack of clear specifications of tolerances is a serious problem in inspection. Tolerances for dimensions of minor importance such as chamfers, corners, radii, and fillets are often not specified at all, but are left to the good judgment and experience of the inspector and the machinist who produces the part. As manufacturing becomes more precise such a practice becomes less acceptable. Many companies today maintain specifications for tolerances on all dimensions regardless of their importance.

Many characteristics of a part cannot be specified quantitatively. Instead the inspector must decide whether or not the part is acceptable based on some qualitative standard. The appearance of a part is such a characteristic. It is desirable, when possible, to express these qualitative characteristics in numerical terms. Surface roughness is a good example of an apparent descriptive characteristic that has been converted into a numerically measured form.

Characteristics such as color and texture cannot easily be put into this form.

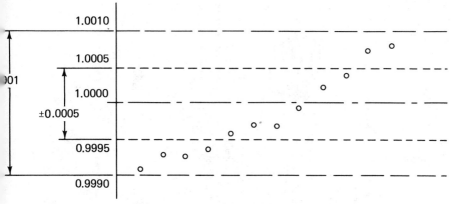

(a) Accuracy acceptable, precision poor

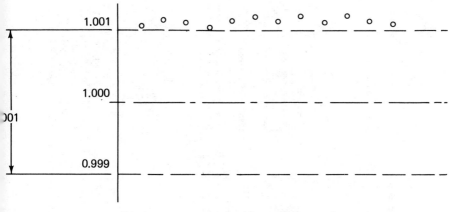

(b) Accuracy not acceptable, precision good

Fig. 12. Illustration of accuracy and precision.

However, visual standards can be developed against which parts may be compared. It is desirable to have a tolerance or acceptable variation specified for all dimensions and characteristics on a part.

Statistical Dimensioning. In many industries it is recognized that 100 percent inspection of all dimensions is not necessary. If sampling is used instead of 100 percent inspection, quality control engineers should establish the size and frequency of the samples to be taken. It is helpful if the inspector understands the reason why he is sampling and the consequences of this sampling. Regardless of the sample size or frequency of sampling (even with 100 percent inspection) some defective parts are likely to pass unnoticed.

In order to control the rate of rejects the practice of statistical dimensioning can be used. This involves the establishment of tolerances compatible with a given sampling procedure to produce a predictable rate of scrap and a predict-

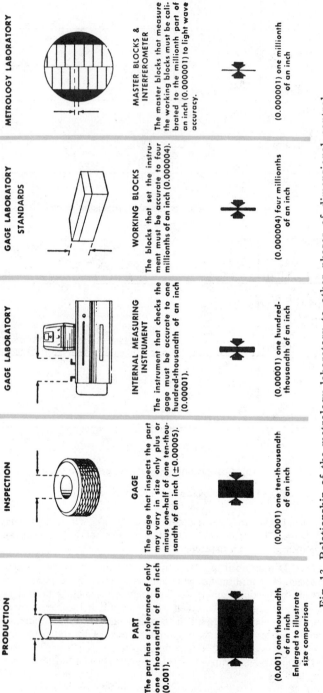

Fig. 13. Relationship of the metrology laboratory to other phases of dimensional control.

PRODUCTION

PART

The part has a tolerance of only one thousandth of an inch (0.001).

(0.001) one thousandth of an inch
Enlarged to illustrate size comparison

INSPECTION

GAGE

The gage that inspects the part may vary in size only plus or minus one-half of one ten-thousandth of an inch (±0.00005).

(0.0001) one ten-thousandth of an inch

GAGE LABORATORY

INTERNAL MEASURING INSTRUMENT

The instrument that checks the gage must be accurate to one hundred-thousandth of an inch (0.00001).

(0.00001) one hundred-thousandth of an inch

GAGE LABORATORY STANDARDS

WORKING BLOCKS

The blocks that set the instrument must be accurate to four millionths of an inch (0.000004).

(0.000004) four millionths of an inch

METROLOGY LABORATORY

MASTER BLOCKS & INTERFEROMETER

The master blocks that measure the working blocks must be calibrated to the millionth part of an inch (0.000001) to light wave accuracy.

(0.000001) one millionth of an inch

able range of dimensions. The inspector is not responsible for establishing such dimensions or sampling procedures, but it is helpful for him to understand the practice.

Calibration. Effective inspection requires (1) that all tolerances be properly specified and interpreted, and (2) that the inspector have some way of measuring or determining the limits of these tolerances. The inspector must know that the equipment and instruments he uses will tell him what he wants to know about a particular dimension. Selection of the proper instrument or piece of equipment is obviously important. Furthermore the instrument must be properly calibrated.

Calibration of measuring instruments and gages consists of comparing them with references which are known to be more accurate than the instrument being calibrated. A gage or measuring device must be more accurate than the workpiece it measures. For very close tolerances it is often necessary to have three and sometimes four levels of gages and references—a working gage, an inspector's gage, a master gage, and a master reference. In many types of work it is necessary that the master reference be calibrated from the standards maintained by the National Bureau of Standards. Figure 13 illustrates the relative range of tolerances which exist among the several levels of gages and references.

FITS AND LIMITS. Terminology, definitions, and specifications for fits and limits of mating parts have been well established for many years. The basic concepts of dimensioning and tolerancing are contained in standard ANSI Y14.5–1966 mentioned earlier in this section.

Definitions. The following terms and their definitions are reproduced from Preferred Limits and Fits for Cylindrical Parts, ANSI B4.1–1967.

Dimension. A dimension is a geometrical characteristic such as diameter, length, angle, or center distance. The term "dimension" is also used for convenience to indicate the size or numerical value of a dimension as specified on the drawing.

Size. Size is a designation of magnitude. When a value is assigned to a dimension it is referred to hereinafter as the size of that dimension.

NOTE: It is recognized that the words "dimension" and "size" are both used at times to convey the meaning of magnitude.

Nominal Size. The nominal size is the designation which is used for the purpose of general identification.

Basic Size. The basic size is that size from which the limits of size are derived by the application of allowances and tolerances.

Reference Size. A reference size is a size without tolerance used only for information purposes and does not govern machining or inspection operations.

Design Size. The design size is the basic size with allowance applied, from which the limits of size are derived by the application of tolerances. If there is no allowance the design size is the same as the basic size.

Actual Size. An actual size is a measured size.

Limits of Size. The limits of size are the applicable maximum and minimum sizes. (See Tolerance Limit.)

Maximum Material Limit. A maximum material limit is that limit of size that provides the maximum amount of material for the part. Normally it is the maximum limit of size of an external dimension or the minimum limit of size of an internal dimension.

Minimum Material Limit. A minimum material limit is that limit of size that provides the minimum amount of material for the part. Normally it is the minimum limit of size of an external dimension or the maximum limit of size of an internal dimension.

NOTE: An example of exceptions: an exterior corner radius where the maximum radius is the minimum material limit and the minimum radius is the maximum material limit.

Allowance. An allowance is a prescribed difference between the maximum-material-limits of mating parts. It is the minimum clearance (positive allowance) or maximum interference (negative allowance) between such parts. (See Fit.)

Tolerance. A tolerance is the total permissible variation of a size. The tolerance is the difference between the limits of size.

NOTE: The plural term "tolerances" is sometimes used to denote the permissible variations from the specified or design size, when the tolerance is expressed bilaterally. In this sense the term is identical to "tolerance limit."

Tolerance Limit. A tolerance limit is the variation, positive or negative, by which a size is permitted to depart from the design size. (See Limits of Size.)

Unilateral Tolerance. A unilateral tolerance is a tolerance in which variation is permitted only in one direction from the design size.

Bilateral Tolerance. A bilateral tolerance is a tolerance in which variation is permitted in both directions from the design size.

Fit. Fit is the general term used to signify the range of tightness or looseness which may result from the application of a specific combination of allowances and tolerances in the design of mating parts.

Actual Fit. The actual fit between two mating parts is the relation existing between them with respect to the amount of clearance or interference that is present when they are assembled.

NOTE: Fits are of these general types: clearance, transition, and interference.

Clearance Fit. A clearance fit is one having limits of size so prescribed that a clearance always results when mating parts are assembled.

Interference Fit. An interference fit is one having limits of size so prescribed that an interference always results when mating parts are assembled.

Transition Fit. A transition fit is one having limits of size so prescribed that either a clearance or an interference may result when mating parts are assembled.

Unilateral Tolerance System. A design plan which uses only unilateral tolerances is known as a unilateral tolerance system.

Bilateral Tolerance System. A design plan which uses only bilateral tolerances is known as a bilateral tolerance system.

Basic Hole System. A basic hole system is a system of fits in which the design size of the hole is the basic size and the allowance, if any, is applied to the shaft.

Basic Shaft System. A basic shaft system is a system of fits in which the design size of the shaft is the basic size and the allowance, if any, is applied to the hole.

Acceptance of Parts. The following terms and standards are also reproduced from ANSI B4.1–1967.

Acceptability. A part shall be dimensionally acceptable if its actual size does not exceed the limits of size specified in numerical values on the drawing or in writing. It does not meet dimensional specification if its actual size exceeds those limits.

Reference Temperature. Limits of size as derived from the tolerances shown herein are the extreme values, within which the actual size of the dimension shall lie, at the standard temperature of 20C or 68F.

Limits and Tolerances. Limits and tolerances are considered to be absolute regardless of the number of decimal places. Limits and tolerances are to be used as if they were continued with zeros beyond the last significant figure.

NOTE: This means that all inaccuracies of size, due to errors, wear, or change in tools, gages, machines, processes or measurement, shall be included within these limits.

Effect of Surface Texture. Parts of necessity are measured over the crests of surface irregularities, yet for moving parts such irregularities soon wear off and clearances are increased. For this reason surface finish is quite critical, especially for the finer grades, and should be specified when considered necessary

Standard Fits. ANSI B4.1–1967 provides for three general types of fits: running fits, locational fits, and force fits. Each of these, in turn, is broken down into several classes of fits which are designated and described as follows.

Designation of Standard Fits. Standard fits are designated by means of the symbols given below to facilitate reference to classes of fits for educational purposes. These symbols are not intended to be shown on manufacturing drawings; instead, sizes should be specified on drawings.

The letter symbols used are as follows:

RC Running or Sliding Clearance Fit
LC Locational Clearance Fit
LT Transition Clearance or Interference Fit
LN Locational Interference Fit
FN Force or Shrink Fit

These letter symbols are used in conjunction with numbers representing the class of fit; thus "FN 4" represents a class 4, force fit.

Running and Sliding Fits. Running and sliding fits are intended to provide a similar running performance, with suitable lubrication allowance, throughout the range of sizes. The clearances for the first two classes, used chiefly as slide fits, increase more slowly with diameter than the other classes, so that accurate location is maintained even at the expense of free relative motion.

These fits may be described briefly as follows:

RC 1 **Close sliding fits** are intended for the accurate location of parts which must assemble without perceptible play.

RC 2 **Sliding fits** are intended for accurate location but with greater maximum clearance than class RC 1. Parts made to this fit move and turn easily but are not intended to run freely, and in the larger sizes may seize with small temperature changes.

RC 3 **Precision running fits** are about the closest fits that can be expected to run freely, and are intended for precision work at slow speeds and light journal pressures, but are not suitable where appreciable temperature differences are likely to be encountered.

RC 4 **Close running fits** are intended chiefly for running fits on accurate machinery with moderate surface speeds and journal pressures, where accurate location and minimum play is desired.

RC 5 **Medium running fits** are intended for higher running speeds, or heavy
RC 6 journal pressures, or both.

RC 7 **Free running fits** are intended for use where accuracy is not essential, or where large temperature variations are likely to be encountered, or under both of these conditions.

RC 8 **Loose running fits** are intended for use where wide commercial tolerances
RC 9 may be necessary, together with an allowance, on the external member.

Locational Fits. Locational fits are fits intended to determine only the location of the mating parts; they may provide rigid or accurate location, as with interference fits, or provide some freedom of location, as with clearance fits. Accordingly they are divided into three groups: clearance fits, transition fits, and interference fits.

These are more fully described as follows:

LC **Locational clearance fits** are intended for parts which are normally stationary, but which can be freely assembled or disassembled. They run from snug fits for parts requiring accuracy of location, through the medium clearance fits for parts such as ball, race and housing, to the looser fastener fits where freedom of assembly is of prime importance.

LT **Locational transition fits** are a compromise between clearance and interference fits, for application where accuracy of location is important, but either a small amount of clearance or interference is permissible.

LN **Locational interference fits** are used where accuracy of location is prime importance and for parts requiring rigidity and alignment with no special requirements for bore pressure. Such fits are not intended for parts designed to transmit frictional loads from one part to another by virtue of the tightness of fit, as these conditions are covered by force fits.

Force Fits. Force or shrink fits constitute a special type of interference fit, normally characterized by maintenance of constant bore pressures throughout the range of sizes. The interference therefore varies almost directly with diameter, and the difference between its minimum and maximum value is small to maintain the resulting pressures within reasonable limits.

These fits may be described briefly as follows:

FN 1   **Light drive fits** are those requiring light assembly pressures and produce more or less permanent assemblies. They are suitable for thin sections or long fits, or in cast-iron external members.

FN 2   **Medium drive fits** are suitable for ordinary steel parts or for shrink fits on light sections. They are about the tightest fits that can be used with high-grade cast-iron external members.

FN 3   **Heavy drive fits** are suitable for heavier steel parts or for shrink fits in medium sections.

FN 4   **Force fits** are suitable for parts which can be highly stressed or for shrink
FN 5   fits where the heavy pressing forces required are impractical.

Preferred Basic Sizes. At one time practically all dimensions used in manufacturing in the United States were based upon the fractional-inch system. That is, all dimensions were expressed as fractions or decimal equivalents of fractions. The designer, the machinist, and the inspector all thought in terms of fractional portions of an inch. Thus, halves, quarters, eighths, sixteenths, thirty-secondths, and sixty-fourths were preferred basic sizes. Even when expressed in decimal form they were thought of as equivalents of fractional denominations.

A strictly decimal system may be used in place of the fractional system. In this system the designer, the machinist, and the inspector think in terms of decimal portions of an inch. Preferred basic sizes in order of priority are then as follows:

1. Whole inches.
2. Even numbered tenths of an inch.
3. Odd numbered tenths of an inch.
4. Even numbered hundredths of an inch.
5. Odd numbered hundredths of an inch.

This system lends itself well to the use of micrometers, verniers, and other decimal-graduated instruments. However the conventional machinists' scale cannot be used and must be replaced by one graduated in decimals, usually 0.02-inch denominations. Terminology associated with the decimal-inch system is illustrated in Fig. 14.

Where the metric system is used, it is used as a decimal system. Therefore, preferred basic sizes are expressed in the same manner as the preferred sizes in the decimal-inch system mentioned above.

Equipment and Techniques of Inspection

ASSIGNMENT OF EQUIPMENT. It is common practice to assign micrometers and other small measuring instruments and equipment to each inspector for permanent use at his work station. Specialized equipment may also be assigned to a particular inspector on a fixed basis if justified by extended usage. Otherwise such equipment should be taken as needed using charge-out methods similar to those used in the production area.

In some companies inspectors are permitted to use their own personal equipment. It is important that this equipment meet the specifications set for the company's own equipment. Personal instruments used for inspection must be examined and calibrated exactly as company-owned equipment. In cases where both personal and company-owned equipment are used, the company-owned equipment must be distinctively marked for identification.

Responsibility for Care and Maintenance. Inspection equipment must be maintained with great care. To see that proper maintenance is given, definite responsibility must be assigned and understood by all inspection personnel.

Typeset[1]	On Drawings[2]	Oral[3]
0.002 inch or 2 mil	.002	two mil or point zero zero two inch
0.012 inch or 12 mil	.012	twelve mil or point zero one two inch
0.02 inch or 20 mil	.02	twenty mil or point zero two inch
0.20 inch	.20	point two inch
2.005 inch	2.005	two inch five mil or two point zero zero 5 inch
2.122 inch	2.122	two point one two two inch
2.00 inch	2.00	two inch
0.000005 inch or 5 × 10⁻⁶ inch or 5 mike	.000005 or 5 MK	five mike
0.00002 inch or 20 mike	.00002 or 20 MK	twenty mike
0.0002 inch	.0002	point two mil or two tenths mil or two hundred mike or point zero zero zero two inch
0.0025 inch	.0025	two point five mil or point zero zero two five inch
2.000005 inch	2.000005	two inch five mike
2.0005 inch	2.0005	two inch point 5 mil or two point zero zero zero five inch

[1] Name of term may be abbreviated, or omitted in tables with adequate column identification.
[2] Longhand or typewritten material may also use this form, followed by the appropriate term or one of its abbreviations.
[3] Where more than one form of expression is shown, the preferred form is stated first.

Manufacturing Engineering and Management

Fig. 14. Terminology used in the decimal–inch system.

The responsibility for the care and maintenance of large fixed pieces of equipment such as comparators, indexing heads, etc., should be assigned to one individual. But, this responsibility is practical only if it is accompanied by authority to use or delegate use of the piece of equipment. If responsibility is assigned to the foreman, the line of his authority already exists. When responsibility is assigned to a non-supervisory inspector it may entail exclusive use of the equipment. If the equipment is needed by other inspectors they must bring their work to the assigned inspector who will perform the inspection procedure on the equipment himself.

While proper care and maintenance is usually within the capabilities of most inspectors, major repairs and adjustments must be performed only by experienced, specially trained technicians. It is important that all inspection equipment be examined and recalibrated periodically in a suitable metrology laboratory especially maintained for this purpose. Military equipment manufacturers are required to calibrate all inspection equipment against the standards established by the National Bureau of Standards of the United States Department of Commerce.

Record Keeping. It is essential that each inspector keep proper records of the work that he does. This is one of his primary responsibilities. The type of records to be kept depend on the part to be inspected, the type of inspection performed, and the uses to which the records will be put. In most cases quality control will determine what records or data are required. Inspectors not only must be given the inspection routine but they must understand the objectives of that particular inspection process.

Data is most accurate when recorded at the time inspection work is performed. Recording information from memory usually results in serious errors and should never be allowed. Comments and other information provided by the inspector should be made as soon as possible. Completed inspection records should be filed with the foreman or with the clerk as soon as the job ends or at the end of each day.

DIMENSIONAL MEASUREMENT. A dimension is the specification of any linear or angular quantity. Dimensional measurement implies either the direct measurement of a dimension or the comparison of the dimension with a given standard.

Equipment for Dimensional Measurement. Proper equipment is a prerequisite for effective inspection. O. W. Ehrhardt (Manufacturing Engineering and Management, vol. 57) provides the following guideline used for the selection of equipment by Giddings and Lewis Machine Tool Company.

1. Economic advantage by reason of expediency, new capability, versatility, or improved performance.
2. Inherent accuracy and precision.
3. Adequate sensitivity, resolution, and response characteristics.
4. Proven principle and functional reliability.
5. Dimensional readout rather than "go" or "not go" gaging.
6. A means for generating a permanent record to facilitate system data or mechanization and for historical purposes.
7. Sufficient physical strength to withstand routine handling in the shop environment.
8. A provision for calibration and/or an accompanying certificate of calibration.
9. Portability.

Dimensional measurement equipment may be classified into three broad categories: direct-measuring devices, comparison devices, and transfer devices. In general, comparison devices are much faster to use than direct measuring devices. Basically a **comparison device** will indicate how much larger or smaller a dimension is than a given standard. Comparison devices incorporate a calibrated motion amplification device. These devices are used as if they were direct measuring instruments, once set to a given standard.

Direct measuring instruments involve one or more of four basic measuring principles.

1. Graduated markings (machinist's scale or protractor)
2. A vernier (vernier calipers)
3. A calibrated thread device (micrometer)
4. The wavelength of a given light source (interferometer)

A **transfer device** is basically a means for transferring surface or dimensional information to a more convenient position. A simple form of a transfer device is a caliper. Some comparison devices, such as dial indicators, are commonly used as transfer devices. Transferring information always increases the possibility of errors and should be avoided if possible.

Low and Medium Resolution Measuring Instruments. Resolution is defined as the smallest quantity of the unit being measured that can be detected by the instrument. It does not refer to the accuracy of the instrument itself, although the instrument must be at least as accurate as the smallest increment to be determined or measured. Assuming the instrument is sufficiently accurate in itself, resolution is largely a matter of the size of increments registered on the instrument. Some refer to this as "readability."

The qualification of "low," "medium" or "high" is quite arbitrary and indefinite. A machinist's steel scale would be classified among the lowest resolution measurement instruments. On the other hand, an instrument which was considered high resolution 10 or 20 years ago might now be considered only medium resolution. An instrument which measures to the nearest 0.0001 inch may be considered a medium resolution instrument.

Probably the most common instrument in this category is the **micrometer**. The conventional micrometer has a screw thread of 40 threads per inch and is calibrated to 0.001 inch. A vernier is sometimes incorporated to extend the calibration to 0.0001 inch. Several modifications of the conventional micrometer are available which are intended to reduce the human error involved in using the micrometer. For example, a micrometer is available with a readout window. Figure 15 shows a micrometer scale arrangement.

Micrometer instruments are available with locking or ratchet devices. The ratchet is intended to produce a uniform pressure on the thimble of the micrometer. Many inspectors, however, prefer not to use the ratchet because it destroys his "sense of feel." A skilled inspector can obtain as consistent results without the ratchet as he can with the ratchet. More sophisticated pressure control devices and amplification devices are available on many micrometer-type instruments.

A **vernier** is often incorporated in the linear or angular scale of an instrument to increase its resolution. Figure 16 demonstrates how a vernier scale is used. In this case the main scale is marked off in 0.025-inch increments. The vernier scale is marked with 25 divisions spaced 0.024 inch apart, 0.001 inch less than those on the main scale. The vernier scale moves along the main scale as the instrument is adjusted to the workpiece dimension being measured. The zero mark on the vernier will appear opposite the main scale reading giving the dimension to the nearest 0.025 inch. The mark on the vernier scale that is exactly opposite any mark on the main scale indicates the number of 0.001-inch increments that must be added to the first reading. In the illustration (Fig. 16) the instrument reads 1.129 inches.

Figure 17 illustrates the construction of a **height-gage size unit**. It consists of a central vertical column upon which there are anywhere from 6 to 24 rings spaced 1.0000 inch apart. It is supported on a micrometer screw with a thimble which is calibrated in 0.0001-inch increments. This can seldom be applied to the workpiece directly but is used with a transfer device. A 0.0001-inch dial

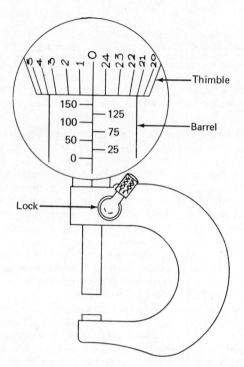

Scherr–Tumice, Inc.

Fig. 15. Micrometer scales. Thimble scale is in increments of 0.001 inch; barrel scale is in increments of 0.025 inch.

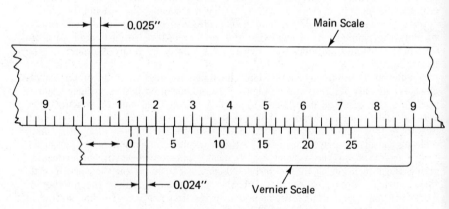

Fig. 16. Principle of the vernier.

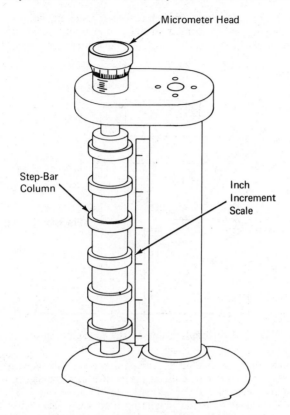

Fig. 17. Typical height-gage size unit.

indicator is commonly used as the transfer device. Busch (Manufacturing Engineering and Management, vol. 54) describes the correct measurement technique for a height-gage size unit:

The first step is to zero the indicator on the workpiece. The indicator is then moved to a position immediately above the closest step on the size unit and the micrometer drum is adjusted until the indicator is again zeroed. Several passes should be made between the size unit and the workpiece to ensure zero readings at both. When the zero readings have been established, the inch steps and the micrometer reading are added to obtain correct workpiece height.

The **autocollimator** is an optical instrument which detects and measures small angular quantities. It involves parallel, or collimated, beams of light as incident light directed against the surface of the workpiece. This surface must be sufficiently smooth so that it will reflect these light beams. If the reflected light beams do not coincide with the incident light beams the surface of the workpiece therefore must not be perpendicular to the incident beams of light. The angle of deviation can be measured on the lens of the instrument.

By attaching plane mirrors to the workpiece the plane of any workpiece surface can be compared with respect to a plane normal to the collimated light beams. This principle can be used to measure parallelism, flatness and squareness. It can also be used to compare a linear dimension on the workpiece with that of a master block or gage by using a plane mirror mounted on pivots as shown in Fig. 18.

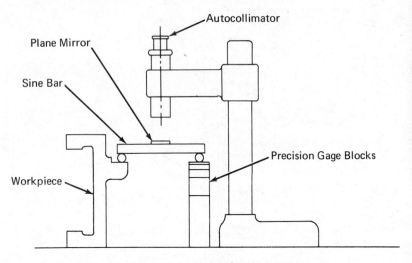

Fig. 18. An application of the autocollimator.

Fixed Gages. A gage is used to compare the workpiece with a standard. This distinguishes a gage from a direct-measuring instrument. A fixed gage is set to a given reference and is not adjusted to the workpiece. The most common types of fixed gages are the plug, ring, and snap gages. Fixed gages are often used in pairs with one "go" gage and one "not go" gage. The "go" and "not go" sizes correspond to the limits of the dimension being inspected. These gages are most often used in routine inspection operations. They permit very rapid inspection and require relatively little skill in using them. On the other hand they do not give much information. They merely indicate if the part is within the prescribed limits or not. Nisbett points out that "go-not go" limit gages are not compatible with statistical methods of quality control, which require quantification of the data being collected. Fixed gages provide only qualitative information.

Earle Buckingham is quoted as having said "when a gage is designed and accepted it becomes the standard of measurement and parts are manufactured not to design print but to acceptance by the gage." This is a very realistic approach but often a hazardous one. It emphasizes the need for great care in maintaining these gages. Plug and ring gages and some types of snap gages are made as integral units. This means when the gage becomes worn beyond acceptable limits the gage becomes useless and usually cannot be repaired. Many types of snap gages are constructed so that they can be readjusted, therefore prolonging their life.

Sight gages and feeler gages also fall in the category of fixed gages.

Deviation-Type Gages. A deviation-type gage is one which is set to a fixed nominal dimension, but also measures deviation from this reference. Deviation from the reference mentioned is detected by a stylus or spindle which rests on the work piece with a light spring tension. Movement of the stylus or spindle is amplified by some type of magnification device and is indicated on a dial or chart. It should be emphasized that the measurement of this deviation is not a measurement of the part itself. Busch (Metalworking News, vol. 9) referring to such a false application states that "this procedure violates a basic principle of reliable measurement in that an indicator is not a direct measurement instrument, it is strictly a tool for making comparative measurements."

Dial indicators are by far the most common type of deviation gage. There are several types of dial indicators also. They are distinguished primarily by the method of amplifying the movement of the stylus. The usual method employs a rack and pinion gear and a gear train as illustrated in Fig. 19. Split reeds and electrical and electronic amplifiers are also used.

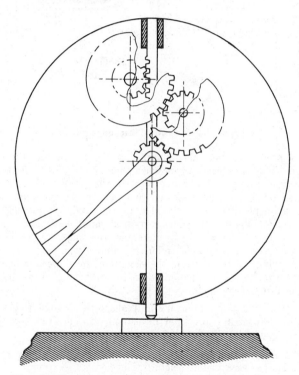

Fig. 19. Basic operating principle of dial indicator.

The optical comparator is usually used as a deviation-type gage also. In the optical comparator a magnified shadow of the part profile is cast upon a screen. A master chart is placed upon the screen and the workpiece, supported on a table, is then moved by a micrometer device thus measuring the deviation from the master chart.

Several types of **gages** employ the use of air pressure to detect the difference between the workpiece and a fixed nominal dimension to which the gage is set. This principle is most commonly applied to plug and ring gages. It can also be applied to snap gages and other special types of gages. Figure 20 illustrates a typical air gage system.

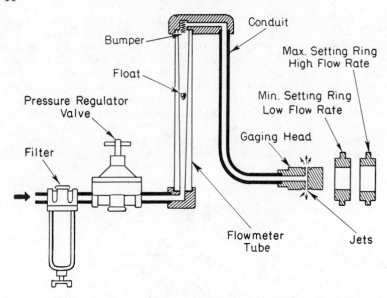

Fig. 20. Flow type of air gage circuit.

Gages that produce a permanent record of the measurement are also used for inspection. This is especially true when there is a desire to study the profile of the dimension concerned. For many years the profiles of gear teeth have been measured on a deviation-type device and recorded on a moving tape or graph. This principle has been extended to the study of circular and other geometric surfaces. Figure 21 shows a typical chart indicating the deviation of a circular surface. The trace does not show true shape or profile but merely indicates the deviation from a given reference circle as the part is turned around. For these recordings to be of value the deviations must be amplified from the actual deviation on the part. One space on the calibrated scales might indicate from 0.0001 inch to 0.00001 inch.

Gage Blocks. Since the First World War precision gage blocks have been popular as a reference for precision measurement. Three classes of precision gage blocks are available:

> **Class AA Master grade gage blocks,** accurate to ± 0.000002 inch per inch of length for blocks over 1 inch in length.
> **Class A Inspection** or **reference gage blocks,** accurate to ± 0.000004 inch per inch of length.
> **Class B Working gage blocks,** accurate to ± 0.000008 inch per inch of length.

Gage blocks are available in several size assortments. One of the most popular is an 81-piece set which consists of four series. The first series has nine blocks ranging from 0.1001 inch to 0.1009 inch in 0.0001-inch increments. The second set has 49 blocks ranging in size from 0.101 inch to 0.149 inch in 0.001-inch increments. The third set has 19 blocks ranging from 0.050 inch to 0.950 inch in 0.050-inch increments. The fourth series has four blocks ranging from 1.000 inch to 4.000 inches in 1-inch increments. The surfaces of gage blocks are so flat that when two blocks are carefully slid or wrung together molecular attraction will hold the blocks together.

Gage blocks are available in either a rectangular shape or a square shape. A variation of gage blocks commonly referred to as **end plugs** are available in larger sizes. These are usually cylindrical in shape with the precision measurement being taken across the ends. The rectangular and square gage blocks are used to establish a reference dimension to which a measuring device or the workpiece itself is compared. End plugs are used for establishing a reference location dimension on optical comparators and other measuring machines and also on some production machines, such as jig borers.

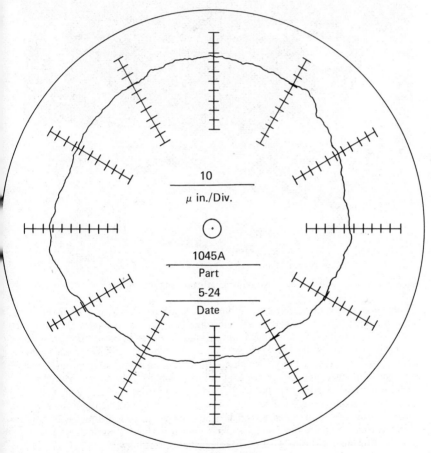

Fig. 21. Typical deviation-from-roundness indicator card.

High Resolution Devices. Any instrument which enables one to measure closer than 0.0001 inch is generally regarded as a high resolution measuring device. High resolution is generally limited to direct measurement and comparison devices. The manipulation of transfer devices involves too many inaccuracies and thus these devices are not considered suitable for high resolution work.

High resolution devices usually involve the same basic principles as low and medium resolution measuring instruments but with added refinements. For example, the amount of amplification and its accuracy is critical in high resolution devices.

Figure 22 illustrates a "super micrometer" or **precision measuring machine** capable of measuring to 0.000010 inch. This machine incorporates a precision micrometer screw with enlarged dials replacing the thimble, and a contact pressure regulating mechanism.

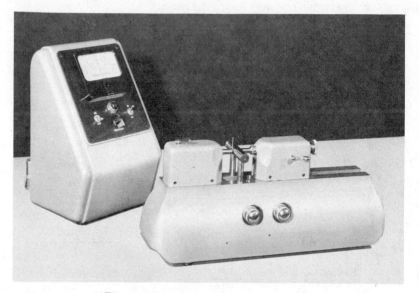

Fig. 22. Horizontal measuring machine.

There are many deviation type indicators available for high resolution work. Most of these involve electric or electronic magnification system. These are generally favored over mechanical amplifying systems because they are faster and more easily adapted to a variety of uses. Electric and electronic systems can be used to trigger relays which in turn may operate sorting mechanisms, start and stop mechanisms, or transfer information to a numerical readout.

In **numerical readout,** numbers representing the dimensions being measured appear in a window or on a screen. The inspector is not required to examine and interpret a scale or a chart to determine the value of the dimension. Figure 23 illustrates a high resolution measuring device with a numerical readout.

O. W. Ehrhardt (Manufacturing Engineering and Management, vol. 57) points out some other advantages of electronic instruments in the metrology field. He states:

The role of electronic instruments in the metrology field will continue to be dramatic, for in addition to the ᵽractical reasons for adopting them they offer definite psychological advantages. . ᴐr example, data collected with these sophisticated devices are not as frequently questioned by other departments' personnel or by customers. This is especially true where the data are in the form of a strip chart recording. Further, these devices can be made part of a complete auto-metrology system. Equipment is available now that permits the programming of a total measuring

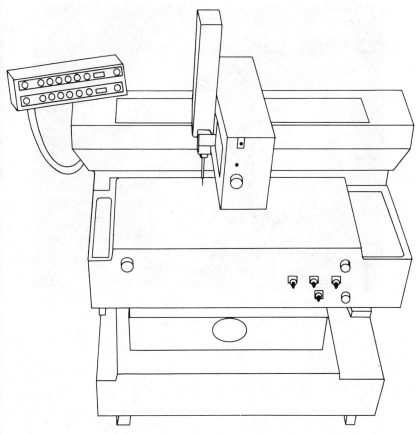

Boice Gages, Inc.

Fig. 23. Coordinate measuring machine with mechanical readout.

process, and the analysis of the result. The system may include a computer which permits immediate processing of data to determine statistical levels of confidence.

The development of the laser provides a new principle for high resolution measurement. Lasers have been used primarily for large dimension measurement. Typical applications include the alignment and inspection of large precision machinery and fixtures. In the measurement of dimensions of ten feet or more an accuracy of ±0.001 inch may be regarded as high resolution measurement. For this type of work the laser has proven to be much more dependable than more conventional methods. A **laser system** using fiber optics (called the Cordax Laser System) is available for use on much smaller parts and dimensions (Fig. 24).

The use of **fiber optics** holds great promise for precision measurement of contoured surfaces as well as recessed or even hidden surfaces. The Hughes Aircraft Co. reported in Metalworking News, vol. 9, the development of an instrument capable of measuring any spot on a contoured surface to "millionths of an inch." Plastic fibers of very small diameter carry beams of light to the workpiece surface. Other fibers pick up the reflected light and carry it back

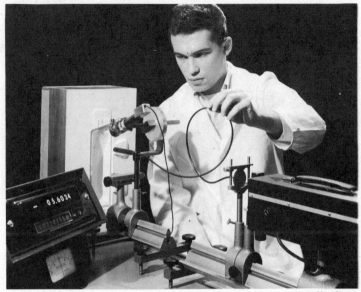

Bendix Corp.

Fig. 24. Cordax Laser System using fiber optics.

to the instrument. The bundle of fibers is moved across the area to be measured in increments of 0.0001 inch.

SURFACE ROUGHNESS TESTS. The measurement and analysis of surfaces has been done for many years. However it becomes of increasing importance as dimensional tolerances are reduced in manufacturing. There is a very significant relationship between a dimension and the conditions of the surfaces which are specified by the dimension. Although standards for the specifications and measurement of surfaces exist, they are not well understood and seldom well utilized.

Units of Surface Roughness. The term "surface roughness" is used very loosely. It is used here in a very broad sense. Technically speaking, roughness is only one of several characteristics of a surface. Terminology accepted by the American National Standards Institute includes roughness, waviness, and lay. These terms are illustrated by Fig. 25. Roughness in a stricter sense refers to

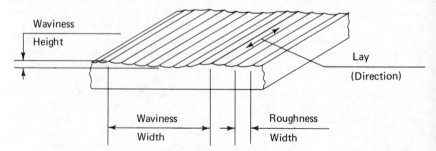

Fig. 25. Terminology used in specifying surface roughness.

the small surface variations that are made by tools on the workpiece and by the structure of the workpiece material itself.

Waviness refers to irregularities of the surface that are of wider spacing than is "roughness," and somewhat repetitive. Waviness commonly results from chatter or vibration causing the tool or the workpiece, or both, to deflect. Waviness height is usually expressed as the peak-to-valley dimension in inches. This value can vary from 0.0002 to 0.0300 inch.

Lay refers to the direction of the roughness pattern. This is the result of the path of the tool used to produce the surface. Lay is not a measured quantity but merely a designated direction with respect to the line or plane in which the symbol is shown. Although lay is often not specified a set of symbols have been designated as follows:

= Parallel to the boundary line
⊥ Perpendicular to the boundary line
x Angular in either or both directions from the boundary line
M Multidirectional or random
C Approximately circular relative to the center of the surface designated
R Approximately radial relative to the center of the surface designated

Another characteristic often ignored yet very significant is the roughness-width cutoff. This refers to the maximum width of irregularity which is to be considered in determining surface roughness. If all irregularities, wide as well as narrow ones, are included a larger roughness height usually results. Wide irregularities are usually amorphous material or waviness and are not considered as representative of roughness. A roughness-width cutoff of 0.030 inch is usually preferred and should be assumed if a value is not specified.

Surface roughness measurement has largely been limited to a measure of roughness height. The numerical measure of surface roughness is expressed in micro-inches and represents the average deviation from a central plane on the surface. This is illustrated in Fig. 26. Originally this average was taken as the

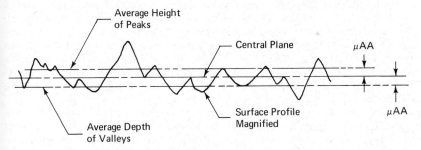

Fig. 26. Interpretation of roughness height.

root mean square (RMS). In 1955 this was changed to arithmetic average (AA), however the difference between the two is very small.

Nisbett, in commenting on some of the new developments affecting inspection and metrology, notes that "on the conceptual level there is the recognition that a quantitative micro-inch finish is not the total description of a surface." As dimensional tolerances become smaller the characteristics of waviness and lay take on much greater significance. Figure 27 shows the ANSI symbols used in specifying roughness (B46.1–1962). All or any portion of these symbols might conceivably be used. Commonly only roughness height is expressed and is designated merely as a check mark with the arithmetic average figure inserted.

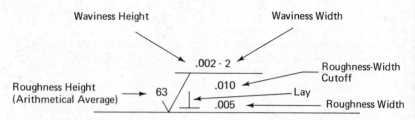

Fig. 27. Standard symbols used in specifying roughness.

Methods of Measurement. Three basic methods are available for determining surface roughness. Each of these methods have some serious limitations.

Interferometry provides an interesting and useful method of measuring surface roughness. In this method a beam of monochromatic light is passed through an optical flat and is reflected by the surface of both the workpiece and the optical flat. Interference of the wavelengths of the reflected light rays give a series of light and dark bands which provide a measure of the variations in the work surface as compared to the surface of the optical flat. This method requires a highly reflective work surface. Interferometry will seldom produce the same information about a surface as other methods, however. Since it depends upon reflectivity of the workpiece surface it will not reveal sharp fluctuations of the surface. On the other hand it shows much more clearly than other methods broad variations, waviness, and lay. If the general contours of the surface are of interest interferometry will give a graphic description of these contours. This method is relatively slow and becomes awkward on many types of work. It is used primarily for the inspection of gages and accurately ground parts.

Another method involves the use of a **profilometer.** The profilometer consists of a stylus mounted on a movable arm, an amplifying mechanism, and a dial or a recording device. The stylus is moved along the surface of the workpiece and variations are noted on the dial or a moving graph. One disadvantage of this method is that a very sharp point is required on the tip of the stylus in order to get good fidelity. A sufficiently sharp point will often scratch the surface and may even cut a path, thus destroying the surface it is intended to measure. Furthermore, this method is a measure of the surface along only a single line on the surface.

A very common method involves the use of standard surface roughness specimens. These standard specimens are in the form of small plates with surfaces typical of those produced by a variety of machining operations. The specimens are available in sets of fairly well standardized increments of roughness height with designations given in micro-inches of roughness height, AA. The common micro-inch designations are 2, 4, 8, 16, 32, 63, 125, 250, 500, 1000, and 2000 micro-inches. The procedure for their use is simple. The standard specimen is held up beside the workpiece, and the workpiece surface finish is compared to the standard specimen. Comparison may be made by any suitable means. Commonly it is done by examining them with a magnifying lens, observing light reflected from each of the two surfaces, or by running a fingernail over the two surfaces. This is obviously a rather crude method and involves many human errors. Nevertheless, it is fast and quite adequate for many inspection operations.

Increased interest in surface measurement has led to the development of other surface measuring devices and modifications of the older methods. Diehl (Manufacturing Engineering and Management, vol. 57) reports the development of an interferometer which combines a high intensity laser light source with precision optics to produce both a visual and photographic readout of surface flatness. This device also includes a camera which provides a permanent record of the surface being checked. The high intensity light source makes it possible to inspect surfaces of much lower reflectivity than is possible with a conventional interferometer. The International Institution for Production Engineering Research (CIRP) reports (Manufacturing Engineering and Management, vol. 53) a European modification which enables an optical measuring device to be used on a machine while the part is in motion.

Ostwald reports (Manufacturing Engineering and Management, vol. 58) the development of an optical device which produces a trace or "optical cut" of the surface being inspected. This device may also incorporate a camera to provide a permanent record.

A workpiece surface may also be analyzed with the use of a replica of the surface made in the form of a metalized plastic film. This method has been used successfully with surface roughnesses of as little as 2 micro-inches AA. This method is discussed by Schellman (Manufacturing Engineering and Management, vol. 58).

For large flat surfaces electronic levels may be used to produce "isolevels" similar to a topographical map.

Matsunaga (as reported in Manufacturing Engineering and Management, vol. 51), has developed a method which utilizes the thin oxide film which covers practically all metal surfaces. The contact resistance of this oxide film is used as a measure of the roughness and character of the surface.

OTHER PRODUCTION TESTS. Inspection involves much more than the dimensional and surface characteristics of a part. In large volume manufacturing, the quantity of a product is important. In many cases various physical properties must also be inspected. Such things as weight, center of balance, density, electrical resistance, or conductivity, and chemical and biological properties may be important characteristics requiring inspection.

Weighing. One of the more common **nondimensional inspections** is that of weighing. Weighing is often used to determine the count of a large quantity of small items. Scales and weighing devices with almost any desired degree of accuracy and precision are available for this purpose.

Much has been said about incorporating the inspection process as a part of the manufacturing operation. Weighing is often used in this manner. Selection of proper weighing equipment and adequate control of the weighing process very often eliminates other inspection operations. Examples include the measurement of plastic materials for preforms in plastic molding processes and the mixing of ingredients for powder metallurgy processes, and abrasive products.

Hardness Testing. Tolerances for hardness can be specified the same way as dimensional tolerances. There are several standard hardness scales in use. Most of them have been developed for some particular type of use, but many have overlapping applications. Conversion from one hardness scale to another is often required. A table of conversions for the most common hardness scales is given in Fig. 28. Many companies have developed their own standards of hardness tolerances, although they are often not specified but merely accepted as common practice.

Brinell		Vickers Diam. Pyramid (50 kg. Load)	Rockwell		Shore
Diam. (mm.) 3000 kg. 10 mm. Carbide Ball	Hardness Number		C Scale 150 kg. Brale	B Scale 100 kg. 1/16″ Ball	
—	767	880	66.5	—	93
2.25	745	840	65.5	—	91
2.30	712	784	64	—	87
2.35	682	737	61.5	—	84
2.40	653	697	60	—	81
2.45	627	667	58.5	—	79
2.50	601	640	57.5	—	77
2.55	578	615	56	—	75
2.60	555	591	54.5	—	73
2.65	534	569	53.5	—	71
2.70	514	547	52	—	70
2.75	495	528	51	—	68
2.80	477	508	49.5	—	66
2.85	461	491	48.5	—	65
2.90	444	472	47	—	63
2.95	429	455	45.5	—	61
3.00	415	440	44.5	—	59
3.05	401	425	43	—	58
3.10	388	410	42	—	56
3.15	375	396	40.5	—	54
3.20	363	383	39	—	52
3.25	352	372	38	110	51
3.30	341	360	36.5	109	50
3.35	331	350	35.5	108.5	48
3.40	321	339	34.5	108	47
3.45	311	328	33	107.5	46
3.50	302	319	32	107	45
3.55	293	309	31	106	43
3.60	285	301	30	105.5	42
3.65	277	292	29	104.5	41
3.70	269	284	27.5	104	40
3.75	262	276	26.5	103	39
3.80	255	269	25.5	102	38
3.85	248	261	24	101	37
3.90	241	253	23	100	36
3.95	235	247	21.5	99	35
4.00	229	241	20.5	98	34
4.05	223	234	19	97	—
4.10	217	228	17.5	96.5	33
4.15	212	222	16	95.5	—
4.20	207	218	15	94.5	32
4.30	197	207	12.5	93	30
4.40	187	196	10	90.5	—
4.50	179	188	8	89	27
4.60	170	178	5	87	26

Fig. 28. Hardness conversion tables.

There are also several methods of measuring hardness. The most common methods involve forcing a hardened ball into the workpiece surface to cause an indentation. The diameter of the indentation is taken as a measure of the hardness of the workpiece. The diameter of the ball and the material, either hardened steel or diamond, must be specified. Another method involves bouncing a ball on the surface of the workpiece. In this case the elasticity of the workpiece material is used as a measure of hardness.

Fluid Properties Tests. Water tightness and pressure capacity are other tests commonly performed. Pneumatically and hydraulically operated and controlled equipment must be inspected to insure proper performance and safety. High pressure, whether in pneumatic or hydraulic lines, receivers, or the like, presents a potential hazard for personnel. For this reason very explicit standards have been developed for high pressure systems, and these standards should be adhered to rigidly. These standards are readily available through the American National Standards Institute and the American Society of Mechanical Engineers.

Nondestructive Testing. Numerous methods are available for detecting internal and subsurface flaws in pressure vessels as well as structural members without destroying the test item.

Subsurface cracks and defects can often be detected by **magnetic particle** inspection. This involves sprinkling fine iron filings on the surface of the workpiece and then placing it in a magnetic field. Any discontinuity in the workpiece, such as a crack, void, bubble, or inclusion, will cause magnetic poles to be established. The filings will be attracted to these poles and will roughly outline the location and shape of the defect.

Penetrant inspection is used to detect cracks and discontinuities which extend to the surface. A fluid dye of low viscosity is brushed or sprayed onto the workpiece surface. Either a **visible** dye or a **fluorescent** dye may be used. These dyes are allowed to penetrate the surface for a period of time then the excess is wiped off. Although the fluorescent dye requires a special light to make the dye visible, it reveals defects too small to be seen with a visible dye.

Magnetic and penetrant inspection require the workpiece to be clean and free of any material or surface condition that would prevent the movement of the filings or dyes.

Radiographic inspection, involving X-rays, is used to detect internal flaws in castings, welded joints, hot and cold worked metal parts, plastics, and so forth. The two types of X-ray testing are referred to as X-ray diffraction and X-ray fluorescence. X-ray diffraction is the more common of the two.

Selection of the optimum voltage for the use of radiographic inspection is important. Too high a voltage becomes costly, and too low a voltage requires a longer exposure time. The recommendations of the equipment manufacturer should be followed.

Ultrasonic energy, sound waves of frequencies from one to ten megacycles, can be used to detect any flaw in the workpiece that affects the propagation of a sound wave. This includes cracks, voids, inclusions, and even variations in density of the material. A sound wave is directed into the workpiece and is reflected by the opposite surface of the workpiece as well as by flaws. The two reflected waves are transmitted to an oscilloscope. An analysis of these waveforms reveals the location and depth of the flaws.

Microwaves are used in a manner similar to ultrasonic energy for testing either metallic or nonmetallic materials. In metallic materials microwaves are satisfactory only for relatively shallow penetration. McCready reports (Product

Engineering, vol. 38) the successful use of microwaves for inspecting rubber and plastic bowling balls.

Eddy current testing uses electromagnetic induction to produce Foucault or eddy currents in conducting objects. These currents in turn produce an additional component of the magnetic field. Measurements of the amplitude of this net field can be made by various means of detection. Analysis of these field signals permit direct measurement of test object geometrical, electrical, and magnetic factors. Other metallurgical, chemical, and mechanical properties of the test material can be correlated to these direct measurements. Material discontinuities can also be detected through eddy current testing.

VISUAL INSPECTION. Most products have some characteristics specified or implied which cannot be measured with instruments. This condition may exist because (1) suitable standards have not been developed, (2) equipment is not available to measure it, or (3) equipment cannot be justified because of time or cost involved. Many of these characteristics serve no functional purpose but are a matter of appearance only. Since appearance is a visual concept, visual inspection is usually adequate. Such characteristics as burrs, scratches, color, texture, etc. are commonly inspected merely by visual examination. Chamfers, radii, and fillets are also often examined visually. In some types of work chamfers, radii and fillets are clearly specified by company, industrial, or military standards. In other cases "burr free" corners and edges are specified but no specific chamfer or radius is given. In the absence of such specifications the inspector depends upon his good judgment or "common practice."

Some industries have developed quite explicit color standards. In most cases it is satisfactory to use a color chart for comparison. However, in some products it is necessary to "measure the color" by the use of light diffraction or even measurement of wavelengths of light emitted from the workpiece surface.

The appearance of a product can be extremely important. This is true not only for consumer goods but for industrial and commercial items as well. One appearance factor of significance is texture. It is common practice to finish exposed surfaces by one of several finishing processes. Thus far however, no satisfactory standards have been developed for specifying texture. Inspection of texture, then, is strictly a matter of judgment, within prescribed guidelines, of course.

Automatic Inspection Methods

ON-LINE INSPECTION. Mass production has brought with it the need for inspection methods which can keep pace with fast production. Sampling, of course, helps to speed up the inspection process, but often this is not fast enough and frequently 100 percent inspection is necessary. As a result, many types of automatic inspection devices have been developed.

Gaging Devices. Automatic gaging machines may involve one or more types of gaging devices. Both fixed gages and deviation-type gages are used. Many machines function merely as sorters. These machines usually consist of some type of conveyor or gravity chute that carries a continuous flow of parts into or through a series of gaging devices. The parts are oriented such that they will pass through the gages in a certain manner. The gaging device then detects whether or not a part will pass through the established limit for that particular dimension. If the part does not fall within the prescribed limits an ejection device will usually remove the part from the conveyor. Those parts that fall within the prescribed limits pass on to the next gage. The rejected parts are ejected into separate pans or boxes. This sorting of defective parts is very helpful to the Materials Review Board.

Dimensions are not the only characteristics that can be inspected on automatic inspection machines. Nondestructive testing methods can also be applied to automatic inspection machines. Ultrasonics and X-rays have been applied to on-line inspection. Automatic weighing and counting are common also.

Automatic inspection as the term is used here is not to be confused with adaptive control. On-line inspection is completely separate from any other process. The automatic inspecting device is placed in sequence along with other fabricating equipment. In practically all cases automatic on-line inspection results in 100 percent inspection.

Numerical Control Inspection Techniques. The terms "numerical control" and "digital readout" have been applied to the many devices developed for measuring coordinate dimensions on a workpiece. The workpiece is held in a fixture and a probe is brought in contact with the work surface to be measured. Either the workpiece or the probe is held on a movable table or arm. The displacement of the probe and the movement of the table or arm are recorded in the readout section of the control device. Two- and three-axis machines are available. A two-axis machine usually registers the vertical displacement of the probe and the horizontal movement of either the workpiece or the probe. A three-axis machine records both of these as well as transverse horizontal motion.

Numerical control inspection is most commonly applied to the inspection of odd shaped contours which cannot easily be measured by other means. Since it is a relatively slow process it is not competitive with automatic gaging devices or other conventional methods for the inspection of easily measured dimensions. Figure 23 illustrates a typical numerical control inspection device.

ADAPTIVE CONTROLS. The prevention of defective parts is only indirectly considered a function of inspection. But if inspection can alert manufacturing immediately after a defective part is produced, corrective measures can prevent defects of a similar nature from being produced in subsequent parts.

In effect this means inspection will control the manufacturing process. This requires on-line monitoring of production and instantaneous feedback of output information. Adaptive controls are designed to serve such a function.

Basic adaptive controls consist of a deviation indicator or sensor, which can be electrical, mechanical, pneumatic, or fluidic; a feedback system, usually electronic or fluidic for speed; and a correcting unit. The deviation indicator monitors the workpieces periodically or continuously and senses whether or not they fall within some preset limits. If the dimension being monitored falls outside these limits, the sensor relays the information, through the feedback system to the correcting unit. The correcting unit then adjusts the position of the workpiece or tool in such a way as to eliminate the defect in future parts. In those cases where corrective action cannot be taken the sensor feeds back into a warning system which stops the machine and/or activates bells or lights.

Adaptive controls can also interface with computer systems which constantly analyze the output of the machine and produce statistical information on the process as it is being performed.

Evaluation of the Inspection Program

EFFICIENCY OF INSPECTION. Efficiency can be defined in terms of the effectiveness of the inspection process in uncovering defective parts or in terms of the physical and mental effort expended during the actual inspection process.

If it was decided to know beforehand the existing number of defective parts in a given lot, then the efficiency of inspection would be the number of defective parts discovered during the inspection process divided by the total number of defective parts in the lot. Since inspection exists for the very reason that the number of defective parts is unknown this measure of inspection efficiency is impossible to attain except under controlled test conditions. **Statistical techniques,** however, are used extensively to predict the number of defective parts that can be expected in a given lot based on a relatively small sample of the lot. But again, there is a measure of risk and uncertainty that the actual number of defective parts differs from the number calculated. The efficiency of the inspection effort may be measured in terms of the actual number of parts inspected. **Motions and methods study** are used in this determination. In this respect the physical layout of the inspection station, the type of equipment used, the location of the work station with respect to production, and many other factors must be considered.

REPORTING TO PRODUCTION CONTROL. Production control is dependent upon inspection in many ways. For example, the production control group must know how much work is being produced on each work order, what percentage of this work is being rejected by inspection, and whether or not this rejected work may be salvaged. Inspection for quantity as well as quality is essential. It is also imperative that the production control group is informed immediately of any change in the status of any work order.

Koepke (Plant Production Control) states "Production control cannot be complete unless the inspection function is included in the plans. Generally this is done by coordinating the quality control group with the dispatch station. This should not be construed as meaning that the dispatcher or the production control department is allowed to tell the inspectors how to sample and inspect. The control is set up to tell them **when** to inspect and **when** to send their findings to the dispatch office." The term "reporting to production control" is used, then, not to mean that the quality control group has any jurisdiction over inspection but rather that the production control group depends upon information from inspection.

The use of inspection tickets discussed earlier in this section lends itself well to this type of report (see Fig. 1). An essential factor in reporting to production control is the matter of time. When the inspection ticket is completed a copy should be sent immediately to the production control group. If inspection tickets as such are not used some other method must be substituted. If the inspection record is kept on the back of the routing sheet a reproduction of the record is sent immediately after inspection to the production control group. In some companies inspection reports must be sent to several different groups or functions within the company. This can result in voluminous paper work and much delay. To avoid separate reports and unnecessary paperwork the inspection ticket or the inspection report should be designed to include information needed by all groups who receive such a report.

COORDINATION WITH QUALITY CONTROL. Inspection is commonly regarded as a **part of the quality control function.** Each function of quality control depends upon the other for information. These functions include establishing specifications for the manufacture of the product, determining manufacturing capabilities, component inspection, and performance testing of the product. The results of the inspection operation are vital to each of these.

Quality control can make use of the same inspection information as production control. In addition, statistical studies made by quality control require detailed records of rejects including the reason for their rejection. Since such data is often of little value to other production functions, separate detailed reports may be prepared for quality control. While the timing of inspection reports is critical for such functions as purchasing and production, it is the thoroughness and accuracy of the reports that are essential if they are to be useful for quality control.

SECTION 8

QUALITY CONTROL AND RELIABILITY

CONTENTS

CONTENTS *(Continued)*

QUALITY CONTROL AND RELIABILITY

The Nature of Quality Control

DEFINING QUALITY. No manufacturer needs to be convinced that reduction of costs is desirable. However, he may doubt that an improvement of quality will reduce costs. If so, he may hold this view because he associates quality chiefly with elaborate design, stylish appearance, extra accessories, and the like. A broader view of quality is shown by the expectations of the buyer, who may complain if a product does not function as intended, does not wear long enough, or does not look attractive. To have these qualities, a product must be sufficient in two respects. The designer must first develop an adequate **pattern,** and the factory must then follow it with adequate **fidelity.** Juran (Quality Control Handbook) terms these concepts "quality of design" and "quality of conformance." Feigenbaum (Total Quality Control) has defined product quality as "the composite product characteristics of engineering and manufacture that determine the degree to which the product in use will meet the expectations of the customer."

Quality of Design. As applied to floor coverings, for example, an ankle-deep carpet has higher quality than a paper-thin one. The former has been designed to use more and better materials in a more complex weave to give superior satisfaction and life. Increases in quality of design normally increase the costs of design and production. This is profitable only to the extent that the higher grade of product leads to a rise of sales income which exceeds the rise in costs.

Quality of Conformance. Two deep, expensive carpets made to the same specification in the same mill may still differ. One may have a uniform, attractive color while another is off-color or blotchy or has loose tufts. These conditions involve lack of conformance to specification.

In general, conformance can be sought either by sorting the good from the bad after manufacture or by taking preventive measures at any source of trouble. Obviously, **sorting after manufacture** is costly. **Prevention,** on the other hand, not only reduces the need for sorting but also reduces the amount of scrap, seconds, and rework; reduces customer complaints; and in many industries avoids much trouble with fitting of spare parts in the field. Thus an increase in conformance can be profitable even without an increase in price. In addition, increased conformance may add to product **reliability, serviceability,** and **maintainability,** all marketable characteristics.

The consumer may not distinguish between quality failures due to design weakness and quality failures due to lack of conformance. If tufts fall out of his rug, he is dissatisfied. He is not interested in whether the designer erred in his specification or the mill failed to conform to it. Nevertheless, the distinction is important to the supplier. The one form of quality improvement may add to costs, the other cuts costs if properly carried out. The choice as to quality of design is made only occasionally, generally at the time a new product is being

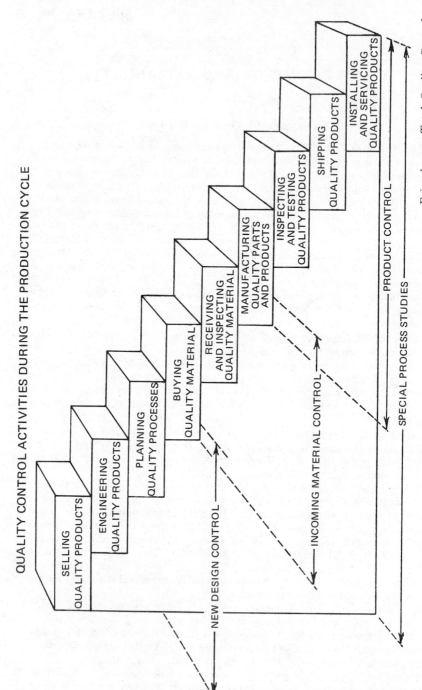

Fig. 1 Activities of quality control

Feigenbaum. Total Quality Control

lanned and launched. The struggle to maintain conformance continues as long
s the product is made.

OBJECTIVES OF QUALITY CONTROL. The objective of quality
ontrol is to provide an effective system for integrating the efforts of those in
he organization responsible for the development, maintenance, and improve-
nent of product quality and to provide for production, distribution, and service
t the most economical levels providing full satisfaction of the customer.

The control endeavor involves:

1. **Establishing standards.** Determining performance, reliability, and cost standards.
2. **Appraising conformance.** Comparing the conformance of the manufactured
   product with the standards.
3. **Corrective action.** Taking corrective action when standards are not met.

Systems Perspective. The absence of objective criteria and real control will
ntroduce cyclic reject rates and cyclic production rates in any industry. This
:yclic behavior is the consequence of the pressures within the firm and between
he firm and the market. If the period of the cycle is too long, management
nay fail to recognize that a cycle, in fact, exists; and if there is recognition, the
.auses may not be readily identified.

The cyclic behavior of reject rates and production may be partially explained
)y the interaction of the sales, manufacturing, and quality functions. It is as-
umed that a constant quality product is manufactured throughout the period
»f this discussion. As sales increase, pressures mount on manufacturing for
ligher production resulting in a general decline in reject rates. At some point,
eedback from the marketplace causes the sales function to bring pressure for
ligher quality. It is reinforced by the desire of the quality function to improve
)roduct quality, with a result that there is a general increase in reject rates.
This condition dominates until pressures for more product again cause a down-
wing. Cyclic behavior continues, usually accompanied by a similar cycle in
he production rate although this is usually out of phase with the reject rate
:ycle. These cycles are accompanied by quality fluctuations caused by day-to-
lay problems in the production process.

Reduced Costs. The total quality control philosophy has only recently be-
ome widely accepted. One of the distinguishing features of this approach is
.he emphasis on the prevention of poor quality and a deemphasis on the screen-
ng or inspection function. Feigenbaum (Total Quality Control) has noted that
:ompetitive conditions present American business management with an op-
)ortunity to improve the quality of many products and practices while at the
;ame time substantially reducing the overall costs of quality. In actual practice,
1umerous companies have found that total quality control programs meet the
:hallenge of achieving both better quality and lower quality cost.

Quality Control Improvement. The improvement effort requires continu-
ng attention to cost, performance, and reliability standards. Disturbing the
existing system may well bring about violent instability for a short period of
:ime.

QUALITY CONTROL ACTIVITIES. Four widely recognized activities of quality control are: new design control (design review), incoming material control, product control, and process study (Fig. 1). Government specifications applicable to most DOD and NASA procurement pertain to the contractor's quality system requirements. The quality system requirements are
:overed in DOD documents MIL–Q–9858A and NASA document NPC 200–2,

and companion inspection system requirements are stated in MIL–I–45208 and NPC 200–3, respectively.

1. New Design Control.

This is the quality control effort for a new product while the marketab] characteristics are being selected. It is during this period that the quality cha acteristics are being identified and design specifications are being establishec At the same time, a manufacturing process is usually being planned, and co calculations are being made. It is necessary to review both product design an process design to eliminate both potential quality problems (including mair tainability and reliability problems) and quality cost problems. The functio of new design control ends when preliminary production runs have indicated tha production is satisfactory in terms of both quality and quantity.

2. Incoming Material Control.

It is necessary to design the procedures that will be used for the acceptanc of components, parts, and materials to be purchased from outside vendors o other units within the company. Incoming material control may be applied t items that are produced in one area of the plant and used in another area o department of the same plant. The establishment of acceptance criteria for a] components, parts, and materials involves the design of sampling plans an procedures to provide protection at the most economical level. The method employed here include vendor rating, evaluation, or certification and acceptanc sampling techniques of various sorts. The design of laboratory tests and tes procedures is also a part of this activity.

3. Product Control.

Product control functions at the production level to evaluate departure fron specification and take corrective action before serious quality problems develop Parts, materials, and processes which contribute to the quality of the produc during the manufacturing operation are all involved. One of the objectives o product control is to see that the product is reliable and that it will perforn satisfactorily under the design-use conditions. Field service related to qualit problems and complaints would fall within this category.

4. Special Process Studies.

These studies are in the form of special investigations, tests, and experiment aimed at identifying major causes of quality problems or defective product The control or elimination of these causes may require product and proces modification which in turn can affect production costs.

Role of Inspection. DOD specification MIL–I–45208A sets forth the "in spections and tests necessary to substantiate product conformance to drawings specifications, and contract requirements and to all inspections and tests re quired by the contract." But complete and reliable screening inspection of al parts and products is expensive; in cases where inspection is destructive, it i impossible. Assuming complete screening, and shipping only products of accept able quality, poor control of product and process variability will result in high scrap losses, as well as interruptions, loss of capacity, costly salvage operations and often selective assembly. The act of screening inspection, even if perfectly performed, can not "inspect" quality into the product; it merely shunts the un acceptable units to one side after manufacture. Planned action must be addec to simple inspection. In contrast, quality control steps in before faulty parts are made.

Feedback of Information. There is strong need for **feedback of inspec- tion information** to production and to those who issue specifications. In a

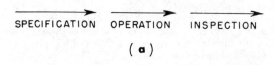

(**a**)

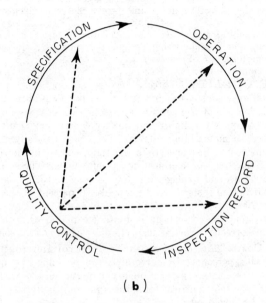

(**b**)

Fig. 2. Introduction of quality control as a coordinator.

startling number of factories the relation between designer, production operators, and inspectors is the coldly formal one illustrated in Fig. 2(a). In his drawings and specifications the designer sets up the quality legislation. The operators make an effort to live within the law but may not know how to go about it. Even worse, they often lack instruments to tell when they are breaking the law. The inspector, in judicial capacity, passes sentence but makes no provision for reform of the guilty. This policing attitude has been widely rejected by modern management.

The modern concept of quality control as a coordinating function is indicated by the completed circuit in Fig. 2(b). Inspection records and special studies are used to identify problems and their causes. Four alternatives are open at this point:

1. **Study and improve** the process.
2. **Change** to a more capable machine or operator.
3. **Revise** the specification.
4. **Sort** the product.

It is quality control's responsibility to get the facts, compare the costs, and find the best way out. Sometimes complex experiments are needed to root out the facts. Sometimes a simple procedure is all that is required, but a systems perspective is always necessary.

Benefits. The benefits of systematic control of quality may be summarized as follows.

1. Improvement in product quality.
2. Reduction of the costs of scrap, rework, and adjustment.
3. Reduction in the costs of the factors of production through random assembly uninterrupted production, and greater utilization of labor and facilities.
4. Reduction in costs of inspection.
5. Improved attainable quality standards, with either higher market values for a given sales volume, or greater volume for a given price.
6. Lower cost designs of products and processes for a given product quality standard.
7. Improved technical knowledge, more reliable engineering data for product development and manufacturing design, and reliable characterization of the attainable performance of processes.
8. Improved employee morale.

COSTS OF QUALITY. Three categories of operating quality cost are prevention costs, appraisal costs, and failure costs. **Prevention costs** are related to the elimination of the causes of defects. These costs include such elements as employee quality training and quality control engineering. **Appraisal costs** are related to maintaining quality. These are typically inspection, test, and quality audit costs. **Failure costs** are those caused by defective materials or by products that do not conform to specifications. These include rework, scrap, field complaints, and spoilage.

Many feel that failure costs represent roughly 70 percent of the quality dollar, appraisal costs 20–25 percent, and prevention costs 10–20 percent. If these figures are even roughly correct, it is quite clear that the quality dollar is usually not spent correctly. Sorting inspection is clearly not the solution.

Prevention Costs. The costs of quality control and engineering work are associated with personnel time required for planning the quality system. It is necessary to translate customer quality requirements as well as design requirements into specific controls for processes and materials. Methods and procedures must be developed.

Process control engineering work costs are those that represent the time of personnel required in the analysis of the process in order to establish the basis for control. In addition, the time required for instructing shop personnel to help them effectively apply and implement quality procedures should be charged here.

The **quality planning cost incurred by other functions** within the organization should be charged as a quality planning cost. Typically, product engineering and design engineering groups will be involved in some pre-production quality analysis and the writing of instructions or operating procedures for test, inspection, and control; and these charges should be reflected as planning cost.

The cost of **developing and presenting quality training programs** throughout the organization should be charged here, if the objective of the program is to train personnel in the use of quality control techniques.

Quality equipment design costs associated with personnel time spent in the design of quality control devices, measurement devices, and instruments should be charged here.

Cost of Appraisal. The cost associated with the time that inspection and testing personnel spend in the **evaluation of quality of purchased materials** should be charged here. This includes supervisory and clerical personnel used in the support of this function. On-site inspection, including travel to vendor locations should also be charged.

Where **laboratory testing** is required or special laboratory services are required for either the evaluation of purchased materials, the calibration of instruments or the test of instruments, these costs should be charged as a laboratory service cost.

The cost associated with the time that the inspection personnel spend in **evaluating quality throughout the manufacturing process** should be charged here. Again, the appropriate supervisory and clerical costs that are incurred in support of this function should also be charged.

Testing costs are those associated with personnel time required to evaluate the technical performance of the product.

Costs associated with the time that manufacturing personnel spend in **checking the quality of their own work** are considered appraisal costs. At some points it is necessary to screen bad lots and to make product and process quality evaluations. These charges would appear as internal failure costs.

From time to time it is necessary to complete **quality audits** and the cost associated with personnel time should be charged here.

Such **miscellaneous inspection and test activities** as setting up equipment to perform inspection and testing, the cost of special materials required for the conduct of tests, and the cost of field testing should be charged here.

Internal Failure Cost. Rework costs represent the extra wages paid to bring a rejected product into specification.

Scrap costs are associated with materials being scrapped. In many companies these costs include the added value of labor to the point of discard.

Engineering costs are those associated with the time required by production engineers to solve quality problems.

If it is necessary to make a **disposition** of a lot rejected during receiving inspection, either by screening or by returning the lot to the vendor, the costs involved are charged to internal failure.

External Failure Cost. The dollar value of items sold and later returned by the customer for nonconformance as part of a **warranty,** including the expense of repair or replacement service should be included here.

The administrative time required to handle customer **complaints** is charged as an external failure cost.

Other Quality Costs. Two other categories of costs are indirect quality cost and equipment quality cost. **Indirect quality costs** are those partially hidden in other cost figures. Indirect quality costs can be reduced by design improvements requiring less material or labor, process improvements reducing labor and material cost, procedure improvements reducing inventories held for test and inspection, as well as the elimination of overstocking to hedge against possible high rejection levels.

Equipment quality costs relate to the cost of equipment purchased to measure product quality. These purchases must be amortized and the equipment maintained thus and adding to the total quality costs.

A Rule of Thumb. Philip B. Crosby (Cutting the Cost of Quality) has used a much less detailed breakdown for cost. He has suggested four cost categories as rework cost, scrap cost, warranty cost, and quality control labor; and it has been his experience that the sum of these costs will be equal to about four percent of sales. In the case where a more detailed identification of cost is maintained, it would be logical to expect this figure to be much greater. These figures will vary widely making it impossible to generalize for all industries.

Analysis of Quality Costs. An operating quality cost report is considered to be a major component of the total quality system. A good, modern cost accounting system will provide for the cost inputs required in most companies;

however, some modifications may be necessary. The analysis usually involves the examination and comparison of each cost element with other cost elements and to the total, and a period-to-period comparison. Bases to which cost comparisons are commonly made include:

1. Direct Labor.
2. Productive Labor.
3. Net Sales Billed.
4. Manufacturing Cost.
5. Standard Labor.
6. Shop Cost (Direct Labor, Material, Indirect Cost).
7. Standard Value of Production.

Reporting. Regular reporting by product line is the most common practice. Reports may be on a weekly, monthly, or quarterly basis as required. Typical graphical displays are illustrated in Fig. 3. The cost or **accounting paper work flow** will vary greatly from firm to firm. Bicking (Industrial Quality Control, Vol. 23) suggests the flow system shown in Fig. 4.

Other Measures. As a measure of **productivity** of the incoming-material appraisal and control function, the following ratio has been suggested:

$$\frac{\text{Direct material dollars}}{\text{Incoming appraisal costs}}$$

As measures of the **effectiveness** of the incoming-material appraisal and control function, two ratios have been suggested:

1. $$\frac{\text{Incoming appraisal costs}}{\text{Manufacturing losses due to vendors}}$$

2. $$\frac{\text{Outside vendor losses recovered}}{\text{Total losses attributable to vendors}}$$

As measures of **timeliness** of defect prevention, the following measures have been suggested:

1. $$\frac{\text{Lots not meeting all requirements}}{\text{Total lots received}}$$

2. $$\frac{\text{Lots per week not processed for use or stock in 24 hours}}{\text{Total lots received at dock per week}}$$

In both cases a reduction in the measure represents an improvement in the timeliness of the activity.

Organization for Quality Control

QUALITY FUNCTIONS. The principal function of the quality control organization is the administration of the activities of individuals and groups concerned with **new design control, incoming material control, product control,** and **process study** as previously defined. In order for the organization to be effective, top management support is essential. Since an important function of the organization is to stimulate all employees to be quality conscious, expert human relations must be practiced. In summary, responsibility for quality is shared as follows.

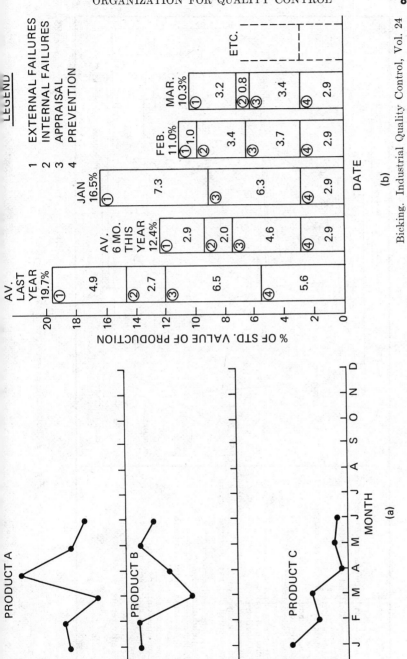

Fig. 3. Graphical methods used in reporting quality costs. (a) Total quality costs. (a) Total quality costs by product lines. (b) Breakdown of total quality costs by elements.

Bicking. Industrial Quality Control, Vol. 24

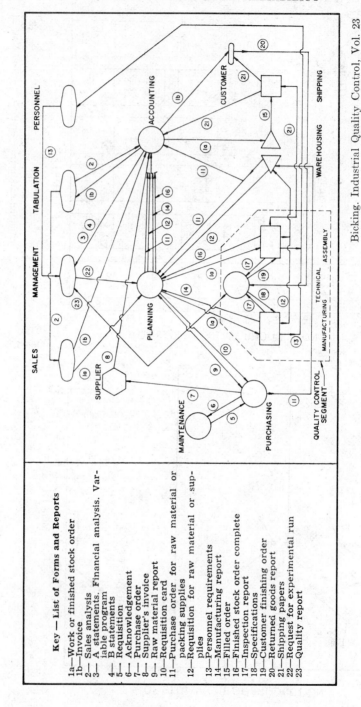

Key — List of Forms and Reports

1a—Work or finished stock order
1b—Invoice
2—Sales analysis. Financial analysis. Variable program
3—A statements. Financial analysis. Variable program
4—B statements
5—Requisition
6—Acknowledgement
7—Purchase order
8—Supplier's invoice
9—Raw material report
10—Requisition card
11—Purchase order for raw material or packing supplies
12—Requisition for raw material or supplies
13—Personnel requirements
14—Manufacturing report
15—Filled order
16—Finished stock order complete
17—Inspection report
18—Specifications
19—Customer finishing order
20—Returned goods report
21—Shipping papers
22—Request for experimental run
23—Quality report

Fig. 4. Accounting paperwork flow system.

1. Top management has the basic responsibility for quality.
2. Each employee has the individual responsibility of producing a quality product.
3. The quality organization is established by top management to serve as a mechanism to integrate and evaluate the activities associated with the four quality control activities.
4. The establishment of a quality organization does not relieve other production functions of the responsibilities for quality.

THE QUALITY CONTROL ORGANIZATION. In many companies little attention has been devoted to the quality organization structure; and as a result, the organization represents an appendage of a quality control function to some existing (and older) "inspection department" or the introduction of some statistical methods into an existing inspection department.

In establishing an effective quality organization to deal with the complex interface between the many functional groups within the company so that the quality objectives are attained, it is necessary that key responsibilities be distributed among the various functional groups. At the same time the quality control function, serving as management's representative, has the overall responsibility for control. Figure 5 shows three subfunctions of the quality control function itself.

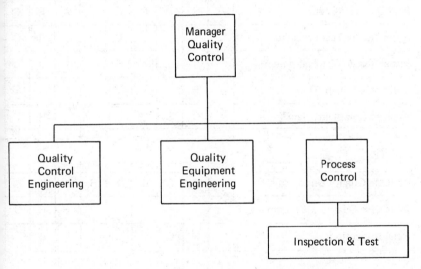

Fig. 5. The subfunctions of the quality function.

Quality Control Engineering is responsible for the overall plan of the quality control system for the company. Quality Equipment Engineering is responsible for the design and development of testing equipment. Process Control (a function to which Inspection and Test reports) is responsible for monitoring the application of the quality system.

Responsibility Relationships. One of the most useful devices for identifying and analyzing quality responsibilities and contributions is called the **Relationship Chart.** An example is shown in Fig. 6.

NEEDS	Gen. Manager	Marketing	Design	Mfg. Manager	Mfg. Engineering	Finance	Materials Control	Quality Control	Process Control
Determine Customer's Needs		R							
Product Design Specs.			R						
Mfg. Process Design			C	C	R		C	C	C
Produce to Design Specs.		C	C	C	C		C	C	R
Determine Process Capabilities				I	C		C	R	C
Vendor Control							R	C	
Plan Quality System	R	C	C	C	C		C	R	C
Inspection & Test Procedures					C		C	R	C
Inspection & Test Equipment					C			R	C
Gather Complaint Data		R							
Analyze Complaint Data		C	C					R	
Compile Quality Costs		C	C	C		R		C	
In-Process Quality Measurement								R	C
Final Inspection			C	C	C			R	

R = Responsible

C = Contributes

I = Informed

Fig. 6. An example of a quality control relationship chart.

Design of the Quality Control Organization. A six-step procedure has been suggested by Feigenbaum (Total Quality Control) to serve as an aid in planning the quality control organization.

1. Define company quality problems.
2. Establish objectives to be achieved by the quality organization if it is to solve the problems.
3. Determine work elements needed to meet the objectives; classify elements into basic functions.
4. Combine basic functions into jobs which meet the following test:
   a. The position must comprise a logical, separate area of responsibility.
   b. The objectives, scope, and purpose of results to be achieved are clearly related to the position.
   c. A single individual should be assigned the job and be fully aware of his responsibilities and how he is to be evaluated.
   d. The functions of the position must be fairly homogeneous.
   e. Authority and responsibility must be balanced.
   f. Communications channels between the position and the rest of the organization must be open.
   g. The holder of the position should be able to supervise the activities of those reporting to him easily.
5. Consolidate the jobs into an organizational component best suited to the particular company.
6. Locate the function in the larger organization where it will produce with maximum effectiveness and a minimum of friction.

Juran (Quality Control Handbook), noted that "looking over the entire picture reveals that one of the chief duties of the quality control department is setting and controlling quality at the most strategic points in the manufacturing process." In regard to the position of the inspection function within the organization, Juran feels that the question is more vital in large organizations than in small; however, he emphasizes that proper placement of quality and inspection functions in the organization is essential and that psychological aspects of the inspection function must be studied. For example, in plants where the operator is paid only for the good products produced, the inspector is in a position to determine the amount of the operator's paycheck. This places the inspector in a difficult position, and Juran feels that the inspection function must therefore be independent of any domination from production in order to maintain specified quality levels. Figure 7 illustrates an organization structure for implementing this philosophy.

The total quality control philosophy is different in that both inspection and testing would be considered a function of manufacturing, so that much of the inspection and test activity would come under the same unit manager as shop operators. This would take the inspector out of the role of policeman, require manufacturing management to assume the responsibility for manufacturing product that satisfies all specifications, and provide for easier working relationships between the quality control organization and the manufacturing organization.

Typical Organizations. Figure 8 (pages 15 and 16) shows typical quality control organization structures for various types of overall organization structure.

QUALITY CONTROL STAFF. The quality control man gives a staff service. Practically all his quality control activities involve advising other persons and coordinating their efforts. He has no direct authority to order specifications changed, to order shop methods changed, or to order an inspector to accept or reject a part. Instead, it is his business to know where the facts can be found: and with whom to check, what to look at, what to measure. He is

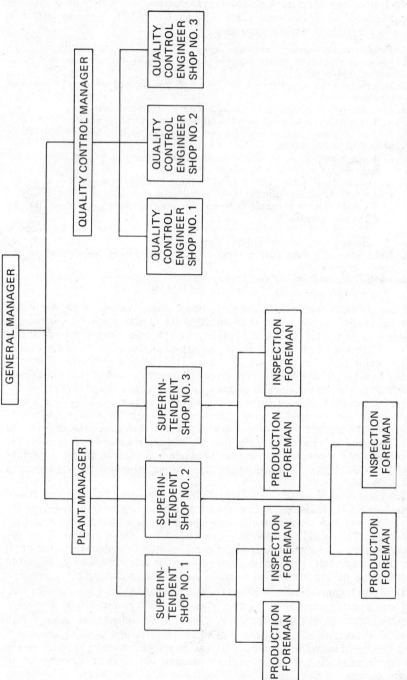

Fig. 7. Typical organization for separation of quality control and manufacturing functions.

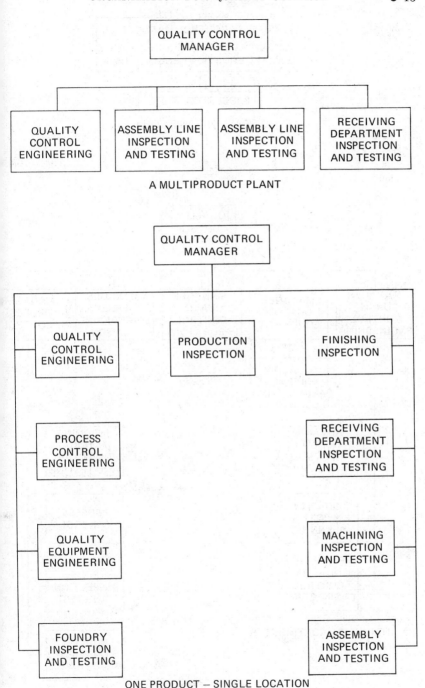

Fig. 8. Typical quality control organizations.

A SMALL ORGANIZATION

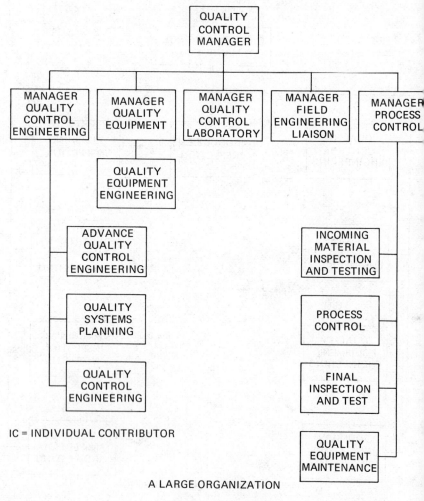

IC = INDIVIDUAL CONTRIBUTOR

A LARGE ORGANIZATION

Fig. 8. Concluded.

authorized to collect facts and to conduct experiments where needed. After he has brought the facts together he must present the results in a way that will achieve cooperation.

Manager of Quality Control. The typical job description for the manager of quality control is as follows:

1. **Formal Training.** Degree or equivalent in engineering. This should include some courses in management and human relations as well as engineering economy, probability and statistics, and cost accounting.

2. **Special Education.** Preferably in quality control, statistics, human relations, and management.

3. **Experience.** This will vary from situation to situation; however, three to five years of work in industry and a background in quality engineering are desirable.

4. **Functional Responsibilities, Authority, and Accountability.** The manager of quality control has the managerial responsibilities for the operation of the quality control department. He is responsible for seeing that quality requirements have been properly defined to permit appropriate quality planning and to see that these requirements have been met. In planning, it is necessary for him to keep informed of the objectives, plans, budgets, and policies of the company and to develop the quality control program including quality objectives, quality policies, quality plans, the quality organization and procedures for company personnel, and to promote the quality control program. As an organizer, he must develop a sound organizational structure for the performance of the quality control activities and the delegation of appropriate responsibility and authority. He must integrate the resources of the quality component to effectively achieved the stated objectives and he must acquaint each individual in the component with his individual authority, responsibilities, and accountabilities.

In addition, he must promote the development of the individual, encouraging each individual to relate personal objectives to job objectives. The manager of quality control must also establish standards to evaluate the performance of his unit and the directors of sub-units. He should analyze and appraise the progress of the quality organization and take any action necessary for improvement. He will:

a. formulate basic policies, programs, plans, standards, and methodologies necessary in carrying out the objectives of the quality control organization;
b. provide adequate facilities for inspection and tests;
c. design and distribute programs to promote quality-mindedness within the organization;
d. maintain relationships with design and engineering units;
e. maintain relationships with marketing and finance organizations;
f. establish a vendor relationship conducive to improving the quality of purchased materials.

The manager of quality control should have full authority to make decisions and to take action required to carry out the responsibilities as assigned. He is accountable for the fulfillment of his responsibilities, and he should not delegate or relinquish any of this accountability. He should be evaluated on:

a. the quality of his leadership in all areas of the quality organization;
b. the quality of his decisions;
c. the attainment of the objectives as stated earlier.

Quality Control Engineering. A listing of the functions of quality control engineering is given on page 18.

1. Recommend realistic company quality objectives and goals.
2. Review new designs for quality, and make recommendations to design engineering which will increase quality and reliability and improve various quality characteristics.
3. Work with marketing and design engineering to establish quality standards covering such characteristics that may readily be perceived by the consumer.
4. Review the results of engineering development work, and analyze prototype performance information.
5. In cooperation with manufacturing engineering, and production management, establish production quality standards.
6. Maintain control over purchased materials and insure adequate definition of quality requirements to all vendors through the purchasing department. Designate which quality characteristics are to be inspected and the procedures for performing quality evaluation. Evaluate vendor facilities as required.
7. Develop procedures for process control and product quality control during manufacturing. The relative importance of the various quality characteristics as well as required quality levels should be stated. Inspection points should be identified, and methods and procedures for measurement and control should be established. Statistical techniques may be applied as needed.
8. Assist manufacturing engineering in the purchase of new production devices which relate directly to, and affect, quality. Establish procedures for adequate preventative maintenance as equipment performance relates to quality products.
9. Conduct process capability studies to determine if manufacturing processes are sufficiently capable of meeting the quality requirements as specified. Recommendations should be made for the improvement of machine and processes to meet quality requirements.
10. Write and provide quality system manuals to satisfy contractural requirements.
11. Develop the information system required to keep management regularly informed of the status of quality control programs as planned.
12. Design and develop quality control orientation programs for operational personnel to insure understanding of quality control processes, programs, objectives, and plans.
13. Conduct an analysis of customer complaints and initiate corrective action.
14. Assist in the development of brochures and publications outlining the quality system and illustrating the advantages to the customer of buying certified quality controlled products.
15. Provide a cost analysis to critically evaluate all elements of quality cost and present periodic reports to the quality control manager.
16. Where there are chronic manufacturing quality problems, provide technical assistance required to diagnose these problems. Progress should be reported to the appropriate management.

Quality Equipment Engineering. The functions of this quality component are:

1. Design and construct inspection tools, fixtures, and test equipment.
2. Provide in-process measuring devices at the point of production.
3. Automate the measurement process where possible, incorporating the measuring devices within the manufacturing equipment.

Process Control. The functions of process control are:

1. Evaluate the effectiveness of the quality plan insofar as it provides for the solution of manufacturing quality problems, economical operations, and the handling of customers' complaints.
2. Review the quality standards to insure that they are adequate from the standpoint of clarity and interpret these standards and their proper use.
3. Interpret the quality plan for all elements of the manufacturing organization as needed.

4. Work with manufacturing personnel to evaluate conformance to inspection and control procedures.
5. Provide troubleshooting service for problems in manufacturing which relate to quality and product performance.
6. Seek out and demonstrate ways to solve scrap and rework costs problems.
7. Design special tests as necessary and arrange for laboratory tests where test procedures are beyond capabilities of quality personnel.
8. Maintain contact with the vendor and his quality control representative, and evaluate his performance, providing vendor rating information to the purchasing department. In addition, determine the position for rejected material and parts.
9. Work with marketing personnel to maintain contact with the customer, helping to interpret standards, specifications, quality requirements. Analyze products returned because of customer complaints. The appropriate organizational components should be advised to provide corrective action.
10. Assure that purchased equipment, including tools, fixtures, and dies meet quality capability specifications.
11. Promote quality-mindedness throughout the entire organization.
12. Provide for the maintenance and calibration of process instrumentation and control devices.

Inspection and Testing. Quality control is responsible for those inspection and testing functions not performed by manufacturing. They include:

1. Schedule inspection and testing in such a manner as to meet overall schedules.
2. Perform the inspection required to insure that only high quality products are accepted from vendors.
3. Perform specific inspection and test operations as necessary during the course of manufacture.
4. Perform quality audits as required.
5. Perform final inspection.
6. Maintain accurate quality records and present the results in such a manner that possible trends can be indicated.
7. Insure that personnel are properly trained for their specific inspection jobs.

The Quality Control Program

PLANNING FOR QUALITY. If efforts to obtain profits through preventing defects are to be successful, they must be made according to a clearly defined plan. A haphazard, uneven attack on the problem may fail to pay for itself. A detailed **master program** for attaining quality should be drawn up at the outset.

Elements of the Quality Program. The development of the plan for the quality system is usually the responsibility of the quality control engineering function. The quality system consisting of all administrative and technical procedures necessary to deliver a product conforming to specifications, must be carefully planned. Some of the **systems and procedures** for which plans must be made are:

1. Preproduction quality evaluation system.
2. Process and product quality control.
3. Purchased material evaluation and quality control:
4. Quality information system—information feedback.
5. Quality information equipment.
6. Training and manpower development.
7. Special process studies.

Special Quality Motivation Programs. There have been numerous promotional programs developed to improve product quality and reliability. A good example is the **Zero-Defects** program which attempts to make each individual in the organization aware of his importance to the product he is helping produce and to the overall endeavor of the organization. The objective is also to obtain management support to the extent that each member of management recognizes the contribution of the employees reporting to him. James F. Halpin (Zero-Defects—A New Dimension in Quality Assurance) noted that "the quality of errors each of us makes is directly proportional to the importance each of us places on the function of the moment." He advises that an organization should carefully consider all of the ramifications of zero-defects before undertaking such a program. An unrealistic or cynical attitude toward the program will likely mean that it will meet with only limited success. Halpin states, "In this day and age of strong competition, no company can stay on top very long without almost absolute quality assurance achieved at a realistic cost. Zero-defects adds that assurance to any organization that will accept it."

Another popular plan is called the **Total Reliability Program.** The objective of a typical total reliability program is quite similar to that of the zero-defects program. One characteristic of both programs is the promotional nature, including slogans, posters, and awards. These programs have been criticized for not having much real substance; however, a number of companies have been sufficiently impressed with their value to continue them over a period of several years.

Military Quality Program Requirements. The military specification for quality program requirements is MIL-Q-9858A. In summary it provides that an effective and economical quality program be planned and developed by the contractor as part of his overall administrative and technical programs. In the words of the specification, "The program shall assure adequate quality throughout all areas of contract performance; for example, design, development, fabrication, processing, assembly, inspection, test, maintenance, packaging, shipping, storage and site location . . . . The program shall include an effective control of purchased materials and subcontracted work."

Physical Control Factors. Juran (Quality Control Handbook) noted that a **quality characteristic** is "a physical or chemical property, a dimension, a temperature, a pressure, or any other characteristic used to define the nature of a product or service." Physical control represents a three stage procedure for measuring this characteristic, comparing it to a standard, and making necessary adjustments (see Fig. 9).

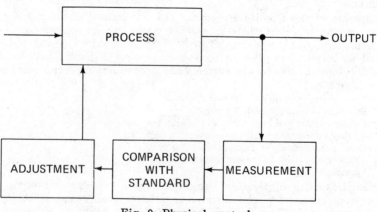

Fig. 9. Physical control.

Statistical Control. The term statistical control refers to the stability of a stochastic process. It implies that the process (distribution), both form and parameters, are constant over time and that independence exists.

Statistical methods are used in the measurement and comparison stages illustrated in Fig. 9. Since the number of product items is very large relative to the number of measurements that could reasonably be made on the items, **sampling** is commonly employed. Values calculated from the sampling process enter into **test statistics** which in turn are used in the comparison procedure. Statistical hypotheses regarding the distribution, its form, parameters, and independence are then either rejected or they are not rejected. If they are not rejected then the process is commonly said to be "in control."

EVALUATING AN EXISTING PROGRAM. In order to judge the kind and extent of quality program that is needed, it is appropriate to begin by taking stock of results with the company's present organization and procedures. A ten-point **checklist** of items to be reviewed in such a quality evaluation is given below:

1. Review of dollar losses due to defectives for:
   a. Amount.
   b. Evidence they are being recovered systematically.
2. Review of specifications for:
   a. Completeness.
   b. Measurability.
   c. Freedom from ambiguity.
   d. Evidence that they describe the properties the customer wants.
3. Review of customer complaints for:
   a. Extent.
   b. Adequacy of corrective action.
   c. Making sure inspection is checking properties complained about.
4. General review of systems for prevention of:
   a. Vendors' defects.
   b. Own process defects.
5. Review of inspection system on such points as:
   a. Organization.
   b. Physical layout.
   c. Gage maintenance.
   d. Number of inspectors.
   e. Attitudes of inspectors.
   f. Standard inspection instructions and operation sheets.
   g. Inspection of components and early operations.
   h. Extent and purpose of 100 percent inspection.
6. Check of attitudes and quality training of production personnel:
   a. Quality taken seriously?
   b. Know what is wanted?
   c. Have means to measure?
   d. Have means to correct?
   e. Attitude toward inspectors?
   f. Training at time of induction?
   g. How is foreman as example?
   h. Posters or other campaign?
   i. Is department's quality record known?
7. Review of organization for:
   a. Quality of top organization.
   b. Clearness of responsibility at firing line (who is responsible for investigation, analysis, action).

8. Review of salvage procedures:
   a. Too loose or too tight?
   b. Information fed back to the shop?
9. Review of reports:
   a. Adequate summaries to produce management action where needed?
   b. Adequate presentation and classification to bring persistent defects to attention of shop, of quality control personnel, etc.?
10. Review of training of inspectors and other quality control personnel on how to make inspection data useful to production for prevention of future defects.

INITIATING THE PROGRAM. Generally speaking, product engineering should be primarily responsible for the preproduction review of its designs; however, a fresh, quality-oriented analysis conducted in the end-use environment is usually necessary. Benefits derived from preproduction evaluation are the identification of important quality characteristics, location and elimination of manufacturing difficulties, and the detection of design details requiring modification.

Quality planning activities, following or coincident with preproduction evaluation, determine which characteristics should be measured, how they are to be measured, where sampling points should be located, who should do the measuring, and the instrumentation required. At this point an information feedback system must also be established. This will likely involve computer and information systems personnel as well as quality control personnel. In many smaller plants, a simplified information system may be employed.

Control procedures for purchased material must be designed. Vendor evaluation techniques as well as receiving inspection must be developed, and the personnel involved must be trained. Channels of communication which provide for immediate corrective action must be established and linked into the total information system of the plant or company.

EFFECTS ON MANUFACTURING. The development of a total quality system usually has a marked effect on the manufacturing function. Their attention is more clearly focused on manufacturing a product meeting all specifications rather than just meeting production quotas.

Improving Product Design and Specifications. One of the most important effects on manufacturing is the improvement in both design and specifications. Frequently, designers use tolerances of a particular nature because of their own bias rather than from a consideration of the functional requirements for the item of product. As a result, tolerances are frequently specified too "tight," that is, the limits of acceptance are very narrow. **Design review,** including tolerance analysis, serves to bring out such difficulties before full scale manufacturing problems as well as morale problems develop. Usually inspection and quality control personnel have no direct control over engineering drawings and written specifications for a product. This is as it should be. Still, rejects cannot be successfully prevented unless all major parties to the specifications (a) have the same understanding as to what they mean and (b) regard the requirements as realistic and necessary.

It is in order for the quality control department to explore the possibility that ambiguous requirements of blueprints are among the significant causes of product rejection. Sometimes obscure language and devious dimensioning are uncovered by a **systematic review of drafting-room standards.** If the uncertainties as to the designer's meaning are really frequent, it may be wise to start revision before launching any new quality maneuvers.

If the prints are only an occasional source of dispute, the troubles can be

ɔinpointed as the shop program goes on. Specific suggestions can then be made
ɔ the designers (or others) who issue and control the drawings and specifi-
:ations. Some of the common flaws to be noted and corrected are given in
Fig. 10.

Cause of Difficulty	Solutions To Consider
Division of engineering work into production design, manufacturing, engineering, metallurgy, etc.	Make designers responsible for issuance of specifications; arrange for others to advise them.
Use of specifications by many people, leading to many interpretations.	Write carefully in standard way using numbers rather than adjectives wherever possible.
Impracticability of expressing all requirements in numbers or unambiguous words. Examples: stains on bulb sockets; colors of rugs; smoothness of finish of paints; scuffs, scratches; fits.	Resort to physical standards. Consider use of "fade-out" or disappearance distance beyond which defect cannot be seen. Specify method of processing so fully as to ensure sufficient uniformity of product, or Specify method of test rather than dimension (etc.) wanted.
Impracticability of specifying against everything undesirable that could possibly occur (e.g. discoloration of molded parts).	Make a reasonable effort; then make a designer arbiter over other requirements when disputes arise; add later restrictions to prints when necessary.
Lack of knowledge of special tolerances needed and of abilities of machines, before going into production.	Pilot runs. Group decisions.
Problem of obtaining perfect conformance and, therefore, the resulting uncertainty as to realism of specifications.	Reject all pieces which do not meet specifications exactly, and Provide a salvage (material review) procedure as a safety valve.

Fig. 10. Improving product specifications.

Revising Production Methods. When specifications are found to be in order and excessive defective material is produced, it is necessary to revise production methods. This may involve equipment replacement, more operator training, or even drastic changes in the method of production.

USE OF STATISTICAL METHODS. Much has been written on the purely statistical aspects of control. These methods are tools which usually must be adapted to the background of each problem. With a product of simple design, exacting specifications, and relatively high unit value (illustrated by the ball bearing), the reclamation afforded by more advanced methods may warrant an expenditure of as much as a third of manufacturing labor-hours in purely statistical control methods. With a product of relatively low precision requirements, assembled from semifinished components of relatively low unit value (illustrated by a lamp base), the simpler and more direct approaches of routine patrol may suffice, with perhaps only a small outlay of control effort warranted.

The essential benefits of the modern analytical approach to quality control derive from an emphasis upon economic and systematic ways of thinking about cause and effect, through observation, interpretation, and action. A fully developed system of statistical quality control is capable of brilliant results, but it is not the initial step in quality work. Statistical thinking starts with the idea that variability is to be expected. It recognizes that the strengths or sizes of parts produced by a process are not constants but constitute a **frequency distribution**. Simple principles derived from this fact are useful long before it is advisable to make a full-dress installation of control charts.

The statistical tools used should be in keeping with the prevailing **level of technical development** in the plant. In an industry where little attention has been given to quality control, first emphasis would rightly center on planning for ordinary inspection, the procedure for marking and charging rejects, the system to be used for salvage, etc. Scarcely any statistics would be involved. Where inspection is already an established function, the first step is to build up the prevention function. This step involves a study of the economics of the situation, problems of coordination and jurisdiction, quality-mindedness, record analysis, and the use of some simple statistical tools. When work of this kind has become familiar, further results can be obtained progressively with control charts, correlation studies, analysis of variance, etc. The more complex tools are usually set up and interpreted by the quality control staff rather than by production personnel, but only at such time as the production personnel are receptive and ready to make real use of the results.

Starting Statistical Quality Control. Programs of statistical quality control should be applied gradually, allowing sufficient time for specific methods to show results and sell themselves. One or two promising men may be encouraged to familiarize themselves with process capability charts, control charts, and other elementary statistical techniques. Applications can then be made to a few problems, perhaps at existing inspection points, in a department having a progressive supervisor. As these applications prove themselves, the techniques can be extended to other problems in the department, using the experience of the original inspectors. Reports on progress can be drafted for the department supervisor's approval and signature. A good record will interest other departments.

Results can be described at a supervisors' round-table conference. A further possibility is designation of a **steering committee for new methods,** with representatives of engineering, production, and inspection. Additional applications can be made as requested in various areas of high scrap loss. Nontechnical talks on what is being accomplished can be set up for operators as well as for foremen and leadmen. Clinics on problems and methods can be organized for such groups as inspectors and engineers.

Correcting Defects

DETERMINING SOURCE OF DEFECTS. One of the first needs in launching a program for the prevention of poor quality is to locate the source of defects. Experience has indicated that most quality losses occur in a small number of places.

Inspection Records. Information on rejection percentages comes primarily from inspection department records. Information on dollar loss comes primarily from cost accounting records. Complete data on **cost of quality losses** is usually the more difficult to obtain; cost systems seldom identify all production factors responsible for the costs.

At the outset it is not unusual to find that the labor and materials charged to a job make no distinction between the base amount and the extra amount required to cover replacement of junked parts, repairs, and the like. This lack adds difficulties for production control as well as for quality control. A good method for obtaining this information is to stipulate that no finished or semifinished parts may go into scrap until a **scrap charge ticket** has been made out. This ticket carries such data as the part number, quantity, operation, reason for junking, and name of operator. Such a system may well put on record large losses not included in inspectors' reports, since in some plants parts junked by an operator may never be reported. Where this has been the case, the unavoidable expenses of scrap (metal shavings, punchings, sawdust, short ends, etc.) are hopelessly merged with the preventable loss due to scrap (defective parts).

Incoming Materials. Analysis of receiving inspection data shows where advice and coordination with vendors are needed. It also guides in the selection of suppliers for future purchases. It is comparatively simple to set up a system for collection of data on rejections of incoming material. For example, the receiving report can show the desired facts on quantity received and disposal of materials in each shipment. Alternatively, a material disposal report or tabulating card can be filled out.

The method of summarizing these original records for action depends on the complexity and volume of the product and on the number of vendors. In many cases an individual ledger card can be prepared for each part number and each vendor (see Fig. 11). A clerk can then enter the facts concerning the part di-

VENDOR'S RECORD

Part No. _113,562_   Part Name _Contact_   Vendor _xx Co._

Receiving Report Data			Inspection Results				Percent Defective Chart					
Date of Receipt	Receiving Rpt. No.	Total Quantity	Quantity Inspected	Rejected	Percent Defective	Disposal of Lot	1	2	3	4	5	6
12/7	27651	5,000	225	3	1.33	Accept						
12/15	27892	8,300	300	6	2.00	Accept						
12/22	27999	6,700	225	2	0.89	Accept						
12/31	28308	5,000	225	4	1.78	Accept						
1/21	29010	10,140	300	4	1.33	Accept						
2/13	29786	4,650	225	6	2.67	Sort						
3/3	30453	7,490	225	13	5.77	Return						
3/25	31224	9,025	300	5	1.67	Accept						
Totals												

Fig. 11. Record of supplier's quality.

rectly on the card. If the quantity and diversity of incoming materials make hand entries too inconvenient and time consuming, automated and computerized systems can be employed.

The increased use of **vendor rating programs** may be attributed in part to their being included as a specified activity in MIL-Q-9858A. The object of these programs is the identification of unreliable vendors and faulty parts or materials, so that suppliers whose shipments do not meet company standards can be dropped. The effort is wasted without the following procedures.

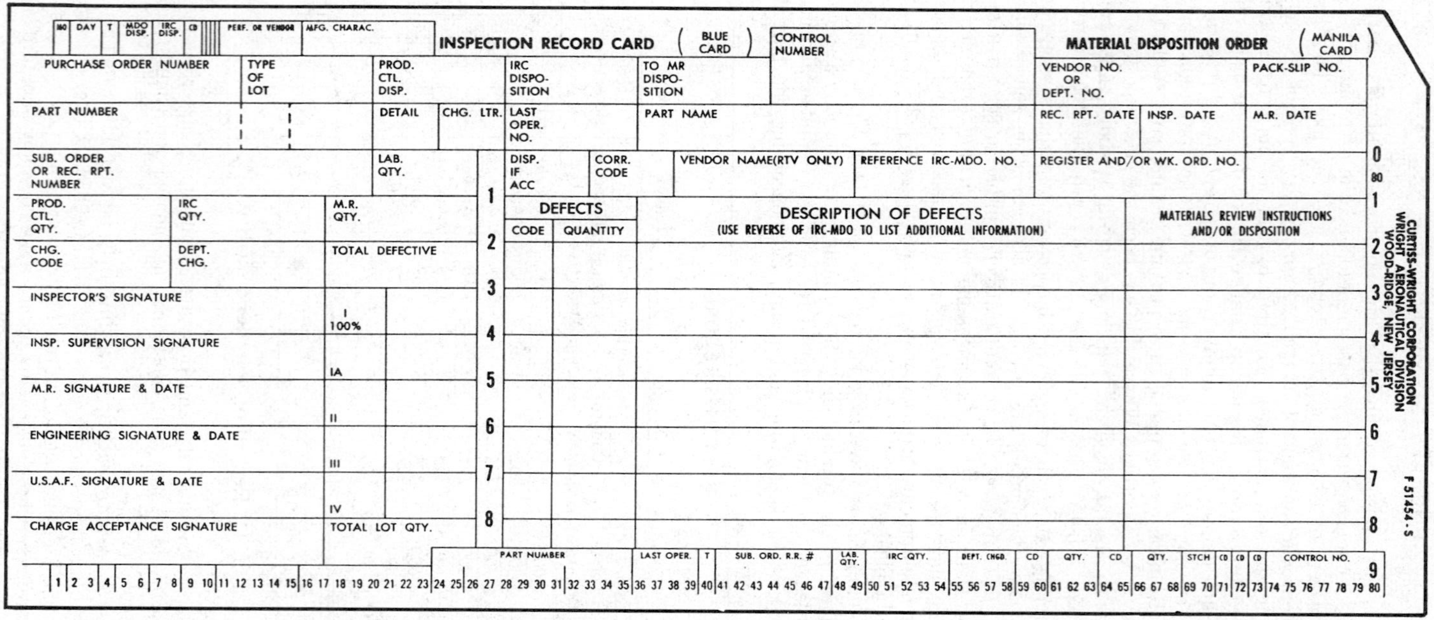

Curtiss-Wright Corp.

Fig. 12. Material disposition form.

1. The quality staff investigates and shows how to correct the trouble.
2. The purchasing department presses the supplier to take corrective action or, failing in this, buys from a reliable supplier.

An early step for the quality men must be a study of the grounds for rejection of any troublesome product. After these grounds have been determined, the cause of trouble must be pinpointed. Some indication of what is taking place at the supplier's factory can be obtained by examining charts of the product measurements. Beyond this it may be necessary to call in the supplier or to send a representative to study the quality picture at the source.

When sizable rejections are being made on all vendors for a particular part or specification, a re-examination of the requirement itself is in order.

In-Process Defects. The procedure in searching out points of loss along the production line is similar to that for collecting data on rejections of incoming material, though it usually puts more emphasis on the dollar losses. Disposition of faulty parts or materials is recorded on forms or tabulating cards similar to the one shown in Fig. 12. Parts scrapped are valued at the material cost plus added direct-labor and overhead costs. If detailed figures are not available, the value can be approximated by multiplying the estimated percentage of completion of the part by the total cost of a finished part. Repairs are valued at the added labor and overhead. These charges are assigned to the department whose work caused the rejection.

The losses thus arrived at can be summarized periodically according to department, operation number, or the like (Fig. 13). The quality control staff then selects from the summary the operations showing the largest losses. The problem of finding the causes of loss still remains. The next step in this direction is to discover the exact nature of the defects observed. The grounds for rejection are seldom specific enough, if previously classified at all.

Quality control may have to prepare a listing of the essential quality characteristics for the operations under study. In some cases this is a separate record (see Fig. 14). As each lot is inspected, the inspector notes the quantity that conforms to blueprint specifications (OK) in that column. Any product that does not conform is identified and classified according to whether it can be reworked (R/W) or is to be submitted for material review (M/R). The counts for these two situations are recorded in their respective columns. In other cases the inspectors are simply provided with a detailed **list of defects** and a code number for each. The appropriate code number is entered by the inspector on the regular tabulating card or disposition order to explain each rejection listed. The code number then appears in a subsequent sheet made up as a monthly summary of scrap.

The result of the foregoing procedure is a focus of attention on the two or three kinds of defects which cause the most frequent rejections of the part or item. All forces can then be brought to bear on the prevention of these defects. Although the above procedure is for evaluation of **items in process,** similar procedures can be applied in analyzing rejections in final inspection and for customer complaints.

ACTION BASED ON INSPECTION RECORDS. The objective of inspection record analysis is to correct quality problems, shifting the emphasis from screening to prevention. The following steps should be taken to **initiate quality improvement:**

1. Call the attention of production supervisors to trouble points by providing periodic reports on losses due to poor quality. The quality cost system should provide data on these "failure costs."

CUT-OFF
RUN-DATE 81/25   MONTHLY RMDO STATISTICAL REPORT

TYPE	CHG. DEPT.	ITEM IDENT NO	DEFECT CODE	DEFECT CODE DOLLARS	DEF CODE QUANTITY	TOTAL DOLLAR PREV. 12 MO
1	0413	3004817 01	H20	1.95	3	.0
		ITEM I.D. TOTALS		1.95	3	.0
		3528539 01	H20	3.60	1	.0
		ITEM I.D. TOTALS		3.60	1	.0
		3616946 00	H20	4.63	1	.0
		ITEM I.D. TOTALS		4.63	1	.0
		3004818 02	H20	9.48	3	6.3
		ITEM I.D. TOTALS		9.48	3	6.3
		3005857 00	H20	11.20	7	.0
		ITEM I.D. TOTALS		11.20	7	.0
		3004624 02	H20	12.60	6	.0
		ITEM I.D. TOTALS		12.60	6	.0
		3005933 01	H20	13.50	5	.0
		ITEM I.D. TOTALS		13.50	5	.0
		3005991 00	H20	16.85	5	.0
		ITEM I.D. TOTALS		16.85	5	.0
		3005462 03	H20	36.40	8	.0
		ITEM I.D. TOTALS		36.40	8	.0
	CHG. DEPT. TOTALS			110.21		6.3

CUT-OFF
RUN-DATE 81/25   MONTHLY RMDO STATISTICAL REPORT

TYPE	CHG. DEPT.	ITEM IDENT NO	DEFECT CODE	DEFECT CODE DOLLARS	DEF CODE QUANTITY	TOTAL DOLLARS PREV. 12 MO.
1	0422	5023826 00	H20	4.32	2	8.28
		ITEM I.D. TOTALS		4.32	2	8.28
		3158980 00	H20	13.04	3	.00
		ITEM I.D. TOTALS		13.04	3	.00
		3159050 00	H20	19.80	9	1.76
		ITEM I.D. TOTALS		19.80	9	1.76
		4094836 00	H20	21.00	6	.00

Fig. 13. Monthly rejected material disposition order.

2. Provide proper instruction to all production personnel and show specimens of defective work to operators.
3. Check operator use of gages and if necessary reinstruct operators in the proper use of such gages.
4. Initiate **process capability study** as necessary.
5. Maintain and replace machinery as necessary to meet specifications.
6. Review tolerances with design personnel.
7. Initiate **station control** at critical work stations.

Part or Assembly No. _N14-C_		Sheet _1_ of _2_		

Description or Name_____ _Jet Nozzle fitting_

Blueprint No. _2157_   File _E_

Date _2/27/_   Shift: D ☑   S ☐   N ☐

Inspector_____ _Dick Herrmann_

Characteristic	OK	R/W	M/R
Operation 38 visual	43	0	1
_I. D._	40	2	2
'' _40 visual_	44	0	0
'' _60 visual_	44	0	0
Toolmark	36	5	3
Radii burrs	30	13	1
Operation 100 visual	41	0	3
(Sheet) Totals	360	53	27

Fig. 14. Form for recording defects.

Follow-up on Quality Changes. Once quality-improvement changes have been made and they are accepted by all personnel, there are several devices that are most useful in maintaining the new quality levels. The control chart discussed later in this section is one such device. Periodic audits may be performed by quality control personnel in order to detect potential problems including a drift away from established quality levels. This should not be a policing activity but rather an audit, with prevention as the major motivation.

Statistical Quality Control

GRAPHICAL ANALYSIS. The analysis discussed up to this point refers to records of the number of good parts or bad parts in a lot. Where size, tensile strength, or other measurements of degree of goodness are made, it is possible to get a better understanding and control of a manufacturing process.

The use of simple charts at selected inspection points often brings out features of the data that would otherwise escape attention. **Bar charts** or **tally sheets** give a picture of the variability of a product and often suggest adjustments in the process that will give better results. These **frequency distribution charts** give much information that is lost in inspection with go-no-go gages. They show the characteristics of the good parts as well as of the bad parts; they emphasize that all parts are not identical; and they are reminders that important decisions should not be made on the basis of only one piece, as, for example, in so-called **first-piece inspection** (see section on Inspection).

In some cases the same information can be obtained in other ways, as from a process capability study or a control chart study. The particular advantages of the bar chart or tally sheet are: (1) it can be applied even if the order of production is unknown (as in receiving inspection); (2) it shows the pattern of quality in the lot of material as a whole; and (3) it is easily made by an inspector.

A typical tally sheet is shown in Fig. 15. The inspector has only to write down the possible readings of his instrument at the left and to put a mark to the right of the correct number of each measurement as he makes it. Figure 16 shows the common appearances and causes of ten types of tallies that occur in practice.

Process Capability Studies. The capability of a machine or manufacturing process is given by Juran (Quality Control Handbook) as "the minimum tolerance to which the machine can possibly be expected to work and produce no defectives under the specified conditions." For many years foremen, dispatchers, and process engineers have used knowledge of equipment capability in assigning work to machines. Their beliefs have usually been derived from years of separate

QUALITY REPORT	**PAGE__of__**	
	NO.	

PART or ASSEMBLY_____ JOB NO._____

PART NO. _____ PRINT NO. _____FILE_____

CUSTOMER_____

ORDER NO._____

INSPECTION RECORD

INSPECTED BY_____DATE_____

INSPECTION METHOD_____

CHARACTERISTIC INSPECTED_____

SPECIFIED LIMITS_____

EQUIVALENT INSPECTION LIMITS_____

WHERE SAMPLE DRAWN_____

DESCRIPTION	TALLY	TOTALS	REMARKS

Fig. 15. Inspection tally sheet.

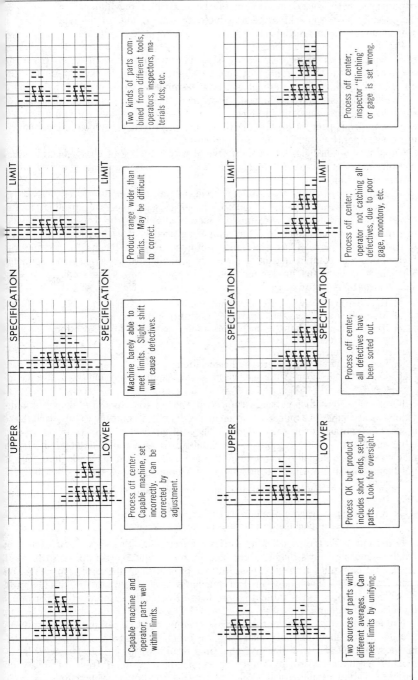

Fig. 16. Interpretations of tallies.

experiences with success or failure of the particular machine on various jobs The results have too seldom been reduced to writing and made available to al who could use them profitably.

Problems of close tolerances met in recent years have encouraged the making of special studies of process capability. The process capability or **machine accuracy study** is an important tool in quality control work. By its use the abilities of production equipment can be measured with reasonable rapidity on a standardized basis and catalogued. Work can then be assigned on the basis of suitability of machines to produce to the tolerances needed. Advantages that can result from process capability studies are improved equipment maintenance improved operator training, and reduced costs.

Capability studies can be made in several ways: by use of charts of individual measurements, by use of bar charts, by use of control charts, or by other statistical techniques.

The most satisfactory way to get an initial impression of process capability is to plot the measurements of 50 to 100 consecutive pieces of product in the exact order in which they are made. Such a **chart of the individual measurements** is a convenient tool for comparing ability of a process with a specification requirement. It is especially useful as a quick means for getting the facts when the competency of a machine is challenged in a clash of opinions. The chart is easy for production people to understand. The steps in making such a **capability study** are:

1. Keep parts strictly in order of manufacture.
2. Measure with a gage that can divide tolerance into tenths.
3. Keep a log of changes in tool settings, materials lots, operators, machine speeds, temperatures, etc.
4. Chart the measurements.
5. Look for significant patterns and associations.
6. Compare the chart with the specification requirements and with the log of changes.

Figure 17 shows five patterns typical of those found in these charts of individual measurements. Since the order of production is known, these charts have a clear advantage over tally sheets. They often reveal tool wear, effects of adjustments, and other reasons for variability. It is not unusual to find that a machine which allegedly "can't hold the tolerance" is actually adequate but is not being operated in the best possible way. Common causes are:

1. Improper setting of the tool, that is, not at the center of tolerance.
2. Inadequate provisions for tool wear.
3. An inadequate gage in the hands of the operator.
4. Resetting of a machine on the basis of one measurement.

Numerical Ratings of Capability. Success of the capability study in trouble shooting shows that it could be even more valuable if applied to prevent trouble. Companies now set up **standard procedures** for rating machines on their capability. The more critical machines are systematically studied and classified. Results, when tabulated, are helpful not only to production personnel but also to designers.

Numerical ratings, to have their greatest value, must be obtained by a uniform procedure and should specify materials and other fundamental factors. An automatic screw machine will not give the same results with mild steel as with hard-machining copper. Also, its ability on diameters may differ from its ability on lengths. Tables of numerical values of process capability or machine accuracy are forecasts of the future performance of a machine. Until it is demonstrated that the performance can be stabilized, it is uncertain that today's performance can be repeated later. Proper forecasts of this type thus should be made only if the machine is in control. It is common practice to establish a condition of control by use of statistical control charts before computing machine capability.

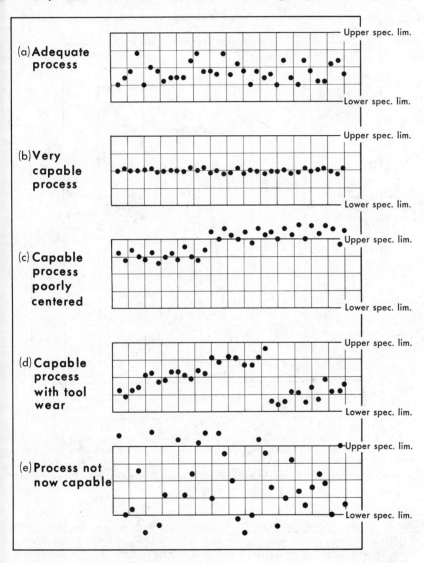

Fig. 17. Charts of individual measurements in the study of capability.

STATISTICAL DETERMINATION OF CAPABILITY. The process capability study is an application of statistical techniques which can be conducted in a short time. The primary purpose of this study is to determine quantitatively the variation inherent in any given process. It can be applied to any process such as machining, heat-treating, mixing, etc., where quantitativ values of variation can be obtained.

There are three sources of variations in any process: the operator, the ma terials, and the machine or process equipment. Figure 18 illustrates the vari

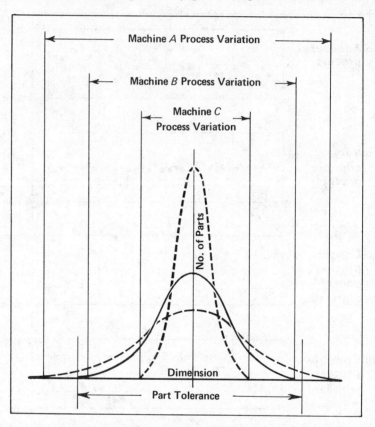

The process variation for machine A exceeds the part tolerance; the variation for machine B is slightly less than the part tolerance; and the variation for machine C is much less than part tolerance.

Fig. 18. Distribution of parts according to one dimension, for three different machines.

ation in process for the same part run on three different machines. Obviously every machine has its particular characteristics with respect to the inherent variation due to chance cause systems. The capability study establishes for each process those variations that are then interpreted as the minimum attainable tolerances in that specific process.

A machine capability study can be conducted in the following manner:

1. Select a part to be processed or machined that is representative of the type of operation performed by the machine.
2. Prepare a data sheet for recording the results of the operation (see Fig. 19).
3. Provide measuring equipment with the capacity to measure beyond the accuracy of the machine being studied.

MACHINE CAPABILITY DATA SHEET

MACHINE TYPE & NO. _Miller # 2157_   OPERATOR _Irv Mehlman_

OPERATION _Slotting_   DEPT. NO. _51_

TOOLING _Cutter # E38_   INSPECTOR _Karl Donero_
Arbor # 112

NOMINAL DIMENSION _0.7500_   TOLERANCE _close as possible_

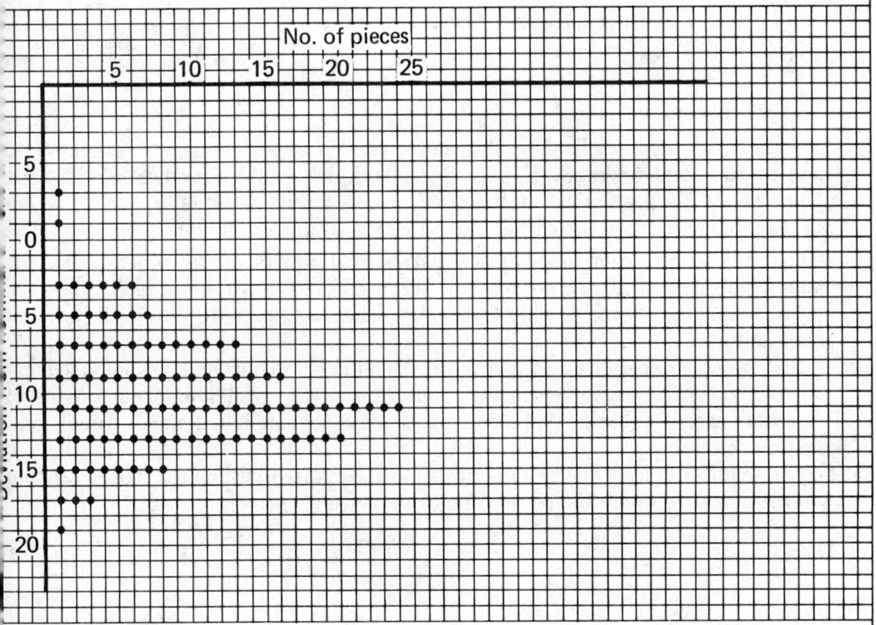

CALCULATIONS:

$$\varepsilon f = 100$$

$$\varepsilon f x = (-3 \times 1) + (-1 \times 1) + (3 \times 6) + (5 \times 7) + (7 \times 13) + (9 \times 16) + (11 \times 24) + (13 \times 20)$$
$$+ (15 \times 8) + (17 \times 3) + (19 \times 1) = 998$$

$$\overline{X} = \frac{998}{100} = 9.98$$

$$\varepsilon f x^2 = (9 \times 1) + (1 \times 1) + (9 \times 6) + (25 \times 7) + (49 \times 13) + (81 \times 16) + (121 \times 24)$$
$$+ (169 \times 20) + (225 \times 8) + (289 \times 3) + (361 \times 1) = 11,484$$

$$\sigma = \left[\frac{\varepsilon f x^2}{n} - \overline{X}^2 \right]^{\frac{1}{2}} = [114.84 - 99.60]^{\frac{1}{2}} = [15.24]^{\frac{1}{2}} = 3.904$$

$$6\sigma = 23.424$$

machine accuracy = 0.0023 in.

STUDY BY _W.S.J_

CHECKED BY _J.P.N._

Fig. 19. Form for machine capability study.

4. Inspect each part as it is produced and record the results in the appropriate place in the data sheet. Inspect at least 100 parts. The number of parts however will vary with the type of machine and the process itself.

The measurements in Fig. 19 are made to ten-thousandths of an inch, and four decimal places are dropped in order to simplify the calculation. These decimal places are of course picked up, as shown, to arrive at the machine capability value. The symbols used in Fig. 19 are defined as follows:

f = frequency of occurrence of each measurement

X = deviation of the dimension from the nominal dimension (in ten-thousandths)

n = total number of measurements ($n \geq 100$)

$$\overline{X} = \frac{\Sigma f X}{n}$$

σ = standard deviation

The machine accuracy is 6σ. Therefore, this machine cannot be expected to hold to any total tolerance (Upper Specification − Lower Specification) smaller than 0.0023. As a practical matter, the most desirable situation would be where the capability value is less than three-fourths of the total part tolerance. The values $\overline{X} \pm 3\sigma$ are sometimes called the **natural limits** of the process, and the $\pm 3\sigma$ values constitute the 6σ spread. With $n \geq 100$, it is highly likely that the natural limits so formed will contain in excess of 99 percent of all product manufactured.

The process capability study may be conducted under highly controlled experimental conditions in which every effort is made to eliminate all sources of variation except machine variation or it may be conducted under usual production conditions.

For small samples, process tolerances must be stated with care. For a random sample of size n, where the population or universe is normal,

$$\overline{X} = \frac{\sum_{i=1}^{n} X_i}{n}$$

and

$$s = \sqrt{\frac{\Sigma X_i{}^2 - n\overline{X}{}^2}{n - 1}}$$

The symbol X_i represents the value of the ith measurement. Natural limits are given by $\overline{X} \pm Ks$ where the factors K are tabulated in the section on Statistical Methods. A probability statement must accompany the tolerance interval.

Confidence Limits in Quality Control. It may be desired to **estimate** the mean and standard deviation of the process from a small or intermediate size sample. Point estimators which are good estimations in a statistical sense are:

Population or Universe	Estimator from Sample
Mean	$\overline{X}$
Standard Deviation	s

In order to deal with the question of the accuracy of the estimators, confidence intervals are formed where it is assumed that the random variable (quality characteristic) X has a normal probability distribution and the sample is a random sample. **Confidence limits on the true mean** are:

$$\overline{X} \pm t_{\alpha/2, n-1} \frac{s}{\sqrt{n}}$$

where $t_{\alpha/2,\ n-1}$ is read from a table of the Student-t distribution with $n-1$ degrees of freedom (see section on Statistical Methods).

Example: Five measurements on Rockwell Hardness of alloy castings yield the following data.

Observation, i	Measurement, Xi
1	50
2	51
3	53
4	47
5	49

$n = 5$

$$\overline{X} = \frac{250}{5} = 50, \qquad s = \sqrt{\frac{20}{4}} = 2.23$$

The value $1 - \alpha$ is called the **confidence level** and it represents the probability that the procedure used will yield an interval that contains the population mean hardness. In the example, let $1 - \alpha = 0.95$ so that $\alpha = 0.05$ and $\alpha/2 = 0.025$. From the table of the Student-t distribution with $n - 1 = 4$ degrees of freedom, $t_{0.025,4} = 2.776$. The interval is thus

$$50 \pm 2.776 \frac{2.23}{\sqrt{5}}$$

or

$$[47.224, \qquad 52.776]$$

Confidence limits on the population standard deviation are given by

$$\left[s\sqrt{\frac{(n-1)}{\chi^2_{\alpha/2,n-1}}}, \qquad s\sqrt{\frac{(n-1)}{\chi^2_{1-\alpha/2,n-1}}} \right]$$

where $\chi^2_{\alpha/2,\ n-1}$ is read from a table of the chi-square distribution with $n-1$ degrees of freedom. The probability is $1 - \alpha$ that the procedure will yield an interval that contains the population standard deviation. In the example, again using $1 - \alpha = 0.95$, we obtain

$$\left[2.23\sqrt{\frac{4}{11.143}}, \qquad 2.23\sqrt{\frac{4}{0.484}} \right]$$

or

$$[1.34, \qquad 6.41]$$

In studying fraction defective where the sample size is large and the number of defectives is counted, the ratio

$$p = \frac{\text{No. defectives in sample}}{\text{Sample size}}$$

is a good estimator of the population fraction defective, and an approximate confidence interval is given by

$$p \pm Z_{\alpha/2} \cdot \sqrt{\frac{p(1-p)}{n}}$$

where $Z_{\alpha/2}$ is read from a table of the normal distribution. Again, $1 - \alpha$ is the confidence level.

Example: A sample of 900 parts is found to contain 300 defectives. The confidence level is 95%.

$$\text{Fraction defective} = \frac{300}{900} = 1/3$$

The confidence interval for this ratio is

$$1/3 \pm 1.96 \sqrt{\frac{(1/3)(2/3)}{900}}$$

$$1/3 \pm \frac{1.96}{90} \sqrt{2}$$

CONTROL CHARTS. In manufacturing, the problem of identifying causes of poor quality occurs again and again. The operator finds that successive parts differ slightly even under good conditions. When is a change substantial enough to justify action such as resetting the machine or looking for faults in the material? Methods which tell the operator exactly when an assignable source of trouble has entered the process are valuable. Not only can they warn of coming difficulty in time to prevent scrap; they also tell when the machine should not be touched because it is producing to the limit of its inherent accuracy.

Ordinary inspection records usually give only general information; for example, "Lathe No. 271 produced more out-of-limit parts yesterday than the day before." The object of the control chart is to be more specific: to warn that there was an unusual increase in the diameter of parts from Lathe No. 271 at 2:30 P.M. today and a return to normal at 3:15 P.M. The control chart can give its alarm within minutes after trouble begins. Investigation can often start before the difficulty has become severe enough to produce parts outside blueprint limits. Also, pinpointing of the time of change permits review of happenings which might have been associated with the change. Was a new piece of bar stock started then? Was there a tool adjustment? A change of operator? A change in appearance of chips and turnings?

While the control chart can distinguish between two broad classes of operating troubles, the great value of the chart is in:

1. Quick reporting of changes.
2. Distinguishing between natural fluctuations and those with assignable causes.
3. Identifying the time at which the change began.

Control Chart Applications. Control charts are used for analyzing data in two different situations. For contrast these are called:

1. **Control with no standard given** (also termed "trial control")—used for testing data for homogeneity, that is, to determine whether sample measurements differ from each other over a period of time by a greater amount than ought to be ascribed to chance. The idea of control is not necessarily related to blueprint tolerances or specifications. Based solely on data from samples, these charts aid in detecting lack of constancy in factors affecting quality. They indicate when any significant change in these factors occurs so that a speedy investigation of the cause can be started. They also indicate whether the machine is making the most uniform product which may reasonably be expected of it with present materials, conditions, and operators. Charts of this kind can thus be used to learn about the natural capabilities of manufacturing processes. It follows that a manufacturing operation may be "in control" in the sense that it is as stable as it is able to be; yet it may not be conforming to specification limits. Charts of this kind may therefore be said to analyze primarily for **stability**.

2. **Control with respect to a given standard**—used to test inspection results for conformance to specifications, that is, to determine whether measurements on samples differ from desired measurements by a greater amount than ought to be ascribed to chance. The **standard values** may be aimed-at levels, such as dimensions set by specifications or drawings. In other cases they may be standard levels selected from experience as practical and economical to adhere to, such as an accepted "normal" percent of defectives. Control charts based on such desired values are used in inspection work to aid in distinguishing the variations expected in the manufacturing process from those which are significantly larger than the standards permit. Charts of this kind may be said to analyze for **conformance.** They are sometimes called **Modified Control Charts** or **Control Charts-Standard Given.**

Usual practice is to make initial control studies on the "trial limit" or no-standard-given basis. This shows what uniformity can be obtained from the process. The chart compares present performance of a machine with its past performance. Thus the limits for action are set from past average behavior of the man-machine-material combination. They are not derived from the engineering specifications for the product. When the trial control limits have been set up and examined they will indicate what should be done next.

Procedure in Applying Control Charts. The process of control charting may be divided into the following steps:

1. Selecting the quality characteristics worthy of control.
2. Analyzing the manufacturing process for probable sources of trouble.
3. Choosing the statistical measures.
4. Planning the selection and grouping of data.
5. Collecting the preliminary data.
6. Establishing trial control limits and starting the charts.
7. Reviewing results and drawing conclusions.
8. Maintaining control.

The starting point for choice of the quality characteristics to be controlled is in the quality records already described. The purpose of control is to reduce defects. The processes to be controlled are those where defects appear in numbers which are serious from the point of view of cost. The first step, then, is to review available inspection records for trouble points.

It does not automatically follow that the control chart should be set up at the existing inspection station. This is particularly true where the output of several machines is mixed together and reaches the inspector as a single stream of product. Best results are usually obtained by working toward the source of this stream. Much of the contamination may occur at one place, and analysis several steps downstream from that point may be ineffective, because the defects will be likely to appear there at random times and in changing patterns which seem inexplicable. For example, in one plant the record showed a high percentage of finish defects on a bearing surface. The parts came from two centerless grinders; the product from both contained the defects. The grinders were supplied with semifinished parts from three lathes. Control studies at the lathes showed that one of the three was much less capable than the others. As a result the replacement of a worn bearing was recommended. After the overhaul the trouble virtually disappeared.

Choice Between $\overline{X}$, R, p, and c Charts. A control chart shows a plot of some characteristic for a series of samples. The characteristic plotted for each sample is most usually one of the following: the average measurement $\overline{X}$ (read as "X bar"); the range of measurements, R; the percentage of defective parts,

$p;$ or the count of defects, $c.$ Each sample, except in the case of the c chart, consists of a number of pieces of product—never one alone.

$\overline{X},$ $p,$ and c charts report on the **prevailing level of quality.** In contrast, the R chart reports on the scatter or nonuniformity of product. R charts warn when a process becomes erratic. Where the greatest ability to diagnose is needed, charts for average, $\overline{X},$ and range, $R,$ are both used. Together they form a sensitive instrument for detecting and correcting **changes in the process.** Charts for average and range require actual measurements on the product, not just results of go-no-go gaging. If measurements are available or can be provided, $\overline{X}$ and R charts will make the most efficient use of the data.

The $\overline{X}$ chart will warn of **changes tending to affect all pieces.** It can detect results of wear of an abrasive wheel, dilution of a chemical solution, a change in heat treatment at a previous operation, improper calibration of a machine for filling and weighing sacks, a decrease of vacuum at the sealing of electronic tubes, and the like. The R chart will warn of a **significant loss (or gain) in uniformity** of the pieces. It can detect results of spotty quality of raw material, unreliability of a worn machine, balky behavior of a weighing scale, inconsistent practices of an operator or inspector, or the like. In general the discovery of erratic quality (from an R chart) requires corrective measures different from the discovery of a shift or trend (from an $\overline{X}$ chart).

CONTROL CHARTS FOR ATTRIBUTES. There are numerous situations where it is desirable to record attributes data rather than variables (measurement) data. In one method of attribute testing each unit is inspected and classified into one of several mutually exclusive classes; the usual practice is to use just two classes, such as **defective** and **satisfactory** but units could also be classified as **defective, satisfactory-first quality, satisfactory-second quality,** etc. Another technique involves counting and recording the number of defects in a constant unit of production.

Attributes testing has the following advantages:

1. If several (or many) quality characteristics are to be observed on each unit, $\overline{X}$ and R charts would have to be maintained on each characteristic. A single chart could be used if each unit is classified as good or bad, or the number of defects is counted.
2. The cost of collecting data and charting is usually much less when dealing with attributes than when measuring variables.
3. Methods are applicable where sensory inspection is involved, and no quantitative measurement is possible.
4. Attribute charts provide data very meaningful to management since they represent a quantitative measure of the effectiveness of production.

Use of p and c Charts. Charts of p and c are not as sensitive as charts of $\overline{X}$ and $R.$ They are not as specific in indicating what is wrong. They require more observations per point plotted and thus may not report the presence of trouble so quickly. Nevertheless these charts have great value to supervisors in telling when to apply pressure for improvement. They can often be set up without any change in existing inspection methods. They distinguish between fluctuations in fraction defective p or count of defects c which warrant action and those probably resulting from ever-present small, unidentifiable causes.

The p chart is used for **control of fraction defective.** It can be applied wherever one can count the possible number of defective articles as well as the actual number of defective articles. The c chart is used for the **control of defect.** (A defect is a flaw of some kind. A single defective article might have several defects.) The c chart is applied chiefly in the special case where the number of defects can be counted but the number of opportunities for defects cannot. Examples are nicks and scratches on silverware, spots on photographic film, punctures in insulating paper, flaws in bearing surfaces. c charts are useful in judging control over complex items, such as radio sets, engines, and watches. The c chart is appropriate only if the total number of possible defects is large (though perhaps not exactly known) and the actual number of defects is relatively small, say less than 10 percent of the possible number.

Selection of Samples. For charts of $\overline{X}$ and R, samples of 5 or 4 are nearly always used. However, for charts of p, much larger samples are needed. The sample should not be so small that the finding of only one defect indicates lack of control. Some authorities advocate a size large enough to lead to an average of at least six defects in a sample. For a p chart this would require the following as an approximate **minimum sample size**:

Average Fraction Defective	Sample Size
0.1	60
0.05	120
0.01	600
0.005	1,200

Charts of c may be used for the count of defects found in a stated number of articles or in an arbitrary (but constant) amount of material, such as a selected length of cable or area of carpet. In general it is desirable to have the number of articles or the area of opportunity of a single specimen sufficient to produce an average count of perhaps **six defects per sample**. It is quite possible that the average count of defects may be large enough in samples of one, since a single specimen may have a number of defects of the same kind and since defects of different kinds may be counted together besides.

Samples for $\overline{X}$ and R charts are generally taken often enough to include from 1 percent to 10 percent of all product made. Samples for p and c charts commonly include 20 percent to 50 percent—sometimes 10 percent—of all product made. The upper percentage is used when the process is unstable or the production rate is small. The lower percentage is used when the process is fairly stable or the production rate is large.

Samples for control purposes should be selected in a way which assures that the specimens in any one sample were produced under the same essential conditions, according to Grant (Statistical Quality Control), so that suspected causes of trouble will operate between one sample and the next. This favors **segregation by source**, that is, forming the sample from the product of one man, machine, batch of raw material, or the like. In the case of $\overline{X}$ and R charts this is usually carried out by taking four or five pieces made consecutively.

Where large samples are necessary, as with p charts, it is sometimes impossible to segregate by source as fully as would be desired. The best that can be done may be to take the sample from the work of one shift, or from one large heat of metal, for example.

CONTROL CHARTS FOR VARIABLES. Variables data depend on the quantitative nature of the measurement being taken. The most common control charts for variables are the $\overline{X}$ and R charts used together; however, the σ chart may be used in place of the R chart. **Cumulative Sum Control Charts** (or cusum charts) are used primarily to maintain current control of a process. Duncan (Quality Control and Industrial Statistics) cites as an important advantage their ability to pick up a sudden and persistent change in the process average more rapidly than other control charts especially if the change is not large.

No. of Observations in Sample	Chart for Averages				Chart for Standard Deviations								Chart for Ranges				
	Factors for Control Limits				Factors for Central Line		Factors for Control Limits						Factor for Central Line	Factors for Control Limits			
n	A	A_1	A_2	A_3	c_2	c_4	B_1	B_2	B_3	B_4	B_5	B_6	d_2	D_1	D_2	D_3	D_4
2	2.121	3.760	1.880	2.659	0.5642	0.7979	0	1.843	0	3.267	0	2.606	1.128	0	3.686	0	3.267
3	1.732	2.394	1.023	1.954	0.7236	0.8862	0	1.858	0	2.568	0	2.276	1.693	0	4.358	0	2.575
4	1.500	1.880	0.729	1.628	0.7979	0.9213	0	1.808	0	2.266	0	2.088	2.059	0	4.698	0	2.282
5	1.342	1.596	0.577	1.427	0.8407	0.9400	0	1.756	0	2.089	0	1.964	2.326	0	4.918	0	2.115
6	1.225	1.410	0.483	1.287	0.8686	0.9515	0.026	1.711	0.030	1.970	0.029	1.874	2.534	0	5.078	0	2.004
7	1.134	1.277	0.419	1.182	0.8882	0.9594	0.105	1.672	0.118	1.882	0.113	1.806	2.704	0.205	5.203	0.076	1.924
8	1.061	1.175	0.373	1.099	0.9027	0.9650	0.167	1.638	0.185	1.815	0.179	1.751	2.847	0.387	5.307	0.136	1.864
9	1.000	1.094	0.337	1.032	0.9139	0.9693	0.219	1.609	0.239	1.761	0.232	1.707	2.970	0.546	5.394	0.184	1.816
10	0.949	1.028	0.308	0.975	0.9227	0.9727	0.262	1.584	0.284	1.716	0.276	1.669	3.078	0.687	5.469	0.223	1.777
11	0.905	0.973	0.285	0.927	0.9300	0.9754	0.299	1.561	0.321	1.679	0.313	1.637	3.173	0.812	5.534	0.256	1.744
12	0.866	0.925	0.266	0.886	0.9359	0.9776	0.331	1.541	0.354	1.646	0.346	1.610	3.258	0.924	5.592	0.284	1.716
13	0.832	0.884	0.249	0.850	0.9410	0.9794	0.359	1.523	0.382	1.618	0.374	1.585	3.336	1.026	5.646	0.308	1.692
14	0.802	0.848	0.235	0.817	0.9453	0.9810	0.384	1.507	0.406	1.594	0.399	1.563	3.407	1.121	5.693	0.329	1.671
15	0.775	0.816	0.223	0.789	0.9490	0.9823	0.406	1.492	0.428	1.572	0.421	1.544	3.472	1.207	5.737	0.348	1.652
16	0.750	0.788	0.212	0.763	0.9523	0.9835	0.427	1.478	0.448	1.552	0.440	1.526	3.532	1.285	5.779	0.364	1.636
17	0.728	0.762	0.203	0.739	0.9551	0.9845	0.445	1.465	0.466	1.534	0.458	1.511	3.588	1.359	5.817	0.379	1.621
18	0.707	0.738	0.194	0.718	0.9576	0.9854	0.461	1.454	0.482	1.518	0.475	1.496	3.640	1.426	5.854	0.392	1.608
19	0.688	0.717	0.187	0.698	0.9599	0.9862	0.477	1.443	0.497	1.503	0.490	1.483	3.689	1.490	5.888	0.404	1.596
20	0.671	0.697	0.180	0.680	0.9619	0.9869	0.491	1.433	0.510	1.490	0.504	1.470	3.735	1.548	5.922	0.414	1.586
21	0.655	0.679	0.173	0.663	0.9638	0.9876	0.504	1.424	0.523	1.477	0.516	1.459	3.778	1.606	5.950	0.425	1.575
22	0.640	0.662	0.167	0.647	0.9655	0.9882	0.516	1.415	0.534	1.466	0.528	1.448	3.819	1.659	5.979	0.434	1.566
23	0.626	0.647	0.162	0.633	0.9670	0.9887	0.527	1.407	0.545	1.455	0.539	1.438	3.858	1.710	6.006	0.443	1.557
24	0.612	0.632	0.157	0.619	0.9684	0.9892	0.538	1.399	0.555	1.445	0.549	1.429	3.895	1.759	6.031	0.452	1.548
25	0.600	0.619	0.153	0.606	0.9696	0.9896	0.548	1.392	0.565	1.435	0.559	1.420	3.931	1.804	6.058	0.459	1.541

Fig. 20. Factors for control charts for variables $\overline{X}$, s, σ, R.

The major advantage to using variables data is that the charts provide the basis for making corrective action as well as other decisions regarding process parameters. It is one thing to know that a part is defective but an entirely different thing to know that a particular dimension is on the average 0.01 inch greater than the nominal value.

Use of $\overline{X}$ and R Charts. Both $\overline{X}$ and R charts are easy to use. After an initial process capability study has been completed, most production employees, with a minimum amount of instruction, can make all necessary calculations at their work stations. The $\overline{X}$ chart is used to make decisions regarding the centering of the process and the R chart is used to make decisions regarding the process spread or dispersion.

ESTABLISHING CONTROL CHART LIMITS. After data have been collected for a preliminary period, the statistics ($\overline{X}$ and/or R, p, or c) are charted together with trial control limits. These initial limits determine whether there is control in the sense of stability. They do not automatically imply control at a suitable level or in sufficient degree to assure acceptable product. Factors and formulas for computing control chart limits and central lines are shown in Figs. 20 and 21. The σ chart has not been explained as it is a seldom used alternative to the R chart. The value σ is calculated as follows:

$$\sigma = \sqrt{\frac{\Sigma X^2}{n} - \overline{X}^2}$$

Trial Limits for $\overline{X}$ and R Charts. Average and range charts are ordinarily set up after collection of about 100 observations (20 or 25 samples). Figure 22 (page 46) gives a typical work sheet, here applied to the study of a machine loading 100-lb. bags of sodium bicarbonate.

The central line of the $\overline{X}$ chart is located at the mean value of all 125 measurements, 101.30 lb. This quantity is termed the grand average, $\overline{\overline{X}}$ (read as "X double bar"). The central line for the range chart is found from the average range, $\overline{R}$ (read as "R bar"). $\overline{R}$ is obtained by adding the 25 separate ranges and dividing by 25. Here $\overline{R}$ is 0.72 lb.

Consulting Fig. 20 for the constants needed, it is found that for samples of five, $A_2 = 0.577$, $D_3 = 0$, $D_4 = 2.115$. The lower control limit for averages as given in Fig. 21 is placed at:

$$LCL_{\overline{X}} = \overline{\overline{X}} - A_2\overline{R} = 101.30 - 0.577(0.72) = 100.88 \text{ lb.}$$

and the upper control limit for averages at:

$$UCL_{\overline{X}} = \overline{\overline{X}} + A_2\overline{R} = 101.30 + 0.577(0.72) = 101.72 \text{ lb.}$$

The 25 averages observed are plotted with these limits in the upper chart of Fig. 23 (page 47).

The central line for the range chart is located at $\overline{R} = 0.72$ lb. The lower and upper limits, respectively, for the range are located at:

$$LCL_R = D_3\overline{R} = 0(0.72) = 0$$
$$UCL_R = D_4\overline{R} = 2.115(0.72) = 1.52 \text{ lb.}$$

The 25 ranges observed are plotted with these limits in the lower chart of Fig. 23.

This process at the time of the study is not in the state of control, as all points do not lie within the control limits. Therefore, the control chart is

Formulas for Variables $\overline{X}$, s, σ, R

Purpose of Chart	Chart for	Central Line	$3 - \sigma$ Control Limits
No Standard Given Used for analyzing past data for control. ($\overline{X}$, $\overline{R}$, $\bar{s}$, $\bar{\sigma}$ are average values for data being analyzed.)	Averages, $\overline{X}$	$\overline{X}$	$\overline{X} \pm A_2\overline{R}$, or $\overline{X} \pm A_3\bar{s}$, or $\overline{X} \pm A_1\bar{\sigma}$
	Ranges, R	$\overline{R}$	$D_3\overline{R}$, $D_4\overline{R}$
	Standard deviations, s	$\bar{s}$	$B_3\bar{s}$, $B_4\bar{s}$
	Standard deviations, σ	$\bar{\sigma}$	$B_3\bar{\sigma}$, $B_4\bar{\sigma}$
Standard Given Used for controlling quality with respect to given standards. ($\overline{X}'$, σ' are given standard values.)	Averages, $\overline{X}$	$\overline{X}'$	$\overline{X}' \pm A\sigma'$
	Ranges, R	$d_2\sigma'$	$D_1\sigma'$, $D_2\sigma'$
	Standard deviation, s	$c_4\sigma'$	$B_5\sigma'$, $B_6\sigma'$
	Standard deviation, σ	$c_2\sigma'$	$B_1\sigma'$, $B_2\sigma'$

Fig. 21. Formulas for

showing that there is an assignable cause of variation in the process. As soon as the assignable source of variability is discovered and eliminated, new trial limits will be calculated to determine if the process is then in control.

If a point lies outside control limits, there is strong evidence that the out-of-control sample was not produced under the same conditions as the others. If a **sample average** is outside, it indicates that a shift affecting all specimens has occurred. The log kept while collecting the data should be consulted to see if there was some known change of operator, tool setting, material, or the like which could account for the out-of-control point. If a **sample range** lies outside the limits, it indicates that the uniformity of the product has changed.

Trial Limits for p Charts. p charts for preliminary data are set up by placing limits around $\bar{p}$ ("p bar"), using the formulas given in Fig. 21. The average fraction defective is

$$\bar{p} = \frac{\text{Total defectives found in all samples}}{\text{Total inspected in all samples}}$$

The limit lines are located above and below $\bar{p}$ by an amount $3\sqrt{\dfrac{\bar{p}(1 - \bar{p})}{n}}$

For example, if the number of pieces in each sample is 200, and the average fraction defective is 0.04:

$$UCL_p = \bar{p} + 3\sqrt{\frac{\bar{p}(1 - \bar{p})}{n}} = 0.04 + 3\sqrt{0.04(0.96)/200} = +0.082$$

$$LCL_p = \bar{p} - 3\sqrt{\frac{\bar{p}(1 - \bar{p})}{n}} = 0.04 + 3\sqrt{0.04(0.96)/200} = -0.002 \text{ (use 0)}$$

If the samples are of different sizes, it becomes necessary to compute separate limits for each point. In this event the quantity

$$3\sqrt{\bar{p}(1 - \bar{p})}$$

Formulas for Attributes p, np, c, u

Purpose of Chart	Chart for	Central Line	$3 - \sigma$ Control Limits
No Standard Given	p	$\bar{p}$	$\bar{p} \pm 3\sqrt{\bar{p}(1 - \bar{p})/n}$
Used for analyzing past data	np	$n\bar{p}$	$n\bar{p} \pm \sqrt{n\bar{p}(1 - \bar{p})}$
for control. ($\bar{p}$, $n\bar{p}$, $\bar{c}$, $\bar{u}$ are	c	$\bar{c}$	$\bar{c} \pm 3\sqrt{\bar{c}}$
average values for data being analyzed.)	u	$\bar{u}$	$\bar{u} \pm 3\sqrt{\bar{u}/n}$
Standard Given	p	p'	$p' \pm 3\sqrt{p'(1 - p')/n}$
Used for controlling quality	np	np'	$np' \pm 3\sqrt{np'(1 - p')}$
with respect to given	c	c'	$c' \pm 3\sqrt{c'}$
standard. (p', np', c', u' are given standard values.)	u	u'	$u' \pm 3\sqrt{u'/n}$

control charts for variables.

is calculated first and then divided by the square root of each sample size in turn. A chart of this type is shown in Fig. 32.

Trial Limits for c Charts. For c charts the sample may consist of a single article, a designated area or length of material, or a designated number of articles. In the discussion that follows it is assumed that the samples are of equal size. If the inspection base changes, a u chart should be used (Duncan, Quality Control and Industrial Statistics).

Control charts for defects are constructed by placing limits around $\bar{c}$. The average count of defects per sample is:

$$\bar{c} = \frac{\text{Total defects counted in all samples}}{\text{Total samples inspected}}$$

The limit lines are placed at a distance of $3\sqrt{\bar{c}}$ above and below this central line.

In a plant making ignition cable, a high-voltage test was applied to all cable. A defect chart was kept on the number of insulation failures. The sample was in each case a 5,000-foot reel of cable. In the trial period 20 counts were recorded as follows: 6, 0, 3, 4, 3, 8, 1, 2, 5, 2, 3, 2, 5, 4, 1, 9, 1, 3, 2, 0. The average number of defects per reel was therefore $\frac{64}{20} = 3.2$, and this value was used for the central line. The resulting c chart is pictured in Fig. 24.

REVIEW AND ACTION. After the data for the initial period have been charted, the first step is to observe whether they indicate lack of control. **Lack of control** is associated with any of the following:

1. One or more points above the upper limit.
2. One or more points below the lower limit.
3. Several consecutive points close to one limit.
4. A marked upward or downward trend among successive points.
5. A sustained period in which the number of points above the central line is quite different from the number below.

The **causes of out-of-control samples** should be tracked down, and the process studied. Notes and recollections concerning changes of material, tool setting, gage, operator, etc., are checked in search of explanations. The chance

Product __Sodium bicarbonate__
Operation __Filling 100-lb. bags__
Machine __#4 Valve packer with preweigh with scale; spout #2__

Specified Limits:
Maximum __None__
Minimum __100.0 (±0.7 tare)__
Unit of Measure __0.1 lb.__

Date __1/30__
Dept. __370__
Plant __2__

SAMPLE No.	1	2	3	4	5	6	7	8	9	10	11	12	13
	101.1	101.0	101.0	101.3	101.2	101.3	100.9	101.3	101.5	101.3	101.5	102.0	101.6
	101.3	100.9	101.3	101.4	101.0	102.0	101.6	101.6	101.3	101.5	101.3	101.3	101.1
	101.5	101.2	100.8	100.9	101.6	101.1	101.7	102.0	100.9	101.9	100.8	101.5	101.2
	101.5	101.2	101.4	101.0	100.9	101.6	101.0	101.4	100.8	101.6	101.2	101.4	101.0
	100.8	101.3	101.3	101.5	101.5	101.5	101.9	101.5	101.2	100.9	101.8	101.2	100.9
TOTAL	506.2	505.6	505.8	506.1	506.0	507.5	507.1	507.8	505.7	507.2	506.6	507.4	505.8
X̄	101.2	101.1	101.2	101.2	101.2	101.5	101.4	101.6	101.1	101.4	101.3	101.5	101.2
R	0.7	0.4	0.6	0.6	0.7	0.9	1.0	0.7	0.7	1.0	1.0	0.8	0.7

SAMPLE No.	14	15	16	17	18	19	20	21	22	23	24	25	26
	100.9	101.0	101.9	101.4	101.6	101.0	101.0	101.2	101.8	101.4	101.0	101.6	
	101.2	101.2	102.0	101.2	101.6	101.1	100.3	101.5	101.7	101.3	101.1	101.2	
	101.3	101.3	101.5	100.8	101.3	100.3	100.5	101.6	102.0	101.2	101.5	101.3	
	101.5	100.9	101.5	101.2	101.0	100.4	100.7	101.6	101.6	101.2	101.8	101.5	
	101.6	101.7	101.5	101.7	101.5	100.5	101.1	101.9	101.6	100.8	101.7	101.0	
TOTAL	506.5	506.1	508.4	506.3	506.7	503.2	503.6	507.8	509.1	505.9	507.1	506.6	
X̄	101.3	101.2	101.7	101.2	101.3	100.6	100.7	101.6	101.8	101.2	101.4	101.3	
R	0.7	0.8	0.5	0.9	0.6	0.7	0.9	0.7	0.4	0.6	0.8	0.6	

No.	TOTALS	R
1	506.2	0.7
2	505.6	0.4
3	505.8	0.6
4	506.1	0.6
5	506.0	0.7
6	507.5	0.9
7	507.1	1.0
8	507.8	0.7
9	505.7	0.7
10	507.2	1.0
11	506.6	1.0
12	507.4	0.8
13	505.8	0.7
14	506.5	0.7
15	506.1	0.8
16	508.4	0.5
17	506.3	0.9
18	506.7	0.6
19	503.2	0.7
20	503.6	0.9
21	507.8	0.7
22	509.1	0.4
23	505.9	0.6
24	507.1	0.8
25	506.6	0.6
TOTAL	12,662.1	18.0

$$\bar{\bar{X}} = \frac{12{,}662.1}{125} \qquad \bar{R} = \frac{18.0}{25}$$

$$\bar{\bar{X}} = 101.30 \qquad \bar{R} = 0.72$$

LIMITS FOR CHART OF AVERAGES:

U.C.L. $= \bar{\bar{X}} + A_2\bar{R} = (101.30) + (0.577)(0.72) = 101.72$

L.C.L. $= \bar{\bar{X}} - A_2\bar{R} = (101.30) - (0.577)(0.72) = 100.88$

LIMITS FOR CHART OF RANGES:

U.C.L. $= D_4\bar{R} = (2.115)(0.72) = 1.52$

L.C.L. $= D_3\bar{R} = (0)(0.72) = 0$

REMARKS:
(18) Following previous practice, operator check-weighed one bag and adjusted automatic scale accordingly.
(20) Operator rechecked one bag after adjustment, found weight too low, set scale up.
(23) Operator reweighed one bag, found weight high, set scale down again.

Fig. 22. Data sheet for $\bar{X}$ and R charts.

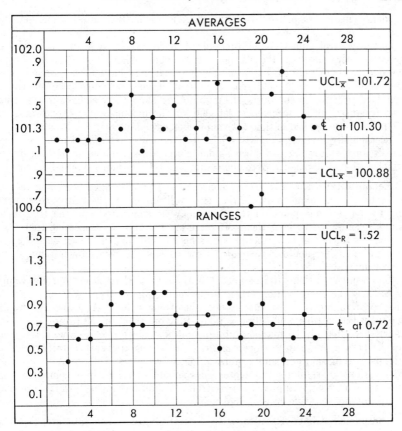

Fig. 23. $\overline{X}$ and R charts for automatic bag loader.

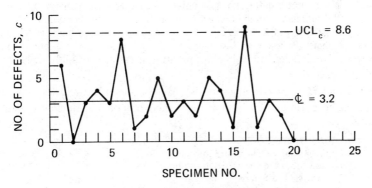

Fig. 24. c chart for cable defects.

of identifying causes of trouble at this stage is obviously greatest while memories are still fresh.

Out-of-control points in the direction of extra-good quality should be investigated as energetically as the others. A new operator may have adopted a better method; a more uniform batch of raw material may have been received. Once the value of any such factor has been seen, it may be possible to change the standard practice to conform. If on the other hand the change was a spurious one, produced by an incorrectly set gage or through inspector carelessness, the chart will have warned in time to avoid customer complaints.

Occasionally investigation will not reveal the cause for an out-of-limits point, and it must simply be put down as one of the three samples in a thousand for which a chance departure from the center line by this much may be expected.

Recomputation of Limits. When causes of trouble have been detected and steps have been taken to assure that they will not recur, the figures corresponding to the relevant point on the chart may be eliminated and the trial-period limits recomputed without them. If there are grounds for suspecting that findable causes for out-of-limits points still remain—and this will often be the case—the initial limits should be extended forward and the periodic sampling should be continued.

The results can thereafter be posted on the control chart soon after any sample is taken. The investigations then being more timely, there will be a greater chance of finding the explanation for out-of-line values. As causes of trouble are identified in turn, and action is taken to avoid them in the future, the points charted will usually be found to group more and more closely around the trend line. There is frequently a decrease in R, reflecting **increased uniformity of pieces** within samples, and counts of defective parts will normally decrease. The various chart limits and center lines should therefore be recomputed from time to time, as the approach to controlled conditions makes available a chart that is a more sensitive decision tool.

Comparing Chart Results with Specifications. The ultimate object of control charts is to secure adequate conformance of product to the design specifications. The control chart is a means, not the end in itself. Thus the quality control engineer must determine not only whether there is control but whether the control is sufficient and at the best level to meet the specification.

In this regard, p and c charts offer the least problem. The question is simply: Is $\bar{p}$ or $\bar{c}$ sufficiently low that the dollar loss through defectives is not excessive? If not, a more detailed study of the process is necessary, possibly including the use of $\bar{X}$ and R charts.

For $\bar{X}$ and R charts the question is: Are the individual articles being made within the **tolerance?** Analysis of the $\bar{X}$ and R data will disclose one of four conditions:

    1. The process is in control
        a. and meets the specification requirements or
        b. does not meet the specification requirements.
    2. The process is out of control
        a. but meets the specification requirements anyway or
        b. does not meet the specification requirements.

If the process is in control, it is easy to forecast its behavior with regard to the specification. The expected **scatter of sizes** of individual parts is estimated from the control chart data. (A symmetrical pattern such as in Fig. 17a or

17b is assumed.) The limits which will include virtually all parts made are stated as:

$$\overline{\overline{X}} \pm 3\overline{R}/d_2$$

with values of d_2 as given in Fig. 20.

Changes To Satisfy Specifications. For the bag-loading operation discussed in connection with Figs. 22 and 23, for example, the limits were revised, and $\overline{\overline{X}}$ was 101.25 and $\overline{R}$ was 0.55 after control was subsequently secured. It could then be forecasted that at prevailing settings individual bag weights would range from $101.25 - 3(0.55)/2.326 = 100.54$ lb. gross weight, to $101.25 + 3(0.55)/2.326 = 101.96$ lb. gross weight.

These limits representing actual performance are then compared with the specification requirements. In the case at hand, allowance for the weight of the empty sack set the lowest permissible gross weight at 100.7 lb. It is evident that some sacks are underweight at the present setting, and an immediate upward readjustment of the scale is needed. Usually such a change of **level** is easily made on a machine.

Sometimes the comparison of actual with intended performance reveals another type of shortcoming not so easy to correct. This is where the range or dispersion of values is too large to satisfy the specification even when the machine is set to give the most favorable average. In the event the **specification cannot be met,** one of the following steps must be taken:

1. Change the job to a more capable machine or operator.
2. Continue as at present but sort the product.
3. Change the specification
   a. to use a material or form more easily made or
   b. to authorize more variation (a wider tolerance).
4. Overhaul or replace machine.

Where the trial limits do not show control, it is not possible to make a firm forecast of future conformance. (Without control over the rudder, one cannot be sure that a ship will hold to its present course in the future.) However, it is common practice even in this case to use the method outlined above, if only to get an idea of what might occur. Depending on the results, it is possible to:

1. Change to a more favorable average level.
2. Improve uniformity by continuing to use the chart until control is obtained.

The procedure described in the preceding sections leads ultimately to control to given standards through the following steps:

1. Set up trial limits.
2. Obtain control.
3. Concurrently, compare actual results with results required by blueprint, etc.
4. Improve level of uniformity if necessary.
5. Finally, adopt the most satisfactory trial central lines and limits as standard values for the future.

Alternate procedures are available under which limits can be derived directly from the specification requirements.

Examples of Statistical Control

AUTOMATIC SCREW MACHINE CAPABILITY. In this example 5,000 bushings (Fig. 25) were scheduled to be made on a four-spindle automatic screw machine. Shortly after the job was started, inspection found that 11 percent of the bushings were outside limits on the 1.000 ± 0.002 diameter. In dis-

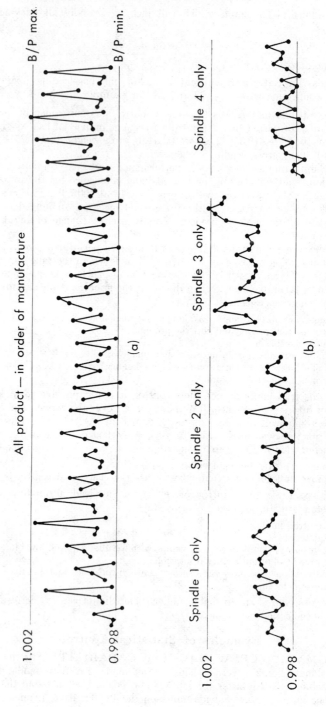

Fig. 25. Capability study on automatic screw machine.

cussing the problem the tool-setter asserted that this was the best that could be done; if this tolerance was to be met, the parts would have to be sorted. The foreman, on the other hand, pointed out that the machine had been used without trouble on ±0.002 work about three months before. The tool-setter retorted that the machine had been in better condition at the time and that inspection had not been so tight.

It was decided to get the facts by means of a **chart of individual measurements.** An inspector was assigned to take bushings in sequence and measure them. In all he measured 100 pieces. A plot of the hundred measurements is shown in Fig. 25(a). At first glance this chart may appear to bear out the tool-setter's contention. However, this is a 4-spindle machine. Bushings 1, 5, 9, 13, etc., are thus made on one spindle; bushings 2, 6, 10, 14, etc., on a second, and so on. Closer examination of the plot shows that pieces coming from spindle No. 3 average about 0.0015 larger than those from spindles 1 and 2. Pieces from spindle No. 4 are usually the smallest in each group of four. To show this more clearly the points have been regrouped, according to source, in Fig. 25(b).

It was immediately evident that an adjustment at spindle No. 3 was necessary to lower the average diameter of bushings made on it. An adjustment at spindle 4 to raise the average diameter of its bushings was also desirable.

A by-product of the study was additional knowledge on **tool wear.** The small upward trend in the diagrams indicates a wear rate of roughly 1/1,000 in. per 200 (total) pieces made. In this short study the product from spindle No. 3 not only had a high average but was the most variable and governed the length of run possible between tool resets. If the machine is initially set with all spindles producing diameters near the lowest dimension permitted, the tool will not have to be reset until about 400 bushings have been turned. If continued study of spindle 3 reduces the variability of its product with that from the other spindles, it may be possible to increase runs without resetting to 700 or 800 bushings.

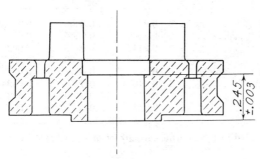

Fig. 26. Molded part for which a capability analysis was applied.

MOLDING MACHINE CAPABILITY. The following discussion by Manuele (Tech Engineering News, vol. 25) suggests useful approaches to control problems:

A good application of statistical analysis is the molded part shown [in Fig. 26]. This part is made in a 16-cavity mold, and considerable difficulty was being experienced in maintaining the 0.245 ± 0.003 in. dimension. Therefore, the first step was to determine whether the part could be made.

Thirty-three pieces were taken from each cavity and carefully measured. From these measurements the average, $\overline{X}$, of the pieces produced by each cavity was calculated. The standard deviation, σ, was also calculated or it can be estimated from $\sigma = \dfrac{\overline{R}}{d_2}$. (See section on Statistical Methods.) These two statistics, for 8 cavities, are shown [in Fig. 27].

Cavity No.	$\overline{X}$	σ	3σ
1	0.2471	0.0009	0.0027
2	0.2535	0.0009	0.0027
3	0.2579	0.0010	0.0030
4	0.2573	0.0006	0.0018
5	0.2559	0.0008	0.0024
6	0.2599	0.0007	0.0021
7	0.2576	0.0005	0.0015
8	0.2542	0.0006	0.0018

Fig. 27. Molding machine capability analysis before correction.

Remembering that according to the normal curve, approximately 99 percent of the parts will be produced within the limits $\overline{X} \pm 3\sigma$, we find that the parts are being produced within the proper tolerances (3σ is smaller than 0.003 in. for each of the cavities), but $\overline{X}$ is too far from the drawing nominal dimension (0.245 in.) in every cavity. This means that the part can be made within the tolerances required, but the individual mold cavities must be adjusted to bring the average of each cavity to 0.245 in.; cavity No. 1 will have to be closed 0.0021 in., cavity No. 2 will have to be closed 0.0085 in., and so on.

Cavity No.	$\overline{X}$	σ	3σ
1	0.2444	0.0006	0.0018
2	0.2446	0.0005	0.0015
3	0.2442	0.0009	0.0027
4	0.2460	0.0009	0.0027
5	0.2454	0.0008	0.0024
6	0.2461	0.0008	0.0024
7	0.2444	0.0009	0.0027
8	0.2442	0.0007	0.0021

Fig. 28. Molding machine capability analysis after correction.

[Figure 28] shows the average and standard deviation for 50 pieces produced in each of 8 cavities after the mold was corrected. This shows that the operation should produce parts which are all satisfactory. A detail inspection of 1,000 parts revealed no rejections; an inspection of 10 percent of a subsequent lot of 6,000 parts also failed to reveal any defective units.

There is a considerable element of approximation in this example because $\overline{X}$ and $(\overline{R}/d_2)$ as calculated here are both random variables, and the sample size of 30 is sufficiently small to be concerned about the statement that $\overline{X} \pm 3(\overline{R}/d_2)$ contains 99 percent of the distribution. In fact, using the values from Table 5 in the Statistical Methods section, the probability is 0.99 that $\overline{X} \pm 3.733(s)$ contains 99 percent of the distribution.

In the example, the estimator $(\overline{R}/d_2)$ is less precise than s. In this light, it is quite possible that the spread for cavities 1, 2, 3 and 5 exceeds that of the specifications; however, after corrections were made and another sample of size 50 was taken, the $\overline{X} \pm 3.4(s)$ interval would contain 99 percent of the distribution with probability 0.99. In this case, it appears that the dispersion is sufficiently small.

WIRE SPRING PRODUCTION. In another case, a fine wire spring was required, as specified in Fig. 29. Note that this specification is rather incomplete, that the right angle with the axis of dimension a is merely implied, that dimension b of the short leg is referred to the somewhat ideal form of the first loop.

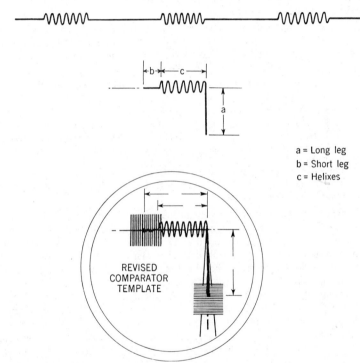

a = Long leg
b = Short leg
c = Helixes

REVISED
COMPARATOR
TEMPLATE

Fig. 29. Spring specification and measurement.

The helixes were first wound intermittently on a small mandrel, with straight-wire connecting links. The connected helixes were then fed off the mandrel one at a time; the projecting long leg of the end helix was bent down and the short leg was cut by a blade positioned from what appeared to be the end loop of the helix. The dimensions were then checked on the template of an optical projection comparator, using hourly samples of four.

As a first step, the historical period was developed in the first 25 samples, plotting points without limits. From this period, average means and ranges were used for computing statistical standards of central lines and control limits.

In the case of short-leg means, the sequence level or grand average was obviously very close to bogey of specification limits, and the specification limit mid-point was therefore used for a central line. This is desirable in such cases. Where the level is appreciably different from bogey, it is better to use a grand mean central line so that **control** can be judged, aside from level, and the more actionable points distinguished from the others. Otherwise, all points would call for action where there may be very few if any causes to investigate, except those creating a systematic displacement of **level**.

The statistical standards of central lines and central limits, computed as above stated, were projected into the then future period of another 25 points, in Fig 30. It is usually held more desirable to have up to 25 points in the historical or qualifying period before projecting a statistical standard. However, in the earlier stages of a problem, a shorter sequence is permissible. As each sample was taken thereafter, the mean and range were easily calculated and plotted on a chart hung near the operator.

Conditions Revealed by Study. When the first statistical standards were computed and also projected back over the past 25 points of the historical period for initial judgment of the process, several conditions were suggested, as follows:

1. The short-leg means were at or near the correct level of specification mid-point, but displayed slight trouble in one point outside limits and in a tendency to trend in points 4 to 10, inclusive.
2. The short-leg ranges were in control but at too high a level: $\overline{R} = 0.0136$, which indicated a dispersion of individual values that would have high scrap.
3. Long-leg means are at a level (0.1441) far below the desired one and show lack of control by a point outside limits and considerable trend in points 4 to 11, inclusive.
4. Long-leg ranges are reasonably in control with an average range of only 0.0333. This suggests possibility of a low rate of scrap if control of means were obtained at proper level.

Corrective Action. The systematically low level of long-leg means was at first puzzling, since a supplementary control investigation of the initial coil-forming operation demonstrated that the **average** spacing between coils on the mandrel was correct. Upon examination of the projection-comparator inspection, it was immediately apparent that the long legs were being bent in an inconsistent manner to angles varying from 45° to 80° instead of 90°. The inspector was measuring only the leg-tip distance perpendicular to the axis of the helix. A tool fixture was designed to remove this effect; i.e., the process was redesigned. Also, both product and inspection specifications were made more objective by requiring angle limits on the long leg, which were in turn placed on the comparator template. Here an assignable cause due to poor specification was removed. Thus, the long-leg systematic difference of means from bogey would be corrected, and with the improvement in the process the degree of variability (level of range) would be reduced. The erratic condition indicated on the chart of means was found to be due partly to spotty wire, which affected the degree of variability in spacing by the coil winder, and partly to inconsistency in operator application. The former factor was worked upon and improved. The latter was due to skill requirements on fine work at the limit of controlled human performance. Use of fixtures and other aids improved the condition.

In the case of the short legs, the cutting operator did not have a well-defined loop from which to reference the cutting blade, and the inspector likewise had difficulty in referencing the measurement. In processing, the end loops were smaller and in some cases not clearly definable. Specification dimensioning was revised, in terms of actual functioning of the part, referencing all dimensions from the apex of the long-leg tolerance angle limits.

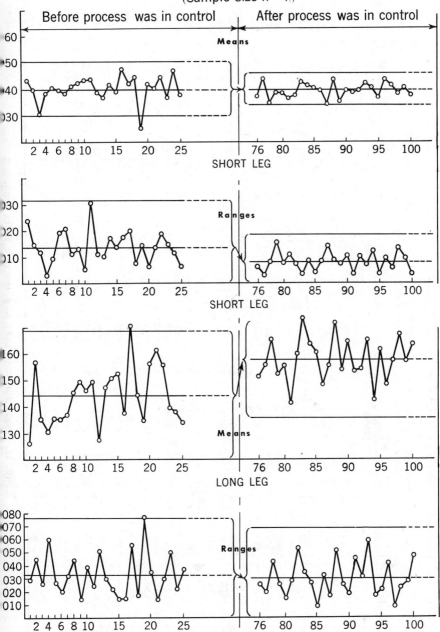

Fig. 30. Control charts used to highlight assignable causes of variation.

PRODUCT: Control Device   DRAWING NO.: D11807   DEPARTMENT: 870·6

AMOUNT INSPECTED: 100%   INSPECTION LOG REFERENCES:

Lot No.	Lot Size N	Gross Def.		A (1 Item)		B (1 Item)		C (6 Items)		D (5 Items)		E (8 Items)		F (Others)	
		r	Fract.	r	Fract.	r	Fract.	r	Fract.	r	Fract.	r	Fract.	r	Fract.
1	2,999	599	.200	103	.0344	131	.0437	73	.0244	67	.0224	49	.0163	176	.0588
2	3,237	537	.166	98	.0303	98	.0219	49	.0151	83	.0256	38	.0117	171	.0512
3	3,156	457	.144	101	.0320	36	.0114	66	.0209	99	.0314	23	.0073	132	.0418
4	3,423	623	.182	156	.0455	62	.0181	68	.0198	142	.0415	45	.0131	150	.0438
5	3,727	727	.195	146	.0391	63	.0169	94	.0252	154	.0413	55	.0147	215	.0577
6	4,263	763	.179	168	.0394	43	.0101	135	.0317	138	.0324	71	.0167	208	.0488
7	3,687	487	.132	80	.0217	38	.0103	41	.0111	79	.0214	24	.0065	225	.0610
8	2,881	676	.234	235	.0815	23	.0080	58	.0201	77	.0267	99	.0344	184	.0638
9	2,262	562	.248	211	.0931	39	.0172	65	.0287	132	.0582	18	.0079	97	.0428
10	2,922	697	.248	234	.0799	94	.0321	96	.0328	90	.0308	31	.0106	152	.0520
11	3,785	875	.231	240	.0634	138	.0364	93	.0246	95	.0251	104	.0274	205	.0542
12	4,814	914	.190	207	.0430	104	.0215	142	.0294	111	.0230	67	.0319	283	.0588
13	2,159	359	.166	84	.0390	44	.0204	55	.0255	50	.0232	20	.0043	106	.0515
14	3,089	484	.157	56	.0189	98	.0317	54	.0175	55	.0178	34	.0110	187	.0605
15	3,156	616	.175	187	.0532	46	.0139	62	.0177	80	.0228	51	.0145	190	.0602
16	2,139	434	.203	110	.0515	36	.0168	45	.0210	56	.0262	12	.0056	175	.0818
17	2,139	503	.194	93	.0359	84	.0324	42	.0162	109	.0421	37	.0143	138	.0533
18	2,510	487	.194	98	.0390	45	.0179	57	.0227	59	.0235	51	.0203	177	.0705
19	4,103	803	.195	197	.0480	121	.0295	92	.0224	150	.0365	39	.0095	204	.0497
20	2,992	547	.183	163	.0545	45	.0150	79	.0264	51	.0170	26	.0087	183	.0612
21	3,545	555	.156	107	.0302	57	.0160	80	.0226	60	.0236	21	.0070	230	.0649
22	1,841	401	.218	65	.0353	40	.0217	27	.0147	132	.0716	18	.0098	119	.0646
23	2,748	418	.156	115	.0419	26	.0095	78	.0284	36	.0131	15	.0054	148	.0539
24	3,924	667	.170	141	.0351	59	.0150	144	.0367	55	.0140	24	.0061	244	.0622
25	2,056	319	.155	82	.0399	29	.0141	21	.0102	20	.0097	16	.0078	151	.0734
26	3,650	474	.130	74	.0203	65	.0178	80	.0219	80	.0219	51	.0140	124	.0340
27	4,001	535	.134	144	.0360	28	.0070	40	.0100	96	.0240	36	.0090	191	.0477
28	2,950	379	.128	88	.0298	38	.0129	74	.0251	50	.0169	24	.0081	105	.0356
29	3,162	550	.173	158	.0499	54	.0171	82	.0259	82	.0259	35	.0110	139	.0439
30	3,827	483	.126	134	.0350	42	.0110	69	.0180	58	.0152	24	.0063	156	.0408

Fig. 31. Inspection data used in a fraction defective chart.

After these steps, i.e., correcting for assignable causes in specifications, operations, and inspection, considerable improvement was obtained. The charts for a
third period, points 76 to 100 inclusive, are shown in Fig. 30. Control was attained at proper levels. The degree of variability was reduced to a point where
very few units of product were outside specification limits. Control inspection
was here used for acceptance inspection as well, and because of ultimate stability of the process, sampling was stretched out to one set of four in 4 hr. for
each operator.

USE OF EXISTING INSPECTION DATA. An illustration of control
from existing work inspection records demonstrates that it is not always necessary to collect new or additional data in order to apply these control techniques.
Frequently, usable data are available and only require interpretation. The data
in Fig. 31 have been summarized from the inspection records of a precision control device. The gross number of devices defective is enumerated for each lot,
together with the fraction it represents. Items A through E are groups of individual characteristics of similar nature for which the group numbers and fractions defective are given. Item F represents all remaining characteristics of the
breakdown.

Control charts prepared for these items are shown in Fig. 32. The central
lines and control limits for the second 15-lot period are based upon the historical
record of the first period. These were projected at the end of the first period to
judge the results of the second, as they appear. Since the sample (lot) size was
quite variable, it was necessary to calculate separate control limits for each
group of samples, according to the relative sample size.

It is apparent that despite the fact that some points in the total fraction chart
do not appear actionable, e.g., points 16 through 19, reference to component
item charts indicates that action is feasible in one direction or another. In this
case consultation of individual entries on the inspector's breakdown led to identification of causes upon which work was done. In other cases additional studies
by variables were necessary to trace removable causes of variability.

It is seen that improvement results during the second period. Therefore, for
the third period beginning with point 31, a new set of central lines and control
limits was computed in most of the component groups and, by summation of
their central line fractions, for the total fraction chart.

Later developments of this problem brought a high degree of control with an
ultimate average fraction defective of about 6 percent in contrast to the original
17 percent. Further reduction of level could be accomplished only by redesign
of a few of the processes for parts, since the residual variability in these cases
was too great for the required specification limits.

PROBLEMS WITH MULTIPLE MACHINES. The preceding illus-
trations did not refer to systematic differences between parallel machines and
their effects upon variability in the common product. In this case, parallel units
(heads in a rotary feed machine) are producing seven-pin bases for computer
plug-in components. Specifications call for the pins to protrude 0.230 in. to
0.260 in. from the plastic base.

The machine producing these parts has 16 heads. Each head indexes around
the machine through all the production and assembly processes. The exposed
pin length is uncontrolled in the common flow of parts to inspection. Thus a
substantial fraction of the parts fell outside specification limits.

Causes of Variations. Possible assignable causes were: improper positioning
of pins in head pockets, misadjustment of heads, improper operation of cams,

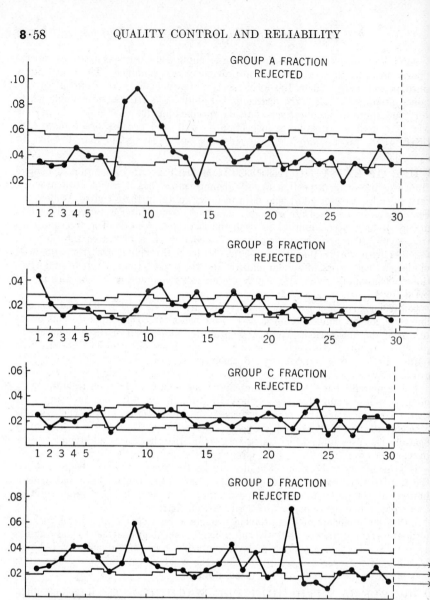

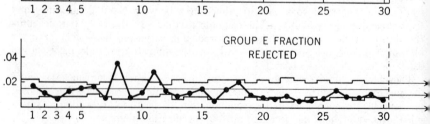

Fig. 32. Fraction defective (*p*) charts prepared from data in Fig. 31.

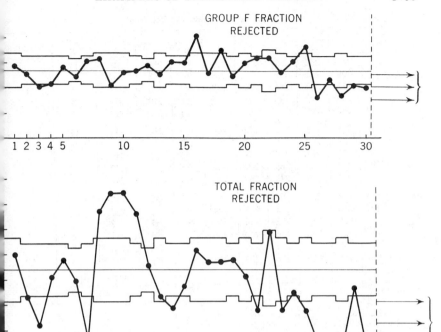

Fig. 32. Concluded.

rollers and linkages in head movements, etc. Since assignable differences were suspected to exist between heads and between each head and specification bogey, one base assembly (consisting of seven pins) was taken each hour from each of the 16 heads. A data sheet indicating the pin length measurements taken during one time period is shown in Fig. 33. Control charts were prepared for means of seven leads for each head. The mean range was computed over a representative chronological period for each head. Control charts for means and ranges were posted on the production floor.

It was soon apparent to the maintenance men that some heads were averaging high, others low. Successive approximations of adjustment quickly brought the heads to average on bogey. In addition, the charts provided prompt indications of the entrance of occasional assignable causes, such as glass particles and fouling in well pockets. The charts for ranges not only indicated such failure in individual pockets, but through improper levels (average range, $\overline{R}$) showed which heads needed cleaning and equalization of the pocket depths by the toolmakers.

Improvements Made. In a relatively short period the control of exposed pin length was improved to a point where extremely few pins were outside the limits; and it was planned to drop 100 percent acceptance inspection in favor of patrol sampling inspection when the stability of control was established. Control charts were continued for supervision, upon the insistence of the production maintenance department.

(Values in Thousandths of One Inch)

Head	Pin Numbers							Mean $\bar{X}$	Range R	Remarks
	1	2	3	4	5	6	7			
1	246	252	252	249	252	249	238	248	14	
2	255	263	265	264	265	265	258	262	10	
3	251	253	251	247	248	247	250	249	6	(1)
4	235	222	246	246	238	250	240	239	28	
5	256	265	261	262	264	261	257	261	9	
6	241	239	240	242	246	248	234	241	14	(2)
7	257	260	260	257	258	255	255	257	5	
8	258	263	260	261	259	260	257	259	4	
9	264	260	265	260	254	256	259	259	11	
10	243	250	245	252	250	239	243	246	8	
11	251	253	254	255	255	251	247	252	8	
12	240	244	234	238	244	247	248	242	14	(3)
13	250	253	252	254	255	252	247	252	8	
14	237	236	236	230	236	238	233	235	8	(4)
15	248	251	254	253	253	255	255	253	7	
16	256	256	258	258	258	257	255	254	3	(3)
Total								4,009	157	
Average								250.6	9.8	

(1) Fouling reported by operator.
(2) Head weight recently adjusted.
(3) Recurrent trouble with cam mechanism on this head.
(4) Recently adjusted.

Fig. 33. Data sheet for control inspection of exposed pin lengths.

Acceptance Sampling

SAMPLING. Sampling for purposes of acceptance, rejection, or other action upon specific lots of material has been an alternative to 100 percent inspection. It can be applied to purchased parts and materials to semifinished items in or passing between different departments or plants of a company, or to final inspection of finished items ready to go to the consumer.

Advantages of Sampling. The benefits of sampling center on the ability to obtain decisions of calculable reliability at least cost. The advantages are:

1. Facts needed for choice of a sampling plan require a conscious consideration of actual quality needs.
2. Less man-hours are alloted to sorting the bad parts from the good so more inspection effort can be applied to prevention of poor quality at its source.
3. More separate quality characteristics can be regularly inspected, as the cost of reaching a decision on any one characteristic is low.
4. A small inspection staff can make decisions on a large volume of material.
5. As the lot size or production rate rises or falls, it is easily possible to adjust the proportion inspected in such a way as to maintain a given assurance of quality.
6. Rejection of an entire lot of material (rather than defectives only) applies strong enough pressure to the supplier to encourage reform.
7. The inspector's care, feeling of importance, and responsibility are increased by the knowledge that his findings from the sample are the basis for an important decision on an entire shipment of material.

8. Sampling inspectors make fewer errors due to monotony, since they usually check fewer parts of the same type before changing to a different measurement or blueprint.
9. Sampling can be used to obtain assurance of quality even where only destructive tests are available.
10. The cost of damage incidental to inspection is reduced because less parts are handled.

Limitations of 100 Percent Inspection. Obviously, 100 percent inspection cannot be applied where tensile strength, life, corrosion resistance, fading, or the like must be measured by destructive tests. Moreover, careful experiments by the Western Electric Company, the Ford Motor Company, and others have shown that inspectors performing 100 percent inspection do not locate all the faulty parts that are present. In general they find only 70 percent to 95 percent of the defects. The greatest difficulty is with large lots containing only small percentages of defective parts. If defective parts are once produced, it is clear that several inspections will have to be made subsequently if defectives are to be reduced to a really low number.

It is true that the most careful inspection of part of a lot still leaves some doubt as to what would be found by inspecting the whole. However, the risk of being misled here can be calculated. With sampling plans an attempt is made to face the difficulties and measure the risks involved. Also, as indicated above, more inspector accuracy is obtained with sampling inspection.

Neither sampling nor screening will guarantee that material accepted is completely free of defects. The best way known to assure 100 percent conformance is to concentrate on control of the manufacturing process. The use of sampling as a measure of results permits a transfer of effort from sorting to prevention of defects.

ACCEPTANCE SAMPLING BY ATTRIBUTES. Where large numbers of parts are involved, the most general type of inspection used in industry is that known as **sampling by attributes.** In this system the inspector merely notes whether the part is accepted or rejected. Actual measurements are not taken, or if they are observed, they are often not recorded. Most acceptance sampling plans use attributes since such data are commonly available. Once a plan has been computed and tabulated, its further use is simple and altogether nonmathematical. This is seldom possible with acceptance plans using measurements.

The earliest published sampling plans for use in civilian industry were jointly developed by Dodge and Romig of the Bell Telephone Laboratories (Dodge and Romig, Sampling Inspection Tables). Other plans originally devised to meet special needs in procurement of material for the armed forces are also in use for civilian goods as well as military. The best known of these is the MIL-STD-105D.

Single, Double, and Multiple Sampling. Acceptance sampling systems may be classified according to the number of samples taken. In single-sampling plans, only one set of specimens is drawn from the lot. The **single-sampling procedure** may be outlined as:

1. Collect and inspect a random sample of n pieces.
2. If the number of defects found is not more than the acceptance number given by the plan, accept the entire lot.
3. If the number of defects found is more than the acceptance number given by the plan, inspect all remaining pieces in the lot, or reject the entire lot.
4. Replace or repair any defective pieces found.

In double-sampling plans a smaller initial set of specimens is taken. If th quality is either very good or very poor, a decision is reached on the basis c this sample alone, with a saving in the amount of inspection. A second set c specimens is taken only if the first sample shows a borderline quality. Doubl sampling usually requires less total inspection than single sampling, by amount ranging from 10 percent to 50 percent. The **procedure in double samplin** is:

1. Collect and inspect a random first sample of n_1 pieces.
2. If the number of defects found is not more than the acceptance number fo the first sample, accept the entire lot.
3. If the number of defects found is equal to or more than the rejection numbe for the first sample, inspect all remaining pieces in the entire lot, or reject th entire lot.
4. If the number of defects found falls between the acceptance and rejectio numbers for the first sample, collect and inspect a random second sample c n_2 pieces.
5. If the total number of defects found in the first and second samples togethe is not more than the acceptance number given for the second sample, accep the entire lot.
6. If the total number of defects found in the first and second samples togethe is equal to or more than the rejection number, inspect all remaining piece in the entire lot or reject the entire lot.
7. Replace or repair any defective pieces found.

In multiple-sampling plans the same line of reasoning is carried further. Pro vision is made for taking a number of samples if necessary to reach a decision For simplicity each of these samples usually has the same size, and this sampl size is very small as compared with the corresponding single-sampling plan The average number of pieces per lot inspected is usually less than required by double sampling by amounts of the order of 30 percent. Balanced against this however, are the costs of administration and training, the necessity to reopen o: rehandle the lot for each additional sample, and the more complex records. The **procedure in multiple-sampling plans** is similar to the one outlined fo: double-sampling plans. Three additional steps equivalent to 4, 5, and 6 above are added for each further sample taken.

Operating Characteristic (OC) Curves. A sampling plan is simply a set of instructions for an inspector. The protection offered by the sampling plan is established by specifying the lot size N, for which it is to be used; the sample size n, to be inspected; and the number of faulty pieces which will cause ac ceptance or rejection of the lot. From this information it is possible to predict the results that will be be obtained when the plan is used in practice. Specifi cally, it is possible to measure the risk that a wrong decision will be made.

Such predictions are made from the **operating-characteristic (OC) curve** of the sampling plan. A typical curve is given in Fig. 34. This particular curve shows the performance that would be obtained if an inspector were instructed to take from a very large lot a sample of 300 pieces and to accept the lot if he found no more than 5 defective pieces. The curve gives the probability that a lot of any stated quality will be accepted if submitted for sampling. For ex ample, if a "good" lot—containing only 1 percent of defective parts—is sub mitted, the probability is 0.92 that it will be accepted. On the other hand, if a "bad" lot—containing 3 percent of defective parts—is submitted, the probability is only 0.12 that it will be accepted.

In general no two sampling plans have identical curves, although their charac teristics may sometimes be very similar. The practical difficulty of printing,

mparing, and using hundreds of curves to show the behaviors of available plans
as led to tabulation of the plans by their responses to a few particularly im-
ortant conditions. The key figures are known as the lot tolerance, the AQL,
nd the AOQL. The first two have to do with the probability of accepting
ated kinds of incoming material. The last relates to the quality of material
at has passed the hurdle of inspection. All are expressed as percentages of
efective parts.

Lot Tolerance Percent Defective (LTPD). Lot tolerance percent de-
ctive is the lowest percent of defectives which a plan is intended to **reject**
gularly. The risk of accepting occasional lots having this **unwanted** quality is
lled the "consumer's risk." In Dodge-Romig plans this risk is arbitrarily set
0.10. Thus, as illustrated in Fig. 34, the lot tolerance is an incoming percent
efective (3.1 percent) which would be rejected 90 percent (accepted 10 per-

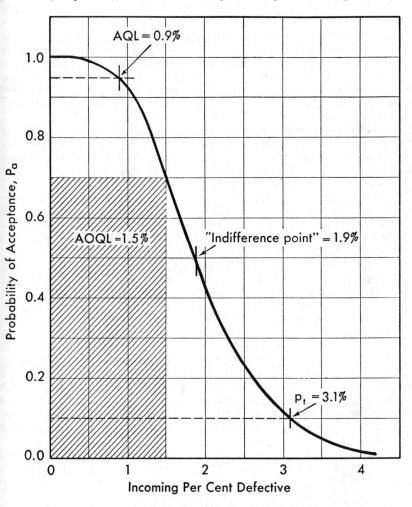

Fig. 34. Typical operating characteristic curve with key points identified.

cent) of the time. The actual quality of accepted material will very rarely b anywhere near this low in the long run. First, a supplier tends to submit lot of varying quality, and the better lots are more likely to be accepted than th poorer ones. Second, lots which are rejected are normally sorted and accepte after the bad parts have been replaced with good. With MIL-STD-105D it i intended that rejected lots are returned to the vendor.

Acceptable Quality Level (AQL). The acceptable quality level is th highest percentage of defectives which the plan is intended to **accept** regularly It may be thought of as the converse of the lot tolerance. To be completel definite, it must be associated with a specific risk that lots having this wantec quality may occasionally be rejected. This is called the "producer's risk." I most plans this risk is set at or near 0.05. Thus, as illustrated in Fig. 34, th AQL is an incoming percent defective (0.9 percent) which would be acceptec 95 percent (rejected 5 percent) of the time. The AQL is most commonly usec in describing military plans.

Average Outgoing Quality Limit (AOQL). The average outgoing qualit limit is the highest percentage of defectives that will be found, on the average, i the flow of product after inspection. The consumer can be assured that this i the worst average fraction of faulty parts he will receive in the long run, n matter what percentage is in the incoming product before inspection. He ma reasonably expect that the actual quality will often be better than the AOQL since this is the limit reached only under adverse conditions.

The AOQL has meaning only when control is exercised over the disposition o rejected lots. In Dodge-Romig AOQL plans it is assumed that any rejected lot: are completely screened and all bad parts are replaced with good. The **outgoing quality** is viewed as a predictable mixture of these perfect lots with the knowr small number of defectives in lots which the plan accepts.

Final inspection and material-in-process inspection usually furnish the con trol needed. The conditions for control may or may not be met in inspection o purchased materials and parts. If rejected lots are returned to the supplier fo sorting and correction, results are doubtful, and plans should be selected on the basis of AQL or LTPD instead.

USE OF STANDARD SAMPLING TABLES. The application of sam pling plans to acceptance inspection can be reduced to a routine procedure in volving no mathematics. The important steps are common to all types of plans.

First, the quality standards (Step 1) must be set. This is normally done by an inspection or quality control executive who can bring together information on cost and value of quality from sales, accounting, production, etc. In some cases the standards are directly imposed by the customer, especially if the gov ernment is the customer. The detailed sampling procedure (Step 2) may be established by the same executive. If routine, it may be established by an in spection supervisor. The actual sampling operation (Step 3) is carried out by inspectors in receiving inspection or between production operations. The subse quent review of results (Step 4) is normally made by the inspection supervisor.

The following **sampling checklist** shows in more detail the questions that must be dealt with in several stages:

1. What degree of conformance is to be set as standard?
   What will constitute an item to be inspected?
   One part?
   A set of parts normally sold or used together?
   What stage of manufacture or receiving inspection?
   What defects are to be looked for?

Are these all of the same severity?

Can more minor than major defects be tolerated?

Will different standards of conformance then apply?

What quality level (AOQL or AQL or LTPD) is appropriate for each class of defect?

Does this defect class involve

Risk to human life?

Risk of injury?

Certain failure of product?

Damage which cannot be repaired in the field?

Shortened life of product?

Damage cheaply repaired?

A flaw that will be revealed in a later stage of manufacture?

Mere annoyance with appearance?

What is the existing record of defects in this factory?

What is the comparable record for competitors? For other vendors?

What is the cost of nonconformance?

How many customer complaints on this defect?

What is the cost of customer returns?

What is the cost of repairs or reworks?

Are extra production operations made necessary?

What is chance and cost of improvement?

Has real effort been made yet?

Is the production process likely to be in control?

Are better methods known to be available?

2. What sampling procedure is to be installed?

How large a lot can be brought together for sampling?

Will such a lot be homogeneous in quality?

If not, can it be subdivided into homogeneous sublots based on date of receipt, time of manufacture, source, or operator?

How is the random sample to be taken?

Will single, double, or multiple sampling be most desirable?

What plan for this is given in the tables?

What data sheets or instructions are needed for the inspector?

3. In operation, which lots are to be accepted?

How many defective items does the sample contain?

Is this a passable number for the defect in question?

Where are the various defects to be listed in the records?

4. What results are being secured?

What process average does sampling show?

Is a different plan indicated for future lots?

Is the process average better than expected?

Worse than expected?

Dodge-Romig Tables. Dodge and Romig (Sampling Inspection Tables) have developed an extensive and valuable set of ready-made tables. Four separate groups of plans are listed:

1. Single-sampling lot-tolerance tables.
2. Single-sampling AOQL tables.
3. Double-sampling lot-tolerance tables.
4. Double-sampling AOQL tables.

The first and third groups of plans are tabulated according to lot tolerance per-cent defective (LTPD) with a consumer's risk of 0.10. The second and fourth groups of plans are tabulated according to the average outgoing quality limit (AOQL) which they assure. Of the four kinds of plans, those of the AOQL double-sampling type have found the widest acceptance. In all cases, the plans have been designed to minimize total inspection, including screening.

Single Sampling (Lot Tolerance Percent Defective = 3%)

Process Average %	0–0.03			0.04–0.30			0.31–0.60			0.61–0.90			0.91–1.20			1.21–1.50		
Lot Size	n	c	AOQL %	n	c	AOQL %	n	c	AOQL %	n	c	AOQL %	n	c	AOQL %	n	c	AOQL %
1–40	All	0	0	All	0	0	All	0	0	All	0	0	All	0	0	All	0	0
41–55	40	0	0.18	40	0	0.18	40	0	0.18	40	0	0.18	40	0	0.18	40	0	0.18
56–100	55	0	.30	55	0	.30	55	0	.30	55	0	.30	55	0	.30	55	0	.30
101–200	65	0	.38	65	0	.38	65	0	.38	65	0	.38	65	0	.38	65	0	.38
201–300	70	0	.40	70	0	.40	70	0	.40	110	1	.48	110	1	.48	110	1	.48
301–400	70	0	.43	70	0	.43	115	1	.52	115	1	.52	115	1	.52	155	2	.54
401–500	70	0	.45	70	0	.45	120	1	.53	120	1	.53	160	2	.58	160	2	.58
501–600	75	0	.43	75	0	.43	120	1	.56	160	2	.63	160	2	.63	200	3	.65
601–800	75	0	.44	125	1	.57	125	1	.57	165	2	.66	205	3	.71	240	4	.74
801–1,000	75	0	.45	125	1	.59	170	2	.67	210	3	.73	250	4	.76	290	5	.78
1,001–2,000	75	0	.47	130	1	.60	175	2	.72	260	4	.85	300	5	.90	380	7	.95
2,001–3,000	75	0	.48	130	1	.62	220	3	.82	300	5	.95	385	7	1.0	460	9	1.1
3,001–4,000	130	1	.63	175	2	.75	220	3	.84	305	5	.96	425	8	1.1	540	11	1.2
4,001–5,000	130	1	.63	175	2	.63	260	4	.91	345	6	1.0	465	9	1.1	620	13	1.2
5,001–7,000	130	1	.63	175	2	.63	265	4	.92	390	7	1.1	505	10	1.2	700	15	1.3
7,001–10,000	130	1	.64	175	2	.64	265	4	.93	390	7	1.1	550	11	1.2	775	17	1.4
10,001–20,000	130	1	.64	175	2	.64	305	5	1.0	430	8	1.2	630	13	1.3	900	20	1.5
20,001–50,000	130	1	.65	225	3	.65	350	6	1.1	520	10	1.2	750	16	1.4	1,090	25	1.6
50,001–100,000	130	1	.65	265	4	.96	390	7	1.1	590	12	1.3	830	18	1.5	1,215	28	1.6

n = size of sample; entry of "All" indicates that each piece in lot is to be inspected.
c = allowable defects for sample (acceptance number).
AOQL = average outgoing quality limit.

Fig. 35. Typical Dodge–Romig table for a single sampling plan.

Lot Tolerance Plans. A constant low consumer's risk is stressed by lot tolerance plans. They give considerable assurance that individual substandard lots will be rejected. Plans from the lot tolerance tables are particularly applicable in cases where lots retain their identity after inspection, as when a lot is shipped as inspected to a consumer in a specific transaction. **AOQL plans** stress the limit on poor quality in the long run but do not maintain uniform assurance that individual low-quality lots will be rejected. Plans from the AOQL tables are particularly applicable when lots are merged in a common supply whose average performance is of interest, as in subsequent operations within the plant or in cases of continual deliveries on large quantity orders. A typical single-sampling lot-tolerance table is shown in Fig. 35. All of the plans in this table afford the same quality protection as measured by the lot tolerance. Plans appropriate to a wide variety of lot sizes are given.

Under acceptance sampling there are two sources of inspection work. For each lot the sample of n items must be inspected. In addition, whenever a rejection occurs, the remaining $N-n$ items in the lot must also be inspected. If the incoming quality is known, it is possible to forecast the total amount of inspection that will result with a given plan. This leads to the possibility of choosing from among several available plans having the same p_t as the one which will involve the least total amount of inspection.

Dodge and Romig have made the extensive calculations needed for this choice. Thus each of the six columns in Fig. 35 lists plans for a specified average value of incoming quality. For example, if the item to be inspected has averaged 0.5 percent defective in the past and no change is foreseen, the least total amount of inspection can be anticipated if a plan is chosen from the column headed 0.31–0.60.

AOQL Plans. Fig. 36 shows a typical double-sampling AOQL table. Once again a whole series of sampling plans is provided to suit varying needs as to lot size and past process average fraction defective. These plans differ considerably from each other as to lot tolerance but have the same AOQL.

MILITARY-STANDARD TABLES. The U.S. Armed Forces have adopted standard tables and procedures for single, double, and multiple sampling. These military-standard plans are designated MIL-STD-105D, Sampling Procedures and Tables for Inspection by Attributes.

Attribute measurement is employed throughout the standard with nonconformance of product expressed either in terms of percent defective or in terms of defects per hundred units.

$$\text{Percent defective} = \frac{\text{No. of defectives}}{\text{No. units inspected}} \times 100$$

$$\text{Defects per hundred units} = \frac{\text{No. defects}}{\text{No. units inspected}} \times 100$$

MIL-STD-105D Terminology. The **process average** is the average percent defective or average number of defects per hundred units (whichever is applicable) of product submitted by the supplier for original inspection. **Original inspection** is the first inspection of a particular quantity of product as distinguished from the inspection of product which has been resubmitted after prior rejection.

Average Outgoing Quality (AOQ) is the average quality of outgoing product including all accepted lots or batches, plus all rejected lots or batches

4.01–6.00						6.01–8.00						8.01–10.00					
Trial 1		Trial 2			LTPD %	Trial 1		Trial 2			LTPD %	Trial 1		Trial 2			LTPD %
n_1	c_1	n_2	$n_1 + n_2$	c_2		n_1	c_1	n_2	$n_1 + n_2$	c_2		n_1	c_1	n_2	$n_1 + n_2$	c_2	
All	0	–	–	–	–	All	0	–	–	–	–	All	0	–	–	–	–
3	0	–	–	–	50.0	3	0	–	–	–	50.0	3	0	–	–	–	50.0
6	0	6	12	2	48.0	6	0	6	12	2	48.0	6	0	6	12	2	48.0
7	0	11	18	3	38.5	7	0	11	18	3	38.5	7	0	16	23	4	36.5
8	0	16	24	4	35.5	13	1	20	33	6	33.5	14	1	24	38	7	32.0
8	0	17	25	4	35.0	14	1	26	40	7	31.5	19	2	29	48	9	31.0
8	0	22	30	5	34.0	15	1	30	45	8	31.0	21	2	44	65	12	29.0
15	1	23	38	6	30.5	16	1	39	55	9	28.5	22	2	53	75	13	27.0
16	1	28	44	7	28.5	22	2	38	60	10	27.5	28	3	52	80	14	26.5
16	1	28	44	7	29.0	22	2	43	65	11	27.0	29	3	56	85	15	26.0
16	1	34	50	8	28.0	24	2	56	80	13	25.5	36	4	69	105	18	24.5
17	1	38	55	9	27.5	24	2	61	85	14	25.0	45	5	95	140	23	23.0
17	1	48	65	10	26.0	33	3	72	105	16	23.0	50	6	115	165	27	22.0
24	2	46	70	11	25.0	41	4	99	140	21	21.5	70	8	150	220	34	20.5
26	2	54	80	12	23.5	44	4	111	155	22	20.0	80	9	195	275	41	19.0
27	2	63	90	13	22.5	50	5	120	170	24	19.5	90	10	240	330	47	18.0
27	2	68	95	14	22.0	60	6	145	205	28	18.5	110	12	265	375	53	17.5
28	2	77	105	15	22.0	70	7	165	235	32	18.0	125	14	320	445	62	17.0
28	2	87	115	17	21.5	80	8	205	285	39	17.5	140	16	355	495	69	16.8
36	3	99	135	20	21.0	85	8	245	330	44	17.0	150	17	390	540	77	16.6

$n_1 =$ size of first sample.
$n_2 =$ size of second sample. Entry of "All" indicates that each piece in lot is to be inspected. The second column under Trial 2 in each case equals $n_1 + n_2$.

Fig. 36. Typical Dodge–Romig

after the rejected lots or batches have been effectively 100 percent inspected and all defectives replaced.

The **Average Outgoing Quality Limit (AOQL)** is the maximum of the AOQs for all possible incoming qualities for a given acceptance sampling plan AOQL values are given in MIL-STD-105D for each of the single sampling plans for normal inspection and tightened inspection.

Average sample size curves for double and multiple sampling show the average sample sizes which may be expected to occur under the various sampling plans for a given process quality. The curves assume no curtailment of inspection and are approximate to the extent that they are based upon the Poisson distribution, and assumed sample sizes for double and multiple sampling of $0.631n$ and $0.25n$ respectively, where n is the equivalent single sample size.

The sampling plans and associated procedures given in the standard were designed for use where the units of product are produced in a continuing series of lots or batches over a period of time. However, if the lot or batch is of an isolated nature, it is desirable to limit the selection of sampling plans to those, associated with a designated AQL value, that provide not less than a specified limiting quality protection. Sampling plans for this purpose can be selected by choosing a **Limiting Quality (LQ)** and a consumer's risk to be associated with it. Tables in the standard give values of LQ for the commonly used consumer's

Process Average % Lot Size	0–0.20 Trial 1 n_1	c_1	Trial 2 n_1+n_2	n_2	c_2	LTPD %	0.21–2.00 Trial 1 n_1	c_1	Trial 2 n_1+n_2	n_2	c_2	LTPD %	2.01–4.00 Trial 1 n_1	c_1	Trial 2 n_1+n_2	n_2	c_2	LTPD %
1–3	All	0	–	–	–	–	All	0	–	–	–	–	All	0	–	–	–	–
4–15	3	0	–	–	–	50.0	3	0	–	–	–	50.0	3	0	–	–	–	50.0
16–50	5	0	3	8	1	53.5	5	0	3	8	1	53.5	5	0	3	8	1	53.5
51–100	5	0	3	8	1	55.0	6	0	8	14	2	43.0	6	0	8	14	2	43.0
101–200	5	0	4	9	1	52.0	7	0	7	14	2	42.0	7	0	12	19	3	38.0
201–300	7	0	7	14	2	42.5	7	0	7	14	2	42.5	7	0	13	20	3	37.0
301–400	7	0	7	14	2	42.5	7	0	7	14	2	42.5	8	0	17	25	4	35.0
401–500	7	0	8	15	2	40.0	7	0	8	15	2	40.0	8	0	18	26	4	34.0
501–600	7	0	8	15	2	40.0	8	0	13	21	3	35.0	8	0	18	26	4	34.0
601–800	7	0	8	15	2	40.5	8	0	13	21	3	35.0	8	0	18	26	4	34.5
801–1,000	7	0	8	15	2	40.5	8	0	13	21	3	35.0	9	0	18	27	4	33.0
1,001–2,000	7	0	8	15	2	40.5	8	0	14	22	3	34.0	9	0	23	32	5	31.0
2,001–3,000	7	0	8	15	2	41.0	8	0	14	22	3	34.0	9	0	24	33	5	30.0
3,001–4,000	7	0	8	15	2	41.0	8	0	14	22	3	34.5	9	0	24	33	5	30.5
4,001–5,000	7	0	8	15	2	41.0	8	0	14	22	3	35.0	10	0	29	39	6	29.5
5,001–7,000	7	0	8	15	2	41.0	9	0	18	27	4	32.5	16	1	29	45	7	28.5
7,001–10,000	7	0	8	15	2	41.0	9	0	18	27	4	32.5	17	1	38	55	8	26.0
10,001–20,000	7	0	8	15	2	41.0	9	0	18	27	4	32.5	17	1	38	55	8	26.0
20,001–50,000	7	0	8	15	2	41.0	9	0	18	27	4	32.5	18	1	42	60	9	25.5
50,001–100,000	8	0	14	22	3	33.5	9	0	25	34	5	30.0	18	1	52	70	10	24.5

c_1 = allowable defects for first sample (acceptance number).
c_2 = allowable defects for first and second samples combined.
●D = lot tolerance percent defective corresponding to a consumer's risk $(pc) = 0.10$.

able for a double sampling plan.

isks of 10 percent and 5 percent, respectively. If a different value of consumer's isk is required, the OC curves and their tabulated values may be used. The ●oncept of LQ may also be useful in specifying the AQL and Inspection Levels ●or a series of lots or batches, thus fixing minimum sample size where there is ●ome reason for avoiding (with more than a given consumer's risk) more than ● limiting proportion of defectives (or defects) in any single lot or batch.

Inspection Level. The inspection level determines the relationship between ●he lot or batch size and the sample size. The inspection level to be used for ●ny particular requirement is prescribed by the responsible authority. Three ●nspection levels: I, II, and III, are given in MIL–STD–105D for general use. ●Unless otherwise specified, Inspection Level II is used. However, Inspection ●Level I may be specified when less discrimination is needed, or Level III may ●e specified for greater discrimination. Four additional special levels: S–1, S–2, ●–3, and S–4, are also given and may be used where relatively small sample ●izes are necessary and large sampling risks can or must be tolerated. In the ●designation of inspection levels S–1 to S–4, care must be exercised to avoid AQLs ●nconsistent with these inspection levels.

Sample sizes are designated by code letters based on the lot or batch size ●nd level of inspection. The AQL and the code letter are used to obtain the ●sampling plan from the appropriate sampling tables (Figs. 37 and 38). When

Lot or Batch Size	Special Inspection Levels				General Inspection Levels		
	S-1	S-2	S-3	S-4	I	II	III
2 to 8	A	A	A	A	A	A	B
9 to 15	A	A	A	A	A	B	C
16 to 25	A	A	B	B	B	C	D
26 to 50	A	B	B	C	C	D	E
51 to 90	B	B	C	C	C	E	F
91 to 150	B	B	C	D	D	F	G
151 to 280	B	C	D	E	E	G	H
281 to 500	B	C	D	E	F	H	J
501 to 1,200	C	C	E	F	G	J	K
1,201 to 3,200	C	D	E	G	H	K	L
3,201 to 10,000	C	D	F	G	J	L	M
10,001 to 35,000	C	D	F	H	K	M	N
35,001 to 150,000	D	E	G	J	L	N	P
150,001 to 500,000	D	E	G	J	M	P	Q
500,001 and over	D	E	H	K	N	Q	R

Fig. 37. Sample size code letters (MIL-STD-105D, Table I)

Acceptable Quality Levels (normal inspection)

Each cell shows **Ac Re** (Acceptance number / Rejection number). ↓ = use first sampling plan below arrow; ↑ = use first sampling plan above arrow.

Sample size code letter	Sample size	0.010	0.015	0.025	0.040	0.065	0.10	0.15	0.25	0.40	0.65	1.0	1.5	2.5	4.0	6.5	10	15	25	40	65	100	150	250	400	650	1000
A	2	↓	↓	↓	↓	↓	↓	↓	↓	↓	↓	↓	↓	↓	↓	↓	↓	0 1	1 2	2 3	3 4	5 6	7 8	10 11	14 15	21 22	30 31
B	3	↓	↓	↓	↓	↓	↓	↓	↓	↓	↓	↓	↓	↓	↓	↓	0 1	1 2	2 3	3 4	5 6	7 8	10 11	14 15	21 22	30 31	44 45
C	5	↓	↓	↓	↓	↓	↓	↓	↓	↓	↓	↓	↓	↓	↓	0 1	1 2	2 3	3 4	5 6	7 8	10 11	14 15	21 22	30 31	44 45	↑
D	8	↓	↓	↓	↓	↓	↓	↓	↓	↓	↓	↓	↓	↓	0 1	1 2	2 3	3 4	5 6	7 8	10 11	14 15	21 22	30 31	44 45	↑	↑
E	13	↓	↓	↓	↓	↓	↓	↓	↓	↓	↓	↓	↓	0 1	1 2	2 3	3 4	5 6	7 8	10 11	14 15	21 22	30 31	44 45	↑	↑	↑
F	20	↓	↓	↓	↓	↓	↓	↓	↓	↓	↓	↓	0 1	1 2	2 3	3 4	5 6	7 8	10 11	14 15	21 22	30 31	44 45	↑	↑	↑	↑
G	32	↓	↓	↓	↓	↓	↓	↓	↓	↓	↓	0 1	1 2	2 3	3 4	5 6	7 8	10 11	14 15	21 22	30 31	44 45	↑	↑	↑	↑	↑
H	50	↓	↓	↓	↓	↓	↓	↓	↓	↓	0 1	1 2	2 3	3 4	5 6	7 8	10 11	14 15	21 22	30 31	44 45	↑	↑	↑	↑	↑	↑
J	80	↓	↓	↓	↓	↓	↓	↓	↓	0 1	1 2	2 3	3 4	5 6	7 8	10 11	14 15	21 22	30 31	44 45	↑	↑	↑	↑	↑	↑	↑
K	125	↓	↓	↓	↓	↓	↓	↓	0 1	1 2	2 3	3 4	5 6	7 8	10 11	14 15	21 22	30 31	44 45	↑	↑	↑	↑	↑	↑	↑	↑
L	200	↓	↓	↓	↓	↓	↓	0 1	1 2	2 3	3 4	5 6	7 8	10 11	14 15	21 22	30 31	44 45	↑	↑	↑	↑	↑	↑	↑	↑	↑
M	315	↓	↓	↓	↓	↓	0 1	1 2	2 3	3 4	5 6	7 8	10 11	14 15	21 22	30 31	44 45	↑	↑	↑	↑	↑	↑	↑	↑	↑	↑
N	500	↓	↓	↓	↓	0 1	1 2	2 3	3 4	5 6	7 8	10 11	14 15	21 22	30 31	44 45	↑	↑	↑	↑	↑	↑	↑	↑	↑	↑	↑
P	800	↓	↓	↓	0 1	1 2	2 3	3 4	5 6	7 8	10 11	14 15	21 22	30 31	44 45	↑	↑	↑	↑	↑	↑	↑	↑	↑	↑	↑	↑
Q	1250	↓	↓	0 1	1 2	2 3	3 4	5 6	7 8	10 11	14 15	21 22	30 31	44 45	↑	↑	↑	↑	↑	↑	↑	↑	↑	↑	↑	↑	↑
R	2000	↓	0 1	1 2	2 3	3 4	5 6	7 8	10 11	14 15	21 22	30 31	44 45	↑	↑	↑	↑	↑	↑	↑	↑	↑	↑	↑	↑	↑	↑

↓ = Use first sampling plan below arrow. If sample size equals, or exceeds, lot or batch size, do 100 percent inspection.
↑ = Use first sampling plan above arrow.
Ac = Acceptance number.
Re = Rejection number.

Fig. 38. Single sampling tables for normal inspection (MIL–STD–105D, Table II–A).

no sampling plan is available for a given combination of AQL and code letter the tables direct the user to a different letter. The sample size to be used i given by the new code letter not the original letter. If this procedure leads to different sample sizes for different classes of defects, the code letter correspond ing to the largest sample size derived may be used for all classes of defect. when designated or approved by the responsible authority. As an alternative to a single sampling plan with an acceptance number of 0, the plan with an ac ceptance number of 1 with its correspondingly larger sample size for a desig nated AQL (where available), may be used when designated or approved by the responsible authority.

Inspection Procedures. Normal, tightened, or reduced inspection continues unchanged for each class of defects or defectives on successive lots or batches except where the switching procedures given below require change. The switch-ing procedures apply to each class of defects or defectives independently.

1. **Normal to Tightened.** When normal inspection is in effect, tightened in-spection is instituted when 2 out of 5 consecutive lots or batches have been rejected on original inspection (i.e., ignoring resubmitted lots or batches for this procedure).
2. **Tightened to Normal.** When tightened inspection is in effect, normal in-spection is instituted when 5 consecutive lots or batches have been considered acceptable on original inspection.
3. **Normal to Reduced.** When normal inspection is in effect, reduced inspection is instituted when all of the following conditions are satisfied:
   a. The preceding 10 lots or batches (or more, as indicated by the sampling table) have been on normal inspection and none has been rejected on original inspection.
   b. The total number of defectives (or defects) in the samples from the pre-ceding 10 lots or batches (or more as in "a" above) is equal to or less than the applicable number given in the sampling table. If double or multiple sampling is in use, all samples inspected are included, not "first" samples only.
   c. Production is at a steady rate.
   d. Reduced inspection is considered desirable by the responsible authority.
4. **Reduced to Normal.** When reduced inspection is in effect, normal inspection is instituted if any of the following occur on original inspection:
   a. A lot or batch is rejected.
   b. A lot or batch is considered acceptable under the procedures.
   c. Production becomes irregular or delayed.
   d. Other conditions warrant normal inspection.

Sampling Plans. A sampling plan indicates the number of units of product from each lot or batch which are to be inspected (sample size or series of sample sizes) and the criteria for determining the acceptability of the lot or batch (acceptance and rejection numbers).

Three types of sampling plans—single, double, and multiple—are given in MIL-STD-105D. When several types of plans are available for a given AQL and code letter, any one may be used. A decision as to type of plan, either single, double, or multiple, when available for a given AQL and code letter, will usually be based upon the comparison between the administrative difficulty and the average sample sizes of the available plans. The average sample size of multiple plans is less than for double (except in the case corresponding to single acceptance number 1) and both of these are always less than a single sample size. Usually the administrative difficulty for single sampling and the cost per unit of the sample are less than for double or multiple.

SAMPLING TABLES FOR PROCESS CONTROL. The Dodge-Romig and MIL-STD-105D tables stress the element of acceptance. They are particularly valuable for checking the quality of incoming parts and materials. In addition, however, they may also be applied in final inspection as well as between departments and operations in the production process.

Plans for Use in Continuous Production. H. F. Dodge (Industrial Quality Control, vol. 7) has formulated several plans for use where articles are manufactured continuously, as on a conveyor. Fig. 39 outlines the operation of one of the simpler plans of this type.

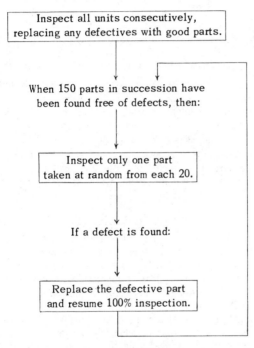

Fig. 39. Operation of a typical Dodge CSP-1 plan for continuous sampling (AOQL = 1%).

ACCEPTANCE SAMPLING BY VARIABLES. The sampling tables so far described are all based on sampling by attributes (presence or absence of a desired characteristic). Great progress has also been made with acceptance sampling systems based on measurements. In general, such plans extract more information from inspection or test of a given number of pieces of product. This means that the same protection for the buyer can be obtained with smaller samples. This possibility is particularly attractive when the property to be tested is one which requires a costly inspection process. Examples are life tests of electron tubes, chemical or metallurgical analyses, tensile or other tests of physical properties, velocity tests for ammunition, and blowing-time tests for fuses.

Military Standard for Variables Sampling. MIL-STD-414 is becoming increasingly popular for some types of product. Plans exist for three primary situations.

1. Variability unknown—sample standard deviation used.
2. Variability unknown—sample divided into subgroups and the average range used.
3. Variability known.

Provisions are made for both tightened and reduced inspection in the estimation of the process average.

To illustrate the application of these plans, consider the variability unknown —sample standard deviation case. Let

$$U = \text{upper specification limit}$$
$$L = \text{lower specification limit}$$
$$\overline{X} = \text{sample mean}$$
$$s = \text{sample standard deviation}$$

A table is provided for AQL conversion. The procedure is to use a master table with lot size and inspection level to determine a sample size code letter If Form 1 of the specification is used, the values of the sample size, n, and the acceptance criterion, k, are read from tables. The sample is taken, $\overline{X}$ and s are calculated, and the test statistic $(U - \overline{X})/s$ or $(\overline{X} - L)/s$, depending on which specification is given, is evaluated and compared to k; if the test statistic is greater than or equal to k, the lot is accepted. Otherwise it is rejected.

Using the specification's Form 2, values $Q_U = (U - \overline{X})/s$ or $Q_L = (\overline{X} -)/s$, or both, are calculated. In the case of a one-sided limit, say U, Q_U is used with n, the sample size, to enter a table for an estimated fraction defective p_u which in turn is compared to a maximum value M read from the same table or the sample size. If $p_u > M$, the lot is rejected.

Reliability and Life Testing

NATURE OF RELIABILITY. The term reliability has been used in a loose manner for many years to represent an attribute of quality. More recently, reliability has been explicitly defined as the probability that a system will perform satisfactorily for at least a given period of time when used under stated conditions. The **reliability function** is this same probability expressed as a function of the time period.

There are several distinct elements of this definition. First, the attribute, reliability, is a probability; second, satisfactory performance or non-performance must be clearly defined; third, a specific time period is associated with a value of reliability; and fourth, the stated conditions are a particular set of environmental conditions associated with the life span over which reliability is measured.

Emphasis in recent years has been shifting towards the quantification of **system effectiveness**. The most used measures incorporate **reliability, design adequacy**, and **operational readiness**. All of these characteristics are measured in terms of the probability scale.

Reliability and Quality Control. The total quality control concept considers reliability as an attribute of quality. Some individuals would delegate reliability activities to the quality function; however, in most aerospace and defense industries a separate reliability function has been established, maintaining close co-ordination with quality control as well as design and production. In these companies, it is not at all uncommon to encounter departments or divisions designated "Reliability and Quality Control" with neither function reporting to the other, but with both reporting to the same manager.

In consumer products industries the reliability engineering efforts were minimal until the 1960s. The principal reason for this was the inability of the consumer o present a united demand for reliability such as that required by government procurement contracts. With increased consumer awareness and direct governmental intervention in some cases, the situation began changing. The impact of federal, state, and local governmental regulations and the establishment of standards of performance has made significant inroads in such industries as home appliance, automotive, and commercial aircraft. The costs associated with replacing failures as well as lost customer good will have been sufficient to motivate the development of reliability functions in many industries.

It is logical to consider reliability as an attribute of quality, and the trend in U.S. industry is to move in this direction.

As the technical competence of quality control groups increases to the point where these groups of individuals can accomplish the more difficult (from an engineering and mathematical standpoint) functions associated with reliability engineering endeavors, the quality control function becomes more capable of implementing the total quality control concept.

Failure. Failure usually implies that the system is in an inoperable condition when needed or it is initially operable but fails during a particular use cycle by not accomplishing its assigned tasks satisfactorily. If we let the symbol t represent the time to failure, and $g(t)$ the overall failure distribution, then the reliability function, $R(t)$ is given by:

$$R(t) = \int_t^\infty g(x)\,dx, \qquad t > 0$$

$$= 1 - g(0), \qquad t = 0$$

Figure 40 (page 76) shows typical failure and reliability distributions. (See section on Statistical Methods for a discussion of probability functions.)

FAILURE PROCESS. When all initial failures are removed we can consider the failure distribution of those remaining. In Fig. 40 this would remove the spike at $t = 0$. Then the function $f(t) = g(t) / \int_t^\infty g(t)dt$, $t \geq 0$ gives the density function of time to failure of the units that do not fail initially. The reliability function is then defined as

$$R(t) = \int_t^\infty f(x)\,dx, \qquad t \geq 0$$

The units that do not fail initially fail eventually due to one of several causes. The most common classification of causes is as follows:

1. Early age failures due to production faults.
2. Random failures.
3. Wearout failures.

Figure 41 (page 77) shows the combination of failure distributions into a single density.

Failure Rate and Hazard Function. The failure rate is defined for a particular interval of time, say t_1 to t_2. It is defined as

$$\frac{R(t_1) - R(t_2)}{R(t_1)(t_2 - t_1)} = \text{FR}\,(t_1, t_2)$$

The units are in failures/hour. A more useful concept is that of the **hazard function**, which is a limiting form of the failure rate as $t_2 \to t_1$. The hazard function is

$$h(t) = \frac{f(t)}{R(t)}$$

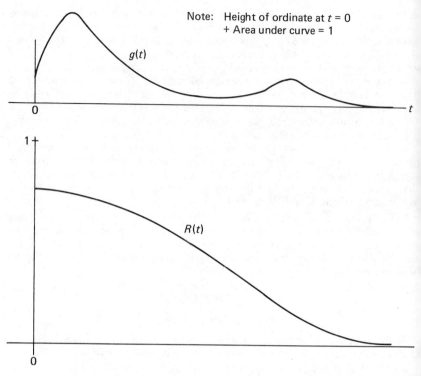

Fig. 40. Typical failure and reliability distributions.

and it may be interpreted as the instantaneous rate of failure at time *t* given survival to time *t*. A typical function $h(t)$ is shown in Fig. 42 (page 78).

Mean Time to Failure. The mean time to failure (MTTF) is the mean of the random variable associated with the combined failure density, $f(t)$, shown in Fig. 41. Theoretically it is calculated as

$$\text{MTTF} = \int_0^\infty t \cdot f(t)\, dt$$

or

$$\text{MTTF} = \int_0^\infty R(t)\, dt$$

LIFE TESTING. In order to estimate such characteristics as $R(t)$, $h(t)$, or MTTF, life tests are conducted. Some units, say n, are placed on test and aged. These tests may be replacement tests or non-replacement tests. In the case of **replacement tests**, when a unit fails, its failure time is recorded, and it is replaced by a new unit. In **non-replacement** tests, the units are not replaced as they fail; however, failure times are noted.

There are two approaches to the problems of estimation of $R(t)$. In the first case, the mathematical form of $R(t)$ is assumed to be unknown, and in this situation it is necessary to operate the test for a time that corresponds to the time for which the reliability estimate is to be made.

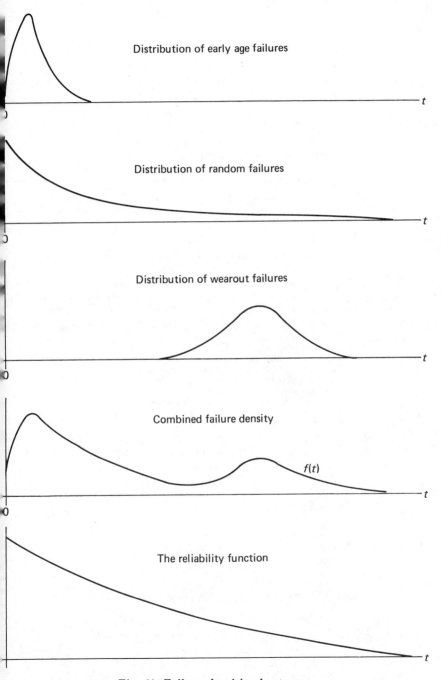

Distribution of early age failures

Distribution of random failures

Distribution of wearout failures

Combined failure density

$f(t)$

The reliability function

Fig. 41. Failure densities by type.

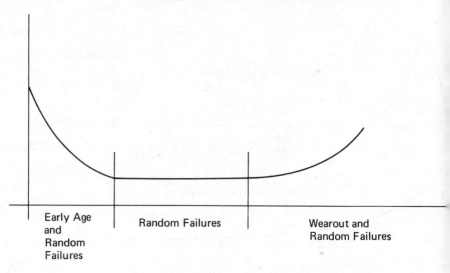

Early Age and Random Failures   Random Failures   Wearout and Random Failures

Fig. 42. Typical hazard function, $h(t)$.

In the case where there is knowledge or assumed knowledge about the form of the function, $R(t)$, then it may not be necessary to operate the test for a time corresponding to the time for which the reliability estimate is to be made.

Tests may be stopped at some pre-selected time or when some prestated number of failures has occurred. Let

n = number of items placed on test at $t = 0$
t^* = time at which the life test is terminated
r = the number of failures occurring to time t^*
r^* = preassigned number of failures
t_i = the operating time on the ith failure by rank in time

Replacement Tests.

$$MTTF = \frac{nt^*}{r}, \qquad \text{test terminated at } t^*$$

$$MTTF = \frac{nt^*}{r^*}, \qquad \text{test terminated at } r^*, R(t) \text{ exponential}$$

Non-Replacement Tests.

$$MTTF = \frac{\sum_{i=1}^{r} t_i + (n - r) \cdot t^*}{r}, \qquad \text{test stopped at } t^*$$

$$MTTF = \frac{\sum_{i=1}^{r^*} t_i + (n - r)t^*}{r^*}, \qquad \text{test stopped at } r^*, R(t) \text{ exponential}$$

Accelerated Life Tests. Due to the large values of MTTF for some units and the difficulty in accumulating failures, accelerated tests are used. In these cases, stress levels of such variables as temperature are increased. In order to successfully use such procedures, however, it is necessary to have knowledge of the relationship between stress level and the characteristic being estimated.

Failure Time Distributions. Many models have been suggested for time-to-failure distributions. The most used model is the exponential,

$$f(t) = \lambda e^{-\lambda t}, \qquad t \geq 0, \qquad \lambda > 0$$
$$= 0, \text{ for all other } t$$

The parameter λ is called the **mean rate of failure.**

The assumption underlying the application of this model is that all failures are random. This implies that either there are no early age failures or burn-in is used to eliminate early age failures and either no wearout failures occur or the time to occurrence for wearout is much longer than the projected unit operating time. The reliability function is

$$R(t) = e^{-\lambda t}, \qquad t \geq 0, \qquad \lambda > 0$$

and the hazard is

$$h(t) = \frac{f(t)}{R(t)} = \lambda, \qquad \text{a constant}$$

Another distribution that has been found to be very useful as an approximation to numerous failure processes is the **Weibull model,**

$$f(t) = \frac{\beta}{\alpha}(t - \gamma)^{\beta-1}e^{-(t-\gamma)^\beta/\alpha}, \qquad t > \gamma$$

$$= 0, \text{ otherwise}$$

so that

$$h(t) = \frac{\beta}{\alpha}(t - \gamma)^{\beta-1}, \qquad t > \gamma$$

and

$$R(t) = e^{-(t-\gamma)^\beta/\alpha}$$

The parameters are γ, the location parameter, β the shape parameter, and α the scale parameter. Both β and α are positive, and frequently $\gamma = 0$ for reliability applications.

In order to estimate α and β it is necessary to estimate $R(t)$ for various values of t. This is done by performing a life test on n units and recording the failure times over the test run. Miller and Freund (Probability and Statistics for Engineers) present a method for obtaining α and β based on a graphical technique. This method makes use of the fact that the Weibull distribution can be transformed into a linear function of $\ln t$ in the form

$$\ln \ln \frac{1}{R(t)} = \beta \cdot \ln t - \ln \alpha$$

Using the method of least squares the parameters α and β can then be determined.

STATISTICAL METHODS

CONTENTS

CONTENTS (*Continued*)

STATISTICAL METHODS

Application of Statistics

STATISTICS IN INDUSTRY. The term "statistics" is used as a general name for a large group of mathematical tools based on the laws of probability which are used to collect, analyze, and interpret numerical data. There are two broad classes of statistical methods: descriptive and inferential. The descriptive methods are used to summarize and describe a set of data while the inferential are used to generalize from a body of data to the population or process from which the data was obtained.

These tools have for many years been indispensable in the various fields of the physical and social sciences and in many areas of engineering and management. Quality control is one of the areas in which they have been adapted and applied with outstanding success. In other areas their application is increasing and will continue to increase as their usefulness to the engineer and administrator becomes more generally known.

The statistical method has been simply described as common sense reduced to calculation. While there is a great deal of truth in the foregoing statement, statistical methods can greatly aid the engineer by quantifying and strengthening the basis for judgments which would otherwise be only intuitive. However, the techniques are not intended to be substitutes for method, reasoning, judgment, common sense, or experience but rather to serve as supplemental tools. More specifically, the **application of statistical techniques** can help in the following ways:

1. In obtaining from a mass of data or observations a concise and understandable summary which will be appropriate for the problem at hand.
2. In determining limits of precision for conclusions reached from the analysis of data; in other words, ascertaining the degree of confidence one may have in information obtained by sampling.
3. In determining the number of observations that must be made or the amount of information that must be gathered and summarized in order to be able to reach conclusions of a specified degree of accuracy.
4. In extracting the maximum amount of useful information from available data and observations. Often, the use of statistical methods can considerably reduce the amount of data that must be collected and analyzed in order to reach conclusions with a required degree of confidence.
5. In appraising in specific terms the uncertainties or variability that is inherent in the procedures, processes, materials, activities, etc., encountered in production.
6. In providing a method for the design of experiments so that only necessary data will be collected, known sources of bias will be eliminated, variability in the factors will be accounted for and the results of the experiment can be analyzed efficiently.
7. In giving evidence of relationships between factors in a process or system that might be unnoticed if a statistical approach were not used, and in determining the mathematical nature and strength of the relationships.

TYPES OF INDUSTRIAL DATA. Many kinds of data are used in industry. From the standpoint of statistical procedures, however, there are two useful ways in which data may be classified. First, they may be classified either as sample data or as population data. When all the individuals or items involved are measured or observed, the observations are **population data.** Population data occurs, for example, when we deal with the measurements on all the items in a production lot. A population is usually a finite collection of objects. When we speak of a population we imply a complete set. When population data are compiled, statistical methods may be of help in summarizing them in a useful way.

In many areas of industrial engineering and management, studies and procedures for control must be based on observations from a sample of items taken from a process that is capable, at least conceptually, of containing an infinite number of items. When we work with such an infinite population we usually have only a sample from the infinite number of items in the population. Time standards, for example, must be determined from time studies of a sample of work cycles for the activity. The number or percentage of errors in the clerical work of a mail-order house may have to be estimated from data collected from selected samples of work. The time to failure of transistors may be estimated from samples of transistors subjected to life tests. When **sample data** are all that are available or that can be justified, statistical procedures may be of much help in drawing conclusions and in making decisions. (See section on Work Measurement and Time Study.)

Data may be classified in another way. They may be classified as **data in variables form** or as **data in attribute form.** When each item is measured by some scale and the measurement itself observed and used, the data are in variables form. Examples of this are time measurements in decimal minutes obtained by a time study, and the weight in ounces of packages leaving a filling machine.

Data are in attribute form when each item has been checked or measured and simply classified as falling in one of two or more possible categories. Checking clerical entries and classifying each as right or wrong, or observing a group of machines and noting whether each machine is or is not running, will give attribute data. One of the most important examples of attribute data arises in quality control when one classifies inspected items as being good or defective.

In some cases one may choose between taking data in variables form or taking them in the form of attributes. An item may be measured, for example, or a "go—no-go" test may be used only to classify the item as good or bad or as above or below standard. When there is a choice, the most satisfactory method to use will depend on the kind of information desired and on the respective costs of obtaining, compiling, and using the data in each of the two forms. Generally, it is more costly to obtain variables data than attribute data, and more work is required for compilation and summary. On the other hand, variables data give more complete information about the items or activities than an equivalent amount of attribute data.

VARIATION IN DATA. Data taken from any product, procedure, system, or activity will almost certainly show some variation from observation to observation, even when an attempt is made to hold all significant factors constant. The reason for this is that it is impractical, too costly, or even impossible to hold all factors strictly constant. Many minor causes of variation are inherent in any system or procedure. Variation due to these inherent causes is frequently referred to as **chance variation.** However, the variation is not due to chance but due to a system of causes that by choice or of necessity have been ignored—ignored because they are relatively unimportant or because their nature cannot easily be ascertained. Also, the method of measurement used in obtaining observations may vary somewhat in its preciseness from measurement to measurement. This, too, will add to the variation in data.

For example, time measurements for a work element—say, for getting a part from a tote box and putting it in a jig—will vary from cycle to cycle, even for the same job and the same operator. Among the inherent causes of such variation may be: (1) variations in distance of reach because parts occupy different positions within the tote box; (2) different amounts of fumbling from cycle to cycle prior to the time the hand selects and grasps a part; (3) different amounts of time spent in positioning the part prior to insertion in the jig, the amount depending on the position of the part in the tote box; (4) differing degrees of operator effort from cycle to cycle; and (5) variations in reading the watch hand from cycle to cycle. Many other inherent causes of variation for this simple activity could be mentioned. Similarly, there may be considerable variation in the time to failure of electronic devices such as transistors or vacuum tubes due to inherent causes of variation in the materials used, the assembly process, or the exact manner in which the devices are used.

In addition to the effect of these inherent chance causes in the product or activity and in the method of measurement, observed variation may also be due to a **change in level** of some important factor. Generally, the purpose of a study is to determine what effect, if any, may be attributed to such a change. In the case of the time requirement for picking up a part and putting it in a jig, for example, a study may be made to determine the effect of a new jig design or of a change to a piece part of a different size or shape.

If the change in level of a factor has an appreciable effect on the activity or product, this may be readily determined by a **comparison of observations** made before the change with observations made after the change. The same remarks also apply, of course, to cases in which the study is not of changes in a factor but of the effects of different levels of a given factor.

If the inherent causes of variation themselves cause an appreciable variation from observation to observation, the effects of a difference in level for some one factor may be difficult to determine. Any effect that a difference in level would have on the observations may be obscured by the variation in data produced by the inherent causes. The effect will be particularly difficult to evaluate if the number of observations available for one or more of the levels is small or if the effect of the change in levels on the observations is small compared to the effects of the inherent system of variation-producing causes.

One of the principal objectives of statistical procedures is to deal with this **variation from inherent causes**—to evaluate the effect it may have in drawing conclusions from study data and to use this evaluation in determining the risk of error in reaching conclusions.

SYSTEMATIC ACCUMULATION OF DATA. Often little thought is given to the mechanics of accumulating data. This is usually a natural result of the engineer being under pressure to get the necessary figures quickly or of his being concerned only about the use of the data and the end results. It is of some importance, however, to plan for the collection of data and to have a systematic procedure for its accumulation. Among the many benefits may be the following:

   1. A saving in time for the person making and recording the observations.
   2. A reduction in the probability of error in recording data, in working it up, or in using it.
   3. The possibility, if the system of accumulation is good, of making some analysis of the recorded data during the course of the study. This analysis

may be helpful in deciding how many more observations will be needed, in determining what other kinds of data may be required, or in other ways.

4. A saving in time in summarizing the data, in posting them to other records, or in using them in other ways.

Probability

DEFINITIONS. The mathematical theory of probability provides the conceptual framework for the understanding and development of statistical methodology. For this reason an understanding of the fundamentals of probability is very important. In addition one often encounters situations involving risk, randomness, and uncertainty in production management. Probability theory provides a method for quantifying and evaluating such risks.

In order to be able to discuss probability unambiguously, use is made of several terms in a special way.

An **experiment** is any process of observation such as the tossing of a coin, the measurement of the diameter of a screw, or the carrying out of an elaborate experiment on a production system.

Each possible result of an experiment is called an **outcome.** For example in the case of tossing a coin there are two outcomes, heads and tails. In the case of measuring the diameter of a screw each possible diameter is an outcome, and hence there are an infinite number of outcomes. In the case of drawing a sample of 50 transistors from a lot of 1000 transistors and testing them, an outcome is a vector of order 50 giving the results of the tests.

The set of all conceivable outcomes of the experiment is called the **sample space,** and outcomes are elements in this set. As in the examples above the sample space could be a finite or an infinite set.

It is often convenient to speak of sets of outcomes which we call **events.** Thus, in the example of the measurement of screw diameters an event of interest might be that the diameters fall within tolerances. This event contains an infinite number of outcomes and if any of the outcomes contained in the event occur we say that the event occurs. Consider another example in which a coin is tossed twice. We denote by H the occurrence of a head and by T the occurrence of a tail. The sample space consists of the outcomes: (H, H), (H, T), (T, H), and (T, T). An event of interest might be that exactly one head occurs. This event is the set of outcomes (H, T) and (T, H). Another event of interest might be that at least one head occurs. This event is the set of outcomes (H, T), (T, H), and (H, H).

Many of the phenomena to which probability theory is applied are essentially numerical. In such cases, the outcomes of the experiment are numbers, and events are sets defined in terms of numbers. The measurement of screw diameters is just such a case. A typical outcome is $D = 1.003$ inches. An event of interest might be the set of diameters $0.995 \leq D \leq 1.005$ inches. In such cases we naturally are dealing with what is called a **random variable.** However, there are many situations in which the outcomes are not simply numbers or they are numbers which we desire to transform or code in some way. Again we will deal with random variables. A random variable is a numerical-valued function defined over all the outcomes in the sample space. Being a function, it is a rule which assigns a number to each outcome in the sample space. An important example is where the outcome is itself a number and the random variable is simply a rule which assigns the same number to the outcome! Again, the screw measurement problem is such a case. However, in the example of tossing two coins the random variable, R, might, for example, assign to an outcome a value equal to the number of heads. Thus $R(H, H) = 2$, $R(H, T) = 1$, $R(T, H) =$ 1, $R(T, T) = 0$. Another random variable, W, might assign to an outcome the number of heads minus the number of tails. Thus $W(H, H) = 2$, $W(H, T) = 0$, $W(T, H) = 0$, $W(T, T) = 2$. Still another, Z, might assign a value of 5 if the first toss is a head and 0 if it is not. Thus $Z(H, H) = 5$, $Z(H, T) = 5$, $Z(T, H) = 0$, $Z(T, T) = 0$. It should be clear that there are many possible random variables that might be used in a given situation. Generally speaking, the random variables encountered in production problems may be classified as being **discrete** or **continuous**. If the random variables take on a continuum of values such as would be the case in the screw measurement problem they are called continuous. On the other hand, if the random variable takes on a finite or countable number of values it is called discrete. The random variable R, W, and Z defined above are examples of discrete random variables.

LAWS OF PROBABILITY. Probability can be defined as a measure of the likelihood of events. The definition of probability in this section will be in mathematical and abstract terms, but it is consistent with the intuitive understanding of natural phenomena, and, in particular, it is consistent with the notion of relative frequency. Suppose a given experiment is performed over and over again under identical conditions. Let A denote some event of interest. Let N denote the number of repetitions of the experiment, and let $N(A)$ denote the number of trials on which the event A occurs. The ratio $N(A)/N$ is called the relative frequency of the event A. It is an empirical fact that many natural, engineering, and economic phenomena exhibit a stabilization of the relative frequency around some value when the number of trials is large. The value of $N(A)/N$ is called the probability of A. Phenomena that exhibit the stabilization of relative frequency provide the most meaningful applications of probability theory and statistical methods.

For a given experiment and sample space, the probability is a function $P\{\cdot\}$, which assigns a real number $P\{E\}$ to each event E in the sample space of the experiment. The fundamental properties of the probability function are:

(a) $0 \leq P\{E\} \leq 1$.

(b) $P\{\text{impossible event}\} = 0$.

(c) $P\{\text{sure event or whole sample space}\} = 1$.

(d) $P\{A \text{ or } B\} = P\{A\} + P\{B\}$, if A and B cannot both occur simultaneously. Such events are called disjoint.

(e) $P\{A \text{ or } B\} = P\{A\} + P\{B\} - P\{A \text{ and } B\}$. This rule is generally true and (d) above is a special case when $P\{A \text{ and } B\} = 0$.

(f) Denoting by A^c the complement of an event A, $P\{A^c\} = 1 - P\{A\}$.

Combinatorial Probability. A special type of experiment of great practical importance is one in which a finite number of outcomes in the sample space each have the same probability assigned to them. That is, each outcome is equally likely. In such situations the probability of an event E can be computed from the ratio,

$$P\{E\} = \frac{\text{Number of outcomes in } E}{\text{Number of outcomes in the sample space}}$$

The main problem in computing probabilities using this model is counting the number of outcomes in the event of interest and in the sample space. An example is the determination of the probability of drawing an ace or king from a deck of cards.

$$P\{\text{ace or king}\} = \frac{\text{Number of aces and kings}}{\text{Number of cards in deck}} = \frac{8}{52} = \frac{2}{13}$$

Of course, this calculation presumes that the deck is well shuffled and the draw is made "at random."

A question which arises frequently in such problems is: In how many ways may k objects be drawn from a set of n objects without regard to either the order of the selection or the arrangement of the k objects? If an experiment I can occur in n_1 ways and another experiment II can occur in n_2 ways the composite experiment consisting of both the experiments I and II can occur in $n_1 \cdot n_2$ ways. Thus n distinct objects can be arranged in $n(n-1) \ldots (2)(1) = n!$ ways and the number of ways k objects can be drawn from n objects is

$$\binom{n}{k} = \frac{n!}{k!(n-k)!}$$

where $\binom{n}{1} = n$ and $\binom{n}{0} = 1$.

For example, a committee of three men can be selected from a club of 5 members in 10 ways since

$$\binom{5}{3} = \frac{5!}{3!2!} = \frac{5 \cdot 4 \cdot 3 \cdot 2 \cdot 1}{(3 \cdot 2 \cdot 1)(2 \cdot 1)} = 10$$

As an application of this counting method to a probability problem, suppose that two transistors are drawn from a box of twelve transistors, four of which are defective. What is the probability that one of those drawn is defective? The number of outcomes in the sample space is the number of ways two items can be drawn from twelve items or $\binom{12}{2}$. The number of outcomes for which one is defective is the product of the number of ways one defective can be selected from the four defectives $\binom{4}{1}$ and the number of ways one good item can be selected from the eight good items $\binom{8}{1}$. Thus,

$$P\{\text{one defective}\} = \frac{\binom{4}{1}\binom{8}{1}}{\binom{12}{2}} = \frac{32}{66} = \frac{16}{33}$$

Conditional Probability and Independence. In some situations partial information about the results of an experiment or process of observation is used to modify a probability. For example, it may be known that it is cloudy at 9:00 A.M., and the probability that it will rain by 5:00 P.M. is desired, or it is known that a screw was produced by a given operator on a certain machine and the probability that it is defective must be determined. The mechanism employed in such cases is that of **conditional probability.** The conditional probability of the event A given that the event B occurs is denoted by $P\{A|B\}$, and is defined as

$$P\{A|B\} = \frac{P\{A \text{ and } B\}}{P\{B\}}$$

when $P\{B\} > 0$. The conditional probability of A given B is not defined when $P\{B\} = 0$. Such a number would have no useful meaning. Consider a combinatorial problem in which the elementary outcomes are equally likely, as in the

following: What is the probability that a card drawn from a deck is a king
given that it is a face card?

$$P\{\text{King}|\text{Face card}\} = \frac{P\{\text{King and Face card}\}}{P\{\text{Face card}\}} = \frac{P\{\text{King}\}}{P\{\text{Face card}\}}$$

$$= \frac{\dfrac{\text{No. of ways King occurs}}{\text{No. of ways to draw a card}}}{\dfrac{\text{No. of ways Face card occurs}}{\text{No. of ways to draw a card}}}$$

$$= \frac{\text{No. of ways King occurs}}{\text{No. of ways Face card occurs}}$$

$$= \frac{4}{12} = \frac{1}{3}$$

In many situations certain events or aspects of an experiment are physically
or logically unrelated or independent. The probability of the occurrence of both
events is the product of the probabilities of each of the events. This suggests
the definition of **probabilistic or statistical independence.** Two events A
and B are independent if

$$P\{A \text{ and } B\} = P\{A\}P\{B\}$$

Furthermore if A and B are independent $P\{A|B\} = P\{A\}$ since if $P\{A\}$ or $P\{B\}$
is not zero,

$$P\{A|B\} = \frac{P\{A \text{ and } B\}}{P\{B\}} = \frac{P\{A\}P\{B\}}{P\{B\}} = P\{A\}$$

$$P\{B|A\} = \frac{P\{A \text{ and } B\}}{P\{A\}} = \frac{P\{A\}P\{B\}}{P\{A\}} = P\{B\}$$

As an example, consider two tosses of a coin. It can usually be assumed that
the two successive tosses are physically independent processes, and hence it can
be assumed that they are probabilistically independent also. Thus to compute
the probability of two heads

$$P\{H \text{ on toss } 1 \text{ and } H \text{ on toss } 2\} = P\{H \text{ on toss } 1\}P\{H \text{ on toss } 2\}$$

Supposing that the coin does not change at all between tosses, and that the coin
is well balanced and fair, $P\{H \text{ on any toss}\} = \frac{1}{2}$. Thus

$$P\{H \text{ on toss } 1 \text{ and } H \text{ on toss } 2\} = \frac{1}{2} \cdot \frac{1}{2} = \frac{1}{4}$$

The notion of independent events also applies to a set of more than two
events. A set of n events $A_1, A_2, \ldots, A_n$ are independent if $P\{A_1 \text{ and } A_2 \text{ and }$
$\ldots \text{ and } A_n\} = P\{A_1\} P\{A_2\} \ldots P\{A_n\}$ and if all subsets of less than n of
the events are also independent. Thus, the probability of throwing three heads
in three independent tosses of a fair coin is

$$\frac{1}{2} \cdot \frac{1}{2} \cdot \frac{1}{2} = \frac{1}{8}$$

DISTRIBUTION FUNCTIONS. In probability and statistics it is convenient to adopt a standard format for representing the probability law associated with a random variable. The standard notation adopted is that of the
cumulative distribution function (C.D.F.) often called simply the distribution
function. If X is the random variable of interest its cumulative distribution
function denoted by $F(t)$ is

$$F(t) \equiv P\{X \leq t\}, \qquad -\infty < t < +\infty$$

In some situations in which one deals with several random variables X, Y, Z etc., it may be convenient to identify the C.D.F.'s of the various random variables by using a subscript notation: $F_x(t)$, $F_y(t)$, $F_z(t)$, etc. The important properties of a distribution function are that it is an increasing function of t, $F(-\infty) = 0$, and $F(+\infty) = 1$. If the distribution function is given, one can compute probabilities for any events of interest associated with the random variable. For example,

$$P\{a < X \leq b\} = F(b) - F(a)$$

In the two important cases of continuous and discrete random variables the distribution function itself may be formed from other functions. In the case of **continuous** random variables the distribution function $F(t)$ is formed from a **density function** $f(t)$ by integration,

$$F(t) = \int_{-\infty}^{t} f(x)\,dx$$

The function $f(t)$ must be non-negative, and it must be the derivative of the distribution function. That is

$$f(t) = \frac{dF(t)}{dt}$$

It is important to note that the density function $f(t)$ is not the probability that X is equal to t. For a continuous random variable this probability is equal to zero. In the case of a **discrete** random variable the distribution function $F(t)$ is formed from a **probability mass function** by summation:

$$F(t) = \sum_{k \leq t} p(k)$$

The probability mass function $p(k)$ is precisely $P\{X = k\}$. The difference between density functions and probability mass functions is an important distinction between continuous and discrete random variables.

DESCRIPTIVE CHARACTERISTICS OF RANDOM VARIABLES. There are many characteristics of random variables which are useful in statistics. Among these are:

Percentiles. The value of t, called t_α such that $P\{X \leq t_\alpha\} = F(t_\alpha) = \alpha$ is called the 100α percentile of the distribution of the random variable X.

Median. The most used of percentiles is the median which is the 50th percentile. It is the value such that with equal probability the random variable is above or below it, and as such it provides a measure of the center of the distribution.

Mode. The mode of the distribution, sometimes called the most probable value, is the value of t such that the density function of a continuous random variable is a maximum, or, if the random variable is discrete, it is the value of t such that the probability mass function is a maximum. The mode need not be unique, and in fact in practice bimodal distributions are sometimes encountered.

Mean or **Expected Value.** The mean or expected value of the distribution of a random variable X is denoted by $E(X)$ and is defined as:

$$E(X) = \int_{-\infty}^{+\infty} tf(t)\,dt, \quad \text{if } X \text{ is a continuous random variable}$$

and

$$E(X) = \sum_{\text{all } t} p(t), \quad \text{if } X \text{ is a discrete random variable}$$

The mean of a distribution is analogous to the center of gravity of a body, and it provides a direct and commonly used measure of the center of a distribution. The connection between the notions of probability and long-run frequency is paralleled by the connection between the notion of the mean and the long run average of a random variable. The long run average of a random variable in a series of repeated trials tends to stabilize near the mean in many actual situations.

Moments. The higher moments of a distribution are expectations of functions of the random variable. In particular, for $k \geq 1$, the kth moment, denoted by $E(X^k)$ is defined as,

$$E(X^k) = \int_{-\infty}^{+\infty} t^k f(t) \, dt, \quad \text{if } X \text{ is continuous}$$

and

$$E(X^k) = \sum_{\text{all } t} t^k p(t), \quad \text{if } X \text{ is discrete}$$

Variance. A particular moment of interest is the variance or second moment about the mean. Denoted by Var X we have

$$\text{Var } (X) = E[X - E(X)]^2 = \begin{cases} \int_{-\infty}^{+\infty} [t - E(X)]^2 f(t) \, dt, & \text{if } X \text{ is continuous} \\ \sum_{\text{all } t} [t - E(X)]^2 p(t), & \text{if } X \text{ is discrete} \end{cases}$$

The variance provides a measure of spread of the distribution and is analogous to the moment of inertia of a physical body. The square root of the variance is called the standard deviation and is denoted by $\sigma(X)$.

SOME IMPORTANT DISTRIBUTIONS. Discrete Random Variables. A random variable that takes on only two values 0 and 1 is said to have a **Bernoulli distribution.** Sampling inspection where 0 denotes acceptable and 1 denotes defective represents an important example of this type of distribution. The probability of a "defective" can be written as

$$p = P\{X = 1\}$$

and the probability of an "acceptable" written as

$$q = P\{X = 0\}$$

where $p + q = 1$. The expectation is p and the variance is pq. The distribution is used in many situations in which a binary classification of an observation is possible.

If n independent samples are made on a Bernoulli random variable a binomial random variable is generated. For example, suppose that n items are produced and each has probability p of being defective. The **binomial probability distribution** gives the probability that k out of the n are defective. Let Y denote the number of occurrences of an event (say production of a defective item) in n independent trials of an experiment where on each trial p is the probability of occurrence of the event. Then

$$P\{Y = k\} = \binom{n}{k} p^k q^{n-k} = \frac{n!}{k!(n-k)!} p^k q^{n-k}, \quad k = 0, 1, \ldots, n$$

where $q = 1 - p$. The mean is Np and the variance is Npq.

The **Poisson distribution** assigns to the event $X = k$ the probability,

$$P\{X = k\} = \frac{e^{-\lambda}\lambda^k}{k!}, \quad k = 0, 1, \ldots$$

where $\lambda > 0$ is a parameter of the distribution which is, in fact, $E(X)$. It is also true that Var $(X) = E(X) = \lambda$. The Poisson distribution occurs often in practice. There are two reasons for this. First, the Poisson distribution provides a good approximation to the binomial distribution for large sample size n and small probability p. It is mathematically a limiting form of the binomial with parameter $\lambda = np$. Second, the Poisson distribution occurs in its own right as a distribution for the number of events of a completely random process occurring in a fixed interval of time t. In this case the parameter is taken as λt and we have

$$P\{\text{exactly } k \text{ events in time } t\} = \frac{e^{-\lambda t}(\lambda t)^k}{k!}, \quad k = 0, 1, \ldots$$

The parameter λ may be interpreted as the mean number of events occurring in a unit of time. This distribution has frequent application in queueing theory and reliability (see the sections on Operations Research and on Quality Control and Reliability).

The **hypergeometric distribution** arises in the experiment of sampling from a finite population. Consider a population (say a given lot of parts to be inspected) consisting of N elements. A sample of n elements is drawn at random without replacement. Suppose that there are M "defective" elements and $(N - M)$ non-defective elements. Let Y be the random variable, the number of defectives in the sample. Then

$$P\{Y = k\} = \frac{\binom{M}{k}\binom{N-M}{n-k}}{\binom{N}{n}}, \quad k = 0, 1, \ldots, n$$

Let $p = M/N$ represent the proportion of defectives. The expectation of Y is np and the variance of Y is $np(1-p)(N-n)/(N-1)$. When the lot size N is large compared to the sample size n the hypergeometric distribution is approximated by the binomial distribution. This is true since sampling without replacement from a large lot is almost the same as sampling with replacement.

Continuous Random Variables. A random variable that can take on any value in the interval $[a, b]$, $a < b$, and for which the probability that it lie in any subinterval is proportional to the length of the subinterval, is said to have a **uniform distribution** on $[a, b]$. The density and distribution functions are:

$$f(t) = \begin{cases} \dfrac{1}{b-a} & a \leq t \leq b \\ 0 & \text{otherwise} \end{cases}$$

$$F(t) = \begin{cases} 0 & t < a \\ t/(b-a) & a \leq t \leq b \\ 1 & t > b \end{cases}$$

The expectation and variance are $(b-a)/2$ and $(b-a)^2/12$, respectively. The most used case is when the interval is $[0, 1]$. An important application of the uniform distribution is in the generation of random numbers.

A continuous random variable taking on only non-negative values with density and distribution functions

$$f(t) = \lambda e^{-\lambda t}$$

$$F(t) = 1 - e^{-\lambda t}$$

is said to have an **exponential distribution**. The expectation and variance are $1/\lambda$ and $1/\lambda^2$. The exponential distribution is used as the distribution of the time between events of a completely random process. It has an unusual property that regardless of how long one has waited since the last event the distribution of the remaining time to the next event has the same exponential distribution with mean $1/\lambda$.

A random variable X which can take on all real numbers from $-\infty$ to $+\infty$ having density and distribution functions;

$$f(t) = \frac{1}{\sigma\sqrt{2\pi}} e^{-\frac{1}{2}(t-\mu/\sigma)^2}$$

$$F(t) = \int_{-\infty}^{t} \frac{1}{\sigma\sqrt{2\pi}} e^{-\frac{1}{2}(x-\mu/\sigma)^2} \, dx$$

is said to have a **normal distribution** with parameters μ and σ. The parameters also are the mean and standard deviation, that is, $E(X) = \mu$ and Var $(X) = \sigma^2$. When $\mu = 0$ and $\sigma = 1$ we have the so-called **standard normal distribution** which has distribution function

$$\Phi(t) = \frac{1}{\sqrt{2\pi}} \int_{-\infty}^{t} e^{-x^2/2} \, dx$$

The function $\Phi(t)$ is tabulated at the end of this section. Computations of probabilities for random variables X with $E(X) = \mu$ and $\text{Var}(X) = \sigma^2$ are made using the relationship,

$$P\{X \leq t\} = F(t) = \Phi\left(\frac{t-\mu}{\sigma}\right).$$

The normal distribution occurs frequently in natural and engineering phenomena, and has the familiar bell-shaped curve for its density function. It arises in part because of the central limit theorem which is discussed elsewhere in this section.

As an example in computing probabilities for a normal random variable, let X be normal with $\mu = 2$ and $\sigma = 4$. Compute $P\{-1 < X \leq 3\}$

$$P\{-1 < X \leq 3\} = F(3) - F(-1) = \Phi\left(\frac{3-2}{4}\right) - \Phi\left(\frac{-1-2}{4}\right)$$

$$= \Phi(0.250) - \Phi(-0.750)$$

The two values of the Φ, found in the table of the normal distribution, are

$$\Phi(0.250) = 0.5987$$

$$\Phi(-0.750) = 1 - \Phi(0.750) = 1.00000 - 0.77337 = 0.22663$$

Thus

$$P\{-1 < X \leq 3\} = 0.59871 - 0.22663 = 0.37208$$

LIMITS FOR A PROBABILITY DISTRIBUTION. For a normal probability distribution, virtually all the observations fall within the range $\mu \pm 3\sigma$. We may note from Table 1 that with $z = 3$—that is, with a limit 3 standard deviations above the mean—the probability is 0.99865, which means that 99.865 percent of the distribution lies below $\mu + 3\sigma$. From Table 1 one may note that 0.135 percent of the distribution will lie below $\mu - 3\sigma$. Thus $99.865 - 0.135$, or 99.73 percent of the observations fall between plus or minus three standard deviations from the mean. Similar analysis will show that $97.725 -$

2.2875, or 95.45 percent of a normal distribution lies within two standard deviations of the mean.

The above remarks apply only for a normal distribution. Limits so computed
will be practical, however, for any distribution that is approximated reasonably
well by the normal distribution. It is possible to compute limits which will be
valid for a distribution of any form by means of an inequality known as **Tcheby-
cheff's inequality.** This inequality states that for any distribution, not more
than $1/z^2$ of the observations will be more than z standard deviations from the
mean. Thus no more than $1/3^2$ or 1/9 of a distribution will lie outside limits that
are plus and minus three standard deviations from the mean. No more than $1/2^2$
or 1/4 of the observations will lie outside limits at plus and minus two standard
deviations.

For non-normal distributions which are known to have only one mode or peak
and whose mode is at or near the value for the mean, an inequality known as
the **Camp-Meidell inequality** may be used to compute limits. This inequality
states that for any distribution meeting the above two requirements, not more
than $1/2.25z^2$ of the observations will be more than z standard deviations from
the mean. Thus no more than $1/2.25 \cdot 3^2$ or 1/20 of a distribution will lie beyond
plus or minus three standard deviations; no more than $1/2.25 \cdot 2^2$ or 1/9 will lie
beyond plus or minus two standard deviations.

Sampling

NEED FOR SAMPLING. In almost all areas of engineering and management, information must be obtained and decisions made by sampling. Working
with sample data entails some risk of misinformation, particularly if the sample
size is small and there is considerable variation in the population or process from
which the sample has been taken.

Sampling is a useful and effective way of obtaining information under many
circumstances. In many cases it is the only way available. The following are
some of the more important circumstances when sampling may be necessary or
desirable:

1. When precise information regarding the population is not necessary.
2. When measuring the entire population would be too costly or not worth the
added accuracy it would give.
3. When information must be obtained quickly.
4. When the entire population is not available for measurement at the time the
study must be made.
5. When measuring or observing an item destroys it or impairs its usefulness.

DRAWING A RANDOM SAMPLE. An important objective in collecting items to make up a sample is to select items that represent without bias the
population from which they have been drawn. In any study some sources that
could introduce bias will be known and others will not. The procedure for
selecting sample items must be designed to circumvent known **sources of bias.**
It must also be designated so as to avoid or minimize sources of bias that may
be present but are not known. If bias cannot be avoided, it is important to
realize that it exists, so that it may be allowed for in the analysis of the results
of the sampling.

Ways of avoiding bias from known sources depend on the nature of the source
itself. Often a reasonable solution is readily apparent. In sampling production
figures in a seasonal industry, samples would be drawn from days throughout
the year, for example. The important thing to do before drawing a sample is
to study carefully the population and the variation that may exist within it so

is to discover all important possibilities that may result in a biased sample. For example, in taking a sample of elapsed times for a work element by a stop watch time study, possible sources of bias are nonstandard working conditions, timing an operator who is not of average ability, unconscious but consistent overreading of the stop watch, and the like.

The best practice for avoiding or minimizing the effect of bias sources that are unknown is to draw or select sample items purely at random. This practice will also, of course, be effective in dealing with known sources. To **select an item at random** simply requires that every item in the population have an equal probability of being selected. While this requirement may be simple to state, it may often be difficult to meet it effectively. Some parts of the population may be more difficult to get at than others. Simply glancing at an item may indicate at once the category into which it will fall. The items may differ in appearance; this may have a subconscious effect on the person selecting the sample. The physical arrangement of the items or figures may also have a subconscious effect.

Ways To Avoid Sampling Bias. The following practices may be of help in avoiding some of the more common sources of bias:

1. Assign a number to each item or the position of each item and select items by number through use of a table of random numbers or simply by drawing numbers out of a hat. This is a particularly useful practice when the nature of an item, such as its size, is apparent before a decision can be made on its selection as a sample item.

2. Avoid any arbitrary scheme for selecting items, such as taking every tenth item in succession. Such an arbitrary scheme might result in taking each observation from the same day of the week, from the same operator, from the same shift, or from the same machine for example.

3. Try to avoid any influence of physical location of items. Do not tend to take most items from the tops of lists, groups, or cartons, for example.

4. When a sample must be taken from some one period of time or on some one date, be sure the time period is representative.

5. Be sure that bias is not introduced into the observations by the method or practice of measurement. In a poll, for example, the response may be influenced considerably by the way questions are worded or by the dress or personal characteristics of the person doing the polling.

6. If a sample is made up only of replies voluntarily returned in a questionnaire or poll, be sure the results will not be biased by those who fail to reply because of pride, lack of interest, fear of a wrong response, or some other reason. On a questionnaire to determine by sampling what proportion of the employees in a company would like to have a lunchroom in the plant, for example, a greater proportion of those sampled who wanted the lunchroom might return their questionnaire than of those who did not. Such forms of bias are very difficult to overcome.

7. If a population is stratified or divided in some arbitrary way, **proportional sampling** may be of help. In this form of sampling, the number of sample items drawn from each segment of the population is in proportion to the number of items or individuals in each segment. In taking a sample of employees or employee records from a plant employing 500 persons on the day shift and 300 persons on the night shift, for example, to make a sample of 40 items, 25 would be selected from the day shift and 15 from the night shift.

RANDOM DIGITS. In simulation experiments or in designing sampling procedures it is sometimes necessary to be able to artificially create a random sample of some distribution. In some cases this can be done mechanically by tossing coins or dice or selecting numbers out of a bowl. Use of a table of random digits eliminates the need for such mechanical procedures. Table 9 provides a set of digits, that is, integers from 0 to 9, which have the appearance of complete randomness. Given one of the digits, or even any sequence of the digits in the table the probability distribution of the next digit in the table being any specific integer 0, 1, 2, 3, 4, 5, 6, 7, 8 or 9 is 0.10. For convenience in use, the integers are arranged as five digit numbers. The use of the table of random digits is illustrated with the following examples.

Selection of a Random Sample Using Random Digits. A shipment of copper consists of 740 coils. A random sample of 5 coils is made for use in an acceptance sampling plan. In order to assure that the sample is random each coil is assigned a number from 1 to 740 and a selection of coils for inspection is made by drawing the numbers of the coils from the table of random digits. Before looking at the coils or at the table of random digits we pick a row and column of the table as the starting point in our sampling. This also could be done by tossing a die or picking numbers from a hat. Suppose we select column 2 and row 4 to start with. We shall read down the 2nd 5-digit column selecting every other 5-digit number for examination. If the last three digits of the number are between 001 and 740 that coil will be inspected, if not the number will be ignored and we select another number. Our procedure generates the following sequence of 5-digit numbers and decisions.

Random Number	Action
93093	Sample coil 93
06907	—
91977	—
36857	—
40961	—
61129	Sample coil 129
12765	—
54092	Sample coil 92
97628	Sample coil 628
58492	Sample coil 492

We are assured that our sample of coils numbered 92, 93, 129, 492, and 628 is a random sample.

Generation of a Random Sample on a Probability Distribution. In a simulation of the effects of changes in maintenance staffing on machine downtime, it was necessary to generate a sequence of observations on a Poisson distribution since the number of breakdowns per shift was approximately described by a Poisson distribution with parameter $\lambda = 2.0$. The cumulative distribution is given below.

k	$P\{x \leq k\}$	Random Digit Range
0	0.135	001–135
1	0.406	136–406
2	0.677	407–677
3	0.857	678–857
4	0.947	858–947
5	0.983	948–983
6	0.995	984–995
7	0.999	996–999
8	1.000	000

A random sample of size 12 on the Poisson distribution with parameter $\lambda = 2$ is generated by selecting an arbitrary starting position in the table of random numbers and proceeding down the column selecting every fifth random number. If the first three digits of the random number fall in the indicated range the number of breakdowns in the shift is taken as the value of k in the above tabulation. Starting with column 3 and row 6 of Table 9 the following sequence of numbers is generated:

Random Number	Number of Breakdowns
11008	0
88231	4
51821	2
85828	4
04839	0
87917	4
27958	1
33362	1
83974	3
02011	0
42751	2
33276	1

The sequence of breakdowns 0, 4, 2, 4, 0, 4, 1, 1, 3, 0, 2, 1 is a random sample on a random variable having a Poisson distribution with parameter $\lambda = 2$.

In using a table of random digits care must be taken to use a different procedure for selecting the random digits each time the table is used in order to avoid bias in the selection of the samples.

FREQUENCY DISTRIBUTIONS. One of the first tasks in dealing with data obtained by sampling, by an experiment, or from historical data is to form a frequency distribution. This is a recording of the measured values together with their frequency of occurrence. The frequency distribution provides a statistical estimate of the probability distribution and hence is very important in most problems.

A simple and generally useful procedure is to summarize the observations or a fairly large sample of the observations (preferably a hundred or more, if available) in the form of a **frequency distribution** as in Fig. 1. It can then be observed whether the frequencies found for each measurement (or range of measurements) correspond approximately to frequencies that would be obtained from one of the standard probability distributions. In what follows we deal with the normal distribution. This distribution, when plotted and connected with a line, gives the familiar **bell-shaped curve.** The most frequently occurring measurements are those close to the average; measurements lying at a greater distance from the average occur less frequently. This distribution is usually the result of **item variation** produced by some system of many chance-acting causes. The distribution is found widely in all the fields of physical and social science as well as in fields of industrial engineering and management. Many of the most commonly used statistical methods are for normal distributions and so their effectiveness is dependent upon the distribution for the measurements under study being approximately the normal distribution.

The normal distribution is illustrated by the following figures for weekly incentive earnings for a group of piece-workers in a shop, all working at the same piece rate.

Incentive Earnings	No. of Employees
$34–36	1
32–34	4
30–32	5
28–30	8
26–28	21
24–26	32
22–24	34
20–22	38
18–20	35
16–18	26
14–16	18
12–14	7
10–12	5
8–10	2
6–8	1
4–6	1
Total	238

These data have been plotted to scale in Fig. 1. A dashed line has been drawn to show the **theoretical normal curve** that the actual frequency distribution approximates.

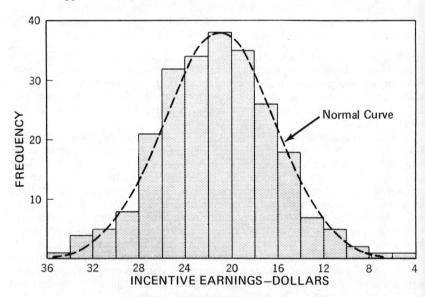

Fig. 1. **Frequency distribution of incentive earnings.**

The frequencies may be **plotted graphically** and connected with a freehand curve, fitted approximately by eye. The shape of the curve may then be observed to see whether or not it exhibits the typical bell-shaped normal form. In cases where a more precise test seems necessary, a more involved **statistical procedure** for comparing observed frequencies with frequencies that could be expected if the distribution were normal will be found in any standard textbook on statistics (chi-square and Kolmogorov-Smirnov goodness of fit tests).

Another method of testing for normality is to plot the cumulative distribution of the sample or population on **normal-probability graph paper** (see Fig. 2). The vertical scale on this paper is so spaced that if the distribution plotted is normal it will plot as a straight line. The plot may also be used to determine the average, to measure dispersion, and to determine the proportion of cases beyond any specified limit. These uses are discussed elsewhere in this section.

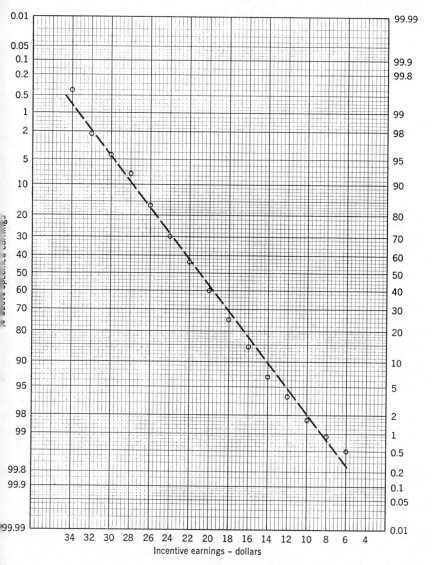

Fig. 2. Cumulative frequency distribution of incentive earnings plotted on normal-probability paper.

The accompanying data illustrate the computation of the cumulative frequency distribution for the incentive earnings example.

Incentive Earning	Frequency	Cumulative Frequency	Cumulative Percentage
$34–36	1	1	0.4%
32–34	4	5	2.1
30–32	5	10	4.2
28–30	8	18	7.6
26–28	21	39	16.4
24–26	32	71	29.8
22–24	34	105	43.7
20–22	38	143	60.0
18–20	35	178	74.6
16–18	26	204	85.5
14–16	18	222	93.2
12–14	7	229	96.2
10–12	5	234	98.3
8–10	2	236	99.1
6–8	1	237	99.5
4–6	1	238	100.0

The cumulative percentages in the last column have been plotted on normal-probability paper in Fig. 2. As the plottings approximate a straight line one may assume the distribution is normal.

Another example of a frequency distribution uses the Poisson distribution. This distribution may be applied to the study of lost-time accidents in industry. For every workman in a plant there are countless moments throughout each day in which a lost-time accident could occur. The probability of one occurring to a man at any one moment is extremely small. The following illustration is from a study of accident figures for a plant in which the average number of lost-time accidents per month has been, over a long period of time, 2.3. Considerable variation from month to month was noted, however, the number ranging from 0 to 5. The meaning of this variation was under question. Use of the Poisson distribution, taking a study period of 60 months, gave the frequency distribution shown in Fig. 3. The distribution showed that the variation in lost-time accidents experienced could be expected by chance from a stable system of accident-producing causes.

OTHER FREQUENCY DISTRIBUTIONS. Occasionally other forms of frequency distributions will be found in industrial engineering and management data. For these special forms the common statistical techniques developed for the normal, the binomial, and the Poisson distributions will ordinarily not apply. For this reason one should be reasonably sure of the nature of the distribution before any such application is made. Usually past experience in the study of similar activities or a knowledge of the variation-producing causes inherent in the activity will indicate the form of the distribution to expect. If there is any doubt, a frequency distribution may be made of a reasonably large sample of observations (at least one hundred) and studied visually to determine its form.

Among the more commonly encountered variations are the bimodal, the screened, the rectangular, and the skewed (see Fig. 4).

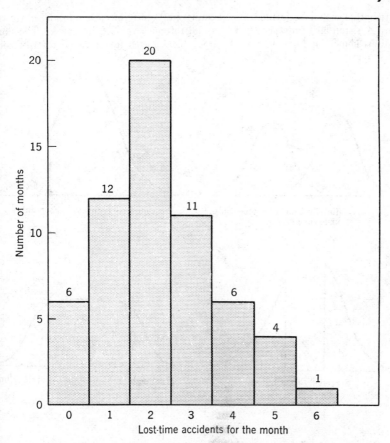

Fig. 3. Frequency distribution of lost-time accidents per month based on the Poisson distribution.

Bimodal. Such distributions are usually formed as a result of combining, in effect, two separate distributions with different averages. This might be the case, for example, for time-study observations of a work element in which a piece part is taken from a container and placed in an assembly jig. Many of the parts may be entangled in the container and when this is the case the grasp and separation may be difficult, giving a longer average time for the element. A distribution of time values obtained when this occurs would give one distribution. A distribution with a lower average value would be formed from time values obtained when parts were not entangled. As one or the other situations occur at random during the study, the resulting distribution will be bimodal and appear as shown in Fig. 4a.

Screened. When data are from an activity or product for which all items or events measuring beyond a specified limit (or between a pair of specified limits) have been removed, a screened distribution is formed. **Out-of-limit items** may have previously been removed by sorting all items, or by use of some procedure or device which automatically or as a matter of routine removed them. Such a distribution would be formed, for example, in a study of cash refunds in a retail store if a standard practice was to pay all refunds over $10 by check. The illustrations shown in Figs. 4b and 4c are typical forms for screened distributions

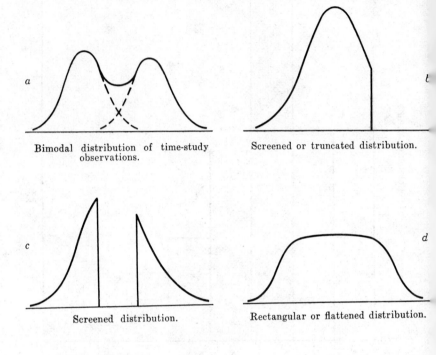

a

Bimodal distribution of time-study observations.

b

Screened or truncated distribution.

c

Screened distribution.

d

Rectangular or flattened distribution.

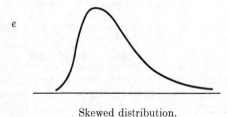

e

Skewed distribution.

Fig. 4. Special types of frequency distributions.

Rectangular. The term "rectangular" is applied to a long, flat distribution such as the one shown in Fig. 4d. Such a distribution may result from a steady drift in one direction during the course of the study of the average tendency of the activity being measured. It may also be a result of several separate frequency distributions. Such might be the case when materials from several different sources, each with different means for the measured characteristic, are mixed. Or a rectangular distribution may result when each outcome within the the range of possibilities is equally likely.

Skewed. Many distributions encountered in production are strongly skewed to the right. Typically service or repair times, times to failure of equipment, and age of items in inventory have skewed distributions (Fig. 4e). The gamma family of probability distributions may often be used to represent skewed distributions.

MEASURES OF CENTRAL VALUE. Presenting data as a frequency distribution will be most useful in some cases. Generally, however, the data may also be summarized by one or both of two measures: (1) the "average" of the measurements—some measure of central value or central tendency—and (2) some measure of the dispersion of measurements about their average.

The use of an average to indicate central tendency makes it possible to:

1. Use one figure to give a useful, meaningful, and concise summary of a body of data.
2. Readily compare one distribution or population with another.
3. Readily use sample observations to obtain data or make inferences about a population.

The Arithmetic Mean. The most generally useful measure of central tendency is the arithmetic mean. The arithmetic mean provides a statistical estimate of the expectation $E(X)$ of a random variable. The arithmetic mean, symbolized by $\overline{X}$, is obtained by taking the sum of the observations and dividing by the number of observations used. In equation form,

$$\overline{X} = \frac{\Sigma X}{n}$$

For example, suppose the arithmetic mean of the following observations of parts per package is required: 15, 19, 16, 15, 18, 14, 15, 18.

$$\overline{X} = \frac{15 + 19 + 16 + 15 + 18 + 14 + 15 + 18}{8} = \frac{130}{8} = 16\tfrac{1}{4}$$

If the observations have been arranged as a frequency distribution, the computation may be more readily made by the following equation:

$$\overline{X} = \frac{\Sigma f X}{\Sigma f}$$

For example, consider the following distribution:

Years of Experience X	No. of Men f	fX
1	7	7
2	16	32
3	22	66
4	9	36
5	5	25
6	1	6
	$\Sigma f = 60$	$\Sigma f X = 172$

$$\overline{X} = \frac{\Sigma f X}{\Sigma f} = \frac{172}{60} = 2.87 \text{ years}$$

In some cases coding the observations will simplify the calculations (Richmond, Statistical Analysis). In the following example, each of the observations has been coded by first subtracting 600 and then dividing by 10.

Temperature X	Coded X	No. of Observations f	fX(Coded)
630	3	8	24
640	4	23	92
650	5	15	75
660	6	4	24
		$\Sigma f = 50$	ΣfX(coded) $= 215$

$$\overline{X} \text{ (coded)} = \frac{215}{50} = 4.3$$

$$\overline{X} \text{ (decoded)} = (4.3 \times 10) + 600 = 643$$

The **advantages** of the arithmetic mean as a measure of central tendency are:

1. It is relatively simple to compute.
2. It is generally the most efficient of all possible measures when using sample data to obtain an estimate of the central tendency of a population.
3. It is the measure most persons assume has been computed when the term "average" is used. Hence when it is used there is relatively little chance of misunderstanding.
4. It is the measure of central tendency required in most statistical tests.

However in some cases particularly when the distribution is quite skewed the arithmetic mean may not be as useful a measure of central tendency as the median. Such is the case, for example, in income or death statistics, or time to failure or breakdown.

Graphical Method for Computing the Mean. If the data have been compiled as a cumulative frequency distribution and plotted on normal-probability paper, a close estimate of the arithmetic mean may be very simply determined. The method is to simply read the value on the horizontal axis of the graph equivalent to the 50 percent point on the vertical axis. This value is an **estimate of the mean.** This method depends on the distribution being normal, and uses the fact that the mean and median are identical. For the incentive-earnings illustration previously plotted in Fig. 2, the mean is read as $21.

The Sample Median. The sample median is the middle value in a group of observations that have been arranged in order of magnitude. It is a statistical estimate of the median of a distribution. If there is an even number of observations, the mean of the two middle values may be taken as the sample median.

For example, suppose the labor cost for a certain operation has been ascertained for nine different plants and that they are as follows (in ascending order of magnitude): $1.08, $1.10, $1.20, $1.22, $1.25, $1.25, $1.30, $1.30, and $1.35. The median cost is the fifth observation, $1.25.

When a more precise measure is not necessary, the sample median may be a useful measure, particularly under the following circumstances:

1. When extreme values must not be given undue weight.
2. When the exact magnitude of extreme items is not known. For such items it is necessary to know only whether they lie above or below the median.
3. When ease of computation is important.
4. When the number of observations is small.
5. When the distribution is quite skewed.

The Mode. The mode for a group of observations is the observation that occurs most frequently. For the observations of parts per package—15, 19, 16, 15, 18, 14, 15, 18—the mode is 15.

The mode is a useful measure of central tendency in many cases, particularly if the number of observations is relatively large. In time study, for example, the modal value for the time observations recorded for a work element serves as a useful measure of the time typically required for performing the element. The general **advantages** of the mode as a measure of central tendency are:

1. It is very easy to determine, particularly if the data have been put in the form of a frequency distribution.
2. Extreme values among the observations do not have undue influence.
3. The exact magnitude of extreme values does not have to be known.
4. It is the most typical value. In many situations this may be the most meaningful measure of central tendency.

MEASURES OF DISPERSION. In many cases the average must be supplemented by some measure of dispersion to summarize or describe a body of data adequately. Also, most statistical tests depend on the use of some measure of **sample dispersion.** The most commonly used measures are briefly described in the following articles.

The Range. A useful measure of dispersion when the number of observations is small is the range. The range, symbolized by R, is simply the difference between the largest and the smallest observation in the group. For example, suppose the following measurements have been obtained: 16.4, 14.7, 17.5, 13.2, 14.5, 16.2, 18.3, 15.8, 14.1, 15.4. Then

$$R = 18.3 - 13.2 = 5.1$$

The range is very easy to compute and is easily understood. However, as only the two extreme values of a group are used in its computation, it is an inefficient measure, particularly when the number of observations available is large. For this reason its use, particularly when working from sample data to estimate the dispersion of a population, should be restricted to applications where the sample size is small.

The Mean Deviation. Another measure of dispersion that is relatively easy to compute is the mean deviation. The mean deviation is the arithmetic mean of the absolute deviation of each observation in the group from the group mean. In terms of an equation,

$$\text{M.D.} = \frac{\Sigma|X - \overline{X}|}{n}$$

For an example, consider the following performance ratings made by a group of eight time-study men during a rating test:

Man	Rating X	$\bar{X}$	$|X - \bar{X}|$
A	115	111.5	3.5
B	112	111.5	0.5
C	120	111.5	8.5
D	107	111.5	4.5
E	110	111.5	1.5
F	110	111.5	1.5
G	100	111.5	11.5
H	118	111.5	6.5
	$\Sigma X = 892$		$\Sigma|X - \bar{X}| = 38.0$

$$\bar{X} = \frac{892}{8} = 111.5$$

$$\text{M.D.} = \frac{38.0}{8} = 4.75$$

The mean deviation is useful as a simple way of describing the dispersion of a group of observations. However, it is not a generally useful measure for statistical procedures. The standard deviation (or the range, in some cases) must be used.

The Standard Deviation. The most generally useful, and therefore the most commonly used, measure of variability is the sample standard deviation. The standard deviation of a probability distribution was defined in the article on probability and is denoted by the Greek lower case sigma (σ).

The sample standard deviation of a set of observations tends to be slightly less than that for the population or distribution from which it is drawn. If the sample standard deviation is computed to serve as an estimate of the **population** standard deviation, this bias may be automatically corrected by dividing the sum of the squared deviations by $n - 1$ instead of by n. Since this is almost always the purpose for computing the sample standard deviation and since almost all data are sample data, the following is the equation for general use:

$$s = \sqrt{\frac{\Sigma(X - \bar{X})^2}{n - 1}}$$

The letter s is used instead of the Greek letter σ to indicate clearly that the measure is obtained from sample data. Thus, s is the estimator of σ. Note that the statistician's notation differs from the notation in the field of statistical quality control where σ is the symbol used for the sample standard deviation and σ' the symbol used for the standard deviation of the population.

For the sample of performance ratings used to illustrate the computation of the mean deviation, the computation of the standard deviation, s, is as follows:

X	$X - \overline{X}$	$(X - \overline{X})^2$
115	3.5	12.25
112	0.5	0.25
120	8.5	72.25
107	− 4.5	20.25
110	− 1.5	2.25
110	− 1.5	2.25
100	−11.5	132.25
118	6.5	42.25

$$\Sigma(X - \overline{X})^2 = 284.00 \qquad n - 1 = 8 - 1 = 7$$

$$s = \sqrt{\frac{284}{7}} = \sqrt{40.6} = 6.4$$

A **short-cut method** for more easily computing the standard deviation, particularly when a calculator or a table of squares is available, is given by the following equation:

$$s = \sqrt{\frac{\Sigma X^2 - \dfrac{(\Sigma X)^2}{n}}{n - 1}}$$

The following is an illustrative application using the data of the previous illustration. Observations have been **coded** by subtracting 110 further to simplify the computations. Note that since the standard deviation involves only differences between each observation and the mean, no decoding step is required. Only if coded numbers are obtained by dividing or multiplying by some constant is decoding of the computed value necessary.

X	Coded X	Coded $(X)^2$
115	5	25
112	2	4
120	10	100
107	− 3	9
110	0	0
110	0	0
100	−10	100
118	8	64
	$\Sigma X = 12$	$\Sigma X^2 = 302$

$$s = \sqrt{\frac{302 - \dfrac{12^2}{8}}{7}} = \sqrt{\frac{284}{7}} = \sqrt{40.6} = 6.4$$

Graphical Method for Determining the Standard Deviation. If the frequency distribution for the population or sample has been plotted on normal-probability paper and a straight line fitted to the points, an estimate of the standard deviation may be made quite simply. The method is to determine first the point on the line above which 84 percent of the distribution lies. Next determine the point on the line below which 16 percent of the distribution lies. The range between these two points, measured on the horizontal scale, is an estimate of twice the standard deviation. For the incentive-earnings illustration previously plotted, the 84 percent point is at $16 and the 16 percent point at $26. This gives a value of ($26 − $16)/2, or $5 for the standard deviation. This method depends on the distribution being normal, that is, the fit on the normal-probability paper must be good.

SAMPLING DISTRIBUTIONS. Due to the effects of chance in the selection of sample items, the characteristics of a sample are likely to differ somewhat from the characteristics of the population from which the sample was taken or differ from the characteristics of another sample, if another were to be drawn. Both the arithmetic mean and the standard deviation for a sample of 25 items, for example, will be somewhat more or somewhat less than the expectation or the standard deviation for the population. Whether one measure is more or less than the other and the exact extent of the difference cannot be known in any single case. However, useful statistical theory and techniques have been developed to use in **estimating the probability** of any specified difference between sample and population values.

The approach is through a study of sampling distributions. To consider a simple illustration, suppose we were to draw many samples of 25 items, each from a population with an arbitrary distribution, a mean, μ, of 200 and a standard deviation, σ, of 15. Suppose, further, that the mean of each sample is computed. It would be found that these means would form a frequency distribution of their own—a distribution with considerably less dispersion than the population distribution, but one that centered about the same mean value as for the population. Sampling distribution theory would show that the expected value for the distribution of sample averages, $\mu_{\bar{x}}$, does coincide with the expected value, μ, of the population from which they have been drawn. That is,

$$\mu_{\bar{x}} = \mu$$

Also, the theory would show that the standard deviation of the distribution of sample averages was equal to the standard deviation for the population divided by the square root of n, the sample size. That is,

$$\sigma_{\bar{x}} = \frac{\sigma}{\sqrt{n}}$$

Finally, sampling distribution theory would show that the distribution of sample averages was approximately a normal distribution. These two distributions are shown in Fig. 5.

Useful conclusions can be drawn with this theory regarding the **reliability of sample data.** In the above illustration, for example, note that if a sample of 25 items is drawn from a population whose standard deviation is 15, one may be rather confident that the mean of the sample will be no more than 9 units away from the mean of the population. More precise techniques based on sampling distribution theory for use in determining confidence limits are described under Statistical Inference in this section. Theoretical knowledge regarding sampling distributions and the sampling-distribution approach also forms a basis for many other statistical procedures.

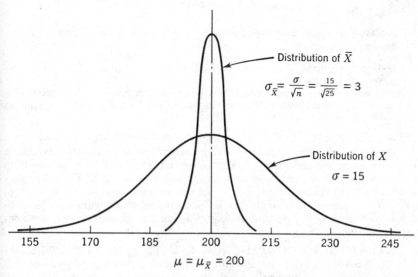

Distribution of $\overline{X}$

$$\sigma_{\overline{X}} = \frac{\sigma}{\sqrt{n}} = \frac{15}{\sqrt{25}} = 3$$

Distribution of X

$\sigma = 15$

| 155 | 170 | 185 | 200 | 215 | 230 | 245 |

$\mu = \mu_{\overline{X}} = 200$

Fig. 5. Sampling distribution of $\overline{X}$.

Statistical Inference

BASIC CONSIDERATIONS. Using a sample of data taken from a given population, such as a lot of tubes, or from a process, such as time-study readings, generalizations can be made about the underlying distribution, its form, the value of its mean or standard deviation, and whether or not it differs from another distribution. The subject which deals with making such generalizations is called **statistical inference** and it can be broken down roughly into two sets of procedures: **estimation** and **hypothesis testing.** Estimation involves estimating the value of parameters of distributions and studying their properties. Hypothesis testing involves making decisions, usually of a yes-or-no variety, about distributions or their parameters. An example of a hypothesis test is the decision that the mean breaking strength of a certain material is or is not above a specified value.

Since only a sample of data is used, the decisions and estimates always involve the possibility of error. The modern theory of statistical decisions studies this aspect of statistics in great detail (Lindgren, Statistical Theory).

ESTIMATION. Usually in working with random variables one does not know the values of the distribution parameters. Instead one estimates the values on the basis of a sample of observations. Suppose that X is a random variable with $X_1, \ldots, X_n$ samples on the distribution of X. One of the parameters, θ, of the distribution of X is unknown. The theoretical problem of estimation is to find that function of the observations, say $g(X_1, \ldots, X_n)$, which "best" estimates θ. There are two types of estimates in common usage, point estimates and interval estimates (Mood and Graybill, Introduction to the Theory of Statistics). Point estimates provide a single number as an estimate of the parameter while interval estimates provide an interval within which the parameter is expected to lie with some specified probability.

A point estimate is

An **unbiased estimate,** if its expectation is equal to the parameter being estimated. In the above example, if $E[g(X_1, \ldots, X_n)] = \theta$, $g(X_1, \ldots, X_n)$, is an unbiased estimate.

A **consistent estimate,** if for very large sample sizes the estimate is equal to the parameter with probability one.

A **minimum variance estimate,** if it has the smallest variance of all estimates of a parameter. Usually it is desirable to have estimates which have minimum variance of all unbiased estimates.

The point estimates $\overline{X}$ of $E(X)$ and s^2 of Var (X) which have already been discussed are unbiased, consistent, and generally minimum variance estimates. However, since the estimates are themselves random variables it is useful to provide more than a point estimate. An interval estimate of $E(X)$ or Var (X) is often desired together with a measure of the confidence one has in the interval estimate. Such interval estimates are called **confidence intervals** and the confidence coefficient is the probability that the random interval covers the nonrandom but unknown parameter.

DETERMINING CONFIDENCE LIMITS FOR THE POPULATION MEAN. A good estimate of the population mean, μ, is given by the sample mean, X. Limits within which the true value for the population mean will lie can be computed for any desired degree of confidence by the following procedures. The procedures are for normal distributions but they will give satisfactory results in cases where a distribution only approximates the normal, as long as the sample size is large. The confidence interval provides an interval estimate of μ.

Population Standard Deviation Is Unknown. The limits between which the population mean, μ, lies are given by:

$$\overline{X} \pm t \frac{s}{\sqrt{n}}$$

where $\overline{X}$ is the mean for the sample, s is the sample standard deviation, and n is the number of items in the sample. The value for t may be obtained from Table 2. For the test, the **degrees of freedom** are $n - 1$ or one less than the number of measurements in the sample. The **probability value** to use in selecting t from the table depends on the desired amount of confidence in the true value being between the computed limits. It should be emphasized that the true but unknown mean is a constant and the computed limits are the random variables. The values in the table heading show the probability of the true value being outside one or the other of the computed limits. The probability of it being beyond a specified one of the limits is one-half the value in the table heading. Thus if a probability of 0.10 is used, the probability is 0.10 or 1 out of 10 that the real mean is outside the limits and 0.90 or 9 out of 10 that it is inside the limits. The probability that it is, say, greater than the upper limit is one-half the value in the table heading or 0.05.

An example may be taken from a plant in which certain parts are made from leather. In studies to determine material cost standards it was found that for a sample run of 30 hides the average number of parts per hide was 137 with a standard deviation of 14. To determine the reliability of this average figure, 95 percent confidence limits were computed as follows:

$$137 \pm 2.042 \frac{14}{\sqrt{30}} \quad \text{or} \quad 137 \pm 5.2 \text{ parts}$$

Population Standard Deviation Is Known. The t-factor in the procedure above allowed for sampling variations in the sample standard deviation. If through past data and computations the standard deviation for the population has been determined, and if it can be assumed this value is good for the current situation, closer confidence limits can be determined. The formula is

$$\bar{X} \pm z \frac{\sigma}{\sqrt{n}}$$

where σ is the population standard deviation. The value for z may be obtained from Table 1. For any value of z this table shows the probability of the population mean being beyond one of the limits. For example, with $z = -1.645$ the probability is 0.05 that the population mean will be above the upper of the computed limits and 0.05 that it will be below the lower. This gives a probability of 0.90 that it lies between the limits.

If in the previous illustration, for example, the population standard deviation for parts per hide was known to be 11, the limits could be computed as follows:

$$137 \pm 1.96 \frac{11}{\sqrt{30}} \quad \text{or} \quad 137 \pm 3.9 \text{ parts}$$

Use of the Range. If the number of observations in the sample is small, a simple alternative procedure using the **sample range,** R, as a measure of dispersion is available. The limits are

$$\bar{X} \pm t_R R$$

The value for t_R may be obtained from Table 3. The probabilities in the table heading are for a value beyond one or the other of the limits. As in the previous procedures, use of $P = 0.10$ gives a probability of 0.10 of the mean lying beyond one limit or the other or a probability of 0.90 that it lies between them. The range is not as efficient as the standard deviation as a measure of dispersion. For this reason the limits for any selected level of confidence will be wider than those determined by the procedure using the sample standard deviation.

Suppose, for example, the man-hours required for an activity have been measured for 8 cycles. The shortest time measured was 43.5 hr. and the longest 46.2 hr. with a mean time for the 8 cycles of 44.7 hr. Confidence limits for the population mean, say at 95 percent confidence, are required. The computations are:

$$44.7 \pm 0.29(2.7) \quad \text{or} \quad 44.7 \pm 0.78$$

The range, 2.7, was obtained by subtracting the smallest measurement from the largest. Table 3 was entered at a probability level of 0.05 to give 95 percent confidence limits.

OTHER LIMITS. Confidence Limits for Population Standard Deviation. The best estimate for the population standard deviation, σ, is given by the **sample standard deviation,** s, computed by the equation

$$s = \sqrt{\frac{\Sigma(X - \bar{X})^2}{n - 1}} \quad \text{or} \quad s = \sqrt{\frac{\Sigma X^2 - \frac{(\Sigma X)^2}{n}}{n - 1}}$$

Limits between which the true value for the population standard deviation will lie can be computed for any desired degree of confidence. The procedure is exact for normal distributions. Any substantial deviations from normality will result in a confidence level somewhat different from that selected.

$$\text{The upper limit for } \sigma = \sqrt{Fs^2}$$

The value of F for the upper limit is obtained directly from Table 6. The probability value to use in selecting F from this table depends on the required degree of confidence that the true value for the standard deviation will lie between the computed limits. The table heading shows the probability of the true value lying above the upper limit. In entering the table for the F-value for this limit, the degrees of freedom for the **greater mean square** are ∞ and for the **smaller mean square** $n - 1$.

$$\text{The lower limit for } \sigma = \sqrt{\frac{s^2}{F}}$$

The value of F for this limit is also obtained from Table 6, but with this difference: the degrees of freedom for the greater mean square are $n - 1$, and the degrees of freedom for the smaller mean square are ∞. The probability value for the table gives the probability of the true value being less than the lower limit. Thus if a probability value of 0.01 is used for each limit, the probability is 0.02 that the true value will be outside the computed limits or 0.98 that it lies between them.

A simple illustration is an application of the study of consistency in an experimental model of a container-filling machine. Twenty-five packages were filled and carefully weighed. The standard deviation, s, of the weights was found to be 0.17 ounces. It was desired to determine with 90 percent confidence the limits between which the true standard deviation will fall. The computations for this case were:

$$\text{Upper limit} = \sqrt{Fs^2} = \sqrt{1.73(0.17)^2} = 0.22$$
$$\text{Lower limit} = \sqrt{\frac{s^2}{F}} = \sqrt{\frac{0.17^2}{1.52}} = 0.14$$

Thus with 90 percent confidence the true standard deviation could be said to lie between 0.14 and 0.22 (with the best estimate being, of course, 0.17).

Confidence Limits for Population Proportion. Under many situations the proportion of times a specified event occurs in a population must be estimated by determining the proportion of times the specified event is found in a sample. The proportion found for the sample is the best estimate available for the proportion for the population. However, the information given by the sample is likely to differ somewhat by chance from the true population figure.

The following procedure gives **approximate limits** to indicate the extent to which the population proportion may differ from the sample proportion. The limits are:

$$p_s \pm z \sqrt{p_s \frac{(1 - p_s)}{n}}$$

where p_s is the proportion of the specified event for the sample and n is the number of items in the sample. The value for z may be obtained from Table 1. For any value of z subtract the table value to obtain the probability of the population proportion being beyond one of the limits. This equation uses the normal distribution as an approximation to the binomial and is accurate only for large values of n. The probability of the population proportion being outside both computed limits is thus twice the probability derived from the table.

For example, in a work-sampling study 300 random observations of an activity were made, and for 39 of them the operation was delayed. This gave 39/300 or 0.13 as the proportion of observations for which a delay was noted. Approximate limits between which one could be 90 percent confident that the population proportion would lie were computed as follows:

$$0.13 \pm 1.645 \sqrt{\frac{0.13(1 - 0.13)}{300}} \quad \text{or} \quad 0.13 \pm 0.032$$

Tolerance Limits for Items in a Population. Often it is necessary to determine the limits between which will lie the measurements for all or some specified proportion of the measurements for individual items in a population. If the expected value, μ, and the standard deviation, σ, of the population are known, and if the distribution is normal, this can be done by use of the table of normal areas by a method that is described elsewhere in this section. Usually, however, these values for the population are not known and the limits must be ascertained by the use of sample data. A simple statistical procedure has been developed for determining limits in such cases with any desired **degree of confidence.** These limits will be somewhat wider than those computed by the use of the population mean and standard deviation because allowance must be made for the fact that the sample mean and standard deviation are likely not to reflect exactly the distribution from which they have been drawn. The limits are determined by the measure

$$\overline{X} \pm Ks$$

where $\overline{X}$ is the mean and s the standard deviation of the sample. Values for K are given in Table 5. For any sample size or number of observations the value for K depends on the level of confidence desired and on the proportion of the population that must lie between the limits.

For the limits to be valid at the selected probability level, the distribution must be normal. If there is any significant departure from normality, the actual probability will differ somewhat from the value used in selecting the factor K.

Suppose, for example, that in **studies to determine estimating time standards** for use in residential air-conditioning installations, time values were obtained for the work of ditching to the cooling tower for 10 jobs. The average time was found to be 12.2 hr., with a standard deviation of 0.7 hr. The computed limits between which the required ditching time for 95 percent of installations can be assumed, with 90 percent confidence, to lie are

$$12.2 \pm 3.02(0.7) \quad \text{or} \quad 12.2 \pm 2.1 \text{ hr.}$$

This variation in the actual time that might be required for this element of work was considered to be so small as to be of no practical significance. Thus, it was decided that in making a cost estimate for a new job, a detailed estimate for ditching need not be made but that an average time figure could be used.

Studies of labor time for sheet-metal and insulation work for ducts for the same 10 installations gave an average of 108 hr. and a standard deviation of 14.8 hr. For this element, 90 percent confidence limits within which the time for 95 percent of the installations could be expected to fall were computed as

$$108 \pm 3.02(14.8) \quad \text{or} \quad 108 \pm 44.7$$

These limits showed that in making a bid for a new job, it would pay to make a detailed estimate for the duct work rather than to use an average figure.

Another typical application would be to use the factors to determine the **specification limits for a manufacturing process.** Suppose, for example, that for a sample of 50 items from a machine, the diameter of each has been measured and the average diameter, X, found to be 2.518 in., with the standard deviation, s, equal to 0.006 in. An estimate is desired of the limits between which virtually all the items produced will measure. A value of K of 3.13 will give limits between which, with 95 percent confidence, 99 percent of the items will lie. The limits are

$$2.518 \pm 3.13(0.006) = 2.499 \text{ and } 2.537$$

DETERMINING SAMPLE SIZE. There is no simple procedure for determining precisely the most appropriate sample size for a study. The **number of observations to obtain** should depend, among other things, on the following:

1. The cost of obtaining a sample observation. If obtaining observations is relatively costly, comparatively few observations may have to suffice.
2. The time and staff available for sampling and for recording data.
3. The value of added confidence in conclusions given by larger sample sizes.
4. The amount of inherent variability in the population being sampled. Generally, the greater the variability in the population, the larger the sample size must be to give conclusions of a specified degree of confidence.
5. The amount of inherent variability in the method of measurement.

A general knowledge of the **relationship between sample size and the degree of confidence** with which conclusions can be drawn from sample data can be obtained by a study of the formulas and tables of factors used in computing confidence limits for averages, standard deviations, and proportions. Consider, for example, the t-factor which can be used to compute confidence limits for the population mean. Note by scanning Table 2 that if the sample size is quite small, say two or three items, a considerable gain in confidence is obtained by increasing the sample size by four or five items. On the other hand, for a sample size of 20 or 30, relatively little gain is obtained by adding four or five more items to the sample. As the sample size is increased, a point of rapidly diminishing return in reliability is quickly reached. Confidence intervals tend to decrease approximately in proportion to the square root of the sample size.

Nomograph for Determining Number of Observations. A nomograph useful in determining the necessary number of observations to reach a satisfactorily precise estimate of the population average is shown in Fig. 6. In this figure, P is the permissible error of estimate, R is the risk that the error is greater than P, and N is the sample size required. Use of the nomograph requires that 10 observations be made and ratio A calculated.

$$A = \frac{\Sigma X^2}{(\Sigma X)^2}$$

Next, the R scale is entered at the decided risk and the intersection of R and the permissible error P is found. From this intersection a line is drawn up to the A value, and then across to the N scale.

For example, say we want to know within 2 percent the average machine time required to finish a part, and we will risk a probability of 0.05 of the error being greater than 2 percent. The first ten values we observe are: 10, 12, 18, 10, 11,

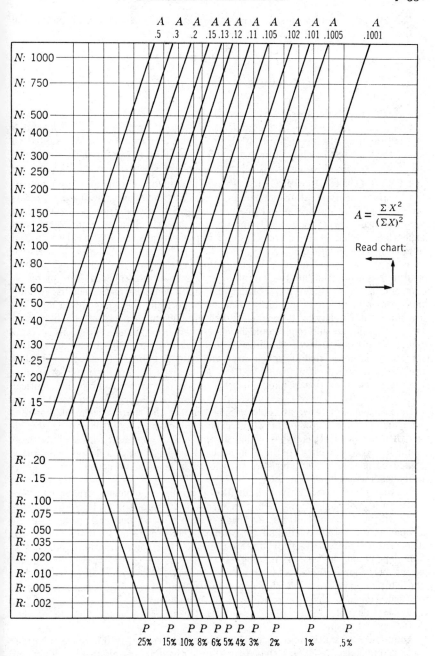

Fig. 6. Nomograph to estimate the number of observations necessary to result in a satisfactorily precise average value.

14, 10, 12, 15, 13. The sum of these values, $\Sigma X = 125$. The sum of the square of these values, $\Sigma X^2 = 1,623$.

$$A = \frac{1,623}{125 \times 125} = 0.104$$

$$P = 2\%$$

$$R = 0.05$$

About 600 observations would be required.

Frequently used values of P and R are 5 percent error and 0.05 risk, and a line is drawn on the nomograph at these values. For example, if we wanted to know the average machine time within 5 percent at a risk of 0.05, we would need about 80 observations.

TESTS OF HYPOTHESIS. In addition to estimation of parameters of distributions or populations one must frequently compare parameters of different groups, or judge whether the parameter of a distribution is greater than a certain value, whether the extreme observation of a data set could have been generated by chance alone, etc. Tests of hypothesis are the statistical tools used in making such judgments.

As a typical problem of this sort, suppose that we are concerned with a determination of whether the sample averages from two different sets of measurements are truly different. For example, they could be measurements of output from two different machines which can perform the same operation. We are interested in having a means of determining and expressing the likelihood that the exhibited difference could have resulted from chance variations in sampling. In other words we should like to be able to say with precision: A difference as large as that obtained from the data is unlikely to have occurred if the samples were drawn from identical populations or distributions; or, a difference as large as this is likely to have occurred between samples of identical distributions.

In situations where we are determining whether the characteristic of the total group is above or below or between certain values, we should like to be able to say with precision: It is unlikely that a sample with characteristics such as these could have come from a population such as we have described; or, it is likely that a sample with characteristics such as these could have come from a population such as we have described.

Making the words **likely** and **unlikely** precise is achieved, statistically, by stating the risk that we are wrong in terms of the probability that our conclusions will be erroneous. If the situation is such that we may risk being in error 1 time in 20, we should intend the word "likely" to mean 19 times out of 20. The **risk of being in error** must be specified for each situation. What it is will depend on the cost of making more measurements and the cost of being wrong. Of course it is possible to err in two ways. First, one can conclude that the samples come from different populations when, in fact, they do not. Second, one can conclude that the samples come from identical distributions when, in fact, they do not. In designing a statistical procedure one must guard against both types of errors.

Differences of **statistical significance** are not necessarily differences of practical significance. The fact that two samples exhibit a statistically significant difference indicates that they were probably drawn from two different distributions. A constant difference will appear more and more statistically significant as larger samples are measured. Whether the difference between the groups is of **practical significance** must be determined by the individual situation. A

ifference so large as to be of practical significance need not necessarily be
tatistically significant; that is, it may be due to chance sampling errors. When
his occurs, additional sampling will show either that the difference is stable and
eal or that the difference will diminish and disappear as more samples are
neasured. Decisions for action should be based on results that are both prac-
ically and statistically significant.

A logical approach to evaluating the statistical significance of differences is
hrough a form of reasoning known as a **hypothesis test**. The general method
akes the following steps:

1. Set up the hypothesis that the samples measured have been drawn from the
   same total group; or alternately, from a described total group.
2. Draw a sample from the population (or each of the populations if several are
   under study).
3. Compute one or more statistics or measures from each of the samples.
4. Using known data regarding sampling distributions for the statistics computed,
   test to see whether the hypothesis is likely to be true; that is, see if the measured samples are likely to have come from the same total group or from the
   described total group.

The following article describes the statistical techniques used in step 4.

Example of *t*-Test. Suppose that a supplier of small purchased parts guarantees that each carton of parts will contain an average of 500 pieces. For the
first shipment the contents of 6 cartons are counted and the average number
per carton found to be 491.2 with a sample standard deviation, *s*, of 5. As
here may be considerable variation in the number of parts from carton to
carton, the significance of the difference between the average number guaranteed
and the average number found for the sample must be questioned. Could the
difference be solely due to chance; that is, is it possible that it just happened
hat the 6 cartons selected for counting tended to be light? Or is the difference
n averages too great for this to be likely?

This difference may be tested by first setting up the hypothesis that the
average for the shipment is 500 pieces per carton. Next, the extent to which
he average for a sample of 6 might deviate from the population average is
determined by consulting a table of the sampling distribution for the statistic
involved (in this case, a table of *t*). This table would show, for example, that
he probability is 0.025 of a sample average being more than $3.163 \times s/\sqrt{n}$ units
ower or higher than the population average purely by chance. (The figure 3.37
s the value for *t* obtained from Table 2.) This is $3.37 \times 5/\sqrt{6}$, or 6.5 units.
The observed difference is $500 - 491.2$, or 8.8 units. As the observed difference
s greater than the difference that could occur by chance at the selected probability level, the hypothesis that the average number per carton for the shipment
s 500 is rejected. The difference is considered significant and thus good evidence
that the supplier's shipment does not meet his guarantee.

While the general approach is as described in the illustration above, the **working procedure** used by most statisticians may differ slightly in detail. In the
above test, for example, the common method of computation is first to determine
a computed *t* by dividing the difference in means by $s/\sqrt{n}$. For this example

$$t = \frac{500 - 491.2}{5/\sqrt{6}} = 4.30$$

The computed value for *t* is then compared with the table value. If it is greater,
the hypothesis is rejected. Note that this procedure is equivalent to the one previously described, differing only in the form of comparison made.

RISKS OF A WRONG CONCLUSION. In using information obtained from a sample of observations to test a hypothesis, there is always some risk of reaching incorrect conclusions. The average for a sample, for example, may deviate considerably from the average of the population from which it has been drawn, particularly if the sample size is small and the population is widely dispersed. The extent of the deviation or the direction of the deviation is not known. In fact, in testing a hypothesis by means of a sample, there are two possible forms of error. One is to conclude that the hypothesis is false when it really is true (a **Type I error**). The other is to conclude that the hypothesis is true when it really is false (a **Type II error**). These two possibilities of error always exist whether statistical tests are used or not. However, the advantage of the statistical approach is that a precise appraisal of these risks, particularly the first, may be made. Knowing the odds of reaching an incorrect conclusion or being able to choose such odds, is of much help in many situations.

The usual procedure in hypothesis testing is to select an appropriate probability figure for the first kind of error mentioned—that of concluding the hypothesis is false when it really is true. This **risk** is measured by the probability level used in working with the tables of sampling distributions such as the table of t and the table of F. In the example above, a probability level of 0.025 was used in getting a value for t to test the hypothesis regarding the population average. The result of using this level was that the risk of concluding the hypothesis was false when it really was true, was 0.025 or 1 in 40 or less. The choice of a probability level for the test depends largely on how costly are the consequences of making an error of this kind. This is an engineering and an economic judgment. In the above example, the level would depend on the costs (both those that may be reduced to money terms and those that are irreducible) of assuming the shipment did not meet the supplier's guarantee and consequently returning it, when in reality it did meet his guarantee.

The second kind of risk, that of assuming the hypothesis is true when it really is false is more difficult to evaluate. Also, it cannot be expressed as a single figure, as the risk depends in part on the extent to which the population actually deviates from the hypothetical value. In the above test, for example, the risk of assuming the average number per carton for the shipment was 500 when it really was not would depend on how much the actual average differed from 500. The extent of this second kind of risk depends also on the sample size and on the probability value selected for the first form of risk. Methods for computing this risk usually require much involved computation. For this reason these risks generally have not been determined as a routine step in statistical applications. Risks for specific tests in the form of **operating characteristic curves** have been developed. It is important to note that the two types of risks and the sample size are linked. The only way to reduce both risks is to increase the sample size. For a fixed sample size, if one of the risks is decreased the other is increased. We do not get something for nothing in statistics. For example the Type I risk can be made zero by the ridiculous expedient of always rejecting the hypothesis. For more discussion on risks and O.C. curves see the section on Quality Control and Reliability.

TESTS FOR DIFFERENCES BETWEEN AVERAGES. In many situations it is necessary to compare a sample mean with a population mean and to evaluate the difference between them. If this difference is not great, and if there is considerable inherent variation among the items being measured, the observed difference may not represent a real difference in population means. Simply by chance the mean for a sample may be greater (or less) than the

:ean of the population from which it was drawn. The following test is based
n an assumption that the population is normal. If the sample size is fairly
arge the assumption is not critical.

**Significance of a Difference Between a Sample Mean and a Population
Mean.** The steps in the test are:

1. Select a suitable probability level for testing the difference for significance.
2. Compute t where

$$t = \frac{\overline{X} - 5}{s/\sqrt{n}}$$

3. Using the selected probability level, determine the value of t from Table 2, using degrees of freedom $n - 1$.
4. Compare the computed and table values of t, ignoring their signs. If the computed t is numerically less than the table value, the observed difference in means is not significant at the probability level selected. If the computed value is numerically greater, the difference may be considered significant.

The probability levels in the table heading give the probability of a computed
being by chance numerically greater than the table t, if the mean of the
population from which the sample has been drawn is the same as the mean μ.
For a test of significance of difference in one direction only, the probability
values in the table headings should be divided by 2. That is, if a table value
with probability of 0.05 is used, the probability is 0.025 of a computed t being
by chance greater than the positive table value. There is an equal probability,
0.025, of its being less than minus the table value.

Sample and Population Means Compared. The average number of batteries per order for a battery manufacturer has, for the past several years, been
270 batteries, although the number on any order may vary considerably from
this figure. A new kind of battery has been introduced which can be stored for
some time without loss of charge. After its introduction a sample of 100 orders
was tallied and the average number per order found to be 306, with a standard
deviation for the sample of 39. It will be helpful to know how much significance
can be attached to this apparent increase in average order size. If the increase
observed from the sample represents a real increase, plans will have to be made
promptly for revised truck routes and schedules, revision of warehouse space,
and the like.

A probability level of 0.025 is selected to test for significance. The old population average, μ, is 270; the sample average, $\overline{X}$, is 306; the sample standard
deviation, s, is 39; the sample size n, is 100. Thus the computed value for t is

$$t = \frac{X - \mu}{s/\sqrt{n}} = \frac{306 - 270}{39/\sqrt{100}} = 9.2$$

Using $n - 1$ or 99 degrees of freedom and a probability level of 0.02, the value
projected is found from Table 2 to be 2.280. As the computed value exceeds
this numerically, the difference is considered significant. The probability is
0.025 of a computed t being numerically greater than 2.280 by chance if the true
average has not changed. The probability of a sample average being greater,
that is of getting a positive t greater than 2.280, is one-half the table probability,
or 0.0125.

Effect of New Tool on Output—Use of t Test. This test may be adapted
as needed to fit the problem at hand. For example, consider a case in which a
costly tool was being studied for use in large numbers in a factory. The vendor

of the tool guaranteed its use would increase the average output from 207 unit.
the present figure, to 260 units per operator per day. **Economy studies** showe
that a changeover to this tool would be economical only if this average increas
of 53 units per operator were to be made. One tool was purchased for tria
Twenty days' use gave daily figures ranging from 242 to 275 units per day, th
average being 261.4 units per operator-day with the standard deviation for th
20 observations being 4.7 units. The significance of these trial figures was que:
tioned. How confident could one be that the average for the group would ir
crease to 260 units? Was this sufficient evidence to warrant making the pur
chase of the additional units needed or should further study be made?

Computation of t gave the following figure:

$$t = \frac{261.4 - 260}{4.7/\sqrt{20}} = 1.33$$

The concern in this case is with a one-sided test. Thus if a risk level of 0.0
is desired the table entries for 0.01 should be used. The table of t gave a value o
approximately 2.86 for t, using a probability level of 0.01. As the computed t wa
not numerically larger than the table value, the difference could not be considere
significant at the probability level selected. Only if the computed t were greate
than 2.86 would this be evidence (at a probability level of 0.01) that the nev
average was 260 or greater. A sample average (for a sample size of 20) coul
by chance be 261.4, with the true average being somewhat below 260. To b
reasonably certain that the average was 260 or greater, an **additional tria
period** was needed. After 160 operator-days were recorded, the average outpu
was 259.2 and the sample standard deviation was 4.0 units. Now,

$$t = \frac{260 - 259.2}{4.0/\sqrt{160}} = 2.53$$

which indicates that if the true average were 260 or greater, it is unlikely tha
we could get 160 operator-days averaging as low as this. The tool should b
rejected.

**Difference Between a Sample Mean and a Population Mean When th
Population Standard Deviation Is Known.** If the standard deviation fo:
the population is known, an alternative test is available. If this alternative tes
can be used, the observed difference between means does not have to be as grea
as in the previous test in order to be significant. The **test procedure** is:

1. Select a suitable probability level for the Type I risk.
2. Compute z where

$$z = \frac{\overline{X} - \mu}{\sigma/\sqrt{n}}$$

3. Using the selected probability level, determine the table value for z from Tabl
1.
4. Compare the computed value for z with the table value. If the computed valu
is less than the table value, the observed difference in means is not significan
at the probability level selected. If the computed value is greater, the difference
may be considered significant.

The probabilities listed in Table 1 give the probability of a computed z being
less than the table value. There is an equal probability of its being more thar
the table value. Thus, for a test of difference in both directions, the table valu
gives one-half the probability of a real difference, either plus or minus, being
indicated in the event that no difference really exists.

Effect of New Battery Ingredient. Suppose that a new ingredient is ʼdded in making the plates for automobile storage batteries in an attempt to ᵻcrease their life. Experience has shown that any change of this kind will have ᵰ effect, if any, only on the average life and not on the dispersion from battery ᵼ battery. Ten batteries are manufactured and put through the relatively ᵰgthy and costly life tests. The increase over the normal life for a battery ᵼ this type averaged 1.6 months per battery. The standard deviation for this ᵼpe battery has been found through past research and mortality studies to be 1.8 ᵼonths. Assuming the standard deviation has not been affected by use of the ᵼw material, does this test indicate the new material has resulted in real imᵼrovement? Can one be 95 percent confident, for example, that the average ᵼifference in life shown by the 10 test batteries indicates a real increase in batᵼry life?

Computation of z gives the following:

$$z = \frac{\overline{X} - \mu}{\sigma/\sqrt{n}} = \frac{1.6}{1.8/\sqrt{10}} = 2.81$$

The table value of z, using a probability level of $1.00 - 0.95$ or 0.05, is -1.645. ᵼs the computed value for z is numerically greater than the table value, the obᵼrved difference is significant at the selected probability level. The probability ᵼ only 5 out of 10 of getting a computed z beyond 1.645 if there had been no inᵼrease in the life average, μ.

Significance of a Difference Between a Sample Mean and a Population Ꭼean Based on the Use of the Range. A simple alternative procedure ᵼhich uses the sample range as a measure of dispersion is available. This test ᵼ a fairly efficient one when the number of observations is small, but it is ᵼnerally not as precise as the t-test using the standard deviation. A greater ᵼbserved difference in averages will be required for an equivalent degree of conᵼdence. The **steps in the test** are:

1. Select a suitable probability level.
2. Compute t_R where

$$t_R = \frac{\overline{X} - \mu}{R}$$

In the above formula, R is the range for the measurements in the sample.
3. Using the selected probability level, determine the table value for t_R from Table 3.
4. Compare the computed value for t_R with the table value. If the computed value is numerically less than the table value, the observed difference in means is not significant at the probability level selected.

The probability values in the table heading give the probability of a computed ᵼalue being numerically greater than the table value by chance, when there is ᵼo real difference between the mean of the population sampled and the mean μ. ᵼor a test of significance of difference in one direction only, the probability ᵼalues in the table heading should be divided by 2.

Effect of New Tool on Output—Use of the Range. This test can be ᵼlustrated by one of the previous examples in which a new type of tool was ᵼested for its effect in improving output. The average for 20 sample observations ᵼas 261.4 units with a range of $275 - 242$, or 33 units. The question was ᵼhether or not the sample average, 261.4, was sufficiently high to be sure the ᵼuaranteed average output for the plant, 260 units, had been met. A probability ᵼevel of 0.02 was used.

The **computation** of t_R is

$$t_R = \frac{\overline{X} - \mu}{R} = \frac{261.4 - 260}{33} = 0.042$$

From Table 3 the table value for t_R is found to be 0.15. As the computed value is not numerically greater, the difference is not significant at the selected level of probability. Note that this is the same conclusion as that reached by the t-test.

Significance of Difference Between Two Sample Means. When two sample averages are compared and the difference between them is found to be small, the significance of the difference may be questionable. This is particularly true if the number of observations is relatively small and if there is considerable variation among the observations within a sample. The difference may be due to chance and not to a real difference in averages between the two populations sampled.

A test for general use is available. This test assumes that the underlying populations are normal. The assumption is not critical for large sample sizes. Further, it is assumed that both populations have the same standard deviation even though it is unknown. (A similar and more precise test for use when the population standard deviation is known is outlined later.) The **steps in the test** for use when the population standard deviations are not known are:

1. Select a suitable probability level for the Type I risk.
2. Compute t where

$$t = \frac{\overline{X}_A - \overline{X}_B}{s_p \sqrt{\dfrac{1}{n_A} + \dfrac{1}{n_B}}}$$

In the above formula $\overline{X}_A$ stands for the arithmetic mean for one sample, $\overline{X}_B$ for the mean of the other. Likewise, n_A is the number of observations in sample A, n_B the number of observations in sample B. The term s_p represents the estimate of the population standard deviation obtained by pooling data from both samples. The following equation may be used to compute s_p:

$$s_p = \sqrt{\frac{\Sigma X_A{}^2 - \dfrac{(\Sigma X_A)^2}{n_A} + \Sigma X_B{}^2 - \dfrac{(\Sigma X_B)^2}{n_B}}{n_A + n_B - 2}}$$

If the standard deviation has been computed for each group of observations, the following alternative formula may be used instead:

$$s_p = \sqrt{\frac{(n_A - 1)s_A{}^2 + (n_B - 1)s_B{}^2}{n_A + n_B - 2}}$$

3. Using the selected probability level, determine the table value for t from Table 2. For this test enter the table with $n_A + n_B - 2$ degrees of freedom.
4. Compare the computed t (ignoring signs) with the table value for t. If the computed t is numerically less than the table value, the observed difference in means is not significant at the probability level selected.

For this test, the probability levels in the t-table heading give the probability of a computed t being numerically larger than the table value if the means of the two populations from which the samples have been drawn do not differ; that is, the probability or the risk of assuming the observed difference in sample means seems significant when it really is not (Type I risk). To test for a change in one direction only, use one-half the probability value given by the table heading.

The above test is exact if the distributions are normal and if the standard deviations of each are equal. For cases that only approximate these requirements, the table probability values will not be exact. However, a test for use when the population standard deviations differ considerably has been developed Duncan, Quality Control and Industrial Statistics).

Comparison of Average Times of Assembly Methods. Consider a case in which two different assembly methods for a product were being studied for possible adoption. A time study was made on a trial setup for each of the possibilities with the following results:

	Method A	Method B
Average Time	4.37 min.	4.09 min.
Standard Deviation	0.28 min.	0.21 min.
Number of Observations	20	25

There was a question as to whether the difference in averages represents a real difference in the time requirements or whether it might simply be a chance difference due to sampling fluctuations. A probability level of 0.10 was selected for the Type I risk.

The computations were:

$$s_p = \sqrt{\frac{(n_A - 1)s_A{}^2 + (n_B - 1)s_B{}^2}{n_A + n_B - 2}} = \sqrt{\frac{19(0.28^2) + 24(0.21^2)}{20 + 25 - 2}}$$

$$= 0.24$$

$$t = \frac{\bar{X}_A - \bar{X}_B}{s_p\sqrt{\frac{1}{n_A} + \frac{1}{n_B}}} = \frac{4.37 - 4.09}{0.24\sqrt{\frac{1}{20} + \frac{1}{25}}}$$

$$= 3.89$$

Using $n_A + n_B - 2$ or 43 degrees of freedom and a probability level of 0.10, the table value for t was found to be 1.68. As the computed value for t was numerically larger, this was accepted as evidence (at the selected probability level) that the true average time requirements for the two methods were not the same. A further conclusion, of course, was that method B was the better of the two.

Comparison of Average Weights of Containers. For another illustration, consider the figures below for the weights of sample containers from two companies.

	Company J	Company K
	124.4 lb.	107.1 lb.
	121.4	107.6
	121.3	108.8
	126.5	115.6
	118.0	108.7
		110.5
		112.4
Average	122.3 lb.	110.1 lb.

The containers from Company J are supposed to weigh at least 10 lb. more on the average, than those from Company K. A question may be raised as to whether or not these sample observations furnish proof that this is the case.

The observed difference is $122.3 - 110.1$, or 12.2 lb. If this difference is sufficiently larger than 10.0 lb. to be significant—that is, if it is so large that it could not be obtained by chance with the true averages just 10 lb. apart—then the observations may be taken as valid evidence that the population average for Company J is at least 10 lb. greater than that for Company K. Suppose that a probability level of 0.01 seems appropriate for the test.

The computations are shown below. Observations have been coded by subtracting 110 lb. from each one so that the computations will be easier to carry out.

Company J		Company K	
X_J	$X_J{}^2$	X_K	$X_K{}^2$
14.4	207.36	-2.9	8.41
11.4	129.96	-2.4	5.76
11.3	127.69	-1.2	1.44
16.5	272.25	5.6	31.36
8.0	64.00	-1.3	1.69
		0.5	0.25
		2.4	5.76

$\Sigma X_J = 61.6 \quad \Sigma X_J{}^2 = 801.26 \qquad \Sigma X_K = 0.7 \quad \Sigma X_K{}^2 = 54.67$

$n_J = 5 \qquad\qquad\qquad\qquad\qquad n_K = 7$

$\overline{X}_J = \dfrac{61.6}{5} = 12.3 \qquad\qquad\qquad X_K = \dfrac{0.7}{7} = 0.1$

$$s_p = \sqrt{\frac{\Sigma X_J{}^2 - \dfrac{(\Sigma X_J)^2}{n_J} + \Sigma X_K{}^2 - \dfrac{(\Sigma X_K)^2}{n_K}}{n_J + n_K - 2}}$$

$$= \sqrt{\frac{801.26 - \dfrac{61.6^2}{5} + 54.67 - \dfrac{0.7^2}{7}}{5 + 7 - 2}}$$

$$= 3.11$$

$$t = \frac{(\overline{X}_J - \overline{X}_K) - 10}{s_p \sqrt{\dfrac{1}{n_J} + \dfrac{1}{n_K}}} = \frac{(12.3 - 0.1) - 10}{3.11 \sqrt{\dfrac{1}{5} + \dfrac{1}{7}}}$$

$$= 6.7$$

To obtain a table value for t the probability values in the table headings are divided in one-half—as this is a test for change in one direction only—for J to be significantly larger than K. Thus with $n_J + n_K - 2$, or 10 degrees of freedom and a selected probability of 0.01 (which requires the use of 0.02 in the table) the table value for t is found to be 2.76. As the computed t is larger, the difference is significant. The tests may be taken as evidence that the average weight per container for Company J will run 10 lb. heavier than for Company K.

Significance of Difference Between Two Sample Means When the Standard Deviation Is Known. If a good estimate of the population standard deviation, σ, has been found, and if it appears that any changes will affect only the population average, an alternative test is available. The **steps in this test** are:

1. Select a suitable probability level.
2. Compute z where

$$z = \frac{\overline{X}_A - \overline{X}_B}{\sigma \sqrt{\dfrac{1}{n_A} + \dfrac{1}{n_B}}}$$

3. Using the selected probability level, determine the table value for z from Table 1, using one-half the selected probability level to enter the table.
4. Compare the computed value for z with the table value. If the computed value for z is numerically less than the table value, the observed difference in means is not significant at the probability level selected.

As for the t-test, the probability of the computed z being numerically greater by chance when no real difference in population means exists is twice the value in the table heading.

For a test in one direction only—for $\overline{X}_A$ to be significantly greater than $\overline{X}_B$, for example—the table of z should be entered at the selected probability level.

Example of Crew Differences in Shrinkage. A cannery has found through extensive studies of shrinkage that, while the average may vary throughout the season or from lot to lot of fruit, the standard deviation remains fairly constant at about 8.6 lb. per day per operator. At the start of a new season it seemed desirable to compare the shrinkage of 2 different crews of 8 women each. Shrinkage was measured over 2 days which gave a sample of 16 woman-days for each crew. One crew had an average shrinkage of 132.5 lb. per woman per day and the other 137.6.

The difference between crews was tested for significance by the following computations (a probability level of 0.01 was used):

$$z = \frac{\overline{X}_A - \overline{X}_B}{\sigma \sqrt{\dfrac{1}{n_A} + \dfrac{1}{n_B}}} = \frac{132.5 - 137.6}{8.6 \sqrt{\dfrac{1}{16} + \dfrac{1}{16}}}$$

$$= 1.68$$

From the table of z, Table 1, a table value of z is found to be -2.58. As the computed value for z was not numerically larger than the table value, the difference was not considered significant. It could have been a chance difference.

Difference Between Two Samples—Use of the Range. A simple alternative procedure using the range instead of the standard deviation as a measure of dispersion is available. It is fairly efficient when the number of observations is small. The number of observations for each sample must be the same. The **test procedure** is:

1. Select a suitable probability level.
2. Compute t_{2R} where

$$t_{2R} = \frac{\overline{X}_A - \overline{X}_B}{\dfrac{1}{2}(R_A + R_B)}$$

In the above formula, R_A is the range for the measurements in sample A and R_B the range for sample B.

3. Using the selected probability level, determine the table value for t_{2R} from Table 4.

4. Compare the computed value for t_{2R} with the table value. If the computed t_{2R} is numerically less than the table value, the observed difference in means is not significant at the probability level selected.

Note that for Table 4 the t_{2R} value is determined by using the sample size and not the number of degrees of freedom. The probability values in the table heading give the probability of a computed t_{2R} being numerically larger than the table value if the means of the two populations from which the samples have been drawn do not differ. To test for a change in one direction only, use one-half the probability value shown by the table.

Test of Machine Attachment. Suppose that a new attachment is tested at a machine for its effect on increasing productivity. Daily production is measured for five days with the device and for five days without the use of the device with the following results:

Units Produced with the Device	Units Produced without the Device
2,214	1,966
1,858	1,707
2,261	1,760
1,918	1,758
1,830	1,581
10,081	8,772

$$\text{Average} = \frac{10,081}{5} = 2,016 \qquad \text{Average} = \frac{8,772}{5} = 1,754$$

$$\text{Range} = 2,261 - 1,830 = 431 \qquad \text{Range} = 1,966 - 1,581 = 385$$

The vendor of the device claimed it would increase average daily output by 200 units. It was desired to know whether this test was valid evidence that his claim would be met over the long run or whether an additional period of testing would be required to be sure. A probability level of 0.025 seemed best for use in this case.

Computation of t_{2R} gave the following:

$$t_{2R} = \frac{(\overline{X}_A - \overline{X}_B) - 200}{\frac{1}{2}(R_A + R_B)} = \frac{(2,016 - 1,754) - 200}{\frac{1}{2}(431 + 385)}$$

$$= 0.15$$

Entering Table 4, the table of t_{2R}, at a probability level of 0.05 (since this test is for a change in one direction only), gives a table value for t_{2R} of 0.61. As the computed value is not numerically larger, the sample observations do not offer proof at the selected probability level that the claim of a 200-unit gain has been met.

TESTS FOR DIFFERENCES BETWEEN PROPORTIONS. In many areas of engineering and management, sample proportions of percentages must be compared for significance of difference. A simple but approximate procedure is available to test such differences. It will give satisfactory results in all applications where the sample size or the proportion observed (or both) are sufficiently large to result in the specified event occurring at least four or five times in each sample.

Differences Between Two Sample Proportions. The steps in the test are as follows:

1. Select a suitable probability level for testing the difference for significance.
2. Compute s_p where

$$s_p = \sqrt{\frac{p_A(1 - p_A)}{n_A} + \frac{p_B(1 - p_B)}{n_B}}$$

and where
p_A = the proportion in sample A
p_B = the proportion in sample B
n_A = the number of items in sample A
n_B = the number of items in sample B

3. Compute d, where

$$d = zs_p$$

The value for z may be determined from Table 1.
4. Compare the difference in sample proportions with the value found for d. If it is numerically smaller than d, the difference in the direction indicated by the sample values is not significant at the probability level used. On the other hand if the difference in sample proportions is numerically larger, the difference in the direction indicated may be considered significant.

Differences in Departmental Turnover. Suppose that in a manufacturing plant the following figures on personnel turnover have been obtained: Department A—out of 810 men hired 109 have quit; Department B—out of 456 men hired 34 have quit. Can these figures be taken as evidence of a real difference between departments? An answer is needed in order to know whether or not to look into working conditions, supervisory practices, and the like for an explanation of the higher proportion of voluntary separations in Department A. A probability level of 0.05 was selected as suitable in applying a test.

For the above figures

$$p_A = \frac{109}{810} = 0.135$$

$$p_B = \frac{34}{456} = 0.075$$

$$s_p = \sqrt{\frac{p_A(1 - p_A)}{n_A} + \frac{p_B(1 - p_B)}{n_B}}$$

$$= \sqrt{\frac{0.135(1 - 0.135)}{810} + \frac{0.075(1 - 0.075)}{456}}$$

$$= 0.0173$$

From Table 1 the value for z is found to be 1.645 for a probability of 0.05. Thus

$$d = zs_p = 1.645(0.0173) = 0.028$$

The observed difference is $0.135 - 0.075$, or 0.060. As this is larger than the computed value for d, the turnover in Department A may be considered significantly larger than the turnover in Department B.

Difference Between a Sample Proportion and a Population Proportion. A variation of the preceding test is available to test the difference between a sample proportion and a population proportion. This variation is also an approximate test, but one that is satisfactory under most circumstances.

The **test procedure** is as follows:

1. Select a suitable probability level for testing the difference for significance.
2. Compute σ_p where

$$\sigma_p = \sqrt{\frac{p(1-p)}{n}}$$

and where

p = population proportion
n = the number of items in the sample

3. Compute d, where

$$d = z\sigma_p$$

The value for z may be determined from Table 1.
4. Compare the difference between the sample proportion and the population proportion with the value for d. If it is numerically smaller than d, the difference in the direction indicated is not significant at the probability level used and could be due to sampling error. If it is numerically larger, the difference may be considered significant.

Change in Clerical Procedures Checked. Suppose that in an area of activity in a mail order house extensive studies have shown the errors in making clerical entries have averaged 0.87 percent of the total entries made. A change in the clerical procedure is made. A sample of 1,000 entries is checked and the weighted proportion defective found to be 0.72 percent. At a probability level of 0.02, could this be taken as evidence of a real reduction in errors, or is it possible that for a sample of 1,000 entries a figure of 0.72 percent could be obtained simply by chance with the population proportion remaining at 0.87 percent? Computation of σ_p gives:

$$\sigma_p = \sqrt{\frac{p(1-p)}{n}} = \sqrt{\frac{0.0087(1-0.0087)}{1,000}}$$

$$= 0.000294, \quad \text{or} \quad 0.0294\%$$

$$\text{Computed } d = z\sigma_p = 2.055 \cdot 0.0294\% = 0.061\%$$

The value for z was obtained from Table 1, using a probability of 0.02.
The observed difference is 0.87 percent $-$ 0.72 percent, or 0.15 percent. As the observed difference is greater than the computed value for d, it may be considered significant.

TEST FOR THE REJECTION OF EXTREME VALUES. Occasionally a group of observations will contain some one value considerably larger (or considerably smaller) than the others. This value may be due to an error in measurement or may be due to some abnormal circumstance. In such cases it may be best to discard it. On the other hand, the extreme value may simply represent an extreme item in the population being measured. If this is so, it may be best to retain it to make the group of data more representative. In any case when it is **known** that the extreme observation is due to a nontypical cause or error it should be rejected outright. The statistical test which follows is appropriate for cases in which there is doubt. However many engineers argue that it is dangerous to ever discard data.

A simple statistical test has been determined to **test for rejection.** The procedure is:

1. Arrange the observations in their order of magnitude.
2. Compute γ.

When the number of values is from 3 to 7,

$$\gamma = \frac{X_2 - X_1}{X_k - X_1}$$

When the number of values is from 8 to 10,

$$\gamma = \frac{X_2 - X_1}{X_{k-1} - X_1}$$

When the number of values is from 11 to 13,

$$\gamma = \frac{X_3 - X_1}{X_{k-1} - X_1}$$

When the number of values is from 14 to 30,

$$\gamma = \frac{X_3 - X_1}{X_{k-2} - X_1}$$

In the equations above, X_1 represents the extreme observation; X_2, the observation nearest to it; X_3, the next in order; and so on to the end of the series, which is represented by X_k. The extreme observation could be either the largest or the smallest observation. X_2 is then the next largest or next smallest value.

3. Select a suitable probability level. The probability values in the heading of Table 7 show the probability of a computed γ being numerically greater than the table value if the extreme observation is not abnormal but is actually a part of the normal population from which the rest of the observations were obtained.

4. Using the selected probability level, determine the table value for γ from Table 7.

5. Compare the computed value for γ with the table value. If the computed value is numerically larger than the table value, the extreme observation may be rejected at the selected probability level.

This test is for observations from a normal distribution. In cases that may only approximate the normal, the table probability values will not be exact.

The table of factors used in the above test may also be used as a test for the rejection of extreme means. This use is discussed in a subsequent article of this section.

Consider the following sample observations obtained in the course of a study:

$$52.4,\ 46.6,\ 49.7,\ 52.7,\ 53.6,\ 51.8,\ 51.6,\ 50.9.$$

The extreme value, 46.6, may be questionable. Arranging the observations in order of magnitude gives 46.6, 49.7, 50.9, 51.6, 51.8, 52.4, 52.7, and 53.6. For the eight values

$$\gamma = \frac{X_2 - X_1}{X_{k-1} - X_1} = \frac{49.7 - 46.6}{52.7 - 46.6} = 0.508$$

Selecting a probability level of 0.05 gives a table value of 0.554 for γ. As the computed value is smaller than the table value, the extreme value may not be rejected at the selected probability level.

Extreme Means. A test similar to the foregoing has been determined to test an extreme mean to see whether or not it is significantly larger (or smaller) than the remainder of the means involved. Such a test might be useful, for example, in determining whether or not to discard an extreme mean as not being representative of the population from which the others have been obtained (perhaps because of an incorrectly drawn sample). Or it may be useful, in some cases, to test for an apparent shift in a population average.

The test for significance of the difference between the **extreme mean** and the other means is:

1. Arrange the means in the order of their magnitude.
2. Compute γ.
   When the number of means is from 3 to 7,

$$\gamma = \frac{\overline{X}_2 - \overline{X}_1}{\overline{X}_k - \overline{X}_1}$$

When the number of means is from 8 to 10,

$$\gamma = \frac{\overline{X}_2 - \overline{X}_1}{\overline{X}_{k-1} - \overline{X}_1}$$

When the number of means is from 11 to 13,

$$\gamma = \frac{\overline{X}_3 - \overline{X}_1}{\overline{X}_{k-1} - \overline{X}_1}$$

When the number of means is from 14 to 30,

$$\gamma = \frac{\overline{X}_3 - \overline{X}_1}{\overline{X}_{k-2} - \overline{X}_1}$$

   In the formula above, $\overline{X}_1$ represents the extreme mean, $\overline{X}_2$ the mean nearest to it in magnitude, $\overline{X}_3$ the next mean in magnitude, and so on to the end of the series which is represented by $\overline{X}_k$.
3. Select a suitable probability level.
4. Using the selected probability level, determine the table value for γ from Table 7. The probability values in the table heading show the probability of a computed γ being greater than the table value for γ because of the extreme mean being obtained by chance from the same population from which the other means were obtained.
5. Compare the computed value for γ with the table value. If the computed value is smaller than the table value, the extreme mean may not be considered significantly different at the probability level selected. If it is larger than the table value, the difference may be considered significant with the probability of error indicated by the probability level selected.

Extreme Work Element Time Retained. Consider a case in which a number of time studies have been made for a work element. It is an element that is considered constant—that is, one whose time requirements, on the average, should not vary from job to job. A number of stop-watch time studies have been made, the average time for the element computed for each study, and a leveling factor applied. The leveled average (in minutes) for each of the studies is as follows: 13.29, 13.27, 11.79, 13.63, 9.49, 12.87, 10.67, 12.33, 11.78 and 10.89. A question of discarding the 9.49-min. study in working up the time for the element has arisen as the value seems to be abnormal. If it is the result of poor leveling, of nonstandard work material, or of some other nonrepresentative condition it should be discarded. On the other hand, if it was at this low figure simply by chance, discarding it before working up the time to allow for the element will not be appropriate.

In order to discover, by testing, whether the value could be obtained by chance from the population from which the other values were obtained, the values are first arranged in their order of magnitude: 9.49, 10.67, 10.89, 11.78, 11.79, 12.33, 12.87, 13.27, 13.29, and 13.63. For the ten means

$$\gamma = \frac{\overline{X}_2 - \overline{X}_1}{\overline{X}_{k-1} - \overline{X}_1} = \frac{10.67 - 9.49}{13.29 - 9.49} = \frac{1.18}{3.80} = 3.11$$

Assuming that a probability level of 0.05 will be appropriate, Table 7 gives a table value for γ of 0.477. As the computed value for γ is smaller than the table value, the extreme mean may not be rejected as significantly different at the probability level selected.

Analysis of Variance

TERMINOLOGY. Variance is a useful measure of dispersion used in many statistical tests. It is the square of the standard deviation. For a **population** or a **distribution** the variance is σ^2 which is defined as

$$\sigma^2 = E[X - E(X)]^2$$

For **sample data** the variance may likewise be defined as the square of the sample standard deviation. The general equation is

$$s^2 = \frac{\text{sum of squares}}{\text{degrees of freedom}}$$

If the data from a **single sample** only are used, the equation is

$$s^2 = \frac{\Sigma(X - \overline{X})^2}{n - 1}$$

As explained elsewhere in this section, the standard deviation for a sample tends to be somewhat smaller than the standard deviation for the population from which it has been drawn. Dividing the sum of squares by $n - 1$ instead of n corrects for this bias and makes s^2 the best possible estimate of σ^2. Computation of s^2 when more than one sample is involved is discussed below.

In the analysis of variance, the general procedure is to set up hypotheses of quite general types concerning either population means or variances. The hypothesis is then tested by comparing sample variances. If the sample variances do not differ by more than an amount possible due to chance or sampling error, the hypothesis is allowed to stand. If they differ more than can be explained as due to chance, the hypothesis is rejected and the difference in sample variances is considered significant.

TEST FOR DIFFERENCE IN VARIANCES. A simple application of the analysis of variance is available to test for the significance of the difference between two sample variances or between a sample variance and a population variance. To consider an example, suppose that experimental changes have been made to a manufacturing procedure to reduce the amount of inherent variability and so produce a more uniform product. The variance for a sample of parts produced and measured after the change is somewhat smaller than the variance for a sample of parts produced and measured before the change. The question may be raised as to whether the reduction in sample variance is good evidence of a real reduction in variability for the process or whether the difference is simply due to chance. It is possible that no real improvement has been made but that, by chance, the items selected for measuring the variation in the old procedure had measurements more widely scattered than those selected for measurement after the change. If putting the experimental changes into effect on the production floor will require considerable investment of money, assurance that the changes have really helped will be required.

Steps in the F-Test. The F-test for significance of difference in a specified direction between two variances requires the following steps.

1. Select a suitable probability level for the Type I risk.
2. Compute the variance separately for each sample. The equation is

$$s^2 = \frac{\Sigma(X - \overline{X})^2}{n - 1}$$

An equivalent equation for ease in computation is

$$s^2 = \frac{\Sigma X^2 - \dfrac{(\Sigma X)^2}{n}}{n - 1}$$

3. Compute F by dividing the larger variance by the smaller; that is,

$$F = \frac{s^2_A}{s^2_B}$$

where s^2_A is the larger of the variances and s^2_B the smaller.
4. Using the selected probability level and degrees of freedom $n_A - 1$ and $n_B - 1$, determine the table value of F from Table 6. If one of the variances being compared is the known variance for the population, the degrees of freedom for this mean square are ∞.
5. Compare the computed value for F with the table value. If the computed value is smaller than the table value, the difference in variances could be a chance happening with no real change in population variance. If the computed value is larger than the table value, the difference could not be assumed to be a chance happening at the selected probability level. The difference in sample variances may be considered an indication of a real difference (in the direction indicated) between population variances.

Application of the F-Test. Suppose that the standard deviation for the tensile strength of paper used as electrical insulation has been stable at about 13 lb. For a new shipment a sample of 10 pieces has been tested and the standard deviation found to be 21 lb. The user wishes to know whether this represents a real increase in variability or whether perhaps by chance the items in the sample were more variable than they typically would be. A probability level of 0.01 is selected as suitable for the test, as it was desirable to take no action in the case unless there was strong evidence of a real change. For this case

$$F = \frac{s^2}{\sigma^2}$$

with σ^2 representing the known variance for the paper as it has been running in the past. Thus,

$$F = \frac{s^2}{\sigma^2} = \frac{21^2}{13^2} = \frac{441}{169} = 2.61$$

The table value of F obtained from Table 6 (using 9 degrees of freedom for the greater mean square and ∞ degrees of freedom for the smaller mean square) is 2.41.

Since the computed value for F is larger than the table value, the sample standard deviation for the new shipment can be considered significantly larger. If the population standard deviation for the new shipment was at the usual value, 13 lb., chances are only 1 in 100 that the value for F for a sample of 10 would be greater than 2.41 purely because of sampling fluctuations.

TEST FOR DIFFERENCE AMONG MEANS. The analysis of variance may be used not only as a test for significance of apparent differences in dispersion but also to test for significant differences among a number of sample means. For example, suppose that time studies for a work element that has been considered constant show some variation in average time from job to job. There is then the question of whether there is a real difference in required time from job to job or whether the observed differences in time-study averages could be due simply to chance or to sampling error.

The general method for the test is first to set up the hypothesis that there is no difference among the population averages from which the samples have been drawn. This hypothesis is then tested by first estimating the variance of the populations by studying the variation of items within each sample. (In using the test it must be assumed that the variances for each of the populations are approximately equal even though the averages may differ, and also that the populations are normal.) The variance of the populations is then estimated again. This estimate is made by analysis of the variation between the sample averages. This second estimate depends on the fact that $\sigma^2 = \sigma_{\bar{X}}^2 n$. The population variance, σ^2, is estimated by computing $s_{\bar{X}}^2$ directly, using the differences between means, and then multiplying by n, the sample size. The two variances are then compared by computing the ratio, F, between the larger and the smaller. If the averages of the populations are all the same, F will tend toward unity. The probabilities that a computed value of the ratio exceeds a tabular value of F due to random effects of the sampling is given in Table 6. If the computed value exceeds the table value, the hypothesis of equal population means may be rejected. Then it cannot be assumed that the difference in variance computed by the two methods is due to chance; part of the difference is probably due to a difference in population averages.

Test Procedures. The steps for the test are as follows:

1. Select a suitable probability level for testing for significance of the differences.

2. **Within-groups variance.** Estimate the population variance by use of the differences between each observation in a sample and the mean of the sample. A **pooled estimate** of the population variance is obtained by pooling the information from each of the samples by using the following equation:

$$s^2 = \frac{\Sigma(X_1 - \bar{X}_1)^2 + \Sigma(X_2 - \bar{X}_2)^2 + \cdots \Sigma(X_m - \bar{X}_m)^2}{n_1 + n_2 + \cdots n_m - m}$$

where X_1 represents an observation in the first sample; $\bar{X}_1$, the arithmetic mean of that sample; n_1 the number of observations making up the sample, and so on; and where m represents the number of samples involved in the test.

An equivalent equation which greatly simplifies the computation of the sum of the squares which makes up the numerator of the equation just given is

$$s^2 = \frac{\Sigma X^2 - \left[\frac{(\Sigma X_1)^2}{n_1} + \frac{(\Sigma X_2)^2}{n_2} + \cdots \frac{\Sigma(X_m)^2}{n_m}\right]}{N - m}$$

where X represents each of the observations making up all the samples, and N is the total number of observations in all samples.

3. **Between-groups variance.** Estimate the population variance by use of the **differences between each group** or sample average and the grand average of all the groups. Such an estimate of the population variance is obtained by the following equation:

$$s^2 = \frac{\Sigma n(\bar{X} - \bar{\bar{X}})^2}{m - 1}$$

where $\overline{X}$ represents the average for a sample and $\overline{\overline{X}}$ the grand average of all the sample averages.

An equivalent equation which greatly simplifies the computation of the sum of the squares in the formula just given is

$$s^2 = \frac{\left(\frac{(\Sigma X_1)^2}{n_1} + \frac{(\Sigma X_2)^2}{n_2} + \cdots \frac{\Sigma(X_m)^2}{n_m}\right) - \frac{(\Sigma X)^2}{N}}{m-1}$$

4. As a check on the above computations, the total sum of squares can be computed, considering the observations in all the samples as one grand sample. The formula is $\Sigma(X - \overline{\overline{X}})^2$. An equivalent formula which simplifies the computations is

$$\text{Sum of squares (numerator)} = \Sigma X^2 - \frac{(\Sigma X)^2}{N}$$

As a routine check on the denominator or degrees of freedom for each of the variance computations, the total number of degrees of freedom can be computed.

$$\text{Degrees of freedom} = N - 1$$

If the computations are correct the sum of squares for within-group variation (obtained in step 2) plus the sum of squares for between-groups variation (obtained in step 3) should equal the total sum of squares (obtained in step 4). Likewise, the degrees of freedom for within-group variation plus the degrees of freedom for between-groups variation should equal the total number of degrees of freedom.

5. Compute F, the variance ratio, by dividing the between-groups variance (step 3) by the within-group variance (step 2).

6. Using the selected confidence level and the degrees of freedom computed in steps 2 and 3, determine the table value for F from Table 6.

7. Compare the computed value for F with the table value. If the computed F is greater than the table value, reject the hypothesis that there is no difference among the averages for the populations sampled and consider the difference among sample averages as significant. If the computed F is less than the table value, accept the hypothesis and assume the observed differences between sample averages to be due to chance. Customary practice is to compute and tabulate the sums of squares and the degrees of freedom separately and to show the results in table form. Also, variance is generally referred to as the **mean square** which is obtained by dividing the **sum of the squares** by the **degrees of freedom**. This practice will be illustrated in the example that follows.

As mentioned previously, for this test to be exact, the standard deviation for each of the populations must be equal (even though their averages may not be), and the distributions must be normal. It has been found, however, that these requirements need be only roughly approximated for the test to be generally satisfactory.

Differences Between Average Incentive Earnings. Consider the following figures for individual weekly incentive earnings for four different crews of women, each crew being made up of five women. A question may be raised as to whether the differences in average weekly earnings among the crews could simply be chance differences, or whether there is evidence that they represent real differences in crew abilities or in conditions of work.

Weekly Earnings

	Crew 1	Crew 2	Crew 3	Crew 4
	$94.00	$82.10	$97.90	$93.20
	88.80	83.90	92.90	91.50
	85.40	88.20	93.60	89.70
	86.00	87.20	94.40	93.20
	89.50	85.80	90.30	91.20
Average	$88.74	$85.44	$93.82	$91.76

The computations from the weekly earnings of the crews are as follows. (Each value has been coded by subtracting $80.00.)

Crew 1		Crew 2		Crew 3		Crew 4	
X_1	X_1^2	X_2	X_2^2	X_3	X_3^2	X_4	X_4^2
14.00	196.00	2.10	4.41	17.90	320.41	13.20	174.24
8.80	77.44	3.90	15.21	12.90	166.41	11.50	132.25
5.40	29.16	8.20	67.24	13.60	184.96	9.70	94.09
6.00	36.00	7.20	51.84	14.40	207.36	13.20	174.24
9.50	90.25	5.80	33.64	10.30	106.09	11.20	125.44

$\Sigma X_1 = 43.70$

$(\Sigma X_1)^2 = 1{,}909.69$

$\Sigma X_2 = 27.20$

$(\Sigma X_2)^2 = 739.84$

$\Sigma X_3 = 69.10$

$(\Sigma X_3)^2 = 4{,}774.81$

$\Sigma X_4 = 58.80$

$(\Sigma X_4)^2 = 3{,}457.44$

$$N = n_1 + n_2 + n_3 + n_4$$
$$= 5 + 5 + 5 + 5$$
$$= 20$$
$$m = 4$$

In the analysis of variance no correction need be made later for coded values if the coding is made simply by subtracting a constant amount from each observation. A probability level of 0.01 was selected as most appropriate for this study. Further computations are:

$$\Sigma X = 14.00 + 8.80 + \cdots 11.20 = 198.80$$
$$(\Sigma X)^2 = 198.80^2 = 39{,}521.44$$
$$\Sigma X^2 = 196.00 + 77.44 + \cdots 125.44 = 2{,}286.68$$

Sum of squares—within-crew variation

$$= \Sigma X^2 - \left[\frac{(\Sigma X_1)^2}{n_1} + \frac{(\Sigma X_2)^2}{n_2} + \cdots \frac{(\Sigma X_m)^2}{n_m} \right]$$

$$= 2{,}286.68 - \left(\frac{1{,}909.69}{5} + \frac{739.84}{5} + \frac{4{,}774.81}{5} + \frac{3{,}457.44}{5} \right)$$

$$= 2{,}286.68 - 2{,}176.36 = 110.32$$

Degrees of freedom—within-crew variation

$$= n_1 + n_2 + \cdots n_m - m = 5 + 5 + 5 + 5 - 4 = 16$$

Sum of squares—between-crew variation

$$= \left[\frac{(\Sigma X_1)^2}{n_1} + \frac{(\Sigma X_2)^2}{n_2} + \cdots \frac{(\Sigma X_m)^2}{n_m}\right] - \frac{(\Sigma X)^2}{N}$$

$$= \left(\frac{1,909.69}{5} + \frac{739.84}{5} + \frac{4,774.81}{5} + \frac{3,457.44}{5}\right) - \frac{39,521.44}{20}$$

$$= 2,176.36 - 1,976.07 = 200.29$$

Degrees of freedom—between-crew variation

$$m - 1 = 4 - 1 = 3$$

Total sum of squares

$$= \Sigma X^2 - \frac{(\Sigma X)^2}{N} = 2,286.68 - \frac{39,521.44}{20} = 310.61$$

Total degrees of freedom

$$N - 1 = 20 - 1 = 19$$

Tabulation of the above values in the following way furnishes a helpful summary:

Source of Variation	Sum of Squares	Degrees of Freedom	Mean Square	F-Ratio
Between Crews	200.29	3	$\frac{200.29}{3} = 66.8$	$F = \frac{66.8}{6.9} = 9.7$
Within Crews	110.32	16	$\frac{110.32}{16} = 6.9$	
Total	310.61	19		

Using 3 as the degrees of freedom for the greater mean square and 16 as the degrees of freedom for the lesser mean square and a probability level of 0.01, interpolation in Table 6 gives a table value for F of 5.3. A comparison of the computed value with this figure shows the computed value to be greater. Thus the observed differences between averages may be considered significant. If the observed differences had simply been due to chance, with the real average earning possibilities being the same in each case, the probability of getting a computed F of more than 5.3 would be one in a hundred or less.

FACTORIAL STUDIES. In some industrial engineering or management studies there may be several factors which must be studied for their effect. A common method of approach is to study the effect of each factor separately, taking one factor at a time. Measurements may be made at a range of levels for the factor while holding all the other factors constant at some one level. This is the traditional experimental method.

In many cases a much more efficient approach may be taken through use of a **factorial design for the study.** In this approach the effects of all the factors are studied simultaneously through use of a planned design for the collection and analysis of data. There are a number of advantages in this approach over the customary method in which only one factor at a time is varied and studied.

1. Much less data and study are required to give conclusions of the required degree of confidence.
2. The data used in evaluating the effects of one factor are based on the full range of values for each of the other factors.
3. Observations will be in a most suitable form for the analysis of variance. Through the analysis of variance the observed effect of varying the level for a factor may be tested for significance. When there is considerable inherent variation in a process, it may be difficult to tell, without the use of statistical techniques, such as the analysis of variance, whether an observed effect of varying a factor represents a real effect, or whether it may be due to chance or experimental error.
4. Observations will be in suitable form for study for possible interactions between one or more of the factors. If there are interactions, the effect of varying one factor will depend on the level of one or more of the other factors. Likewise, the results of varying two factors together will not be the same as the sum of the effects that would be found if they were to be varied separately.

Factorial Design Procedure. In a factorial design for a study, each level for each factor is combined with each possible combination of levels for each of the other factors. For each combination of levels one or more observations are obtained. The general plan for collection and arrangement of the data for analysis is indicated by the following illustration. A similar plan may be developed for any number of factors and any number of levels for each. Also, the number of observations obtained for each combination of levels may be any number desired. In this plan three will be obtained. The number will generally depend on the amount of inherent variation in the procedure, the importance of the study, the precision required for the conclusions, etc. The assignment of factors to lines and columns may follow any convenient pattern.

Factors:

$$\begin{array}{ll} \text{Operators:} & A, B, \text{ and } C \\ \text{Materials:} & X \text{ and } Y \\ \text{Tools:} & 1 \text{ and } 2 \end{array}$$

The effect, if any, of each of these factors singly or in combination on output is to be determined.

	Material X		Material Y	
	Tool 1	Tool 2	Tool 1	Tool 2
Operator A	$O_A\ T_1\ M_X$	$O_A\ T_2\ M_X$	$O_A\ T_1\ M_Y$	$O_A\ T_2\ M_Y$
	$O_A\ T_1\ M_X$	$O_A\ T_2\ M_X$	$O_A\ T_1\ M_Y$	$O_A\ T_2\ M_Y$
	$O_A\ T_1\ M_X$	$O_A\ T_2\ M_X$	$O_A\ T_1\ M_Y$	$O_A\ T_2\ M_Y$
Operator B	$O_B\ T_1\ M_X$	$O_B\ T_2\ M_X$	$O_B\ T_1\ M_Y$	$O_B\ T_2\ M_Y$
	$O_B\ T_1\ M_X$	$O_B\ T_2\ M_X$	$O_B\ T_1\ M_Y$	$O_B\ T_2\ M_Y$
	$O_B\ T_1\ M_X$	$O_B\ T_2\ M_X$	$O_B\ T_1\ M_Y$	$O_B\ T_2\ M_Y$
Operator C	$O_C\ T_1\ M_X$	$O_C\ T_2\ M_X$	$O_C\ T_1\ M_Y$	$O_C\ T_2\ M_Y$
	$O_C\ T_1\ M_X$	$O_C\ T_2\ M_X$	$O_C\ T_1\ M_Y$	$O_C\ T_2\ M_Y$
	$O_C\ T_1\ M_X$	$O_C\ T_2\ M_X$	$O_C\ T_1\ M_Y$	$O_C\ T_2\ M_Y$

This design is for 36 observations, 3 for each of the 12 combinations of levels for the three factors. The observations may be taken in any convenient order. However, if there is any possibility of changes in other factors while data are being collected, observations should be collected in random order. This will minimize the possibility of such a change showing as an effect of changing one of the factors under study.

The effect of changing each factor is determined by simply computing the average for all the readings for each level and then comparing these averages. Any difference among the averages for a factor that might seem to be due simply to chance may be tested for significance by the analysis of variance. This procedure and the procedure for testing for interactions will be described in the discussion that follows.

ANALYSIS OF VARIANCE IN FACTORIAL STUDIES. There may be a question of whether the observed difference among averages for each of the levels for the factor represent real differences or whether they might be only chance differences arising from the inherent variation in the activity. When such a question does arise, the analysis of variance technique can be used to test the observed differences for significance.

The averages for each level of each factor may also be compared by simple observation to determine what interactions between factors, if any, exist and to determine their importance. In cases where an interaction seems to be present but where there is some question as to whether the variations noted in the observations may simply be chance variations, the analysis of variance may again be used as a test for significance.

Steps in the Procedure. In the analysis of variance for a factorial study, the total sum of squares and the total number of degrees of freedom for all the observations are separated into:

1. The sum of squares and the degrees of freedom for each factor that may be assigned to changes in levels for that factor.

The sum of squares for between-level variations for a factor may be determined by the following equation:

$$\text{Sum of squares} = \left[\frac{(\Sigma X_1)^2}{n_1} + \frac{(\Sigma X_2)^2}{n_2} + \cdots \frac{(\Sigma X_m)^2}{n_m} \right] - \frac{(\Sigma X)^2}{N}$$

where ΣX_1 represents the sum of all the observations at level 1 for the factor being studied and n_1 the number of observations at that level. Likewise, ΣX_2 represents the sum of all observations at level 2, n_2 the number of observations at that level, and so on through all levels. The expression ΣX represents the sum of all the observations for the entire study and N the total number of observations.

The degrees of freedom for between-level variation are $m - 1$ or one less than the number of levels for the factor.

2. The sum of squares and the degrees of freedom that may be assigned to each of the possible interactions between factors.

The sum of squares for variation due to **interaction between any two factors** may be determined by first computing the sum of the squares, using the sum of the observations for each of the combinations of levels for the two factors, and then subtracting the sum of squares for between-level variation (previously computed) for each factor. The equation may be written:

$$\text{Sum of squares} = \left[\frac{(\Sigma X_1)^2}{n_1} + \frac{(\Sigma X_2)^2}{n_2} + \cdots \frac{(\Sigma X_m)^2}{n_m} \right] - \frac{(\Sigma X)^2}{N} - \text{sum of squares for be-}$$

tween-level variation for one factor − sum of squares for between-level variation for the other factor

In the preceding equation, ΣX_1 represents the sum of the observations for one combination of levels for the two factors, ΣX_2 the sum of observations for another combination of levels, and so on until all possible combinations have been included. The value for n_1 is the number of observations for the first combination of levels, and so on.

The sum of squares for variation due to **interaction between three factors** may be determined by first computing the sum of squares, using the sum of the observations for each of the combinations of levels for the three factors, and then subtracting the sum of squares for between-level variation (previously computed) for each factor and in addition subtracting the sum of squares for each of the two-factor interactions as previously computed.

The number of degrees of freedom for any interaction is the product of the degrees of freedom for between-level variation for each of the factors in the interaction.

3. The sum of squares and the number of degrees of freedom that may be assigned to the within-group or inherent variation in the activity under study.

This sum of squares and degrees of freedom may be obtained by subtracting all the previously computed sums of squares from the total sum of squares and all the previously computed degrees of freedom from the total degrees of freedom. An alternative procedure is to compute these measures directly, using the following equations:

$$\text{Sum of squares} = \Sigma X^2 - \left(\frac{(\Sigma X_1)^2}{n_1} + \frac{(\Sigma X_2)^2}{n_2} + \cdots \frac{(\Sigma X_m)^2}{n_m} \right)$$

$$\text{Degrees of freedom} = n_1 + n_2 + \cdots n_m - m$$

In the above equations, ΣX_1 represents the sum of the observations for one combination of levels for the factors; n_1 represents the number of observations for that combination; ΣX_2 represents the sum of observations for the next combination of levels, and so on.

4. The total sum of squares and the total number of degrees of freedom may be computed. The equations are:

$$\text{Total sum of squares} = \Sigma(X^2) - \frac{(\Sigma X)^2}{N}$$

$$\text{Total degrees of freedom} = N - 1$$

After this breakdown of the total sum of squares and total number of degrees of freedom has been made, the results may be posted to an **analysis-of-variance table** for easy reference. Next, the mean-square values for each source of variation may be computed by dividing the sum of squares for that source by its degrees of freedom.

The F-ratio for each source of variation may next be determined. The observed value for F may then be compared with the table value at the selected probability level. If the computed value is greater than the table value, the observed difference in means may be considered significant.

The interaction of the three factors should be tested first, using the within-group, or residual, mean square to compute F. If the F-test shows the interaction to be significant, this interaction mean square should be used as the denominator to compute F for each of the two-factor interactions. If this is done, the F-test will test for a two-factor interaction effect over and above the three-factor effect. If the three-factor interaction is found not to be significant, its degrees of freedom and sum of squares may be added to the within-group, or residual, sum of squares and degrees of freedom. The resulting mean square from the pooled data may be used as the denominator to compute F for each of the two-factor interactions.

If one of the two-factor interactions is found to have a significant effect, the mean-square value for this interaction should be used as the denominator in computing F for the between-level variation for each of the two factors involved. This will test for an independent effect of changing the level of the factor over and above the interaction effect of the factor. If the F-test for the interaction does not show significance, its sum of squares and degrees of freedom may also be pooled with the residual sum of squares and degrees of freedom to give a revised mean square. This new mean square may be used as the denominator to compute F for between-level variation for each of the separate factors.

A Factorial Design Study with Analysis of Variance. A simple example of a factorial design and of variance analysis is given by the following productivity study. There are two crews doing the work under study, one supervised by Foreman A, the other by Foreman B. Both crews are made up of male and female operators. Productivity figures have been obtained for a ten-day period for the crew under each foreman, and for each foreman, separately for male and for female operators. These figures are shown below. Three questions have been asked. (1) Is the apparent difference in productivity between foremen significant? (2) Is there evidence of a real difference in productivity for this work between male and female operators? (3) Is there an interaction between foreman and sex of operator; that is, does the productivity of a specified sex depend on which of the two foremen is the supervisor?

The following are the computations that were made. Actual productivity figures have been coded by subtracting 100 to simplify the arithmetic.

Foreman A		Foreman B	
Male	Female	Male	Female
4	4	4	4
4	4	3	4
5	2	1	0
5	2	2	2
3	2	2	3
6	3	2	0
4	2	3	2
4	5	3	2
4	1	4	3
6	2	1	3
$\Sigma = 45$	$\Sigma = 27$	$\Sigma = 25$	$\Sigma = 23$

$$\text{Average, Foreman } A = \frac{\Sigma X}{n} = \frac{45 + 27}{10 + 10} = \frac{72}{20} = 3.6 \text{ or } 103.6$$

$$\text{Average, Foreman } B = \frac{\Sigma X}{n} = \frac{25 + 23}{10 + 10} = \frac{48}{20} = 2.4 \text{ or } 102.4$$

$$\text{Average, male operators} = \frac{\Sigma X}{n} = \frac{45 + 25}{10 + 10} = \frac{70}{20} = 3.5 \text{ or } 103.5$$

$$\text{Average, female operators} = \frac{\Sigma X}{n} = \frac{27 + 23}{10 + 10} = \frac{50}{20} = 2.5 \text{ or } 102.5$$

$$N = 10 + 10 + 10 + 10 = 40 \qquad \Sigma X = 45 + 27 + 25 + 23 = 120$$
$$\Sigma X^2 = 4^2 + 4^2 + 5^2 + \cdots 3^2 + 3^2 = 442$$

Between foremen:

$$\text{Sum of squares} = \left(\frac{72^2}{20} + \frac{48^2}{20}\right) - \frac{120^2}{40} = 374.4 - 360.0 = 14.4$$

Degrees of freedom = 2 − 1 = 1

Between sexes:

$$\text{Sum of squares} = \left(\frac{70^2}{20} + \frac{50^2}{20}\right) - \frac{120^2}{40} = 370.0 - 360.0 = 10.0$$

Degrees of freedom = 2 − 1 = 1

Interaction:

$$\text{Sum of squares} = \left(\frac{45^2}{10} + \frac{27^2}{10} + \frac{25^2}{10} + \frac{23^2}{10}\right) - \frac{120^2}{40} - 14.4 - 10.0 = 6.4$$

Degrees of freedom = 1 × 1 = 1

Within-group, or residual:

$$\text{Sum of squares} = 442 - \left(\frac{45^2}{10} + \frac{27^2}{10} + \frac{25^2}{10} + \frac{23^2}{10}\right) = 442 - 390.8 = 51.2$$

Degrees of freedom = 10 + 10 + 10 + 10 − 4 = 36

These values are summarized below. The mean square for each source of variation has been computed as shown. The F-values taken from Table 6 are for a probability level of 0.05.

Total:

$$\text{Sum of squares} = 442 - \frac{120^2}{40} = 442.0 - 360.0 = 82.0$$

Degrees of freedom = 40 − 1 = 39

Source of Variation	Sum of Squares	Degrees of Freedom	Mean Square	F-Ratio	$F=$ Table F Value (Probability = 0.05)
Between Foremen	14.4	1	$\frac{14.4}{1} = 14.4$	$\frac{14.4}{6.4} = 2.25$	161
Between Sexes	10.0	1	$\frac{10.0}{1} = 10.0$	$\frac{10.0}{6.4} = 1.56$	161
Interaction	6.4	1	$\frac{6.4}{1} = 6.4$	$\frac{6.4}{1.42} = 4.50$	4.08
Within Group	51.2	36	$\frac{51.2}{36} = 1.42$		
Total	82.0	39			

The analysis shows the foreman-sex interaction to be significant. Accordingly, the plant was rearranged so that all men could be placed under Foreman A and all women under Foreman *B*. The analysis of variance indicated no significance for the difference in means between foremen or between sexes independent of the interaction effect. This does not mean that workers under Foreman *A* may not really be more productive than workers under Foreman *B* or that male operators are not more productive than female. It does mean that the observations for this study give no evidence (at the selected probability level) of an effect of these factors that is independent of the interaction effect.

Regression

RELATIONSHIPS BETWEEN VARIABLES. Regression techniques are useful in estimating the value of one variable when another variable is known. Knowing the relationship between two or more variables is useful in production management applications when: (1) an important variable is difficut to measure and control and it is related to one that is easier to measure and control; (2) the output or some other characteristic of a process is related to a variable which is possible to control or predict; (3) future values of variables are related to known present values of other variables.

For an example of the first case, it would be difficult to make a time study of the exact standard hours' worth of work performed by a longshoreman crew handling thousands of different items, but if it is known that a relationship exists between tonnage handled and standard hours, the crew performance may be evaluated on the basis of tonnage handled. At least two aspects of the relationship are desired: what is the relationship of the average value of the dependent variable to the value of the independent variable (how many standard hours per ton), and how precise or variable is the relationship—how wrong one is likely to be in making estimates or predictions. The strength of the relationship can be thought of as the relative importance, in the example, of tonnage compared to the importance of factors such as bulkiness or shape of the goods moved or chance factors such as weather in determining time required.

Consider also the control of packaged net weight of a semifluid product such as peanut butter. It would be difficult to empty and weigh the contents of many jars, but if it is known that net weight per jar (for a given volume of fill) is related to the viscosity of the contents of the filling hopper, control can be kept on an easily read viscosity measurement. In this example also, we have to know what the relationship is and how reliably fill weight is related to viscosity—how important viscosity is compared to gasket leaks, pressure variations, and other factors.

Still another example is the knowledge of the relationship between hours flown and standard hours of repair required by an aircraft, used in forecasting and scheduling workload requirements in aircraft overhaul and repair stations.

Standard time for new products can be predicted by knowing the relation between measurable product dimensions and standard time.

Regression techniques are designed to reveal:

1. The **nature** of the relationship between two variables, that is, what formula can be used to determine one from the other.
2. The **strength** of the relationship, that is, how widely deviant are the real values likely to be from the predicted values.
3. How well the observed sample of values indicate the real nature of the relationship; that is, whether a large enough sample has been observed so that the **real relationship** will not be distorted by one factor varying independently of the other.

GRAPHIC METHODS. The simplest approach, and one that is entirely satisfactory in many cases, is to plot the available data on graph paper. It may then be examined for **evidence of a relationship.** If there is good evidence, a curve or line may be fitted readily by eye. This line can then be used directly as a device to predict the value for one variable, given some value of the other, or the line may be used to derive an equation for use in making such predictions.

If the points fit some line closely, placing the line by eye will usually give satisfactory results. If the points have a **linear trend** so that the relationship may be represented by a straight line, some accuracy may be gained in placing the line by the following procedure: (1) compute the mean for the observations for each of the variables; (2) plot a point on the graph representing these two averages; (3) draw the line (fitting it by eye) so as to pass through this point.

The following data was used in a study to determine the relationship between the total labor hours required to assemble an airframe subassembly and the total number of "hole operations" in the subassembly. (Drilling a hole, reaming a hole, placing a rivet, etc., constitute hole operations.)

Study	Total No. of Hole Operations X	Total Labor Hours for Assembly Y
A	236	5.1
B	80	1.7
C	127	3.3
D	445	6.0
E	180	2.9
F	343	5.9
G	305	7.0
H	488	9.4
I	170	4.8

Computation of the averages gives the following:

$$\Sigma X = 2{,}374 \qquad \Sigma Y = 46.1$$

$$\overline{X} = \frac{2{,}374}{9} = 264 \qquad \overline{Y} = \frac{46.1}{9} = 5.1$$

Points representing the number of hole operations and the labor hours for each study together with a point at $\overline{X}, \overline{Y}$, have been plotted in Fig. 7. A straight line has been fitted by eye, drawing it so as to pass through $\overline{X}, \overline{Y}$.

Some notion of the precision of predictions made from the plotted line may be obtained by observing the scatter of observed values about the line. Notice that while most values are within one hour of the line, one observation is more than two hours off. The entire graphical procedure is very useful, but it should be remembered that no statistical properties should be attributed to it.

METHOD OF LEAST SQUARES. Fitting a line by eye depends on the judgment of the person doing it. No two persons will place the line in exactly the same place. Also, a line placed by eye may be placed far from the most suitable position, particularly if there are few points and they are widely scattered. Even though there is a relationship between two variables, it may be obscured by the inherent variation in the process or system measured.

A relatively simple procedure is available for using sample data to make an estimate of where to place the line for its intended use. This procedure is referred to as the method of least squares. Specifically, the method of least squares produces a straight line passing through the data which minimizes the sum of the squared deviations from the line in the direction of the dependent variable.

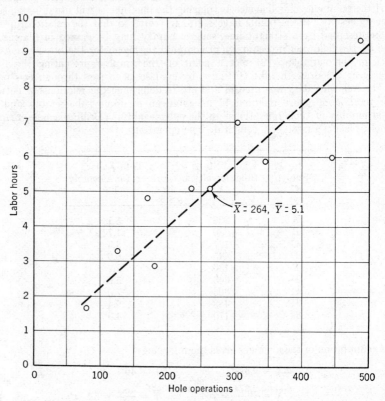

Fig. 7. Line fitted by eye showing the relationship between hole operations and labor hours.

Although the method of least squares gives us a "line of best fit," this line is not unique. For example, we could obtain another least squares line by reversing the roles of dependent and independent variables. It is not clear which of these two lines is best. It is important for engineers to recognize that of itself the method of least squares has no particular statistical properties. When one makes some special statistical assumptions about the variables and the nature of the relation the method of least squares provides best estimators and predictors in a statistical sense. These assumptions are discussed under Regression Analysis elsewhere in this section. Least squares of itself is simply a curve fitting procedure.

In the statistical procedures it is customary to use the Y-axis for the dependent variable and the X-axis for the independent variable. The line is referred to as "a fit of Y on X." For **linear** (straight line) relations, the equation of the line may be represented by

$$Y = a + bX$$

where Y is a point on the line for any given value of X. The symbol a is, of course, the value for the Y-intercept for the line, and the symbol b represents the slope of the line. When the values for a and b have been determined, this equation may be used to give an estimate for Y for any given value for X. An alternative is to use the computed values for a and b to draw the line on the graph and then to use the line for making estimates.

The value for a, the Y-intercept, may be determined by the following equation:

$$a = \frac{\Sigma X^2 \Sigma Y - \Sigma X \Sigma XY}{n \Sigma X^2 - (\Sigma X)^2}$$

where n is the number of pairs of observations.

The value for b, the slope of the line, may be determined by the following equation:

$$b = \frac{n \Sigma XY - \Sigma X \Sigma Y}{n \Sigma X^2 - (\Sigma X)^2}$$

An alternative procedure for finding a and b is available. It is to solve the following simultaneous equations:

$$\Sigma Y = na + b \Sigma X$$

$$\Sigma XY = a \Sigma X + b \Sigma X^2$$

The point $(\overline{X}, \overline{Y})$ discussed above lies on the line. These values may be computed and plotted to serve as a check on the values determined by the formulas.

Curvilinear Relationships. The relationship may be more closely approximated by some curved line rather than by a straight line in some studies. When the points fall closely along what seems to be a well-defined line, the line may be placed by eye reasonably well. When this is not the case, a more accurate method may be necessary.

One curvilinear relationship frequently observed is that of a parabolic line. The formula for this line is:

$$Y = a + bX + cX^2$$

The following simultaneous equations may be used to compute the values for the constants a, b, and c:

$$\Sigma Y = na + b \Sigma X + c \Sigma X^2$$

$$\Sigma XY = a \Sigma X + b \Sigma X^2 + c \Sigma X^3$$

$$\Sigma X^2 Y = a \Sigma X^2 + b \Sigma X^3 + c \Sigma X^4$$

Time Studies Using Least Squares. The time study data for the airframe subassembly illustration, discussed under Graphic Methods are used to determine the least squares line (the Y^2 values have been computed for use in the discussion on Regression Analysis which follows).

Study	Hole Operations X	Labor Hr. Y	X^2	Y^2	XY
A	236	5.1	55,696	26.01	1,203.6
B	80	1.7	6,400	2.89	136.0
C	127	3.3	16,129	10.89	419.1
D	445	6.0	198,025	36.00	2,670.0
E	180	2.9	32,400	8.41	522.0
F	343	5.9	117,649	34.81	2,023.7
G	305	7.0	93,025	49.00	2,135.0
H	488	9.4	238,144	88.36	4,587.2
I	170	4.8	28,900	23.04	816.0

$$\Sigma X = 2{,}374 \quad \Sigma Y = 46.1 \quad \Sigma X^2 = 786{,}368 \quad \Sigma Y^2 = 279.41 \quad \Sigma XY = 14{,}512.6$$

$$\overline{X} = \frac{\Sigma X}{n} = \frac{2{,}374}{9} = 264 \qquad \overline{Y} = \frac{\Sigma Y}{n} = \frac{46.1}{9} = 5.1$$

$$a = \frac{\Sigma X^2 \Sigma Y - \Sigma X \Sigma XY}{n\Sigma X^2 - (\Sigma X)^2} = \frac{786{,}368(46.1) - 2{,}374(14{,}512.6)}{9(786{,}368) - (2{,}374)^2}$$

$$= 1.245 \text{ hr.}$$

$$b = \frac{n\Sigma XY - \Sigma X \Sigma Y}{n\Sigma X^2 - (\Sigma X)^2} = \frac{9(14{,}512.6) - 2{,}374(46.1)}{9(786{,}368) - (2{,}374)^2}$$

$$= 0.0147$$

The equation for the least squares line thus becomes

$$Y = 1.245 + 0.0147X$$

This line has been plotted in Fig. 8. As a check, $(\overline{X}, \overline{Y})$ has been plotted to see that it falls on the computed line.

REGRESSION ANALYSIS. In certain situations the line fitted by least squares has special statistical properties. This is not to say that when the conditions which specify the statistical properties do not hold that least squares does not provide a useful engineering tool. Rather, in certain cases the engineer may place increased confidence in the results of the least squares procedure.

In simple linear regression with one independent and one dependent variable the following assumptions are made:

1. The independent variable X is deterministic and not random. This variable is considered completely under control and its values known without any significant experimental error.
2. The dependent variable Y is a function of X, $Y(X)$, in that its expectation $E(Y(X))$ depends linearly on X, that is

$$E(Y(X)) = \alpha + \beta X$$

and the variance Var $(Y(X)) = \sigma^2$ is independent of X, in fact, being constant throughout the entire range of X.
3. The series of observations used to fit the line are denoted by (X_1, Y_1), (X_2, Y_2), . . . , (X_n, Y_n) and it is assumed that the Y_i are themselves uncorrelated random variables.

When the above conditions hold, the least squares line provides a "best" fit to the data in that it provides an estimator a of the parameter α, and an estimator

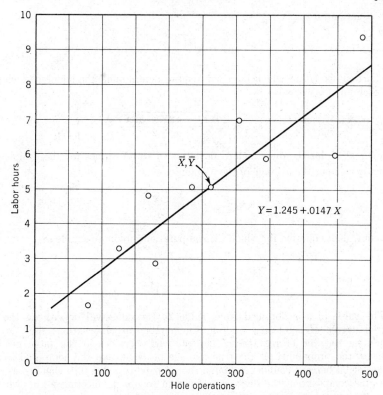

Fig. 8. Line computed by the method of least squares.

b of the parameter β which are minimum variance unbiased linear estimators of the parameters. The calculated regression line $a + bX$ can be thought of as an estimate of the true relationship between the expected value of Y and X, that is, the true regression line $\alpha + \beta X$. The estimators a and b also have the property that for very large samples they will be arbitrarily close to α and β.

CONFIDENCE LIMITS. If one can make the additional assumption that the deviation in Y about the regression has a normal distribution it is possible to compute confidence intervals and conduct tests of hypotheses on the least squares estimates and the regression line. The estimates a and b are functions of the sampling data and hence are themselves random variables. If the Y's are normally distributed so are a and b. This fact is used in forming confidence intervals. The confidence intervals should be interpreted in a similar fashion as those discussed previously. The confidence limits for the expected value of Y given a specified value of X, $\mu_{y \cdot x}$, is given by the following equation:

$$Y_{r_x} \pm t s_{y \cdot x} \sqrt{\frac{1}{n} + \frac{(X - \overline{X})^2}{(n-1)s_x{}^2}}$$

where $Y_{r_x} =$ the value for the average Y given by the least-squares line or equation
$s_{y \cdot x} =$ the estimate of the standard deviation of the Y-values obtained from the sample observations about the line of regression

The equation for $s_{y \cdot x}$ is

$$s_{y \cdot x} = \sqrt{\frac{\Sigma(Y - Y_r)^2}{n - 2}}$$

A short-cut equation which does not require determining the squared deviation for each point is

$$s_{y \cdot x} = \sqrt{\frac{\Sigma Y^2 - a\Sigma Y - b\Sigma XY}{n - 2}}$$

The equation for S_x, the standard deviation of the X-values for the sample observations about the average of the X-values, $\overline{X}$, is

$$s_x = \sqrt{\frac{\Sigma(X - \overline{X})^2}{n - 1}}$$

An equivalent equation for which the computations are more simple is

$$s_x = \sqrt{\frac{\Sigma X^2 - \dfrac{(\Sigma X)^2}{n}}{n - 1}}$$

The value of t is obtained from Table 2. Its value will depend on the required degree of confidence and on the number of observations used in determining the line of regression. The probability values in the table heading indicate the probability of the true average being outside the computed limits. Thus for 90 percent confidence limits, the probability value 0.10 should be used. For this computation the degrees of freedom to use in selecting t are two less than the number of pairs of observations, or $n - 2$.

Computation of a Confidence Limit. Suppose, for example, that one wishes to know with 90 percent confidence the limits between which the true average time for the assembly of airframes with 300 hole operations will lie. The computations are as follows:

$$s_{y \cdot x} = \sqrt{\frac{\Sigma Y^2 - a\Sigma Y - b\Sigma XY}{n - 2}}$$

$$= \sqrt{\frac{279.41 - 1.245(46.1) - 0.0147(14,512.6)}{9 - 2}}$$

$$= \sqrt{\frac{8.58}{7}} = 1.07$$

$$s_x = \sqrt{\frac{\Sigma X^2 - \dfrac{(\Sigma X)^2}{n}}{n - 1}} = \sqrt{\frac{786,368 - \dfrac{2,374^2}{9}}{9 - 1}}$$

$$= \sqrt{\frac{160,160}{8}} = 141$$

$$Y_{r_x} = 1.245 + 0.0147X$$

$$= 1.245 + 0.0147(300) = 5.66$$

Confidence limits for $\mu_{y \cdot x}$ are

$$Y_{r_x} \pm t s_{y \cdot x} \sqrt{\frac{1}{n} + \frac{(X - \overline{X})^2}{(n-1)s_x{}^2}}$$

$$5.66 \pm 1.90(1.07) \sqrt{\frac{1}{9} + \frac{(300 - 264)^2}{(9-1)141^2}}$$

$$5.66 \pm 0.70 \text{ hr.}$$

This means that for units with 300 hole operations one can be 90 percent confident that the average assembly time required will be between 5.66 ± 0.70 hr. or between 4.96 hr. and 6.36 hr.

Other Confidence Limit Formulas. Confidence limits for individual values for the dependent variable can also be determined to indicate how far off one is likely to be in any individual prediction. In general, the surer the true relationship, the less will be the deviation of any one value; and the greater the importance of the independent variable in the relationship, the less likely will be a deviation.

The formula for limits for **individual values** is

$$Y_{r_x} \pm t s_{y \cdot x} \sqrt{1 + \frac{1}{n} + \frac{(X - \overline{X})^2}{(n-1)s_x{}^2}}$$

For the above illustration, 90 percent confidence limits for Y when X is 300 are

$$5.66 \pm 1.90(1.07) \sqrt{1 + \frac{1}{9} + \frac{(300 - 264)^2}{(9-1)141^2}}$$

$$5.66 \pm 2.16 \text{ hr.}$$

This means that for any one assembly with 300 hole operations, one can be 90 percent confident that its labor time will be between 5.66 ± 2.16 hr. or between 3.50 and 7.82 hr. Notice that these limits are much wider than those for the average value.

Confidence limits for the **true slope** of the line of regression can also be computed. The formula is

$$b \pm t \frac{s_{y \cdot x}}{s_x \sqrt{n-1}}$$

As in the two tests above, the number of degrees of freedom to use in getting the value of t from Table 2 are $n - 2$.

For the illustration above, 90 percent confidence limits are

$$0.0147 \pm 1.90 \left(\frac{1.07}{141\sqrt{9-1}} \right)$$

$$0.0147 \pm 0.0051$$

This computation shows that one can be 90 percent confident that the real slope of the line of regression lies between 0.0147 ± 0.0051 or between 0.0096 and 0.0198.

The slope of the regression line provides a basis for a statistical test of the significance of the regression. One sets up a test with a null hypothesis that the slope $\beta = 0$. The test may be carried out at a given level of risk by computing the confidence interval and rejecting the hypothesis if the interval does not include zero. In the example, the regression is significant at the 90 percent level as the interval (0.0096 to 0.0198) does not include zero.

CORRELATION. Regression analysis determines the relationship between a **random variable**, Y, and a **deterministic variable**, X. Correlation analysis examines the relationships between random variables Y and X. In many production situations it is not realistic to assume, as was done in the discussion of regression, that X is completely controllable. For example, suppose that the relation, if any, between the amount of a chemical additive X and the breaking strength Y of a yarn must be determined. Measurements of X may be subject to random error as are measurements of Y. Thus the two random variables must be related. The fundamental statistical measure of the relationship between X and Y is the covariance

$$\text{Cov}(X, Y) = E[(X - E(X))(Y - E(Y))]$$

The covariance is a parameter of the joint probability distribution of X and Y and will be large and positive if there is a positive linear relationship between X and Y, large and negative if there is a negative linear relationship between X and Y, and zero if X and Y are independent random variables. The **correlation coefficient** is a scaled version of the covariance given by the equation

$$\rho_{XY} = \frac{\text{Cov}(X, Y)}{\sigma_X \sigma_Y}$$

It is a more useful measure of the relationship between X and Y since

1. If $\rho_{XY} = +1$ the variables are positively linearly related. It is also true that if the variables are positively linearly related $\rho_{XY} = +1$.
2. If $\rho_{XY} = -1$ the variables are negatively linearly related. It is also true that if the variables are negatively linearly related $\rho_{XY} = -1$.
3. If the variables are independent $\rho_{XY} = 0$. The **converse is not true,** for it is possible that X and Y can be related and still have $\rho_{XY} = 0$.

Estimation of the Correlation Coefficient. The correlation coefficient ρ_{XY} may be estimated by the sample correlation coefficient r_{XY} by using n pairs of observations of X and Y, $(X_1, Y_1), \ldots, (X_n, Y_n)$ in the equation

$$r_{XY} = \frac{\sum_{i=1}^{n} (X_i - \bar{X})(Y_i - \bar{Y})}{\left\{ \left[\sum_{i=1}^{n} (X_i - \bar{X})^2 \right] \left[\sum_{i=1}^{n} (Y_i - \bar{Y})^2 \right] \right\}^{1/2}}$$

An alternative equation which is more suitable for computation is

$$r_{XY} = \frac{n \sum_{i=1}^{n} X_i Y_i - \left(\sum_{i=1}^{n} X_i \right) \left(\sum_{i=1}^{n} Y_i \right)}{\left\{ \left[n \sum_{i=1}^{n} X_i^2 - \left(\sum_{i=1}^{n} X_i \right)^2 \right] \left[n \sum_{i=1}^{n} Y_i^2 - \left(\sum_{i=1}^{n} Y_i \right)^2 \right] \right\}^{1/2}}$$

In using this equation only ΣX, ΣY, ΣXY, ΣX^2 and ΣY^2 need be computed from the original data. Whichever equation is used, r_{XY} provides an unbiased

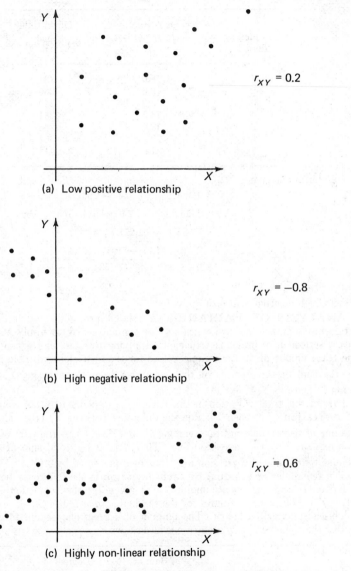

Fig. 9. Data sets and sample correlation coefficients.

estimate ρ_{XY}. Figure 9 shows several data sets with their associated values of r_{XY} and interpretations.

Correlation Between Material Defects and Down-time Cost. Defects in the metal rod used in the manufacture of wire can cause breakage·of the wire during the drawing process. An investigation was made of the relationship between the average number of defects found per lot of rod in sampling inspection and costs due to machine down-time. The available data were

Production Period	Average Number of Defects per Lot X	Down-time Cost Y
1	8.0	131
2	9.6	266
3	13.0	240
4	14.3	251
5	18.0	258
6	20.2	367
7	21.3	418

Computations give

$$\Sigma X = 104.4 \qquad \Sigma Y = 1931$$
$$\Sigma X^2 = 1715.38 \qquad \Sigma Y^2 = 584,757$$
$$\Sigma XY = 31,271.7$$
$$r_{XY} = \frac{(7)(31,271.7) - (1931)(104.4)}{\sqrt{[7(1715.38) - (104.4)^2][(7)(584,757) - (1931)^2]}}$$
$$= 0.86$$

Thus a high positive relationship exists.

ANALYSIS OF VARIANCE. A related procedure for testing the relationship between the two variables for significance is to apply the F-test, a test described in a previous article. In applying this test in regression analysis, the **total variation** in the sample values for the dependent variable is measured by the sum of the squares of the deviations from the average, $\overline{Y}$. This total sum of squares is $\Sigma(Y - \overline{Y})^2$.

Part of this total variation in the Y-values is explained by the computed **line of regression.** For each Y-value the difference between Y_r (the value given by the line of regression) and the average, $\overline{Y}$, or $(Y_r - \overline{Y})$, is the part of the deviation explained by the regression line. That part of the total sum of squares explained by the line of regression may thus be represented by $\Sigma(Y_r - \overline{Y})^2$.

The remainder or **residual of each deviation** is the difference between each value of Y and the corresponding value given by the regression line Y_r, or $(Y - Y_r)$. The sum of squares for the residual is thus $\Sigma(Y - Y_r)^2$.

This is summarized below. The number of degrees of freedom to use in computing the mean square in each case is also shown.

Source of Variation	Sum of Squares	Degrees of Freedom	Mean Square
Variation explained by the line of regression	$\Sigma(Y_r - \overline{Y})^2$	1	$\dfrac{\Sigma(Y_r - \overline{Y})^2}{1}$
The residual	$\Sigma(Y - Y_r)^2$	$n - 2$	$\dfrac{\Sigma(Y - Y_r)^2}{n - 2}$
Total	$\Sigma(Y - \overline{Y})^2$	$n - 1$	

If there were no actual relationship between the variables, the two mean-square values should be approximately the same. Any difference would be due only to the effect of chance on the sample observations. However, if the **residual mean square** (which measures the dispersion of the observations about the line of regression) is significantly smaller than the other, one may assume that the line of regression represents a real relationship.

F-Test Procedure and Formulas. The steps in applying the F-test are as follows:

1. Select a suitable probability to test for significance.
2. Compute F, where

$$F = \frac{\text{regression mean square}}{\text{residual mean square}}$$

3. Using the selected probability level, determine the table value for F from Table 6.
4. Compare the computed value of F with the table value. If the computed value for F is larger than the table value, the line of regression may be considered significant.

Equations that may be used for easy computation of the sum of squares are:

$$\Sigma(Y_r - \overline{Y})^2 = b\left(\Sigma XY - \frac{\Sigma X \Sigma Y}{n}\right)$$

$$\Sigma(Y - Y_r)^2 = \Sigma Y^2 - a\Sigma Y - b\Sigma XY$$

$$\Sigma(Y - \overline{Y})^2 = \Sigma Y^2 - \frac{(\Sigma Y)^2}{n}$$

Application of this test to the previous time-study example shows, using a probability level of 0.05, that the relationship may be considered significant. The computations made in reaching this conclusion are

$$\Sigma(Y_r - \overline{Y})^2 = 0.0147\left(14{,}512.6 - \frac{2{,}374(46.1)}{9}\right)$$

$$= 34.6$$

$$\Sigma(Y - Y_r)^2 = 279.41 - 1.245(46.1) - 0.0147(14{,}512.6)$$

$$= 8.6$$

$$\Sigma(Y - \overline{Y})^2 = 279.41 - \frac{46.1^2}{9} = 43.2$$

Source of Variation	Sum of Squares	Degrees of Freedom	Mean Square	F-ratio
Variation explained by the line of regression	34.6	1	$\frac{34.6}{1} = 34.6$	$F = \frac{34.6}{1.2} = 29$
The residual	8.6	7	$\frac{8.6}{7} = 1.2$	
Total	43.2	8		

From Table 6 the table value of F for a probability of 0.05 is found to be 5.59. Only five times out of a hundred could a computed F larger than this value be obtained by chance if there were no relationship between the variables. As the computed F is 29, one assumes a real relationship does exist.

Statistical Tables

PURPOSE. The following tables are among those most frequently used in statistical analysis and are necessary for the application of many of the statistical procedures described in this section.

Table 1. Values of z, Normal Distribution Areas

	0.00	0.01	0.02	0.03	0.04	0.05	0.06	0.07	0.08	0.09	
0.0	0.500 00	0.503 99	0.507 98	0.511 97	0.515 95	0.519 94	0.523 92	0.527 90	0.531 88	0.535 86	0.0
0.1	0.539 83	0.543 79	0.547 76	0.551 72	0.555 67	0.559 62	0.563 56	0.567 49	0.571 42	0.575 34	0.1
0.2	0.579 26	0.583 17	0.587 06	0.590 95	0.594 83	0.598 71	0.602 57	0.606 42	0.610 26	0.614 09	0.2
0.3	0.617 91	0.621 72	0.625 51	0.629 30	0.633 07	0.636 83	0.640 58	0.644 31	0.648 03	0.651 73	0.3
0.4	0.655 42	0.659 10	0.662 76	0.666 40	0.670 03	0.673 64	0.677 24	0.680 82	0.684 38	0.687 93	0.4
0.5	0.691 46	0.694 97	0.698 47	0.701 94	0.705 40	0.708 84	0.712 26	0.715 66	0.719 04	0.722 40	0.5
0.6	0.725 75	0.729 07	0.732 37	0.735 65	0.738 91	0.742 15	0.745 37	0.748 57	0.751 75	0.754 90	0.6
0.7	0.758 03	0.761 15	0.764 24	0.767 30	0.770 35	0.773 37	0.776 37	0.779 35	0.782 30	0.785 23	0.7
0.8	0.788 14	0.791 03	0.793 89	0.796 73	0.799 54	0.802 34	0.805 10	0.807 85	0.810 57	0.813 27	0.8
0.9	0.815 94	0.818 59	0.821 21	0.823 81	0.826 39	0.828 94	0.831 47	0.833 97	0.836 46	0.838 91	0.9
1.0	0.841 34	0.843 75	0.846 13	0.848 49	0.850 83	0.853 14	0.855 43	0.857 69	0.859 93	0.862 14	1.0
1.1	0.864 33	0.866 50	0.868 64	0.870 76	0.872 85	0.874 93	0.876 97	0.879 00	0.881 00	0.882 97	1.1
1.2	0.884 93	0.886 86	0.888 77	0.890 65	0.892 51	0.894 35	0.896 16	0.897 96	0.899 73	0.901 47	1.2
1.3	0.903 20	0.904 90	0.906 58	0.908 24	0.909 88	0.911 49	0.913 08	0.914 65	0.916 21	0.917 73	1.3
1.4	0.919 24	0.920 73	0.922 19	0.923 64	0.925 06	0.926 47	0.927 85	0.929 22	0.930 56	0.931 89	1.4
1.5	0.933 19	0.934 48	0.935 74	0.936 99	0.938 22	0.939 43	0.940 62	0.941 79	0.942 95	0.944 08	1.5
1.6	0.945 20	0.946 30	0.947 38	0.948 45	0.949 50	0.950 53	0.951 54	0.952 54	0.953 52	0.954 48	1.6
1.7	0.955 43	0.956 37	0.957 28	0.958 18	0.959 07	0.959 94	0.960 80	0.961 64	0.962 46	0.963 27	1.7
1.8	0.964 07	0.964 85	0.965 62	0.966 37	0.967 11	0.967 84	0.968 56	0.969 26	0.969 95	0.970 62	1.8
1.9	0.971 28	0.971 93	0.972 57	0.973 20	0.973 81	0.974 41	0.975 00	0.975 58	0.976 15	0.976 70	1.9
2.0	0.977 25	0.977 78	0.978 31	0.978 82	0.979 32	0.979 82	0.980 30	0.980 77	0.981 24	0.981 69	2.0
2.1	0.982 14	0.982 57	0.983 00	0.983 41	0.983 82	0.984 22	0.984 61	0.985 00	0.985 37	0.985 74	2.1
2.2	0.986 10	0.986 45	0.986 79	0.987 13	0.987 45	0.987 78	0.988 09	0.988 40	0.988 70	0.988 99	2.2
2.3	0.989 28	0.989 56	0.989 83	0.990 10	0.990 36	0.990 61	0.990 86	0.991 11	0.991 34	0.991 58	2.3
2.4	0.991 80	0.992 02	0.992 24	0.992 45	0.992 66	0.992 86	0.993 05	0.993 24	0.993 43	0.993 61	2.4
2.5	0.993 79	0.993 96	0.994 13	0.994 30	0.994 46	0.994 61	0.994 77	0.994 92	0.995 06	0.995 20	2.5
2.6	0.995 34	0.995 47	0.995 60	0.995 73	0.995 85	0.995 98	0.995 09	0.996 21	0.996 32	0.996 43	2.6
2.7	0.996 53	0.996 64	0.996 74	0.996 83	0.996 93	0.997 02	0.997 11	0.997 20	0.997 28	0.997 36	2.7
2.8	0.997 44	0.997 52	0.997 60	0.997 67	0.997 74	0.997 81	0.997 88	0.997 95	0.998 01	0.998 07	2.8
2.9	0.998 13	0.998 19	0.998 25	0.998 31	0.998 36	0.998 41	0.998 46	0.998 51	0.998 56	0.998 61	2.9
3.0	0.998 65	0.998 69	0.998 74	0.998 78	0.998 82	0.998 86	0.998 89	0.998 93	0.998 97	0.999 00	3.0
3.1	0.999 03	0.999 06	0.999 10	0.999 13	0.999 16	0.999 18	0.999 21	0.999 24	0.999 26	0.999 29	3.1
3.2	0.999 31	0.999 34	0.999 36	0.999 38	0.999 40	0.999 42	0.999 44	0.999 46	0.999 48	0.999 50	3.2
3.3	0.999 52	0.999 53	0.999 55	0.999 57	0.999 58	0.999 60	0.999 61	0.999 62	0.999 64	0.999 65	3.3
3.4	0.999 66	0.999 68	0.999 69	0.999 70	0.999 71	0.999 72	0.999 73	0.999 74	0.999 75	0.999 76	3.4
3.5	0.999 77	0.999 78	0.999 78	0.999 79	0.999 80	0.999 81	0.999 81	0.999 82	9.999 83	0.999 83	3.5
3.6	0.999 84	0.999 85	0.999 85	0.999 86	0.999 86	0.999 87	0.999 87	0.999 88	0.999 88	0.999 89	3.6
3.7	0.999 89	0.999 90	0.999 90	0.999 90	0.999 91	0.999 91	0.999 92	0.999 92	0.999 92	0.999 92	3.7
3.8	0.999 93	0.999 93	0.999 93	0.999 94	0.999 94	0.999 94	0.999 94	0.999 95	0.999 92	0.999 95	3.8

Table 2. Values of t

DF	0.90	0.80	0.70	0.60	0.50	0.25	0.10	0.05	0.025	0.010	0.005
					P						
1	0.158	0.325	0.510	0.727	1.000	2.414	6.314	12.71	25.45	63.66	127.3
2	0.142	0.289	0.445	0.617	0.817	1.604	2.920	4.303	6.205	9.925	14.09
3	0.137	0.277	0.424	0.584	0.765	1.423	2.353	3.183	4.177	5.841	7.453
4	0.134	0.271	0.414	0.569	0.741	1.344	2.132	2.776	3.495	4.604	5.598
5	0.132	0.267	0.408	0.559	0.727	1.301	2.015	2.571	3.163	4.032	4.773
6	0.131	0.265	0.404	0.553	0.718	1.273	1.943	2.447	2.969	3.707	4.317
7	0.130	0.263	0.402	0.549	0.711	1.254	1.895	2.365	2.841	3.500	4.029
8	0.130	0.262	0.399	0.546	0.706	1.240	1.860	2.306	2.752	3.355	3.833
9	0.129	0.261	0.398	0.543	0.703	1.230	1.833	2.262	2.685	3.250	3.690
10	0.129	0.260	0.397	0.542	0.700	1.221	1.813	2.228	2.634	3.169	3.581
11	0.129	0.260	0.396	0.540	0.697	1.215	1.796	2.201	2.593	3.106	3.500
12	0.128	0.259	0.395	0.539	0.695	1.209	1.782	2.179	2.560	3.055	3.428
13	0.128	0.259	0.394	0.538	0.694	1.204	1.771	2.160	2.533	3.012	3.373
14	0.128	0.258	0.393	0.537	0.692	1.200	1.761	2.145	2.510	2.977	3.326
15	0.128	0.258	0.393	0.536	0.691	1.197	1.753	2.132	2.490	2.947	3.286
20	0.127	0.257	0.391	0.533	0.687	1.185	1.725	2.086	2.423	2.845	3.153
25	0.127	0.256	0.390	0.531	0.684	1.178	1.708	2.060	2.385	2.787	3.078
30	0.127	0.256	0.389	0.530	0.683	1.173	1.697	2.042	2.360	2.750	3.030
40	0.126	0.255	0.388	0.529	0.681	1.167	1.684	2.021	2.329	2.705	2.971
60	0.126	0.254	0.387	0.527	0.679	1.162	1.671	2.000	2.299	2.660	2.915
120	0.126	0.254	0.386	0.526	0.677	1.156	1.658	1.980	2.270	2.617	2.860
∞	0.126	0.253	0.385	0.524	0.674	1.150	1.645	1.960	2.241	2.576	2.807

* Two-tailed probabilities P for given degrees of freedom, DF.

Table 3. Table of t_R

Sample Size	Probability			
	0.10	0.05	0.02	0.01
2	3.16	6.35	15.9	31.8
3	0.89	1.30	2.11	3.01
4	0.53	0.72	1.02	1.32
5	0.39	0.51	0.69	0.84
6	0.31	0.40	0.52	0.63
7	0.26	0.33	0.43	0.51
8	0.23	0.29	0.37	0.43
9	0.21	0.26	0.32	0.37
10	0.19	0.23	0.29	0.33
11	0.17	0.21	0.26	0.30
12	0.16	0.19	0.24	0.28
13	0.15	0.18	0.22	0.26
14	0.14	0.17	0.21	0.24
15	0.13	0.16	0.20	0.22
16	0.12	0.15	0.19	0.21
17	0.12	0.14	0.18	0.20
18	0.11	0.14	0.17	0.19
19	0.10	0.13	0.16	0.18
20	0.10	0.13	0.15	0.18

Table 4. Table of t_{2R}

Sample Size	Probability			
	0.10	0.05	0.02	0.01
2	2.32	3.43	5.55	7.92
3	0.97	1.27	1.72	2.09
4	0.64	0.81	1.05	1.24
5	0.49	0.61	0.77	0.90
6	0.41	0.50	0.62	0.71
7	0.35	0.43	0.53	0.60
8	0.31	0.37	0.46	0.52
9	0.28	0.33	0.41	0.46
10	0.25	0.30	0.37	0.42
11	0.23	0.28	0.34	0.38
12	0.21	0.26	0.32	0.36
13	0.20	0.24	0.29	0.33
14	0.19	0.23	0.28	0.31
15	0.18	0.22	0.26	0.29
16	0.17	0.21	0.25	0.28
17	0.16	0.20	0.24	0.26
18	0.16	0.19	0.23	0.25
19	0.15	0.18	0.22	0.24
20	0.14	0.17	0.21	0.23

Table 5. Tolerance Factors* for Normal Distributions

n	90% Confidence that percentage of population between limits is			95% Confidence that percentage of population between limits is			99% Confidence that percentage of population between limits is		
	90%	95%	99%	90%	95%	99%	90%	95%	99%
2	15.98	18.80	24.17	32.02	37.67	48.43	160.2	188.5	242.3
3	5.847	6.919	8.974	8.380	9.916	12.86	18.93	22.40	29.06
4	4.166	4.943	6.440	5.369	6.370	8.299	9.398	11.15	14.53
5	3.494	4.152	5.423	4.275	5.079	6.634	6.612	7.855	10.26
6	3.131	3.723	4.870	3.712	4.414	5.775	5.337	6.345	8.301
7	2.902	3.452	4.521	3.369	4.007	5.248	4.613	5.488	7.187
8	2.743	3.264	4.278	3.136	3.732	4.891	4.147	4.936	6.468
9	2.626	3.125	4.098	2.967	3.532	4.631	3.822	4.550	5.966
10	2.535	3.018	3.959	2.839	3.379	4.433	3.582	4.265	5.594
11	2.463	2.933	3.849	2.737	3.259	4.277	3.397	4.045	5.308
12	2.404	2.863	3.758	2.655	3.162	4.150	3.250	3.870	5.079
13	2.355	2.805	3.682	2.587	3.081	4.044	3.130	3.727	4.893
14	2.314	2.756	3.618	2.529	3.012	3.955	3.029	3.608	4.737
15	2.278	2.713	3.562	2.480	2.954	3.878	2.945	3.507	4.605
16	2.246	2.676	3.514	2.437	2.903	3.812	2.872	3.421	4.492
17	2.219	2.643	3.471	2.400	2.858	3.754	2.808	3.345	4.393
18	2.194	2.614	3.433	2.366	2.819	3.702	2.753	3.279	4.307
19	2.172	2.588	3.399	2.337	2.784	3.656	2.703	3.221	4.230
20	2.152	2.564	3.368	2.310	2.752	3.615	2.659	3.168	4.161
21	2.135	2.543	3.340	2.286	2.723	3.577	2.620	3.121	4.100
22	2.118	2.524	3.315	2.264	2.697	3.543	2.584	3.078	4.044
23	2.103	2.506	3.292	2.244	2.673	3.512	2.551	3.040	3.993
24	2.089	2.489	3.270	2.225	2.651	3.483	2.522	3.004	3.947
25	2.077	2.474	3.251	2.208	2.631	3.457	2.494	2.972	3.904
26	2.065	2.460	3.232	2.193	2.612	3.432	2.469	2.941	3.865
27	2.054	2.447	3.215	2.178	2.595	3.409	2.446	2.914	3.828
28	2.044	2.435	3.199	2.164	2.579	3.388	2.424	2.888	3.794
29	2.034	2.424	3.184	2.152	2.554	3.368	2.404	2.864	3.763
30	2.025	2.413	3.170	2.140	2.549	3.350	2.385	2.841	3.733
35	1.988	2.368	3.112	2.090	2.490	3.272	2.306	2.748	3.611
40	1.959	2.334	3.066	2.052	2.445	3.213	2.247	2.677	3.518
50	1.916	2.284	3.001	1.996	2.379	3.126	2.162	2.576	3.385
60	1.887	2.248	2.955	1.958	2.333	3.066	2.103	2.506	3.293
80	1.848	2.202	2.894	1.907	2.272	2.986	2.026	2.414	3.173
100	1.822	2.172	2.854	1.874	2.233	2.934	1.977	2.355	3.096
200	1.764	2.102	2.762	1.798	2.143	2.816	1.865	2.222	2.921
500	1.717	2.046	2.689	1.737	2.070	2.721	1.777	2.117	2.783
1000	1.695	2.019	2.654	1.709	2.036	2.676	1.736	2.068	2.718
∞	1.645	1.960	2.576	1.645	1.960	2.576	1.645	1.960	2.576

* Values of K for the limits $\overline{X} \pm Ks$.

Table 6.　Table of the F Distribution: $P = 0.10$.　Values of F for right tail of the distribution

n_1	\begin{array}{c}n_2\\1\end{array}	2	3	4	5	6	7	8	9	10	15	20	24	30	40	60	120	∞
1	39.86	49.50	53.59	55.83	57.24	58.20	58.91	59.44	59.86	60.20	61.22	61.74	62.00	62.26	62.53	62.79	63.06	63.83
2	8.53	9.00	9.16	9.24	9.29	9.33	9.35	9.37	9.38	9.39	9.42	9.44	9.45	9.46	9.47	9.47	9.48	9.49
3	5.54	5.46	5.39	5.34	5.31	5.28	5.27	5.25	5.24	5.23	5.20	5.18	5.18	5.17	5.16	5.15	5.14	5.13
4	4.54	4.32	4.19	4.11	4.05	4.01	3.98	3.95	3.94	3.92	3.87	3.84	3.83	3.82	3.80	3.79	3.78	3.76
5	4.06	3.78	3.62	3.52	3.45	3.40	3.37	3.34	3.32	3.30	3.24	3.21	3.19	3.17	3.16	3.14	3.12	3.10
6	3.78	3.46	3.29	3.18	3.11	3.05	3.01	2.98	2.96	2.94	2.87	2.84	2.82	2.80	2.78	2.76	2.74	2.72
7	3.59	3.26	3.07	2.96	2.88	2.83	2.78	2.75	2.72	2.70	2.63	2.59	2.58	2.56	2.54	2.51	2.49	2.47
8	3.46	3.11	2.92	2.81	2.73	2.67	2.62	2.59	2.56	2.54	2.46	2.42	2.40	2.38	2.36	2.34	2.32	2.29
9	3.36	3.01	2.81	2.69	2.61	2.55	2.51	2.47	2.44	2.42	2.34	2.30	2.28	2.25	2.23	2.21	2.18	2.16
10	3.28	2.92	2.73	2.61	2.52	2.46	2.41	2.38	2.35	2.32	2.24	2.20	2.18	2.16	2.13	2.11	2.08	2.06
11	3.23	2.86	2.66	2.54	2.45	2.39	2.34	2.30	2.27	2.25	2.17	2.12	2.10	2.08	2.05	2.03	2.00	1.97
12	3.13	2.81	2.61	2.48	2.39	2.33	2.28	2.24	2.21	2.19	2.10	2.06	2.04	2.01	1.99	1.96	1.93	1.90
13	3.14	2.76	2.56	2.43	2.35	2.28	2.23	2.20	2.16	2.14	2.05	2.01	1.98	1.96	1.93	1.90	1.88	1.85
14	3.10	2.73	2.52	2.39	2.31	2.24	2.19	2.15	2.12	2.10	2.01	1.96	1.94	1.91	1.89	1.86	1.83	1.80
15	3.07	2.70	2.49	2.36	2.27	2.21	2.16	2.12	2.09	2.06	1.97	1.92	1.90	1.87	1.85	1.82	1.79	1.76
16	3.05	2.67	2.46	2.33	2.24	2.18	2.13	2.09	2.06	2.03	1.94	1.89	1.87	1.84	1.81	1.78	1.75	1.72
17	3.03	2.64	2.44	2.31	2.22	2.15	2.10	2.06	2.03	2.00	1.91	1.86	1.84	1.81	1.78	1.75	1.72	1.69
18	3.01	2.62	2.42	2.29	2.20	2.13	2.08	2.04	2.00	1.98	1.89	1.84	1.81	1.78	1.75	1.72	1.69	1.66
19	2.99	2.61	2.40	2.27	2.18	2.11	2.06	2.02	1.98	1.96	1.86	1.81	1.79	1.76	1.73	1.70	1.67	1.63
20	2.97	2.59	2.38	2.25	2.16	2.09	2.04	2.00	1.96	1.94	1.84	1.79	1.77	1.74	1.71	1.68	1.64	1.61
21	2.96	2.57	2.36	2.23	2.14	2.08	2.02	1.98	1.95	1.92	1.83	1.78	1.75	1.72	1.69	1.66	1.62	1.59
22	2.95	2.56	2.35	2.22	2.13	2.06	2.01	1.97	1.93	1.90	1.81	1.76	1.73	1.70	1.67	1.64	1.60	1.57
23	2.94	2.55	2.34	2.21	2.11	2.05	1.99	1.95	1.92	1.89	1.80	1.74	1.72	1.69	1.66	1.62	1.59	1.55
24	2.93	2.54	2.33	2.19	2.10	2.04	1.98	1.94	1.91	1.88	1.78	1.73	1.70	1.67	1.64	1.61	1.57	1.53
25	2.92	2.53	2.32	2.18	2.09	2.02	1.97	1.93	1.89	1.87	1.77	1.72	1.69	1.66	1.63	1.59	1.56	1.52
26	2.91	2.52	2.31	2.17	2.08	2.01	1.96	1.92	1.88	1.86	1.76	1.71	1.68	1.65	1.61	1.58	1.54	1.50
27	2.90	2.51	2.30	2.17	2.07	2.00	1.95	1.91	1.87	1.85	1.75	1.70	1.67	1.64	1.60	1.57	1.53	1.49
28	2.89	2.50	2.29	2.16	2.06	2.00	1.94	1.90	1.87	1.84	1.74	1.69	1.66	1.63	1.59	1.56	1.52	1.48
29	2.89	2.50	2.28	2.15	2.06	1.99	1.93	1.89	1.86	1.83	1.73	1.68	1.65	1.62	1.58	1.55	1.51	1.47
30	2.88	2.49	2.28	2.14	2.05	1.98	1.93	1.88	1.85	1.82	1.72	1.67	1.64	1.61	1.57	1.54	1.50	1.46
40	2.84	2.44	2.23	2.09	2.00	1.93	1.87	1.83	1.79	1.76	1.66	1.61	1.57	1.54	1.51	1.47	1.42	1.38
60	2.79	2.39	2.18	2.04	1.95	1.87	1.82	1.77	1.74	1.71	1.60	1.54	1.51	1.48	1.44	1.40	1.35	1.29
120	2.75	2.35	2.13	1.99	1.90	1.82	1.77	1.72	1.68	1.65	1.54	1.48	1.45	1.41	1.37	1.32	1.26	1.19
∞	2.71	2.30	2.08	1.94	1.85	1.77	1.72	1.67	1.63	1.60	1.49	1.42	1.33	1.34	1.30	1.24	1.17	1.00

Table 6 (Continued). Table of the F Distribution: $P = 0.05$. Values of F for right tail of the distribution

n_2 / n_1	1	2	3	4	5	6	7	8	9	10	15	20	24	30	40	60	120	∞
1	161.45	199.50	215.71	224.58	230.16	233.99	236.77	238.88	240.54	241.88	245.95	248.01	249.05	250.09	251.14	252.20	253.25	254.32
2	18.51	19.00	19.16	19.25	19.30	19.33	19.35	19.37	19.38	19.40	19.43	19.45	19.45	19.46	19.47	19.48	19.49	19.50
3	10.13	9.55	9.28	9.12	9.01	8.94	8.89	8.85	8.81	8.76	8.70	8.66	8.64	8.62	8.59	8.57	8.55	8.53
4	7.71	6.94	6.59	6.39	6.26	6.16	6.09	6.04	6.00	5.96	5.86	5.80	5.77	5.75	5.72	5.69	5.66	5.63
5	6.61	5.79	5.41	5.19	5.05	4.95	4.88	4.82	4.77	4.74	4.62	4.56	4.53	4.50	4.46	4.43	4.40	4.36
6	5.99	5.14	4.76	4.53	4.39	4.28	4.21	4.15	4.10	4.06	3.94	3.87	3.84	3.81	3.77	3.74	3.70	3.67
7	5.59	4.74	4.35	4.12	3.97	3.87	3.79	3.73	3.68	3.64	3.51	3.44	3.41	3.38	3.34	3.30	3.27	3.23
8	5.32	4.46	4.07	3.84	3.69	3.58	3.50	3.44	3.39	3.35	3.22	3.15	3.12	3.08	3.04	3.01	2.97	2.93
9	5.12	4.26	3.86	3.63	3.48	3.37	3.29	3.23	3.18	3.14	3.01	2.94	2.90	2.86	2.83	2.79	2.75	2.71
10	4.96	4.10	3.71	3.48	3.33	3.22	3.14	3.07	3.02	2.98	2.84	2.77	2.74	2.70	2.66	2.62	2.58	2.54
11	4.84	3.98	3.59	3.36	3.20	3.09	3.01	2.95	2.90	2.85	2.72	2.65	2.61	2.57	2.53	2.49	2.45	2.40
12	4.75	3.89	3.49	3.26	3.11	3.00	2.91	2.85	2.80	2.75	2.62	2.54	2.51	2.47	2.43	2.38	2.34	2.30
13	4.67	3.81	3.41	3.18	3.03	2.92	2.83	2.77	2.71	2.67	2.53	2.46	2.42	2.38	2.34	2.30	2.25	2.21
14	4.60	3.74	3.34	3.11	2.96	2.85	2.76	2.70	2.65	2.60	2.46	2.39	2.35	2.31	2.27	2.22	2.18	2.13
15	4.54	3.68	3.29	3.06	2.90	2.79	2.71	2.64	2.59	2.54	2.40	2.33	2.29	2.25	2.20	2.16	2.11	2.07
16	4.49	3.63	3.24	3.01	2.85	2.74	2.66	2.59	2.54	2.49	2.35	2.28	2.24	2.19	2.15	2.11	2.06	2.01
17	4.45	3.59	3.20	2.96	2.81	2.70	2.61	2.55	2.49	2.45	2.31	2.23	2.19	2.15	2.10	2.06	2.01	1.96
18	4.41	3.55	3.16	2.93	2.77	2.66	2.58	2.51	2.46	2.41	2.27	2.19	2.15	2.11	2.06	2.02	1.97	1.92
19	4.38	3.52	3.13	2.90	2.74	2.63	2.54	2.48	2.42	2.38	2.23	2.16	2.11	2.07	2.03	1.98	1.93	1.88
20	4.35	3.49	3.10	2.87	2.71	2.60	2.51	2.45	2.39	2.35	2.20	2.12	2.08	2.04	1.99	1.95	1.90	1.84
21	4.32	3.47	3.07	2.84	2.68	2.57	2.49	2.42	2.37	2.32	2.18	2.10	2.05	2.01	1.96	1.92	1.87	1.81
22	4.30	3.44	3.05	2.82	2.66	2.55	2.46	2.40	2.34	2.30	2.15	2.07	2.03	1.98	1.94	1.89	1.84	1.78
23	4.28	3.42	3.03	2.80	2.64	2.53	2.44	2.37	2.32	2.27	2.13	2.05	2.00	1.96	1.91	1.86	1.81	1.76
24	4.26	3.40	3.01	2.78	2.62	2.51	2.42	2.36	2.30	2.25	2.11	2.03	1.98	1.94	1.89	1.84	1.79	1.73
25	4.24	3.39	2.99	2.76	2.60	2.49	2.40	2.34	2.28	2.24	2.09	2.01	1.96	1.92	1.87	1.82	1.77	1.71
26	4.23	3.37	2.98	2.74	2.59	2.47	2.39	2.32	2.27	2.22	2.07	1.99	1.95	1.90	1.85	1.80	1.75	1.69
27	4.21	3.35	2.96	2.73	2.57	2.46	2.37	2.31	2.25	2.20	2.06	1.97	1.93	1.88	1.84	1.79	1.73	1.67
28	4.20	3.34	2.95	2.71	2.56	2.45	2.36	2.29	2.24	2.19	2.04	1.96	1.91	1.87	1.82	1.77	1.71	1.65
29	4.18	3.33	2.93	2.70	2.55	2.43	2.35	2.28	2.22	2.18	2.03	1.94	1.90	1.85	1.81	1.75	1.70	1.64
30	4.17	3.32	2.92	2.69	2.53	2.42	2.33	2.27	2.21	2.16	2.01	1.93	1.89	1.84	1.79	1.74	1.68	1.62
40	4.08	3.23	2.84	2.61	2.45	2.34	2.25	2.18	2.12	2.08	1.92	1.84	1.79	1.74	1.69	1.64	1.58	1.51
60	4.00	3.15	2.76	2.53	2.37	2.25	2.17	2.10	2.04	1.99	1.84	1.75	1.70	1.65	1.59	1.53	1.47	1.39
120	3.92	3.07	2.68	2.45	2.29	2.17	2.09	2.02	1.96	1.91	1.75	1.66	1.61	1.55	1.50	1.43	1.35	1.25
∞	3.84	3.00	2.60	2.37	2.21	2.10	2.01	1.94	1.88	1.83	1.67	1.57	1.52	1.46	1.39	1.31	1.22	1.00

Table 6 (Concluded). Table of the F Distribution: $P = 0.01$. Values of F for right tail of the distribution

n_2 \ n_1	1	2	3	4	5	6	7	8	9	10	15	20	24	30	40	60	120	∞
1	4052.	4999.	5403.	5625.	5764.	5859.	5928.	5982.	6022.	6056.	6157.	6209.	6235.	6261.	6287.	6313.	6339.	6366.
2	98.50	99.00	99.17	99.25	99.30	99.33	99.36	99.37	99.39	99.40	99.43	99.45	99.46	99.47	99.47	99.48	99.49	99.50
3	34.12	30.82	29.46	28.71	28.24	27.91	27.67	27.49	27.34	27.23	26.87	26.69	26.60	26.50	26.41	26.32	26.22	26.12
4	21.20	18.00	16.69	15.98	15.52	15.21	14.98	14.80	14.66	14.55	14.20	14.02	13.93	13.84	13.74	13.65	13.56	13.46
5	16.26	13.27	12.06	11.39	10.97	10.67	10.46	10.29	10.16	10.05	9.72	9.55	9.47	9.38	9.29	9.20	9.11	9.02
6	13.74	10.92	9.78	9.15	8.75	8.47	8.26	8.10	7.98	7.87	7.56	7.40	7.31	7.23	7.14	7.06	6.97	6.88
7	12.25	9.55	8.45	7.85	7.46	7.19	6.99	6.84	6.72	6.62	6.31	6.16	6.07	5.99	5.91	5.82	5.74	5.65
8	11.26	8.65	7.59	7.01	6.63	6.37	6.18	6.03	5.91	5.81	5.52	5.36	5.28	5.20	5.12	5.03	4.95	4.86
9	10.56	8.02	6.99	6.42	6.06	5.80	5.61	5.47	5.35	5.26	4.96	4.81	4.73	4.65	4.57	4.48	4.40	4.31
10	10.04	7.56	6.55	5.99	5.64	5.39	5.20	5.06	4.94	4.85	4.56	4.41	4.33	4.25	4.17	4.08	4.00	3.91
11	9.65	7.21	6.22	5.67	5.32	5.07	4.89	4.74	4.63	4.54	4.25	4.10	4.02	3.94	3.86	3.78	3.69	3.60
12	9.33	6.93	5.95	5.41	5.06	4.82	4.64	4.50	4.39	4.30	4.01	3.86	3.78	3.70	3.62	3.54	3.45	3.36
13	9.07	6.70	5.74	5.21	4.86	4.62	4.44	4.30	4.19	4.10	3.82	3.66	3.59	3.51	3.43	3.34	3.25	3.17
14	8.86	6.51	5.56	5.04	4.69	4.46	4.28	4.14	4.03	3.94	3.66	3.51	3.43	3.35	3.27	3.18	3.09	3.00
15	8.68	6.36	5.42	4.89	4.56	4.32	4.14	4.00	3.89	3.80	3.52	3.37	3.29	3.21	3.13	3.05	2.96	2.87
16	8.53	6.23	5.29	4.77	4.44	4.20	4.03	3.89	3.78	3.69	3.41	3.26	3.18	3.10	3.02	2.93	2.84	2.75
17	8.40	6.11	5.18	4.67	4.34	4.10	3.93	3.79	3.68	3.59	3.31	3.16	3.08	3.00	2.92	2.83	2.75	2.65
18	8.29	6.01	5.09	4.58	4.25	4.01	3.84	3.71	3.60	3.51	3.23	3.08	3.00	2.92	2.84	2.75	2.66	2.57
19	8.18	5.93	5.01	4.50	4.17	3.94	3.77	3.63	3.52	3.43	3.15	3.00	2.92	2.84	2.76	2.67	2.58	2.49
20	8.10	5.85	4.94	4.43	4.10	3.87	3.70	3.56	3.46	3.37	3.09	2.94	2.86	2.78	2.69	2.61	2.52	2.42
21	8.02	5.78	4.87	4.37	4.04	3.81	3.64	3.51	3.40	3.31	3.03	2.88	2.80	2.72	2.64	2.55	2.46	2.36
22	7.95	5.72	4.82	4.31	3.99	3.76	3.59	3.45	3.35	3.26	2.98	2.83	2.75	2.67	2.58	2.50	2.40	2.31
23	7.88	5.66	4.76	4.26	3.94	3.71	3.54	3.41	3.30	3.21	2.93	2.78	2.70	2.62	2.54	2.45	2.35	2.26
24	7.82	5.61	4.72	4.22	3.90	3.67	3.50	3.36	3.26	3.17	2.89	2.74	2.66	2.58	2.49	2.40	2.31	2.21
25	7.77	5.57	4.68	4.18	3.85	3.63	3.46	3.32	3.22	3.13	2.85	2.70	2.62	2.54	2.45	2.36	2.27	2.17
26	7.72	5.53	4.64	4.14	3.82	3.59	3.42	3.29	3.18	3.09	2.81	2.66	2.58	2.50	2.42	2.33	2.23	2.13
27	7.68	5.49	4.60	4.11	3.78	3.56	3.39	3.26	3.15	3.06	2.78	2.63	2.55	2.47	2.38	2.29	2.20	2.10
28	7.64	5.45	4.57	4.07	3.75	3.53	3.36	3.23	3.12	3.03	2.75	2.60	2.52	2.44	2.35	2.26	2.17	2.06
29	7.60	5.42	4.54	4.04	3.73	3.50	3.33	3.20	3.09	3.00	2.73	2.57	2.49	2.41	2.33	2.23	2.14	2.03
30	7.56	5.39	4.51	4.02	3.70	3.47	3.30	3.17	3.07	2.98	2.70	2.55	2.47	2.39	2.30	2.21	2.11	2.01
40	7.31	5.18	4.31	3.83	3.51	3.29	3.12	2.99	2.89	2.80	2.52	2.37	2.29	2.20	2.11	2.02	1.92	1.80
60	7.08	4.98	4.13	3.65	3.34	3.12	2.95	2.82	2.72	2.63	2.35	2.20	2.12	2.03	1.94	1.84	1.73	1.60
120	6.85	4.79	3.95	3.48	3.17	2.96	2.79	2.66	2.56	2.47	2.19	2.03	1.95	1.86	1.76	1.66	1.53	1.38
∞	6.63	4.61	3.78	3.32	3.02	2.80	2.64	2.51	2.41	2.32	2.04	1.88	1.79	1.70	1.59	1.47	1.32	1.00

Table 7. Table of γ—Criteria for Testing Extreme Value or Mean

Statistic	Number of Means, k	Probability	
		0.05	0.01
$\gamma = \dfrac{\overline{X}_2 - \overline{X}_1}{\overline{X}_k - \overline{X}_1}$	3	0.941	0.988
	4	0.765	0.889
	5	0.642	0.780
	6	0.560	0.698
	7	0.507	0.637
$\gamma = \dfrac{\overline{X}_2 - \overline{X}_1}{\overline{X}_{k-1} - \overline{X}_1}$	8	0.554	0.683
	9	0.512	0.635
	10	0.477	0.597
$\gamma = \dfrac{\overline{X}_3 - \overline{X}_1}{\overline{X}_{k-1} - \overline{X}_1}$	11	0.576	0.679
	12	0.546	0.642
	13	0.521	0.615
	14	0.546	0.641
	15	0.525	0.616
$\gamma = \dfrac{\overline{X}_3 - \overline{X}_1}{\overline{X}_{k-2} - \overline{X}_1}$	16	0.507	0.595
	17	0.490	0.577
	18	0.475	0.561
	19	0.462	0.547
	20	0.450	0.535
	21	0.440	0.524
	22	0.430	0.514
	23	0.421	0.505
	24	0.413	0.497
	25	0.406	0.489
	26	0.399	0.486
	27	0.393	0.475
	28	0.387	0.469
	29	0.381	0.463
	30	0.376	0.457

Table 8. Cumulative Poisson Distribution*

				λ				
x	0.01	0.05	0.10	0.20	0.30	0.40	0.50	0.60
0	0.990	0.951	0.904	0.818	0.740	0.670	0.606	0.548
1	0.999	0.998	0.995	0.982	0.963	0.938	0.909	0.878
2		0.999	0.999	0.998	0.996	0.992	0.985	0.976
3				0.999	0.999	0.999	0.998	0.996
4					0.999	0.999	0.999	0.999
5							0.999	0.999

				λ				
x	0.70	0.80	0.90	1.00	1.10	1.20	1.30	1.40
0	0.496	0.449	0.406	0.367	0.332	0.301	0.272	0.246
1	0.844	0.808	0.772	0.735	0.699	0.662	0.626	0.591
2	0.965	0.952	0.937	0.919	0.900	0.879	0.857	0 83
3	0.994	0.990	0.986	0.981	0.974	0.966	0.956	0.946
4	0.999	0.998	0.997	0.996	0.994	0.992	0.989	0.985
5	0.999	0.999	0.999	0.999	0.999	0.998	0.997	0.996
6		0.999	0.999	0.999	0.999	0.999	0.999	0.999
7				0.999	0.999	0.999	0.999	0.999
8							0.999	0.999

				λ				
x	1.50	1.60	1.70	1.80	1.90	2.00	2.10	2.20
0	0.223	0.201	0.182	0.165	0.149	0.135	0.122	0.110
1	0.557	0.524	0.493	0.462	0.433	0.406	0.379	0.354
2	0.808	0.783	0.757	0.730	0.703	0.676	0.649	0.622
3	0.934	0.921	0.906	0.891	0.874	0.857	0.838	0.819
4	0.981	0.976	0.970	0.963	0.955	0.947	0.937	0.927
5	0.995	0.993	0.992	0.989	0.986	0.983	0.979	0.975
6	0.999	0.998	0.998	0.997	0.996	0.995	0.994	0.992
7	0.999	0.999	0.999	0.999	0.999	0.998	0.998	0.998
8	0.999	0.999	0.999	0.999	0.999	0.999	0.999	0.999
9			0.999	0.999	0.999	0.999	0.999	0.999
10							0.999	0.999

* Entries in the table are values of $F(x) = P(C \leq x) = \sum_{c=0}^{x} e^{-\lambda}\lambda^c/c!$. Blank spaces below the last entry in any column may be read as 1.0; blank spaces above the first entry in any column may be read as 0.0.

Table 8 (Continued). Cumulative Poisson Distribution

				λ				
x	**2.30**	**2.40**	**2.50**	**2.60**	**2.70**	**2.80**	**2.90**	**3.00**
0	0.100	0.090	0.082	0.074	0.067	0.060	0.055	0.049
1	0.330	0.308	0.287	0.267	0.248	0.231	0.214	0.199
2	0.596	0.569	0.543	0.518	0.493	0.469	0.445	0.423
3	0.799	0.778	0.757	0.736	0.714	0.691	0.669	0.647
4	0.916	0.904	0.891	0.877	0.862	0.847	0.831	0.815
5	0.970	0.964	0.957	0.950	0.943	0.934	0.925	0.916
6	0.990	0.988	0.985	0.982	0.979	0.975	0.971	0.966
7	0.997	0.996	0.995	0.994	0.993	0.991	0.990	0.988
8	0.999	0.999	0.998	0.998	0.998	0.997	0.996	0.996
9	0.999	0.999	0.999	0.999	0.999	0.999	0.999	0.998
10	0.999	0.999	0.999	0.999	0.999	0.999	0.999	0.999
11			0.999	0.999	0.999	0.999	0.999	0.999
12							0.999	0.999

				λ				
x	**3.50**	**4.00**	**4.50**	**5.00**	**5.50**	**6.00**	**6.50**	**7.00**
0	0.030	0.018	0.011	0.006	0.004	0.002	0.001	0.000
1	0.135	0.091	0.061	0.040	0.026	0.017	0.011	0.007
2	0.320	0.238	0.173	0.124	0.088	0.061	0.043	0.029
3	0.536	0.433	0.342	0.265	0.201	0.151	0.111	0.081
4	0.725	0.628	0.532	0.440	0.357	0.285	0.223	0.172
5	0.857	0.785	0.702	0.615	0.528	0.445	0.369	0.300
6	0.934	0.889	0.831	0.762	0.686	0.606	0.526	0.449
7	0.973	0.948	0.913	0.866	0.809	0.743	0.672	0.598
8	0.990	0.978	0.959	0.931	0.894	0.847	0.791	0.729
9	0.996	0.991	0.982	0.968	0.946	0.916	0.877	0.830
10	0.998	0.997	0.993	0.986	0.974	0.957	0.933	0.901
11	0.999	0.999	0.997	0.994	0.989	0.979	0.966	0.946
12	0.999	0.999	0.999	0.997	0.995	0.991	0.983	0.973
13	0.999	0.999	0.999	0.999	0.998	0.996	0.992	0.987
14		0.999	0.999	0.999	0.999	0.998	0.997	0.994
15			0.999	0.999	0.999	0.999	0.998	0.997
16				0.999	0.999	0.999	0.999	0.999
17					0.999	0.999	0.999	0.999
18						0.999	0.999	0.999
19							0.999	0.999
20								0.999

Table 8 (Concluded). Cumulative Poisson Distribution

				λ				
x	7.50	8.00	8.50	9.00	9.50	10.0	15.0	20.0
0	0.000	0.000	0.000	0.000	0.000	0.000	0.000	0.000
1	0.004	0.003	0.001	0.001	0.000	0.000	0.000	0.000
2	0.020	0.013	0.009	0.006	0.004	0.002	0.000	0.000
3	0.059	0.042	0.030	0.021	0.014	0.010	0.000	0.000
4	0.132	0.099	0.074	0.054	0.040	0.029	0.000	0.000
5	0.241	0.191	0.149	0.115	0.088	0.067	0.002	0.000
6	0.378	0.313	0.256	0.206	0.164	0.130	0.007	0.000
7	0.524	0.452	0.385	0.323	0.268	0.220	0.018	0.000
8	0.661	0.592	0.523	0.455	0.391	0.332	0.037	0.002
9	0.776	0.716	0.652	0.587	0.521	0.457	0.069	0.005
10	0.862	0.815	0.763	0.705	0.645	0.583	0.118	0.010
11	0.920	0.888	0.848	0.803	0.751	0.696	0.184	0.021
12	0.957	0.936	0.909	0.875	0.836	0.791	0.267	0.039
13	0.978	0.965	0.948	0.926	0.898	0.864	0.363	0.066
14	0.989	0.982	0.972	0.958	0.940	0.916	0.465	0.104
15	0.995	0.991	0.986	0.977	0.966	0.951	0.568	0.156
16	0.998	0.996	0.993	0.988	0.982	0.972	0.664	0.221
17	0.999	0.998	0.997	0.994	0.991	0.985	0.748	0.297
18	0.999	0.999	0.998	0.997	0.995	0.992	0.819	0.381
19	0.999	0.999	0.999	0.998	0.998	0.996	0.875	0.470
20	0.999	0.999	0.999	0.999	0.999	0.998	0.917	0.559
21	0.999	0.999	0.999	0.999	0.999	0.999	0.946	0.643
22		0.999	0.999	0.999	0.999	0.999	0.967	0.720
23			0.999	0.999	0.999	0.999	0.980	0.787
24					0.999	0.999	0.988	0.843
25						0.999	0.993	0.887
26							0.996	0.922
27							0.998	0.947
28							0.999	0.965
29							0.999	0.978
30							0.999	0.986
31							0.999	0.991
32							0.999	0.995
33							0.999	0.997
34								0.998

Table 9. Table of Random Digits

10480	15011	01536	02011	81647	91646	69179	14194	62590
22368	46573	25595	85393	30995	89198	27982	53402	93965
24130	48360	22527	97265	76393	64809	15179	24830	49340
42167	93093	06243	61680	07856	16376	39440	53537	71341
37570	39975	81837	16656	06121	91782	60468	81305	49684
77921	06907	11008	42751	27756	53498	18602	70659	90655
99562	72905	56420	69994	98872	31016	71194	18738	44013
96301	91977	05463	07972	18876	20922	94595	56869	69014
89579	14342	63661	10281	17453	18103	57740	84378	25331
85475	36857	53342	53988	53060	59533	38867	62300	08158
28918	69578	88231	33276	70997	79936	56865	05859	90106
63553	40961	48235	03427	49626	69445	18663	72695	52180
09429	93969	52636	92737	88974	33488	36320	17617	30015
10365	61129	87529	85689	48237	52267	67689	93394	01511
07119	97336	71048	08178	77233	13916	47564	81056	97735
51085	12765	51821	51259	77452	16308	60756	92144	49442
02368	21382	52404	60268	89368	19885	55322	44819	01188
01011	54092	33362	94904	31273	04146	18594	29852	71585
52162	53916	46369	58586	23216	14513	83149	98736	23495
07056	97628	33787	09998	42698	06691	76988	13602	51851
48663	91245	85828	14346	09172	30168	90229	04734	59193
54164	58492	22421	74103	47070	25306	76468	26384	58151
32639	32363	05597	24200	13363	38005	94342	28728	35806
29334	27001	87637	87308	58731	00256	45834	15398	46557
02488	33062	28834	07351	19731	92420	60952	61280	50001
81525	72295	04839	96423	24878	82651	66566	14778	76797
29676	20591	68086	26432	46901	20849	89768	81536	86645
00742	57392	39064	66432	84673	40027	32832	61362	98947
05366	04213	25669	26422	44407	44048	37937	63904	45766
91921	26418	64117	94305	26766	25940	39972	22209	71500
00582	04711	87917	77341	42206	35126	74087	99547	81817
00725	69884	62797	56170	86324	88072	76222	36086	84637
69011	65795	95876	55293	18988	27354	26575	08625	40801
25976	57948	29888	88604	67917	48708	18912	82271	65424
09763	83473	73577	12908	30883	18317	28290	35797	05998
91567	42595	27958	30134	04024	86385	29880	99730	55536
17955	56349	90999	49127	20044	59931	06115	20542	18059
46503	18584	18845	49618	02304	51038	20655	58727	28168
92157	89634	94824	78171	84610	82834	09922	25417	44137
14577	62765	35605	81263	39667	47358	56873	56307	61607
98427	07523	33362	64270	01638	92477	66969.	98420	04880
34914	63976	88720	82765	34476	17032	87589	40836	32427
70060	28277	39475	46473	23219	53416	94970	25832	69975
53976	54914	06990	67245	68350	82948	11398	42878	80287
76072	29515	40980	07391	58745	25774	22987	80059	39911
90725	52210	83974	29992	65831	38857	50490	83765	55657
64364	67412	33339	31926	14883	24413	59744	92351	97473
08962	00358	31662	25388	61642	34072	81249	35648	56891
95012	68379	93526	70765	10592	04542	76463	54328	02349
15664	10493	20492	38391	91132	21999	59516	81652	27195

OPERATIONS RESEARCH

CONTENTS

CONTENTS (*Continued*)

OPERATIONS RESEARCH

Nature of Operations Research

DEFINITION. The Committee on Operations Research of the National Research Council (Operations Research with Special Reference to Non-Military Applications) defines Operations Research as ". . . the application of the scientific method to the study of the operations of large complex organizations or activities." The Committee notes that its objective "is to provide top-level administrators with a quantitative basis for decisions that will increase the effectiveness of such organizations in carrying out their basic purposes."

ASSUMPTIONS. This definition of operations research serves as a basis for a characterization of operations research as it has developed to date in American industry. Operations research is, in short, research on the operations of business, governmental, or military organizations. Business has for some time recognized the importance and validity of the application of experimental research techniques to the study of equipment, processes, and products. New methods of psychology have been applied to the study of individuals, their relationships, and the structure of the organization. The **thesis** of operations research is that the operations of an organization exhibit basic patterns of orderly behavior and that the combination of men, equipment, organization, and technology at work toward an economic or social goal is, therefore, a subject for fruitful application of the techniques of experimental research. The subject matter studied is not the equipment used, the capabilities of participants, or the physical properties of the output: It is the combination of these as a **total economic process.** This combination is subjected to types of quantitative analysis associated with experimental research, or research in the physical sciences.

OBJECTIVES. There has been a great deal of disagreement concerning the place of operations research in the American industrial scene—disagreement caused by the diverse background of many of the men now engaged in it. Operations research is a field of **applied research;** its objective is intensely practical. From the point of view of operations research, an **operation** is viewed as a set of decisions or strategies, with an outcome associated with each. The objective of operations research is to clarify the relations between the several actions and their outcomes, to indicate the action whose outcome is most nearly consistent with the purposes of the managing executive, and thereby to assist him in choosing his decision or course of action intelligently.

The operations studied in industry may be those of a department, a plant, a division, or a company as a whole. The important point is that, whatever the organizational unit, it is studied as a unit. Where the unit studied is in some sense subsidiary, certain problems arise in the statement of objectives. These problems are noted in the discussions that follow.

The Scientific Approach to Industrial Problems

SCIENTIFIC METHOD. Scientific research differs markedly from the ordinary operation of a business. These differences give it its power in the

particular areas in which it is used. Conant (On Understanding Science) re
marks that a man who has been a successful investigator in any field of experi
mental science approaches a problem in science with a special point of view
even when the problem lies outside his particular area of knowledge. This poin
of view, independent of the particular facts or techniques in the area, will almos
always, in Dr. Conant's opinion, be missed by even the highly educated an
intelligent citizen without research experience. The layman will fail to grasp th
essentials in a discussion which takes place between scientists, not because of hi
lack of knowledge or his failure to comprehend the technical jargon, but in larg
degree because of his fundamental ignorance of what science can accomplish ir
the course of planning a scientific investigation. This, in essence, is the reasor
why operations research has been so rewarding and fruitful in areas which normally have not been explored by the scientific method. The core of science o
scientific research is not so much a body of fact related to a particular area a
it is a method of thinking related to the use of **experiment, conception, and
hypothesis**—a way of thinking which has been shown to be exceedingly powerful in areas where it is applicable, though by no means the only legitimate one
or the only sound method of approaching any particular industrial problem.

An important characteristic of the scientific method is its careful examination
of observable facts. Facts become the raw material to which the thought processes of scientific research are applied, principally the processes of **induction**
and **inverse deduction.** In the former, the scientist reasons from the facts to
the mechanism; in the latter, he attempts to set up assumed mechanisms from
which he can deduce the equivalent phenomena to check against observed facts.
These methods of thought, together with observed information, are used to arrive at an approximation to the underlying system.

BUSINESS APPROACH. It is true, obviously, that all businessmen are
used to absorbing and using facts. Modern executives use facts and figures to
control their operations, and these are interpreted in the light of the objectives
which are maintained in any company. The questions asked of the figures stem
from the objectives, and the executive is primarily concerned with the results—
only secondarily with causes. In examining figures related to production or
financial results, the first figure the business executive usually looks at is net
profit. When an executive looks at sales figures, he does it in terms of the success of his sales campaign and its effect on profit. When the analyst looks at
figures and facts, however, he seeks in them a clue to the fundamental behavior
pattern underlying the figures, the **fundamental system** which produced this
particular set or combination of figures and facts. For example, when he studies
the same sales figures, he seeks light on the behavior pattern of the customers
who produced these figures. Thus, he is often able to improve the sales program
and the total business obtained from a given sales area. This difference of approach, in essence, is the basic reason for the success of operations research even
in organizations which are normally regarded as well managed.

TECHNIQUES OF THE SCIENTIFIC APPROACH. The process
of scientific research involves the following steps:

Observation. The observation of fact is the first and probably the basic step
in the scientific process. The process of observation is an essential and continuing process. It is one of the most common methods for collecting data.

Hypothesis. From observation comes hypothesis, and the development of
methods for the making and establishing of fruitful hypotheses' has probably
been the greatest strength of scientific research. Hypotheses must generate new

concepts of underlying causal relationships. These concepts lead in turn to the development of theory or models to describe explicitly how the concept works.

Model or Theory. A model, conception, or theory, if it is to hold up, must be valid, efficient, and fruitful. For a model to be **valid,** the facts on which the model is based must be reproducible. A model, or conception, is **efficient** if it reduces the amount of data required to explain observed fact. A concept is **fruitful** if it supplies new insight into the observed facts or suggests new lines of experiment toward understanding these facts. A model should be something beyond a method for obtaining statistical reduction of data.

Prediction. To be valid, a concept or model must predict. Its predictions are tested through experiment. A great deal of testing and theorizing goes on in business every day, and each of us in our daily lives is forced to try things out and come to conclusions. However, techniques of experimentation in the field of scientific research have developed to such a degree beyond everyday testing as to reach an essential difference in kind.

Research experimentation is done, first, to learn facts, and second, to test theories. Experiments in the fact-finding category are aimed at developing the raw material for the creation of hypothesis and theory. One of the great advantages of the conceptual or model-making method of the scientist is the opportunity to use a **critical experiment** to test the validity of a theory by relatively simple and elegant means. If the theory holds up in these critical experiments, then security in accepting conclusions from the theory may be considerably strengthened, where other experimental verification is difficult or unwieldy.

In business operations, trials often go on under the names of "tests" or "experiments" which are, in fact, **pilot trials** of working policies. This testing amounts to trying out conceivable operating policies which might be substituted for existing policies on the large scale if pilot trials showed them to be superior. It is important to distinguish such pilot runs from information-gathering or critical theory-testing experiments.

An important strength aspect of scientific research is the **interplay of experiment and theory.** The processes of observation, hypothesizing, and model-building, repeated time and again, are used to arrive at the scientist's best approximation to the system underlying observed fact. The **test of the validity** of this approximation is simple: Does the assumed mechanism or systematic organization act enough like nature? The word "enough" is noteworthy. The scientist knows his analogue to nature will never be perfect; his objective is to make it sufficiently accurate to suit the particular purposes at hand. A satisfactory analogue to nature should be quantitative in order that it can be predictive—the only accepted fundamental test of being physically meaningful.

ROLE OF OPERATIONS RESEARCH. Recent decades have seen a tremendous growth in the intricacy, diversity, and sheer size of business operations. The increasing efficiency of conversion processes, with rising direct labor productivity and reduction in direct labor content of material consumed, has been brought about by the intensive technological research and development implemented by extended investment in physical processing facilities. Size, increased productivity, and specialization have brought their own associated problems. Further, automation has made it necessary for management to be right when equipment is purchased, since the total capital outlay involved is often a sizable percentage of the capital available for fixed investment purposes. Errors here would mean the financial end of the business. Operations research, therefore, is

PAST	PRESENT	FUTURE
Internal expansion (new products)	Acquisition	Mergers—growth analysis
		Optimize expansion pattern
	Design computer system	Schedule and analysis—computer system
	Corporate financial model	
	Cash flow analysis	
	New venture analysis	
	Tax model	
	Credit policy analysis	
Developing procedures	Manpower planning and control (training middle managers)	Division coordination analysis
		Employee motivation system
		Top management resource development
		Top management effectiveness
Facility design	Advertising effectiveness	Urban development and social systems
		R & D effectiveness
		Analysis of competition—social behavior
Inventory control	Inventory control	Inventory—scheduling (joint, expanded)
Production scheduling	Production scheduling	
Production control	Production control	
	Facility location	
	Fleet size	
Quality control	Distribution	World-wide distribution
	Corporate planning (short-run)	Corporate planning (long-run)
	Capital investment	
Sub-information system—information system (short-run)	Forecasting (short-run)	Management information system
		Forecasting (long-run)
Cost accounting	Profit centers	Industry stimulation (simulation)
Plant simulation	Corporate simulation	
Portfolio analysis		Portfolio analysis (simulation)
		Growth—expansion

a logical tool to be employed to increase the accuracy of management decisions and to improve management's ability to predict the future.

TYPICAL PROJECTS HANDLED. The industrial operations research team has wide scope. In questions related to **marketing,** it helps to determine the most profitable size and use of an advertising budget, to set up means for judging branch sales and expense, and to set up a more profitable price structure. In **production** areas, the operations research unit helps set up quality standards for component parts consistent with company goals and designs scheduling and order flow techniques to achieve a desirable balance among production, inventory, and manpower. **Distribution** problems include location and capacity of branch plants, warehouses, and distribution points, and the analysis of the operation of transport fleets. Operations research teams assist in the design and installation of central data-handling and information systems, and work with research and engineering groups in designing experiments, in guiding product development through analysis of anticipated cost and performance, and in running field and production-line test programs to discover causes of product and process failure.

Industrial problems have become more critical and difficult to solve due to the increasing technological and operating complexity of modern business, but many of the problems are not new. Nor are the basic methods and concepts of operations research new. They go back to the origins of experimental science. What is new is the organized and systematic application of these methods and concepts, which have been supported by business in product and process work since the beginning of the twentieth century, to the study of **operations as an entity,** accepting existing products, skills, and technology, rather than the study of products, equipment, or technology.

Past, Present, and Future Activities. In a survey of the 107 largest corporations in the United States, Turban (Industrial Engineering, vol. 1) recorded about 500 operations research projects ranging from the comparatively simple inventory control problems to complex projects involving simulation of entire industrial and social systems. Some of the more frequent projects are shown in Fig. 1 together with trends brought out by questioning the management on its past uses and future plans for applying operations research. The trend is clearly in the direction of the more sophisticated projects usually encompassing a much broader area than in previous applications.

Operations Research Concepts

THE MODEL. The most frequently encountered concept in operations research is the notion of the model. The operations research model is a simplified representation of an operation, containing only those aspects which are of primary importance to the problem under study. This concept has been of great use in facilitating the investigation of operations.

Types of Models. Models are, in most cases, abstractions of the actual system under study. An example of an engineering model as given by Buffa (Models for Production and Operations Management) is the aerodynamicists' model used in conjunction with wind tunnels. Since the engineer's primary interest is in aerodynamic performance, those characteristics, such as shape, that influence the performance are carefully duplicated in the model. Other factors such as weight, strength of individual parts, interior design are ignored. Using the model the engineer can make measurements more easily, manipulate variables at will, and make changes in the model all at a relatively low cost.

Another kind of model is the **accounting model.** This is a simplified representation, on paper in the form of accounts and ledgers, of the flow of goods and services through a business enterprise, for the purpose of providing measures of the rate of flow, the values produced, and the performances achieved. It is adequate to these purposes, although it is hardly a complete representation of a business in detail.

Many types of physical models are used in physics and chemistry, often for visual demonstrations.

Operations Research Models. The operations research model is a model of an operation. It is usually not physical, although it might be. For example, a model of waiting lines or queueing processes has been set up using radioactive sources and selective counters to simulate the arrival and servicing of customers under a variety of possible assumptions. The operations research model comes in a variety of forms. The two common characteristics are its explicit and quantitative nature.

A particular operations research model can be described first by the manner of representation used. Most operations research models are **mathematical in form,** being a set of equations relating significant variables in the operation to the outcome. Another type of model frequently used is the **punched-card model,** where components of the operation are represented by individual punched cards and masses of these are manipulated on standard punched card equipment. For example, in a study of a sales distribution problem, each customer, of thousands served by the company, was represented by a punched card containing significant information about his location, type of business, frequency of purchase, and average rate of business. The punched cards representing the customers could then be subjected to assumed promotional treatments, with the effects of the promotions punched into the cards, resulting business calculated, and the punched card model thereby used to test out alternative sales promotion campaigns. The steelmaking functions of the Inland Steel Company, Chicago are represented in financial terms in a set of models. The models are mathematical descriptions of the metallurgical, physical, and financial characteristics of the key manufacturing operations (Industrial Engineering, vol. 2). Finally, there is the **physical model,** such as the wing described above.

Exact and Probabilistic Models. Operations research models can also be differentiated as exact or probabilistic models. An **exact model** is used in operations or processes where chance plays a small role, where the effect of a given action will be reasonably closely determined. Exact models can be used, for example, in long-range production scheduling problems in the face of known or committed demand. The exact model is sufficiently accurate, since it can be assumed that over the long run, planned and actual production will coincide reasonably closely, barring a major catastrophe.

Probabilistic models, on the other hand, contain explicit recognition of uncertainty. Probabilistic models are highly useful in the analysis of quality control problems, where the uncertainty of material and processing is a major consideration. Probabilistic models make extensive use of the highly developed theory of probability which has come to be of such great value in the physical sciences. Methods developed for problems involving mass behavior under random conditions can be applied with great facility and value to operating problems.

The model is a major goal of the operations research analyst. In one sense, the construction of the model, or a faithful representation of the operation, is the analyst's primary job. In the **construction** of the model is contained the development of a theory to explain the observed characteristics of the operation.

The remaining task is the interpretation of this theory through the **manipulation** of the model, whether mathematical or physical.

Model for Setting Time Standards. The evaluation of machine operating characteristics for the purpose of setting time standards illustrates the combined use of experimental and a priori information in building a model. In one such case, a model was built which was based in form on the physically known facts related to the strengths of materials used. The actual numerical value of certain constants in the model was derived by comparison of the model with observational data.

The machine being observed is used for winding various types of protective tapes on steel cable. The cable is pulled horizontally through a "taping head," which carries the roll of tape and revolves around the cable. As the "taping head" revolves, the tape is unwound and wrapped around the cable. Metal, rubber, cloth, and paper tapes are used to wrap a wide range of diameters of cable. A number of machines of different designs, purchased at different times, were being used, and time study had failed to yield adequate time standards due to the nature and variability of operators' jobs. Statistical analyses of work records failed to explain the time variations found.

Preliminary investigation indicated that setup time was about the same for all jobs, and the run-speed was set by experience to keep tape breakage down. Investigation indicated tension on the tape was proportional to speed, and tensile strength was proportional to tape width. The time needed to wrap L feet of cable is given by

$$T = T_0 + \frac{L}{nl}$$

where T_0 is the setup time, n is the number of revolutions per minute the machine makes, and l is the "lay" of the tape, the length of cable covered in one

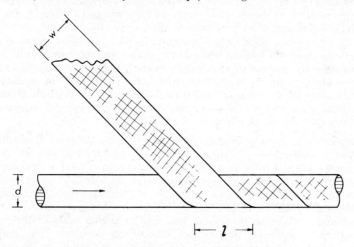

Fig. 2. Variables in a tape-winding operation.

revolution. Figure 2 shows the geometrical relationship between the width and lay of the tape, the diameter of the cable, and the velocity and tensile strength of the tape.

The amount of tape applied in one revolution is $\sqrt{\pi^2 d^2 + l^2}$. The speed at which the tapes moves is $n\sqrt{\pi^2 d^2 + l^2}$. Since the maximum tension the tape will take is proportional to its width, and the actual tension is proportional to the tape speed, the maximum machine speed n is

$$n = \frac{Kw}{\sqrt{\pi^2 d^2 + l^2}}$$

where K is related to the tensile strength of the particular material.

In operating the process, the product specifications give the diameter, d, and lay, l, and the appropriate width, w, is chosen. The required width, as shown in Fig. 2, is

$$w = l \sin \theta = \frac{l \pi d}{\sqrt{\pi^2 d^2 + l^2}}$$

and the maximum machine speed is

$$n = \frac{K \pi l d}{\pi^2 d^2 + l^2}$$

The time required to cover a length, L, of cable of diameter d at lay l is given by

$$T = T_0 + \frac{L}{Kl} \frac{\pi^2 d^2 + l^2}{\pi l d}$$

This formula was found to predict reported times accurately enough for use in setting standard times. It was applicable to all types of taping machines, and showed that the apparent differences among machines were due to different mixes of product assigned.

THE MEASURE OF EFFECTIVENESS. Fundamental to the concept of the model or theory of operation is the notion of a measure of effectiveness, or an explicit measure of the extent to which an operation is attaining its goal. For example, one common, overall measure of effectiveness in industrial operations is return on investment. Another one frequently found is net dollar profit. A measure of effectiveness which might be used in a smaller-scale or more detailed operation might be the number of customers serviced per hour, or the ratio of productive to total hours of a machine operation. The ability to construct an explicit measure of effectiveness implies that there exists an **explicit statement of the goal** toward which the operation is directed. Around this measure, and toward this goal, the model is built.

Inconsistent Company Goals. The goal underlying the measure of effectiveness must be self-consistent. Inconsistent goals appear frequently in management discussions, and while they may be useful administrative techniques, they can often be interpreted as danger signs, and frequently lead to management conflicts and frustrations. The ideal example of an inconsistent set of goals was the statement by one executive that his objective was maximum dollar profits with maximum return on investment and with maximum share of the market.

Operations research on **production scheduling** problems has frequently brought to light inconsistencies in company goals. Very often, for example, the object of the production scheduling has been stated as scheduling production to meet sales forecasts, with minimum production costs and minimum inventory investment and without customer service failure. It is apparently not at all obvious that minimizing inventory investment typically requires the use of very expensive production plans resulting in excessive production costs, or that elimination of the chance of inability to meet customer demand requires huge inventories in the face of fluctuating and at least partially unpredictable demand.

Explicit and Consistent Goals. A consistent statement of the fundamental goals of the operation is an essential counterpart of the mathematical logic of the model. The goals may be complex, but they must be explicit and consistent. Just as the model cannot make 2 and 2 add up to 5, so it is impossible to make fundamentally inconsistent objectives consistent and meaningful. A frequent contribution of operations research to management problems is the **forced clarification** of operating goals. Goals which appear inconsistent can often be combined into a unified and consistent goal. For example, in the production scheduling problem, the several goals of customer service, production economy, and investment minimization can often be expressed in terms of costs—the cost of inefficient production, the cost of inability to meet a customer's demand, and the cost of investment in inventory. While the last two costs are primarily policy costs, experience has shown them to be reasonably well determinable. Thus the cost of investment in inventory may be the rate of interest the treasurer wishes to charge to conserve his funds, or perhaps the return on investment which can be earned through alternative uses of the available funds. Hence the three apparently inconsistent goals by the **process of sublimation** can be expressed as a single, unified, and consistent goal of minimizing the total scheduling cost made up of the three components—production, investment, and service costs.

The statement of a complete and wholly consistent goal of company operations must be recognized as an ideal. Business goals are very complex, and to catch the full flavor of the objectives of a complex business operation is difficult in any simple, explicit statement. Many business goals remain, and probably always will remain, at least in part, intangible. These include such objectives as the improvement of employee morale, or contribution to public welfare. Thus, to the extent that **intangibles** enter into the formulation of a company goal, the objective of operations research must be more modest than the construction of a complete model and the measurement of the extent to which the operation is attaining the complete set of goals established for it. Rather, the operations research objective then becomes the clarification of the interdependencies of those aspects of the operation and company goals which are measurable, as a guide, but only a guide, to executive decision.

Ultimate and Subordinate Goals. In many problems in large organizations, the ultimate goal of the organization (for example, "maximum long-run profit") is too remote from the particular operating situation studied to be useful. Some subordinate goal must be chosen more directly connected with the operation studied. While this sounds easy, experience shows that choice of these subgoals needs care. Maximum accounting profit vs. investment for a division, minimum cost for a producing department, maximum sales for a sales district, and similar goals can be and have in some cases been found to be inconsistent with broader company goals. An operations research unit has two responsibilities in this connection: first, to choose objectives or effectiveness measures in its work which are, at least to a close approximation, consistent with broader goals; second, to help management analyze subordinate goals and incentives for consistency.

THE ROLE OF DECISION MAKING. A third concept fundamental to operations research is that of decisions and decision making. The element common to all true operations research problems is the existence of **alternative courses of action,** with a choice to be made among them. Without the requirement or opportunity for choice among alternative courses of action or alternative decisions, the study of an operation becomes academic or theoretical. Operations research, on the other hand, is applied research. In those problems where there is no decision to be made, or where the choice of decisions is severely limited, research is powerless to help the executive choose a sound course. He has no choice at all. From the point of view of operations research, an operation is viewed as a set of decisions or strategies, with an outcome associated with each. The **objective of operations research** is to clarify the relations between the several actions and the outcome, as expressed by the measure of effectiveness, to indicate the action whose outcome maximizes the measure of effectiveness used, and thereby to assist the executive in choosing his decision or course of action intelligently. In every case, however, the ultimate choice of the course of action lies with the executive.

Management Function. Buffa (Models for Production and Operations Management) sees management's **primary function** as the making of decisions that determine the future course of action for the organization over the short and long term. The decisions cover the entire range of the company's problems. Operational as well as policy decisions deal with marketing, finance, personnel, material, equipment, and the like. Usually decisions cut across formal functional lines. As new managerial problems arise, they are faced by management with the only tools it has—experience, judgment, and even intuition. The decisions may become formalized and become company policy thus establishing guidelines for future decisions. In some cases decision models are derived thus providing automatic decision guidelines. The flow of problems and their effect on the decision-making process is shown in Fig. 3 (based on Buffa, Models for Production and Operations Management).

The development of an inventory control system illustrates the role of **executive decision.** The analyst's job is to design an efficient control system and to show the relationship among, for example, the average inventory investment, the degree of fluctuation in production level, and the level of customer service provided by the system. The executive must weigh his goals to decide what alternative available under the system best meets these. Perhaps the analyst can help by measuring the cost of production fluctuations and even, perhaps, the cost of service failures. However, to express the system objective completely in terms of cost means a cost must be assigned to inventory, in particular to investment in inventory. This cost, the desired return on investment, characteristically depends very directly on basic financial policies of the company. Management must set these policies; then the analyst can adjust the control system to work with maximum efficiency under them.

THE ROLE OF EXPERIMENTATION. Another important concept of operations research is that of the role of experimentation. Operations research is the application of experimental science to the study of operations. The theory, or model, is generally built up from observed data or experience, although in some cases the model development may depend heavily on a priori or external information. The theory describing the operation must be verifiable experimentally. This means two things: first, verification must be possible; second, the theory and observation must check.

Experiments for Information and Prediction. Two kinds of experiments are important in the field of operations research. The first is simply to get information, and sometimes it takes the form of an apparently rather impractical test. For example, in one case the operations analysts directed advertising to-

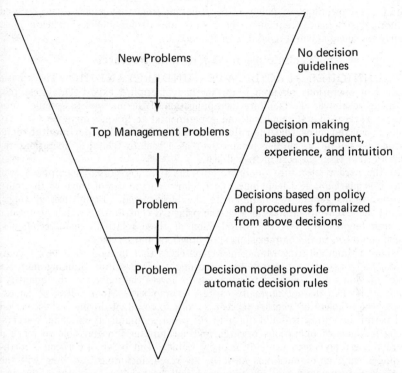

New Problems — No decision guidelines

Top Management Problems — Decision making based on judgment, experience, and intuition

Problem — Decisions based on policy and procedures formalized from above decisions

Problem — Decision models provide automatic decision rules

Fig. 3. The development of problems into policy and automatic decision models.

ward potential customers the company considered not worth promoting, and refrained from promoting customers the company typically sought. The reason for this, as expressed by the operations research group, was simple. There was plenty of evidence indicating what happened when advertising was directed toward those normally promoted, and not toward those normally not promoted. To evaluate the effectiveness of the advertising, therefore, it seemed necessary in this case to find out what happened to those normally promoted when they weren't promoted, and what happened to those normally not promoted when they were.

Nadler (Work Design) discusses the desirability of quantitative measurement and the collection of data for structuring a model. Experiments with actual operations, and in laboratory settings, may be required. In a study of the receiving, checking, and marking activities in a department store, cited by Nadler, it was necessary to establish an experiment to determine times for putting certain types of information on the tickets. Comparison was required with the time for omitting that information. Both procedures were tried in a small section, and time measurements were made as the two processes were used. The second type of experiment is the critical type, or the experiment designed to test the validity of conclusions. Again the forms of experimentation used frequently appear to be rather impractical. Sometimes the most sensitive experiment of this type, designed to test the validity of the theory or model, may require the use of a test technique or policy rather different from that which would be used in normal practice. Frequently the results of extreme policies indicate most clearly the adequacy and reliability of the model.

Techniques in Operations Research

TECHNIQUES FACILITATE UNDERSTANDING. The basic method of operations research is the method of applied experimental research, as described above. To facilitate the application of this method to specific problems, a variety of mathematical and experimental techniques may be used. It is important, however, that the **role of mathematical and experimental techniques** be put in its proper perspective: as a tool to facilitate operations research rather than the core of the subject.

In the earlier literature on operations research, considerable emphasis was placed on mathematical techniques, particularly on the use of some of the more complex forms generally unfamiliar to the businessman. From this, at times, sprang the erroneous impression that the principal contribution which operations research might offer was the introduction of these relatively new techniques, which could be grafted onto existing staff functions and services.

Moore (Manufacturing Management) observes that in modern industry computers and operations research keep pushing managers into thinking more in terms of numbers. The need to use numbers forces the managers to "quantify" things that they do not normally express as numbers. Moore cites, as an example, the problem of whether to keep a work force intact during a slack period and permit inventory to build up or to cut production and lay off some workers. Using methods of operations research and a computer it is possible to reach a decision based on consideration of many variables. But before a decision is forthcoming a number of questions must be answered, including how long will the layoff last, how many laid off employees will be able to find other jobs and thus not collect unemployment insurance, how much additional unemployment tax will be paid since some of the employees will not find jobs, how many experienced workers will return and how many will be lost when production is resumed, how much will training and lost production cost until qualified replacements are developed, how big will the inventory get before it starts going down, how much will it cost to carry this inventory, etc.? In addition there are many intangibles such as the effect of a layoff on community and personnel relations, union demands in the area of security, retirement, vacations, and the like.

By its very nature, operations research requires a rethinking of many problems in quantitative terms so that the variables involved are more clearly identified, their relationships more easily presented, and their behavior and performance more accurately predicted. Whether or not operations research is used, a decision must be made. Operations research merely makes more use of the known facts and less of intuition.

MATHEMATICS IN OPERATIONS RESEARCH. Mathematics in operations research serves basically as a language in which the findings from investigations can be expressed and the consequences of these investigated. Its advantages lie in the conciseness, rigor, and lack of ambiguity of statement which it permits. As a mathematical argument develops in complexity, short cuts in the form of more sophisticated techniques may be fruitfully introduced. In many problems, particularly problems which are relatively new or in fresh areas, the investigator cannot rely on the existence of suitable well-developed echniques; he must rely on his ability to think broadly, constructively, and creatively in mathematical terms.

Role of Mathematical Techniques. The practitioner of operations research must possess a sound basic grasp of the concepts of **analysis.** Many reasons may be cited. In particular:

1. Application of more complex or highly developed methods of mathematical or statistical analysis generally requires that we make **restrictive and simplifying assumptions** about the problem being investigated. For example, use even of the methods of integral calculus to determine the area under a curve requires that the curve be defined throughout the interval. Use of many devices of mathematical statistics requires the assumption that the distribution underlying the data conform, e.g., to the normal distribution. Linear programming methods (described below) are based on assumptions that the cost or other objectives and the restrictions on the solutions can be expressed as linear functions of the variables to be chosen. The investigator must have enough of basic grasp of analysis to understand the logic and assumptions on which such methods are built, to test these assumptions for suitability in the particular circumstances under study, to estimate the influence of possible differences between assumption and fact on conclusions, and to construct suitable arguments or methods of attack where existing procedures do not apply well.

2. Many problems in operations research today require basically a **numerical attack.** The conditions which must be satisfied and the complexity of interaction among variables which may exist may force the investigator into the use of the most basic methods of mathematical analysis. This does not mean that the investigator need know only the methods of arithmetic. Frequently, an appropriate line of attack or analysis using numerical techniques can be found only after clearing up conceptual questions in a "rough cut" using more complex techniques. Understanding of the more complex methods is required to define the equivalent numerical operations and correctly carry them out.

Techniques Used Most Often. There are a number of **quantitative techniques** applicable to a wide range of production, management, and financial situations as well as some developed for specific problems. When the variables in the mathematical model representing the situation are controllable, the model is called **deterministic.** As noted previously many processes are not controllable and their outcomes are not individually predictable. These processes, called **probabilistic** or **stochastic** or **chance** processes, can be described in terms of the relative frequencies (probabilities) of the various outcomes. Some operations research techniques are called **heuristic,** meaning that they solve a given problem by making a search for reasonable answers. They are applied when a situation consists of such a great number of alternatives that it would be impossible to list and evaluate all of them. Thus heuristic techniques may not lead to the best or optimum solution but to one that with a high probability is suitable for the desired purposes. Among other methods used are queueing (waiting line) theory and the Monte Carlo method. Figure 4 shows the most frequently used techniques in the largest U.S. corporations as brought out by Turban's survey (Industrial Engineering, vol. 1). Note that statistical analysis, simulation, and linear programming represent almost 75 percent of all problem-solving methods used.

Probability and Statistics. **Probability** is one of the most useful branches of mathematics in operations research. This is because most real world situ-

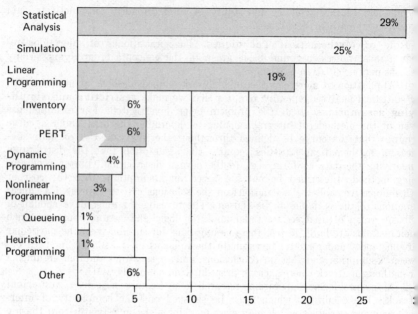

Fig. 4. **Most often used operations research techniques.**

ations in industry consist of outcomes that are not individually predictable with certainty. As stated by Richmond (Operations Research for Management Decisions) probability theory enables one to assign a numerical value to the **degree of confidence** that a specified outcome will occur. That is, when a probability value is assigned to a real world phenomenon, it describes, in a quantitative way the degree of belief or intensity of conviction that a particular event will occur (see section on Statistical Methods).

Much of probability is used in conjunction with statistics or statistical analysis. Richmond (Statistical Analysis) explains that the purpose of statistics and statistical analyses in business was usually to provide the information on which to base decisions. The traditional view was that of numerical data assembled and presented for the purposes of describing some process or phenomenon These numbers (or statistics) may be nothing more than simple sums, such as the total number of rejects produced by a machine during a given period. This is the so-called **descriptive statistics.** In recent years, statistics, in the form of **statistical inference,** had its role expanded to encompass even more of the managerial process. The basic problem of statistical inference is that of forming general conclusions about data that is not (and usually cannot be) examined completely. Richmond (Statistical Analysis) cites as examples of both of these processes the case where it is known that all members of the labor force in a particular area are engaged in manufacturing. We can then deduce that any sample of, say ten, workers will consist exclusively of manufacturing workers. If, however, no previous knowledge about the distribution of workers was known then a sample of ten workers, if they were all manufacturing workers, would tell us nothing more definitely than that there are at least ten manufacturing workers in the area. To say more about the workers in general in the area would require forming a conclusion from the sample taken, and this is the basic roblem of statistical inference. The techniques and reasoning used in this analysis is covered in the section on Statistical Methods.

Inventory Theory

DEFINITION OF COST ELEMENTS. Inventory theory is concerned with making decisions related to inventories which minimize the associated costs of production and inventory. However, not all costs in the production and inventory area are affected by inventory decisions. Magee and Boodman (Production Planning and Inventory Control) cite two criteria for the costs which should be considered in inventory problems:

1. The costs shall represent out-of-pocket expenditures, i.e., cash actually paid out or foregone opportunities for profit.
2. The costs shall represent only those out-of-pocket expenditures or foregone opportunities for profit whose magnitude is affected by the decision in question.

The particular elements of cost which need to be included in a specific inventory problem depend on the operating conditions associated with that problem. For example, the cost of warehouse space is generally considered to be one of the components of inventory carrying cost, but under some conditions warehouse space cost does not vary with inventory level and can be ignored. If a company owns its own warehouse, the warehouse space cost is a fixed cost and is not affected by short-range inventory decisions.

The determination of the cost elements to include in a problem analysis is the first step in the solution of an inventory problem. There are three general categories of costs associated with inventory problems: production costs, inventory carrying costs, and shortage costs. The actual component cost elements of each of these cost categories depends, as pointed out above, on the operating environment in the inventory problem.

Production costs are those costs incurred in the production operation. There are many components to this cost, most of which exist in the typical manufacturing company. Setup cost is the cost most frequently affected by inventory decision problems. Other non-routine or abnormal costs such as hiring and lay-off costs, under-capacity costs, overtime costs and inefficiency costs may also be affected by inventory decisions.

Inventory costs include the cost of handling the product in and out of inventory; storage costs such as rent, heat, insurance, and taxes; obsolescence and shrinkage, which may occur in a variety of forms; and the costs of capital invested in inventory. However, these must be the out-of-pocket or marginal costs, in the same way as the manufacturing cost.

Shortage costs refer to the costs incurred as a result of being out of stock of particular products when these are ordered by customers. Shortage costs include the costs incurred in expediting production in order to fill a backorder; the cost of lost profits when sales are lost as a result of being out of stock; and the cost of lost goodwill, which represents future potential sales.

ECONOMIC ORDER QUANTITY. Probably the most common inventory decision problem in the production area is concerned with the determination of the economic order quantity. The production manager is faced with two conflicting objectives:

1. To minimize setup costs by extending run lengths and reducing the frequency of runs.
2. To minimize inventory levels by reducing run lengths and increasing the frequency of runs.

The best **strategy** is to select a run quantity which is neither extremely long nor extremely short but an intermediate run quantity which minimizes the combined cost of setup and inventory.

There are many different formulas for calculating the economic order quantity. Some of these variations in the formula arise as a result of changes in the assumptions on which the **cost model** is based. Different model assumptions mean that the formula is applicable to different conditions. The simple, classical economic order quantity formula is applicable to a highly specialized situation.

Other variations in the economic order quantity formulas result from defining the variables in the formulas in different ways. For example, inventory carrying cost can be defined in at least three different ways:

1. A single term expressed in dollars per piece per year.
2. A single term expressed as a fraction of the unit cost.
3. Two terms: one representing the cost of capital expressed as a fraction of the unit cost; the other term representing storage cost expressed in dollars per piece per year.

Each of these definitions results in a slightly different EOQ formula. However, as long as the parameter values which are substituted in the formulas are equivalent, the results obtained from these variations of the formula are identical.

In the application of economic order quantity formulas, it is quite important that the model used correspond to the actual situation. Use of the wrong model will give erroneous results; use of erroneous results leads to ordering non-optimal quantities and cost penalties. A more important consequence is the possible rejection of all scientific management techniques by the production management because of the lack of confidence in its results.

Simple Economic Order Quantity. The simple economic order quantity (EOQ) formula has been well known for many years. Mennell (APICS Quarterly Bulletin, vol. 2) traced the history of this formula as far back as 1917 in the literature of inventory control.

The parameters used in the simple EOQ model are the setup cost, the inventory carrying cost, and the sales forecast:

1. The setup cost includes all those cost elements, direct and indirect, which are incurred at the beginning or end of a production run and are independent of the run quantity or length. Inefficiency and under-capacity costs are included if they are applicable. These costs have been discussed in the previous article.
2. The inventory carrying cost includes all cost elements incurred in maintaining inventory which are affected by the level of inventory. This may include storage costs, obsolescence and spoilage, cost of capital and handling, if appropriate.
3. The sales forecast is an estimate of the expected production requirement for the coming period. A period of one year is usually used, although any time period may be used. The time period for the sales forecast must correspond to the time period on which the inventory carrying cost is based.

The assumptions made in the simple economic order quantity model are as follows:

1. Replenishment occurs instantaneously. Therefore, no sales occur during replenishment.
2. There are no interactions in setup cost with other products.
3. There is no interference with other products in getting production completed and capacity is adequate so that lead times for receipt of production runs are reasonably constant.
4. The unit cost of production is constant, i.e., unit cost is independent of run quantity.

5. The run quantity has no effect on (is independent of) the safety stock.
6. Demand is constant and continuous over time. Therefore, inventory depletion results from a continuous stream of very small orders relative to the production quantity.

The basic problem is to determine the production quantity which minimizes the total cost of production setups and of carrying inventory over some period of time corresponding to a sales forecast.

An **example of the calculation** of a simple EOQ is as follows:

S = pieces per time period = 20,000 pieces per year
A = setup cost in dollars per run = $50 per run
I = inventory carrying cost in dollars per piece per time period
  = $2 per piece per year

The formula used is

$$Q = \sqrt{\frac{2AS}{I}}, \quad \text{where } Q = \text{EOQ}$$

$$Q = \sqrt{\frac{2 \times \$50 \times 20,000}{\$2}} = 1,000 \text{ pieces}$$

$$C_m = \sqrt{2ASI}, \quad \text{where } C_m = \text{minimum total cost}$$

$$C_m = \sqrt{2 \times \$50 \times 20,000 \times \$2} = \$2,000$$

The economic order quantity is 1,000 pieces and the minimum total annual cost for setups and carrying inventory is $2,000.

Typically, neither the cost factors nor the sales forecast are known very accurately. Furthermore, it is not always convenient or practical to produce exactly the economic order quantity. Fortunately, the total cost function is quite flat near the minimum as is shown in Fig. 5. Moderate errors in the input parameter values and moderate differences between the actual order quantity and the economic order quantity produce very small increases in total cost. The formula for determining the relative increase in total cost for a given relative error in setup cost, inventory carrying cost, or sales forecast is as follows (relative errors are expressed as decimals, e.g., $33\% = 0.33$):

c = ratio of actual total cost to minimum cost
m = relative error in sales forecast, in setup cost, or
  in inventory carrying cost

$$c = \frac{1}{2}\left(\sqrt{1 + m} + \frac{1}{\sqrt{1 + m}}\right)$$

If in the previously mentioned example, a value of $100 was used for setup cost instead of the actual setup cost of $50, an erroneous "economic order quantity" would be calculated. The error in setup cost is $+100\%$. Using the above equation to determine the relative increase in cost resulting from the use of the incorrect setup cost gives the following result:

$$c = \frac{1}{2}\left(\sqrt{1 + 1.00} + \frac{1}{\sqrt{1 + 1.00}}\right) = 1.06$$

That is, the total cost incurred as a result of using an EOQ calculated with a value for setup cost which is 100 percent greater than the actual setup cost is only 6 percent higher than the minimum cost.

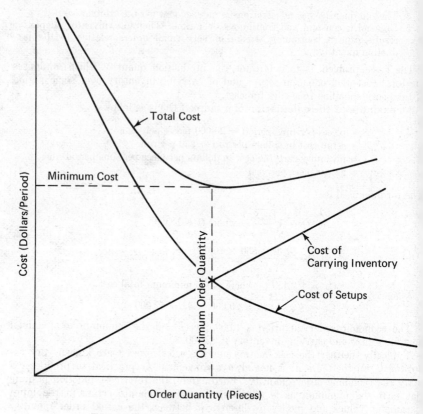

Fig. 5. Costs affecting economic order quantity.

The formula for determining the relative increase in total cost for a given percent error in order quantity is as follows:

q = percent error in order quantity relative to the economic order quantity

$$c = \frac{1}{2}\left[(1 + q) + \frac{1}{(1 + q)}\right]$$

In the previous example, if it is desired to express the run quantity in whole shifts of production and the production rate is 300 pieces per shift, the run quantity should be rounded to 3 shifts (1,000/300 = 3⅓ shifts). This means that the actual run quantity is 900 pieces instead of the "economic" order quantity of 1,000 pieces:

$$q = \frac{-100}{1,000} = -0.10$$

$$c = \frac{1}{2}\left[1 + (-0.1) + \frac{1}{1 + (-0.1)}\right] = 1.0055$$

Decreasing the run quantity to 900 pieces in order to run in whole shifts increases total annual cost by only about one-half of one percent.

Economic Order Quantity with Gradual Replenishment. In most manufacturing situations the assumption that replenishment of inventory occurs instantaneously is not valid. It is usually found that replenishment occurs gradually over a period of time during which sales or inventory depletions also occur. It is necessary to use a slightly different cost model than the simple economic order quantity model to accurately represent this condition. The inventory level behaves as shown in Fig. 4, under this condition. Replenishment occurs gradually, with the rate of inventory buildup equal to the difference between the production rate and the sales rate. For the same order quantity, the peak inventory level and the average inventory level are lower when there is gradual replenishment than when there is instantaneous replenishment because sales (depletions) occur during replenishment. Thus, use of the gradual replenishment model will give larger economic order quantities than the simple economic order quantity model for the same cost and forecast values.

The only change introduced in the total cost equation is in the average inventory level which results from the inventory behavior shown in Fig. 6.

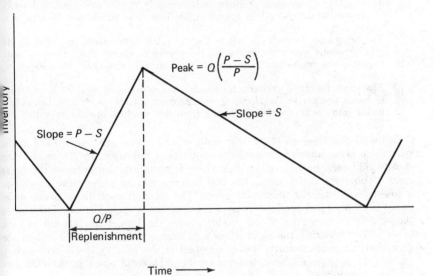

Fig. 6. **Inventory level with gradual replenishment.**

This leads to the following economic order quantity formula:

$$Q = \sqrt{\frac{2AS}{I\left(1 - \dfrac{S}{P}\right)}}$$

where P = production rate in pieces per year

If the forecast sales rate is very small relative to the production rate so that S/P approaches zero, then the above formula reduces to the simple EOQ formula. This is logical since a very high production rate relative to the sales rate means that inventory replenishment occurs essentially instantaneously.

As the forecast sales rate approaches the production rate, the economic order quantity approaches infinity. This simply means that the product must be produced continuously in order to be able to satisfy the forecast sales requirement Obviously, a forecast sales rate higher than the production rate is an impossib situation since this means that requirements exceed capacity.

An example shows how the formula is used. Assume:

$$S = 20,000 \text{ pieces per year}$$
$$A = \$50.00 \text{ per run}$$
$$I = \$2.00 \text{ per piece per year}$$
$$P = 25,000 \text{ pieces per year}$$

$$Q = \sqrt{\frac{2 \times 50 \times 20,000}{2[1 - (20,000/25,000)]}} = 2,236$$

Run Quantities for Cycled Products. There are three conditions, unde one or more of which it is desirable to cycle or follow a fixed product sequenc in production:

1. The total requirements for products produced on a facility are very close t its capacity. Controlling products independently results in a random backlo of production orders and an unpredictable lead time which may become ver; long.
2. The setup cost for each product in a set of products normally produced on th same machine varies depending on the previous product run. There is som sequence for producing these products which will minimize the total setu; costs.
3. There is a family of products for which a major setup is incurred in switchin; to run a product in the family from a product outside the family, but only minor setup is incurred in switching from one product to another within th family.

In the first of these conditions the assumption made in the simple EOQ mode that there is no interference between products is not true. In the second an third conditions, the assumption that there is no interaction between the setu; costs of products is not true. The best way to operate under these condition; is to establish a regular production cycle. The calculation of the economic orde: quantities for cycled products must recognize that a cycle is to be followed s(that the order quantities of the products have the proper relationship to eacl other. The previously run quantity of a product should be depleted just at th(time that the product starts into production in the next cycle (safety stock i; maintained to protect against forecast errors). This will result in the minimun average inventory.

One model can be formulated which can be applied to all three types of pro-duction cycling conditions by using the resulting formula in slightly different ways. The model is set up for the case of a product family with a major family setup cost, A, and minor setup costs, a_j, for each product in the family. The total setup cost for one complete cycle of the family is the major setup cost plus the minor setup costs for each item in the family $(A + \Sigma_j a_j)$. Of course, it may be desirable to not run some of the low demand products in the family every cycle. These products could be run every second, third or fourth cycle in a quantity correspondingly larger to achieve a more economical balance be-tween setup and inventory carrying costs. In the model, a parameter, k_j, is used to indicate for each product in the family, the multiple of the basic cycle at which it should be run. (k_j is limited to integer values.) The basic pro-duction cycle, T, is defined as the interval from the start of one cycle (i.e., the time a major family setup is incurred) to the start of the next cycle expressed as a fraction of the sales forecast time period. The total cost over the sales recast time period for a product family is given by the following equation in which the adjustment for gradual replenishment is included:

$$C_T = \frac{A}{T} + \Sigma_j \frac{a_j}{k_j T} + \Sigma_j \left[\frac{k_j T S_j}{2} \left(1 - \frac{S_j}{P}\right) I_j \right]$$

The product cycle multipliers, k_j, interact with the production cycle time, T, so it is not possible to solve for independent formulas for these parameters. Brown (Decision Rules for Inventory Management) gives an iterative procedure for solving the above equation with the following steps:

1. Calculate an initial estimate of the family cycle time assuming all k_j equal one—

$$T_0' = \sqrt{\frac{2(A + \Sigma a_j)}{\Sigma S_j[1 - (S_j/P)]I_j}}$$

where the summations include all n products in the family.

2. Calculate an approximation for each product multiplier—

$$k_j' = \sqrt{\frac{2a_j}{T^2 S_j[1 - (S_j/P)]I_j}}$$

3. Using Fig. 7, select the integer value for k_j corresponding to the non-integer value for k_j'.

Range of k_j'	Round to Integer Value
0 to 1.414	1
1.414 to 2.449	2
2.449 to 3.464	3
3.464 to 4.472	4
4.472 to 5.477	5
5.477 to 6.480	6
6.480 to 7.483	7

For larger values of k_j', use the integral part of $(k_j' + 0.52)$.

Fig. 7. Table for rounding-off tentative product multiplier k_j' to an integer.

4. Recompute the family cycle time, using the set of k_j values obtained in step 3 (unless all $k_j = 1$)—

$$T_0 = \sqrt{\frac{2(A + \Sigma(a_j/k_j))}{\Sigma k_j S_j[1 - (S_j/P)]I_j}}$$

5. Repeat steps 2 through 4 until either the set of k_j or T_0 are the same on two consecutive calculations. (From a practical standpoint, it is unlikely that the values will change enough to make a significant difference in the total cost function after the first iteration.)

The economic order quantity for each item is obtained by multiplying the production cycle interval for the product by its sales forecast:

$$Q_j = k_j T_0 S_j$$

If this procedure is used for a set of products for which there is no family setup cost, then the parameter A should be set equal to zero and the procedure followed in the regular manner. The first condition given for cycling in which a group of products are cycled to avoid the problems of interference and fluctuating lead times, would be handled by using a zero value for the family setup cost. If there is a physical constraint in the production process, such as might be found in chemical process industries, which requires that all products must be produced every cycle (i.e., all $k_j = 1$), then the economic cycle interval i obtained in the first step and the other steps can be ignored.

When there is a preferred production sequence which minimizes setup cost for a group of products, the setup cost incurred in this preferred sequence should be used for the a_j value for each product in step 1 of the above procedure. A pointed out above, if all products must be produced every cycle because of an operating constraint, then step 1 gives the optimum production cycle. If it is possible for low demand products in the group to be produced less frequently than the basic cycle of the majority of products and it is found in steps 2 and 3 that this is desirable for some products, then the individual item setup costs a_j, should be adjusted before recomputing the family cycle time in step 4. The adjusted value of a_j should be a weighted average reflecting the frequency and changeover costs for the likely sequences resulting from the set of product cycle multipliers obtained in step 3. This adjustment is only necessary if the differences in setup cost incurred using the preferred sequence and other sequences are quite large, e.g., greater than a ratio of two to one. It was shown for the simple EOQ model, that results are not sensitive to minor errors in setup cost.

Economic Order Quantity for Slow Moving Items. All of the previous inventory cost models have included the assumption that inventory depletions are continuous and constant. For some items, demand is characterized by a low frequency of depletion orders. The orders for some of these infrequently ordered items are typically for a small quantity of the basic sales unit and for others are typically for large quantities. The assumption of continuous and constant demand is obviously not valid. The problem in this kind of situation is to accurately represent the average inventory level.

One approach to this problem proposed by Hanssmann (Operations Research in Production and Inventory Control) is to assume that:

1. The interval between demands is constant.
2. The demand quantity is a constant amount, n.
3. Replenishment always takes place at one of the demand points in time.

The result after formulating a cost model based on these assumptions and solving for the optimum order quantity for that model can be simply stated. The optimum order quantity is obtained by solving for Q using the most appropriate of the previous economic order quantity formulas (with the assumption of continuous and constant demand) and then rounding off the result to the nearest integer multiple of the constant demand quantity, n. For example, if the constant demand quantity is one dozen and the calculated EOQ is 75 pieces, the order quantity should be rounded to 72 pieces (6 dozen).

Queueing approaches have also been suggested for this problem (Hanssmann, Operations Research in Production and Inventory Control) but these approaches are not considered in the area of inventory theory. Moreover, these approaches have not proven to be useful for practical inventory control systems.

ORDER POINT. The order point is a replenishment trigger level. When it is detected that the actual inventory level (available inventory level which includes on-order is used in some inventory control systems) is below the order point, a replenishment order is initiated.

Lead Time. The order point represents the **maximum reasonable demand** ver the lead time. For a fixed interval inventory control system, i.e., a system a which the inventory levels are reviewed at periodic intervals such as weekly r monthly, the lead time equals one review interval plus the procurement lead me. The review interval is included because if actual inventory is above the rder point at one review time, the earliest that replenishment could occur would e the result of an order initiated at the next review time. This is illustrated in 'ig. 8a.

For a continuous review inventory control system, i.e. a system in which inentory is reviewed after each transaction, the lead time is simply equal to the rocurement lead time. Replenishment is initiated as soon as the actual inventory level penetrates the order point. This is illustrated in Fig. 8b.

Safety Stock. The order point is made up of two components: one which epresents the forecast or expected demand over the lead time and the other which protects against the uncertainty in the actual demand over the lead time. Jncertainty arises in two forms. The rate of demand or requirements is usually ot known exactly and the procurement lead time is also usually subject to ariation. If there was no uncertainty, the order point would simply be set qual to the expected demand over the lead time. The other component of the rder point would be zero.

In most situations there is uncertainty; therefore, it is necessary to carry additional inventory which is commonly called safety stock. Safety stock is maintained to protect against running out of stock before replenishment is received. The cost of carrying this extra inventory must be balanced against the value of he protection obtained. As the amount of safety stock is increased, the chance f a stockout is decreased. However, typically the decrease in the chance of a tockout diminishes as the safety stock increases. In other words, each additional unit of safety stock buys less and less protection. Therefore, it is usually not economical to carry enough safety stock to avoid any chance of a stockout. The question is how much safety stock will minimize the combined costs f carrying inventory and stockouts.

In addition to depending on the stockout cost and inventory carrying cost, he amount of safety stock required also depends on the magnitude of the forecast errors over the lead time. It is necessary to know the probability that demand over the lead time will exceed the forecast by more than various levels. The **statistical distribution** of forecast errors gives the probability of an error of any specific size. While the exact shape of the distribution will be different for each product and forecasting technique, Brown (Statistical Forecasting for Inventory Control) states that "the normal distribution is a good enough description of the actual characteristics of forecast errors."

The standard deviation is a measure of the magnitude of the forecast errors. The common practice is to calculate safety stock as a multiple of the standard deviation of the forecast errors over the lead time, i.e.:

$$SS = k\sigma$$

where SS = safety stock for the lead time
   σ = standard deviation of forecast errors over the lead time
   k = safety factor

Determination of the Safety Factor When Stockout Costs Are Known. The value of the safety factor is ideally determined so as to minimize the combined costs of carrying safety stock inventory and stockouts. While stockout

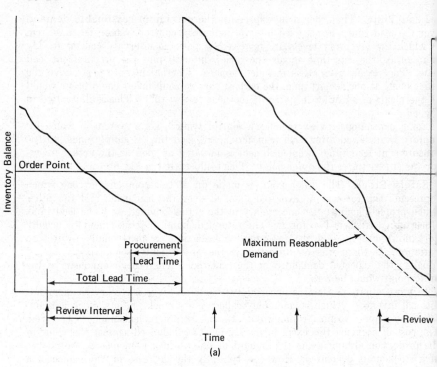

(a)

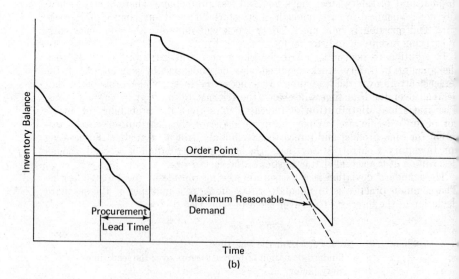

(b)

Fig. 8. Inventory control systems. (a) Fixed interval. (b) Continuous revi osts may arise in many ways, there are two ways of expressing stock costs which are frequently encountered:

1. A fixed cost per stockout which is independent of the length of the stockout or the number of units of demand involved.
2. A fixed cost per unit of demand backordered or lost. In this form, the cost roughly approximates a fixed cost per order backordered or a fixed cost per unit of time out of stock for which cost models for determining the safety factor cannot be easily formulated.

The total cost equation using the first form of stockout cost is as follows:

$$C_1 = k\sigma I + \frac{S}{Q} B F(k)$$

where
- k = safety factor
- σ = standard deviation of forecast errors over the lead time
- I = inventory carrying cost in dollars per piece per time period
- S = forecast sales in pieces per time period
- Q = order quantity per cycle in pieces
- B = shortage cost in dollars per shortage
- $F(k)$ = probability of a forecast error over the lead time larger than $k\sigma$; i.e., the probability of a stockout
- $p(x)$ = density function of a unit normal variable which represents the distribution of forecast errors normalized by its standard deviation

$$p(x) = \frac{1}{\sqrt{2\pi}} \exp\left(-\frac{x^2}{2}\right)$$

The optimum value for the safety factor is

$$p(k) = \frac{IQ\sigma}{BS}$$

The value of k corresponding to $p(k)$ can be easily obtained using a table of probabilities. A few selected values for $p(k)$ are given in Fig. 9 from Brown (Decision Rules for Inventory Management).

Safety Factor, k	$p(k)$	$F(k)$	$E(k)$
0.00	0.3989	0.5000	0.3989
0.50	0.3521	0.3085	0.1978
1.00	0.2420	0.1587	0.0833
1.50	0.1295	0.0668	0.0293
2.00	0.0540	0.0227	0.0085
2.50	0.0175	0.0062	0.0020
3.00	0.0044	0.0013	0.0004

Fig. 9. Safety factor parameters for the normal probability distribution.

The use of the above equation can be demonstrated with an example. Assume:

$$Q = 4,000 \text{ pieces per run}$$
$$\sigma = 1,000 \text{ pieces}$$
$$I = \$2 \text{ per piece per year}$$
$$S = 20,000 \text{ pieces per year}$$
$$B = \$1,500 \text{ per stockout}$$

$$p(k) = \frac{2 \times 4,000 \times 1,000}{15,000 \times 20,000} = 0.2667$$

$k \simeq 0.89$ (by interpolation from Fig. 9)

The total cost equation for the second form of expressing stockout cost (dollars per unit of demand backordered) is as follows:

$$C_2 = k\sigma I + \frac{S}{Q} b\sigma E(k)$$

where all variables are the same as in the previous cost equation, except:

b = stockout cost per unit of demand backordered
$\sigma E(k)$ = expected demand backordered per cycle

$$\sigma E(k) = \int_k^\infty (x - k)\sigma\, p(x)\, dx$$

The equation for C_2 can also be solved to determine the optimum value of the safety factor:

$$F(k) = \frac{IQ}{bS}$$

The value of k for some representative values of $F(k)$ are given in Fig. 9.

The use of the optimum safety factor equation is illustrated by the following example. Assume:

Q = 4,000 pieces per run
S = 20,000 pieces per year
I = \$2 per piece per year
b = \$10 per piece short

$$F(k) = \frac{2 \times 4,000}{10 \times 20,000} = 0.0400$$

$k \simeq 1.70$ (by interpolation from Fig. 9)

Determination of the Safety Factor When Stockout Costs Are Unknown. It is frequently the case that the cost of a stockout includes many intangibles which cannot be quantified. When the cost of a stockout cannot be quantified, it is necessary to use judgment to select the best service objective. The two most common forms for expressing the service objective correspond to the two previously mentioned forms for expressing stockout cost:

1. The probability of no stockout per replenishment cycle.
2. The fraction of demand satisfied directly from inventory. (This is the complement of the fraction of demand backordered or lost sales.)

The first form of the service objective corresponds exactly to $1 - F(k)$. Thus

$$F(k) = 1 - Z_1$$

where Z_1 = service objective expressed as the probability of no stockout per replenishment cycle. That is, if the service objective is given as a 95-percent probability of no stockout per replenishment cycle, then the required value of the safety factor is the k corresponding to an $F(k) = 0.05$, $k = 1.64$. In other words, the $F(k)$ represents the probability of a stockout per replenishment cycle for a safety factor k.

The second form of the service objective can be represented with the following equation:

$$Z_2 = 1 - \frac{\sigma E(k)}{Q}$$

where Z_2 = service objective expressed as the fraction of demand
satisfied without backorder or lost sales
$\sigma E(k)$ = expected shortage per cycle for safety factor k
Q = order quantity per cycle

Thus

$$E(k) = \frac{Q}{\sigma}(1 - Z_2)$$

Selective values for $E(k)$ are given in Fig. 9.

For example, if the desired service objective is to satisfy 95 percent of demand from inventory, the run quantity is 4,000 pieces and the standard deviation is 1,000 pieces:

$$E(k) = \frac{4,000}{1,000}(1 - 0.95) = 0.20$$

$$k \simeq 1.66$$

Calculation of the Order Point. The order point is calculated by adding the required safety stock to the expected demand (forecast of the average) over the lead time. Thus

$$OP = S_l + SS_l$$

where OP = order point
S_l = forecast sales over the lead time
SS_l = safety stock for the lead time

The lead time was previously explained. The safety stock is calculated using the safety stock equation on page 23. The safety factor used to calculate the safety stock is determined by using equations for $p(k)$, $F(k)$ or Z_2, depending on whether stockout costs are known and the service definition.

An Example of a Complete Inventory Calculation. As an example, assume that a fixed interval inventory control system is used with a review interval of one week. The procurement lead time is assumed to be three weeks for an item for which it is desired to calculate the order point. Thus the total lead time for the order point is four weeks. Further assume that:

Q = 4,000 pieces per run
S = 20,000 pieces per year
I = \$2 per piece per year
S_l = 1,540 pieces for 4 week lead time (i.e., 20,000/13)
σ_l = 400 pieces
b = \$10 per piece short

Since stockout cost is expressed on a "per unit short" basis, the equation for $F(k)$ is used to determine the safety factor:

$$F(k) = \frac{IQ}{bS} = \frac{2 \times 4,000}{10 \times 20,000} = 0.0400$$

Using Fig. 9 and interpolating between values of 0.0668 and 0.227 for $F(k$ gives

$$k = 1.70$$

Safety stock is next calculated:

$$SS_l = k\sigma_l = 1.70 \times 400 = 680 \text{ pieces}$$

Using the order point equation,

$$OP = S_l + SS_l = 1,540 + 680 = 2,220 \text{ pieces}$$

If the shortage cost was not known, and the service objective was expresse as "99 percent of demand should be satisfied without backorder $(Z_2 = 0.99)$, the calculation of the safety factor would use the equation for $E(k)$:

$$E(k) = \frac{Q}{\sigma_l}(1 - Z_2) - \frac{4,000}{400}(1 - 0.99) = 0.1$$

By interpolation from Fig. 9,

$$k = 0.92$$

The safety stock and order point are then calculated:

$$SS_l = k\sigma_l = 0.92 \times 400 = 368$$

$$OP = S_l + SS_l = 1,540 + 368 = 1,908$$

INTERACTION OF THE ECONOMIC ORDER QUANTITY AN‍ ORDER POINT. In the previous discussions in which formulas for the eco nomic order quantity and the order point are given under various conditions, ‍ has been assumed that these two parameters are independent of each other. Th‍ assumption is obviously not valid but is made to simplify the formulas for pract‍ cal application in inventory control systems. The justification for this assump‍ tion is that the potential savings which would be achieved by using more sophist‍ cated procedures would be more than offset by additional costs incurred i‍ implementing and operating the more complex inventory control system. Further more, there is also a question as to whether the cost and forecast values used i‍ the formulas are accurate enough to make the more precise answers meaningfu‍ However, in some situations, particularly when the forecast errors are large rela‍ tive to the average sales rate so that safety stock is relatively large, there ma‍ be good reason to use a procedure for calculating the economic order quantit‍ and order point which recognizes their interaction.

Hadley and Whitin (Analysis of Inventory Systems) give some approximat‍ formulas and a heuristic procedure for solving these formulas which give th‍ economic order quantity and order point considering the interaction betwee‍ them. The approximation used in this formulation is that backorders or los‍ sales have no effect on the average inventory level. Since the level of back‍ orders or lost sales is zero most of the time and is expected to be very small th‍ few times that they occur, this assumption should have very little effect on th‍ results.

Linear Programming

METHOD OF RESOURCE ALLOCATION. Linear programming i‍ one of the best developed and most widely applied disciplines of operations re‍ search. It concerns itself with the **optimum allocation of limited resource‍** among competing activities, under the constraints imposed by the problem bein‍

nalyzed. Those constraints could be of a financial, technological, marketing, organizational, or any other nature. In broad terms, linear programming can be defined as a mathematical representation aimed at programming or planning the best possible allocation of scarce resources, where the **mathematical model** used in that representation is characterized by allowing exclusively linear functions to be used.

Although the roots of linear programming could be traced back to the origins of the first optimization problems, it was not until 1947 when the general linear programming problem was formulated and solved by George B. Dantzig, as part of the effort of project SCOOP (Scientific Computation of Optimum Programs), a research group of the U.S. Air Force. At that time, Dantzig developed the **simplex method** for solving the general linear programming problem. The extraordinary computational efficiency of the simplex method, together with the availability of highly powerful digital computers, have made possible the application of linear programming to an enormous class of military, governmental, industrial, and business problems.

APPLICATIONS OF LINEAR PROGRAMMING. Buffa (Models for Production and Operations Management) considers linear programming one of the most far reaching developments in management science technology. The following list (based on Buffa) gives some general and specific problems to which linear programming has been applied:

1. Distribution of products from a set number of origins to a number of destinations, with a minimum of transportation costs and within a number of supply and demand constraints (see discussion of the Transportation Problem).
2. Distribution of products from plants to warehouses, with a minimum of production and distribution costs.
3. Multiple plant location studies in which production and distribution costs for the entire system are minimized.
4. Production studies for multiple plants in which production costs are minimized for the entire system for a given total demand.
5. Redistribution of empty freight cars with a minimum of transportation costs.
6. Allocation of a limited amount of raw material among a number of different products to meet demand and maximize profit.
7. Allocation of production facilities when alternate routings are available to minimize costs and maximize profit.
8. Blending problems in which the total cost of materials is minimized.
9. Maximizing material utilization from a given standard amount of raw material.
10. Production program for seasonal demand that will minimize inventory and production costs while meeting the demand.
11. Product mix problems the objective of which is to meet product demand within certain cost and profit constraints.
12. Long-range planning involving production capacity, demand forecasts, costs, and operations.

Basic Theory. Springer, Herlihy, and Beggs (Advanced Methods and Models, Vol. 2) define a **linear program** as a mathematical program in which the constraints and the objective function involve the choice variables in a mathematically linear way (that is, the exponents of the variables are 1). The **choice variables** are those variables chosen by the analyst as reflecting the available management decision choices. **Constraints** are the relationships among the variables, usually in the form of linear inequalities, that restrict the values that may be assigned to the choice variables. The **objective function** is the mathematical expression whose value may be computed when the values of the choice variables are given.

Linear programming is the process of developing a linear program that w reflect some business or production problem. The **optimum solution** of linear program is that set of values for the choice variables that maximizes (minimizes) the value of the objective function. As an example: A and B a two choice variables; A cannot be more than twice B and B cannot exceed 5 find the maximum value of $A + B$.

Using the above definitions,

Choice variables: A and B

Constraints: $A \leq 2B$
$B \leq 5$

Objective function: $A + B$

The objective function is maximized subject to the constraints when $A = 1$ and $B = 5$.

MATHEMATICAL FORMULATION OF LINEAR PROGRAM MING.

In mathematical terms, the linear programming model can be expresse as the maximization of an objective (or **effectiveness**) function, subject to given set of linear constraints. Specifically, the linear programming problem can be described as finding the values of n variables, $x_1, x_2, \ldots, x_n$, such tha they maximize an objective function z

$$z = c_1 x_1 + c_2 x_2 + \ldots + c_n x_n \tag{1}$$

subject to the following constraints

$$
\left.
\begin{aligned}
a_{11}x_1 + a_{12}x_2 + \ldots + a_{1n}x_n &\leq b_1 \\
a_{21}x_1 + a_{22}x_2 + \ldots + a_{2n}x_n &\leq b_2 \\
\vdots \\
a_{m1}x_1 + a_{m2}x_2 + \ldots + a_{mn}x_n &\leq b_m
\end{aligned}
\right\} \tag{2}
$$

$$
\left.
\begin{aligned}
x_1 &\geq 0 \\
x_2 &\geq 0 \\
\vdots \\
x_n &\geq 0
\end{aligned}
\right\} \tag{3}
$$

where c_j, a_{ij} and b_i are given constants.

It is easy to provide an immediate interpretation of the general linear programming problem just stated in terms of a production problem. For instance we could assume that in a given production facility there are n possible products we may manufacture, whose production quantities we wan! to determine and we designate by $x_1, x_2, \ldots, x_n$. In addition, there are m limited resources for which these products compete. These resources could be manpower availability, machine capacities, product demand, working capital, etc., and are designated by $b_1, b_2, \ldots, b_m$. Let a_{ij} be the amount of resource i required by product j and let c_j be the unit profit of product j. Then the linear programming model seeks to determine the production quantity of each product in such a way as to maximize the total resulting profit z (Eq. 1), given that the available resources should not be exceeded (Eq. 2) and that we can only produce positive or zero amounts of each product (Eq. 3).

Linear programming is not restricted to the structure of the problem presented above. First, it is perfectly possible to minimize rather than maximize the ob-

ective function. In addition, "greater than equal" inequalities or equalities can
be handled simultaneously with the "less than equal" constraints presented in
Eq. (2). Finally, some or all the variables can be set to be unrestricted in sign,
although in practice most of the variables assume the nonnegative constraints
presented in Eq. (3).

PRACTICAL FORMULATION OF LINEAR PROGRAMMING
PROBLEMS. To illustrate the rationale that should be used in formulating
linear programming problems, highly simplified practical problems are described
here together with their corresponding linear programming formulations. The
statements of these problems were developed by Professor Robert M. Thrall of
the University of Michigan.

Use of Raw Materials. A producer of aluminum makes a special alloy
which he guarantees contains 90 percent or more aluminum, between 5 percent
and 8 percent copper, and the remainder of other metals. The demand for this
alloy is very uncertain so he does not keep a supply on hand. He has just received an order for 1,000 lbs. at $0.45/lb. He must make the alloy from two
batches of scrap, pure copper and pure aluminum. The analysis of the scrap
shows that they contain:

	Al	Cu	Other
Scrap I	95%	3%	2%
Scrap II	85%	1%	14%

Their costs are:

	$/lb.
Scrap I	$0.15
Scrap II	0.05
Al	0.50
Cu	0.60

It costs $0.05/lb. to melt the metal. He has more than 1,000 lbs. of each type of
metal on hand. How should the producer charge his furnace so that he will
maximize his profit?

Program formulation:
The first step in formulating a linear program is to define the **decision variables** of the problem. In the present example, these variables are simple to
identify, and correspond to the number of pounds of Scrap I, Scrap II, aluminum, and copper to be used in the production of the alloy. Specifically, the decision variables are denoted as follows:

$$x_1 = \text{pounds of Scrap I}$$
$$x_2 = \text{pounds of Scrap II}$$
$$x_3 = \text{pounds of aluminum}$$
$$x_4 = \text{pounds of copper}$$

The next step in the formulation of the program is to determine the objective
function. In this case, the total profit resulting from the production of 1,000
pounds of the special alloy is to be maximized. Since exactly 1,000 pounds of the
alloy are to be produced, the total income will be the selling price per pound
times 1,000 pounds. That is:

$$\text{Total Income} = 0.45 \times 1{,}000 = \$450$$

To determine the total cost incurred in the production of the alloy we should add the melting cost of $0.05/lb. to the corresponding cost of each metal used. Thus, the relevant unit cost, in dollars per pound, is:

Scrap I	$0.15 + $0.05 =	$0.20
Scrap II	0.05 + 0.05 =	0.10
Al	0.50 + 0.05 =	0.55
Cu	0.60 + 0.05 =	0.65

Therefore, the total cost becomes:

$$\text{Total Cost} = 0.20x_1 + 0.10x_2 + 0.55x_3 + 0.65x_4 \tag{4}$$

and the total profit we want to maximize is determined by the expression:

$$\text{Total Profit} = \text{Total Income} - \text{Total Cost}$$

Thus,

$$\text{Total Profit} = 450 - 0.20x_1 - 0.10x_2 - 0.55x_3 - 0.65x_4 \tag{5}$$

It is worthwhile noticing in this example that since the amount of special alloy to be produced was fixed in advance (1,000 pounds), the maximization of the total profit, given by Eq. (5), becomes completely equivalent to the minimization of the total cost, given by Eq. (4).

The constraints in the problem should now be defined. First, since the producer does not want to keep any supply of the alloy on hand, the total amount to be produced is exactly 1,000 pounds, i.e.:

$$x_1 + x_2 + x_3 + x_4 = 1,000$$

Next, the alloy should contain at least 90 percent aluminum. This restriction can be expressed as follows:

$$\frac{0.95x_1 + 0.85x_2 + x_3}{1,000} \geqq 0.90$$

The numerator of the left-hand side of this constraint indicates the total number of pounds of aluminum in the alloy. By dividing it by 1,000 we get the proportion of aluminum in the alloy, which must be at least 0.90 or 90 percent.

Similarly, the restrictions regarding copper content in the alloy can be represented by the following inequalities:

$$\frac{0.03x_1 + 0.01x_2 + x_4}{1,000} \geqq 0.05$$

$$\frac{0.03x_1 + 0.01x_2 + x_4}{1,000} \leqq 0.08$$

The first inequality establishes the minimum copper content in the alloy at 5 percent, while the second inequality sets the maximum copper content at 8 percent.

Finally, the obvious nonnegativity constraints are given by

$$x_i \geqq 0, \quad i = 1, 2, 3, 4.$$

If the total cost is to be minimized, and if the previous constraints are multiplied by 1,000, the resulting linear programming problem can be expressed as follows:

$$\text{Minimize } z = 0.20x_1 + 0.10x_2 + 0.55x_3 + 0.65x_4$$

subject to the constraints:

$$x_1 + x_2 + x_3 + x_4 \quad\quad\quad = 1{,}000$$
$$0.95x_1 + 0.85x_2 + x_3 \quad\quad \geq \quad 900$$
$$0.03x_1 + 0.01x_2 + x_4 \quad\quad \geq \quad\ 50$$
$$0.03x_1 + 0.01x_2 + x_4 \quad\quad \leq \quad\ 80$$
$$x_1 \geq 0,\ x_2 \geq 0,\ x_3 \geq 0,\ x_4 \geq 0$$

Product Mix for Maximum Profit. An electronics company manufactures three lines of products for sale to the government: transistors, micromodules, and circuit assemblies. It has four process areas: transistor production; circuit printing and assembly; transistor and module quality control; and circuit assembly test and packing.

Production of one transistor requires 0.1 standard hour of transistor production area capacity, 0.5 standard hour of transistor quality control area capacity and $0.70 in direct costs.

Production of one micromodule requires 0.4 standard hour of the circuit printing and assembly area capacity, 0.5 standard hour of the quality control area capacity, 3 transistors, and $0.50 in direct costs.

Production of one circuit assembly requires 0.1 standard hour of the capacity of the circuit printing area, 0.5 standard hour of the test and packing area, 1 transistor, 3 micromodules, and $2.00 in direct costs.

Any of the three products may be sold in unlimited quantities at prices of $2.00, $8.00, and $25.00 each, respectively. If there are 200 hours of production time open in each of the four process areas in the coming month, what products should be manufactured, and in what quantities, to yield the most profit?

Program formulation:

As previously indicated, the first decision in formulating a linear programming model is the selection of the proper variable to represent the problem under consideration. In the present example there are two different sets of variables that can be chosen as decision variables. Let x and y be defined as follows:

x_1 = total number of transistors produced
x_2 = total number of micromodules produced
x_3 = total number of circuit assemblies produced

and,

y_1 = total number of transistors sold as transistors
y_2 = total number of micromodules sold as micromodules
y_3 = total number of circuit assemblies sold as circuit assemblies

Since each micromodule needs three transistors, and each circuit assembly needs one transistor, the total number of transistors sold as transistors is given by:

$$y_1 = x_1 - 3x_2 - x_3$$

Similarly, for the total number of micromodules sold as micromodules:

$$y_2 = x_2 - 3x_3$$

Finally, since all the circuit assemblies are sold as such:

$$y_3 = x_3$$

Expressing the x's in terms of the y's:

$$x_1 = y_1 + 3y_2 + 10y_3$$
$$x_2 = y_2 + 3y_3$$
$$x_3 = y_3$$

The linear program can be formulated using either the x's or y's as decision variables. First the formulation is shown in terms of the x's, the production variables.

The objective function is stated by using the information provided in the statement of the problem with respect to the selling prices and the direct cost of each of the units produced. The total profit is given by:

$$\text{Total Profit} = 2.0y_1 + 8.0y_2 + 25.0y_3 - 0.7x_1 - 0.5x_2 - 2.0x_3$$

The relationships between the x's and y's allow us to express this total profit in terms of the x variables alone. After performing the transformation:

$$\text{Total Profit} = 1.3x_1 + 1.5x_2 - 3x_3$$

Now the restrictions due to the maximum availability of 200 hours of production time in each of the four process areas must be introduced.

The transistors are the only ones requiring processing time in the transistor production area, and they consume 0.1 standard hour of production per transistor, so that:

$$0.1x_1 \leq 200$$

Similarly for the other three areas:

$$0.4x_2 + 0.1x_3 \leq 200$$
$$0.5x_1 + 0.5x_2 \leq 200$$
$$0.5x_3 \leq 200$$

In addition to these constraints the number of transistors, micromodules, and circuit assemblies sold as such (i.e., the y variables) are nonnegative. Therefore, the following constraints must be added:

$$x_1 - 3x_2 + x_3 \geq 0$$
$$x_2 - 3x_3 \geq 0$$
$$x_3 \geq 0$$

Finally, the trivial nonnegativity conditions are given by:

$$x_1 \geq 0 \qquad x_2 \geq 0 \qquad x_3 \geq 0$$

Notice that besides the nonnegativity of all the variables, which is a condition always implicit in the methods for solving linear programming, this form of stating the problem has generated seven constraints. If the problem is expressed in terms of the y variables, however, the nonnegativity of the x variables is automatically guaranteed by the nonnegativity of the y variables; the x's expressed in terms of the y's are the sum of nonnegative variables, and therefore are always nonnegative.

By performing the proper changes of variables, it is easy to see that the linear programming formulation in terms of the y variables is given by:

$$\text{Maximize } z = 1.3y_1 + 5.4y_2 + 14.5y_3$$

subject to:

$$0.1y_1 + 0.3y_2 + y_3 \leq 200$$
$$0.4y_2 + 1.3y_3 \leq 200$$
$$0.5y_1 + 2.0y_2 + 6.5y_3 \leq 200$$
$$0.5y_3 \leq 200$$

$$y_1 \geq 0 \qquad y_2 \geq 0 \qquad y_3 \geq 0$$

Since the computation time required to solve a linear programming problem increases roughly with the cube of the number of rows of the problem, in this example the y's constitute a better choice of decision variables than the x's.

Cargo Space Allocation. A ship has three cargo holds, forward, aft, and center. The capacity limits are:

Forward	2,000 tons	100,000 cubic feet
Center	3,000 tons	135,000 cubic feet
Aft	1,500 tons	30,000 cubic feet

The following cargoes are offered; the shipowners may accept all or any part of each commodity:

Commodity	Amount (tons)	Volume per Ton (cu ft)	Profit per Ton ($)
A	6,000	60	6
B	4,000	50	8
C	2,000	25	5

In order to preserve the trim of the ship, the weight in each hold must be proportional to the capacity in tons. How should the cargo be distributed so as to maximize profits?

Program formulation:

To determine the decision variables of this problem, it is important to realize that what is really required is the amount of each commodity to be allocated in each of the three cargo holds. Thus, let the decision variables be:

	Commodity (in tons)	Location
x_{AF}	A	Forward hold
x_{AC}	A	Center hold
x_{AA}	A	Aft hold
x_{BF}	B	Forward hold
x_{BC}	B	Center hold
x_{BA}	B	Aft hold
x_{CF}	C	Forward hold
x_{CC}	C	Center hold
x_{CA}	C	Aft hold

The objective function can be defined as the maximization of the total profit given by

$$\text{Total Profit} = 6(x_{AF} + x_{AC} + x_{AA}) + 8(x_{BF} + x_{BC} + x_{BA}) + 5(x_{CF} + x_{CC} + x_{CA})$$

where, for instance $(x_{AF} + x_{AC} + x_{AA})$ represents the total tons of commodity A being shipped and 6 is the profit per ton of that commodity.

The first set of constraints has to do with the total capacity, in tons, existing in each of the cargo holds. The 2,000-ton limit of the forward hold is represented by

$$x_{AF} + x_{BF} + x_{CF} \leqq 2,000$$

Similarly for the center hold

$$x_{AC} + x_{BC} + x_{CC} \leqq 3,000$$

and for the aft hold

$$x_{AA} + x_{BA} + x_{CA} \leqq 1,500$$

In order to express the cubic feet capacity limit we have to convert tons into cubic feet, by multiplying the tons shipped of each commodity by the corresponding cubic feet per ton. Thus the cubic feet capacity constraints for forward, center, and aft holds are, respectively:

$$60x_{AF} + 50x_{BF} + 25x_{CF} \leqq 100,000$$
$$60x_{AC} + 50x_{BC} + 25x_{CC} \leqq 135,000$$
$$60x_{AA} + 50x_{BA} + 25x_{CA} \leqq 30,000$$

A third set of restrictions has to do with the maximum availability of each commodity. Only 6,000 tons of commodity A are available, thus:

$$x_{AF} + x_{AC} + x_{AA} \leqq 6,000$$

Similarly, only 4,000 tons of commodity B are available

$$x_{BF} + x_{BC} + x_{BA} \leqq 4,000$$

and finally, only 2,000 tons of commodity C are available

$$x_{CF} + x_{CC} + x_{CA} \leqq 2,000$$

The fourth set of constraints refers to the need to keep the weight in each hold proportional to its capacity in tons. That is to say:

$$\frac{x_{AF} + x_{BF} + x_{CF}}{2,000} = \frac{x_{AC} + x_{BC} + x_{CC}}{3,000} = \frac{x_{AA} + x_{AB} + x_{AC}}{1,500}$$

The numerators of these fractions represent the total weight, in tons, allocated in the forward, center, and aft holds, respectively. The denominator gives the corresponding capacities, in tons, of those cargo holds.

The last expression can be incorporated in the linear programming model by the following two equations, which represent the proportionality between forward and center, and center and aft, respectively.

$$\frac{x_{AF} + x_{BF} + x_{CF}}{2,000} - \frac{x_{AC} + x_{BC} + x_{CC}}{3,000} = 0$$

$$\frac{x_{AC} + x_{BC} + x_{CC}}{3,000} - \frac{x_{AA} + x_{BA} + x_{CA}}{1,500} = 0$$

Finally, the obvious nonnegativity constraints of all the variables involved are:

$$x_{AF} \geqq 0 \qquad x_{AC} \geqq 0 \qquad x_{AA} \geqq 0$$
$$x_{BF} \geqq 0 \qquad x_{BC} \geqq 0 \qquad x_{BA} \geqq 0$$
$$x_{CF} \geqq 0 \qquad x_{CC} \geqq 0 \qquad x_{CA} \geqq 0$$

THE GEOMETRY OF LINEAR PROGRAMS. Although the actual computational procedures for solving linear programs will not be discussed, some insight into those procedures can be gained by looking at the geometry of a simple example. Consider a variation of an example given in Driebeek (Applied Linear Programming). The problem is limited to two decision variables, so that only simple graphs will be required.

A Production Scheduling Problem. A plant manager has one machine which can be used to produce two types of insulating material, say B and R. Because of the difference in the densities of the two materials he can produce $16\frac{2}{3}$ carloads of B per day but only $8\frac{1}{3}$ carloads of R per day. He can produce any combination of these two activities, as long as the total does not exceed an equivalent of $16\frac{2}{3}$ carloads of B per day, where the manufacturing of 1 carload of R is equivalent to 2 carloads of B. Further, special raw materials for B can-ot be made available in quantity so that production of B is limited to 8 carloads per day. Finally, there are only 10 boxcars made available per day for all the production. If the net profit to the plant is $200 per carload for B and $300 per carload for R, what should the manager's production schedule be in order to maximize net profits?

Problem formulation:

First the decision variables are defined. For this problem the number of carloads of each type of insulation to produce each day is desired. Let

$$x_B = \text{number of carloads of B produced per day}$$
$$x_R = \text{number of carloads of R produced per day}$$

Next, the objective function is easy to establish since net profits are to be maximized:

$$\text{Net Profit} = 200x_B + 300x_R$$

The constraints can be written next. Since one half as much per day of insulation type R as insulation type B can be produced, the constraint imposed by limited machine capacity in units of carloads of insulation type B is

$$x_B + 2x_R \leqq 16.67$$

Since only 10 boxcars are made available each day:

$$x_B + x_R \leqq 10$$

Finally, the available raw materials limit the amount of insulation type B that can be produced

$$x_B \leqq 8$$

With the usual nonnegativity constraints on the division variables

$$x_B \geqq 0 \qquad x_R \geqq 0$$

the following linear program is obtained:

$$\text{Maximize P} = 200x_B + 300x_R$$

Subject to:
$$x_B + 2x_R \leqq 16.67$$
$$x_B + x_R \leqq 10$$
$$x_B \qquad \leqq 8$$
$$x_B \geqq 0 x_R \geqq 0$$

Graphical Representation of the Decision Space. If just the constraints of the above linear programming problem are examined all the values of x_B and x_R satisfying these constraints can be geometrically represented by the cross-hatched area given in Fig. 10. Notice that each line in this figure is represented by a constraint expressed as an equality. The arrows associated with each line show the direction indicated by the inequality sign in each constraint. The set of values of x_B and x_R satisfying simultaneously all the constraints (indicated by the cross-hatched area) are all the feasible production possibilities. Among these feasible production alternatives is the one which maximizes the resulting net profit.

The Optimal Solution. The optimal solution may be found by starting at some feasible solution (i.e., a point within the cross-hatched area), say $x_B = 0$, $x_R = 0$, and looking for an improvement. Since producing R is worth $300/car-

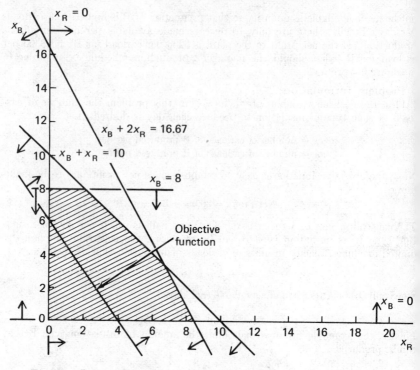

Fig. 10. Graphical representation of feasible production alternatives.

load while producing B is only worth \$200/carload, x_R is increased as much as possible. Then $x_B = 0$, $x_R = 8\frac{1}{3}$ and the profit P = \$2,500. Now x_B cannot be increased without decreasing x_R. If x_B is increased one unit the return is \$200/carload, however x_R must be reduced one-half carload to accommodate the increase in x_B and this causes a loss of \$300/carload or \$150 for each carload of B added. The net effect is $200 - (\frac{1}{2})(300) = 50$; hence x_B should be increased as much as possible while decreasing x_R appropriately. Then $x_B = 3\frac{1}{3}$, $x_R = 6\frac{2}{3}$ and the profit P = \$2,667. If x_B is increased further another carload of x_R must be balanced by a decrease in x_R of one carload. The net effect is $(1)(200) - (1)(300) = -100$ which is a decrease in profit of \$100. Therefore x_B should not be increased. Note that any point in the interior of the feasible region cannot be the optimal solution since more profit could be made by increasing both x_B and x_R simultaneously.

The optimal solution is shown in Fig. 11. The line through the point marked optimal solution is the equation of net profit at the optimal solution.

Fractional Solutions. In linear programming the decision space is continuous in the sense that fractional answers are always allowed. This is contrasted with discrete or integer programming where integer answers are required for some or all variables. In this example if all ten carloads must be shipped each day and mixed carloads are not permitted then an integer solution is necessary. The initial reaction is to round off the fractional quantities:

$$x_R = 7, \qquad x_B = 3$$

However, this solution is infeasible since the production capacity constraint would be exceeded. Hence, rounding off is not always possible. Rounding down yields a feasible solution

$$x_R = 6, \ x_B = 3 \text{ and } P = \$2,400$$

but it is easy to see that

$$x_R = 6, \ x_B = 4 \text{ and } P = \$2,600$$

is also a feasible solution and it is clearly preferable since it has a higher net profit. Basically the integer programming problem is inherently difficult to solve and falls in the domain of combinatorial analysis rather than simple linear programming. Often what seem to be integer variables can be interpreted as rates and then the integer difficulty disappears. In this example, if it is unnecessary to ship all ten carloads each day then the fractional carloads not completed by the end of the day are merely completed first the following day. Thus the rate of output will still be ten carloads per day. Finally, when the numbers involved are large, rounding to a feasible solution usually results in a good approximation.

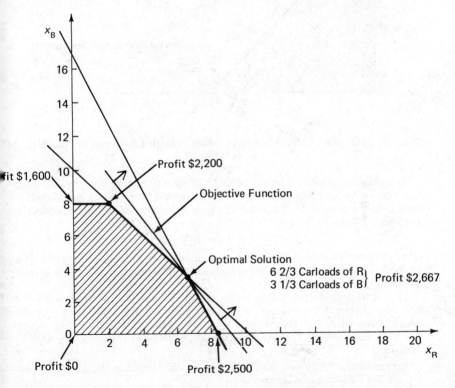

Fig. 11. Optimal solution to production scheduling problem of Fig. 10.

Uniqueness of the Optimal Solution. Very often it happens that the optimal solution to a linear programming problem is not unique. The problem may have more than one solution with the same value of the objective function. This results from the objective function being "parallel" to one of the constraints (Fig. 12).

If the objective function in the scheduling problem were changed to

$$P' = 2x_B + 4x_R$$

then the optimal solution is no longer unique.

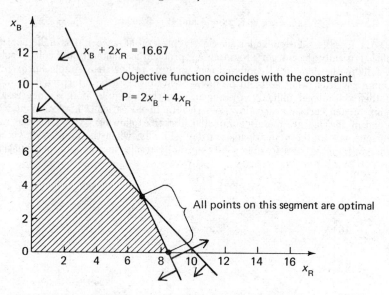

Fig. 12. Multiple solutions obtained when objective function is "parallel" to constraints.

Starting Solutions. The null activity of $x_R = 0$, $x_B = 0$ was a feasible starting solution to the problem as stated. However if a new item is considered, say minimum production to keep the machine operating efficiently, the situation changes. Suppose

$$x_B + 2x_R \geq 10$$

This constraint means that there is a minimum rate at which the machine can be used to produce the two types of insulation. The new feasible region is shown in Fig. 13.

Since a feasible starting solution may not be obvious, the initial objective is to find one. Hence it is necessary to perform what is called a Phase I procedure which shows whether a feasible solution exists. Once an initial starting solution has been obtained, then Phase II proceeds to optimize the economic objective function.

Infeasibility. Suppose that in typing in the data for the additional constraint, the following mistake was made:

$$x_B + 2x_R \geq 20$$

The graphical representation of this error is shown in Fig. 14.

There are no points that satisfy all the constraints simultaneously and the problem is therefore **infeasible.** Computer routines for solving linear programming problems indicate when an error of this type has occurred. Infeasibility is presented, in general, whenever a set of constraints is so restrictive that it produces an impossible solution.

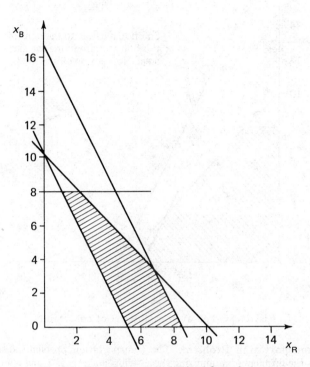

Fig. 13. Feasible region limited from below.

Unboundedness. Suppose the control message for the original problem was typed incorrectly so that a solution for the following problem is attempted instead:

$$\text{Maximize: } P = 300x_B + 200x_R$$
$$\text{Subject to:} \qquad x_B + 2x_R \geqq 16.67$$
$$x_B + x_R \geqq 10$$
$$x_B \geqq 8$$
$$x_B \geqq 0 \quad x_R \geqq 0$$

Clearly the maximum of the objective function is now infinity (Fig. 15). Linear programming solution techniques also indicate when this kind of error has been made. In practice, an unbounded solution is always an indication that the problem has been ill-formulated since no real situation produces an unlimited value of the objective function.

SPECIAL LINEAR PROGRAMMING MODELS. There are several kinds of linear programming models which present a special structure and for which efficient and practical solution algorithms have been designed, addressed to take advantage of the particular structure of these models. Some of these special linear programs will be reviewed because they have several significant practical applications.

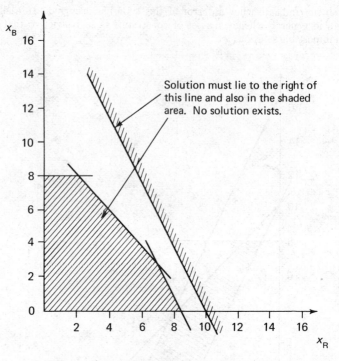

Fig. 14. **Graph of infeasible set of constraints.**

The Transportation Problem. The transportation problem is usually formulated as the problem of minimizing the distribution cost of a single homogeneous product from m sources (plants or warehouses) to n destinations (warehouses or consumer markets), where the total number of units available at each source and the total number of units required at each destination are known, constant numbers.

To precisely formulate the problem, the following terms are defined:

a_i = number of units available at source i ($i = 1, \ldots, m$)
b_j = number of units required at destination j ($j = 1, \ldots, n$)

The total product availability is assumed to be equal to the total product requirements, that is:

$$\sum_{i=1}^{m} a_i = \sum_{j=1}^{n} b_j$$

However, it is trivial to relax this condition by introducing an additional source or destination to provide for the difference. In addition the following are defined:

c_{ij} = unit transportation cost from source i to destination j
    ($i = 1, \ldots, m; j = 1, \ldots, n$)
x_{ij} = number of units to be distributed from source i to location j
    ($i = 1, \ldots, m; j = 1, \ldots, n$)

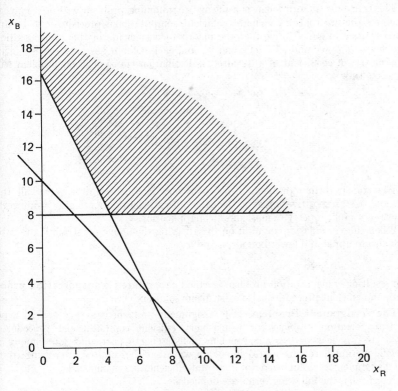

Fig. 15. Graph showing unbounded "solution" situation.

The transportation problem can be formulated as follows:

$$\text{Min } z = \sum_{i=1}^{m} \sum_{j=1}^{n} c_{ij}x_{ij}$$

subject to:

$$\sum_{j=1}^{n} x_{ij} = a_i \qquad i = 1, \ldots, m$$

$$\sum_{i=1}^{m} x_{ij} = b_j \qquad j = 1, \ldots, n$$

$$x_{ij} \geqq 0 \qquad \begin{cases} i = 1, \ldots, m \\ j = 1, \ldots, n \end{cases}$$

This is similar to the earlier examples. Here, total distribution cost z is subject to the following constraints:

1. The total amount shipped from the ith source to all destinations is equal to the amount available there, a_i.
2. The total amount shipped to destination j is the total requirement, b_j, at that destination.
3. All variables are nonnegative.

The transportation problem is a linear programming problem with a very particular structure that allows highly efficient computational procedures to be applied in its solution. Notice that the matrix of coefficients of the transportation problem is only composed by 1's and 0's and each column has only two 1's in it. The matrix of coefficient of a 3-source, 3-destination transportation problem will look like this:

x_{11}	x_{12}	x_{13}	x_{21}	x_{22}	x_{23}	x_{31}	x_{32}	x_{33}	
1	1	1							$= a_1$
			1	1	1				$= a_2$
						1	1	1	$= a_3$
1			1			1			$= b_1$
	1			1			1		$= b_2$
		1			1			1	$= b_3$

The transportation problem has also a very particular property: if all the a_i's and b_j's are positive integers, then the optimal solution x_{ij} is also in integers. This is a useful property in most practical applications.

When the variables x_{ij}, in addition to satisfying the above constraints, are subject also to upper and lower bounds, u and l,

$$l_{ij} \leq x_{ij} \leq u_{ij}$$

the resulting transportation problem is called **capacitated transportation problem,** and efficient procedures also exist for its solution.

The Assignment Problem. The assignment problem is a very special type of transportation problem for which very efficient computational procedures exist. It can be stated as the problem of assigning n particular jobs among n specific workmen (or machines) in such a way that the total cost resulting from this assignment be minimized (or the total efficiency be maximized).

Specifically, the following terms are defined:

c_{ij} = cost of assigning job i to workman j $(i = 1, \ldots, n; j = 1, \ldots, n)$

$$x_{ij} = \begin{cases} 1 \text{ if job } i \text{ is assigned to workman } j \\ 0 \text{ if job } i \text{ is not assigned to workman } j \end{cases}$$

With this notation, the assignment problem is defined as follows:

$$\text{Minimize } z = \sum_{i=1}^{n} \sum_{j=1}^{n} c_{ij} x_{ij}$$

subject to:

$$\sum_{j=1}^{n} x_{ij} = 1 \qquad i = 1, \ldots, n$$

$$\sum_{i=1}^{n} x_{ij} = 1 \qquad j = 1, \ldots, n$$

$$x_{ij} \geq 0 \qquad \begin{cases} i = 1, \ldots, n \\ j = 1, \ldots, n \end{cases}$$

The first equation is the total assignment cost to be minimized.

The next two equations show that each job i has to be assigned to one and only one workman j (x_{ij} is either $+1$ or 0) and that one and only one job should be assigned to each workman j. Finally the usual nonnegativity restrictions apply to the variables involved.

The assignment problem is merely a transportation problem in which $a_i = b_j = 1$ for all the sources and destinations. This immediately implies that x_{ij} will be an integer number. (As in the transportation problem, the x_{ij} are all integral whenever a_i and b_j are integers.) In the assignment problem, the constraints allow the x_{ij} only the integer values of 0 or 1. This property makes it unnecessary to introduce that requirement into the assignment model, allowing avoidance of integer or nonlinear representations of the problem.

As stated, the assignment model assumes there are the same number of jobs as there are workmen available. This may not necessarily be the case. However, the difficulty can be solved by introducing "dummy" jobs, if the number of workmen is in excess of the number of jobs, or by introducing "dummy" workmen, if the opposite case is true.

The Generalized Transportation Problem. Ferguson and Dantzig (Management Science, vol. 3) have developed an efficient procedure to solve a class of problems they called **generalized transportation** problems.

The generalized transportation model can be formulated as a machine assignment problem where m products should be processed in n machines, each one having a different efficiency in making the various products. As usual, the allocation should be performed in such a way as to minimize the resulting cost of the assignment. Let:

a_i = number of units of product i to be produced, $i = 1, \ldots, m$
b_j = time available on machine j, $j = 1, \ldots, n$
P_{ij} = time required to process one unit of product i on machine j
$\quad (i = 1, \ldots, m; j = 1, \ldots, n)$
c_{ij} = cost of producing one unit of product i in machine j
$\quad (i = 1, \ldots, m; j = 1, \ldots, n)$
x_{ij} = number of units of product i to be made on machine j
$\quad (i = 1, \ldots, m; j = 1, \ldots, n)$

Then the generalized transportation problem can be described as follows:

$$\text{Minimize } z = \sum_{i=1}^{m} \sum_{j=1}^{n} c_{ij} x_{ij}$$

subject to:

$$\sum_{j=1}^{n} x_{ij} = a_i, \qquad i = 1, \ldots, m$$

$$\sum_{i=1}^{m} P_{ij} x_{ij} = b_j, \qquad j = 1, \ldots, n$$

$$x_{ij} \geq 0 \qquad \begin{cases} i = 1, \ldots, m \\ j = 1, \ldots, n \end{cases}$$

Obviously, the generalized transportation problem becomes a regular transportation problem whenever $P_{ij} = 1$ for all i and j.

APPLICATION TO PRODUCTION PLANNING. One of the most used applications of linear programming in production is in the planning of aggregate production. The plant and demand characteristics which make this approach most appropriate are **limited plant capacity** and **seasonal demand**. The programming problem is: find the plan of aggregate production which accumulates seasonal stock to meet the demand with the least cost of inventory and production. The primary decision variables are the amount of overtime to use each period and the allocation of the production time among the items to be produced. In addition, one may reduce regular time production during the off-peak season if increasing overtime during the peak is preferable to carrying greater inventories.

Generally, the objective is the minimization of the total cost of production (regular time and overtime) and inventory carrying during the planning period. That planning period is generally the time until the next peak season. In some activities there are two peaks, and the planning cycle needs to be carried through both of them in order to show the optimum plan of seasonal inventory accumulation.

The simplest form of this problem usually requires the following decision variables:

r_{it} = regular production hours of the ith product type in the tth period
s_{it} = overtime production hours of the ith product type in the tth period

To describe the objective function, the following cost factors are required:

f_i = cost of one regular production hour of the ith type
g_i = cost of one overtime production hour of the ith type
h_i = cost to carry one unit of the ith type in inventory for one time period
I_{it} = inventory level of the ith type at the end of the tth period

Our objective is to minimize the cost

$$C = \sum_i^n \sum_t^T (f_i r_{it} + g_i s_{it} + h_i I_{it})$$

(where there are n product types and T time periods) subject to the constraints

$$\sum_i^n r_{it} \le R_t \qquad \text{(total regular production time in the } t\text{th period)}$$

$$\sum_i^n s_{it} \le S_t \qquad \text{(total overtime production in the } t\text{th period)}$$

$$
\begin{array}{ll}
r_{it} \ge 0 & \text{all } i,\, t \\
s_{it} \ge 0 & \text{all } i,\, t \\
I_{it} \ge 0 & \text{all } i,\, t
\end{array}
\left.\begin{array}{l}\\ \\ \\ \end{array}\right\}
\begin{array}{l}
\text{Regular and overtime production hours} \\
\text{and ending inventories must be non-} \\
\text{negative}
\end{array}
$$

$$I_{it-1} + p_i(r_{it} + s_{it}) - I_{it} = D_{it}$$

where

p_i is the production rate of the ith type in units per hour
D_{it} is the requirement (demand) for the ith type (in units) during the tth period

This last constraint is a material balance equation which says the ending inventory (I_{it}) is equal to the beginning inventory (I_{it-1}) increased by the period production $[p_i(r_{it} + s_{it})]$ and decreased by the demand (D_{it}). The effect of this constraint is to assure that all demands are met by current and previous production.

The result of the optimization is the set of decision variables, the regular and overtime production to be devoted to each type during each time period, which yields the least total cost of production and inventory carrying.

The problem described here is one of the simpler forms of the production planning problem. It can readily be extended to include multiple production points and, in that way, assist in solving a complex seasonal planning and plant balancing problem, including interplant shipment costs.

Constraints can be added to the problem to deal with some extensions. For example, limits on total production could be established by product type or by group of products. Total inventory limits can be set, length of time in storage can be limited, etc.

A particular characteristic of this approach is its use as a routine monthly planning system. By updating the starting inventory position and forecasts of future requirements each month, an economic sequence of production plans is developed, adapting smoothly to forecast errors and changes in requirements estimates.

SOLUTION STAGES OF A LINEAR PROGRAMMING PROBLEM. Formulating the Model. The first step in the application of linear programming to a practical situation is the development of the linear programming model. The basic aspects of model formulation are:

1. **Definition of the decision variables.** Care and ingenuity should be applied in selecting the decision variables to describe the problem being examined. In some instances, it is possible to decrease drastically the number of constraints or to transform an apparent nonlinear problem into a linear one by merely defining in a different way the decision variables to be used in the model formulation.

2. **Selection of the objective function.** Once the decision variables are established, the objective function (measure of performance or effectiveness) to be minimized or maximized should be described in terms of the decision variables. The measure of performance provides selection criteria used to evaluate the various courses of action that are available in the situation being investigated. The most common index of performance selected in production applications is dollar value, thus defining the objective function as the minimization of cost or the maximization of profit. However, other objective functions are more relevant to the decision in some instances. Examples of alternative objectives are:

Maximize total production, in units
Minimize production time
Maximize share of market for all or some products
Maximize total sales, in dollars or units
Minimize changes of production pattern
Minimize the use of a particular scarce (or expensive) commodity

The definition of an acceptable objective function is difficult in some situations, especially when social and political problems are involved. In addition, there may be conflicting objectives, each one important in its own right, that the decision maker wants to fulfill. In these situations it is usually helpful to define multiple objective functions and to solve the problem with respect to each one of them separately, observing the values that all the objective functions assume in each solution. If no one of these solutions appears to be acceptable, additional constraints representing the minimum acceptable performance level of each of the objective functions can be introduced and the problem solved, again having as an objective the most important of the objective functions being considered. Sequential tests and sensitivity analysis can be quite valuable in obtaining satisfactory answers to this problem.

3. **Description of the constraints.** Finally, all the constraints that characterize the problem under consideration must be described in terms of linear equations in the decision variables. Often, some vital constraints are overlooked or

some errors are introduced in the initial model description, leading to unacceptable solutions. The linear programming solution, however, provides enough information to assist in the detection of these errors and their prompt correction. The problem has then to be reformulated, repeating the process just described.

Gathering the Data. Having defined the model, all the data required for its solution must be collected. The data required include the objective function coefficients, the matrix coefficients and the constraints (the so-called "right-hand side") of the linear program. This stage usually represents one of the most time consuming and costly efforts required in solving the problem.

Obtaining the Optimum Solution. Because of the lengthy calculations that are required to obtain the optimum solution of a linear program, a digital computer is invariably used in solving applied linear programming problems. In fact, almost all the computer manufacturers offer highly efficient linear programming codes which are available to be used without charge. These codes can presently handle general linear programming problems of up to 5,000 rows (constraints). Instructions to use these codes are provided by the manufacturers and vary slightly from one computer firm to another.

When dealing with large linear programs, it is best to write small computer programs to prepare the required input data and instructions. Those programs called **matrix generators**, are designed to cover specific applications. Similarly, computer codes are often written to translate the linear programming output, usually very technical in nature, into managerial reports ready to be used by middle and top managers.

Applying Sensitivity Analysis. One of the most useful characteristics of the linear programming codes is their capability to perform sensitivity analyses on the optimum solutions obtained for the problem originally formulated. These post-optimum analyses are important for several reasons:

1. **Data uncertainty.** Much of the information used in formulating a linear program is uncertain to some degree. Future production capacities and product demand, product specifications and requirements, cost and profit data, etc. are usually evaluated through projections and average patterns and cannot be known with complete accuracy. Therefore, it is often important to determine how sensitive the optimum solution is to changes in those quantities, and how the optimum solution varies when actual experience deviates from the values used in the original model.

2. **Dynamic considerations.** Even if the data were known with complete certainty, a sensitivity analysis of the optimum solution should be performed to find out how the recommended courses of action should be modified after some time, when changes have taken place in the original specification of the problem.

3. **Input errors.** The optimum solution should be tested to show the effect of errors in the original formulation of the problem. It is important to investigate changes in the objective function coefficients, in the right-hand side elements, and in the matrix coefficients. Sometimes it is important to know the impact on the objective function of introducing new variables or new constraints into the problem. Although it is often impossible to assess all of these changes simultaneously, the good linear programming codes provide several means for obtaining pertinent information about the impact of such changes or permit evaluation of new optimum solutions with a minimum of additional computational effort.

Most codes provide the following:

1. Specification of the values of those variables which are in the solution (i.e., those decision variables which have non-zero values). These are called **basic variables.**

2. Cost range information associated with each basic variable, indicating the lower and upper value that the cost coefficient of a given basic variable can take, without affecting the value of that variable in the optimum solution.
3. Marginal costs associated with each nonbasic variable (i.e., those variables which have a zero value in the optimum solution). These marginal costs represent the change in the magnitude of the objective function if the value of the nonbasic variable is changed from 0 to 1. Marginal costs are extremely useful in determining the impact in the overall system of the availability of additional resources.
4. Dual variables (or shadow prices) associated with each constraint. The shadow prices are also marginal "costs" and represent the change in the objective function as a consequence of changing a constraint by one unit.
5. Constraint range information, showing upper and lower bounds within which the constraint could vary without affecting the optimum solution.

In addition to this information, most codes allow for the use of multiple right-and sides (sets of constraints) and multiple objective functions. This permits efficient exploration of the changes in the optimum solution with changes in the available resources and the costs or profits. Some codes permit parametric changes, by specifying the resulting optimum solution when either constraints or objective function coefficients are continuously varied from a set of prescribed values.

Testing and Implementing the Solution. The solution should be fully tested to make sure the model clearly represents the real situation. Sensitivity analyses are an important part of this testing effort. Should the solution be unacceptable, new refinements have to be incorporated into the model and new solutions obtained until the linear programming model is an adequate representation of the real situation.

When testing is complete, the model can be implemented. Implementation usually means the operation of the model with real data to arrive at a decision or set of decisions. If the model is to be used continually or repetitively (e.g., in production planning or blending) as opposed to a single or one-time use (such as a plant location problem), routine operating systems should be established to provide input data to the linear programming model from other information systems and to transform the output into routine operating instructions. Care must be taken to insure that changes which take place in the real operating system are reflected in changes in the model.

CPM, Simulation, Queueing, and Other Techniques

CRITICAL PATH METHOD (CPM). Horowitz (Critical Path Scheduling) describes the critical path method as a system for **planning, scheduling, and controlling a project.** The operations involved in the project are represented graphically in a diagram called a **network.** The sequence of operations and the interrelationships of the various tasks are indicated on this network. Once the network has been completed, the planner must estimate the time it will take to do each operation. Since the network will usually consist of a number of parallel paths and some operations will be done simultaneously with others, the total time for the project will not be a sum of the estimated operations times. The key to the network is, in fact, a relatively small number of operations that do control the project's completion time. These operations, called **critical operations,** produce a path through the network called the **critical path.** It is knowledge about this critical path that gives the method its greatest value in planning and controlling the project.

Projects Suitable for CPM. In order to use CPM effectively, a project must have a definite beginning and a definite end. As noted above it must also be capable of being divided into a sequence of operations which are performed in a logical, orderly sequence. Horowitz (Critical Path Scheduling) cites construction projects as prime examples of projects lending themselves to CPM. Other types of projects suitable for CPM analysis include:

1. Setting up a new department.
2. Introducing a new product.
3. Research and development projects.
4. Engineering or architectural design.
5. Assembling a large piece of machinery or aircraft.

Steps in Using CPM. The **basic elements** of the CPM approach to project planning and scheduling are given by Horowitz as follows:

1. Analyze the project. Determine the individual tasks or operations that are required.
2. Show the sequence of these operations on a network chart.
3. Estimate the time it takes to perform each operation.
4. Compute the critical path.
5. Use this information to develop the most economical and efficient schedule.
6. Use the schedule to control and monitor job progress.
7. Revise and update the schedule frequently throughout the execution of the project.

The CPM approach can be extended to include the cost of each operation thus providing a means for determining an **optimum time** for completion of the project with a **minimum of total cost.** The method relating operation times with costs is called **Least Cost Scheduling.**

Computing the Critical Path. An example based on Horowitz (Critical Path Scheduling) will be used to demonstrate the technique and analysis required to determine a critical path. Figure 16 shows a network on which are

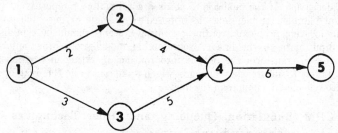

Fig. 16. Network diagram for CPM study.

shown the operations (indicated by arrows) and their duration (the number next to the arrows). The numbers in the circles, called **events,** indicate the beginning and end of each operation. An operation begins at the tail of the arrow and ends at the arrowhead. Event 1 is the beginning of the overall project and event 5 is the end of the project. Since the project will begin at time 0 the earliest start time for all operations beginning with event 1 will be 0. The earliest finish time for an operation is the earliest start time plus the time it takes to do the operation. Of course the finish times for the initial operations become the start times for the subsequent operations. The network diagram shows that operation 4-5 cannot start until two operations, 2-4 and 3-4 are completed. Since 2-4 is completed on day 6 and 3-4 is completed on day 8

| Operation | | Description | Duration | Earliest | | Latest | | Total Float |
i	j			Start	Finish	Start	Finish	
1	2		2	0	2	2	4	2
1	3		3	0	3	0	3	0
2	4		4	2	6	4	8	2
3	4		5	3	8	3	8	0
4	5		6	8	14	8	14	0

Fig. 17. Worksheet for CPM calculations.

the earliest start for operation 4-5 is day 8. Operation 4-5 takes 6 days so that its earliest finish is day 14, which is also the earliest finish date for the project represented by the network. A worksheet is used to record the results of all calculations (Fig. 17).

To find the latest times for starting each operation, it is necessary to work backwards. The object is to find the latest time an operation can start and still produce the shortest overall time for the project. Thus the earliest finish time (14) is the starting point for these calculations. To find the latest start time for operation 4-5 subtract the time it takes to do the operation. In this case, $14 - 6 = 8$. If operation 4-5 is to begin on day 8 the preceding operation must finish on day 8. Two operations precede operation 4-5. The start time for each is calculated by subtracting their duration from 8:

Operation 2-4: $8 - 4 = 4$ Operation 3-4: $8 - 5 = 3$

These are latest start times for the respective operations. Continuing back from these operations and subtracting the durations for each of the operations, the latest start time for operation 1-3 is day 0, and that for operation 1-2 is day 2.

The amount of **leeway** or **delay** that could occur in any operation without affecting the overall project time is called **float**. **Total float** is the **maximum time** that any operation can be delayed. It is the time available for an operation if all preceding operations are started as early as possible. Total float for the preceding problem is calculated and recorded in the worksheet. In each case the total float is found by subtracting the earliest from the latest finish (subtracting the earliest from the latest start will produce the same results).

To find the **critical path** the critical operations must be determined, since any delay in the critical operation will delay the project completion. Critical operations always have zero total float. From Fig. 17 it is seen that operations 1-3, 3-4, and 4-5 have no total float. The original network is redrawn, showing these operations in heavy black lines (Fig. 18). The continuous path indicated by these heavy lines comprises the critical path.

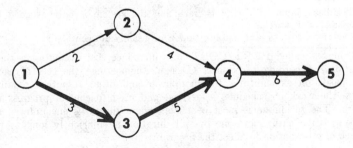

Fig. 18. The critical path.

It is possible to have more than one critical path. Paths may divide or branch out but in order to be a critical path there must be continuity from the beginning to the end of the network, that is, there must be no breaks or interruptions in the path. Every operation in the critical path must be a critical operation, but this does not necessarily imply any degree of difficulty in performing the particular operation. In fact, in most projects only about 10 to 20 percent of all operations are critical, that is, their operation directly controls the overall completion time of the project.

PROGRAM EVALUATION AND REVIEW TECHNIQUE (PERT). In using CPM it was assumed that the amount of time it took to perform an operation was known with some degree of certainty. For many projects this is a realistic approach. Experience and a tremendous amount of data in past projects provide the background for making fairly accurate time estimates.

In some projects, however, estimates of this type are difficult if not impossible to make. Research and development work, the design of new products, unusual construction projects, and other instances where new or untried techniques or materials are being used involve operations for which realistic time estimates are not possible. PERT was designed to deal with this type of project. Both PERT and CPM involve network diagrams, events, operations (called **activities** in PERT), and a critical path. The basic difference in the two, however, is the manner in which the estimated times are determined.

PERT requires estimating three times for performing an activity. The **optimistic time** is the shortest possible time in which the activity could possibly be completed. The **pessimistic time** is the longest time the activity could reasonably require. The **most likely time** is the time the activity would take most often if it were repeated over and over again under the same conditions. The three time estimates are used to determine an **expected time** which is based on the recognition of the probabilistic nature of the activities involved.

Advantages of PERT. Abramowitz (Production Management) lists the following benefits derived from the use of PERT in managerial decision making:

1. The planning required to create a valid network helps define and control a complex problem.
2. The critical path reveals interdependence and problem areas and indicates where resources are needed.
3. It allows for the orderly presentation of large amounts of data.
4. It leads to improved planning on all levels.
5. It allows for the introduction of probability reasoning in scheduling activities.

The **disadvantages** of PERT are:

1. The basic input of PERT is time, and the system is only as good as the estimates of time.
2. Factors of cost, capital, and manpower are usually not considered.

A technique called **PERT/COST** does introduce cost estimates to the time calculations. According to Vidosic (Design Engineering) the cost of each activity, based on expenditures for manpower and other resources required to perform the work on schedule, is derived and used to control progress in the network. The technique is most beneficial where the tasks encountered are of a non-repetitive nature and the overall duration is sufficiently long to permit feedback of information to affect the program.

GRAPHICAL EVALUATION AND REVIEW TECHNIQUE (GERT). A generalized network technique called GERT was developed by Pritsker (GERT: Graphical Evaluation and Review Technique) for the **analysis of stochastic networks** which have the following characteristics:

1. Each network consists of modes denoting logical operations and directed branches.
2. A branch has associated with it a probability that the activity represented by the network will be performed.
3. Other parameters describe the activities which the branches represent. Parameters may be additive (time, for example) or multiplicative (reliability, for example).
4. A realization of a network is a particular set of branches and modes which describes the network in one experiment.
5. If the time associated with a branch is a random variable, then a realization also implies that a fixed time has been selected for each branch.

Basically, GERT is a procedure for the formulation and evaluation of systems using a graphical approach to problem solving. It utilizes the following steps (Whitehouse and Pritsker, AIIE Transactions, vol. 1):

1. Convert a qualitative description of a system or problem to a model in stochastic form.
2. Collect data needed to describe the functions denoting the network branches.
3. Combine branch functions (network components) into an equivalent function (or functions) which describe the network in terms of one branch. (A FORTRAN computer program is available for performing this step.)
4. Convert the equivalent functions into performance measures for studying the system or solving the problem for which the network was formulated.
5. Make inferences based on the performance measure developed in step 4.

GERT has been applied to production problems involving inventory, reliability, queueing, and maintenance.

SIMULATION MODELS. A simulation model is one that is used to describe a system for the purpose of determining the value of the output for many alternative arrangements of the input variables so that a course of action leading to an optimum solution may be found.

The Nature of Simulation. The process of **simulation** involves developing a model that represents some aspect of a company's operations and using the model to understand the operation, predict its behavior, and prescribe courses of action to produce desired effects. The models used in business and production are characterized by the joining of two or more subsystems into an overall system.

In an actual system the inputs and outputs of the various subsystems are so interrelated that an analytic solution is very difficult and often impossible. Even in cases where the problem lends itself to representation with mathematical models there may be no known way of determining the optimum solution. Simulation calls for an **experimental approach** involving numerous trials. The advent of the high-speed electronic computer, however, makes the massive number of repetitive computations required feasible.

An Example of Simulation. Skeith, Curry, and Hairston (Industrial Engineering, vol. 1) describe a machine interference problem studied through simulation. The simulation involves a repairman servicing 40 machines. The machine running times between breakdowns are considered exponentially distributed with a mean time for all machines of 3,200 units (each unit being 0.1 min.). Machines are repaired on a first-in-first-out basis; repair times for all machines can be described by the exponential distribution with a mean time of 60 units. A computer simulation program of 18 statements serves to represent the primary characteristics of most machine interference problems. These characteristics cover the following.

1. The limitations on the total number of possible events that may occur in the system represented (in this case, 40).
2. The arrival rate representing the machine breakdown rate. A computer statement serves to delay each machine a specific time interval according to the exponential distribution. The machine arrival rate itself is a Poisson distribution.
3. The limitations on number of service channels. In this case it is one, representing the single repairman.
4. The service rate, or the time it takes the repairman to service the machine and return it to its previous condition. The time of each repair is exponentially distributed with a mean time of 60 units.
5. The order of service (in this case first-in-first-out, meaning the machines join the queue and depart from it in the same order).

The results of the simulation problem give the average percentage of machines operating, average number of machines waiting for repair, maximum number of machines waiting for repair, average time to become operational after breakdown, percentage utilization of repairman, and the probability of a machine having to wait for service.

With minor modifications the program can be used to analyze other models, for example, a last-in-first-out order of repair, a random selection of machines to be repaired, two repairmen with different work rates, or two machines classes and one repairman. The computer language used in the above simulation was General Purpose Systems Simulator (GPSS).

MONTE CARLO METHOD. The term "Monte Carlo method" has come to be applied frequently to the use of simulations of gaming methods in operations research. In many problems, an analytic model may be too complex to manipulate. It may be possible, however, to construct a **numerical model**, with tabulated probability distributions, stated basic functional relationships, specified transition probabilities (probability of making a change from one condition to another), and the like. A large number of trials of operating "on paper," with random draws from probability distributions, or computer generated random numbers repeated with functions and distributions adjusted for changes in controllable parameters, can be used to trace out the dependence of outcome on parameter values.

A Problem in Manpower Assignment. Moore (Manufacturing Management) cites the following example of a problem lending itself to Monte Carlo simulation. In a small manufacturing plant, setup men do machine repairing in addition to setting up jobs. Machine running times, setup times, and repair times vary from machine to machine and are as given in Fig. 19. There are 10 machines. Management wants to know how many men to assign to the setup/

Length of Job		Setup or Repair Time		Likelihood of Repair During Job	
Hours	Per Cent of Instances	Hours	Per Cent of Instances	Length of Job Hours	Probability of Repair Per Cent
4	20	2	50	4	5
8	30	4	20	8	5
16	20	8	15	16	5
24	20	12	10	24	10
40	10	20	5	40	10
				No repair	65

Fig. 19. **Running, setup, and repair times for ten machines in Monte Carlo simulation.**

repair job. Since the Monte Carlo method depends on trials, or random selection, a means must be devised to effect this randomness. A **table of random numbers** can be used for this purpose (see section on Statistical Methods). The following system will be used to select job lengths randomly from the table: numbers 01 through 20 will represent a job length of 4 hr.; 21 through 50, 8 hr.; 51 through 70, 16 hr.; 71 through 90, 24 hr.; and 91 through 00, 40 hr. The solution is begun by assuming the jobs have all been set up. The length of job for each machine must now be determined. To do this the table is entered randomly and ten consecutive two-digit numbers are read. Each number represents the time it takes to do a job on each of the ten machines according to the previously established system.

The random table is again used to determine likelihood of repair. Here numbers 01 through 05 represent 4 hr., 06 through 10, 8 hr.; 11 through 15, 16 hr.; etc. Ten 2-digit numbers are selected and each is compared with the length of the job for each machine. If the two times match, then breakdown occurs while the machine is operating. Another system is needed to represent setup or repair times in the table of random numbers. In this case 01 through 50 can represent 2 hr., 51 through 70, 4 hr.; and so on. Random number selection for the ten machines completes the first run. The machines are then put through another run beginning with setup times, then length of jobs, and repair time (if any). The data for each run is recorded. The runs are repeated until each machine has accumulated 160 hr. of time. A bar graph is developed to show the results for each machine (Fig. 20). The graph shows the results beginning

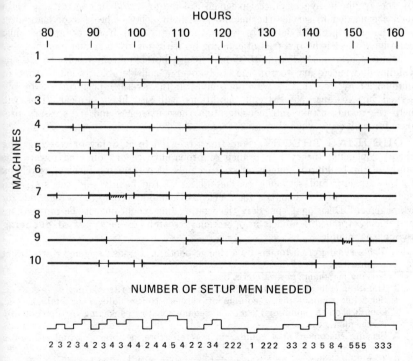

Fig. 20. Machine times based on Monte Carlo simulation. Heavy lines indicate machine running time, thin lines represent setup time, and shaded lines indicate repair time.

with the third week (80 hr.) and extending to the end of the test (160 hr.) This was done so that the initial start-up process (at time 0) would not unduly influence the steady-state figures. From the 40 two-hour blocks of time the following manpower requirements are indicated:

No. of Men	Number of Instances
1	1
2	13
3	11
4	9
5	5
6	0
7	0
8	1

The manpower requirement based on this table is approximately 3.25 men at all times so that four setup men will be used. If the setup men do no other work except machine repair and setup jobs there will be the equivalent of three-fourths of a man idle all the time.

Note that the quantitative part of the Monte Carlo technique ended with the calculation of the required manpower. The manager, however, is left with the task of analyzing the effects of the decision indicated. For example, in the above problem, if four men are used to service the ten machines there will be considerable idle machine time as indicated by the bar graph. As shown also by the graph, in five instances five men are required and in one instance eight men are needed to service the machines without delay. The lack of manpower will therefore produce delays in repairs and setup and from the analysis just completed there is no prospect of catching up with this lost machine time.

Monte Carlo method is quite simple to use but to obtain results with some validity it is necessary to run the study over a considerable period of time or number of cycles. In this way not only will the average performance be observed but instances of extreme conditions will also likely appear. This will help the manager take into consideration these extremes and if necessary prepare for such cases when they arise.

QUEUEING THEORY. Sometimes referred to as **delay** or **waiting line** theory, queueing theory is a branch of probability theory directed toward the study of the buildup of queues or waiting lines at a servicing facility, as related to the capacity and servicing characteristics of the facility and the statistical characteristics of the demands for service. Richmond (Operations Research for Management Decisions) considers the waiting line model one of the most useful and widely applicable models in operations research. Some **types of queueing processes** include:

1. The servicing of customers by a change booth in a transit system or at a ticket window.
2. Serving customers in a cafeteria line.
3. Handling calls on a telephone exchange or over a set of trunk lines.
4. Servicing semiautomatic equipment subject to breakdown or stalling, e.g., spindles on a spinning loom or semiautomatic packaging equipment, by servicemen.
5. Serving customers at a counter in a retail store, a library, a tool crib, or a stockroom.
6. Handling shipments on a transportation system with limited facilities.
7. Handling papers in a clerical station or operation, such as processing orders.
8. Setting up a balanced assembly line, where the rates of output of individual

stations may be subject to fluctuation due to equipment breakdown or quality rejection.

Basic Characteristics of Queueing. Queueing theory is not a single set of mathematical formulas but an expanding collection of methods and concepts based on a variety of assumptions. Some of the basic characteristics which may vary from problem to problem include:

1. The **size of the group** being serviced, whether it is finite or "infinite," in effect whether it is small enough so that the size of the waiting line influences the rate of demand for new calls.
2. Whether the "customers" or elements requiring service are "patient" or not; i.e., whether they will stay in line indefinitely or not.
3. The distribution of holding or **servicing times**, the two most commonly encountered being the constant and exponential servicing time assumptions. Under the latter, the probability a unit requires service lasting longer than a time t is given by e^{-at}, where $1/a$ is the average service time. A distribution of servicing times approximating this distribution has been encountered in a wide variety of circumstances, such as times to service equipment, length of telephone calls, time for a truck to deliver an order and return, or time for a clerk to wait on a customer, including answering inquiries, at a retail counter. One (not the only) set of physical conditions generating this distribution would be where the amount of time previously spent in service would have no bearing on the amount remaining to be done. It has been remarked that a woman in a comfortable telephone booth is the ideal example of this condition.
4. The **characteristics of arrivals**, i.e., whether demands for service are generated in a uniform, regular pattern, or at random around some mean rate.
5. The number of servicing units.

Servicing Defective Parts—A Single-Channel System. A single-channel servicing system is characterized by a number of elements arriving at a service facility, waiting in a line, if necessary, receiving service, and leaving the facility. The **number of elements in the system** is the sum of the elements in the waiting line plus the number of elements being serviced. As an example of a queueing problem in production consider the following: The inspection department of a parts manufacturer detects on the average of 2 defective parts per hour. These parts are turned over to a special service man who corrects the defect and forwards the part to finished products. The service man is able to correct on the average of 3 parts per hour. Management is interested in predicting the behavior of this system. The arrival of defective parts at the service station and the length of time for adjusting each part are random variables. Traditionally the random arrivals are described by the Poisson distribution and the servicing times by the exponential distribution. The average arrival time rate, λ, means that, on the average, λ units per unit time will be expected to arrive at the service station. In the above problem $\lambda = 2$. The service time, μ, is the number of units that can be serviced in a unit time and sent out of the system. In the above problem $\mu = 3$. Single-channel system analysis is feasible only if $\dfrac{\lambda}{\mu} < 1$. In the above problem $\dfrac{\lambda}{\mu} = \frac{2}{3}$.

The average number of parts in the system, $\bar{n}$, is given by the formula

$$\bar{n} = \frac{\dfrac{\lambda}{\mu}}{1 - \dfrac{\lambda}{\mu}}$$

$$\bar{n} = \frac{\frac{2}{3}}{1 - \frac{2}{3}} = 2$$

Thus 2 parts will be in the system on the average at any given time.

The average number of parts waiting to be serviced at any given time, n_q is found from the formula:

$$\bar{n}_q = \frac{\lambda}{\mu}\,\bar{n}$$

$$= \frac{2}{3} \cdot 2 = 1\tfrac{1}{3} \text{ parts}$$

The average time spent by a part in the system, W, is

$$W = \frac{1}{\mu - \lambda}$$

$$= \frac{1}{3 - 2} = 1 \text{ hour}$$

The average wait before a part is serviced, W_q, is

$$W_q = \frac{\lambda}{\mu} \cdot W$$

$$= \frac{2}{3} \cdot 1 = \frac{2}{3} \text{ hour or 40 minutes}$$

Percentage of time that the service man is idle, $P(0)$, that is, no parts have been turned over to him for servicing, is given by

$$P(0) = 1 - \frac{\lambda}{\mu}$$

$$= 1 - \frac{2}{3} = \frac{1}{3} \text{ or } 0.333$$

The percentage of time in which exactly one part will be in the system, $P(1)$ is

$$P(1) = \left(\frac{\lambda}{\mu}\right)\left(1 - \frac{\lambda}{\mu}\right)$$

$$= \left(\frac{2}{3}\right)\left(\frac{1}{3}\right) = \frac{2}{9} \text{ or } 0.222$$

The percentage of time in which exactly two parts will be in the system, $P(2)$, is

$$P(2) = \left(\frac{\lambda}{\mu}\right)^2\left(1 - \frac{\lambda}{\mu}\right)$$

$$= \left(\frac{4}{9}\right)\left(\frac{1}{3}\right) = \frac{4}{27} \text{ or } 0.148$$

It is apparent that the ratio $\dfrac{\lambda}{\mu}$ is very important in the single-channel system analysis. This ratio, sometimes called the **utilization factor,** is given the symbol ρ and expressed in units called "Erlangs" in honor of A. K. Erlang who did pioneering work in queueing theory.

As λ approaches μ the average number of units in the system increases rapidly, as does the average time spent by a unit in the system (which includes waiting as well as servicing times, as noted previously).

The relationship between $\bar{n}$ and the utilization factor is shown in Fig. 21 (Richmond: Operations Research for Management Decisions). Notice the ex-

treme sensitivity of the system as a result of small changes in the average arrival
rate or the servicing rate, when the two values are almost equal.

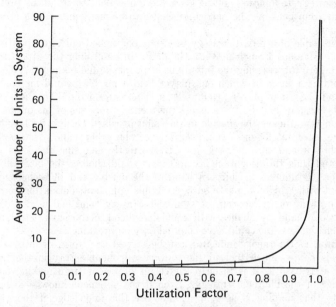

Fig. 21. Relationship between average number of units in single-channel
system and utilization factor.

GAME THEORY. Game theory, first discussed by J. von Neumann and
O. Morgenstern (The Theory of Games and Economic Behavior) is the analysis
of **choice of strategies** in a competitive situation. Problems arising in competitive games, warfare, and business competition are equally open to analysis
using this general theory of strategy. Whereas probability theory received early
stimulus from problems due to chance in games and gambling, game theory
assumes that the chance elements of the contest, and the rules of probability
governing them, are understood. The essential problem in game theory is to
choose a strategy in the face of a conscious antagonist which will in some sense
be "optimum."

A basic assumption underlying game theory is that the antagonist is intelligent
and equally able to choose a "good" strategy. Part of the objective in choosing
a strategy, therefore, is to choose it in such a way that the opponent cannot tell
what the complete strategy is, if this can help him; and a "good" strategy is
defined by the theory as one which will permit the player to win at least a specified amount—or, alternatively, to lose no more than a specified amount—no
matter what the opponent's strategy may be. Three important concepts of the
theory flow from this: the "mixed" strategy, the role of bluffing, and the "minimax principle."

Basic Concepts of Game Theory. A **mixed strategy** is one in which some
or all of the steps taken are chosen by chance. The purpose of such a strategy
is to prevent the opponent from using knowledge he may have, to employ some
damaging counterstrategy. For example, in matching pennies, the accepted

strategy is to toss the coin—i.e., for both contestants to select heads or tails with equal probability. Von Neumann and Morgenstern show that this is the appropriate strategy to follow—not to keep the game "honest," but to prevent the opponent from guessing the particular choice to be made and taking appropriate counteraction.

The principle of a mixed strategy need not be applied only in repetitive situations. The famous French novelist Gaboriau, in describing the capacities of his detective-hero for assessing the intentions of people, noted his remarkable abilities as a child at a game in which one player held a marble behind his back in one hand, and the other player attempted to choose this hand. The hero was successful because of his ability to assess the mental processes of his opponent: the simpleton would leave the marble in the hand in which he first held it; the next higher level of intelligence would shift it to the other hand; the next higher would shift it and shift it back, etc. By correctly assessing the level of intelligence and thus the deception his opponent could choose, the boy was able to determine the number of shifts of hand of the marble and thus choose the correct hand with a high degree of success. If his opponent, no matter how simple, had been aware of the concept of a "mixed strategy" and had used a device for choosing one hand by chance with equal likelihood, Gaboriau's detective would have been helpless—he could have counted only on breaking even.

Bluffing, as the theory indicates, can be viewed as a form of mixed strategy. Its role is not so much to lead the unwary opponent astray in the particular occasion it is used as it is to keep the clever opponent from making use of his knowledge of the bluffer's strategy. Even if the opponent knows—as he should find out—that bluffing is used, he cannot use this knowledge if the bluffing is properly done, since he does not know in any particular case whether a bluff is on or not.

The "good" strategy in game theory is the one based on the **minimax principle**; that is, the one which makes the player's maximum expected losses a minimum. Employing this principle, game theory leads essentially to a conservative strategy. It has been criticized, on occasion, as being too conservative, as not permitting the strategist to take advantage of opponents' mistakes. The answer is, of course, that one never knows when the "mistake" may be the bluff of a clever opponent.

Game theory to date is still largely a field of pure theory; applications have been very sparse. Only very **limited applications** have been made to, e.g., some simple military combat problems. Limited application is due to the unmanageable complexities that arise in contests more extensive than the "two-man, zero-sum" variety (two opponents, the losses of one equalling the winnings of the other). Some applications have been attempted in cases of planning in the face of uncertainty, using the concept of a "game against nature" (i.e., taking "nature" as the opponent with future uncertainty as its strategy). However, the applicability of game theory here, the validity of the concept of "nature" as an intelligent, malignant force, is open to question.

Management of Operations Research

AREAS FOR MANAGEMENT DECISION. Development and management of an operations research unit in business involves decisions and action in three areas:

1. Selection of a place to begin.
2. Selection of personnel to do the work.
3. Establishment of a permanent form and program for the unit.

Experience indicates that there are no pat answers to questions in these areas; they must be answered by each firm with particular reference to its own circumstances and needs. Timing of decisions within these three areas will also vary from company to company. Some companies begin with selection of the problem area, and follow this with organization of the group to attack it; others first find the personnel, then the most likely way for them to begin.

SELECTION OF PROBLEM AREA. Two bases are generally used in the selection of initial problem areas. In some companies the areas selected have been **troublesome** areas in which conventional techniques have failed to provide adequate help, resulting in a desire on the part of the company for additional help and a fresh attack. In other companies, where interest in operations research has stimulated a desire to test out its value without particular reference to an immediately pressing problem area, search has been made for **manageable problems,** where there exists a possibility of testing and evaluating the solutions within a reasonably short period of time.

These have been typically complex problems to which the answers were neither obvious nor trivial, and for which present aids and guides were substantially incomplete. The problems have been in important areas where the investment in knowledge could be expected to pay off manyfold through use in later assignments. A study of retail store operating costs for a company that maintains an extensive retail chain yielded useful measures for control of operating costs and for spotting developing weaknesses in operations of individual units. The resulting knowledge about the operations and associated costs provided part of a significant and far-reaching analysis of chain merchandising policies.

Review of industrial problems where operations research has been successful indicates one characteristic of overwhelming importance, **executive interest and cooperation.** The area and specific question chosen must be one on which the executive wants and is prepared to take action, and where the problem is significant enough to him for him to want assistance and to be willing at least to listen to the research unit's findings. Figure 22 based on the Turban survey (Industrial Engineering, vol. 1) shows the distribution of projects among the three basic organizational levels.

Problems need not be well defined and quantitative—at least at first—to be useful subjects for operations research. On the other hand, it is important that the executive and the team have some common and clear understanding of how the research can lead to decisions, what kinds of decisions can be made, and what limitations are imposed by resources or higher policies, in order to prevent the research unit's work from drifting into "blue sky" or "study" projects.

Situation for Initial Research. Within these general qualifications, the following characteristics are important, particularly in initial work:

1. **Opportunity for decision.**
2. **Possibility of quantitative study and measurement.** The area should be at least in part quantitative and should lend itself to quantitative interpretation and description. At least some aspects of the problem should be measurable, and at least a minimum of data related to the operations should be available. A preliminary study to provide bases for the prediction of the acceptance of fabric styles was quickly dropped in one case because of the inability to construct within a reasonable period an adequate quantitative description of the complexities of fabric, style, pattern, and color.
3. **Chance for experimentation and data collection.** Disruption of operations for special experiments is by no means always necessary, but opportunity for collection of special data is frequently required. In one problem, analysis of accounts receivable extending back over two years yielded the key to a knotty

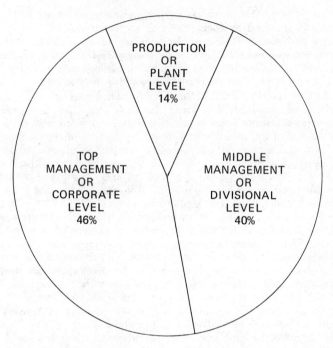

Fig. 22. **Distribution of operations research projects among the three basic organizational levels.**

marketing problem. In another case, an initial study of maintenance problems was found to be uneconomical because of lack of available records showing maintenance and breakdown histories on equipment to be studied. Recommendations were made, however, for the collection of necessary data to provide a basis for investigation at a later time when information of a quantitative nature was available.

4. **Ready evaluation of results.** Neither the analyst nor the most enthusiastic executive can expect operations research activities to be supported on the basis of faith alone. Ready ability to measure effects of changes provides bases for evaluating the research, in order to build up an attitude of mutual confidence. These problems, furthermore, have not been so large in scope as to be utterly indefinite. There should be some aspect which can be tackled readily, an understanding of which would be a contribution to the development of an approach to the whole area under investigation.

The final selection of the specific problem to be studied is best made in cooperation with the research team. Executives have found it useful to map out the area and types of problems in advance; the research group can then comment on those aspects most amenable to study, indicating where the problem can be clearly formulated, and where progress can be expected with reasonable effort. A problem can then be selected which meets the requirements of both the executive (for importance and use) and the research group (for suitability of existing data for quantitative study).

ACHIEVING SUCCESS IN OPERATIONS RESEARCH. Much frustration and dissatisfaction can be avoided if the research team and the executive keep in mind each other's needs. While it is necessary for the research team to arrive at a sufficiently understandable statement of the problem and method of attack to provide the executive with confidence in giving initial support, the executive must also recognize that this activity is research, and part of any research program is typically the **explicit formulation of the problem.** Advance specification of a detailed program, scope, and goals is frequently difficult and usually meaningless.

The executive dealing with an operations research unit will obtain value from it in proportion to the discretion he can use in controlling the group's activities. On one hand, he must recognize that the unit's work is a form of research, and research takes time. Too much insistence on "progress reports" or short-term progress can stifle the group's work. On the other hand, the executive has a responsibility to himself and to the unit to exercise enough control over its activities to be sure first, that its work is practically oriented, and second, that the unit proves its value in the application of its results. Steps in the successful **executive control** of operations research, like any research, are agreement on an initial area of investigation, provision for the group to have access to the data and the people to help formulate the specific problems to be studied, and continued contact to guide and redirect the work along the lines of greatest value as the work develops.

PERSONNEL AND ORGANIZATION. Operations research, as the foregoing discussion indicates, is a technical field. To be used successfully, it requires persons with some experience in, and training for, experimental research. It is not simply applying mathematical techniques to clearly defined business problems; the investigative and experimental aspects are vital. Individuals with the training needed may be obtained from universities or they may be recruited internally, for example, from the physical research laboratory. In either case, it is important to look first to the **experimental research capacities** and **training** of the individuals rather than to their knowledge of the business or business experience in general.

Most operations research departments are very small. A professional staff of five, supplemented by about one or two clerical or secretarial staff members, is considered average. The professional staff is characteristically young and highly educated. Most researchers are under 35 and more than half have advanced degrees. More than three-quarters of the managers of operations research departments have advanced degrees. At the bachelor's degree level studies in engineering, mathematics, and statistics dominate. At the advanced degree level business and management science studies dominate. The operations research department is sometimes used as a training ground for top management and staff positions. For this reason and others there is a high turnover of personnel in the department.

Access to Management. Operations research does have some claim to executive attention. In particular, it must be established at or report to some executive level where, within its area of work, it is in a position to:

1. Talk directly with the part of operating management concerned about problems it may have or the unit may see.
2. Get access to data without generating interdivisional jealousy.
3. Work directly with operating management on the use of results.

Direct contact between people with operating questions and problems and the operations research unit is essential. The unit cannot get its questions secondhand. Often questions are not framed initially in the right terms. Only by give-and-take between the unit and the operating man concerned can the problem be clarified or stated in the right terms for study. If successful, the operations research unit should be given an opportunity to expand the area of its studies as it proves its case, and should not be shut off from or restricted to specific and limited operating areas. Furthermore, the group needs the interest and encouragement of top management if it is to achieve its maximum level of usefulness within the company. Turban's survey (Industrial Engineering, vol 1) shows that in a cross section of some of the largest U.S. corporations most of the operations research departments report directly to a vice president and in some cases, the director of the department is a vice president himself.

LONG-TERM PROGRAM. If a company undertakes operations research, it should generally do so with the intention of continuing and expanding its use. The investment in knowledge and method, no matter how low the initial cost, is too valuable to be thrown away. The problems first recognized and attacked have, experience has shown, generally not been the most productive problems the group could study. An industrial company should undertake operations research, therefore, with a full recognition of the implications—that when operations research is properly carried out it is not a toy or an idle pastime, and that it is hardly suited to sporadic, offhand use.

Considerations governing choice of problems in the long run are not too different from those governing the initial few problems. Certainly, the **basic criterion** that the question be in an area where management can and will take action of some kind (even if only objecting to higher management) remains important. There is, perhaps, less need to concentrate on problems with an immediate possibility of measurable return, and there is more reason as the unit proves itself to gamble part of the unit's time on some difficult assignments where there is a good-chance of inconclusive results.

The **value** of an operations research unit will depend in part on its ability to answer questions raised by management, and in part on its ability to formulate and raise new questions of its own. These questions can be found and properly formulated only if the group has some time to spend on projects of its own which it need not justify. On the other hand, a full diet of such problems could easily turn a unit into a "study" group, adding nothing to the company but volumes in its archives. Specific-application projects (many of which should come out of the longer-range studies) are important to help the group justify its work by directly measurable results, to help it keep its whole program practically oriented by requiring it to take partial responsibility for getting its findings applied, and to keep it in touch with the practical workings of the business. In a long-run program, a balance of about one-third of the group's total effort devoted to **longer-range projects** and two-thirds to **specific-application projects** appears desirable.

Evaluation of Operations Research

BASIS FOR SOUND DECISIONS. Case histories show that operations research provides a basis for arriving at an **integrated and objective analysis** of operating problems. An integrated, broad view is fundamental to the method, and one characteristic of operations research studies is a tendency toward expansion in viewpoint during the investigation. Objectivity is likewise fundamental to the method, partly because the mathematical force of the model and techniques limits opportunity for biases as a result of personal viewpoints. Moreover, since the method is experimental, the views and biases which may be introduced must check with fact. The end-products of operations research studies are quantitative results. These provide an opportunity for sound estimates in quantitative terms of requirements, objectives, and goals, and the basis for more precise planning and decision-making, leading to more nearly optimal action.

Contributions from Existing Data. The contributions of operations research to **business analysis and planning** have been important and substantial. One of the most valuable of these has been the application of organized thinking to data already existing within a company. Frequently a major operations research contribution has been the location, collection, and ordering of existing data scattered through widely separated branches of the company. In one study, an operations research team found fundamentally the same problem cropping up under various guises in a number of different parts of the company. Each division or section had its own point of view toward the problem, and each had data and significant information to contribute. Unfortunately, however, even in this company, well known for its sound and progressive management, failure to recognize the fundamental and pervasive character of this problem and to draw together data existing in diverse parts of the organization had up to then prevented a unified and comprehensive attack upon it.

Contributions Through New Methods. Another contribution has been the introduction of new concepts and methods of analysis. Some of these concepts, such as information theory, control theory, and certain aspects of statistical mechanics, can be carried over from other fields. The physical sciences, and in particular modern physics, have been a very fruitful source of transplanted analytical techniques. The **conscious search** characteristic of scientific research for formalization and generalization of concepts and techniques has led to the development of concepts and methods specific to operations research. The notion of **measure of effectiveness** is simple enough, but the introduction of this concept has often been exceedingly helpful in uncovering and eliminating policy conflicts and inconsistencies in such problems, for example, as production scheduling, or the establishment of incentive compensation plans. The theories of search and of allocation of effort were conscious generalizations of specific military research studies, and the translation of these theories into industrial terms is under way. Within industrial operations research, theories of clerical organization and consumer behavior and inventory control systems illustrate the opportunity for the development of powerful tools for attacking important business problems. The opportunity to explore the effects of alternate decisions, in advance of commitment, through the concept of the model, provides clarification of the interdependencies among variables and the connections between actions and results. This explicit statement of the connection between action and result, together with the quantitative prediction of outcome, is the source of the power of operations research methods in organizing bases for executive action.

LIMITATIONS OF OPERATIONS RESEARCH. Operations research is not a cure-all for every business ill, nor is it a source of automatic decisions. As a quantitative field of analysis, it is limited to the study of tangible, measurable factors; yet many important factors affecting business decisions remain intangible or qualitative. The responsibility rests with the operations research group, therefore, to frame its analysis and statement of conclusions in recognition of this limit. On the other hand, the executive must be prepared to adjust or modify the conclusions drawn from the quantitative analysis to adjust for the impact of qualitative, nonmeasurable factors. The limitation of operations research to quantitative and measurable concepts, however, has often forced successful efforts to find ways and means of measuring factors which had been assumed or presumed to be "intangible" or qualitative.

Executive and Research Roles. The need for decisions and decision-making is by no means removed by operations research. The **research responsibility** is one of analysis: to use existing or experimental factual data to the extent possible; to indicate the assumptions underlying analysis; to indicate the implications of alternative assumptions; to interpret the results of the analysis as they bear on the necessary decisions. The **executive responsibility** remains: to examine the assumptions and evaluate them; to consider factors which may have been neglected, or questions of policy which may have been assumed; to reach, and take responsibility for, decisions.

The distinction between the analytical and executive responsibility is clearly brought out by the following example. A series of conferences were called at a plant to implement the results of a long and major operations research investigation. The results of this study were based in part on the assumption that the volume of output of the plant in question could be increased substantially at the existing level of efficiency. The executive responsible for the operation of the plant agreed with this assumption and felt that it could be matched in practice. The official responsible for the ultimate decision, however, expressed the fear that increases in volume would lead to slackening of control and resulting efficiency losses. He decided to follow a more conservative modification of the course of action suggested by the study—a modification based primarily on his estimate of the psychology of the plant personnel as affected by changes in the level of plant operations.

Management Control of Operations Research. The characterization of operations research as an applied science implies certain limitations on its usefulness. Analyses are always incomplete, with results following from successive attacks and successive approximations. Each level of approximation has its own value and limits, but this sequential method of attack and **successive approximation** creates problems in control for the executive in determining whether extension of the study is practical, or whether limitation of the study will smother potentially useful results. The characterization of operations research as scientific, rather than expert, opinion means that while certain advantages may exist in the power and usefulness of results, corresponding disadvantages arise in the time required to achieve useful conclusions—more time than is required for normal engineering analyses.

As an **applied science,** the work is torn between two objectives. As "applied" it strives for practical and useful work; as "science" it seeks increasing understanding of the operation, even when the usefulness of this information is not immediately clear. The executive who plans to support research work of this character must be fairly warned of the duality in application versus research and of the need for restraint in controlling the research or analytical activities. The natural tendency to require that the studies or analyses be "practical" can result in loss of substantial benefits, if enforced too rigidly. Finally, as a form of research, the results of studies of this type are necessarily somewhat speculative. When operations research is purchased, neither the specific program to be followed, the precise questions to be answered, nor the successful achievement of results can be guaranteed. Recognition of this difference between the nature of operations research and that of more conventional engineering methods is essential to the satisfaction of both the controlling executive and the research analyst.

PROCESS CHARTS

CONTENTS

PROCESS CHARTS

Nature of Process Charts

DEFINITION. A process chart is defined (Operation and Flow Process Charts, ASME, Standard 101) as a "graphic representation of events and information pertaining thereto occurring during a series of actions or operations." Such charts display in a compact form a great deal of information regarding such activities or operations as those involved in manufacturing a part in a plant or in processing forms in an office. F. B. and L. M. Gilbreth, who developed the process chart technique, state (ASME Transactions, vol. 43):

The process chart lends itself equally well to the routines of production, selling, accounting, and finance. It presents both simple and complicated problems easily and successfully; it provides records that are comparable; it assists in solving problems of notification and interdepartmental discrepancies, and it makes possible the more efficient utilization of similarities in different kinds of work and in the transfer of skill.

As for the types of people in the organization who can use process charting, Barnes (Motion and Time Study) states that, "The process chart may be made profitably by almost any one in an organization. The foreman, supervisor, process and layout engineer, as well as the industrial engineer should be familiar with the process chart and should be able to use it."

PURPOSE. Process charts can present a picture of a given process so clearly that an understanding of its every step may be gained by those who study the chart. Krick (Methods Engineering) states that, "For purposes of specification, information recording and presentation, visualization, explanation, improvement of method, and the like, the methods design process is supplemented by a number of special techniques for description and communication of work methods." Process charting is such a technique and only one tool among many available for the study of present and proposed industrial and business processes.

Like statistical methods or charting and graphing procedures, process charts may be effective in analysis and may help in detecting inefficiencies. Their special merit is that they encourage an overall, yet analytical, **view of processes.** They do not provide any final answers in themselves, however. These are to be found in such **fundamental investigatory methods** as are described for the use of the individual analyst in the section on Motion and Methods Study and for the use of groups analyzing and improving methods in the section on Work Simplification.

TYPES OF PROCESS CHARTS. The process charts discussed in this section include **operation and flow process charts,** operator process charts, and multiple activity charts. The flow process chart can be a man, product, or material chart. Flow diagrams and procedure flow charts are also discussed. More detailed analysis charts such as Simo and Therblig charts are covered in the section on Motion and Methods Study. While the major process charts presented here have often been subdivided into many specialized types, Shaw (The

Purpose and Practice of Motion Study) recommends that the detailed type of chart should not be followed rigidly, since,

. . . they are all fundamentally process charts and the same theory and method of analysis applies to all. If each chart is made with the definite object of "visualizing a process as a means of improving it," it will be built up so that the controlling factors of the work are brought into prominence by the form of the chart. The result may resemble other charts and fall roughly into one of these categories or it may be something quite new and original though appropriate to the presentation of the facts it records. It is unwise to set out to make a chart conforming to a particular type. A suitable shape and form should emerge in each case if a careful attempt is made to set out the material logically and vividly.

Repetitiveness of Work. Any work performance may be classified in one of three ways:

1. Work (or an operation) which is **repetitive,** that is, repeated over and over on identical parts, forms, or other units.
2. Work which **varies slightly** from cycle to cycle, such as the work of stock handlers, inspectors, lead men, or clerical workers.
3. Work which is **nonrepetitive.** Supervisory personnel and those doing creative and imaginative work are performing nonrepetitive activities. Other types of such activities include maintenance and the work of production control personnel, foremen, or engineers.

Selecting the Chart or Analysis Technique. The process chart or technique to be used depends on four factors:

1. Whether an analysis of sequence or of methods is desired.
2. Characteristics of the process or work situation.
3. The repetitiveness of the work.
4. The procedure for collecting the information.

These factors are shown with their interrelationships in Fig. 1. Many of the chart techniques are overlapping. Some of them vary on the basis of how much detail is obtained. In some situations, the selection of the chart depends upon the detail required and the time allotted to the project. All the techniques in Fig. 1 are not used for every problem. They are like the tools in a carpenter's kit. The carpenter knows how to use each of his tools, but it is his judgment which determines the tool to be used for a particular problem. An analyst, who is familiar with each chart and knows where and how it can be used best, should select the charts to be used for a problem according to his judgment of the situation.

Basic Process Charts

OPERATION PROCESS CHARTS. The basic process chart, called an operation process chart, is defined (ASME Standard 101) as "a graphic representation of the points at which materials are introduced into the process, and of the sequence of inspections and all operations except those involved in materials handling. It includes information considered desirable for analysis such as time required and location." A typical operation process chart is shown in Fig. 2.

Use of Charts. As step-by-step accounts of exactly what is done to materials, operation process charts can be made properly only by actually observing what transpires and listing the details in the order in which they occur. They graphically portray every raw material which is purchased and each operation and inspection performed in a productive process. Such charts visualize the entire procession of a single product from beginning to end. They show the

rocess Chart Technique	When Used (Characteristics)	Repetitive-ness	Procedure for Gathering Data
	Analysis of Sequence		
ation process art	When information about process operations or inspections is desired.	R, V	O, D
erial or product w process chart	When information about the process used to manufacture or work with a material or product is desired.	R, V	O, D
edure flow chart	When information about the work with, and handling of, forms or other paper work is desired.	R, some-times V	D, some-times O
	Analysis of Method		
flow process art	Man goes from one workplace to another to accomplish his operation.	R, V	O, D, some-times M
t- and left-hand operation chart	Work done at one workplace without time-controlling equipment.	R, V, some-times NR	O, D, M
-machine chart	Work with one or more time-controlling pieces of equipment.	R, V, some-times NR	O, M, some-times D
ti-man chart	Two or more men working coordinately, with or without time-controlling equipment.	R, some-times V	M best, sometimes O, D
rblig chart	Work done at one workplace. Less time and money available for analysis than needed for Simo chart.	R, V	O, D
chart	Work done at one workplace. Usually long-run, skilled, or finger operations.	R	M

R = Repetitive. D = Discussion.
V = Slightly varied cycles. M = Motion pictures.
O = Observation. NR = Nonrepetitive.

Fig. 1. Factors involved in the selection of process chart technique.

operations to be performed on each part in their proper sequence as well as the order of part fabrication and assembly. In addition they indicate the complexity of components, the nature of the materials used, and the point at which each material, part, or subassembly enters the process. They may be used by accounting and purchasing to check on raw material costs and to schedule purchases.

Operation process charts can serve as **labor cost estimate sheets** or as checklists for **facilities needed,** such as workspaces and machines. (See the description of the use of these charts in the sections on Production Planning; Production Control; and Plant Layout and Facilities Planning.) They are helpful in planning a **new product** and coordinating the efforts of those involved in putting it into production. These charts can be made on drop curtains, projection slides, or display boards for use in conference rooms where **project teams** may work on

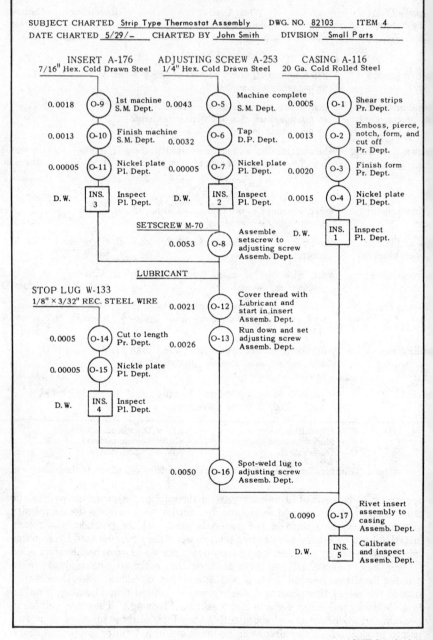

Fig. 2. Typical operation process chart.

improving small portions of the whole. They are also useful as a training aid for teaching new technical personnel, salesmen, and servicemen. They may be varied readily to suit specific needs. Busy executives can keep informed on what operations are being discussed by reference to these overall charts.

PROCESS CHART SYMBOLS. Many types of symbols and variations of them have been used as process charts developed through study of different factory and office processes and through application to different industries. The ASME standard has grouped the actions that occur during a given process into five **activity classifications.** These are known as operations, transportations, inspections, delays, and storages. Symbols for these actions have been established by the ASME. Fig. 3 shows these standard symbols for, and the definitions of, the basic activity classifications which will be found applicable under the majority of conditions encountered in process charting work. Certain other commonly used symbols and their definitions are also shown in Fig. 3 as given by Nadler (Motion and Time Study).

When unusual situations outside the range of these definitions are encountered, the intent of the definitions summarized in the following tabulation will enable the analyst to make the proper classifications.

Classification	Predominant Result
Operation	Produces or accomplishes
Transportation	Moves
Inspection	Verifies
Delay	Interferes
Storage	Keeps

As shown in Fig. 3 and as was originally used by Gilbreth, a small circle is quite commonly used to represent the transportation activity. Sometimes the storage symbol, an inverted triangle, is used to cover both delay and storage; it may also be used to represent **temporary storage,** and one triangle inside another may represent **controlled storage** (or vice versa), when it is desirable to distinguish these two types of storage. It is not so important that everyone use the same symbols as it is that the symbols should be used uniformly and the actions defined accurately and consistently in making a given process chart or developing a related series of charts. A number of other symbols for specialized types of process charts are shown in this section and in the section on Motion and Methods Study, and anyone who uses a given kind of process chart frequently, or becomes involved in a very thorough analysis of an operation, may find it convenient or necessary to develop special symbols for his own purposes.

Construction Procedures. A graphic representation of the principle of operation process chart construction is given in Fig. 4.

The **sequence** in which the events depicted on the chart must be performed is represented by the arrangement of the process chart symbols on vertical flow lines. **Material,** either purchased or upon which work is performed during the process, is shown by horizontal material lines feeding into the vertical flow lines (see Fig. 4). One of the parts making up the completed product is selected for charting first, usually the component on which the greatest number of operations is performed or, if the chart is to be used for laying out a progressive assembly line, the part having the greatest bulk to which the smaller parts are assembled.

When the component to be charted first has been chosen, a **horizontal material line** is drawn in the upper right-hand portion of the chart, and a description of the material is recorded directly above this line (see Figs. 2 and 4). A **vertical flow line** is then drawn down from the right-hand end of the horizontal material

Process Chart Symbols		Names of Activities	Activities Represented on Process Charts
ASME Symbols	Other Symbols		Definitions of Activities
○	Ⓛ Ⓜ	Operation / Labor operation Modification operation	An **operation** occurs when an object is intentionally changed in any of its physical or chemical characteristics, is assembled or disassembled from another object, or is arranged or prepared for another operation, transportation, inspection, or storage. An operation also occurs when information is given or received or when planning or calculating takes place. / Expenditure of labor or cost on product at one workplace which does not add value to the product. Modification (changing shape or size, machining, permanent assembly or disassembly, etc.) of product at one workplace. (Modification may be accomplished by machines and/or labor expenditure.)
⇧	○	Transportation / Move	A **transportation** occurs when an object is moved from one place to another, except when such movements are a part of the operation or are caused by the operator at the work station during an operation or an inspection. / Change in location of product from one workplace to another workplace.
□	◇	Inspection / Verification	An **inspection** occurs when an object is examined for identification or is verified for quality or quantity in any of its characteristics. / Comparison of product with a standard of quantity or quality at one workplace (a specialized labor operation). A control point established by management action.
D	▷	Delay / *Temporary storage	A **delay** occurs to an object when conditions except those which intentionally change the physical or chemical characteristics of the object, do not permit or require immediate performance of the next planned step. / Delay, waiting, or banking of product when no special order or requisition is required to perform next activity.
▽	▷▷	Storage / *Controlled storage	A **storage** occurs when an object is kept and protected against unauthorized removal, shown by inverted triangle. / Delay, waiting, or banking of product when a special order or requisition is required to perform the next activity.
 ⊙ (circle in square)		Combined activity	When it is desired to show activities performed concurrently or by the same operator at the same work station, the symbols for those activities are *combined*, as shown by the circle placed within the square to represent a combined operation and inspection.

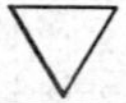

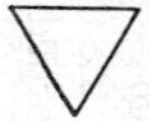

*In much motion and time study literature, the temporary storage symbol is a double triangle, and the controlled, or permanent, storage symbol is a single triangle. Because the temporary storage symbol is used overwhelmingly more than the controlled storage symbol, the principles of methods design (see section on Motion and Methods Study) indicate that it is better to use the symbols as indicated here, i.e.. single triangle for temporary storage. double triangle for controlled storage.

Fig. 3. Types of symbols and definitions for process charts

STEPS OF PROCESS ARRANGED IN CHRONOLOGICAL ORDER

ASME Standard 101

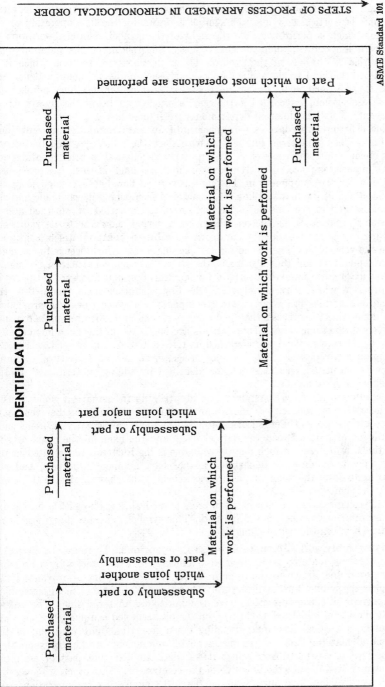

Fig. 4. Principles of operation process chart construction.

line, and, approximately ¼ in. from the intersection of the horizontal material line and the vertical flow line, the symbol is drawn for the first operation or inspection which is performed. To the right of the symbol, a brief description of the event is recorded, while the time allowed for performing the required work is recorded to the left of the symbol. Other pertinent information which it is considered will add to the value of the chart, such as department in which the work is performed, male or female operator, cost center, machine number, or labor classification, is recorded to the right of the symbol below the description of the event. Thus, the different departments are entered in Fig. 2.

This charting procedure is continued until an **additional component** joins the first. Then a material line is drawn to show the point at which the second component enters the process. If it is purchased material, a brief identification of the material is placed directly above the material line. If work has previously been done on the component in the plant, a vertical flow line is erected from the left-hand end of the material line. The material from which it was made and the operations and inspections performed on it are then charted as indicated above. As each component joins the one shown on a vertical flow line to its right, the charting of the events which occur to the **combined components** is continued along the vertical flow line to the right. The final event which occurs to the completed apparatus will thus appear in the lower right-hand portion of the chart.

Operations are numbered serially for identification and reference purposes in the order in which they are charted; the first operation is numbered 0–1, the second 0–2, and so on. When another component on which work has previously been done joins the process, the operations performed upon it are numbered successively in the same series. **Inspections** are numbered in the same manner in a series of their own. They are identified as INS–1, INS–2, and so on (see Fig. 2). No number is repeated. If it becomes necessary to add an activity after numbering is completed, then it should be identified by adding subscript "a" to the preceding activity number.

According to the ASME standard, the **identifying information** which is usually necessary on such charts consists of the subject charted, whether the chart represents the present method or a proposed method, any identifying items such as drawing or part number, the date charted, and the name of the chart maker. Additional information which may be valuable is the **location** represented in the chart (plant, building, or department), the chart number, approvals, and the **chart span** from the point in the process at which the chart begins to the point at which it ends.

If for some reason it is necessary to cross a vertical flow line with a horizontal line, a cross-over symbol, Fig. 5, is used. This symbol indicates that there is no juncture between the two crossing lines.

Disassembly and Alternate Process Conventions. To represent disassembly operations, material is represented as flowing from the process by a horizontal material line, drawn to the right from the vertical flow line approximately ¼ in. below the symbol for the disassembly operation. The name of the disassembled **component** is shown directly above the horizontal material line. The subsequent operations which are performed on the disassembled component, if any, are shown on a vertical flow line extending down from the right-hand end of the horizontal material line. If the disassembled component is later **reassembled** to the part or assembly from which it was disassembled, that part or assembly is shown as feeding into the flow line of the component. This practice moves the major vertical flow line always to the right. In numbering the operations, the perations performed on the disassembled component, after disassembly, are umbered before numbering the operations on the part from which it was disssembled. Then if the part later rejoins the disassembled component, the conentional numbering practices may be followed. This practice also applies to inspections.

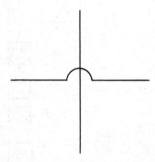

Fig. 5. Cross-over symbol used to show no juncture when horizontal material line crosses a vertical flow line.

Figure 6 shows several other conventions used in constructing operation process charts. When two or more **alternate courses** may be followed during part of the process, a horizontal line is drawn below the vertical flow line, the central point of the line being at the intersection of the vertical flow line and the horizontal line. Vertical flow lines are then dropped from the horizontal line for each alternative which it is desired to show. If no operations or inspections are performed during one alternative, a vertical flow line only is shown. In all cases, operation and inspection symbols are added in the conventional manner. They are numbered serially, beginning with the first unused number in the operation or inspection series, with the symbols on the flow line furthest to the left numbered first. When all of the alternative paths have been charted, a horizontal line is drawn connecting the lower ends of all the alternate flow lines. From the mid-point of this line, a vertical flow line is dropped and the balance of the process is charted in the conventional manner.

In some cases, the unit shown by the chart changes as the process progresses. The lines "piece 52 in. long," "piece 12 in. long," etc., in Fig. 6 indicate the convention for showing such a **change in the unit**. Figure 6 also illustrates the convention for **combined operations** and **inspections**.

Method Summaries. When a proposed method is to be presented by an operation process chart, it is often desirable to show the advantages it offers over the present method. This may be done by including with the information shown on the chart a summary of the important differences between the two methods. The summary may show the number of operations and inspections, the time of these for both the present and the proposed methods, and the difference between them; in addition, it may compare the unit cost for direct labor, the total yearly saving for direct labor, the installation cost of the proposed method, and the estimated saving the first year. It should be placed in a prominent location on the chart. On an 8½ in. × 11 in. chart it will usually be placed in the lower left-hand corner.

Analysis of Operation Process Charts. During the observation and after the charts of the present methods have been completed, every activity should

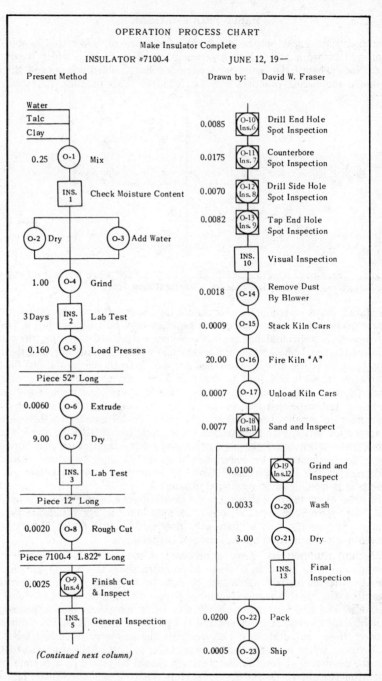

Fig. 6. Operation process chart showing conventions for indicating unit on which operations and inspections are performed, alternate routes, and combinations.

11·10

e carefully analyzed to determine if there are any ways in which these activities may be discarded or improved. Niebel (Motion and Time Study) suggests that he analyst should

. . review each operation and inspection from the standpoint of the primary approaches to **operation analysis**. In particular the following approaches apply when tudying the operation process chart:

1. Purpose of operation.
2. Design of part.
3. Tolerances and specifications.
4. Materials.
5. Process of manufacture.
6. Setup and tools.
7. Working conditions.
8. Plant layout.

The procedure is for the analyst to adopt the questioning attitude on each of the above criteria that influence the **cost and output of the product under study.** Typical questions that should be asked are:

1. Why is this operation necessary?
2. Why is this operation performed in this manner?
3. Why are the tolerances this close?
4. Why has this material been specified?

These lead to other questions such as: What is the purpose of the operation? How can the operation be better performed? Who can best perform the operation? Where could the operation be performed at lower cost? When should the operation be performed to reduce handling?

Morrow (Motion Economy and Work Measurement) indicates that during the construction and study of operation process charts the following questions should be kept constantly in mind:

1. Can any of the operations or inspections shown on the charts be **eliminated?** Sometimes operations are performed which are not necessary at all. For example, the backs of metal-plated trays were being buffed at one factory, while at another factory of the same company, the backs of the trays were not buffed. The buffing operation was eliminated at the first factory as unnecessary, thereby saving time and materials.
2. Can one operation be **combined** with another operation? A common example of this is the manufacture of a part involving press operations of perforate, form, and blank. Formerly the part was sold in small quantities, the volume so small that the three operations were done separately. When production volume increased, it was worthwhile to make a progressive die and combine the three operations in one.
3. Can a better **sequence** of operations or inspections be followed? For example, inspections might be made to better advantage at other points in the manufacture of a part. In the manufacture of an amplifier where many connections were soldered, it was found advisable to have an inspection on the assembly line rather than to wait until the amplifier was completely wired.
4. Can operations be **simplified?** Most operations can be simplified by the correct application of the principles of work analysis. In the case of operations in a job shop, the work analysis may be only a visual analysis of the operation. An engineer trained and experienced in motion economy can see all the motions and recognize those which may be eliminated to simplify the operation. If the operation is highly repetitive and the volume of work is large, it may be worthwhile to go to full micromotion study, using slow motion pictures and simo-motion charts. (See section on Motion and Methods Study.)

Such questions may also be useful in analyzing any other types of process charts.

Flow Process Charts

DEFINITION. David B. Porter (ASME Standard 101) defines a more highly developed, but still basic, type of process chart—the flow process chart —as

. . . a graphic representation of the sequence of all operations, transportations, inspections, delays, and storages occurring during a process or procedure, and includes information considered desirable for analysis such as time required and distance moved.

Since the flow process chart shows more activities than the operation process chart, it is used primarily on portions of the process. It is especially useful in detecting hidden, non-productive costs such as delays, temporary storages, and distances traveled.

There are two types of flow process charts:

1. The **material** or **product type** presents the process in terms of the events which occur to the material.
2. The **man type** presents the process in terms of· the activities of the man.

The material or product process chart is the symbolic and systematic presentation of the procedure used to modify and/or work on a product. It assists in gathering and presenting information concerning the sequence of work on, and handling of, the product, and in planning future production sequences and lay outs. The man process chart is the symbolic and systematic presentation of the method of work performed by a man when his work requires him to move from workplace to workplace. The chart depicting work at one workplace is known as a right- and left-hand chart, or operator process chart, and shows the method of work performed by the hands (and other body members, if used) when the work is at one workplace, without the use of cycle-time controlling equipment. Right- and left-hand charts will be discussed later in this section.

It will be noted that flow process charts are distinguished from operation process charts in that the flow process charts show all the basic actions, including transportation, delays, and storages. The latter are not shown on the operation process charts, which are limited to operations and inspections. Flow process charts analyze activities more ·completely than do operation process charts and may be applied to any processes or operations, whether in plants or offices. Figure 7 is a flow process chart of the material type with preprinted headings and captions, showing the activities involved in a requisition for supplies.

Prepared Forms and Symbols. In Fig. 8, Goodwin (Lake Placid Work Simplification Conferences) shows a flow process chart of the material type, depicting what happens to a special order form, and set up on a preprinted sheet using ASME symbols. In constructing the chart, the symbol for an activity is connected with the symbol for the next activity occurring on the next line, and so on. This form features the summary in the upper left-hand corner; it carries the descriptive details of the present or proposed method being analyzed, the columns for the five preprinted symbols for activities, columns for distance quantity, and time, and columns to check and comment on possibilities for improving the methods. Such preprinted forms and symbols are sometimes recommended on the basis that they do not require drawing freehand or with a template. Furthermore, untrained personnel can use them effectively, since they are self-explanatory in arrangement and simple in use.

An example of a more complete development of a **prepared chart form** of the flow process type consists of a 17 in. × 22 in. sheet which is folded twice into

PROCESS CHART

NAME OF PART OR PRODUCT		Requisition for Supplies — Rush Job			

Chart Begins at Machine Shop Foreman's Desk, Ends on Typist's Desk in Pur. Dept. | CHART NO.

ORDER NO.   LOT SIZE   DEPT. | SHEET 1

CHARTED BY   C.H.H.   DATE CHARTED   7/28/–   BLDG.   M. E. Lab. | OF 1 SHEETS

Travel in ft.	Time in min.	Symbol	Operations	Remarks
		(1)	Written longhand by foreman.	
		[1]	On foreman's desk (awaiting messenger).	
1,000		▷1	By messenger to secretary of head of department.	
		[2]	On secretary's desk (awaiting typing).	
		(2)	Typed.	
15		▷2	By messenger to head of department.	
		[3]	On head of department's desk (awaiting approval).	
		O-3 INS-1	Examined, approved, and coded (signed and code stamped).	
		[4]	On head of department's desk (awaiting messenger).	
2,000		▷3	To purchasing department.	
		[5]	On purchasing agent's desk (awaiting approval).	
		▢2	Examined and approved.	
		[6]	On purchasing agent's desk (awaiting messenger).	
25		▷4	To typist's desk.	
		[7]	On typist's desk (awaiting typing of purchase order).	

SUMMARY

Number of operations   3

Number of delays   7

Number of inspections   2

Number of transportations . . .   4

Total travel in feet   . . .   3,040

ASME Standard 101

Fig. 7. Flow process chart of material type.

11·13

ANALYSIS
WHY?
WHAT? WHERE? WHEN? WHO? HOW?

QUESTION EACH DETAIL

FLOW PROCESS CHART

NO. 1
PAGE 1 OF 1

SUMMARY

	PRESENT NO.	PRESENT TIME	PROPOSED NO.	PROPOSED TIME	DIFFERENCE NO.	DIFFERENCE TIME
○ OPERATIONS	12					
⇨ TRANSPORTATIONS	4					
□ INSPECTIONS	3					
D DELAYS	5					
▽ STORAGES	–					
DISTANCE TRAVELED	140 FT.		FT.		FT.	

JOB Special Will Call & Mail Orders while in general office

☐ MAN OR ☑ MATERIAL The Order Form

CHART BEGINS At receptionist's desk

CHART ENDS In mail chute

CHARTED BY H.F.G. DATE

	DETAILS OF PRESENT ~~PROPOSED~~ METHOD	OPERATION	TRANSPORT	INSPECTION	DELAY	STORAGE	DISTANCE IN FEET	QUANTITY	TIME	POSSIBILITIES: ELIMINATE	CHANGE: COMBINE	SEQUE.	PLACE	PERSON	IMPROVE	NOTES
1	Waited in box at reception				D											
2	Picked up by confid. clerk	○													✓	Use wire basket
3	Taken to desk at A		⇨				30'						✓		✓	To files instead-shorter distance
4	Examined (for information)			□									✓			At files
5	Waited (procure info.)				D					✓						Not necessary if taken to files
6	Prices written on order	●											✓			At files
7	Taken to post. clerk at B		⇨				40'								✓	Shorter distance
8	Placed in desk tray	○														
9	Waited for clerk				D											
10	Picked up	○														
11	Examined (for information)			□												
12	Prices added (machine)	○														
13	Total written on order	●														
14	Waited (clerk gets ledger)				D											
15	Tot. transferred to ledger	○														
16	Placed in special out box	○								✓						Taken directly to mail chute
17	Waited for routing clerk				D					✓						Not necessary
18	Picked up	○								✓						Not necessary
19	Taken to desk (C)		⇨				40'			✓						Not necessary
20	Examined (determine route)			□									✓	✓		By B
21	Placed in envelope	○								✓						Not necessary—save cost of envel.
22	Addressed to proper dept.	●								✓						Have B route & drop in mail chute
23	Taken to recept. desk		⇨				30'							✓	✓	By B—shorter distance
24	Placed in mail chute	○														

Fig. 8. Material flow process chart with preprinted symbols.

ᴉn 8½ × 11 in. size for filing. Features of this form are: the provision for showing both present and proposed methods; prepared symbols; a summary showing cost data and savings; and more complete information under the identification heading. The reverse side is arranged to provide a **cross section background** for a sketch of a layout or flow diagram. The two center pages are for a detailed explanation of **proposed changes,** and the back cover provides for recording action taken on the proposed method. Figure 9 shows a man-type flow process chart on this form, presenting both the present and the proposed methods and comparing them in the summary in the upper right-hand corner.

PRINCIPLES OF FLOW PROCESS CHARTING. Certain principles and precautions can be helpful in the study of a work process or operation and the construction of a flow process chart representing it.

1. Determine the **subject** which will be followed. Either a person (man chart) or an object (material or product chart) may be chosen as the subject of a flow process chart. The determination is made on the basis of which will portray the flow of activity most effectively. An object (material) is usually charted when several people successively handle the same object during the sequence of events (see Figs. 7 and 8). Otherwise, it would require different charts of the activity of each person to tell the whole story. If one person performs all the work, then that person logically becomes the subject of the chart. Everything he does is then recorded (see Fig. 9). Once the chart has been started in following either an object or a person, it should continue to follow that object or person throughout the activity.

2. Determine the **starting** and **ending points.** If the chart is to be as specific and complete as intended, the starting and ending points of the activity should be selected and recorded. Since most projects involve a portion of an overall procedure, this indication is helpful in defining the limits of the intended study.

3. Record the appropriate **symbol** and a brief **description** of each detail. Every step or activity which occurs should be listed with an accurate description, no matter how small or insignificant it may seem to be. Every time something happens, such as a move, delay, or inspection, it should be recorded exactly the way it is observed.

4. Shade in the **do symbols.** Since most **make-ready** and **put-away** details depend on the do activities, it is important to segregate these key details graphically so that they may be challenged first (see Fig. 8). If there is any doubt about whether the detail actually is a do according to definition, the customary procedure is to shade it in anyway, since it is the use of the chart which counts rather than its exactness.

5. Note **transportation distance.** Transportation and handling between operations is expensive, nonproductive work. A significant amount of this type of activity is a challenge to the ingenuity to reduce or eliminate it entirely. Estimates by **pacing** are considered adequate if more accurate measurements are not obtainable easily.

6. Record the **quantities handled.** Often the subject of a flow process chart is handled in varying quantities during the complete sequence charted. Material may start off in a roll, be cut into bundles, and wind up as individual pieces. Such information is pertinent to the formation of a clear picture of exactly what is happening.

7. Indicate **time consumed** or **production rates,** if possible. An idea of the performance attained at each step becomes useful when comparisons are being made between different methods and summaries of savings are being worked up.

IDENTIFICATION		**FLOW**

SUBJECT CHARTED _STAMP AND PACK WELDING ROD_
DRAWING NO. _____ PART _3/16" DIA. WELD. ROD_
POINT AT WHICH CHART BEGINS _GET SHIPPING BOX_
LOCATION _EDGE OF WORKPLACE_
POINT AT WHICH CHART ENDS _COMPLETE PACKING WITH BOX OUT OF_
THE WAY LOCATION _TRUCK OR CONVEYOR_

QUANTITY INFORMATION

CHART NO. _15-43_
TYPE OF CHART _MAN_
SHEET NO. _1_ OF _1_ S
CHARTED BY _WALKER B. W_
DATE _MAY 5, 19 —_
APPROVED BY _H.W. CHAP_
DATE _MAY 9, 19 —_

YEARLY PRODUCTION _21,000_
COST UNIT _100-Pound_ ?

PRESENT METHOD

QUANTITY UNIT CHARTED	SYMBOLS	DESCRIPTION OF EVENT	DIST MOVED IN FEET	UNIT OPER. TIME IN MIN.	UNIT TRANSP. TIME IN MIN.	UNIT INSPECT TIME IN MIN.	DELAY TIME IN ___	S
1 BOX	①⇨□D▽	PICK UP WOODEN WELDING ROD BOX		.103				
1 BOX	○⇨□D▽	MOVE BOX TO STAMPING TABLE	6		.117			
1 BOX	②⇨□D▽	POSITION BOX TO STENCIL		.030				
1 BOX	③⇨□D▽	STENCIL BOX		.200				
1 BOX	○⇨□D▽	MOVE BOX AND LID TO SCALE	5		.093			
1 BOX	④⇨□D▽	WEIGH BOX AND LID AND MARK		.079				
1 BOX	⑤⇨□D▽	REMOVE LID FROM BOX		.051				
1 BOX	○⇨□D▽	MOVE BOX TO TRUCK	3		.070			
1 BOX	⑥⇨□D▽	POSITION BOX ON TRUCK		.059				
1 BOX	⑦⇨□D▽	PUT PATENT SLIP IN BOX		.047				
100 LB.	○⇨□D▽	GO TO STOCK BIN	12		.131			
100 LB.	⑧⇨□D▽	REMOVE WELDING ROD FROM STOCK		.755				
100 LB.	○⇨□D▽	MOVE WELDING ROD TO STAMPING MACH.	12		.154			
100 LB.	⑨⇨□D▽	SIT DOWN AT STAMPING MACHINE		.039				
100 LB.	⑩⇨□D▽	STAMP ONE PLACE ON WELDING ROD		7.420				
100 LB.	○⇨⑪D▽	INSPECT STAMP ON ROD				.039		
100 LB.	○⇨□D▽	WALK TO SCALE	7		.132			
100 LB.	⑪⇨□D▽	WEIGH AND PLACE 100 POUNDS IN BOX		.868				
1 BOX	⑫⇨□D▽	PLACE LID ON BOX		.127				
1 BOX	⑬⇨□D▽	POSITION BOX TO STRAP		.095				
1 BOX	⑭⇨□D▽	PICK UP STRAP AND STRAPPER		.073				
1 BOX	⑮⇨□D▽	STRAP BOX IN 3 PLACES		1.055				
1 BOX	⑯⇨□D▽	PUT DOWN STRAPPER		.051				
1 BOX	○⇨□D▽	GET BOX OF NAILS, TACKS + HAMMER	10		.047			
1 BOX	⑰⇨□D▽	NAIL LID ON BOX		1.317				
1 BOX	⑱⇨□D▽	LAY 2 LABELS ON BOX		.094				
1 BOX	⑲⇨□D▽	START 10 TACKS WITH THUMB		.563				
1 BOX	⑳⇨□D▽	DRIVE 10 TACKS WITH HAMMER (LABELS)		.172				
1 BOX	○⇨□D▽	PUT AWAY NAILS, TACKS AND HAMMER	10		.046			
1 BOX	○⇨□D▽	MOVE BOX TO TRUCK	8		.123			
	○⇨□D▽							
	○⇨□D▽							

Fig. 9.   A preprinted flow process

SS CHART — SUMMARY

TAL YEARLY SAVING—DIRECT LABOR $1,635.00	(LABOR RATE = $1.80/HOUR)	PRESENT METHOD	PROPOSED METHOD	DIFFERENCE
	UNIT COST DIRECT LABOR & INSP	$.4245	$.3467	$.078
	DISTANCE TRAVELED IN FEET	73	52	21

INSTALLATION COST OF PROPOSED METHOD $375.00		NO.	TIME IN MIN	NO.	TIME IN MIN	NO.	TIME IN MIN
	◯ OPERATIONS	20	13.198	21	10.758		2.440
	⇨ TRANSPORTATIONS	9	.913	9	.760		.153
ESTIMATED NET SAVING—FIRST YEAR $1,260.00	☐ INSPECTIONS	1	.039	1	.039		0
	D DELAYS						
	▽ STORAGES						

PROPOSED METHOD

QANTITY UNIT HARTED	SYMBOLS	DESCRIPTION OF EVENT	DIST MOVED IN FEET	UNIT OPER. TIME IN MIN	UNIT TRANSP. TIME IN MIN	UNIT INSPECT TIME IN MIN	DELAY TIME IN	STORAGE TIME IN
Box	① ⇨ ☐ D▽	PICK UP WOODEN WELDING ROD BOX		.103				
Box	◯ ① ☐ D▽	MOVE BOX TO STAMPING TABLE	6		.117			
Box	② ⇨ ☐ D▽	POSITION BOX TO STENCIL		.030				
Box	③ ⇨ ☐ D▽	STENCIL BOX		.200				
Box	◯ ② ☐ D▽	MOVE BOX AND LID TO SCALE	5		.093			
Box	④ ⇨ ☐ D▽	WEIGH BOX AND LID AND MARK		.079				
Box	◯ ③ ☐ D▽	MOVE BOX AND LID TO RACK	4		.076			
Box	⑤ ⇨ ☐ D▽	PLACE BOX AND LID ON RACK		.033				
o LB.	◯ ④ ☐ D▽	GO TO STOCK BIN	12		.131			
o LB.	⑥ ⇨ ☐ D▽	REMOVE W.R. FROM STOCK TO SCALE		.755				
o LB.	◯ ⑤ ☐ D▽	GO FROM STOCK TO STAMPING MACH.	12		.154			
o LB.	⑦ ⇨ ☐ D▽	SIT DOWN AT STAMPING MACH.		.039				
Box	⑧ ⇨ ☐ D▽	REMOVE LID FROM BOX		.051				
Box	⑨ ⇨ ☐ D▽	PICK UP BOX		.031				
Box	◯ ⑥ ☐ D▽	MOVE BOX TO CHUTE	3		.053			
Box	⑩ ⇨ ☐ D▽	POSITION BOX AT CHUTE		.068				
Box	⑪ ⇨ ☐ D▽	DROP CHUTE ON BOX		.039				
Box	⑫ ⇨ ☐ D▽	PUT PATENT SLIP ON BOX		.047				
o LB.	⑬ ⇨ ☐ D▽	STAMP ONE PLACE ON WELDING ROD		7.470				
o LB.	◯ ⇨ ① D▽	INSPECT STAMP ON ROD				.039		
Box	⑭ ⇨ ☐ D▽	RAISE CHUTE		.053				
Box	◯ ⑦ ☐ D▽	SLIDE BOX FROM UNDER CHUTE	2		.043			
Box	⑮ ⇨ ☐ D▽	PUT LID IN BOX (INSERT)		.056				
Box	◯ ⑧ ☐ D▽	SLIDE BOX AWAY ON ROLLERS	4		.046			
Box	⑯ ⇨ ☐ D▽	POSITION BOX TO STRAP		.095				
Box	⑰ ⇨ ☐ D▽	PICK UP STRAP AND STRAPPER		.073				
Box	⑱ ⇨ ☐ D▽	STRAP BOX IN 3 PLACES		1.055				
Box	⑲ ⇨ ☐ D▽	PUT DOWN STRAPPER		.051				
Box	⑳ ⇨ ☐ D▽	LAY 2 LABELS ON BOX		.094				
Box	㉑ ⇨ ☐ D▽	GLUE 2 LABELS ON BOX		.386				
Box	◯ ⑨ ☐ D▽	SLIDE BOX AWAY ON ROLLERS	4		.047			
	◯ ⇨ ☐ D▽							

Adapted from ASME Standard 101

chart used for comparing methods.

It is also significant in determining which details are the most costly, and, there fore, should receive the most attention. A convenient way of expressing th time is by noting the production record in units per hour or day.

8. Prepare **summaries.** Total each type of detail, the distance traveled, an the time consumed. The summary is the chart feature in which higher authori ties will be most interested. It may be used to make comparisons with the pro posed method chart and indicate the extent of improvement attained.

Construction Procedures. The same kind of identifying information i placed on a flow process chart as on an operation process chart, with the additiou that the **chart type,** whether material or man, is usually identified. With chart: drawn on plain paper where more than one item is to be charted, the same ar rangement and conventions which are used for the operation process chart ar likewise used for the flow process chart. However, in the man type of flow process chart, there are no horizontal lines representing the entrance of materia into a process, and the storage symbol is not used. Flow process charts ar usually drawn on plain paper of sufficient size to accommodate them when the; portray the events occurring to more than one item of material, the activities o: more than one person, or the alternate routes or procedures followed by mate rial or men.

When a flow process chart is made about a single item of material or a single person, however, usually only one column of symbols is needed, and no horizonta. line is used to introduce material (see Fig. 7). The symbols are usually drawr with the aid of a **symbol template** for efficiency and uniformity. There is enough similarity among flow process charts of single items to permit the use o prepared forms which are both convenient and time-saving. The simplest o: these forms, as shown in Fig. 7, provides headings and identification spaces, but no vertical or horizontal rulings.

Simple, Converging, and Diverging Charts. Three types of material flow process charts can be made. The first is the simple or **straight line chart,** shown in Fig. 10a (Nadler, Motion and Time Study), using the types of symbols illustrated and defined in Fig. 3. This charts one material or product alone, re gardless of what might be added to or taken away from it, and it is usually a part of one of the more complex chart types. The second type is the converging chart (Fig. 10b), made for most **manufacturing** and **assembly work.** The activity charted utilizes different materials which come together to form a final product. The third type is the diverging chart (Fig. 10c), which would be made in industries where a single material is divided into **several final products,** as in the food processing, metal-working, and meat industries. The converging and diverging types of flow process charts are a series of simple charts placed together in proper relationships.

In Fig. 11, Nadler (Motion and Time Study) illustrates a complete material flow process chart of the converging type, using the common symbols defined in Fig. 3, with minimum detail for a repetitive switch box assembly. The assembly of the boxes is the basic activity, so the principal straight line on the left ana lyzes this. The covers and the cartons converge onto the box assembly at the proper time from the right-hand lines. Figure 11 is a planning chart, analyzing future work. Before starting production, the industrial engineer, the foreman of the department involved, and the production manager met to discuss the process to be used for assembly of the switch boxes. They knew the floor space available for the assembly, developed a flow diagram for the process, and decided

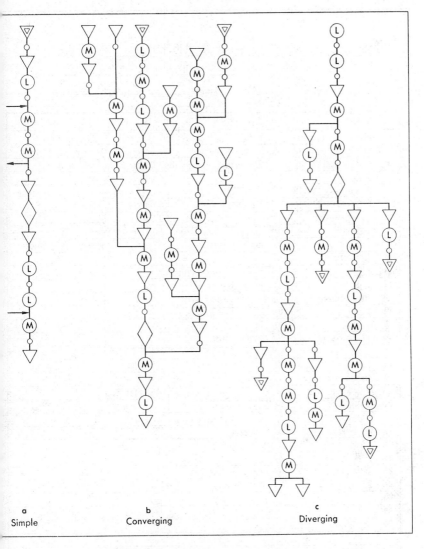

Fig. 10. Types of flow process charts.

to make up the chart shown in Fig. 11 for the procedure that seemed likely to be the most effective.

Defining the Workplace. Every chart, including the flow process chart, has to be conceived in a flexible manner. There is no one "correct" chart. The key to a good flow process chart is the definition of the workplace for the process. In Fig. 12 (Nadler, Motion and Time Study), a lathe, a supply table, and a finish table are shown. The operator takes a part and prepares it on the supply table. After machining the part, the operator unloads it onto the finish table.

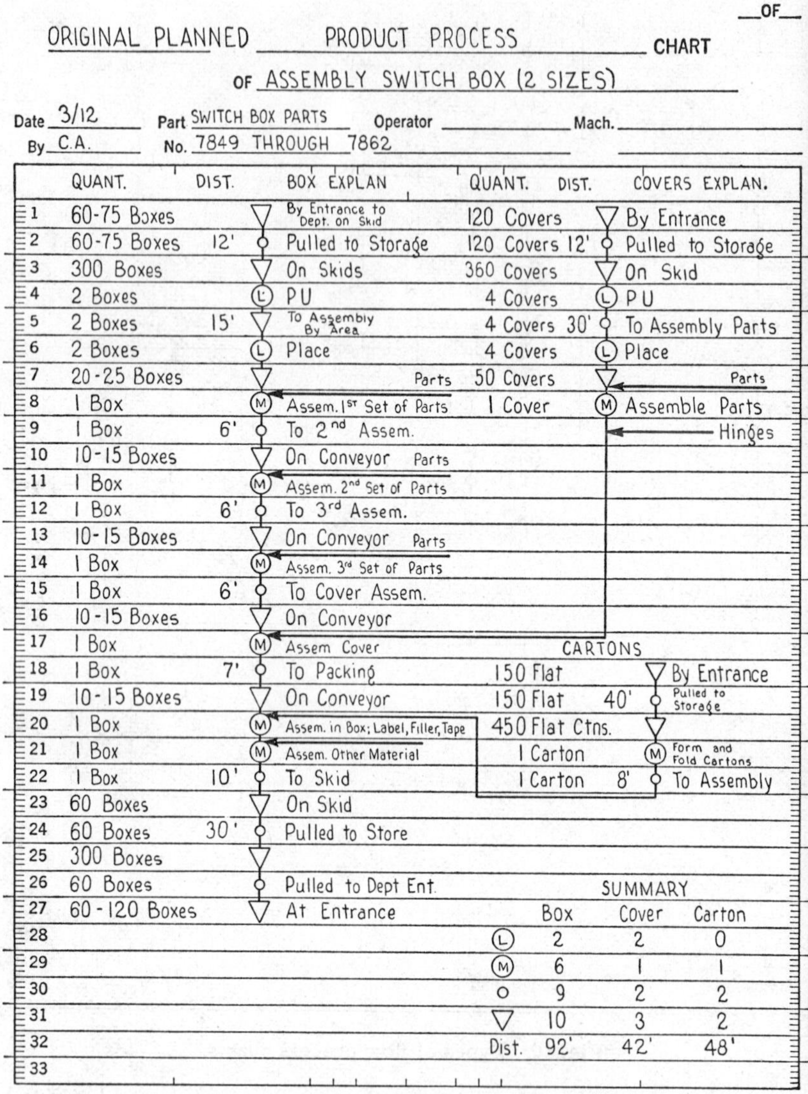

ORIGINAL PLANNED _____ PRODUCT PROCESS _____ CHART ___OF___

OF ASSEMBLY SWITCH BOX (2 SIZES)

Date 3/12   Part SWITCH BOX PARTS   Operator _____ Mach. _____
By C.A.   No. 7849 THROUGH 7862

#	QUANT.	DIST.	BOX EXPLAN.	QUANT.	DIST.	COVERS EXPLAN.
1	60-75 Boxes		▽ By Entrance to Dept. on Skid	120 Covers		▽ By Entrance
2	60-75 Boxes	12'	○ Pulled to Storage	120 Covers	12'	○ Pulled to Storage
3	300 Boxes		▽ On Skids	360 Covers		▽ On Skid
4	2 Boxes		(L) PU	4 Covers		(L) PU
5	2 Boxes	15'	▽ To Assembly By Area	4 Covers	30'	○ To Assembly Parts
6	2 Boxes		(L) Place	4 Covers		(L) Place
7	20-25 Boxes		▽ Parts	50 Covers		▽ Parts
8	1 Box		(M) Assem. 1st Set of Parts	1 Cover		(M) Assemble Parts
9	1 Box	6'	○ To 2nd Assem.			◀── Hinges
10	10-15 Boxes		▽ On Conveyor Parts			
11	1 Box		(M) Assem. 2nd Set of Parts			
12	1 Box	6'	○ To 3rd Assem.			
13	10-15 Boxes		▽ On Conveyor Parts			
14	1 Box		(M) Assem. 3rd Set of Parts			
15	1 Box	6'	○ To Cover Assem.			
16	10-15 Boxes		▽ On Conveyor			
17	1 Box		(M) Assem Cover			CARTONS
18	1 Box	7'	○ To Packing	150 Flat		▽ By Entrance
19	10-15 Boxes		▽ On Conveyor	150 Flat	40'	○ Pulled to Storage
20	1 Box		(M) Assem. in Box; Label, Filler, Tape	450 Flat Ctns.		
21	1 Box		(M) Assem. Other Material	1 Carton		(M) Form and Fold Cartons
22	1 Box	10'	○ To Skid	1 Carton	8'	○ To Assembly
23	60 Boxes		▽ On Skid			
24	60 Boxes	30'	○ Pulled to Store			
25	300 Boxes		▽			
26	60 Boxes		○ Pulled to Dept Ent.			SUMMARY
27	60-120 Boxes		▽ At Entrance	Box	Cover	Carton
28				(L) 2	2	0
29				(M) 6	1	1
30				○ 9	2	2
31				▽ 10	3	2
32				Dist. 92'	42'	48'
33						

Fig. 11. Converging flow process chart.

Notice that work is being performed at the supply table, the lathe, and at the finish table. Depending upon the definition of a workplace, the analyst can make different types of charts. In Fig. 12a, the analyst assumed that all the **work stations** were within one **workplace**, contained within the dotted lines. In this case, the minimum detailed material flow process chart for the activity at this workplace would be like Fig. 12b. To obtain greater detail with this definition of workplace, the analyst could make a chart like that shown in Fig. 12c.

Figure 13a represents another way of defining workplaces. Here three differ-
¬t workplaces are shown, divided by the dotted lines. The minimum chart for
is workplace concept would be that shown in Fig. 13b. For greater detail, the
¬art shown in Fig. 13c could be made. There are, then, various ways of
¬arting the same activity, depending upon the determination of the workplace
¬d the detail desired. A consistent pattern of workplace definition should be
llowed on any given chart.

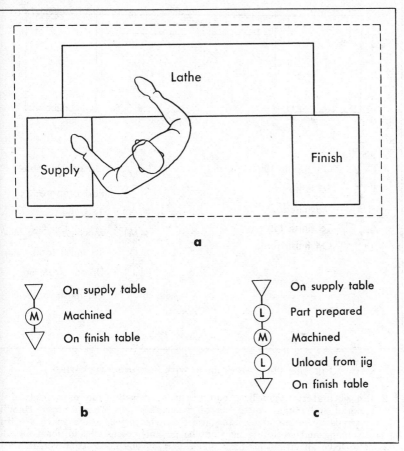

Fig. 12. One workplace with flow process charts.

ADVANTAGES OF FLOW PROCESS CHARTS. Morrow (Motion
¬conomy and Work Measurement) summarizes the improvements which may
¬esult from the development and use of flow process charts and flow diagrams
¬s follows:

1. **Reduction in distances** that the work travels. The distance a part may travel
   through a factory in the course of manufacture, not only back and forth on
   each floor, but up and down in elevators, is almost unbelievable. Savings in
   travel time can be made by relocation of machines or departments, changes
   which usually pay for themselves many times over.

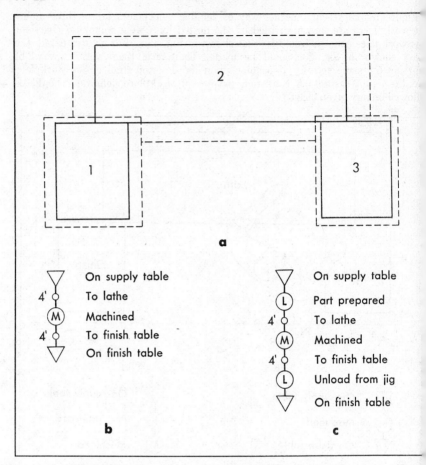

On supply table
4' To lathe
(M) Machined
4' To finish table
On finish table

b

On supply table
(L) Part prepared
4' To lathe
(M) Machined
4' To finish table
(L) Unload from jig
On finish table

c

Fig. 13. Three workplaces with flow process charts.

2. Use of **material handling equipment** to expedite the movement of the material and obtain better use of floor space and storage areas. Handling materials by means of elevators usually involves long waits which are time consuming and costly. It may not be possible to install additional elevators or to have all operations for a single part on one floor. However, relocation of machines, and handling materials between floors by conveyors or chutes will help to solve this problem.

3. Reduction of number of **periods of temporary storage** of materials between operations or elimination of such storage entirely. This cuts the work in process and saves floor space.

4. Reduction of the number of **inspections** needed, or relocation of inspection points.

THE FLOW DIAGRAM. The flow diagram in its simplest form shows a rough view of the space in which the activity being studied occurs and the location and extent of the work areas, machines, or desks, with a connecting series of arrows and lines to indicate the route of travel. Flow diagrams are often made up in conjunction with process charts, since they are tremendously helpful in

isualizing the process and conceiving changes in it. Flow diagrams may be made
p into more elaborate **engineering drawings** if desired, or **three-dimensional
models** with colored strings based on flow diagrams may be very effective, par-
icularly with the less technically trained levels of supervision and personnel.

Figure 14a and b (The Standard Register Co.) shows the flow diagrams of a
production control process before and after the process was improved. The
low diagram of the old method (Fig. 14a) reveals much wasted time and energy.
The solid lines show how each order once traveled 162 ft., while the broken lines
ndicate how auxiliary forms traveled 80 ft. As shown in the flow diagram of the
ew method (Fig. 14b), proper arrangement of the desks, consistent with the
low of paperwork, corrected the first situation. The total floor space was re-
luced from 960 to 460 sq. ft., and both the order and the auxiliary form travel
nly 35 ft.

The flow diagram should identify each activity near where it occurs by the
ccompanying symbol and number shown on the flow process chart. Niebel
Motion and Time Study) notes: "It can be seen that the flow diagram is a
elpful supplement to the flow process chart because it shows up backtracking
nd areas of possible traffic congestion, and facilitates making an ideal plant
ayout."

PROCEDURE FLOW CHARTS. Sometimes also known as the form
process chart, the procedure flow chart is a symbolic and systematic presentation
f the procedure used to modify, work on, and handle a form or forms. It may
be thought of as a specialized type of a flow process chart of the product or
material kind. The usual activity charted on a procedure flow chart is a **system
or procedure.** This refers to the flow of paperwork, since the form process
chart is usually made for a complete procedure, like purchasing or employment.
Hence it is sometimes called a **paper-work flow chart.** However, the procedure flow chart can be made for just one form. Many factory procedures can
be analyzed with these charts, such as production control, time records, inventory
control, or incentive earnings processing. Paperwork has become so complicated
with multipartite forms and new data-gathering needs that those specializing in
it have designed this type of chart which they feel more nearly meets their
needs. Procedure flow charts are sometimes done on a **horizontal charting
basis,** which, of course, can readily be done with any flow process chart.

Construction Procedures. The symbols commonly used in procedure flow
charts are shown in Fig. 15. The symbol for Information Transmission may be
modified by a letter or letters above the arrow to indicate source of information
other than another form (MD = Man's Decision, T = Telephone). These are
similar to those used in the regular flow process chart, it will be noted, although
additional symbols are used to take care of information on forms, which is the
main product, and the symbols are named and defined somewhat differently.
Using the symbols shown in Fig. 15, Nadler (Motion and Time Study) shows in
Fig. 16 a procedure flow chart made in a company which was having an extensive cost reduction program which included a review of systems and procedures.
The specific goal for the order form was to reduce the time and money spent in
working on it.

The columns on the procedure flow chart represent the departments or other
units of the organization through which the forms move. Within these columns,
the symbols are indicated, with an explanation of them, and with the distances
covered by the movement of the forms. **Equipment** used can also be listed when
it is an important factor in the procedures being analyzed. The procedure flow
chart is started by noting the symbol for the activity first observed at the point

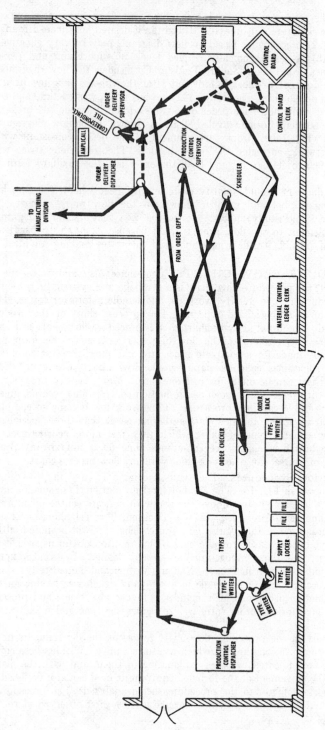

Fig. 14a. Flow diagram of procedure before improvement.

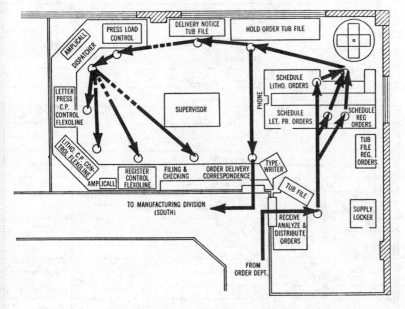

Fig. 14b. Flow diagram of procedure after improvement.

Geometric Symbol	Name	Activity Represented
②	Origination	Form being made out at one workplace. (Number in center represents number of copies made.)
◯	Operation	Modification of, or addition to, form at one workplace.
○	Move	Change in location of form from one workplace to another.
▽	Temporary storage	Delay or waiting of form where no special order is required to perform next activity (e.g., in desk basket).
▽	Controlled storage	Delay or waiting of form where a special order is required to perform next activity (e.g., in file cabinet).
◇	Verification	Comparison of form with other information to ascertain correctness of form.
┽	Information transmission	Reading or removal of information on form for use by someone or some machine.
⊠	Disposal	Form destroyed.

Fig. 15. Procedure flow chart symbols.

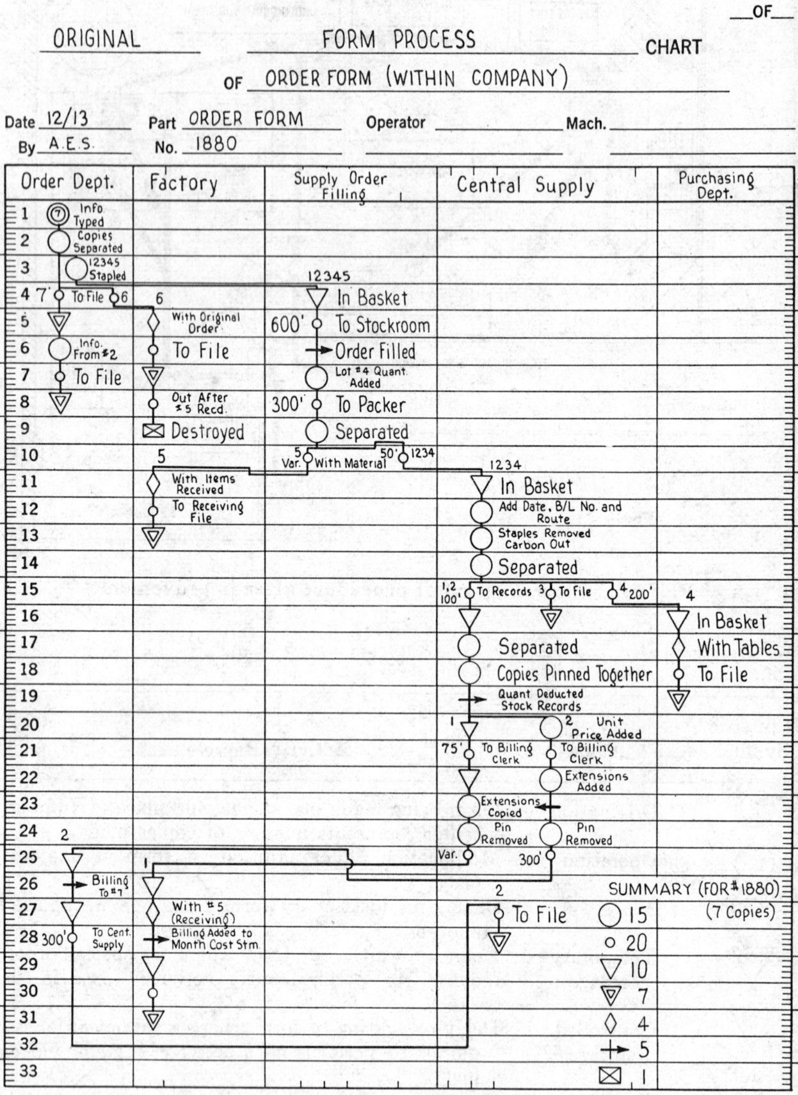

Fig. 16. Procedure flow chart for order form.

where the analysis is to begin. The next activity is charted on the next line, and so on.

As is the case with the other types of charts, **specialized symbols** have been developed also for procedure flow charts, particularly when applied to office work. Some of these are illustrated in Fig. 17 (The Standard Register Co.) which shows a flow diagram of an office procedure.

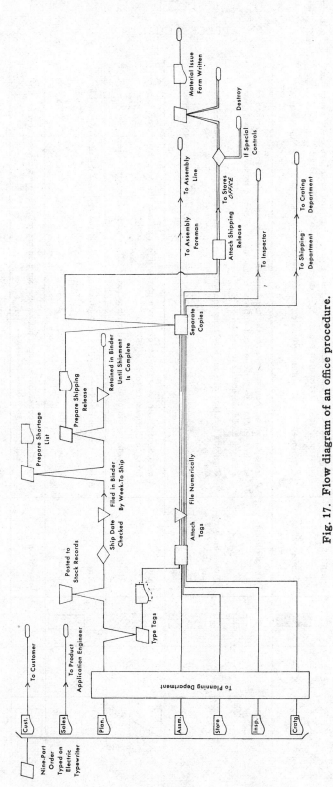

Fig. 17. Flow diagram of an office procedure.

Right- and Left-Hand Chart Symbols				
ASME Symbols	Other Symbols	Time Symbols	Names of Activities	Hand Activities Represented on Right- and Left-Hand Charts — Definitions of Activities
○	○	■ (black rectangle)	Operation	An **operation** occurs whenever the hand picks up something, drops, or lays down something, positions, uses, assembles something — each activity a separate detail.
			Sub-operation (1, 2, 3)*	A sub-operation occurs in performing something at one work area in a workplace.
 ⇧ (arrow)	○	▦ (cross-hatched)	Transportation	A **transportation** occurs whenever the hand moves from one location in the workplace to another, in order to perform such actions as moving from center to supply of work, or bringing tool from holder to fixture.
	⊘ (circle with line)	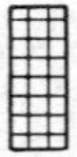	Movement without load (2)*	Changing location without load from one work area to another work area in a workplace.
		 ▨ (cross-hatched)	Movement with load	Changing location with load from one work area to another work area in a workplace.
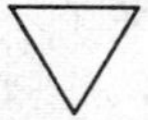 ▽	▷	▨ (hatched)	Hold	A **hold** occurs whenever the hand holds a part or a tool so that the other hand may do something to that part or tool.
			Hold	Maintaining an object in a fixed orientation to allow work on or with object
 ⌂ (D shape)	▷	▭ (open rectangle)	Delay	A **delay** occurs when the hand is motionless or is not in the performance of an operation or transportation detail.
	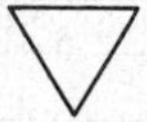		Delay (1, 2, 3)*	Waiting or idle.
☐			Inspection	**Inspection** is occasionally used, but rarely found on pre-printed forms since it occurs so infrequently as a hand motion.

*Used also for (1) Feet (2) Eyes. and (3) Knees.

Fig. 18. Types of symbols and definitions used for operator process charts.

Complex Process Charts

OPERATOR PROCESS CHARTS. Sometimes known as the right- and left-hand process chart, the operator process chart is the symbolic and systematic presentation of the method of work performed by the hands and perhaps other body members, when the work is performed at one workplace, without the use of cycle-time controlling equipment. It is a useful tool in motion study. The purpose of this chart is to present the operation in sufficient detail so that by analysis and application of the principles of motion economy, the operation can be improved. According to Niebel (Motion and Time Study), "This chart will facilitate changing a method so a balanced two-handed operation can be achieved and ineffective motion either reduced or eliminated. The result will be a smoother, more rhythmic cycle which will keep both delays and operator fatigue to a minimum." An operator process chart will show in detail what each hand of the operator is doing at each stage of the operation, whether picking up, laying down, moving from one location to another, or holding a part or tool. Similar activities of the feet or knees of the operator can be analyzed when they are involved in the work.

Construction Procedures. The operator process chart is set up in two columns, one representing the activities of the left hand and the other those of the right hand. The symbols for these activities are inserted down the center of the chart and the time may be shown beside the symbols. This chart is drawn to scale with an appropriate unit of time represented by each ¼ inch of vertical space on the chart. The basic ASME and other symbols as well as time symbols are shown in Fig. 18, with the definitions of the activities represented by the symbols. The motions of the hands may be indicated by using combinations of some of the fundamental motions. These motions and their symbols are:

Reach	....Re	Grasp	......G	Move	......M	Use	........U
Position	...P	Release	....Rl	Delay	......D	Hold	.......H

Getting a part would be a combination of Reach and Grasp. Placing that part could include Move and Position. Dispose would consist of Move and Release.

The title of the chart should be Operator Process Chart with other identifying information such as part number, drawing number, operation description, present or proposed method, date, department, and the name of the person preparing the chart. A sketch of the workplace should always be included, drawn to scale. Figure 19 (Niebel, Motion and Time Study) shows a simple Operator Process Chart, with cross-section lines to facilitate sketching. In making the chart it is important to establish the proper time scale to fit the chart. It is best to chart the movements of one hand completely, and then the other. The normal starting point should be after the release of the completed item. A summary of cycle time, pieces per cycle, and time per piece may be shown at the bottom of the chart.

Figure 20 (Nadler, Motion and Time Study) shows a right- and left-hand chart of the operation of placing a glue compound at the end of a case filled with hard powder. Two operators were doing this work. A time column is shown for each hand, the approximate times being obtained with a sweep-second wrist watch. A group of actions were timed first, and the time was then prorated as indicated on the chart. The chart also illustrates how repetitions of the same work in a given cycle are shown. A more detailed and exact analysis can be made in the right- and left-hand **time chart**, using time symbols, when motion pictures of the work are available.

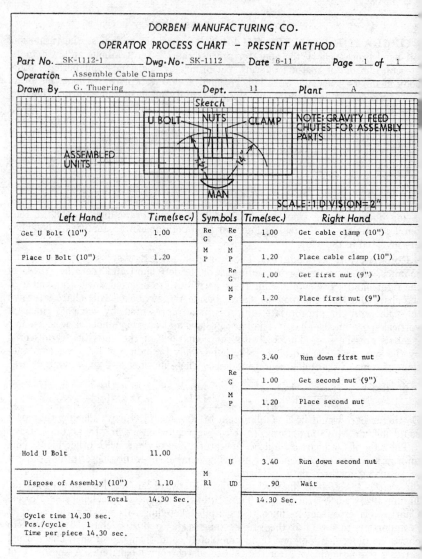

DORBEN MANUFACTURING CO.

OPERATOR PROCESS CHART - PRESENT METHOD

Part No. __SK-1112-1__   Dwg·No·__SK-1112__   Date __6-11__   Page __1__ of __1__

Operation __Assemble Cable Clamps__

Drawn By __G. Thuering__   Dept. ____11____   Plant ____A____

Left Hand	Time(sec.)	Symbols		Time(sec.)	Right Hand
Get U Bolt (10")	1.00	Re G	Re G	1.00	Get cable clamp (10")
Place U Bolt (10")	1.20	M P	M P	1.20	Place cable clamp (10")
			Re G	1.00	Get first nut (9")
			M P	1.20	Place first nut (9")
			U	3.40	Run down first nut
			Re G	1.00	Get second nut (9")
			M P	1.20	Place second nut
Hold U Bolt	11.00		U	3.40	Run down second nut
Dispose of Assembly (10")	1.10	M R1	UD	.90	Wait
Total	14.30 Sec.			14.30 Sec.	

Cycle time 14.30 sec.
Pcs./cycle   1
Time per piece 14.30 sec.

Fig. 19. Operator process chart of assembly of cable clamps.

Defining the Work Area. The degree of refinement to which motions should be analyzed in making these charts is a question of judgment. Some simple breakdown of the job is necessary in order to effect any improvement, but it is also possible to go to the other extreme of refinement in subdivision of the job beyond that which is economically feasible. Workplace diagrams or layouts to show the various work areas significant in the chart are often made up and sometimes shown on the chart itself. These diagrams do not usually show the lines of movement of the body members. If this were done, the diagram

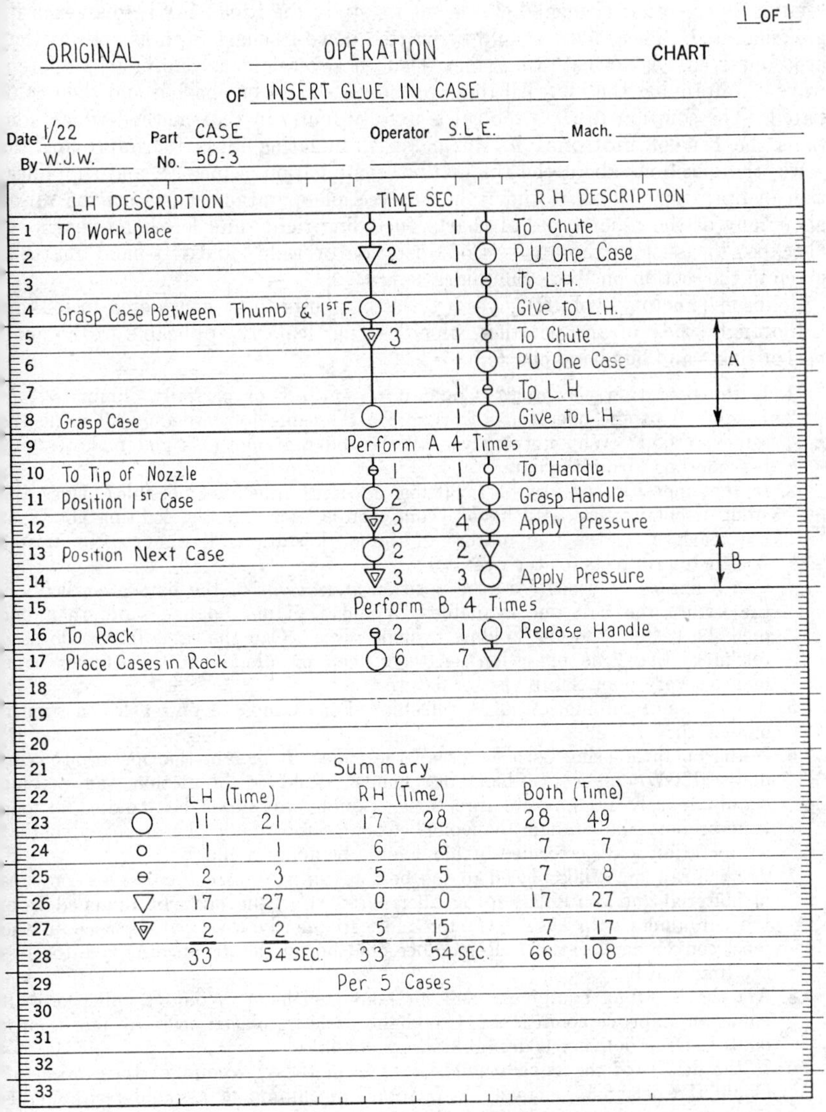

1 OF 1

ORIGINAL _____ _____ OPERATION _____ CHART

OF INSERT GLUE IN CASE _____

Date 1/22 ____ Part CASE _____ Operator S.L.E. _____ Mach. _____

By W.J.W. ____ No. 50-3

LH DESCRIPTION	TIME SEC	R H DESCRIPTION
1 To Work Place	○ 1 1 ○	To Chute
2	▽ 2 1 ○	P U One Case
3	1 ○	To L.H.
4 Grasp Case Between Thumb & 1ST F. ○ 1	1 ○	Give to L.H.
5	▽ 3 1 ○	To Chute
6	1 ○	PU One Case A
7	1 ○	To L.H
8 Grasp Case ○ 1	1 ○	Give to L.H.
9	Perform A 4 Times	
10 To Tip of Nozzle	○ 1 1 ○	To Handle
11 Position 1ST Case	○ 2 1 ○	Grasp Handle
12	▽ 3 4 ○	Apply Pressure
13 Position Next Case	○ 2 2 ▽	B
14	▽ 3 3 ○	Apply Pressure
15	Perform B 4 Times	
16 To Rack	○ 2 1 ○	Release Handle
17 Place Cases in Rack	○ 6 7 ▽	
18		
19		
20		
21	Summary	
22	LH (Time) R H (Time) Both (Time)	
23 ○	11 21 17 28 28 49	
24 ○	1 1 6 6 7 7	
25 ⊖	2 3 5 5 7 8	
26 ▽	17 27 0 0 17 27	
27 ▽	2 2 5 15 7 17	
28	33 54 SEC. 33 54 SEC. 66 108	
29	Per 5 Cases	
30		
31		
32		
33		

Fig. 20. Right- and left-hand chart with time columns.

would become too complex and hard to understand. It is important that this diagram show all tools, jigs, bins, parts, and supplies in their proper relationship to one another at the workplace. A three-dimensional sketch or photograph is often helpful in defining these relationships.

Analysis of the Charts. The act of making the chart helps the observer to acquire an intimate knowledge of the details of the job, and the chart serves as the means whereby he may study each element in the operation by itself and it relation to the other elements. From such a study the **ideas for improvement** are generated. These ideas should be written down in chart form as soon as they are conceived. A normal study may lead to several ideas embracing different ways of improving the job. All these proposals should be charted and then com pared. The solution to the problem is usually found in that method which con tains the **fewest motions**. As an aid in formulating ideas for improving the work, the questions about the elimination, combination, sequence, and simplifica tion of operations detailed under the analysis of operation charts can be raised about any of the other types of charts, including right- and left-hand charts. A checklist to assist in the analysis of operation or right- and left-hand charts i given in the section on Work Simplification.

Johnson (Factory, vol. 92) gives a series of points to be considered in making a thorough study of an operation, which are particularly applicable to the anal ysis of right- and left-hand charts:

1. Is the operation necessary? Could it be entirely or partially eliminated?
2. If more than one operator is working on the same job, are they all using the same method? Why not analyze all the different methods and make a "one best method" from the data?
3. Is the operator comfortable? Sitting down as much as possible? Does the stool or chair being used have a comfortable back? Is the lighting good but not glaring? Is the temperature of the work station all right? No drafts? Are there arm rests for the operator?
4. Can a fixture be used? Are the position and height of the fixture correct? I the fixture the best one available? Would a fixture holding more than one piece be better than one holding a single piece? Can the same fixture be used for more than one operation? Always keep in mind that the human hand makes a very poor clamp vise, or fixture.
5. Are any semi-automatic tools applicable? For example, a power-driven wrench or screwdriver.
6. Is the operator using both hands all the time? If so, are the operations sym- metrical? Wherever possible, both hands should be in motion and moving simultaneously in opposite directions. Could two pieces be handled at one time to better advantage than one? Could a foot device be arranged so that an operation now performed by hand could be done by foot?
7. Are the raw materials placed to the best advantage? Are there racks for pans of material and containers for smaller parts? Can the parts be removed from the containers with ease? Are the most frequently used parts placed in the most convenient location? Remember, the shorter the distance moved, the less the time will be.
8. Are the handling equipment and methods sufficient? Would a roller or belt conveyor improve conditions? Could the parts be placed aside by means of a chute? Drop delivery is desirable where possible.
9. Is the design of the apparatus the best from the viewpoint of the workman? Could the design be changed to facilitate machining or assembly without af- fecting the mechanical or electrical qualities of the apparatus?
10. Is the job on the proper machine? Are the correct feeds and speeds being used? Are the specified tolerances all right for the use to which the part is to be put? Is the material being used the best for the job? Could one operator run two or more machines?

Multiple Activity Charts

TYPES OF MULTIPLE ACTIVITY CHARTS. The multiple activity chart is the symbolic and systematic presentation of the method of work per- formed by a man when his work is coordinated with one or more **cycle-time**

controlling devices, such as another man, a machine, a process, or several machines. Since a great many activities are of this type, multiple activity charts are not uncommon. Their primary value is in gathering together and properly presenting information about these activities. This in turn aids in the development of new and improved procedures for the most efficient utilization of men and machines. A checklist to assist in the analysis of multiple activity charts is given in the section on Work Simplification.

Any of the following types of multiple activity charts may be made.

1. Man and machine chart.
2. Man and multi-machine chart. One man is working with two or more time-controlling mechanisms.
3. Multi-man chart. Two or more men are working coordinately on the same work. It is not sufficient to have two or more men working in an area, a department, or side by side. The men must be involved in a given work activity, like moving 20-ft. lengths of steel pipe or carrying a prefabricated house section, and must be dependent on each other in performing each activity.
4. Multi-man and machine chart.
5. Multi-man and multi-machine chart.

The analyst has to determine the characteristics of the work in relation to the number of men and machines, and then work up the appropriate chart.

Construction Procedures. Multiple activity charts may be drawn using flow process chart symbols or operator process chart symbols. The man–flow process chart breakdown is used when the individual or crew members move from one workplace to another, or when the number or sequence of men or machines is changed without changing the actual method used in each of the individual activities. The usual man–machine chart is concerned with one work station and more often with the number of machines one individual can operate.

Motion pictures have been used more and more for gathering information about multiple activity operations. (See section on Motion and Methods Study.) This is especially true of work which has seldom been analyzed before, like jobs with large crews numbering 12 to 15 men. It would be virtually impossible to have an analyst watch all these men and make a meaningful chart. However, a motion picture of the same activities permits the analyst to review the same cycle over and over until he has obtained information about each person involved. This provides a well-interrelated chart. The motion picture **speeds** most frequently used for such purposes are 1 FPS or 100 FPM (memomotion study).

Every multiple activity chart should be accompanied by a workplace diagram or a flow diagram, depending upon the work situation and activity being analyzed. A flow diagram will normally be more useful for a multiple activity flow process chart, and a workplace diagram for a multiple activity right- and left-hand chart.

Man-Machine Charts. Only two symbols are used for a machine in any type of multiple activity chart. These are the operation and the delay symbols. The **operation symbol** is used when the machine, equipment, or apparatus controls the time. It is not necessary for the machine to modify the product, but merely to be controlling the time. The **delay symbol** is used for the machine when it is not controlling the time. In some cases, the machine being used does not always control the time when it is doing work, but it may restrict the operator in some way.

In Fig. 21, Niebel (Motion and Time Study) illustrates a man-machine chart representing one operator running two milling machines. This chart is drawn to a suitable time scale like the operator process chart. The usual symbol for

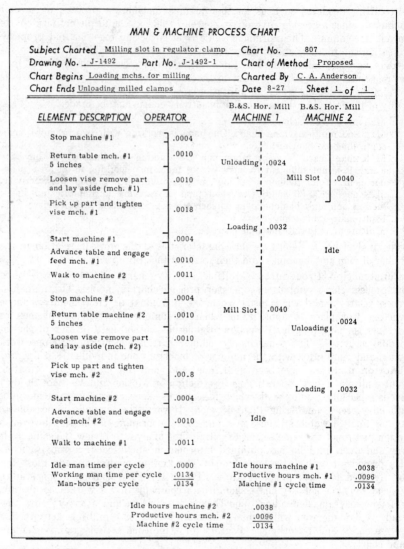

MAN & MACHINE PROCESS CHART

Subject Charted Milling slot in regulator clamp   *Chart No.*   807

Drawing No. J-1492   *Part No.* J-1492-1   *Chart of Method* Proposed

Chart Begins Loading mchs. for milling   *Charted By* C. A. Anderson

Chart Ends Unloading milled clamps   *Date* 8-27   *Sheet* 1 of 1

ELEMENT DESCRIPTION	OPERATOR	B.&S. Hor. Mill MACHINE 1	B.&S. Hor. Mill MACHINE 2
Stop machine #1	.0004		
Return table mch. #1 5 inches	.0010	Unloading .0024	
Loosen vise remove part and lay aside (mch. #1)	.0010		Mill Slot .0040
Pick up part and tighten vise mch. #1	.0018		
		Loading .0032	
Start machine #1	.0004		
Advance table and engage feed mch. #1	.0010		Idle
Walk to machine #2	.0011		
Stop machine #2	.0004		
Return table machine #2 5 inches	.0010	Mill Slot .0040	Unloading .0024
Loosen vise remove part and lay aside (mch. #2)	.0010		
Pick up part and tighten vise mch. #2	.00.8		
			Loading .0032
Start machine #2	.0004		
Advance table and engage feed mch. #2	.0010	Idle	
Walk to machine #1	.0011		

Idle man time per cycle	.0000	Idle hours machine #1	.0038
Working man time per cycle	.0134	Productive hours mch. #1	.0096
Man-hours per cycle	.0134	Machine #1 cycle time	.0134

Idle hours machine #2   .0038
Productive hours mch. #2   .0096
Machine #2 cycle time   .0134

Fig. 21. Man and machine process chart for milling machine operation.

working time of both men and machines is a solid vertical line. A break in this line is used to indicate idle time for man or machine. The machine may be prevented from productive operation by such actions as loading or unloading. Such a state is indicated on the chart with a dotted or broken vertical line. The chart is of sufficient length to show the complete cycle of the man and all machines being charted. Total time for the man must equal the total productive and idle time of each machine charted. The determination of the precise details of an improvement, however, requires a knowledge of the times taken for each activity of the man and machine. These times may be obtained by stop-watch observation or by means of motion pictures with a chronometer in the field of the camera. (See sections on Motion and Methods Study and on Work Measurement and Time Study.) The times for each activity of man and machines are charted on a **vertical time scale** (as in Fig. 21). By this means the beginning, duration and ending of each activity are clearly set forth in their proper time sequence relative to the other activities.

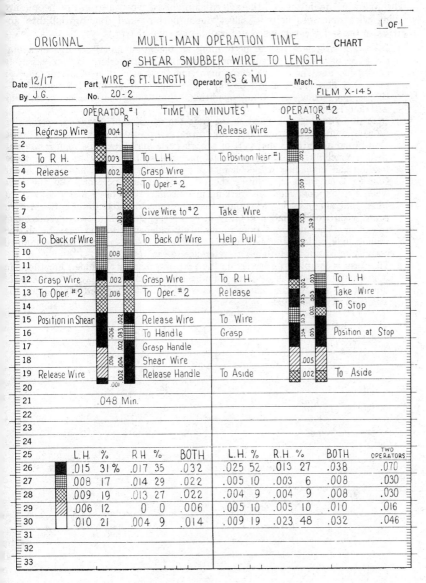

Fig. 22. Multi-man operation time chart.

Multi-Man Charts. This type of chart, sometimes called a **Gang Process Chart**, can be used most effectively in showing the coordination of work within a group of people with or without machines. It serves as a means of making a graphical record of an intricate set of relationships so that they can be more easily visualized and studied for improvement. Figure 22 (Nadler, Motion and Time Study) is a multi-man time chart of the process of shearing snubber wire to length, using the time symbols given in Fig. 18.

The following definitions of the activities involved when a group of people are working with machines help to select the important and **distinguishable activities** so that they can be used in formulating such charts:

1. **Man working** is a person performing an operation independently of a machine or of another person. For example, a man working at an assembly operation; also, in a work cycle involving the use of a machine or another man, that part of the man's time during which he is working independently—that is, not setting up, loading, operating, or unloading the machine or working with another man.

2. **Machine running** is defined as the time the machine is operating, performing its work without requiring attention, so that the operator is free for other work.

3. **Combined activity** is that part of a work cycle during which a man is working with a machine or another man. It includes such activity as setting tools, loading and unloading work where this ties up the machine, and the machine running time when it requires the attention of an operator; also the time when two or more men are working in unison. It has been found convenient when analyzing a cycle of man and machine time to differentiate between their times when working independently of each other and when one depends upon the other. The blocks of time representing independent work may be shifted around independently of each other, whereas the blocks representing combined activity must not be shifted with reference to each other.

4. **Idleness** is complete inactivity on the part of a man or machine. For the man it is usually during the machine running time when there may be nothing for him to do, and for the machine it is when it is stopped and waiting for the attention of an operator.

WORK MEASUREMENT AND TIME STUDY

CONTENTS

CONTENTS (*Continued*)

CONTENTS (*Continued*)

WORK MEASUREMENT AND TIME STUDY

Work Design and Measurement

CONCEPTS OF WORK DESIGN. The field of work methods and measurement encompasses all aspects of the design of work systems. The generic term **work design** is used to describe the design of work systems taking into account men, machines, materials, sequence of operation, and the appropriate working facilities (Journal of Industrial Engineering, vol. 16). The process technology along with the physiological and behavioral characteristics of man are taken into consideration. Individual areas of study may include analysis and simplification of manual motion components; design of jigs, fixtures, and tooling; man-machine system analysis and design; or the analysis of gang or crew work. G. Nadler (Work Design) uses the term work design to define the systematic investigation of contemplated and present work systems in order to formulate, through the ideal system concept, the easiest and most effective systems and methods for achieving the necessary functions.

In both definitions, the concept of analysis and design of the total system has replaced the traditional approach to detailed study of individual operations or fragmented segments of a process. Work design includes techniques of analysis such as operation analysis (methods engineering, methods study), work study, motion study, motion analysis, motion economy, work simplification, and job design. Although these terms are often used interchangeably, each describes a specific aspect of work design as outlined in the following definitions.

Operation Analysis (also called **methods study, methods engineering**). A study which encompasses all those procedures concerned with the design or improvement of production, the purpose of the operation or other operations, inspection requirements, material used and the manner of handling material, setup, tool equipment, working conditions, and methods used.

Work Study. The technique of methods study and work measurement employed to ensure the best possible use of human and material resources in carrying out a specific activity.

Motion Study (also **motion analysis**). Detailed study of the manual and/or body motions used in a work task or at one work area, often involving comparative analysis of right- and left-hand motions.

Motion Economy. Use of the basic principles of the manner in which body motions are performed to simplify and reduce work content.

Work Simplification. Improvements in work methods or work flow initiated and developed by supervisors or the workers on the job as a result of methods training and/or economic incentives.

Job Design. The establishment of job content using the concept of enlarging or broadening the scope, variety, and responsibility of a job in order to improve personal job satisfaction and increase output quality and quantity.

PURPOSE OF WORK DESIGN AND WORK MEASUREMENT. The purpose of work design and related techniques is to determine the best or most effective method of accomplishing a necessary operation or function. The criteria for best could be an increase in job satisfaction and personnel morale; reduction in physiological fatigue; decrease in the number of accidents and personal injuries; minimization of material usage, tool breakage, or usage of consumable supplies; increase in productivity by reduction of performance time, etc. These measures, or a combination of these measures, demonstrate that every operation contains to a certain degree **mechanical, physiological, psychological, and sociological factors.** The purpose of work measurement is to quantify these factors.

Traditionally, **productivity** as measured by performance time has been used as the sole criterion for evaluation of a given work design situation. In recent years there has been an emphasis on consideration of the **physiological** aspects of human work, especially in the determination of fatigue allowances (Journal of Industrial Engineering, vol. 17). Also, there has been a growing concern about the **psychological** and **sociological** aspects of job design starting with the work of L. E. Davis (Journal of Industrial Engineering, vol. 8), and F. Herzberg (Work and the Nature of Man). Consideration of the two latter factors usually is relegated to the disciplines of industrial psychology and sociology; they will become an important aspect of work measurement for the evaluation of the evolving automated work systems.

Thus the purpose of work design is to find the best way of completing a given task while work measurement is concerned with the evaluation of the resulting design in terms of what the designer has decided to minimize or maximize. Traditionally, performance time has been used as the measure of goodness and consequently measurement has become synonymous with **time study.** However, other measures such as employee satisfaction, physiological demand, product quality, and operating costs should be considered in certain situations.

OPERATION ANALYSIS PROCEDURE. Work measurement, and more specifically time study, is considered as an integral part of a formal operation analysis. The overall procedure of operation analysis, including the design or improvement of an operation, work measurement, and job standardization, is divided into the following steps:

1. **Preliminary analysis.** Analysis and recording of existing conditions or proposed specifications as to the process, equipment, materials, material handling, workplace layout, methods, volume of required output, and costs, as well as of limitations in the availability of resources such as capital, space, and labor.
2. **Description of operation.** Breaking the operation down into arbitrarily defined functional or motion elements and a systematic charting of these elements to describe the existing or proposed method.
3. **Detailed analysis.** Critical study of the elements of the operation to determine which ones may be eliminated or improved. This step includes the analysis of methods, tools, equipment, materials, and workplace layout, as well as of the motion pattern used by the operator.
4. **Work measurement.** Analysis of the time consumed by productive and nonproductive elements of the existing or proposed operation.
5. **Synthesis.** Integration of improved elements into alternative operation designs.
6. **Evaluation.** Selection of the best alternative in terms of such specific criteria as unit time, cost, and space required.
7. **Job standardization.** Standardization of methods, job conditions, and unit production times by reducing them to some form of written standard practice. Preparation of detailed instructions covering tools necessary, and explanation of the elements of the method in detail and in proper sequence, with the time allowed for each element.

8. **Installation.** Planning and executing the introduction and installation of new methods, including the training of the operator to perform the task in the manner and time specified.

9. **Control.** Maintenance of standardized working conditions, equipment, methods, material quality and supply, and standard times during the life of the operation prior to a formal redesign of the method. Included in control are the means of detecting deviations between standard and actual performance.

DISTINCTION BETWEEN MOTION STUDY AND TIME STUDY.

The contemporary practice of motion and time study, which is a part of the operation analysis procedure, resulted from the integration of concepts and practices developed by Frederick W. Taylor and by Frank B. and Lillian M. Gilbreth. Taylor stressed the procedure of time study, while the Gilbreths developed motion study as the core of their work. Both time study and motion study were concerned with the systematic analysis and improvement of manually controlled work situations. Both groups employed the stop watch and motion analysis as components of their studies. However, motion study developed into a part of methods study and time study became a technique of work measurement.

Motion study is usually a qualitative analysis of a work situation leading to the design or improvement of an operation. Time study is a quantitative analysis leading to the establishment of a time standard, regardless of the ultimate use of the results. Suggestions for improvements of a method may arise during the course of a formal time study, but it is not a substitute for a motion or methods study. Similarly, through the use of motion films and basic-motion time data, a time standard may be an important by-product of a given motion study.

Figure 1 illustrates the relationship between motion and time study as a part of the total work design procedure.

BASIC TERMINOLOGY OF WORK MEASUREMENT.

The basic terms associated with work measurement are described in the following paragraphs.

Work Measurement. The procedure involved in measuring or forecasting the rate of output of an existing or newly designed operation, as well as in determining how much time is consumed for various productive and nonproductive activities of a process, operation, or job, is known as work measurement. Also involved is the determination of **standard times** which represent the allowable time for the performance of work. The term work measurement is used generically and pertains to all techniques of time measurement of work systems.

Standard Time. The results of the application of a work measurement technique may be expressed in terms of a standard time which represents "that gross time required by a normal operator, working under normal conditions, and with normal skill and pace to complete a unit of work of satisfactory quality." This subjective criterion of normal or standard performance leaves considerable range for interpretation.

Other terms used to express the idea of standard are "a fair task," or "a fair day's work." Work measurement is more than measurement of time. This has been true since its inception by Taylor, who said (ASME Trans., vol. 34): "Mere statistics as to time which a man takes to do a given piece of work do not constitute a time study. Time study involves careful study of the time in which work ought to be done."

The determination of what a worker "ought" to produce in a day is a problem potentially fraught with difficulties. Both **operator method** and **pace** are in-

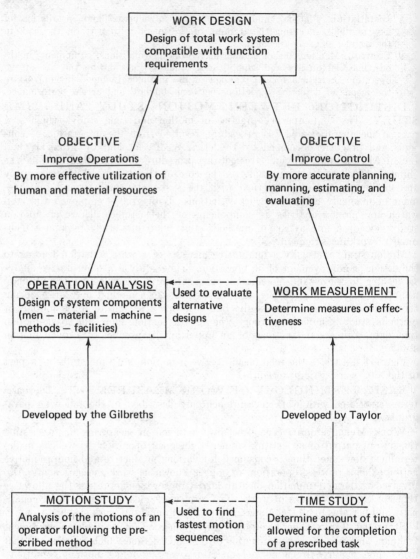

Fig. 1. Relationship of motion and time study to work design.

volved. As a formal procedure, time study may contribute to the development of a "best" method, and this may be done with a certain amount of engineering objectivity. But standard pace, or how fast a worker should produce, is a matter of judgment, at least with present knowledge of human capacities. An objective procedure for determining an **optimal worker pace** in terms of physiological and psychological criteria may be developed as our knowledge through research improves.

Proper time study usually locates the task at a level which is not detrimental to the worker's health. Rates of production or tasks resulting from inadequate time studies are sometimes so high as to meet, at first, with incredulity and opposition from operators. It is desirable to all concerned that tasks be measured according to standards which are fair. In a wage incentive program, management must be prepared to train the operator in the best method, to guarantee his wages during a reasonable period of training, and to guarantee standard conditions of work upon which the study was based.

Job Standardization. Recording the exact method on an **instruction card** together with the **time** for each element of the operation constitutes job standardization. Standard job conditions, material specifications, tools, and workplace layout should be carefully recorded.

Once established, time standards should not be changed so long as the method or conditions remain unchanged. When new tools, fixtures, or machines are designed for a standardized job and the standard conditions controlling the job have been changed, the original standard time can no longer apply. A new time standard must then be established.

FUNCTIONS OF WORK MEASUREMENT. Work measurement results are used by management as an analytical technique for evaluation of work methods, determination of standard time values for given tasks, cost analysis and comparisons, and for development of standard time data systems as described in the following paragraphs.

Evaluation of Work Methods. One function of work measurement is to determine where and how time is employed in a present work method or proposed operation. In an ordinary industrial operation a **stop-watch time study** measures the various elements of the productive cycle of an operation, and the results can be used to indicate the sources of delays as well as areas where time may be reduced.

In highly repetitive, short duration operations that move at a speed too rapid for satisfactory observations, a **micromotion study** using a motion picture camera (960 frames per min., or faster) can be performed in order to evaluate the motion patterns. Lengthy events, group activities, or processes that do not move rapidly can be analyzed by a memomotion study which records events at 50 or 100 frames per minute. Extremely long operations which might occur at irregular intervals can be studied over a protracted period of time by means of **work sampling.**

Regardless of the method involved, the results from the work measurement procedure will indicate areas of inefficiency that exist in the process or operation. Also, time standards, expressed in terms of monetary units, are often used to make production operators and supervisors motion-minded in order to stimulate participation in a work simplification program as illustrated in Fig. 2.

Determination of a Standard or Task Time. When an operation has been improved and standardized as to method, the primary function of work measurement is to establish a standard time. This standard reflects the time required for a unit of output. Depending on the level to which it is set, the standard time can become the basis for:

1. Individual or group wage incentives.
2. Evaluating the actual performance of an individual or group.
3. Scheduling work to an operation or process.
4. Determining the number of operators required for a specific volume of output.

TIME IS MONEY. BE MOTION MINDED,

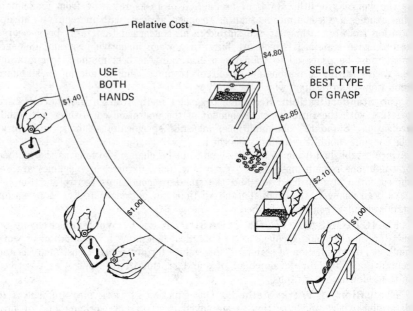

Fig. 2. Making production operators and supervisors motion-minded.

5. Programming work to production or service processes and operations.
6. Estimating cost of jobs performed at the operation.
7. Compiling standard labor and overhead costs and budgets.

It should be noted that the interpretation of standard time will vary according to the ultimate use of the standard. A standard time for incentive purposes might not be the time used for scheduling work or making a cost estimate. That is, the time that management considers representative of **standard performance** may not represent the **expected time in actual practice.** For instance, the "standard time" for incentive plans may have such a relationship with expected performance that the average operator will be expected to earn a bonus.

Synthesis of Standard Time Data. The third major function of direct work measurement is to provide initial data which may be expanded into a system of standard time data. Such standard data constitute an inventory of elemental times which may be synthesized to provide a standard for a job, making direct timing methods unnecessary. Standard data may be expressed in terms of functional or motion elements. During the stages of designing a process or operation, standard data are usually the only practical basis for evaluating **alternative designs** in terms of time, as well as for estimating a standard time for the selected design prior to installation.

Elements of an Operation

TYPES OF ELEMENTS. An operation may be described by functional elements or motion elements. **Functional elements** are arbitrary descriptive subdivisions of a work cycle readily identified and defined. In a given sequence these elements designate what is doné to alter material properties or to provide a service by means of a combination of manual movements and mechanical actions. Examples are: pick up part, place part in jig, close jig, drill ⅜-in. hole to ½-in. depth, and open jig. These elements are also referred to as **time study elements** because they are capable of being timed with a stop watch. A time study element which is manually controlled will consist of groups of motion elements.

Motion elements are basic motions or simple motion combinations used to describe the sensory-motor activity in an operation. The Gilbreths named the 17 fundamental hand motions **therbligs** and this term is often used as a synonym for motion elements. There are a number of **predetermined motion-time** systems used today in which time values are associated with precisely defined motion elements found in such predetermined time systems. They are:

Reach	Assemble	Foot Motion
Move	Disassemble	Bend
Grasp	Pre-position	Stoop
Turn	Position	Kneel
Apply Pressure	Release	Sit
Eye Focus	Disengage	Arise
Eye Travel	Leg Motion	Stand
Walk	Sidestep	Turn Body

METHODS OF WORK MEASUREMENT. There are five commonly used methods or techniques of work measurement used to obtain time standards, namely: time estimation, historical records, direct time study, predetermined time systems, and work sampling.

Time Estimation. This is not a true measurement technique but rather is a means of obtaining time values based on the personal judgment of the estimator to ascertain how the work should be performed and the time required to perform the work. The accuracy of the time estimates depends upon the experience of the estimator with similar work projects. Rather than forcing a single point estimate, the expected time value may be calculated by a weighted average of the most **optimistic** time estimate, the most **pessimistic** time estimate and the **most likely** time estimate using the relationship

$$\text{Expected time} = \frac{\text{Optimistic} + 4\,(\text{Most Likely}) + \text{Pessimistic}}{6}$$

B. W. Niebel (Motion and Time Study) points out that time estimates used for labor standards have an average deviation of about 25 percent from the values obtained from correctly conducted time studies. However, time estimates are very useful for project planning purposes where the deviations in the estimates can be adjusted as the project progresses toward completion.

Historical Records. Time standards might be calculated from production records. Standards based on historical data are usually more accurate than time estimates. However, they merely reflect how the work has been done in the past and not necessarily how the work should be done. However the opportunity to make an analytical study of the work situation by more direct work measurement techniques may be lost.

Direct Time Study. This is the procedure by which an analyst using a **stop watch** or some other suitable timing instrument measures the elapsed time for a number of complete cycles. The **average performance time** is calculated

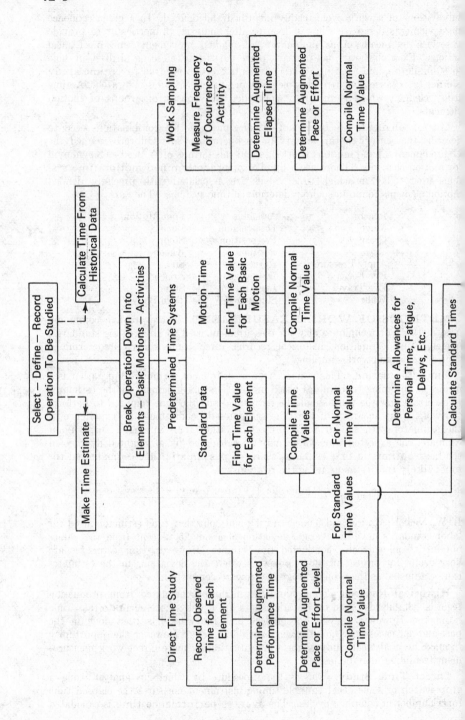

from an adequate sample of individual cycle times. The average time is adjusted for any observed variance from the time-study analyst's concept of normal effort or pace to obtain the **normal time** (also called **base time** or **leveled time**). The normal time is converted to **standard time** by the addition of allowances for personal, fatigue, and delay factors (pfd) plus any other special allowances such as learning or training effects. If the task is of sufficient length, it is broken down into short, relatively homogeneous **work elements**, each of which is treated separately and then compiled into a standard time for the entire cycle.

Predetermined Time Systems. This is the generic term used to describe the work measurement procedure of assigning time values to elements or basic motions of an operation from a collection of values obtained from previous measurements of tasks in similar job situations.

Standard Data Systems usually consist of normal time values for similar elements determined by direct time studies. Because they are based on time study results conducted in the organization using the data, allowances may be included in the time values; thus resulting in a system of standard time data.

Predetermined Motion Time Systems consist of normal time values for basic manual or body motions. Motion time data have a wider, or more universal, application than the standard data systems based on time study elements. This is because normal time values must be supplemented by a company's allowance system in order to obtain standard times.

Work Sampling. This technique consists of intermittent, instantaneous observations which may be taken at regular spaced intervals or at random intervals, distributed over the entire duration of the study or over several strata of the observation period. Regardless of the method of sampling, the observations are made stochastically independent of the work cycles under study. The primary objective of a work sampling study is to determine the percentage of time spent on the various activities or delays. If the observations include rating the operator's performance in terms of pace or effort and the average time spent on the various activities, normal time values can be established for the activities.

SELECTING THE APPROPRIATE METHOD. Selection of the appropriate work measurement method should be consistent with the degree of precision and accuracy required for the ultimate use of the data. Economic considerations usually determine the technique selected. Each technique relies on the experience of the time-study analyst and the data used to compile the results. Figure 3 summarizes the various techniques that may be used to establish standard time values.

Applicability of a Standard Time. The predictive value of a time measurement depends on the similarity between the parent population and the extended population for which the prediction is made. The causes for variation must remain essentially constant over the protracted time period. This is basically a problem of standardization and control of job factors both during the study and in the subsequent performance of the job.

When it is apparent, during the life of the operation, that unusual causes for variation have arisen, they must be corrected to bring the operation back into control, or a new time study must be made to account for them in the standard time.

Accuracy. One of the sources of inaccuracy is bias in the time study procedure. A result may be inaccurate due to biases, as follows.

1. Error in reading the stop watch.
2. Error in interpreting the elements being measured.
3. Influences injected in the method and pace by the operator.
4. Variations from standard in materials, methods, and conditions during the study.
5. Failure to include nonrepetitive or cyclical sources of variation by limiting the length of the study.

Like applicability, accuracy depends on the degree of standardization and control exerted during the study. Taylor recognized this in his early studies and therefore stressed the necessity of standardizing and controlling **job variables** during the actual time study. Much of the early unpopularity of time study, from labor's point of view, can be traced to the failure of practitioners to give sufficient attention to this matter.

The fundamental task of methods and job standardization is to remove and stabilize the causes of variation to the extent that the remaining variability is due to constant chance causes. That is, the **accepted variability** in the time data is essentially random within some specified and predictable limits.

During the formal time study the observer is required to use his judgment as to when standard methods, motions, and conditions prevail, even though the standards may be specified carefully in writing. This is manifested in his refusal to time a job if the specifications are not followed or in his deleting from the study individual readings which represent nonstandard situations.

Avoiding Biases. Work measurement deals with the measurement of a phenomenon which in itself is sensitive to the measuring procedure. In manually controlled operations variation and change can be intentionally initiated and controlled by the operator. The observer can also introduce bias, particularly in the rating procedure.

The purpose of an effective work measurement training program is to convey knowledge and provide sufficient practice to prevent, or at least minimize, human and procedural biases. The time study practitioner should be selected on the basis of his ability and background to fully understand the operations he is asked to evaluate. He should be well trained in the mechanics of data collection, statistical analysis of data, a sound system of evaluation of variances in operator performance levels, and an internally consistent procedure for determination of allowances. Appropriate training should be provided in the use of a given predetermined time system and for conducting valid work sampling studies.

Time Recording and Accessory Equipment

TIME RECORDING DEVICES. The most widely used device for measuring elapsed time is the stop watch with the motion picture camera ranking second. Special purpose electromechanical time recorders are also available for recording time study data and machine running times. Finally audio and video tape recorders are used effectively in work measurement.

STOP WATCHES. Basically a stop watch consists of a good watch movement and a method for controlling the movement of the watch's hands. Sequence timing watches use the winding stem for controlling the timing mechanism. Successive depressions of the stem cause the hand to start, stop, and return to zero. The watch used for **continuous** and snap-back timing (Fig. 4) has a slide A to start and stop the timing mechanism and a stem B to return the hands to zero. The hand starts to move as soon as the depressed stem is released.

Decimal-Minute Stop Watch. The decimal-minute watch (Fig. 4) has a large hand making one revolution per minute and a small hand making one revolution in 30 minutes. The watch continues to run for periods longer than 30 minutes but the number of 30-minute periods must be recorded elsewhere.

Fig. 4. Decimal stop watch.

Decimal-Hour Stop Watch. The decimal-hour stop watch is similar to the decimal-minute watch in design and operation. Each of the 100 subdivisions for the large hand represents 0.0001 hour, and each of the 30 subdivisions for the small hand represents 0.01 hour. The large hand makes 100 revolutions per hour, representing 0.01 hour for each completed revolution. The small hand makes 3⅓ complete revolutions per hour.

The chief advantage of the decimal-hour watch over the decimal-minute watch is that the readings are represented directly in fractions of an hour; however, the four-place data is more difficult to record than the two decimal places obtained with the decimal-minute watch.

Split-Hand Stop Watch. The split-hand watch (Fig. 5) has two large hands. By successive presses of stop B, both hands will start together, stop, and return to zero. By pressing stop A, the lower hand is held wherever it is while the upper hand continues its progress. This is done when a delay occurs so that its time may be read without interrupting the overall timing. At a second depression of stop A, the lower hand instantly catches up with the upper hand and continues with it. This watch is not adapted for snap-back recording.

Another variation is a series of numbers about a dial which indicate the number of times any elapsed time is repeated per hour. This feature is more confusing than helpful.

Accuracy and Calibration of Watches. G. G. Wiley and J. W. Heuer (Instruments & Control System, vol. 33) in their review of manual timing techniques, state that a worthy timepiece manufacturer can guarantee precision within one beat in 10,000 (approximately ⅓ second per hour of timing). Watches should be calibrated from time to time by operating them for a suf-

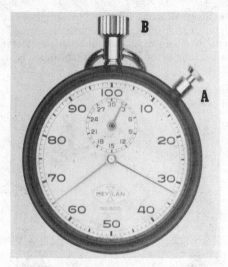

Fig. 5. Split-hand stop watch.

ficient period of time and comparing their readings with a regular watch or chronometer of known reliability.

MOTION PICTURE CAMERAS. The motion picture is an accurate record of both the method employed and time taken during a particular study (see section on Motion and Methods Study).

Synchronous Motor-Driven Cameras. If the camera is driven by a synchronous motor, the time for individual motions, elements, or cycles may be determined by converting frame counts to time units. The usual camera speed is 1,000 frames per minute which is equivalent to 0.001 minute per frame. Camera speeds greater than 1,000 frames per minute may be used for analysis of extremely fast motions. A **time-lapse drive** may be attached to the camera to take pictures at slow speeds such as 1 frame per second or 100 frames per minute for the analysis of long cycle operations as in memomotion studies (M. E. Mundel, Motion and Time Study: Principles and Practice). **Random observations** can be recorded on film by attaching an external triggering device to the camera, using a method devised by Gambrell and Barany (The Office, vol. 50).

Spring-Wound Cameras. If the camera uses a spring drive or if the projector lacks a frame counter, a special timing device such as a **microchronometer** or **wink counter** should be placed in the subject field in order to provide direct time readings on each frame. H. G. Thuesen (The Development of a Memo-Activity Camera) has designed a special purpose camera which records the absolute time at which each frame is taken by photographing the dial of a miniature, battery-powered timer built into the camera itself.

Microchronometer. A specially designed clock driven by a synchronous motor is used with spring-driven cameras. The dial of the clock is divided into 100 equal divisions. The large hand makes 20 revolutions per minute, and the small hand makes 2 revolutions per minute. Therefore, each subdivision on the clock represents 0.005 minute.

Wink Counter. Instead of a clock for recording time, a motor-driven counter resembling a speedometer may be used. The device, called a wink counter, has three revolving numbered discs from which time can be read accurately to 0.005 minute with a helix on the drive shaft that may be used for closer timing. The counter may be used in conjunction with a camera or may be read by the observer at the same instant he observes the motion of the operator, in a manner similar to reading a stop watch.

Film Costs. The major disadvantage of using a motion picture as a timing device is the cost of obtaining an adequate sample size. Most industrial applications require 16-mm. film containing approximately 40 frames per foot. Therefore, a micromotion study conducted at 1,000 frames per minute would consume a 100-foot roll of film in about 4 minutes. In order to get an adequate sample size, a relatively large number of cycles might be required. The cost of processing and analyzing the film might make this technique of timing impractical.

ELECTROMECHANICAL TIME RECORDERS. A number of special purpose electromechanical recorders have been developed for timing manual operations and recording machine running times. A kymograph, time recorder, marstochron, and various tape recorders are discussed below.

Kymograph. The kymograph consists of a synchronous-motor-driven paper tape on which lines are drawn by a series of solenoid-operated pens. The solenoids are actuated by switches or photocells situated so that they can intercept or monitor activities at the work station. When there is no activity, the pens produce continuous parallel lines on the tape. When a solenoid is activated the pen to which it is connected is jogged and its line is offset slightly. The line is returned to its original position again by a switch or photocell. Time values can be determined by measuring the length of the offset line. Several electronic kymographs have been designed to record the lapsed times on punched cards or tape suitable for tabulation with a digital computer. The kymograph is capable of measuring time to the nearest 0.00001 minute.

Time Recorders. Mechanical or electrical time recorders may be used to record productive machine time. A **Servis** recorder is a mechanical timer consisting of a graduated circular chart revolving once in 8 or 10 hours. The device, attached to the piece of equipment being studied, records activity time by means of a vibration-sensitive pen which produces a jagged line on the chart paper when the equipment is running and a smooth line when it is stopped. Delay causes are noted by the operator or some other person. Other devices used to record running times utilize the electrical supply to the machine for pen activation. Connections may be made to a limit switch placed at an appropriate place for equipment that runs continuously. Similar principles of operation are used in time totalizer mechanisms that give elapsed run-time and down-time, thus removing the necessity for chart analysis.

Marstochron. This device was designed to record time intervals too short to be measured with a stop watch (for example, 0.01 minute or less). The marstochron consisted of a small box through which a **scaled paper tape** was drawn at a uniform speed by means of an electric motor. The beginning and end of each element were recorded by pressing one or two control keys mounted on the case. Pressure on a key depressed a type bar which made a mark on the tape. Elements were identified by use of the keys in different arrangements, such as pressing both keys or double pressing one or the other. The tape moved at a rate of 10 to 20 inches per minute, depending upon the motor drive used. At 10 inches per minute, 1 inch of tape represents 0.1 minute and 1/10 inch equals 0.01 minute.

The principal advantage of the marstochron was that the observer could keep his eyes on the operator and was not required to make any written observations Its major disadvantage was that there was no means of making notes during the study or of handling the many foreign elements and interruptions during the study.

Audio Tape Recorder. Both audio and video tape recorders may be used for timing purposes. J. M. Kminek (Journal of Industrial Engineering, vol. 12) developed a device to place a time-based pulse signal on the tape in a portable audio recorder. Element break points are recorded along with description of the activity. Time per observation is obtained by counting the pulses after completion of the recording session. The technique is used for timing short elements (less than 0.04 minute) and for studies where there is inadequate lighting for reading a stop watch. Use of the time-based pulses removes the need for expensive high, constant speed recording equipment.

Video Tape Recorder. Apparatus designed to record television signals on magnetic tape is sufficiently small and mobile to make it practical for use in work design studies. The equipment is capable of collecting, transmitting, recording, and reproducing the audio-visual information simultaneously, instantaneously, and continuously. The major advantages are that the tape is reusable, requires no special processing, and can record relatively long operations (60–180 minutes). Stop-action and slow-motion techniques are possible with video equipment.

COMPARISON OF TIMER ACCURACY. Comparisons have been made of the accuracy of time values recorded by the use of a stop watch, wink counter, and marstochron. For equal confidence in results, 2 to 5 times as many observations are necessary with the stop watch as with the marstochron. With the marstochron, an element of 0.01 minute can be read as readily as 0.10 minute with the stop watch. For the wink counter, the smallest recommended element is 0.025 minute in comparison to 0.04 minute for the stop watch.

ACCESSORY TIME STUDY EQUIPMENT. A time study board for holding the time study observation sheet and the stop watch is required for ease of recording data. The **time observation sheet** should provide space for recording all pertinent information required for setting a standard time.

Time Study Board. The data or observation sheets, and usually the watch, should be held on a specially designed board. A stop watch has a more delicate movement than an ordinary watch, and it should be handled carefully. If the analyst holds it in his hand, he is sure to lay it down occasionally, often on vibrating machines or benches where it can be damaged. Moreover, it should be in a fixed position relative to the observation sheet. Therefore, the watch should be mounted on the board.

The time study board shown in Fig. 6 (Meylan Stopwatch Co.) combines the features needed for general time study work. It provides a convenient writing surface for holding the observation sheet and holds the stop watch in position to be operated by the left hand while the left arm supports the board, leaving the right hand free to record observations.

Time Observation Sheet. There is considerable variation in time observation sheets, depending upon the technique used by the analyst in taking observations. The essential feature of a well-designed observation sheet is that it provides space for all necessary data. Both sides should be used, one side for time study observations and the other for all additional data of identification, analysis tools, etc.

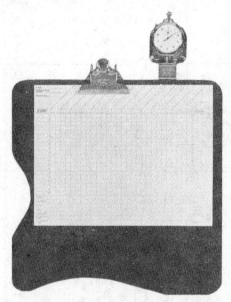

Fig. 6. Time study board.

For a non-repetitive operation, the sheet shown in Fig. 7 may be used. Two columns are used for writing down each element, or its code letter, as it occurs. This sheet is more adaptable to **indirect labor operations** where the sequence of elements is not fixed. Typical data for such an operation are recorded in Fig. 7. Columns for rating each element, for clock reading, and for element time are included.

The sheet shown in Fig. 8 is designed for studying a **repetitive operation.** The details of the work cycle are recorded in sequence. Space is provided in the middle of the sheet for elemental times for ten cycles. The reverse side of the sheet can be used for sketches of workplace layout and information on jigs, fixtures, tools, etc.

Stop Watch Time Study Procedure

OVERALL PROCEDURE. The major steps included in a formal time study are as follows:

1. The objective of the study is clearly defined. This includes a definition of the use of the results, the precision required, and the desired confidence in the estimation of the time parameters.
2. The purpose of the study and necessary preparations are reviewed with the supervisor of the operating department.
3. A preliminary analysis is made of the operation to determine whether standard methods and conditions prevail and whether the operator is properly trained. If necessary a request may be made for a formal methods study and further operator training.
4. Where there is more than one operator, the operator to be studied is chosen.
5. The selected operator is oriented as to the purposes and, if necessary, the method of the study.
6. A detailed record is made of the standard method and conditions, either in writing or by motion picture.

TIME STUDY OBSERVATION SHEET   N⁰ 1331

OBSERVER J. O. Moore   DATE 3/4/   SHEET NO. 2 OF 8 SHEETS

ELEMENT	RTG.	READ'G	EXTN.	ELEMENT	RTG.	READ'G	EXTN.
Unlock door, turn on lights Rm. 101		10	.10	Move bucket & applic. 10'		81	12
Move materials into room		26	.16	A - 400 sq. ft.		26 81	8.00
Walk 30 ft.		37	11	D - 1 chair		85	.04

Element Code
A - Wax clear area
B - Buff clear area
C - Move desk or table
D - Move chair

OPERATION AND PRODUCT

Wax and buff office floors – Rm. 101

| START 5:30 | STOP 9:30 | ELAPSED TIME 4 |
| PRODUCTION | STD. HRS. EARNED | OVERALL EFF. |

SKILL POOR FAIR AVER. GOOD [✓] EXCEL SUPER EXCES.
EFFORT POOR FAIR [✓]

OVER-   R & D   TOT. 95
ALL %   %   RTG.

OPERATOR'S NAME L. Brown
M   OR   F
REG. NO. 1268

Fig. 7. Time study observation sheet for nonrepetitive operations.

TIME OBSERVATION SHEET

FACT. NO.	18	DATE	5-21-	PART NO.	62 B 67
DEPT NO.	22	MODEL NO.	Std.	SCHEDULE NO.	T4920
PART NAME	Bearing Cap			CANCELS SCH. NO.	T 1151
OPERATION NO.	20A	OPERATION	Drill 2 (19/32) Holes	GROUP NO.	none
MACH. NO.	681	MACHINE	24" Cinn. Dr. Press	PAGE	2 OF 3

		NO.	DETAILS OF WORK CYCLE	ELAPSED TIME IN DECIMAL MINUTES										ALLOWED TIME		
SKETCH (OVER)	✓														IDLE TIME/CYCLE	.60
REMARKS (OVER)	✓	1	Place 1 part in fixture (27")	.03	.04	.04A	.04	.05	.03	.07	.03	.04	.05	.04	IDLE TIME/PIECE	.60
SAFETY	OK	2	Clamp 1 part (socket wrench & nut)	.08	.07	.09	.09	.08	.10	.09	.10	.08	.07	.06	DELAY/PIECE	✓
LUBRICANT	Sol. Oil	3	Lower drill head (3")	.03	.03	.03	.02	.03	.04	.03	.04	.02	.03	.02	MACH. CAP'TY/HR.	57
DRIVE	Elect.	4	Drill 2 holes (19/32 x 2)	.68	.66	.65	.69	.67	.66B	.68	.67	.65	.65	65	OPER. CAP'TY/HR.	57
LOT SIZE	✓	5	Mach. feed trip, raise head	.02	.04	.03	.02	.02	.02	.03	.04	.02	.03	.02	PIECES/CYCLE	1
LOTS/YEAR	✓	6	Release 1 pt. (socket wrench & nut)	.05	.05	.06	.07	.08	.05	.06	.07	.06	.08	.05	RE'QD. PROD./HR.	50
CONT'US PROD.	Yes	7	Remove 1 part (16")	.03	.03	.04	.05	.04	.03	.03	.05	.04	.05	.03	NO. PCS. STUDIED	10
CONE STEP US'D	✓	8	Blow out fixture (Hose)	.05	.05	.06	.05	.05	.05	.04	.05	.06	.04	.05	NO. OF OPERATORS	1
FEED LEVER	.010	9													NO. OF HELPERS	✓
SPEED LEVER	✓	10													NO. OF MACHINES	1
MATERIAL		11													FORMER BASE RATE	2.20
Mall Iron		12													NEW BASE RATE	2.20
		13													FORMER STD. TIME	.0160
DELAYS		X	4A - go to next spindle to spotface												NEW STD. TIME	.0175
A (12).08		X	4B - spotface 2 holes	see oper. #20B											FORMER PROD./HR.	62.5
Examine fixture		X	4C - return to position												NEW PROD./HR.	57
B		17													% INCREASE PROD.	✓
Change drill .74		18													% DECREASE PROD.	10
C		19													% INCREASE TIME	9.4
		20													% DECREASE TIME	✓
D		21													FORMER COST/PC.	.0352
		22													NEW COST/PC.	.0386

ITEM #	4	WORK CYCLE TIME PER PIECE	.920	
CUTTING DIA.	.693	MULTIPLE PCS./WORK CYCLE, PCS.; PRORATE/PC.	✓	DEMONSTRATION
CUTTER DIA.	"	" MACH./ " " ; MACH., " /MACH.	✓	GOOD ☑ FAIR ☐ POOR ☐
NO. OF TEETH	2	" OPERS./ " " ; OPER.; " /OPER.	✓	
CUTTER MATL.	HSS	HANDLE SUPPLIES MIN. FOR /PCS.; " / "	✓	APPROVED BY:
R. P. M. WORK/TOOL	360	" SCRAP " " / " ; " / "	✓	ANALYST
F. P. M. WORK/TOOL	56	" STOCK TO WORK AREA .30 " " 15 / " ; " / "	.020	
TYPE OF CUT	Dr.	" " FROM " " .36 " " 15 / " ; " / "	.024	CHECKER
FEED/REV. WORK/TOOL	010	MACH. SET UP TIME ALLOWANCE ✓ %; MIN./SET UP; " ; " / "	✓	
FEED/MIN. WORK/TOOL	3.6	TOOL CHANGE TIME ALLOWANCE " ; 1 MIN./CHANGE; 200 " ; " / "	.005	FOREMAN
FEED/TOOTH WORK/TOOL	005	REST OR DELAY ALLOWANCE ✓ " MIN./DAY, " ; " / "	✓	
LENGTH OF CUT	2¼	MACHINE ALLOWANCE ✓ " ; 19 " / " ; " ; " / "	✓	GEN FOREMAN
WIDTH OF CUT	✓	SPECIAL ALLOWANCE ✓ " ; " / " ; " ; " / "	✓	
DEPTH OF CUT	2"	PERSONAL ALLOWANCE ✓ " ; 21 " / " ; " ; " / "	✓	EFFICIENCY ENG.
LIMITS	✓	TOTAL ALLOWANCE 9 " ; 40 " / " ; " ; " / "	.083	
REASON FOR CHANGE Holes were 17/32 dia		TOTAL TIME ALLOWED PER PIECE	1.052	SUPERINTENDENT

Fig. 8. Time study observation sheet for repetitive operations.

7. The operation is divided into time study elements.
8. A number of time study observations are taken. This provides the basis fc estimating certain statistical parameters which are used to determine the tot; number of observations to be taken. Also, the length of the study period ma be specified.
9. The actual measurement is made with the requisite number of observation over the specified time period.
10. The operator is rated during the study by comparison to the performance leve of the normal operator.
11. Allowances for various job interruptions and delays are determined by polic. or by independent measurement.
12. The standard time is derived from the observed times, the rating factor, an allowances.

PRELIMINARY INVESTIGATION. Prior to the actual measuremen of net cycle times the **objective** or purpose of the study should be clearly under stood. The objective determines the ultimate use of the standard time. With clarified objective, a rational determination of allowable error and number o observations to take can be made. For instance, if the standard is to be used a a basis for **wage incentive**, the error that will be tolerated in the measuremen may be less than if the standard is to be employed, along with others, in ap proximating an aggregate **production program** or a **labor budget.** In th latter cases accuracy and precision may be sacrificed if operation and stud controls are relaxed and the number of observations are limited.

A preliminary investigation should disclose the extent to which method analysis has been previously employed and whether further effort in method improvement and job standardization is desirable. Thus, the time measurement may be deferred until a **methods study** is concluded. The lack of origina operation design, standardization, and control may be incompatible with the ob jectives of the time study, including the accuracy and precision desired. I should be the company policy to permit the time study engineer to recommen a formal methods study when it is apparent that the time study requirement demand it. In certain cases, for expediency, a methods study may be exclude in favor of the immediate setting of a standard time. The standard may the be designated as a **temporary standard** subject to revision after a prope methods study has been made.

Careful **liaison** must be established between the time study analyst and th supervisor of the operating department. Further, this liaison must be extende to the foreman and the operator himself. Reasons for the study and the neces sary preparation to insure a valid study must be handled with subtlety an care by the analyst in charge. Without rapport between supervisor, operator and analyst, the entire study may be either inadvertently or wilfully sabotaged

STANDARDIZATION OF METHODS AND CONDITIONS. The method and job conditions should be standardized in writing, because the stand ard time is relevant only to a specific method and set of conditions. After a stan ard time has been set, considerable difficulties may arise if various interpretation of standard methods and conditions are possible. This is particularly true i the standard time is to be used as a basis for wage payment and an unexpecte discrepancy between standard and actual performance may be the result pri marily of a methods change. Such a discrepancy in a highly repetitive long-rur operation may lead to a loss of confidence in both methods and time study techniques.

Standardization by Motion Pattern. In highly repetitive operations where there is more than one operator a motion pattern may be difficult to standardiz as the one "best" method. Each operator may employ variations in motion which are peculiar to his own aptitudes. The analyst may have to employ careful judgment in standardizing a given motion pattern which he thinks should be representative for a number of operators.

Changes in the standard method, and particularly of the motion pattern, may be initiated by the operator after the standard has been set, with a resulting gain in performance level. These changes, where minor, may not be readily differentiated from skillful performance or good pace and it may be inadvisable to attempt to change the standard by retiming the job. But where the variation in method is enough to cause a significant change in performance, the job can be retimed. A **significant performance change** is one which permits the operator to have earnings well above the average, or one which can contribute to a specified shift in net cycle times as indicated on a control chart.

A basic problem here is what criteria may be employed to differentiate standard from nonstandard situations. An **operation description** in all of its details may suffice. This refers to sketches of workplace layout and a complete description of such items as the types and relative locations of equipment, motion pattern, machine feeds and speeds, and specifications of material quality. The analyst should allow for a detailed record of these factors to accompany the observation data sheets and should take the actual time readings only under these standard conditions. Where the operation is highly repetitive with short elements, a motion picture can be used to provide an accurate record of the standard method or motion pattern in detail.

Standardization by Control Chart. A more objective criterion of standardized methods and conditions is a control chart of the means of small samples of net cycle times. If the measurements are consistent or in statistical control, that is, taken from a constant chance cause system, the means (averages) of successive and randomly selected small samples would lie within statistically determined **limits.** The distribution of these means and their limits may be depicted on a control chart. These charts can be constructed for each element of the study or for overall cycle times. (See section on Quality Control and Reliability also.)

SELECTION OF OPERATOR FOR STUDY. The selection of the worker for the study depends somewhat upon the procedure used but is a very important factor in the success of the study. Micromotion and time study techniques should make use of the best workers. For timing purposes, it is advantageous to select the operator whose motions are performed with ostensible automaticity or without conscious effort. This condition is partially evident when **ballistic movements** are used, as opposed to restricted or controlled movements, and where **rhythm** is present in the repetitive portions of the work.

The foreman and the existing production records can assist the analyst in selecting the worker to study. In general, intelligence and ability to learn are as important as dexterity and ability to maintain high production. The worker should be one who has the respect and confidence of the workers in the same job classification, for he or she interprets the work of the analyst to them and can do much to help or hinder the general acceptance of the standard time.

The analyst should make selection of the worker part of the preliminary investigation. During the preliminary investigation he can learn who are the best operators and which ones are most likely to cooperate and lead. Sometimes the choice may be limited to a few or there may be only a single operator.

ATTITUDE OF ANALYST. The analyst should work in the fullest cooperation with foreman, union steward, and worker. The foreman should intro-

duce the analyst to the operator, and the latter should be guaranteed no loss o
earnings during the study. The analyst should treat the worker as one sharing
in the investigation and should endeavor to win his interest and cooperation
Generally there is little difficulty with intelligent workers who want to avoi
delays caused by poor material, irregular serving, bad work on preceding opera
tions, or improper care of equipment, for these delays limit their earnings and
cause irritation. The stop watch should be shown to the operator, its use ex
plained, and the idea of measuring explained as against the idea of driving
The operator should be put at ease and instructed to work at his **normal rate**
The time study man should avoid standing in front of or directly behind the
operator and should select a position from 4 to 6 ft. away at one side. It i
reassuring to the operator if he can look around at the analyst occasionally, an
furthermore the analyst may need to ask questions now and then.

PREPARATION FOR STUDY. The analyst should so arrange an
simplify his own routine that he can give his whole attention to the job being
studied. He can make himself comfortable but should not lounge or relax hi
attention. He must keep off outside interference and try to put himself int
the worker's situation. He must be on the alert to catch all the **variables** con
sisting of those having to do with the worker, with the surroundings, and with
motion. Motions may be classified as frequent and infrequent. The latter are
the ones most likely to be overlooked—for instance, starting new stock, oiling
machines, or changing tools. In many industries there is **seasonal variability**
changes from light to heavy materials, etc. No variable likely to affect motion
and time should be ignored.

Time observation sheets should be prepared and all **general data covering
conditions** recorded, such as date, time of starting and stopping, operation
work in process, operator, machine or workplace, temperature, humidity, light
and sound. Figures 8 and 9 show complete forms for an extensive study of
machine operations where considerable supporting data must be recorded. In a
particular plant the form can be simplified to contain only the information
essential to the products and conditions therein.

DETERMINATION OF TIME STUDY ELEMENTS. The operation is broken down into time study elements and timed accordingly.

Element Breakdown. There are a number of criteria employed to divide
the operation into a **pattern of elements.**

1. **Manually performed or hand elements** should be separated from **machine-controlled elements.** For the latter type the **machining time** may be
calculated on the basis of the proper tool or table speeds, feeds, depth of cuts,
types of material, etc. This may be done whether the machine feed is automatic
or hand fed. Machine elements are not rated against a normal.

2. **Constant elements** should be differentiated from **variable elements** for
manually controlled portions of the job. In this case variable does not refer to
the inherent variability of individual observed times. It refers rather to the fact
that certain elements common to many jobs are relatively constant in different
jobs performed at the operation, while other element times vary with the job,
according to some factor such as part size or shape, tolerances required, or location of materials. For instance, the time to close and open a jig, lower drill to
work, or rotate jig may be relatively constant over a range of jobs, whereas the
time to pick up part, place part in jig, etc., may vary with different parts or jig
characteristics. With this sort of breakdown, standard data may be formulated
from a small number of studies to apply to a range of similar but not identical

TIME OBSERVATION SHEET

DATE 5-21--   MODEL NO. Std.

FACT. NO.	18
DEPT. NO.	22
PART NAME	Bearing Cap
OPERATION NO.	20B OPERATION Spotface (2) 1 3/4 Holes
MACH. NO.	682 MACHINE 24" Conn. Dr. Press

PART NO.	62.B67
SCHEDULE NO.	T-4920
CANCELS SCH. NO.	T-1151
GROUP NO.	none
PAGE	3 OF 3

Details of Work Cycle — Elapsed Time in Decimal Minutes

No.	Details of Work Cycle	Cont	Indv	Cont	Indv	Cont	Indv	Cont	Indv	Cont	Indv	Allowed Time
1												
2												
3	Place part under spotfaces (18")	.02	.02	.60	.03	.94	.02	1.57	.02	1.94	.04	.02
4	Lower spindle (4")	.04	.02	.63	.03	.96	.02	1.59	.02	1.96	.02	.02
5	Spotface (1 3/4 x 1/16)	.20	.16	.74	.11	1.06	.10	1.72	.13	2.07	.11	.11
6	Raise job, turn part end for	.23	.03	.77	.03	1.10	.04	1.75	.03	2.10	.03	.03
7	end & lower spd.											
8	Spotface (1 3/4 x 1/16)	.42	.19	.87	.10	1.508	.11	1.85	.10	2.22	.12	.11
9	Raise spindle	.44	.02	.90	.03	1.52	.02	1.87	.02	2.24	.02	.02
10	Remove part (22")	.57A	.03	.92	.02	1.55	.03	1.90	.03	2.26	.04	.03
11	MATERIAL Mall. Iron											
12–22	DELAYS											.34

Delays:
A — Deep & pick up (.13)
B — Move stock .40
C
D

SKETCH (OVER)	✓
REMARKS (OVER)	O.K.
SAFETY	Lat. Out.
LUBRICANT	Elect.
DRIVE	✓
LOT SIZE	✓
LOTS/YEAR	✓
CONT'US PROD.	Yes
CONE STEP US'D	✓
FEED LEVER	✓
SPEED LEVER	✓

Summary

IDLE TIME/CYCLE	none
IDLE TIME/PIECE	none
DELAY/PIECE	none
MACH. CAP'TY/HR.	162
OPER. CAP'TY/HR.	162
PIECES/CYCLE	
RE'C'D. PROD./HR.	50
NO. PCS. STUDIED	5
NO. OF OPERATORS	1
NO. OF HELPERS	1
NO. OF MACHINES	1
FORMER BASE RATE	2.26
NEW BASE RATE	2.20
FORMER STD. TIME	.0059
NEW STD. TIME	.0062
FORMER PROD./HR.	175
NEW PROD./HR.	162
% INCREASE PROD.	
% DECREASE PROD.	8
% INCREASE TIME	5
% DECREASE TIME	
FORMER COST/PC.	.0125
NEW COST/PC.	.0136

Work Cycle Time Per Piece

	Pcs.	Prorate/Pc.
MULTIPLE PCS./WORK CYCLE:	1	✓
" MACH.	1	✓ /MACH.
" OPERS./	1	✓ /OPER.
HANDLE SUPPLIES MIN. FOR PCS.:	1	✓
" SCRAP	1	✓
" STOCK TO WORK AREA	1	✓
" FROM "	1	✓
MACH. SET-UP TIME ALLOWANCE ✓ %:	1	✓ MIN./SET-UP:
TOOL CHANGE TIME ALLOWANCE ✓ ":	3 MIN./CHANGE, 400 ":	.007
REST OR DELAY ALLOWANCE ✓ ":	MIN./DAY:	1
MACHINE ALLOWANCE	✓ :	1
SPECIAL ALLOWANCE	✓ : 10	1
PERSONAL ALLOWANCE	✓ : .21	1
TOTAL ALLOWANCE	7 ": .31	.024
TOTAL TIME ALLOWED PER PIECE		.371

Machine / Tool Data

ITEM #	5-8
CUTTING DIA.	1 3/4
CUTTER DIA.	"
NO. OF TEETH	4
CUTTER MATL.	H.S.S
R.P.M.	220
F.P.M.	69
TYPE OF CUT	S.F.
FEED/REV. TOOL WORK	HD
FEED/MIN. TOOL WORK	"
FEED/TOOTH TOOL WORK	
LENGTH OF CUT	1/16
WIDTH OF CUT	"
DEPTH OF CUT	1/16
LIMITS	1/64
REASON FOR CHANGE:	Was 1" dia.

Demonstration

DEMONSTRATION   GOOD ☑   FAIR ☐   POOR ☐

APPROVED BY:
ANALYST
CHECKER
FOREMAN
GEN. FOREMAN
EFFICIENCY ENG.
SUPERINTENDENT

Fig. 9. Method of recording continuous time study.

parts of jobs performed at the operation in question. Thus the normal time for a job may be determined from the standard data by knowing the elements necessary to complete a cycle of work, the characteristics of the part, and any other factors influencing the selection of the element time.

3. The choice of elements may be influenced by the ease with which the **break points,** points separating successive elements, can be observed or discerned. Thus, sound or some other factor affecting the observer's senses can be employed to designate the end of one element and the start of another with accuracy. This is important where the elements are short.

4. Elements should be made **short** within the limits of accuracy of the timing device. This is more important where methods or motion study is involved than in obtaining a standard time for a well-controlled operation. It also facilitates a detailed description of the standard method. With the stop watch, elements less than 0.04 minutes are difficult to time, especially if they are not separated by longer elements.

It is said that breaking the job down into small elements tends to increase "accuracy" due to the canceling out of errors in the elements. This does not have much foundation, in that it assumes that the sources of error lack consistency or randomly affect the elemental readings. It is more likely that biases may be consistent. For instance, if the snap-back method of timing is used, the **consistent error** in this method would be magnified by small elements.

5. Elements should consist of **homogeneous groups of therbligs.** The elements should consist of a "natural" subdivision of work, such as a series of motions with a single part, inspections, or use of a tool.

6. **Irregular elements** should be noted. These are elements which do not occur regularly with each cycle of output but are nevertheless productive elements and must be included in the observed data and prorated over the regular times.

Element Recording. When time study elements are in a fixed order, it is better to write them down on each form before beginning the study of time. If the order cannot be fixed, the elements may be listed at the bottom of the sheet and given **element symbols,** *a, b, c,* etc. These symbols can then be used in any sequence which may develop. There is some danger of miscopying, but it is not always possible to fix the sequence completely. Clearness as to just what constitutes an element is important. In such divisions as "adjust tool in post, tighten tool, position," there is danger of being indefinite as to exactly where one ends and the next begins.

Break points should be definitely given as:

1. From reaching for tool to end of tool adjustment in post. The method should be described on the back of the sheet or on simple motion study.
2. From end of adjustment in post to end of tightening. The motions of tightening should be described as above.
3. From end of tightening to starting machine. The method should be described as above.

These break points must be fixed in mind, and it is safer to write them in advance of timing the job. In establishing these various points, use of judgment to cut out unnecessary detail will avoid too much writing.

RECORDING TIME VALUES. During the recording portion of the study, minutes need not be repeated with every decimal reading, but they should be put down frequently to prevent doubt. Usually each element is timed except n overall timing. Snap-back, continuous, and sequence timing are the three most commonly used techniques.

Overall Timing. This is **timing of cycles** only, without reference to elements. This will not permit any detailed analysis of the operation nor will it indicate minor delays. Where there is some assurance that the operation is normally under control, overall timing may be used to collect sample data for use in control charts.

Snap-back Timing. At the beginning of each element the watch hand starts from zero. At the end of each element the watch is read, the hands are snapped back to zero, and observed time is recorded (see Fig. 9). No computations are necessary to obtain element times, as these are recorded. Thus the clerical work in computing the study is less than in the continuous method.

A major **disadvantage** of the snap-back method is the bias caused by time taken by the observer to manipulate the watch at the end of each element reading. Also, the observer may anticipate individual readings after a number of readings have been taken. An extensive controlled study by Lazarus (Advanced Management, vol. 15) gave the following results: The standard deviation of errors in both snap-back and continuous types of readings was 0.081 minute. The average error in the continuous method was +0.000097 minute and for the snap-back method −0.0008 minute. Lazarus concluded also that the absolute magnitude of the error is independent of the size of the element.

Except for highly repetitive, short-cycle operations, the error in the watch reading may be minor in view of the distribution of time study cycle times resulting from the variation in the subject being studied.

Continuous Timing. This procedure gives the most satisfactory results in general on most operations. Elements are recorded in sequence without stopping the watch. The observer keeps the watch going continuously during the period of study, making a mental note of the time as shown on the watch at the instant each elementary operation is completed and recording that time on the sheet opposite its name or appropriate symbol. He should do all this with sufficient speed and concentration to be free to note and write the time of completion of the next elementary operation. As reading of the watch is practically instantaneous, there is no necessity for stopping the hand.

The advantages of the continuous method are that it gives not only the time for each element as a distinct entity, but also the times of all elements in the order of their performance; furthermore, it charges every minute of time for the duration of the study either to some necessary element, called a **productive** element, or to an unnecessary one, called a **nonproductive** element; it also eliminates any danger of omitting delays. The split-hand watch greatly facilitates the use of this procedure.

Cumulative readings only are recorded during the run. Elapsed times are derived by subtraction later and entered beside the corresponding cumulative readings (Fig. 9). As a check, **the sum of individual times** should equal the final cumulative reading. The individual times should not be entered too far from the original readings. A simple way is to enter the cumulative reading with hard pencil on one-half of each square and the extensions with softer or colored pencil on the other half. This method keeps the two sets of figures together and yet allows either set to be scanned without confusion.

Sequence Timing. Sequence timing can be performed by two watches, three watches, or split-hand watches. In **three-watch sequencing,** the watches are mounted next to each other and are connected by a lever mechanism in suc' a manner that one watch is in motion for timing the present element, the nex watch is set at zero for starting to time the next element when the lever i depressed, and the last watch is stopped for recording the time associated wit] the preceding element. The sequence is repeated each time the lever is presse with a given watch running, stopping, and being reset to zero. The advantage of the method are that the watches are read while the hands are stopped and the elapsed time data makes subtraction unnecessary.

LENGTH OF STUDY. The length of the study determines the size of th parent population or universe from which samples are taken. The study lengt] affects accuracy where bias may result if the period of study is too short t include all significant sources of variation.

There may be **irregular work elements** that appear at given intervals o production within or between hours. These include such tasks as replenishing supply bins, filling out a production ticket, and inspecting every twentieth uni produced. If such tasks are to be included as productive elements rather thar delays, they must be recognized, timed, and **prorated** over units produced in th interval between their appearances.

One factor to consider in the study length is the influence of **fatigue** on the net cycle time. When fatigue results in a between-hour variation of cycle times the study should be extended over the working day to include such variation However, it is conceivable that in most light industrial operations fatigue may have little influence on net cycle times. Any fatigue that is not remedied by res periods may show up primarily in an increase in the number and length of mino delays rather than in a change in pace. These factors are then considered in the **measurement of delay allowances** rather than of net cycle times.

Traditional time study practice usually limits the study to observing net cycle times over a relatively short time period to include essentially between-cycle variation. Consideration is usually given to between-hour and between-day varia- tion by taking the observations during, say, the middle of the morning work period and during the second or third day of the work week, when it is assumed that **production rate** is average as far as these sources of variation are con- cerned. The actual length of the study is the time necessary to take observations of a consecutive number of units of output equal to the required sample size.

A second method is to extend the study over the period of a day or longer. Observations are then taken randomly, in small sub-sample sizes, over the specified extended time period. This is the recommended procedure where there is reason to believe that net cycle time variation is affected by factors operating irregularly and cyclically over this period.

An intelligent selection of **optimal study length** may be difficult. The eco- nomics of the situation, in which costs of a long study are weighed against costs of having some results of inferior accuracy, may favor a short study period. Also, past experience may be reflected in standards of study length set as a matter of policy.

NUMBER OF OBSERVATIONS. When the criterion of statistical con- trol is employed to determine standardized conditions, initial observations may be necessary merely to discover any need for further methods study, training, or control before setting the actual standard. It is possible to **pool** some of the initial observations with those taken in the formal study, provided the initial times represent standard conditions. Experience and judgment may be the sole basis for deciding the number of readings to take.

Statistical Determination. A number of techniques have been developed to determine the appropriate sample size required to provide a given degree of confidence that the observed average time value for a given element or cycle will not deviate more than a specified amount from the true population time value (accuracy). The theoretical aspects of these N' formulas are discussed under Statistical Considerations in Time Study.

FORMULATION OF STANDARD TIME. The major steps in arriving at standard time from stop-watch time data may be diagrammed as in Fig. 10.

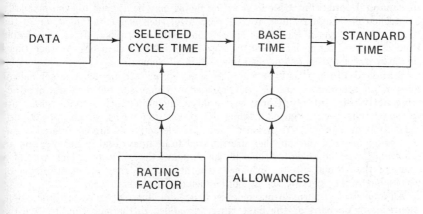

Fig. 10. **Steps in determining standard times.**

Selected Cycle Time. The selected cycle time is the sum of averages of selected elemental times. This includes those elements which appear regularly in every cycle and those which are irregular, whose average times are prorated over the number of cycles intervening between their occurrences. An example of an **irregular element** is, "Inspect every 20th part." The cycle time for this element is equal to the average time for the inspection operation divided by 20.

Prior to determining each elemental average time, certain individual **abnormal elemental times** may be removed from the study by circling the values on the observation sheet. This removal is based on the judgment of the observer, who must decide whether the abnormally high or low value is nonrepresentative of standard or good performance.

The removal of abnormal times should be substantiated with evidence that the times are not indicative of the normal method or conditions. For instance, in an assembly operation the operator may have difficulty in positioning two components during a cycle because of a burr on one of the parts. This may be a periodic occurrence and considered as representative of normal. If it is nonstandard, the time may be removed from the study; however, the causes for the burring should be discovered and remedied, if possible.

Certain abnormal elements of work which appear during the study may be separated from the normal sequence and designated as **foreign elements.** These elements are listed on the side of the observation sheet as they occur, along with the time for their performance. They should be carefully scrutinized to determine that they are not a normal part of the work. If their causes are unavoidable by the operator, they may be considered as part of the productive activity of the job or they may be designated as interruptions to be considered under allowances.

Base Time. The selected cycle time is an estimate of the actual performance level of the operator observed. This performance may differ from that which would be expected of a **normal operator.** If this is the case, the selected time is adjusted by a **rating factor** to obtain the **base time.** Other terms used to denote the base time are **normal, rated, or leveled time.** The method employed, first to determine a concept of normal, and second to adjust the observed data to normal, is called **rating or leveling.**

The operator's performance is rated, in relation to normal, at the time the observations are taken. Only those portions of the job over which the operator has **manual control** are rated. A rating factor may be applied to each manually controlled elemental reading, to the average of such elemental times, or to the selected cycle time if the job is entirely manually controlled. It is conceivable that a separate rating factor can be applied to different elemental average times if the operator does not show the same skill in all elements.

Standard Time. The base time represents "pure" production time per unit of output for a normal operator. It is exclusive of time consumed in **nonproductive activities, interruptions, and delays,** including such activities as setup, personal time, instruction, and filling out production forms, which appear periodically over a protracted period of time. If the activities are considered essential components of the job, they are referred to as **unavoidable delays,** and an allowance must be made for them in the standard time. Delays which are not a part of the normal job performance, whether or not they are controllable or avoidable by the operator, should not be included in the standard.

Allowances for normal unavoidable delays or interruptions to production, figured as a **percent of the base time,** are added to the base time to give the standard time.

The selected net cycle time represents expected performance for the operator observed. When allowances for interruptions and delays are added directly to this time, without reference to rating, the resultant time may be used for forecasting future actual performance. This can be the basis for estimating, scheduling, and budgeting, where **predicted actual** rather than **hypothetical standard performance** is pertinent. However, if an incentive plan is to be installed based on the time study, the actual future performance may deviate significantly from the observed performance due to the influence of the incentive.

Standard Study Report. The write-up of every job studied should be a formal and precise report including a permanent record of all conditions under which the study was made and a detailed description as to the one best way developed. A typical report may take the following form:

1. **Identification.** At the top of the report data necessary for classifying and identifying the study should be entered. The results of the study should be summarized and entered here also.

2. **Analysis.** This should include a detailed description of the method of performing the operation and all details necessary to set up the job in standard fashion at a future time. As part of analysis, elementary operations are listed in the sequence in which they are performed by all personnel involved in the study. Each detailed operation may be assigned a symbol for convenience in reference and shown with the time values determined by the study. The description of each operation should be so detailed and complete as to leave no doubt about the series of motions used.

3. **Synthesis.** Under this heading are put together the detailed operations to form a complete cycle. Explanation of the method of putting the details together are given here. If any mathematical formula is used, it is stated by means of the symbols previously assigned in algebraic form. At this point allowances for rest and delay may be entered.

4. **Calculation of Standard.** Space is provided for the calculation of time standards and, where necessary, piece price determination.

Although the report may be made on plain paper using the sequence recommended above, it is sometimes convenient to use prepared forms emphasizing those points important in a particular plant. All reports should be signed by the person making the study and checked. A space should also be provided for the signature of the executive to make the study results effective.

Instruction Card. The instruction card is an important part of the write-up. Its function is to convey to the worker complete information required for doing the standardized task, covering:

1. Equipment, tools, fixtures, and gages to use, and drawings required, if any. If the same set of tools is used on several different jobs, the list may be made out on a separate card so that it can be used with the instruction card for any job.
2. Feeds and speeds to be used, if they are under the worker's control.
3. What to do and how to do it, including elementary operations listed in their final sequence and in detail, sketches of setup if necessary for clearness, and standardized time allowed for each elementary operation and for the whole operation so that the worker may know in advance all that is expected.
4. Pay, and incentive, if any, allowed for performance of task.

In general, information on the instruction card will follow closely that on the observation sheet covering the final or perfected stage of study, with such additional data on equipment and method as the production worker may require. It forms a complete and permanent record of the task as standardized. With the tools called for and the best motions described, a properly qualified worker, with a reasonable amount of training and practice, should be able to perform the task in the time set.

Methods of Performance Rating

NEED FOR RATING. An individual operator's rate of production or speed of performance will show variation from cycle to cycle, hour to hour, and, perhaps, from day to day. Such variation is indicated in the precision of the study and the number of observations to take. The variability in performance among operators is considered in the **rating function** in which each operator is related to a **hypothetical normal operator.**

In order to obtain a standard time, rating is essential unless normal performance is defined as the statistical average of the rate of output of a sample of different operators performing the same operation. However, this may be an inadequate standard of good performance, and in many cases the number of operators on the same job is small. If the number is large, it may be uneconomical to sample more than a single operator.

THE RATING PROBLEM. There are two basic factors to be considered in the rating function. The first is the definition and determination of a concept of **normal performance** for a given type of work. The second is the adjustment of the observed time data to "normal" by some **numerical index.** These problems are not independent and the method of determining "normal" may be related to the method of arithmetically adjusting the observed data.

All rating demands some sort of judgment on the observer's part, and therefore the rating procedure has been subjected to considerable controversy throughout the history of time study. Much of the controversy has resulted from the claims of some that rating is a "scientific procedure." As pointed out by Abruzzi (Work, Workers, and Work Measurement), "Since the rating component of time study performs an evaluation function, it cannot possibly be a scientific process.

The claims of scientific validity, however, have been so well advertised that almost everyone in this area has examined rating as though it were a scientific activity."

Concept of Normal. In order to provide a reference standard for normal performance, the following definitions of **normal effort** are listed in the Industrial Engineering Terminology Manual (Journal of Industrial Engineering, vol. 16):

1. The effort expended in manual work by the average experienced operator working with average skill and application.
2. Performance of an average operator over an eight-hour day under Measured Daywork conditions.
3. Performance of an average operator working at an efficient pace over an eight-hour day without undue fatigue or without experiencing cumulative fatigue the following day.
4. A generally accepted industry norm arrived at by mutual agreement of labor and management.
5. Performance with a steady exertion of reasonable effort by one who has mastered the standard method. Normal pace or normal performance comes from the concept of the expected output from an average experienced operator, without the stimulus of an incentive wage plan. (It is the level of performance for which properly evaluated base rate wages are paid).

From these definitions, it is evident that the concept of normal may be defined on the basis of collective bargaining. Most incentive programs in this country provide for an average bonus of approximately 30 percent. Therefore, normal effort may be defined as **100/130** of **maximum incentive performance** for the **qualified** piecework operator. Also, certain **benchmarks** may be used to define normal, such as walking on level ground at 3 miles per hour. A given company may adopt the norms established on their own rating films or those produced by the Society for the Advancement of Management (SAM).

Accuracy. In rating, accuracy has a questionable objective meaning. There is no true objective standard, independent of human judgment, by which to evaluate a given operator's performance. Since judgment is the essential feature, accuracy has meaning only in the sense of a statistical average of the numerical rating indices made by a number of independent observers of a given operator's performance. **Precision** in rating, then, would be related to the variability of the indices around this average. Generally, it is uneconomical to take a large number of independent ratings of a given operation. However, a statistical norm may be determined for **sample basic operations** which are deemed indicative of the types of operations encountered in a given plant. That is, in a given plant films may be taken of basic operations including a sample of different operator paces. These can be rated independently by time study engineers and an agreement made as to what constitutes normal performance for these operations. These standards become the basis for training time study engineers in the concept of normal.

SAM Rating Films. The Society for the Advancement of Management has conducted extensive studies, based on existing practices in industry, to establish a concept of **a fair day's work** and to develop films as a medium for **training in rating** (Rating of Time Studies). Films were made of 24 simple industrial, clerical, and laboratory operations. These were rated by 1,800 men from 200 companies. Norms for the operations were subsequently established and widely circulated for training purposes.

An analysis of the SAM procedure was made by William Gomberg (A Trade Union Analysis of Time Study) in which he stated a number of weaknesses in the study. Two fundamental points brought out were: (1) Since the ratings made by each rater were from single estimates of single workers on a specific operation at each pace rated, both within-day and between-day rater variation for a given pace were eliminated. (2) There was a lack of proper statistical treatment of the rating data obtained from the study.

Localized Rating. Opposed to the notion of a universal standard of performance is the idea that effective norms can be established only at the local level of the individual plant and that, within the plant, each operation presents an **individual rating problem.** Coupled with this is the belief that experience on the rater's part in assessing the particular situation is all-important.

It is conceivable that the second problem of rating, that of making the numerical adjustment, may be avoided by having the operator work only at **normal pace** and **skill.** It must be noted that the observer still exercises judgment as to what normal performance is and instructs the operator accordingly. This method is basically unsound, due to the problem of attempting to make an operator perform at some pace or skill level which is unnatural to him.

The performance rating procedure may be avoided at the time of a particular time study by employing **standard time data** which are already leveled or rated. However, worker reactions to the fairness of the standard may not be eliminated, since the standard data were rated at the time of formulation. Also, when the observer applies motion-standards, he must use his judgment in determining the best method and in interpreting this method in terms of the motion-time system employed.

Consistency. The predominant requirement of good rating practice seems to be consistency. There may be a significant range within which the absolute rating index is acceptable to everyone concerned. But where the rating is inconsistent, as indicated by significant **standard time variation** between like operations performed in the plant, considerable objection will result from the operators concerned.

Simplicity. Another necessary characteristic of a good rating procedure is simplicity. The operator is more likely to accept a standard time if the rating procedure is not too complex for him to understand.

Machine elements are not generally rated. If the machine elements are automatically controlled by the machine, the elemental base times can be determined from engineering data or from a number of stop-watch readings. If the pace of the machine element can be controlled by the operator, an **allowed element time** can be calculated from empirical data. For instance, the time to drill a ¾-in. hole to a depth of 1 in. in 120 steel can be determined from tables of drill speeds and feeds for the particular material and drill size. The time to lower and raise the drill spindle to and from the material are manually controlled elements which are timed with the watch.

PERFORMANCE RATING. In performance rating the speed of the operator's motions or his general pace is rated against a concept of normal.

Percentage Basis. One widely used method of speed rating is to rate performance on a percentage basis. The **normal** operator is rated at 100 percent. A slower operator would be rated at a percentage less than 100, and for a faster operator the rating would be over 100 percent. For example, data from a time study would be rated as follows.

Average Time per Piece

a. Pick up piece, insert in fixture........................... 0.056 min.
b. DRILL... 0.107 min.
c. Remove piece from fixture to tote box.................... 0.038 min.

Base Time

a. 0.056 rating at 90%, or 0.056 × 0.90........................... 0.050
b. 0.107 rating at 100%, or 0.107 × 1.00........................... 0.107
c. 0.038 rating at 90%, or 0.038 × 0.90........................... 0.034

Point Basis. Rating an operator's time for speed alone on a point basis has the advantage of simplicity, but it also is dependent on judgment. Usually each element is individually rated. Shumard (Primer of Time Study) suggests the use of the following rating table:

Symbol	Speed	Symbol	Speed
100	Superfast	65	Good −
95	Fast +	60	Normal
90	Fast	55	Fair +
85	Fast −	50	Fair
80	Excellent	45	Fair −
75	Good +	40	Poor
70	Good		

Data from a time study on an operator working "good +" for elements a and c, and "normal" for element b, is as follows:

Average Time per Piece

a. Pick up piece, place on bench............................ 0.050 min.
b. BURR piece with file.................................... 0.270 min.
c. Place piece in tray..................................... 0.030 min.

Rated Time

a. 0.050 time with 75 rating, or 0.050 × 75/60...................... 0.063
b. 0.270 time with 60 rating, or 0.270 × 60/60...................... 0.270
c. 0.030 time with 75 rating, or 0.030 × 75/60...................... 0.038

Elements a and c were judged to be performed 75/60 or 25 percent faster than normal. The operator performing at 60 on element b is producing at normal speed equivalent to **unstimulated day work effort,** and the time does not have to be adjusted.

Shumard says, "From compiled industrial data covering many plants that have attractive wage payment incentive plans which result in excellent quality of product at low unit costs, it was learned that group average speed of the operators in each plant was 33⅓ percent faster than normal. . . . Consequently we may safely set up 80 as the ideal effort-goal in manufacturing . . . ," the 80 being 133⅓ percent of 60, normal. This condition, however, will not hold true in all cases.

LEVELING PLAN. Leveling is a method of rating in which the causes producing differences in performance are analyzed according to **factors of skill, effort, conditions, and consistency** (Lowry, Maynard, and Stegemerten, Time and Motion Study). Each of these factors is graded with a numerical index associated with each grade as shown in the following table.

	Skill			Effort	
+0.15 +0.13	A1 A2	Superskill	+0.13 +0.12	A1 A2	Excessive
+0.11 +0.08	B1 B2	Excellent	+0.10 +0.08	B1 B2	Excellent
+0.06 +0.03	C1 C2	Good	+0.05 +0.02	C1 C2	Good
0.00	D	Average	0.00	D	Average
−0.05 −0.10	E1 E2	Fair	−0.04 −0.08	E1 E2	Fair
−0.16 −0.22	F1 F2	Poor	−0.12 −0.17	F1 F2	Poor
	Conditions			Consistency	
+0.06	A	Ideal	+0.04	A	Perfect
+0.04	B	Excellent	+0.03	B	Excellent
+0.02	C	Good	+0.01	C	Good
0.00	D	Average	0.00	D	Average
−0.03	E	Fair	−0.02	E	Fair
−0.07	F	Poor	−0.04	F	Poor

Skill is defined as the ability to follow a given method. **Effort** is "the will to work." **Conditions** are those factors which affect the operator rather than the operation, such as lighting, temperature, and minor variations from normal conditions. **Consistency** takes into consideration minor variations not detected as skill, effort, or conditions.

With reference to both skill and effort, characteristics of the six grades are given qualitative description. For example, **good effort** is described as follows:

1. Little or no lost time.
2. Takes an interest in the work.
3. Takes no notice of the time study man.
4. Works at best pace suited for endurance.
5. Follows a set sequence.
6. Conscientious about his work.
7. Has faith in time study man.
8. Encourages advice and suggestions and makes suggestions.
9. Well prepared for job and has workplace in good order.
10. Steady and reliable.

The observer judges the skill, effort, and consistency of the operator and the conditions, checking off each item on the leveling chart. For instance, if skill is at level B2 (0.08), effort B2 (0.08), conditions C (0.02), and consistency C (0.01), the sum of these, 0.19, would be the **leveling factor** applied to the average times for each element. All elements are not necessarily given the same grading, for the observer may decide to grade certain elements differently from the general grading applied to the study. In using leveling, it should be borne in mind that the accuracy of the final standard is no greater than the accuracy of the observer's judgment. One aid to the observer's judgment of skill is the record of the operator's length of experience on the class of work or the particular job. The operator's diligent application to work at hand with no unnecessary delays or interruptions is an indication of good effort, the value of which must be judged by the observer. Variations from conditions previously standardized for the job are allowed for in the leveling. Consistency of the operator may be determined while the study is in progress or later from the observations. It should be remembered that use of high or low figures on the **leveling scale** introduces greater chance of error, and in those cases it is better to restudy the job on another operator.

SYNTHETIC LEVELING. Another method of leveling is to compare the actual time obtained for certain elements with predetermined standard elemental times. According to Morrow (Motion Economy and Work Measurement), the method of applying synthetic leveling is to take the time study as usual, break it down into elements, and then compare as many of these elements as possible with the **predetermined standard times.** The percentage of variation of the actual times from the standards is arrived at, and the same percentage applied to parts of the study for which predetermined standard times are not available. Machine times are kept entirely separate from handling time and are rated at 100 percent.

OBJECTIVE RATING. Mundel (Motion and Time Study) reduces the rating procedure to two functions—rating of **observed pace** and of **job difficulty.** As he says, "The actual pace of performance observed must be understood to be a function of the skill, aptitude, and exertion of the operator, but these variables are neither separately identifiable nor is a separate appraisal pertinent." His conclusion is that all three of these factors are reflected in either **pace** or **job difficulty** or both.

In the typical time study procedure, the time study observer first judges job difficulty, in order to form a concept of the appearance of adequate performance for the job (as required by the definition of standard time he is using), and then judges observed pace, against this imagined concept. Mundel proposes a two-step procedure which is a reverse of the conventional order. The procedure consists of the following steps:

1. The rating of observed pace against an objective pace-standard which is the same for all jobs. In this rating, no attention whatsoever is paid to job difficulty and its limiting effect on possible pace; hence, a single pace-standard may be used instead of a multiplicity of mental concepts.
2. The use of a difficulty adjustment, consisting of a percentage increment, added after the application of the numerical appraisal from step 1 has been used to adjust the original observed data.

Mundel has developed adjustment tables from experimental data for amount of body used, foot pedals, bimanualness, eye-hand coordination, handling requirements, and weight or resistance.

The rating factor with the secondary adjustment percentages expressed in decimal form would be:

$$\text{Rating factor} = (\text{Pace rating})(1 + \Sigma \text{ Sec. Adj.})$$

Nadler (Work Design) lists eight factors for adjustment of performance due to job restrictions. His percentage values are used to find the Rating Factor as follows:

$$\text{Rating factor} = \frac{\text{Pace rating}}{1 - (\Sigma \text{ Sec. Adj.})/100}$$

The adjustment values are given in terms of the factor occurring in 100% of the element. Therefore, the values may be prorated according to the percent of element in which a given factor occurs.

Statistical Considerations in Time Study

STATISTICS IN TIME STUDY. Time study is concerned with measuring certain time properties of a dynamic system. The system, such as an operation and its environment, is dynamic, in that variation and change occur in one or more of the operational factors over a period of time. This variation or change presents a measurement problem as well as a problem of operation control. The observer must not only reckon with variation at the time of measurement but must consider it in setting standard times which are used to make forecasts of future performance. If the analysis of variation and change and their causes is inadequate, there may be a critical discrepancy between the estimated or predicted performance and the subsequent actual results. This discrepancy should not be confused with an expected difference between standard performance and actual performance. The latter difference may be predicted by the level to which the standard is set.

Nature of Variation and Change. The dynamic nature of the operation is reflected in variation in the time taken for successive units of output of a repetitive operation or in the relative time consumed in successive irregular productive or nonproductive job activities. It is also reflected in changes in certain average time parameters over a protracted time period. These parameters might be, for example, the average cycle time for a repetitive operation, the average percent time consumed in handling materials on a job, or the average amount of variation in the time for successively produced units. Since variation or change in an operation may be attributed to a complex of mechanical, physiological, psychological, and sociological factors, it is suggested that work measurement deals primarily in the areas of variable and constant chance cause systems (see Statistical Methods section).

Population Sampling. Time study is essentially a sampling procedure. A relatively small number of observations are taken, at random or otherwise, from a parent population or universe whose time parameters are to be inferred. Sampling may be avoided by observing the entire population, but this is usually uneconomical and unnecessary.

The parent population consists of all possible time units from which sample observations or measurements may be selected. The sample is randomly selected if all possible time units in the population have an equal likelihood of being selected for inclusion in the sample. For a repetitive operation the parent population may be defined as the cycle times for all units produced in a given day, for example. From this population a sample of N observations or cycle time measurements may be taken.

The only valid inference that can be made from the sample is to the parent population. It is customary, however, to assume that future performance will have the same characteristics as the parent population. Thus, the inference is in practice extended to future performance as well.

The population time parameters of usual interest are a measure of central tendency, such as arithmetic mean or average, and a measure of dispersion or variation, such as range or standard deviation. Both parameters refer to whatever time dimension is being measured, such as net cycle time for a unit of output, time required for a given motion, or percent time consumed in setting up a machine.

If the population is a constant chance cause system, the time dimensions for successive units of output or activity will have a constant pattern of dispersion as well as a stable average. Also, the successive unit times will be randomly dispersed around the central tendency. The population is then **statistically stable** or homogeneous. If the population is a variable chance cause system, there will be a lack of stability in the pattern of dispersion or in the average or both. Thus, a statistically stable operation is one in which the known causes for time variation are controlled so that the variation which does exist is due essentially to constant chance causes. **Change** is reflected in shifts in the value of the population parameters, either in the measure of central tendency or in the measure of dispersion.

Time study involves three basic **sampling procedure problems:**

1. The sample estimates of particular time parameters may be used to predict future performance. The **applicability** of the sample estimate to future situations depends on the similarity between the parent population from which the sample was extracted and the extended population over the protracted period of time for which the forecast is made.
2. The estimate should be **accurate** in that it is an unbiased representation of the true value of the parent population. **Accuracy** is a measure of the degree by which the estimated average value of a set of measurements differs from the true value of the quantity being measured. An inaccurate estimate results from sources of bias built into the measuring or sampling procedure.
3. The estimate should conform to a required **precision. Precision** is a measure of the reproducibility of the measured value of the quantity in question. A precise estimate would be reflected in a high degree of similarity between the estimates derived from successive samples. Thus, a sample estimate may be precise but lacking in accuracy.

ESTIMATION OF PARAMETERS. In the determination of statistical control the first problem of actual measurement is the estimation of the **average net cycle time** ($\overline{X}$), which is a measure of central tendency. This time represents the productive time necessary to complete one cycle of output. The second problem is the estimation of the standard deviation (σ'), which is a measure of dispersion and is defined as the root-mean-square (rms) of the deviations of the individual observations about their average.

A simple statistical model employed to estimate the parameters of net cycle time is as follows:

Let the time study elements be numbered $1, 2, \ldots, i, \ldots M$. Example: pick up piece and place in jig, close jig, etc.

$C_j =$ the jth observed cycle of output of a successive number of units produced.
$j = 1, 2, \ldots N$

$x_{ij} =$ the observed time of the ith element in the jth cycle of output

$\overline{x}_i =$ estimate of average time for the ith element

$$= \frac{\displaystyle\sum_{j=1}^{N} x_{ij}}{N}$$

X_j = the jth cycle time

$$= \sum_{i=1}^{M} x_{ij}$$

$\overline{X}_j$ = estimate of average net cycle time for the operation

Where N is a number of units, the time study takes the form of a cycle as
shown below.

| Element | Observed cycle, j | | | | | | Averages |
i	1	2	·	·	·	N	
1	x_{11}	x_{12}	·	·	·	x_{1N}	$\overline{x}_1$
2	x_{21}	x_{22}	·	·	·	x_{2N}	$\overline{x}_2$
·	·	·	·	·	·	·	·
·	·	·	·	·	·	·	·
·	·	·	·	·	·	·	·
M	x_{M1}	x_{M2}	·	·	·	x_{MN}	$\overline{x}_M$
Cycle time	X_1	X_2	X_j	·	·	X_N	$\overline{X}_j$

The **standard deviation** of the population of net cycle times is estimated as
follows, using the cycle times:

σ_{X_j} = estimate of the standard deviation of the population of net cycle times

$$= \sqrt{\frac{\sum_{j=1}^{N} (X_j - \overline{X}_j)^2}{N}}, \quad \text{if} \quad N > 30$$

or

$$= \sqrt{\frac{\sum_{j=1}^{N} (X_j - \overline{X}_j)^2}{N - 1}}, \quad \text{if} \quad N < 30$$

It should be pointed out again that the parameters $\overline{X}$, and σ_{X_j} refer to the
measured or **observed cycle times** and not to base or standard times.

In the above model each cycle time (X_j) represents an observation from units
of output. There are disadvantages in confining the length of the observation
period to that which is included by N consecutive readings. The observation
period will probably be too short unless N is extremely large, and the estimation
of the population standard deviation may have little meaning because of a lack
of independence between successive observations. This can be remedied by
taking random subgroups of observations over a longer period of time, such as
a day or two, and analyzing the subgroups for statistical control.

Each subgroup will consist of a small number of consecutive observations. In
this case the following notations apply:

Let

C_{jk} = the jth observed cycle of output in the kth subgroup

$j = 1, 2, \ldots n; k = 1, 2, \ldots K$

$$N = \sum_{k=1}^{K} n$$

$$K = N/n$$

$\overline{X}_{jk}$ = the average cycle time for the kth subgroup

The average subgroup cycle time may be determined from elemental readings of n successive units of output:

$\overline{\overline{X}}_j$ = estimate of average net cycle time for the operation

$$= \frac{\sum\limits_{k=1}^{K} \overline{X}_{jk}}{K}$$

and

$$\sigma_{x_j} = \sqrt{\frac{\sum\limits_{j=1}^{N} (X_j - \overline{\overline{X}}_j)^2}{N}}, \quad \text{if} \quad N > 30$$

$$= \sqrt{\frac{\sum\limits_{j=1}^{N} (X_j - \overline{\overline{X}}_j)^2}{N - 1}}, \quad \text{if} \quad N < 30$$

STATISTICAL DETERMINATION OF SAMPLE SIZE. The sample size necessary to give a certain precision may be determined statistically.

Assume that the engineer desires to have an **assurance** of 95 in 100 that the population average cycle time lies within ±2.5 percent of the estimated observed value $\overline{X}_j$ from a sample of size N. (The symbols used here are explained above under Estimation of Parameters.) In other words he is willing to select a sample size N such that there is a risk of 5/100 that the population value will lie outside the range ±2.5 percent $\overline{X}_j$. The problem might be to determine N for an individual elemental time, but the approach is the same as that employed to determine N for cycle times.

The **validity** of the approach is based on three fundamental statistical laws. First, the mean of the distribution of sample means is the mean of the population of individual values from which the sample is taken. Second, the standard deviation of the distribution of sample means equals $1/\sqrt{N}$ times the standard deviation of the population of individual values. Third, the form of the distribution of sample means approaches the form of a normal probability distribution as the size of the sample is increased. Now let:

$\sigma_{\overline{X}_j}$ = standard deviation of the distribution of sample means

Then

$$2\sigma_{x_j} = 0.025\overline{X}_j$$

(For a precise determination of the coefficient of σ, see the section on Statistical Methods.)

In this case $\overline{X}_j$ is the average of the cycle times from a sample of size N, $\overline{\overline{X}}_j$ may be substituted for $\overline{X}_j$ if the average is taken from K subsamples of size n.

Since

$$\sigma_{x_j} = \frac{\sigma'_{x_j}}{\sqrt{N}} \text{ where } \sigma'_{x_j} \text{ is the population standard deviation}$$

hen

$$\frac{\sigma'_{x_j}}{\sqrt{N}} = \frac{0.025\overline{X}_j}{2}$$

$$N = \left(\frac{2\sigma'_{x_j}}{0.025\overline{X}_j}\right)^2$$

Since $\overline{X}_j$ and σ'_{x_j} are usually unknown at the start of the study, an initial number of observations N' must be taken in order to estimate these parameters. Now assume that the analyst takes an **initial sample** of cycle times, $N = 16$. The values observed (X_j), in minutes, are:

1.00	0.95	0.81	0.97
0.95	0.89	0.92	0.96
0.80	1.01	0.87	1.11
1.10	1.04	1.01	1.03

The population standard deviation (σ'_{x_j}) can be estimated by the formula for $N < 30$, by means of the average subgroup range, or by the moving range technique. C. R. Hicks and H. H. Young (Journal of Industrial Engineering, vol. 3) compared several methods for determination of sample size and concluded that they are quite consistent, except that the range techniques may result, on the average, in slightly larger sample sizes. However, the necessary computations for the range techniques are much easier than for the standard deviation formula.

Use of Standard Deviation Formula. The required sample size (N') can be found from the previous data by estimating the standard deviation of the population as follows:

$$\sigma'_{x_j} = \sqrt{\frac{\sum_{i=1}^{N}(X_i - \overline{\overline{X}}_j)^2}{N-1}} = \sqrt{\frac{N\sum_{i=1}^{N}(X_j{}^2) - \left(\sum_{i=1}^{N}X_j\right)^2}{N(N-1)}}$$

Substituting in the values of the initial observations

$$\sigma'_{x_j} = \sqrt{\frac{16(14.9838) - 237.7764}{16(15)}} = 0.09047$$

and

$$\overline{X}_j = \frac{\sum_{i=1}^{N}X_j}{N} = \frac{15.42}{16} = 0.96375$$

$$N' = \left(\frac{2\sigma'_{x_j}}{0.025\overline{X}_j}\right)^2 = \left(\frac{80(0.09047)}{0.96375}\right)^2 = 56.3 \approx 57 \text{ obs. required}$$

Therefore, an additional 41 observations are necessary in order to meet the specified confidence and accuracy limits with the amount of variation observed in the initial 16 observations. After gathering the additional data points, the entire sample of 57 observations would be used to recompute N' to determine if the appropriate sample size has been obtained.

Use of Average Subgroup Range. The average subgroup range can b
used to estimate the population standard deviation using the statistical qualit
control chart relationship

$$\sigma'_{X_i} = \frac{\overline{R}}{d_2}$$

where R = the high time study value minus the low time study value in each sub-grou

   $\overline{R}$ = average range for k sub-groups

   d_2 = estimating coefficient based on the number of readings in each sub-grou

Analysis of the data would be as follows:

Observation		Subgroup		
	1	2	3	4 = Â
1	1.00	0.95	0.81	0.97
2	0.95	0.89	0.92	0.96
3	0.80	1.01	0.87	1.11
4	1.10	1.04	1.01	1.03
Subgroup Range	0.30	0.15	0.20	0.15

$$\overline{R} = \frac{\sum_{k=1}^{K} R_k}{K} = \frac{0.80}{4} = 0.200$$

d_2 for subgroups of 4 = 2.059

$$\sigma'_{X_i} = \frac{\overline{R}}{d_2} = \frac{0.200}{2.059} = 0.097$$

$$N' = \frac{(80\sigma'_{X_i})^2}{\overline{X}_i} = \left(\frac{80(0.097)}{0.96375}\right)^2 = 65.0 \approx 65 \text{ obs.}$$

Use of Average Moving Range. The average moving range (range be
tween adjacent readings) can be used to estimate the population standard devi
ation. To use this technique the order in which the data were taken must b
kept so that the moving range can be calculated. Analysis of the previous 1
data points would be as follows:

	1	2	3	4	Observation 5	6	7	8	9
Data	1.00	0.95	0.80	1.10	0.95	0.89	1.01	1.04	0.81
R_m		0.05	0.15	0.30	0.15	0.06	0.12	0.03	0.23

	9	10	11	12	Observation 13	14	15	16
Data	0.81	0.92	0.87	1.01	0.97	0.96	1.11	1.03
R_m		0.11	0.05	0.14	0.04	0.01	0.15	0.08

$$\text{Avg. Moving Range} = \overline{R}_m = \frac{\sum_{m=1}^{N-1} R_m}{N-1} = \frac{1.67}{15} = 0.1113$$

$$\sigma'_{X_i} = \frac{\overline{R}_m}{d_2} = \frac{0.1113}{1.128} = 0.0987$$

where d_2 is for n equal to 2 because two adjacent readings are used to find each R_m value.

$$N' = \left(\frac{80\sigma'_{X_i}}{\overline{X}_i}\right)^2 = \left(\frac{80(0.0987)}{0.96375}\right)^2 = 67.08 \approx 68 \text{ obs.}$$

It should be emphasized that in a given study only one method of estimating N' would be used. The discrepancy among the sample sizes found by the three methods are due to individual sampling error. If a given method is selected prior to analysis of the data, the average outcome of the N' formula over all possible values of N observations selected from the population will be the same except for the restrictions found by Hicks and Young, noted earlier.

The required sample size (N') may be used to determine if the process is in statistical control and ready for setting a time standard. An extremely large sample size would indicate a lack of internal consistency.

CONTROL CHARTS FOR STANDARDIZATION. In order to test whether or not the observed time data are consistent, that is, whether the operation is in **statistical control**, a control chart for sample means of cycle times can be constructed. In this case the entire number of observations (N) is broken down into successive subgroups of size n and the means of these subgroups plotted on the control chart.

For example, assume that, at random times over a two-day interval, samples of four observations each are taken of cycle times of an operation. The cycle times may represent the **sum of elemental times** or they may be **overall times** without reference to an elemental breakdown. The results of the observations in either event are given in Fig. 11.

Assuming the subgroups of 4 were taken at random, the validity of the control chart is based on the three basic laws previously mentioned.

Thus

$$\overline{\overline{X}}_j = \text{estimate of population average cycle time}$$

$$= \frac{\sum_{k=1}^{K} X_{jK}}{K}$$

and

σ_{X_i} = estimate of the standard deviation of the universe of individual cycle times
$\sigma'\overline{X}_i$ = standard deviation of the distribution of sample means

$$= \frac{\sigma'_{X_i}}{\sqrt{n}}$$

If the operation is in a state of statistical control, with respect to mean cycle times, the probability of an individual sample mean $\overline{X}_j$ falling outside the limits $\pm 3\sigma_{\overline{X}_j}$ is approximately 3 in 1,000.

Sample No.	Time of Day Randomly Selected	Successive Observed Cycle Time Values X_j				Sample Average (Mean) $\overline{X}_j$
1	8:25	1.00,	0.95,	0.98,	1.01	0.985
2	9:15	1.15,	0.98,	1.00,	1.17	1.075
3	10:00	0.95,	1.30,	1.02,	0.92	1.047
4	11:00	0.80,	0.75,	0.85,	1.08	0.870
5	1:20	0.94,	1.11,	1.25,	1.01	1.077
6	2:30	1.20,	1.17,	1.45,	1.30	1.280
7	3:10	1.12,	0.86,	0.79,	0.82	0.897
8	3:45	0.82,	0.98,	0.97,	1.13	0.975
9	8:30	0.72,	0.63,	0.87,	0.90	0.780
10	9:10	1.10,	1.28,	0.90,	1.13	1.102
11	10:05	0.99,	0.82,	0.86,	0.95	0.905
12	11:30	0.98,	1.17,	1.29,	0.69	1.032
13	1:30	1.05,	1.12,	0.66,	0.93	0.940
14	2:15	0.91,	0.80,	1.18,	1.18	1.017
15	3:25	0.76,	0.91,	1.06,	0.82	0.888
16	4:00	0.75,	1.11,	0.69,	1.17	0.930

n = Subgroup size = 4
N = Total number of observations = 64

Fig. 11. Tabulated data for use on a control chart.

Using the above results, the following parameters are calculated:

$$\overline{X}_j = \frac{\sum_{1}^{16} \overline{X}_j}{.16} = 0.988$$

$$\sigma'_{X_i} = \sqrt{\frac{\sum_{N=1}^{64} (X_j - \overline{X}_j)^2}{N - 1}} = 0.179$$

$$\sigma_{X_i} = \frac{\sigma'_{X_i}}{\sqrt{n}} = \frac{0.179}{\sqrt{4}} = 0.0895$$

Upper control limit = $\overline{X}_j + 3\sigma_{\overline{X}_i} = 1.256$
Lower control limit = $\overline{X}_j - 3\sigma_{\overline{X}_i} = 0.720$

The control chart with individual sample means plotted in succession is shown in Fig. 12. It may be seen that the mean of sample No. 6 lies outside the upper control limit, suggesting an absence of statistical control.

Lack of Statistical Control. A number of conclusions may be drawn from the above example concerning the **out-of-control point.**

1. The sample in question (No. 6, Fig. 12) reflects **nonstandard performance** by the operator in unusual pace or method. This presents the fundamental problem of variation which is controllable by the operator. Theoretically, if the cycle times are in a state of statistical control, the manual elements are performed habitually and without conscious effort on the operator's part. Where

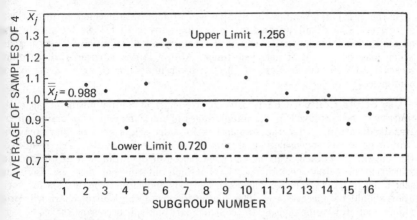

Fig. 12. Control chart for sample means of net cycle times.

there is a **conscious effort** on the operator's part to change pace or method, a state of statistical control would probably not exist, as indicated by sample means falling out of control on the chart or by a lack of randomness in the variability of the sample means even if they are within limits.

2. The sample may reflect **nonstandard conditions,** uncontrollable by the operator, such as a change in material quality, tool dimensions, or part locations. Such conditions should be checked and causes removed before making a new study.

3. It is possible that the sample (No. 6) reflects a **cyclical productive activity** which is nevertheless a part of the standard method. That is, the sample may represent the handling of the last few parts in a tote pan, or the initial cycles after a rest period. In these situations it may be desirable to include the out-of-control points as a part of the estimate. Factors which operate cyclically during the work day will be weighted in the normal random observation period so that no points fall out of control. But this phenomenon will show up on control charts by a lack of randomness of sample means about the grand average, by groupings of successive sample points above or below the average line, or by runs of successive readings.

4. Out-of-control points may reflect an irregular or **intermittent element,** which may or may not be an allowable part of the method. The advantage of elemental timing is that such minor delays or interruptions may be noted and cause for their being out of control assigned to them. In some cases, these irregular elements may be studied independently of the repetitive elements and their times prorated over the average number of parts completed between their appearances.

Subgroup Size and Frequency of Sampling. In general, small subgroups are **advantageous** because of the economics of sampling and making the calculations. Where the subgroup size is small and the samples are taken frequently over a day, shifts in the population mean may be detected more readily than with larger samples taken less frequently. That is, changes in the population average may be masked in the larger sample size. The objective then is to make each subgroup as homogeneous as possible (to represent a single population), to maximize between sample variation, and to minimize within sample variation.

In the previous example the sample size of four was selected to meet thes objectives. The **disadvantage** of small subgroups is that their means are no normally distributed, particularly if the population distribution is non-normal which may be typical of net cycle times. Also, a larger subgroup will be mor sensitive to variation because of the narrower limits for a specified degree o confidence.

Use of the Control Chart. The control chart of sample means of net cycl times may be employed in the initial stages of measurement as a **criterion o standardization.** After the standard is formally set, a control chart may b kept over an extended period of time as a **control device.** The results o random samples taken over the extended period of operation will give a curren indication of actual performance levels. Such information may be useful i scheduling work, appraising changes in methods, indicating when causes for vari ation should be checked, etc. The continued use of the control chart will pro vide a pool of data for revising the control chart limits, if necessary.

CONTROL CHART OF VARIATION. The control chart of means indi cates the variability in central tendency. This may be supplemented with control chart of **sample variability**—in terms either of sample ranges or o sample standard deviations. The control chart of sample variation indicates th consistency of individual time values.

USE OF LINEAR PROGRAMMING AND MULTIPLE REGRES SION IN SETTING STANDARDS. Situations exist in work measuremen studies in which the number of work units exceeds the number of data sets avail able. The data sets consist of information concerning work counts of the worl units together with the total work time for all outputs in the data set. Thus, us of simple simultaneous equations is not possible.

Mundel (Motion and Time Study) examines the use of linear programmin and multiple regression techniques in situations in which the following four condi tions exist:

1. The output data and the associated time data are available only as a series o data sets with each set containing work counts of more than one work unit.
2. The work count of each work unit in each data set is separated from the worl counts of other work units.
3. The time for each kind of work unit is not separated in each data set; th work time is available only as an aggregate for each data set.
4. The number of different work units for which work counts are available exceed the number of data sets available; the reverse may also be true.

As Mundel points out, these conditions are frequently encountered when usin historical production and time study data. In recording the data at the time th work was being studied, the work unit may not have been clearly defined. Thu there was no breakdown within the total times recorded.

Mathematical Model. A partial model may be used to represent the situ ation described above, as follows:

$$BV + WC_11 \times ST_1 + WC_21 \times ST_2 + \ldots + WC_i1 \times ST_i + S_1 = \Sigma MH_1$$
$$BV + WC_12 \times ST_1 + WC_22 \times ST_2 + \ldots + WC_i2 \times ST_i + S_2 = \Sigma MH_2$$
$$\cdot$$
$$\cdot$$
$$BV + WC_1N \times ST_1 + WC_2N \times ST_2 + \ldots + WC_iN \times ST_i + S_N = \Sigma MH_N$$

where BV = a basic value associated with any level of operation (unknown).

$WC_i1, WC_21, WC_i1, \ldots WC_i2, \ldots WC_iN$ = work count of work unit $1, 2, \ldots, i$ in data set $1, 2, \ldots, N$ (known).

$ST_1, ST_2, \ldots, ST_i$ = standard time for work unit $1, 2, \ldots, i$ (unknown).

$S_1, S_2, \ldots, S_N$ = slack time in data set $1, 2, \ldots, N$ (that is, some amount of time that cannot be explained by the indicated relationships) (unknown).

$\Sigma MH_1, \Sigma MH_2, \ldots, \Sigma MH_N$ = sum of man-hours (or any other appropriate time reference) of work time associated with data set $1, 2, \ldots, N$ (known).

Solution of Model. Mundel (Motion and Time Study) outlines two **mathematical routines** for finding the unknown values in the above equations.

Linear programming method:

1. $S_1 + S_2 + \ldots + S_N =$ a minimum value, referred to as **residual slack.**
2. $BV, S_1, S_2, \ldots, S_N \geqq 0.$
3. $ST_1, ST_2, \ldots ST_i \geqq 0.$
4. All work units need not be assigned a standard time other than 0.
5. It is assumed that any work count multiplied by the appropriate standard time is part of a linear function (that is, a work count of 30 takes three times as long as a work count of 10, and one-half as long as a work count of 60, etc.)

Multiple regression method:

1. $S_1^2 + S_2^2 + \ldots + S_N^2 =$ a minimum value, referred to as **residual variance.**
2. $BV, S_1, S_2, \ldots, S_N \gtrless 0.$
3. $ST_1, ST_2, \ldots, ST_i \gtrless 0.$
4. All work units need not be assigned a standard time other than 0.
5. The condition of linearity given as condition 5 of the linear programming method also applies here.

Obviously these techniques cannot be applied blindly. The solutions obtained must be examined carefully and their validity verified with comparison against other applicable criteria. The solutions for ST_1, ST_2, etc., are obviously different for the linear programming and multiple regression methods.

Hand computation of either method is impractical when many work units and data sets are involved. In such cases, use of the **computer and appropriate programs** is called for. Limitations in capacity often limit problems to 50 data sets and 200 work counts even with large computers.

Allowances

TYPES OF ALLOWANCES. The **base time** represents the net production time per unit of output for the normal operator. The **standard time** is a gross unit time at normal performance. This gross time will be greater than the net production time due to interruptions or delays reducing production time, and to certain factors such as fatigue, adverse or extreme job conditions, and rework whose influences on production time were not considered in the measurement or rating stages of the study. The base time is therefore adjusted by percent allowances to account for these factors.

Interruptions. As pointed out in the time study procedure, certain interruptions to the productive cycle, if minor, may be classed as irregular elements of the time study and included in the base time as **prorated items.** Interruptions of a large magnitude and periodic nature may be classed as **separate jobs** or operations and given an independent standard time. Allowances, then, are added to account for those interruptions and delays which are residual and not included in base time or separately standardized. **Allowances** include time necessary for the following.

1. Personal needs and rest periods.
2. Unavoidable delays, such as:
   a. Setup, tear-down, and cleanup.
   b. Tool and equipment maintenance.
   c. Getting supplies and raw material.
   d. Receiving instructions from the supervisor.
   e. Filling out production or other forms.
   f. Periodic inspection of material or parts.
   g. Reading blueprints or instruction cards.

Personal and Delay Allowances. Personal allowances cover the time that must be allowed every worker for his personal necessities. They may range from 3 to 5 percent, varying with jobs but tending to be higher when the work is heavy or unpleasant.

Delay allowances may be avoidable or unavoidable. **Avoidable delays** are not included in the standard time. While avoidable delays usually refer to those under the direct control of the operator, certain delays classified in this manner may be avoided by better shop management or supervision. Although incidental interruptions and minor delays from tool breakage, machine stoppages, variations in material, and supervisory contacts should be kept to a minimum, they do take place. These are **unavoidable delays** which are covered by an allowance determined by **all-day time studies** or **work sampling studies.**

The interruptive tasks, which are considered to be normal constituents of the job and are referred to as unavoidable, are susceptible to proper design or improvement as a part of operation analysis. Prior to determining their time for allowance purposes, effort should be made to analyze and standardize these elements of the job just as with productive operations.

Fatigue Allowance. Fatigue in varying degrees is a normal part of every manually controlled operation. Fatigue is manifested during the course of the day in the gradual reduction in rate of output. This is reflected in an increase both in mean cycle time and in the number and intensity of delays. Lengthening of the study in conjunction with a sufficient number of observations may make very little direct allowance necessary for a light operation.

Other Allowances. Certain allowances besides those for delays may be added to base time to adjust for circumstances peculiar to the operation in question such as:

1. **Small production lot sizes.** Sometimes the actual production period is too short to allow the operator to reach a pace in line with his usual performance. Where a large number of small-lot jobs are worked on during the day with various setups included, an allowance as high as 15 percent may be required to allow the normal operator to make normal earnings.
2. **Training.** During a training period an allowance may be added to give the trainee an opportunity to gain reasonable earnings. This may be a "sliding" allowance which is progressively decreased during the training period until it reaches zero at a predetermined time. If the company has sufficient data to show the effect of learning on production rates this training allowance may be correlated with the learning curve.
3. **Extreme job conditions.** Such conditions as unfavorable weather on outside jobs, and extreme heat, noise, or fumes on inside operations may be allowed for.
4. **Spoilage and rework.** On certain operations experience may indicate that a certain percentage of parts are spoiled due to factors beyond the operator's control. The time required to rework these parts may be included as an allowance.

DETERMINATION OF ALLOWANCES. The direct measurement of any of the factors which are to be included as allowances is difficult. In many cases an allowance may be the result of company policy, judgment based on long experience, precedence, or tradition. In certain instances, allowances for specific types of jobs may appear in the union contract. In still other cases, the individual time study engineer may resort to his own judgment in setting the amount.

For allowances such as those for abnormal job conditions and small production lot sizes, the analysis of company data may lead to more objective results. Special studies employing statistical methods may be made to determine the need for and the amount of allowances for the various conditions in question.

INTERRUPTION STUDIES. Interruption studies are sometimes referred to as **production** or **all-day studies.** The general purpose of the interruption study is to determine the various activities carried out by an operator on a job during the normal work day and the time consumed in each activity. The stop watch is usually used as the measuring device. The results of the study may be employed to:

1. Determine allowances for fatigue, and for personal and unavoidable delays.
2. Check the results of a time study.
3. Indicate areas of inefficiency in the operation or job.

Whatever the use of the study, the method of taking the observations is the same. Although the study is very simple to take, it is rather tedious. The method requires a critical observation of the operation or job and a careful record of the activities. The observer starts at the beginning of the work day and keeps a record of the time for each and every delay that occurs, at the same time recording the productive time separately.

Data Obtained. In Fig. 13, Morrow (Motion Economy and Work Measurement) shows a hypothetical example of an interruption study on a drilling operation. The interruption study is shown from 7 A.M. to 8:30 A.M., and it is continued throughout the day in a similar manner. The stop watch is allowed to run continuously and, in addition, every half-hour the time is recorded from an ordinary watch. Observed readings are noted, as shown in the column headed "Ob." It will be noted that symbols have been used for parts of the operation which are frequently repeated:

D—Drilling time.
D–p—Part of drilling time.
H—Handling time, from the end of drilling time to start of drilling again.
H–p—Part of handling time.

In the case illustrated, the **productive time** has been divided into machine and handling time. Ordinarily this separation would not be necessary and the productive time would be kept as one item. The number of pieces produced should be recorded. Symbols can be used for delays if their nature is known beforehand. In this case delays are described, viz., talk, belt broke, adjust oil pump, etc. Production time is recorded as shown. After the study has been completed, the differences between the recorded times are calculated to obtain the time for each element in the study. Figure 14 shows the **interruption study summary sheet.** The observation sheet shown in Fig. 7 may be effectively used for the study itself. The study may extend over one or more shifts and possibly for more than a day if it is advisable.

Computation of Allowances from Interruption Study. Considering the above example, assume that during a previous time study on the same operation the **selected net cycle time** was determined to be 2.04 minutes per unit.

INTERRUPTION STUDY

OPERATION: Drill
ARTICLE: G18M Slide
TIME: START: 7 A.M.   STOP: 4 P.M.   DEPARTMENT: Drilling   ELAPSED: 8 hours (1 hour out)   UNITS PRODUCED: 135
OPERATOR: #260

Ob — Observations   Df — Differences

	Ob	Df
Start....	0.00	— 7 A.M.
Get work ticket..	4.01	4.01
Talk..	5.26	1.25
Grind tools and adjust..	20.10	14.84
Arrange work bench..	22.00	1.90
Talk..	24.50	2.50
Place tote box..	24.63	.13
H..	25.98	1.35
D..	26.34	.36
H..	28.32	1.98
D..	28.70	.38
H..	30.45	1.75 7:30 A.M.
D..	30.82	.37
H..	2.44	1.62
D..	2.83	.39
H..	4.41	1.58

	Ob	Df
D..	4.79	0.38
H..	6.51	1.72
D..	6.88	.37
H..	8.29	1.41
D..	8.60	.39
H..	10.18	1.50
D..	10.56	.38
H..	12.08	1.52
D..	12.45	.37
H..	14.33	1.88
D-p..	14.53	.20
Belt broke..	24.78	10.25
D-p..	25.06	.28
H..	26.69	1.63
D..	27.05	.36
H..	28.58	1.53 8 A.M.

	Ob	Df
Adjust oil pump....	1.76	3.18
D..	2.13	.37
H-p..	2.43	.30
Talk..	4.58	2.15
H-p..	5.84	1.26
D..	6.19	.35
H..	7.77	1.58
D..	8.15	.38
H..	10.01	1.86
D..	10.38	.37
H..	11.99	1.61
D..	12.38	.39
H..	13.94	1.56
D..	14.31	.37
H-p..	14.59	.28
Adjust Machine..	17.74	3.15

	Ob	Df
H-p..	19.06	1.32
D..	19.42	.36
H..	21.00	1.58
D..	21.39	.39
H-p..	21.95	.56
Talk..	22.80	.85
H-p..	24.05	1.25
D..	24.42	.37
H..	25.91	1.49
D-p..	26.17	.26
Belt off..	27.23	1.06
Adjust oil pump..	29.90	2.67
D-p..	.20	.30 8:30 A.M
H..	1.87	1.67
D..	2.25	.38
Personal..	5.00	2.75

Rest of day not shown. Refer to summary.

Fig. 13. Observation and analysis sheet for an interruption study.

INTERRUPTION STUDY—SUMMARY

Operation	Productive Time (min.)	Preparation Time (min.)	Delay Time (min.)	
			Unavoidable (and personal)	Avoidable
Drilling time, 135 pieces.	52.30			
Handling time.........	240.50			
Grind and adjust tools..		25.60		
Change work ticket.....		8.00		8.61
Place tote box, arrange work bench.........		2.95		
Talk.................				20.62
Belt broke, slipped off, etc................			5.00	25.68
Waiting for work.......				35.81
Pump adjustment, packing, etc.............				31.34
Adjust machine........			9.71	
Personal.............			13.88	
Total time 480.00.......	292.80	36.55	28.59	122.06

Total time, excluding avoidable delays = (sum of 292.80, 36.55, and 28.59) = 357.94 min. *Time per piece*, 2.64 min. excluding avoidable delays.

Fig. 14. Summary sheet of an interruption study.

Productive time from interruption study .... 292.80 min.
Productive time using independent time study results
(135 units × 2.04 min./unit) .... 275.40 min.
Difference .... 17.40 min.

The difference may be assumed to be caused by fatigue. This is based on the assumption that the independent time study was taken at a time of day before the results of fatigue were noticeable. However, the difference could partially be due to sources of bias in the time study, including particularly a change in operator performance level.

Percentage allowance for fatigue .... $\frac{17.40}{275.40} = 6.3\%$

Personal time .... 13.88 min.

Percentage allowance for personal time .... $\frac{13.88}{275.40} = 5.1\%$

Unavoidable delays .... 14.71 min.

Percentage allowance for unavoidable delays .... $\frac{14.71}{275.40} = 5.3\%$

Total Allowances 16.7%

In order to determine standard time, the 16.7 percent of base time is added to the base time per unit. The method of computing delay and personal time allowances shown above is biased toward giving a high value. This is in comparison with use of 292.80 minutes as the base rather than 275.40 minutes. It should also be noted that **avoidable delays** are removed from the study before calculating allowances. It is assumed that these avoidable delays will largely be absent in actual practice due to closer controls. Also, it is assumed that the time consumed in avoidable delays, 122.06 minutes, will be utilized in production time, personal time, and delays in about the same proportion as observed in the interruption study. The allowances determined from the interruption study may be biased toward the high side if the amount of avoidable delay time in the study is high. That is, the more the time devoted to avoidable delays the smaller becomes the production time which is used as a base for calculating the percent allowances.

If the period needed to study an operation must be long, the interruption study is a relatively uneconomical method of observation. This is particularly true if the purpose of the study is merely to measure **delay times.** A more economical method in this case may be to use work sampling. If the purpose of the all-day study is to gather data for **methods improvement** or for more efficient programming of the operation, there may be no substitute for continuous observation (see the section on Motion and Methods Study).

Work Sampling

DEFINITION OF METHOD. The work sampling procedure involves the taking of qualitative observations of a work system randomly over a protracted time period. The state of the work system is classified by **type of activities** such as production, setup, and idle for repairs. At a given instant for a given machine or operator, these activities are mutually exclusive. An instantaneous **sample observation** is taken to determine the state of the operation in terms of the class of activity. A ratio of the number of observations recorded for a particular activity to the total number of observations made in the sample gives an estimate of the **percent time spent in that activity.** When the study is made of a group of similar objects the results may be expressed as a percent of the group in a given activity if each item in the group is **independent.**

USES OF WORK SAMPLING. The basic purpose of work sampling is to estimate the percentage of a protracted time period consumed by various activity states of a resource, such as equipment, machines, or operators. The technique has broad application in both **manufacturing** and **service** endeavors. It may be used to:

1. Determine allowances for inclusion in standard times.
2. Indicate the nature of the distribution of work activities within a gang operation.
3. Estimate the percent utilization of groups of similar machines or equipment.
4. Indicate how material handling equipment is being used.
5. Provide the basis for indirect labor time standards.
6. Indicate areas where methods study may profitably be used.
7. Determine the productive and nonproductive utilization of clerical operations.
8. Determine a standard time for a repetitive operation (as an alternative to the stop watch method).

PROCEDURE IN WORK SAMPLING. Just as in time study, the accuracy of an estimate obtained from work sampling is dependent on the control and execution of the **study procedure,** whereas precision is controlled by the **number of observations** taken.

Defining the Parent Population. The objective of the study must be clearly defined. This is tantamount to defining carefully the parent population to be sampled. This population should be **homogeneous** in the sense of containing items with similar operating or other characteristics. For instance, assume that he purpose of the study in a printing plant is to determine percent downtime n the presses. If there are a number of different types of presses with different operating characteristics, then a number of independent work sampling studies hould be taken, each with a population of like presses. On the other hand, if he purpose of the study is to determine time spent for personal reasons by the press operators, the population may cut across dissimilar presses and include all press operators.

Where unlike objects are grouped to constitute a **heterogeneous population,** he sample estimates have only an aggregate meaning, and no real conclusions may be drawn about the characteristics of the unlike individuals in the population.

Activity Classification. The classes of activities for a given population must exhaust all possibilities and be **mutually exclusive.** Each type of activity or each state of the entity being measured must be so clearly defined as to its content that a proper **discrimination between classes** can be made at the moment of observation. There should be little opportunity for indecision regarding the activity to which a particular sample observation is assigned. In the example of the three machines, assume that the classification includes, among others, the categories "downtime for adjustment" and "downtime for maintenance." On a given observation the analyst may find a machine undergoing a repair while it is also being adjusted. This leaves a question regarding the state the machine is in.

Length of Study. The period over which observations are taken determines the **size of the parent population.** Just as with time study, the work sampling study should be of such length that all activities which are a normal part of the operation have a chance of being observed and that cyclical influences can also be covered. Conway (Journal of Industrial Engineering, vol. 8) states:

In general, the period should be at least as long as the longest period of any cyclical behavior of the characteristic being studied. Such behavior could depend on clock time, calendar time, physical characteristics of the activity, or on policy or procedural requirements such as the frequency of inspection or maintenance activity. For example, if there is any a priori basis for suspecting that the pattern of activity is different on different days of the week, the period should obviously not be less than a full week.

If the study is made short for economic reasons, the extended population should be tested at certain intervals for homogeneity. This may be done with the use of the control chart (see Fig. 16). If there are reasons to believe that there are present in the extended population significant causes for variation that were not present in the parent population, then subsequent data may be **pooled** with the original data to provide a new estimate.

Random Observations. A single observation consists of an **instantaneous** or **snapshot** look at the operation in order to determine the state of activity. Such an observation should be taken at a randomly selected point in time.

A **random selection** of observation times is made in the following manner. The study period, or parent population, is divided into specific time units. For instance, the length of the study may be designated as 10 days (4,800 minutes). This period is broken down into units of, for example, 1/10 hour (6 minutes). The parent population contains 800 units, from which a sample of N readings may be selected. The number of observations, N, should be less than 800, unless there is more than one observable entity in the population and a number of observations can be made in a single trip. The randomization can be done with the use of a table of random numbers. If the length of the study is short and the number of required observations large, the trips may be so frequent as to warrant the use of an interruption or all-day time study instead of work sampling.

Nonrandom **systematic sampling** may be employed under certain conditions By this method sample observations are taken at regular intervals during the course of the study. This has the advantage of greater ease in planning the observer's work. Also, it facilitates the programming of several different studies to be conducted over the same period. That is, several jobs may be observed during a trip.

The use of systematic sampling is recommended only when there is no regular pattern or **cyclical behavior** in the elements or activities of the job being studied. If these cyclical conditions exist, then systematic sampling may introduce a significant bias into the estimate, and the models employed to determine precision or sample size will be meaningless.

Use of Random Digits. In work sampling, the times of observation for any given sample must be selected without **bias.** That is, these observations must be completely random with respect to time. Therefore, the problem is one of determining the times at which observations should be made. The number of observations (sample size) has been determined for a given confidence level, and the specific observation times must be distributed randomly through the study period.

This can be accomplished in several ways. The most effective way, especially where there are multiple studies involved, is through the use of a table of random digits (see section on Statistical Methods).

For example, assume 15 observations are required for a given study over an eight-hour work day. The work day begins at 8:00 A.M., lunch is between 12 noon and 1:00 P.M., and the work day ends at 5:00 P.M. The morning can be divided into five-minute segments, numbered 1 through 48; the afternoon can be broken into five-minute segments, numbered 49 through 96. The table of random digits is then used to select fifteen two-digit numbers, making sure to reject all numbers over 96 and all duplications. Each valid number selected will identify the five-minute segment during which an observation should be made. Thus 37 represents the 37th five-minute segment (11:05 A.M.) and 63 represents the 63rd five-minute segment (this is the 15th segment in the afternoon or 2:15 P.M.).

This specific pattern need not be followed, but there should be some specific pattern agreed upon in order for the end result to be a truly random selection.

OBSERVATION PROCEDURES. The observer passes the operation, takes an instantaneous look, and places a check on the observation sheet next to the class of activity observed. Fig. 15 shows a sample observation sheet used for the study of clerical operations, with frequency based on the table of random digits.

Since the observation is instantaneous the activities should be clearly defined as to context and demarcation points to avoid indecision in classification. The classification may be difficult to make instantaneously in some cases, and the observer may have to delay the observation momentarily to check the activity.

Independent Studies. Different operations may be observed during a single trip by the observer. In this case the observer's time is concentrated on making observations for an extended time. Analysis of the data and formulation of standards can be performed subsequently, without interruptions for making observations. Where more than one study is conducted, the time for starting the

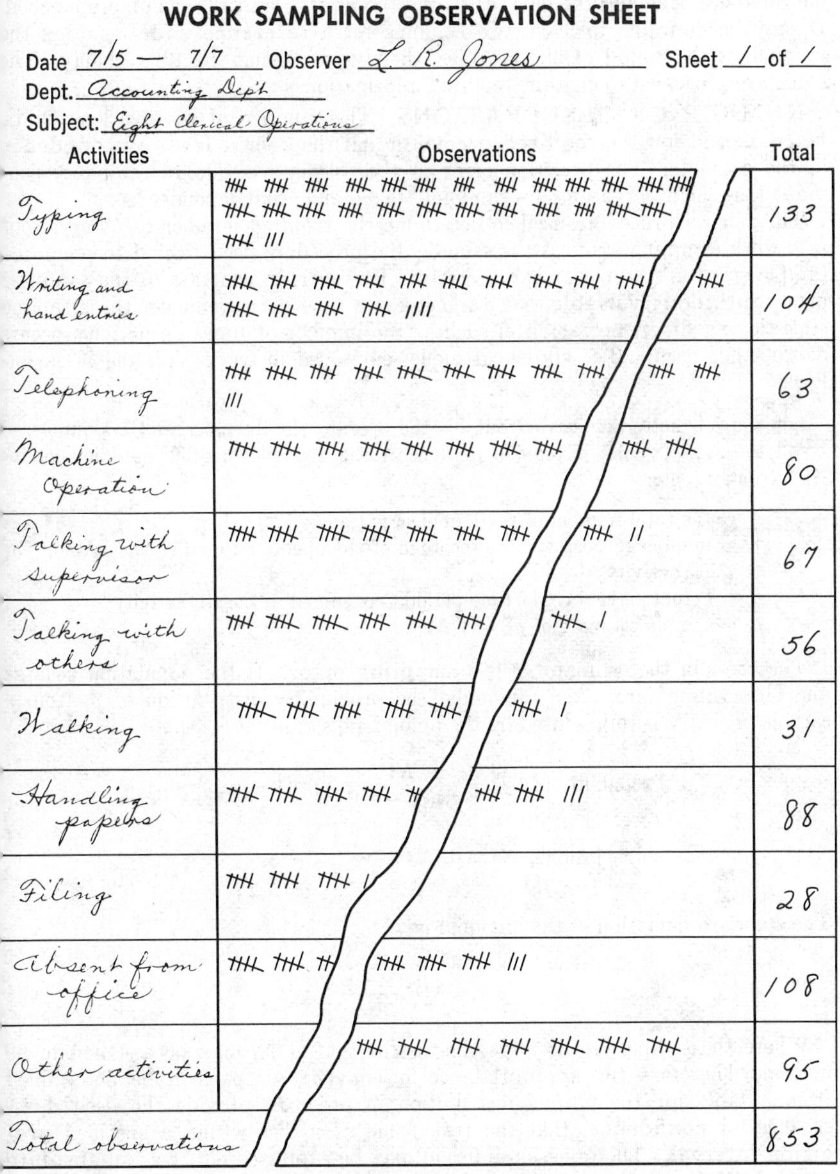

WORK SAMPLING OBSERVATION SHEET

Date *7/5 — 7/7* Observer *L. R. Jones* Sheet *1* of *1*

Dept. *Accounting Dep't*

Subject: *Eight Clerical Operations*

Activities	Observations	Total
Typing	~~THL~~ ///	133
Writing and hand entries	~~THL~~ ~~THL~~ ~~THL~~ ~~THL~~ ~~THL~~ ~~THL~~ ~~THL~~ ~~THL~~ ~~THL~~ ~~THL~~ ~~THL~~ ~~THL~~ ~~THL~~ ~~THL~~ ~~THL~~ ~~THL~~ ~~THL~~ ////	104
Telephoning	~~THL~~ ~~THL~~ ~~THL~~ ~~THL~~ ~~THL~~ ~~THL~~ ~~THL~~ ~~THL~~ ~~THL~~ ~~THL~~ ~~THL~~ ///	63
Machine Operation	~~THL~~ ~~THL~~ ~~THL~~ ~~THL~~ ~~THL~~ ~~THL~~ ~~THL~~ ~~THL~~ ~~THL~~ ~~THL~~	80
Talking with supervisor	~~THL~~ ~~THL~~ ~~THL~~ ~~THL~~ ~~THL~~ ~~THL~~ ~~THL~~ ~~THL~~ //	67
Talking with others	~~THL~~ ~~THL~~ ~~THL~~ ~~THL~~ ~~THL~~ ~~THL~~ ~~THL~~ /	56
Walking	~~THL~~ ~~THL~~ ~~THL~~ ~~THL~~ ~~THL~~ ~~THL~~ /	31
Handling papers	~~THL~~ ~~THL~~ ~~THL~~ ~~THL~~ ~~THL~~ ~~THL~~ ~~THL~~ ///	88
Filing	~~THL~~ ~~THL~~ ~~THL~~ /	28
Absent from office	~~THL~~ ~~THL~~ ~~THL~~ ~~THL~~ ~~THL~~ ~~THL~~ ///	108
Other activities	~~THL~~ ~~THL~~ ~~THL~~ ~~THL~~ ~~THL~~ ~~THL~~ ~~THL~~	95
Total observations		853

Fig. 15. Observation sheet for work sampling study.

trip and the operation to be observed at the start of the trip are randomized. The direction of the trip may also be altered.

When the number of observations and amount of data required are large, the clerical procedure may be reduced by using a **mark-sensitive** tabulating card system. A common method of making the observations is to use a card for each observation. Other information required for the analysis, such as department or file reference code and the date, can either be written on the card or prepunched It may be desirable to leave two columns for a **reference code** denoting the number of the round of observations, thereby signifying the time of day. The cards are processed to give the required information.

NUMBER OF OBSERVATIONS. The number of observations to be taken depends on the **required precision** and the desired **level of confidence** in the estimate. The precision error in the estimate is due to sampling, provided that the observations are randomly taken and are independent.

The general procedure used to determine the required number of observations in a work sampling study is the same as the procedure used to find the required sample size for a stop watch time study. However, in contrast to the sampling of a **continuous variable,** such as net cycle time measurements, the sampling statistic is a **discrete variable,** such as the number of times an activity occurs or does not occur. The statistical model employed in work sampling is as follows:

Let the job activities carried out by the operator in time period t be numbered $1, 2, \ldots, i, \ldots M$. Example: produce at the machine, handle material, get instructions, etc.

N = the total number of random observations

r_i = number of observations recorded of the operator found in the state of the ith activity

p_i = actual percentage of time period t consumed in the ith activity

$\bar{p}_i$ = sample estimate of $p_i = r_i/N$

The error in the estimator $\bar{p}$ is a **sampling error.** If the population is large and observations are taken randomly, the probability distribution of $\bar{p}_i$ from a sample of size N is represented by the binomial distribution:

$$\text{Probability } (r_i/N) = \frac{N!}{r_i!(N - r_i)!} \, (p_i)^{r_i}(1 - p_i)^{N-r_i}$$

$$\sum_{r=0}^{N} \text{Prob } (r_i/N) = 1$$

The standard deviation of the binomial is:

$$\sigma_{p_i} = \sqrt{\frac{p_i(1 - p_i)}{N}}$$

Where the sample size is large the distribution of $\bar{p}_i$, may be assumed to be normal. Therefore the appropriate coefficient (Z) for σ_{p_i} can be determined from a table for the normal distribution in order to provide the desired assurance, or **confidence,** that the true value of p_i lies within a specified **precision interval.** The precision interval may be expressed either as an **absolute error,** $\pm X\%$, or as a percentage of the estimated p_i **relative error** in terms of $\pm X\% \ p_i$. The formula used to determine the required sample size for an absolute error of $\pm X\%$ would be:

$$\pm Z\sigma_{\bar{p}_i} = \pm Z \sqrt{\frac{\bar{p}_i(1 - \bar{p}_i)}{N}} = \pm X$$

$$N = \left[\frac{Z}{X}\right]^2 \bar{p}_i(1 - \bar{p}_i)$$

Whereas the formula for a relative error of $\pm X\%$ p_i would be:

$$\pm Z\sigma\bar{p}_i = \pm Z\sqrt{\frac{\bar{p}_i(1-\bar{p}_i)}{N}} = \pm X\bar{p}_i$$

$$N = \left[\frac{Z}{X}\right]^2 \frac{(1-\bar{p}_i)}{\bar{p}_i}$$

For a given error of $X\%$, the number of observations required using the relative measure is greater than those required for an absolute measure, assuming the confidence level is constant. If the analyst desires to be 95% confident ($Z \approx \pm2$) that the results of his study are within $\pm5\%$ of p_i, the required sample sizes and corresponding precision intervals for selected values of p_i are:

Percent of Occurrence (p_i)	Sample Size $N = \left[\dfrac{2}{0.05}\right]^2 \dfrac{(1-p_i)}{p_i}$	Precision Interval $\pm X\% \, p_i$
50	1,600	±2.5
40	2,400	±2.0
30	3,730	±1.5
20	6,400	±1.0
10	14,400	±0.5
1	158,400	±0.05

The procedure would allow for an error of $\pm2.5\%$ in estimating activities that occur 50% of the time but would permit an error of only 0.05% for estimating activities that occur 1% of the time. Obviously, the high degree of precision resulting from the large sample sizes would not be warranted in most situations because of economic considerations. Specifying an absolute error of $\pm2.5\%$ for all values of p_i would result in smaller sample sizes as shown below:

Percent of Occurrence (p_i)	Sample Size $N = \left[\dfrac{2}{0.025}\right]^2 p_i(1-p_i)$	Precision Interval $\pm2.5\%$
50	1,600	±2.5
40	1,536	±2.5
30	1,344	±2.5
20	1,024	±2.5
10	576	±2.5
1	64	±2.5

The desired level of confidence and the degree of precision used to determine the required number of observations should be chosen in accordance with economic considerations of the use of the final results of the study. For example, assume that the problem is to determine the percent idle time for a group of three similar machines employed in a certain department. A work sampling study might result in the following data.

Time of Observation (Randomly Selected)	Number of Observations per Trip	Number of Occurrences of Idle Machine
8:20 A.M.	3	3
8:32 A.M.	3	1
9:00 A.M.	3	2
.	.	.
.	.	.
.	.	.
4:50 P.M.	3	2

Totals: 80 Trips   240 Observations   72 Idle Machine Observations
Estimate of percent idle time $= 72/240 = 0.30$
Estimate of idle machines $= 0.30 \times 3 = 0.90$

The analyst could state that he is 95% confident that the machines are idle 30% of the time based on his 240 observations. The accuracy of this statement would be:

$$X = Z \sqrt{\frac{p_i(1 - p_i)}{N}} = 2 \sqrt{\frac{(0.30)(0.70)}{240}} = 0.059$$

That is, 95 times out of 100 the true idle time per machine will be 30% ± 5.9%, or between 24.1% and 35.9%. The analyst must decide whether these estimates are sufficiently accurate to meet his needs for the data. If he requires a higher degree of precision he could increase the number of observations or reduce his confidence in the results.

CONTROL CHART ANALYSIS. A control chart of p taken from sub-samples of size n can be constructed to indicate **consistency** in the study data as well as to detect shifts in the extended population parameters. For instance, assume the results shown in Fig. 16 of a work sampling study made on the three machines.

Date	Number of Observations	Number of Occurrences of Machine Idle	Estimate of $\bar{p}$ from Sub-sample
6/9 A.M.	33	11	.33
P.M.	39	7	.18
6/10 A.M.	36	14	.39
P.M.	30	6	.20
6/11 A.M.	39	10	.26
P.M.	30	7	.23
6/12 A.M.	30	8	.27
6/14 A.M.	33	17	.52
P.M.	39	16	.41
6/15 A.M.	36	7	.19
P.M.	36	11	.31
TOTALS	381	114	.30

Fig. 16. Results of a work sampling study made on three machines.

The limits for the control chart are calculated at $\bar{p} \pm 3\sigma_p$. The subsample size, $\bar{n}$, is set at 35, the average of the number of observations taken at each period. The subsample size affects the width of the control limits. In general, the average sample size, $\overline{n}$ can be used if the individual subsample sizes do not vary more than 30% from $\overline{N}$.

$$3\sigma_p = 3\sqrt{\frac{\overline{\overline{p}}(1-\overline{\overline{p}})}{\overline{n}}}$$

$$= 3\sqrt{\frac{0.30 \times 0.70}{35}}$$

$$= 0.24$$

The results are shown in Fig. 17.

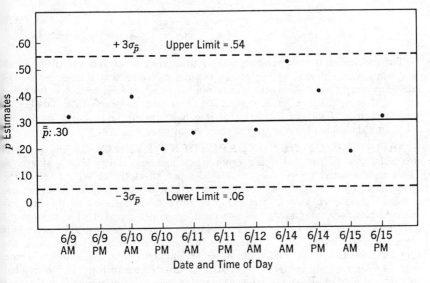

Fig. 17. Control chart for percent machine idle time.

STRATIFIED SAMPLING. The use of the simple binomial model implies that the probability of finding the operation in question in a particular activity state is the same throughout the period of study. This is not the true state of affairs in many applications where a given activity may predominate at the beginning, middle, or end of the day. For instance, more delays may be encountered at the beginning or end of the day or, in an operation such as welding, setup time may predominate during the first part of the day.

In these situations the population may be stratified or divided into periods. Conway (Journal of Industrial Engineering, vol. 8) suggests that such a **stratification study** take the following form:

f_i is the fraction of the total length of study represented by the ith period or stratum.

p_i is the probability of finding an observation in the ith stratum in the particular activity state under consideration.

n_i is the number of observations taken in the ith stratum.

r_i is the number of observations in the ith stratum for which the activity was in the particular state under consideration.

The overall proportion of time represented by the particular state of activity in question is given by a weighted average of the **strata probabilities**.

$$p = \sum_i f_i p_i$$

An unbiased estimate of this overall proportion is:

$$\bar{p} = \sum_i f_i \frac{r_i}{n_i}$$

The variance of this estimator is given by:

$$\sigma_p^2 = \sum_i \frac{f_1^2 p_i (1 - p_i)}{n_i}$$

The conclusion is that **proportionally allocated stratified sampling** is always at least as precise as the simple random sampling that is appropriate for the binomial model. When the p_i are not all equal, the estimate based on the stratified sample has a smaller variance (assuming the same total number of observations). This means either that it is more precise or that it is capable of obtaining the same precision with fewer observations.

SAMPLING OF NON-INDEPENDENT EVENTS. In the previous examples, one of the underlying assumptions has been independence of events. That is, the activity level of one unit does not affect the activity level of the remaining units. Consequently, three simultaneous observations could be recorded during each visit to the group of three machines used as the first example. However, the activity of one machine or one member of a crew might influence the activity of the other machines or crew members. The erroneous practice of recording these situations as independent events and then using the binomial model usually will result in an understatement of the standard error, σ_p. The appropriate model is referred to as **cluster sampling**. If the number of units under observation is a constant for each visitation made during the study, the following procedure can be used to estimate σ_{p_i} (W. G. Cochran, Sampling Techniques):

m is the number of elements, machines or men, in a given group.

n is the total number of observations recorded.

v is the number of visits made to the group and is equal to n/m.

r_{ij} is the number of units observed in the ith activity during the jth visit. The term takes on the values of 0, 1, 2, ..., m.

p_{ij} is the proportion of units engaged in the ith activity during the jth visit.

$$p_{ij} = \frac{r_{ij}}{m}$$

The overall proportion of time represented by the particular state of activity in question is the unweighted mean of quantities p_{ij}.

$$p_i = \frac{\sum_{j=1}^{v} p_{ij}}{v}$$

An unbiased estimate of this overall proportion is:

$$p_i = \frac{\sum\limits_{j=1}^{v} r_{ij}/m}{n/m} = \frac{\sum\limits_{j=1}^{v} r_{ij}}{n}$$

The variance of this estimator is given by:

$$\sigma^2_{p_i} = \frac{m^2}{n^2} \sum_{j=1}^{v} (p_{ij} - p_i)^2$$

If the number of elements is not constant for all groups, which often is the case with work crews, the term m_j must be added to the model to reflect the number of elements in the group during the jth visit. The addition of this factor changes the estimating formulas to the following:

$$p_{ij} = \frac{r_{ij}}{m_j}$$

$$p_i = \frac{\sum\limits_{j=1}^{v} r_{ij}}{\sum\limits_{j=1}^{v} m_j}$$

$$\sigma^2_{p_i} = \frac{v}{\left(\sum\limits_{j=1}^{v} m_j\right)^2 (v-1)} \left(\sum_{j=1}^{v} r_{ij}^2 - 2p_i \sum_{j=1}^{v} r_{ij} m_j + p_i \sum_{j=1}^{v} m_j^2\right)$$

Halsey (Journal of Industrial Engineering, vol. 11) converted the formula for σ_{p_i} into terms of coefficients of variation and illustrates the use of sample ranges for estimating these components, thus simplifying the computational procedures.

STANDARD TIME FROM WORK SAMPLING. In contrast to the use of stop watch time study, work sampling may be used to determine a standard time for a repetitive production operation.

Barnes and Andrews (Journal of Industrial Engineering, vol. 6) made extensive studies of industrial operations in which work sampling and time studies were compared. From their studies of 13 operations in 8 different companies they concluded: "Our studies seem to indicate that work sampling will give time standards for repetitive standardized manual operations which are substantially the same as standards obtained by time study."

The **procedure** employed to obtain a standard time from work sampling is as follows:

1. A work sampling study is made of the operation in which production activity, or measured work, is observed, as well as various classes of delays and interruptions. The production activity is differentiated into **manually controlled** and **machine-controlled** categories. During the study the operator's performance is rated at each observation of manually controlled production activity.

2. At the end of the study the individual rating indices are averaged to obtain an **overall rating index.**

3. A **physical count of parts** produced over the study period is taken.

4. The actual time per unit of output is calculated by dividing the total clock-hours consumed in production or measured work by the number of parts pro-

duced. This is equivalent to the **net cycle time** obtained from a stop-watch time study. This unit time is broken down into manually and machine-controlled elements in proportion to the observations made in these two categories.

5. The **base time** is obtained by multiplying the manually controlled portion of observed unit time by the rating index and adding the machine-controlled element, which may be independently calculated from empirical standard data.

6. The **standard time** is obtained by adding allowances obtained from the study or from other sources. It should be noted here that percent delay time obtained from work sampling is normally related to total time. The percent allowances added to base time are **percentages of base time**. Therefore, the total percent allowances for inclusion in the standard time should be obtained by taking a ratio of the total percent delay obtained from the work sampling to 100 minus this percent.

For example, assume that a five-day work sampling study is taken of a repetitive operation with the following results:

Total number of observations............................		1,000
Observations of production activity.......................		900
Manually controlled elements..........................	650	
Machine-controlled elements...........................	250	
Percent production time, 900/1,000.......................		90%
Total clock time over the study,		
5 days × 8 hr./day × 60 min./hr......................		2,400 min.
Total production time, 2,400 × 0.90......................		2,160 min.
Total units produced.....................................		3,860
Actual time per unit, 2,160/3,860........................		0.56 min.
Manually controlled elements, 0.50 × 650/900..........	0.404 min.	
Machine-controlled elements, 0.560 − 0.404............	0.156 min.	
Rating index, average of individual ratings................		112%
Base time per unit, (1.12 × 0.404) + 0.156		0.610 min.
Observations of unavoidable delays.......................		80
Percent unavoidable delay time, 80/1,000.................		8%
Allowance percent for unavoidable delays, 0.08/0.92.........		8.7%
Standard time, 0.610 + (0.087 × 0.610)..................		0.663 min.

Schmid (14th Annual Conference Proceedings of A.I.I.E.) developed a systematic sampling technique termed **Work Measurement Sampling** for the study of long cycle jobs. The **essential features** of this technique are:

1. Determination of time standards for a large number of job positions (10 to 50 jobs may be studied simultaneously).
2. Reduction of the observations to punched cards preserving the sequence of observations and identification of the time of day of each observation.
3. Use of a computer to summarize the data in order to estimate the maximum duration of an activity, the minimum duration, the median time value, and a confidence interval established for the average time allowed for an activity.

GUIDELINES FOR WORK SAMPLING. The following points should be kept in mind in considering the use of work sampling:

1. Only homogeneous groups should be combined—such as delays on similar operations performed on similar kinds of machines, or delays of operators on work of a similar nature.
2. A large number of observations are recommended. Studies are best adapted to large groups of machines or operators.
3. Results from a few hundred observations may be useful if p is not too small. The use of the normal approximation to the binominal distribution is adequate if $pN \geq 5$.

4. The accuracy of the results depends on the amount of bias introduced in the study by the sampling procedure.
5. As percentage of delay time increases, more observations are necessary for a given absolute degree of precision.
6. Observations must be taken at random intervals and distributed over a long period of time.
7. Work sampling data provides a basis for evaluating various departmental operations or processes.
8. Two or more studies may be conducted simultaneously during the observation period.
9. Work sampling may produce fewer complaints from operators being studied than continuous or all-day stop-watch studies.
10. The technique is applicable to a wide variety of situations in manufacturing, distribution, or service industries.
11. The cost of work sampling studies is considered to be about a third that of production or interruption studies.

Standard Time Data

STANDARD DATA METHOD. Standard time data consists of an inventory or file of elemental times accumulated from time studies made in the plant or from special studies. These times are expressed in terms of **job variables.** When it is desired to set a standard on a new job, elemental times are extracted from tables, charts, or graphs, and then synthesized in logical sequence to provide the base time for the job. In such cases the elements and variables of the job are similar to or identical with those from which the original data were formulated.

The elements referred to here are **time study** or **functional elements** and are usually peculiar to the experience of a given plant. Standard data based on motion elements, which are more or less universally applicable, are discussed later. The two forms of standard data are often referred to as **macroscopic** (time study standard data) and **microscopic** (motion-time standard data).

The element times comprising standard data are usually **base times** because the standard data may be used on jobs with varying allowance requirements. Allowances must be added at the time the data are used to obtain a standard time.

A particular element time used in the standard data may be the average of a large number of independent time studies made on operations which include the element in question. Or the element time can be a single value removed from a data curve. This curve may be the result of interpolation or extrapolation of two or more original times.

The standard data method can be traced back to studies of Taylor and Merrick. It was planned to compile a "dictionary" of fundamental elements necessary to perform work, with times that skilled operators should take for these elements.

ADVANTAGES OF THE METHOD. There are a number of advantages in compiling and using standard time data as opposed to taking direct time studies on every new operation.

1. The **cost** of setting standard times using standard data is less than taking a direct study if the potential requirement for standards in the plant is large.
2. Standard data contributes to a **consistency** in time standards for operations employing the same elements.
3. Standard times can be **calculated** before the operation or job is activated. This is an obvious advantage in planning a production process and in estimating labor requirements and costs.

4. Certain **indirect labor** operations, such as material handling, cleaning and janitorial work, and maintenance, can be timed economically only through the use of standard data.

5. Standard data can be employed to evaluate **methods designs** and to balance production or assembly lines in the design stages.

ACCURACY OF STANDARD DATA. Accuracy is often claimed as an advantage of standard data. But there are basic features of the standard data method which make claims of great accuracy questionable. The first factor is that standard data are collected from time studies and therefore are susceptible to some of the sources of **bias** present in the original timing. This is important where the standard data may represent essentially single studies on different jobs which contribute points on a curve. Also, standard data are based on **rated times** and therefore are subject to criticisms of accuracy applied to the rating procedure. This is less important where times are averages of a large number of independent ratings.

A second factor concerns the **additivity** of element times. Are elements so independent that their times may be extracted from an original job context and resynthesized perhaps in some different order in another context involving a different machine or operator? Research in this area seems to indicate that independence of this sort is not present. In practice the problem can be avoided only by transplanting groups of elements from one job context to another, where possible. This is admittedly difficult in most situations, especially where operations are predominantly manually controlled and subject to variation in method and motions of operators.

A third factor, related to these, is that the elements constituting the standard data may not be similar or comparable to the elements employed in the operation for which a standard time is set. Again, this refers to variation in methods. The only way of reducing this source of error is to ensure that the **similarity of elements** is great enough to justify the use of the data.

A final factor is that there may be sources of error built into the data due to failure to reduce the data in terms of the correct variables or to correctly determine the nature of the variation.

DERIVING STANDARD DATA. Standard data may be derived by the use of a number of procedures and aids.

Grouping Like Operations. Standard data is relevant to similar operations or jobs. Like operations should be grouped for **spot studies.** In some cases the standard data may pertain to like or similar products on a given type of, or identical, machines.

Preliminary Studies. Time studies are taken of a representative number of parts on a given **machine grouping,** including a distribution of spot studies made on a range of product weights, shapes, or other significant variables of the work. Care should be taken to obtain studies on high and low **limits** of the variable in question. For example, where the weight of the part is an important factor, both the heaviest and the lightest parts should be studied. The data will be good for further use between these limits. The results of each individual time study should be summarized on an instruction sheet showing the base time for the various elements.

During the time study, particular care must be taken to segregate the variable elements from the constant and manually controlled from machine elements. Also, elements involved in manipulating tools, jigs, or fixtures should be segregated.

Time Study File. The accumulation of time studies taken over an extended period of time for the purpose of setting time standards is commonly referred to as the time study file. The **advantage** in using this file of time studies for deriving standard data is that there is no cost involved. The **disadvantage** is that, because the data has been collected over a long period, it is quite vulnerable to changes in methods, equipment, quality, and conditions, as well as in elemental definitions. Also, there are numerous instances where failure to record certain information when the time studies were taken makes the file an unusable or inferior source of time data.

Comparison Sheet. The results of the initial time studies are compiled on a comparison sheet. The **basic elements** are listed in the left column. **Elemental times** from the various studies are entered in succeeding columns and then compared. At the top of each column representing a given study, **job** or **product variables** are noted.

Morrow (Motion Economy and Work Measurement) gives a comparison sheet illustrating data from drill press operations (Fig. 18). On this sheet the elemental base times are recorded for each study. It is to be noted that the elements in the master list in the left-hand column are not necessarily in the order in which they are performed. For the element "piece up on table," the first four final base times indicated are 0.24, 0.09, 0.27, and 0.09. Comparing these with the weight of the piece for the first four studies—35 lb., 10 lb., 42 lb., and 14 lb., respectively—it is quite evident that the lighter the piece the shorter the time. The same relationship applies for the element "piece off and aside." From these data **standard curves** can now be constructed and the standard (base) times determined for the elements for any piece of which the weight is known.

The times for the element "tighten one strap" are 0.16, 0.14, 0.15, 0.16. Evidently a slight variation in time is due to causes other than the weight of the piece, but the time is considered to be a constant of 0.15 minutes.

The comparison sheet can be expanded in complexity to include other factors that will contribute to variability. Its essential purpose is to determine the nature of variability between jobs.

INTERPOLATION AND EXTRAPOLATION OF BASE TIMES.

In timing similar operations on a range of sizes of the same product, it is usually possible to determine some of the times by interpolation. A few sizes are selected at suitable intervals and their base times plotted. If there is no variation in conditions, an even curve can be drawn through the points, and base times for intervening sizes may be safely interpolated. This principle can be used to determine by extrapolation the time for sizes smaller or larger than those studied, though with much less confidence. The danger in both cases is that with certain sizes sudden changes of condition may be made, which change the operation, for instance, the introduction of crane service. Such factors alter time abruptly. When this change of condition is known, sizes just above and below should be studied and discontinuity determined (see Figs. 19 and 20).

SETUP STANDARDS. The method of determining setup standards is similar to that described for production operation standards. Usually it is better to have the two types of standards separate, although in some cases setup time may be prorated over the number of pieces in the lot, and the operation standard will then include setup time. The latter method is more convenient for **shop records,** but it will not be accurate unless the size of lots is uniform or setup time is very short in proportion to operation time.

COMPARISON SHEET

HANDLING UPRIGHT DRILL PRESS

	Rate	R1830	C1790	C1820	R1842	R1843	C1860	R1858	B1863	B1870	R1907	R1902	C1793	C1794	R1906
Study number		R1830	C1790	C1820	R1842	R1843	C1860	R1858	B1863	B1870	R1907	R1902	C1793	C1794	R1906
Part number		12132	14011	11121	17473	10110	12186	10407	15003	14115	15165	10123	12501	16657	10132
Machine name		Barnes	Barnes	Edlund	Sipp	Sipp	Barnes	Buffalo	Barnes	Barnes	Buffalo	Sipp	Edlund	Buffalo	Sipp
Shop number		166	115	92	403	114	121	118	166	115	123	114	92	123	114
Operator		Walsh	Short	Jones	Smith	Walsh	Black	Smith	Jones	Nott	Short	Walsh	Brown	Brown	Walsh
Material		C1	Brz 2	C1	C1	Brz Steel	Brz 1	C1	C1	C1	C1	Brz 2	Brz Steel	C1	C1
Weight piece		35	10	42	14	3	6	26	12	21	7½	26	21	52	26
Jig number		714	731		512	598	285		786		377				
Weight jig		18	4		3	2	3		5		2				
Rate	Standard	70	70	75	65	75	75	60	65	25	80	65	55	60	65
Piece up on table	Curve 1	.24	.09	.27	.09	.05	.06	.16	.11	.17	.07	.17	.14	.38	.15
Piece off and aside	Curve 1	.15	.04	.20	.05	.03	.04	.09	.05	.09	.04	.10	.09	.28	.11
Piece in jig Class 1	Curve 2	.05				.02			.03		.02				
Piece out jig Class 1	Curve 2	.04				.01			.02		.02				
Piece in jig Class 2	Curve 2		.02		.03		.02								
Piece out jig Class 2	Curve 2		.02		.02		.03								
Locate under spindle	Curve 4	.07	.04	.06	.04	.04	.05	.05	.05	.07	.04	.05	.05	.07	.06
Move hole to hole	Curve 4	.04	.03	.04	.02	.02	.02	.03	.03	.04	.02	.04	.04	.05	.03
Move spindle to spindle	Curve 4			.05							.03			.06	
Tighten one screw	Chart 6				.05	.04	.06								
Loosen one screw	Chart 6				.03	.03	.04								
Tighten one strap	.15	.16	.14						.15		.16				
Loosen one strap	.11	.10	.11						.10		.09				
Tighten vise	.04			.08								.04	.06	.08	
Loosen vise	.03			.05								.04	.04	.03	
Tighten chuck	.06							.07							.06
Loosen chuck	.05							.05							.05

Fig. 18. Time study comparison sheet used for compiling standard data.

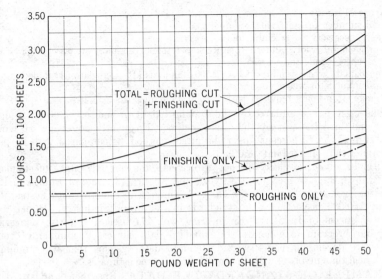

Fig. 19. Curve for interpolating times on sheet copper work.

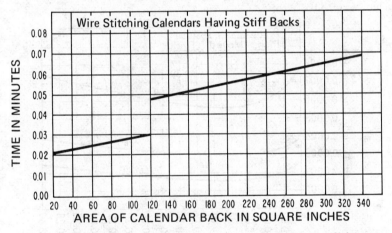

Fig. 20. Curve showing sudden break due to change in conditions.

COMPARATIVE COSTS. Although the indirect or standard data method ˍs slower at the beginning, because no standards are set until many time studies have been taken and computations are made, the cost of time study work by this method is far less than by the direct method. Carroll (S.A.M. Journal, vol. 4) takes a hypothetical case requiring 2,500 standards and shows comparative time and cost for each method.

Direct Method:

Total number of standards required 2,500
Average number of standards set per week per man 25
Number of time study men 3
Approximate weeks required to set 2,500 standards 33
Average Man-hours required to set each standard 1.584

Indirect Method:

Total number of standards required 2,500
Weeks necessary to assemble data 4
Average number of standards per week per man after data are assembled 125
Number of time study men 3
Approximate weeks to set 2,500 standards after data are assembled 7
Total number of weeks to assemble and set standards 11
Average Man-hours required to set each standard 0.528

Thus **cost per standard** by the standard data or indirect method is only about one-third the cost by the direct method. Frequently **coverage of work** by **standards** is 40 percent under the direct method, as compared with 95 percent under the standard data method.

With the advent of electronic data processing, the cost advantage of standard data will continue to increase. Motycka and Auburn (Journal of Industrial Engineering, vol. 8) report the use of a computerized standard data system for setting 100,000 inspection time standards in the aircraft industry. Jelinck and Steffy (Journal of Industrial Engineering, vol. 17) used a stepwise regression computer program to generate the standard data curves. A number of companies are developing computerized standard data systems that will minimize the clerical time involved with the calculation of time estimates.

EXAMPLE OF STANDARD TIME DATA. Specimen standard data times for a part handled by hand are given in Figs. 21, 22, and 23 (American Machinist, vol. 82).

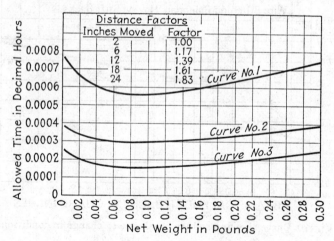

Curve 1 shows allowance for time to pick up a part and place it in a jig or fixture. Curve 2 is read both for time allowed to pick up a part and place it on a table or in a vise, and for time allowed to remove the part from the jig or fixture and lay it aside. Time allowed for laying a part aside is read from curve 3. Distance factor is to be applied in all cases.

Fig. 21. Time curves for finger, wrist, and arm movements.

The data in the figures have been compiled with the object of standardizing part handling times in all departments of a factory, thus making the information of general value. They are compiled from studies on actual production jobs in several shop sections, some studies being taken with ideal setups to show how various factors affect the time.

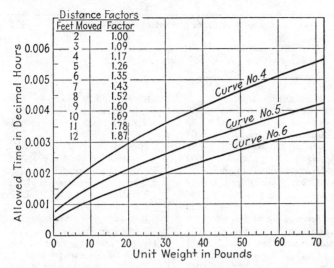

Time allowed to pick up a part and place it in a jig or fixture is read from curve 4. Curve 5 shows both time allowed to pick up a part and place it on a table or in a vise, and time allowed to remove a part from a jig or a fixture and lay it aside. Curve 6 is read for time allowed to lay a part aside. Each reading must be multiplied by the proper distance factor.

Fig. 22. Time curves for finger, wrist, arm, and body movements.

Predetermined Motion-Data Systems

THE METHOD. A motion-time standard data system consists of **pre-determined times** associated with basic or **fundamental motions,** as the elements of the system. The basic motions employed are similar to therbligs in concept and magnitude. There are numbers of such systems in use, some of which have been developed and are used by only one company, and others which have been publicized for universal application.

These systems are developed on the basis of two assumptions. The first is that all manual work can be divided into basic units or elements which are universally descriptive and applicable. The second is that an average time in terms of job variables can be associated with each qualitative element. These times also have some degree of universal applicability. Included in the time system is the recognition of **work variables** which influence the time taken to perform certain elements of motions. Thus, the predominant factor of utility in these systems, as compared to time study standard data, is the universality of their application to a wide variety of manual operations within a given plant or within a group of plants in general.

In these plans a time standard for a particular job is synthesized from the motion-time values found in **predetermined time tables.** The data in these tables are the results of studies made in the laboratory or in factories. Such studies include, in principle:

1. The derivation of a system of basic motions including a qualitative definition of the motions.
2. The study of a representative number of operations in the laboratory or factory. This usually involves the recording of the operation method on film.

DESCRIPTION OF OPERATION ELEMENT

Description	Allowed Time Std. (hours)	Description	Allowed Time Std. (hours)
Open and close cover of jig or fixture when hinged	.0008	Tighten and release two short locating screws by hand	.0023
Close cover of jig or fixture when hinged	.0003	Tighten two short locating screws by hand	.0013
Open cover of jig or fixture when hinged	.0005	Release two short locating screws by hand	.0010
Put on and remove cover of jig or fixture when not hinged	.0029	Tighten and release one long locating screw by hand	.0037
Put on cover of jig or fixture when not hinged	.0020	Tighten one long locating screw by hand	.0021
Remove cover of jig or fixture when not hinged	.0009	Release one long locating screw by hand	.0016
Tighten and release wing nut	.0012	Tighten and release two long locating screws by hand	.0047
Tighten and release thumb screw by hand	.0013	Tighten two long locating screws by hand	.0027
Tighten thumb screw by hand	.0008	Release two long locating screws by hand	.0020
Release thumb screw by hand	.0005	Tighten and release locating screws with a wrench	.0053
Tighten and release one short locating screw by hand	.0018	Tighten locating screw with a wrench	.0035
Tighten one short locating screw by hand	.0010	Release locating screw with a wrench	.0018
Release one short locating screw by hand	.0008		

Fig. 23. Sample time data for parts handling.

3. The analysis of individual motions employed in the various operations t determine variables which influence each motion time.
4. A description of the motion system in terms of the critical variables, fo instance, the influence of part weight or distance moved on the motion em ployed to reach with the hand or move a part by hand.
5. The determination of base times for the motions at different values of th critical variables.
6. The presentation of the final data on charts, tables, or curves.

Motion-time standard data should be applied only by **analysts** who have ha experience and training in motion and time study and in the particular plan the are going to use. When an inexperienced, untrained person attempts to use thes systems, the danger is that he will overlook motions which are quite evident t the experienced observer and that he will classify the motions incorrectly.

USES AND ADVANTAGES OF MOTION-TIME DATA. A num ber of uses and advantages of the motion-time data method are proposed:

1. **Operation design.** One of the major uses of such data is in determinin elemental or standard times for operations in the design stages prior to installa tion or production. This is advantageous for two reasons. The first is tha alternative method designs may be evaluated in terms of time without having t set up mock-up operations for actual testing of the methods. Also, the table indicate the factors which influence time for given motions. The second reaso

that a standard time can be set for the chosen method for use in estimating osts, planning labor requirements, etc.

2. **Methods improvement.** The data can be used to detect gradual changes 1 the methods used in existing operations. If the original method is defined in ɔrms of motion standard times, changes can be detected and their effects evalu- ted. In the setting of a new standard by using standard data, the method and iotion pattern must be analyzed and specified in detail.

3. **Product design.** The influence of alternative product designs on produc- on or assembly times may be determined through the study of motion times. f the standard data indicate the influence of product variables such as weight nd shape on the time to perform a motion, then **critical product design ariables** can be directly assessed during the design stage. Similarly, the in- uence of product design changes on an existing job standard time can be de- ɛrmined.

4. **Tool and equipment design.** Tools and equipment should be designed to ɪaximize the manual efficiency of their operation. Alternative tool and equip- ɪent designs can be assessed in terms of motion times in order to minimize perating or handling time.

5. **Establishing a standard time.** A standard time can be calculated for an xisting job without resorting to a stop watch study. The standard can be used ɪr wage payment purposes or for any use to which time standards are put.

6. **Developing time formulas.** Time formulas are developed to aid the ɪgineer in establishing standards for specific types of work. Motion-time data an be resynthesized for inclusion in such formulas to apply to specific jobs ʻithin a given plant. For a given formula, motion times can supplement time tudy standard data accumulated by the plant in question.

VALIDITY OF MOTION-TIME DATA. The validity and accuracy f any system of motion-time standards must be considered by the analyst efore he uses it. In general it can be concluded that most companies using a niversal system must make periodic adjustments in the data to fit their own onditions. It is advisable that a firm continually **audit standards** which have een set from motion-time standard data, particularly to determine the influence f job variables which have been ignored in the original data system.

1. The **universality** of the data depends on the similarity between the popu- ation from which the data were derived and the population to which the data re applied. The times recorded in tables represent sample averages of data aken from laboratory studies or from studies of actual industrial operations. ʼhe population on which studies were originally conducted must of economic ɪecessity be limited.

2. The data presented in tables are usually rated or leveled times. The con- ept of normal or standard performance may vary between different systems nd between a given system and the plant in which the data are applied.

3. The nature of the job or method variables which affect motion times and he interaction between individual motions must be considered. A valid system hould allow for the effect of variables influencing motion times. Although those hat have little effect may be ignored, it should be the policy of the investigator o include all variables which may possibly influence the validity of the data.

4. The question as to **additivity** relates to motion-time as well as to time tudy standard data. The argument of nonadditivity is based on certain evi- lence that the time for a given motion depends on the preceding and succeeding ɪotions.

The validity of motion-time data in general and a given system specifically is difficult to assess, and considerable research is needed before many of the basic questions can be answered.

APPLICATION ERRORS. Besides possible errors in the data themselves, the manner in which the data are applied may cause an inaccurate standard time to be set for a particular job.

For a new operation the analyst must decide which **motion pattern** represents the best method. In the case of an existing operation the analyst must decide which motion pattern is representative of the best method or of expected performance, depending on what use will be made of the standard. Any difference between the motion sequence considered best by the analyst and the sequence which is optimal from the operator's point of view introduces an error into the standard time.

The analyst must interpret the selected method in terms of the motion system employed. This points out the necessity of having clearly **defined motion elements,** including description of situations where each motion applies or does not apply. Failure to select the proper motion element or to omit motions will introduce error.

The analyst must correctly recognize and measure the variables affecting given motion times. This includes those variables which are a part of the system used as well as those variables peculiar to the operation in question and perhaps ignored in the system.

MOTION-TIME ANALYSIS. One of the original systems of motion-time data was proposed by Segur and called motion-time analysis. Over a period of 25 years Segur accumulated data on times required for motions of experts when putting forth their best efforts. An **expert** is defined as:

. . . a person in good health whose physical, sensory, and mental characteristics adapt themselves to any method involved, and who is able to perform that method automatically, and at a pace which can be maintained at operating efficiency.

The times are synthesized by means of formulas which take into account the conditions and methods involved. Here times are calculated to the fifth decimal of a minute.

Segur does not claim that any worker will maintain these **ideal times** continually. On a short cycle he is likely to attain such times once out of three or four performances. A percentage allowance is therefore made for fatigue, fumbles, and errors, and other minor delays. This percentage varies with the number of body members, senses, and mental actions which may be operating simultaneously in performing an operation. The percentage is lowest on such operations as foundry work, where heavy loads are handled and the amount of attention is at a minimum, and highest on such operations as intricate assembly of machine parts, especially where sensory and mental attention is required by the operator. An analysis is made of any job under consideration to determine what motions are used and how they are combined. Times are then calculated as selected from the formulas and their sums checked with a stop watch to make sure that actual work habits have been fully recorded.

These actual motions are then compared with ideal motions and improvements made according to the rules of the **correct human motion.** As permitted by operation and production limitations, one or more synthetic improved methods are set up. The times for these synthetic methods are calculated from predetermined formulas and entered on the analysis. The best practical method is selected and used as a basis for instruction of operators. The selected method is described in sufficient detail to permit accurate and specific instruction to perators, and the allowances are added to the time so that the job is rated before any instruction begins.

The data secured on the original analyses form the basis for further improvement of work or rate-setting on similar allied operations. Since the time values and the analysis are based primarily on conditions rather than on performance, much closer check can be kept on methods and operating conditions than is possible on time study of operation performance alone.

PREDETERMINED TIME VALUES. An approach to the use of "predetermined time values" was first developed by Engstrom and his associates at the General Electric Company. This system confines itself to certain types of jobs, chiefly those that can be done in the normal individual work area, either sitting or standing. Time values are assigned according to the size of the part, that is, whether it can be handled by one hand, two hands, two fingers, or three fingers, and the values are classified according to the size of the part and also the conditions of getting and placing the part. The **get values** include the motions of transport empty and grasp; the **place values** are for the motions of transport loaded, pre-position, position, and release load. By using the Gilbreth right-hand left-hand analysis and breaking the motions into the get-and-place groups, values are assigned according to the size of the part and the condition for getting and placing specific parts.

The most effective use of this system is for developing standard data sheets for each type of work (see Fig. 24). The circled items in Fig. 24 represent those which apply to the particular activities with a Hardinge hand-screw machine being analyzed. The times for the machining operations are derived from a chart. This system has its limitations in that it does not attempt to cover the entire gamut of man's work, such as from heavy foundry operations up to light bench assemblies. It is limited, as previously described, to particular sized types in normal work areas. It's advantages are that it is simple to use, it has more than 25 years of usage, and has proven itself to be reasonably correct.

WORK-FACTOR PLAN. This is a method of analyzing the **factors involved in work** developed by The WOFAC Co. It is a system of time values for individual finger, hand, arm, leg, or body motion which a human being can make. These time values have been set into **moving time tables** and rules have been written for their use.

Work-Factor time data is an outgrowth of research led by Joseph H. Quick involving 17,000 motion times measured by means of special watches, motion picture cameras, photoelectric relays, and stroboscopic lighting units. Subsequently, Quick et al. (Work-Factor Time Standards) modified and improved the original methods based on their experience with actual rate-setting problems encountered in factories and offices.

Work Factors. Although there are dozens of variable factors involved in motions, for practical purposes, the following major ones must be recognized in measuring them for rate-setting purposes.

1. The body member used [finger (F), hand (H), arm (A), forearm swivel (FS), trunk (T), foot (FT), leg (L), or head turn (HT)].
2. The length of the motion measured as a straight line between the starting and stopping points of the motion (5 in., 10 in., etc.).
3. Basic motion, which is the simplest and fastest.
4. The kind of motion involving work-factor.
   a. Does it carry weight or overcome friction? (W)
   b. Does it require control to steer it? (S)
   c. Does it require care or precaution? (P)

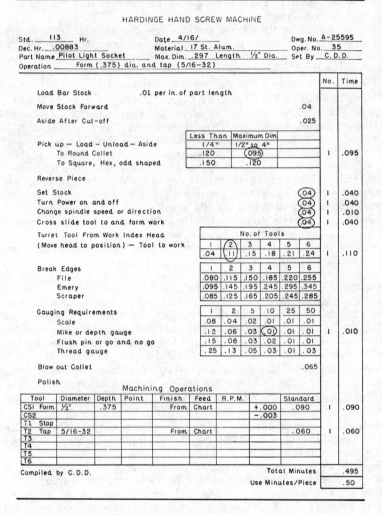

HARDINGE HAND SCREW MACHINE

Std. __113__ Hr. Date __4/16/__ Dwg. No. __A-25595__
Dec. Hr. __.00883__ Material __17 St. Alum.__ Oper. No. __35__
Part Name __Pilot Light Socket__ Max. Dim. __.297__ Length __½″ Dia.__ Set By __C.D.D.__
Operation _____ Form (.375) dia. and tap (5/16-32) _____

			No.	Time
Load Bar Stock	.01 per in. of part length			
Move Stock Forward		.04		
Aside After Cut-off		.025		

Pick up — Load — Unload — Aside	Less Than	Maximum Dim		No.	Time
	1/4″	1/2″ to 4″			
To Round Collet	.120	(095)		1	.095
To Square, Hex, odd shaped	.150	.120			

Reverse Piece

		No.	Time
Set Stock	(04)	1	.040
Turn Power on and off	(04)	1	.040
Change spindle speed or direction	(04)	1	.010
Cross slide tool to and form work	(04)	1	.040

Turret Tool From Work Index Head — No. of Tools
(Move head to position) — Tool to work

1	2	3	4	5	6	No.	Time
.04	(11)	.15	.18	.21	.24	1	.110

Break Edges

	1	2	3	4	5	6
File	.080	.115	.150	.185	.220	.255
Emery	.095	.145	.195	.245	.295	.345
Scraper	.085	.125	.165	.205	.245	.285

Gauging Requirements

	1	2	5	10	25	50	No.	Time
Scale	.08	.04	.02	.01	.01	.01		
Mike or depth gauge	.12	.06	.03	(01)	.01	.01	1	.010
Flush pin or go and no go	.15	.08	.03	.02	.01	.01		
Thread gauge	.25	.13	.05	.03	.01	.03		

Blow out Collet		.065

Polish

Machining Operations

Tool	Diameter	Depth	Point	Finish	Feed	R.P.M.		Standard	No.	Time
CSI Form	½″	.375		From	Chart		+.000	.090	1	.090
CS2							-.003			
T1 Stop										
T2 Tap	5/16-32			From	Chart			.060	1	.060
T3										
T4										
T5										
T6										

Compiled by C.D.D.

Total Minutes	.495
Use Minutes/Piece	.50

Fig. 24. Standard data sheet for an operation.

d. Does the motion change direction? (U)
e. Must it end at a definite stop point? (D)

If each of these factors is known about a given motion, that motion can b[e] identified according to the amount of time required to make it. One of th[e] fastest and simplest motions which an operator can make is tossing an objec[t.] It is evident that the factors noted above, such as weight or resistance, steerin[g] to exact location, moving with care or precaution, making a change in directio[n] or making a definite stop (controlled by the operator), tend to slow down [a] motion.

Standard Elements. In the Work-Factor system, all work is divided int[o] standard elements in order to analyze an operation in an orderly sequence an[d]

> provide a practical method of **cataloging** time values so that once an element has been analyzed, its established time value can be used whenever the element recurs. The standard elements are:

Transport (Reach or Move)	R or M
Grasp	Gr
Release	Rl
Pre-position	PP
Assemble	Asy
Use	Use
Dissemble	Bsy
Mental Process (Inspect, etc.)	MP

Symbols designating body members are

A	Arm
L	Leg
T	Trunk
F	Finger
H	Hand
Ft	Foot

Symbols for work factors include

W	Weight or resistance
S	Directional Control (steer)
P	Care (precaution)
U	Change Direction
D	Definite Stop

Examples of Work-Factor Usage. Fig. 25 shows some examples of motions classified according to work-factor analysis and their time values.

No.	Description of Motion	Work Factor Classification	Work Factor Motion Time (Min.)
1	Move finger 1 in.	F* 1	0.0016
2	Move hand 10 in. (arm)	A 10	0.0042
3	Move hand 20 in.	A 20	0.0058
4	Move hand 20 in. to wrench	A 20 D**	0.0080
5	Move hand 20 in. carrying wrench (3 lb.) to place aside	A 20 WD	0.0102
6	Move hand 20 in. carrying wrench (3 lb.) to place on nut	A 20 WSD	0.0124
7	Move hand 20 in. carrying wrench (3 lb.) around and behind fixture to place on nut	A 20 WSUD	0.0144
8	Move foot to depress machine pedal (leg) (10 in.)	L 10 D	0.0070

* The symbol before the number designates the body member.
** Each symbol used after the number represents a work factor.

Fig. 25. Typical motions classified according to work-factor analysis.

All values shown on the work-factor table of motion times are known as "select" times. When allowances for personal time, fatigue, and delays are added to "select" time, the result is the time which the normal skilled operator working at an incentive pace should consistently maintain. This is assumed to be equivalent to 25 percent above a base rate level of output. The followin allowances are recommended for use in conjunction with work-factor.

For an incentive operation:

Assume select time from table is.........................	1.00 min.
Incentive allowance = 25%.............................	0.25 min.
Personal, fatigue and delay allowances = 18%	
of normal skilled time..............................	0.22 min.
Standard time is.................................	1.47 min.

On this basis, the normal incentive worker should earn 25 percent above hi base rate (or "day" when the base and day rate are the same).

In addition to the **Detailed** Work-Factor analysis, three other systems hav been developed. The **Simplified** system is based on average values of the De tailed for use in estimating values with an accuracy of 0 to +5% in the origina data. An **Abbreviated** system has been developed for rapid estimating pur poses using 0.005-minute time units rather than the 0.0001-minute units. **Read** Work-Factor was developed to fill the need for a predetermined time systen that can be taught readily to persons not engaged in, or skilled in work study

METHODS-TIME MEASUREMENT. Maynard *et al.* (Methods-Tim Measurement) describe the development of MTM. A methods engineering re search project was carried on over a period of many months to arrive at "methods formula" which would enable determination of effective methods be fore these methods were introduced into the shop. The investigation was limite at first to sensitive drill-press operations. However, it was found that the data which had been compiled for drill-press work applied equally well and with very satisfactory degree of accuracy to all classes of work involving manual mo tions. Hence, instead of a methods formula applying merely to sensitive drill press work, it was recognized that truly basic methods-time data had been de veloped. While the present data must be regarded as incomplete, they do appl to the majority of industrial operations.

Uses of Methods-Time Measurement. MTM has been used for:

1. Developing effective methods in advance of beginning production.
2. Improving existing methods.
3. Establishing time formulas or standard data.
4. Estimating.
5. Guiding product design.
6. Developing effective tool designs.
7. Establishing time standards.
8. Selecting effective equipment.
9. Training supervisors to become highly methods conscious.
10. Settling grievances.
11. Research, particularly in connection with methods, learning time, and per formance rating.

Methods-Time Data Tables and Application. Tables of time data have been developed for the following motions:

1. Reach. (R)
2. Move. (M)
3. Turn and apply pressure. (T and AP)
4. Grasp. (G)
5. Position. (P)
6. Release. (Rl)
7. Disengage. (D)
8. Eye Travel Time and Eye Focus. (ET and EF)

9. Body, Leg, and Foot Motions.
10. Simultaneous Motions.

Fig. 26 (MTM Association for Standards and Research) shows sample data for some of these motions.

When the data are applied with an understanding of the characteristics of the motions covered, it is possible to establish with a certain degree of accuracy the time required to perform the vast majority of industrial manual operations.

In these tables the unit of measurement is a TMU (Time Measurement Unit). 1 TMU = 0.00001 hour = 0.0006 minute = 0.036 second. Thus, if the time required to perform a given series of motions is found from the tables to be 325 TMU, the time in decimal hours is 325 × 0.00001 or 0.00325 hour.

The tables are set up in terms of **leveled TMU.** They show the time required by the operator of normal skill, working with a normal effort to make the motion under normal conditions. Conventions for recording motion classification are shown in Fig. 27, on page 76.

BASIC MOTION-TIME STUDY.
The BMT system for studying motion times was developed by the staff of J. D. Woods & Gordon, Ltd., of Toronto. The system and times were derived from laboratory experiments and were carefully checked against a variety of factory operations before being accepted for general use. The laboratory experiments were set up in such a manner that the variable factors which influence motion times could be introduced either one at a time or in combination and in carefully controlled degrees.

BMT data are based on **basic motions.** A basic motion is considered to be a single complete movement of a body member. A basic motion occurs every time a body member which is at rest moves and again comes to rest. The action of knocking on a door, for example, requires two basic motions for every knock— one to draw the hand back and another to move the hand forward and knock. It can be noted that this system avoids describing a motion in terms of its purpose or in terms of a description of the object being handled. For instance, a grasp is specified by listing the **component motions** as such, and not by describing the shape, size, condition, or location of the object being grasped.

Analysis of Basic Motions. Basic motions are classified as: finger, hand, and arm motions; foot and leg motions; and miscellaneous body motions.

The first factor considered in **finger, hand, and arm motions** is the role of muscular control in stopping the motions.

1. A motion can be stopped by impact with a solid object, as in the downstroke of a hammer or in pushing a flat sheet of metal against a stop in metal shearing. These are identified as **Class A motions.**
2. A motion can be stopped in mid-air by muscular control without coming in contact with any object. Motions like this are used for the upstroke in hammering and for tossing objects aside. These are **Class B motions.**
3. A motion can be stopped by grasping or placing an object. Muscular effort is used here to slow down the motion before the object is grasped or placed in position. This is illustrated by a motion used to reach and grasp a desk pad or to carry and place the desk pad on top of the desk. **These are Class C motions.**

In summary, Class A motions are stopped without muscular control by impact with an object; Class B motions are stopped entirely by the use of muscular control; and Class C motions are stopped by the use of muscular control both to slow down the motion and to end it in grasping or placing action.

There are no separate times for **grasping** or **releasing.** Any grasp that is accomplished as part of the initial contact with the object being grasped is

REACH (R)

Distance Moved (in.)	Time TMU				Hand in Motion	
	A	B	C or D	E	A	B
¾ or less	2.0	2.0	2.0	2.0	1.6	1.6
1	2.5	2.5	3.6	2.4	2.3	2.3
2	4.0	4.0	5.9	3.8	3.5	2.7
3	5.3	5.3	7.3	5.3	4.5	3.6
4	6.1	6.4	8.4	6.8	4.9	4.3
5	6.5	7.8	9.4	7.4	5.3	5.0
6	7.0	8.6	10.1	8.0	5.7	5.7
7	7.4	9.3	10.8	8.7	6.1	6.6
8	7.9	10.1	11.5	9.3	6.5	7.2
9	8.3	10.8	12.2	9.9	6.9	7.9
10	8.7	11.5	12.9	10.5	7.3	8.6
20	13.1	18.6	19.8	16.7	11.3	15.8
30	17.5	25.8	26.7	22.9	15.3	23.2

Case and Description

A   Reach to object in fixed location, or to object in other hand, or on which othe hand rests.

B   Reach to single object in location which may vary slightly from cycle to cycle

C   Reach to object jumbled with other objects in a group so that search and selec occur.

D   Reach to a very small object or where accurate grasp is required.

E   Reach to indefinite location to get hand in position for body balance, or next motior or out of way.

Fig. 26. Examples of methods-tim

really part of the motion to the object. The motion and grasp terminate at th same instant—at the instant of contact with the object. The grasping actior may be more complex, such as getting a single object from a large number o picking up a part from a flat surface. Here **extra finger motions** are needed which must be measured separately. Releasing is treated similarly. The re leasing action is part of the motion away from the object being released.

Motion-Time Determination. Other factors influencing motion times are

1. **Distance moved.** The data are classified in terms of distance moved i inches.

2. **Visual attention.** If the eyes move to the ending point of a motion as i is taking place, the motion time is greater than if there were no eye movement Class B and Class C motions, which use muscular control in the stopping actior may or may not be accomplished by a movement of the eyes to the endin; point. When an eye movement is needed to complete these motions, they ar

MOVE (*M*)

| Distance Moved (in.) | Time TMU | | | Hand in Motion |
	A	B	C	B
¾ or less	2.0	2.0	2.0	1.7
1	2.5	2.9	3.4	2.3
2	3.6	4.6	5.2	2.9
3	4.9	5.7	6.7	3.6
4	6.1	6.9	8.0	4.3
5	7.3	8.0	9.2	5.0
6	8.1	8.9	10.3	5.7
7	8.9	9.7	11.1	6.5
8	9.7	10.6	11.8	7.2
9	10.5	11.5	12.7	7.9
10	11.3	12.2	13.5	8.6
20	19.2	18.2	22.1	15.6
30	27.1	24.3	30.7	22.7

Weight Allowance

Wt. (lb.) up to	2.5	7.5	12.5	17.5	22.5	27.5	32.5	37.5	42.5	47.5
Factor	1.00	1.06	1.11	1.17	1.22	1.28	1.33	1.39	1.44	1.50
Constant TMU	0	2.2	3.9	5.6	7.4	9.1	10.8	12.5	14.3	16.0

Case and Description

A   Move object to other hand or against stop.

B   Move object to approximate or indefinite location.

C   Move object to exact location.

measurement time data.

aid to be visually directed. They are then identified as **Class *B* visually directed** (*BV*) or **Class *C* visually directed** (*CV*) motions. *BV* and *CV* class motions occur only when the eyes move with the hand. If the eyes can be fixed on the ending point of the motion before it starts, the basic arm motion is not delayed and no allowance is necessary for visual direction.

3. **Precision requirements.** Precision is the term applied to the **extra muscular control** required where a motion ends in grasping a small object or in placing an object in an exact location. The degree of precision needed in any motion situation can be stated in quite definite terms. In the case of motions that end with a grasp, this is done by determining the limits within which the fingertips must be located in order to make a satisfactory grasp.

4. **Weight.** Whenever a heavy object must be handled or when friction must be overcome, added muscular effort is required. BMT defines the use of this extra effort as **force.** In a motion that involves handling a heavy object, force may be introduced in three phases.

Table	Example	Significance
I	R8C	Reach, 8 inches, case C
	R12Am	Reach, 12 inches, case A, hand in motion
	R14C	Reach, 14 inches, case C
II	M6A	Move, 6 inches, case A, object weighs less than 5 pound
	M16B15#	Move, 16 inches, case B, object weighs 15 pounds
III	T30°S	Turn 30°, small part
	T90°L	Turn 90°, large part
	AP	Apply pressure
IV	G1A	Grasp, case 1A
V	P1NSD	Position class 1 fit, non-symmetrical part, difficult t« handle
VI	RL1	Release, case 1
VII	D2E	Disengage, class 2 fit, easy to handle

Fig. 27. MTM conventions for recording motion classifications.

   a. To apply pressure in grasping the object in order to gain control of the weight That is, the fingers are pressed firmly against the object so that it will not sli» through the fingers during the move.

   b. When the weight is brought under control by this squeezing action, to apply force to start the weight in motion.

   c. As the weight is brought near its destination, to apply force in the opposite direction to slow up and stop the weight.

BMT deals with the effect of weight on an individual motion time by firs recognizing the separate presence of these different phases of the force factor Then appropriate allowances are added to the basic motion time. The allow ance requirements are determined by both the weight of the object and the length of the motion.

5. Simultaneous arm motions. If either of the two motions that are occur ring simultaneously can be completed without the use of the eyes to direct it neither motion will take longer to complete than it would as a single motion However, when both motions must be visually directed, the eyes are directed first to one ending point and then to the other. This delays completion of the second motion and an allowance must be added. The amount of the allowance depends on both the distance between the end points of the two motions and the precision needed to end the motions.

Tables of BMT Data. Fig. 28 shows a table of BMT data expressed in ten-thousandths of a minute. The "Reach or Move" table supplies the data for the five classes of motion. To assist in describing the activity, the letters R and M are used to indicate when the motion is for the purpose of reaching (R) to an object or of moving (M) an object to a new location. A 10-in. Class C reach would be coded R10C.

Arm movements that consist of rotating the forearm, as in using a screwdriver are called **turns** and are set out in a separate table. Turn motions are described under the same five classes as used in dealing with reaches and moves. A Class C turn that rotates the hand through 90 degrees is coded T90C.

Separate tables headed "Precision," "Simultaneous Motions," "Turn," and "Force" set out the **time allowances** that must be added to the basic motion times when a motion involves these conditions. It will be noted that the allow ance for precision varies with two factors: the **tolerance limits** at the ending

REACH OR MOVE

nches	½	1	2	3	4	5	6	7	8	9	10	12	14	16	18	20	22	24	26	28	30
A	27	30	36	39	42	45	47	50	52	54	56	60	64	68	72	76	80	84	88	92	96
B	32	36	42	46	49	52	55	58	60	62	64	68	72	76	80	84	88	92	96	100	104
BV	36	42	48	53	57	60	63	66	68	70	73	77	81	85	89	93	97	101	105	109	113
C	41	48	55	60	64	68	71	74	77	79	81	86	90	94	98	102	107	111	115	119	123
CV	45	54	62	67	72	76	79	82	85	87	90	95	99	104	108	112	116	120	124	128	132

PRECISION

nches	1	2	3	4	5	6	7	8	9	10	12	14	16	18	20	22	24	26	28	30
1/2″ tol.	3	4	6	7	8	9	10	11	12	13	14	16	17	18	19	20	21	22	23	24
1/4″ tol.	13	16	18	21	23	25	27	29	31	32	36	39	42	45	48	51	53	55	57	59
1/8″ tol.	33	37	41	45	48	52	55	58	60	62	67	72	76	80	83	87	91	94	98	101
1/16″ tol.	60	65	69	73	76	80	83	87	90	93	98	103	107	112	115	119	123	127	131	135
1/32″ tol.	90	97	102	106	110	114	117	120	123	126	131	135	139	143	147	150	153	157	161	165

Fig. 28. Examples of basic motion-time values (expressed in ten-thousandths of a minute).

point, and the **distance** covered by the motion. The allowance for **simultaneous motions** that require visual direction also varies with two factors: the tolerance limits at the end points of the motions and the distance separating the two end points. The simultaneous motion allowances increase as the distance between the end points of the motions increase.

Multiple Machine Operations

MULTIPLE MACHINE ASSIGNMENTS. When automatic or semiautomatic machines are used and the operator handles two or more machines, computation of standard times becomes complicated by interference between machines and variations in assigned work loads. The solution to multiple machine assignments involves the calculation of **machine interference idleness,** the time that a machine is idle because the operator is servicing another machine in the group, and **operator idleness,** the time the operator is idle because all the machines are running. These man-machine relationships can be classified as:

Completely systematic situations, involving machines with a known time interval between requests for attention with a known and highly constant operator service time. Such problems are solved by means of **man-machine** charting.

Complete random situations, existing when machine service requests and service times are unknown. Average values and distributions for service rates and times are available, permitting use of **probability** and **queueing** theory.

Systematic-Random situations, characterized by random machine service requests with regular service times. Here the distribution has an average service time with low variation. These problems are handled by using algebra and probability distributions such as the **binomial expansion.**

MAN-MACHINE CHART ANALYSIS. Plastic-molding presses, gear-hobbing machines, and automatic screw machines are a few examples of operations having regular, systematic servicing demands. The operator may place a part or piece in the machine, clamp it in place, start the automatic feed and have nothing further to do until the cut or machine run is finished and the piece is ready to be removed. In some operations, **incidental** work such as inspection and minor tool adjustments may have to be done during the cutting time. This fact should be taken into account in making machine assignments.

Although cutting or machine time for operations on machines of the types mentioned is taken from a time study, this time should be checked with calculations based on **approved cutting feeds and speeds.** Usually a 5 percent allowance is added to computed machine time for variations. The percent allowance will depend upon kind and condition of machine and upon accuracy of its adjustment.

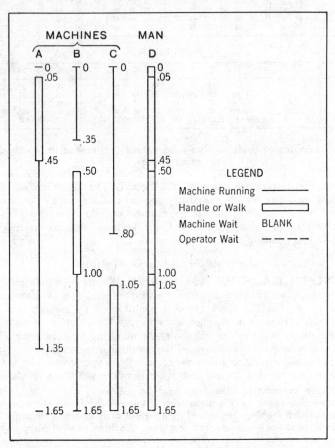

Fig. 29. **Man-machine chart for automatic machine cycle.**

Machine Idleness. Charts such as that shown in Fig. 29, with time indicated on the vertical scales, are used to show machine running, operator handling, walking, interference, and man-waiting times. *A, B,* and *C* are the charts for three machines, and *D* is a man chart. The figures are the **indicated times.** Charts may be constructed commencing with machines at any desired stage of operation. The chart illustrated indicates the man as walking to *A,* having just started *C,* which is now running, while *B* is partly through its run. While the man is handling machine *A,* machine *B* stops at 0.35 minute and must wait until the operator finishes handling *A* and has walked to *B.* At this point, **interference time** is from 0.35 to 0.45, or 0.10 minute. If the operator is assigned too many machines, interference time becomes excessive and cost increase becomes uneconomic.

The man chart D in Fig. 29 shows the operator working or walking 100 percent of the time, and computation of **standard times** can be based on the walking and handling times alone. Allowances for personal time, fatigue, and delays are made as usual.

EXAMPLE OF MAN-MACHINE ANALYSIS
(Time in minutes)

Factor	Machine A	Machine B	Machine C
Machine time..........................	0.900	1.000	0.800
Handling time..........................	0.400	0.500	0.600
Walking time...........................	0.050	0.050	0.050
Handling and walking..................	0.450	0.550	0.650
Personal, fatigue, delay allowance of 15%..	0.068	0.083	0.098
Working cycle.........................	0.518	0.633	0.748

Preparation, or time for getting tote box of work and removing finished pieces, takes 2.00 minutes per 100 pieces, to which an allowance of 20 percent is added. Preparation on each machine is $(2.00 + 20\%) \div 100 = 0.024$ minute per piece.

EXAMPLE OF PREPARATION ALLOWANCES IN MAN-MACHINE ANALYSIS
(Time in minutes)

Factor	Machine A	Machine B	Machine C
Working cycle.........................	0.518	0.633	0.748
Preparation...........................	0.024	0.024	0.024
Standard time per piece in minutes......	0.542	0.657	0.772

The allowance on handling time for personal needs, fatigue, and delays will vary for different jobs and should be determined from studies.

Operator Idleness. The machine time may be so long that the machine will not have completed its run by the time the operator returns to the machine after handling the other machine or machines. This occurrence is shown in Fig. 30, a chart of **two-machine operation.** After the man has finished handling machine A, he may walk to B, arriving there at 0.60 minute, where he may wait until it is time to handle B. This wait is also shown on man chart C by the dotted line. Other man-waits occur throughout the cycle.

The charts are extended for handling and running machine A three times, and for handling and running machine B four times. How far to carry out the chart is determined by cut-and-try methods.

The **standard time** for this cycle may be computed from the man-machine chart as follows.

Total handling and walking times 0.60 + 0.55 + 1.10 + 0.50 + 0.55 + 0.50 (shown by double lines, chart C)............................. 3.800 min.
Personal, fatigue, and delay allowance, 15%......................... 0.570 min.
Man-wait 0.20 + 0.45 + 0.65 + 0.80 (no added allowance) (shown by dotted lines on chart C).. 2.100 min.

Total time... 6.470 min.

Operator's handling or walking time:
Machine *A*, 1.65/3.80 . . . .   43.5%
Machine *B*, 2.15/3.80 . . . .   56.5%
Machine *A* produces 3 pieces:
Standard time per piece, (43.5% × 6.470)/3 . . . .   0.938 min
Machine *B* produces 4 pieces:
Standard time per piece, (56.5% × 6.470)/4 . . . .   0.914 min

In doing his work, the operator may have to bring tote boxes of work to hi
machine and remove finished parts. If this moving is merely a lift from floor t
bench taking 0.10 minute, it may be done during any waiting time the operato

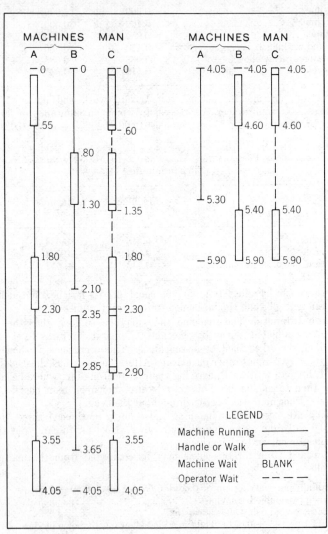

Fig. 30. Man-machine chart with operator waiting for automatics to finish
operations.

as available and no allowance need be made in the computations. If the work must be obtained from a distance, taking a longer time (2.00 minutes, for instance), allowance must be made, because both machines will have stopped running during the 2.00 minutes.

Machine waiting time or delay due to interference decreases productive machine time, eventually a point being reached where it is uneconomic to assign the operator more machines. It is important, therefore, to know the interference time. Graphic charts may be used in cases similar to the foregoing. **Interruption sheets** are sometimes used to record the observations of interruptions.

Another method, with an experienced operator willing to give a good study on the machines, is to take a production or ratio delay study, recording machine running, "down" due to handling or "down" due to interference, together with other usual data. Interference is then obtained as a percentage of the handling time, or as a percentage of handling plus running times.

Interference may be determined from formulas developed to cover particular machines and operating conditions. Such formulas are recommended when their application and limitations are clearly defined.

MACHINE INTERFERENCE FORMULAS. Theoretical evaluation of machine interference loss has received considerable attention. Working from a formula originally developed by Fry (Probability and Its Engineering Uses), Wright (Mechanical Engineering, vol. 58) adapted it in terms of machine interference as follows:

$$I = 50\{\sqrt{(1 + X - N)^2 + 2N} - (1 + X - N)\}$$

where I = interference in percentage of attention time
  X = ratio of running time to attention time
  N = number of machine units assigned to one operator

The formula was checked with the analysis of more than 1,100 hours of actual shop observations during which interference had been measured and recorded. These studies covered the operation of eight entirely different kinds of machines and were therefore considered entirely general. It was found that the formula checked accurately with these actual shop studies for assignments of six or more machine units per operator, but did not agree when the assignment was less than six.

Wright recommended using a set of **empirical curves** (Fig. 31) when the number of machines is six or fewer than six. The curves for two to five machines give interference values which are less than values obtained by solving the formula. For six machines, the chart and the formula give the same results, except at the upper and lower ends of the curve.

Wright's adaptation of Fry's formula and several other formulas were checked and compared with synthetic production studies. Wright's adapted formula checked more closely with production studies than any of the others. A simple formula to apply, it is most widely used in industry.

Following Wright's work, a number of mathematical solutions to the problem of machine interference have been proposed. Ashcraft (Journal of the Royal Statistical Society, vol. 12) applied the rules of probability to determine the extent of reduction in productivity when one operator tends several machines if the probability of a particular machine requiring service within a short period of time is independent of both the state of the remaining machines and the elapsed time since the last servicing. Tables were prepared for the case of equal repair times for each failure.

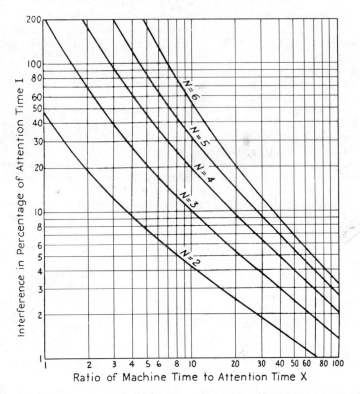

Fig. 31. Interference in percentage of attention time, operator running six or fewer machines.

The problem of a battery of machines being attended by a team of operators was solved for an exponential distribution of service times by Palm (Journal of Industrial Engineering, vol. 9). Feller (Introduction to Probability Theory and Its Applications), following Palm's analysis showed the problem of machine interference to be a special case of steady-state birth and death processes. Fetter (Journal of Industrial Engineering, vol. 6) continued Palm's work and derived a table for the case of collaborating servicemen.

Smith (Work Study, vol. 14) modified Wright's formula as follows:

$$I = 50\{\sqrt{(N - 1\frac{1}{2} - X)^2 + 2(N - 1)} + (N - 1\frac{1}{2} - X)\}$$

A comparison of Smith's **Universal formula** with the exact solutions proposed by Ashcraft and Palm proved it more accurate than previous solutions by Wright and others, eliminating the need for the corrections given in Fig. 31.

Work Assignments

COMPLETELY SYSTEMATIC OPERATIONS. Machine assignments for operations involving man-machine chart analysis with constant running times and service requirements are made on the basis of the cost of each idle machine and the hourly rate of the operator. Niebel (Motion and Time Study) gives the following quantitative procedure for establishing the best assignment of machines and operators.

First estimate the number of facilities the operator should be assigned by establishing the lowest whole number from the equation:

$$N_1 \leqq \frac{l + m}{l + w}$$

where N_1 = lowest whole number
  l = normal loading and unloading time of the facility in hours
  m = machine time in hours (power feed)
  w = normal walking time to the next facility in hours

It can be seen from the above, the cycle time with the operator servicing N_1 machines is $l + m$ since in this case, the operator will not be busy the whole cycle, while the facilities he is servicing will be occupied during the entire cycle. Using N_1, the total expected cost may be computed as follows:

$$C_{N_1} = \frac{K_1(l + m) + N_1 K_2(l + m)}{N_1}$$

$$= \frac{(l + m)(K_1 + N_1 K_2)}{N_1}$$

where C = cost of production per cycle from one machine
  K_1 = operator rate in dollars per hour
  K_2 = cost of machine in dollars per hour

After this cost is computed, a cost should be calculated with $N_1 + 1$ machines assigned to the operator. In this case the cycle time will be governed by the working cycle of the operator, since there will be some idle machine time. The cycle time now will be $(N_1 + 1)(l + w)$. Let $N_2 = N_1 + 1$. Then the total expected cost with N_2 facilities is:

$$C_{N_2} = \frac{(K_1)(N_1 + 1)(l + w) + (K_2)(N_1 + 1)(l + w)}{(N_1 + 1)}$$

$$= [(l + w)][K_1 + K_2(N_1 + 1)].$$

The number of machines to be assigned will depend upon whether N_1 or $N_1 + 1$ gives the lowest total expected cost per piece.

COMPLETELY RANDOM OPERATIONS. A further application of machine interference formulas is the development of procedure to obtain the most economic number of machine units assigned to an operator. The various investigators cited previously have determined **economic assignments** pertaining to their specific analytical approach to machine interference.

For example, Fig. 32 represents Wright's chart for determining the economic assignment of machines. Application of the procedure is illustrated for the situation where:

  Machine running time = 25 min. per lb.
  Attention time = 1 min. per lb.
  Cost chargeable to machine = $0.50 per unit per hr. ($M$)
  Direct labor cost = $2.00 per operator per hr. (L)

In Fig. 32, line $X = \dfrac{25}{1} = 25$ and line $M/L = \dfrac{0.50}{2.00} = 0.25$ intersect at $N = 20$, the economic number of machine units to assign to the operator.

To arrive at the standard time, it is necessary to compute the interference:

$$I = 50[\sqrt{(1 + 25 - 20)^2 + 2(20)} - (1 + 25 - 20)]$$
$$= 136\% \text{ of the attention time, or 1.36 min. per lb.}$$

Machine running time... 25.00 min
Attention time, including allowances............................... 1.00 min
Interference.. 1.36 min
                                                                   ————
Total running, attention, interference time........................ 27.36 min
Personal allowance 3%... 0.82 min
                                                                   ————
Standard time for 20 lb... 28.18 min

As one pound is produced on each of the 20 machines handled, then:
Standard time per 100 lb.. 2.34 hr.

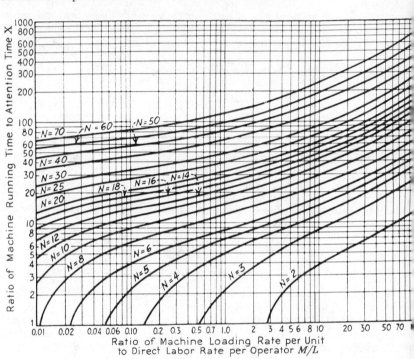

Fig. 32. Diagram for determining the number of machines to assign to an
operator.

SYSTEMATIC–RANDOM OPERATIONS. Use of the binomial expansion for determining the appropriate work assignment was illustrated in an
early study by Bernstein (Factory, vol. 99). The procedure illustrates how the
correct number of machines may be determined from a delay study on any
number of machines that might be available for observations.

In an actual case, where one operator ran 3 and another ran 4 machines, a
delay study was made, resulting in the analysis in Fig. 33.
Bernstein states:

For purposes of multiple-machine basis, only the stocking-up item need be considered in relation to the running time, because on other items the adjuster and not
the operator does the work; i.e., on tool or machine breaks. The stocking-up time

vas 290.05 min. and the running time 4,339.65 min. Therefore, considering only these two items, the stocking-up time, when the operator is occupied, is 6.2% of the total, and the running time is 93.8%. Machine and tool delays amount to 269.90 min., and personal delays are arbitrarily set at 30 man-min. per working day. Machine and tool delays amount to 5.8% of total time.

When one machine is down, there is no loss in efficiency on a multiple basis over a unit basis. When two machines are down, interference occurs—one machine must wait while the operator attends the other. When three machines are down, two machines wait while the first is being attended, and one machine waits while the second is attended, a total of 3 **machine waits**.

								Total Min.
Machine number ..	3411	3437	3436	4196	3432	3434	3409	
R.P.M.	115	115	127	119	122	122	135	
Personal delay	1.60	.35	2.50				.65	5.10
Broken tool	111.70			9.90			26.75	148.35
Stocking up [1]	34.45	39.45	53.30	39.50	36.35	43.85	43.15	290.05
Adjust machine	78.85		9.50			33.20		121.55
Machine down [2] ...	5.40	5.60	11.65	12.00	8.90	5.15	13.30	62.00
Wait for work containers [3]			30.85	10.80	10.00	10.65	11.00	73.30
Total delays	232.00	45.40	107.80	72.20	55.25	92.85	94.85	700.35
Time on study [4]....	720.00	720.00	720.00	720.00	720.00	720.00	720.00	5,040.00
Running time	488.00	674.60	612.20	647.80	664.75	627.15	625.15	4,339.65
Efficiency percent..	68	93.5	85	90	92.5	87	87	

[1] One machine down only—no loss in efficiency over a one-machine basis.
[2] Machine down while operator is occupied on another machine that is down.
[3] This item will be eliminated—more containers purchased.
[4] Not continuous—three different days.

Fig. 33. Delay analysis (seven machines—two operators).

On an eight-machine basis, average interference per machine is determined by using the **binomial expansion** for 6.2 percent stocking up time and 93.8 percent running time. Interference is determined by multiplying probability times the average machines down. For instance, for four machines down the average is 6 machine waits divided by 4, or 1.5.

Binomial Expansion	Machines Down At One Time	Interference
$(0.938)^8$ = 0.5992	0	0
$8(0.938)^7(0.062)$ = 0.3170	1	0
$28(0.938)^6(0.062)^2$ = 0.0733	2	0.0733×0.5 = 0.0366
$56(0.938)^5(0.062)^3$ = 0.0097	3	0.0097×1.0 = 0.0097
$70(0.938)^4(0.062)^4$ = 0.0008	4	0.0008×1.5 = 0.0012
$56(0.938)^3(0.062)^5$ = 0.0000	5	0
	Total	0.0475

On an eight-machine basis, the **efficiency per machine** would be:

$$\text{Efficiency on one-machine basis} = 0.9380$$
$$\text{Interference or machine waits} = 0.0475$$
$$\text{Efficiency on eight-machine basis} = \overline{0.8905}$$

This is running efficiency. However, the operator is allowed 30 man-minutes out of 480 for personal delays, so that the actual minutes per hour is 56.2 and not 60. The overall efficiency is therefore $0.8905 \times 56.2/60.0 = 0.835$.

Study Procedure. The time study man first makes a delay study on one machine of the group, with the operator attending the number of machines which is historically normal. He writes down delays which occur on this one selected machine, differentiating between machine waits for the operator when more than one machine is down at a time, and delays where the operator is attending to that machine. The first named delays, machine waits, are not used in computing the efficiency because that is taken care of in the binomial expansion. Also, at the end of the study, the actual production should be checked against the theoretical in order to see if any loss through nonfeeding or any other shrinkage has occurred. The **shrinkage** should be added to delays in attending the machine in order to compute the one-machine running efficiency to be used in the binomial expansion.

In addition, the operator is attending to gaging work, filling the hopper, etc. while the machines are running. The minutes consumed this way should be added to the delay time to see that the operator is not theoretically occupied more than 100 percent of his available time. If the operator is occupied 6 percent of his time on each machine while it is running, then 48 percent of his available time is used up in this manner. Added to this 48 percent are his attendance times on machines when they are down. For an eight-machine basis these times are as follows:

Machines Down	Attendance Time, Etc.
1	$0.3170 \times 1 = 0.3170$
2	$0.0733 \times 2 = 0.1466$
3	$0.0097 \times 3 = 0.0291$
4	$0.0008 \times 4 = 0.0032$
5	0.0000
over 5	0.0000
Total	0.4959

Thus, 50 percent plus 48 percent adds up to 98 percent of operator's available time.

Finally, the labor costs for various bases must be determined, to see that the law of diminishing returns has not set in. At certain times, when machine tools are scarce or when production is more important than labor costs, the basis may have to be lowered so that full use can be made of machine time available.

When the operator does his own adjusting, the number of machines handled should be fewer. In the first example, the restocking, adjusting, and attention time will be 11.4 percent and machine running time will be 88.6 percent, computed from Fig. 37, on page 90.

Hence, for four machines the man is occupied slightly under 50 percent of his time, which is a fair basis, as $11.4 \times 4 = 45.6$ percent.

Binomial Expansion	Machines Down At One Time	Interference
$(0.886)^4$ $= 0.6162$	0	0
$(0.886)^3 \times (0.114) \times 4 = 0.3171$	1	0
$(0.886)^2$ $(0.114)^2 \times 6 = 0.0612$	2	$0.0612 \times 0.5 = 0.0306$
$(0.886)^1$ $(0.114)^3 \times 4 = 0.0053$	3	$0.0053 \times 1 = 0.0053$
$(0.114)^4$ $= 0.0002$	4	$0.0002 \times 1.5 = 0.0003$
		0.0362

Analysis	Percent
One machine down (normal)....................	11.40
Machines down while operator is occupied........	3.62
Personal delays..............................	5.00
	20.02
Efficiency....................................	79.98

Considering only labor cost, would it be better to have an operator-adjuster handle four machines or have an operator on eight machines? The task is computed by the formula $RPM \times RE \times MH \times 8$, where RE equals running efficiency and MH equals minute-hour.

Task for 8 machines = $120 \times 0.8905 \times 56.2 \times 8 = 48{,}044$
Task for 4 machines = $120 \times 0.7998 \times 56.2 \times 4 = 21{,}575$

Comparative labor costs would be:

No. of Machines	Output at Task for 8 Hr.	Labor Cost for 8 Hr.	Labor Cost per Hundred
8	48,044	$28.80	$0.0607
4	21,575	$24.00	0.111

The eight-machine basis gives 46 percent lower labor cost and only 10 percent lower production per machine.

For a group of machines on which different components are being run and restocking varies greatly, it is a good practice to rate each machine individually, and let the operator run as many machines as he can look after. Suppose earnings per hour for the correct number of machines is set at $1.00. In a group of ten machines, assume five were being run as shown in Fig. 34.

To bring his task earnings up to $1.00, he should run an additional machine like No. 30022 or No. 29334.

This method works satisfactorily when the efficiency of operation is approximately the same in the new combination of machines as in the combinations for which the rates were originally set. Otherwise, corrections should be made for the changed efficiencies.

AUTOMATIC SCREW MACHINES. A method for computing hourly production and standard times for Brown and Sharpe automatic screw machines is given by Varga of the Neptune Meter Co.

$$\text{Standard minutes per piece} = \left[\frac{T}{60} + \frac{.50}{\frac{(L \times 12) - K}{l + l'}} \right] \left[1 + \begin{array}{l} \text{Percentage for} \\ \text{tool,} \\ \text{oil, and} \\ \text{general} \\ \text{allowances} \end{array} \right]$$

where T = cam time in seconds per piece, from blueprint
L = length of bar stock, in feet
K = scrap, in inches
l = length of one unit produced, in inches

Machine Number	Percent Restocking	Correct Machine Basis	Task Earning per Machine
29323	8.8	5	$0.200
29331	6.7	6	0.166
30022	4.7	10	0.100
30021	16.2	3	0.333
29334	5.4	9	0.111
Operator's task earnings...			$0.910

Fig. 34. Machine assignment when restocking varies greatly.

The tabulations in Figs. 35 and 36 give basic data for the factors K and l' and tool allowances.

Procedure for Setting a Standard. In setting a standard on a job of this kind, the following procedure is used:

1. From operation sheets obtain cam time to produce piece and the class of machine job is run on.
2. Apply formula. Refer to tables for values of l' and K. Compute time to produce piece, including the 0.50 minutes for loading the rod. Compute the rod time, which equals the product of the time per piece times the number of pieces in the rod.
3. List tools required to machine part. Refer to table of tool allowances and list corresponding percentage tool allowances for given material. List general allowances. Refer to the chart in Fig. 37, and obtain percent allowance for rod running out and also overall allowance on tool adjustments. Total these allowances.

Diam. of Material (inches)	Mach. No.	Values for l' (inches)		K Length of Scrap (inches)
		Metallic Material Without Hole	Metallic Material With Hole	
0.062 to 0.312...........	00	0.055	0.055	1¾
0.343 to 0.500...........	0	0.072	0.072	2⅜
0.531 to 0.875...........	2	0.080	0.100	2¹⁵⁄₁₆
0.875 to 1.000...........	2	0.100	0.100	2¹⁵⁄₁₆
1.032 to 2.000...........	—	0.125	0.125	2¹⁵⁄₁₆

Note: When facing tool is used, add .01 to values of l'. Rod lengths: Brass = 12 ft. All others = 10 ft.

Fig. 35. Factors K and l' for automatic screw-machine calculations.

Tool	Metal Being Machined		
	Brass	Steel	Monel
Cutoff	.490	1.850	2.600
Form	1.400	2.070	2.740
Position	0.035	0.035	0.035
Spot face	0.029	0.043	0.049
Drill	0.390	0.660	0.930
Reamer	0.160	0.205	0.250
Box	1.040	1.540	1.750
Die	0.240	0.243	0.246
Tap	0.080	0.119	0.135
Swing	2.500	3.700	4.200
Slotter	0.033	0.048	0.055
General	8.000	8.000	8.000
Oil	2.000		
Inspect	0.950		

Fig. 36. Tool allowances, automatic screw machines.

4. Increase time to produce piece by these allowances.
5. Convert standard time into production per hour.

$$\frac{60}{\text{Standard time}} = \text{Pieces per hour}$$

Illustration: Part: Steel hand shaft. 9/16 in. diameter, steel. Run on No. 2 B. & S. screw machine. Given: $T = 35$ sec., $l = 2\frac{1}{4}$ in., $L = 10$ ft.

$$\frac{T}{60} + \frac{.50}{\dfrac{(L \times 12) - K}{l + l'}} = \frac{35}{60} + \frac{.50}{\dfrac{120 - 2^{15}\!/_{16}}{2.25 + 0.080}}$$

$$= 0.583 + \frac{0.50}{50}$$

$$= 0.583 + 0.01 = 0.593 \text{ min.}$$

Rod time $= 50 \times 0.593 = 29.65$ min.

Allowances	Percentage	
Cutoff.............................	1.85	
Form.............................	2.07	
Position..........................	0.035	
Two box..........................	3.08	
Die..............................	0.243	
Oil...............................	2.00	
Inspect...........................	0.95	
General...........................	8.00	
Tool adjustment..................	0.465	(from Fig. 39)
Rod..............................	6.50	(from Fig. 39)
(Includes Interference)		
	25.193	

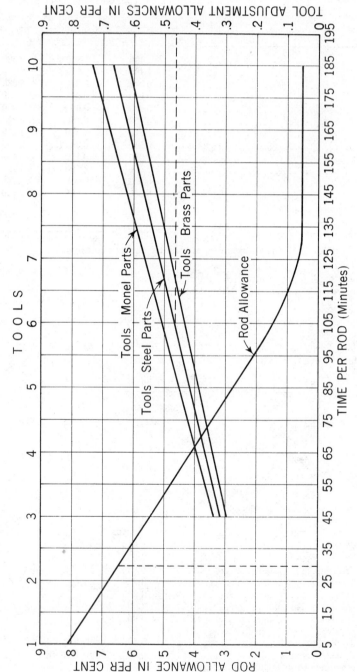

Fig. 37. Rod (bar stock) and tool adjustment allowances for automatic screw-machine work.

Standard time = 0.593 × 1.2519 = 0.742 min. per piece

Production per hour for 1 machine = 81 pieces
Production per hour for 2 machines = 162 pieces
Production per hour for 3 machines = 243 pieces

TEAM OR GROUP WORK. Where two or more operators work together on the same unit of production, there must be close balancing, that is, **synchronizing**, or one operator will be idle part of the time. Synchronizing is difficult but not a matter of refinement. When the best possible synchronization has been established, the group time is limited by the longest subdivision or by the slowest individual. Timing should be done on a single individual because it is seldom that all workers involved start and stop at the same point in the operation. There is some overlapping, and consequently the true time of the cycle is measured between corresponding points on the cycle of one worker.

Where it is feasible to measure the output of an individual, individual tasks are to be preferred to group tasks, both because of the limitation mentioned above and because of the greater strength of individual incentives. There are cases, for instance emergency repair, in which a group of employees work on the same job and where it is difficult to measure the output of the individual: and there are cases, particularly in the automobile industry, where models change frequently, high output is demanded, and there is insufficient time to study and set individual tasks.

When group jobs are reasonably stable, each individual job in the group should be studied separately, or better yet, the time should be synthesized from unit data and time allowances figured. Time allowances for all individual jobs constituting a group job should be made to reapportion elementary operations in such a manner that the individual times would be approximately equal. In cases where reapportionment of work between members of the group is impossible, a study of the longest part is all that is necessary. In such cases the output of this portion will be the output of the group. It is sometimes possible to place a second employee on the longest subdivision of the job and apportion the balance of the work so that the time required of each of the others is approximately half the time required for the longest subdivision.

Measurement of Indirect Labor Operations

TYPES OF INDIRECT LABOR. Indirect labor operations include, among others, material handling, cleaning or janitorial tasks, maintenance, and clerical jobs. These operation types present an **obstacle to measurement** and standardization for a number of reasons:

1. There is usually an ostensible lack of repetitive pattern to the operational method.
2. The operations may involve a group activity.
3. The unit of output may be difficult to define.
4. The job may be made up of numerous suboperations.
5. A cycle of output is long.
6. The operation may be constantly changing in geographic location.

Besides measurement difficulties, certain **economic factors** must be considered. The period of measurement is usually long, being a matter of weeks or even months. In addition to actual measurement time, considerable engineering hours must be consumed in analysis of data and in compilation and classification of standard data for use in setting standard times for individual jobs. Also, the failure of an indirect labor incentive plan due to improper or hasty measurement and planning can be costly to correct.

NEED FOR MEASUREMENT. The measurement, standardization, and control of these types of jobs should not be neglected. There are a number of **reasons for measurement,** among which are the following:

1. The proportion of indirect labor cost to total labor cost is high, being as much as 75 percent in some plants.
2. With automation the proportion of indirect costs to total costs increases.
3. The use of incentive plans on direct labor has resulted in a need to place incentives on indirect labor in order to have some equity in wage-earning opportunities.

INDIRECT LABOR MEASUREMENT. Both time study and work sampling techniques are applicable to measuring indirect labor jobs. The following applications suggest approaches to the derivation of **standard data** for indirect operations using the stop watch.

Janitorial Work. The following steps for measuring the operation of dry-mopping office floors and cleaning desk tops include those necessary to compile basic standard data for the job.

1. Set up procedures covering equipment and methods employed in carrying out the job. Sufficient attention should be given to methods design or improvement. Put these in written standard practice form.
2. Determine the important elemental activities of the operation. In this case these activities might include:
   a. Unlock and open office door and turn on lights.
   b. Get cleaning material out of hand truck.
   c. Mop clear floor area.
   d. Move objects in the room.
   e. Clean desk and table tops.
   f. Mop under objects.
   g. Remove dust from mops and cloths.
3. Time study the full range of operations, with the janitors following the standard methods. Snap-back timing is an economical method where the purpose may be to build up suboperation times. Times can be recorded directly on a floor plan blueprint.
4. Analyze the data, considering the variables and problem of applying the data to obtain standards. The important variables in this case will be:
   a. Area to be cleaned in terms of square feet.
   b. Indirect time necessary to get ready for cleaning.
   c. Shape of the area to be cleaned.
   d. Congestion of the space.
   e. Movable objects.
   f. Walking time.
5. Establish elemental standard data.
   a. **Time to mop floor.**
   b. **Time to clean desk and table tops.**
   c. **Time to move objects and walk.**

Now let:

A = total clear floor area in square feet.

a = time to mop 1 square foot of floor as determined from curve.

b = initial preparation time per room, determined from curve or from separate time studies.

b' = preparation time necessary for each subarea. This may be specified as some fraction of b.

N = number of subareas.

d = time to clean 1 square foot of table or desk top.

A' = table- or desk-top area.

C = number of tables or desks.

e = unit time to move an object in the room.

M = number of movable objects.
MTM = MTM standard walking time per foot.
D = number of feet walked.
T = total standard, time required to dry-mop an office floor and clean table and desk tops.

$$T = aA + b + b'(N - 1) + dA'C + eM + MTM \ (D)$$

The standard data in this form provide the basis for planning a program or schedule of janitorial work necessary for a given building or group of offices. If the standard data and a **standard schedule** are to be used as the basis for incentive wages the incentive earnings must be built into the base rate of pay. This assumes that the schedule, representing good performance, will be adhered to. Provisions must be made to insure quality and to evaluate complaints concerning nonstandard job conditions.

Maintenance Operations. An attempt to measure and standardize various maintenance crafts or operations will prove uneconomical if the number of maintenance operators in the plant is small. If the maintenance labor force is large, particularly for a given craft, measurement can be profitable.

PREDETERMINED STANDARD DATA SYSTEMS. In order to avoid the high costs of development of standard data systems for indirect labor as outlined in the previous examples, commercially available predetermined time systems have been developed by a number of management consulting firms. **Universal Maintenance Standards** (Factory, vol. 113) was one of the earliest attempts to classify groups of jobs for the various crafts according to duration. The technique applies work measurement to a job by comparing it with a **library** of standard tasks of similar content. The standard time selected for a given job is that of a task whose range of operations is very similar to the job in question. **Engineered Performance Standards,** developed by the U.S. Navy's Bureau of Yards and Docks (Factory, vol. 120) is an example of commercially available data for maintenance standards.

The use of a computer to calculate time standards from standard data in the computer's memory system is the basis for a procedure termed **Autorate** developed by the Service Bureau Corporation, an IBM subsidiary, (Factory, vol. 120). **Universal Office Controls** (Maynard, et al., Practical Control of Office Costs) and **Master Clerical Data** (Birn, et al., Measurement and Control of Office Costs) are two examples of systems developed for clerical operations.

Fatigue Measurement and Reduction

NATURE OF FATIGUE. The measurement and reduction of **fatigue** in industrial operations may be considered an adjunct problem to work design and measurement. Unfortunately, the term "fatigue" is used rather indiscriminately to apply to various phenomena present in both **muscular** and **sedentary** or mental tasks. These phenomena may be described and studied from both physiological and psychological points of view.

Ryan (Work and Effort) pointed out the difficulty of definition and research in the area of fatigue. "One of the primary difficulties is that the term 'fatigue' is not clearly enough defined to aid us in a search for a measure. Not only is the term vague as it appears in common usage, but it has been difficult for scientists to agree upon a common meaning of the term which would put their research upon a straightforward basis."

From the industrial point of view, fatigue has been defined as the effect of work upon an individual's mind and body which tends to lower his rate or grade of quality in production, or both, from his optimum performance.

Fatigue is manifested in a number of ways, some lending themselves to objective measurement and others more subjective. The manifestations of fatigue include:

1. A decrement in performance. A reduction in the rate and quality of output over a specified time period—**industrial fatigue.**
2. Reduction in the capacity to do work, such as **impairment** of a muscle or inability of the nervous system to give stimulus—**physiological fatigue.**
3. Subjective reports from workers concerning feelings of tiredness, physical discomfort, or localized pain—**perceived** or **psychological fatigue.**
4. Changes in other activities and capacities. These include changes in physiological functions or changes in ability to perform psychological activities other than the job itself—**functional fatigue.**

Modern manufacturing methods are eliminating heavy muscular activity from industrial jobs. While these types of jobs are being eliminated, sedentary tasks introduce their own particular problems of fatigue. **Localized** and **static muscular fatigue** are associated with this type of job. Also involved are the kindred problems of boredom, monotony, and nervous or emotional strain.

Rohmert (The Production Engineer, vol. 45) lists the following factors that induce or favor fatigue on workers when performing industrial tasks:

a. Lengthy durations of physical work.
b. Uniform work with monotonous surroundings.
c. Hot and damp surroundings.
d. Demands on concentration, on mental alertness, and on skill.
e. Noisy surroundings.
f. Unfavorable lighting conditions.
g. High degree of responsibility and care.
h. Psychic feelings of aversion (worries, lack of interest, etc.).
i. States of illness and pain.
j. Conditions of nutrition.

MUSCULAR FATIGUE. Muscular work consists of those activities in which the primary aim is the development of mechanical force. Included in these activities are tasks which make heavy demands upon the whole muscular system, such as running, walking, carrying large loads, shoveling, and scrubbing. It would also include those tasks which involve the development of force by a restricted muscle group. Ryan (Work and Effort) described some of the types of muscular activity as:

1. Rapid expenditure of a large amount of energy ("dromal" tasks), e.g., running a competitive race.
2. Steady grind, e.g., a long day, or walking under a load.
3. Repetitive local task with high local energy expenditure, e.g., lifting a weight with one finger.
4. Postural restrictions ("static work"), e.g., maintaining an awkward position; holding a weight in outstretched hand.

Laboratory Experiments in Muscular Fatigue. Physiological and psychological studies have been made on muscular activity. Classic experiments upon localized muscular activity have been conducted using the **ergograph.** This instrument graphically records the nature of contraction of specific or local muscles in an intact organism. In a typical ergograph study a weight is attached by a cord to the finger and the subject is asked to raise and lower the weight at specific times until fatigue makes contraction impossible. These studies are useful in attempting to determine such things as: the parts of the organic system—muscle, peripheral nerves, or central nervous system—responsible for loss in capacity for further performance; the nature of the spread of fatigue from a locally stimulated muscle to those surrounding; and the optimal load for a fixed number of lifts of a weight.

Muscular activity also produces changes in the blood pressure, heart rate, cardiac output, pulmonary ventilation, body temperature, rate of sweating, oxygen consumption, and chemical composition of the blood. Measurement of these **metabolic** factors may be used to quantify the physiological cost of doing a certain amount of work.

Studies such as these are important in determining **optimum work loads** for various jobs and under various working conditions. The suitability of a person to handle long continuous heavy work can be determined, for instance, by the capacity for **absorption of oxygen.** Also, depending on whether the absorption of oxygen is great or small, certain conclusions can be drawn with regard to the degree of **strain** in a given stage of work for certain constant conditions.

Industrial Experiments in Muscular Fatigue. Studies by Christensen (Symposium on Fatigue, Floyd and Welford, eds.) were made to determine a physiological valuation of work in the Nykroppa Iron Works in Sweden. Measurements were made of oxygen consumption, pulse rate, maximum body temperature, and maximum fluid discharge for various types of work in the plant.

The **oxygen consumption** is an expression of the intensity of consumption in the body which the work entails; the heavier the work, the greater the amount of oxygen consumed. On the basis of oxygen consumption the production of **calories** can be calculated. A small proportion of these calories, estimated at 10 percent, is converted into mechanical work; the larger part degenerates into heat and must be conveyed away from the body.

The **pulse frequency** per minute is determined in order to check that the intensity of work at the time of determination of oxygen consumption was representative.

The **body temperature** can give valuable information about the heaviness of the work. The body temperature normally adjusts itself at a constant level after about one hour's work to a value which depends on the work load. The greater the load, that is, the greater the rate of development of energy, the higher the temperature.

The **loss of fluid** during the shift can also give information about the combined effect of work and heat. A certain amount of perspiration is normal, but it is increased very much if the temperature of the air or the intensity of heat radiation is high. The time of exposure is also a factor determining the amount of fluid lost.

Figure 38 (adapted from Christensen, Symposium on Fatigue, Floyd and Welford, eds.) shows the results of the tests in terms of a ranking of each job on each factor measured, as well as an overall grading figure.

Passmore and Durnin (Physiological Reviews, vol. 35) collected a number of studies in various human activities as determined by physiological studies. The data, expressed in terms of gross energy expenditure for the "average" human male or female, was converted from units of Kcal/min. to Btu/hr. by McKarns and Brief (Heating, Piping and Air Conditioning, vol. 38) as shown in Fig. 39. These values can be used to estimate heat stress loads and rest periods for jobs of similar physical exertion.

Belding and Hatch (Heating, Piping and Air Conditioning, vol. 27) list the estimates of energy expenditures in Fig. 40 for light, moderate, and heavy work activities.

Work Task	O₂ Litres Per Min. Mean Value	Pulse Rate Per Min. Mean Value Under O₂ Uptake	Pulse Rate Per Min. Mean Value by Shifts	Maximum Pulse Values	Maximum Body Temperature	Litres, Fluid Discharge Exclusive of Urine, Mean Value	Grading
Open hearth:							
Slag removal	1	4	—	8	—	7	5
Dolomite shovelling.	2	3	—	4	—		3
Tipping the moulds.	9	10	9	11	7	—	12
Cogging mill:							
Tending the heating furnace	3	6	2	4	3	2	2
Tending the soaking pit	10	—	5	9	3	9	11
Hand rolling	5	1	7	9	4	6	7
Tending the saw pits	8	5	3	3	5	8	6
Wire rod mill:							
Roughing	6	8	4	2	1	4	4
Finishing	10	—	13	13	6	5	13
Wire bundling	3	2	1	1	2	6	1
14-inch merchant mill:							
Merchant mill rolling	4	9	8	6	4	3	8
Large hammer:							
Forging	7	7	$\{\begin{smallmatrix}6\\10\end{smallmatrix}\}$	7	8	(1)	$\{\begin{smallmatrix}9\\10\end{smallmatrix}\}$
Chipping:							
Chipping	10	11	11	10	9	10	14
Overhead crane:							
Operator	11	—	12	12	4	—	15

Fig. 38. Results of experiments in muscular fatigue.

Reduction of Muscular Fatigue. Brouha (Advanced Management, vol. 19) points out that each worker owns a certain amount of "physiological capital" which enables him to have a given amount of "physiological credit." As he works, he contracts a "physiological debt" which varies with the nature of the job. The amount of physiological debt determines the degree of muscular fatigue. In general, the average human male can withstand a rate of energy expenditure of approximately 5 Kcal/min. (Murrel, Human Performance in Industry) corresponding to a heart rate of approximately 120 beats per minute. Brouha (Physiology in Industry) lists the following recommendations for reducing the amount of physiological fatigue:

1. Reduce the work load by means of work design.
2. Reduce the heat load by maintaining temperature and humidity as low as ventilation cost permits.
3. Organize adequate rest periods based on physiological considerations.

Activity	M, Btu/hr	Activity	M, Btu/hr
Typing, electrical*	270–330	Bricklayer	950
Typing, mechanical*	300–375	Timbering	975–2140
Lying at ease	335–360	Plastering walls	975
Sitting at ease	380–395	Machine fitting	1000
Standing at ease	405–450	Mixing cement	1115
Draftsman	430	Walking on job	1165–1610
Drilling, machine	430	Pushing wheelbarrow	1190–1660
Light assembly line	430	Chiseling wood	1355
Armature winding	525	Shoveling	1285–2495
Light machine work	570	Loading mixer	1425
Machine wood sawing	570	Forging	1520–1595
Measuring wood	570	Tending heating furnaces	1595–3850
Medium assembly work	640	Working with axe	1640–5730
Driving a car	670	Drilling wood, hand	1665
Sheet metal worker	715	Cross cutting with bucksaw	1780–2500
Machinist	740	Climbing stairs or ladder	1830–3140
Drilling rock	880–2255	Planing wood	1925–2160

* Women

Fig. 39. Energy expenditures, *M*, for various industrial activities (data may be converted to Kcal/min by multiplication of Btu/hr figures by 0.0042).

	Activity	M, Btu/hr
Light Work	Sitting quietly	400
	Sitting, moderate arm and trunk movement	450–550
	Sitting, moderate arm and leg movement	550–650
	Standing, light work at machine or bench	550–650
Moderate Work	Sitting, heavy arm and leg movement	650–800
	Standing, light work and some walking	650–750
	Standing, moderate work and some walking	750–1000
	Walking, moderate lifting or pushing	1000–1400
Heavy Work	Intermittent, heavy lifting, pushing, or pulling	1500–2000
	Hardest sustained work	2000–2400

Fig. 40. Energy expenditures, *M*, for light, moderate, and heavy work activities.

4. Organize adequate work teams, balancing the amount of physiological stress among the crew members.
5. Provide for an adequate water and salt supply for workers in hot environments.
6. Select workers based on their natural physical fitness and degree of training for specific activities requiring heavy energy expenditures.

In the area of reduction of work load by means of workplace design, Dacrus (Symposium on Fatigue, Floyd and Welford, eds.) suggests the following factors in work design to be considered in relation to tasks requiring muscular exertion.

First, optimum use must be made of the muscle power available. The position of the control to be used should be such that it can be operated with the maximum force and that the load on the control should be the smallest proportion of that which can be exerted upon it. . . .

The part of the body used will depend on the nature of the work; for instance, the force required, the speed, the amplitude and the direction of movement. . . .

Much can be done when forces are required by providing suitable counter-pressure for the body. Certain devices which fix the body more rigidly, such as backrests, relieve the fixator muscles and increase the efficiency of the prime mover as well as reducing unnecessary displacements of the body.

A reduction should be made in total static muscular activity. A considerable degree of useless static work can be imposed by the design and positioning of controls. . . .

A further improvement could be made by ensuring that any control that has to be manipulated is placed so that the body can assume a posture which can be maintained with the minimum muscular effort. If the body is at all "off-balance," muscular activity has to be increased in order to preserve equilibrium.

If possible the load should be shared around the body—a division of labor instituted. . . . Improvements might also be effected by allowing the same movement to be produced by different muscle groups; . . . If the task is particularly heavy the operator should have the opportunity of moving from one process to another in which different groups of muscles are used.

Unnecessary body movement can also be reduced by designing the control so that it can be moved with the minimum shift of body position when it is moved through its required range.

FATIGUE IN SEDENTARY WORK. The term "sedentary work" is preferred to "mental work" for the tasks considered since they are truly body activities rather than activities that take place in the mind. These tasks include situations where control, timing, skill and directions of activities are more important than muscular force. Ryan (Work and Effort) described various types of sedentary work:

1. **Problem solving** (with minimum muscular and sensory involvement). Examples are calculation, solution of mathematical problems, composing, planning, directing the work of others, supervising, and socialized tasks such as selling.
2. **Continued sensory adjustment** (primary visual tasks). Examples are proofreading, visual inspecting, radio code reception, piano tuning, and reading under difficult conditions or when long continued.
3. **Motor skill** (with patterning and accuracy of movement as the central core, and force only an accessory feature). Examples are typing, drawing, various machine operations, assembling machines, woodworking, sewing, acting, and speech-making.
4. **Sedentary muscular** (light muscular tasks with little skill or control involved). Examples are the work of a watchman, crossing guard, or machine feeder, or any task where the main requirement is that the proper movements be made at the proper time. The movements themselves involve few elements of skill or of force.

In many industries these task types predominate over heavier muscular types of jobs. The types of fatigue associated with these jobs must be discerned and reduced where possible. However, the fatigue present in sedentary operations is characteristically difficult to measure and evaluate. It is manifested in **deterioration of performance** or in subjective reports from the worker concerning his feelings of fatigue.

Subjective reports from workers are poor indicators of fatigue if remedial action is to follow. They are usually delayed reactions to the onset of fatigue and come too late for immediate remedies.

Output Decrement in Sedentary Work. An hourly production curve is often presented as typical of output of factory operations. Such curves tend to show a composite of a number of factors which obscure the real influence or presence of fatigue. Some of these factors, such as a daily task or production goal set by the worker, the tendency of rhythmic operations to be more constant in rate of output, or the length of the day, may have an influence on when production will fall off.

Often it is difficult to find clear-cut changes in performance which accompany the progress of fatigue. There are few results among the studies of mental work which are clear enough to be of value in understanding the problem.

Industrial Experiments in Sedentary Work. A now classic study in fatigue was conducted at the Hawthorne Plant of the Western Electric Company (Roethlisberger and Dickson, Management and the Worker). One part of the study is referred to as the Relay Assembly Test Room. Six girl workers were selected and placed in a separate room where experiments were made with different kinds of working conditions. The operators chosen were assemblers of telephone relays, neither inexperienced nor expert. The experiment was to provide the answers to questions concerning fatigue.

In thirteen different test periods spread out over a number of years, various factors were altered, including length of work period, introduction of food during rest periods, number and duration of rest periods, and methods of payment. In general the daily rate of output of the group continued to increase during the overall experimental period. A statistical analysis of the results showed no simple correlation between output changes and changes in the various working conditions. Nor was there any correlation between output and other physical factors of the job for which records were kept, such as temperature, humidity, hours of rest, and changes of relay type.

The significant result of the experiment was the increased output over the entire length of the study. This was considered to be due to psychological and sociological factors. Statements of the subjects indicated the important reasons for increased productivity. In conclusion a company report pointed out that: "Upon analysis, only one thing seemed to show a continuous relationship with this improved output. This was the mental attitude of the operators. From their conversations with each other and their comments to the test observers, it was not only clear that their attitudes were improving but it was evident that this area of employee reactions and feelings was a fruitful field for industrial research."

ENVIRONMENTAL CONDITIONS AND FATIGUE. Besides the physical and movement requirements of a job, certain other factors, both inherent in the nature of the job and environmental to it, introduce fatigue. Some of these factors are noise, illumination, duration and time of work periods, and effective temperature of the work environment.

Noise of Work Environment. Noise has been increasing in factory work as in all modern life. Fortunately, public opinion and governmental regulations (Dept. of Labor Safety and Health Standards, Occupational Noise Exposure) are bringing some relief. The first step in abatement of noise nuisance is one of measurement. A unit of measurement called the decibel has been standardized and an instrument made to register sounds in that unit. The **decibel**, abbreviation db, is defined as the smallest change which the ear can detect in the level of sound. The decibel scale is a logarithmic scale, on which a difference of intensity level of 1 decibel corresponds to a ratio of intensities whose logarithm is 0.1.

Compared with the scale running from the threshold of human hearing (0 decibels) to the threshold of painful sound (130 decibels), Hunt (Standard Handbook for Mechanical Engineers, Baumeister, ed.) lists the data in Fig. 41 as typical sound levels for illustrative purposes.

Decibels			Decibels	
	120	Threshold of feeling	60	Average factory
		Thunder, artillery		Noisy home
Deafening	110	Nearby riveter	Moderate 50	Average office
		Elevated train		Average conversation
	100	Boiler factory	40	Quiet radio
		Loud street noise		Quiet home or private office
Very Loud	90	Noisy factory	Faint 30	Average auditorium
		Truck unmuffled		Quiet conversation
	80	Police whistle	20	Rustle of leaves
		Noisy office		Whisper
Loud	70	Average street noise	Very Faint 10	Soundproof room
		Average radio		Threshold of audibility
			0	

Fig. 41. Typical sound levels.

The relative intensity of sound alone does not account for **pitch** or **quality** and does not show **regularity** or **irregularity**, all of which make a difference in the tiring effect due to noise. Most factory work is between 90 and 40 db. Some idea of this range may be gained from Hunt's table describing average radio music as equivalent to 70 db.

McCormick (Human Factors Engineering) makes the following recommendations concerning noise:

1. There is abundant evidence that continuous exposure to high noise levels contributes to hearing loss, although there is not much known about the degree and character of hearing loss caused by different noise conditions There is evidence, however, to support certain general conclusions regarding degree of hearing loss, as follows: (a) hearing loss is related to level of noise to which exposed; (b) hearing loss is related to exposure time for high exposure intensities, though to a limited extent, or not at all, for low exposure intensities; and (c) hearing loss usually is greater in the 4000-Hz range than in the 1000- and 2000-Hz ranges.

   Continuous and extensive exposure to noise levels above 80, 85, or 90 db is generally considered to bring about hearing loss.

   Impact noise, such as that from drop forges, and *impulsive* noise, such as that from gun blasts generally bring about hearing loss more quickly than exposure to continuous noise.

2. While there is no systematic evidence that noise *generally* brings about degradation of work performance, there is accumulating evidence that it affects performance under certain circumstances, such as where sustained performance is required on tasks that are not intrinsically challenging.

   People have considerable capacity to adapt to the annoying characteristics of noise, but this adaptation probably is not complete.

   The characteristics of noises that cause them to be annoying seem to be high intensities, high frequencies, intermittency, and reverberation effects.

3. Noise control can be accomplished in various ways, depending upon the circumstance. Some methods are (a) control of noise at the source, such as

through machine design, proper maintenance and lubrication, or mounting equipment on rubber; (b) isolation of noise with enclosures, rooms, barriers, etc.; (c) use of baffles and sound absorbers; (d) acoustical treatment; (e) use of ear-protection devices; and (f) taking advantage of the acoustic reflex in impulse-type noise.

Illumination of Workplace. In general, conditions of lighting which are suboptimal for particular visual tasks accelerate the onset of fatigue.

A study by Luckiesh (Light, Vision, and Seeing) was made to determine the effect of illumination upon a number of factors (Fig. 42). **Acuity** increases

Type of Test	Intensity of Illumination (foot-candles)		
	1	10	100
Visual acuity	100%	130%	170%
Contrast sensitivity	100%	280%	450%
Muscular tension (key pressure) while reading	63 grams	54 grams	43 grams
Change in frequency of blinking after reading one hour	100%	77%	65%
Decrease in heart rate while reading one hour	10%		2%
Decrease in convergence reserve of ocular muscles after reading one hour	23%		7%

Fig. 42. Influence of level of illumination.

rapidly as light passes through the lower levels of intensity and more and more slowly as higher levels of intensity are attained. The increases are proportional to the ratio of increase of light rather than to absolute increase. The gain from added intensity is relatively small beyond 10–20 footcandles.

The Committee on Recommendations for Quality and Quantity of Illumination of the Illuminating Engineering Society has presented a number of recommendations concerning the **appropriate levels of lighting** in modern practice (see Plant Layout and Facilities Planning section).

Temperature and Ventilation. Much factory work is done under poor air conditions. The effects of temperature and ventilation, or lack of ventilation, is still not thoroughly understood. Woodson and Conover (Human Engineering Guide for Equipment Designers) note that certain extreme temperatures have proven detrimental to work efficiency. Moderately complex tasks such as problem solving, hand coordination, or visual attention without physical exertion are possible in temperatures as high as 85°F. By increasing the complexity or adding physical or mental strain, however, this maximum is lowered slightly. The relationship between temperature and working efficiency is given by Woodson and Conover in Fig. 43.

Excessively hot and damp working environments may not only reduce the efficiency of a given worker but can result in physiological harm due to excessive heat stress. Hertig and Belding (in Temperature: Its Measurement and Control in Science and Industry, Vol. 3, Herzfeld, ed.) discuss a procedure to balance the sum of the components of the heat load with the physiological and physical capacities of the individual to dissipate the excess heat without suffering a significant

120°F	Tolerable for about 1 hour, but is far above physical or mental activity range (160°F for ½ hour).
85°F	Mental activities slow down—slow response, errors begin.
75°F	Physical fatigue begins.
65°F	Optimum condition.
50°F	Physical stiffness of extremities begins.

$\begin{cases} 65° \text{ to } 75°F \\ 63° \text{ to } 71°F \end{cases}$ Summer comfort zone. $\Big\}$
           Winter comfort zone.

Humidities between 30 and 70 percent have been found comfortable by most people.

Fig. 43. Relationship between temperature and working efficiency.

rise in body temperature. The following five measurements must be obtained at the work site:

    a. Ambient dry bulb temperature.
    b. Ambient wet bulb temperature.
    c. Estimate of the mean radiant temperature of the solid surroundings.
    d. Air velocity.
    e. Bodily heat production.

These measures can be used with a set of nomographs prepared by McKarns and Brief (Heating, Piping and Air Conditioning, vol. 38) to predict the appropriate amount of rest breaks to provide the appropriate allowable exposure times and recovery times.

REST PERIODS. Because of the cumulative nature of fatigue, it is natural for workers to seek relief in pauses of various kinds. If there is incidental work to be done, that is, work that is legitimate but not directly productive, they will do this intermittently and sometimes prolong it.

The basic question is whether or not **formally authorized** rest periods are better than indiscriminate and unauthorized rest, and if so, what sort of rest program is best.

In general, most researchers agree that there is nothing lost by use of regular rest periods; however, the optimum frequency and distributions of the pauses has not been established.

In heavy, **muscular tasks**, efforts have been made to establish rest pauses based upon physiological data similar to the efforts of Hertig and Belding to determine heat stress allowances. The total rest period could be determined from the formula

$$R = \frac{T(K - 5.0)}{K - 1.5}$$

where R = rest period in minutes.
       T = total working time in minutes.
      K = energy expenditure for the task in Kcal/min.

Müller (Quarterly Journal of Experimental Physiology, vol. 38) uses this basic relationship to estimate the distribution of rest pauses in order to obtain optimum output and minimum fatigue. Brouha (Physiology in Industry) uses pulse rate measurements to determine the recommended levels of rest pauses.

The determination of rest pauses for **sedentary work** is more obscure than the procedure used for heavy tasks.

An interesting study of employee morale was conducted by Herzberg and his colleagues (The Motivation to Work). Various classes of employees were asked to "think of a time when they felt exceptionally bad about their jobs" and then asked to describe what happened and why they felt as they did. The factors which made people satisfied with their jobs were not the same (nor opposite) as those which made them dissatisfied. Apparently, the presence of so-called "satisfiers" would act to increase an individual's satisfaction, but their absence would not make him actively dissatisfied, only apathetic. Similarly, the presence of so-called dissatisfiers made people feel they had a "bad" job, but the absence of dissatisfiers did not make a "good" job.

The satisfiers were: achievement, recognition, the work itself, responsibility and advancement. These forms of satisfaction arise out of the job itself. The dissatisfiers were: interpersonal relations (both with one's superiors and peers), the technical ability of the supervisor, company policy and administration, physical working conditions, and the individual's personal life off the job. The dissatisfiers pertained to the environment within which the job is performed. Salary ranked both as a satisfier and a dissatisfier.

Herzberg suggests that the presence of satisfiers lead to higher productivity (and for this reason they are called "motivators"). Dissatisfiers, on the other hand, do not lead to lower production nor does their elimination tend to raise production. Their elimination may reduce active resistance to the job, but promote only a passive acceptance; therefore, dissatisfiers are called "hygienic factors" since they are used to avoid trouble.

MONOTONY AND BOREDOM OF THE JOB. Ryan (Work and Effort) distinguishes fatigue and boredom as follows: "Fatigue reduces the capacity for performance, while boredom reduces the effort level . . . it is a reduced level of motivation of the worker, involving a distaste for the work, a desire to cease work." He states: "The general problem is therefore twofold—to discover the job factors which tend to reduce the general level of interest of the workers in that occupation, and to determine the factors which make an individual especially susceptible to boredom in a given type of work."

Certain characteristics of modern industrial tasks may contribute to boredom, for example, the **repetitive performance** of simple standardized tasks providing little opportunity for varied and capacity performance of skills and development of aptitudes. However, these job characteristics have a different influence on the performance of different individuals. That is, the **susceptibility** to boredom is an individual characteristic.

Boredom is a personal experience and its existence must be verified from reports by the worker. However, the results of boredom may be seen in decreased productivity, increased variability of output, as well as overt manifestations of worker dissatisfaction, such as absenteeism and requests for job changes. These are not reliable indicators of boredom, however, because of the many factors involved in a particular situation.

Boredom and Repetitive Work. Wyatt and Langdon conducted a study for the British Industrial Health Research Board to determine the nature of boredom (Fatigue and Boredom in Repetitive Work). The workers included in the study were employed in four factories situated in widely different parts of the country. Results were obtained from 355 workers engaged in fairly simple forms of repetitive work. **Boredom assessments,** based on the replies to a number of carefully prepared questions, were obtained for each worker. The more important results are summarized as follows.

1. Boredom was a fairly common experience among the operatives included in the study. There were marked individual differences in susceptibility to boredom. Some workers were seldom free from boredom while others showed a definite preference for monotonous work.
2. The amount of boredom experienced seems to be related to (a) intelligence (b) inability to mechanize simple manual processes, (c) temperamental tendencies satisfied in active contact with external world (extrovert) rather than in phantasy (introvert), and (d) a desire for creative rather than repetitive work.
3. Boredom and discontent are related to the type of work. Slight differences between one job and another may have widely different effects on the worker. Efficiency and contentment may be increased by giving the beginner a short trial on different types of work resulting in assignment to the job which is liked best.
4. The most frequent causes of discontent (as revealed by the number of objections recorded to particular items on a given list) were waiting for work, faulty material, atmospheric conditions, fatigue, monotony, and noise. It seems that boredom increases sensitivity to objectionable features associated with conditions of work.
5. The proportion of workers who complained of fatigue was high (49 percent) but it is significant that only 3 percent referred the fatigue to the parts of the body actually used in the performance of work. The fatigue experienced was mainly static, or concerned with the maintenance of posture, rather than dynamic, or due to bodily movement.

EMOTIONAL FACTORS IN FATIGUE. Emotional strain may have a deleterious effect on the performance of an operator regardless of the type of work. Emotional stress may be traced to poor human relations within the plant or to situations without the plant, such as in the home. As pointed out by the Committee on Work in Industry of the National Research Council, "Social contacts in industry, often close and constant, often unavoidable and formed without choice, are accompanied by emotional strain which may not only interfere with effective collaboration but lead to a marked decline in individual efficiency, and even to incapacity for work." Regardless of the causes of emotional illness, whether they be traceable to relationships within or without the plant, management should make a concerted effort to alleviate them if possible. As the Committee notes, "Neither the direct observation of the worker on the job nor the interview directed to reveal working conditions is likely to be carried out by the doctor unless someone in daily contact with the worker suggests where the trouble may originate."

SECTION 13

MOTION AND METHODS STUDY

CONTENTS

MOTION AND METHODS STUDY

Nature and Scope

DEFINITIONS. The area of motion and methods study has many names, such as methods analysis, methods engineering, methods research, work study, and work analysis. This section covers the general material usually referred to by any one of these terms.

F. B. and L. M. Gilbreth (Primer of Scientific Management) defined **motion study** as "the science of eliminating wastefulness resulting from using unnecessary, ill-directed, and inefficient motions. The aim of motion study is to find and perpetuate the scheme of least-waste methods of labor." Another definition by the same authors (Applied Motion Study) indicates a procedure: "Motion study consists of dividing work into the most fundamental elements possible; studying these elements separately and in relation to one another; and from these studied elements when timed, building methods of least waste."

As practiced by the Gilbreths, **motion study** included an analysis of the flow and processing of material and paperwork and the movements of men as well as the study of the fundamental elements of each worker's job. However, their development of **micromotion study** and **therbligs** has in many cases indicated to others that motion study concerned only the study of activities within an operation. For the purposes of this section, motion study will be generally confined to the analysis of such individual operations. The concept of studying the whole is contained in the definition of methods study.

Methods study will be considered to be synonymous with such terms as work study, methods engineering, and methods analysis. **Methods study** is defined as a systematic procedure for the analysis of work; it is a critical analysis of the movements made by men, materials, and machines in performing any work. Because this definition includes the study of all facets of work and all factors affecting the work, motion study is considered a part of methods study. Henceforth, there will be no differentiation between these two terms, because motion study techniques will be integrated into the procedures for methods study. Work simplification, on the other hand, refers to the educational and group procedures which have been developed for studying the whole job or work situation and for improving its efficiency. Time and motion study techniques are often applied after work simplification has indicated that the work is necessary and can be improved.

Therefore, since the four areas—work measurement and time study, motion and methods study, work simplification, and process charts—are not entirely independent, an examination of the four sections dealing with these subjects should be carried on simultaneously, in order to attain the proper perspective. These four areas cover rather completely that phase of industrial engineering concerned with effective production and therefore their application contributes the bulk of the gain to the productivity and income of both employee and employer.

SCOPE OF MOTION AND METHODS STUDIES. Work is defined as any physical or mental activity. Motion and methods studies influence the overall work procedures to make the most efficient use of human effort. There is no work to which motion and methods studies cannot be applied. The tools and techniques of analysis and design are universal in application. The analysis o problems should not be looked upon as a series of techniques, but rather as a philosophy. All problems should be approached with the idea that improvements are possible.

Motion and method study analyzes the effect of work with respect to:

1. Raw material.
2. Product design.
3. Work order and work processes.
4. Work area, equipment, and machines.
5. Motion pattern of individual worker performing the work.

This analysis is then used to:

1. Eliminate all unnecessary work.
2. Provide the best order of work.
3. Provide the best design conditions for work.
4. Standardize the work.
5. Train workers in the standard method.

Modern industrial systems invest a good deal of money in equipment and training of personnel. Motion and methods study is concerned with making the most effective use of this investment by developing a smoothly operating man-machine system.

BASIC PROCEDURES. Motion and method study is the analysis of the work necessary to perform the job. The job may be an operation, a process, or an activity. It may involve a product or a service. Even though the use of motion and methods study has grown rapidly, the total range of applications is not fully appreciated. In general its **objectives** are:

1. Cost reduction of operations.
2. Increased effectiveness of operations.
3. Improvement of customer service.
4. Increased productivity by elimination of waste (time, energy, materials).
5. Development of a climate receptive to change.

Steps in Making Study. The following general approach is used in making a motion and methods study:

1. Document the present method, or gather and organize necessary data for a new method.
2. Question every aspect of the job, develop an effective sequence of doing the job.
3. Generate alternative approaches to the job, and evaluate the feasible alternatives.
4. Devise a new method with the following considerations.
   a. Selection of specific criteria that need to be optimized.
   b. Prediction of performance for each alternative. Performance being measured in terms of time, fatigue, monotony, effort, learning cost, and job satisfaction.
   c. Conversion of analysis into monetary terms.
   d. Weighing the effect of non-quantitative factors.
5. Prepare specification of proposed method.
6. Implement and follow-up proposed method.

Obstacles in Making Study. The two situations which are most often encountered are: the job currently exists or a new job must be designed. In both cases there is a great degree of interaction among the individuals involved.

Obstacles to the successful application of motion and methods study come from three groups of people.

1. People performing the job. A worker might have several different reactions when told the method for his work is being analyzed. He might resent outside criticism of analyst, or he may fear loss of job content, obsolescence, and economic insecurity.
2. People managing the job. A manager might have many of the reactions that workers have. He may feel a greater fear and resentment for outside help. There is also the problem of inertia and uncertainty.
3. People making the study. An analyst can be his own worst enemy. He must recognize that no matter how much technical merit his proposal may have, he has to convince and sell his ideas to all concerned. A tactless approach, inopportune timings, lack of clarity in presentation of proposal can forestall success. If he is overbearing and egocentric, his recommendations may be doomed to failure.

In a nutshell, it can be said that success of motion and methods analysis is a measure of the ability to understand relationships and effective communications. In general, the major obstacle is **resistance to change.** Management leadership in assuring job and economic security can help develop an atmosphere receptive to change and progress.

APPLICATIONS. Motion and methods study is a pervasive art and science. It can be applied to all industrial enterprises, past, present, or future. Jobs which have been done in the past and may recur can be studied in retrospect, and improvements can be made in procedures before the task reaches the plant again. Motion and methods study has been applied in department stores, on the farm, in the home, in process industries, in hospitals, and in assembly plants. It is very effective in correcting office procedure.

One of the important contributions of motion and methods study is to produce **methods consciousness** in the minds of all operators, supervisors, and managers in an industrial plant or a commercial enterprise. Such methods consciousness permits all personnel to analyze their work habits as they proceed with their daily assignments. While not all people are analytical enough by nature to make self-improvements, many are, and others may be able to make suggestions which will improve the work of those who are not analytically minded. Motion and methods study will pay for itself in the creation of motion-mindedness alone.

Effective communication and understanding of interacting relationships is a must for improvements and progress. Hence a **systems approach** is very essential and desirable. Because systems provide a framework for visualizing relationships, they permit an effective analysis of the change and its impact.

The classical use for motion study is to determine the "one best way" for performing a given task. This was Gilbreth's original principle, and it has been followed by all of those who have studied and used his methods. The term "one best way" indicates that there is probably no better way of performing a given job under given conditions. It does not presuppose, however, that this is the best way for all time, since one of the implications of modern motion and methods study is continual re-evaluation of what may now be the "one best way" of doing a task.

Analysis of Work

ASPECTS OF ANALYSIS. The basic question answered by motion and methods analysis is "How should the job be done?" This is the qualitative analysis which precedes the quantitative question answered by time study, "How much time should the job take when done according to the recommendations of the motion and methods analysis?"

There are several types of changes that could be recommended after a motion and methods analysis.

1. Change of product design.
2. Change of raw material.
3. Change of production sequence.
4. Change of tools used.
5. Change of motion pattern and movements of the operator.

Each of these changes may or may not influence any of the others. For example, a change of raw material may change product design; however, a change in the product design may not change the raw material.

The extent of change in any given situation will be determined by:

1. Volume of production.
2. Length of production run.
3. Cost of change
   a. retraining
   b. new equipment
   c. savings
4. Time available for making change.
5. Effect of change on organization policy and sales.

The above factors should be weighed in terms of both present and future demands.

DETERMINATION OF GOAL. In order to process a given problem with the techniques and principles of motion and method study, a specific set of objectives should be stated, and the limits of the problem defined. A methodical approach to this goal determination could be the following.

General Goal. The management objectives for the plant, area, or department in which motion and methods activities are to take place should be explicitly recognized. In most cases the analyst will be given the general goal for the particular department or area, although sometimes he must, himself, determine the goal, and he will find that it varies from area to area. If there is no other general goal, the analyst can, at least, use the standard goals of **cost reduction** and **increased productivity.**

Other general goals might be: to reduce the cost for a specific product or department; to design the best layout to produce a product for an anticipated volume; to reduce the scrap; to make the work easier; to reduce indirect labor costs; to increase productivity; to change the design of the product for lower cost; or to reduce the cost of material handling. The general goal will remain the same for the area or department until such time as it has been met satisfactorily.

Specific Problem. The most difficult part of goal determination is the selection of the first work activity to study. It is usually essential to limit the initial problem to be worked on to as small a unit of the **work sequence** as possible. Generally, the specific problem will be one operation. In some cases, it may not be possible to limit the study to one operation. Then the specific problem should be the smallest possible **group of operations.** The smaller the problem unit, the more readily the savings may be computed and the faster a solution may be found.

Although the overall problem is broken down into small units in this step, the procedure of gathering information in the analysis of the work will help relate the small problem to the whole situation. In this way, the whole work activity ill remain in perspective while one problem is worked on. After the initial problem is solved, it is necessary to return to this part and select another problem. In this way, all the problems (operations) in the area or department can be solved (proper methods design) to meet the general goal.

Specific Goal. Types of specific goal statements are:

Eliminate time spent in obtaining materials and tools.
Eliminate the bottleneck holding up an operation.
Revise an operation method where an operator looks uncomfortable at his work.
Rearrange a disorganized workplace.
Provide proper equipment for an operation.
Eliminate some make-ready time.
Eliminate some put-away time.
Reduce the effort required.

The specific goal will frequently include **limiting statements** concerning other conditions in the area. Thus, volume has an important effect and floor space limitations are important.

General Approach. Some additional preparations must be made for reaching the specific goal for the specific problem. In some cases, a **preliminary survey** is required to decide whether or not the operation is necessary. This question is one of the most important in all parts of the motion and methods study procedure. If the operation is not necessary and that fact can be determined at this point, much analysis time and effort will be saved.

If the operation is necessary, the preliminary survey determines the approach to the problem and the **economic limits** within which the analyst must work. Typical questions to be answered are:

What authority should be established before the project is worked on?
How should the workers be approached?
How should the supervisory or management staff be approached?
How much time is available to solve this problem?
What are the possible savings?
What amount of money can be spent to meet the goal?
What effect will quality requirements for the operation have on the problem?

With the answers to these questions, the analyst can plan the type of analysis to use and the amount of detail for all the following steps of the motion and methods study procedure. The extent of the analysis to be made must be determined, usually on economic grounds.

GATHERING INFORMATION. To make an analysis, data must be available. Techniques presented later will show how the data can be organized to obtain the most useful information. However, first it is necessary to determine the means whereby the information can be collected. Each procedure for obtaining information is not exclusive of the others. Frequently, **combinations** of all the techniques may be used for any particular problem.

Observation. This is the most common procedure used for gathering information. In most cases it is necessary for the investigator to trace and verify every step in the process when analyzing present work. This cannot usually be done by sitting at a desk and consulting route sheets in an office. The analyst must make actual observations where the work is occurring and frequently combine this procedure with discussion.

A good analyst must have the ability to notice motions and activities which are usually unobserved by the layman. This is part of the **methods consciousness** required.

Discussion. Discussion with those who do or supervise the work can frequently provide information not obtainable by observation. The discussion procedure is used most widely where special or **irregular work** is performed Likewise, discussion is valuable when trying to analyze **past work** in order to improve efficiency in performing future work.

Discussion also helps the analyst make contact with people. Discussion is frequently used even where observation by itself may accomplish the purpose of gathering data. This can help to develop good human relations.

Records. Even if the work being studied was never done before, much valuable information can be obtained from records of all sorts. These include production, costs, time, inventory, and invoice records. Sampling or analysis of records of this type can save much time in predicting or emphasizing future activities.

Motion Pictures. Taking motion pictures represents one of the most detailed and accurate procedures for gathering information. It is an excellent technique for communication to all levels of an organization. Motion pictures can focus attention at any point or motion in an operation. They help "sell" improved methods. For some work, motion pictures are the only way of getting a good analysis. Work which requires **large crews** is almost impossible to analyze in any other way. Work which requires **care** and **dexterity** is analyzed much better with motion pictures. There are many advantages of motion pictures in addition to the basic purpose of gathering information.

The general **advantages** of motion pictures in motion and methods study are:

1. Permits great detail.
2. Permits review of details in quiet surroundings.
3. Convenient for study.
4. Enlists cooperation of all concerned.
5. Accurate record of times.
6. Good training aid.
7. Positive record.
8. Permits evaluation of methods changes before the change.
9. Accurate portrayal of simultaneity.

Motion pictures can be taken at various speeds. They can be run forward and backward to obtain the exact relationships among work activities. Proper amount of detail is determined by the speed. The **speeds** of motion pictures are measured in frames per second (FPS) or frames per minute (FPM). Normal speeds are 16 FPS or 1,000 FPM. Speeds below or above these have special uses. Studies at speeds slower than the usual 16 FPS are frequently called memomotion studies (usually taken at 1 FPS or 100 FPM), and studies at normal and frequently higher speeds are sometimes called micromotion studies.

Micromotion Study. Micromotion study was developed by the Gilbreths. At the same time, they formulated the concept of therbligs which divide work into fine detail. These therbligs are described later as used in a technique for analysis of work. Micromotion study employs motion pictures with a timing device to obtain times for each motion or therblig. Generally, micromotion studies are made with 16 FPS or 1,000 FPM motion pictures. However, it is possible to use higher speed if required. In most cases, higher speeds, like 64 FPS or 4,000 FPM, are used for research in motion study and not so much for industrial activities. Because motion pictures with a timing device can be used to make many different types of analysis, the general usage of this term is now applied to gathering information with motion pictures at 16 FPS or more.

Memomotion Study. This study requires motion pictures at usually 1 FPS or 100 FPM, but if slightly more detail is needed the pictures could be taken

t 8 FPS or 500 FPM. The detail is less than in micromotion study. Memomo-
ion study is generally for analyzing **longer cycles of work,** and for analyzing
work which requires three, four, or more people **working coordinately.** Uses of
memomotion study will be discussed later under the appropriate technique.

PRESENTING INFORMATION. After the data have been collected,
hey need to be put together in an organized manner for the purpose of analysis,
esign, improvement, and study. A graphical presentation or chart is one of the
most effective ways of doing this. Such charts are also good for "selling" the
deas, and for information to all concerned. These charts include a logical
escription of events and activities as they occur, diagrammatic sketches, sym-
ols, and standard motion pattern descriptions.

In the section on Process Charts several types of such charts are illustrated
nd discussed. The therblig and simo charts, which represent a specialized appli-
ation of motion and methods study analysis are discussed in this section. Both
f these charts are primarily concerned with a workplace activity and the motion
attern of the operator.

THERBLIG CHART. The therblig chart may be defined as a detailed
ystematic analysis of the work performed by the body members of a man,
usually when the work is performed at one workplace. This chart provides the
reatest amount of qualitative detail available. Although mainly devised for
wo-hand analysis, it is usable for the analysis of any type of activity.

Uses. Because the therblig breakdown provides a fine detail analysis of
ctivity it is most often used when there is a large volume of work. If the
ands and fingers perform relatively skilled motions, it may be difficult to
nalyze the work completely with other techniques which provide less descriptive
etail.

The therblig chart is very useful in planning work for future operations. With
uch a chart, it is easier to analyze different methods before the work is done.
This approach also develops an overall method consciousness.

Definition of Therbligs. The Gilbreths in their quest for methods improvement established a system in which all work is composed of seventeen basic
motions. To these basic motions the Gilbreths assigned the term "therblig"
Gilbreth spelled backwards—with the last two letters transposed).

It is recognized that not all basic motions are performed in every job; some
nvolve only a few while other jobs may involve many. Furthermore, the manner
n which the basic motions are combined will vary greatly from job to job. It is
his variation of combinations of basic motions which should be kept in mind
when defining each job motion.

The definition of a basic motion consists of three parts:

1. When the motion begins.
2. The nature of the motion.
3. When the motion ends.

If one can define, in his own words, each basic motion in terms of the hands
o that the definition is always applicable, then one can readily and more
horoughly analyze a job.

Barnes (Motion and Time Study) lists the definitions of the seventeen
herbligs as follows (one of Gilbreth's original therbligs (find) is omitted be-
ause it is really the end point of another therblig and actually represents a
mental reaction rather than a physical motion):

1. *Search* (Sh): that part of the cycle during which the eyes or the hands are
unting or groping for the object. Search begins when the eyes or hands begin
o hunt for the object, and ends when the object has been found.

2. *Select* (St): the choice of one object from among several. In many cases is difficult if not impossible to determine where the boundaries lie between searc and select. For this reason it is often the practice to combine them, referring t both as the one therblig *select*.

Using this broader definition, select then refers to the hunting and locating of one object from among several. Select begins when the eyes or hands begin t hunt for the object, and ends when the desired object has been located.

EXAMPLE. Locating a particular pencil in a box containing pencils, pens, an miscellaneous articles.

3. *Grasp* (G): taking hold of an object, closing the fingers around it preparator to picking it up, holding it or manipulating it. Grasp begins when the hand of fingers first make contact with the object, and ends when the hand has obtaine control of it.

EXAMPLE. Closing the fingers around the pen on the desk.

4. *Transport empty* (TE): moving the empty hand in reaching for an object. I is assumed that the hand moves without resistance toward or away from the objec Transport empty begins when the hand begins to move without load or resistance and ends when the hand stops moving.

EXAMPLE. Moving the empty hand to grasp a pen on the desk.

5. *Transport loaded* (TL): moving an object from one place to another. Th object may be carried in the hands or fingers, or it may be moved from one plac to another by sliding, dragging, or pushing it along. Transport loaded also refers t moving the empty hand against resistance. Transport loaded begins when the han begins to move an object or encounter resistance, and ends when the hand stop moving.

EXAMPLE. Carrying the pen from the desk set to the letter to be signed.

6. *Hold* (H): retention of an object after it has been grasped, no movement of the object taking place. Hold begins when the movement of the object stops, an ends with the start of the next therblig.

EXAMPLE. Holding bolt in one hand while assembling a washer onto it with th other.

7. *Release load* (RL): letting go of the object. Release load begins when th object starts to leave the hand, and ends when the object has been completel separated from the hand or fingers.

EXAMPLE. Letting go of the pen after it has been placed on the desk.

8. *Position* (P): turning or locating an object in such a way that it will b properly oriented to fit into the location for which it is intended. It is possible t position an object during the motion *transport loaded*. The carpenter, for ex ample, may turn the nail into position for using while he is carrying it to the boar into which it will be driven. Position begins when the hand begins to turn of locate the object, and ends when the object has been placed in the desired positio or location.

EXAMPLE. Lining up a door key preparatory to inserting it in the keyhole.

9. *Pre-position* (PP): locating an object in a predetermined place, or locating i in the correct position for some subsequent motion. Pre-position is the same a *position* except that the object is located in the approximate position that wi be needed later. Usually a holder, bracket, or special container of some kind hold the object in a way that permits it to be grasped easily in the position in which i will be used. Pre-position is the abbreviated term used for *pre-position for th next operation*.

EXAMPLE. Locating or lining up the pen above the desk-set holder before re leasing it. The pen may then be grasped in approximately the correct position fo writing. This eliminates the therblig position that would be required to turn th pen to the correct writing position if it were resting flat on the desk when graspec

10. *Inspect* (I): examining an object to determine whether or not it complies with standard size, shape, color, or other qualities previously determined. The inspection may employ sight, hearing, touch, odor, or taste. Inspect is predominantly mental reaction and may occur simultaneously with other therbligs. Inspect begins when the eyes or other parts of the body begin to examine the object, and ends when the examination has been completed.

EXAMPLE. Visual examination of pearl buttons in the final sorting operation.

11. *Assemble* (A): placing one object into or on another object with which it becomes an integral part. Assemble begins as the hand starts to move the part into its place in the assembly, and ends when the hand has completed the assembly.

EXAMPLE. Placing cap on mechanical pencil.

12. *Disassemble* (DA): separating one object from another object of which it is an integral part. Disassemble begins when the hand starts to remove one part from the assembly, and ends when the hand has separated the part completely from the remainder of the assembly.

EXAMPLE. Removing cap from mechanical pencil.

13. *Use* (U): manipulating a tool, device, or piece of apparatus for the purpose for which it was intended. Use may refer to an almost infinite number of particular uses. It represents the motion for which the preceding motions have been more or less preparatory and for which the ones that follow are supplementary. Use begins when the hand starts to manipulate the tool or device, and ends when the hand ceases the application.

EXAMPLE. Writing one's signature in signing a letter (use pen), or painting an object with spray gun (use spray gun).

14. *Unavoidable delay* (UD): a delay beyond the control of the operator. Unavoidable delay may result from either of the following causes: (*a*) a failure or interruption in the process; (*b*) an arrangement of the operation that prevents one part of the body from working while other body members are busy. Unavoidable delay begins when the hand stops its activity, and ends when activity is resumed.

EXAMPLE. If the left hand made a long transport motion to the left and the right hand simultaneously made a very short transport motion to the right, an unavoidable delay would occur at the end of the right-hand transport in order to bring the two hands into balance.

15. *Avoidable delay* (AD): any delay of the operator for which he is responsible and over which he has control. It refers to delays which the operator may avoid if he wishes. Avoidable delay begins when the prescribed sequence of motions is interrupted, and ends when the standard work method is resumed.

EXAMPLE. The operator stops all hand motions.

16. *Plan* (Pn): a mental reaction which precedes the physical movement, that is, deciding how to proceed with the job. Plan begins at the point where the operator begins to work out the next step of the operation, and ends when the procedure to be followed has been determined.

EXAMPLE. An operator assembling a complex mechanism, deciding which part should be assembled next.

17. *Rest for overcoming fatigue* (R): a fatigue or delay factor or allowance provided to permit the worker to recover from the fatigue incurred by his work. Rest begins when the operator stops working, and ends when work is resumed.

Data for a Therblig Chart. The data for preparing a therblig chart come from visualizing the motion pattern of a new job or by observing an existing job either directly or through films. It is important to have some quantitative measures of performance, since the degree of detail in qualitative descriptions (the therbligs) is so small. Without a quantitative measure, the therblig chart would be simply an operation chart (or left- and right-hand chart) with motion patterns described in terms of therbligs (see section on Process Charts).

SIMO CHARTS. The simo chart (short for the simultaneous motion cycle chart, developed by the Gilbreths) is really a therblig time chart. The simo chart is the detailed symbolic, systematic, time presentation of the method of work, as recorded by motion pictures, performed by the body members of worker usually performing his job at one workplace. The simo chart is not made often, and the volume of work must be rather large before it is worth while. However, it is an excellent **training device** because it forces the person to record each detail of the motions and accompanying times for both hands. The simo chart was the graphical presentation developed by the Gilbreths when they originated the therbligs. Therbligs and simo charts are generally considered a part of **micromotion study.**

Types of Simo Charts. The simo chart may be made for hand and/or body members, or for fingers. The **finger simo chart** gives much more detail than the hand simo chart. Ordinarily, the **hand simo chart** is made with fine detail therbligs, and the finger simo chart with basic therbligs.

Making a Simo Chart. The simo chart could be made directly from the film without making a record of the analysis of the activity. However, this is more difficult than following the usual procedure. The time values and the type of therblig are usually determined in an intermediate step on the form shown in Fig. 1. On this **film analysis sheet** the analyst records, in sequence for one hand, the therbligs and the clock time of the frame in which the therblig first appears (or projector counter if no clock is in the picture). After one hand has been charted, the other hand's activity would be charted. Subtraction is used to obtain the time for a given therblig. Making the simo chart from the film analysis becomes a simple step for a clerk. For almost all cases a **workplace diagram** should be made with a simo chart. A **summary** is sometimes difficult to make. If a clear presentation of the number of symbols can be made, a summary is helpful.

Part of a regular fine detail simo chart is shown in Fig. 2 (page 12), based on the film analysis shown in Fig. 1.

Although therbligs provide fine detail, an analyst soon learns even therbligs are inadequate. This presents a problem, because trying to make finer divisions makes therbligs less useful. And unless fineness is further developed, it is frequently not possible to obtain sufficient detail. Research is continuing to try to find solutions to this problem.

When new work methods are to be developed micromotion analysis is not possible since films of the method do not exist. In such cases predetermined time systems are used (see section on Work Measurement and Time Study).

NONREPETITIVE WORK ANALYSIS. All the techniques presented to this point are used mainly for analyzing repetitive or slightly varied cycle work. However much work in industry is of the nonrepetitive type. Motion and methods study personnel must, through new techniques, take into consideration an analysis of nonrepetitive work if a complete job is to be done. Some of the **advantages** to be found in performing such analyses are:

1. Activities to which work simplification can be applied are pinpointed.
2. Information for action is provided to supervisors and management.
3. Facts are provided for discussions with workers. Frequently opinion has been the major basis for discussions.
4. Areas where training is required are pinpointed.
5. A basis for evaluating new programs can be developed.

There are four techniques of analysis in general use, and many variations upon them to suit the tasks at hand.

FILM ANALYSIS DATA

Operation _Cap Assembly on Mailing Tube_ Page _1_ of _1_

Operation No. _30_ Dept. _Mailing_ Specif. No. _____

Parts _Tube, Cap, Glue_ Part No's. _____ Draw. No. _63_

Machine _____ Mach. No. _____ Fixt. No. _____

Operator & No. _D. Jones #3809_ Film No. _112_ Film Date _Oct._

Analysis by _R. Nanda_ Analysis Date _Nov. 1_ Time Unit _1/1000 min._

Film count	Elapse time	Motion	Description Left Hand	Film count	Elapse time	Motion	Description Right Hand	Film count	Elapse time	Motion	Notes
0	8	TE	To box of tubes	0	9	TE	To glue pot				
8	6	G	Tube	9	5	G	Glue Brush				
14	12	TL	To work area	14	12	TL	To work area				
26	40	H	Hold Tube	26	40	U	Apply glue				
66	6	TL	Turn Tube	66	6	UD	Wait				
72	40	H	Hold Tube	72	40	U	Apply Glue				
112	44	UD	Wait	112	10	TL	Brush to Glue Pot				
156	1	RL	Tube	122	2	P	Brush in pot				
157	10	UD	Idle	124	1	RL	Brush				
167			To Box of Tubes	125	8	TE	To box of caps				
				133	4	G	Cap				
				137	10	TL	To work area				
				147	2	P	Cap to tube				
				149	3	A	Cap on Tube				
				152	1	RL	Cap				
				153	4	G	Tube				
				157	9	TL	To mailing box				
				166	1	RL	Tube.				
				167		TE	To Glue Pot				

Fig. 1. Film analysis sheet.

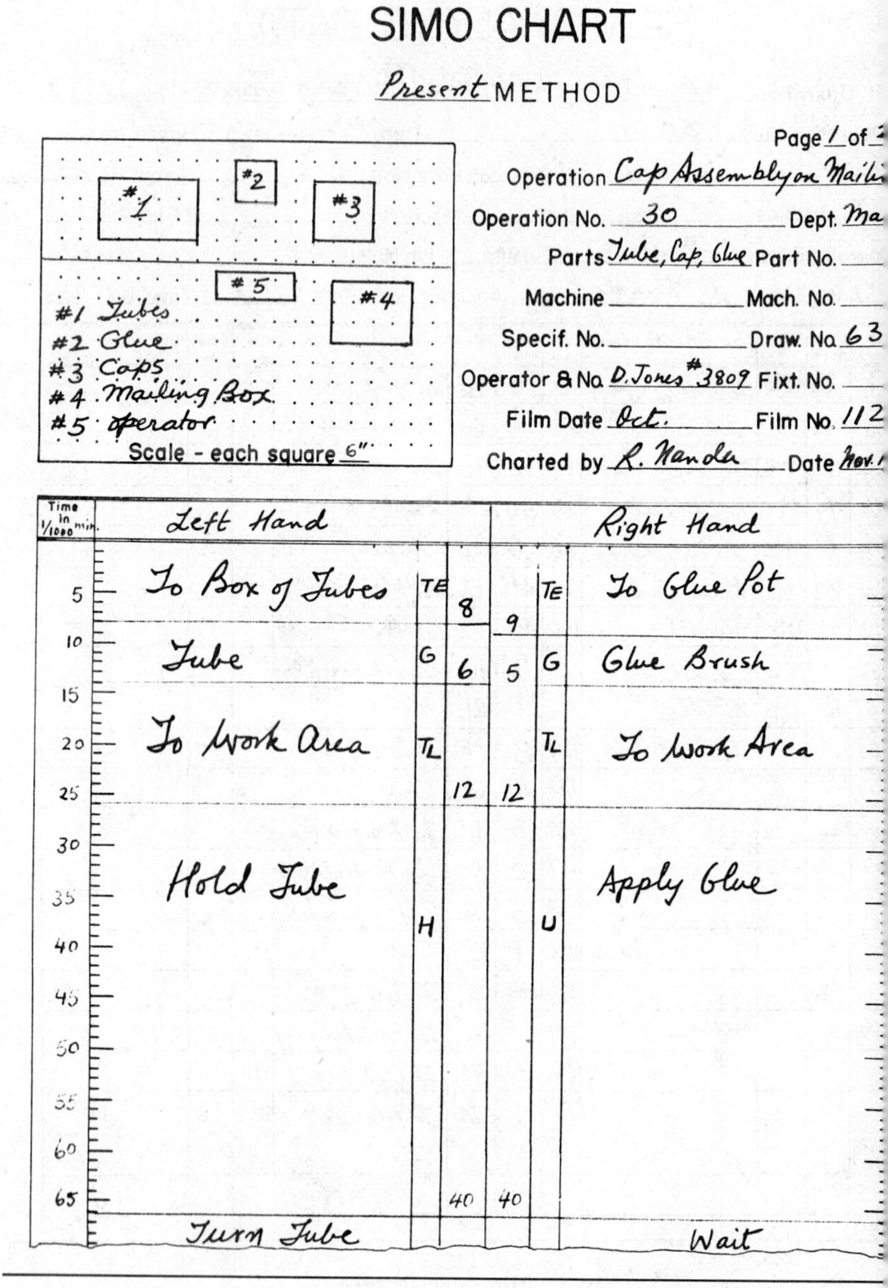

SIMO CHART

Present METHOD

Page _1_ of _2_

Operation _Cap Assembly on Maili..._

Operation No. _30_   Dept. _Ma_

Parts _Tube, Cap, Glue_ Part No. ____

Machine ____ Mach. No. ____

Specif. No. ____ Draw. No. _63_

Operator & No. _D. Jones #3809_ Fixt. No. ____

Film Date _Oct._ Film No. _112_

Charted by _R. Nandu_ Date _Nov. 1_

#1 Tubes
#2 Glue
#3 Caps
#4 Mailing Box
#5 operator

Scale - each square _6"_

Time in 1/1000 min.	Left Hand				Right Hand
5	To Box of Tubes	TE	8	9 TE	To Glue Pot
10	Tube	G	6	5 G	Glue Brush
15					
20	To Work Area	TL		TL	To Work Area
25			12	12	
30					
35	Hold Tube				Apply Glue
40		H		U	
45					
50					
55					
60					
65			40	40	
	Turn Tube				Wait

Fig. 2. Partial simo chart based on Fig. 1.

Observer with Worker. In this procedure, an analyst is assigned to a given perator or group, and he follows the workers through all activities. The nalyst records the time and activity for every **work situation** in the day. The ecord shows what was done and how much time was spent doing it.

This procedure is used mainly when people go over many routes and cover reat distances in performing their work. The **advantages** of this technique are hat it gives information not otherwise obtainable. However, there are several lisadvantages. Ordinarily, the analyst cannot spend too many days following he worker. This becomes too expensive. Therefore, the information gathered is ot usually **representative** of all the individual worker might do. This procedure oes not result in completely accurate data, because the analyst may be an utside influence on the worker. Only one worker or one small group is followed y this technique, whereas many people or groups may be performing the same ctivity. A question arises about the similarity between the one person or group nd the others doing the same work.

Because the procedure is simple, it has been and will be used frequently in nalyzing nonrepetitive work. However, it must be judged in relation to the ther techniques to determine its usefulness.

Memomotion Study. This procedure is closely related to the one above. Iowever, the motion picture camera is virtually substituted for the analyst. The film provides the same type of information as gathered above. Motion ictures provide an excellent record of what has gone on and good information bout time. If the photographer is the analyst, then analyzing the film will be asier; 1 FPS (100 FPM) pictures can be shown again and again to review what appened. The technique has certain limitations. A photographer must stay vith the camera, and there is the additional cost of analyzing the film later. This technique is not usable unless the work is restricted to a general area vhich can be covered by a motion picture camera. If the person moves from lace to place, it would be difficult to use. Although there is a restriction on the eriod of time covered by motion pictures, it is not so severe as the first tech-ique above. It is possible to use the camera as a **sampling device** to obtain nformation from different periods of time.

Work Sampling. The work sampling study determines the percentage of time pent by an individual or a group of individuals in various activities by means f random observations. In effect, the analyst samples the activity of the in-lividual to obtain the information for the study.

The work sampling study has many **advantages** that the other techniques do ot have. It can be made for a short period of time or for as long as desired. t is simple and can be made by almost anyone. It also results in fairly accurate nformation. Work sampling also permits the analyst to do other work while he s making the study. And work sampling can be made on a large group of eople doing similar nonrepetitive work as well as on one person or crew. (The rocedure for making work sampling studies is described in detail in the section n Work Measurement and Time Study.)

Record Keeping by Worker. Frequently, work is performed by people who nove not only throughout the whole plant but from plant to plant, area to area, nd even city to city. The easiest way to handle this problem is to provide an nalysis form with a list of activities down the left and the times or periods luring the day across the top so the man can check off his activities throughout he day at, say 15-minute intervals. It is unlikely that great accuracy will be obtained this way, but at least some valuable information will be available abou the type and frequency of activities the man performs.

Techniques and Equipment of Work Study

MOTION PICTURES. Since motion pictures are such an importar method of gathering information for motion and method study the analyst shoul be familiar with all phases of their use. The information given here abou motion pictures refers to 16 mm. film and equipment. Home movies are fre quently 8 mm. (or Super 8). Commercial motion picture theaters usually hav 35 mm. film. The latter is too expensive in industry or business, and 16 mn film is usually used instead of 8 mm. because it gives a larger picture. Most ir dustrial rental film is 16 mm.; sound pictures, and a wide range of movie ac cessories and equipment are available in 16 mm.

Cameras and Lighting. Most 16 mm. cameras are equipped to take picture at different speeds, ranging from 8 FPS to 64 FPS. These cameras usually hav an electric battery operated motor. The motor speed is accurately regulatec With electric motor drives the normal running speed can be adjusted to 100 FPM instead of 16 FPS, permitting a decimal (0.001 minute) time interva between picture frames. Standard film magazines on cameras are of 100 foo capacity; however, interchangeable film magazines up to a capacity of 40 feet are also available as accessories with some cameras.

The lens is the most important factor in determining picture quality. Th camera should be capable of taking **interchangeable lenses** with focusing mounts There should be at least a wide angle lens of 15 mm. focal length for takin pictures in cramped quarters and a 1-in. lens for normal conditions. The 2-in. o larger telephoto lens is convenient for getting close-up views. If a camera is t be used without auxiliary floodlighting, the lens should be "fast," that is, witl maximum **diaphragm openings** of f. 1.9 or f. 1.5. On the other hand, if ade quate auxiliary illumination can be provided, a less expensive, slower lens (smal openings), f. 3.5, can be used. Also, when pictures are taken with smalle diaphragm openings, the **depth of focus** is increased, which means that th range over which objects will be in sharp focus is increased. This factor i especially important considering the relatively short range of speeds and lightin, within which most motion pictures are taken. The **view finder** should show th exact field of the lens being used. If an eye-level finder is on the camera, thi finder should be adjustable for all focus distances, to compensate for the parallaɔ between the optical axes of the lens and that of the view finder.

For motion study purposes sharp pictures are necessary, and for this purpos the available sensitive or "fast" **films** are needed, sometimes aided by **auxiliary lighting.** Color pictures are desirable for "selling," but the cost is usually ex cessive for general purposes. Color films are also slower than black and white films. If auxiliary lighting is needed, **light stands** may be necessary. Reflector are placed on top of the stand to floodlight the area.

Timing Equipment. Some sort of **timing device** should be used with motior pictures to obtain a more complete analysis. In most cases the camera is suffi ciently accurate to be used as a timing device. For other situations a separat timing device should be included. When taking pictures at 1 FPS or 100 FPM a special **drive mechanism**, sometimes included on motor drives, is needed. A timing device in the picture is not needed for the usual memomotion study situation.

Auxiliary Equipment. The following additional equipment or facilities are recommended for making motion pictures and analyzing them for motion and methods study.

1. Camera tripod stand.
2. An exposure meter for determining lens settings.
3. A measuring tape—this is especially critical for close-ups.
4. Camera drive for memomotion analysis.
5. Camera drives for speeds other than normal for micromotion analysis, e.g. 64 FPS or 4000 FPM.
6. A movie projector with a frame counter and remote controls for stopping and running at speeds from 0 to 1500 FPM.
7. A screen for projecting films.
8. An analysis room or booth equipped with desk or writing space and auxiliary lighting.
9. Film analysis forms (see Fig. 1).
10. Film editing and splicing unit.
11. Film storage cabinets.
12. Auxiliary lenses and filters for cameras.

Making Motion Pictures. Taking motion pictures is a simple procedure. Some common errors to be avoided are insufficient lighting, background color too similar to the subject, making panoramic moves too fast, poor camera angle, improper distance, and wrong focus. The procedure to follow in taking motion pictures for motion and methods study is:

1. **Obtain the permission** of the foreman or supervisor and union steward, if any, and the cooperation of the operators. When the people involved are informed of what is going on they are more apt to give their cooperation.

2. **Prepare all the equipment for use.** Load the camera, make certain the photofloods work, the tripod is ready, etc.

3. **Locate all the equipment in the proper places.** The camera should be placed to obtain desired detail. Usually the camera is placed in front of and somewhat above the subject to be photographed. Sometimes a shot over the shoulder of the operator is desirable. Lights should be placed to illuminate the work area properly. They should not be pointed into the camera lens or at the operator and must not be too far away from the workplace.

4. **Decide on the camera speed.** Figure 3 adapted from Nadler (Motion and Time Study) summarizes the uses of motion pictures related to camera speeds.

5. **Make all necessary camera adjustments.** Camera speed, aperture opening, and distance from the workplace must be set. The aperture opening is determined by the amount of light on the work area and the film speed. The exposure meter is used to determine the amount of light on the subject. The camera should be placed to obtain maximum detail of the work itself, avoiding general background views.

6. **Keep records of what is done.** A motion picture record form used to record data during a "shooting" is shown in Fig. 4. Full information is thus available to future photographers.

7. **Make film analysis.** To collect the information, the analyst should record the data on a film analysis sheet (see Fig. 1). The sheet has columns for the clock reading (or frame count), subtracted time, and description of the activities of each hand.

Limitations. The analyst must use his judgment concerning the **adaptability** of motion pictures to a specific situation. Motion pictures take time to be developed, and if information is needed rapidly, this may cause difficulties. Fre-

Camera Speeds

Standard (FPS)	Decimal (FPM)	Situations	Special Advantages
1	100	1. Three or more in a coordinate crew. 2. Nonrepetitive work in one area. 3. Long-cycle repetitive work. 4. Gross body motion operations. 5. Need for times in work load determination (for operations with characteristics of 1 through 4 above).	1. Low film cost. 2. Extended time coverage
8, 10	500	1. Medium (0.75 to 2 min.) cycles. 2. Short-cycle nonrepetitive work. 3. Need for times in work load determination.	1. Low film cost. 2. More cycles than above 3. Projects as relatively continuous motions.
16	1,000	1. Short-cycle work. 2. Need for measurement in work load determination. 3. Skilled operation. 4. Transfer operation to another plant. 5. Comparison of methods. 6. Finger or hand operations. 7. Evaluation of equipment.	1. Normal projection speed. 2. Can be used as training film.
24, 32, 64	1,500, 2,000, 4,000	1. Research. 2. Especially skilled or complex operation. 3. Evaluation of complex equipment usage.	1. Fine detail 2. Complex and skilled operations can be stopped and viewed one frame at a time. 3. Can be used for training in skills.

Fig. 3. Uses and advantages of motion pictures.

quently, the **time lag** between the exposure and development of film can be reduced by using more rapid means of transportation to processors. Films require added costs of analysis. There is also the additional time for the photographer and the attendant cost, regardless of the speed at which the film is taken. Although 1 FPS motion pictures (memomotion study) are a **film-cost saver,** it is still necessary to take the pictures and then analyze them, both of which add cost to the analysis. A more detailed analysis of film will be more costly than a less detailed analysis. Motion pictures are not a "cure-all" for motion or method study problems. Figure 3 should be used as a guide for determining when motion pictures should be taken, in conjunction with a full consideration of the situation (human relations, technical difficulties, etc.) and the disadvantages and economics of motion pictures.

MISCELLANEOUS TECHNIQUES AND EQUIPMENT. Video Tape Recorders. Closed circuit television used in conjunction with video tape recorders offer several advantages over motion pictures as a means of gathering

MOTION PICTURE RECORD

Page_____of_____

Operation_____Date_____

Operation No._____Dept._____

Parts_____ Part Nos. _____ Dwg. No._____

Machine_____ Mach. No._____

Operator & No._____

Photographer_____ Assisted by _____

Equipment:

Camera type & No._____Lens _f_____F.L._____

Camera drive type & No._____

Exposure meter type & No._____

Lighting type & No._____

_____

_____

Film type_____ B & W_____Color_____

    Exposure index_____Reel length_____

Other equipment_____

_____

Lens opening_____Camera speed _____

Distance to subject _____

Show relationship of operator, workplace, camera, lights, timing device, etc. Indicate scale

   Symbols: ☐→ camera   (→ light   Ⓣ timing device   ○ operator

Scale =

Fig. 4. Form for recording filming data.

information. They are as versatile as motion pictures in almost all aspects. The main difference perhaps is in the initial cost of equipment. When the overall systems are considered video tape recorders become competitive in price to motion picture equipment. Some of the attractive features of video tape recorders are as follows:

1. The recording tape can be reused.
2. Instant playback is possible. No time needed to develop film as with motion pictures.
3. Electronic mixing permits simultaneous recording of close up and wide angle views on one frame, or recording of two different angles simultaneously on one frame.
4. Relatively less illumination necessary.
5. Simultaneous audio recording and commentary is relatively easy.

Chronocyclegraph. This technique was developed by the Gilbreths. A light is attached to the middle finger of each hand. A still camera is used. A relay in the circuit of the lights flashes them on and off, with a variable amount of electricity while the light is on. This forms a **pear-shaped dot** on the exposed film. In this way, a record of the motion and its relative speed is obtained. This is more a "selling" device than an effective tool for motion analysis. It is primarily used to make presentations in which the "before" and "after" relationships are dramatized.

The **cyclegraph** is almost the same as the chronocyclegraph except the lights do not flash. Only continuous white lights (or lines) are recorded on the film.

Predetermined Time Systems. A general extension of the therblig concept was predetermined time systems. These systems contain time values for basic motions; hence, it is possible to plan and compare methods designs by predicting the performance times. These systems provide useful information for both analyzing existing methods and developing new ones. They are treated in greater detail in the section on Work Measurement and Time Study.

Methods Engineering

GUIDELINES FOR IMPROVING METHODS. Through the years every field tends to build up a set of principles and concepts which can be used as guides for designing or improving activities in the area and this is true of motion and methods study. Principles have been developed through applications, research, and discussions. Of course, every principle is not applicable to every problem.

A questioning attitude should be developed by all involved in methods analysis. Questions pertaining to material, design, sequence, equipment, and method should be raised. Too often the methods analyst is brought into the picture only after problems arise, often after the product has been designed and manufactured or purchased. In such cases the methods analyst is not totally effective; he can only develop or improve the motion pattern of the operator. In a proper systems approach the methods analyst should be involved in all decisions regarding the activities to meet the goal.

A checklist of the factors involved in any methods study is given in Fig. 5 (pages 20–22). Although the list is by no means complete it does provide guidelines along the major areas to be analyzed.

PRINCIPLES OF MOTION ECONOMY. The rules of motion economy and efficiency which referred to hand motions of operators were developed by Gilbreth. From time to time other investigators have added to the list. The principles of motion economy are today divided into three groups:

1. Effective use of the operator.
2. Arrangement of workplace.
3. Tools and equipment.

Figure 6 (page 23) is a table of twenty-two principles as developed by Barnes (Motion and Time Study). Some of the important principles are discussed in subsequent paragraphs.

Simultaneous and Symmetrical Hand Patterns. The first three principles refer directly to the symmetrical and simultaneous motion pattern for the operator's hands. This symmetrical pattern will generally cause less fatigue to the operator, because the hands and arms balance one another. Due to some increased demand on coordination, simultaneous operations will take more time than one-handed operations. However, the time increase is not double the time of one-handed work although twice the production is obtained. Thus the application of these principles will not only produce less fatigue and mental effort but will increase productivity.

Normal Work Area. Principle 11 in effect defines a normal work area. The operator works in areas bounded by the limits of his arms and hands. **Normal spheres** can be generated by rotating arms around center points represented by the elbow and shoulder joints. There is also a **maximum working area** defined by the arcs made by the hands when the arm is outstretched and pivoted at the shoulder. Layouts should be designed for normal rather than maximum work areas. Figure 7 (General Motors Engineering Journal, vol. 2) shows the normal and maximum areas for male and female operators as well as other information relative to principles 16 and 17. (See page 24.)

Mechanical Controls. The best foot pedal design is with the fulcrum under the heel, with the resistance overcome under the ball of the foot. The dimensions for cranks, levers, handles, hammers, pulls, and knobs depend on the physical measurements of the human being. Although a valve handle may be 10 in. in diameter, it is not wise to assume that one hand can operate it; two hands may be needed. For cranks, 4 in. is about the best diameter.

Some other principles pertaining to equipment design are listed by Krick (Methods Engineering):

1. Use color, shape, or size coding to maximize speed and minimize errors in finding controls.
2. Use simple on/off, either/or indicators whenever possible. If this does not suffice use a qualitative type indicator if adequate.
3. When adequate, controls should be of the simple on/off, either/or type. If this does not suffice use a control with a limited number of discrete settings if adequate.
4. Direction of motion switches, levers, handwheels, knobs, and other controls should conform to stereotyped reactions.

An analyst should not assume that this list is complete. New applications in industry and research are adding to these principles.

Applying Principles. The principles of methods design are used as a stimulus to the imagination to find as many ideas as possible concerning probable good methods design for the work. Even though procedures will be given for helping this process, the analyst should always work with other people in the organization. Foremen, operators, supervisors, and persons in other departments of the organization, like purchasing and engineering, are good sounding boards for the application of the principles. It is wise to ask these persons many of the principal questions. Frequently, they come up with good ideas.

Important aspects of the five basic factors of methods design. Consideration should be given to their simplification, rearrangement, combination, and elimination.

A Activities at Workplace	B Material Handling	C Storage	D Holding Activities	E Delays	F Inspection
1. Materials Received					
Type of materials used (scrap, new material)	Type of material	Quantity shipped	Dimensions of part	Finish of part	Acceptance sampling
Packaging	Physical characteristics of material quantity	Nature of load (palletized, bulk containerized, etc.)	Weight of part	Weight of part	Prepacking to specification
Type and number of parts (standardization)		Quantity packaged			
Purchased vs. manufactured parts					
Physical dimensions of part					
Positioning devices					
Physical characteristics of material (weight, strength, gage, etc.)					
Quality of incoming material					
Auxiliary materials (lubricants, coolants, etc.)					
Finish and shape of part					
2. Product Design					
Amount of material tolerances	Weight of product	Self-stacking features	Jigs and fixtures	Container	Requirements for finished product
Automation	Quantity of parts			Part symmetry method of shipping	Sampling inspection
Physical characteristics (weight, strength, gage, etc.)	Handling and positioning devices (eyelets, hooks, handles, flats, etc.)				Go-no-go inspection specifications
Processing technique					
Finish					
Interchangeability					
Standardization					
Number of parts assembled					
Packaging					

3. Sequence of Production

Order of operations	Quantity moved	Number of steps	Number of operations in each step	Number of steps in process	Number of operations before inspection
Jigs and fixtures	Location of interrelated operations	Method of shipping	Number of steps		Number inspections grouped together
Physical processing techniques	Number of steps in process	Number of steps			Inspection included with production operation
Balancing work	Number of operations	Amount of handling			Number of controls
Lot size		In-process inventory			Specifications
Multimachine operation		Location of storage			Electronic inspection equipment
Location of workplace					
Training					
Reprocessing					
Requirements for succeeding operations					

4. Equipment, Tools, Machinery, and Workplace

Physical processing technique	Quantity per pickup	Housekeeping procedures	Jigs and fixtures	Lighting	Methods of counting
Jigs and fixtures	Moving distances	Conveyors	Positioning devices	Type of furniture	Automation
Furniture—type and positioning	In-plant transport: hoists, cranes, etc.	Stacking cranes	Foot controls	Noise levels	Electronic measuring devices
Material handling	Pneumatic, hydraulic systems	Skids	Magnetic and vacuum devices	Files	Go-no-go gages
Automation		Hoppers, bins	Clamps and vises	Adjustments of furniture	Lighting
Lighting					
Feeds					
Space-work area					
Location of services and facilities					

Fig. 5. Checklist of five major factors of methods design.

4. Equipment, Tools, Machinery, and Workplace (Continued)

A Activities at Workplace	B Material Handling	C Storage	D Holding Activities	E Delays	F Inspection
Location of tools, jigs, and fixtures Conveyors Visibility Noise control Safety features	Location of switches and controls Trucks, dollies, etc.				

5. Work Methods and Motion and Body Patterns

A Activities at Workplace	B Material Handling	C Storage	D Holding Activities	E Delays	F Inspection
Performance in normal work area Manpower utilization Training and instruction Body movements Part movements Location and positions of parts Area of performance Handling at workplace Part identification Energy utilization Tools, jigs, and fixtures Machine utilization	Distances traveled Multi-operations Type of movements Eye movements Body movements	Housekeeping techniques Stacking	Pressure Comfort Balance	Eye movement Simultaneous symmetrical patterns Balance Training Rotation of operators Manpower utilization	Part movements Eye movements Lighting Counting methods Part identification Sampling techniques

Fig. 5. Concluded.

Use of the Human Body

1. The two hands should begin as well as complete their motions at the same time.
2. The two hands should not be idle at the same time except during rest periods.
3. Motions of the arms should be made in opposite and symmetrical directions, and should be made simultaneously.
4. Hand and body motions should be confined to the lowest classification with which it is possible to perform the work satisfactorily.
5. Momentum should be employed to assist the worker wherever possible, and it should be reduced to a minimum if it must be overcome by muscular effort.
6. Smooth, continuous curved motions of the hands are preferable to straight-line motions involving sudden and sharp changes in direction.
7. Ballistic movements are faster, easier, and more accurate than restricted (fixation) or "controlled" movements.
8. Work should be arranged to permit easy and natural rhythm wherever possible.
9. Eye fixations should be as few and as close together as possible.

Arrangement of the Workplace

10. There should be a definite and fixed place for all tools and materials.
11. Tools, materials, and controls should be located close to the point of use.
12. Gravity feed bins and containers should be used to deliver material close to the point of use.
13. Drop deliveries should be used wherever possible.
14. Materials and tools should be located to permit the best sequence of motions.
15. Provisions should be made for adequate conditions for seeing. Good illumination is the first requirement for satisfactory visual perception.
16. The height of the workplace and the chair should preferably be arranged so that alternate sitting and standing at work are easily possible.
17. A chair of the type and height to permit good posture should be provided for every worker.

Design of Tools and Equipment

18. The hands should be relieved of all work that can be done more advantageously by a jig, a fixture, or a foot-operated device.
19. Two or more tools should be combined wherever possible.
20. Tools and materials should be pre-positioned whenever possible.
21. Where each finger performs some specific movement, such as in typewriting, the load should be distributed in accordance with the inherent capacities of the fingers.
22. Levers, crossbars, and hand wheels should be located in such positions that the operator can manipulate them with the least change in body position and with the greatest mechanical advantage.

Fig. 6. Principles of motion economy.

Time is required in using the check lists and principles of motion economy, and frequently the analyst does not have enough time to use them all. In these cases, a **mental application** is feasible as long as the basic principles or concepts are applied. The more experience an analyst has, the more likely it is that he can apply the principles mentally.

Generally, the principles for analyzing material, design, and sequence should be used first, and then the principles for improving the equipment and method should be analyzed.

PRESENTING SUGGESTIONS. Suggestion Lists. A suggestion list is the written presentation of the ideas evolved for meeting the goal of the work study. The list should be formally presented giving all pertinent information concerning the job or operation under study. The suggestions should be specific

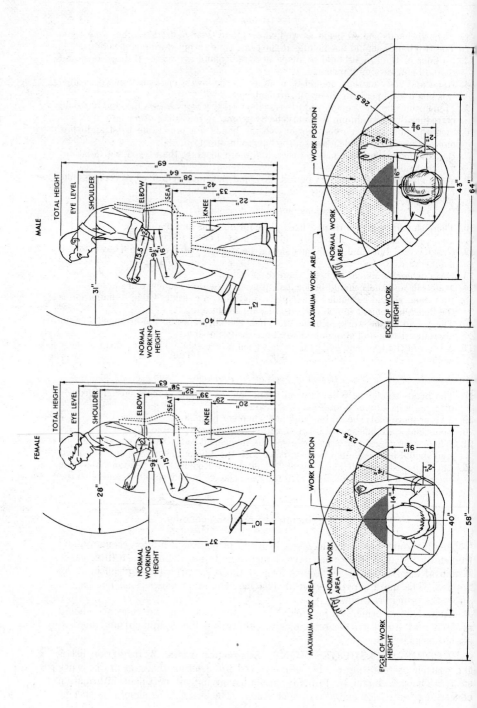

and if possible given in the order in which the job is performed. It is important that the objectives of the suggestions be clearly stated since most operations must compromise one or another basic principle in actual situations. Thus the objective "reduce operator fatigue" will probably lead to increased production but the primary aim may be to alter the workplace to make the operator more comfortable.

An analyst may get many ideas as he works with the principles. It is important that, regardless of the step in the procedure, all ideas should be recorded. The ideas may come to mind even though the application step has not been reached or has been passed. Ideas should be recorded regardless of how silly or ridiculous or costly they may seem. But, by recording all ideas, it is possible for the analyst to review them later, perhaps to get other, more feasible ideas sparked by the costly or silly suggestion.

Generally, the suggestion list is made with three columns. The first column orders the suggestions numerically. The second column gives the number for the most complex principle involved: (1) material is the most complex; (2) design, second most complex, etc. This provides an indication of the area covered and the authority needed to get the change made (more authority is needed to make more complex changes). The third column lists the specific idea. Each idea should be specifically spelled out. For example, an idea for a specific job would not be "use a simultaneous symmetrical hand pattern." It is better to indicate "two jigs for holding parts at assembly area, and two chutes for disposing of the assembly."

Suggestion Guide. If time is available, a suggestion guide may be made to give details about each suggestion on the suggestion list. For example, a suggestion may involve a material change. The suggestion guide has five columns on it, one each for material, design, sequence, equipment, and method. The idea for the change in material would detail what would be changed in each of the five areas if the change in material went into effect. This permits the analyst to get some insight into what the suggestion entails and its feasibility. It also permits some of the ideas to be combined to make still better suggestions.

SELECTING THE OPTIMUM WORK METHOD. Factors in Selection. With all the ideas collected, the analyst next determines which one or group of ideas is best for the circumstances. In determining which is the "best" the analyst must review certain factors which are common to the solution of all problems.

If the same job is performed in different plants or companies, there is nothing to indicate that the same method should be used in each case, because circumstances vary. However, certain **standard factors** have been found to affect the selection of feasible solutions, regardless of where the work situation may occur. There are four of these which usually need to be evaluated when selecting a feasible solution, namely, the economic, safety, control, and psychological factors —the economic factor being the most important in most situations.

Evaluation Worksheet. To make certain all information about a suggestion in each of these four factors is recorded properly, the use of an evaluation worksheet, such as that shown in Fig. 8, is recommended. Enough space is provided so that each of the factors may be adequately stated. The number of suggestion columns will depend on the available alternatives. Only those suggestions that appear to be feasible need be evaluated on the worksheet. In the usual case, most of the space on the worksheet will be devoted to the evaluation of the economic factors.

Page ____ of ____

EVALUATION WORKSHEET

Suggestions for (operation) ______________________________

______________________________ Department

Objectives: ______________________________

____

Analyst ______________

Date ______________

Factors	Suggestions		
1. Economic			
2. Safety			
3. Control			
4. Psychological			

Fig. 8. Typical evaluation worksheet.

Economic Factors. Generally two economic questions must be asked about any suggestion under consideration: "How much will it cost?" and "How much will it save?" There are many factors determining the net cost for a particular suggestion. The following are some of the factors to be considered.

1. The cost of **purchasing and installing new machines.** This refers particularly to major pieces of equipment purchased from outside vendors.
2. The total charge for **new tools, jigs, and fixtures** if purchased from outside sources. If made within the company, material and labor would be entered separately. Companies have their own policies with regard to overhead charges and these policies should be followed when making charges for the new devices.
3. In some companies, charges for **engineering services** are made if new designs are necessary. Engineering time is almost always required to make design changes in machines and equipment or even of products and materials.
4. **Labor and material charges** for installing the suggested method. Some companies include the cost of installing new machines, mentioned above.
5. **Overhead** may or may not be added as a cost of a suggestion. In some cases a company will want to charge overhead for the service functions of the organization to a particular project. In other cases, the overhead may include interest on any money borrowed in connection with the suggestion.
6. **Losses** due to a change in materials or methods. These represent out-of-pocket costs to make a change and therefore the new method must support any losses of this nature. However any equipment or machines that are made obsolete by a proposed change cannot be charged against the new method. Grant and Ireson (Principles of Engineering Economy) state, "The consequences of any decision regarding a course of action for the future cannot start before the moment of decision. Whatever has happened up to date has already happened and cannot be changed by any choice among alternatives for the future. This applies to past receipts and disbursements as well as to other matters in the past."
7. Other charges may be included in a **miscellaneous** category. In some cases, the amount of **lost production** might be charged to a particular suggestion.

These charges give a gross cost for a suggestion, from which any **salvage value** must be subtracted to give the net total cost. Naturally each factor will not be used in each case and often other factors are added, depending on the company and the circumstances.

Monetary savings are based on the present operating cost as compared to the proposed operating cost for the suggestion. Operating costs are made up of direct labor, material, overhead, and machine rate. In some cases machine rate and overhead are in the same category. It is possible to calculate savings on an effort or a reduced skill basis, but these savings must be given in terms of their monetary value. Usually, **material savings** are relatively simple to calculate. At least, this information is available from many sources. In a similar fashion, it is proper to use the **overhead calculations** of the company, whatever they may be. There are many problems in establishing overhead figures, but the analyst should use the present procedures of his company.

Labor costs are dependent upon the amount of time it takes to perform an operation. The biggest aid to estimating the time for the proposed method is to make a **synthetic chart** for the activity with the suggestion included. If the chart has enough detail, it is possible to compare directly the number of symbols and other activities to obtain an estimate of savings. It should be emphasized that this is only an estimate, and this procedure cannot always be utilized with every type of chart. If a time or simo chart was made of the original method, it is easier to estimate the proposed time because some relationships of time and method are available. **Predetermined time values** are one of the best ways of estimating time for proposed methods. Although these time values are not precise, their consistent application will result in a reliable estimate. Other ways of obtaining estimated times are to make a **mock-up,** have the analyst go through the hand pattern without the actual setup, use previous time study data or make an out-and-out guess as a last resort. The time information obtained should be converted to dollars and cents for use on the evaluation worksheet.

Safety Factors. Industry should always try to make work safer for people. The purpose of evaluating this factor is to make certain the suggestion does not include any factors which make the work less safe. Every industry tends to have a general **safety level,** but hazards should be decreased regardless of the safety level of the plant. Most important, hazards should not fall below this level for an individual activity.

On the evaluation worksheet, comments are made concerning the analyst's thoughts and knowledge about the safety of the activity with the suggestion included. He should frankly note if he thinks the suggestion increases hazard to the operator so this can be taken into consideration when making the final selection.

Control Factors. Almost every activity has certain characteristics of **quality** or **quantity control.** Making a suggestion for an activity can change these characteristics. Basically, this factor determines how the suggested change affects the specifications of quantity or quality control, and whether the suggestion makes for better product quantity or quality control.

This does not mean it is necessary to increase or decrease quality or quantity control levels. The analyst should evaluate this type of change so a proper decision about a feasible solution can be made.

Psychological Factors. Depending upon the circumstances, this factor can be one of the most important of all. It involves both the feelings and attitudes of supervisors of work as well as those performing the work. In most cases, it may be better to use an employee's or supervisor's suggestion, though technically not the "best," because a well-motivated employee or supervisor using his own fairly good method can frequently outproduce a poorly motivated employee or supervisor using the "best" method.

There is space on the evaluation worksheet to permit the analyst to record some of these factors as they may affect the various solutions. The analyst should be perfectly honest with himself in describing the problems which may arise through people. This should help select the "best" and easiest method for the operation.

Evaluating suggestions for nonrepetitive work is much more difficult. Judgment plays a more important role in the final decisions, because frequently the volume or number of times the activity will be performed is not known. The "guesses" can be more reliable if more people are concerned with the decision.

FORMULATION OF PROPOSED METHOD DESIGN. This might be called the actual design step. Although a synthetic chart may have been made, it is more important now to design all the details regarding the proposed method. In many cases, this involves charting the proposal and making the designs of equipment and workplace. There are cases where only a mental formulation may be possible, for example, where a time study must be made, and there is no other opportunity to make wholesale changes; or where the volume may be low, and not much time can be spent in developing a better method. In all cases, however, some degree of design or formulation of the details is necessary for the proposal. The basic purpose is to prepare the information and details to assure proper performance of the proposed method. Without such information, many proposed methods will not be properly performed. All possible errors should be eliminated. Also, the formulation is a guide to those who will be performing and supervising the work.

Three processes are usually necessary, although there can be different combinations or additions. Because of the mental process involved, there are no fine lines between the three.

1. **Detailed visualization of methods.** Although visualization was employed in the selection of the feasible solution, it is even more important here to visualize the method so that proper details can be designed.
2. **Design of tools and equipment.** The proposal may involve new equipment or material specifications. This step involves the design of all factors changed in the material, design, equipment, and sometimes, sequence areas. The exact changes must be specified to permit the proposed method to be installed. The analyst does not have to be an accomplished draftsman, but he should be able to sketch his ideas for a tool designer who may make the drawings or even the tool. **Flow diagrams** should also be made as an aid for proper layout, if changes are made, and **workplace diagrams** should be made to describe completely the activity to the supervisors. The material entering the plant and the design of the product leaving the plant must be considered. The analyst may not make drawings for the changes in these areas, but he should be capable of sketching and drawing these changes for submission to the proper authorities.
3. **Design of the actual work methods.** Visualization usually includes two concepts, one concerning the design of mechanical aspects, and the other concerning the hand or motion pattern to be used. The general way of recording hand or motion patterns is to make a proposed chart of the same type made in the analysis-of-work step.

After the proposed method has been designed, these designs frequently must be submitted for approval. When this is done, it is wise to have a **proposal sheet** to point out the actual changes to be made, the advantages of the change, the savings anticipated, a recapitulation and comparison of the charts made for the original and proposed methods, and the references attached. Some companies have standard forms for summarizing a methods proposal, but the above items are the basic ingredients.

Review and Testing of Methods Design

REVIEW PROCEDURES. Because human beings design methods and equipment, there are possible errors included in the formulation step. There are other problems that may have been overlooked which should be solved. These problems are the reason for the **review step.** It is a "stop, look, and listen" step to make certain all details are completed. There are three parts of the review step.

Adequacy of Design. There are two key aspects of good methods design. One concerns the **total possible savings,** and the other concerns the **reaction of the people** involved with the change or new design. The total possible savings should reflect the maximum return of benefits to employer and employee. To the employee it should reflect job satisfaction, less fatigue, no drastic reduction in skill level, and economic security. To the employer it should appeal in terms of overall effective utilization of resources towards a goal, namely higher productivity, better employee morale, and cooperation.

Details for Proper Functioning. The analyst must check the design of all new **equipment parts** and **products** to make certain they will work as desired. This is economically and psychologically important. The analyst cannot be embarrassed more than when a designed part, product, or jig goes into production and will not work because some design detail has been overlooked. Other people are important in reviewing the designs and checking all details, since they may spot better methods.

The enthusiasm of the analyst can contribute to the improper design of a method. He may overlook small parts of the product or jig design which can make the design inoperative. In addition, if he takes more time in looking over the design, he may be able to make the jig or fixture applicable to more than just the one operation he studied. The supervisors of the activity can help outline potential uses of a general design.

Planning for Implementation. Supposedly the analyst has been working carefully with the people involved. He has obtained their ideas about the work, and what might be done about it. He has explained to them the purpose and value of the analysis techniques, and the approach he is using. But even with such good preparation, the analyst should take time to plan still further the approach to be used in introducing a new method to the people.

People are an important ingredient of the review step. Even if the analyst decided not to use other people in any way in the preceding steps, the review should force him to do so. A proper time schedule of the proposed changes should be presented to all the personnel involved in the installation of the change. This schedule should include the training adjustment periods, wage changes, revision of supervision, duties, and any other aspect of departmental activities.

TESTING TECHNIQUES. Testing is used to catch errors before actual usage. It is a way to determine whether a proposed method will work. It demonstrates whether the operator can do what is required by the proposed method, whether the mechanical devices will perform as specified, and whether the parts will perform as expected. Not every new method proposal must be tested. This step is designed as a **checkpoint** for those problems which cannot be solved in the formulation and review steps. If the test step finds a method unworkable, the analyst returns to the selection step to start with another proposal.

Testing can be done more often if the organization has a **methods laboratory.** Many companies have such laboratories where tests can be performed prior to installation. Even where a laboratory is not available, some area where the methods analyst can review and test some of the more complicated proposals is desirable. Various ways for making a test of proposed methods changes are discussed below.

Model or Pilot Plant. Frequently a plant layout or material handling problem, or a large portion of the production sequence, forms the basis for a problem being studied. For such gross activities, the only testing procedure is frequently a pilot setup, or a scale model of the involved area. A **model** is the reduced-scale version of the actual situation and a **pilot plant** is a full-scale version of the layout or area but with a somewhat smaller scope than would be the case in the actual situation.

Workplace Mock-Up. One of the most common tests involves making an actual workplace out of inexpensive materials. This permits the analyst to perform some of the components of the work method at a low cost. This test is about the closest possible approach to getting actual information. Ordinarily, a mock-up is made out of wood and other available scrap material to save the expense of using expensive tool steels and other pieces of equipment. Of course, its use is restricted to activities where there is a great savings potential.

Trial Run. If the volume is sufficiently large, it may be desirable to make an actual workplace or jig setup and perform a trial run, just as if the method were going to be used. Another condition where this test would be performed is where there are several operators doing exactly the same thing and where making one of the actual workplaces would be a small part of the total cost of installing.

In some cases, a new proposal requires **commercial equipment.** It may be advisable to purchase this commercial equipment and test the method when it is known that the commercial equipment can be used in other areas in the plant if it fails to work at the suggested point. This type of test is the best for this situation because it will be performed in exactly the way the proposal suggests.

Simple Methods Changes. This procedure is a good way of testing under actual conditions of production. Frequently, a proposal consists of making some simple changes in present layout or methods. For example, it might require moving a table several feet, or moving a bin several inches or feet. This type of change can be made, tested, and reviewed to see if the desired results are obtained. If the expected results are not forthcoming, it is easy to return the equipment to its previous location.

New Machinery Trials. If the proposed method requires new machinery, it may be possible to test, depending on the type of machinery involved. If the machinery can be borrowed, so much the better. If not, the machinery to be purchased should be "tested" by obtaining as much knowledge as possible about its **operating characteristics.** This information can be obtained from manufacturers or by visiting other plants where similar machinery is used. These are not real tests, but they do provide for some information that would not otherwise be available.

Installation of Method

INSTALLATION ACTIVITIES. In some companies, the analyst may be completely responsible for all the activities needed in installation, and in others he may not have any relationship with these activities at all. There are many possibilities between these two extremes. Some of the other aspects have already been discussed under Planning for Implementation. A good tool for the overall coordination is a PERT chart.

Co-ordination with Production Procedures. The installation of a change on the production line should involve a minimum of disruption in normal activities. The installation could take place on offtime or weekends, and should be preceded by a retraining program for the personnel involved, and a complete testing of equipment and method. The impact of operations preceding and following the changed operation should also be co-ordinated. The problem of co-ordination is directly related to the nature and size of the change.

Installation of Equipment and Facilities. This would involve procedures similar to those discussed above. In many cases the equipment and facilities are made to order and require greater co-ordination and cooperation during installation and development.

Training Operators and Supervisors. Those concerned with using the new method must be given the opportunity to learn that method. The general way for providing this learning opportunity is through a training program, which can

Bus Duct Assembly (¼ in. x 3 in.)

(With or Without End Protectors, Two-Man Method)

1st Man	2d Man
1. Go to covers, grasp cover at one end, and pick up with help of 2d man. Help carry cover to squeezer jig and place in position.	Go to covers, grasp cover at one end, and pick up with help of 1st man. Help carry cover to squeezer jig and place in position.
2. Pick up rabbet and two screws. Place and align rabbet and start 1st and 2d screws.	Pick up four insulators and place one in each of the four insulator brackets.
3. Go to copper bars. Grasp one bar at end and slide it toward near end of truck. Regrasp bar at center and pick up. Carry bar to cover and push through all four insulators.	Go to guide, pick it up, and place in position over insulators.
4. Same as step 3.	Go to truckload of housings, grasp one at its center, and lift it over onto horses.
5. Same as step 4.	Pick up one coupling and four screws. Place and align coupling at end of housing and start each screw in its position.
6. Go to center of cover, pick up guide and set it aside.	Go to labels, pick up, and place in water can to soak. Remove labels from water, bring them to housing, and put in place.*
7. Pick up four supports. Go to 1st and 2d positions and place 1st and 2d supports. Close the nearest two squeezers.	Go to end of cover and pick up rabbet and two screws. Place and align rabbet and start 1st and 2d screws.†
8. Go to 3d and 4th positions and place 3d and 4th supports. Close the remaining two squeezers.†	
9. Go to end of cover and pick up length of rope. Double the rope and thread it through holes in ends of bar. Tie the two ends of the rope together.	Same as the 1st man.
10. Go to horses and grasp one end of housing. Pick up the housing with help of 2d man. Help carry housing to cover and place it on the cover so that the holes are aligned.	Same as the 1st man.
11. Pick up four screws and get power screwdriver from overhead hook. Run down the screws at the end of the duct. Place and run up one screw at each of the four screw positions. Replace screwdriver on hook.	Same as the 1st man.
12. Open all four squeezers.	
13. Go to one end of the assembled duct. Grasp and pick up with help of 2d man. Help carry duct to truck and place on truck.	Same as the 1st man.

* If wooden end protectors are used, place them on at this point in the cycle in the following manner:

  1st man: Pick up end protector and four screws. Hold the end protector in position and start the screws.

  2d man: Pick up end protector and four screws. Hold end protector in position and insert one screw, lock washer, and nut in each position. Tighten nut by hand until it is caught by the washer.

† If four copper bars are used, the 2d man puts on two supports and tightens two squeezers at this point.

Fig. 9. Written standard procedure (WSP) for assembly operation.

range from just a few words to the operator to extensive training programs in the plant. The people doing the training range all the way from the analyst himself to a company **training instructor.** The amount of time necessary for training ranges from a few minutes to many hours, depending on the skills required in the work. Some of the data developed under analysis (simo charts, motion pictures, video tapes, etc.) may be useful in training the personnel involved.

Written Instructions for Procedures. Documentation of the entire analysis, design, development, and installation should be maintained as a record. These records prove to be a valuable source of information for future work and for training of new personnel. Each documented piece of information should be prepared to serve specific purposes. For training operators and supervisory personnel, a convenient type of document is **written standard procedure** (WSP).

In general, three **classes of information** are required on the written standard procedures. The first is the written statement of the general conditions, equipment, and tools needed for the work. The second is the description of the desired method. The third class is the workplace or flow diagram for the operation.

Figure 9 shows the written standard procedure for a rather short-run operation, which occurred off and on in a plant. The workplace diagram is not shown. Other WSP's have been made with minute details of work, showing what each hand and finger performs. Some of these include drawings of where the fingers should be placed.

In some plants, the WSP is enclosed in a plastic cover and posted conspicuously at the workplace so that the operator can check his method at any time.

Follow-Up. Even though all the mechanical changes have been made, and the training has been given, there is no assurance that the method will be used the way it was designed. This requires the analyst to follow up the new method. The follow-up is probably most important when operations are not continually running. When the analyst checks on his proposed method, it is somewhat like insurance on his investment of time and effort. The most important purpose of follow-up is to act as a "loop closer" in the system approach to methods analysis. A closed-loop system will permit a re-evaluation of the initial problem in contrast to the implemented changes. If the change falls short of the initial goal subsequent and further changes and evaluations may be undertaken.

WORK SIMPLIFICATION

CONTENTS

WORK SIMPLIFICATION

Philosophy of Work Simplification

DEFINITION. The term "work simplification" was coined by the late Professor Erwin H. Schell of Massachusetts Institute of Technology to portray the approach to methods improvement developed by Allan H. Mogensen. He used the principles of motion study established by Frank B. and Lillian M. Gilbreth to structure a program in which every member of an organization might participate. His definition for this procedure was "The organized use of common sense to find easier and better ways of doing work."

Work simplification as originally developed by Mogensen was essentially a training program. Since the 1940's it has become more widely used as an overall improvement program and has evolved a broader and deeper meaning. The emphasis has shifted to the human relations involved in the improvement process. A **basic philosophy** of **friendliness, understanding,** and **teamwork** pervades the program. Profitable growth and overall effectiveness is enhanced. The tools and techniques of finding improvements are more group-oriented. Much attention is being given to effective application and use of work simplification by organizing a continuous and self-perpetuating program of improvement within the enterprise. Year by year, new avenues of communication are being opened for reporting progress, recognizing participants, and introducing new material. The **objective** of work simplification is an increased rate of improvement as a result of the coordinated efforts of everyone in the organization to achieve that goal and the accelerated personal development of the men and women who participate in the program. Work simplification emphasizes the development of people as the most important asset of any company. It is a vehicle for personal development inasmuch as everyone is expected to improve his own activity first, then join in a team effort to improve the entire enterprise. People learn by doing—by solving today's actual problems. The experience sets a pattern for coping with similar problems in the future.

PEG BOARD EXPERIMENT. The philosophy of the work simplification approach can be exemplified in any group meeting by a simple experiment known as "peg board demonstration." The peg board, originated by Professor Ralph Barnes of the University of California, Los Angeles, as a motion study illustration, was first used to develop the work simplification philosophy by Harold Dunlap of H. P. Hood & Sons, Inc. It has since become almost the universal method of introducing the basic concept of work simplification to those unfamiliar with it. In this experiment, one member of the group is requested to assemble some pegs on a board and remove them just as if he were performing an industrial or office operation. Another participant records how long it takes to do it. Spontaneous discussion and numerous trials of different methods of doing the operation follow, in which the leader is often forgotten. Between 10 and 20 people in a group is the most desirable number of participants, although the experiment has been carried on with several hundred by utilizing subgroups, each with one or more boards. Normally, final cycles take only a fraction of the time of initial attempts. The effort expended is quite apparently less. The job is easier and less fatiguing. **Productivity is increased.**

Conclusions from Experiment. The **group's discussion** and **actions** can readily be summarized in the four simple statements which follow.

1. **Participation is a motivator.** What is done and said, the enthusiasm and laughter which are so apparent, indicate that the group is enjoying the experience. An analysis of why this happens leads to reflections on the importance of teamwork. Most people like to work in groups. Several heads are better than one. When people work together in an atmosphere of friendliness and cooperation, there is usually a much greater feeling of accomplishment, satisfaction, and pleasure as a result of the effort. The most important factor in the group approach is that the people involved come up with the ideas themselves. The common-sense procedure is developed by the group. Ideas are not pressed upon them by someone else. When attitudes toward improvement become more positive, conditions for motivation develop spontaneously.

2. **Improvement is natural.** People like to improve when given the opportunity—unless inhibited. If the leader has taken care not to mention the words "method," "way," "improvement," "better," etc., then the group has demonstrated a spontaneous urge to improve.

It is usually agreed that fear is the universal inhibitor. Loss of security and loss of acceptance by one's friends or fellow workers rank high on the list. Most people hold back if they feel afraid. Work simplification seeks to replace these fears with understanding, mutual confidence, and teamwork. The natural desire to improve becomes available for effective use when the people begin to realize that improvement is in their own best interest. They begin to have confidence in their leadership.

3. **It is the method that counts.** Observations by the group include such fundamental concepts as, "No one worked harder or faster, yet productivity was increased." "The group worked smarter, not harder, by finding an easier and better method." It is the method that determines productivity. Likewise, the method of approach is what determines acceptance. Good methods of managing, of leading, of considering the feelings and points of view of other people, are all demonstrated effectively in this experiment. The group is organized in such a way as to utilize the potential of all the people. They respond as a cooperative team. It takes a little longer, but the results are more satisfactory and, therefore, more acceptable to all involved.

4. **Communications are easier and more open.** Work simplification opens up new avenues of communication. It also sets up standards of courtesy and consideration which establish the approach as one of friendliness and understanding. This point too can be demonstrated in the peg board experiment. The man who first performed the operation readily agrees that he would have resented being "told" he was using a poor method. The rest admit that they would have been inclined to take his side if any remarks had been made in that direction. The sear of sarcasm would have made the resentment still deeper.

Typical ways of getting results through other people are to: "tell them, sell them, or involve them." The **telling** approach is resented unless complete confidence exists. The **selling** approach is more desirable but often insincere. The **involvement** approach is generally the most acceptable, since it recognizes the ability of the people to make contributions. It is basically more friendly. If there is a proper balance among the three, with the trend toward the last whenever possible, then, when it becomes necessary in an emergency to "tell," people will usually respond more willingly. **Confidence in leadership** is a quality established through considerate, patient action in an atmosphere of understanding and friendliness.

MOTIVATION. The success of any human activity depends upon the effectiveness of the motives which stimulate it. These can be positive, whereby there is something to be gained, or they can be negative, in which case there is a defense against loss.

Positive Motives. The work simplification philosophy emphasizes the positive motives for improvement. "What's in it for me?" "What is my own objective?" "Is there a common goal toward which we all can work as a team which will help to fulfill the hopes of each as an individual?" A clearer understanding of the **interdependency** of everyone in a free economy is the foundation of progress. Labor, management, owner, and consumer all look to the improvement process in order to achieve a higher standard of living. Henry Ford said that this objective could be attained by "making the best possible product at the lowest possible cost while paying the highest wages possible."

The **improvement motive** is broad, positive in nature, and easily developed in open discussion with practically any group of people. It can be reduced to specifics in the case of most individuals as, for example, the following statements:

We want to belong.

We want to work for and with people who are understanding, human, and appreciate us as individuals.

We want to believe our own feeling of importance in the scheme of things.

We want confirmation by others of our right to enjoy our self-respect in the form of recognition, esteem, and honor.

We want to associate with people more important than ourselves.

We want any feeling of increased importance to manifest itself in increased earnings.

We want to share in the tangible results of our efforts.

We want an opportunity to enhance our own personal status.

We want satisfaction and a sense of achievement from our efforts.

The work simplification philosophy is built on an understanding of these human aspirations. It seeks to provide an opportunity for a higher degree of fulfillment.

Negative Motives. The negative motives which may retard or completely stifle improvements of any nature are almost entirely associated with the fear of loss of a job and the related potential changes in social status or the fear of losing one's feeling of importance and self-esteem. The responsibility of recognizing these defensive stimuli rests in the upper levels of management. A **consistent policy of remedial action** in this regard is basic to the success of work simplification (see discussion of Security). Open discussion of this problem among groups within any specific organization will usually resolve the situations which need attention.

CHANGE. People appear to resist change. History repeats itself over and over in accounts of the unwillingness of people to accept something new. Iron ships, steamboats, automobiles, airplanes, telephones, and television are just a few examples of improvements which most people claimed were impractical, would never work, and could not possibly be economically useful.

Resistance to Change. The work simplification philosophy recognizes this phenomenon in human behavior as one of the greatest **obstacles to the improvement process.** It attempts to stimulate a desire for a better understanding of why this resistance exists. Participants are urged to test their powers of analysis with respect to the behavior of others by first analyzing themselves. Why do they appear to resist change? What is the frame of reference that dominates their own thinking? What biases and fixed ideas do they have themselves? How do others feel on the same issues? Why?

When the notions which arise as answers to such questions are pursued, it becomes clear, as a common-sense concept, that people do not resist changes toward which their attitude is favorable. In a positive environment people like to change. The problem thus becomes one of influencing attitudes. Professor Leo Moore of M.I.T. said, "You don't change behavior, you change attitudes." There is no one best method of doing this; rather, it is a combination of several approaches that usually brings success. The atmosphere of the work simplification approach is of itself one of participation, understanding, and friendliness. People are recognized for their knowledge of the job and are given the opportunity to contribute toward its improvement (see treatment below of A Program of Improvement). **Communication channels** are opened and utilized to keep everyone informed of activity which affects them. Information is freely available, and **open discussion** of the facts leads to understanding and confidence rather than mistrust and suspicion. The net result is a new outlook and attitude toward change as a result of teamwork and cooperative effort. A positive environment develops within which change is commonplace, sought after, and accepted as desirable.

CRITICISM. The implication of any proposed improvement is criticism of the existing procedure. It is next to impossible to criticize a method without criticizing the person doing the job, the one responsible for it, or the person who prescribed that method in the first place. People usually resent criticism. It is necessary to recognize this situation and deal with it effectively if we expect improvements to be acceptable. The inference of all change is criticism and we cannot have improvement without change. What appears to be resistance to change is often unexpressed resentment of criticism or fear of loss of the job or job status (see Security below).

Here again, an open discussion of this phenomenon by groups within the organization exemplifies the work simplification philosophy. A sincere effort to realize why people resent criticism reveals a surprising insight by most and leads to better understanding of the problem and a new outlook. People admit that they resent criticism because they would rather be told they are right. We like to have our own good opinion of ourselves verified. Anyone worthwhile has a high degree of pride in his work and naturally resents being told that it is not as good as it could be. A realization that people are trying to help, not to criticize, is a good first step. The **team approach** also does much to alleviate ill effects. If each man knows that his turn is likely to be next, he may be a little more considerate of the others. The very fact that it is a mutual effort and each understands the feelings of the rest is a major part of the answer.

Symbolic Aids to Discussion. Numerous symbolic devices have become almost universally used as aids in dramatizing these points. The **traffic light** is one such symbol. If the green light is on, the mind is open to new ideas and cognizant of the fact that in order for improvement to take place, obsolete ideas must be discarded. The red light symbolizes resistance, resentment, and the closed mind. A desk ornament composed of a ball which is half red and half green serves the same purpose. Its presence is a license to start a discussion over again, if it appears to one of those present that he has run into resistance or resentment. It works well, particularly in cases where there are two levels of the organization involved. The symbol makes the effort impartial. The **parachute** is another such symbol: "The mind is like the parachute. It functions only when open, and you can't go back and get another if it doesn't work."

SECURITY. Fear is without doubt the greatest barrier to improvement and progress. Much of the resistance to change is fear manifested in a manner acceptable to the ego of the individual in question. It is only natural for a man to find reasons why something won't work or can't be done if the doing of it seems to him to pose a threat to his importance, his job, or his future. The work simplification philosophy recognizes that the opinions and feelings of the people involved are as real as the facts of a situation and must be treated with just as much respect. A worthy objective is to try to get opinion and fact to coincide. No amount of telling, selling, or convincing will accomplish this end if action belies the words. Moreover, the conduct of a competitor or just a neighbor within any industrial or business community may have as profound an effect on personnel as if it were occurring in their own company. Be sure you are aware of the influences beyond your control—not the least of which is everyday news interpretation by columnists, commentators, and politicians.

Solving the Fear Problem. The area of solution to the problem of fear lies first of all in the **sincerity** of the management leadership. If management fails to accept a responsibility in this regard, there is no reason to expect employees to be anything but wary. What happens to people whose jobs have been removed by improvement or progress is of vital concern to the people involved and likewise to the success of any improvement program. An open discussion of the implications of a hypothetical change which makes it possible for one person to complete the same number of units in a given time as were formerly completed by three people invariably produces the same answers whether the group be executives, supervisors, or hourly employees. The change is progress. It raises the standard of living. It creates employment in the long run. It is a desirable thing, if it doesn't happen too fast, and if management makes provision for **retraining** and **relocating** those displaced.

Figure 1 lists **implications** of a **hypothetical change** elicited from the three different groups indicated. It is interesting to note that all three started on a note of fear that someone would lose his job and in each case concluded with the assurance that no one had to lose his job.

A few companies have gone so far as to guarantee that no one will lose his employment as a result of a work simplification improvement. This may be dangerous as a written policy because of misinterpretations and activity beyond management control. Usually, it is reasonably easy to carry out such a policy.

Improvement and Turnover. The **labor turnover rate** is usually several times the **improvement rate.** It is rare indeed when the rate of displacement of people as a result of improvement is faster than the natural attrition through resignation and retirement. Many firms publicize their turnover rate, since it shows that no turnover has been caused by technological improvements. It is unfortunate that some managers who have been condoning certain methods as satisfactory for 20 years insist on a change overnight once they find out about its possibilities, no matter what the cost in human feelings. A policy of concern expressed in a sincere effort to "do something about it" gains the most respect. The alternative to an **effective reputation for appropriate action** is mistrust, featherbedding, and a slower improvement rate. Conversely, understanding, confidence, and enthusiasm can add new acceleration to the improvement rate anywhere.

AUTOMATION. An open discussion of the implications of automation, both in the factory and in the office, with groups of executives, supervisors, or workers alike produces thoughts like the following.

As Seen by a High-Level Group from a Hospital	As Seen by a Supervisory Group in a Plastic Molding Plant	As Seen by a Worker Group from an Appliance Plant
Someone out of a job	Two out of a job	Two people not necessarily out of work
Insecurity	Lower cost	Save space
Lower morale	Save space	Save dollars
Increased efficiency	Increase capital	Training job must be done
Mistrust	What to do with other two?	Start with less output
Personnel policy problems	Operator more skilled	Sell for less
Lower cost	Shorter delivery	Produce product at lesser cost
Space gained	More profit	Sell more
Transfer problem (other departments may not want)	Improved method	More help needed to sell it
Capacity increased	Better product	More people needed to produce product
Re-education problems	Better working conditions	Cost money to change
Transition time (time for retraining)	More breakdown hazard	More work for builders of equipment
Possible higher capital requirements (initial costs)	New maintenance problems	More work for everybody
Possible higher maintenance costs	Could cost be higher?	Easier to do the work
May need more people	Less handling	
May have to pay more for higher skills	Less setup	
	How would worker feel?	
	What happens to pay and thoughts of pay?	
	Is this the right time to change?	
	Two people transferred to other jobs	
	Retraining	
	Must be able to keep earnings up	
	New markets	
	New products	

Fig. 1. Implications of a change.

This is just another type of improvement. It causes more employment before it even goes into production. It requires people with greater brainpower at higher pay rates. It will make certain products on a high-volume basis which will lower costs. If costs go down we will sell more goods or people will have more money to spend on something else. Either of these events will create more jobs. It will produce certain special products which man has been unable to fashion economically. It will be a fine thing for humanity, if pressure groups don't try to exploit it without regard for the people involved.

The **effective initiation and use of automation** in industry will depend largely on the methods used in installing it. The work simplification philosophy outlined here will be no less helpful in this area than in any improvement procedure. Costly and difficult-to-change errors may be avoided if the people involved have an opportunity to participate. The contribution that can be made by people not skilled in the engineering specialties required for most automation may appear nebulous, yet it is nonetheless important in obtaining enthusiastic cooperation.

COMPETITION. A proper question to ask is, "What if we don't improve?" If competition succeeds in making improvements that we could make and do not, he will grow and prosper by beating us competitively in the marketplace and we will have fewer jobs as a result. In answer to the question, "Who is our unknown competitor?" a three-hour discussion at the top executive level came up with this definition: "Our unknown competitor is anyone, anywhere with an idea we don't have that could be used in our job or our company."

Tools and Techniques

PROCEDURE. The tendency to make so-called "flash decisions" or "snap judgments" is overcome by a scientific step-by-step work simplification procedure. This scientific approach began with the early pioneers: Taylor, Gilbreth, and Gantt followed by Porter, Barnes, Mogensen, and many others who have made substantial contributions to its growth. What has evolved is a logical, orderly pattern which itself exemplifies simplicity. The problem-solving pattern is described in the paragraphs that follow.

Step 1. Select a Job, Activity, or Situation To Improve. Too much emphasis cannot be placed on the importance of a good beginning. In selecting a situation which is an opportunity for improvement, it should be borne in mind that the end result is a real improvement enthusiastically utilized by the people involved. This, then, is the time to utilize the concepts of the work simplification philosophy previously outlined. The situation is a dynamic one depending on the current attitude toward improvement in any given organization. If an improvement program were well established, the procedure might be quite different from that which would obtain if one were just being launched.

At the outset, the analyst should be urged to select a job, or some part of a job, that he performs himself. In this way, initial resistance to change or any implied "criticism" may be avoided or minimized; moreover, the job a man does himself will be the job he knows best. Jobs that he supervises or is directly concerned with which have high costs, are bottlenecks, or are giving trouble, are appropriate situations for selection. Any activity that entails much walking for materials, tools, or supplies, any place where waste of time, energy, or materials is suspected, is a good possibility. Sometimes the activities which need improving must be hunted out. Groups with common interests but **different frames of reference** from within an organization can generate long lists of such opportunities if they are requested to do so in a "green light" atmosphere (see the discussion of creativity and brainstorming under Step 3). These lists may then be **segregated** by departmental or individual responsibilities, and priority ratings given to those that seem most important.

Practically any activity is a potential project for improvement, but the objective should be to have everyone select activities that are commensurate with his level of responsibility, as well as those which are more important. Several projects can be under consideration at all times, so that unavoidable delays in the progress of one will allow continual activity on others. While first experimenting with work simplification, the smaller, more easily completed projects are desirable because of the need for experiencing accomplishment during the learning stages. Later, larger projects may be selected for investigation.

Step 2. Get the Facts. In order to improve an activity, procedure, or operation, the facts involved must be assembled in detail in such a way as to render them available for analysis in an orderly and scientific manner. One way of doing this is to make a process chart. The process chart is a universal tool of work simplification. It is often the means of applying the work simplification pattern. Once the job to be improved has been selected, a process chart becomes a record of the facts and the work sheet for applying subsequent steps in the procedure. Depending on the type of process chart made, the important features are graphically portrayed in such a way that they can be easily recognized and readily challenged in Step 3. All forms of process charts, including the more complex and special forms, are fully described in the section on Process Charts.

The tendency in work simplification is to de-emphasize technical perfection in the charting process, which is only Step 2 of the pattern. In order to encourage participation at all levels of the organization, technicians are often assigned to make the charts for a project team. Simple informal documentation of the present procedure or pertinent facts about the problem is often all that is needed to lead to substantial possibilities for improvement.

Step 3. Challenge Every Detail—List the Possibilities. This is the creative stage of the work simplification pattern in which every detail of the process or procedure is challenged and possibilities for improvement or alternatives that might be better are listed. The green light must be on, the minds must be open, resistance to change and resentment of criticism must be forgotten. The attitude of the people involved must reflect understanding, the basic work simplification philosophy, if this step is to be effective. This is **organized creative thinking** and **brainstorming** with "possibilities for improvement" the objective, catalyzed by a simple but thorough questioning procedure. Judicial thinking and evaluation of the possibilities should be postponed until ideas have been developed. Creative thinking begins with the simple mental procedure of wondering. The ability to ask questions leads to possible answers. If we ask certain specific questions, we are guided in the direction of improvement possibilities.

What is being done? Why? What is the purpose of doing it? Why should it be done at all? Can it be eliminated? What are possible alternate methods of accomplishing the same result? The elimination of a detail is the ultimate in simplification. If we can eliminate certain details, other dependent ones may automatically be eliminated. Hence, the "do" details are challenged first, since "make-ready" and "put-away" details depend on them. A chain reaction is often set up. A **"do"** detail is one that actually accomplishes work, brings about a change of condition, or adds value to the product. **"Make-ready" details** such as setup, pickup, and load-machine, prepare for the "do." Cleanup, toss-aside, or replace-supplies are **"put-away" details.**

Where is the detail done? Why is it done there? Where else could it be done better, easier, or with less time and energy? Could it be relocated nearer to the next operation? Could it be combined with another detail at a different location?

When is the detail done? Why is it done then? Could it be done at another time? Could it be combined with another detail? Could it be done in a different sequence, thus improving some of the make-ready and put-away details?

Who does it? Why does that person do it? Who else could do it? Can someone less skilled do the detail? Can it be combined with another detail done by someone else? Who else is qualified?

How is it performed? Why is it performed that way? What other ways are there to do it? Are there easier or safer ways? Can it be economically mechanized?

Answers to these questions on all of the "do" details are noted on the **process chart** along with all possibilities for improvement. The questions are then repeated for the make-ready and put-away details, which are next questioned in the light of the other possibilities as well. It will be noticed that the order of asking questions is organized to follow a descending order of importance in terms of the "steamshovel" improvements first and the "teaspoon" improvements later.

Elimination is the biggest improvement that can be made. In the "Elimination Approach" Spinanger asks, "If it were not for what key point this cost would be unnecessary?" If we can eliminate this one most important element the rest could therefore also be eliminated.

Smaller increments are made by improving minute "make-ready" and "put-away" details, but they are important nonetheless. This procedure can be carried out by groups of average people with unusually good results. They catalyze each other as a result of diverse experience and different points of view. One idea sets up a chain reaction of several ideas. Group activity and participation really develop into fun in this step. It sets the stage for the steps that follow. The trend in work simplification is more and more toward the task-force or **project-team approach.**

Step 4. Develop Improved Methods. The task now becomes one of organizing the possibilities into the best conceivable procedure under expected conditions and circumstances. This is the **judicial step.** The group can be organized at this point to check all people concerned who might not yet have become involved. Their participation now may be the key to their eventual acceptance of the improvement. Engineering, safety, sales—all may have an interest, yet not have been represented on the project team.

Alternate possibilities must be evaluated, and many will have to be discarded simply because they conflict with each other. Wherever practical, trials of proposed or alternate methods are desirable. A laboratory mock-up model is useful in many instances. Comparative cost estimates should be prepared by the appropriate staff people with the help of the project team. The best available information with regard to expected results should be gathered for the final decision-making process.

The following three-part **improvement classification** may be useful:

1. **Conservative:** Those improvements which can be installed with the tools, materials, equipment, and authority available at the level where the study was made.
2. **Radical:** Those improvements which require major changes involving much higher authority and large sums of money.

3. **Intermediate**: Those improvements which lie somewhere between the two extremes.

It is apparent that these classifications are variable, depending on what level of authority is involved in the project. A conservative possibility for the general manager might be a radical one for a setup man or an assistant foreman.

It is desirable at this point to impress upon those making the decision that in many cases, two or more improved methods should be developed. A conservative improvement may often be installed right away and pay for itself many times over before intermediate or radical improvements can be prepared for installation. Frequently, a more radical change cannot be justified on the basis of savings to be realized when compared to a good conservative one. This alone is usually reason enough for making most first improvements conservative.

A new way of doing something is not easier and better unless the man who is going to use the new method thinks it is better and willingly accepts it. Participation on his part during these steps in the pattern assures more effective application in the next step. A process chart may be made of each proposed method so that those involved in the installation will have a basis for instructing others in the new procedure. **Proposed method charts** often serve to catalyze still further improvements, as well as becoming a convenient method of reporting the details of the change.

Step 5. Install Improvements. If the preceding four steps have been conducted within the framework of the work simplification philosophy, actual application of the improvements developed often is a routine matter. Approval must be obtained from those in appropriate positions of responsibility and authority. Their previous participation to a certain extent in some of or all the steps has set the stage. Trial or **pilot runs** are a usual procedure for most changes. These afford opportunity for still further refinement as they proceed. Assumptions may be checked during these experimental runs, and estimates of results compared and validated. The training of those who are going to perform the new method is part of the installation procedure. The process charts that were made up as part of the previous step may thus become instruction sheets. In short, careful planning is a requisite of successful implementation.

The **retraining** of people who have previously done the job in a different way needs the concentrated attention of supervisor and worker alike. If both are familiar with the new method through participation in its development, this attention is practically automatic. Productivity and output may even be less during the early stages of the retraining period, since new habit patterns and thought processes must be used. Frustration or discouragement may often retard or even prevent learning of the new method if this point is not understood and anticipated by all concerned.

Step 6. Feedback and Review. A major part of the installation procedure is the **final report** on results. This information completes the circuit initiated when alternate methods were discussed in Step 4 and estimates of savings were used to decide on the acceptable improvement. Comparison of the actual with the estimated costs in the report of the final results is a good way of developing confidence. "Did we get the expected results?" "How can we go at it better the next time?" are the big questions. The executive can notice accuracy and conservatism in such estimates and therefore be more inclined to accept future estimates. He is also in possession of information on which he may comment favorably as a form of recognition to those involved in the project.

PROCESS CHARTS. The three most common and **basic process charts** are the flow or product process chart, the multiple activity chart, and the operator or right- and left-hand chart. Of these, the flow process chart has the most general application. It is universally taught as the fundamental fact-gathering tool in the educational phase of a work simplification program. In its simplest form it is easily understood by all levels within the organization and is readily used with the work simplification pattern previously discussed.

The making of the flow process chart is a simple procedure of recording information.

1. Indicate the job or activity under study.
2. Identify the subject (man or material) to be followed.
3. Indicate the starting and ending points of the chart.
4. Record a brief description of each detail.
5. Assign the proper symbol.

Additional instructions to note in making a process chart are:

1. Shade in the "do" symbols.
2. Record distances of transportation.
3. Record quantities handled.
4. Indicate time consumed, or production rate.
5. Prepare a summary.

All the above steps are explained in detail in the section on Process Charts. **Special-purpose charts** include the simultaneous motion chart and the procedure flow chart. All process charts have the common objective of portraying facts in a complete, orderly, and graphic fashion. Each is handled the same way in the work simplification pattern. The special features of those mentioned above are discussed in appropriate detail in the section on Process Charts. The ways in which they can be used in work simplification groups are presented below.

The computer has spawned countless other special purpose techniques. Some, such as CPM and PERT, are outgrowths of the flow process chart. Others, such as the large number of decision-making and optimization techniques, fall within the framework of portions of the work simplification pattern as they are applied (see Improvement Management, later in this section, as well as the section on Operations Research).

Plant Layout. Use of **process charts** with accompanying **flow diagrams** and the work simplification pattern inevitably leads project groups to propose minor changes in layout or even a complete **rearrangement of facilities.** The work simplification approach brings about closer teamwork between the line organization and the plant layout engineer. Preliminary to a major project involving plant layout, a group discussion by those involved of the fundamentals of different **layout theories** sets the stage. The advantages and disadvantages of each are listed on a general basis, along with the broad factors which influence the eventual decision. After discussion of them, the specific case in point may be reviewed in detail and the actual policy of procedure established (see section on Plant Layout and Facilities Planning).

The objectives of the discussion are, first, to understand the fundamentals behind the decision, and second, to make the decision which will give the most desirable combination of the advantages of both types of layout with the fewest disadvantages. Understanding by all concerned of the factors that underlie the decision leads to more effective application, once the layout is installed. This is particularly true of new products being put into production for the first time.

Workspace Layout and Motion Economy. One of the most important factors in the development of skill is the continued execution of an unhurried, repetitive sequence of motion which tends to form a smooth, effortless **habit path.** Those workers at the highest levels of productivity appear to expend little or no extra effort. This is the real difference between **speed-up** and **high-speed** work. The former is a hurrying of all motions, both the essential and the nonessential. The latter is attained by eliminating or reducing to a minimum all nonproductive motions and actually making more time available for a better quality performance of the "do."

Individuals left to their own resources rarely use the easiest method without special training or specific instruction. Their own efforts to increase output, particularly under the influence of a financial incentive, will usually result in "speed-up." It is easy to see how this results in increased fatigue and, therefore, lower morale and resentment. Work simplification proceeds on the basis that most people are working at a reasonable level now and that any increase in productivity from their effort must mean a more effective utilization of that effort. Their participation in the development of the improvements, their understanding of the basic principles involved, and their gratitude for being given the opportunity to participate in improvements should combine to raise morale, productivity, and cooperation.

The workspace and the motions of the individual at work within it constitute the area of activity in which the worker is most likely to be capable of making a contribution. This is his area of experience. He should be in a position to know more about it than anyone else. The fundamental principles of the workspace are based on the fact that there are only five different classifications of motion:

1. Finger motions.
2. Motions involving fingers and wrists.
3. Motions involving fingers, wrists, and forearm.
4. Motions involving fingers, wrists, forearm, and upper arm.
5. Motions which include all of these with the addition of a body motion or a change of posture.

A glance at the normal and maximum working areas diagrammed in Fig. 2 shows that the easiest motions are arcs or circles, which vary within the limits established by the dimension of the human body. A full analysis is given in the section on Motion and Methods Study.

Some basic **principles for effective motions** can be stated as follows:

1. Motions should be as **productive** as possible. The elimination of nonproductive motions is the first step.

2. Required motions should be performed within the **lowest classification** possible. Reaching can be reduced to a minimum by locating supplies and tools as near as possible to the normal working area. Any motion beyond the maximum or fourth class should be avoided if at all possible. The shorter the motion, the less time and effort it will take to perform it.

3. Motions should be **smooth and rhythmical.** Simultaneous motions of both hands in equal and opposite directions are natural, smooth, and easy. Different kinds of motions performed at the same time will require more concentration, with resulting increase in fatigue, and reduce the time for the most rapid to that of the slowest. Abrupt changes in direction should be avoided by smoothing them to circular motions if possible. This point is easily demonstrated by having people try to draw circles, then squares, then triangles. The circle,

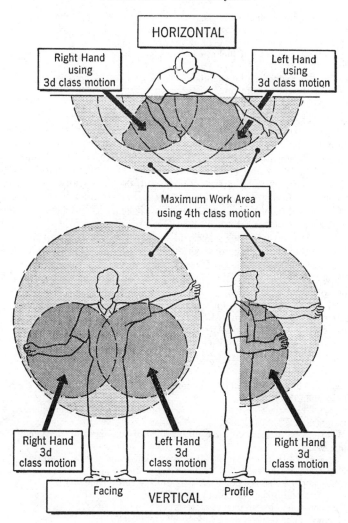

A. H. Mogensen. Work Simplification Conference
Fig. 2. Normal and maximum working areas for hand motions.

of course, is easiest. Attempts to draw a circle and a square simultaneously demonstrate the previous point.

4. **Tools and materials should be pre-positioned whenever possible:** Pre-positioning of tools and materials not only aids in consistently reducing the length of the motion, but minimizes the mental energy necessary. Habit makes the thought process virtually automatic.

Additional principles may be developed by discussion groups as they study their own workspace and motions. They should be encouraged to develop their own checklist. Different wordings arise from specific cases. A typical **workspace** and **motion checklist** follows.

1. Supply stations at proper levels.
2. Leverage at the best mechanical advantage.
3. Holding fixtures wherever possible.
4. Sliding grasps instead of pickups.
5. Pre-position for the next operation.
6. Foot pedals may often be useful.
7. Operator should be comfortable.
8. Workspace should be well lighted.

See the Right- and Left-Hand Checklist for detailed questions.

Multiple Activity Checklist. The multiple activity chart is a graphic representation of the coordinated activity of men, or machines, or both, in any combination. It may be used to represent the detailed activity of one man and one machine, one man and several machines, or just several men working as a team. The introduction of the time relationship of one activity coordinated with another becomes important in establishing the shortest path to completion. The information presented in the multiple activity chart is subjected to the same searching group attack as that in the flow or product process chart, through the work simplification pattern. Checklists made by each group to fit general conditions in their own organization arouse enthusiasm and catalyze specific ideas. One such follows:

1. Checklist for **"do" details.** (Details themselves cannot be generalized because they are specific for each machine.)
   a. Is this the right machine for the job?
   b. Are we using the most up-to-date attachments and tools for the job?
   c. Are machine speeds and feeds at a maximum?
   d. Are the tolerances too tight or too loose?

2. Checklist for **make-ready details.** (Generalized list; could apply to most machines; aimed at greater machine utilization.)
   a. Can we have tools, materials, or blueprints brought out to the machine beforehand?
   b. Can we have instructions accompany blueprints or specifications and see that they are faultlessly clear?
   c. Where possible, can we have duplicate tools prepared or sharpened by someone else?
   d. Can better planning reduce setups by making longer runs possible?
   e. Can oiling be done on "off" shift or during lunch periods?
   f. Can someone else lay out workspace during setup time?
   g. Can machine be left running with safety?
   h. Can someone else prepare materials?
   i. Can a hopper feed be used to load machine?
   j. Are we taking full advantage of guides and stops?

3. Checklist for **put-away details.** (Generalized list; could apply to most machines; aimed at greater machine utilization.)
   a. Can machine be left running safely?
   b. Is a limit switch practical?
   c. Can we use a drop delivery?
   d. Can someone else inspect work?
   e. Will a chute do the transporting of parts to the next job?
   f. Can cleanup of scrap be done automatically during the "do"?
   g. Why is so much paper work necessary?
   h. Will longer runs result in fewer change-overs?
   i. Can someone else return tools and excess materials?

Right- and Left-Hand Checklist. The right- and left-hand or operator chart is an analytic description of the motions of the two hands of an individual operator. Like the multiple activity chart, it provides a space for the parts sketch and the workspace layout. In effect, it is a multiple activity chart of the two hands of an operator, although in its simplest form no time scale is included. Rather, the activity of the two hands is coordinated by simply recording what each hand is doing at any instant. Whenever a detail performed by one hand is recorded, that which is being done by the other hand at the same time is noted in the other column.

The same Steps 3, 4, and 5 of the work simplification pattern are again applied just as with the other charts. Checklists developed by each group to fit conditions in their own plant are also helpful. A typical one follows:

1. Challenging the **"do" details:**

WHAT?

    a. Is it necessary?
    b. Is there some other way of getting the same results?

WHERE?

    a. Can the "do" be performed in a fixture or holding device?
    b. Is the "do" performed in the most convenient location?

WHEN?

    a. Can the "do" be performed on several units at once?
    b. Can the "do" be reversed?

WHO?

    Are both hands performing "do" details?

HOW?

    a. Are parts designed for easy performance of "do" details?
    b. Will simple mechanical aids be helpful?

2. Challenging the **make-ready** and **put-away details:**

WHAT?

    a. Is it necessary?
    b. Can an ejector be used on jigs and fixtures?

WHERE?

    a. Will gravity-feed hoppers bring materials closer to point of "do"?
    b. Are materials and tools located within the normal grasp areas?
    c. Are tools and materials pre-positioned whenever possible?
    d. Is the workspace laid out to facilitate the best sequence of motions?
    e. Are supply stations at proper working levels?
    f. Can the workspace be arranged so as to insure a rhythmical habit path of motion?

WHEN?

    a. Can two parts be picked up at the same time with one hand?
    b. Can a tool and a part be picked up in the same trip?
    c. Can parts be picked up ahead of time and be palmed?
    d. Can several pieces be put away at the same time?

WHO?

    a. Can each hand perform a complete cycle?
    b. Can alternate motions be used?
    c. Will a foot-pedal-operated device assist the hands?
    d. Can certain skilled details be simplified to the extent that the left hand can easily perform them?

HOW?

    a. Can a drop delivery be used?
    b. Can parts be slid instead of picked up?

c. Is the lowest possible classification of motion being used?
d. Can a foot pedal simplify the detail?
e. Can a change in leverage be made to advantage?

MOTION PICTURES. The motion picture is an integral part of work simplification activity. Films become teaching media and vehicles of recognition and reporting, as well as records of progress. They greatly stimulate the interest of those who are in them, and concentrate thinking in one area in a way that no other technique seems to do. Groups of people can walk by an operation day after day in the office or on the production floor and not see possibilities for improvement. When the group is shown a two-minute movie of the same job, ideas explode like magic. Sound motion pictures add other opportunities for the people on the job to participate. They not only see themselves, but hear themselves tell of the virtues of working more easily and economically.

Early films customarily showed the old method followed by the new. This technique has been replaced by the **documentary** type of film which tells the story of the improvement. The emphasis is on the people involved and the things they tried and did. The failures along the way are no less important than the successes as part of the story. The mock-ups, the trial runs, and the transferred people are all put into the picture. The result is a human story of progress rather than a technical illustration of methods. Both 8 mm and 16 mm film are in common use.

VIDEO TAPE RECORDING. Compact, low cost video tape recording systems are replacing motion pictures in many work simplification applications. One of the advantages of tape recording is its ability to instantly replay what has just been recorded. This permits analysis and discussions immediately after a particular process has been recorded. With motion pictures the time between the actual filming, processing, and setup before projection often reduces the impact of the demonstration.

Using closed circuit television a group may in fact assemble at some remote point to view an operation as it is being performed. Instant replay permits going back and viewing significant parts of the operation for group discussion. If a permanent record is not needed the video tape can be erased and used again.

CONFERENCE LEADERSHIP, PROBLEM SOLVING, AND DECISION MAKING. Since work simplification is group-oriented and emphasizes the organized team approach, effective leadership of the project group is a necessary adjunct to education in work simplification. Supervisors, as well as some hourly-rated employees, are being regularly included among those who receive instruction in this area while participating in the project's activities. Skills in conference leadership, problem solving and decision making are absolute requirements.

Nelson, Moore, Trego and many others (Work Simplification Conferences) recommend a step-by-step pattern for **practicing conference leadership,** which is extensively used in improvement projects where process charts are not practical, as follows:

1. The situation: What is unsatisfactory? Why be concerned?
2. The problem(s): define it specifically.
3. Possible solutions: to be arrived at by creativity and brainstorming.
4. Best solution: consolidation of best combination.
5. Summary, program of action: Who does what? When?

The similarity of these steps to the work simplification pattern is quite apparent, and they could be considered as merely a different wording. The important point is that they have the same thread of organization and participation running through them. They have been adopted as part of work simplification now for several years.

A Problem-Solving Procedure. Dunlap has designed a unique form for implementing the problem-solving procedure (Figs. 3 and 4). This is the result of many years of development by the participants in their own work simplification program. Many companies find that similar forms designed by their own personnel to fit their special needs are most effective.

WORK SIMPLIFICATION AND DESIGN. Project teams operating within the work simplification framework make improvements in the **design of new products,** as well as in already existing ones, by systematic review of each part in terms of its possibilities. Many companies schedule regular reviews of every product by specially organized task forces under the leadership of the engineering personnel responsible for the design. Such task forces include all levels of production people. Even those whose jobs are completely foreign to the problems of design make contributions. The questioning attitude in a "green light" atmosphere reveals numerous possibilities for such changes as elimination of parts, redesign for easier assembly, better end use of product or equipment, reduction in the cost of materials, and improved styling.

WORK SIMPLIFICATION AND STANDARDIZATION. Standardization of tools and equipment, component parts, and procedures can effectively be promoted by the work simplification approach. This is, perhaps, one of the most important vehicles for involving the engineer and technicians in the work simplification program at their own levels. Standards, when established and energetically enforced, provide a variety of economies. Project teams at all levels respond in this area when given the opportunity.

MORE EFFECTIVE TRAINING. Improvement means change, which requires retraining in the proposed methods. The process charts of proposed methods become effective **instruction sheets** useful to both supervisor and operator. Perhaps the greatest reduction in the tremendous nonproductive cost of retraining results from attitude changes on the part of those involved. Time invested in reviewing the work simplification philosophy and developing a cooperative atmosphere reduces the learning time considerably. The operator chart in a simplified form, without symbols and with several details grouped into word descriptions, becomes the best instruction sheet available for individual operations. Here is a simple example of five details of a right- and left-hand chart simplified for an instruction sheet.

Normal Detail	Condensed Statement
→ To nut O Pick up nut → Nut to fixture O Place on bolt O Tighten loosely with fingers	Get nut; assemble to bolt; tighten loosely with fingers.

Time study analysts can cooperate with the work simplification program by using this simplified format in their work. Then charts are already made if project teams desire to review them for improvement or if the supervisor wants to instruct his people. The very act of making the breakdown for instruction or

LICK THE PROBLEM	DATE	DIVISION Vehicle	LOCATION	DEPT. Garage	BY

OBJECTIVE	To cut down on pickup trips for parts

WHAT IS THE TROUBLE?	CAUSES	CORRECTIVE ACTION	RESPONSIBILITY FOR ACTION ASSIGNED TO	DUE DATE	DATE COMPLETED
Excessive trips into town to pick up material.	Shop men don't list all parts needed.	Foreman or assistant check list to see that it is complete	Foreman	1/15	
Three or four trips made on some days.					
Should be one trip only.	Outside locations not anticipating needs.	Schedule time for sending requisitions weekly.	Branch Foreman	1/15	
	Unnecessary "emergency" calls for parts.	Try to get parts locally first, then call in as early as possible.	Branch Foreman	1/15	
	Stockroom inventory incomplete.	Improve inventory.	Foreman	3/15	
	Incorrect part from supplier.	Closer check by pickup man.	Foreman	1/15	
	Wholesaler out of part.	Change source of supply.	Foreman	1/15	

Fig. 3. Form to implement problem solving.

H. P. Hood & Sons, Inc.

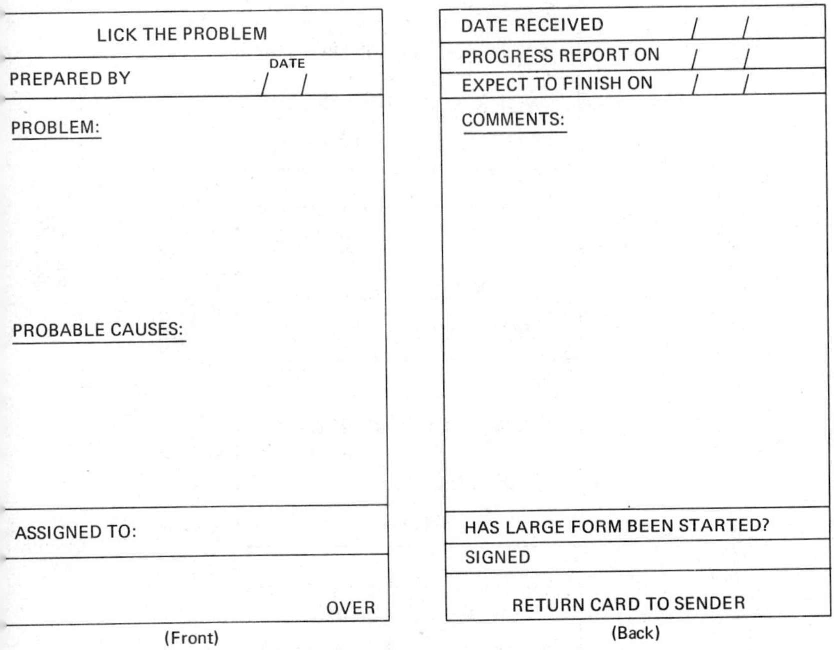

LICK THE PROBLEM	DATE RECEIVED / /
PREPARED BY DATE / /	PROGRESS REPORT ON / /
	EXPECT TO FINISH ON / /
PROBLEM:	COMMENTS:
PROBABLE CAUSES:	
ASSIGNED TO:	HAS LARGE FORM BEEN STARTED?
	SIGNED
OVER	RETURN CARD TO SENDER
(Front)	(Back)

H. P. Hood & Sons, Inc.

Fig. 4. Pocket card version of Dunlap form.

time study purposes in the form of a right- and left-hand chart may catalyze an improvement project.

WORK SIMPLIFICATION AND INCENTIVE SYSTEMS. It is difficult for a work simplification program to operate at all levels in a plant where incentives exist unless supervisors and workers alike have confidence in the management and the related systems. This is usually a problem of education. Time invested in developing understanding in this area will pay off not only in a better improvement program but in a better incentive system as well. The power of **financial incentives** is considered in many circles to be waning. Companies which have financial incentives often experience great difficulty if efforts are made to eliminate them. A better understanding of incentives and the time studies associated with them enhances an atmosphere of acceptance of their use. Improvements in existing systems and evolutionary changes toward "team-oriented" incentives are more readily made in such an environment. The simple fact is that everyone lives by the clock. All of us are time study men and women as we go about our daily lives. The plant could not run unless it was known how long it should take to do each task. Schedules, machine loads, delivery promises, labor requirements, purchasing rates, and capital requirements are just a few of the things which depend on time and expected performance. A simple graphic representation such as that shown in Fig. 5 is often helpful. The answers to each question, in Fig. 5, involve separate techniques, but they are related to the extent that each must build on the previous group, in descending order. In plants where an incentive program exists, the work simplifi-

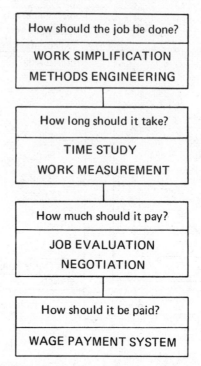

Fig. 5. A graphic representation of job questions and related techniques for
solutions.

cation activity may be enhanced if people at all levels understand the difference
between all of these techniques and the way they depend on each other.

WORK SIMPLIFICATION AND SUGGESTION SYSTEMS. In
the early days of work simplification, a suggestion plan was encouraged as part
of a program. Because work simplification is now group-oriented and suggestion
programs are still for the most part directed at the individual, fewer new suggestion plans are being installed. Here, again, is a financial incentive which is
waning in the face of the incentives of recognition, belonging and participation
and personal satisfaction which develop through work simplification. Where
both programs now exist, they complement each other quite well. Increases in
the improvement rate through work simplification stimulate individual initiative.
The rate of suggestions submitted usually increases. Rarely do people hold back
an idea in a group meeting so they can slip it in the suggestion box later. It
usually becomes quite apparent that someone else will think of it anyway. Few
companies pay for suggestions made by members of work simplification project
teams who leave the job for the specific assignment of working on a project.
Most people are so pleased with the opportunity to participate that the subject
does not even arise.

The philosophy that everyone is responsible for improvement on the job he
manages leaves little doubt that, in the long run, each must benefit in proportion
to the progress of the company.

A Program of Improvement

CRITERIA. Goodwin (Advanced Management, vol. 22) states that work mplification becomes an effective "program of improvement" within any organization when:

1. Costs are lower and the quality of product or service is higher as a result of the coordinated efforts of everyone in the organization to achieve these results.
2. The men and women who make up the organization begin to develop personally at an accelerated rate as a result of the improvement process in which they are participating.

The errors which originally prevented the almost universal acceptance of ıotion study appear to be the same that plague scientific and technical progress »day. So many of us assume that people will accept something new and useful ı its merits alone, or that a simple "order" will put the results of new technical nowledge into profitable practice. In actual fact, it becomes increasingly evient that the development of methods for motivating people to understand and se new techniques and to develop a positive attitude toward the application of ıe results, and the organization of the entire work force to participate to the xtent of their ability are by far the most important and the most difficult parts f our task.

A work simplification program is fun. It provides the satisfaction of being art of a team. It furnishes an opportunity for each person to participate to ıe extent of his ability. It satisfies man's natural desire to improve. It deelops an organization which understands and believes that in the method of oing anything lies the secret of progress, and which applies this axiom not only ɔ technical methods and the physical motions of doing a job, but also to the ıethods of management, leadership, and participation which bring about the enhusiastic use of any improvement. All results are achieved through a medium of ommunication which is acceptable to and utilized by the entire organization. 'his is a program of improvement within a framework of executive, supervisory, nd worker development, based on a simple philosophy of understanding and eamwork.

PHASES IN DEVELOPMENT. Goodwin identifies three distinct phases hrough which such a program develops:

Appreciation. In the appreciation phase, the entire organization becomes ,ware of what work simplification is and understands its implications as it may nvolve them as individuals and eventual participants in the program. This beomes the key phase in the continuation of the program, by completing the ircuit of communications and by constantly adding new grist to the mill.

Education. In the educational phase, the philosophy, the tools and techıiques, and the procedures for application are developed internally by small ;roups using the conference method. The research point of view and the desire ɔ experiment with methodology is the order of the day. Such an approach orients the thinking at the levels of the ability and responsibility of the people n any group. (This phase has been erroneously considered by many as the only ɔhase and thus in many instances has been used as training alone. So employed, ts effect is usually meager and short-lived compared to the constantly growing benefits from a complete program.)

Application. In the application phase, organized and continuous use of common-sense tools and techniques is made in order to develop and apply improvements at every level of the organization and in all activities. This is done

through appropriate project teams within a policy framework created fro
within to fit the particular organization and circumstances.

EFFECTIVE COORDINATION OF THE PROGRAM. Goodw
presents these three phases graphically in the accompanying chart (Fig. 6
Few organizations have done all of it in one program but many are well on tl
way. Others intentionally are interested only in parts of the program, or tal
an entirely different approach. The chart is fully explained in the discussic
below.

Yes or No Man. Every organization has one man who can say "yes" or "nc
and mean it. This may be the president, the treasurer, the chairman of tl
board, or a little old lady who owns 51 percent of the stock but never put
in an appearance at the place of business. In large corporations with man
divisions, the **division head** may be the man who can say "yes" or "no," pro
vided his operation stays appropriately in the black. He may be able to wor
quite independently of other divisions or of any central policy. Some companie
are so organized that a **small group** is expected to make final yes or no de
cisions. Even so, one individual is often likely to dominate the group.

Whatever the specific situation, the so-called "yes or no man" must some
how find out about work simplification. He must become interested at least t
the extent that he wants to learn more about its potential and how to go abou
setting up a program. This may occur in a host of ways. He may hear of i
through someone in his own organization or read of it in a book or periodica
He may learn of it through a friend at a social gathering. He may listen to
convincing address on the subject or sit in on a management seminar where it i
discussed. He may become acutely aware of work simplification through it
successful use by one of his competitors. Appreciation must eventually permeat
the entire organization, but it must begin at the top. This happens when th
"yes or no man" wants more information and takes action to get it.

The Top Executive Group. The executive leadership of the organization a
represented by the heads of each major activity usually make up the clos
advisors of the "yes or no man." They should have an opportunity to influence
him in his decision with respect to the advisability of proceeding with worl
simplification and should participate in any preliminary review and evaluatio
of it.

Complete Management Organization. Initial **appreciation sessions** fo
the entire management organization are usually limited to a two-hour informa
tive presentation and are often scheduled as an evening dinner meeting. Large
organizations often find it easier to arrange the presentation during working
hours by planning for two or more sessions. One portion of the management
group attends, while the other covers the supervisory needs of the operations;
then, vice versa. An evening dinner meeting, however, usually arouses more
interest and attaches greater importance to the new activity. Executive prefer
ence usually decides this point.

The importance of the initial meeting lies in the fact that it exemplifies the
work simplification approach. It is preferably opened by a few words from the
"yes or no man." All hear about the proposed program at the same time.
"Grapevine" interpretations of unknown activities are therefore minimized—a
most important point. Interest and enthusiasm are aroused as much as possible.
If a group is small enough, most of the basic work simplification philosophy can
be developed during this meeting. Everyone hears that the top executives have
participated in an appreciation session and were impressed to the extent that

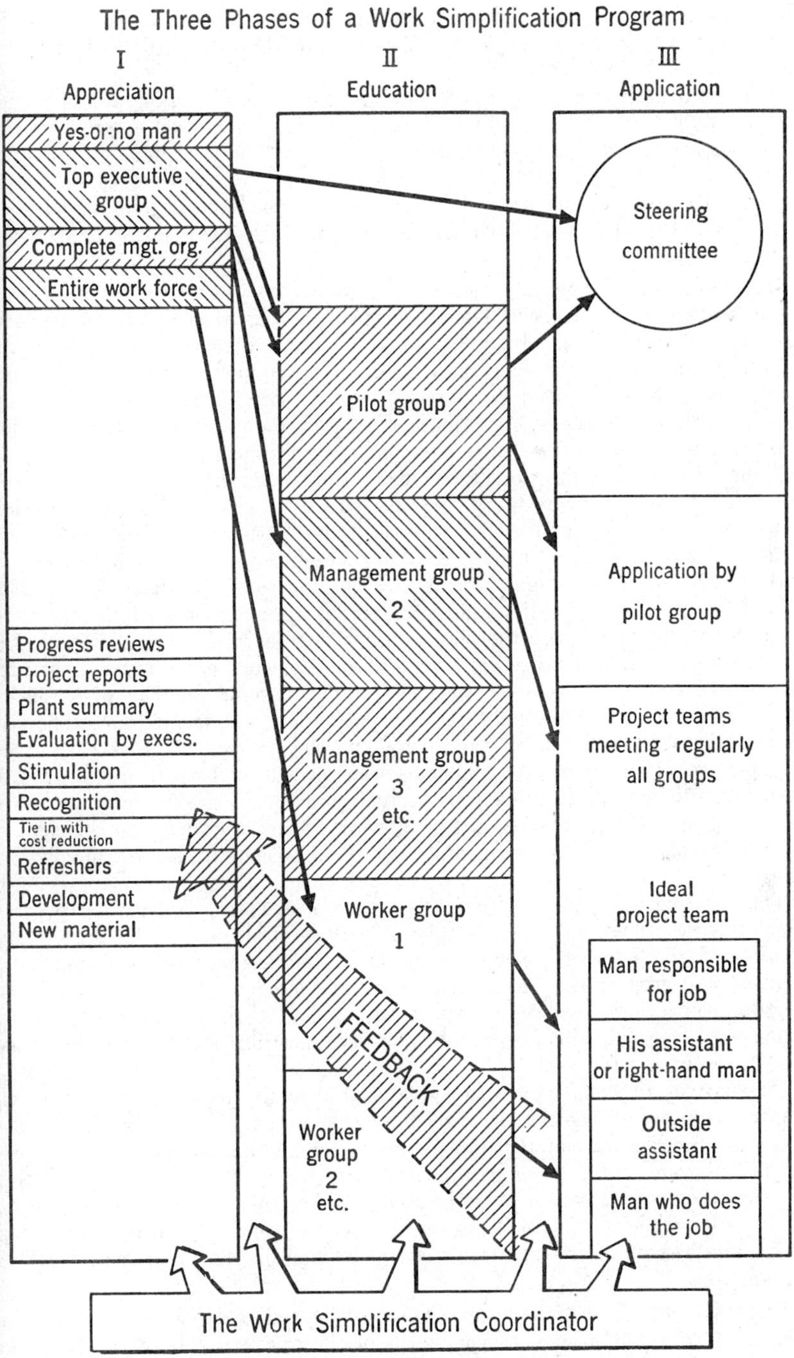

The Three Phases of a Work Simplification Program

I — Appreciation   II — Education   III — Application

Yes-or-no man

Top executive group

Complete mgt. org.

Entire work force

Progress reviews

Project reports

Plant summary

Evaluation by execs.

Stimulation

Recognition

Tie in with cost reduction

Refreshers

Development

New material

Steering committee

Pilot group

Management group 2

Management group 3 etc.

Worker group 1

Worker group 2 etc.

FEEDBACK

Application by pilot group

Project teams meeting regularly all groups

Ideal project team

Man responsible for job

His assistant or right-hand man

Outside assistant

Man who does the job

The Work Simplification Coordinator

Fig. 6. Phases of a complete work simplification program.

they have approved **experimentation with a pilot group.** This group w begin regular discussions soon and will be followed by other groups if its discu sions prove to be successful.

The group hears that progress of the pilot group and the program as a who will be publicized to the greatest extent possible. A motion picture may shown as an illustration, and the fact that films will be used may be announce at this time. This group should be notified of the appreciation session for th entire work force if one has been scheduled. Union officers and stewards a often invited to this session as a preview of the work-force session or in lieu such a session.

Entire Work Force. Appreciation by the entire work force takes on a di ferent degree of importance. The confidence that this group needs in the ma agement in order to supply understanding and cooperation may well deper on the willingness of the "yes or no man" to schedule this meeting or arran; appropriate publicity. People invariably mistrust what they do not understan and rumor can sometimes twist the facts so that they are unrecognizable. Th is particularly true if a few people have unaired grievances.

A meeting of the entire work force at one time or by departments or shifts fo about a half-hour appears to be the most effective means of informing worke of the program and soliciting their interest, participation, and cooperation. Suc a meeting would review a little of the work simplification philosophy, what it i and equally important, what it isn't. All get the same first-hand story. The hear of the previous executive and management appreciation sessions and of th decision to experiment with the pilot group. They learn that work simplificatio is being tried as a long-range improvement and development program that mu; benefit everyone if it is to be successful. They become familiar with the o portunities they will eventually have to participate in the program. They ar assured that they will hear of the progress of the program as it develops. Mc tion pictures, bulletin boards, and house organs will all become useful **supple ments to direct communication.** Actual educational sessions may be sched uled later on. Movies may be taken of workers and they may be asked to si in on some of the early experimental project discussions if they are involved i the activity.

This meeting is the step in the early stages of a work simplification program which is most often omitted. It is quite difficult to implement in the larger plant because of the numbers of people involved and, sometimes, inadequacies o facilities and geography. The cost is often considered too high this early in th program, since there has been no major return from work simplification thus far When work simplification becomes established, meetings are occasionally sched uled to report on progress.

Alternatives to such a meeting are available in the form of **reports** in the plant newspaper. Sometimes a special issue is prepared to announce the new program. Since work simplification has been set up thus far only as an experi mental program and is being tried out only by the pilot group, some executive feel that they should hold back any publicity other than that fact alone. Inade quate early publicity will usually place a heavy responsibility on the supervisor staff. The success of the program may well hinge on the effectiveness of thei explanations.

Pilot Group. The "pilot group," usually limited to from 12 to 16 men, i preferably composed of department heads or the **operating leadership** of eacl area of activity. It should include at least one or two of the top executives ir

rder to maintain a continuity between the original appreciation sessions and
ie pilot experiments.

This group reviews in detail all of the **educational material** which may be
sed in subsequent sessions by other groups. It participates in the development
f the work simplification philosophy, experiments with the use of the tools and
echniques, and reviews typical programs of application as developed by other
ompanies. Its schedule typically consists of 15 to 20 meetings of approximately
 hours duration spaced about a week apart. Some companies are experimenting
ith fewer but longer sessions spaced farther apart.

The principal responsibility of the pilot group is to analyze critically the ma-
erial presented with a view toward redesigning it to fit their own organization.
o that each participant may have an opportunity to "try out" work simplifi-
ation as a part of this procedure, each is requested to select five or more
perations or activities which seem to need improving or appear to him suitable
or experimentation. Subsequently, each member of the group is expected to
ractice on one of these, thus actually becoming involved in the improvement
rocess at the same time that he is learning about it.

During this exercise, each individual collects all of the pertinent facts about
is own project and presents them to the rest of the group. In so doing, he uses
he work simplification approach and those tools of analysis which the group has
hus far discussed. These usually include the work simplification and problem-
olving patterns, the flow process chart and flow diagram, films, video tapes, and
xisting reports of any kind which are already available in the particular organi-
ation involved. After the data thus collected have been reviewed by the group
as a whole, it selects three or four of the projects for further study. **Three-
or four-man teams** then concentrate on these projects during the remaining ses-
ions. Actual situations thus become the vehicle for discussing each new tool or
echnique, and involve the group in the human relations problems incident to
he use of these procedures as well. The projects not selected become a backlog
or future study.

The teams of three or four also meet apart from the main group between each
scheduled session, and a portion of each scheduled session is devoted to progress
reports, review of any difficulties encountered, and creative suggestions by the
entire group. Films of these small group meetings along with those of the actual
activity are effective in adding interest and enthusiasm, and in addition provide
catalysts for creative thinking and a means for documenting the progress. Sound
supplied extemporaneously by the people appearing in the pictures adds a most
effective measure of participation and recognition.

How to proceed most effectively becomes the main subject of discussions as the
participants begin to orient their thinking toward the problems involved in useful application of the results of their efforts. This leads to their final session,
which naturally becomes a problem-solving conference directed at organizing
themselves into the application phase, establishing **committees** to reorient material to be presented to succeeding groups, and scheduling **progress review
meetings,** at which time the pilot group will reconvene to check on its own
effectiveness.

Experiences of other companies are often brought into the discussion as guides
or possible alternate approaches. The group is normally quite willing to accept
one of these as an experimental procedure. Members usually feel that they will
be in a much better position at a later date to establish their own specific policy.
This is a very desirable situation, since it produces a condition of flexibility

requiring constant review. This effectively leads the way to subsequent organ
zation of the **steering committee.**

Application by the Pilot Group. Application by the pilot group usuall
involves experimentation with a few more projects from the list of those prev
ously charted but not studied during the educational phase. Project tea
groupings become a matter of personal preference rather than the "ideal projec
team," since often those in the pilot group may not be directly responsible fc
the activities they select to study. This very fact usually emphasizes the nee
of educating the entire organization, meanwhile informally involving everyon
concerned with any particular project.

These new project teams meet on their own at times they schedule themselve:
It is desirable to set up a procedure for **circulating minutes** of such meeting
to all members of the pilot group and to require the scheduling of the nex
meeting before adjourning. This policy provides excellent control over dupli
cation, establishing good liaison between all members of the group, and reducin
the likelihood of a few being "just too busy to get at it."

Approximately once a month, the group may reassemble as a whole to hear th
progress of each team, review any new films they have taken of the activitie
under study, and have a round of creative discussion on all aspects of eac
project. Such meetings lead the way to regular review sessions, with reports o
progress becoming the agenda of the meeting. The implications of the review
session take on added importance as the second management group (see Ad
ditional Management Groups, below) nears the completion of its educatio
phase and prepares to join the pilot group in application. Need for new plan:
for organizing the application activity becomes apparent as this situation de
velops.

Additional Management Groups. The second management group is usuall
made up of the right-hand men of those participating in the pilot group. Often
they are selected by the pilot group as it finishes its own educational sessions
They participate in a similar series of **educational discussions,** which incorpo
rate whatever revisions have been proposed by their predecessors.

There is rarely any major change in the subject-matter of the discussions
Rather, the recommendations usually call for more or less weight on certain
topics or sections as they appear to affect the particular situation. Increased
emphasis is quite regularly requested in the area of human relations, particularly
among the upper levels of management. The way is generally left open for the
groups that follow to continue to edit the material. This is highly worthwhile
on the basis that it gives every group an opportunity to participate in the de
velopment of the program. Since there is new material appearing all the time,
it becomes a very convenient arrangement from the point of view of adding new
grist to the mill and setting up the need for later refresher sessions.

All members of the second group, and of ensuing ones, are requested to select
five or more jobs or activities which appear to them to need improving. These
they use in the same manner as the pilot group did. Each member chooses one
from his own list as a **practice subject** about which to accumulate all the facts.
A few of these subjects are then selected by the experimental project teams. The
remainder become a backlog as the group joins the previous one in the appli
cation phase. These experimental project teams have the advantage of **counsel**
from members of prior groups. They are encouraged to seek assistance in this
area, since it initiates the eventual liaison necessary for transition to the ap
plication phase.

In essence, each group receives the same treatment as those preceding, except
that their final discussions can now be developed out of what has been previously established. The subject of the final session is always the identical theme,
"How can this group best merge with the previous ones in an organized program
of application?" Once again, there is actual participation on the part of the
people who are close to the job. Since the programming procedure remains
flexible, each group has a chance to influence the direction in which it will move.
Policies set up by previous groups and approved by the steering committee become a base line. Major changes are rarely recommended, but it would be still
more unusual for any group to fail to produce some useful innovation in the
procedure.

It may be desirable to bring the pilot group or the steering committee into
session to review these recommendations if they are at all radical in nature, or if
it appears appropriate in order to give some recognition to those who are responsible for them. This can add a generous measure of team spirit for all who
have participated in the development of the program thus far.

This same format is used with each succeeding management group until everyone in the organization has participated in the basic educational phase. Their
backlog of "activities which need improvement" becomes a regular source of new
project material as they join previous groups in the application phase.

Application by Later Groups in Organized Project Teams. As each
new group moves from the educational phase to the application phase, a new
opportunity to reorganize for the most effective procedure becomes available.
New project teams can be immediately set up to include the new men. Since
these are often assistants to previous group members, the ideal project teams
can begin to take shape in some instances. With teams set up on a **vertical
basis** within the organization, it becomes easier for upper levels of management
to delegate the more detailed work to those who now join them.

Since it is expected that innovation will continue even within the improvement procedure, the atmosphere of research and experimentation is maintained
constantly. The questioning attitude and the creative approach are exemplified.
New lines of communication are more readily established and old ones are freed
for more effective use. As the number of people involved increases, the need for
progress reporting and liaison between teams becomes obvious. Published **lists
of projects** being studied and the participants on each team keep everyone
abreast of activity. It affords executives an opportunity to pass along a word
of advice with respect to the importance of one project over another or to add
suggestions of their own. They are often in possession of information which
may well affect the activity under study, and which they might not normally
communicate to lower levels unless they knew of the work through publication
of these activity lists. It also allows executives to participate more regularly
by giving them the opportunity to add an item or two to the list if they choose
to do so.

As **organized application** begins in earnest, it rapidly becomes apparent that
the people actually doing the job have an important role. They can most effectively be drawn into the program early in the application phase on an informal basis. Films taken of the activities under study afford a convenient opportunity to do this. Portions of the project teams' discussions may be held
right on the job so as to include the workmen involved, or the workmen might
be brought into the meetings occasionally. Some means of including them must
be found, since a more formal introduction to work simplification must naturally

wait until all of the supervisory personnel have participated and the regula worker program can be set up in turn.

Steering Committee. The steering committee is a natural outgrowth of th pilot group's activities. It fills the need for establishing broad policies of pr cedure and for keeping the top executives informed early in the program. Th committee is appropriately composed of two or three of the top executive grou preferably including the "yes or no man," in addition to two or three membe of the pilot group. The most likely candidates are those who have shown th most enthusiasm and by their actions have indicated a desire to see the progra effectively utilized.

This committee now becomes a small, effective team of executives whose r sponsibility it is to see that the company gets the greatest benefit possible. The should meet regularly to **review** the overall progress of the work simplificatio program, **evaluate** its effectiveness, and **clarify** policies with respect to lon range aspects. In the early stages of the program they may wish to meet qui often. Later, once or twice a year is a typical meeting schedule; or they ma meet only when requested to by the staff member who has the full responsibilit of leading work simplification within the company. Since members of the steerin committee are high-level policy-makers, they back the program with prestige an authority and render needed executive recognition. It is through this group tha the feedback procedure is initially set up. They also supply needed top-lev counsel if any difficulties should arise that involve interdepartmental relation

Feedback. The feedback portion of the work simplification program con pletes the **circuit of communication** which began with the "yes or no man when he first indicated his interest. At regular intervals, reports of progres statements of dollar or man-hour savings, or summaries of accomplishments a necessary if interest is to be continued. Of equal importance is the need f top-level recognition of the results obtained.

One company holds a monthly progress-review dinner which is attended regu larly by each department head. If he cannot be present, he sends a represent tive in his place. (This group is essentially the original pilot group, or those the same positions.) All of the top executive group are invited and at least tw put in an appearance at each meeting. All persons, including workers, who hav participated in projects which have been completed during the last month a also invited. Appearance at the dinner becomes a form of recognition. Th after-dinner schedule includes an introduction of each individual who is prese for the first time, as well as a word or two from the top executives. This followed by a review of completed projects with those participating in eac telling the story. Any films which may have been taken are shown as part of th presentation. An open discussion usually follows the review of each projec which invariably leads to more projects' being added to the list. A **progres summary** for the year to date is passed out to all those present and circulate to all regular members and the top executive group. This report lists th projects by department, the number undertaken, those completed or discarde and the man-hour or dollar savings.

There are many other approaches used by various companies. The plant new paper or magazine is used in some instances. One big department store has "President's Luncheon" similar to the dinner meeting described above. Th president himself always attends. Some feel that less fanfare is more effectiv and written reports are simply circulated to interested parties. All successf programs have one thing in common—a realization at the top that the circu ust have complete continuity, that the lines of communication must be open, ad that new activity must be stimulated by top management recognition.

Annual **refresher sessions** are scheduled in numerous instances, even though ey are not always necessary if everyone is continually participating in the ogram on an active basis. One company has every person who has ever articipated in the educational phase return to the conference room for one full ay each year. These persons bring their reference material up to date, discuss e latest developments in the field, and lay plans for their activity in the coming ar. This is a constant appreciation procedure.

As programs develop, many of them are integrated into the routine procedures the plant operation. If a cost-reduction program and/or a suggestion plan ist, most companies tie work simplification in with them. Sometimes companies even initiate such programs to stimulate activity.

The Ideal Project Team. The ideal situation is one where everyone is constantly participating in the improvement process, each to the extent of his ability. he initiative should be in the hands of the man in charge of the activity, and he ould be using the most powerful forces for improvement available to him, that e economically appropriate for the particular project. The line organization is aphasized on the basis that staff service functions are desirable and should be quested by the line to the fullest extent available. The ideal project team ould then consist of the man responsible for the activity, his right-hand man assistant, outside assistance, and the man who does the job.

For example, a supervisor wants to improve the production flow through several operations in his department. His logical team might include, besides himlf, his assistant foreman, one of the industrial engineers, and the men who are tually performing the operations. The work simplification philosophy brings e foreman to realize that, in addition to the contributions of his own right-hand an, the **different frame of reference** of someone else like the industrial engiier plus that of the men on the job should provide better results and more thusiastic acceptance. If a new piece of equipment is being considered, the itside point of view might be supplied by a sales engineer of the equipment anufacturer or a technical expert from the mechanical engineering department. The bigger and more important the project, the higher the level of brains thin the organization that should be brought to bear on it. An open mind d experience with creativity among all people lead the man in charge to seek t ideas from everyone who could help. He should make sure that he gives all opportunity to participate who should be interested in doing so, and he should ke the responsibility on himself to create that interest.

It is not always possible to have the best possible team work on any specific oject. Often a team is not necessary at all, as in the case of a mechanic arnging his own workspace. An understanding of the need for priority and a nstant review of project lists develops into a policy of **delegating** more and ore activity down the line as the organization improves its ability and warrants ore confidence from above. The project teams change their character as the pacity of each individual becomes more fully utilized. The trend is one of tting closer to the actual doing of the job.

Worker Groups. Some companies balk at the idea of the hourly work force articipating in the educational phase on a formal basis. There is no doubt that effective job can be done informally by enlightened and enthusiastic first-line pervision. But we rarely find this type of supervision where executives refuse allow experimentation at least with a small portion of the work force.

Formal worker participation in the educational phase, while not normal as productive in tangible dollars saved as participation at higher levels of infl ence, nonetheless has almost always produced some gratifying results. The ve fact that management has considered workers important enough to include the program has had a marked effect on attitudes and feelings toward the com pany. Improvement in this area comes with understanding, which is the bas of the work simplification philosophy.

Worker series vary greatly in their content from plant to plant. In sor companies, it would appear quite natural that because of their narrower sphe of influence the hourly-paid employees would be given a much narrower versi of the basic work simplification material. In others, top management insists th receive exactly the same treatment as management people.

People who read the same papers, listen to the same radio programs, vote f the same representatives in government, can and do understand the simp philosophy of work simplification. If enthusiastically led to apply it on the own jobs, they will do so, if not inhibited by fears of insecurity or criticis from their fellow workers. This is neither allowing people on the job to run th business, nor asking permission of employees before making changes and in provements. It is part of an organized procedure of utilizing all of the brai available.

Workers who have participated in a formal educational series in work simp fication for the most part find new interest in their jobs. They look forward participating in improvement activity which involves them. The successful a plication of whatever improvement is made depends greatly upon their willin ness to make it work.

At Texas Instruments a Work Simplification Task Force was developed involve all members of a work group in the review and improvement of the total operations. This has resulted in the kind of job enlargement illustrated Fig. 7.

The Work Simplification Coordinator. The work simplification coord nator is the man who has the detailed operation of the program as his full-tin responsibility. He is preferably selected from within the organization. **Qual fying prerequisites** are his interests, his previous training, his knowledge the plant and its people, and his desire to do the job. He may need speci training to prepare him for all phases of the activity, but this can often gained by attending any of a number of courses offered by many colleges ar universities or by specialists in the field.

The need for this man becomes acutely apparent early in the appreciatic phase, when the "yes or no man" is just getting interested or the top manag ment group is discussing how to proceed with the pilot group. Often a delay necessary at this stage until the man selected has time to prepare himself lead such a group. In some cases, professional assistance is used during th early stages and the coordinator is trained along with the pilot and secor management groups, and then assumes control.

The coordinator's **responsibilities** include such things as leadership of th educational conferences, scheduling project group meetings, taking motion pi tures, compiling or editing progress reports, maintaining projects lists, and con piling plant-wide summaries for evaluation by the executive group. He is member of the steering committee and keeps that group informed of the sta of the program. He **reports** in many instances to personnel, industrial eng neering, the comptroller, the vice-president, or the president. The importar point is not where he reports in the organization but the implication of th

HORIZONTAL

• Assemblers on a transformer assembly line each performed a single operation as the assembly moved by on the conveyor belt. Jobs were enlarged horizontally by setting up work stations to permit each operator to assemble the entire unit. Operations now performed by each operator include cabling, upending, winding, soldering, laminating, and symbolizing.

• A similar transformer assembly line provides horizontal job enlargement when assemblers are taught how to perform all operations and are rotated to a different operation each day.

VERTICAL

• Assemblers on a radar assembly line are given information on customer contract commitments in terms of price, quality specifications, delivery schedules, and company data on material and personnel costs, breakeven performance, and potential profit margins. Assemblers and engineers work together in methods and design improvements. Assemblers inspect, adjust, and repair their own work, help test completed units, and receive copies of customer inspection reports.

• Female electronic assemblers involved in intricate assembling, bonding, soldering, and welding operations are given training in methods improvement and were encouraged to make suggestions for improving manufacturing processes. Natural work groups of five to 25 assemblers each elect a "team captain" for a term of six months. In addition to performing her regular operations, the team captain collects work improvement ideas from members of her team, describes them on a standard form, credits the suggestors, presents the recommendations to their supervisor and superintendent at the end of the week, and gives the team feedback on idea utilization. Though most job operations remain the same, vertical job enlargement is achieved by providing increased opportunity for planning, reorganizing and controlling their work, and earning recognition.

HORIZONTAL PLUS VERTICAL

• Jobs are enlarged horizontally in a clad metal rolling mill by qualifying operators to work interchangeably on breakdown rolling, finishing rolling, slitter, pickler, and abrader operations. After giving the operators training in methods improvement and basic metallurgy, jobs are enlarged vertically by involving them with engineering and supervisory personnel in problem-solving, goal-setting sessions for increasing production yields.

• Jobs in a large employee insurance section are enlarged horizontally by qualifying insurance clerks to work interchangeably in filing claims, mailing checks, enrolling and orienting new employees, checking premium and enrollment reports, adjusting payroll deductions, and interpreting policies to employees. Vertical enlargement involves clerks in insurance program planning meetings with personnel directors and carrier representatives, authorizes them to sign disbursement requests, permits them to attend a paper-work systems conference, and enables them to recommend equipment replacements and to rearrange their work layout.

Fig. 7. Examples of horizontal and vertical job enlargement.

H. F. Goodwin

the ART — PHILOSOPHY
Human Considerations

ATTITUDE — MOTIVATION

PARTICIPATION (everyone)
- Fun
- Natural
- Method
- Communications

YOU (Ourselves)
- In It For Me
- Personal Objectives
- Desire
- Motivation

PEOPLE (Most Important Asset)
- Resistance
- Resentment — WHY ?
- Fear
- Desire
- Tell, Sell, Involve

EMPLOYMENT (Economics)
- Implications of Change
- Automation
- Jobs and Company at Stake
- What if We Don't Improve?
- (The Facts of Life)

ACCELERATIVE CHANGE
- Obsolescence
- Your Unknown Competitor

the SCIENCE — TOOLS AND TECHNIQUES
Engineering — Organized Approach

SYSTEM — CREATIVE TEAMWORK

THE PROBLEM SOLVING PATTERN
1. The Job, Activity or Situation
   (Selection)
2. The Facts — Problems/Opportunities
   (Information Gathering)
3. Possible Solutions, Alternatives
   (Creative Thinking & Analysis)
4. Preferred Solutions
   (Practical Evaluation & Decision Making)
5. Installation
   (Plan of Action)
6. Feedback & Review

BASIC TOOLS
- Process Charts
- Flow Diagrams
- Layouts

ADVANCED TECHNIQUES
- Time Study & Work Measurement
- Plant Layout & Material Handling
- Planning, Scheduling & Forecasting
- Statistical Control, Analysis & Sampling
- Value Analysis, Engineering & Design
- Data Gathering; Processing & Analysis
- Operations Research & Mathematical
  Models

the PLAN — PROGRAM OF ACTION
Deliberate Planning & Scheduling of Improvement Activities

RESULTS — DOING THE RIGHT THINGS

EDUCATIONAL PROGRAMS
- Executive Development
- Supervisory Training
  (Prod., Office, Sales, Eng., etc.)
- Worker Sessions

MANAGEMENT OF IMPROVEMENT
- Executive Leadership
- Company Objectives
- Long Range Planning
- Policy, Procedure, Control
- Communications
- Department Goals & Programs
  (Line Responsibility)
- Planning
- Commitment (target dates)
- Communication & Scheduling
- Progress Review
- Measurement & Evaluation
- Replanning
  (What's Important?)
- Task Forces
- Special Projects
- Interdepartmental Teams
- Recognition (Individual)
- The Personal Inventory
  (Strengths & Weaknesses)
- Promotions & Advancement

ct in terms of how broad a field he is allowed to cover effectively. If he works
it of the industrial engineering department or manufacturing office and this
oes not limit him to the factory alone, it can work quite well. On the other
and, if any major functional line in the organization avoids representation in
ie pilot group or succeeding groups, it may be necessary to transfer the co-
:dinator's official reporting to the president's office or to the comptroller in
:der to give him the needed initial prestige.

In large plants employing several thousands, there may be several men on the
ork simplification staff, since many groups may be working simultaneously.
his activity has been used by numerous companies as an **executive develop-
ent area,** since it exemplifies improvement-mindedness in an atmosphere of
amwork. In smaller plants, the work simplification coordinator may even take
1 other duties as the program moves into the application phase and essentially
l of the personnel have participated in the educational phase.

Work simplification is an effective program of improvement because it functions
 an environment of friendliness. It develops a philosophy of improvement
irough teamwork and understanding together with the personal development of
ie individual. It gives each man an opportunity to participate to the extent of
is ability. What may be the proper procedure in one organization may not be
 in another. Each organization must find out by actually studying its own
tuation.

IMPROVEMENT MANAGEMENT. Improvement must be managed.
oodwin states, "Since it is virtually impossible for managers or engineers alike
 be knowledgeable and skilled to a high level of proficiency in all or even a
ajor portion of the improvement techniques, it is increasingly obvious that
ecialists are here to stay and must become members of the team. The wise
ader uses all of his people to generate and implement results. He succeeds
ecause of the way he goes at it—the way he manages the process. The ef-
:ctive innovator puts people and technique together in a powerful pattern of
rogress to achieve the desired results."

This framework for managing improvement (Fig. 8) was developed through
ork simplification and becomes a useful way of visualizing the total work
mplification effort.

Work simplification is a coordinated team effort of "do-it-yourself" improve-
ent at all levels. A positive environment of change is developed as a result of
veryone trying to improve their own effectiveness by improving the things they
o. **Attitude** and **motivation** are the mainsprings of the philosophy of human
onsiderations. **System** and **creative teamwork** are the patterns of success in
sing all of the improvement techniques. Doing **the right things** to get re-
ults is fundamental to an effective program of action. Understanding, confi-
ence, respect, and teamwork enhance the profitable growth of the company by
veryone working toward common goals.

SECTION 15

INDUSTRIAL RESEARCH AND DEVELOPMENT

CONTENTS

INDUSTRIAL RESEARCH AND DEVELOPMENT

Nature of Research and Development

DEFINITION AND SCOPE. Research and development, a commonly used phrase, describes a wide variety of distinctly different technical endeavors among different industries and companies.

Research is defined by Webster as "studious inquiry; usually, critical and exhaustive investigation or experimentation having for its aim the revision of accepted conclusions in the light of newly discovered facts." Research, in this broad sense, is thus applicable to all human affairs and all fields of endeavor. In industry, the term research has come to mean scientific investigations, early engineering development, preliminary pilot-plant studies, product and process evaluation, and formalized study of operations and functions.

Equally, **development** refers to the technical activity involved in converting the results of research into usable products and processes, and usually requires design and test of prototypes and pilot plants, material selection, environmental and fatigue testing, and a major effort to achieve favorable economics.

OBJECTIVES OF RESEARCH AND DEVELOPMENT. In a profit-making organization the objectives of research and development can normally be stated in terms of:

1. Technical information required by management.
2. The solution of a specific problem.
3. Formulation of a product or process concept.
4. Elaboration of a technical concept to its full potential (development) within time and cost constraints.

These objectives are usually further translated into detailed product specifications, research project objectives, and specific schedule and cost objectives for (1) the technical manpower required to solve the problem or achieve the product and (2) the resulting product itself.

ROLE IN INDUSTRY. The overall role of the research and development function in a business is to apply scientific knowledge to a product, process or service which an organization with existing or future resources can make or sell to achieve the organization's objectives. In a commercial organization this objective is usually the making of a profit. Research and development can fill many different roles, depending upon the need of the company since each industry has its own unique requirements. In some industries, state-of-the-art development in electronics, materials, and information processing is normally required. Therefore, the research and development role in the aerospace industry, for example, is radically different from that of research and development in consumer packaging or construction materials—where evolutionary, economically dominated development is required. These differences are reflected in the widely varying amounts committed to research and development by different industries as a percentage of sales. While not an ideal measure, the gross differences in level of spending indicate the difference in importance of research and development to each industry.

15·1

Research and development may be prompted by a variety of needs. Some o these are:

1. Improvement of existing products.
2. Development of new products.
3. Improvement of existing production processes.
4. Development of new, more efficient processes.
5. Solution of current production problems.
6. Reduction of production costs.
7. Development of new uses for existing materials, processes, or devices.
8. Adaptation of products to new markets.
9. Improvement of customer and public relations.
10. Exploration of diversification or expansion opportunities.
11. Assurance of sources of materials supply.
12. Assistance to management planning.
13. Abatement of dangers and nuisances.
14. Assistance in standardization.
15. Development of prestige and acceptance for the company and its product.

Several specific objectives may exist in a company at any one time, depending upon the company's competitive position, availability of resources, and historica technical success. Therefore, most research and development programs are usually a compromise among objectives, many of which compete for scarce technica and financial resources.

It is not easy to assess precisely the contribution of research and developmen' to a company. That such contribution exists, however, cannot be questioned either internationally among countries or within a single industry. As stated by J. Servan-Schrieber (The American Challenge), the United States has beer notably more successful in translating scientific knowledge into useful enc products and processes than have countries in the European Common Market Within an industry, leading companies who have achieved their position basec upon technical performance frequently can be identified. As described by Morrison and Neuschel (Harvard Business Review, vol. 40) industry leaders perform approximately twice as well as the poorer competitors. In the technica industries, research and development management is a major factor in this distinction.

STRUCTURE OF RESEARCH AND DEVELOPMENT. The technical and managerial activity which takes place in the research and development function of a business may be described in many ways. One way is to describe this activity as a sequence of steps:

1. Fundamental or basic research
2. Applied research
3. Development
4. Product design
5. Technical field support/customer service

Not all companies participate in all phases of research and development Each company has its own definition of the technical process and its own philosophy for assigning responsibility for various phases of this process (see discussion under Organization of the Department).

Fundamental or Basic Research. The term fundamental research has different meanings to different people and its imprecise use leads to much confusion. In a pure sense, **fundamental research** means scientific inquiry conducted solely out of curiosity, without a practical or utilitarian objective. It is sometimes described as "research for research's sake." Universities, some governmental laboratories, and institutions dedicated to learning conduct such research. Some industrial corporations encourage research personnel to spend part of their time pursuing inquiries into new fields, without commercial objectives defined or contemplated.

Of course, many companies conduct exploratory research into areas where it is believed new knowledge might be uncovered that could lead to industrial values. Such research, called **directed basic research** by some research administrators, contemplates a useful end, even though no specific end product can be defined. Industry spends millions of dollars yearly on exploratory investigations of this nature, and these investigations, while not meeting a pure definition of fundamental research, produce a large portion of the additions to scientific knowledge. In this section, the term "fundamental research" will be used in the absolute sense only. When directed basic research is meant, it will be called by that name or by the shortened term "basic research."

Applied Research. The bulk of the research conducted by both industry and government in the United States is applied research. Broadly, applied research includes all technical undertakings aimed at **solving problems of practical significance.** By this definition, applied research includes those activities frequently distinguished as development. It also includes that portion of a directed basic research that is continued once an end use or a practical value is foreseen.

While applied research aims at solving problems, the problem is not necessarily of a troublesome nature. It may arise from the desire for progress. Finding a new product that can lead to increased corporate prosperity and expansion is solved through applied research, just as a problem in cutting production costs or eliminating a product deficiency may be solved.

Development. While there is no sharp distinction between applied research and development, development generally takes a known technical solution from research and completes the supporting mechanical, electronic, chemical, and other disciplinary information needed to move the solution beyond the laboratory bench. It is at this stage that laboratory-proved fact is translated into **pilot plants or prototypes** and the problems of use and production are solved.

It is possible for development work to be conducted without preceding laboratory study. Information gained from experience in the operation of a process or in the use of a product may be the basis for **evolutionary development** in industries where technology evolves slowly.

Development includes the effort to improve existing products within known technology, as companies meet the demand for change and improvement in their products. Although it is difficult or impossible to ascertain how much of the country's technical effort is expended on the improvement of products, there is no doubt that this type of research constitutes a substantial portion of the total.

New-product development means new horizons for an enterprise—expansion, progress, and profits beyond those of the past. Often the most successful industrial firms are those that have been able to use research to produce a constant stream of new products. The Minnesota Mining and Manufacturing Company is an excellent example, as are General Electric, Westinghouse, and du Pont.

In one survey, reporting companies rated the development of new products as the most important aim of their research programs. Forty percent stated that the development of new products was their principal objective; the same percentage stated that product improvement was their principal objective.

A study of 191 companies (Dearborn, et al. (eds.), Spending for Industrial Research) reveals that 50 percent of research expenditures were directed to the improvement of present products or processes, 42 percent to the development of new products or processes, and 8 percent to unspecified objectives.

Most industries today cannot depend on the fortuitous inventions that produced many important products in the past. A **systematic exploration program** is usually needed to find new products for manufacture. This requires a company, intent upon growth and expansion, to conduct considerable directed basic research in its fields of interest. Larger corporations with greater resources are more able to take the risks inherent in this type of research. For this reason, new products that alter an industry's economics usually come from the development laboratories of the large corporations. In newer technologies, this is much less true.

In the process industries, research and development is directed toward **improving existing processes** or developing new ones. Such research may deal with process steps or with the complete manufacturing method.

Improvement or innovation in processes may be desirable either to improve the quality of the product or to reduce production costs. Frequently, process research is aimed at devising an economical method of manufacturing a product that has not been produced commercially before. An example of the latter would be the development of a commercial process for the manufacture of a new chemical or a previously unused metal. Obviously, process research may involve the most difficult of technical problems and may require the highest type of scientific work.

Process improvement receives more attention in industry than the development of new processes. To maximize the return on invested capital, it is desirable to get the highest efficiency from existing processes, particularly in those industries which have a high invested capital per dollar of sales.

Process research is extremely important in the primary metals and chemical industries and increasingly important in electronics, where solid-state processing and circuit design are interdependent.

Product Design. Much of the engineering effort in many companies is devoted to this last step in the technical process—just prior to release of technical results to manufacturing. Although this phase of the activity is often included in development work it deserves separate mention. Product design has the following **functions**:

1. Designing parts and processes so they are producible in the company's manufacturing facilities to as large an extent as possible.
2. Reducing the cost of the product by selecting common parts and multiple supplier sources, and conducting value engineering.
3. Ensuring that proper tolerances, materials, and dimensions are specified, so that the product can be mass produced.

During this phase **production drawings** are completed and **parts lists** are compiled. Little or no conceptual work is done in this last step. If technical problems are encountered, product design engineers normally request assistance from the development engineers familiar with the product.

Standard test specifications are written and the documentation required is normally provided by this function. While usually of low rank in the technical departments, this group can have a significant effect upon the cost of manufacture through drafting practice, dimensional decisions, and selection of standard parts.

Technical Support. After a design has been released for procurement and manufacture, supporting technical service is often required by manufacturing, procurement, or customers. This is particularly true for highly complex products. The functions provided in support engineering are:

In support of manufacturing or procurement:
1. Interpretation of the drawings for preproduction and tooling engineers.
2. Writing, evaluating, and processing engineering changes to permit manufacture or reduce the cost of manufacture.
3. Answering questions concerning the process which although successful at the pilot-plant level has new problems at the full-scale level.

For marketing or customers:
1. Helping solve customer problems, or identifying and defining problems relating to product or process, in the customers' plant.
2. Defining new products or processes required to compete effectively and translating them into product specifications which may be used for planning the research and development program.

Organization of Research and Development

BASIC ORGANIZATION. A skeleton chart of organization of the research function is shown in Fig. 1. The chart represents an idealized situation and would probably rarely exist in practice. As shown, the chief research executive has vice-presidential status and is directly responsible to the president. Through the presidential office he has contact with production, sales, finance, and other executive branches of the company. He has **direct executive responsibility** for the technical sections, but at the same time, all details of day-by-day management need not clear through him, but may go to his assistants.

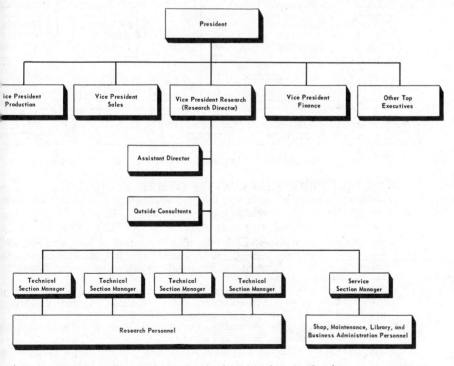

Fig. 1. Chart of an idealized research organization.

He is also responsible for broad policy in the management of the service and business units that supplement his technical sections.

In this **idealized scheme,** quality control and other laboratories have been deliberately omitted from the responsibility of the research director. In theory these are functions of production and merely dilute the activities of the research department. Many companies, however, put all laboratories under the executive responsibility of the research director. Practical considerations, including the nature of the company's product, the size of the operation, and the number of research personnel employed, make deviations from the idealized scheme expedient.

Bachman (Research for Profit) shows a modified organization chart (Fig. 2) for a chemical company emphasizing the operations with which research is most concerned.

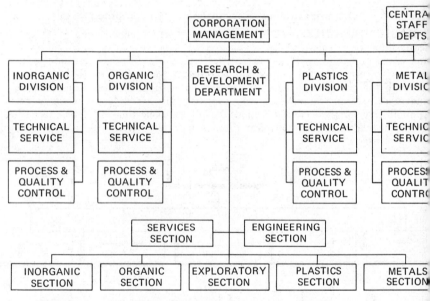

Fig. 2. Modified organization chart for a chemical plant.

INTERRELATIONSHIPS WITH OTHER FUNCTIONS. One of the major objectives of research and development management is the establishment of effective working relationships with the corporate staff, marketing manufacturing, finance, and other units of the company. The reasons are:

1. Even though the research and development department may be well run, and its program executed properly, the program may not be directed to the critical needs of the business, so that results are not of major significance to the corporation as a whole.
2. The original product design may be well-conceived and executed from a technical point of view, but customer-prompted changes under crash conditions introduced late in the design cycle are very expensive.
3. A new product may be technically excellent, but its effect on manufacturing and marketing causes an unusually high introduction cost.

Corporate Staff. Research and development is one of the major avenues for implementing corporate strategy and, as described by Ansoff (Corporate Strat-

egy) is the fundamental alternative to acquisition of a product line or competence. As such, the success of research and development has a major influence
upon the breadth and nature of the product line.

Another major responsibility is determining the technical feasibility of diversi-
fying the product line. This helps management decide whether a specific market
should be entered, the resources required, and the rate of investment. It also
helps identify major obstacles. Technical groups sometimes find such diversification analyses difficult to perform with confidence, since the technology is unfamiliar and technical problems are not apparent. The result is that a company
may enter a new technical field without fully understanding the difficulties to
be encountered.

The research and development department is frequently called upon by
corporate management to advise on the timing of a research and development
venture. This is particularly essential if a large market introduction cost or
capital equipment investment is required, in order to commit minimum financial
resources at an early date, and not be at a competitive disadvantage due to
the delay in investment.

The research and development department works with corporate planning or
marketing staffs to provide **competitive intelligence.** This includes general
technical forecasting, but more specifically, an analysis of the direction and
amount of resources competitors are devoting to specific areas of technology.

The corporate staff/research and development interaction is usually most
formal during the **annual planning process.** At that time the direction and
rate of product/process investigation is analyzed in the light of the company's
overall objectives.

Top management's support of research is needed since **divisional and departmental management** can greatly influence the profits resulting from the
investment in the research. Bachman (Research for Profit) cites as an example
a research project that had been well conceived, evaluated, and accepted as
worthwhile. The research proved successful and the forecasted economics of a
commercial project was acceptable. The project was subjected to pilot-plant
verification and the results scaled up for commercial operation. Again the economics of the project were found acceptable and the project was considered for
commercialization. At this point the divisional management decided not to go
through with commercialization and the research department was left carrying
the costs of the unused research.

Marketing. In most technical industries a good relationship with marketing
is required in order to plan product lines well and to define product-line and
individual product specifications. The selection of a product line includes specifying:

1. Product families within the line.
2. Number of products in each family.
3. Levels of technical performance.
4. Level of reliability.
5. Cost/price levels within the line.
6. Maintainability.
7. Relation to competitive products.

Following specification of the line, specification of individual products is required. Based upon marketing requirements, specifications are normally written
in the research and development department and sometimes reviewed by technical members of the marketing department.

Agreement is also required concerning **reliability of product.** In most complex electromechanical products, high reliability may be achieved, if a company is willing to accept the cost of design and test. Against this design cost must be balanced the cost of service in the field, if a lesser degree of reliability is accepted.

Another relationship to be managed is that between research and development and service to customers in the field. The design concept itself often determines which parts will require servicing, the technical sophistication required by the servicemen, the frequency of product malfunction, and the simplicity with which a malfunction may be corrected. In some companies, these parameters are translated into technical service specifications by the field services group in marketing headquarters, and become a design requirement.

Another significant problem for many manufacturers is **spare parts.** Often overlooked or given a low priority in both production and engineering planning, new product introductions are often tainted by the unavailability of spare parts at a depot or location where repairs will be made. The problem seems to exist in most companies. Production planning and separate scheduling of spare parts, coupled with separately assigned design responsibility, often eliminate this major irritant in new product introduction.

Manufacturing. The research and development department, in a large measure, sets the cost levels at which manufacturing will work, by designing into product cost the material requirements, machining tolerances, and sequence of assembly. Use of large numbers of nonstandard individual parts increases the overhead cost in manufacturing support groups, such as production control, inspection, purchasing, expediting, and receiving.

The effect of engineering on manufacturing is receiving increased attention partly due to **value engineering/value analysis.** The Society of Manufacturing Engineers (Value Engineering in Manufacturing) describes value engineering as an organized effort to attain **optimum value** in a product, system, or service, by providing the necessary functions at the lowest cost. Its importance in research and development stems from the increasing isolation of the research personnel from the rapidly expanding shop technology. The young researcher may never have been in the shop while the senior engineer may have experience that has been made technically obsolete by recent developments. Although companies have had varying success with the technique, there is little doubt that substantial savings may be gained by its use.

Research and development has a major effect on the purchasing function as well. In many complex products, two-thirds of the product cost is purchased outside the company. Steps to reduce this cost include:

1. Locating purchasing analysts in the research and development design section to assist designers in selecting low-cost, standard parts.
2. Using purchase analysis to identify high-volume parts whose redesign might effect major savings.
3. Eliminating sole-source suppliers as much as possible.

Another major relationship between research and development and manufacturing concerns **equipment and facilities.** In the electronics field, as the technology has changed from eletromechanical to vacuum tubes to transistors to integrated circuits, major changes in manufacturing facilities have been required. To a lesser degree, every product designer affects his company's ability to utilize existing facilities. Often the designer determines the "make-buy" decision by designing for special equipment. Temporary assignment of design engineers to manufacturing departments helps to alleviate this condition.

A fourth major relationship between research and development and manufacturing concerns **engineering changes.** Release of a design, including a parts list and test specifications, is normally regarded as the interface between development and manufacturing. However, after release of design, a steady stream of engineering changes can seriously affect manufacturing. While relatively few changes are critical by themselves, large numbers overload manufacturing administration channels, obsolete material, and disrupt manufacturing schedules and inspection. The normal administrative answer to such a problem is a strong **engineering change control** function headed by an engineering change board. The board is responsible for approving the change, establishing when the change is to be made, stopping production, and determining whether tooling and material should be reworked or scrapped.

Finance. Research and development departments normally work on a budget, as do most departments in a company. Often, however, a budget is not particularly meaningful to research and development managers for the following reasons:

1. The most important product cost decisions are made when the designer or development engineer sits at the drafting board, a function not reflected in most accounting or control systems.
2. The cost of the product is far more important than the budget of the engineering department. For example, a research and development budget of 5 percent of sales is high for many industries, although the "designed-in" unit cost of the product may be 70 to 80 percent of sales. These differences in importance are not properly understood or reflected in budgets.
3. The cost that the manufacturing department will incur due to design is not reflected in accounting costs until it is far too late for the engineer to make other than minor changes.

This raises a basic problem concerning the relationship between financial and engineering groups. Unless financial functions account for commitments rather than expenditures early in the design process, during development and selection of vendors, there is little financial control exerted during the design process.

External Factors. Few corporations can afford to investigate all of the technologies that are pertinent to their business. This requires selecting those areas of technology that are most important for detailed study, and simply maintaining an awareness of other sources of technology to ensure that other developments are not a threat.

Often informal, systems of maintaining awareness about external technology development include:

1. Consulting arrangements with university professors knowledgeable in particular disciplines.
2. Review of all doctoral theses in the field.
3. Attendance at technical conferences and seminars.
4. Continual scanning of journals and papers.

ORGANIZATION OF THE DEPARTMENT.

ORGANIZATION OF THE DEPARTMENT. Sound organization is one of the requirements for a smoothly operating research and development function. However, structure alone does not solve major problems. The management process is at least as important as organizational structure.

As elaborated upon by Sherman (It All Depends) organizing is "the process of grouping activities and responsibilities, and establishing relationships (formal and informal) that will enable people to work together most effectively in determining and accomplishing the objectives of an enterprise."

Alternative Organizational Structures. The basic forms of organization available to a research and development department to accomplish its objectives are:

1. A project organization.
2. A specialist or discipline organization.
3. Phasing of the research and development process, i.e., separate research, development, and field support.

Further, these three forms may be either centralized or decentralized. Each form has its advantages. The advantages and disadvantages of centralization and decentralization were summarized well by Sherman (It All Depends):

Centralization permits
1. Greater development of technical expertise in a community of similar skills.
2. Easier scheduling of scarce resources.
3. Development of techniques for managing technical people.
4. Efficient use of services such as laboratories, test cells, pilot plants, and drafting.
5. Improved managerial control over a difficult-to-control function.

Decentralization permits
1. Closer coupling of the technical function to manufacturing and marketing at lower levels.
2. Decision making closer to the business.
3. Development of broader managers earlier in their careers.
4. Differences in organization and style appropriate to the individual product lines.

Organizing the research and development department by phase of the research process permits separation of the research function from the applied research function, and both from development thus bringing together people with skills and technical responsibility, and separating dissimilar skills.

In research, one man or a small team normally investigates a problem or phenomenon. Their job is to generate information, understanding, and even alternative ways to achieve the same end. **Researchers** tend to be self-motivated, expert in their particular discipline, and more concerned with technical solutions than with time and cost considerations.

In contrast, a **development engineer** is significantly concerned with time and cost. While the product must have technical integrity, the engineer works within time and cost constraints. His job is not to generate additional conceptual alternatives, but rather to select the one that offers the best solution to a problem, and to move it quickly to production. Managing skills as different as these is extremely difficult when both groups are managed by the same individual, by the same procedural rules, and within the same structure.

Many companies organize by discipline or technical specialty, such as metallurgy or circuit design. The significant advantage of this approach is that it concentrates specialty or discipline expertise in a single group which can result in improved sophistication within the discipline—sometimes at the expense of individual projects. It permits subspecialization within the disciplines, and allows managers to form project teams to solve a particular problem within the discipline more readily. On the other hand, this organization tends to complicate the job of schedule and cost control since all disciplines must be closely woven together on a major technical project, in support of the rest of the company.

The fourth major organizational orientation is that of **project management** which is practiced extensively in the aerospace and construction industries. Appropriate numbers of applied researchers, development engineers, and discipline specialists are gathered within the aegis of the project manager, who is responsible for results-oriented functional performance. This tends to get the job done on time and increases cost control, but often at the sacrifice of sophistication within a discipline. That is, less effort is normally devoted to improvement of knowledge within the discipline and a much greater effort is devoted toward obtaining a particular project objective. Such an approach can create friction between the project manager and the discipline manager when a project form of organization is used.

MANAGEMENT PROCESSES. While organization of the research and development department is important, the processes by which it is managed are critical. These management processes determine the quality and quantity of output, the direction of the technical effort, the extent to which corporate objectives are met, and the capacity to expand beyond the current level of achievement. The major processes are:

1. **Project selection:** deciding upon the work to be done to meet corporate objectives and marketing requirements.
2. **Personnel management:** hiring, evaluating, and rewarding technical personnel.
3. **Work assignment:** selecting technical group leaders or project leaders within research and development.
4. **Review:** checking on the quality of work and seeing that the original objectives are met.
5. **Allocation of resources:** budgeting and scheduling of projects and planning facilities.

Project selection, review of projects, and allocation of resources will be treated in more detail later in this section. The processes concerning personnel are described in the following paragraphs.

PERSONNEL MANAGEMENT. Success in research is largely dependent upon the people conducting the research. A good research staff is able to carry an idea to fruition at a minimum cost, while an inferior staff will vacillate, waste time, and frequently will not achieve the desired objective. The qualities for research personnel, as defined by Engstrom and Alexander (Proceedings, Institute of Radio Engineers, vol. 40) typify the thinking of modern management. According to them, industry expects its research workers to:

1. Possess originality and, even if in varying degrees, creativeness (inventiveness).
2. Be well trained in the fundamentals of science and continue to improve their knowledge and skills throughout their research careers.
3. Be reliable and of good character.
4. Be energetic and have drive in the conduct of research.
5. Be scientifically inquisitive in the fields assigned and for the objectives established for the research program.
6. Have a practical outlook—a common-sense approach to their work.
7. Have wholesome attitudes toward their work and toward the persons with whom they come in contact, and understand and practice cooperation.
8. Make appropriate use of the freedoms of the research environment.
9. Be, insofar as possible, young in mental outlook.

Williams, ed. (Proceedings, Eighth Annual Conference on the Administration of Research) sums up the **requirements for the research man** as follows:

1. Creativeness, or the ability to develop ideas from new associations of information.
2. Scientific integrity, or absolute reliability in the conduct of experiments and the interpretation of results.
3. Cooperativeness, or the ability to get along with others and work harmoniously on the research team.
4. Drive, or the practiced will to get the job done.

5. Knowledge of fundamentals and of research methods.
6. Practicality, including economic sense.
7. Salesmanship, meaning the ability to organize and present ideas so as to communicate to others and to motivate them.

Creativeness or **inventiveness** is important in the research worker, because the research process is essentially a process of creation or invention. This is an intangible capacity, difficult to measure, and usually evidenced only after the worker has been on the job for some time.

A **good understanding of scientific and engineering principles** is basic. While training is no substitute for inventiveness, in the present state of technology it is a necessary prerequisite for solving problems. Industrial research personnel frequently have advanced university training, and the doctor's degree is commonplace in many laboratories.

Curiosity is a basic trait of the scientist and an important adjunct to the creative spirit. Dissatisfaction with the status quo frequently inspires new ideas, but dissatisfaction per se is no indication of research ability. The research worker should understand that the laboratory exists to promote the earnings of the company and should be able to identify his own motivation with this function. Industry usually strives to promote the research worker's interest in the total activity of the company, as well as his understanding of the part played by his own group.

Previous experience, scholastic records, and various aptitude and other psychological tests are used in singling out competent personnel. Recommendations of teachers and previous employers are given weight. Interviews with supervisory personnel on the job are helpful in evaluating intangible qualities. Most research administrators feel it is wiser to recruit from many educational institutions, rather than from just a few.

While many of these aspects are equally true of other functional departments within the corporation, some are unique to the research and development man. Since requirements of individual companies vary for research or development, the mix of characteristics required for the man must vary also. Unfortunately, hiring programs do not always identify these requirements accurately.

Research and development people, as a group, are extremely mobile. The ease with which they can change companies, either throughout the country, or, indeed, throughout the world, has no precedent. At the same time, many professional engineers and scientists have developed an allegiance to their work which is stronger than their allegiance to a particular corporation or institution. They work for a particular corporation only as long as the environment is proper for them, and leave when the challenge of the work or environment elsewhere is more hospitable. Increasingly, companies must understand what competitive environments are or suffer continual turnover. One good study of the conditions leading to staff stability, by Deutsch, Shea, and Evans Inc. (Motivating Factors in Engineer Employment), indicates that autonomy, engineer-management communications, advancement, personal development, recognition, competent technical supervision, interpersonal relations, and work assignments rank highly with research and development people, while personal benefits, working conditions, status, and prestige and even salary are much the lesser contributors to stability.

Needs of the Research Engineer. Managing a technical group differs from managing other functional departments, such as manufacturing or marketing, in degree, if not in kind. It is generally agreed that a research and development management must provide a technical challenge to good department members, challenge which is tangible and stands a reasonably good chance of solution. The challenge of the work plays a large part in the technical man's job decision.

The mix of assignments which the technical man receives is equally important. While he frequently becomes absorbed in a single project or assignment, sooner or later the project ends. At this time, the assignment has either contributed to his development or not, and he is usually aware of the result. Many companies program a series of assignments for individual technical people to ensure that their mix of experience meets personal criteria and, at least in a general way, forces a broadening of capability.

Research people normally require a considerable amount of discretion in their work. This places an increased burden on the research and development manager to define the results required or the problem requiring solution and a more extensive delegation of the "how" of the job than for most functions.

A major issue is whether a good specialist can also be a good manager. Companies have chosen to go both ways, but it appears that the prevailing route is to provide equal measure of internal success to both the specialist and the manager, but require that the specialist eventually manage at some level. The IBM Fellow program or the GE specialist category are good examples of the high rank that specialists can attain without becoming line managers.

Incentives and Motivating Factors. Managing a research and development department usually requires an understanding of the following aspects of **personnel management** within the field:

1. Pay scales
2. Bonus practices
3. Stock options
4. Facilities and working environment
5. Educational and training allowances
6. Promotion practices
7. External recognition
8. Educational rotation or training tracks
9. Motivation

It is not possible to establish an absolute ranking of the importance of each of these, for they vary by man and by company. It is possible, however, to describe the range of significance of each.

Promotion is usually the strongest motivation for engineers, although not necessarily so for research men. Described by Arch Patton (Men, Money and Motivation), promotion takes precedence over pay in motivating men. In the research environment, promotion normally means greater freedom, greater discretion concerning the choice of project to be worked upon, and therefore greater challenges. It is less often hierarchical progress. In development, however, actual promotion within the organizational hierarchy is a requirement in order to retain talented development engineers.

Salary level is a major consideration. Many companies use surveys, such as that from Deutsch, Shea, and Evans, Inc. and the Engineers Joint Council to determine normal ranges of pay for engineers of varying disciplines often based upon years-since-degree. Engineering salaries tend to rise rapidly, then to level out unless a man moves on into management, or in companies which provide specialist positions, rise to a senior specialist position. **Pay scales and pay grades** are commonly used in development, but are less precise for research workers (Engineering Management Commission of the Engineers Joint Council, Professional Income of Engineers).

The objective of a **bonus** is to tie a man's compensation to his performance. In many cases, this can be done where the man's personal contribution is discernible and measurable during the course of the pay review period. If his contribution cannot be sharply isolated or is reflected only over a several-year period, a bonus system rarely adds to motivation.

An element of compensation used increasingly for technical people is **stock options, or stock participation.** Prompted by the success of technical entrepreneurs near Boston, San Francisco, and Los Angeles, who have offered stock options to attract bright people from leading companies, corporations are under pressure to include more of their technical people in the stock option plans, including the use of "phantom stock option."

Facilities and working environment are a further consideration. Much has been written concerning the quality of facilities available to technical personnel, but there is far-from-unanimous agreement on its importance. It is generally agreed that research people dislike poor equipment or laboratory services, and engineers expect to have modern test equipment and computer capability. The need for specific types of housing and offices is much less clear. In any industry, engineering departments may be observed in widely differing physical facilities, yet little correlation exists between success of the department and its physical facilities. Except for grossly inadequate facilities, housing was not considered an important factor unless other aspects of the job were already an irritant to the technical man.

The Deutsch, Shea & Evans report, however, reveals that location has become one of the more influential factors in job selection. Among material attraction it runs a close second to salary. Although climate seems to be the most important feature, the report cites other factors such as availability of educational facilities, and professional and cultural opportunities. For the married engineer one of the main concerns is finding an excellent environment for his family.

Educational and training allowances are finding increasing favor, and are increasingly required. The research and development manager must be able to meet competitors in this regard.

External recognition is an important aspect of technical personnel administration, particularly in research. Most companies permit technical people to attend conferences and seminars and to present papers, except when competitive information might be released. To prevent this, speeches concerning technical matter must be cleared by a senior technical manager who can judge whether important information is being disclosed. Control over attendance at seminars ranges from highly structured (certain levels to a certain number) to complete discretion by the individual. Most organizations require some justification for attendance and limit the number at any individual conference.

Finally, rotation of technical people through a series of positions, both inside the technical department and outside, is increasing. One company, a leader in its industry, requires that all technical managers work in both marketing and manufacturing before being promoted into senior positions of the corporation. Rotation for training is practiced by many companies; the best known has been General Electric's Engineering Test Program, where engineers worked for short periods on problems in various departments, then were assigned more permanent jobs of much longer duration. Even then, however, managers were transferred among functions to give them technical and managerial breadth.

Successful Project Manager. Steiner and Ryan (Industrial Project Management) have studied the methods leading to successful project management in the aerospace industries. Among their recommendations are:

1. **Leadership of Staff.** The project manager should be not only a coordinator of actions needed to achieve the objectives of the project, but he should be an architect, a leader of men in the highest sense of the term, and a superb organizer.
2. **Authority.** The project manager should have broad authority over all elements of the project. . . . He should have appropriate authority in design and in the making of technical decisions in development.
3. **Staffing.** The project manager should hold the size of his central project staff over which he exercises direct control to about ten people, if possible, but no more than thirty.

Resources and Facilities

THE TECHNICAL POSITION OF THE COMPANY. The importance of technical position is obvious. The company in a technological industry which loses position normally loses its profit and eventually its ability to reinvest in research and development. The loss of position often leads to crash programs for introducing products, or the introduction of products too late to capture a significant share of the market. As a company's competitive position deteriorates it becomes increasingly difficult to attract first-rate technical men which leads to further deterioration of position.

Maintaining Technical Position. In those industries which depend upon the technology inherent in their product or process for competitive survival, it is important to maintain a superior technical position. Dominance in its field normally permits management the freedom to take competitive moves at the company's own choosing.

There are several alternatives open to the research and development and corporate manager concerning technical position. The company may invest its research and development resources across a broad front of technology. To do this well requires a large expenditure. Or the company may choose to concentrate upon a relatively narrow technology. When commitment is made to a relatively narrow technology, some effort is frequently devoted to competing or supporting technology, so that if breakthroughs occur, the company can move to maintain a competitive position. Increasingly, companies structure **listening posts** on technology in areas which are potential threats to them. An individual assigned listening post responsibility is required to know what competitors are doing in relevant facets of technology, and to interpret the information for research and development management.

Factors Influencing Choice of Program. Technical position depends upon the research and development program, and is almost always a compromise among the following factors:

1. **Program balance versus resources available.** Research and development programs would be better balanced, more inclusive, and reduce the risk of failure if they were larger, more comprehensive, and better planned. However, most companies work with limited resources for their research and development effort and cannot afford the breadth desirable or the administrative expense to provide elaborate controls.
2. **Timing of market entry versus risk of competition.** Most research and development programs run some risk that the technical effort will not be timely enough to anticipate competitor market entry. The alternative, however, is to greatly increase technical effort in one area at the expense of others, or markedly increase overall resources committed to research and development. Although this quandary is normal, the risk is often not made explicit by marketing and technical personnel to corporate management.

In the final analysis, a technical position is adequate if the results are adequate. Some corporations appoint one member of the board of directors, usually technically competent to oversee the technical program. His function is to review, with the director of research and development, the extent to which the overall program is likely to meet the opportunities and threats seen by the board. In lieu of involvement by the board of directors, a scientific council may be appointed to advise the president and/or board of directors concerning technical position. Technically competent senior scientists and engineers from outside the company are asked to review the technical program, including individual projects, to determine whether the direction, rate of movement, and sophistication of the program are adequate to the corporation's needs.

Sources of Technology. To obtain technology, management has several alternatives. They are:

1. Develop the technology within the company.
2. Buy a company that has the technology.
3. Hire experts in the technology for a limited period of time.
4. Participate in trade association research and development.
5. Conduct research jointly with another company or organization, for example, a university.

Buying the technology through acquisition provides technology that is already developed, requires modest additional investment, and immediately establishes a position for the corporation. At the same time, however, the weaknesses of the other company are inherited whether they are technical, managerial, or financial. If the operating philosophies of the two companies differ, new friction may be created; what were two successful independent laboratories may become a joint failure with a resultant loss of position.

Mace and Montgomery (Management Problems of Corporate Acquisition) in their in-depth study of the problems, advantages, and disadvantages of acquisitions, advocate this route to obtaining technical know-how. They state: "To neglect the developments of others through an unwillingness to consider the acquisition of technical know-how created in other laboratories is to deprive the company of a significant opportunity to strengthen its own position. . . . Most larger companies can augment their own research and development programs by searching out and acquiring the know-how developed elsewhere. The smaller organizations may need and want the resources of a larger, financially capable, and established company."

A second major way of obtaining the technology is from independent research facilities. Although this practice dates back many years, it has been growing rapidly since the end of World War II.

It is sometimes referred to as sponsoring or "farming out" research, and is essentially the buying of a **research service.** A number of organizations specialize in **contract research.**

There are good reasons for using contract research. Companies are spared capital costs for laboratories and the expenses of maintaining the facilities and a staff. Unnecessary involvement in the details of the research program is obviated. Equipment and personnel, beyond the ability of the company to provide for itself, are available for use as needed in executing the program. Use of contract research enables the company to use more of its research funds for actual execution of research, rather than tying up the capital in building and equipment. The contract research organization is a **central laboratory** that serves various companies, rather than various divisions or subsidiaries.

INDEPENDENT RESEARCH FACILITIES. Independent research nstitutes, university research foundations, commercial consulting and research aboratories, and the laboratories of other companies are sources for contract research.

Independent Research Institutes. The independent research institutes are organizations set up for the specific purpose of helping industry solve technological problems. These serve on a not-for-profit basis, and a few are backed oy endowments. Some have nominal ties with universities, although their operations are usually independent of the university management. Because of the gradual merging of character, it is sometimes difficult to tell where the line of demarcation between independent research institutes and university research foundations can be drawn.

Research institutes conduct **diversified research** and often are experienced n many phases of industry and technology. They conduct research in confidence and give the sponsoring company exclusive right to all information, developments, and patents. Research institutes do not normally engage in engineering consultation work, analytical or testing work, or the evaluation and comparison of products for trade purposes.

Commercial Laboratories. Laboratories that operate for profit and do consulting, testing, and research work are commonly grouped under the designation, commercial laboratories. Consulting and testing laboratories are the oldest form of technical assistance to industry. They are outgrowths of private consulting services and the chemist's assaying shop. With the expansion of private laboratories in industry, the types of work referred to consulting and testing laboratories have changed, and the latter have had to adapt their facilities and interests to serve companies that have no research laboratories of their own. Many have evolved into organizations engaging in research as much as in testing, analysis, or consultation. Some of these laboratories resemble the large research institutes in their handling of broad research programs in many areas of technology. Consulting laboratories are situated in most of the larger industrial centers.

Cooperative Research With Universities. University laboratories can be helpful to companies wishing to conduct research, particularly for the conduct of **directed basic investigations.**

Research arrangements between universities and companies are "cooperative" in that both parties achieve benefits they could not have working alone. Funds made available by a company to a university supplement the regular university research budget. This makes possible an expansion of the university program, and the supplementary income going to its faculty members enables the university to hold scientific talent on its teaching staff. At the same time, the results of the research are useful to the company financing the work.

Research is placed in universities through **fellowships, grants,** or **contracts.** Fellowships and grants imply the donation of a fixed sum, to be expended, usually as the university sees fit, toward the solution of a technical problem. This may be done as the basis for a master's or doctoral thesis or for a scientific paper, and the actual work may be performed by a faculty member or by graduate students under his supervision. Many colleges and universities offer such research services to industry. Frequently, the college or school of engineering of the university, the engineering experiment station, or departments of physics or chemistry provide services that are directly analogous to those performed by the separately organized university foundations. In most instances, faculty members and graduate students conduct the research.

University Research Foundations. A number of universities have established research foundations, where work is done on a contract basis by faculty members or, in some cases, by full-time research personnel having no other duties. In all such cases, the sponsoring company's **rights to discoveries or patents** are defined by the agreement and may vary from no rights whatsoever to complete rights of ownership, depending upon the policies of the particular university. University research foundations are very similar to independent research institutes, the chief points of difference being affiliation with a parent university, or a close tie with a university to the extent that faculty members and graduate students may be used on the research staff. Many, however, employ full-time research personnel who have no teaching assignments.

The typical university research foundation seeks and negotiates contracts and provides overhead control, financial management, and patent assistance in connection with **industrial contracts.** Some have separate facilities, while others use the regular university facilities. The distribution of these research organizations throughout the country makes contract research facilities convenient to most industrial firms.

Trade Association Research. Trade association research is a method of solving problems that are common to a whole industry. To obtain the benefits of trade association research, an individual company associates itself with the one or more trade organizations serving its industry. In addition to trade associations, which have other functions as well as research, special research associations or institutes serve some industries.

Associations conduct industrial research and development on **common industry problems** in several ways. Some maintain and operate their own facilities. Many, however, employ the technical and scientific equipment and personnel available at universities, independent research institutes, and commercial laboratories. In some instances, the laboratories of member companies are used. Some employ a combination of facilities.

Altogether, there are some 2,000 trade associations of national and regional scope in the United States. Most were formed for trade or legislative functions, but more and more are embracing technical research in their programs.

Research financed by trade associations is usually paid out of the association's general funds, although there are instances when special assessment of members is practiced. Committees, set up for the purpose, usually manage the trade association research program.

Trade associations have been very helpful in advancing the technical progress of industries and in protecting them from the onslaughts of competitive technologies. The efforts of associations in advancing the industry can be very fruitful if such efforts are well-founded and executed. If, however, industry-wide research is initiated only after a competitive problem has become acute, it may be too late to realize more than token values from the program.

Technical Consultants. Technical problems not involving great amounts of laboratory work may also be solved through the use of technical and engineering consultants. Numerous firms supply advisory and consultation services, and specialists are available in almost all fields of industrial endeavor. Fees are charged for the service; and if actual laboratory or pilot-plant investigation is required, the work is done on a contract basis. The dividing line between a true consultation service and a commercial research laboratory is not sharp, one merging into the other. Many university professors also engage in consultation work in addition to their academic work.

Research and Development Procedures

PROJECT SELECTION. Project selection is the means by which the company translates research and development objectives, and corporate objectives, into actual work. These objectives help to establish criteria for project selection in order to choose among the projects available.

Most technically oriented companies have a research and development program of some kind. Suggestions for projects are either developed during the planning process or arise from other functions in the company because of:

1. Lack of corporate growth or too low a growth rate.
2. Diminishing earnings.
3. Competitive difficulties, reduced sales, increased sales resistance.
4. Excessive labor and material costs.
5. Production difficulties, excessive "rejects," lack of uniformity in products, or process failures.
6. Saturated markets, vanishing markets, or the threat of a saturated and vanishing market.
7. High incidence of customer complaint; "bugs" in the product.
8. Waste disposal and other environmental problems.
9. Lack of sales appeal in the product; product obsolescence in terms of utility or customer taste and fancy.
10. Shrinking prestige of the company.

Sources of Research Ideas. Every research project is based on an idea, which is a strong argument for performing basic research.

Ideas, of course, do not come from the research staff alone. Some companies encourage everyone to submit ideas for research, and substantial rewards may be granted for those that result in savings or profit. Suggestions relating to existing products and processes arise from the production, operating, sales, and executive departments. Sales people, coming as they do in contact with the market, are frequently excellent sources of suggestions for product improvement or new products. In one large chemical company, it was found that 50 percent of the ideas offered came from the research and development department; 25 percent came from the sales department; 15 percent from production; and 10 percent from management, market research, and other groups. Of the ideas offered, 65 percent of those from the research and development personnel were accepted and 35 percent of those from nontechnical sources.

Selection Criteria. Projects are usually selected against criteria which reflect objectives, needs, and relative attractiveness of projects. Figure 3 lists a set of criteria for one research and development program, as described by John S. Harris of the Monsanto Company.

Several related projects may often be an identifiable subset of the total research and development program. If, for example, a product-line position has been eroded by competitor improvements, it is likely that several small or large development projects may be required to revitalize a company's market position. Interrelationships such as these are most important, even though they complicate the project selection task.

PROJECT PLANNING. Only a limited amount of discretion in project selection exists in each planning period. While any project may be discontinued at any time, continued stopping and starting of programs without acceptable continuity is so disruptive to the technical departments that output will decline and good people will depart.

		R & D PRODUCT PROFILE CHART			
		Minus		Plus	
		-2	-1	+1	+
FINANCIAL ASPECTS	• ROI*				
	• Annual sales*				
	• Time to reach estimated sales volume*				
	• Risk**				
PRODUCTION AND ENGINEERING ASPECTS	• Required corporate character*				
	• Equipment*				
	• Raw materials**				
	• Process familiarity**				
	• Risk**				
R & D ASPECTS	• Research know-how**				
	• R & D time investment*				
	• Patent status**				
	• Risk**				
MARKETING AND PRODUCT ASPECTS	• Marketability to present customers**				
	• Suitability of present sales force**				
	• Product advantage**				
	• Length of product life*				
	• Market trend**				
	• Market development**				
	• Risk**				

*Quantitative evaluation   **Qualitative evaluation

Fig. 3. Criteria for evaluating research and development projects and their ranking.

Cetron and Monahan (in Technological Forecasting for Industry and Government, J. R. Bright, ed.) provide a research and development planning sequence (Fig. 4) which shows that technology, including technological forecasting and threat and policy determinations play a major role in developing operational objectives. Ultimately the technical program is developed and the research and development tasks executed. The crucial point in this planning system is the **feedback concept** in which technical information generated in the research and development tasks is fed back into threat and policy from which it may in turn affect the operational objectives. As stated by Cetron and Monahan, the planning sequence is iterative, and dynamic technological forecasting is not only an ingredient but also a catalyst.

There is increasing recognition that an annual cycle is not desirable for planning research and development activity. The projects do not neatly correspond to annual increments of technical effort. Increasingly, companies use a rolling portfolio, with quarterly review of progress. Each quarter, the least desirable

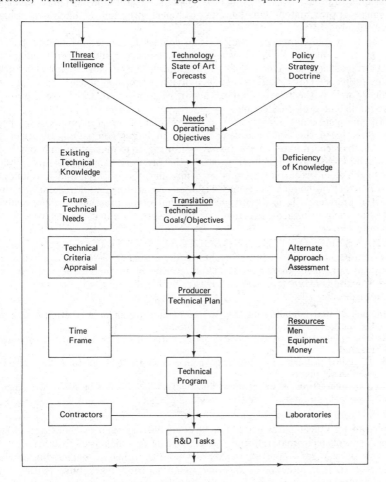

Fig. 4. An example of a research and development planning sequence.

projects are dropped from the active portfolio and new projects are substituted. "Roll-over" planning more closely approximates reality in gathering of information, assessment of the likelihood of technical success, and prediction of implications, should the project be successful.

Many companies insist that new product introductions be planned so that a separate function, such as the planning department, can monitor progress against the plan, in order to indicate to the president when a development is not on schedule. Since it is often difficult for such a group to interpret technical progress accurately, friction may be created between planning and the technical department involved as to whether or not the project is on schedule.

Program Balance. Program balance is an important objective in every company having a research and development activity. Figure 5 is an example of a desirable program balance, derived by an assessment of the funding or manpower that should be devoted to the function. Such an approach introduces a discipline for the research and development director and corporate manager to gauge approximately whether the technical program is likely to achieve corporate objectives.

As a means of comparing projects one against the other, Monsanto has developed the ranking system shown in Fig. 3. Such a ranking forces judgments on specific aspects or elements of the program, leading to an improved comparison of projects. During discussions of individual projects, a selection committee normally weighs the relative importance of each criterion, which contributes to more intelligent project selection.

In many companies, the research and development director proposes his portfolio directly to the chief executive officer. In larger companies, the planning director may coordinate such a review—ensuring that marketing and other departments have had ample time to study the proposed portfolio before submission to the chief executive. In smaller companies, the research and development portfolio for the coming year may be discussed informally in regular meetings with the president. It is usually desirable that other functions review, analyze, and offer their informed assessment of the program's objectives relative to the company's needs.

TRANSFER OF TECHNOLOGICAL RESULTS. Despite excellent management of a project in the technical departments, a project may fail due to improper transfer of results to manufacturing or to the customer. The **reasons for ineffective transfer** are many, both subjective and technical. They are, for example:

1. Lack of understanding on the part of manufacturing or marketing concerning the limitations or complexities of the product.
2. Incomplete technical data, particularly parts lists, specifications, or operating manuals.
3. Technical problems in scaling up from a laboratory bench to a pilot plant to a full-scale plant.
4. Inadequate transfer of technical knowledge which cannot be easily written.
5. Resistance to further technical change.

The effects of such problems can be substantial. In manufacturing, for example, an imperfect parts list plus many engineering changes soon after design release can obsolete material and tooling and delay schedules substantially. An analysis of changes, reported weekly or monthly, plus an engineering change review board (composed of engineering, manufacturing, marketing, and accounting) to approve each change helps to forestall adverse schedule or cost effects.

TECHNICAL RESOURCE ALLOCATION GRID

R & D RESOURCE		TECHNOLOGICAL SOURCES			
		PRESENT PRODUCTS	FORESEEABLE NEW PRODUCTS		ENTIRELY NEW PRODUCTS
FUNCTIONS	ALLOCATION		LICENSE AND ACQUISITION	INTERNAL DEVELOPMENT	
OFFENSIVE					
Generation of new areas	10%		1%	8%	1%
Fundamental research	10				10
Development research	10			10	
DEFENSIVE					
Monitoring	5		3	1	1
Applications engineering	15	10%	2	3	
Product improvement	25	25			
Cost reduction and process improvement	25	25			
TOTAL	100%	60%	6%	22%	12%

Fig. 5. Program balance among technological product sources.

Poor transfer of customer need or product requirements from marketing to the technical function can result in development of products which do not meet the needs of the marketplace. A simple but formal review of product specifications early in design by marketing and development often avoids this commonplace problem.

Spare parts are another frequent and aggravating problem. As manufacturing "stepchildren" during the introduction of a new product, spare parts requirements are often overlooked. Appointment of a separate **spare parts coordinator,** either in production control or in marketing, helps to alleviate this problem.

Corporations use several means to manage transfer. Some techniques are as follows.

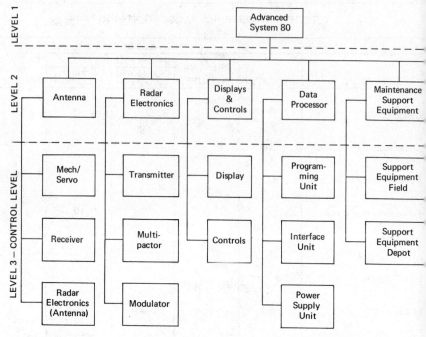

Fig. 6. Work breakdown structure

1. Transfer personnel "downstream" from development to manufacturing or marketing to assist these two functions during introduction of the product.
2. Move manufacturing personnel "upstream" into development during the design phase to convey the difficulties that particular design approaches will incur for manufacturing.
3. Schedule all interfunctional relationships through a **new product coordinator** for all functions.

Management of Research and Development

PROJECT MANAGEMENT. Once a project is selected, the objectives are essentially the same in both research and development—technical success, on time, within funds.

In research, the technical objectives are often difficult to define sharply. Because the technical objective cannot be stated as precisely in research, time and cost cannot be estimated as accurately. Often, the total amount of resources committed to a project is relatively modest. And the cost of a project is less important than the achievement of technical objectives, which tends to repay many times over the cost of the research.

In contrast, the development project usually has very specific technical goals. Because development normally works within the state of the art, time and cost can be planned. In managing development projects, an important early step is to plan manpower. Although in research, numbers may be planned approximately and modified by adding relatively few men to a small team, in development, a more rigorous approach must be taken. Frequently, manpower scheduling by month and by discipline may be done early in the project, so that manpower may be added to the project during its lifetime according to plan.

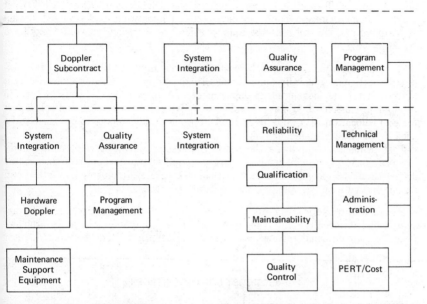

from **PERT/Cost System.**

Defining the technical work to be done is the most critical step in managing either a research or a development project. Specifications may be verbal, but complex objectives or a large project usually require written specifications.

Hill (Journal of Industrial Engineering, vol. 18) in commenting on improved methods for estimating and controlling research and development tasks states:

Effective managerial control requires that planning at all stages be accepted by both researchers and management. If a **laissez-faire** attitude exists within the research organization, even a properly conceived control system can hardly be expected to function correctly. Certain of the traditional approaches to recording and analysis of technological development programs are satisfactory when used, although indications are that in many cases they are not too effectively applied. However, complementary techniques are seriously needed to accommodate the unique characteristics of research activities. In the technological development area, it is especially important for control purposes that the complex of program objectives and subobjectives be clearly identified and that uncertainty be dealt with in an explicit fashion.

Tools of Planning and Control. Research projects tend to be scheduled in approximate fashion, since the milestones of work are not clear at the beginning of the project or, if clear, cannot be scheduled with precision. This is not the case in development. Since the early 1960's development project networks have become standard in engineering. Development departments use PERT, CPM, and CPS, among others, and powerful computer programs to handle calculations. Networks sequence activities in a valuable way when complexity is great. Bar charts are used where complexity is low.

Steiner and Ryan (Industrial Project Management) consider PERT an effective technique in the research and development stage of a project. In their study of the aerospace industry they found that when project managers had the opportunity, PERT was generally used, but most used it for planning and not control. Following the publication of the DOD and NASA PERT/Cost Guide, Hughes Aircraft Company developed an integrated computerized system of planning and control centered upon a PERT network and a work breakdown structure (Fig. 6).

Cost Problems. In research, cost objectives may be difficult to state explicitly although generally understood. In development, however, precise cost objectives for the development department and precise cost objectives for the product are essential. Figure 7 indicates a typical project cost control report for cost allocation of a large development project. The report is designed to be used by the project manager.

Cost problems seem to be the most troublesome during development, partly because substantial management effort is required but is not available. Clear cost assignment, therefore, to the individual engineer, monitored monthly can be established with the cost breakdown described above.

Review and Reporting on Projects. Review of the projects may be accomplished informally, and is in most nondefense companies. In a research department, an informal review may be quite appropriate. The research director

PROJECT COST CONTROL REPORT

WORK PACKAGE DESCRIPTION: SUBASSEMBLY A		PLANNED COST			EXPENDITURES TO DATE		COMMITMENTS TO DATE		TO DA
PROJECT UNIT OR PART NO.	DESCRIPTION	ORIGINAL	APPROVED REVISE	REVISED	AMOUNT	(O)/U* PLANNED COST	AMOUNT	(O)/U PLANNED COST	(O)/ PLA
38686	Corner	1,300	200	1,500	640	100	35	(15)	8
38687	Panel	3,350	–	3,350	1,000	(350)	690	0	(35
TOTAL		14,000	1,600	15,600	6,200	(850)	1,600	100	(75

*Over/Under.

Fig. 7. Project cost control report.

is often familiar with the project, has a clear idea of the objectives required and, with long experience in the technology, can assess whether or not the investigator is doing a good job.

In a very large research project or in a company with many research projects, each of which is complex, a formal **management reporting system** is required to determine whether projects are behind schedule, whether technical decisions are reasonable, and how well the project manager is performing.

In development projects, a history of data is essential to predict whether costs will underrun or overrun the original estimate. Unfortunately, most accounting systems do not provide such a trail of data until long after the project has been completed, so that data must be informally maintained in the technical departments. The cost structure described above assists in such accounting.

Reviews, whether held monthly, weekly, or semiannually, normally are for identifying problems.

Technical problems arise most often—a product or process will not perform as expected—and tend to be well handled in most corporations. Schedule and cost problems are frequently not well handled. Determining whether a project is behind schedule is difficult if milestones have not been planned. And there may be a tendency to accept schedule delay without analyzing alternatives, some of which would give acceptable technical performance without delay.

EVALUATION OF RESEARCH AND DEVELOPMENT EFFORT.
There are few ways to evaluate research and development effort with great precision, but because of the large cost involved, most companies make some attempt to do so. The **basic causes for imprecision** are:

1. Results of research and development often do not affect profits for several years.
2. The original development objective often changes as market knowledge becomes more accurate. This complicates measuring actual performance to objective.
3. Competitive position of individual products, after development is complete, is often difficult to ascertain when they are part of a broader product line.
4. No economic technique is available to determine a reasonable level of research and development investment.

For these reasons, evaluation tends to be highly judgmental and primarily based on performance against plan, rather than against an external standard.

Companies use different techniques to evaluate their research and development. Some of these are:

1. **Product introduction review.** Periodically new product introductions are evaluated by an interfunctional committee. The costs, time, and technical performance are reviewed to determine whether new products are being introduced in a reasonably short time with good technical and cost performance. This is more typical in companies with a large number of new product introductions, since it takes considerable skill to collect and review the data intelligently and to make the judgments.
2. **Technical councils.** As subcommittees of the board of directors or operating committees which advise the president, these councils review the results of the research and development program a few times each year. They may consist of division research directors in a multidivision company, or they may be individuals of recognized stature from outside the corporation.
3. **Technical planning meetings.** In planning meetings, each division's technical plan and results are reviewed for the quarter, half year, or the year. Meetings concentrate on the achievement of technical objectives relative to schedules and costs.

4. **Corporate new program introduction manager or marketing program manager.** He reviews each of the projects to determine whether or not the technical effort met the objectives of the marketing department.

Techniques for measurement are not exact, and the numbers which they develop are only approximations of what has happened. **Schedule evaluation** is often made against milestones. However, it is not always easy to determine whether the content of the work described by the milestone has been completed. For example, "release blueprints" is a frequent milestone on many new product introductions. Much less frequently defined, however, is whether blueprint release includes parts lists as well as the prints themselves. Lack of a parts list, of course, greatly complicates manufacturing and data processing efforts.

Many corporations, when reviewing a research and development program, reduce their concern to two questions:

1. What alternatives do we have for the existing research and development program, and are they better?
2. Are our competitors improving their market position based upon performance of their technological departments?

PLANT LAYOUT AND FACILITIES PLANNING

CONTENTS

CONTENTS (*Continued*)

PLANT LAYOUT AND FACILITIES PLANNING

Plant Layout

CLASSIFICATION OF MANUFACTURING PROCESSES. Manufacturing industries may be classified according to the nature of the process performed—continuous process, repetitive process, and intermittent process industries. Generally speaking, a **continuous process** industry is one in which production is carried on 24 hours per day; a **repetitive process** industry is one in which production of similar products is broken into lots or batches; an **intermittent process** industry is one in which production proceeds when and as orders are received. In the repetitive process industry, the lots may follow each other with such regularity as to create a situation analogous to a continuous process, except that production need not be carried on 24 hours a day.

Mallick and Gaudreau (Plant Layout Planning and Practice) have presented this classification very clearly in diagrammatic form as shown in Fig. 1. Owing to plant site, time of construction, caliber of management, nature of process, and many other factors, no two plants have identical layouts, even though their operations may be the same.

RESPONSIBILITY FOR PLANT LAYOUT. The responsibility for layout planning will vary among organizations depending on the size of the plant, the product manufactured and the importance of the layout function itself. According to a survey made by the International Material Management Society plant engineering departments and industrial engineering departments are each responsible for approximately 26% of all layout planning; process engineering departments account for another 13%. This survey indicated that the remaining work was being accomplished by either operating managers or other various groups.

Scope of the Plant Layout Function. The specific activities assigned to the plant layout unit are dependent upon the same factors used to determine the initial responsibility for the layout planning function. Apple (Plant Layout and Materials Handling) has listed the following most common duties which may be performed by the layout planning group.

1. Design workplace layouts.
2. Plan workplace interrelationships.
3. Plan material handling.
4. Draw layout plans.
5. Prepare construction drawings.
6. Design auxiliary equipment.
7. Supervise construction for new layouts or changes in layouts.
8. Study completed layouts for further improvement.
9. Keep files of layouts.
10. Determine costs of setting up layouts and savings from the improvement of layouts.
11. Determine processing procedures with the aid of methods engineers.
12. Establish time standards with the aid of the time and motion study department.

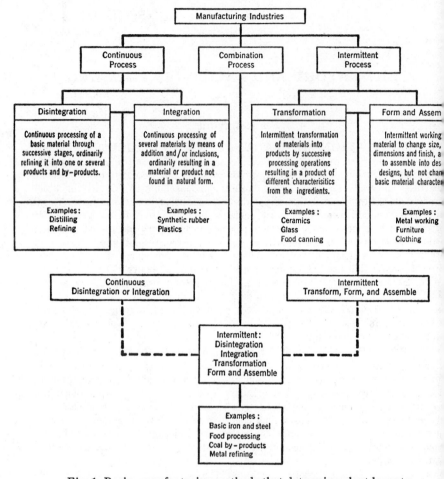

Fig. 1. Basic manufacturing methods that determine plant layouts.

13. Determine numbers of machines required for the respective processes.
14. Determine work methods with the aid of time and motion study engineers.
15. Design building to house manufacturing and other layouts.
16. Plan non-productive operations associated with the manufacturing processes.

Types of plant layout problems are classified into the following four categories by Moore (Plant Layout and Design) in order of frequency:

1. Minor changes in present layouts.
2. Existing layout rearrangement.
3. Relocating into existing facilities.
4. Building a new plant.

These classes of layout problems are similar in some respects but differ in others.

In most plants **minor layout changes** are made quite frequently because of improvements in an operation, new types of inspection, the introduction of a new (but similar) product into a department, the development of a new process

r the like. These problems are usually the easiest to handle, since they often
equire less planning and manpower to devise a workable solution. This class
f problem is the one that is most frequently met in practice.

Industries that are involved with products that require frequent redesign face
he problem of **relayout of existing facilities.** A complete rearrangement per-
nits the introduction of the latest methods, equipment and procedures. A com-
)any is more likely to abandon obsolete processes and methods when a whole
lepartment is being replanned than it is if only minor changes are being made.
The layout man is usually restricted to the area presently occupied by the de-
)artment in question. Although he may be given a free hand within the depart-
nent, he must stay within the presently allotted space.

Moving to a new location presents an opportunity to reevaluate existing
nethods and processes in order to bring them up to date with a minimum of
xpense. Improving processes by making minor changes can be costly in the
ong run, but when the opportunity to make a new layout is presented it is
ssential to consider the very latest methods, machines, and processes to avoid
apid obsolescence.

The **design of a completely new plant** requires a large amount of manpower
nd is the most complicated of the four classes of layout problems. All facets of
he production process must be considered as well as all auxiliary functions
necessary to make the plant a complete and integrated operation. Once this is
lone the shell—that is, the structure enclosing these facilities—is planned so that
t aids rather than hinders the production processes, as is often the case when
aying out existing buildings.

The above classes of problems are caused by various contributing factors.
Moore (Plant Layout and Design) graphically depicts (Fig. 2) the relationships
)etween the stimulating cause and the resulting type of layout problem. In this
liagram the common reasons for layout problems are shown across the top while
he four classes of layout problems are at the bottom.

IMPORTANCE OF EFFICIENT LAYOUT. In effect plant layout is
he master plan that physically integrates and co-ordinates the five basic factors
)f industrial management: men, materials, money, machinery, and markets.
Mallick and Gaudreau (Plant Layout Planning and Practice) have called plant
ayout the **blueprint of management.**

Because of its comprehensive scope, plant layout is the joint product of the
various fields of engineering and management. It presents in visual form the
)uildings and machinery needed to turn out, within a specified investment ex-
)enditure and from the materials specified for the product, the volume of
)roduction required for the delivery schedule established by the accepted sales
orecast.

The importance of an effective plant layout cannot be overestimated. Reed
(Plant Layout: Factors, Principles, and Techniques) outlines some of the advantages of a good plant layout:

1. Improved methods and control of the manufacturing process resulting from:
   a. Elimination or reduction of delays through better work balance between
      machines or operators.
   b. Smoother materials flow into, through, and from the process. A major
      factor in this area is the possibility of incorporating the techniques of
      automation or automatic material handling, thereby reducing flow fluctu-
      ation.
   c. In-process identifying, counting, and inspecting of goods.

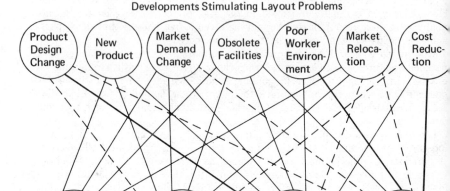

——— Cause and result occur very frequently
——— Cause and result occur less frequently
— — — Cause and result occur occasionally
(no line) Cause and result occur very seldom

Developments Stimulating Layout Problems

Product Design Change · New Product · Market Demand Change · Obsolete Facilities · Poor Worker Environment · Market Relocation · Cost Reduction

Build New Plant · Move to Existing Plant · Rearrange Existing Layout · Minor Changes

Classes of Layout Problems

Fig. 2. Graphical correlation between the development stimulating layouts and the classes of plant layout problems that result.

2. Improved quality control due to the consideration of quality factors, and their incorporation in the most effective manner for maximum control and minimum cost.
3. Improved material handling due to better location of equipment, reduced handling distances, and better co-ordination of the entire handling activity. The application of the principle of standardization to material handling reduces the variety of handling units and equipment, permitting greater flexibility without sacrificing efficiency. Standardization also may reduce the investment required for material handling.
4. Reduced equipment investment through planned machine balance and location, minimum load handling distances, and a resulting reduction in idle or partially loaded units in the production areas. This reduction of equipment investment applies to service and maintenance equipment, material handling equipment and office equipment, as well as production machines.
5. Effective use of available area resulting from the location of equipment and services so that they can perform multiple functions, the development of up-to-date work areas, and operator job assignments leading to full utilization of the labor force.
6. Improved utilization of labor through the proper design of individual operations, the manufacturing process, the work flow, and material handling. Balancing of labor to production needs and machine requirements, the layout

of equipment for ease of maintenance, improved handling, and the design of the process to facilitate control may lead to a reduction in production and office personnel requirements.

7. Improved employee morale may be obtained by providing for employee convenience and comfort in the layout. Adequate lighting, proper heat and ventilation, noise and vibration control, and sufficient and convenient rest rooms, locker rooms, and lunch facilities improve job performance and reduce idle time.

8. Improved efficiency in plant services results from the design of new utility distribution systems having maximum flexibility and capacity for future rearrangement or expansion.

Developing the Layout

BASIS OF LAYOUT. Layout of a plant, both of departments and of machines, should be the expression of a purpose. To this end, the processes through which materials pass, their sequence of flow, the machines and equipment required for the anticipated volume, and the location of the auxiliary departments (receiving, shipping, maintenance, toolroom) and services (lavatories, cafeteria, dispensary, and others) are vital. But the practical and psychological aspects of other factors, the building structure, heating, ventilating, lighting, noise control, etc., must receive thorough consideration.

Although essentials of plant layout are theoretically the same for all industries, in application the results will vary depending upon the type of plant and its size, the variety of output, and building and financial limitations imposed.

Mallick and Gaudreau (Plant Layout Planning and Practice) suggest the following fundamental concepts be considered in developing a layout:

1. The major part of production work is not processing, as is usually supposed, but material handling.
2. The speed of production in a plant is determined primarily by the adequacy of its material-handling facilities.
3. A good plant layout is designed to provide the proper facilities for material handling as well as processing.
4. The factory building is altered or constructed around the prescribed plant layout design.
5. The production efficiency of a plant is determined by the limitations of its layout.

GATHERING DATA. Before the actual assignments of space can be made in a layout, certain basic data must be collected and investigated. Inquiries are made concerning the product, material handling facilities, processing, raw material storage. Deficiencies in current operations such as bottlenecks, high percentage of rejects, and operator inefficiencies are also analyzed.

A comprehensive table suggested by Reed (Plant Layout: Factors, Principles and Techniques) listing basic data requirements for plant layout planning is shown as Fig. 3.

Analysis of Production. After the necessary basic data has been gathered the process engineer will:

1. Plan the operations necessary to process the rough stock into the finished part.
2. Decide on proper sequence of the necessary operations.
3. Select machines and equipment required to perform each operation.
4. Specify or sketch necessary auxiliary equipment such as tools, jigs, and fixtures.
5. Procure production standards, usually in terms of hours per piece, from the time-study records.
6. Write up these data as a production routing or operations list.

Problem	Data Required	Representative Data Sources
1. Product	Type of product, production quantity, specific designs.	Specifications, sales and market forecasts.
2. Production	Preferred and alternative order of operations for manufacture.	Parts lists, assembly charts, operation sheets.
3. Control	Quality and quantity control points and methods.	Operation process charts, operating procedures.
4. Personnel	Number, job and area assignments of, and facility requirements for employees.	Job descriptions and specifications, man and machine charts, organization, procedures, personnel policies.
5. Maintenance	Maintenance activity volume.	Individual machine and area requirements and records, general plantkeeping requirements, office and shop requirements.
6. Storage	Requirements for continuous plant operation without materials delays.	Production schedules, sales schedules, inventory control and purchasing policies, in-process goods estimates.
7. Material handling	Volume and type of activity, equipment type and quantity.	Materials and parts specifications, production rates and quantities, flow patterns.
8. Services	Requirements for personnel, equipment, plant, and general business.	Power, light, heat, ventilation, maintenance, and noise control requirements. Personnel services, plant and general offices, receiving, shipping, and warehousing requirements.
9. Costs	Initial investment and operating costs.	Company records, depreciation and investment recovery policies, sales and market forecasts, engineering economy studies.
10. Legal restrictions	Limiting design requirements.	Building codes, labor laws, safety codes, tax policies, waste disposal restrictions.
11. Space limits	Building or plant area limitations.	Property plots, building plans, deed restrictions, city or county planning commission regulations.
12. Appearance	Public and worker acceptance.	Approved practices in design, opinion surveys.
13. Future alteration or expansion	Estimates of growth.	Sales forecasts, industry trends, flow flexibility, building design, building to plot relations.

Fig. 3. Design data for plant layout.

In order to provide a benchmark to evaluate the final proposed layout, a necessary step to be taken at the beginning of the study is to ascertain the present performance of the existing layout. Mallick and Gaudreau (Plant Layout Planning and Practice) suggest a format for this performance analysis (Fig. 4). A further analysis of production can be made by breaking down total output into type of products.

Description of Data	Last 12 Months (1)	Average per Month (2)	Highest Month (3)	Latest Month (4)
OUTPUT				
Number of units assembled	6,543	545	659	455
STANDARD LABOR HOURS ALLOWED				
Net allowed hours	69,520	5,793	6,861	5,413
Setup hours	1,777	148	148	79
Extra allowance hours	16,767	1,397	1,570	1,383
Expense hours	6,177	515	393	497
Non-saleable product	11,716	976	935	442
Miscellaneous hours	212	18	29	26
Total allowed hours	106,169	8,847	9,936	7,840
ELAPSED HOURS				
Total elapsed hours	88,259	7,355	8,336	5,916
EFFICIENCY				
Percent efficiency	120%	120%	119%	133%
AVERAGE HOURS PER UNIT ASSEMBLED				
Net allowed hours	16.2	16.2	15.1	17.2
Elapsed hours	13.5	13.5	12.6	13.0
NUMBER OF WORKERS				
Assembly crew, 1st shift	24	24	23	19
Assembly crew, 2nd shift	18	18	18	16
Operator, 3rd shift	1	1	1	1
Foreman	1	1	1	1
Inspectors	3	3	3	3
Lapping machine operators	2	2	2	2
General laborers	2	2	2	2
Packer	1	1	1	1
Time-study man	1	1	1	1
Shop clerks	4	4	4	4
Total force in department	57	57	56	50

Fig. 4. Example of actual performance on existing layout.

Aside from the economic data indicating layout deficiencies, a number of inefficient operating conditions will provide evidence of an inadequate layout. Mallick and Gaudreau list the more common of these indicators:

1. Receiving department.
   a. Congestion of materials.
   b. Complaints of delays at trucking lines.
   c. Recurrent demurrage.

    d. Damage to materials by exposure to the elements.
    e. Necessity for material handlers to work outside in inclement weather.
    f. Difficult manual handling and rehandling operations.
  2. Storerooms.
    a. Congestion of stock storage.
    b. Damage to materials in storage.
    c. Frequent loss of material.
    d. Poor control of inventories.
    e. High ratios of storeroom clerks and material handlers to productive operators.
    f. Frequent rehandling and restorage of materials before processing.
  3. Process departments.
    a. Necessity for skilled operators to handle materials.
    b. Presence of large quantities of materials on the floor, not under production control.
    c. Poor quality of work in process.
    d. Complaints from foremen and production supervisors regarding the lack of floor space while overhead space remains unused.
    e. Excess length of the manufacturing cycle over actual processing time.
    f. Presence of congestion and hazards in narrow or crooked traffic aisles.
    g. Over 15 percent of the total floor space in the plant taken up by traffic aisles.
    h. Difficulty in the material-handling service to the productive equipment.
    i. Excessive maintenance costs.
    j. Necessity for frequent rearrangements of equipment.
    k. Periodic requests for additions to equipment or to the working area.
  4. Productive operators.
The existence of any of the following excesses in material-handling work performed by productive operators:

Material-Handling Condition	Male Operator	Female Operator
a. Lifting frequently from floor level to	A point overhead	Chest height
b. Lifting above knee level when weight exceeds	75 lb.	35 lb.
c. Lifting or moving with the assistance of	1 helper	1 helper
d. Handling constantly for longer than	30 min.	30 min.
e. Handling similar materials daily for	Several wks.	Several wks.
f. Moving heavy loads over more than	50 ft.	50 ft.

  5. Employee morale.
    a. Complaints from operators regarding:

      (1) Heating   (4) Rest rooms
      (2) Lighting   (5) Congestion
      (3) Ventilation   (6) Hazards

    b. High accident rate.
    c. High labor turnover.
  6. Shipping department.
    a. Delays in shipping even though materials are ready for shipment.

Coordination of Basic Data. After the basic data have been accumulated it has been found helpful to coordinate it by means of two graphical techniques. These are the assembly chart or graphical parts list (see section on Production Control) and the operation process chart (see section on Process Charts).

An **assembly chart** shows in simple graphical form:

  1. The parts making up the total assembly.
  2. The subassemblies and their components.

3. The order in which parts are assembled.
4. The flow of parts and subassemblies into the assembly.
5. The first concept of the production flow pattern.

Apple (Plant Layout and Materials Handling) illustrates a typical assembly chart as shown in Fig. 5.

As a further aid in coordinating the basic data, the operation process chart can prove helpful. Such a chart has the following uses and advantages:

1. Coordinates information from parts list and routings.
2. Gives an idea of the complexity of the production problem.
3. Indicates roughly the relative importance of production lines for various parts.
4. Presents a rough picture of the desired flow pattern.
5. Gives an overall picture of all necessary production operations.
6. Shows relationships between various parts.

Determination of Equipment Requirements. A basic factor in developing an effective layout is the determination of proper equipment requirements. In order to do this correctly it is necessary to obtain estimates of the following information:

1. Production volume
2. Scrap estimates
3. Standard times for operations

The method of calculating machine requirements will depend upon whether the layout is to be **product** or **process** oriented. As **an example,** a product-oriented layout is assumed to have the following characteristics:

1. Annual desired output of 100,000 good units.
2. Standard 40-hr week for 50 weeks per yr. (2,000 hrs. per yr.).
3. Scrap loss equals 5%.
4. Machine production efficiency equals 90%.
5. Machine capacity is 31 pieces per hour. The machine requirements can be determined by the following calculation:

$$\frac{100{,}000 \text{ good units}}{2{,}000 \text{ hr. per year}} = 50 \text{ good units per hour}$$

$$\frac{50 \text{ good units}}{(1.00-0.05) \text{ total units per hr.}} = 52.6 \text{ pieces per hour (for material requirements)}$$

$$\frac{52.6 \text{ total units per hr. (desired)}}{31 \text{ pieces per hr. (Machine capacity)}} = 1.69 \text{ machines required at 100\% efficiency}$$

$$\frac{1.69 \text{ machines}}{0.90 \text{ production efficiency}} = 1.88 \text{ machines}$$

Since less than a whole machine cannot be purchased, two machines would be needed in this case. If when computing machine requirements, the answer should indicate, for example, a need for 1.15 machines, the decision to purchase 1 or 2 would depend upon possible methods improvement, overtime considerations, use of alternate equipment to absorb access requirements, and the opportunity for rescheduling production demands.

If a **process** layout is being planned where more than one product can be processed by a given machine, machine-loading techniques must be utilized to determine machine requirements. Moore (Plant Layout and Design) presents a form (Fig. 6) to aid in determining machine requirements in this situation. A computation similar to the one depicted must be made for each machine, or production center, to be included in the layout plan.

Material Handling Plan. Since handling of material accounts, on the average, for 25–35 percent of all production costs, it is important that thoughtful

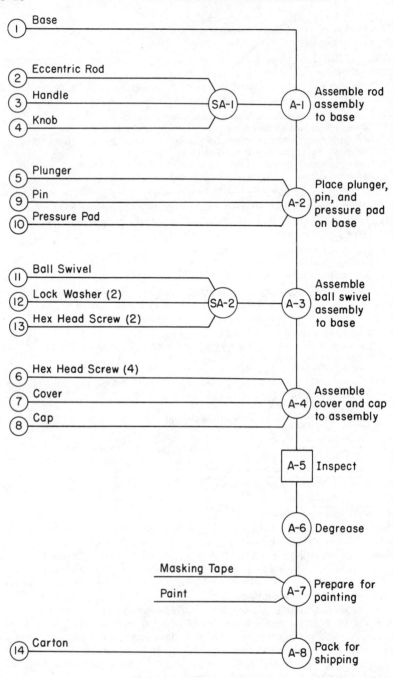

Fig. 5. Typical assembly chart.

MACHINE REQUIREMENTS – PROCESS LAYOUT

Machine: _HCT Hardinge Lathe_　　　Machine No: _70342_

Department: _Adding Machine_　　　Operating Hours per _mo.:_ _240_

Part No.	Name (2)	Operation (3)	Time S.U.* (4)	Time per pc. (5)	Volume (6)	No. of S.U. per mo. (7)	Hrs./mo S.U. (8)	Hrs./mo Prod. (9)	Capacity Req'd (10)
1	Cover handle	Face both sides	.30	.010	5500	4	1.2	55.0	56.2
3	Main shaft	Cntr drill + turn	.25	.035	8000	5	1.2	280.0	281.2
4	Frame	Face bottom	.85	.082	2000	3	2.6	169.0	166.6
									504.0
			$\frac{504}{240} = 2.1$						

No. of Machines Required	Theoretical 2.1	Actual 3

S.U. = Set-up of machine tools

Fig. 6. Machine requirements worksheet.

consideration be given to this problem. In determining the material handling plan it is well to remember that time required and distance traveled in the course of manufacturing a product directly affect its cost of production (see section on Material Handling).

At this point, then, some decision should be made as to the general method or system of handling which will be used. It may be tentative, but at least it will serve as a guide in succeeding steps.

Construction Factors in Production and Material Handling. Although the primary purpose of a building is to provide protection for the plant facilities, it can and should contribute to the effectiveness of the plant layout it contains. Naturally, if a new building is to be constructed it is possible to plan the layout and then design the structure around the layout and its process. In preparing a relayout plan the layout engineer is constrained by the physical limitations of the existing structure.

In general, **relayout projects** require, as a first step, a complete survey of existing conditions. Moore (Plant Layout and Design) recommends that the engineer have up-to-date building plans showing all permanent structural features of the building, such as windows, columns, beams, doors, shafts, stairways, and the like. In addition all permanently installed service facilities, such as piping and conduit buried in walls and floors should be accurately located. The engineer should carefully note floor loadings, especially in those cases where there is a possibility of heavy machinery or materials being installed or stored. Foundations, footings, walls, and roof decks should be examined for signs of weakness and those places needing repairs noted so that the repairs may be made when the new layout is installed.

Layouts for new plants present quite different situations to the engineer. In the case of a new layout, departmental arrangements, production layouts, auxiliary equipment and facilities, and services are integrated so as to achieve optimum efficiency. The building is then designed to house the layout. Column spacings, floor loadings, and provisions for horizontal and vertical movement into and within the building can then be determined. The initial design must provide for sufficient flexibility so that future changes and relayouts can be effected smoothly and efficiently within the existing structure.

Muther (Systematic Layout Planning) suggests a general guide for industrial building features (Fig. 7) which can be a valuable aid in relating construction considerations to production and material handling requirements.

PLOT PLAN. The plot plan is a diagrammatic representation of the building outline, showing its location in the property and the location of external transportation facilities. It may show such items as yards, roads, railroad tracks, rivers, tanks, storage areas, fire hydrants, recreation areas, and parking lots. Aerial photos are excellent aids in developing plot plans.

The plot plan is the key to the actual layout and will determine much of what goes on inside the plant (see Fig. 8). The plant should, if possible, be oriented on the property to take advantage of favorable sun, wind, and weather conditions, as well as of the existing or planned transportation facilities, utilities, and parking, and future plant expansion. The location of shipping and receiving areas in the plant will depend, of course, on external transportation facilities, whether existing or to be constructed. Therefore, these two facilities should be spotted on the sketch of the plot plan first.

Receiving and Shipping. It may be convenient and economical to operate the receiving and shipping departments adjacent to one another or combined. A shipping department located at one end of a plant and a receiving department located at the other end is often an effective arrangement providing for straight-line flow of work in certain kinds of production. Trackage alongside a plant may suggest unloading near one end and loading near the other. Solutions to the placement of receiving, shipping, and other auxiliary departments are to be found in various examples of layout presented in this section.

Parking. Never underestimate the size and the need for parking facilities. Too often plant layout engineers from metropolitan areas who are used to most of their employees arriving by public transportation overlook this factor, particularly with suburban site planning.

GENERAL FLOW PATTERN. One of the primary objectives of plant layout is to provide an arrangement of equipment which will facilitate the manufacturing process. In an ideal situation, the manufacturing equipment would be laid out to suit the process, and the plant would then be designed around the layout. However, in most cases, a workable flow pattern must be fitted into an existing area and around existing facilities such as elevators, loading docks, and railroad sidings.

The **flow of material** is the core of plant layout. Since the analysis of this flow pattern is the heart of successful layout planning, it is essential that the analysis be made through carefully guided planning efforts. An aid to such efforts has been prepared by Apple (Plant Layout and Materials Handling), through development of various flow planning principles. Although certainly all of the principles listed cannot be followed in every case, their consideration should aid in arriving at the most effective flow pattern. These principles are:

1. Plan for movement of materials in as direct a path as possible through the plant.
2. Minimize backtracking.
3. Use the line-production principle wherever feasible.
4. Plan for incoming materials to be delivered directly to the work areas when practicable.
5. Install material handling equipment which will permit production employees to spend full time on production.
6. Use mechanical handling equipment to assure a constant rate of production.
7. Combine operations whenever possible to eliminate handling between them.
8. Eliminate "rehandling."
9. Combine processing with transportation whenever practicable.
10. Plan for storage of a minimum of material in the work area.
11. Minimize walking required of production operators.
12. Reduce manual handling to a minimum.
13. Plan for each operator to dispose of a part in a convenient location for the next operator to pick it up.
14. Use gravity to move materials.
15. Place related activities near each other.
16. Plan for processes involving heavy materials to be located near the receiving area.

Factors Affecting the Flow Pattern. In deciding upon a general flow pattern, there are many factors to be taken into consideration, such as:

1. External transportation facilities.
2. Number of parts to be handled.
3. Number of operations on each part.
4. Number of subassemblies made up ahead of assembly line.
5. Number of units to be processed.
6. Amount and shape of space available.
7. Necessary flow between work areas.

Flow Analysis Through Charting. Flow-of-material analysis can be the most important part of layout planning when the movement of materials is a significant factor in the process. There are a number of different methods of analyzing the flow of materials and the choice of analysis will be dependent upon the process conditions. As a general guide:

1. In a product oriented layout use **operation process charts.**
2. For a process layout use **travel charts.**

Operation process charts represent schematically the sequence of all operations in a process except those involved in material handling. The objective in using these charts is to develop a clear picture of the flow process and thus facilitate the actual planning of the layout. A typical example of an operations process chart is shown in Fig. 9.

Where multiple flow patterns exist, and the operation process chart is not capable of adequate analysis of the flow pattern, the **travel chart** or **cross chart**

Use general-purpose or multipurpose building when the following are important or predominant:
    Initial cost
    Speed of getting the layout into production
    Probability of selling the building later for:
        Profit
        A better location
        Foreclosure
    Frequency of changes in:
        Products or materials
        Machinery and equipment
        Processes or methods
        Volume or output

Use single-story construction, possibly including balconies and/or basement, when the following conditions exist:
    Product is large or heavy
    Weight of equipment causes heavy floor loads
    Large, more-or-less unobstructed space is needed
    Land value is low
    Land is available for expansion
    Product is not adapted to handling by gravity
    Erection time is limited
    Frequent changes in layout are anticipated

Use a relatively square building where there are:
    Frequent changes in product design
    Frequent improvements in process
    Frequent rearrangements of layout
    Restrictions on building material availability or substantial savings desired in amount of materials used

Use other shapes or separate buildings when there are:
    Limitations in physiography of the land
    Property lines at awkward angles
    Operations that cause dirt, odors, noise, vibration
    Operations not part of production
    Operations susceptible to fire, explosion, contamination

Use a basement if these features can be obtained:
    Ample headroom   Ample lighting
    Good ventilation   Waterproofed walls
    Sound foundations   Freedom from seepage or flooding

Use balconies for these typical situations:
    Light subassembly above final assembly on ground level
    Assembly operations with heavier forming machinery below
    Light-machine operations with heavier machines below
    Treating operations with forming operations and assembly of bulky units on ground level
    Supporting activities that can be kept off the production floor—storage, wash, and locker rooms, production offices, packing, auxiliary equipment, and the like
    Operating or servicing upper parts of tall, high machinery
    Liquid or bulk material storage and preparation areas, involving mixing, aging, blending, and the like

Fig. 7. Industrial building features guide.

Use no windows generally in cases when:
    Work is affected by changes in temperature, humidity, light
    Work is subject to dust, dirt, contamination
    Workers or work are affected by external noise
    Artificial light and power is inexpensive
    Seeing things outside is not necessary
    Windows would get dirty fast

Use these desired floor characteristics where practical:
    Floors of various buildings at same level
    Strong enough to carry machines and equipment
    Made from inexpensive materials
    Not too expensive to install
    Ready for use quickly after laying down
    Easily and quickly repaired, removed, and replaced
    Resistant to shock, abrasion, heat, vibration
    Not slippery under any condition
    Noiseless and sound absorbing
    Attractive to eye and with numerous colors available
    Unaffected by changes in temperature and humidity, or by oils, acids, alkalies, salts,
        solvents, or water
    Odorless and sanitary
    Resilient enough to seem soft underfoot and to minimize damage to articles dropped
        on it
    Easy to fasten machines and equipment to
    Will dissipate static electricity and not cause sparks
    Easily kept clean

Use these roof and ceiling features when applicable:
    Overhead space and height clearance for:
        Production machines
        Process equipment—treating vats, drying ovens, etc.
        Handling equipment—cranes, conveyors, etc.
        Elevated traffic ways
        Sprinkler (with 24 inches clear underneath)
        Electrical distribution
        Heating and ventilating
        Air circulation
        Washroom and toilets, service and storage areas built off the production floor
    Strength for underside (or above) support of:
        Machinery and process equipment
        Handling equipment for material or machinery
        Heating and ventilating systems
        Elevated traffic ways, storage or service areas
    Light: Roof lighting independent of walls or expansion plans
    Heat conductor for:
        Heat losses in winter
        Effect on personnel in summer
    False ceiling:
        Dust accumulation and dust drop-off
        Appearance

Fig. 7. Concluded.

Ford Motor Co., Research and Engineering Center

Fig. 8. Plot plan of existing buildings.

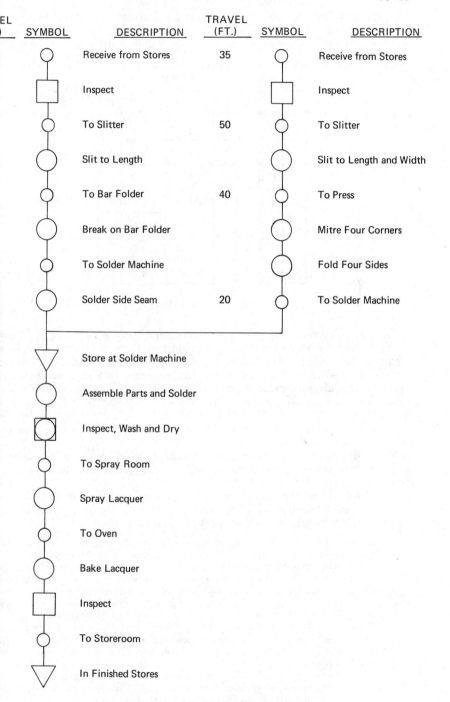

VEL .)	SYMBOL	DESCRIPTION	TRAVEL (FT.)	SYMBOL	DESCRIPTION
	○	Receive from Stores	35	○	Receive from Stores
	□	Inspect		□	Inspect
	○	To Slitter	50	○	To Slitter
	○	Slit to Length		○	Slit to Length and Width
	○	To Bar Folder	40	○	To Press
	○	Break on Bar Folder		○	Mitre Four Corners
	○	To Solder Machine		○	Fold Four Sides
	○	Solder Side Seam	20	○	To Solder Machine
	▽	Store at Solder Machine			
	○	Assemble Parts and Solder			
	⊡	Inspect, Wash and Dry			
	○	To Spray Room			
	○	Spray Lacquer			
	○	To Oven			
	○	Bake Lacquer			
	□	Inspect			
	○	To Storeroom			
	▽	In Finished Stores			

Fig. 9. Typical operation process chart.

has proved helpful. A travel chart is a matrix, similar to the mileage chart on a road map, which has been adapted to the analysis of movements of materials to and from a number of locations (see Fig. 10). Numbers in the squares represent

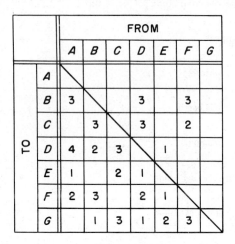

TO \ FROM	A	B	C	D	E	F	G
A							
B	3			3		3	
C		3		3		2	
D	4	2	3		1		
E	1		2	1			
F	2	3		2	1		
G		1	3	1	2	3	

Fig. 10. Travel chart.

the number of moves to and from plant areas identified as A, B, C, D, E, F, and G, as dictated by sequences shown on production routings or operations lists. Upon completion of the travel chart, a **preliminary layout** is prepared using the volume of movements as a guide to the flow sequence. Following the design of this layout, a **distance-volume chart** is prepared in the format of the travel chart, but indicating the product of the number of movements and the distance moved for each location. The efficiency of the layout is determined by the sum of the rows or columns on the distance-travel chart. This sum represents the total distance-volume handled.

By developing the schematic layout, checking the efficiency of the layout, preparing new travel charts, again plotting and analyzing them, the most effective plant layout can be determined.

In general **flow pattern** analysis should not be used as the sole basis for layout planning. According to Muther (Systematic Layout Planning) analysis bearing on relationships should be considered for the following reasons:

1. The supporting services must integrate with the flow in an organized way. This integration results from total analysis—analysis of the reasons underlying why certain supporting activities should be close to certain producing or operating areas. The maintenance crib, the superintendent's office, the locker and rest room, and the transformer bank, all have a relatively preferred closeness to each of the producing areas. They are all part of the layout; they must be planned into it, yet they are not part of the flow of materials.
2. Frequently, flow of material is relatively unimportant. In some electronic and jewelry plants only a few pounds of material will be transported during an entire day. In other industries, materials are piped, or one skid-load lasts a worker all week.
3. In completely service industries, office areas, or maintenance-and-repair shops, there is often no real or definite flow of material. Therefore, any general rule must offer us a way of relating areas to each other without being tied to

material flow. And this is true even if we recognize that we will substitute paper work, equipment, or even people, as the "material" that will flow.

4. Additionally, in heavy material-movement plants, where the influence of material flow will dominate the layout planning, flow will not be the sole basis for arranging the process operations and equipment. Basically, we chart flow to determine the sequence of operations or which departments should be near each other. But flow of materials is only one reason for this closeness. There are many others. And these may conflict with or at least cause adjustments in the closeness as based on the analysis of flow. For example, the routing may call for the sequence: form, trim, treat, subassembly, assembly, and pack. For best flow of material, treating should lie between trimming and subassembly. But treating is both a very dirty and a dangerous operation. Therefore, it should be kept away from the delicate subassembly area and its high concentration of workers. The effect of factors such as these—or the distribution of utilities, the cost of controlling quality, the contamination of the product and the like—must be compared with the importance of material flow, and adjustments made as practical.

When circumstances such as the above are encountered the **relationship chart** (Fig. 11) is an ideal vehicle in aiding the layout planning process. The relationship chart is a form on which the relationship between each activity and all other activities can be depicted. The basis of the relationship chart has been clearly described by Muther (Systematic Layout Planning) and is shown in Fig. 12.

Once the flow of materials or activity relationship analysis has been made it may be helpful to diagram this information. A useful suggestion when **diagraming flow of materials** is to keep the diagram as simple as possible in order to maintain maximum clarity. A method of **diagraming relationships** is presented by Muther (Fig. 13). The procedure involves connecting the activities by a number-of-lines code. The shape of each symbol indicates the type of activity; the number inside is the activity identification; the number of connecting lines indicates the rated closeness.

Types of Flow Patterns. Since every layout problem is different, there is no one best way to lay out a flow pattern. There are, however, several basic types which have come into favor, and which can be adopted to many situations. They can be divided into two groups: (1) flow patterns designed for **production lines** and (2) flow patterns required for **assembly lines.** Although both may be incorporated into the overall layout, their characteristics differ somewhat.

Reed (Plant Layout, Factors, Principles, and Techniques) lists and diagrams (Fig. 14) the basic types of production-line flow patterns as:

1. Straight-line.
2. U-shaped.
3. S-shaped.
4. Convoluted.

The theory behind the **straight-line** flow pattern is that so long as the distance between machines is a minimum, total handling distance will be minimal. It is not always possible to arrange a facility to provide for straight-line production. As a result it is often advantageous to use a flow path resembling a **U-shape.** This pattern has the basic advantage of allowing shipping and receiving on the same side of the building. If the line is relatively long compared with the building, additional loops may be placed in the pattern. The addition of one loop results in the **S-shaped,** while more loops result in the **convoluted** arrangement.

Plant_____		Project_____			
Reference_____		Date_____	Charted by_____		

Dept.	Sign	"OK"	Dept.	Sign	"OK"

Code	REASON
1	
2	
3	
4	
5	
6	
7	
8	
9	
10	
11	
12	
13	
14	
15	

Value	CLOSENESS	Color
A	Absolutely Necessary	Red
E	Especially Important	Or.-Yel.
I	Important	Green
O	Ordinary Close.	Blue
U	Unimportant	U
X	Undesirable	Brown

Fig. 11. Relationship chart.

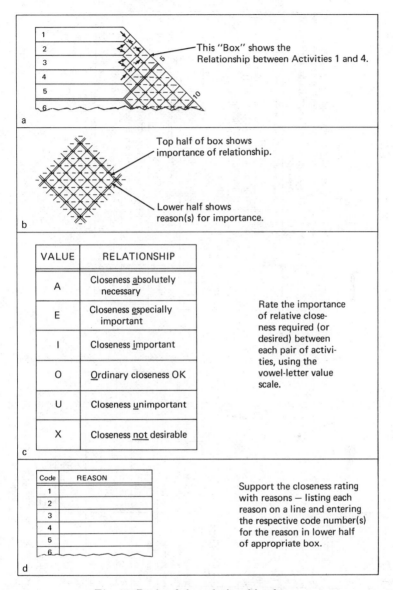

Fig. 12. Basis of the relationship chart.

Basic flow patterns for assembly lines are charted by Reed as in Fig. 15.

In the **comb** pattern the main assembly line is fed from a series of subassembly or parts' lines, all originating from the same side of the main assembly line. In the **tree** pattern, the main line is fed by subassembly or parts' lines from each side. The **dendritic** flow pattern is more irregular than either the comb or tree pattern. Each part progresses along its production line until production is completed to the point of assembly. The **overhead** pattern is distinguished by

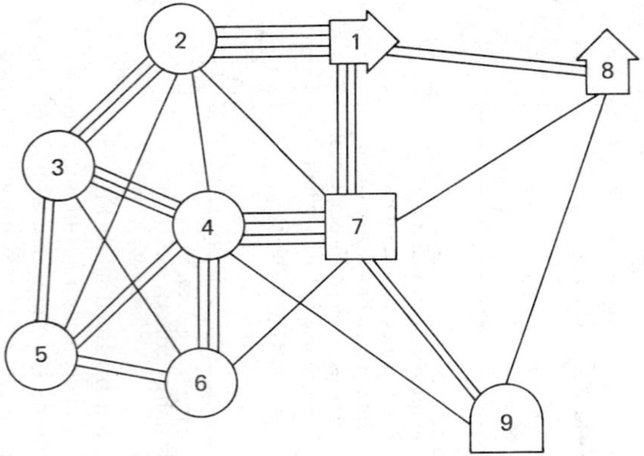

ACTIVITY IDENTIFICATION

Symbol	Color*	Type of Activity, Area, or Equipment
○	Red	Operation or Production (Subassembly and Assembly)
○	Green	Operation or Production (Processing or Fabrication)
⇧	Orange-Yellow	Transport-related activities (Receiving, Shipping, Rail siding)
▽	Orange-Yellow	Storage
□	Blue	Inspection, Test, Check
D	Blue	Services (Maintenance, Utilities, Personnel Services)
⇦	Brown	Office areas or activities not directly part of the main area or its support

Note: Activity number is inserted inside the symbol when diagramming
*Use is optional

CLOSENESS CODING

Rating	Closeness	Color*	No. of Lines
A	Absolutely necessary	Red	4 Straight
E	Especially important	Orange-Yellow	3 Straight
I	Important	Green	2 Straight
O	Ordinary Closeness	Blue	1 Straight
U	Unimportant	- - - -	0
X	Undesirable	Brown	1 Wiggly
XX*	Extremely undesirable	Black	2 Wiggly

Note: Minus sign behind letter indicates half a degree of closeness value. It is translated to a dotted line in color or in number-of-lines coding.
*Use is optional

Fig. 13. Relationship diagram indicating closeness rating of activities.

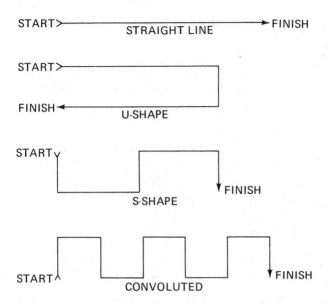

Fig. 14. Basic production flow patterns.

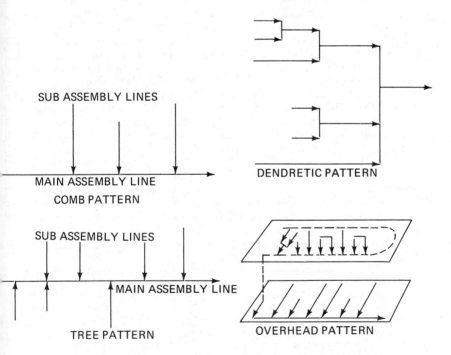

Fig. 15. Basic assembly-line patterns.

parts, materials, or subassemblies moving between various levels of manufacturing activities.

PROCESS LAYOUT. There are two general classes of layout. One is based on the kind of product being made. The other is based on the kinds of manufacturing processes the products must undergo. A process layout is set up where the product is not, and cannot be, standardized, or where the volume of like work produced is low.

Advantages of Process Layout. The main advantages of the process oriented layout are summarized as follows:

1. Higher degree of machine utilization by producing more than one product.
2. Greater flexibility of production process.
3. More efficient supervision due to subdivision of process.
4. Less interruption in work flow resulting from machine breakdown.
5. Lower equipment investment since each machine can process more than one type of product.
6. Better control of total manufacturing costs.
7. Higher level of individual operator performance through continuous performance on machine.

Disadvantages of Process Layout. The several disadvantages inherent in the process layout are:

1. Production is not always continuous.
2. Cost of transporting product between process departments may be high.
3. Damage may occur to fragile goods due to the extra movement required with this layout.
4. Additional man-hours must be spent on checking, sorting, and counting as product moves through various processing departments.

PRODUCT LAYOUT. A product layout is set up for a standardized product that is to be manufactured in large quantities for a considerable period of time or indefinitely. A product layout may also be used for shorter runs if another product could be produced with only a few changes in equipment. In general, product layout is best adapted to mass production industries. The **automotive industry** is an outstanding example of what may be done in production by "straight-lining." But the principles of straight-lining can be, and have been, applied in many other industries, in either permanent or flexible setups. In a product layout of machines, all emphasis is placed on the product. The machines required in the processing of the product are brought together within a department and set up in accordance with the necessary sequence of operations. Such a line is set up for a specific production rate, and the machine for each operation selected and geared to a suitable speed with that requirement in mind. Ideally the product will flow smoothly along with little or no delay.

Advantages of Product Layout. Some of the principal advantages of the product oriented layout are:

1. Lower material handling costs since transportation between departments is minimized.
2. Lower unit production time.
3. Less work-in-process.
4. Less total floor space required per unit of production.
5. Less record keeping and simpler production control.
6. Less total inspection.
7. Greater utilization of unskilled personnel.

Combination of Product and Process Layout. Frequently, the best solution may be obtained by combining the basic features of both product and process layout types. In effect, each process is set up as a unit or department, and these units are arranged into a product layout. This is typical of many "job shops" and semi-mass-production situations.

PLANNING FOR FLEXIBILITY. One of the objectives of plant layout is to provide flexibility of arrangement and of operation. Future **equipment rearrangement** may be necessary to:

1. Alter the process to fit a change in the design of a part.
2. Allow for change in production rate.
3. Permit the addition of a new part or product within the present plant.

Flexibility can be obtained in **existing structures,** by the following means:

1. Good general lighting in plant.
2. Electrical facilities which allow for plug-in connections to machines.
3. Mobility through use of self-contained machinery.
4. Portable conveyor units.
5. Extensive use of small tools.
6. Portable jigs and fixtures.
7. Standardization of equipment.
8. Mounting machinery on casters or skids.

In planning for flexibility in the design of a **new plant,** the following factors are to be considered:

1. Possible extension of production lines.
2. Use of mezzanines and auxiliary departments.
3. Shape of buildings.
4. Design of plants into manufacturing units.
5. Large unobstructed floor areas.
6. Sufficiently strong floors to stand the possible relocation of heavy machinery.
7. Design area to permit access at more than one point or side.

Planning Individual Work Stations. After the flow pattern has been established, the direction of materials flow through each processing area and each workspace can be determined. A check of the flow pattern when planning an individual work station or production center is necessary so that each operation will fit properly into the overall plan.

Work stations are defined and discussed by Mallick and Gaudreau (Plant Layout Planning and Practice) as follows:

The floor area occupied by the worker and the machine or group of machines which he operates is designated as the workplace, or work station. It constitutes the smallest indivisible space unit on a layout and includes the following amount of floor space:

1. The rectangular space occupied by the length and width of the machine or group of machines operated by one worker, or a group of operators working as one unit, in a given area. This space is expanded to allow for travel of moving parts and for projecting machine parts, such as shafts, levers, pulleys, and doors.
2. Floor space for the machine's own motor or power source when placed on the floor or within the working area.
3. Working space for the operator.
4. Clearance for feeding the work on and off the machine.
5. Space for the skids, tote pans, racks, conveyor stations, etc., which either contain the work to be processed or receive the work after it has been processed on the machine.
6. Space for whatever tool racks, work benches, and auxiliary equipment each particular machine may need for its operation.
7. Portion of the aisle space or conveyor space immediately adjacent to the operator, the machine or the group of machines he operates.

Fig. 16 from Apple (Plant Layout and Materials Handling) shows a typical work station or workplace. The complete work station requires much more space and equipment than the machine itself would indicate. With the use of accurately designed two-dimensional templates the **clearance area** can be utilized for effective layout. The space in which the operator is standing also serves as clearance for the bench drawer. The tool locker is located adjacent to the machine in a space often overlooked when working with block templates.

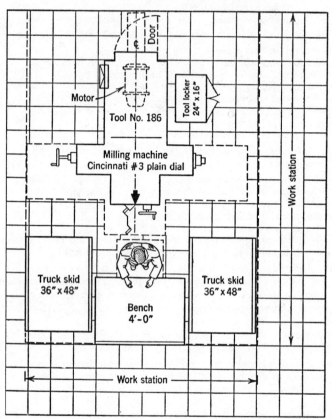

Fig. 16. Work-station layout.

Of great importance in effective operation planning are the principles of motion economy. By following these principles in planning equipment layouts the planning engineer may bring about important operating savings, which is a major objective of his activities (see section on Motion and Methods Study).

At the work station, men, materials, and machines are brought together to perform an operation. The basic approaches to bringing the three together are:

1. Stationing the man at a given point or machine with his tools and bringing the material to him. The automobile assembly line is an example of this approach.
2. Bringing the man and his tools and machines to the material. This approach is used to manufacture very large products that are too bulky or heavy to move through a production line easily.

3. Bringing the man and the materials to the machine. This is common in many plants where the machines are large and immobile. The man can move from machine to machine with the materials to perform the various operations.

Operator Cycling or Multiple Machine Operation. Where semi-automatic and automatic machines are in use, one operator can frequently be assigned to more than one machine. After loading and starting one machine, the operator can service a second and perhaps a third machine, or even more, before the first machine finishes its operation and is ready to be serviced again. (See section on Work Measurement and Time Study.) To prevent excessive walking by the operator, machines tended by one man should be placed adjacent to one another as much as possible, and material handling equipment should be designed to bring the work-in-process close to the work station. This is readily accomplished with process layout where all machines are similar and where line flow is not maintained. A screw machine department, for instance, where machines are loaded at relatively long intervals offers a good illustration of this principle.

The assignment of more than one machine to an operator is somewhat more difficult to apply with product layout because of the need for maintaining line flow and the fact that an operator must run different types of machines. Even here, however, as a general rule, it can be accomplished, often through grouping machines. While idle machine time may be increased by grouping machines, if properly planned it will be more than compensated for by savings in direct labor. The delays caused by machine idleness, adjustments, breakdowns, or difficulty with the material, are classified as **machine interference.** (See section on Work Measurement and Time Study.)

Number of Machines an Operator Can Tend. The number of machines an operator can tend depends upon the standard time for the operation, the hourly production required from the machine, and the coordination of man and machine time. When machines are to be placed in a group, they should be located as near one another as possible to save walking time and effort of the operator. In a product layout this arrangement depends upon maintaining the line of flow without excessive travel of material. **Line bends** or **loops** often may bring the product back into an area through which it has already passed in the course of operations (see Fig. 17). The improved layout shown in Fig. 17 is designed to lessen walking time and effort in operating and tending several machines.

Machine Breakdowns. As machine breakdowns are inevitable, provision must be made to keep the maximum amount of the plant working regardless of individual machines which may be out of order. In the process type of layout, operations of subsequent departments are not dependent solely upon flow of work in large volume along a definite production line under any one manufacturing order, nor in successive lots, so a machine breakdown does not cause a general plant stoppage. Since similar machines are grouped together, urgent operations can be shifted readily to another machine.

In the product type of layout a single machine breakdown can stop the entire line. Two provisions can be made to minimize such plant tieups: (1) immediate replacement of the crippled machine, and (2) providing a bank of work ahead of each operator. These procedures require additional investment in capital equipment and work-in-process, respectively. The automobile industry, for the most part, replaces the crippled machine, or sets up a temporary machine through which it shunts the parts while the machine in the line is being repaired. A decision as to which method to use must compare the cost factors of one method against those of the other. If a **work bank** is to be used it should be large

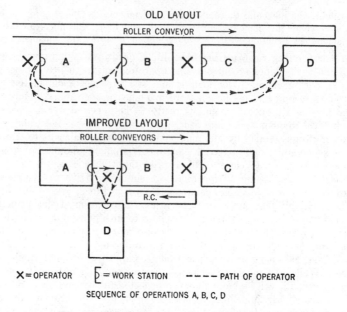

OLD LAYOUT

IMPROVED LAYOUT

X = OPERATOR   ⊃ = WORK STATION   – – – – PATH OF OPERATOR

SEQUENCE OF OPERATIONS A, B, C, D

Fig. 17. Old and improved machine layouts.

enough to keep each operator busy for the length of time taken to make the average repair. If the machine is to be replaced, or temporary machines are to be set up, aisle space or crane service must be available.

Maintenance and **machine repair** must also be considered when positioning a machine so that it is completely accessible for normal maintenance and repair (see section on Plant Maintenance).

Selecting Specific Material Handling Equipment. After the individual work stations have been planned, or possibly as the planning is being done, it is wise to go back to a consideration of material handling equipment. A preliminary consideration must be given to material handling equipment before the flow pattern is "roughed in." The work stations are then planned on the basis of this general selection of material handling equipment and the flow pattern (see section on Material Handling).

STORAGE SPACE REQUIREMENTS. The storage of materials in a plant is a major concern to the layout engineer. Storage space is necessary for raw materials, finished products, and for finished or partly finished materials in process. The size of the plant and the variety of manufacturing processes and products influence the degree of centralization of these areas (see sections on Materials Management and Material Handling).

Reed (Plant Layout: Factors, Principles, and Techniques) offers a list of questions which must be answered before the storage area layout is prepared:

1. What is to be stored?
2. What is the mix which can be expected?
3. What are the normal inventory levels of individual goods and supplies for which storage will be required?
4. What is the condition and packing of goods at the time of arrival?
5. What is the condition and packing of goods at the time of dispatching?
   a. What changes are made in the goods prior to storage?

b. What changes are necessary to prepare the goods for issue or dispatch after the receipt of an order?
6. What is the handling method by which goods will be received?
7. What is the handling method by which goods will be dispatched?

Raw materials are usually located close to a receiving railroad siding or platform and adjacent to the manufacturing departments which will perform the initial operation on the material. Separate storerooms may be advisable for different classes of raw material, as for example paper, steel stocks, tool steel, and supplies. Size of plant, location of departments, available trackage, and character of the raw materials to be stored are determining factors. An outdoor storage yard is practical for material not injured by weather or made more costly to handle or use. Materials having special characteristics, such as very high value, or light weight; materials requiring special handling; or material used exclusively by one department may be stored in, and under the control of, the department concerned.

In-process storage is best located adjacent to or near those departments which require the material next. **Finished products** are usually stored near shipping departments, convenient for the packing and loading operations.

In many cases, a few simple calculations will solve the major problems in **determining storage area requirements.** The following illustration is offered by Apple (Plant Layout and Materials Handling):

Assume that production requirements call for the storage of two weeks' supply of a certain casting. Production is at the rate of 50 units per hour, or 4,000 units for two 40-hr. weeks. If one casting is approximately 6 x 6 x 6 in., then 8 castings will require one cubic foot. A 3 x 5 x 2½ ft. container will hold:

$$3 \text{ ft. x 5 ft. x 2½ ft. x 8 per cu. ft.} = 300 \text{ units}$$

Then 4,000 can be contained in:

$$\frac{4,000}{300} = 13\frac{1}{3}, \text{ or 14 containers}$$

If containers can be stacked only 3 high, because of the floor-load or ceiling-height limitation, then:

$$\frac{14}{3} = 4\frac{2}{3}, \text{ or 5 stacks}$$

will be required.

As each stack requires 15 sq. ft. of floor space, then 75 sq. ft. of floor space will be required to store a two-week supply of the castings.

The actual allocation of storage space will be determined by process location. Some storage space for raw materials will be needed at the beginning of the process, some along the fabrication line for parts in process, and some at the end, for finished parts. It may be advisable to construct a chart or tabulation of **storage space requirements** so as to be sure that space is provided. Such a tabulation is shown in Fig. 18.

FLOW DIAGRAMS FOR PRODUCTION CENTERS. The flow diagram should show, to scale, all the machines, stock containers, benches, and other auxiliary equipment necessary to the proper functioning of the particular department or production area.

In making these sketches, attention should be paid to the general flow pattern and the work-station plans developed previously. The flow diagrams may or may not be detailed, but enough information should be included to show fairly accurately the total amount of floor area needed by each production center. It is these flow diagrams which will later be molded into the final layout around a

STORAGE REQUIREMENTS WORKSHEET

☐ Incoming materials
☐ Work-in-process
☐ Finished components
☐ Finished products

ITEM							QUANTITY		RECEIPT		HANDLING UNIT							SPACE REQUIREMENTS			
Part No.	Description	Dimensions					Max. Invent'y	Normal. Received	Freq.	Type Carrier	Type	L	W	Ht.	Wt.	Items/ Hdlg. Unit	Hdlg. Unit/ Max. Invent.	Cu. Ft. Req'd.	Type Handling Unit	Type Storage Space	Storage Location
		L	W	Ht.	Wt.																

Fig. 18. Storage requirements worksheet.

network of aisles and columns, with the necessary allowances included for service areas.

Line Production. Layout by product and line production are generally considered synonymous phrases, and the application is of interest at this stage. The advantages of product layout have already been cited, but there are **prerequisites** to its use, which are given by Apple (Plant Layout and Materials Handling) as follows:

1. There must be sufficient quantity to justify the product type of machine arrangement.
2. The product must be of such a nature that the operations to be performed on the line can be broken down into units sufficiently small:
   a. To permit each to be learned in a relatively short time by a relatively unskilled operator, thus facilitating the shifting of personnel necessitated when production increases or decreases.
   b. To permit the elements of work to be recombined into time units of about equal length for each operator in the line.
3. Jigs and fixtures must be used to make sure that each operation is performed in exactly the same manner on each part or assembly.
4. The line, after it is set up, must not be so inflexible as to prevent minor alterations which might be required by design, model, or methods changes on the part or product. In spite of the application of the production-line technique to the "nth" degree to automobile assembly lines, one manufacturer turns out 20,000 variations of his product on one line.
5. Materials must be continuously supplied to the line at the required places so that a materials shortage will not cause the line to shut down. Production control and materials control must be properly worked out and must function well.
6. There must be enough operations to be done on the part or product to warrant the line.
7. The production of each kind of unit must extend over a sufficient period of time. A line cannot be set up if a month's supply can be turned out in a few hours, since machine shutdown and changeover is time consuming and probably expensive.
8. The line must run a sufficient portion of the working time to be economical.
9. The job must last long enough to justify setting up the line.
10. The design must be fairly well "frozen" or standardized so that changes will not disrupt the line too often.
11. Parts must be interchangeable.

Balancing Production Lines. Although this activity falls primarily into the area of operations rather than plant layout, it is an obvious factor to be considered when developing the layout scheme. Line balancing ensures that each work station will carry an equal work load in order that a continuous flow of production can be maintained. Failure to adhere to this principle will result in bottlenecks, downtime, reduced employee efficiency, and lower profits.

In order that each work station receives, processes and feeds to the following station the material at the required manufacturing rate, it is necessary to determine the **line speed** of the production stations. If that speed does not conform to the required production rate, then adjustments must be made in the line and the layout should reflect these adjustments.

Various methods have been developed to solve the line balance problem, including:

1. Utilization of an operator at adjacent work stations.
2. Sub-dividing the task to be performed into multiple elements of equal length.
3. Determining a better way of performing the operation (improved methods or operator training).

4. Assigning additional sub-assembly tasks to operators with idle time.
5. Stationing inventories of work in process along the line to pick up slack resulting from slower stations "up the line" or equipment breakdowns.

Each of these techniques influence the plant layout directly.

Subassemblies. The problem presented by subassemblies should be thoroughly explored before planning the assembly area in detail. Apple (Plant Layout and Materials Handling) lists some of the reasons for using subassemblies:

1. To facilitate the handling of smaller or larger parts at the assembly point.
2. To shorten the general assembly line, where the subassembly requires much space.
3. To reduce final assembly time if a subassembly requires a greater amount of time than can conveniently be fitted into the operations planned for the line.
4. To separate from the line any equipment, the nature of which would interfere with the line.
5. To reduce complications on the line when a part must be machined after it is assembled, and the processing equipment could not well be fitted into the line.
6. To provide for required testing of the subassembled unit before it becomes a part of the final assembly.
7. When the subassembled part may become a part of any one of several different products.

These factors will influence the layout of not only subassembly areas but also the assembly areas into which they feed. The beginning and ending locations of subassembly lines should be planned carefully. Usually it is desired to have a subassembly line end at a specific point on the final assembly line, and let the beginning fall as appropriate.

Developing the Plot Plan

ALLOCATION OF PRODUCTION CENTERS ON PLOT PLAN.
After the flow diagrams have been made for individual production centers and it is known how much space each will require, area allocation is necessary to fit them into the building outline.

Area allocation diagrams are generally made to a smaller scale than that for the layout. Moore (Plant Layout and Design) illustrates such a diagram as in Fig. 19. This shows building lines, internal partitions, and area allocations, but not machinery, equipment or facilities. They indicate that such a diagram is generally used only for reference in analyzing area arrangement and not for plant layout drawings.

Aisles. Aisles provide a means of communication and transportation between various sectors of a facility. Adequately designed aisles facilitate the movement of workers, materials, and equipment. Poorly placed aisles can result in greater expenses due to excessive handling requirements; too much space allocated to aisles will result in reduced production space, while insufficient aisle space can cause severe bottlenecks and unsafe conditions.

The typical industrial plant will have two types of aisles: **main aisles** and **departmental aisles.** Main aisles are utilized for interdepartmental movements or for transporting men or materials into or out of the plant. Departmental aisles permit movement within a given department.

Suggestions for planning aisles are as follows:

1. Locate main aisles first; the larger the plant, the farther apart they can be.
2. Plan at least one main "back-bone" aisle at or near the center of the building, 10 to 20 ft. wide, depending on plant size.

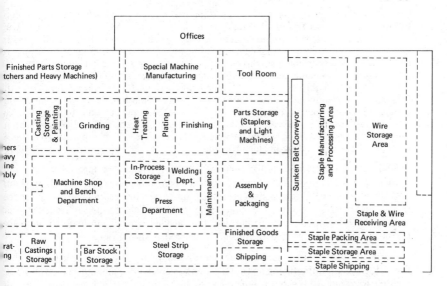

Fig. 19. Area allocation diagram for a manufacturing facility.

3. Start main aisles at an outside entrance and run them as straight as possible.
4. Use interior aisles of less width for "feeder" purposes, such as movement of personnel, materials, supplies, scrap, service, and fire access.
5. Consider unit-load sizes and truck-turning radii when determining aisle width.
6. Plan for loaded industrial trucks to pass each other, at least in selected spots, along aisles.

Structural Elements. Reed (Plant Layout: Factors, Principles, and Techniques) presents the following discussion on the influence of building structural elements on plant layout:

Various building features necessitated by structural engineering or architectural requirements tend to restrict or limit the flow patterns which may be used in manufacturing or production area. The most common of these are the columns used to support the roof structure and around which the flow must be designed. The flow arrangement should be designed in such a manner that, to the greatest extent possible, columns either are located between machine areas or are located at the side of aisles between machines in such a manner that they effect minimum interference with the smooth and most desirable production process flow. The second structural feature that may be a limiting factor is structural **beams.** Where columns create a floor plan interference, beams represent vertical limitation as to the utilization of overhead space and must be taken into consideration in planning flow or future expansion. Thought must be given in the initial design of the building to the maximum height requirements for manufacturing equipment or manufacturing process. Clear ceiling heights—taking into consideration the clearance required for utilities and sprinkler distribution systems—must be designated to meet these requirements. If a limited number of machines or other types of equipment require higher ceilings than the bulk of the equipment, these taller machines may be placed in separate bays or buildings with the necessary high ceilings. Building areas for normal equipment can then be constructed with somewhat lower ceilings. Scattering tall machines throughout the manufacturing area will increase the total cost of building above that which is economical, due to wasted wall and columnar heights.

The third major type of building structural limitation is **floor loading.** Floor loading is particularly important when multiple-story buildings are being considered,

because of the hazards created on the lower floors from overloading an upper floo
Before finalizing the layout, it should be determined by conference with the struc
tural engineer and the architect that no dangerous floor-loading conditions ar
created by the desired layout. Furthermore, when heavy equipment is being re
located during expansion or relayout of existing plants, it is paramount that th
determination be made as to whether floor capacity is sufficient to support th
material and/or equipment with a satisfactory safety factor during use. For th
reason, it is normally advantageous to locate heavy equipment on the ground floo
Structural alterations or designs to provide special supports and foundations ca
then be incorporated only at the points necessitating them rather than throughou
the entire structure. Special foundations for large presses in a separate area are a
example of the application of this principle.

SERVICE AREAS. While service areas are frequently located where roor
is available after manufacturing space has been allocated, their location shoul
not be chosen arbitrarily. Service areas must be integrated into the overa
layout on the basis of their functional relationship with the manufacturing area
Failure to develop a sound plan for the service areas can lead to increase
operating expenditures due primarily to wasted man-hours.

General Office. In small and medium-sized plants, general offices and thos
for principal manufacturing executives, engineers, and other special groups, ar
located together at the main entrance to the plant. This arrangement provide
maximum convenience for visitors and keeps them out of the manufacturin
buildings and grounds. The general offices should be close to the various plar
departments and buildings. In larger concerns a separate building and locatio
for **factory administration offices** may be preferable so as to bring togethe
various phases of factory management, to afford closer contacts for executive
both with the departments they supervise and with each other, and to provid
a better work environment.

Factory Office. The factory office from which the manufacturing procedure
are coordinated should be located as close as practicable to the actual productio
areas. Frequent personal communication with key men and personal observa
tion of operations are often desirable; hence the need for close proximity. Nois
dirt, and fumes from manufacturing operations must be excluded and a pleasan
office atmosphere maintained. Sometimes the ideal location for factory offices i
on balconies or **mezzanine floors.**

Tool Cribs and Toolrooms. Tool cribs, in which tools, fixtures, dies, etc
are kept, should be located conveniently close to the manufacturing areas servec
A common practice is to have a central toolroom or department where too!
making, if any, is carried on, where expensive and less frequently used tools ar
kept, where major repairing is done, and tool records are kept (see section o
Tools, Jigs, and Fixtures). This central department is supplemented by
number of smaller tool cribs located about the plant, where needed, for issuin
tools to setup men and workers, and where minor tool repairs may be made.

Powerhouse. If the plant generates its own power, the powerhouse or engin
room should be located so as to facilitate fuel delivery and waste disposa
Where large quantities of coal are used, outside storage areas will usually b
necessary and **car dumpers** or **receiving hoppers,** sometimes served by conveyo
systems for yard storage and later reclaiming of coal, may have to be installec
If oil is used, similar receiving and storage accommodations must be made
Equipment and **distribution systems** for electric power, gas, compressed ai

:eam, hot water, etc., must be planned as carefully as those for manufacture of roducts. In some locales it is mandatory, and in most locales it is desirable, 1at the powerhouse be so situated as to keep air and water pollution at a 1inimum.

Locker Rooms, Washrooms, and Toilets. Locker rooms should be located 1 or near the areas in which employees work, and preferably adjacent to uilding entrances through which the employees pass as they enter or leave 1e plant. **Time clocks** should be placed near entrances so that the fewest teps are taken in checking in and out. Passageways in these areas should rovide ample space for peak traffic. To avoid interference with production,)cker rooms, washrooms, and toilets are sometimes placed in wings attached to a uilding, or in central **service areas,** along with elevators and stairways. An rrangement employed in one-story buildings with high ceilings, where only men 'ork, is to place toilets on a mezzanine floor above the working area, reached by pen metal stairways. In multistory buildings, washrooms and toilets are placed ne above the other on all floors to facilitate piping runs.

Ordinarily it is not desirable to require employees to go from one building ɔ another, or from one floor to another, to reach lockers, washrooms, and toilets. 'he time lost, the extra supervision required to prevent unnecessary idleness, nd the possibility of falls in going up and down stairs from floor to floor entail a onsiderable and unnecessary loss of work time, especially in the case of women orkers.

Where it seems advisable to have one washroom and toilet room serve two oors in a multistory building, it may be placed between floors where the mployees need go down or up only half a flight of stairs. In any case, it is well o have service facilities sufficiently **decentralized** to avoid congestion.

Personnel Department. The personnel department is preferably located on he ground floor near the street entrance through which applicants for jobs may nter and leave conveniently. It is not desirable to have them go through he plant to reach the personnel office. Since this department must have direct ontact with the manufacturing departments, keep employee records, and may ave hospital and dispensary facilities, it should be located close to the manu- acturing area.

Dispensary. The dispensary should be located so that it is convenient to all nanufacturing departments and can be reached in a minimum of time in an mergency. It should preferably be near an outside driveway for **ambulance ervice.** The dispensary and the personnel departments are often found adjacent ɪr co-located in one department. This arrangement permits greater ease in onducting personnel activities, making physical examinations, and attending to ɪrst-aid patients.

Cafeterias. Restaurants and cafeterias should be centrally located for the onvenience of the majority of the employees and with a minimum of travel. Although not necessarily profit-making departments, they should contribute to he well-being of the employee, his feeling of satisfaction, and his volume of ɪroduction. As a supplement to central service, **branch cafeterias** may be •perated to supply hot foods, sandwiches, fruit, coffee, and milk. Employees hould be discouraged from eating at their workplaces for sanitary reasons as vell as possible damage to equipment or product. Therefore attractive lunch- 'oom facilities, in areas especially set aside for the purpose, should be available or those employees who bring their own lunch. Comfortable chairs and tables, together with a limited number of vending machines for dispensing coffee, milk candy, etc., should be provided.

A comparison of food service methods is charted by Moore (Plant Layout and Design) as shown in Fig. 20.

	Cafeteria	Snack Bar	Lunch Wagons	Vending Machines
Food:				
Quality	Variable	Variable	Uniform	Uniform
Variety	Unlimited	Somewhat limited	Limited	Limited
Sanitation	Difficult	Easier	Difficult	Easier
Price	Low but subsidized	Low but subsidized	Medium	Low
Cost to Company:				
Installation	High	Moderate	Low	Low
Space	Large	Moderate	Very little	Little
Subsidy	High	Moderate	Moderate	Low
Clean-up	High	High	Moderate	Moderate
Administration	Many employees	Few employees	Some employees	Janitor only
Service:				
Speed	Slow	Faster	Fast	Fast
Availability	Limited	Scheduled	Scheduled	All times
Distance from work area	Remote	Remote	At scene	Nearby
Personal touch	Yes	Yes	Yes	No
Reliability	Good	Good	Good	Occasional malfunction

Fig. 20. Comparison of food service methods.

Maintenance Department. A **central** maintenance department should be located in an area that is accessible to the various production activities of the plant. When a number of **decentralized** maintenance areas are to be provided their location is determined by the desire to place them at or near the center of the departments which they will serve. A combination approach to maintenance department location is sometimes used (see section on Plant Maintenance).

Other Departments. As with the departments just mentioned, other miscellaneous departments should be located with regard to the function they perform and their inherent requirements.

Facilities for employee recreational, educational, and social activities should not be overlooked, and should be considered wherever space, convenience, and the nature of the activity permits. In large industrial complexes meeting halls and auditoriums, as well as baseball fields and basketball courts, may be located within the plant grounds. In smaller plants, cafeterias, lunchrooms, and large office areas can be converted when necessary into auditoriums, meeting rooms etc., for both company and non-company affairs. More effective use can thus be made of places, such as cafeterias, which are otherwise used for only short peak periods during the work day.

Completing the Layout

EVALUATION OF ALTERNATE LAYOUTS. When the preceding work has progressed step by step through area allocation and assigning of service areas to their locations on a sketch, the plant layout engineer usually is faced with a number of layout **alternatives** that appear to satisfy his requirements. The next step therefore is to **evaluate** these alternatives in order to arrive at an optimum solution.

Muther (Systematic Layout Planning) suggests a method of analysis that considers the important or significant factors in deciding which layout to select. The procedure for tabulating and evaluating alternatives by **factor analysis** is shown in Fig. 21 and as follows:

1. **Identify the plans to be evaluated**
   a. Select the layout plans that are to be evaluated.
   b. Have a visual plan or sketch of each layout in front of each rater, and clearly understood by him, during the evaluation process.

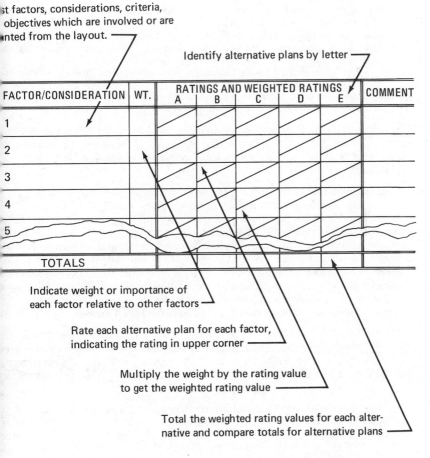

Fig 21. Rating sheet for factor analysis of alternate layouts.

    c. Identify each visual plan by letter—A, B, C, etc. Also give it a brief three-to-five-word description.

2. **Establish the factors or considerations**
   a. Establish what factors, considerations, criteria or objectives are involved or are wanted from the layout.
   b. Define the factors so they are clearly understood. Avoid duplication between terms and confusion as to meaning.

3. **Arrange a rating sheet**
   (See Fig. 21.)

4. **Determine the relative importance of each factor**
   a. Determine a weight or importance value of each factor relative to the other factors.
   b. Record by whom the weight values were determined.

5. **Rate each alternative plan by factors**
   a. Establish a rating code or system.
   b. Rate each alternative on the extent to which it achieves or affords the end represented by the factor in question—rating each layout exactly as it is planned.
   c. Rate all plans for one factor; then take next factor.
   d. Enter rating symbol above slant line on rating sheet.
   e. Record by whom the ratings were made.

6. **Calculate weighted values and total**
   a. Translate the rating symbols into numerical values and multiply by the weight value.
   b. Total the weighted rating values for each alternative plan by adding the respective columns.
   c. Record by whom the extension and tally were made.
   d. Take action as appropriate, based on the totals.

A listing of the most frequently identified factors which may affect the selection of the most suitable plan is shown in Fig. 22 (based on Muther). A definition of each factor is given, in addition to the key points to consider in making the evaluation.

Cost Comparisons. In addition to the subjective factors analysis approach, a study of the **relative economics** of competing plans should also be made. Two methods are available in developing the cost analysis. One considers **total** (annual costs if appropriate) investment and operating expenses associated with each alternative, while the other deals only with **incremental** differences in cost between the various plans. Whatever the method selected, the valuator should be sure to consider all the investment charges associated with the various plans and develop the annual operating expenses to be experienced. Then, based upon engineering economics techniques (discussed in the section on Capital Investment Analysis) he can determine which plan presents the lowest costs.

CONSTRUCTING THE MASTER LAYOUT. The master plan is the finished product. It represents the work of the plant layout engineer and others and it should be done well. There are four common methods of **representing the final layout.** According to Apple (Plant Layout and Materials Handling) they are:

1. Drawn in the conventional drafting manner on drawing or tracing paper.
2. Constructed with two-dimensional templates mounted on a suitable base or grid.
3. Constructed with three-dimensional scale models in place of templates.
4. Constructed with scale models and templates—for ease of reproduction.

1. Ease of Future Expansion (the simplicity of increasing the space employed):

   a. Tie-in with long-range potential use of the space.
   b. Ability to spread out to adjacent areas.
   c. Freedom from fixed or permanent building features.
   d. Regularity of allocated space amounts.
   e. The amount of disruption or re-arrangement of areas other than the ones specifically being expanded.
   f. Ease of contracting the lay-out economically, to cut down the size if necessary.

2. Adaptability and Versatility (the ease of accommodating changes (normal or emergency) in items like the following):

   a. Product, materials, or items.
   b. Quantity or volume.
   c. Frequency of delivery.
   d. Process equipment.
   e. Operation sequence.
   f. Working methods and operating time.
   g. Handling or storing methods.
   h. Utilities or auxiliaries.
   i. Test runs, pilot lots, experimental engineering.
   j. Type or classification of employees.
   k. Time-keeping system.
   l. Hours of work.
   m. Material dispatching procedure.
   n. Inspection controls.
   o. Rework procedures.
   p. Standby equipment.
   q. Additional space for stock.
   r. Alternate routes.

3. Flexibility of Layout (the ease of physically rearranging the layout to accommodate changes):

   a. Mobility of machinery and equipment.
   b. Relative size of equipment.
   c. Standardization of equipment, containers, workplaces.
   d. Freedom from fixed building features or walls, unmatching floor levels, other barriers.
   e. Overly dense saturation of space.
   f. Independence or self-sufficiency of facilities (not dependent on central coordination or centralized service tie-in).
   g. Ready accessibility of service lines, piping, power distribution, heating and ventilating, manholes, etc.
   h. Access to the area laid out at more than one point or side.

4. Materials Handling Effectiveness (the ease or simplicity of the handling system, equipment, and containers):

   a. Ease of tie-in with external handling methods and equipment: rail line, docks, highway, and other accessways.
   b. Necessity for re-handling, extra handling, delays, awkward positioning, undue physical effort, undue dependence on frequency or urgency of moves, undue amount of temporary connections on non-integrated equipment.
   c. Traffic congestion and interferences other than due to flow pattern.
   d. Balanced variety of handling systems, equipment and containers.
   e. High utilization of handling equipment and containers.
   f. Simplicity of handling devices.
   g. Equipment integrated for multiple use.
   h. Dependence of materials handling equipment on maintenance, repair, replacement parts.
   i. Avoidance of synchronizing two or more people at same time or place.
   j. Ability to move completely around buildings on company property.
   k. Use of gravity for moving materials.
   l. Combined purposes of handling equipment for storing, pacing, sequencing, inspecting, workholding, weighing, and the like, as well as moving.

Fig. 22. Factors or considerations in selecting the layout.

5. Flow or Movement Effectiveness (the effectiveness of sequenced working operations or steps):

a. Greatest flow intensities with minimum distances.
b. Basic regularity or consistency of flow patterns.
c. Proximity of related areas to each other where movement of material, people, or major paper work is involved, or where frequent, urgent or significant personal contact takes place.
d. Access to, away from and between major areas (like receiving, shipping, key operating areas).
e. Flow of auxiliary or service materials: supplies, tools, scrap or waste, and other service materials.
f. Accessibility for delivery and pick-up, visitors, or employed non-company service personnel.

6. Storage Effectiveness (the effectiveness of holding required stocks of materials, parts, products, service items):

a. Inclusion of all storage—raw, in-process, finished goods, supplies, tools, scrap or waste, trash and equipment or materials not in current use.
b. Accessibility of items stored.
c. Ease of locating or identifying items stored.
d. Easy stock, inventory control.
e. Ability to make stored items available according to urgency of demand.
f. Protection of material (fire, moisture, dust, dirt, heat, cold, pilferage, deterioration, spoilage).
g. Adequacy of storage spaces.
h. Closeness to points of delivery and use.

7. Space Utilization (the degree to which floor area and cubic space is put to use):

a. Conservation of floor space, property, or land.
b. Utilization of overhead space in terms of cubic density.
c. Ability to share or exchange space among similar activities, and balancing of areas with seasonally complimentary space requirements.
d. Effectiveness of aisle space.
e. Waste or idle space, caused by split, divided, cornered, scattered or otherwise honey-combed structures, too-close columns, too-frequent partitions or walls.
f. Less desirable or out-of-way space utilized for slow, dead areas; convenient space for fast, active areas.

8. Effectiveness of Supporting Service Integration (the way supporting areas are arranged so as to serve the operating areas):

a. Ability of existing (or planned) systems, procedures, and controls to work well with layout.
b. Ability of the layout to integrate with desired or effective pay plans, performance measure, cost reports, lot size, order quantities.
c. Physical closeness of service areas according to each area's need for the service (actual versus desired relationships).
d. Ability of the utilities, auxiliary service lines, and central distribution or collection systems to serve the layout.
e. Convenience of scrap collection and waste control areas and equipment.
f. Ability of engineering groups and technical advisors to support the layout effectively.

9. Safety and Housekeeping (the effect of the layout and its features on accidents or damage to employees and facilities, and on the general cleanliness of the areas involved):

a. Basic regularity of the aisles and work areas, and degree of freedom from equipment protruding into aisles or work areas, congestion, blind corners.
b. Degree to which all safety codes and regulations are satisfied.
c. Risk of danger to people or equipment.
d. Availability of adequate exits and clear escapeways.
e. Access to first-aid facilities and fire extinguishers.
f. Floors unobstructed, tidy, and not overly congested.
g. Adequate protection or segregation for dangerous or unsightly operations.
h. Workers not located too near moving parts, unguarded equipment, or other hazards.
i. Workers able to get benefit from special safety devices or guards.
j. Effectiveness of housekeeping methods and equipment.

Fig. 22. Continued.

16·40

10. Working Conditions and Employee Satisfaction (the extent to which the layout contributes to making the areas pleasant places in which to work):

a. Effect of layout on attitude, performance, or general morale of employees.
b. Working conditions suitable to the type of operation.
c. Suitability of the layout's arrangement and allocated space to the personnel.
d. Convenience for employees.
e. Freedom from features causing workers to feel afraid, hemmed-in, embarrassed, discouraged, discriminated against.
f. Noise, distractions, or undue heat, cold, drafts, dirt, glare, or vibrations.
g. Utilization of employee know-how and skills.
h. Balanced manpower allocations.

11. Ease of Supervision and Control (the ease with which supervisors and managers are able to direct and control operations):

a. Ability to see the area fully and easily.
b. Ability to get around the area conveniently.
c. Ease of controlling quality, quantity counts, schedules, inventories in process.
d. Ease of controlling pilferage, waste time, lost materials, or supplies.
e. Ease of moving or reassigning personnel to other work.

12. Appearance, Promotional Value, Public or Community Relations (the ability of the layout to promote the company name or reputation in the community):

a. Attractiveness of external or viewable features, yards, main structure, out buildings.
b. Ability to serve as a show-place or reflect reliability, progressiveness, or other company qualities.
c. Regularity, symmetry, clean lines, and organized appearance.
d. Fit with community appearance, tradition, character.
e. Effects on neighbors (benefits and irritants).

13. Quality of Product or Material (the extent to which the layout affects quality of the product, material, or their workmanship):

a. Damage or risk to materials caused by nature of the layout or its transport facilities.
b. Contamination, corrosion, spoilage, or other detriments to the product's nature or condition as caused by the layout.
c. Convenience and inter-relationship of quality control activities.

14. Maintenance Problems (the extent to which the layout will benefit or hinder maintenance work):

a. Adequacy of facilities for maintenance and repair work.
b. Sufficiency of space for access to machinery and equipment to be lubricated, checked, cleaned, adjusted, on-spot repaired, etc.
c. Appropriate janitorial facilities.

15. Adaptability with Company Organization Structure (the degree to which the layout matches the organizational structure):

a. Eliminates or streamlines supervision and improves the alignment of managerial personnel.
b. Areas having the same supervisory responsibility are adjacent or convenient to each other.
c. Staffing or manning of layout fits with job classifications and salary schedules.

16. Equipment Utilization (the extent to which machinery and equipment is used):

a. Degree of utilizing all equipment.
b. Necessity for duplicating equipment caused by layout versus use of common equipment and services.
c. Overcapacity equipment necessitated by the layout.
d. Man-machine efficiency planned into the layout.

Fig. 22. Continued.

16·41

17. Utilization of Natural Conditions, Building, or Surroundings (the extent to which the layout takes advantage of the natural conditions of the site, physical surroundings, building structure, or neighboring areas):

   a. Slope, topography, foundation, drainage.
   b. Direction of sun, prevailing wind.
   c. Rail line, highway, waterway, bridges, accessways, crossings.
   d. Building features, structure, shape, height, construction,

   docks, door locations, elevators, windows, walls, columns.
   e. Zoning of site and restrictions of community or neighborhood.
   f. Fit of the areas laid out onto the natural site or into the existing building or area allocated.

18. Ability to Meet Production Requirements (how well the layout actually meets the planned needs or desired output):

   a. The right products or materials, properly meeting specifications.
   b. The right quantities of each variety or item in the operating

   time planned, without overtime or premium pay.
   c. The right yield in terms of projected quantities and qualities of product.

19. Investment or Capital Required (the amount of money actually required to get the layout installed and operating):

   a. Cost to construct, to erect, to modify, or otherwise prepare the space and/or site.
   b. Cost to actually install or rearrange machinery, equipment, utilities, and services.
   c. One-time expenditures or other expenses occasioned by the installation.
   d. Subsequent investment costs re-

   quired to make the layout perform properly.
   e. Ability to finance.
   f. Investment consistent with the age or suitability of the building and the anticipated life of the layout.
   g. Ease of step-at-a-time, piece-meal, or programmed financial outlay for installation.

20. Savings, Payout, Return, Profitability (the extent of savings offered by one layout over other alternatives and by comparison with the investment capital required):

   a. Total operating expenses, one layout versus another.
   b. Difference in annual cost and

   the allocated or amortized portion of investment.
   c. Future or latent savings or economies.

Fig. 22. Concluded.

A guide to the use of drawings, templates and models is suggested by Muther (Systematic Layout Planning) as shown in Figure 23.

Drawings. Drawings are probably the easiest, most practical and most frequently used method of presenting a layout. Used generally to present planning concepts, they are quite handy to rough out various possible arrangements for further review before advancing to more detailed templates or models.

Layout Boards. The master layout is usually mounted on a backing board. If the project is small, the work can be done on a **drawing board** or drawing table. In the larger plants, special **layout tables** made up of removable sections, about 2 x 4 ft. in size, are used, to permit removal of any section to be studied, worked on, or discussed.

Boards, or sections of the layout, may also be stored or filed in a horizontal position in a rack, with each board pulling out like a drawer.

Sometimes layouts are mounted in a vertical position. They may be placed on walls, stands, or built into **vertical filing racks** of one sort or another. They could be mounted like a window sash, so they could be lowered into viewing position from a "file" above or mounted on larger boards, about 7 x 8 ft., and brought into position by sliding on overhead tracks.

DRAWINGS	1. When "paper and pencil" are at hand, and time does not warrant or allow getting templates or models. 2. When roughing out an arrangement idea prior to moving templates or models. 3. When you have no experience with templates or models and the project will not justify or allow time to gain it. 4. When you want a quick replica for someone while away from your layout planning office. 5. When you are refining your template or model layout and recording suggestions or capturing new ideas on a print of the alternative or proposed layout.
TEMPLATES	1. When planning detail layouts for a large overall area or for repeated relayouts. 2. When template materials are readily available and materials or facilities for making drawings are not at hand. 3. When you have had prior experience with templates. 4. When project involves many alternatives — especially when preliminary to preparing 3–D model. 5. When equipment for print reproduction is readily available.
MODELS	1. When the layout involves complex interferences and diagonally-oriented vertical equipment — that is, a 3–D problem. 2. When the layout involves new or changed processes, products, methods or procedures, or a completely different type of layout than previously familiar to the company. 3. When planning a layout involving substantial capital investment or a long-term (relatively permanent) layout where full detail planning makes good "insurance" — especially when the cost of floor space is high. 4. When layout planners and operating and support people have little ability or experience in layout planning so that review and check of layout by others is significant. 5. When the layout will involve many outsiders during the construction and/or installation planning, many problems in selling management or directors, and many new employees and supervisors to be trained.

Fig. 23. Guide to the use of drawings, templates, and models.

Scale. The most common scale in use in industry is 1/4 in. = 1 ft. for th
layout itself. Smaller scales (1/8, 1/16, 1/32 in., etc.) are used for plot plans an
preliminary sketches of large areas. Larger scales (3/8 or 1/2 in.) are use
when smaller areas are being laid out, or when considerable detail must be shown

Templates. Templates are the most flexible method of visualizing layout plans
Templates are pre-drawn or pre-printed scaled replicas of the equipment asso
ciated with the layout being considered. Equipment outlines are those which
would be made around the periphery of a piece of equipment, although projection
such as wheels, handles, etc., are separately indicated as details not in th
periphery. Templates are usually not used to represent storage materials
service facilities or similar units. Templates can be constructed of plastic, heavy
cardboard, sheet metal, wood, or paper.

Several kinds of two-dimensional templates commercially available are printe
on translucent plastic and some are fitted with **permanent magnets.** Layout
are made on metal-surfaced boards, over a **translucent grid,** and a print i
made directly from the actual layout, with no drawing necessary.

The Johns Mansville Corporation (Plant Layout Manual) suggests the follow
ing guidelines for making templates:

a. Letter templates before cutting out so that they will be ready to use whe
separated, and the small ones will not be mislaid. Letter templates so tha
they can be read with the observer standing on either one long side or shor
side of the board.

b. Make templates in duplicate, one set to be labelled and filed away. Th
duplicate set will facilitate making changes on a spare board and allow com
parison with the present layout.

When using transparent grid sheets, a second or revised layout may be pre
pared quickly by placing a new set of grid charts over the first layout an
locating all equipment which is to be moved by matching the lower charts

c. Templates should not be copied from other templates. If an error was mad
in laying out the template, it will be repeated. Only actual measurements are
to be used in laying out templates.

As each template is laid out and lettered it should be checked so that wher
cut out and fastened to the layout board, the lettering will be readable and ir
the correct direction; that is, the lettering will not be upside down.

Care should be taken so that no equipment, machinery, areas, aisles, etc. wil
be missed. Check equipment lists and flow charts prepared in advance of th
layout.

d. Show only enough detail necessary for layout plan. Be sure that the templat
looks the same as the equipment, areas, etc. would appear if viewed from above

Do not be misled when measuring equipment, check all projections above th
floor line. This is especially true of planers, milling machines, etc. that hav
movable tables. These areas should be shown as dotted lines extending to th
full travel so that other equipment or areas will not be located in space that i
not available.

When the template does not appear clear with just the outline, only enough
detail should be added to make the template clear to an observer. The menta
picture that the observer has in his mind should be confirmed when he look
at the template.

e. The scale used in making up all templates should conform to the building out
line and cross section scale.

f. It may be necessary, because of size limitations, to use letter or number
designations instead of complete names to identify templates.

Do not use abbreviations which will not be familiar to other observers. If letters are used, a separate legend should be made up with the letters and the name of the equipment clearly marked. If abbreviations are used they should also be a part of this legend.

Three-Dimensional Scale Models. Three-dimensional scale representations are made from wood, metal, cardboard, or plastic. Details included are those sufficient to identify the model and show important structural and operating features for ready identification. To further aid identification, the name, number and size of the equipment are given in abbreviated form somewhere on the model.

Models are usually the clearest device for presenting layout planning. They enable non-technical people to study and evaluate the layout, allow for quick rearrangement of alternate schemes, make employee orientation and training easier and readily show overhead details and clearances. Due to their cost, in normal practice, models are only used for large capital expenditure projects or those involving problems of vertical complexity.

Flow Lines. Flow lines are shown on the finished layout to indicate the path to be followed by the material as it progresses through the plant. Such lines should follow on the drawings, as nearly as possible, the same path the part will follow. Flow lines, if indicated accurately, should present a pattern similar to that planned in the flow diagrams developed earlier. They will show, to some extent, how well the principles of good plant layout have been followed. Flow lines are usually made with narrow, color-coded, pressure-sensitive tapes or drawn directly on the layout.

Presentation of Layout. It is often desirable to make reproductions of the layout for use in the plant by millwrights, plumbers, electricians, and other service men. The more common methods of **producing copies** of the layout board are:

1. Drawing a tracing from which prints can be made.
2. Photographing the layout.
3. Photostating the layout.
4. Making direct prints from translucent layouts.

Two other items of interest at this point are the legend on the layout and the use of color. If the layout is to be meaningful to those observing it, certain data should be shown on the drawing or in a **legend.**

Items suggested by Mallick and Gaudreau (Plant Layout Planning and Practice) are:

1. Title and scale of the drawing.
2. Sketch number.
3. Approved space.
4. Detail notations for use of plant engineer.
5. Construction notations for architects and detail engineers.
6. Notations on heating, lighting, ventilation, and communication facilities.
7. Directional locations and orientation on plant site.
8. Transportation and handling facilities.
9. Overhead obstructions and obstructions below floor levels where foundations or superstructures are necessary.
10. Machine-tool and equipment identification numbers.
11. Column numbers.
12. Service outlets for gas, electricity, water, air, and other utilities.
13. Fire protection and other safety apparatus notes.
14. Signature and date.

Color is useful in emphasizing certain items on the master layout. It is also an aid in "selling" the layout to others. Color may be used as follows.

1. Indicate certain types of equipment, such as: salmon = production equipment yellow = material handling equipment; red = storage equipment; green = office and service facilities.
2. Emphasis flow lines.
3. Differentiate between present and proposed equipment.
4. Indicate equipment used on different parts.

The color can be applied to templates by either making templates out of colored stock or marking on uniform colored templates with crayons or pencils.

OFFICIAL REVIEW OF LAYOUT. During the development of the layout the layout engineer should contact all supervisory and managerial personnel affected by the proposal. After completing the total layout, approval must be obtained for the final product. It is wise to have at least the following persons sign their approval in an appropriate place on the layout:

1. Department foreman.
2. General foreman.
3. Methods engineer.
4. Production control manager.
5. Chief plant layout engineer.
6. Manager of quality control.
7. Safety engineer.
8. Chief industrial engineer.
9. Personnel manager (if personnel facilities are involved).
10. Plant manager.

These approvals, as with the consultation mentioned above, assure that the proper persons have seen the layout, studied it, and given approval from their own particular points of view. Each person who has had a part in planning or approving the layout becomes a supporter of the ideas contained therein, when the layout is put into effect. Artful compromise is sometimes the most successful tactic of the layout engineer.

FINAL REVIEW AND APPROVAL. The final decision as to whether or not to invest in a new or revised plant layout rests with the management of the plant. The report to management is a presentation of the proposed solution to the problem. Data presented may vary considerably from company to company but the following outline could be effective:

1. Presentation of problem.
2. Summation of proposals.
3. Statement of costs of installing proposals.
4. Cost analysis and engineering economy study; summary comparison with present conditions.
5. Schedules of installations; procedures for relayout or rearrangement of plant area.
6. Recommendations for improvements in related areas.
7. Appendix:
   a. Drawings and blueprints of present and proposed plans.
   b. Detailed calculations supporting costs and savings presented in report.
   c. Specifications for all equipment, supplies, tools, and other items which must be purchased or made.
   d. Explanation of any proposed plans that may need amplification.
   e. Summary of plans or proposals that were studied and discarded, together with reasons for their discard.

Installation of the Plant Layout

MAKING THE INSTALLATION. The finished plant layout is only the beginning.

The following is suggested as a means of making changes with a minimum amount of difficulty:

1. Locate and prepare all utility and service outlets (electrical, telephone, gas, oil, air, water, etc.).
2. Lay out exact position of machines on the floor.
3. Prepare a list of all machines to be moved and include the name and number of machine as well as its old and new locations.
4. Tag each machine. Include name and number of machine on tag and old and new locations. Allow spaces on tag for inspection check, adjustment check, and final OK.
5. Prepare a schedule for the movement of each of the machines. Determine truck loadings and estimate times for disconnecting and moving machine to shipping platform, if necessary. Schedule departure times for each machine.
6. Issue a move order for the movement of each machine. This will serve as a work guide for the moving crews.
7. Maintain a daily record of moves and installations until the job is completed.
8. Make final check and adjustment of all machines before releasing them to production. Make sure each machine is properly leveled at new location.

Automation and Plant Layout

SYSTEMS APPROACH. Automated production adds another dimension to the plant layout problem. The use of closely integrated and interconnected equipment modules limits the amount of flexibility available to the layout engineer. Large units such as transfer machines (see section on Manufacturing Processes and Materials) often dominate a production area and their location presents a fundamental space problem upon which all other space considerations must depend. More than ever, the layout problem becomes a subproblem within an overall production system. Harris and Smith (Journal of Industrial Engineering, vol. 19) advise using cost-effectiveness techniques as a means of analyzing the system. They recommend as a first step in developing a systems approach to layout the restating of the layout problem in the form of a set of criteria that must be satisfied by the actual layout. The purpose of this restatement is to place the facilities layout problem within the realm of systems analysis. Quantification of the various criteria involved interacting with other subsystems such as scheduling, process design, man-machine interface, job design, and management philosophy can lead to a quantitative evaluation of the plant layout alternatives.

ANALYZING THE PRODUCTION PROCESS. Automation requires that a **complete review** of products and processes be made for profitability. This is because the high capital and fixed operating costs of automation require a substantially large production demand for the goods in question. Since automation also requires **standardization** it is primarily adaptable to those processes which tend to remain unchanged over a period of time. As a result processes which appear to offer a degree of stability of design should be first investigated for automation possibilities.

Wilburn (Automation, vol. 13) suggests as profitable areas for automation manual assembly lines with the following characteristics:

1. High volume operations with a relatively stable product design not subject to radical change over a period of time.
2. Routine assembly jobs where identical operations are performed for many products.
3. High labor content operations, particularly those in which operators are not required to inspect or use selection judgment in assembling parts.

4. Highly skilled operations where normal operators cause high reject rates.
5. Operations where new methods are needed to cut product costs.
6. Hazardous operations where safety for either the operator or the product is problem.
7. Operations where increased production capacity is needed. Automatic assembl: systems are high production systems and can be operated on one, two, o three shifts to provide variable capacity requirements.
8. New product operations where no previous manual assembly line exists.

Basically, each automatic assembly system is composed of the following sub systems:

1. **Holding system** to collect and secure parts as they are added to the as sembly.
2. One or more **parts feeding systems** to deliver parts at the proper time, in a desired sequence, and in an oriented position to elements of the holdin; system.
3. For multi-station systems, a **transport system** to move and accurately locate the elements of the holding system as they are carried through various as sembly, processing, and inspection positions.
4. **Work-station tooling systems** that act where the holding system and the feeding system come together.
5. **Drive system** to provide controlled motions for the actuation of the trans port system and the station tooling.

Getting the product into and out of each unit of the system is a consideration that must come in the early stages of layout planning. Since it is not uncommon for the layout of an automated system to consist of numerous machines, from many different manufacturers, the intake and discharge mechanisms of each machine are often based on different design concepts. As a result, considerable modification is frequently necessary to combine the modules into a system.

INTRODUCING NEW EQUIPMENT. Ideally, installation of an auto mated system consists of merely uncrating the machine and connecting the power source. Usually, however, many preparations must be made to ac commodate the new equipment. Proper preparation calls for adequate planning before the automatic equipment is built and then necessary steps after the equipment is assembled to bring the completed machines to a state of satisfactory operating performance.

If new equipment is introduced where none existed previously a space problem is generated. This often requires shifting storage areas and aisle space rather than the more expensive alternative of shifting machines. An analysis of specific areas and requirements is necessary to properly locate new machinery.

Replacing old machines with newer automated machines may create a prob lem in material flow but usually produces a saving in space. Leone (Production Automation and Numerical Control) cites the example of an automotive com ponents manufacturer with five conventional machines, including millers, dril presses, and a special purpose machine. These machines were replaced with a single numerically controlled machining center with capabilities in excess of those represented by the five machines it replaced. The machining center occupied a nominal floor space of about 200 square feet including the space for the operator; the five replaced machines required 250 square feet of machine space plus an additional 200 square feet for aisles and passageways between machines, space for the operators, and materials access. A saving of more than 50 percent of the original floor space was realized for the same production capability.

The introduction of new equipment and machinery usually requires a layout study be made of the existing facility including the availability of services

Templates of the new equipment should be prepared and the layout procedures outlined elsewhere in this section followed. If a new machine is introduced in an active production area moving and installation schedules must provide for a minimum of disruption and downtime. Thought must be given to the effects of the new machine on line balance since the simple act of adding increased manufacturing capacity can influence space allocations for storage of final products and work in process.

In a case discussed by Keller (Unit-Load and Package Conveyors) a manufacturer desired to double production by doubling the number of machines in one manufacturing area. It was not possible to increase the size of the plant and the manufacturer wished to keep his present location. The solution was to install belt conveyors over the machines and eliminate the large aisles formerly used to truck material. The belt conveyors served the machines with raw material and carried the finished product to an inspection station. Vertical conveyors then elevated the product to upper floors for further processing and packing. The packed material was then lowered by conveyors to the shipping dock.

STARTING UP NEW PRODUCTION LINES. Starting-up operations require active participation by the layout engineer, production engineer, and plant engineer who may have to maintain existing production while at the same time activating the new automatic process. The start-up phase usually includes limited tryouts, up to a point where a production run is attempted. Debugging of new equipment is almost unavoidable since not every factor can be anticipated in the design and analysis stage. **Typical problems** that might be encountered could be in the areas of:

1. Handling and feeding of parts in the feeding operation.
2. Deviations of parts from their specifications. (This can become a particularly difficult problem when a part is transferred from a hand-processing operation to an automated process where tolerances are more critical.)
3. Injection of foreign matter (chips, filings, etc.) into automatic equipment causing machine jamming.
4. Manufacturing defects by equipment supplier.
5. Failure to provide adequate auxiliary services (electricity, steam, water, compressed air, etc.).
6. Difficulties associated not with the automatic machinery, but with a new process or a new material.

It is good practice to designate a **project leader** to be responsible for starting up new automated production lines from the time the project is started until it is formally accepted by the production department.

USING AUTOMATED MATERIAL HANDLING. Every improvement in material handling directly or indirectly improves production. This is because the actual manufacturing process usually takes only a small fraction of the total production time. The major amount of time is expended in handling materials rather than processing them. Leone (Production Automation and Numerical Control) describes an analysis made by a manufacturer which indicated that the **material handling cost** for a group of representative parts produced in his plant accounted for approximately 6 to 20 percent of the product cost, depending on the part. For all produced items, material handling costs averaged 16 percent of the product cost. A switch to automated equipment led to a 25 percent saving in handling or an average reduction of 4 percent in the overall product cost.

Automation in material handling has been accomplished in part by combining handling and machining operations into an integrated sequence. The **transfer** machine is a composite of a number of standarized machine unit stations. Each machine station performs a set of operations and at the conclusion of a work cycle, the part is transferred to the following station in the machine. Transfer machines usually save space and result in greater production (see section on Manufacturing Processes and Materials).

Automated material handling is also used to bring the parts in production through the plant. Separate machines or assembly units are placed at various points along the material handling conveyor. Flexibility is the major advantage in this method since the work stations can be positioned in many arrangements about the conveyor. Automobile production is based on this system.

Some industries use automatic conveyors for both product and operator. Keller (Unit-Load and Package Conveyors) cites the ceramic industry as using a triple conveyor system. In this industry unfired ware is placed in a sagger manually. The saggers advance on a belt conveyor while the operator stands on another belt conveyor synchronized with the sagger belt. A third conveyor in back of the operator keeps a supply of unfired ware moving with him. This type of system is usually arranged so that two assembly lines (two sets of three conveyors) form a closed loop. In this way the operator is again performing work while being returned to the starting point.

PROCEDURES FOR LAYING OUT AN AUTOMATED LINE. The degree of automation selected, the area of application, and the design of the system will primarily be the responsibility of the systems engineer and the components designer. The layout planner must, however, consider such things as the arrangement of related automated equipment with respect to each other using basic layout techniques discussed earlier, and the optimum location of maintenance facilities to service the automated line. In this respect the automated line presents no special layout problem not previously discussed.

Plant Facilities: Heating, Ventilation, and Air Conditioning

EFFECTS ON HEALTH AND PRODUCTION. Air tempering systems are used to control such factors as dust, pollen, bacteria, odors, and toxic gases in addition to air temperature and humidity.

There are two basic kinds of heat release that will affect the comfort of personnel. One is a hot-dry heat which allows some relief by means of evaporation of body perspiration. The other type is warm-moist heat which prevents evaporation of perspiration. The latter presents the greater hazard to worker health and comfort. The introduction of outside air with low relative humidity will contribute greatly to worker comfort in warm moist areas. The physiological and psychological effects of a **controlled environment** upon workers, and the effects of controlled environment on the quality of products must often be considered. In some instances only a control of temperature is necessary, whereas in others, simultaneous control of temperature, humidity, ventilation, and air cleanliness may be required.

A systematic and logical procedure should be followed in each situation where the design and selection of a heating and ventilating system is involved; the most important items to be considered should be evaluated, and a system should be chosen which will be adequate to provide the degree of control desired without exceeding the economic limitations of initial investment or operation.

HEATING SYSTEMS. All heating systems consist of certain basic components when reduced to their most elementary form:

1. **A heat source,** which may be the burning fuel in a boiler or furnace; electric energy; or the earth, water, or air in heat pump systems.

2. **A conveying or distribution system** used to transfer a heating medium from the heat source to the space to be heated, e.g., steam piping, hot water piping, or air ducts.
3. **Heat disseminators** located within the space to be heated, e.g., convectors or radiators in steam and hot water systems, registers and grilles in air systems.

The variations in the actual systems which include these basic components are numerous, and no attempt will be made to describe in detail all possible combinations. Some classifications are important enough, however, to be discussed for purposes of comparison.

Combined or Split Systems. A combined system of heating and ventilation is designed to provide for both the heating requirements and the ventilation requirements by air which is supplied to the building. The entire heating requirements of the building may be supplied by heated air without the use of auxiliary devices such as radiators or convectors, or a part only of the **total heating load** may be supplied by the air-handling system. In any event, sufficient fresh air for ventilation purposes is possible along with a distribution system for air movement throughout the heated space. The distinguishing characteristic of this system is that the temperature of the **supply air** is raised to a level high enough to provide either part or all of the heating requirements.

The **spilt system** makes use of auxiliary radiation or convection devices to provide for building heat losses and supplies tempered air at approximately room temperature to satisfy ventilation requirements. When ventilation (implying the supply of outdoor air) is not required at times, such as overnight or during periods of low occupancy, operating cost may be reduced by supplying building-heating requirements without the use of the ventilating system.

A similar operational procedure is commonly used with the combined system, making use of dampers to allow for **recirculation** of room air in order to maintain temperatures without providing ventilation.

Central or Unit Systems. A central air-handling system consists of one or more main supply fans with heating coils, filters, heat generation equipment, etc., usually located in a **mechanical equipment room.** A **plenum chamber** is ordinarily provided for supplying air to a system of ductwork which serves the entire area to be heated. Return air ducts, if required, bring recirculated room air back to the equipment room where it is mixed with a supply of fresh outdoor air for ventilation. Individual control of temperature by areas or zones may be accomplished by dampers located in the plenum chamber.

A **unit system** differs from the central system in that the fans, filters, heating coils, and ductwork, including ventilation air supply, are combined in separate units, usually located in or adjacent to the spaces to be heated. Supply and return ductwork may be incorporated with unit systems, but direct discharge of supply air through grilles attached to the housing enclosing the equipemnt is commonly used. The units may have steam or hot-water heating coils or may be direct-fired by gas or oil. Units of this type range from small unit heaters with no ductwork or ventilation air provisions, to completely self-contained units including filters, ventilation provisions, a mixing chamber with automatically controlled dampers for multiple-zone temperature control, and a complete system of ductwork.

Heat-of-Light Systems. With increasing light intensities and the related increase in by-product heat, lighting heat becomes an important energy source to be recovered and used rather than rejected. Approximately 55 percent of the electric energy to a recessed fluorescent luminaire is directed upward into the plenum as convected, conducted, or radiated energy. The remaining 45 percent enters the occupied space as light and radiant energy.

Transfer and control of lighting heat are essential. Transfer of the lighting heat must be employed to prevent excessive heat buildup, which will adversely affect lamp and ballast performance. Heat-of-light systems use either an air or water media to absorb the heat of the luminaire and transfer it to the area requiring heat. An important consideration in selecting a heat recovery system is the time–amplitude function of the interior heat gains and exterior heat losses. If the gains and losses are of equal amplitude and concurrent, the design becomes a matter of control and transfer. If the functions are not equal, some form of supplemental energy will be required.

SELECTION OF A SYSTEM. The classifications of systems outlined are only the more general ones into which heating and ventilating systems can be grouped; the designer is afforded many more choices than are noted here. Types of heat exchange elements, air distribution devices, and methods of capacity and temperature control may vary widely. In general, there is no one clearly defined "best" system; there may be several available ways by which the same end result can be achieved. Economic considerations will often be the most important factor in final selection, but individual application studies should be made to determine which type of system offers the most return for the investment required. In order to provide the best system for a specific application, a designer should have a thorough understanding of the fundamentals involved in the design of heating and ventilating systems; he should be familiar with the types of equipment available and should be capable of incorporating a group of individual components into an **integrated system** which will function satisfactorily and require a minimum of maintenance.

VENTILATION SYSTEMS. Ventilation may be provided for industrial buildings by either **natural means,** making use of wind forces and indoor-outdoor temperature difference, or by a **mechanical system** employing either powered exhaust ventilators, supply fans and ductwork, or a combination of both. A disadvantage encountered with natural ventilation systems is that both the air quantity and distribution are variable with little control of air movement afforded. Positive distribution of specified quantities is more likely if a mechanical system is used.

Due to the many variables involved, there are no precisely defined rules for general ventilation practice; in many instances **state** or **local codes** prescribe minimum requirements which must be met. These requirements are usually stated in terms of the **quantity** of outdoor air required per occupant or the number of air changes required per hour within a given space.

Although insufficient attention is often devoted to it, the problem of satisfactory **ventilating-air distribution** is as important as the quantity required. The designer should provide for adequate control of air movement in all areas which are to be ventilated, with sufficient quantities of fresh air to eliminate or minimize undesirable conditions arising from heat and moisture loads, odors, dust, dirt, toxic gases, vapors, or other air contaminants due to occupants or industrial processes.

Frequently, the heating or cooling system can provide adequate amounts of outside air for ventilation, but where severe problems of contamination exist, exhaust hoods or other localized ventilating systems may be required (see section on Industrial Safety). In all instances, the designer must provide a sufficient quantity of outdoor air to replace that which is exhausted.

AIR CONDITIONING SYSTEMS. By popular usage, the term "air conditioning" is often used to describe systems which provide cooling of the atmosphere within a building structure. Technically, however, air conditioning refers to the **simultaneous control** of air temperature, humidity, motion, distribution, pressure, and purity, including such factors as dust, odors, bacteria, and toxic gases. A true air-conditioning system must be able to provide control of these factors during both the heating and cooling seasons.

Air-conditioning systems are frequently classified as **comfort** air conditioning, the providing of an atmospheric environment considered desirable for human comfort, and **industrial** air conditioning, the maintenance of an environment for the purposes of manufacture or storage of a product.

In many applications, both end results are achieved, although the emphasis in industrial air conditioning is on product control.

AIR CONDITIONING EQUIPMENT. There are certain major types of air-conditioning systems and equipment used industrially which should be considered and compared in determining the appropriate design of air conditioning under specific conditions.

Unit and Central Systems. For applications where cooling loads are relatively small, the use of package or unit conditioners is common. Unit conditioners are ordinarily factory assembled and have the advantage of simplicity of installation and operation. Condensing equipment may be air- or water-cooled; service connections are relatively simple to install and for applications where suitable, unit equipment is to be preferred to built-up systems.

For cooling loads of perhaps 10 tons of refrigeration (1 ton of refrigeration = 12,000 Btu. per hr.), or less, **unit equipment** is ordinarily considered more desirable than a **built-up system.** As the size of the load increases, system design usually increases in complexity and built-up systems become more desirable. Obviously, no general rule can be given for selecting one system in preference to another, but advantages of each type must be compared before the best selection for any given application may be made.

Reciprocating and Centrifugal Compressors. In smaller central systems, reciprocating compressors are ordinarily used. In the range of from 10 to 100 tons capacity, single or multiple reciprocating machines with water-cooled condensing equipment are most common.

Although centrifugal machines at one time were not considered for applications where the cooling load was less than 100 tons, machines are available which may offer advantages well worth considering for the 50 to 100 ton range. In this size range, the initial cost of centrifugal equipment will probably be somewhat higher than the reciprocating, but simplicity of operation, reduced maintenance, and reduced operational noise and vibration may overcome the higher cost.

Capacity control for reduced-load operation can be achieved with cylinder-unloading devices on reciprocating compressors or by the use of multiple machine installations. The provisions for capacity regulation are more complex than for centrifugal machines, however, and the degree of regulation is ordinarily much more limited.

Direct-Expansion and Chilled-Water Systems. Direct-expansion cooling coils, using refrigerant supplied from condensing equipment and returned to the compressor are widely used for both unit systems and built-up systems with reciprocating compressors. Mixing dampers or face-and-by-pass dampers may be used for close regulation of air temperatures.

Chilled-water systems come with factory-assembled **water chillers** incorporating compressor, condenser, and capacity-control equipment. Installation is frequently simplified appreciably by the use of such equipment, resulting in savings in field installation costs. Coil costs may also be reduced by the use of water-type cooling coils, helping to offset the increased first cost of the package equipment.

Condensing Equipment. Although air-cooled condensers are used on some of the smaller unit air conditioners, water-cooled condensers are common on equipment above approximately one ton in capacity. State and local regulations governing the use of water often require recirculating systems; cooling towers or evaporative condensers are commonly required for installations above perhaps 10 tons capacity.

Heat Pumps. The main advantages of the heat pump are that the same items of equipment may be used for cooling during the summer as are used for heating during the winter, and equipment required for direct burning of fuels is eliminated.

Successful and economical operation of the heat pump is dependent upon the existence of **heat sources** and **sinks** which are satisfactory from the standpoint of temperature, adequate in capacity, and reliable. The principal heat sources (excluding those which may be available as a result of certain industrial processes) are the air, the earth, and water. **Solar energy** is also an important and practical source of energy for heat pump applications.

Although frequently described as a "reversed" refrigeration cycle, the **heat pump cycle** is thermodynamically identical to a conventional refrigeration cycle. The difference is merely in terms of the desired effect; in the **refrigeration cycle**, the desired effect is the absorption of heat from the air in a room by the refrigerant which boils in an evaporator. The heat equivalent of the compression work along with the heat absorbed in the evaporator is rejected to a sink and wasted, as in a water-cooled condenser.

In the heat pump, however, the heat rejected from the condenser may be used to warm the air inside a building, hence the desired effect. The heat absorption by the refrigerant in the evaporator may be from outdoor air, the earth, or from water. In order to utilize this heat, compression work is required so that the temperature can be raised to the level required for heating the building.

When conditions are suitable, the heat pump offers an attractive possibility for providing for the heating and cooling requirements of a building. Very careful analysis is required, however, to insure that application is practical. **Auxiliary heating** is frequently required and may in some instances make operational costs prohibitive. Reliability of operation as well as of heat sources must be established.

Plant Facilities: Lighting

EFFECTS OF LIGHTING ON PRODUCTION. With some important exceptions, most manufacturing functions involve seeing tasks. In order to perform these seeing tasks adequate lighting must be provided at the work station. For safety and security proper lighting is also necessary in those non-production areas normally frequented by personnel. Illumination may be natural daylight coming through plant windows or artificial lighting.

Good illumination affects production in the following ways:

1. It permits greater accuracy and precision in workmanship thus improving product quality and decreasing spoilage.
2. Safety and security are increased in and around the plant.
3. Reduces employee eye strain, discomfort, and fatigue.

4. Acts as an incentive to maintain clean and neat work areas.

5. Pleasant surroundings improve employee morale and reduce labor turnover.

Factors of Good Illumination. Good illumination is made up of more than simply a proper level or quantity of illumination. Proper quality, which includes the color of light, its direction, diffusion, brightness, steadiness, and absence of glare, in many cases is as important as proper quantity. Consideration of the many factors involved in good illumination is a complex problem.

These factors have been defined and described (Benjamin Electric Mfg. Co.) as follows:

1. **Quantity.** The term "quantity of light" refers to the amount of light which produces the brightness of the task and the surroundings. The minimum quantity of light for seeing efficiently varies greatly with the different kinds of work.

2. **Quality.** The factors involved include amount of glare, diffusion, distribution direction, color, and brightness. The quality of the light is extremely important because moderate differences are not easily detected. However, the resulting conditions often produce a material loss of seeing efficiency and undue fatigue.

Diffusion, Shadow, and Distribution. The light source and the type of lighting equipment determine the duration and the amount or concentration of light provided. High specular reflecting surfaces require highly diffused light. Difficult-to-see, minute objects may require high concentration of light without regard to its specular quality.

Distribution refers to the amount of light provided to the various parts of the entire working area. Even distribution permits flexibility and rearrangement of equipment, and elimination of dangerous light and dark areas and spotty lighting.

Color. While variations in the color of light have little or no effect on visual acuity, there are certain operations involving color discrimination where the color of the light is a tremendously important factor.

Quantity of Light. Quantity of light supplied from whatever the source, either natural or artificial, should be determined by the work to be done. In general, the more trying the seeing task and the higher the degree of accuracy or fineness of detail required, the greater should be the illumination. Investigations in field and laboratory have proved that as the illumination on the task is increased, the ease and speed with which the task can be accomplished are increased. These tests have not yet established an upper limit, but the harmful effects of low foot-candle values are well known.

A table of recommended lighting levels for various types of industrial activities is given in Fig. 24 (Practice for Industrial Lighting, ANSI Std. A11.1—1965).

GENERAL LIGHTING SYSTEMS. Luminaires can be classified into the following groupings (Illuminating Engineering Society, Lighting Handbook):

1. Direct lighting
2. Semi-direct lighting
3. General diffuse lighting
4. Semi-indirect lighting
5. Indirect lighting

Basically, the direct-lighting fixture directs nearly all of its output downward. This system produces a highest level of illumination; it also tends to result in shadows, glare or spotty illumination. Semi-direct fixtures direct less of their light downward. The general diffuse fixture distributes light equally in all directions. With the semi-direct fixture most of the light is directed upward

Area and Seeing Task	Foot-candles[a]	Area and Seeing Task	Foot-candles[a]
Airplane manufacturing		Building surrounds	1
Stock parts		**Chemical works**	
Production	100	Hand furnaces, boiling	
Inspection	200	tanks, stationary driers,	
Parts manufacturing		stationary and gravity	
Drilling, riveting, screw		crystallizers	30
fastening	70	Mechanical furnaces, gener-	
Spray booths	100	ators and stills, mechanical	
Sheet aluminum layout		driers, evaporators, filtration,	
and template work,		mechanical crystallizers,	
shaping and smoothing		bleaching	30
of small parts for		Tanks for cooking,	
fuselage, wing sections,		extractors, percolators,	
cowling, etc..........	100	nitrators, electrolytic	
Welding		cells	30
General illumination ...	50	**Clay products and cements**	
Precision manual arc		Grinding, filter presses,	
welding	1000[b]	kiln rooms	30
Subassembly		Molding, pressing, clean-	
Landing gear, fuselage,		ing, trimming	30
wing sections, cowling		Enameling	100
and other large units ..	100	Color and glazing—rough	
Final assembly		work	100
Placing of motors,		Color and glazing—fine	
propellers, wing sec-		work	300[b]
tions, landing gear	100	**Cloth products**	
Inspection of assembled		Cloth inspection	2000[b]
ship and its equipment .	100	Cutting	300[b]
Assembly		Sewing	500[b]
Rough easy seeing	30	Pressing	300[b]
Rough difficult seeing ...	50	**Electrical equipment**	
Medium	100	**manufacturing**	
Fine	500[b]	Impregnating	50
Extra fine	1000[b]	Insulating: coil winding ..	100
Automobile manufacturing		Testing	100
Frame assembly	50	**Explosives**	
Chassis assembly line	100	Hand furnaces, boiling	
Final assembly, inspection		tanks, stationary driers,	
line	200	stationary and gravity	
Body manufacturing		crystallizers	30
Parts	70	Mechanical furnace, gener-	
Assembly	100	ators and stills, mechani-	
Finishing and inspecting.	200	cal driers, evaporators,	
Bookbinding		filtration, mechanical	
Folding, assembling,		crystallizers	30
pasting, etc.	70	Tanks for cooking, ex-	
Cutting, punching,		tractors, percolators,	
stitching	70	nitrators	30
Embossing and inspection	200	**Flour mills**	
Building exteriors		Rolling, sifting, purifying .	50
Entrances		Packing	30
Active (pedestrian and/or		Product control	100
conveyance)	5	Cleaning, screens, man	
Inactive (normally locked,		lifts, aisleways and	
infrequently used)	1	walkways, bin checking .	30
Vital locations or			
structures	5		

[a]Minimum recommended.

[b]Combination of general lighting and special supplementary lighting.

Fig. 24. Recommended lighting levels.

Area and Seeing Task	Foot-candles[a]
Forge shops	50
Foundries	
Annealing (furnaces)	30
Cleaning	30
Core making	
Fine	100
Medium	50
Grinding and chipping	100
Inspection	
Fine	500[b]
Medium	100
Molding	
Medium	100
Large	50
Pouring	50
Sorting	50
Cupola	20
Shakeout	30
Garages—automobile and truck	
Service garages	
Repairs	100
Active traffic areas	20
Parking garages	
Entrance	50
Traffic lanes	10
Storage	5
Glass works	
Mix and furnace rooms, pressing and lehr, glass-blowing machines	30
Grinding, cutting glass to size, silvering	50
Fine grinding, beveling, polishing	100
Inspection, etching and decorating	200
Inspection	
Ordinary	50
Difficult	100
Highly difficult	200
Very difficult	500[b]
Most difficult	1000[b]
Jewelry and watch manufacturing	500[b]
Leather manufacturing	
Cleaning, tanning and stretching, vats	30
Cutting, fleshing and stuffing	50
Finishing and scarfing	100
Loading and unloading platforms	20

Area and Seeing Task	Foot-candles[a]
Machine shops	
Rough bench and machine work	50
Medium bench and machine work, ordinary automatic machines, rough grinding, medium buffing and polishing	100
Fine bench and machine work, fine automatic machines medium grinding, fine buffing and polishing	500[b]
Extra-fine bench and machine work, grinding, fine work	1000[b]
Material handling	
Wrapping, packing, labeling	50
Picking stock, classifying	30
Loading, trucking	20
Inside truck bodies and freight cars	10
Offices	
Cartography, designing, detailed drafting	200
Accounting, auditing, tabulating, bookkeeping, business machine operation, reading poor reproductions, rough layout drafting	150
Regular office work, reading good reproductions, reading or transcribing handwriting in hard pencil or on poor paper, active filing, index references, mail sorting	100
Reading or transcribing handwriting in ink or medium pencil on good quality paper, intermittent filing	70
Reading high-contrast or well-printed material, tasks and areas not involving critical or prolonged seeing such as conferring, interviewing, inactive files, washrooms	30
Corridors, elevators, escalators, stairways	20
Paint manufacturing	
General	30
Comparing mix with standard	200[b]

[a]Minimum recommended.

[b]Combination of general lighting and special supplementary lighting.

Fig. 24. Continued.

Area and Seeing Task	Foot-candles[a]	Area and Seeing Task	Foot-candles[a]
Paint shops		**Final inspection**	
Dipping, simple spraying, firing	50	Tube, casing	200[b]
Rubbing, ordinary hand painting and finishing art, stencil and special spraying	50	Wrapping	50
Fine hand painting and finishing	100	**Service space**	
Extra-fine hand painting and finishing	300[b]	Stairways	20
		Elevators, freight and passenger	20
Paper-box manufacturing		Corridors	20
General manufacturing area	50	Storage (see Storage rooms)	
		Toilets and wash rooms	30
Paper manufacturing			
Beaters, grinding, calendering	30	**Sheet metal works**	
Finishing, cutting, trimming, papermaking machines	50	Miscellaneous machines, ordinary bench work	50
Hand counting, wet end of paper machine	70	Presses, shears, stamps, spinning, medium bench work	50
Paper machine reel, paper inspection and laboratories	100	Punches	50
Rewinder	150	Tin plate inspection, galvanized	200
Plating	30	Scribing	200
Polishing and burnishing	100		
		Shoe manufacturing—leather	
Rubber goods—mechanical		Cutting and stitching	
Stock preparation		Cutting tables	300[b]
Plasticating, milling, Banbury	30	Marking, buttonholing, skiving, sorting, vamping, counting	300[b]
Calendering	50	Stitching	
Fabric preparation		Dark materials	300[b]
Stock cutting, hose looms	50	Making and finishing	
Extruded products	50	Nailers, sole layers, welt beaters and scarfers, trimmers, welters, lasters, edge setters, sluggers, randers, wheelers, treers, cleaning, spraying, buffing, polishing, embossing	200
Molded products and curing	50		
Inspection	200[b]		
Rubber tire and tube manufacturing		**Shoe manufacturing—rubber**	
Stock preparation		Washing, coating, mill run compounding	30
Plasticating, milling, Banbury	30	Varnishing, vulcanizing, calendering, upper and sole cutting	50
Calendering	50	Sole rolling, lining, making and finishing processes	100
Fabric preparation			
Stock cutting, bead building	50	**Soap manufacturing**	
Tube and tread tubing machines	50	Kettle houses, cutting, soap chip and powder	30
Tire building		Stamping, wrapping and packing, filling and packing soap powder	50
Solid tires	30		
Pneumatic tires	50	**Storage battery manufacturing**	
Curing department		Molding of grids	50
Tube and casing	70		

[a]Minimum recommended.

[b]Combination of general lighting and special supplementary lighting.

Fig. 24. Continued.

Area and Seeing Task	Foot-candles[a]	Area and Seeing Task	Foot-candles[a]
Storage rooms or warehouses		Drawing	
Inactive	5	White	50
Active		Colored	100
Rough bulky	10	Warping	
Medium	20	White	100
Fine	50	White (at reed)	100
Storage yards		Spinning (frame)	
Active	20	White	50
Inactive	1	Colored	100
Testing		Spinning (mule)	
General	50	White	50
Extra-fine instruments,		Colored	100
scales, etc.	200[b]	Twisting	
Textile mills—cotton		White	50
Opening, mixing, picking	30	Colored	100
Carding and drawing	50	Colored (at reed)	300[b]
Slubbing, roving, spinning,		Weaving	
spooling	50	White	100
Beaming and splashing on		Colored	200
comb		Gray-goods room	
Gray goods	50	Burling	150
Denims	150	Sewing	300[b]
Inspection		Folding	70
Gray goods (hand turning)	100	Wet finishing	
Denims (rapidly moving).	500[b]	Fulling	50
Automatic tying-in	150	Scouring	50
Weaving	100	Crabbing	50
Drawing-in by hand	200	Drying	50
Textile mills—silk and		Dyeing	100[b]
synthetics		Dry finishing	
Manufacturing		Napping	70
Soaking, fugitive tinting,		Shearing	100
and conditioning or setting		Conditioning	70
of twist	30	Pressing	70
Winding, twisting, rewinding		Inspecting (perching)	2000[b]
and coning, quilling,		Folding	70
slashing			
Light thread	50	Welding	
Dark thread	200	General illumination	50
Warping (silk or cotton		Precision manual arc	
system)		welding	1000[b]
On creel, on running ends,			
on reel, on beam, on warp		Woodworking	
at beaming	100		
Drawing-in on heddles and		Rough sawing and bench	
reed	200	work	30
Weaving	100	Sizing, planing, rough	
Textile mills—woolen and		sanding, medium quality	
worsted		machine and bench work,	
		gluing, veneering,	
Opening, blending, picking	30	cooperage	50
Grading	100[b]	Fine bench and machine	
Carding, combine, recombing and gilling	50	work, fine sanding and finishing	100

[a]Minimum recommended.

[b]Combination of general lighting and special supplementary lighting.

Fig. 24. Concluded.

to be reflected by the ceiling and upper walls. This system practically eliminates shadows and glare. The indirect fixture distributes almost all of its light upward, thereby removing shadows and glare.

Fig. 25 (Benjamin Electric Mfg. Co.) provides a quick picture of the **characteristics and applications** of three common light sources available for industrial lighting. Modern practice is to combine different light sources in various arrangements when special lighting problems warrant.

Light Source	Applications	Special Characteristics
Incandescent	In all classes of plants, at low, medium, or high mounting heights. Used alone or in combination with mercury. Also used for supplementary lighting.	Smallest investment cost. Easily installed. Flexibility in size, layout, and operation. High heat output. High brightness of source requires adequate shielding. Relatively large number of lamps needed for usual industrial and commercial lighting levels.
Fluorescent	In all classes of plants, at low and medium mounting heights; in large assembly plants at high mounting (high output lamps). Also used for supplementary lighting at benches and machines.	Relatively low brightness, high efficiency, choice of color, lack of heat, offer possibility of increasing illumination without a complete wiring job being necessary. Long lamplife. Requires auxiliary ballast or transformer.
Mercury	At medium to high mounting heights. Street and parking lot lighting.	Blue-green color combines well with yellow-orange of filament lamps to give a cool appearance. Lowest cost of lighting on overall basis. Higher brightness means higher reflected glare from shiny surfaces. Very long lamp life. Requires auxiliary ballast. Slow starting; needs warmup period.

Fig 25. Characteristics and applications of common light sources.

The **design of any lighting installation** is dependent on many factors, among which is the provision of the proper quantity of illumination. This is accomplished by first analyzing the seeing task and its particular illumination requirements. It then becomes possible to select the most desirable type of lighting equipment and calculate the number of luminaires which would give the average illumination level sought after.

Location of Luminaires. For general lighting, luminaires are usually located symmetrically within the floor area or in each bay or pair of bays, although this practice is often modified by the position of existing ceiling outlets, the general architecture of the area, the type of luminaire, or the nature and location of the seeing tasks to be performed. Where working positions are fixed, it is sometimes desirable to position the luminaires with reference to the particular area where high intensities are necessary. In offices or other areas where work is performed throughout the room, a high degree of uniformity is desirable.

In rooms relatively wide and long in proportion to the height (for example, where the width is greater than three times the height), a **spread type of direct**

lighting equipment is usually appropriate. Good quality illumination in this case would be provided by fluorescent luminaires mounted in continuous, properly spaced rows. A **grid pattern** of fluorescent units provides even a higher level of illumination, and at the same time reduces shadows. In areas where the width is equal or less than the height, direct lighting equipment which concentrates the light is usually used. Incandescent or mercury vapor sources are recommended in such instances. When facilities which contain high-bays and also extreme length and width are encountered, it is appropriate to use spread types of direct lighting equipment with a large number of luminaires, each contributing a small amount of light to a given point.

Outdoor Lighting. Lighting of buildings and grounds at night is used for:

1. Pedestrian and vehicular traffic.
2. Exterior work, such as platform loading and night construction work.
3. Plant security.
4. Advertising.

Supply lines to exterior lights should be concealed underground or in walls. **Switches** should be available only to authorized persons, key-operated, or automatically actuated by timers or photoelectric relays.

Other Environmental Features

NOISE CONTROL. The problem of sound attenuation should be considered by the layout engineer, although in most cases his contribution will be limited to methods involved with the arrangement of facilities.

Federal, state, and local laws in many cases limit or regulate the level of the noise environment for industrial plants (Fig. 26, Division of Industrial Safety

Frequency Band, Hz	Octave Band Sound Pressure Level, Decibels (Re: 0.0002 dyne/sq cm), for Following Amount of Exposure per Day		
	Over 5 hours	Less than $2\frac{1}{2}$ hours	Less than $1\frac{1}{4}$ hours
20– 75	110	113	116
75– 150	102	105	108
150– 300	97	100	103
300– 600	95	98	101
600– 1200	95	98	101
1200– 2400	95	98	101
2400– 4800	95	98	101
4800–10000	95	98	101

Fig. 26. California standards for noise control.

for California). There is some disagreement among experts in this field concerning the noise-level danger point; however, all agree that continuous exposure to high noise levels can affect one's health causing temporary or permanent hearing loss, nervous disorders, and frequent headaches. Obviously, all of these conditions can contribute to the reduction of employee effectiveness and production efficiency (see section on Work Measurement and Time Study).

Controlling Noise Levels. Primary attention should be directed to the source of noise to ascertain if the sound level can be reduced there. If this is unsuccessful, it is then necessary to resort to isolation or sound absorbers to contain the noise.

The **reduction of noise** in a plant may be brought about by the following methods:

1. Elimination or reduction of the noise at its prime source.
2. Isolation of the noise through relayout.
3. Use of sound absorbers or baffles.
4. Acoustical treatment of the noise source.
5. Use of ear-protection devices by individuals.

Planning for Noise Control. Floyd (Mechanical Engineering, vol. 90) believes that new plant construction provides an ideal planning situation for noise control. The best results obviously can be obtained by considering potential noise problems very early in the plant design stage. Unfortunately, he feels, this is often not considered until much of the layout has been established. Many of the noise problems produced by plant machinery can be anticipated, based on data obtained on existing equipment or from installations in other plants. Geographical planning should be included in any plant layout survey since the location of the plant itself can contribute to the noise control problem. Landscaping, such as trees and shrubs, have little acoustical effect for industrial noise conditions but they do have a psychological effect on residents of a neighboring community since they do screen the plant from sight and thus make the noise seem less disturbing.

AESTHETIC FACTORS. Plant design has moved away from the sterile architecture of the past when the sole purpose was to enclose the production process. Modern architects are given wide latitude in planning well-landscaped, functional buildings reflecting favorably the company's image, significantly adding to employee morale, and enhancing resale opportunities.

The architect's attention to such factors as column spacing, window expanse, and low maintenance exterior and interior design, all affect the layout planning function.

Use of Color. Modern human factors engineering recognizes the important effect of color on work safety and efficiency. A number of purposes can be served by selection of the proper color scheme. Among these are greater safety, better housekeeping, improved visibility, increased morale, and greater work efficiency. Also, the company's public image can be enhanced by proper selection of color on the plant's exterior.

The intelligent use of color for safety is recognized as an important element in the modern plant. Widely accepted is the code developed for this purpose by the American National Standards Institute (Safety Color Code for Marking Physical Hazards and the Identification of Certain Equipment, ANSI Z53.1—1967) which is summarized below:

Red Fire protection and apparatus; danger; stop.

Orange Dangerous parts of machinery or energized equipment which may cut, crush, shock, or otherwise injure.

Yellow Caution; for marking physical hazards, such as stumbling, falling, tripping and "caught in between"; frequently striped with black to demand attention.

Green Safety equipment; location of fire-aid equipment (other than fire-fighting equipment).

Blue Caution, limited to warning against starting, the use of, or the movement of equipment under repair or being worked upon.

Purple Radiation hazards (ANSI endorses use of three-bladed propeller as a radiation symbol, in purple on a yellow background, as adopted by the Atomic Energy Commission).

Black and White Traffic and housekeeping markings.

Color can also form the basis of improvements in the performance of visual tasks in plants and offices (Pittsburgh Color Dynamics, Pittsburgh Plate Glass Industries, Inc.). Eye fatigue is caused by unnecessary travel of the eye over ill-defined areas of about the same color, and constant adjusting when changing from light surfaces to dark surfaces. **Focal colors** center the worker's attention on working points. **Receding colors** on surrounding machine parts cause these to drop back and relax the eye. A light green is suitable for most applications, but in certain cases (for example, food and drug plants) white surfaces produce a more sanitary appearance. Double contrast between the work itself and the immediate work area is necessary for more accurate distinction. The work area color therefore must give satisfactory contrast with both material and surrounding machine parts. In addition, cool colors offset the psychological effects of unavoidable high temperatures in workplaces, and warm colors similarly offset the effects of cold workplaces.

Walls, Floors, and Aisles. Attention to machine colors must be accompanied by corresponding care regarding walls, columns, partitions, doors, work or tool cabinets, ceilings, roof trusses, overhead cranes or conveyors, piping, factory floors, aisles, and other objects or areas within the workers' field of vision. **Walls and column** should have approximately the same general tone of color (not necessarily the same color) as machines and work areas. A light green is good for walls within the workers' vision as they glance up from their work, and brighter colors, such as yellow, for other walls to gain the benefit of simulated sunlight. If work in a department is done on colored materials, the machines and walls should be painted in the complementary color to prevent the image of the work from appearing momentarily in the workers' vision when they look up. When there are numerous roof trusses or wires, pipes, etc., overhead, they should be made to recede into the background by proper painting. If the **ceiling** has little reflection value, a lighter blue can be used to produce the blending effect. For indirect lighting, of course, colors with a high reflecting value are necessary.

Floors should be a light color with good reflectivity. Where good visibility is necessary below as well as above work, as for example in many assembly operations done on the floor, light gray is often used for the floor. **Tops of benches** should be light for bench assembly and related work. Even where it is not necessary to have light floors under and around equipment it is best to paint the **aisles** light and to band them along each side to form traffic lanes, in a distinctive color, both to mark them off as traffic limits for the safety of workers at machines, and to keep them clear of materials, finished products, tools, boxes, benches, and other objects. Platforms, ladders, steps, large assembly jigs, and other **operations auxiliaries,** such as in aircraft plants, should be painted a distinctive bright color to mark them off from the surroundings and for safety in use. Display areas of the machine (control meters, switches, instruction plate) should differ in color from that of the main area, but strong contrasts, strong colors and reflected glare are to be avoided. Strong contrasts should be reserved for dial scales and printed instructions.

Color codes for **safety devices** and other color standardized items should be overriding. Inside of safety guards must be strongly colored so that they show up if guards are left off.

Moving Equipment. Industrial trucks, cranes, conveyors, etc., are a source of accident hazard which can be cut down by proper distinctive painting. Trucks may be painted yellow all around, including any side projection, because this color has high visibility. **Boxes, racks,** etc., used for moving work from place to place may be painted green to distinguish them from trucks. If the insides of such receptacles are painted in light colors, it is easier to see how much material they contain. **Overhead cranes and crane hooks** painted yellow (sometimes striped in places with black) have high visibility and thus are noticed as they move over areas in which employees are operating. There is less likelihood of accidents from crane operation when this practice is adopted.

SPECIAL ENVIRONMENTAL SITUATIONS. Clean Rooms. Clean rooms can be divided into two major types, based upon the type of air flow designed for the facility. One category is designed with turbulent air conditions present within the facility. This type of room is usually referred to as a **conventional clean room.** The turbulent flow patterns within the conventional clean room serve to provide temperature and humidity distribution. A second category of clean rooms is known as the **laminar flow clean room** as a result of the air pattern in these facilities.

These areas are specially designed and laid out to exclude contamination which might enter in the air, on bodies and clothing, and on material. Circulating air is cleansed of dust through use of electrostatic and mechanical filtration. Possibly all services in the assembly area are brought up through the floor to eliminate the need for overhead ducting and walls; floors and ceilings can have smooth non-dust-catching surfaces. Use of scavenging lobbies, in which dust is removed from the outer clothing of operators by subjecting them to an air shower as they enter the building, can be very helpful. In addition, special gratings to remove mud and dirt from footwear is desirable. Once in the plant the employee can be required to wear special clothing.

Water Use and Treatment. A supply of clean, wholesome, and safe drinking water should be provided from a source approved by the local health authorities. In situations where such water is unavailable, aid should be acquired for rendering the available water safe for human consumption. Water taken from unapproved sources and used for industrial processes or fire protection should have adequate notices conspicuously posted indicating that such water is unsafe for drinking. There should never be a pipe connection between a system furnishing drinking water and one supplying water unsuitable for drinking.

Process water uses are as diverse as the processes or products of industry and, consequently various qualities are desirable or acceptable. It is not uncommon for intake water to be split into several systems, each given a different degree of treatment and use. A single large petrochemical plant complex may have separate piping systems for raw water, clarified water, filtered water, distilled water, boiler feed water, domestic water, and so forth.

Plant process or production personnel may specify many different minimum qualities of water for various functions: product mix, heat exchange, fire protection, sanitation, boiler makeup, etc. There are few formal quality standards for an entire industry or for that matter for even specific products. Small plants without restrictive quality requirements may be able to buy potable water from local utilities, but most industrial water is self-supplied. Deciding what treatment to give which portions of the water intake is an engineering and economic decision. By far the most common problems are turbidity and hardness, and both are amenable to relatively inexpensive treatment by conventional methods.

Air Pollution. The potential dangers to the health of plant workers are directly related to the degree of atmospheric pollution to which they are exposed. The greater the concentration of contamination, the greater the hazard. Methods commonly employed to control pollution at its source are:

1. Local exhaust ventilation.
2. Good housekeeping.

Exhausting an operation or process in order to trap and carry off the dusts, fumes, or vapors given off is perhaps the most important method of controlling atmospheric contaminants at their source. The disposal of the air exhausted from the production process also presents various problems of no little importance to the layout engineer. Since merely discharging the exhausted air outside the building may serve only to transfer the hazard to another area special precautions are necessary: (1) to ensure that the plant pollution problem does not become a community problem and (2) to be certain that air currents do not carry the contaminant back into the plant building through windows or other openings. Preventing the latter situation usually means construction of high stacks, use of special equipment to clean the exhaust, and even redesign of the process causing the pollution. The purpose is to substantially reduce the amount of pollution falling in neighboring communities. Locating plants to take advantage of wind patterns certainly should be a factor in developing the master plot plan. The general desirability of installing various types of collecting equipment whose function is to separate contaminants from the air exhausted must also be investigated.

Housekeeping is also a very good air pollution control method. Dust allowed to remain on floors and machinery can be disseminated to various parts of the plant. A regular schedule of vacuum cleaning is a very important factor in maintaining an effective program for controlling in-plant pollution. Immediate cleaning of spilled material is also a method of removing the contaminant at its source. The interior design of the plant and the layout should aid housekeeping functions.

Radioactive Contamination. Exposure to radioactive materials above the maximum permissible levels can be avoided by such measures as (1) maintaining exposure to external radiation as low as feasible and (2) by minimizing entry of radionuclides into the body by ingestion, inhalation, absorption, or through open wounds when unconfined radioactive material is handled. The successful accomplishment of these objectives requires planning well beyond the usual level taken when working with other materials When involved with radioactive materials it is absolutely necessary to analyze the hazards of each job, provide safeguards against each possible accident, and to use satisfactory protective devices and planned emergency procedures in the event accidents do occur.

Ayres (ed., Decontamination of Nuclear Reactors and Equipment) defines decontamination as removal of unwanted radionuclides from surfaces, and cleaning (mechanical or chemical) as the removal of nonradioactive contamination, such as dirt and grease, from a surface. The removal of 99% of the nonradioactive contaminants from an article usually makes it suitable for reuse. In decontamination the removal of contaminants must be increased by a factor of 10, 100, or even more. Articles that appear bright and clean after an initial decontamination may have sufficient radioactivity on their surfaces to warrant further decontamination. The main criterion for decontamination is the amount of residual radioactivity on the surface.

The layout engineer is faced with the problem of providing special cor taminated areas and decontamination rooms. Waste from these areas must b given special handling.

Physical safeguards to control exposure to external radiation make use c the fact that radiation intensity diminishes with distance from the source. first line of control includes physical enclosures to isolate the radioactive mate rial, and effective ventilation patterns throughout the facility. **Shielding** serve to minimize the external radiation reaching the body. It should be placed, possible, near the source. If adjacent rooms (both horizontal and vertical) ar occupied then plans should be made to protect these areas. In constructin shields care should be taken to ensure that cracks do not extend through th shield. This is particularly true if the construction material is lead brick.

For low-level radiation work, an ordinary fume hood may be used to contai the radioactive materials and prevent contamination of the work and th laboratory. In general, the ventilation system should be designed to cause ai flow to direct any radioactive material away from the worker or other material Basically the air flow should be from a noncontaminated area toward the cor taminated one. The good ventilation system will confine the contaminant, ex haust it through suitable duct work and then pass this material through collector or scrubber before releasing it to the atmosphere. In addition, th ventilation system should be under negative pressure so that leakage in the duc system will be into the ducts rather than the working area. As with any anti pollution system the discharge should be high enough above the building roof t prevent the fumes from being carried back into the facility through air intake or other openings. A common design for facilities housing radioactive materia provides for clean ventilating air to be admitted to offices and corridors firs then exhausted through rooms of low- and high-level radiation work, in order, b control of static pressures.

Areas in which radioactive materials are handled or stored should be s marked and isolated with a physical barricade to prevent unauthorized entry The type of barricade is a function of the level of contamination hazard and ca include chain and posts, permanent fences, or separate rooms and vaults. Area where radiation levels are significant should be provided with controlled entry

Within the work area surfaces of floors, benches, hoods and other facilitie should be impervious or capable of economic replacement. For example, un coated wood, concrete and plain rolled metals absorb radioactive materials. Til is commonly used for work surfaces because it is easily replaced in small section once contaminated. Ordinary paints, varnishes and lacquers are not recom mended for wear surfaces. Dust collecting surfaces should be eliminated a much as possible; for this reason lights should be recessed, pipes and shelves en closed, and corners between walls and floors should be coved to aid house keeping. Special facilities for disposal of liquid radioactive wastes should b provided. Separate sewer systems permit better radiation control and enabl construction of more economical waste treatment facilities where necessary. Al sinks and floor drains used for radioactive waste should be so marked and han wash basins should have knee or foot operated faucets.

Waste and Sewage Disposal Facilities. Every industrial facility generate wastes and consequently the layout engineer must concern himself with it nature, source, and method of removal. The simplest solution to the wast and sewage problem is to dispose of it through a municipal sewage system However, in larger facilities it is quite often necessary to install separate wast treatment and disposal facilities. In many instances more than one system ma

e required. When planning for the disposal of wastes, the layout engineer is
ot normally involved with the design of the specific system. He should,
owever, be consulted during the preparation of specific design objectives.

Safety Considerations. Every layout should be reviewed with a con-
ideration for the safety and welfare of the employee (see section on Industrial
afety). Any potentially hazardous situations should be corrected before imple-
nentation of the final plan. Some factors to be considered in this area include:

1. Provision for fire extinguishing and detection devices.
2. Adequate clearance along all traffic lanes.
3. Isolation of toxic fumes.
4. Clear marking of all personnel and handling aisles.
5. Safety railings on all stairwells, ledges, and catwalks.
6. Adequate lighting in all corridors and stairwells.
7. Provision for at least two emergency exits from each plant area.

Applications of Quantitative Techniques

NATURE OF PROBLEMS. Certain types of problems which frequently
ccur in plant layout and facilities planning are capable of solution by quantita-
ive methods. Considerable attention has been given to the formulation of models
or such problems as developing a plant layout, determining the relative location
f machines, evaluating equipment and product flow, and assembly line balanc-
ng. The models which have been developed frequently require a minor amount
f modification before they are applicable to a specific problem. However, con-
iderable time can be saved if one is familiar with them.

DECISION-MAKING PROCESS. All problems for which a decision
nust be made involve certain factors which may be determined explicitly for a
leliberate decision process or determined instinctively if the problem situation
rises suddenly. These factors are:

1. Elements subject to choice.
2. Criteria by which to evaluate the alternatives.
3. Relating the elements to the criteria.
4. Method for making a choice.

Elements subject to choice are called **variables.** As an example, a recurrent
problem that is found in a machine shop is the scheduling of work, over a time
pan, to men and machines. The variables would then include the allocation of
vork to machines, men, and time periods. The **alternative** schedules are the
particular assignments that might be made.

A decision involves the **selection of a particular alternative** based on the
onsequences resulting from the decision's implementation. The **selection of
he appropriate criteria** requires that the criteria be relevant to both the ob-
ectives and the variables.

Objectives. A large number of objectives are typically used in plant layout
nd facilities planning, including:

1. Maximize flow capacity.
2. Maximize on-time deliveries.
3. Minimize trip time.
4. Minimize capital investment.
5. Minimize operating costs.
6. Minimize idle time.

It should be noted that more than one objective may be applicable to a given
problem, but may not be dimensionally comparable. Furthermore, the selection

of differing criteria to be evaluated for otherwise identical problems may resul in different operating decisions.

Steps in Solution. The procedure widely used in problem solving can be de scribed by the following steps:

1. Defining the problem.
2. Gathering data on factors affecting the results.
3. Constructing a mathematical model to represent the problem under study.
4. Obtaining a solution from the model.
5. Testing the solution.
6. Putting the solution to work.

Certain quantitative techniques have found wide application as tools for mak ing decisions for the types of recurrent problems present in plant layout and fa cilities planning. The techniques, their characteristics or conditions, as well a their application to classes of problems, are discussed in the paragraphs tha follow.

QUANTITATIVE TECHNIQUES. The quantitative techniques mos often used to solve the problems that confront the facilities planning enginee can be conveniently grouped as follows:

1. Allocation models.
2. Waiting line models.
3. Simulation models.
4. Heuristic methods.

One of the real problems for the facilities planning engineer is to recognize tha his problem may be solved by one of the above techniques. For further discus sion of these techniques, see the section on Operations Research.

Allocation Models. Allocation models are used to solve three basic types o facilities problems:

1. With the activities and resources specified, allocate the resources to the ac tivities, optimizing the objective, i.e., the overall measure of effectiveness.
2. With the resources specified, determine the mixture of activities that will opti mize the overall measure of effectiveness.
3. With the activities specified, determine the mixture of resources that will opti mize the overall measure of effectiveness.

Allocation problems are solved most frequently by **linear programming.** Th technique requires that the variables of a problem be fairly constant and thei values known. The variables must represent a linear or straight-line relationshi to one another. The technique is thus applicable mainly for deterministic condi tions. The **simplex method** is most commonly used to solve linear programs The mathematical computations are relatively simple and can easily be accom plished manually for small or medium problems.

Waiting Line (Queueing) Models. Waiting line or queueing theory, technique based heavily on probability theory, attempts to solve **bottlenec problems** dealing with four basic situations:

1. **Facilities**—determining the number of service facilities.
2. **Scheduling**—determining the schedule of arrivals.
3. **Sequencing**—determining the order in which items are to be serviced.
4. **Line balancing**—combining operations so as to minimize the service time.

Waiting line models are extremely valuable tools for the facilities plannin engineer, for he may use it to solve the problems associated with random flov systems. The mathematics of queueing theory is somewhat complicated, thoug most problems can be solved manually.

Simulation Models. Simulation models are used as an alternative to direct analysis of problems. Simulation permits various courses of action to be tested experimentally under conditions similar to the real situation, without requiring costly and disruptive changes to actual operations. Conclusions reached from the simulated situation can then be applied to the actual operation with fairly predictable consequences. The experimentation is accomplished quickly on paper or computers, utilizing data and relationships corresponding to the real world under study. Another advantage of simulation is that it can represent a great deal of experience in a short time. The understanding gained is not truly equivalent to the years of experience, but should provide a useful insight into the behavior of a system.

Simulation has been used to attack a broad range of plant layout and facilities problems including:

1. **Congestion problems** involving cranes, conveyors, and material handling systems in general, as well as service facilities with variable demand.
2. **Sequencing of production operations** for job shops, to minimize in-process times, and planning for peak capacity.
3. **Simulation of entire production and distribution systems** to aid and evaluate management decisions concerning short- and long-range planning.

As is true for most mathematical models, simulation of complex situations may prove to be an expensive process because of the requirements for large amounts of data concerning the pertinent activities and their relationships.

Heuristic Methods. Heuristic models use principles or rules that simplify the environment of the decision maker by reducing the search in problem-solving activity. The advantage gained through heuristic programming is not that they give good decisions (as compared to optimal decisions for algorithms), but rather that they limit search by reducing the number of alternatives for combinational decision problems to a manageable size. Heuristic models typically require much less effort to use, even for complex situations, than the use of an algorithm. The most notable **applications of heuristic programming** within the area of plant layout and facilities planning have been to assembly line balancing, job shop scheduling, and plant layout analysis.

APPLICATION TO PROBLEM AREAS. The quantitative techniques discussed earlier, as well as other mathematical techniques, find wide application in problems associated with plant layout and facilities planning. Chief among the areas of application are problems associated with:

1. Workplace design.
2. Assembly or production line balance.
3. Material handling.
4. Transportation and distribution.
5. Utility system analysis.
6. Material flow.
7. Allocation (or layout).
8. Machine or facility location (or layout).
9. Warehouse layout.
10. Office layout.
11. Packaging or packing requirements.
12. Alternative layouts.
13. Long-range planning.
14. Plant location.
15. Warehouse locations.

The remaining portion of this section is devoted to several specific classes of problems relevant to the facilities planning engineer. Other applications are discussed in the section on Operations Research.

Relative Location of Facilities—Facility Layout. The use of trave charts, operation process charts, and the relationship charts as aids in layou planning have been described earlier in this section and in the section on Ma terial Handling for both product and process layouts. The layout of a product oriented facility involves following the flow of the product, which is relativel simple. The layout problem in a process or job shop facility requires followin; multiple product flow paths—a problem which can become virtually unmanage able for the above charts when the number of departments becomes large, unles the flow has a dominant pattern. Five heuristic **computer-aided program** have been developed to assist the engineer in planning job shop or multi-depart ment layouts. They are CRAFT (**C**omputerized **R**elative **A**llocation of **F**acili ties **T**echnique) developed by E. Buffa, U.C.L.A., T. Vollman, and G. Armou Indiana University; CORELAP and Interactive CORELAP (**CO**mputerize **RE**lationship **LA**yout **P**lanning), developed in the Industrial Engineering De partment at Northeastern University headed by J. Moore; ALDEP (**A**uto mated **L**ayout **DE**sign **P**rogram), developed by I.B.M.; and RMA Comp. I developed by Richard Muther and Associates.

Figure 27 compares the input requirements, method of operation, and outpu format for each of these techniques. Computer-aided layout planning is still a an early stage of development and must be conducted in conjunction with soun(manual planning such as is built into the procedure of Interactive CORELAP.

Relative Location of Facilities—Machine Location. Machine locatio determination also comes under the problem classification covering the relativ location of facilities. Two **types of machine location determination prob lems** exist:

1. The location of additional machines (or other facilities) in an existing layout.
2. The relative location of machines in a new layout.

The location of new machines in an existing layout usually has a limite number of possibilities. With the assumption that a pre-existing layout is to b modified, there is non-interdependence between new machines to be added Moore (Journal of Industrial Engineering, vol. 12) has applied the **assignmen technique** (a variant of the linear programming technique) to achieve optima solutions for the locations considered. The machine assignment problem is state(as given n machines (or other facilities) and n potential locations (jobs, assign ments, etc.) and given the effectiveness of each machine for each potential loca tion, assign each machine to one and only one job to optimize the measure o effectiveness.

For example, a machine shop desires to add three new machines (A, B, and C to their existing layout. Engineers have three potential locations (1, 2, and 3) fo the new machines. (The availability of more than three potential locations woul(require a change in the solution procedures.) The objective is to minimize th(amount of travel required between the new and existing machines. The measur(of effectiveness will be pound-feet per day.

The first step is to determine the production in pounds that will flow betwee the new and existing machines per week and the distances between the potentia locations and existing machines. Using matrix multiplication, the following effec tiveness matrix is obtained from Figs. 28 and 29.

$$
\begin{bmatrix} 50 & 16 & 8 & 0 & 60 \\ 0 & 14 & 20 & 24 & 16 \\ 16 & 10 & 120 & 0 & 32 \end{bmatrix}
\begin{bmatrix} 4 & 2 & 12 \\ 4 & 6 & 8 \\ 8 & 6 & 4 \\ 12 & 14 & 4 \\ 18 & 12 & 6 \end{bmatrix}
=
\begin{bmatrix} 1408 & 964 & 1120 \\ 832 & 732 & 384 \\ 1640 & 836 & 944 \end{bmatrix}
$$

	Input Requirements	Operating Procedure	Output
CRAFT	Material flow between two activity areas. Space requirements in the form of a proposed or existing layout (location of activity can be fixed). Cost per unit moved per unit distance.	Cost calculated as a product of flow, cost, and distances between centroids of activities in the initial layout. Interchanges locations based on the greatest cost reduction (two-way and three-way interchanges are considered). Process is continued until no significant cost reduction can be found.	Layout is printed in basic rectangular form. Irregular activity shapes may be manually adjusted into more practical dimensions. Latest total cost is calculated and the differences between it and the initial total cost is shown as savings.
CORELAP	Relationship data as determined by a relationship chart. Space requirements for each activity. Maximum building length-to-width ratio.	After summing each activity's closeness ratings (TCR) with other activities, the activity with the highest TCR (called the *Winner*) is selected and placed first in the layout. The relationship chart is searched to find activities that have an A (absolutely necessary) closeness rating with the Winner. If one is found, this activity (called the *Victor*) is placed next to the Winner. If an A activity can be found for the Victor, the Victor becomes the new Winner and the procedure is repeated. Continuing the process, if no Victor can be found for the Winner, a search is made for an E (especially important) rating, then an I (important) rating, finally an O (ordinary closeness) rating, until a valid relationship is found and a potential new Winner is determined. Two other closeness ratings are also used: U (unimportant) and X (undesirable). The above process is continued until all activities have been placed in the layout.	Layout printed in an irregular format on a total and activity basis is manually adjusted into a workable layout.

Fig. 27. Comparison of computerized approaches to the relative location of facilities.

	Input Requirements	Operating Procedure	Output
Interactive CORELAP (Designed for a Time-Shared Computer System)	Relationship data as determined by a relationship chart. Space requirements for each activity.	Similar to CORELAP with the provision that the heuristic logic of CORELAP can be interrupted by the user to make adjustments (template shuffling) or by the logic itself when it gets into difficulty. In either case, the user may make some adjustments or push the heuristic re-start button. The Interactive CORELAP routine contains a scoring routine providing the user with the cost involved in moving the template, considering all pairs of relationships in his solution.	As the process progresses, partial irregular layouts may be printed, scored, modified by the user, the rescored with the final shape of the total and individual layout dependent upon the user's manipulations.
ALDEP	Relationship data as determined by a relationship chart. Space requirements for each activity. Building description including size and location of fixed facilities (aisles, etc.). Locations to be pre-assigned.	A modified random selection technique is used to generate alternate layouts with the user specifying the number of layouts wanted which must satisfy a minimum score. The first activity is selected and located at random. The relationship matrix is searched to find an activity highly related to the first activity placing this activity adjacent to the first. If a second activity is not found, a second activity is selected at random and placed next to the first. The above procedures are repeated until all activities are placed. Other layouts are generated by repeating the above procedure.	Layout is printed in the format of the building description with activities irregularly shaped to be manually adjusted into a workable layout.
RMA Comp. 1	Relationship data as determined by a relationship chart. Space requirements for each activity.	The activity with the largest TCR is placed in the center of the layout. Subsequent activities are placed in the layout considering both relationships to activities placed or to be placed. All activities are placed without regard for the area requirements of the activity.	Layout is printed as a space relationship diagram or as blocks (roughly to scale) located by best fit of all desired closeness relationships. The diagram must then be manually adjusted into a block layout.

Fig. 27. Concluded.

	Existing Machines				
New Machines A	50	16	8	0	60
B	0	14	20	24	16
C	16	10	120	0	32

Fig. 28. **Flow between new and existing machines; numbers in matrix represent pounds × 10.**

	Existing Machines				
Potential 1	4	4	8	12	18
Locations 2	2	6	6	14	12
3	12	8	4	4	6

Fig. 29. **Distance between new and existing machines; numbers in matrix represent feet × 10.**

The steps required to solve the assignment problem by the **Hungarian method** are summarized below:

1. Subtract the minimum of each row from all elements of the row in the effectiveness matrix. In the resulting matrix, subtract the minimum element of each column from all the elements of the column, obtaining the first reduced matrix.
2. Determine the optimal assignment. If the optimal assignment is a feasible solution, the procedure terminates here. If not, step 3 must be executed.
3. Cover all zeros with the minimum number of straight lines. Find the smallest element not covered by a line. Subtract this minimum element from all elements not covered by lines; then add it to all elements lying at the intersection of two lines, which reveals the second reduced matrix.
4. Repeat steps 2 and 3 until a completely feasible solution is obtained.

	Potential Locations				Potential Locations				Potential Locations		
	1	2	3		1	2	3		1	2	3
New A	1408	964	1120		0	0	156		[0]	⍉	156
achines B	832	732	384		4	348	0		4	348	[0]
C	1640	836	944		360	0	108		360	[0]	108
	Effectiveness Matrix				Step 1				Step 2		

Fig. 30. **Matrices for the machine location problem.**

The matrices for the machine location problem are shown in Fig. 30. The optimal solution is to:

1. Assign machine A to location 1.
2. Assign machine B to location 3.
3. Assign machine C to location 2.

For the assignment of a single machine, the effectiveness matrix would produce the answer for multiple potential locations.

Queueing Model—Waiting for an Elevator. Queueing models offer a means of reducing the amount of judgment required on capacity decisions under changing demands. Waiting line problems occur when units arrive at one or more facilities for service, with the arrival and service times subject to random changes. (For more details, see section on Operations Research.)

For example, stock trucks arrive at a freight elevator randomly at the rate of 20 per hour. The trip to the third floor averages 2 minutes. The production engineer wants to determine the average number of trucks waiting for the elevator, the average length of time a truck spends waiting for the elevator, and the increase in the elevator service rate required to lower the waiting time to an average of 1 minute. Trucks are served on a first-come, first-served basis. In all of the queueing problems discussed, it will be assumed that arrival–service times are such that the Poisson-exponential distributions are applicable.

If

$$\lambda = \text{average arrival rate} = 20 \text{ per hour}$$
$$\mu = \text{average service rate} = 30 \text{ per hour}$$

then

Average length of the non-empty waiting line $= \dfrac{\mu}{\mu - \lambda} = 3$ trucks

Average waiting time of an arriving truck $= \dfrac{\lambda}{\mu(\mu - \lambda)} = 0.067$ hour or 4 minutes

Increase in service rate needed to lower waiting time to 1-minute range $= \dfrac{\lambda}{\mu'(\mu' - \lambda)} = 0.0167$ hour or

$$\mu' = 43.2 \text{ trips per hour.}$$

Queueing Model—Determining the Number of Truck Docks. During the peak period of the day, trucks arrive at the receiving dock at the rate of 19 per hour on a random basis. The service time is found to have a mean of 30 minutes. Management wants to know how the number of truck docks will affect the waiting time of the drivers, in order to provide enough doors so that 75 per cent of the drivers will get immediate service. Saaty (Elements of Queueing Theory with Applications) gives a summary of formulas useful for a variety of waiting line problems:

$$\text{Average waiting time of an arrival} = \frac{\mu(\lambda/\mu)^k}{(k - 1)!(k\mu - \lambda)^2} P_0$$

where k = number of truck docks and

$$P_0 = \frac{1}{\left[\sum\limits_{n=0}^{k-1} \dfrac{1}{n!} \left(\dfrac{\lambda}{\mu}\right)^n\right] + \dfrac{1}{k!} \left(\dfrac{\lambda}{\mu}\right)^k \dfrac{k\mu}{k\mu - \lambda}}$$

where P_0 = probability of zero in systems.

Peck and Hazelwood (Finite Queueing Tables) present tables applicable to waiting line calculations. Use of these tables leads to the information shown in Fig. 31. Based on the study, 13 truck docks would be recommended.

Number of Dock Doors	Average Time a Truck Must Wait for Service	Trucks Which Will Not Have to Wait	Maximum Wait for 95 Per Cent of Trucks
10	51 minutes	15%	170 minutes
11	11	45	48
12	4½	62	24
13	2	77	13

Fig. 31. Truck dock information.

Queueing Model—Tool Crib Service. The tool crib problem lends itself to waiting line theory to determine the number of clerks needed to service the facility under uncertainty.

Mechanics arrive at the tool crib randomly with an average of 50 seconds between arrivals. They are served in an average of 1 minute. The idle time cost for the mechanic is $7.50 per hour; for the clerk it is $3.00 per hour. The problem is to determine the number of clerks that should be assigned to the tool crib in order to minimize cost.

As in the truck dock example, this problem involves a multi-server queueing model with $\lambda = 3,600/50 = 72$ and $\mu = 3,600/60 = 60$. We know that we must have more than one clerk or an infinite waiting line would develop. The total cost of clerks per hour at $3.00 each per hour would be:

Clerks	2	3	4
Cost/hour	$6	$9	$12

The expected idle time of the mechanics can be determined from the expected queue length. Evaluating the waiting line formula for the expected queue length with 2, 3, and 4 clerks, the results obtained are as shown in Fig. 32. The most economical choice appears to be the assignment of 3 clerks to the tool crib.

Clerks	Expected Waiting Line Length	Cost of Waiting Line/Hour	Total Expected Cost/Hour
2	1.875	$13.12	$19.12
3	1.294	9.06	18.06
4	1.216	8.51	20.51

Fig. 32. Tool crib service problem.

Recognition of facility problems which can be formulated as waiting line problems requires a certain degree of ingenuity and insight. The example above could as well have been concerned with how many mechanics to assign to a given group (see queueing problems in section on Operations Research).

General Applications of Queueing Models in Facilities Planning. Hillier (Management Science, vol. 10) gives the following examples of facilities planning problems that lend themselves to solution using queueing theory:

1. How many semiautomatic or automatic machines should be assigned to each operator?
2. How many repairmen should be assigned to a given group of machines?
3. What type and what quantity of material handling equipment should be used in a given factory area?
4. How many inspectors should be assigned to an in-line assembly inspection department?
5. In a process layout, how many machines (or other operating units) of a given type should be provided?
6. How many electronic computers of a given type should be provided?
7. What quantities of various types of auxiliary service lines (water, electricity, steam, oil, and the like) should be provided in a given factory area?
8. How many specialists should be provided for performing a specific service on call, for example, typists, draftsmen, programmers, artists, or clerks in such service areas as reproduction, documents, library, tool crib, and so forth?

9. What should be the size of a crew for jointly performing a certain task, for example, maintenance work, loading and unloading operations, inspection work, and setup of machines?

10. How many employee facilities of a given type, for example, rest rooms, first-aid centers, drinking fountains, and vending machines, should be distributed throughout a given area?

11. How much space should be provided for in-process storage at a given location?

Assembly Line Balancing Techniques. An assembly line may be characterized as a moving conveyor that passes a series of work stations in a uniform time interval called the **cycle time.** At each work station, work is performed on the product, either by adding parts or by completing assembly operations. The work performed at each work station is comprised of a number of **work units.** The work units are considered indivisible since the elements cannot be split between two or more operators without paying a penalty for extra time. The problem is then to determine the most efficient operating conditions required to balance a given assembly line by either:

1. Finding the production line of minimum length for a given cycle time, or,
2. Finding the production line of minimum cycle time for a given line length.

The optimal solution to 1 and 2 becomes the minimization of the **balance delay,** defined as the amount of idle time on the line due to the imperfect division of tasks among the work stations. Several common assumptions have been used in balancing procedures:

1. Precedence relations exist which restrict the order in which tasks can be performed.
2. There is no product mix on the assembly line.
3. The work unit times are deterministic and known.
4. With few exceptions, zoning restrictions (concerned with the position of the object being assembled with relation to the operator or operators) do not exist.

Because of these restrictions, the distinction between optimal and non-optimal balances tends to fade with the possibility of redefining tasks, precedence, zoning, and other problems.

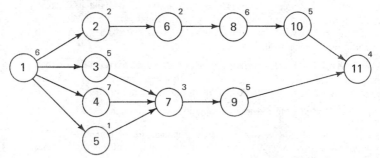

Numbers in circles are work element numb[

Numbers outside circles are times to comp
the particular work elements.

Fig. 33. **Assembly precedence diagram for assembly process requiring eleven work elements.**

Figure 33 is an example of an assembly precedence diagram. The numbers inside the circles identify the work units; the numbers outside the circles refer to the corresponding time durations. The arrows show the precedence relations.

Jackson (Management Science, vol. 2) has developed an exhaustive and enumerative algorithm that will find an optimal solution when carried to completion. The procedure will minimize the number of stations for a given cycle time (T) and is good for manual solution of relatively small problems. The procedure is to:

1. Construct all feasible first work stations.
2. For each first work station, construct all feasible second work stations.
3. Construct all feasible third stations for each first–second combination, etc.
4. At some point, it will be found that one or more of the balances have assigned all the tasks, thereby minimizing the number of work stations for the given cycle time.

Jackson's computational procedure is illustrated in Fig. 34 with maximum cycle time $T = 10$.

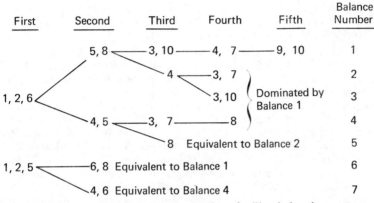

Fig. 34. Computational procedure for line balancing.

Several heuristic approaches to the line balancing problem have been developed which are suitable for manual problem solving. Helgeson and Birnie (Journal of Industrial Engineering, vol. 12) have developed a **positional weight method** in which the positional weight of a work unit is its time plus the time of all other work units which must follow it. The procedure is to:

1. Calculate the positional weights by summing the element times following and including each work element on a particular network path. For instance, the positional weight for work element Z in Fig. 33 is found by adding the time for element Z (Z units) to the times for all following work elements on the particular network path (elements 6, 8, 10, 11). The positional weight $= 2 + 2 + 6 + 5 + 4 = 19$. Note that for element 1, all other work elements on all network paths are follower elements.
2. Rank the work units with the largest weight first.
3. Assign work units to stations in order of ranking.
   a. If the work unit is longer than the time remaining, or would violate precedence, continue to the next work unit.
   b. Continue until no work units can be assigned to a station.
   c. Assign work units to the next station, beginning with the units passed over.

Using the example problem of Fig. 33, with $T = 10$, the positional weights and preceders are:

Work Unit	1	2	4	3	6	8	5	7	9	10	11
Positional Weight	46	19	19	17	17	15	13	12	9	9	—
Immediate Preceders	—	1	1	1	2	6	1	3,4,5	7	8	9,10

The resulting balance is obtained by taking the work units in order of positional weight and assigning them to work stations such that the combined times do not exceed the station time. Jumps in order may be utilized as long as precedence restrictions are met.

$$1, 2, 6 \mid 4, 5 \mid 3, 7 \mid 8 \mid 9, 10 \mid 11$$

Note that with this approximate technique, 6 instead of 5 work stations were required.

When the number of work units per station is large, the heuristic technique of Kilbridge and Webster (Journal of Industrial Engineering, vol. 12) becomes applicable. Their technique attempts to find the perfect balance where the total time would be the product of the work stations and the cycle time. The procedure requires the addition of column labels to the precedence diagram. Then two properties of the work units in the diagram are exploited:

1. The order in which tasks in the same column are performed is unimportant.
2. Work units can be transferred between columns.

A procedure is developed based on a column rule. The main feature is the grouping of work units into columns to guide the selection of work units. Each work element is identified by two column numbers denoting the first and last columns in which the work unit may be selected. The solution is achieved when the smallest number of work stations is found that will achieve perfect balance.

Moodie and Young (Journal of Industrial Engineering, vol. 16) have also developed a heuristic procedure with a **two-phase operating procedure.** Phase I assigns the longest of those work units that will fit the current station, with Phase II rearranging the work elements between the stations. Manual computation is possible for the procedure, which also allows variable performance times.

Other heuristic and exact solution procedures have been developed, but are more suitable for computer than for manual solving of realistic problems. Mastor (Management Science, vol. 16) has conducted an experimental investigation to show that there are significant differences among the effectiveness results of line-balancing techniques. For the larger balancing problems, however, the computing time difference may warrant more important consideration than the effectiveness results.

MATERIAL HANDLING

CONTENTS

SECTION 17

MATERIAL HANDLING

———

The Material Handling Function

DEFINITION. The Materials Handling Division of the American Society of Mechanical Engineers has approved the following short, simple, and all-inclusive definition: **Materials handling is the art and science involving the moving, packaging, and storing of substances in any form.** This is but one of many ways by which material handling has been defined yet no single definition has received universal acceptance. Other definitions include:

Creation of time and place utility.
Movement and storage of material at the lowest possible cost through the use of proper methods and equipment.
Lifting, shifting, and placing of material which effect a saving in money, time, and place.
Art and science of conveying, elevating, positioning, transporting, packaging, and storing of material.

SCOPE OF MATERIAL HANDLING. The scope of material handling activity within a specific industrial enterprise depends on the type of company, the product manufactured, the size of the company, the value of the product, the value of the activity being performed, the relative importance of material handling to the enterprise, the personalities of individuals involved, and the organizational structure of the enterprise. Figure 1 shows the material flow cycle for a typical production organization. Note that each path implies some form of handling takes place.

In a somewhat chronological sense material handling activities in a company go through three stages: the traditional point of view; plant-wide concern for the overall flow of material; and the systems point of view.

In the **traditional** interpretation of material handling, primary emphasis is on the movement of material from one location to another, more likely than not, within the confines of the individual plant. The concern of the material handling engineer is merely to find the best way to move something from point A to point B. His interest is usually in individual, isolated, independent material handling problem situations. More likely than not the analyst is busy jumping from crisis to crisis in order to meet the immediate needs of those who call for his services, and most of his attention is given to those who shout the loudest. Very little attention is given, nor is there much interest in, the possible interrelationships between the various individual handling situations with which the analyst is involved. Material handling functions operating in this fashion are neither very efficient nor progressive.

Plant-wide concern centers the attention of the analyst on the overall flow of material in the enterprise. The engineer is concerned with the interrelationships between all handling problems and the possibility of establishing a general overall material handling plan. His overall effort is directed toward typing each problem solution into all others, in order to obtain a totally integrated material han-

17·1

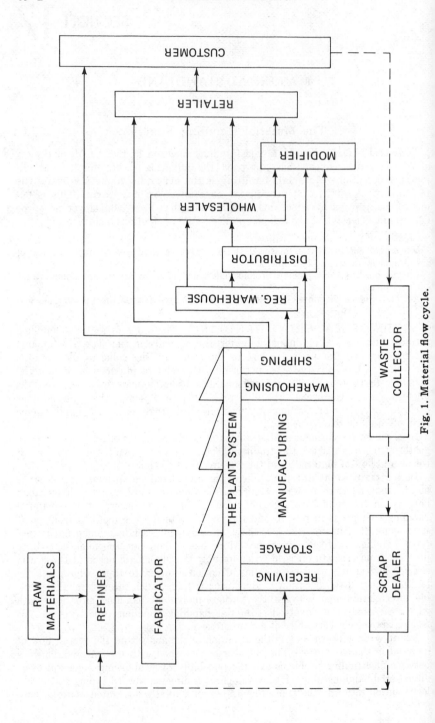

Fig. 1. Material flow cycle.

dling plan. This point of view is fairly common in the typical manufacturing enterprise.

The **systems** point of view of material handling requires that the analyst visualize the material handling problems, the physical distribution activities, and all closely related functions as one, all-encompassing system. This point of view involves a much broader consideration of all material handling activities involving the movements of all materials from all sources of supply, all handling activities within and around the plant itself, and the handling activities involved in the distribution of finished goods to all customers of the enterprise. This broader, more inclusive, point of view has the theoretical goal of conceptualizing a total solution to the overall handling problem in terms of a theoretical ideal system. The material handling engineer must design as much of the total system as is currently practicable and implement those portions which are feasible to install. He would continue to work on other portions of the "theoretical" system and over a period of time design and implement as many as possible as means become available and their installation is practical and/or economical.

OBJECTIVES. The following outline suggests those objectives or goals which the material handling engineer strives to accomplish:

1. **Reduce costs**
   a. Minimize handling
   b. Reduce material handling effort
   c. Reduce handling by direct labor
   d. Reduce related indirect labor
   e. Reduce waste
   f. Reduce paper work
   g. Reduce the total amount of material in the system
   h. Use less subsidiary materials
   i. Control pilferage
   j. Lower inventory and production control costs
2. **Increase capacity**
   a. Maintain constant rate of production
   b. Coordinate handling system with production
   c. Revise plant layout, to:
      (1) Reduce travel distances
      (2) Smooth work flow
      (3) Improve space utilization
   d. Increase utilization of manpower
   e. Obtain higher equipment utilization
   f. Handle larger loads
   g. Speed loading and unloading of carriers
   h. Reduce overall production cycle
   i. Eliminate production bottlenecks
3. **Improve working conditions**
   a. Provide safe working environment
   b. Reduce job effort
   c. Reduce worker fatigue
   d. Improve personal comfort
   e. Upgrade employees to more productive and desirable work
4. **Improve customer service**
   a. Speed delivery
   b. Reduce product damage
   c. Improve handling methods

In working with as many people as he does in carrying out his everyday activities, the material handling engineer is in an outstanding position to reduce costs in many phases of the plant operation.

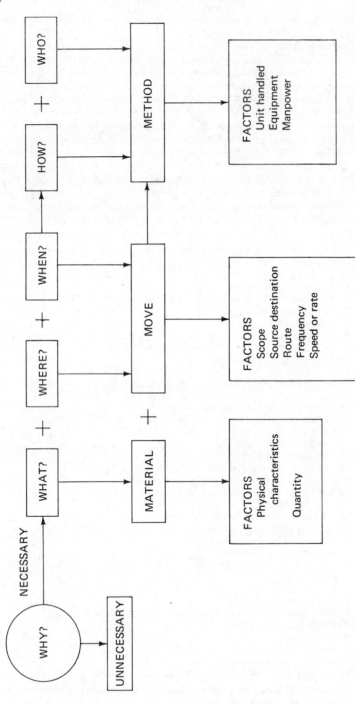

Fig. 2. Material handling equation.

BASIS FOR ANALYSIS. Past approaches to the solution of material handling problems have frequently led to alleviating the problems, rather than solving them. Very likely the viewpoint from which the problem was analyzed was entirely too narrow, and the background thinking of the analyst was very likely to be equipment (hardware) oriented. This frequently resulted in a rather inadequate solution.

Modern production technology is entirely too complex for a piecemeal approach, and complex handling tasks must be examined in an organized fashion. This begins with a proper recognition by management of the material handling activity, and involves the development of a planned program of analysis, implemented with carefully selected procedures and techniques to assist in obtaining the best solution to the problem.

The organized approach should begin with a device for visualizing the many factors inherent in a material handling problem. One such device, the material handling equation (Fig. 2), indicates that there are six major questions to be answered in finding the solution to a handling problem. The question "Why?" implies a careful look at the problem situation to be sure that the problem is properly identified and defined. The analysis then asks "What?", in order to properly acquaint the analyst with the **material** to be moved. The next two questions, "Where?" and "When?" will assist in identifying and specifying the **move** that needs to be made. The last two questions, "How?" and "Who?" pertain to the **method** to be used and the personnel involved. A careful examination of the equation will show that one of the major causes of poor solutions to handling problems in the past has been the temptation of the analyst to jump from the "What?" to the "How?" in his rush to find a quick answer to the problem.

After having divided the problem into its three major phases—material, move, and method, it is necessary to look deeper into the implications of the equation to uncover the many factors involved in each phase.

MATERIAL HANDLING ENGINEERING. Some fundamental practices have been developed over a period of years, representing the accumulated experience of many past material handling practitioners. They represent ideas that have been tested and findings which have been observed, tried out, and found worthwhile. Over a period of time, many of these have been written down and have come to be known formally as **Principles of Material Handling**. The following is based on principles proposed by the College-Industry Committee on Material Handling Education:

Related to Planning

1. Planning Principle. All handling activities should be planned.
2. Systems Principle. Plan a **system** integrating as many handling activities as is practical and coordinating the full scope of operations.
3. Material Flow Principle. Plan an operation sequence and equipment arrangement optimizing material flow.
4. Simplification Principle. Reduce or eliminate unnecessary movements and/or equipment.
5. Gravity Principle. Utilize gravity to move material whenever practicable.
6. Space Utilization. Make optimum utilization of building cube.
7. Unit Size Principle. Increase quantity, size, and weight of load handled.
8. Safety Principle. Provide for safe handling methods and equipment.

Related to Equipment

1. Mechanization/Automation Principle. Use mechanized or automated handling equipment when practicable.

2. Equipment Selection Principle. In selecting handling equipment, consider all aspects of the **material** to be handled, the **move** to be made, and the **methods** to be utilized—and in terms of the lowest overall cost.
3. Standardization Principle. Standardize methods as well as types and sizes of handling equipment.
4. Flexibility Principle. Use methods and equipment that can perform a variety of tasks and applications.
5. Dead-Weight Principle. Reduce the ratio of equipment dead-weight to pay load.
6. Motion Principle. Equipment designed to transport material should be kept in motion.
7. Idle Time Principle. Reduce idle or unproductive time of both handling equipment and manpower.
8. Maintenance Principle. Plan for preventive maintenance and scheduled repair of all handling equipment.
9. Obsolescence Principle. Replace obsolete handling methods and equipment when newer methods or equipment will pay off in a reasonable time.

Related to Operations

1. Control Principle. Use material handling equipment to improve production control, inventory control, and order handling.
2. Capacity Principle. Use handling equipment to help achieve full production capacity.
3. Performance Efficiency Principle. Determine efficiency of handling performance in terms of expense per unit handled.

Survey of Operations. While the above list represents the general directions to be taken in analyzing material handling problems, there are many specific areas that require investigation. These areas are covered in the statements on the Preliminary Material Handling Survey Checklist (Fig. 3). The questions on the checklist are stated in the negative in order to emphasize their value as **indicators of ineffective material handling** practices.

Material Handling and Plant Layout

BUILDING DESIGN. To keep handling at a minimum the functional approach to facility planning should be adopted. This involves the development of an efficient layout around a preplanned flow pattern. Thus the **flow of material** becomes the basis for the layout and the production processes are performed in their respective locations along the flow lines.

Floors. Floor loading capacities will determine the advisability of utilizing industrial trucks, floor-supported conveyors and cranes, powered hand trucks, or the like. The type of floor will have an important bearing on the specifications of any rolling equipment used. Power requirements and wheel design are two of the equipment factors dependent upon floor material and grades. **Aisle space** must be reserved to facilitate movement of various types of industrial vehicles. However, the amount of space assigned to aisles necessarily decreases the amount of space available for storage and production or services. Thus the ultimate assignment of space is generally a compromise based on space cost, handling equipment and labor expense, and the physical requirements of the move. State and local industrial regulations and equipment specifications often establish minimum requirements for aisle space.

Structure. The mounting height of lighting fixtures, pipes, ductwork, sprinkler systems, unit heaters, and other ceiling and wall-mounted equipment must be carefully coordinated with the movement of supplies and material in the plant. A desirable arrangement is to install all such accessories above the bottom chord of the roof truss or the bottom flange of the roof or main beams. The objective is to keep a uniform unobstructed operating clearance, as high as practical, throughout the plant.

Spacing of supporting columns may be a limiting factor on the type of handling equipment used or the size of the unit loads permitted. The National Wooden Pallet and Container Association (Pallets and Palletization) recommends uninterrupted spans of 50 to 60 feet for modern warehouse structures. The design of **columns and pilasters** will be influenced by the use of wall- and column-mounted equipment such as cranes and craneways. Provision for this equipment in the original design can usually save the expense of auxiliary structures in the future. Typical bay widths should be designed with due consideration of craneway operating areas.

Anticipation of future conversion from warehousing to manufacturing occasionally requires the overdesign of warehouse structures to permit installation of handling and production equipment.

Other important **building details,** such as doors, loading docks, ramps, elevators, fire walls, and such permanent installations as toilets, offices, and other service facilities should be given careful consideration in light of their possible effect on the handling function. Door sizes, elevator sizes and capacities, fire walls, and service facility locations should be developed after the material flow and production layout have been defined.

Multistory Design. Although single-story plants have been in vogue for a considerable period of time, there is an argument for the use of multistory plants in situations where vertical material handling may be less costly than horizontal handling. For example, developments in vertical handling equipment make it considerably easier to move material automatically and at less cost between floors, than would be required if the handling were done on elevators or even on a single floor, but over a greater distance.

Plant Location. Many factors must be considered when choosing a site for a manufacturing plant, distribution warehouse, or other industrial facility. The value of real estate, the plant's effect on the environment, the availability of labor, community acceptance, the relationship to market and resources, all must be considered. (See section on Plant Layout and Facilities Planning.) It is now recognized that the material handling aspects of site selection must also be considered.

The **nature of the product** has a great deal of bearing on the nature of the material handling problem. It is logical to assume that such heavy industries as steel, aluminum, glass, heavy chemicals, etc. would select locations which would give them a favorable relationship to their source of raw material. They would probably seek water or rail transportation facilities and perhaps, at some future date, cross-country conveyors. On the other hand, industries which depend upon a large volume, low unit value, and rapidly distributed product, would probably seek locations which would be easily accessible to their market. It is unlikely that a brewery, dairy, or newspaper plant would be located very far outside of a large population center.

These **transportation considerations** relate themselves to material handling in the plant through the nature of unloading or shipping requirements. Plants which use only one type of transportation have far simpler handling problems than those which use several types of carriers. The variations between rail-car and truck docks and the differences in loading and unloading problems have a real bearing on the design of the facility; thus, traffic problems relate not only to the choice of the site, but to the economics of the plant itself.

Plant _____ Building _____

Area _____

Prepared by _____ Date _____

Indicators of Ineffective Material Handling (if checked in YES column, INVESTIGATE!)	Check		Comments, Suggestions for Improvements, etc.
	YES	NO	

General

1. Crowded conditions ....
2. Empty floor space ....
3. Poor housekeeping ....
4. Excessive temporary storage ......
5. Material piled directly on floor ....
6. Wasted cube...................

Material

1. Characteristics of material causes handling problems ......
2. Quantity justifies mechanical handling ....
3. Too much/too little on hand ......
4. Damaged materials.............
5. Excessive scrap ....

Move

1. Scope of move beyond area under investigation ....
2. Building characteristics restrict move ....
3. Carrier characteristics restrict move ....
4. Move appears too long ....
5. Move not in direct path ....
6. Flow pattern complicates material handling ....
7. Zig-zag, or crooked path ....
8. Backtracking ....
9. Lack of alternate paths ....
10. Cross-traffic impedes flow ....
11. Too much distance between operations ....
12. Obstacles in material flow ....
13. Related work scattered ....
14. Traffic jams ....

Method of Handling
  —General—

1. Moving one item at a time ....
2. Excess manual handling ....
3. Inadequate scrap removal.........
4. Excess storage at workplace ......
5. Insufficient storage at workplace ...
6. Poor flow between work areas ....
7. Material piled on floor ....
8. Scheduling difficulties ....
9. Improper location of feeder or sub-assembly lines ....

Fig. 3. Preliminary material handling survey checklist.

Indicators of Ineffective Material Handling (if checked in YES column, INVESTIGATE!)	Check		Comments, Suggestions for Improvements, etc.
	YES	NO	

Method of Handling
  —General— (cont.)

10.	Inspection not integrated with production	
11.	Hard, hazardous work done by hand	
12.	Safety hazards	
13.	Two-man lifting job	
14.	Rehandling	
15.	Unplanned material handling methods	
16.	Frequent, short, repetitive moves by hand	
17.	Makeshift methods	
18.	High load/unload time	
19.	Difficult handling	
20.	Overmechanized handling	

—Unit Handled—

1.	Items not moved in unit loads	
2.	Unit received not utilized in subsequent moves	
3.	Unit received inefficient for handling	

—Equipment Used—

1.	Idle equipment	
2.	Excessive equipment repairs	
3.	Overloaded equipment	
4.	Underloaded equipment	
5.	Equipment obsolete	
6.	Unsafe equipment	
7.	Shortage of equipment — no spares .	
8.	Operating over/under rated speed ..	
9.	Inadequate maintenance/repairs ...	

—Containers—

10.	No container used in move	
11.	Frequent change of containers	
12.	Heavy container	
13.	Non-standardized containers	
14.	Shortage of containers	
15.	Container not suitable for mechanized handling	

—Utilization of Manpower—

1.	Excessive injuries	
2.	Large no. of men doing handling ...	
3.	Handling done by direct labor	
4.	Men walking for material	
5.	Operators waiting for material	
6.	Heavy physical exertion	

—Costs—

1.	High overhead costs	
2.	High indirect labor cost	
3.	High unit handling costs	

Fig. 3. Concluded.

Real estate values also have a double-barreled effect on site selection. The recognition of handling cost reductions which result from single-story construction has tended to move industry into areas of lower real estate values. However, if other factors dictate the choice of high-value real estate and multistory construction, material handling costs influence the selection of the site and design of the building. The accessibility of in-town plants to truck and rail traffic is usually more complicated than in a rural installation. The internal handling may also be complicated by the vertical travel element.

The selection of **distribution facilities,** such as warehouses and retail outlets, is usually closely related to their position in regard to the market, purchasing policies among the customers, and the inventory policy of the company involved. However, the increasing recognition of the high cost of handling material in the distribution phases of industry has led to a more careful selection of distribution facility locations. The same problems which are described above for manufacturing plants apply equally well to the choice of sites for distribution facilities.

In cases where outdoor operations are necessary, climate enters into the material handling picture. Many types of material handling equipment are considerably less efficient in wet or icy environments.

STORAGE AND WAREHOUSING. The term storage as used here deals with pre-production and in-process holding of material and supplies awaiting use in the manufacturing operation; warehousing as used here implies the holding of finished goods awaiting shipment. While this distinction may not be universally accepted, it is necessary to separate these two concepts of storage activity to distinguish their particular objectives.

Apple (Plant Layout and Materials Handling) states that both the storage and warehousing functions involve (1) the determination of what goods to store, (2) the quantity to store, (3) the amount of space required, (4) control of the overall activity, and (5) a system for coordinating the operations and facilities. Storage and warehousing usually consist of the following operations:

1. Receiving
2. Identification and sorting
3. Dispatching to storage
4. Placing in storage
5. Storage
6. Order picking
7. Order consolidation
8. Packing
9. Loading
10. Shipping
11. Record keeping

This list also provides a basis for considering the possible mechanization or automation of individual functions discussed elsewhere in this section.

General Objectives. In carrying out the above activities, the general objectives should be maximum use of space, effective utilization of labor and equipment, ready access to all items, efficient movement of goods, maximum protection of items, and good housekeeping. These objectives can only be carried out properly with careful planning of the overall storage and warehousing facilities and operations. Details on some of the activities listed above are presented below.

Receiving. This activity is concerned with all operations involved in accepting deliveries. It usually involves screening all documents for information on unusual situations, such as special handling, relocation of existing stock, consideration of first-in, first-out, and the like; analyzing documents for planning purposes; scheduling and controlling receipts and work loads; spotting carriers; and unloading carriers.

Identification and Sorting. This activity is concerned primarily with the determination of what is received and the decision as to where it should be stored. The process usually involves initiating a receiving report, recording receipt of goods in inbound shipment register, physically checking merchandise against documents, checking goods against purchase order, determining quantity received, and separating goods if there is any question of acceptance.

Dispatching to Storage. This activity involves the movement of goods away from the point of receipt and to their desired or required locations. This move is normally into a storage area, although it could be to an inspection or testing location.

Placing in Storage. This is considered here as a separate activity to emphasize its importance. In too many cases, the goods to be stored are brought to the storage area and put down, rather than being immediately placed into storage. The distinction is also made because there is frequently a dividing point between the two in a mechanized or automated storage activity. When practical the dispatching and placing operations should be combined.

Storage. Storage is the actual holding of material and supplies. While in storage the material must be protected against physical damage, loss from theft, and abnormal deterioration. The items in storage must be easily accessible and conveniently located in the receiving and shipping flow patterns. The factors that go into making an efficient storage and warehouse system are discussed below under Storage Planning and Layout.

Order Picking. One of the basic activities performed in the warehouse or storage area is order picking or order selection. For that reason the entire warehouse should be designed primarily to facilitate the withdrawal of items. In some storage situations, the entire facility is one big picking area while in others a separate area is set aside from which "broken case" quantities are picked. Briggs (Warehouse Operations: Planning and Management) suggests the following criteria for use of a separate area:

1. Less than case lots are generally issued.
2. Full case issues are a small volume compared to total.
3. It is impractical to issue small quantities from bulk stock.
4. Counter or customer pickup service is desired.
5. Items are fed to a production line.

Whenever it is difficult to resolve several criteria, a composite solution can frequently be worked out (see Fig. 4). Figure 5 shows a segregated order picking arrangement for shelf stores. Order picking is performed by removing items by hand from picking storage, placing the items on a tray resting on a bench or gravity wheel conveyor along the main conveyor, and placing the tray containing the completed order on the main conveyor for delivery to shipping. Various coding devices may be used to indicate the shipping method. Colored trays or different positions on the main conveyor can denote the various types of shipping, such as parcel post, private trucker, and the like. Stock is replenished by moving items from the reserve storage area to the picking area using hand trucks. Shelving open at both sides is used in the picking area so that first-in, first-out selection is possible.

In Fig. 6 an order picking system for cartons is shown in which roller storage is utilized. Pickers on each side of the conveyor remove cartons from the pallets and place them on the conveyor for delivery to shipping. For large orders, lift trucks can take entire palletized loads to shipping or directly to the carrier. Picking storage is replenished by lift trucks which bring pallet loads from the reserve area to picking area.

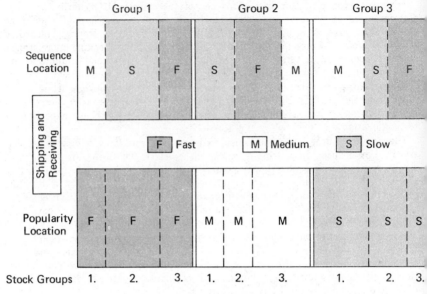

Fig. 4. Composite order picking area.

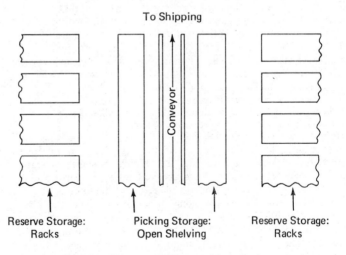

Fig. 5. Order picking of shelf stores.

Order Consolidation. After an order has been picked, it must be assembled with and separated from other orders which may have been picked at the same time. This process is known as order consolidation and involves checking the items against the original customer order. Consolidation can be accomplished by:

1. Conveyor systems traveling from picking to packing areas.
2. Manually loaded carts pushed to a central location for consolidation.
3. Carts pushed to nearest tow-type conveyor.
4. Carts, cartons, or containers mechanically or electronically coded for automatic identification and consolidation.

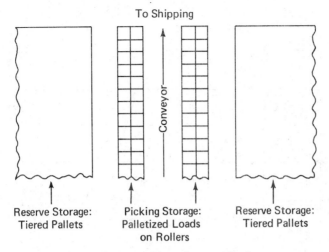

Fig. 6. Order picking of carton loads.

Packing. After the order has been consolidated, it must be packed to provide protection during the shipping process. The requirements, regulations, and practices vary in accordance with the type of carrier, type of commodity, etc. In general, the packing operation involves proper attention to freight classifications, tariffs and regulations; the selection of appropriate containers and packing materials; packing; and the selection and use of appropriate equipment for the necessary packing operations.

Loading. After the goods have been packed, they are usually placed in a marshalling area to await loading. The area involved is frequently divided into sections identified by overhead signs or floor markings according to customers, geographical locations, specific carriers, and the like. It may also contain racks for smaller shipments. When an item is removed from this area it must be checked against the documentation to insure that the customer receives the proper goods.

Shipping. Shipping is not only the last phase of the actual storage or warehousing cycle, but it is the last link in the chain between supplier and customer. In planning for shipping operations, the following factors should be considered:

1. Quantities to be shipped
2. Weight or volume to be shipped
3. Number of shipping points
4. Distances involved
5. Modes of transportation
6. Delivery date promised
7. Documentation

Record Keeping. Paperwork or record keeping activities occur throughout the storage and warehousing phase. This includes the proper accumulation and analysis of the data and information required for efficient operation of the total storage or warehousing function (see sections on Inventory Control, and Materials Management).

STORAGE PLANNING AND LAYOUT. The key to efficient warehouse operations is careful planning and layout of the storage area. Briggs (Warehouse Operations: Planning and Management) cites the following commodity and space factors as being of primary importance:

1. **Commodity factors**
   a. **Similarity**—Goods which are usually ordered, shipped, received and inventoried together should be stored together.

WAREHOUSE PLANNING WORK SHEET   Page 1 of 35   STOCK GROUP ___ 43 ___

| | INVENTORY DATA | | | RETAIL DATA | | | POP. | BULK DATA | | | | | | | |
STOCK NO. (1)	ON HAND (2)	DUE IN (3)	TOTAL (4)	NUMBER RETAIL ISSUES 90 DAYS (5)	TOTAL QUANTITY ISSUE 90 DAYS (6)	SIZE BIN REQUIRED (7)	RELATIVE ACTIVITY (8)	SIZE PALLET (9)	UNITS PER PALLET (10)	HEIGHT OF MATERIAL ON PALLET (11)	NUMBER OF PALLETS REQUIRED (12)	NUMBER OF PALLETS HIGH (13)	RACK SPACE REQUIRED (14)	NUMBER OF BASE PALLETS (15)	TOTAL NET SQ. FT REQUIRED (16)
G43-B-13650	300	1200	1500	90	580	#24	F	40" x 48"	4500	36"	1	1	1	0	0
G43-B-13658	150	1000	1150	75	200	#12	F	40" x 48"	1200	36"	1	1	1	0	0
G43-B-13688	200	1500	1700	60	140	#2	F	40" x 48"	1920	30"	1	1	1	0	0
G43-B-14222	800	0	800	30	80	#1	M	40" x 48"	1500	30"	1	1	1	0	0
G43-B-14248	750	0	750	28	48	#1	M	40" x 48"	800	30"	1	1	1	0	0
G43-B-14368	390	0	390	12	36	#2	M	40" x 48"	240	30"	2	2	0	1	16
G43-B-8009	5000	15000	20000	150	2000	#1	F	40" x 48"	24000	15"	1	1	1	0	
G43-B-8091	250	0	250	3	15	#1	S	40" x 48"	750	30"	1	1	1	0	
G43-B-8093	175	0	175	2	10	#1	S	40" x 48"	500	30"	1	1	1		
G43-B-8200	500	5000	5500	80	350	#1	F	40" x 48"	10000	24"	1	1	1		
G43-B-8213	6000	0	6000	60	325	#1	F	40" x 48"	2500	24"	2	2	1		
G43-B-8215	3000	0	3000	50	320	#1	M	40" x 48"							
G43-B-8302	2500	0	2500	25											
G43-B-8304	5000	0	5000	1											
G43-B-8314	1500	0	1500												
G43-B-8318	5														
G43-B-8324															
G43-B-8332															
G43-B-8336															
G43-B-															
G43-															
G4															

SIMULATED TOTALS → 3600 → 1600 → 1000

Fig. 7. Warehouse planning worksheet.

 b. **Popularity**—Turnover is one of the most important factors for consideration for selecting the storage location. Fast movers should be near the "front"; slow movers in more distant locations.
 c. **Size**—The relative size of individual items is important in determining both storage locations and space requirements.
 d. **Characteristics**—Hazardous commodities, perishable items, high-value items, sensitive materials, etc., require special consideration.
2. **Space factors**
 a. **Size** of space—Volume.
 b. **Nature** of space—Suitability for storing specific items.
 c. **Location**—In relation to other activities
 d. **Availability.**
 e. **Building characteristics.**

Floor load capacity	Column spacing, size,
Doors	number
Loading and unloading	Clear stacking height
facilities	Elevators, ramps, etc.

 f. **Area required for auxiliary functions**

Handling equipment maintenance,	Employee facilities
repair, and storage	Offices
Fueling or battery-charging areas	Fire protection facilities

 g. **Space required for aisles**

Space Analysis. Before space planning can begin it is necessary to accumulate such data as:

Stock quantities	Issue volume
Inventory policies	Storage area type
Replenishment practices	Methods of handling
Issue units	Equipment capabilities

After accumulating the information, Briggs (Warehouse Operations: Planning and Management) suggests the use of a **Warehouse Planning Worksheet** as shown in Fig. 7. Once the worksheet has been completed and basic space requirements have been established Briggs further suggests the use of a **Warehouse Planning Analysis Sheet** (Fig. 8).

In making the space layout, there are other items to be given consideration, including:

1. Commodity size	7. Column spacing (bay size)
2. Pallet size	8. Building size and shape
3. Equipment to be used	9. Desired location of receiving and shipping
4. Aisle width	10. Aisle location
5. Pallet spacing	11. Required service areas, their location and
6. Pallet rack spacing	size

Based on the preceding analysis, an approximation of the total space requirements can be made. Auxiliary service activities must also be identified along with estimates of their space requirements. After a consideration of the balance of the above factors, it will be possible to sketch a proposed floor plan.

Stock Location. Once the layout is developed, space may be assigned to specific stock classes, commodities, and functions. Generally speaking, commodity factors will determine the type of storage space for each item. Other suggestions to aid in selecting stock locations are:

1. Store by commodity factors.
2. Assign areas according to size of lots.
3. Use high areas for goods that can be safely and efficiently stored at maximum height available.
4. Store heavy bulky items on strongest floor and nearest to shipping area.

WAREHOUSE PLANNING—ANALYSIS

STOCK CLASS	CLASS DESCRIPTION	BUILDING	FLOOR	LOAD PER SQ. FT.	STORAGE HEIGHT	Date
43	Bolts, Nuts, Rivets, Screws, Washers	42	2	450	7'-6"	

A. RETAIL BIN REQUIREMENTS (Equipment and Space)

1 SHELVING REQUIREMENTS

			2. SHELVES 36" x 18" x 12"	3. BOXES	
a.	No. 1	3350	+ 10% x 0.333 =	1228	x 3	3684
b.	No. 2	350	+ 10% x 0.666 =	257	x 3	771
c.	No. 3	0	+ 10% =	0	x 3	0
d.	No. 3x	50	+ 10% =	55		
e.	No. 4	0	+ 10% x 1.333 =	0	x 3	0
f.	No. 5	0	+ 10% x 1.666 =	0	x 3	0
g.	No. 6	0	+ 10% x 2. =	0	x 3	0
h.	No. 6x	0	+ 10% x 2. =	0		
i.	No. 12	1700	+ 10% ÷ 12. =	156	x 12	1870
j.	No. 24	800	+ 10% ÷ 24. =	37	x 12	444
k.			TOTALS	1733		

4. Total Line 1733 ÷ 7 or 8 = 217 ....TOTAL
   A. 1. k. Col. 2   18" x 36" x (87" or 99") shelf sections required

5. Total Lines A. 1. a. b. c. e. f. g. Col. 3   4455 ....QUANTITY
   large shelf boxes required

6. Quantity A. 1. i. Col. 3   1870 .... = TOTAL
   one compartment shelf boxes required

7. Quantity A. 1. j. Col. 3   444 .... = TOTAL
   two compartment shelf boxes required

8. Total A. 1. i. and j. Col. 3   193 .... = NUMBER
   intermediate shelves required for small shelf boxes

9. Answer A. 4   217 .... x 10.5 Sq. Ft. (Area occupied
   by one section plus aisle requirements) = 2279 .... Gr. Sq. Ft.
   required for Retail Section

B. BULK REQUIREMENTS (Equipment and Space)

10. RACK REQUIREMENTS

a. Total Col. 14 (W.S.) 1600 ÷ 2 or 1 = 800   Total Racks Required

b. Total Racks 800   x 16 sq. ft. = 12,800   Net Rack Space Required
   Required per Rack

11. PALLET REQUIREMENTS

a. Total Col. 12 (W.S.) 3600   =   Pallets Required

b. Total Col. 15 (W.S.)   =   16,000   Net Sq. Ft. Required
   x 16 Sq. Ft. per Pallet   Pallet Storage

12. SPACE REQUIREMENTS

a. 8. 10. b.   B. 11. b.   (Operating space–ship'g.   Total Gross
   rec'g aisles, etc.)   Required for Bulk

   12,800 + 16,000 + 10,000 = 38,800

b. B. 12. a.   A. 9.

   38,800 + 2279 = 41,079   Total Gross Space Required for Retail and Bulk Operations

NOTES:

11a – 1600 Pallets for racks
     2000 Pallets for bulk
     3600 Total pallets required

11b – 2000 Pallets of 11a for pallet storage
     2000 ÷ 2 (storage height)
     = 1000 base pallets x 16
     = 16,000 (net sq. ft. required for pallet storage)

RECAPITULATION

COLUMN CENTERS		
N/S 20'	E/W 20'	

13. SPACE
Net   28,800
Gross   41,079

14. SHELVING
7 Ft.   0
8 Ft.   217

15. SHELF BOXES
Size
1   4455
12   1870
24   444

16. RACKS
2   800
Level
3   0
Level

17. PALLETS
40" x 48"   3600
40" x 40"   0

18. OTHER

Fig. 8. Warehouse planning analysis sheet.

5. Store light items on limited floor load areas, mezzanines, etc.
6. Locate items as near as possible to similar items already stored.
7. Use remote locations for inactive goods or for small, light, easy-to-handle items.
8. Store slow-movers farther from receiving and shipping and at higher levels.
9. Place fast-movers near shipping and in lower locations.
10. Locate service areas in low-ceiling areas.
11. Use outdoor space for selected items.

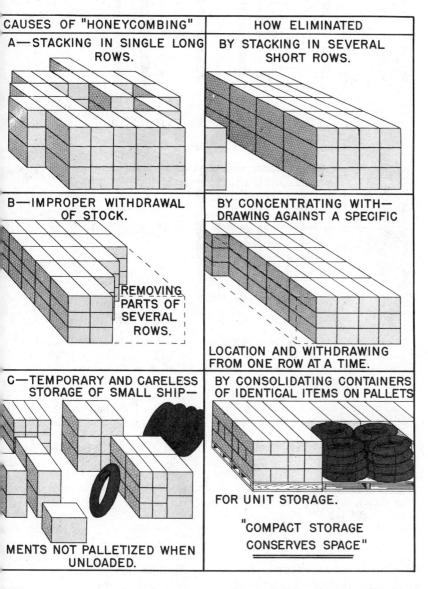

Warehouse Operations Handbook

Fig. 9. Honeycombing: its cause and cure.

Figures 9 and 10 indicate some of the common problems encountered in stock location and selection and indicate suggestions for more efficient operation.

STORAGE BY SEQUENCE

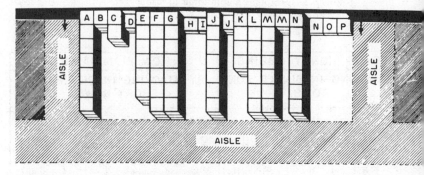

STORAGE BY SPACE

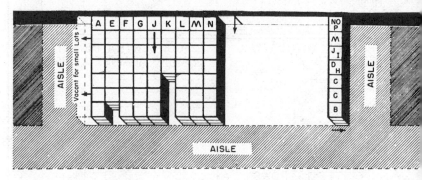

Fig. 10. Typical storage arrangements.

Stock Location Systems. One method of simplifying stock location and order picking is to assign each space in the warehouse or storage area an **identification or code number.** Briggs (Warehouse Operations: Planning and Management) suggests a system that would be useful for large multi-building organizations though it could be modified for use in any size operation. Basically the system requires that each building, floor, row, stack, and level in each storage area be given a number. A given space can then be identified by a code number that lists, in order, each of these numbers. The resulting number may be five or more digits long. For convenience in reading the number, the digits may be segregated into groups. Figure 11 shows a cutaway view of a storage area with the various numbers specified. The shaded space would carry the identification number 324-112-123 which would be translated as Building 32, 4th floor, Row 112, Stack 12, 3rd level. In systems of this type confusion is avoided by maintaining the same number of digits for all code numbers. Thus if the above space were in Building 7, and Row 9, the complete number would read 074-009-123, with zeros making up the required nine-digit code.

Storage and Warehousing Equipment. The above discussion has implied the use of various kinds of equipment in the storage and warehousing function.

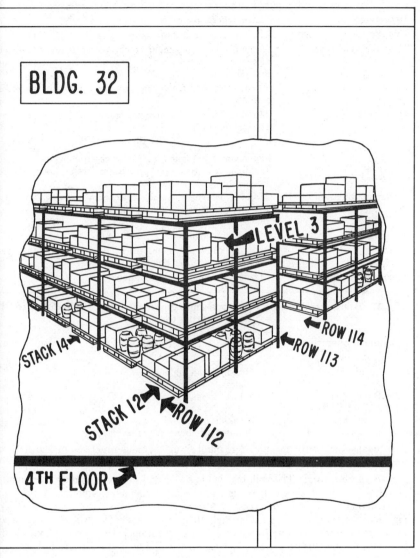

Fig. 11. An example of a coded storage location.

In general this equipment can be divided into three categories: storage, handling, and picking.

In many cases it is difficult to place a piece of equipment in a particular classification. Figure 12 lists the most common equipment used in storage and warehousing operations. For specific information on the different types listed see the discussion of material handling equipment elsewhere in this section.

Storage Structures. The storage structure is primarily a means of protecting goods from the elements. However, it must be designed for efficient and

Receiving	Carts, trucks, conveyors
Identification and sorting	Manual, mechanical, electrical
Dispatch to storage	Manual, conveyor, hand truck, power truck, lift truck, tow-line, tractor–trailer train, crane, hoist, stacker crane, special device
Storage	Floor, shelf, bin, conveyor, rack (fixed flow), pallet—floor and rack (fixed flow)
Order picking	Manual, lift truck, conveyor, crane, hoist, stacker crane, storage machinery

To Shipping

	Conveyor, hand truck, power truck, tractor–trailer train, tow-line, lift truck, pickup station (crane, hoist, stacker crane, storage machinery)
Order consolidation	Manual, conveyor, hand truck, power truck, tow-line, tractor–trailer train, crane, hoist, stacker crane, storage machinery
Packing	Manual, mechanized
Loading	Manual, conveyor, hand truck, power truck, lift truck, crane, hoist, stacker crane, special device
Record keeping	Manual, mechanical, automated

Fig. 12. Basic warehousing functions and equipment types commonly used in implementing them.

economical operation, and planned with proper consideration given to such items as:

Location	Handling methods	Doors
Plot size	Aisles	Docks
Building size	Service areas	Construction methods
Layout	Column spacing	Lighting
Flexibility	Stacking height	Ramps
Expansion	Number of floors	Elevators

In connection with the above, a number of common practices in the field of warehouse building design should be reviewed.

1. Site planning. The building should be located so as to minimize the amount of fill required. It should be oriented so that a minimum amount of roadway construction is required for access to streets and highways. Railroad and trucking facilities should be situated in such a way as to allow for future expansion of these facilities without disruption of existing operations. In general, plans should be made for the future growth and expansion of the facility in all aspects.

2. Building planning. Square buildings are usually more economical than other shapes in both space utilization and construction costs. Square bays (and pallets) however are less flexible than some other shapes. As much as practical use the building height rather than the horizontal building area to increase the use of available space. Up to 30 feet clear stacking height is common where goods are stored for a relatively long period (the clear height should allow for maximum extension of lift truck fork carriages or masts). Adequate floor loadings should be designed especially where very heavy loads are to be stored (loaded electric trucks will weigh up to four times their capacity; gasoline trucks up to twice their load capacity). The floor (6-in. concrete) should be poured on compact fill. The roof should be supported wherever feasible by building columns, not walls. The outer walls of the original construction should be designed so that they can be used as fire walls after the building is expanded. If panel construction is used for the walls, the same panels can be reused for the expansion.

3. Dock area planning. Allow 15 ft. of dock space and 65 ft. of apron space for truck widths and lengths. Dock heights of 48 in. (with mechanical leveling dock board) to 52 in. (without mechanical dock board) are required for high-way trucks. Pick-up and city delivery trucks require lower dock heights. Canopies should be provided above the dock areas so that operations can continue despite inclement weather. In some cases, the nature of the product and the normal climatic conditions may require the use of total enclosure dock seals. Dock bumpers and wheel chocks should be provided at all truck docking spaces.

4. Facility planning. All obstructions to a clear floor and ceiling area should be kept to a minimum. Where permitted, down spouts, conduits, water lines, and the like should be run within the flanges of the building steel (I-beams). Horizontal runs should be run through bar joists. Lighting fixtures should be mounted between joists with as high a clearance as possible. When necessary, overhead obstructions (unit heaters, for example) should be located in main traffic aisles rather than in storage areas. Sprinkler system heads should be given adequate clearance to permit full pallet stacking while providing proper sprinkler coverage. Panel or rolling doors with vertical interior guide rails are preferred to outside opening doors. Minimum door dimensions should be 9 ft. wide and 8 ft. high but some provision must be made if tall or very wide equipment must be removed for servicing. Flexible interior doors are required to facilitate movement of lift trucks. Minimum dimensions for these doors are 8 ft. high and 7 ft. wide. Maximum utilization should be made of all marginal areas such as mezzanines and balconies.

Organization for Material Handling

NEED FOR ORGANIZATION. The scope of handling covers such a wide range of activity that it is necessary for the overall function to be carefully organized. This will insure proper coordination and control of the handling efforts. It will also be remembered that a successful handling system is built-in to the overall production activity, and the integration of two such important functions can hardly be accomplished without proper guidance. In general this is true, regardless of the size of the organization.

MATERIAL HANDLING ORGANIZATION. The material handling activity may be organized as a line or staff function, a specialist group, a committee, or a combination of any of these.

If material handling activities are to be delegated to line supervision, then material handling planning as well as operations will be coordinated by foremen, department heads, supervisors, and the like. In many cases material handling planning and engineering activities are assigned to a separate staff department, or it may be a part of either the industrial engineering or plant engineering function. The person or persons planning the handling activities will then serve as staff assistants to the line organization.

Another alternative is for the planning and engineering activities to be performed by staff specialists, operating out of the company central office or headquarters, and serving the individual plants when problems arise, or when their services are requested.

Still another alternative would be to provide for a plant or corporation-wide committee on material handling, which would oversee all handling activities, consider problems as they arose, and then request assignment of specific task through the regular line or staff functions.

It is also possible that the handling function could be carried out by a combination of the above methods. For example, a central office staff might serve in a consulting capacity, with a **division staff** serving as an advisory link between the central office and the individual plants. The actual material handling activities would be under the jurisdiction of a material handling department serving the line organization directly.

Place in the Corporate Structure. The relationship of material handling to the overall structure of the company is dependent upon the basic objectives of the company. For example, warehouse operations of a mail order firm would probably require considerably more attention to material handling than warehouse operations of a typical manufacturer.

Similarly the nature of the products involved in the business affect the place of the material handling function in the company. Manufacturers of **high-unit-value** products such as machine tools, automobiles, watches, jewelry, and aircraft usually relegate the material handling function to a staff or secondary line executive in the organization. In the manufacture of **low-unit-value** products, such as lumber, ore, coal, and food, the material handling function is more often a senior line operation or a function of top management with the support of top-level staff aid. The high-unit-value products normally require that a much lower percentage of their total production costs be devoted to material handling whereas in the low-unit-value products a much higher percentage of the total production cost must be expended on handling. Fragility and perishability of the product also require that a higher percentage of the total production cost be devoted to material handling.

Apple (Lesson Guide Outline on Material Handling Education) presents an overall organization chart which shows a dozen possible locations for the material handling function within the corporate structure (Fig. 13).

An outside consultant is frequently valuable in either organizing the handling function or analyzing handling activity. The broad experience which he can bring to bear can be extremely helpful in a plant where material handling experience is limited.

EQUIPMENT POOLS. Some plants make use of a pool-type material handling organization in order to get maximum utility from automotive equipment, fork trucks, railroad, and tractor-trailer equipment. This **pool-type operation** can be based upon a variety of management techniques such as:

1. Routing and scheduling of vehicular equipment on a predetermined network of trips to create an itinerant transportation system. In this type of operation the production or service departments schedule their work to coordinate with the pickup or delivery schedule of the itinerant vehicular equipment.
2. Assigned work-center-type use of vehicular equipment involves the assignment of fork trucks and other equipment to specific operating departments. In this case, the daily operation of the equipment comes under the jurisdiction of the department being serviced while training, maintenance, and technical activities are supervised by the head of the transportation pool. This type of operation permits a better distribution of equipment, which can be shared between adjacent departments and permits the development of uniform techniques and trained personnel.
3. The dispatching-type vehicular system uses call boxes, two-way radio, or depot dispatching as a control technique. In this case, vehicular equipment is assigned on a job basis and controlled on a plant-wide level.

Many large plants find that the pool technique alone is not applicable to their operation. It is not uncommon to find some departments operating on an **assignment basis,** with several scheduled tractor-trailer or truck routes running through the plant, and a central dispatching operation controlling a flexible equipment pool.

MULTIPLANT OPERATIONS. Multiplant operations bring to the material handling function some additional problems not usually found in the intra-plant handling system. The selection of the carrier for moving material between plants is directly related to the distance to be traversed, climatic conditions, and the nature of the product.

It can be seen that an inter-plant shipment on a nationwide basis bears very little resemblance to an intra-plant handling system. It is similar to and a part of the traffic operations of the organization involved. On the other hand, multiple-plant operations within the confines of a city or a company-owned compound can be handled on the same basis as intra-plant handling as long as traffic regulations and climatic conditions permit.

Intra-plant operations can often be controlled by two-way radio. For inter-plant operations within the range of medium-power radio systems, a radio-dispatched operation is usually highly desirable. The use of radio in long-range operations is not normally economical.

Inter-plant handling often involves the storage of material en route, either in company warehouses or depots or, in some cases, in public warehouses. The same considerations with regard to inventory control, storage techniques, and the storage bank apply here as in the intra-plant operation.

In large corporations where decentralized management of the various plants has been accomplished and where inter-plant movement is a factor, it is advantageous to have central coordination of these inter-plant movements. Each plant manager must of necessity be responsible for the proper scheduling of shipments and their safe handling.

In situations where a main plant operates as a center for several feeder plants whose schedules are controlled with relationship to the "main" plant, a centralized management of material handling and inventory control can have a very favorable effect upon the overall scheduling and operation of the system. The decentralization of these functions can result in scheduling malfunctions and confusion if the teamwork between the various plant managers is not perfect.

TRAINING IN MATERIAL HANDLING. Meaningful attention to handling problems and costs depends upon knowledge and information on the subject of material handling. Material handling training can be helpful in this regard at all levels in the organization. For instance, **top management** personnel might be given an "appreciation" course in material handling to increase their awareness of the extent, significance, and cost of the activity. For the same reasons, **middle management** should be provided with an opportunity to learn enough about handling so that they will understand the functioning of the material handling staff. A third group requiring training in material handling is the **operating personnel** themselves. This would include job-related training for those who are actually performing handling operations, such as lift truck operators, equipment operators, material handlers, and the like. The educational program would also include the material handling **planning and engineering staffs,** primarily to keep them up to date with new equipment and analytical approaches. Finally the **production personnel** should know enough about handling, to better appreciate the importance and cost of handling activities.

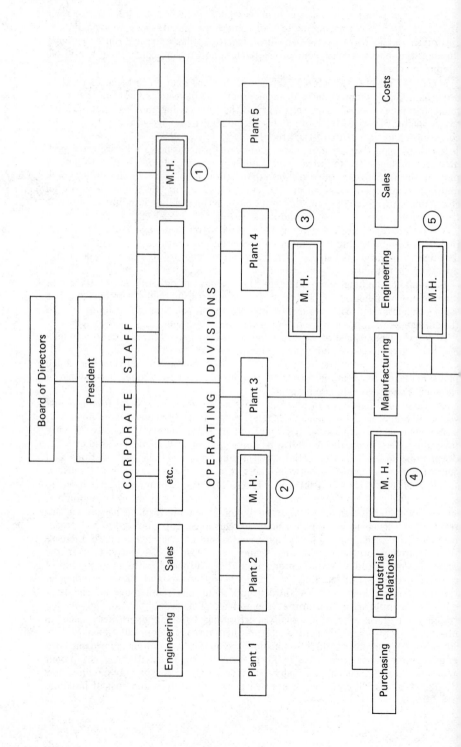

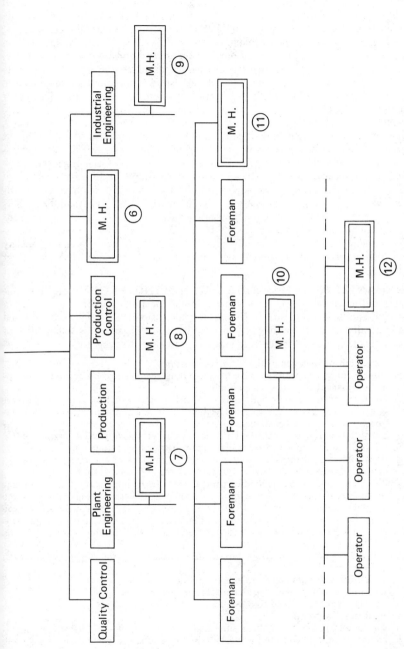

Fig. 13. Possible locations of the material handling function in the corporate structure.

Responsibility for Training and Maintenance. The maintenance of material handling equipment and the training of personnel in the operation and maintenance of this machinery are found in various places in the organization. The maintenance of fixed equipment, such as conveyors, feeders, hoppers, chutes, piping, and the sundry other plant facilities utilized for handling material in process and storage, is usually a part of the **plant engineer's** responsibility. In most cases he trains the maintenance personnel, plans and operates any preventive maintenance programs, and initiates the establishment of maintenance cost control and analysis records (see section on Plant Maintenance). This latter function is sometimes shared by the staff **material handling engineer.**

Automotive and mobile equipment, such as fork lift trucks, tractors, trailers, cranes, hand trucks, and handling tools, is sometimes maintained by the plant engineer, but more often, when a pool-type operation exists, automotive equipment maintenance becomes a part of the responsibility of the **material handling department.** An example of this can be found in the maintenance of delivery truck fleets under the supervision of the **delivery department manager,** who in turn reports to the executive in charge of material handling. This arrangement is particularly common where the top-level material handling executive mentioned above is in control of both the operating and planning of material handling functions. The advantage of this relationship stems primarily from the enforcement of preventive maintenance techniques through driver training.

ORGANIZING A MATERIAL HANDLING PROGRAM. The following outline is suggested as a general guide for initiating and developing a material handling program. It is not intended that this procedure be used automatically or in a routine fashion, rather it should serve as a checklist of those items that constitute the foundations for an effective program.

1. **Establish objectives** of the handling program.
2. **Check company policies** related to handling, such as inventory, wages, equipment replacement, project payoff, etc.
3. **Outline scope of program** so as to identify the breadth of functional coverage, the proposed organizational structure, and the organizational interrelationships with other functions.
4. **Establish a list of activities** required to meet the objectives stated above.
5. **Divide the activities into logical groupings** or functional assignments.
6. **Establish a preliminary budget** based on estimates of the manpower, equipment, and other costs necessary to carry out the proposed activities.
7. **Select a number of desirable projects** for early implementation. It might be suggested that the earlier projects should deal with obvious trouble spots in which the implementation can be accomplished in a relatively short time and with a fairly positive assurance of success. In this way, the problem solution will stand out and serve as an "advertisement" of the potential of the handling program.
8. **Estimate possible savings potential** or other goals or objectives for each project.
9. **Establish a sequence** of projects for action.
10. **Determine manpower requirements** for each project.
11. **Set up a project schedule.**
12. **Assign personnel.**
13. **Carry out projects,** making sure that a procedure is established for implementing each, establishing a mechanism for evaluating progress, and planning for the periodic evaluation of each project.
14. **Document and report all project savings** as a basis for periodic reports.

Interorganizational Relationships

INDUSTRIAL ENGINEERING. Material handling activities are frequently considered as a function of industrial engineering. Other production
functions often under the jurisdiction of industrial engineering and intimately
related to material handling in their day-to-day activities are process engineering, methods planning, work measurement and standards, and plant layout.

Process Engineering. This function involves the **basic design of the
manufacturing process** which in turn establishes the number of machines,
the operation sequence, and the degree of mechanization and automation necessary to produce the product. Line balancing and in-process handling and storage
operations are affected by the decisions made at this point.

The process engineer should consider the following suggestions designed to
facilitate handling operations:

1. Use **mechanized handling** for
   a. Large quantities or volumes of material.
   b. Frequent, repetitive moves, even though short.
   c. Long moves.
   d. High-effort, hazardous, difficult moves.
   e. Replacing excessive manual handling.
   f. Replacing excessive handling labor.
   g. Feeding and removing material from machines.
   h. Reducing handling time.
   i. Scrap removal.
   j. Providing a continuous, uniform, maximum controlled rate of movement.
   k. Moving material while it is being processed or inspected.
2. Do not overmechanize.
3. Design or select containers suitable for mechanical handling.
4. Use automated equipment when practicable.
5. Consider mechanizing the movement of personnel.
6. Use mechanization for the movement of equipment, tools, jigs, fixtures, and
   the like.
7. Use modern communications techniques to facilitate material movement.
8. Use automatic couplings, switches, transfers, etc.

Methods Planning. This function is concerned with the design of individual
workplaces, the motion and methods used in performing the operations, and the
interrelationships between the individual workplaces and the overall material
flow. The methods engineer can facilitate handling operations by considering the
following suggestions:

1. Avoid placing material directly on the floor; use pallets or other supports.
2. Provide adequate storage space at the workplace for the proper amount of
   material, both ahead of, and following the operation.
3. Use the same container throughout the system; avoid frequent changes.
4. Provide adequate clearances in and around each workplace for maneuvering
   handling equipment and material.
5. Place product on packing base (pallet, skid, etc.) as early in the production
   process as practical.
6. Properly locate the material supply and disposal areas.
7. Plan an effective scrap removal procedure.
8. Plan for productive operations and inspections to be done during transportation.
9. Keep intermediate handling and walking to a minimum.

Work Measurement and Standards. Work standards have been applied
to production operations for many years. Their use in measuring and evaluating

material handling methods is less common but nevertheless valid. The material handling staff should cooperate with the work standards personnel in an attempt to establish work standards for material handling operations. In many cases such standards can be used as the basis for an incentive plan for compensating material handling personnel.

Plant Layout. This function is generally responsible for developing the physical arrangement of the facilities in the plant. Since the layout is frequently built around the material flow pattern, it is extremely important that layout and material handling personnel work closely on the plan. For additional details on the plant layout function, see the section on Plant Layout and Facilities Planning.

The plant layout engineer can facilitate material handling by considering the following suggestions:

1. Avoid crowded conditions.
2. Eliminate obstacles from materials flow.
3. Carefully observe building and carrier restrictions.
4. Plan movement in a direct path (avoid back-tracking, zig-zag flow, crooked paths).
5. Arrange for alternate paths, in case of difficulty.
6. Be aware of cross-traffic and take necessary precautions; avoid traffic jams.
7. Keep related work areas close together.
8. Use product-type layout when possible.
9. Plan proper location of sub-assembly and feeder lines.
10. Combine operations to eliminate intermediate handling.
11. Plan for definite pick-up and delivery areas.
12. Minimize moves between floors and buildings.
13. Process heavy and bulky materials close to receiving.
14. Move the greatest bulk and/or weight the least distance.

PRODUCTION GROUPS. This general grouping includes all functions directly related to producing the product or performing the service of the company. Specific functions include production control, inventory control, material management, quality control, and purchasing. Good communications with these functions is essential to good material handling. The Material Handling Institute (Booklet No. 3) states that

The materials handling activity must know production and shipping schedule plans far enough in advance to schedule its labor and equipment to provide the service required. If such changes in plans represent greater than normal fluctuations, changes in handling methods and facilities may be necessary, or desirable.

Changes in normal inventories of materials should be made known to the materials handling function to plan changes in storage. The materials handling activity should be advised of changes in scheduled receipt of materials. . . .

Materials handling must be in a position to advise production and materials control of its ability to provide service required by a given schedule. That requires knowledge and data of the potential capacity of all segments of the materials handling system. Also, it requires an understanding of the degree of flexibility in the system—changes that can be made and how long it takes to make the changes.

The materials handling function should make known the cost effect of various schedules. It should suggest changes that may effect economies in handling costs.

Purchasing and Material Management. A large portion of in-plant material handling is attributable to the receipt and maintenance of raw materials and supplies. At the very beginning of the production process an opportunity exists for the purchasing function to help reduce handling costs. No sizable quantity of material should be ordered without checking with material handling on the best method or form for receipt of the material. Suppliers and carriers should be involved in developing efficient techniques for shipping, unloading, and otherwise handling purchased material.

In many cases, purchase price considerations appear to have a major effect on quantity, often to the exclusion of the effect of quantity on handling costs. Since the scheduling activity established by production control normally determines **withdrawal quantities** it also affects order characteristics.

If the raw material and supply problem is studied in its entirety with full consideration to all relationships involved, it may be found that the controlling cost factor is the material handling expense, which may have been given the least amount of attention.

The control of material in storage is traditionally governed by accounting requirements, as this relationship is critical in raw material inventory. The result is that accounting procedures frequently influence handling practices in storage operations. In such cases, it may be found desirable to remember that material costs more to handle than does paper work and therefore adjusting the accounting procedure to permit the most economical material handling activity, may be worth considering.

It is precisely the situation implied here that fostered the development of the **materials management** concept which combines a number of previously independent, but intimately related, functions within the industrial organization (see section on Materials Management).

Production Control. The requirements of production control can frequently result in the addition of unnecessary handling operations to the manufacturing sequence, as well as to in-process storage. Whenever two operations producing parts at unequal speeds follow one another, a **balancing operation** must take place. In some cases this can be accomplished by multiple machines on one or the other of the operations, but more often it is necessary to create a production bank or float to level off the inequalities in production rates.

Production banks are also necessary for the storage of materials and semi-finished parts which must be scheduled into production at a specific place in the sequence. The storage of standard parts for multiple products and the storage of semi-finished parts which can be finished into various products can also be classified as a form of production bank.

The handling of material from one production operation to another or through a production bank to a production or inspection operation with an unequal rate can be costly and unnecessary. Analysis of scheduling often uncovers the cause of these apparently unnecessary handling operations.

Some of the specific ways in which material handling can cooperate with production control are:

1. Provide direct, mechanical paths for material movement.
2. Move material in lots, batches, or containers of a predetermined quantity or size.
3. Store or pack material in containers holding a specific number of pieces.
4. Use containers that facilitate identification of their contents.
5. Utilize two-way radio or closed circuit television to expedite movement and control of material.
6. Make optimum use of mechanical handling in order-picking, accumulation, loading.
7. Pace production with mechanical handling equipment.
8. Build production, inventory, and accounting control features into the material handling system.
9. Move material on a schedule and in lots to match production and to avoid rush delivery, partial loads, or duplicate moves.

10. Use move tags or orders as authority for all moves—avoid verbal orders.

11. Coordinate handling schedule with purchasing, manufacturing.

Conveyorized handling systems have a favorable effect on minimizing the complexity of production scheduling and control methods. This is due to the fact that no paper work is required when the product is moving between operations and travelling over a fixed path at a uniform rate of speed. In contrast however, a truck or trailer system is unrestricted in its routing, handles individual units, and as a result, loads may become misdirected.

Conveyor storage of material in process is often adequate as a production bank between operations of unequal rate. If sufficient material can be stored in transit on conveyors, the difference in the rate of the sequential operations can be cushioned by the material on the conveyor. The unit load system, although far more flexible, usually requires the addition of stacking and unstacking operations to permit its use in a production bank situation.

Quality Control. The care with which a product or material is handled in the plant plays an important role in its quality. The characteristics of the product, such as size, weight, finish, etc., will determine just how much care is needed. Quality is affected during material handling principally by physical damage due to the handling equipment or the surroundings while the product is in transit. As part of its responsibility, quality control must investigate any abnormal increase in rejects due to defects resulting from material handling. Coordination with product engineering and material handling can result in a change in product characteristics or handling methods. The solution may involve the design of special handling equipment to eliminate a particular problem such as vibration, abnormal stresses, or the like.

SALES AND DISTRIBUTION. The nature of the sale can frequently affect handling practices within the plant. Wholesale houses and mail-order organizations are good examples of shipping operations which require the picking of diverse orders from **multiple-item picking lines.** This type of operation usually results in nonstandard shipping practices which limits the use of unitization and can make conveyor handling difficult. Most of the organizations which have this type of shipping problem unitize their product at the end of the production line, store in bulk, and advance unit loads into the picking line for full-package picking (see Order Picking elsewhere in this section).

In addition to the nature of the sale, **rate structures** often affect material handling practices. Unitization of low-unit-value products for rail or truck shipment frequently results in a prohibitive dunnage and return charge if pallets are used. The shipment of unitized loads of unassembled products for assembly by the purchaser or distributor can, on the other hand, result in a favorable effect on freight rates and frequently overcomes the cost of dunnage and unitizing as well as returning a handling-cost saving.

The **destination** of the product must also be taken into consideration when determining handling methods in the warehouse and shipping department. Merchandise destined for overseas shipment must often be crated and can be handled by different methods from those used for the usual carton-packed domestic shipments. Crate handling attachments for lift trucks eliminate the need for skids and pallets in handling and storing many types of crates. Merchandise which is being shipped short distances in intra-city trucks can often be unit-loaded without suffering serious dunnage or pallet return costs. A distant destination would limit this procedure.

Effects of Customer Requirements. The customer frequently dictates delivery schedules and loading methods. In many industries, notably the automobile industry, vendors deliver unitized or specially packed products to the customer's plant on a precisely predetermined schedule. The products are shipped in large, reusable compartmented cartons mounted on pallets or skids, which permit fork lift truck handling, while "drop" sides permit easy production line unpacking.

Some plants receive products directly into conveyor systems while some unitize the product in the carrier or at the tail gate. In most cases, a packaged product can be shipped either in bulk or unitized to a firm which unloads by conveyor, without making any inroads on their receiving cost. However, when a customer receives material into a unit-load system, it is often desirable to ship to him in unit loads which are suitable to his operations. Customers sometimes require this type of shipment. As a result, many suppliers are forced to use a multitude of unit-load sizes in their warehousing and shipping operations, in order to satisfy the customer's desire for unit unloading without double-handling the product from their own pallet to that of the customer.

SAFETY. This activity is important in every production function, but possibly more so in the material handling function, since two-thirds of the general causes of industrial accidents, as listed by the safety engineers, are directly related to material handling. For this reason the material handling engineer will probably find himself working closely with the safety engineer in an attempt to design safety into both material handling methods and equipment.

These two functions can cooperate in assuring safe handling operations by consideration of the following:

1. Install adequate guards and safety devices on handling equipment.
2. Keep handling equipment in good operating condition.
3. Furnish mechanical handling equipment for difficult, hazardous handling activities.
4. Do not permit handling equipment or devices to be overloaded or operated beyond rated capacity.
5. Examine two-man lifting jobs for possible use of mechanical equipment.
6. Use mirrors at busy aisle intersections to permit seeing around the corners.
7. Keep aisles clear and uncluttered.
8. Install adequate lighting.
9. Maintain floors in good condition.
10. Avoid crowded conditions.
11. Provide good housekeeping.
12. Stack materials carefully.
13. Be sure operators are properly instructed in method and use of equipment.
14. Provide mechanized part feeding and removal devices.
15. Use remote emergency switches and controls.
16. Plan for removal of undesirable dust, fumes, smoke, etc.
17. Isolate inherently dangerous equipment and operations.
18. Allow for liberal factors of safety.
19. Use bright colors or moving lights to highlight handling hazards or danger areas.
20. Install "dead-man" switches on equipment requiring operators.

Selection of Material Handling Equipment

SYSTEMS AND EQUIPMENT. One of the major problems facing the handling engineer is the vast quantity and variety of available handling equipment. There are over four hundred types of conveyors, trucks, vehicles, cranes, hoists, containers, racks, and items of auxiliary equipment. In fact, it is unlikely that the material handling engineer will have sufficient in-depth knowledge of these equipment types to make more than a limited choice.

METHOD	10.		MANPOWER REQUIRED				
	9.	Equipment	EQUIPMENT INDICATED				
			MANUAL				
			NONE				
	8.	Load Handled	DISPOSAL				
			HOW CARRIED				
			LOADS / TOTAL QUANTITY				
			WEIGHT OF LOAD				
			ITEMS/LOAD (container)				
			TARE				
			CONSTRUCTION				
			SIZE				
			TYPE				
MOVE	7.		RATE				
			SPEED				
	6.		FREQUENCY				
	5.	Route	CROSS TRAFFIC				
			OPERATIONS IN TRANSIT				
			LOCATION				
			LEVEL				
			DIRECTION / PLANE				
			COURSE				
			PATH				
			AREA COVERED				
			DISTANCE				
	4.	Source & Destination	DESTINATION				
			SOURCE				
	3.		SCOPE				
MATERIAL	2.	Quantity	MAXIMUM INVENTORY				
			QUANTITY/DELIVERY				
			ANNUAL QUANTITY				
	1.	Physical Characteristics	WEIGHT / ITEM				
			DIMENSIONS				
			HOW RECEIVED				
			ITEM DESCRIPTION (part no., form, type, properties, etc)				
			MOVE NUMBER (from PROCESS CHART)				

Fig. 14. Material handling analysis recap form.

17·32

The **selection process** requires that the characteristics of the material and the specifications of the move be properly matched with the equipment capabilities. Ideally the handling method or equipment should be selected by function. That is, equipment should be selected to best implement each move in the process with the end solution an integration of several equipment types. In practice other factors enter into the selection decision. For example, equipment might be selected from a maintenance standpoint, suggesting the desirability of standardization about one type of method, equipment, or system. Similarly in a particular flow pattern, a conveyor might be best for some moves, a crane for others, and a truck for still others but the small volume of handling involved may not justify investment in so diversified a group of equipment. In that case the engineer may be forced to choose the best or most versatile method to do a wide range of diversified jobs.

Objectives. The principal objective of the equipment or method selected is to optimize the flow of material in the plant or warehouse. It should do so in the most economical manner possible and with a minimum amount of complexity in terms of equipment or technique. Wherever possible the handling operation should be combined with other operations such as production, storage, inspection, packing and the like. If the equipment or system is being introduced in an existing system it should be compatible with that system. Versatility is important where product lines and models are changed frequently or where methods of manufacture are often being revised. To prevent or work around bottlenecks the system should provide for alternate handling methods. Overdesign can represent as great a loss as underdesign in that both are costly in terms of investment and space requirements. The handling system should be compatible with the limitations of the facility, with consideration given to future expansion. Ideally the equipment chosen should require little maintenance, consume as little power or fuel as possible, and have a long, productive life. In most cases the final choice will represent a number of compromises with the above objectives.

Factors Influencing Selection. In analyzing a material handling problem the engineer must consider the many parameters connected with the material, move, and method. The principal **material parameters** connected with **bulk moves** are size, density, chemical behavior, viscosity, flowability, friability, temperature, angle of repose, abrasiveness, and moisture content. With **unit loads** these parameters include size, weight, crush resistance, shape, load supports, and temperature.

The **move parameters** are distance, rate per hour, frequency, loading and unloading levels, slope, aisle and path dimensions, range and area, cross traffic, sequence, running surface, plane, location, speed, and origin and destination.

The **method and equipment parameters** involve cost of floor space, headroom, congestion, and for a lift truck, such items as capacity, load center, maximum lift, lift speed, and the like.

EQUIPMENT SELECTION PROCEDURE. Assemble all factors pertinent to the problem. The material handling equation can serve as a basis for accumulating necessary data. The Material Handling Analysis Recap Form shown in Fig. 14 will be helpful in recording and correlating this information. Each line represents a move step in the process. Filling out the form will force the analyst to obtain information that might otherwise have been overlooked. As it is filled out, alternative material handling methods or equipment types may suggest themselves to the analyst and should be indicated in columns 9 and 10.

TASK and EQUIPMENT CHARACTERISTICS	EQUIPMENT TYPE → CONVEYORS Moving uniform loads continuously from point to point over fixed paths where primary function is transporting	CRANES and HOISTS Moving varying loads intermittently to any point within a fixed area	INDUSTRIAL TRUCKS Moving mixed or uniform loads intermittently over various paths with suitable surfaces where primary function is maneuvering
MATERIAL — Volume	high	low, medium	low, medium, relatively high
Type	individual item, unit load, bulk	indiv. item, unit load, variety	indiv. item, unit load, variety
Shape	regular, uniform, irregular	irregular	regular, uniform
Size	uniform	mixed, variable	mixed, or uniform
Weight	low, medium, heavy, uniform	heavy	medium, heavy
MOVE — Distance	any, relatively unlimited	moderate, within area	moderate, 250–300 ft.
Rate, Speed	uniform, variable	variable, irregular	variable
Frequency	continuous	intermittent, irregular	intermittent
Origin, Destination	fixed	variable	may vary
Area covered	point to point	confined to area within rails	variable
Sequence	fixed	may vary	may vary
Path	mechanical, fixed pt. to fixed pt.	may vary	may vary
Route	fixed, area to area	variable, no path	variable, but over defined path
Location	indoors, outdoors	indoors, outdoors	indoors, outdoors
Cross traffic	problems in by-passing	can by-pass, no effect	can by-pass, maneuver, no effect
Primary function	transport, process/store in move	lift & carry, position	stack, maneuver, carry, load, unload
% Transport in operation	should be high	should be low	should be low
METHOD — Load support method	none, or in containers	suspension; pallet, skid, none	from beneath; pallet, skid, container
Load/unload characteristics	automatic, manual, designated points	manual, self, any point	self; any point on available path
Oper. accompany load	no	may or may not, usually does	usually does; may be remote
BUILDING CHARACT. — Cost of floor space	low, medium	high	medium, high
Clear height	if enough, conv. can go overhead	high	low, medium, high
Floor load capacity	depends on type conv. & mat'l.	depends on activity	medium, high
Running surfaces	not applicable	not applicable	must be suitable
Aisles	not applicable	not applicable	must be suitable
Congested areas	fair	good	poor

Fig. 15. General characteristics of ...

Tentatively select equipment type. The form (Fig. 14) will focus attention on the **problem characteristics** so that the analyst can match them with the general capabilities of the major handling equipment types (see Fig. 15). In using the form, the column containing the largest number of appropriate characteristics will indicate the **basic equipment type** for further consideration.

Narrow the choice. It is now necessary to further investigate the specific equipment types within the category previously selected. This will require a review of the analyst's knowledge of equipment and/or discussion with equipment salesmen. It cannot be overemphasized that the proper selection of handling equipment depends largely on the experience of the material handling engineer and/or his ability to obtain accurate information from appropriate equipment representatives.

Evaluate alternatives. After the alternatives have been narrowed down to a relatively few pieces of equipment, it is necessary to determine which one appears to be the best solution to the problem. The evaluation process requires the determination of the cost of each alternative, and then a comparison of both the cost and the intangible aspects of each alternative (see section on Capital Investment Analysis). Consideration must be given to the **cost factors** shown in Fig. 16.

Besides cost factors, there are a number of **intangible factors** that in the ordinary sense defy quantification or calculation of a dollar value and cannot be included as items in a cost comparison, although in many cases they may outweigh the cost factors. These are also shown in Fig. 16.

The cost factors may be evaluated in terms of common practice in engineering economy, with proper consideration given to company policies, procedures, etc. Intangible factors can be ranked, weighted, evaluated, and taken into consideration as appropriate.

Check selection for compatibility. Attention should be given to the compatibility of the equipment being considered with the equipment already in use or also being considered. Each segment of the system should be carefully fitted into the whole. In general, the best material handling system is the one that smoothly ties together all interrelated handling situations.

Select specific piece of equipment. The **final selection** must be based on the facts available to the analyst, the equipment types and categories applicable to the problem, and the experience of the analyst, along with the advice of specialists in matching problem requirements with equipment capabilities.

Procure equipment. Detailed specifications should be written carefully and in sufficient detail to assure comparable bids. Sketches and design specifications should be prepared, if appropriate, but not so rigid as to preclude the possibility of several suppliers working together on components of a complex system. Requests for quotations should be issued, accompanied by the appropriate specifications, drawings, etc. The engineer should carefully evaluate the quotations, and if appropriate, make comparison of the specifications and details of the several proposals. After a careful analysis of the information contained in the quotations, either a vendor should be selected and a purchase order issued or a work order issued to have the company's own shops fabricate the necessary equipment.

INTEGRATION INTO EXISTING PLANT SYSTEMS. Selection of material handling equipment in any existing application is usually tempered by the nature of the systems already in use in the plant. The interfaces between

1. Direct. Commonly associated with the operation of a piece of equipment, and including such items as:

FIXED COSTS	VARIABLE COSTS
Depreciation	Operating personnel
Interest on investment	Fuel and power
Taxes	Lubrication
Insurance	Maintenance parts and supplies
Supervisory personnel	Outside maintenance labor
Plant maintenance personnel	

2. Indirect. Associated with the investment or operation, but not usually in a direct relationship. These will include such items as:

EQUIPMENT AND METHOD RELATED COSTS	MANAGEMENT RELATED COSTS
	Re-layout
Space occupied	Personnel training
Taxes	Travel involved in investigation
Inventory	Damaged equipment, etc.
Repair parts	
Downtime	

3. Indeterminate. Cannot be precisely determined or fixed, or may be vague, or frequently not known in advance, or which do not lend themselves to the determination of a definite cost figure. This category might include some of the following frequently overlooked costs:

EQUIPMENT AND METHOD RELATED COSTS	MANAGEMENT RELATED COSTS
Space lost or gained	Production lost due to delay in installation
Changes in overhead	Percent of time equipment is utilized
Inventory control savings	Turnover of work in process
Production control savings	Business volume
Changes in product or material quality	Equipment trends
	Paperwork
EQUIPMENT AND/OR METHOD RELATED FACTORS	MANAGEMENT RELATED FACTORS
Quality of equipment	Financial policy
Flexibility	Plans for expansion
Adaptability	Effect on morale
Safety	Improved customer service
Manufacturer's reputation	
Availability of equipment	
Availability of repair parts	
Quality of service	

Fig. 16. Factors for evaluating equipment alternatives.

existing facilities and new systems must be seriously considered when selecting new material handling methods. For example, the cost of converting production from a conveyor system to a unit-load storage and shipping system might seriously affect the economics of the unit-load handling system. In this case an automatic pallet-loader or a central pallet-loading-station setup could probably be worked out to permit smooth transition from one handling method to the other. However, the location of the transition point might also have serious economic implications. If the unit loads were made too early, uneconomical long-haul trucking might result; if too late, high investment in conveyor systems and serious space loss might result. In some cases, the location of a transition point might also be affected by supervision problems and labor relations considerations.

Whenever material handling changes are considered, the relationship of the change to adjacent functions and the overall facilities must be taken into consideration.

EQUIPMENT CLASSIFICATIONS. The three **basic categories of handling equipment** are: conveyors, cranes and hoists, and industrial vehicles. Their characteristics are tabulated in Fig. 15 and described briefly in the paragraphs that follow. There are also several categories of auxiliary equipment.

There is rarely a clear-cut distinction among handling equipment best suited to perform any specific task. For this reason, and others, there have been developed a number of methods of classifying handling equipment designed to distinguish the various groups, types, and items available. One such classification is as follows:

1. Conveyors
2. Cranes, derricks, hoists, winches
3. Industrial vehicles (trucks, trailers, stackers)
4. Storage facilities (containers, supports)
5. Positioning, weighing, control equipment
6. Elevating devices
7. Shipping containers
8. Packaging materials, supplies, and equipment
9. Transportation equipment (highway, railway, marine, air)
10. Miscellaneous (portable equipment; building, dock and yard equipment; handling and storage aids)

TYPES OF HANDLING SYSTEMS. In addition to the several methods of classifying handling equipment, handling activities are frequently classified in terms of types of **handling systems.** It should be pointed out that the word system, as used here, applies to a related group of handling devices and methods commonly used together. It does not mean the same as the word systems in the **systems approach,** discussed elsewhere in this section.

Handling systems are as varied in their composition as are the operations to which they may be applied. Different plants in the same industry will often use different systems for the same purpose. Differences in plant layout, processes, volume of production, receiving procedures, and in many other factors may call for variations in the handling system. The systems described below, however, are rather basic and will often be used exclusively in a plant, or as components of an integrated plant material handling system. The equipment types referred to are described later in this section.

Equipment-Oriented Systems. These are commonly described in terms of the three basic groups previously suggested: conveyors, cranes and hoists, and industrial vehicles.

Industrial truck systems. Platform trucks and skids constitute one system. The high lift platform truck and the low lift truck will pick up, transport, and set down skid-loaded materials without manual rehandling. With the low lift truck primarily used for moving, the high lift truck is used for stacking, maneuvering, positioning, and the like.

The principle of the **fork truck and pallet system** is the same as that of the platform truck and skid system. The forks require less clearance than the platform, making it possible to use pallets, which are shallower than skids, and which save space in tiering. Double-faced pallets make possible wider load distribution, another space saving feature. The powered hand truck may be used with either the skid or pallet system.

The **tractor-trailer** system is economical for hauling large quantities of materials for distances over 300 feet. The cost per ton for handling materials in this way is very low, as one tractor can move many loaded trailers at one time. Loading and unloading may be done by cranes, hoists, platform lift, or fork trucks, or even by hand.

Conveyor systems. Conveyors and conveyor systems are commonly found in operations where material or items of uniform size and shape are handled over the same path repeatedly or for long periods of time.

Conveyor systems, in general, are more adaptable to mass movement than the unit system. When the conditions are right, the application of control devices, programming systems, and careful layout can often reduce the time and manpower used in material handling to a negligible quantity. Conveyors become uneconomical when they must be loaded and unloaded frequently or when complicated installations must be changed frequently. It should be remembered however that material handling requires careful application engineering including a thorough economic study. Each system must be tailor-made to the situation in which it will operate and no basic principles or rules of thumb will apply all the time.

Conveyor systems are very useful in the **warehousing operations** of plants that produce large volumes in reasonably long runs. For example, several packaging lines may feed into main-line power conveyors controlled by a central panel. These lines in turn could feed into a battery of pallet loaders in a warehouse some distance from the production operation. A control system might schedule the lines so that the pallet-loader receives only full unit-load quantities. The loaded pallet could be delivered by conveyor and automatic pallet elevator to the floor on which it would be stored. From this point it might be picked up by a fork lift truck and placed in storage. It would then be removed from storage and delivered to a truck or car by fork lift truck for shipment.

Overhead systems. In some cases overhead cranes and monorail equipment are used in operations where floor space or product characteristics make the use of fork lift trucks or conveyors undesirable, and where travel distances and paths are reasonably restricted. For example, most rug cutting and storing operations use monorail systems from which racks are suspended. These racks support rolls of carpet in a horizontal position, one above the other. Several racks are mounted side by side with the rails running perpendicular to the aisle or cutting floor. For cutting operations, the rack can be moved out and the carpet unrolled without removal from the racks. This same system is utilized for handling other types of rolled materials.

Material (Load)-Oriented Systems. These are commonly identified as: unit handling systems, bulk handling systems, and liquid handling systems. Here again it can be seen that certain types of equipment fit better into one category than another, while some are entirely impractical in a particular category.

Unit load systems. A **unit** usually refers to an individual piece such as a box, bale, roll of material, piece of lumber, sheet of glass, machine, or part of a machine, either as individual items or grouped into a unit load.

A **unit load** has been defined as a quantity of items or bulk material arranged or restrained in such a way as to permit their being picked up and moved as a single object, and upon being released will retain their original arrangement. A unit load is assumed to be too large for manual handling; single objects too large for manual handling can be regarded as unit loads.

This unit-load concept—handling and storing a number of objects or a large mass as one item—has had a profound impact on material handling activities. It is probably best illustrated by a pallet upon which a number of cartons, bags or other material are carefully stacked, with the entire load being handled by a lift truck; unit loads can also be handled by such methods as:

1. A lifting surface under the mass (pallet or skid)
2. Inserting the lifting element into the body of the unit load (spearing such items as coils, rolls, tiles, blocks, etc.)
3. Squeezing the load between two lifting surfaces (bales, cartons, rolls, etc.)
4. Suspending the load (hooking, slings, magnets, vacuum, etc.)

Generally speaking the unit-load system is more flexible and requires less investment than any other approach to material handling cost reduction. It can usually be applied in any size organization with some degree of success. It is particularly applicable to operations where non-repetitive handling sequences are found, or where a variety of products and material are handled. It can also be applied in highly repetitive, large volume operations with considerable success. Many types of specialized apparatus and attachments have been developed for use in such operations.

In storage areas, unit loads are arranged and stacked to permit direct access to every item. Identification and location of items in storage is usually based upon a grid locator system. Storage in unit loads provides for good housekeeping and simplified inventory control. Storage and warehousing are discussed in greater detail elsewhere in this section.

The development of palletless handling has made it possible to eliminate pallets and skids. Such devices as carton clamps, slip sheet attachments, shrink film systems, etc., have increased the capabilities of unit load operations through the elimination of the cost of purchasing and handling pallets and skids, strapping operations, and the like.

Bulk handling systems. Typical bulk materials are coal, grain, ore, gravel, stone, powdered materials, and the like. Bulk systems handle a flow of material on a continuous basis by means of various types of conveyors, power shovels, scoops, cranes, drag lines, and construction equipment. Fork lift trucks are also used to handle bulk material through the use of bulk handling containers. Front-end loaders and scoop attachments on fork lift trucks are common for some handling operations, such as the loading and unloading of bulk materials from freight cars.

Bulk material by its very nature is shapeless and often confined in storage and transit by containers, such as sacks, tubs, tanks, bins, hoppers, skid boxes, barrels, and many others. Once the bulk material is confined, the handling problem is really no different than in the case of any other unit load. Many types of bulk materials are also handled by pneumatic conveying equipment.

Liquid material systems. These are in a class almost by themselves, and require rather specialized handling equipment. The extreme differences in material characteristics and handling equipment involved in the handling of liquids, make it impractical to cover them in this brief section. It should be noted, however, that the most common handling methods are piping systems and a wide variety of tanks.

Method (Production)-Oriented Systems. These systems are defined in terms of the types of production in which they are used, for example, manual, mechanical, automated, mass production, or job shop. The **manual system** uses manual handling methods because of the nature of the operations, such as low volume, wide variety, extreme fragility, etc. Such situations would normally indicate that anything other than manual handling would be undesirable.

Mechanized or **automated systems** are used where greater volume or standardized products are involved and therefore the use of increasingly more sophisticated, complex, or mechanized equipment is justified as the situation may warrant. Such systems make extensive use of conveyors, automatic controls, transfer machines, and other methods of mechanized or automated handling between production operations.

Mass production handling systems are used for high volume or mass production. Complex material handling machinery becomes economically justifiable in such cases. In mass production the basic machinery is no different from that which is used in other applications. The advantages of mass production are derived from the ability to control and manage repetitive handling operations automatically by the use of mechanical, electrical, electronic, photo-electric, and magnetically controlled equipment. High volume flow usually results in fixed paths of material movement and minimum unit handling costs. High density storage operations and mass movement activities are possible with quantity production. Warehouses can be laid out with storage depths based on full truck or full car shipments. Automatic pallet loaders and unloaders are used to best advantage in mass production situations.

A typical **mass production unit handling system** can be found in the manufacture of appliances, where major components are delivered to a conveyor-mounted assembly line from subassembly and fabricating shops by various types of conveyors. Small components are delivered to the feeder lines and main line work stations in various types of unit loads. The final work station delivers a finished, tested, and packed appliance to a collecting conveyor. Fork lift trucks with special carton handling attachments may stack the packed appliances in a storage area at this point. The appliances may proceed by an automatic elevator to an overhead long distance conveyor for transportation to a warehouse where they are stored with the same type of lift truck equipment. Appliances are stacked twenty-five to thirty feet high in deep storage blocks with a minimum of aisle space loss. Shipments are usually in truck or car load quantities and the product is handled from storage to the carrier with specially equipped lift trucks.

This type of integrated system demonstrates the value of mass production techniques in simplifying complex handling operations. Some of the components are handled to the main assembly line by an overhead chain conveyor, and some by unit loads. The assembly line moves on a slat conveyor, and the finished product is on roller and belt conveyors. Warehousing is based on industrial truck methods. The timing and synchronizing of the various components of the systems are the results of careful planning, good management, and effective design. Mass production permits the optimum in integrated handling techniques.

A **mass production bulk handling system** is typified by a coal burning power plant at the mouth of a mine. In this case, mining machinery develops the product (lump coal) in the mine, and loaders deliver it to belt conveying equipment in the mine. It is automatically handled through cleaning and sizing operations and delivered by conveyor to stockpiles. When required for fueling the boilers, the coal is collected by power shovels for conveyor type loaders. It is conveyed to pulverizing machinery, ground to powder consistency, and blown into the boiler. This system with a multiple of variations can be found in many electric utility installations. The uniformity and high volume of the product handled gives it mass production characteristics. Automatic controls are very precise at the boiler end of the system, and much less so at the mine end of the system. The stockpile permits adjustment between rates of supply and demand.

Job shop handling systems typically involve metal working operations on a small volume basis. However, many semi-production operations also fall into this category. The manufacture of machine tools, heavy aircraft, glass tableware, pottery, furniture and many other everyday items resembles job shop operations. In this case an order for a machine may last from one hour to one month, but it is seldom long enough to justify large investments in automatic handling equipment. The unit-load system and portable conveyors or sectional conveyor units are usually sufficient. The use of numerical control machines has resulted in complete automation of much between-operations handling. In some cases, efficient in-process inventory management can build up handling quantities to a level approaching small run mass production. Thus, the handling efficiency of a job shop is more dependent upon good management than on equipment applications.

In general **bulk** handling operations incorporate several types of handling equipment into an integrated system. As in the case of job shop unit handling systems, management is more critical than in mass production applications.

Function-Oriented Systems. This method of classification, which appears useful from a problem-solving point of view, is outlined by Haynes (Distribution Worldwide, vol. 52) as follows:

1. **Transportation systems.** Horizontal motion over fixed or variable, level or nearly level routes, by pulling or pushing, on surface riding vehicles.
2. **Elevating systems.** Vertical motion over fixed vertical or steeply inclined routes with continuous or intermittent motion.
3. **Conveying systems.** Horizontal, inclined or declined motions over fixed routes by gravity or power.
4. **Transferring systems.** Horizontal, or vertical compound motions, through the air, over fixed routes or limited areas with intermittent motion.
5. **Self-loading systems.** Intermittent motion with machines that pick up, move horizontally, set down and, in some cases, tier loads without other handling (also known as unit load systems).

RELIABILITY OF EQUIPMENT. Within each classification or type of material handling equipment, there are varying degrees of ruggedness of design. A conveyor with a given capacity might be completely adequate in a clean sheltered installation, but might fail if used outdoors or in dusty atmospheres with the same or less load requirement. Likewise, industrial automotive equipment with gasoline power plants might be the best selection in some installations, whereas under other operating conditions, battery electric power might be mandatory.

The reliability of material handling equipment has a direct bearing on production costs. Selection of equipment with suitable design characteristics to meet the severity of service anticipated in a particular installation is as important as selection of the right type of equipment in the first place. However, reliability cannot be 100-percent designed into the machinery. Even the most rugged installation requires preventive or at least remedial maintenance.

Measuring Material Handling Operations

IMPORTANCE OF MEASUREMENTS. The measurement of material handling operations is important in determining machine and manpower requirements, planning manufacturing schedules, and establishing productivity standards.

Prior to any investment in capital facilities, it is necessary to estimate equipment and manpower requirements, with possible substantial errors in judgment because of the lack of practical experience. With some capital facilities available, it is possible to analyze actual operations in detail, resulting in maximum accuracy in the evaluation of additional equipment and manpower requirements.

In planning a new plant, it is essential to analyze the flow of raw materials, work-in-process, and finished products throughout the plant (see section on Plant Layout and Facilities Planning). Based upon this analysis, minimum requirements of material handling equipment and manpower may be established.

Types of Measurements. After a plant is in operation, and minimum material handling facilities have been provided, it is relatively simple to determine additional facility requirements with a high degree of accuracy. The techniques of **time studies, standard time data,** and **work sampling** are well suited to the evaluation of additional equipment and manpower requirements.

Keys to the proper determination of material handling equipment utilization include standard time data for the equipment in use, and a reasonable approximation of the number and character of material handling moves involved, related to a given production schedule.

Depending upon the specific type of manufacturing involved, any given production schedule may be evaluated into a certain frequency and character of material handling **moves.** Application of standard time data to the moves will develop standard hours of work for the various types of material handling equipment employed. Dividing standard hours of work by actual hours of work will develop percent utilization of the equipment.

In the development of standard time data, it is essential to consider the same factors of leveling, base time, personal time, fatigue, incidental delays, and incentive allowances, as are included in the development of standard time data for the more conventional manufacturing operations (see section on Work Measurement and Time Study).

Material handling costs provide another means for measuring and evaluating operations.

MATERIAL HANDLING COSTS. Material handling costs have traditionally been surrounded by misunderstanding, confusion, and mystery. There are a number of reasons for this. First, material handling functions, activities, and duties normally are not under one authority. As a result there are varying concepts of what actually constitutes material handling. This makes it difficult to accurately identify the relevant costs. In addition, material handling costs are frequently mixed with or inherent in other costs and are therefore difficult to separate. In fact, much of what is commonly classified as **direct labor** is actually material handling.

Nevertheless material handling cost, however it is defined, is too large a segment of the production dollar to be overlooked. Instead it should be considered as a prime target for cost reduction. In general, the process of accumulating and determining material handling costs consists of:

1. Determining what costs are wanted
2. Identifying cost elements
3. Assigning costs to each element
4. Totaling elemental costs

Cost Factors. Fundamental to the discussion of handling costs is an understanding of the various factors connected with them. **Direct costs** are those commonly associated with the operation of material handling equipment. Both equipment and manpower costs are included in this category. **Indirect costs** are those associated with the investment or operation of a piece of equipment but not in a direct relationship. Costs of training personnel in the operation of a particular type of equipment would fall into this classification. **Indeterminate costs,** although related to indirect costs, cannot be precisely determined or fixed, are often vague, frequently not known in advance, or do not lend themselves to definite cost figures. **Intangible factors,** in the ordinary sense, defy quantification or dollar-evaluation and therefore cannot be included as items in a cost comparison. Typical items which might be included in each of these categories are shown in Fig. 16.

Material handling costs may be classified further according to the **character** of the costs.

Equipment costs are those related to the actual capital investment. They may be determined with the aid of an **Equipment Operating Cost Determination** worksheet (Fig. 17). After having determined investment costs and added typical fixed and variable charges for a predetermined time period, the same worksheet is used to calculate **operating costs,** that is, those costs directly related to the operation of the equipment.

Activity costs are those expenditures required by a specific handling activity. A careful analysis of the activity under consideration is required to make this determination. First it is necessary to identify and isolate each **move** (or **transportation**) in the activity. This is best done with the aid of a process chart (Fig. 18) which shows the number of moves involved in the activity. Each individual move is then analyzed to identify the manpower, equipment, and other costs required to perform it. A **Cost Determination** worksheet (Fig. 19, page 46) will assist in this analysis. It provides space for calculating manpower, handling equipment, and auxiliary equipment costs related to each specific activity. The four right-hand columns are a re-cap of the three cost items and the resulting total cost. The sum of the items in colunm S represents the overall total annual cost involved in carrying out the move portion of the activity depicted on the process chart.

Function costs are concerned with such organizational units as receiving, warehousing, and the like. In general their determination involves little more than a combination of the costs outlined above, with traditional accounting concepts.

COST REDUCTION. It is commonly acknowledged that material handling activities account for up to 35% of all production costs. Since the handling process itself adds no value to the product, it represents a prime target for a cost reduction program. Ideally all material handling should be eliminated, but practically this is not possible. Apple (Plant Layout and Materials Handling) outlines the following suggestions for obtaining **economical material handling:**

1. Keep all handling at a minimum.
2. Handle as many pieces in one unit as is practical.
3. Provide large, heavy, or bulky objects with handling aids such as eyelets, hooks, lips, etc.
4. Coordinate production planning, scheduling, and dispatching.
5. Always consider handling problems in analyzing and designing the overall production system.
6. Modify existing building facilities to permit full use of available capacity.
7. Whenever possible manufacturing processing should take place while the product is being moved.
8. In general, move the greatest weight or bulk the least possible distance.

ITEM	ALTERNATIVE NO. 1		ALTERNATIVE NO.	
EQUIPMENT DATA				
Make				
Type				
Model				
Capacity				
Accessories				
Attachments				
Operating Characteristics				
INVESTMENT				
Invoice Price				
Installation charges				
Maintenance facilities				
Fueling &/or power facilities				
Alterations to present facilities				
Freight &/or transportation				
Design Work				
Supplies				
Other charges				
Credits				
TOTAL INVESTMENT COST				
FIXED CHARGES	8 hours	16 hours	8 hours	16 hours
Depreciation				
Interest on investment				
Taxes				
Insurance				
Supervision				
Clerical				
Maintenance Personnel				
Other				
TOTAL FIXED COST				
VARIABLE CHARGES				
Operating Personnel (% x $ wage)				
Power &/or fuel costs				
Lubricants				
Maintenance parts & materials				
Maintenance labor				
Other				
TOTAL VARIABLE COST				
OTHER OVERHEAD (% x $)				
TOTAL ANNUAL COST				
OPERATING HOURS PER YEAR				
COST PER HOUR OF OPERATION				

Fig. 17. Equipment operating cost determination.

PROCESS CHART   Page 1 of 1

	SUMMARY	
		NO.
	○ OPERATIONS	1
	⇨ TRANSPORTATIONS	6
	☐ INSPECTIONS	0
	D DELAYS	0
	▽ STORAGES	6
	TOTAL STEPS	13
	DISTANCE TRAVELED	96'

ART NAME ____ Item No. 10 — Wax

ROCESS DESCRIPTION ____ Move wax from R.R. car to Melt

EPARTMENT ____ Production

LANT ____ Aerosol Products Co.

ECORDED ____ A.M.J. ____ DATE _ 9/20

STEP	Operations	Transport	Inspect	Delay	Storage	DESCRIPTION OF PRESENT METHOD	Dist.	How Moved	Est. Hrs./ Load	Hrs. per Bag
1	○	⇨	☐	D	▽	in R.R. car — (10,000 100# bags)				
2	○	1	☐	D	▽	stack onto pallet (25/pallet)	20 ft.	hand	.01	.0100
3	○	⇨	☐	D	1 ▽	on pallet at dock				
4	○	2	☐	D	▽	pallet from Dock to Receiving Stores (25 b./p.)	30'	F.T.	.08	.0032
5	○	⇨	☐	D	3 ▽	in Receiving Stores on floor				
6	○	3	☐	D	▽	Pallet to Mixing Area (25 bags/pallet)	30'	F.T.	.10	.0040
7	○	⇨	☐	D	4 ▽	On floor in Mixing Area				
8	○	4	☐	D	▽	Pallet to Mixing Level (25 bags/pallet)	10'	F.T.	.12	.004S
9	○	⇨	☐	D	5 ▽	Pallets on floor at Mixing Level				

Fig. 18. Process chart for one component (wax).

Material Handling Equipment

MAJOR EQUIPMENT GROUPS. As noted previously, material handling equipment can generally be grouped into **three major categories:** conveyors, cranes and hoists, and industrial vehicles. Several categories of auxiliary equipment are also in general use. There are, however, few clear-cut choices of equipment to best solve a particular handling problem. In fact, most decisions involve compromises and qualifications of varying degrees. While it is obvious that equipment classification systems will not in themselves solve a handling problem or select a piece of equipment, they are nevertheless helpful in distinguishing the many different types available for purposes of comparison.

CONVEYORS. This classification includes all equipment for transporting material between two fixed points with continuous or intermittent forward movement and continuous drive, whether the equipment is fixed or portable. This definition, with some exceptions, is close to that given by the Conveyor Equipment Manufacturers' Association (Conveyor Terms and Definitions).

Keller (Unit-Load and Package Conveyors) further classifies conveyors according to the **type of medium** used to convey the material. The three natural mediums listed by Keller are gravity, air, and water. Other mediums include rollers, wheels, chutes, slides, belts, chains, cables, and screws.

Belt Conveyors. A belt conveyor consists of a power driven endless belt, with terminal pulleys and idlers or slider beds carrying the loaded and empty

COMPANY_____ PLANT_____ DEPARTMENT_____ PROJECT NO_____ ANALYZED BY_____ DATE_____

MOVE IDENTIFICATION					MANPOWER COST				HANDLING EQUIPMENT COST					AUXILIARY EQUIPMENT COST (containers, pallets, racks, etc.) DO NOT enter more than once if equipment stays with material						TOTAL COST			
Move No.	Move Description	Yearly Quan. A	Item per Load B	Loads per Year (A÷B) C	Time per Load Hrs. D	Hrs. per Year (CxD) E	Wage per Hour F	Cost per Year (ExF) G	Equipment Identification	Time/Load in Hrs. H	Hrs. per Year (CxH) J	Cost per Hr. K	Cost per Yr. (JxK) L	Equip. Ident.	No. Reqd. M	Cost per Item N	Amort. Rate P	% Used This Move Q	Cost per Yr. (MNPQ) R	Man-pow. Cost/Yr. G	Equip. Cost per Yr. L	Aux. Cost per Yr. R	Total Cost per Yr. S

Fig. 19. Cost determination worksheet.

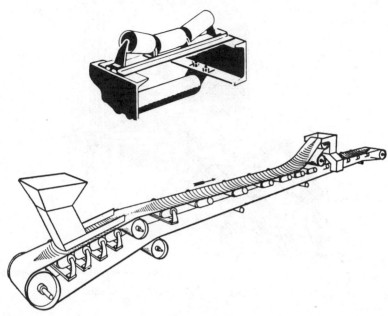

Rex Chainbelt Inc.

Fig. 20. Typical troughed belt conveyor with feeder and tripper and showing idler.

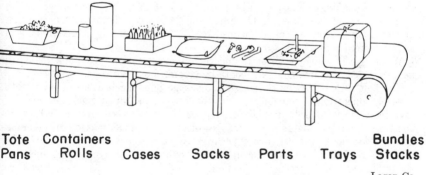

Tote   Containers   **Bundles**
Pans   Rolls   Cases   Sacks   Parts   Trays   Stacks

Logan Co.

Fig. 21. Applications of typical belt conveyor.

strands of the belt (Fig. 20). Its first cost is moderate and its maintenance cost low (barring accident to the belt).

Belt conveyor systems are very economical of power and are used for handling constant and intermittent flows of materials over long or short distances. They can transport goods horizontally or up or down an incline, and can be used to restore grade in gravity systems or control travel rate down an incline.

Belt conveyors may be either fixed (Fig. 21) or portable (Fig. 22). The portable variety can be adjusted to move goods horizontally or up or down an

Fig. 22. Typical portable belt application.

incline, as is needed. This versatility is an extremely valuable feature. Conveyors of the so-called booster type belong to this group.

Idlers are a series of rollers on which the endless belt travels. For light loads up to 50 lb. per sq. ft., a steel or wood trough or slider bed may be substituted for the idlers. Belts are often troughed to reduce spillage in handling bulk materials. Troughing idlers normally are inclined at 20°. Molded-rubber, rubber disc, spiral, and flexible-shaft troughing rolls are used. Thinner belts are used and troughing idlers are often inclined to 45° for handling light materials, such as wood chips or grain.

While a belt conveyor most frequently discharges at its end, bulk loads can be discharged at points along its length by **trippers** or by **plows**.

The belt conveyor is highly suited to package handling because of its smooth, noiseless operation, reversibility, and the ease with which packages may be diverted. More efficient at high speeds than any other continuous carrier, its continuous surface adapts it to packages of even the smallest size. In manufacturing operations belt conveyors perform two-way service by carrying loaded boxes of work-in-process on the upper side and returning empty containers on the underneath or return side of the belt.

The customary angle of inclination is 15° for flat-belt conveyor installations conveying miscellaneous materials up or down. This incline can be increased to 30° by using a rough-surface belt. Cleats fastened to the belting permit handling of packages on even greater inclines.

A **wire-mesh conveyor** belt is often used in light-weight production on washing, draining, or other operations involving immersion of parts in, or their separation from, liquids. Metal-mesh belts are also used where fire is applied through the mesh.

A special form of the belt conveyor is the closed belt **"zipper" conveyor**. This conveyor is constructed of a piece of flat conveyor belt upon which are mounted two pieces of rubber with teeth molded in one edge. When forced together by two rollers bearing against the bead upon which the teeth are molded,

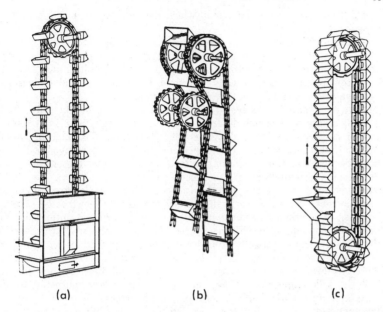

(a)   (b)   (c)

Rex Chainbelt Inc.

Fig. 23. Bucket elevators.

the teeth lock to form a closed joint, with the result that the conveyor becomes the equivalent of a moving pipe.

The conveyor is able to negotiate any slope up to the vertical and can accomplish numerous changes in direction. It is particularly suited to handling materials which are dusty or those which must not be contaminated, and where degradation must be kept to a minimum.

Elevating Conveyors. The most widely used means of elevating material in a restricted horizontal distance is the **bucket elevator.** It is made up of buckets mounted on either belt or chains running over terminal wheels. Such elevators are built in a number of types (see Fig. 23).

Centrifugal discharge bucket elevators (Fig. 23a) may be vertical or inclined, with buckets mounted at intervals on a single or double strand of chain or on a belt. This type of elevator is used to handle bulk materials which can be picked up by spaced buckets as they pass under the footwheel and are discharged by centrifugal force as the buckets pass over the headwheel.

Positive discharge elevators (Fig. 23b), operate at slow speeds, and are made up of spaced buckets mounted between two strands of chain. The material is scooped up by the buckets passing under the footwheels and discharged at the head when the buckets are inverted over the discharge opening by means of knuckle wheels. This type is recommended for handling light, fluffy, or sticky materials.

Continuous bucket elevators (Fig. 23c) may be vertical or inclined, with buckets mounted continuously on single or double strands of chain or on a belt. The material is usually fed to the buckets through a loading leg and is discharged over the face of the preceding bucket in passing around the headwheel. These elevators may be used to handle the same kinds of material as the centrifugal type, but are particularly adapted to handling materials that are difficult to pick up or are friable due to the manner in which they are fed to and discharged by the elevator.

Super-capacity elevators are continuous bucket elevators in which the buckets are mounted between two strands of chain and project backward in toward the center of the elevator, thus carrying larger capacities and permitting larger lumps to be handled. They are used for handling friable, heavy, or abrasive materials ranging from fines to heavy lumps.

The **power** required to drive an elevator is an uncertain quantity, largely because all of the load is not discharged at the discharge chute, and some recirculation takes place. Many designers select an elevator capable of handling the nominal load with the buckets 75 percent full. Fifty percent is then added to the horsepower computed to lift this load. The resulting motor will be adequate to handle recirculation and other losses.

Arm elevators are the simplest type of equipment for elevating or lowering packages at steep angles or vertically. They are usually constructed of one or two strands of continuous chain with projections or arms at intervals. When such elevators are equipped with finger-arm carriers or solid tray arms, packages can be picked up automatically from loading fingers or stations on the up side and discharged over the top. When loaded on the down side, fragile packages can be safely lowered to any desired level. As lowering devices, these elevators have fairly wide application. They find extensive use in multistory storage buildings and in industrial plants where space is important.

The **suspended-tray elevator**, also known as the pivoted-tray elevator, consists of a series of pivoted suspended trays attached to two strands of endless chain running over top and bottom sprockets. Because of their design, the weight of the load is always centered well below the suspension points and holds the tray level as it passes over the sprockets. Suspended tray elevator-lowerers may be divided into two general classes: the simple swing-tray machine with solid or specially constructed trays, which are loaded and unloaded wholly or partly by hand; and the highly developed automatic loading and discharging machine.

Reciprocating elevators are those equipped with an electric hoist which raises or lowers a counter-weighted car or platform, operating along suitable guides, by means of a chain or cable over a sprocket or sheave. The construction is simpler than that of the suspended-tray elevator; hence the cost is less. Such an elevator handles packages, cartons, barrels, boxes, pallets, or trays but is used where great capacity is not required.

Keller (Unit-Load and Package Conveyors) describes the en-masse conveyor (Fig. 24) as consisting of a stationary casing or conduit with a moving articulated conveying medium fitted with flights. As the flights move through the casing, the material is conveyed and is protected against contamination or spilling from the conveyor. The moving element is usually a chain with the flights formed or cast integrally with the chain links. Cable or solid V-belts are also used in place of chain. En-masse conveyors are frequently used as self-feeders in which case lumpy materials should be avoided. The unit is very compact and flexible, lending itself to various paths of travel. It has wide application and will handle most materials satisfactorily handled by bucket elevators, in addition to finely ground, dry, nonabrasive materials, for which it is ideally suited.

Chain and Cable Conveyors. According to Apple (Plant Layout and Materials Handling) this type of conveyor contains one or more strands of endless chain operating on suitable tracks or guides. Load is placed on and carried

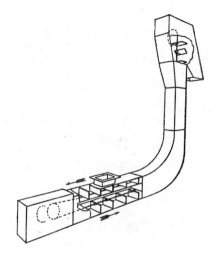

Fig. 24. En-masse conveyor–elevator.

by the chain surface or by special carriers fastened to the chain. In addition to the moving chain propelling the load, projections mounted on the chain may assist in pushing the load. Chain and cable conveyors are often classified according to the type of load carrying arrangement used.

The **slat conveyor** (Fig. 25) consists of steel or wood slats attached at their ends to two strands of chain, running in or on parallel steel tracks to form a nearly continuous traveling platform. The platform may be horizontal, inclined, or both. With blocks, cleats, or brackets to prevent materials from shifting, a slat conveyor can be used on inclines up to 45° or more. It is normally used where the loads lack a surface suitable for live-roll conveyors or where belting would be damaged. It is used for conveying heavy objects or packages, such as cartons, boxes, barrels, and bales weighing as much as 500 lb. Slat conveyors are frequently installed flush with the floor to permit cross traffic.

The **apron conveyor** is a modified slat conveyor with the "slats" overlapping to form a continuous jointed and leakproof tightly moving bed. They are primarily known in the bulk field, but are also found in many heavy-duty-unit-type installations such as foundries, quarries, and steel plants. This type of conveyor is designed to withstand severe loading conditions and high temperatures, and

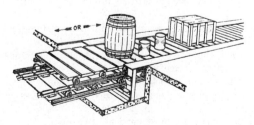

Fig. 25. Slat conveyor installation at floor level.

to deliver a uniform flow. On the basis of handling 50 lb. of material per cu. ft., at 100 ft. per min., capacities of this type of equipment vary from 80 tons per hr. for an apron 18 in. wide to 290 tons per hr. for an apron 60 in. wide. Such conveyors are usually operated at speeds ranging from 60 to 100 ft. per min. and can negotiate inclines up to 25°.

The simplest of the chain types is that consisting of a single strand of chain riding in a steel track and serving as a carrier for the object to be conveyed. This equipment, sometimes referred to as a **drag chain** or **carrier chain conveyor,** is designed to convey light, small, and uniform-size objects such as bottled beverages, but can be designed to carry heavy steel drums and logs. Multiple-strand conveyors are also available. These conveyors can turn 90° corners with most loads. The chain usually is guided in a steel track or wheel at turns.

Car-type conveyors are used in foundries to move flasks from molding to pouring, to shake out empty flasks, etc. Other varieties of **pallet** conveyors are used in assembly operations where the work is not suited to flat-belt conveyors, and pallets designed for holding specific parts carry them through progressive assembly operations. These conveyors are often intermittent in movement and operations are performed without removing the work from the pallet.

The **trolley conveyor** consists of an endless-circuit overhead track beneath which moves an endless chain connected to trolleys at regular intervals. Two-wheel trolleys carry the load by means of hooks, racks, special carriers, etc., and the chain pulls them along. Line of travel may include 90° and 180° bends on the horizontal plane, and inclines up to 45° on the vertical plane, depending upon the spacing of trolleys and loads carried.

The various types of carrier attachments distinguish trolley conveyor systems. In many systems, the attachments are no more than hooks on which materials are hung for transportation and storage. There are many varieties, and some producers design and manufacture overhead chain conveyor attachments as their principal business. Special installations have been made where the load is also carried vertically as well as horizontally. The most common application for this type of equipment is feeding from subassembly lines to assembly conveyors, or for carrying parts through spraying, plating, baking, painting, and other operations. It is widely used in the automobile and meat-packing industries. **Carrier attachments** can be pivoted on the chain pins or swivels to create rotary or other motions of the load, thus increasing their usefulness for electroplating baths, cleaning baths, etc.

The conveyor also acts as a **storage medium** because if an object is not removed it will recirculate over the system repeatedly until taken off at the point where it is needed. By utilizing the overhead space for transportation, floor space is saved and the vertical or inclined line of travel permits easy unloading at working height.

Horizontal turns are made by means of multiple-roller curves—useful because of low friction resistance and large radii—and by traction wheels. The latter method is best when the conveyor passes through temperatures of 400° or more because of its simple lubrication.

Drives are of the corner-sprocket or the caterpillar type. The former is used except where the corner radius is too large. Drives should be at the point of maximum pull and near a down curve, if any exists, to take chain slack away from the drive. Radii of vertical dips should be as large as practicable. If the conveyor has more than four horizontal turns, it is necessary to calculate the pull on the chain progressively and to add 5 percent for friction loss at each turn.

In some instances a trolley conveyor uses a cable in place of the chain to transmit the power. The general principles of operation of the cable type are similar to those with chain. There are some differences in sprocket design and long downhill runs usually must be avoided to prevent "piling up" of the carriers as a result of the lack of rigidity of the cable.

A variation of the trolley conveyor is the **tow conveyor.** It can be installed overhead—as well as in the floor. In either case, a floor truck or cart is pulled by attaching it to the chain.

Among the many applications of the tow conveyor system is the order-picking line of a warehouse. In such cases a warehouse trailer or hand truck is pushed or towed to a point above or below the chain, and a connection is made by a pin (through the floor) or a hook (to the overhead chain). The truck is then pulled to its destination without further manual effort. An automatic disconnecting device can be installed at the point of delivery.

These systems are also useful for delivering orders from the picking area of the warehouse to the shipping dock or for maintaining continuous motion of the order-picking truck while the warehouseman removes merchandise from shelves or pallets to build the order. They can also replace tractor-trailer systems in some instances.

Haulage Conveyors. According to the Materials Handling Handbook (Bolz and Hagemann, eds.) this group includes drag conveyors, flight conveyors, and tow conveyors. Drag and flight conveyors involve dragging or pushing material by means of a chain, or chains, traveling against the material, making use in some cases of flights and in other cases of the surfaces innate in the chains themselves. Tow conveyors, on the other hand, actually tow trucks, dollies, or cars on which are placed materials or packages, as discussed above. The tow power is again obtained from a chain or cable.

A **floor-type system** frequently used in assembly lines and warehouses consists of a chain or other linkage mounted near or flush with the floor and guided by tracks. Pulling or pushing attachments on the chain connect to the unit in motion and, in the case of assembly lines, tracks often guide the wheels supporting the work unit. In most instances, these conveyors are timed to pace the assembly or machining operations and maintain continuous motion. In some instances periodic moves are used.

Drag conveyors consist largely of one or more endless propelling mediums, usually chains or cables, which drag material in a trough, or along a defined path. The materials being carried rest on the chain, and in the trough.

Cross-bar conveyors consist of two strands of chain with cross-bars mounted between the chains at regular intervals and moving above a stationary bed. The cross-bars push the loads, such as packages or boxes, horizontally or up an incline. This type of conveyor is also sometimes known as a booster or pusher-bar elevator (Fig. 26) and is used extensively as a grade retriever in gravity conveyor systems. It is also used for elevating or lowering packages from one floor to another and is suited for a fairly uniform range of packages, or for practically any object of sufficient solidity and shape to slide on a runway at inclines of 30° to 60°. Capacities range from 500 to 1,200 packages per hr. at a chain speed of 60 to 90 ft. per min.

In the **flight conveyor** (Fig. 27) scrapers or flights replace the pusher bars described above and serve to push bulk materials along a stationary trough. For

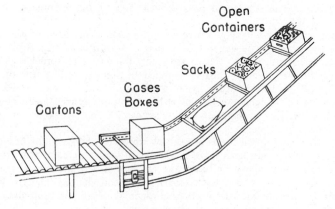

Logan Co.

Fig. 26. Pusher-bar elevator.

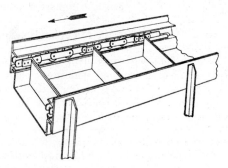

Rex Chainbelt Inc.

Fig. 27. Scraper flight conveyor.

light conveyors a single strand of chain is sometimes used, while two strands are used for heavier units.

The principal advantages of flight conveyors are their ability to negotiate slopes up to 45° and to deliver material at intermediate points. The disadvantages are high power consumption and rapid wear on trough bottoms and sides. This susceptibility to wear makes it unsuitable for use with the more abrasive materials. A lesser disadvantage is the build-up of sticky material on the scrapers when such material is being handled. The power requirements for flight conveyors are dependent upon the material being handled and the design of the equipment.

Roller and Wheel Conveyors. In general, the roller conveyor consists of two laterally braced channels with equally spaced holes punched in the webs; tubular steel rollers supported by anti-friction bearings are usually suspended between these channels. The plane of rolling can be either above or below the frame, depending upon whether or not the frame is to be used as a guide. Roller conveyors are available in standard widths from 6 in. to 36 in. and on special order can be made almost any width. The size of the roll depends upon the weight of the object to be conveyed and whether or not the roll will be subject to shock from loads dropped on the conveyor.

The wheel conveyor is similar to the roller conveyor except that wheels replace the rollers. Single-rail conveyor units are often used in pairs or sets for special handling operations. A gravity conveyor is used in both portable and permanent installations. The supports vary with the requirements. Wheel conveyors are available with steel, aluminum, magnesium, or stainless steel frames; steel, aluminum, and plastic wheels; closed and open tops; and in various capacities. The wheel conveyor is usually considered most practical for smooth-bottom unit loads such as cartons, boxes, trays, and bundles. It is not usually desirable for installations where irregular surfaces might be encountered and is not capable of withstanding much impact or cross loading. It is the least expensive, most easily repaired type of gravity equipment for many package-handling installations. In most cases a grade of 5 in. per 10 ft. of conveyor will give a free flow for smooth, hard-bottomed packages.

A roller conveyor is usually more rugged and versatile than a wheel conveyor. Roller conveyors can be built to carry heavy loads and to withstand severe impact and cross loading. They are occasionally built in a "V" or "U" troughed form to handle cylindrical objects, such as coils of steel or beer kegs. Roller conveyors can handle items from light-weight packaged materials to heavy castings, molds, and machine parts.

Roller and wheel conveyors are available in adjustable length designs known as "accordion" and telescopic conveyors. This equipment is commonly used in situations where the work station is changed frequently or where the conveyor must be retracted when not in use to permit cross traffic.

Directional changes are made by switching devices of all types and curved track sections. In complex installations, the gravity conveyor is often fed by and delivers to the power conveyor. Switches, junctions, and various other devices utilize gravity where applicable. A power conveyor is often used to elevate or regain height for long gravity runs.

Live roller and wheel conveyors are powered by a belt snubbed against the underside of the carrying rolls or wheels. The result is a powered conveyor which is ideally suited for automatic accumulation or temporary storage. With this type of conveyor the load can be moved up or down grades and can be stopped without undue friction. Roller-chain and sprocket drives are also used to drive the rollers. Loads should have rigid riding surfaces, but need not be of uniform weight.

Gravity Conveyors. Chutes are a common handling device for both bulk and unit materials. They rely on gravity as the primary motive power. However, they are likely to require more field alteration than any other equipment when used for bulk handling. Material behavior is greatly influenced by size, consistency, and moisture content. Chutes are commonly constructed of galvanized steel, stainless steel and wood. Teflon coatings are often applied to the sliding surfaces to reduce friction and sticking of certain material.

Flumes are special chutes in which solids mixed with a liquid, generally water, are transported. Since the mixture of solids and water is often the result of a process which controls the percentage of solids, the conveyability may not be ideal.

Screw Conveyors. Keller (Unit-Load and Package Conveyors) describes screw conveyors as constructed of formed or rolled flights fastened to a shaft. The flights form a continuous spiral that moves the material along a trough or

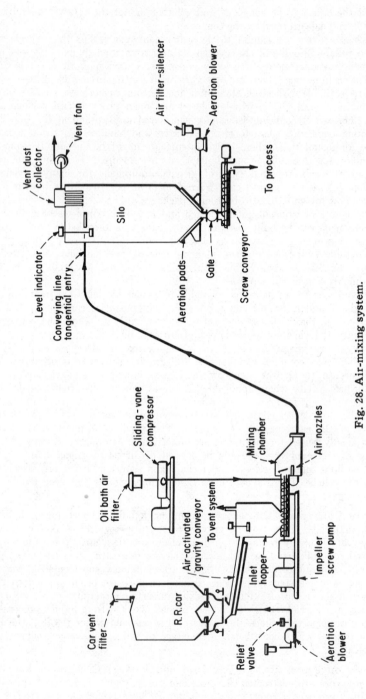

Fig. 28. Air-mixing system.

casing as the shaft turns. When abrasive materials are conveyed the screw and casing are usually made of cast metal.

The screw conveyor has no return strand or other outside parts and consequently requires the least amount of space of any form of conveyor, and readily lends itself to handling materials which require a completely sealed conveyor. It is best adapted to conveying loose bulk materials which fall in one of the following general classifications, are not very sticky, and are of a maximum size not greater than ¼ the diameter of the spiral.

Class 1. Non-abrasive (bituminous coal, crushed asphalt, pebble lime, etc.)
Class 2. Moderately abrasive (portland cement, crushed gypsum, phosphate rock, etc.)
Class 3. Very abrasive (coke, cinders, cement clinker, etc.)

When damp or sticky materials are to be conveyed, the flights are mounted outward of the shaft to reduce material buildup on the shaft.

Capacities of spiral conveyors depend upon the nature of the material to be handled, since both the maximum cross section of the load and the maximum speed of the conveyor are affected.

Pneumatic Conveyors. According to Kraus (Pneumatic Conveying of Bulk Materials) three types of pneumatic conveying systems are available commercially. These are (1) the so-called "pressure systems" in which the material enters a stream of air under either negative or positive pressure, or is induced into the stream of air by vacuum, (2) systems in which air and the material are intermixed simultaneously at the entrance of the conveying line in a special type of feeder (Fig. 28), and (3) the pressure-tank or blow-tank systems in which air enters a mass of material and causes it to flow. Variations of the pressure systems include the combination vacuum-pressure (V-P) and loop systems.

Pneumatic conveying systems are used to transport particles of solid materials, solid fluidized materials, and small unit loads. The techniques are used in loading and unloading bulk materials into freight cars, ships, barges and the like, and for handling bulk materials in storage and warehousing facilities.

Vibrating Conveyors. Vibrating conveyors transport any granular material which is not sticky or tacky through the application of strokes which cause the material to hop along the conveyor. The conveyor consists of a flexibly supported trough or tube; vibration may be induced **electrically** with a balanced electric power unit or **mechanically** by an eccentric shaft connected to the deck or trough. The material, being free to move, does not return with the backward movement of the deck, but falls under the force of gravity until it is intercepted by the next forward and upward stroke. Thus, while in appearance the movement of the material is that of a uniform flowing stream, it is, in reality, a continuous series of rapid, short forward hops which are imperceptible to the eye. Abrasive wear on the deck is negligible, and for this reason, on large tonnages material can be handled most economically.

Vibrating conveyors, both mechanical and electrical, are generally preferred when dealing with high temperatures, abrasives, extreme fines, dust, or easily degraded materials. Fine, dusty materials are handled by covered decks or tubes. Transfers from one conveyor to another can be made with drops of the order of an inch; consequently degradation at points of transfer is the lowest obtainable with any conveyor. Vibrators in operation in many industries handle over 400 different materials, ranging in density from 5 lb. per cu. ft. to 400 lb. per cu. ft., and ranging in unit size from powder to 5-ft. cubes. It is obvious that neither the electric nor mechanical vibrator competes with an elevating or an apron conveyor, if the material must be raised. Likewise, these conveyors cannot compete with a belt conveyor in first cost or operating costs when the material must be transported over long distances. Vibrators do, however, furnish an efficient type of conveyor which, when properly selected and applied, is best and most economical for its purposes.

CRANES, ELEVATORS, AND HOISTS. This group includes equipment whose primary function is to raise and lower material either as a primary task or as an adjunct to the lateral movement of the material. The distinguishing feature of these devices is the use of a drum windup or a plunger drive. Continuous drive elevating conveyors would not fall into this category. Pneumatic and hydraulic plunger hoists, as well as capstans and winches, may be included in the crane, elevator, and hoist grouping.

Cranes are versatile lifting and transporting mechanisms that are made adaptable to a wide variety of jobs by the attachment of an almost infinite number of specially designed grabs. This category of equipment may be subdivided broadly into three groups: fixed cranes, traveling cranes, and portable cranes. The first of these groups is generally associated with limited area handling. The second group is capable of travel along a fixed track, and the last is capable of considerable mobility. **Basic crane designs** include jib cranes, gantry cranes, bridge cranes, cableways, towers, and derricks or boom cranes. Each of these designs may be applied to one or more of the three groups described above. The more important design types are treated below.

Jib Cranes. There are several types of jib cranes. They are generally classified according to the type of construction into the five basic groups of bracket jib, cantilever jib, interlocking jib, pillar jib, and walking jib. Jib cranes may be fixed, traveling, or portable and may be used as integral parts of, or as supplements to, many other types of equipment. They range in size from small manual units to loading towers with capacities exceeding 300 tons. The basic design is illustrated in Fig. 29.

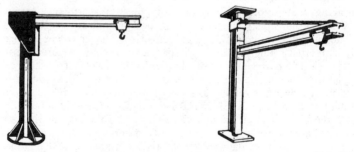

Flow Directory

Fig. 29. Typical jib cranes.

Gantry Cranes. Gantry cranes (Fig. 30) are basically bridge cranes which have girders supported at either or both ends by gantry legs which may be fixed, wheeled to travel on rails, or mounted on casters for portability. They are widely used outdoors in steel yards, shipyards, etc., or in buildings which will not support a bridge crane structure. There are several variations of the gantry crane. The standard gantry operates on the same principle as the bridge crane. The cantilever type is designed to permit the bridge crane to travel

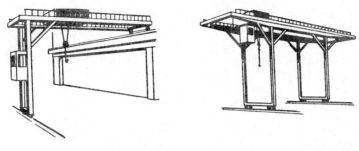

Flow Directory

Fig. 30. Gantry cranes.

beyond the legs. The semi-gantry is so called because one end of the bridge girders are supported by a building or runway attached to a building and the other by a gantry leg.

Bridge Cranes. One of the more common crane systems is the bridge crane. Bridge cranes vary in complexity from a simple chain hoist on a girder spanning two beams in a building structure to a highly complex multiple-hoist power-operated system with interlocking bridges and cab controls (Fig. 31). Ca-

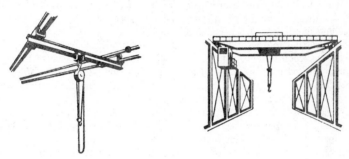

Flow Directory

Fig. 31. Traveling overhead bridge cranes.

pacities can range up to 1000 tons. The selection of bridge crane equipment for a particular application is complex and usually requires the assistance of crane engineering specialists.

Monorail Cranes. Overhead traveling hoists using a rigid single track are often called monorail cranes or conveyors. The monorail, being suspended from the ceiling or other overhead portion of a building, may be installed and used without interfering with operations on the floor underneath.

Monorail systems may be floor-operated traveling hoists or can be run by an operator who rides in a cab suspended on trolleys. Practically any kind of power-operated hoist can be used with a monorail system. Slings or skips are often used to handle unit materials, or buckets for bulk, loose, or wet materials.

Switches and turntables in a monorail system can be manually or electrically operated. Photoelectric cells or indicator pins can be used to actuate the electric mechanisms so that the systems are fully automatic.

Capacities of monorail hoists run as high as 6 tons or more. Speeds of operation run up to 35 ft. per min. or higher, depending on the number of switches, curves, and other characteristics of the system.

Mobile Cranes. Mobile or portable cranes are built in many sizes and capacities. Booms and others of the basic designs referred to above are given broad operating ranges when mounted on a variety of vehicles. The basic types, categorized by their motive power and running gear, include:

1. Crawler-mounted cranes of the track-laying or caterpillar type. They have two parallel continuous crawler or tread belts.
2. Locomotive-mounted cranes, in which the revolving superstructure is mounted on the deck of a railroad car on standard railway trucks, permitting movement on railway trackage.
3. Truck-mounted cranes, which may be operated by a power take-off from the propelling motor or by a separate engine. This type of mounting permits highway travel and provides maximum mobility in both speed and distance.
4. Wheel-mounted cranes, cranes mounted on a wheeled chassis which may be propelled by the same engine which drives the mechanism on the superstructure. This type is suited for operations that require constant and relatively rapid movement of the crane around the job, such as in a factory or storage yard.

Elevators. According to the Materials Handling Handbook (Bolz and Hagemann, eds.), a well-planned freight elevator installation in a factory, warehouse, or multistory terminal can be integrated with almost every method of horizontal material handling—manual or mechanized—to speed the vertical flow of raw materials and in-process or finished goods.

In all cases the characteristics of the elevators, either passenger or freight, depend on the requirements of the individual installation, because of the wide variation in the weight and bulk of articles to be carried. Where factories produce and handle small articles of considerable weight, the inside dimensions of the elevator platform should be approximately 6 ft. by 6 ft. or larger. When larger articles are to be carried, the platforms may be 8 or 9 ft. in width by 10 to 20 ft. in depth, or—the width and depth dimensions can be reversed, since a wide elevator is easier to load and unload.

In most buildings, a variety of industrial trucks and trailers are used and in many cases must be carried from floor to floor on elevators. Detailed information on the characteristics of such equipment must be obtained before platform dimensions can be determined.

Freight elevator car speeds vary with the height of the buildings and may range from 100 to 300 ft. per min. in industrial buildings and up to 800 ft. per min. in commercial buildings.

The traction-type electric elevator has proved far safer and more efficient than the drum-type elevator machine and is in general use. Traction elevators of the geared type have a worm and gear combination and are used for car speeds up to 350 ft. per min. For higher speeds the elevator machines are gearless with slow-speed motors, to permit faster car speeds and greater efficiency. Gearless machines provide car speeds up to 1,200 ft. per min. for intensive passenger service in tall buildings. Hydraulic electric elevators are used in low-rise buildings primarily for freight.

Portable elevators are of great importance in industry for various lifting operations which are intermittent and not in a fixed location. Such units can be used to lift maintenance men to high repair jobs, to stack heavy objects, to lift heavy parts onto machines, to handle dies, and many other diverse jobs.

A **skip hoist** consists of a large steel bucket equipped with flanged wheels, running on vertical or inclined guides of steel channels or T-rail tracks. The bucket is loaded manually or automatically, hoisted up to a discharge position over a storage bin or silo by winch and cable, and automatically dumped. The skip hoist is a unit complete in itself for receiving, elevating, and discharging batches of materials, usually abrasive, corrosive, or containing large, sharp lumps, or when exceptionally high lift is needed. It is used largely for handling coal or ashes and capacities vary up to 250 tons of coal per hr. Size of buckets varies up to 150 cu. ft. and height of lifts ranges up to 160 ft.

INDUSTRIAL VEHICLES. According to the Materials Handling Handbook this classification covers all types of material handling vehicles used in industry, excluding motor highway vehicles and public-carrier railroad cars, and includes industrial trucks, rail cars, trailers, tractors, excavating and grading equipment, off-highway carriers and agricultural vehicles used in material handling.

Fork Lift Trucks. The fork lift truck (Fig. 32), with electric, gasoline, diesel or propane gas power plant, is the most important type of **industrial truck** and is probably the best known mobile material handling device. It is built in a range of capacities up to 100,000 lb. The fork lift truck is well suited for combined horizontal and vertical movement of material. The normal industrial range of capacities is from 1,000 to 6,000 lb., measured in the form of a 48-in. homogeneous cube. This hypothetical load is normally used as a basis for rating lift trucks.

The fork lift truck is usually equipped with tapered steel forks to handle loads on pallets or skids, or in containers. However, it is possible to equip these trucks with various attachments for specialized applications. Paper roll clamps, carton clamps, cotton, wool, and scrap paper bale clamps, rotators, side shifters, scoops and end dumping devices are typical of the attachments available.

Approximately 90 percent of the fork trucks in use are within the 6,000-lb. capacity range. The standard sizes within this range are 1,000 lb. at 24 in., 2,000 lb. at 24 in., 4,000 lb. at 24 in., and 6,000 lb. at 24 in. These trucks are available with either pneumatic or semi-pneumatic tires, and usually must operate on relatively smooth areas. Figure 32 gives the general overall dimensions, together with turning radii for some typical trucks.

Fork lift trucks are built in a range of standard overall heights and maximum fork elevation heights. It is customary to use a lower load capacity for trucks where the load is elevated in excess of 144 in. The amount of reduction in load-carrying capacity for increased height of lift is largely dependent upon the type of truck, the wheel base of the truck, and operating conditions. It is recommended that the manufacturer be consulted on problems involving exceptionally high lifts. Fork trucks are available for high tiering up to 30 ft. Such applications require careful attention to the maintenance of the warehouse, plant, or area, excellent floors, and extensive driver training.

A major modification of the fork truck, equipped with outriggers to stabilize loads, is known as the **outrigger fork truck** and is intended for load carrying capacities up to 4,000 lb. It is especially suited for warehousing uniform loads and is designed for operations in a minimum aisle width. The aisle width for operation of this type of truck varies with the size of the load but is normally between 6 ft. and 7 ft. This truck is advantageous in warehouses or storerooms where selective or rack storage is required.

The outrigger type stacker is particularly useful in multistory warehouses where a low floor load is a factor. Because of small diameter wheels on the

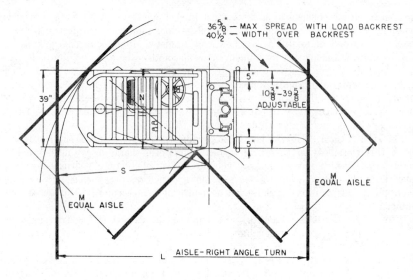

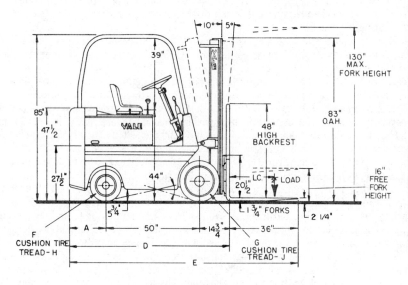

Fuel—Gas, Liquid Propane, or Diesel
Speed—Up to 10 mph
Capacity—4,000 lbs.
Load Center (LC)—24″

Dimensions (in inches):

A—$19\frac{1}{4}$	G—$21 \times 6 \times 15$	M—$67\frac{1}{2}$
D—84	H—29	N—$38\frac{3}{4}$
E—120	J—$32\frac{1}{2}$	S—$77\frac{3}{4}$
F—$16\frac{1}{4} \times 5 \times 11\frac{1}{4}$	L—$128\frac{1}{2}$	

Eaton Corp., Industrial Truck Div.

Fig. 32. Typical fork lift truck specifications.

outriggers (4 in. to permit the operation with wing type pallets) this type of truck is intended for smooth floors and should not be operated on ramps or dock plates. It is essential that the size of pallet be standardized for the operation of this type of vehicle.

Platform Trucks. Fixed, or low and high lift platform trucks are another important class of **industrial truck.** Some of these trucks are equipped with four-wheel steer and end control, making them extremely maneuverable. These trucks are widely used in handling material on skids and should be considered when quantities of large area low strength material, such as plaster board, wallboard, or thin sheet stock, are to be handled without palletizing or other dunnage. These trucks are frequently tailor-made for the specific applications such as ingot handling, foundry operations, warp beam handling, rug storage, and others.

Platform trucks are generally battery powered, but like other battery-powered units, they are occasionally equipped with gasoline-driven generator power packages.

Tractors and Trailers. The prime advantage of a tractor-trailer system lies in the combination of large pay loads and flexible routing. Individual loads can be detached en route in a distribution-type operation or they can be added in the same manner for collection-type operations. The individual trailers can be moved manually for short distances and in some applications are arranged to permit stacking by fork lift trucks. When used in combination, fork lift trucks and tractor-trailer systems tend to increase efficiency by increasing loads for long hauls and eliminating long "dead-head" lift truck runs.

Industrial tractors are available with electric or internal combustion power plants and with two-, three-, and four-wheel construction. They are built in pedestrian and rider types. Maximum draw-bar pulls range from 2,000 to 6,000 lb., the latter being sufficient to tow a trailing load of 150 tons. Normally one tractor can serve three trailer sets—one being loaded, one being unloaded, and one being moved.

The type of trailer that is used is dependent upon the nature of the load to be moved. Most loads permit the use of standard platform trucks. These can be supplied with various superstructures but the most common have removable end and side racks. The trailers can have four or six wheels. Caster wheels are preferred over rigid wheels since they provide good trailing qualities and allow maneuvering close to walls and other obstructions. Metal or rubber tired wheels can be used depending upon the load to be transported and the surface over which the trailers must travel. Trailers can be coupled automatically using self-couplers or manually with hand-operated couplers.

Dump bodies are frequently used for bulk materials to facilitate unloading. Dollies or heavy-duty flat beds are used to transport barrels and other heavy objects too heavy for the standard platform trailer.

Radio and photoelectric-cell-controlled tractors requiring no operator are most frequently applicable where the trailer train follows a preplanned repetitive route. **Radio-control systems** employ a wire buried in the floor, along the path of travel. Control signals are transmitted through the wire to a receiver in the tractor. **Photoelectric-control systems** use a line painted on the floor. A light sensitive device in the tractor guides it along the path by responding to the light reflected off the painted line. Coded controls (either mechanical or electrical) can be used to stop and start the trailer. Operatorless trailers are particularly useful for long hauls from production to storage, storage to picking, and picking to shipping.

Straddle Carriers. Straddle carriers are four-wheel inverted frame trucks designed to straddle the material to be handled (Fig. 33). The load is often handled on skid-type platforms called "bolsters." Long loads can also be handled without bolsters on lifting arms. These self-loading vehicles are used extensively to handle long and heavy materials, and are capable of highway travel.

Fig. 33. Straddle carrier.

The straddle carrier is used in a wide variety of material handling functions. Its ability to load itself with different types of cargo, to travel on highways, and to operate over rough terrain as well as paved areas has made it useful in widely differing applications. This machine, which was originally designed for handling lumber, is now frequently used in agricultural and industrial operations. **Specialized containers** have made the straddle carrier applicable to handling bulk materials and unpacked products in large loads. One example is the containerization system for transporting freight, which takes advantage of the most economical aspects of trucking and railroad transportation. This shipping and storage method can be used with truck, rail, marine, or air transportation, or in combined movements. Automatic locking devices permit the containers to be fastened securely to the bed of a flatcar, a flatbed highway truck, a ship's hold or deck, or an aircraft floor.

Walkie-Type Trucks. When small volumes, short or infrequent hauls, space limitations, and structural restrictions do not warrant or permit rider-type trucks, a pedestrian-controlled powered vehicle, or walkie, is often effective. Walkie models of many of the trucks described above are available including fork trucks, platform trucks, pallet trucks, tractors, and specialized equipment for specific applications.

Walkie fork trucks are built in capacities of 1,000, 1,500, 2,000 and 3,000 lb., all at 24-in. load center. They can be used effectively in narrow aisles and on upper floors where floor loading is an important consideration. Because of their overall size and weight they will frequently fit into existing freight elevators, permitting the use of a single unit on several floors.

Walkie-type pallet and platform trucks (Fig. 34) are very effective for quick short movements of unit loads. They are intended for horizontal movement over relatively short distances (usually less than 100 ft.). Materials to be handled should be on pallets or skids. Loose parts or bulk materials can be handled in skid or pallet bins. These units travel at normal walking speeds and are usually battery powered. Powered hand trucks are manufactured in a variety of fork lengths and widths and usually require chamfered bottom boards on the pallets with a minimum of 8 in. opening in the bottom face to permit the retractable load wheels to pass through to the floor. The steerable drive

Flow Directory

Fig. 34. Powered walkie skid and pallet trucks.

wheel is controlled from a handle and steered in the same manner as a toy wagon.

Hand trucks are built for the same purpose and with very similar characteristics as the powered trucks discussed above. They are usually impractical for loads in excess of 4,000 lb., on ramps, for distances over 100 ft., and for highly repetitive moves. These trucks should be considered a warehousing or handling tool rather than a vehicle.

Various types of hand trucks are used in manufacturing and warehousing operations (Fig. 35). Four-wheel trucks with a variety of wheel arrangements includ-

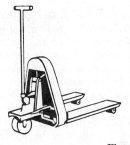

Flow Directory

Fig. 35. Hand skid and pallet trucks.

ing castered wheels, fifth-wheel arrangements, double end steering, etc., are available in capacities and sizes to meet almost any requirement. These trucks are used as trailers in tractor-trailer trains, with drag line installations, and for manual shifting and moving functions. The more common sizes have capacities up to about 6,000 lb. In some instances very large units are available.

The common two-wheel warehouse truck is also available in special designs for handling barrels, home appliances, crates, acid carboys, bales, etc. They are made of wood, steel, aluminum, and magnesium, and are equipped with iron, aluminum, rubber, and plastic treads. Few plants are completely independent of these hand trucks. They are very useful as maintenance tools and in some operating departments. Mechanical handling systems have reduced their significance, however, and they should be avoided wherever possible because of their small unit load capacity.

CONTAINERS AND SUPPORTS. Containers may be defined as special carriers designed to consolidate a large quantity of an item or items as one unit. A wide variety of standard and custom-made containers are available, designed to carry in-process and finished parts, assemblies, or products through all phases of the manufacturing cycle, including shipment of the end product. This classification covers all types of pressure, tight, loose, and open-top containers; also platform and coil supports, and all types of securements such as strapping, cinches, bulkheads, dunnage, etc.

Pallets and Skids. Among the most widely used supports are skids and pallets of which there are many varieties. **Skids** (Fig. 36) are platforms made

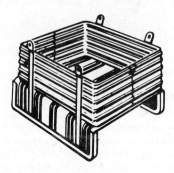

Fig. 36. Typical skids.

of wood or metal with enough underclearance to admit the platform of a platform lift truck. They can, of course, also be handled by fork trucks.

Skid bins permit handling heavy loads of small parts by platform lift truck or fork lift truck. Some are designed for tiering. The corrugated steel skid bins usually have sling eyes for crane handling. Sectional sided bins provide adjustable capacity and easier access to the contents in the bottom. **Skid racks** hold stacked loads of parts and offer better accessibility than bins.

Pallets are elevated platforms with lower underclearance than skids. Single-faced pallets are simple wood platforms on runners. High stacking or tiering of single-faced pallets may be feasible if the top of the load can withstand the concentrated forces imposed by the runners. Double-faced wooden pallets are the most widely used. The lower face often provides clearance for the wheels of a hand pallet truck. When pallets admit forks on all four sides, they are known as four-way pallets. When they can be entered from all four corners as well, they are called eight-way pallets. Two-way double-faced pallets have the largest load-bearing area on the lower face. This is the principal advantage of the two-way double-faced pallet. Double wing-type stevedore pallets provide for bar slings or spreaders to hoist unit loads by ship tackle or pier cranes. Plywood can be used to make pallets light and strong.

Steel pallets are used to withstand hard usage such as occurs in the metal-fabricating industries. Several types are available. Steel wire mesh pallets are light in weight, but the bearing area is comparatively small. In some applications the fact that they do not absorb odors is important. They are also used where drainage of liquids is a factor. Aluminum pallets combine strength and lightness, and usually avoid the corrosion problem.

Expendable paper pallets were developed to solve the problem of the expense for returning pallets accompanying shipments. Moisture resistant paper pallets withstand static loads up to 18,000 lb. Prescribed stacking patterns must be followed. Another type paper pallet rests on three runners consisting of square, diagonally braced paper tubes. Expendable pallets may also be made of wood or metal. One type is of wire-bound wood construction.

Pallet frames and other devices are often used for handling and storing irregularly shaped or easily crushed merchandise. They are made in a wide variety of designs.

The National Wooden Pallet and Container Association has established standard nomenclature and a conventional dimensioning system which is generally accepted in industry today.

Pallet Loaders and Unloaders. Pallet loaders are capable of unitizing cartons and bagged goods in a variety of pallet patterns. Modern automatic pallet-loading equipment is capable, through an electronic programming device, of handling many different pallet patterns through a single machine. The pattern can be changed by an operator or by a predetermined schedule.

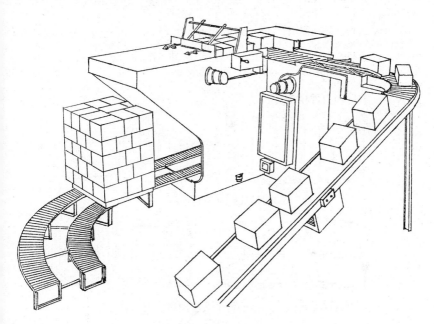

Fig. 37. Pallet loader.

The **pallet-loading machine** (Fig. 37) consists of a pallet-feeding device into which a lift truck places a stack of empty pallets; a pattern-forming device which places the cartons or bags in the proper arrangement for the pattern; a tier-loading mechanism which places the tiers one upon the other and lowers the pallet to accommodate the next tier; and a conveyor removal system for delivering the loaded pallet into a conveyor system or to a fork lift truck pickup point.

A **semi-automatic pallet loader** consists of a ball table or power-operated pattern table upon which a man forms the tier pattern. Each tier is then automatically loaded onto the pallet in the same manner as in the automatic machine. Semi-automatic bag palletizers handle 600 to 1,200 bags per hour, while automatic pallet-loading equipment handles 1,500 to 1,800 cartons or cases per hour, and semi-automatic carton- and case-palletizing machinery falls some-where in between, depending upon the weight and size of the package.

Automatic **pallet unloaders** reverse the process. A pallet load of cartons or cases is placed in the machine and the machine unloads it, aligns the packages, and feeds them onto a conveyor. This unit is particularly helpful in food pack-ing and bottling plants for delivering empty bottles or cans in cartons or cases to the filling line.

The trend toward development of palletless unit-load handling methods has resulted in the development of modified automatic pallet-loading machines capable of building unit loads without pallets. These machines can be used with carton clamp and other palletless handling lift-truck attachments.

The question of the economic justification of unitizing or pallet-loading equip-ment depends upon the economics of the particular installation. In most cases a high rate of production is a prerequisite to economical use of this type of equipment.

Containers. In addition to cartons, barrels, drums, and pallets, this classifi-cation includes a variety of large containers for unit-load shipment of packages, parts, and bulk materials. Basic types as illustrated by Dean and Norton (Mechanical Engineering, vol. 78) are shown in Fig. 38. The collapsible bag

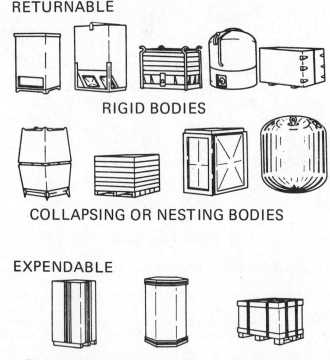

Fig. 38. Basic types of unit-load containers for shipping bulk products.

(balloon) handling method is an effective unit bulk handling technique. This technique permits shipment of bulk products in boxcars or open gondola cars in units of up to 400 cu. ft. In most cases these bags are handled into and out of the cars with cranes or by fork lift trucks with crane attachments. The containers are stored on the ground and are placed on steel frame supports with standard, flexible-type, bin valves and squeezing devices to convert them into service hoppers to feed the material directly into the manufacturing process. The bag, usually made of Neoprene, is collapsed for the return trip to the supplier. This method is particularly adaptable for such hard-to-handle materials as lamp black.

Racks, Shelving, and Bins. In many instances, storage operations require auxiliary equipment which permits high cube utilization with small lots while maintaining selectivity. Pallet racks, shelving, bins, and stackable bins, fall into this category.

Pallet racks are simply shelves for pallets which make it possible to remove the unit load at the bottom of the stack without disturbing those above it. They are used for irregularly shaped loads, short lots, and picking lines. Fig. 39 shows a typical example of a pallet rack in which short lots of various-shaped items are stored.

When open-stock items must be handled, storage equipment is usually required. Large quantities of such items can be handled in stackable pallet boxes from

Fig. 39. Pallet racks for stacking of uneven loads.

which single pieces can be withdrawn through an opening at the end even when other boxes are stacked on top. Smaller quantities are often handled in bin boxes of similar design. Small quantities are often stored in permanent shelf units, drawers, or rotary bins. There are many types of specialized racks such as barrel racks, reel racks, etc.

The flow, or live, rack is widely used as a convenient storage medium (Fig. 40). A series of roller or skate wheel conveyors are placed one above the other

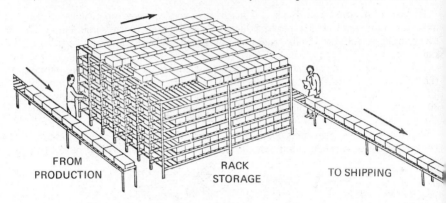

FROM PRODUCTION   RACK STORAGE   TO SHIPPING

Fig. 40. Flow rack.

in side by side stacks 10 to 20 ft. or more in length. These units slope slightly from the input (back) to the output (front) end. They are quite useful for accumulating packages or tote pans into unit loads at a sorting station and for maintaining a first-in, first-out process inventory between operations. The box or tote pan is placed in the proper slot (conveyor) on its arrival at the storage point and it takes its place in an automatically scheduled moving storage for delivery to the next operation. This system can be very useful as a scheduling device for job machine shops and other diverse operations. It is also widely used for manual order picking and as a basis for many automated warehousing installations, with automatic release of cartons to a moving belt.

The **live floor** storage system is another method which is gaining acceptance. It operates on the same principle as the conveyor rack described above, except that it is a single-level operation. This principle is used in the car slot storage system in warehousing. Live floor storage is also quite useful in manufacturing operations which require the handling of bulky products in a continuous flow. Appliance plants are particularly adaptable to this system. It can also be used as a machine-loading device whereby pallet loads of tote pans or in-process materials can be loaded on a conveyor leading to a production operation and thus provide an automatic scheduling and storage function.

Material Handling at the Workplace

DEFINITION AND SCOPE. It is sometimes thought that all material handling projects are extensive; all handling equipment is large; and any handling activity requiring less than the systems approach is of little or no concern to the material handling engineer. In practice this is far from the truth.

Material handling at the workplace is defined as that handling which must be done after material has been delivered and set down for use at the workplace, and before it is picked up and moved to the next operation. It can normally be considered to cover no more than three or four feet in any direction from the center of the machine, bench, or workplace. Yet it constitutes one of the greatest opportunities for reducing handling costs. For convenience of analysis, workplace material handling can be divided into five individual phases:

1. **Preparatory handling** of material at the workplace
2. Handling material **into the workplace** or machine
3. Manipulation of material **within the workplace**
4. **Removal of material** from the workplace
5. **Transfer of material** to the next workplace (this portion of the activity may be considered traditional material handling activity)

CHARACTERISTICS OF WORKPLACE HANDLING. The following characteristics of material handling at the workplace suggest the uniqueness that separates it from traditional material handling:

1. Individual handling distances are relatively short.
2. The number of moves is usually high, and is a function of the number of cycles performed.
3. The greatest portion of the operation cycle time at an individual workplace is taken up by material handling.
4. Much of workplace handling is commonly classified as direct labor.
5. Much of the handling activity at the workplace is subject to improvement by traditional material handling techniques and devices.

Effect on Costs. Workplace handling operations present another area for investigation and identification of cost improvement possibilities. In a typical production operation, as shown in Fig. 41, it is found that the actual **machine**

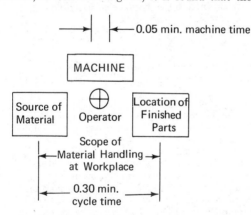

Fig. 41. Cycle time for production operations.

time portion of the total cycle is relatively small. In terms of the relationships shown in the figure, 25/30ths of the total cycle time (83⅓%) is devoted to handling at the workplace.

Planning for Effective Handling. In planning an efficient workplace, the material handling engineer should give due consideration to the principles of workplace layout, as stated by Apple (Plant Layout and Materials Handling):

1. Tools, gages, material, and machine controls should be located close to and in front of the operator to minimize motions and effort.
2. Provide a definite place for tools, gages, and material.
3. Use gravity when possible to feed and remove material.

4. Pre-position material and tools within the workplace.
5. Deliver material directly to point of use.
6. Plan prompt and efficient removal of material from workplace.
7. Locate materials within the workplace so they can be reached with the most efficient sequence of motions.
8. Plan proper height relationships between material supply, point-of-use, and disposal.
9. Coordinate each workplace relative to preceding and following operations.
10. Leave sufficient space at the workplace for efficient delivery and removal of material.
11. Plan for storing a practical minimum of incoming material and finished work awaiting removal at the workplace.
12. Be sure handling equipment at the workplace is properly integrated into the overall handling system.

Implementing the Plan. The following suggestions will be helpful in implementing the above principles:

1. **Preparatory handling of materials at the workplace**
   a. Use gravity chutes, conveyors, etc., to move material to the point-of-use.
   b. Request unit packs from vendors in place of individual cartons, packages, etc., for convenience at workplace.
   c. Have material delivered to the correct location and position on the first delivery, not merely to the general area.
   d. Plan for mechanical delivery, if practical, to assure a uniform, constant, regulated, uninterrupted flow.
   e. Allow sufficient space for convenient handling of required material.
2. **Handling material into the workplace or machine**
   a. Use mechanical means, if practicable.
   b. Use feeders for individual pieces, sheets, coils, bars, etc., to increase both efficiency and safety in loading.
   c. Provide for a uniform flow or supply.
3. **Manipulation of materials within the workplace**
   a. Review principles of motion economy (see section on Motion and Methods Study).
   b. Review principles of workplace layout.
   c. Use "convenience" devices such as magnetic sheet separators, lift tables, vibratory hoppers and feeders, containers with counterbalanced bottoms, special trays and racks, etc.
   d. Use indexing devices and similar aids.
4. **Removing materials from the workplace or machine**
   a. Allow space for storing a normal amount of finished materials.
   b. Plan for removal at proper intervals to avoid congestion.
   c. Arrange for mechanical removal of materials.
   d. Plan for prompt removal of chips, turnings, scrap, etc.
5. **Transferring material to the next workplace or machine**
   a. Use mechanical devices or aids.
   b. Use gravity where practical.
   c. Move material promptly to avoid accumulation.
   d. Move material in unit loads where practical.

Material Handling Equipment at the Workplace. An important aspect of the whole concept of material handling at the workplace, is the availability of a variety of material handling devices which can be used to eliminate, reduce, or simplify such handling activities. These devices, singly or in combination, along with custom designed devices of a similar nature, and conveyors, can be extremely helpful in minimizing material handling at the workplace. This **equipment** may be classified as follows:

1. **Feeding.** Delivers material closer to, or into the work area.
2. **Positioning.** Places or facilitates the placement of material into the proper position for the operator or machine.
3. **Holding.** Maintains the desired relative position between the material and the operator or equipment.
4. **Indexing.** Sequentially positions material into working position.
5. **Cycling.** Automatically (or semi-automatically) initiates next work cycle.
6. **Transferring.** Moves material from one workplace to the next.
7. **Miscellaneous.** Not otherwise classified.

Trade magazines and advertising literature contain information on a wide variety of devices that are normally overlooked by the material handling engineer. An in-depth investigation of the workplace operations may reveal new areas for cost savings in material handling.

The Systems Concept in Material Handling

ADVANTAGES OF SYSTEMS APPROACH. If the typical material handling situation is represented by the traditional definition given at the beginning of this section, then the systems approach is merely a larger scale, results-oriented procedure that is much broader in scope, greater in depth, and interdisciplinary in nature.

Hare (Systems Analysis: A Diagnostic Approach) states that the **systems concept** is distinguished by its disregard of the traditional rigid boundaries of functions and departments. By cutting across these lines it attempts to discover and sort out the tangle of factors involved in a complex situation and find relevant and reliable interrelationships between significant factors. This more detailed investigation and analysis makes it easier to diagnose, conceptualize, design, and evaluate complex combinations of sophisticated hardware, organizational interrelationships, and information flow. It usually results in an integrated composite of facilities, activities, and information flow and encompasses as much of the total problem environment as is feasible and economical.

Objective. The overall objective of the systems approach to a material handling problem is to conceptualize a total solution to the entire problem, and then to implement those portions that are technologically possible and economically feasible. The total solution could involve—in the implementation of some of its phases—some aspects that would be either impossible or unfeasible to implement at the present. Therefore, the application of the systems approach should provide for a continuing investigation, with the ultimate goal of implementing additional phases as the means become justifiable.

BASIC PRINCIPLES OF SYSTEMS DESIGN. The systems approach involves a more sophisticated treatment of a material handling problem situation requiring a more thorough and rigorous investigation. As a result, the solution should be of a higher calibre than if the problem was approached from the traditional point of view. In fact, Nadler (Work Design) states that "If your guide for planning a system is a perfect system, your final system will be better than if your guide was something else."

It is aiming toward the perfect system that also forces the interdisciplinary approach, thereby resulting in the expanded scope and greater depth of the problem investigation. The following basic principles for **designing the ideal system** are based on Nadler (Work Design):

1. **Eliminate the need for the function.** This is a basic principle applicable to almost all functions, especially at the conceptual stage. As the system level gets higher this principle will be applied less frequently.

2. **Specify one low-cost input and one low-cost output.** The number of criteria for specifying inputs and outputs is quite high and by minimizing them a corresponding reduction or elimination of other elements in the system may be effected. This in turn can reduce the cost of the individual elements and the overall system. Standardization is an important result of this principle.

3. **Automate.** Automation begins at the input and continues through to the output. Make certain all handling is free of human effort. If it is absolutely necessary for a workman to handle material at the input stage, once it is in the process, no further human effort should be necessary. In addition automatic data processing should be used to eliminate transferral, review, and record keeping by production and clerical personnel.

4. **Use adaptive control.** The control elements should respond automatically to any deviation from the norm and correct these deviations without human intervention. Quick transient response will prevent overreactions in other parts of the system. The ideal system, of course, would not require any correction.

5. **Utilize personnel skills to their utmost.** Whenever human effort is required, the skill of the individual should be used to its maximum, theoretically 100 percent of the time. The principles of methods and motion study should be applied to their fullest.

6. **Design systems for normal conditions before dealing with exceptions.** The ideal system is always designed along the lines of the norm. The normal conditions are assumed to occur 100 percent of the time. Once the norm has been identified, the exceptions and deviations may be equated with the norm and proper action taken. The norm is not always a regular cyclic pattern but may be some recurring routine. The ideal system would then be built around the most prevalent routine.

SYSTEMS PROCEDURE. Although the procedure for analyzing a handling problem from the systems point of view is more comprehensive than that used in solving traditional material handling problems, some of the steps are necessarily similar or identical. The systems approach includes the phases and steps discussed in the paragraphs that follow. It should be mentioned, that it is not necessary for the procedure to be followed in the sequence outlined. While some problems require the entire procedural approach, many will not because of the nature of the problem. It may be necessary to perform some of the steps out of sequence due to the problem requirements or limitations. However, it is wise to check each step, in order to be sure that no important item or factor is overlooked during the analysis.

Problem Definition. Identify the problem. One of the most important steps in analyzing a handling problem is that of identifying the problem. One way of approaching this is with the use of a "Preliminary Material Handling Survey Checklist" (Fig. 3). This form represents the equivalent of many of the principles of material handling and their corollaries in abbreviated form. They are restated as indicators in order to imply questions to be asked while observing each plant area in question. Rather than use it on too large an area, it is better to use two or more separate sheets. Each indicator observed should be checked in the "yes" column, and each check mark further investigated in an effort to properly identify material handling problem areas.

Determine the scope of the problem. The next step is to determine the complete scope of the problem in terms of the philosophies previously expressed. It is important that the problem should not be solved out of context, with significant activities or related situations either overlooked or ignored. The entire scope must be identified so that the complete problem can be both defined and solved within its total framework and in proper perspective to related activities.

Define the problem. Only after the problem has been properly identified and its complete scope established, can the **total problem** be accurately defined. The definition will serve as a review of previous thinking and help to establish parameters or boundaries for the investigation.

Establish objectives. Having defined the problem, it is easier to establish the objectives. They should be clearly stated in terms that will make it possible to check on the degree to which they are achieved in the proposed solution. A clear statement here will also facilitate auditing the problem solution, after it has been installed.

Problem Investigation. Determine what data to collect. Just as important as collecting the data, is the need for establishing what data should be collected. The Basic Data form (Fig. 42) should be useful at this point.

This form carries out the theme of the **material handling equation,** and is based on the three major phases of the problem, plus the ten primary factors and their subdivisions. It should be emphasized, however, that the form will not solve a problem! It will only serve as a guide to the organized collection of the necessary data.

Establish work plan and schedule. Depending on the scope and complexity of the problem, this step might involve one or more of the following:

1. Meeting with those persons who will be concerned with the problem and discussing it with them.
2. Grouping similar and/or related materials, problems, or activity areas for convenient treatment as a single problem.
3. Breaking the overall problem down into smaller segments to facilitate their analysis as separate handling problems.
4. Preparing plans to study the problem, including determination of who is going to do what, what additional data are required, etc.
5. Establishing a detailed work schedule for each major activity.
6. Approving the work schedule.
7. Assigning responsibilities for each project phase.

Collect data. The collection of the data will probably require the analyst to:

1. Review sources of data.
2. Establish project relationships with other company functions and activities.
3. Carefully observe the activities being analyzed to make sure that they are thoroughly understood; or carefully synthesize the problem situation in the event that no situation exists for analysis.
4. Obtain complete data on the material, move, and method (actual or proposed) and record it in a convenient form.
5. Get supplementary data such as schedules, layouts, building drawings, equipment details, etc.
6. Obtain a layout of the area under study and be sure the layout is complete and accurate.
7. Obtain data on the material flow, in order to cover the move phase of the material handling equation. Some of the following sources of information or analytical techniques may be useful at this point:
   a. Bill of materials
   b. Assembly chart
   c. Production routings
   d. Operation process chart
   e. Multiproduct process chart
   f. Process chart
   g. Flow process chart
   h. Flow diagram
   i. From-to chart
   j. Activity relationship chart
   k. Activity relationship diagram
   l. String diagram
   m. Memomotion pictures

MATERIAL HANDLING ANALYSIS—BASIC DATA

Plant _____ Building _____

Area _____

Problem _____

_____

_____ compiled by _____ Date _____

(Attach Process Chart and/or Flow Diagram)

(Check and/or fill in, as applicable)	Remarks, Explanation, etc.

MATERIAL — Part No(s). _____

    Description _____

1. Physical Characteristics

    Form: solid _____ liquid _____ gas _____ other_____
    Type of material, load:
        individual item _____ unit load _____
        packaged _____ bulk
    How Received _____
                    (carton, pallet, drum, bag, etc.)

    Nature: fragile _____ sturdy _____ bulky_____
    Properties: dimensions_____
        shape_____ wt./item _____
        density_____

2. Quantity   annual quant._____
               quant./deliver_____ max. inventory _____
               lot size _____

MOVE

3. Scope (check each phase applicable)

☐ Packaging by vendor   ☐ Interdepartmental
☐ Packing by vendor   handling
☐ Loading by vendor   ☐ Service & auxiliary
☐ Common carrier   operations
☐ External transportation ☐ Quality control
☐ Unloading   activities
☐ Receiving   ☐ Packaging
☐ Material storage   ☐ Packing
☐ Material issue   ☐ Finished goods
☐ Production activities   warehousing
☐ Intradepartmental   ☐ Stock picking
   handling   ☐ Order assembly
                ☐ Loading operation
☐ Workplace handling   ☐ Shipping operation
                ☐ Carrier-from plant
☐ In-process storage   ☐ Intraplant handling

4. Source & Destination

      Source   Destination

☐ vendor   ☐ receiving
☐ carrier   ☐ storage
☐ storage area   ☐ point-of-use
☐ work station   ☐ work station
☐ fixed   ☐ fixed
☐ variable   ☐ variable
☐ other   ☐ other_____

Fig. 42. Basic data required for analysis.

17·76

4. Source & Destination (continued)

Remarks, Explanation, etc.

Building Characteristics

Source		Destination
_____	aisle width	_____
_____	column height	_____
_____	truss height	_____
_____	truss capacity	_____
_____	floor construction	_____
_____	floor load capacity	_____
_____	floor condition	_____
_____	power	_____
_____	openings(no.,size)	_____
_____	dock	_____
_____	duct work	_____
_____	sprinklers	_____
_____	elevators	_____
_____	other	_____

Common Carrier Characteristics

Type_____ Volume_____
Dimensions _____ Load capacity_____
Door size _____

Storage Area & Volume

_____ Length x _____Width x _____Height

5. Route

Distance_____
Area Covered: size _____
☐ fixed   ☐ variable
Path
☐ straight   ☐ curved
☐ fixed   ☐ obstacles
☐ variable   ☐ combination

Course
☐ fixed pt.–fixed pt.   ☐ var. pt.–fix. pt.
☐ fixed pt.–var. pt.   ☐ two-way
☐ var. pt.–var. pt.   ☐ other_____

Direction/Plane

☐ horizontal   ☐ multi-level
☐ vertical   ☐ combination
☐ incline   ☐ ramp____Length x____Width
☐ single level   ☐ ramp_____ % grade

Level

☐ on floor   ☐ overhead
☐ working ht.   ☐ level/level

Location

☐ inside   ☐ between bldgs.
☐ outside   ☐ beyond bldgs.

☐ Operations in Transit   ☐ Cross Traffic

6. Frequency

☐ Regular   ☐ Continuous
☐ Irregular   ☐ Intermittent
☐ Unpredictable   ☐ Reciprocating

Fig. 42. Continued.

17·77

	Remarks, Explanation, etc.

7. Speed or Rate

Speed

☐ uniform ☐ synchronized
☐ variable ☐ other ______

Rate – ______ items/ ______ (time period)

☐ fixed ☐ pounds/hr. ______
☐ variable ☐ ft./min. ______

METHOD

8. Load Handled ☐ uniform ☐ variable

	Alternative 1	Alternative 2	Alternative 3
Type			
Size			
Construction			
Tare Weight			
Items/load			
Weight of Load			
Loads/total quant.			
How carried			
Disposal			
Cost	$______	$______	$______
Pro-rated cost/load	$______	$______	$______

9. Equipment

Desired Characteristics

☐ powered ☐ self-loading
☐ operator req'd ☐ tiering/stacking
☐ mobile ☐ elevate/lower
☐ control ☐ positioning
 ☐ manual ☐ transferring
 ☐ automatic
 ☐ remote

Equipment indicated

☐ none ☐ combination
☐ manual handling ☐ comm. carrier
☐ conveyor ☐ rack
☐ crane/hoist ☐ other ______
☐ truck

Cost/hour $______

Time/move ______ equip. ______ men ______

% Capacity required for this move:

weight capacity ______ % time capacity ______ %

10. Manpower (load, move, unload)

Time/load ______ Hourly rate $ ______

Cost/load $ ______

Fig. 42. Concluded.

8. Obtain information on each item of existing material handling equipment.
9. Procure data on the manpower requirements of the activity.
10. Tabulate all information pertaining to the storage aspects of the problem.
11. Investigate and analyze procedures and techniques concerning the communications and control aspects of the problem.

Develop, weigh, and analyze data. This step is concerned with the evaluation of the information on hand. It might involve such activities as:

1. Sorting and classifying information into major aspects of the problem.
2. Checking information for accuracy and completeness.
3. Summarizing data on each phase of the problem.
4. Determining practicable ranges of data.
5. Averaging, weighting, or otherwise treating data to assure that the information is in the most usable form.
6. Developing charts, graphs, etc.
7. Checking for inconsistencies, omissions, errors, irrelevant data, etc.
8. Summarizing the data for use in developing improvements, or solutions to the problem.

System Synthesis. Conceptualize system possibilities. One of the distinguishing features of the systems approach is that it requires the analyst to stretch his imagination far beyond the boundaries within which he is used to working. This provides a greater opportunity for conceptualizing or creating alternative possibilities for the solution of the problem at hand. Conceptualizing is frequently done in the form of a flow chart (Fig. 43), where each of the symbols represents a major activity or function of the potential system.

If the total system appears too large to be manageable as a single system, it may be necessary to divide the total system into subsystems. This will very likely require a new flow chart for each subsystem, and probably additional flow charts for the several alternative possibilities for implementing each alternative or subsystem.

Structure alternative systems. At this step the systems approach becomes more pronounced. Computer simulation of the proposed solutions or alternatives may be involved. In fact the flow chart is the most common preliminary step in the structuring of a problem solution for possible computer simulation. The procedure is as follows:

1. Sketch a preliminary flow chart of the entire system or systems.
2. For each step, function, or activity shown on the flow chart, determine the parameters, characteristics, factors for consideration, etc.
3. Establish the constraints to be met or complied with by the system.
4. Develop a "block diagram" of the system logic. A typical block diagram is shown in Fig. 44.
5. Convert "model" to computer language. The block diagram (Fig. 44) is frequently referred to as a "model." When it is complete in terms of the **logic flow**, its meaning is converted into a **computer language**. One of the more common computer languages useful in the material handling field is called "General Purpose Systems Simulation," commonly referred to as GPSS. Figure 45 represents the conversion of several steps of the preceding block diagram into the GPSS language, through the use of punched cards.
6. Run the model on the computer and carry out the steps called for. This is the way the model is manipulated, or put through its paces to check the reasoning, parameters, constraints, and the results of the operation, as a means of testing out the proposed system. Several sets of data would be utilized in making the runs.
7. Analyze results. The analysis of the computer runs would normally include a comparison of the simulations of several alternative systems or subsystems.

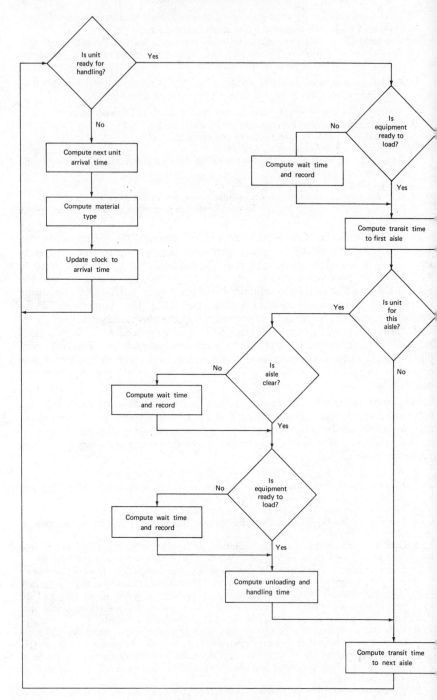

Fig. 43. Flow chart.

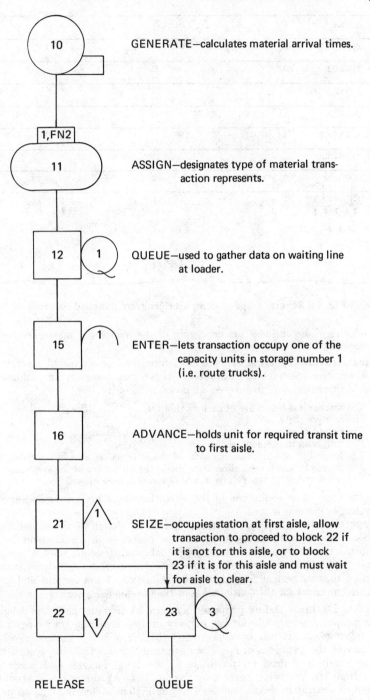

Fig. 44. Portion of block diagram.

21	SEIZE	1			FN	3		
2	QTABLE	2	0	1	100			
20	QUEUE	2			21			
16	ADVANCE				20		2	
15	ENTER	1			16			
1	QTABLE	1	0		100			
12	QUEUE	1			15			
11	ASSIGN	1	FN2		12			
10	GENERATE				11		1	FN1

Fig. 45. Several steps of computer program punched on cards.

This step also includes the debugging of the model and making corrections necessary in the program.

Select feasible systems. Having structured and/or simulated several possible alternative solutions to the problem, it becomes necessary to evaluate the available alternatives. This involves an investigation of:

1. The mechanical feasibility of each alternative.
2. The relative costs.
3. The potential savings.
4. The return on investment.
5. The feasibility of implementing some of the alternatives at a future date.
6. The degree to which each alternative meets the objectives of the system.
7. The relative ease of supervising, maintaining, etc., each alternative.

On the basis of an evaluation of the above factors, a preliminary selection of the system or systems is made.

Prepare preliminary justification. Before the design of the actual system can take place, it is advisable to present the proposals to management for approval. At this stage, a cost estimate is probably not feasible, since it would be difficult to specify any of the equipment, etc. However, enough detail must be presented to management to permit their evaluation of the concept and to obtain their approval for the continuation of the development work.

System Design. Define proposed system. A true and complete definition of the proposed system cannot be properly drafted until the preceding actions have taken place. It will be remembered that Step 3 was concerned with a definition of the **problem area.** The distinction here is that the system must also be carefully defined on the basis of the facts, factors, and conclusions drawn from the preceding portion of the analysis. At this stage it should be possible to accurately define and describe the system in terms of its scope, its redefined objectives, and its expected results.

Establish functional specifications. Having defined the proposed system, it is now necessary to establish general specifications for the hardware required. It is possible at this point that some of the equipment items will be fairly well identified, so that their specifications may be recorded. It is also necessary to conceptualize the plans for the information system and to begin preliminary work on the design of the records, forms, etc., to be used in the operation of the system. In addition, it will be necessary to prepare preliminary plans for the design of the control aspects of the system. This will include the integration of the information phases of the system with the plant's electronic data processing activity.

Develop preliminary budget for implementation. Having structured the proposed system in terms of functional specifications, the next step is to select the potential suppliers of equipment or services, and submit specifications and a request for a quotation. Upon receipt of bids from the interested suppliers, it will be necessary to study them carefully and evaluate them relative to the specifications. After bids for all key items have been evaluated, it will be possible to prepare at least a **preliminary** budget for the proposed system.

Develop and design system components. After the preliminary budget has been approved and the project released by management, it is possible to proceed with the detailed design of system components. This will involve the functional design, in preparation for fabrication or procurement of all major segments of the system. It will be necessary to establish the responsibilities of the user, the vendors, and the consultant, if one is assisting in the design process. An important phase of the functional design is the establishment of performance requirements to be used as guides by the suppliers as well as bases for measuring the effectiveness of the components.

It may be necessary to prepare prototypes of selected portions of the system, and to specify the terms and conditions under which pilot operations may be necessary. It may be desirable to measure the system's performance in terms of labor content, costs, or other unit. All of the above applies not only to the hardware aspects of the proposed system, but to the information phases and the control devices as well.

Evaluate development progress. It is good practice to evaluate progress periodically during the design of the complex system. It is therefore desirable to have a planned evaluation process as a part of the total systems design procedure. This provides a definite place to measure progress against the original objectives and time schedule.

Prepare justification report and/or presentation. Preparation of the necessary justification of the proposal for management review and approval can proceed as outlined below:

1. Compare the costs of the present and proposed methods of operation.
2. Summarize capital investment required for the proposed methods.
3. Determine the rate of return on the investment, according to company policy.
4. Review the applicability and importance of intangible gains.
5. Review the proposed solution in terms of the original problem definition and objectives.
6. Present the justification clearly and accurately.

Company policy will usually dictate whether the problem solution should be presented in the form of a report, an oral presentation, or a combination of the two. In any case, the preparation will involve:

1. Definition of the problem.
2. Brief statement of proposed solution or solutions.

3. Statement of installation cost.
4. Statement of expected savings in operating expenses.
5. Detailed cost and economic analysis, with a comparison of the present method to the proposed alternative or alternatives.
6. Procedure for putting the proposal into effect.
7. Recommendation of any other potential improvements uncovered by this study.
8. Appendix, which might include such items as:
   a. Drawings and blueprints.
   b. Detailed calculations.
   c. Specifications for equipment.
   d. Detailed manpower requirement.
   e. Further explanation of any items needing clarification over and above content of report proper.
   f. If advisable, reference to alternative plans and/or equipment considered but not accepted.
   g. Selected exhibits of data and information.

The contents will vary, of course, with the individual problem, as well as with company practice. If desired, plans should be made for an oral presentation of the report, using appropriate visual aids such as slides, photos, charts, mock-ups, models, etc.

Obtain approvals. Approval is not only necessary but usually advisable as a means of creating an awareness of the problem among higher levels of management. In fact, acceptance as well as successful operation of the proposed plan may hinge on the fact that management personnel affected by the proposed plans were given an opportunity to be a part of the decision-making group.

Revise as necessary. In many cases immediate approval of the original proposal may be withheld pending a further investigation of several items, or questions by the review group. If such changes become necessary they should be made, approved, and incorporated into the problem solution.

System Implementation. Organize for procurement. The procurement of a system which may cost hundreds of thousands or even millions of dollars, requires organization prior to the actual purchasing operations. This involves making detailed plans for the investigation of each major segment of the system, arranging contacts with the several parties who might be involved in the procurement process, both inside and outside of the buying organization, and properly scheduling all procurement efforts to assure proper coordination of the parties. It is important that all assignments be made with one individual in overall control of the procurement process.

An important aspect of the procurement problem is that of establishing the responsibilities of the user, vendor, consultant, and architect. The more complex the system, the more important it is that all individuals know their specific responsibilities in the implementation of the system. These should be spelled out in sufficient detail that they can be adequately followed up during the implementation process. When applicable it would appear wise to establish a schedule to insure, as well as facilitate, the coordination of the several individuals and parties concerned.

Procure equipment. After the preliminary work, it is time to actually procure the equipment required for the proposed system. Activities connected with obtaining the required equipment include the following:

1. Obtain firm quotations and delivery dates on all items to be purchased.
2. Tabulate information on quotations received to assist in the evaluation of details and the comparison of vendors.
3. Select vendors for all components and issue purchase orders (or take necessary steps to initiate fabrication of components).

4. Determine company manpower requirements and budget both time and money for their contribution to the installation.
5. Establish a time schedule for the installation, preferably in the form of a CPM network, a Gantt chart, or both.

Any prototype or pilot installation must be built, run, checked out, and revised as necessary, prior to the procurement activity.

It may also be wise in some cases to include, as a part of the purchase order, a note regarding vendor followup at predetermined intervals, or requiring proper operation for a specified period of time—without breakdowns—before payment. It may also be desirable to spell out customer and supplier responsibilities as a part of the purchase order.

Plan for training of operators and managers. Since a complex new system might be considerably different from previous methods of operation, proper training plans must be developed in detail for all those who will be concerned with its operation. Orientation sessions must also be prepared for all management personnel who will be associated with the installation.

Prepare for employee and public relations situations. Plans must be prepared to deal with such problems as reductions in personnel or changing classifications of personnel. It may also be necessary to consider any public relations problems created by the proposed installation. In either situation, the consequences of the installation may be either negative or positive. If the reactions are likely to be negative, steps should be taken to counteract their effects. However, in a good many cases it will be found advisable to capitalize on the prestige value of a new, expensive, and complex installation. Those aspects worthy of promotion should be promulgated among operating personnel, management personnel, customers, vendors, the community in general, and stockholders to take advantage of the publicity value of the installation.

Supervise installation of equipment. A number of activities and problem areas must be considered in conjunction with installation of equipment. Allan Harvey (Proceedings of the AIIE Annual Conference) suggests the following:

1. Delivery of equipment
   a. Changes in delivery schedule
   b. Availability of building resources
   c. Equipment protection
   d. Responsibilities for:

Construction equipment	Building permits
Cranes	Building codes
Utilities	Resolving problems of changes
Insurance	

2. Coordination and supervision
   a. With architects and contractors
   b. Coordinating and resolving differences among trades
3. Inspection at predetermined stages of the installation
4. Establishing the roles of the several parties during the installation process
5. Integrating the information system with the physical system
6. Planning the move sequence
   a. Parallel operations
   b. Phasing-out procedures

If possible, the installation should be made with guidance of a PERT network as an aid in scheduling the various phases of the project and the specific items of equipment. If the building is new, it will be necessary to insure that the site will be ready to receive equipment when it is scheduled to arrive. On the other hand, if the installation is being made in an existing building, it will be necessary to develop plans so that production operations can continue as much as is possible during the installation.

Start-up and debug. With the initial operation of any complex system, it will be found necessary to establish debugging procedures, schedules, and responsibilities. Certain difficulties can be anticipated and steps taken to alleviate negative consequences. The various types of problems and downtime situations should be logged, as an aid to their evaluation and subsequent correction. It should be remembered that acceptance testing is a normal part of the start-up operation and previously prepared tests should be performed early in the period as well as at scheduled intervals to assure that design criteria and objectives are achieved. Agreement will have to be reached between parties on any point where performance does not meet expectations.

Audit and follow-up. In the period following the acceptance of the installation it will be necessary to continue the follow-up process in order to assure uninterrupted operation of the system in accordance with original goals and objectives. In order to assure an orderly follow-up procedure, it is wise to establish responsibilities for specific segments of the system and assign personnel to each. In making observations, evaluations, etc., the follow-up personnel should pay particular attention to equipment warranties. They should also watch closely for such unforeseeable events, as items or components requiring excessive maintenance, safety hazards, material build-up on components, and jams and conditions which are subject to change after the installation has been made.

During the audit of the new system, it will again be necessary to pay close attention to the responsibilities of the several parties involved. Some areas of concern might be the following:

1. The customer should carefully review all contractual arrangements in order to be sure that all parties have carried out their respective duties and responsibilities.
2. Since the supplier is at a disadvantage, he will very likely be motivated to correct any difficulties, in order to gain acceptance and preserve his reputation.
3. The customer can lose a considerable amount of money if the system does not operate properly, therefore, problems of contingent liability should be carefully explored and settled.
4. The vendor could be liable for unsafe conditions.
5. All back-up systems should be checked out to make sure that they will operate properly in case of a breakdown.
6. A continual watch should be made in order to assure that the original problem analysis was complete, and that some oversight or neglected aspect of the problem has not resulted in a suboptimal solution to the problem.
7. In order to assure adequate follow-up over a reasonable period of time, personnel should be specifically assigned to perform necessary inspections and evaluations, with reports to appropriate personnel.

The last aspect of the system audit should concern an **evaluation of the accomplishments** of the installation. Some factors worth considering are:

1. Degree of performance versus objectives.
2. Indications of overmechanization.
3. Pay-off not up to expectations until operating at a higher level of output than that for which it was designed, or that obtained during break-in.
4. Need for recalculation of data to obtain real facts as a basis for evaluation, rather than data on which design was based.
5. Consideration of the fact that some justification for the installation may be in the intangible aspects.

MECHANIZATION AND AUTOMATION. Almost nowhere has the stop-and-go aspect of traditional handling methods interfered with the goal of continuous flow more than in the attempt to mechanize or automate a production activity. As pointed out earlier in this section, material handling can be interpreted to include any movement of material occurring between the source of raw material and the destination of the finished product. While this may appear to be an ambitious task, it is the challenge of the handling engineer to mechanize and automate as much of the handling involved in the total cycle as mechanically feasible and economically justifiable. And in many cases this opens up as much as 90 percent of the total cycle for re-analysis by the engineer.

In a manufacturing facility the beginning of the total handling task is the point where the incoming material is unloaded from a carrier. The end of the handling cycle in the facility would be at the point where the finished product is loaded into a carrier. This defines the field of examination for the material handling engineer as any handling activity occurring between the receiving and shipping locations. Within these bounds all handling activities are prospects for mechanization and automation. The theoretical goal would be complete automation of all activities from receiving through shipping. Such an overall system may seem somewhat impossible to achieve, however, by setting the highest possible goal the desired ultimate solution to the total handling problem will more likely be realized. Conceptually, the engineer should think in terms of one continuous move between receiving and shipping, during which all necessary operations would be performed. This concept complies with the systems approach discussed earlier and the search for the theoretical ideal system.

It should be pointed out that the typical handling analysis is much more likely to begin at the output end of a particular operation and become overly involved in the relatively insignificant task of transporting the material to the input location for the succeeding operation. This viewpoint will make it difficult to achieve total mechanization or automation of the process. Nevertheless, because this viewpoint is so common, and the theoretical ideal system appears so remote, handling technology has concentrated its attention on subsystems, frequently resulting in small, unrelated islands of automation.

An Approach to Total Automation. Before proceeding with the development of an automation concept, automation will be defined as **mechanization plus control,** with the implication that the control system enables the mechanized process to be self-regulating.

In his research study, Bright (Automation and Management) suggests the seventeen levels of mechanization. Fig. 46 shows a similar delineation of the levels of **handling mechanization** in manufacturing as developed by Apple and Bazaraa at the Georgia Institute of Technology. Based on this concept, Fig. 47 illustrates the concept of one move between source and consumer, with the move implemented at the appropriate level of mechanization, and integrated with the other functions (fabrication, assembly, test, store, and control) at their appropriate levels of mechanization.

In the concept presented here, material handling is the integrating force, with a goal no less than a carefully and thoroughly researched solution to the handling aspects of the total material flow cycle. The desired result would consist of an integrated composite of facilities, activities, and information flow encompassing as much of the total problem scope and environment as is feasible and economical.

Mechanized Warehousing. One of the major subsystems of the total material flow cycle is the storage function. The importance of inventory value

Classification		Level	Description	Examples	Characteristics
Manual Control	Manual Power	1	Hand	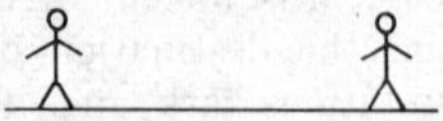	Man carries load.
Manual Control	Manual Power	2	Hand Equipment	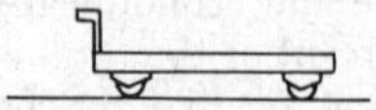	Equipment carries load.
Manual Control	Manual Power	3	Mechanized Hand Equipment	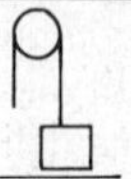	Uses mechanical advantage.
Manual Control	Gravity	4	Gravity Equipment	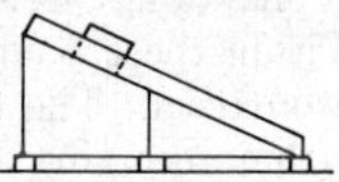	Positive control of object.
Manual Control	External Power	5	Power Equipment, Hand Control	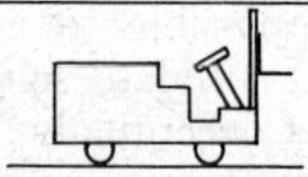	Power does work; man controls power.
Manual Control	External Power	6	Power Equipment, Remote Hand Control	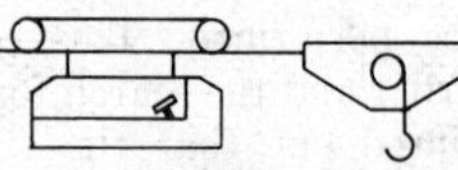	Control remote from load.
Automatic Control	External Power	7	Power Equipment, Program Control	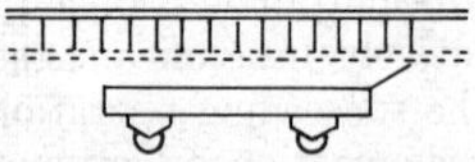	Control according to program.
Automatic Control	External Power	8	Power Equipment, Feedback Control	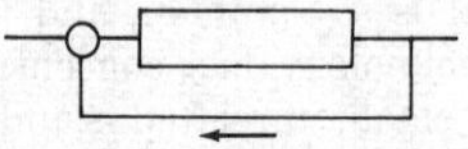	Automatic correction, according to signal.
Automatic Control	External Power	9	Adaptive System Equipment	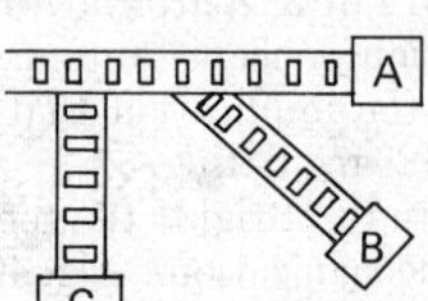	Integrated system of signals and actions.
Automatic Control	External Power	10	Fully Automated System Equipment		

Fig. 46. Summary of the levels of mechanization of material handling equipment.

and turnover has caused much interest in the possibility of mechanizing or automating the storage function, in order to control and minimize inventory, and reduce storage or warehousing cost. Here again it should be stated that the complete automation of the function would involve the entire cycle of activities from the receipt of goods (from either vendors or production) to the dispatching of goods to shipping. However, common practice usually limits automated warehousing installations to the following.

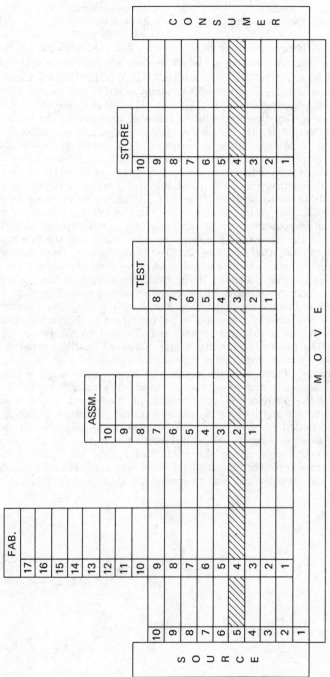

Fig. 47. The one-move concept in material handling.

1. Identification of goods
2. Sorting of items
3. Dispatching to storage
   area

4. Storage
5. Order selection
6. Order consolidation
7. Record keeping

Most installations are either built around a flow rack with conveyorized feeding and/or discharge, or they are based on some adaptation of the stacker crane.

MATERIAL HANDLING AND THE COMPUTER. Direct linkage of the manufacturing function with handling activities can be achieved through the use of a computer. IBM (Material Handling Engineering, vol. 24) has developed a system that works in the following manner: At the beginning of a shift each worker inserts his identification tag into a data transmitter in his department. The transmitter is connected to a computer which has stored within it all pertinent information concerning each employees skills. The computer sends to each department manager a report on the skills available in his department for that shift; the plant manager receives a report on the total skills available in the plant for that shift. A tag is also prepared for all parts and materials received in the plant. The tag can call forth from the computer all information concerning the storage, production, and shipping of that part or material. In the receiving department the tag is inserted in the data transmitter and the computer, after verifying the identity of the part, sends instructions on where to store them. A mechanized handling system then directs the material to its proper storage location. If the clerk has dispatched the material incorrectly, the computer prints out instructions to correct him. The plant inventory records are automatically updated by the computer also. The computer alerts the clerk when material is required in production. The material is automatically retrieved, put on a conveyor system, and delivered to the proper production activity. The production activity is also monitored by the computer. For example, the operator might get a printout telling him to stop the operation for a tool change. Throughout the production process everyone is kept informed about all activities that affect them. An assembler might receive a printout telling him that there is a machine breakdown. Each work station is monitored by the computer and its output measured against production schedules. Supervisors are alerted for downtimes and shortages in the schedule. At the end of the shift a complete report of the shift's activities is automatically prepared for each production area. Using this method of control paperwork is kept to a minimum or completely eliminated for some functions. Material handling, production, and the attendant data processing are all intimately linked in this type of system.

MANUFACTURING PROCESSES AND MATERIALS

CONTENTS

CONTENTS *(Continued)*

CONTENTS *(Continued)*

MANUFACTURING PROCESSES AND MATERIALS

Manufacturing Processes

DEFINITION AND CLASSIFICATION. A knowledge of tools, machines, and processes is imperative in working toward the goals of improved production methods and lower manufacturing costs. Manufacturing processes are the **primary processes** used in the fabrication of engineering materials and must be kept in mind when considering alternative production procedures and attempting to avoid production problems.

Most manufacturing processes can be grouped into three basic categories: forming, machining, and assembly. The objective is the same in each case, that is, changing the shape or physical characteristics of the original material. Each of the categories can be further divided into functions and processes.

Forming. Includes casting, forging, stamping, embossing, spinning, and other processes, the major purpose of which is to change the shape of the workpiece without necessarily removing or adding material.

Machining. The basic process of metal removal. This group includes all the well-known machine tool operations such as turning, milling, drilling, grinding, etc. It also includes the chipless processes such as electrodischarge and electrochemical machining, chemical milling, laser drilling, etc.

Assembly. Involves the joining of components or pieces to make a single functioning piece. The processes in this group include welding, brazing, soldering, riveting, bolting, etc.

SELECTION OF A PROCESS. The choice of an operation is influenced by several factors, including the product quality desired, the cost of labor needed, and the number of units to be produced. Although many items can be produced by several methods, there is usually one method that is best for a given set of variables. In the following, each process is defined and described and its major variations presented, with emphasis upon applications of the process and its advantages and limitations. The major financial considerations in selecting machines and equipment for a manufacturing process are analyzed in the section on Capital Investment Analysis.

Casting Methods

CASTING OF METALS. The casting process consists of the pouring or forcing of molten metal into a mold and allowing sufficient time for the metal to solidify and retain the shape of the mold cavity. Because the process is so versatile practically all metals can be cast. The ferrous metals and alloys that are cast in the greatest tonnage are steel, gray iron, and malleable iron. Of the nonferrous group, alloys of tin, copper, aluminum, magnesium, lead, and zinc are the most important.

Methods of Melting. Metals may be melted in a variety of different furnaces depending on the metal and the quantity of it to be cast. Pig iron is melted in the **blast furnace** and various grades of steel for casting are melted in **open-hearth** or **electric furnaces.** Cast irons may be melted in cupolas, air furnaces, or electric furnaces. These are all high-tonnage furnaces and are used in large ferrous metal foundries. Smaller melts of these ferrous alloys are normally made in induction furnaces or crucible furnaces as are most non-ferrous metals.

Induction furnaces operate by passing high-frequency current through a set of coils surrounding the crucible containing the metal. Eddy currents are induced in the metal and it is these induced currents, coupled with the inherent resistance of the metal, that heats the charge. Extremely high temperatures can be obtained in this manner making it possible to melt very refractory metals in induction equipment. **Crucible furnaces** may be oil or gas-fired, the crucible containing the melt being heated directly by this flame. Temperature limitations due to the type of fuel mixture restrict the use of this type of furnace to temperatures under 2,000°F. Electric furnaces operate either by maintaining an arc between graphite electrodes and the molten metal, a process which produces very high temperatures, or by resistance heating of coiled elements, a method suitable only for lower temperatures (less than 2,000°F).

Since many metals are subject to oxidation, atmospheric control during melting is essential. This may be provided by blanketing the melt either with a protective atmosphere of an inert gas or with a mixture of reducing gases. In some cases vacuum melting is preferred to avoid all contaminants and to insure the quality of the melt. Gaseous atmospheres can be used with most furnace equipment, but vacuum melting is generally used with induction furnaces.

Sand Casting. Most metals can be cast to shape in a sand mold, as represented in Fig. 1. The molds are easy to prepare and have few limitations as to size and shape of casting to be produced. In sand casting, sand is packed around a pattern after which the pattern is removed leaving the cavity to be filled with molten metal.

Although the pattern is a model of the part to be cast it is not the same shape and size as the finished product. Metal shrinks when it solidifies. Therefore the pattern must be made larger in all dimensions than the final casting by an amount called a **shrinkage allowance**. This allowance ranges from ⅛ to 3/16 inch per foot depending on the metal. In order to remove a pattern from the mold without damaging the mold, the pattern must have **draft** or a taper parallel to the direction in which the pattern is withdrawn. A minimum draft of one degree is usually recommended. Interior surfaces require more draft than exterior surfaces. A machining or **finishing** allowance, usually about ⅛ inch, must be added to those surfaces of the casting that are to be machined. Patterns for very large castings also contain allowances for **shake** and **distortion.**

A cavity in a casting is produced by using a body of sand known as a **core** in the mold. Cores are generally made of dry sand mixed with a **binder,** such as molasses or linseed oil, and then baked or air dried. They are held in place in a mold by impressions called **core prints** made in the sand during the molding operation, or by chaplets. **Chaplets** are small metal supports made from low melting point alloys. Cores must be refractory to withstand the molten metal; the binding material must burn out or break down and the chaplets melt after a short interval of time in order to minimize the stresses set up and allow easy removal of the core.

Molds, classified according to the types of materials used, are:

1. **Green sand molds.** The material most commonly used for molds is a damp mixture of sand and clay.

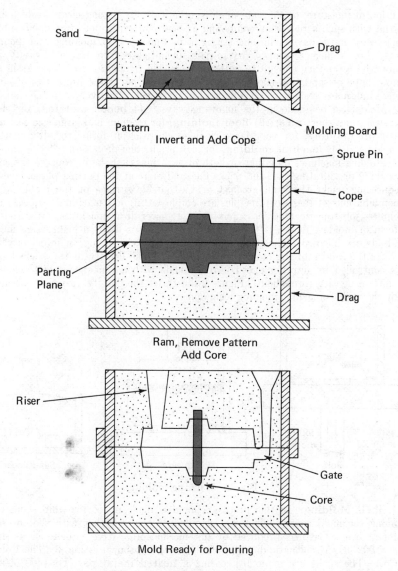

Invert and Add Cope

Ram, Remove Pattern
Add Core

Mold Ready for Pouring

Fig. 1. Sand molding procedure.

2. **Skin-dried molds.** In this process the surface of the cavity is treated with a special wash or binder which is dried by heat prior to pouring. The mold itself is made of green sand.

3. **Dry-sand molds.** Similar to cores, these molds are made from clean sand and a binder, and are then baked or air dried.

4. **Loam molds.** Made from a loam mortar, this type of mold is used only for very large castings.

Green sand is a mixture of silica sand grains, a clay bond material, and moisture. Foundry sands are used either as found in a natural state with the addition of moisture, or they are made synthetically by mixing clean washed silica sand with such a pure clay compound as bentonite with moisture added. Molding sand must have sufficient strength to undergo the molding and pouring operation, sufficient permeability to pass air and gases from the mold cavity, and must be refractory enough to withstand pouring temperatures. Since grain size and shape of sand, and clay and moisture content affect the properties of molding sand, many tests have been devised to predict its effectiveness in the foundry.

Molds can be made in the following ways: (1) **bench molding,** for small castings and short runs; (2) **floor molding,** for medium and large size castings; (3) **pit molding,** for very large castings where a pit forms the drag part of mold; and (4) **machine molding,** for high production processes.

Molding machines do either or both of two things: shake or squeeze the sand. Shaking or vibrating the sand packs the sand more at the parting plane, whereas squeezing packs the sand greatest at the outside surface of the mold. Many combinations of the two principles are employed in: jolt machines; squeeze machines, jolt-squeeze machines; jolt-squeeze power-draw machines, etc. Another form of molding device is the **sandslinger** which is used to make large molds. Molds are filled and packed in one operation by employing an impeller which slings the sand into the mold with high speed. The density of the packed sand is controlled by impeller speed. Molds made with a sandslinger have close to uniform density throughout. Figure 2 illustrates the density variations obtained by the various machine molding techniques.

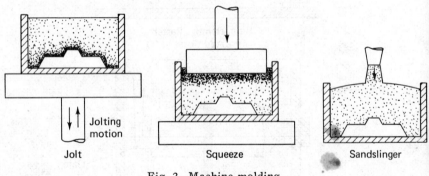

Fig. 2. Machine molding.

Shell Molding. In this process a mold is made up of two thin shells outlining the mold cavity, which are clamped or held together with shot or sand for pouring. Castings produced by this method often have a tolerance as small as 0.002 in. A schematic diagram of the process is shown in Fig. 3 (The Borden Co.). The shells are made by coating a **heated metal pattern** (400–560°F) that has been sprayed with a **release agent** with a mixture of sand and phenolic resin. A shell forms over the pattern due to the heating of the resin; excess sand-resin material does not adhere and falls away. The shell is baked or cured for 35–65 seconds at temperatures of 600–1,400°F and is then removed from the pattern. Metals cast by this process include brass, bronze, cast iron, low carbon steel, and the aluminum alloys.

Die Casting. Die castings are made by forcing metal under pressure into a metallic cavity or die. The pressures ranging from a few pounds per square inch to approximately 50,000 psi are maintained until solidification is complete.

The **principal advantages** of die castings are: rapid production, excellent surface finish, close tolerances of from 0.002 to 0.001 in. possible, dense structure, and suitable for casting thin sections. Since the cost of dies and molds is high, the method is not used for short-run jobs. Metals commonly die cast include alloys of zinc, aluminum, copper, tin, lead, and magnesium.

The two basic types of die casting machines are the hot-chamber and the cold-chamber machines. A melting pot is an integral part of the **hot-chamber** machine, and the injection cylinder remains immersed in the molten metal. Inasmuch as the melting pot and injection equipment are subject to very high temperatures, only the lower melting point alloys are cast in this type of machine. Care must always be taken to be certain that the molten casting alloy does not dissolve the steel equipment.

The **cold-chamber** machine is not equipped with a melting pot, but must be furnished with a separate melting furnace. The molten metal is placed in the injection system by hand ladle or mechanical means. Metals that have higher melting temperatures and require greater pressures for sound castings are cast in this machine. If these metals were not heated in a separate pot, the life of the pot and the parts of the injection system would be short. Alloys commonly used in die casting and some of their advantages are shown in Fig. 4 (page 8).

Permanent-Mold Casting. Permanent molds are made from metal and their life depends on the temperature of the metal being cast. The mold is coated with a wash before each pouring to protect the mold and to facilitate the removal of the casting. A permanent mold offers cleaner, smoother, more accurate, gas-free, and shrink-free castings than sand molds. All metals can be cast by this method, but the cost of using this method increases rapidly as higher melting point alloys are used. Permanent-mold casting is rarely used for large castings or short runs and where tolerances ranging from 0.0025 to 0.010 in. are required.

The method consists of pouring the metal into a permanent mold and allowing only gravity and hydrostatic pressure to force the metal to fill the mold cavity, unlike die casting where pressure is applied to the molten metal. The two halves of the mold are normally hinged to facilitate opening and removing the casting. The mold is usually heated at the start of each run to avoid excessive chilling and surface defects.

Centrifugal Castings. Centrifugal castings are produced by pouring a molten metal into a **rotating mold,** thus utilizing centrifugal force as a means of forcing the metal into the mold cavity. The centrifugal force insures greater surface detail, a denser and stronger structure, and freedom from impurities. Metals of all kinds can be cast by this process, depending, of course, on the mold material. There are three types of centrifugal castings.

True centrifugal castings are those made by rotating the mold about a horizontal or vertical axis, to form a cylindrical cavity on the inside.

Semi-centrifugal castings are those made by rotating the mold about its vertical axis, but with the central axis full of metal. The less dense metal and the impurities are in the center of the mold which is normally machined out in a later operation.

Centrifuge castings are made in molds having several cavities around the outer areas of the mold, the metal being fed to these cavities by radial gates from the center of the mold.

Molds for centrifugal casting may be either the **permanent metal** type or the **sand rammed** type. According to Jones (Centrifugal Molding and Casting)

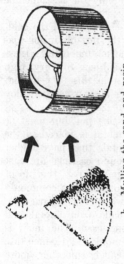

b. Mulling the sand and resin

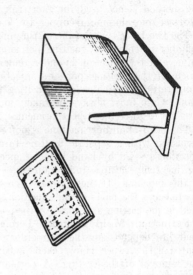

d. Excess resin-sand material falls back in dump box

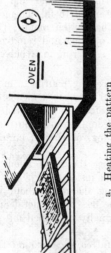

a. Heating the pattern

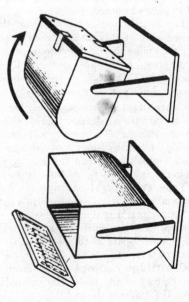

c. Resin-sand mixture applied to heated pattern

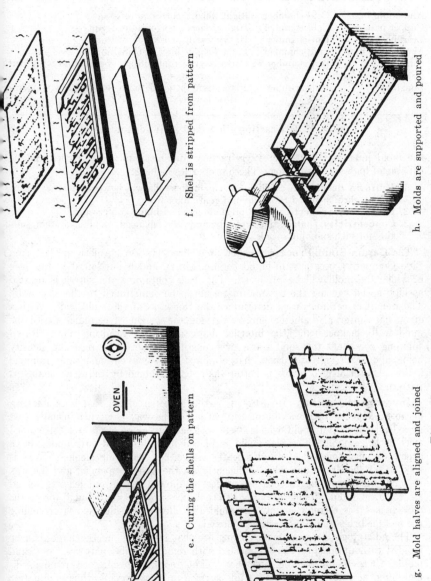

f. Shell is stripped from pattern

e. Curing the shells on pattern

OVEN

h. Molds are supported and poured

g. Mold halves are aligned and joined

Fig. 3. Schematic of the shell molding process.

Alloy	Machine	Advantages of Alloy
Aluminum base	Cold-chamber	Light weight, corrosion resistant
Copper base	Cold-chamber	High strength, corrosion resistant
Lead base	Hot-chamber	Heavy weight, corrosion resistant
Magnesium base	Cold-chamber	Light weight, high machinability
Tin base	Hot-chamber	Corrosion resistant, suitable for contact with food and beverages
Zinc base	Hot-chamber	High strength, good finish, low casting temperature, low cost

Fig. 4. Die casting alloys and their advantages.

the mold must have satisfactory refractory and mechanical properties and be capable of high speed spinning. These properties include:

1. **Uniform density** to avoid localized mold swells which would restrict longitudinal shrinkage and cause circumferential hot tears.
2. **Refractory properties** which preclude mold scabbing and erosion.
3. **Concentricity** contributing to a dynamically balanced molding system suitable for high spinning speeds.

The **Ceram-Spun Process** (Acipco Steel Products Div., American Cast Iron Pipe Company) uses a semi-liquid ceramic slurry and a horizontal metal flask in which the mold will be produced. The flask containing the slurry is rotated at high speed causing the ceramic materials to be centrifuged to the flask wall. The weight of ceramics used determines the thickness of the mold wall. With a given flask internal diameter, simply varying the weight of ceramics varies the mold wall thickness and the internal diameter of the resulting mold. Rapid spinning compacts the solid refractory constituents with the resulting density being about 20% greater than that obtained with conventional sand rammed molds. Uniformity of molds is better than that obtained by ramming against a pattern.

Investment Casting. Investment casting is often called **lost-wax casting** because of its use of a wax or plastic pattern. The wax pattern is usually cast in a lead-alloy split mold which itself was cast using an accurately machined steel or brass mold. The wax mold is placed in a flask and surrounded by a slurry of very fine silica, ethyl silicate, and either water or alcohol. Heat is used to set the slurry and when sufficiently set the flask is upended and the wax that has melted during the heating process is allowed to run out leaving behind an accurate mold. Using gravity, external pressure, or centrifugal force, the molten metal is introduced into the mold and allowed to solidify. After cooling the mold is broken and the casting removed.

The advantages of investment casting are that intricate, precise shapes can be molded without the problems connected with removing solid patterns from sand molds, there is no parting line that might require finishing to remove, dimensional accuracy is good and a fine surface finish is possible thus eliminating the need for further machining. Jewelry is often cast by this method.

The **Mercast** process is very similar to the lost wax process in that a pattern is produced by freezing mercury in a metal mold or die. The mercury pattern is then coated with a ceramic slurry which forms a shell around the pattern. The ceramic mold is allowed to set and the mercury return to its normal liquid state; the mercury is then poured out of the shell leaving an accurate mold. Any metal with a melting point under 3,000°F can be accurately cast to a tolerance of approximately 0.002 in. by this process. The Mercast technique is relatively expensive to use.

CASTING OF PLASTICS. Casting processes for plastics are similar in many respects to those used with metals. Plastic materials are melted, blended with the necessary catalysts and filler materials or hardening agents, and then poured into molds. Most plastic materials including acrylics, polyesters, ethyl cellulose, and epoxies can be cast.

Plastic materials are also formed into finished products by compression, transfer, or injection molding; extrusion; vacuum forming; and blowing. Although some plastics can be processed in several ways, many of them are best adapted to one procedure. Of the methods given above, **compression molding** is one of the most widely used.

Behavior of Plastics Under Heat. Plastics include a large group of synthetic or natural organic materials that become pliable with heat and are formed into shape by pressure. Their reaction to heat once they have hardened depends on whether they are thermosetting or thermoplastic. **Thermosetting** materials are first softened by heat and then shaped; as additional heat and pressure are applied, the material sets permanently by a chemical process known as polymerization. After setting the material cannot be remelted by the addition of heat. **Thermoplastic** materials are similarly formed by heat and pressure but do not undergo any chemical change. They remain hard at normal temperatures but upon application of heat they are softened and can be reshaped.

Conventional Casting. As in the casting of metals, **mold construction** necessitates consideration of draft and shrinkage. However, since plastics are usually lighter and have lower melting points, the molds are lighter in construction and less expensive to make than those used for metals. Frequently, open molds for short rods and simple shapes are formed by dipping a tapered steel mandrel into molten lead and stripping the shell from the sides of the mandrel after it solidifies. The wall thickness of such mold is from ⅛ to 3/16 in. **Lead split molds,** produced in a die casting machine, are used for more involved shapes having undercut. Molds of a **rubber latex** or an **elastomeric plastic** substance can be used for some materials. The elastic material is applied on the object to be reproduced by dipping or spraying until sufficient thickness is obtained. It is then removed by cutting at a desired parting line and is backed with plaster of Paris to give it rigidity. Molds of **glass** or **plaster of Paris** can also be used. The curing and hardening phase of the process takes place in the mold and varies considerably according to the material used. It is often done in closed ovens under controlled conditions as most materials require heat in the curing process, though a few materials can be cured in open molds.

Advantages of the casting process include low tooling costs and the ability to cast large sections and stock shapes. Stock shapes, such as short rods, tubes, and plates, are often used as material for machining operations. Costume jewelry, lenses, clock cases, handles, and drilling jigs are usually cast. Large punches and dies used in the aircraft industry for drawing, forming, and stretching operations are also cast from plastics.

Compression Molding. Compression molding (Fig. 5) is used principally for thermosetting plastics such as phenolics, alkyds, melamines, ureas and for materials that are molded cold. The **molding cycle of operations** is as follows.

1. Material in the form of powder or special preformed shapes is placed into the mold.
2. The mold is closed while heat and pressure are gradually applied.
3. When the maximum temperature is reached and material becomes fluid, the maximum pressure is applied and maintained as the material hardens.
4. The mold is opened and the formed part is removed.

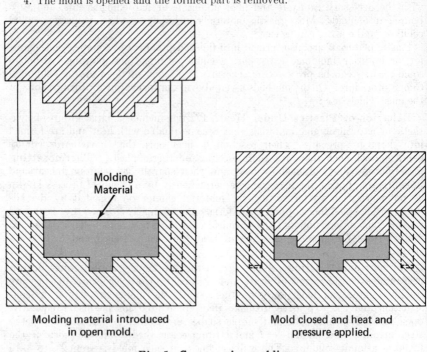

Molding material introduced in open mold.

Mold closed and heat and pressure applied.

Fig. 5. Compression molding.

Thermoplastic materials require that the mold be cooled in order to harden the liquid mass before the part is removed; otherwise, distortion is apt to result.

Preforms consisting of small pellets or shapes of plastic material facilitate the loading of molds, prevent the wasting of material, control the weight of the charge, and permit preheating. They are used only in compression and transfer molding.

A large variety of hydraulic presses, ranging from hand-operated to completely automatic, are available for compression molding. The press is the main functional unit in the operation as it furnishes both pressure and heat for the process. Heat, transferred from heated platens or applied directly to the metal mold, can be applied by steam, electrical resistance, high frequency current, or heated fluids. In most cases the powder or preforms are preheated before entering the mold. Presses for compression molding range in size up to 3,000 tons capacity.

Transfer Molding. Transfer molding is a process of molding thermosetting compounds in a closed mold, but differs from compression molding in that the powder is first placed in a pressure chamber adjacent to the mold cavity. Here it is heated to a liquid state and then injected into the mold under pressure where it is cured and hardened. The mold is closed by the upward movement of a press platen. The operator drops the material into an opening in the center of the transfer chamber plate where it is plasticized. The plunger supplies the necessary pressure and forces the liquid molding material into the mold where it cools and hardens. The cycle differs from compression molding in that the mold is closed before any material enters.

Transfer molding is desirable for producing parts requiring metal inserts since the location of the inserts is not disturbed by the liquid material entering the mold. It is also used for articles having intricate sections and in cases where there is a large variation in thickness section. With this process, more uniform density is obtained and close tolerances can be maintained. The cost of the mold is high and there is some loss of material in the transfer chamber and sprue.

Injection Molding. The operation of an injection molding machine for thermoplastics is very similar to a cold-chamber die casting machine. Material for the machine is placed in an overhead hopper where it is fed by gravity to a metering device and then to a circular heating chamber as shown by Begeman and Amstead (Manufacturing Processes) in Fig. 6. Here it is compressed by a

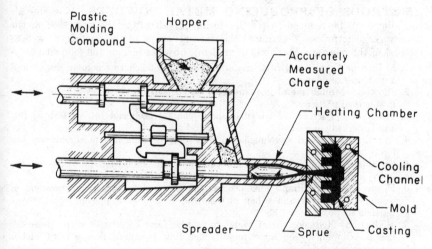

Fig. 6. Operation of plastic injection molding machine.

plunger, softened, and finally injected into the relatively cold mold under considerable pressure. Since the mold temperature is maintained below the melting temperature, the molded parts harden rapidly. The mold then opens and the parts are ejected. Most injection presses are fully automatic, the presses having one hydraulic unit for feeding the material to the mold and another for opening and closing the mold.

Some of the **common thermoplastic materials** used in injection molding are cellulose acetate, nylon, polystyrene, methyl methacrylate, and various vinyl resins. The use of thermoplastic materials in injection molding equipment permits a much faster cycle operation than in compression molding since the mold does not have to be alternately heated and cooled. As a rule, the mold is held at a fairly constant temperature by circulating water, permitting a production cycle of several shots per minute. Because of the rapid operation of this machine, fewer die cavities are necessary to maintain equivalent production than by other possible methods. Material loss when using thermoplastic materials is low, as sprues, gates, and defective parts can be reused. A limitation of the process is the high equipment cost and accompanying maintenance and overhead charges.

Powder Metallurgy

BASIC POWDER PROCESS. Producing commercial products from metal powders represents one of the few processes where the product is not made from a homogeneous solid material. The operation appears simple, namely, mixing the powders in suitable proportions, pressing a given amount of the mixture to shape, and finally heating the compressed powder to the proper temperature. The heating or **sintering** results in binding the solid particles together by molecular forces. The investment in equipment for this process is relatively high and requires parts to be produced in large quantities. It is unique in that it is the only way that some parts can be made. It permits combining different powders for products that must have a controlled density. Although there is some limitation as to size and shape, parts made by this technique have a close tolerance which frequently eliminates the necessity for further processing.

METHODS OF PRODUCING METAL POWDERS. Most metals and alloys can be reduced to powder form, but all cannot be processed in the same way. The process determines the size and shape of the particle as well as the cost of the powder. **Methods of making powder** include the following:

1. Milling or Crushing. Brittle materials can be reduced to irregular shapes and any degree of fineness.
2. Shotting. Molten metal is poured through a sieve or orifice resulting largely in spherical particles.
3. Machining. Particles of irregular shape can be produced in this fashion but size is difficult to control.
4. Atomization. Metal spraying is suitable for producing fine powders from low temperature alloys.
5. Granulation. A process which depends upon the formation of oxides on the individual particles during the stirring operation.
6. Electrolytic Deposition. This process can be used for producing powders of iron, copper, silver and other metals, and results in a powder of dendritic particle shape.
7. Reduction. Reducing metal oxides to powder form by contact with a reducing gas is economical for some metals including tungsten, copper, and nickel. Irregularly shaped particles are produced.

Other methods include precipitation, condensation, and various chemical processes.

The **powder characteristics** of flowability and compressibility are dependent upon the shape of the powder particles. Likewise particle size distribution has considerable influence on the apparent density and final porosity of the product. Other properties of metal powders which must be considered in the processing operation include fineness, flowability, and sintering ability. Procedures for testing the important characteristics of the powder have been prepared by ASTM.

METHOD OF MANUFACTURE. Several processes are usually involved in manufacturing metal powder products.

Selection of Powder. Since the particle fineness, shape, and size distribution in a powder all influence the density, proper selection requires experience and a thorough knowledge of the process.

Blending. Mixing or blending of powders is often necessary to obtain desired flowability, and density or particle size distribution. Blending is also necessary when non-metallics are added. Many powders have lubricants (stearic acid, lithium stearate, powdered graphite, etc.) added to reduce wall friction in the dies and to aid in the ejection of the compact.

Pressing. Metal parts are produced from powders by pressing them to shape under high pressure in suitable steel dies as shown in Fig. 7 (Delco Moraine

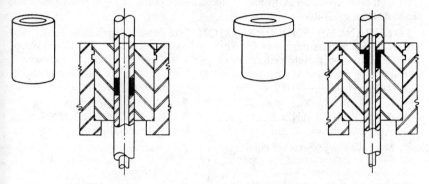

Fig. 7. **Briquetting punch and die for forming hollow cylindrical parts.**

Div., General Motors). Because soft particles can be pressed together quite readily, powders that are plastic do not require so high a pressure as the harder powders to obtain adequate density. Quite obviously the hardness and density increase with pressure up to some optimum pressure above which little change in properties takes place. The product resulting from the pressing operation is known as a **cold or green compact.** In a few cases sufficient strength is obtained by this operation but usually subsequent heating is necessary.

Many of the conventional mechanical and hydraulic presses can be used for compacting of the powdered metal. Pressures for different powders vary greatly, ranging from 5 to 5,000 tons per sq. in. Pressure of from 1 to 2 tons is sufficient for hot pressing.

Sintering. Heating a green compact to some elevated temperature below the melting point of the principal powder is known as sintering. It is done in a protective atmosphere soon after the compact is made. The protective atmosphere depends on the metal: metals which are readily oxidized require a reducing atmosphere; oxide ceramics require an oxidizing atmosphere. Both batch- and continuous-type furnaces are used in this operation. In all cases, temperature and time are closely controlled. The application of heat to the compact, at temperatures above the **recrystallization temperature** of the metal, increases the plasticity and mechanical interlocking of the particles, presses them into more intimate contact, and facilitates the bonding of the solid particles by atomic forces. When the sintering is carried out at a temperature above the melting temperature of one of the components, the process is called **liquid-phase sintering.** Here the solid particles dissolve slightly in the liquid phase where **diffusion** of atoms is easier and bonding between solid particles is established more rapidly than in normal sintering processes.

Hot Pressing. In some cases, pressing and sintering are combined to provide parts with improved strength and hardness. This procedure requires proper atmospheric control, method of applying heat, and selection of suitable dies to resist high temperature, wear, and creep.

Sizing and Finishing. Since there is always some size change or distortion after sintering, a sizing or coining operation is often necessary. This provides close tolerance to the part and improves its strength. All pressed metal products can be heat treated, but the best results are obtained with those having dense structures. Strength and density can also be increased by dipping the sintered parts in molten metal so as to fill up the external pores. Electroplating of these products is also possible.

TOLERANCES AND PRECISION. Tolerances are influenced by the size and shape of the part, and the kind of powder used. Vertical tolerances of ±0.002 in. can be maintained on short parts but as the length is increased a corresponding increase in tolerance must be made. Radial or side-to-side tolerances can be held to ±0.001 to 0.002 in. Slightly closer tolerances can be obtained by sizing.

ADVANTAGES AND LIMITATIONS OF POWDERED METALS. The major **advantages** in manufacturing products with metal powders are:

1. Machining operations are eliminated.
2. There is little waste of material.
3. Labor cost is low.
4. Products of extreme purity can be made.
5. A wide range of physical properties is possible with a given material.
6. Structure and porosity can be controlled.
7. Ceramics and refractory materials which cannot be cast or are difficult to machine can be used to produce parts.

The principal **limitations** in the use of powder metallurgy are:

1. Large or irregularly shaped products are difficult to produce.
2. Cost of dies and equipment is high.
3. Cost of metal powder is high.
4. A completely dense product is not possible.
5. Some powders may present explosion hazards.
6. It is difficult to sinter low-melting powder.
7. Protective atmospheres are essential.

Design Considerations. For successful operation parts and dies for use with metal powder must be designed and constructed with extreme care. Some of the **design requirements** are as follows:

1. Avoid designing holes in the part that are at an angle to the punch since they cannot be produced directly but must be machined separately.
2. Provide fillets on the inside corners of the part to permit better punch design.
3. Outside bevels on the part should be made as flat as possible to avoid feather edges on the punch.
4. Undercuts such as on threads should be avoided since they cannot be produced directly.
5. Punch and die surfaces in contact with the powder must be very smooth since metal powders lack the ability to flow readily.
6. The die material should be sufficiently hard so that scoring of the die surfaces by the powder is minimized.
7. Dies should be provided with a slight draft to facilitate ejection of the part.

APPLICATIONS OF POWDER METALLURGY. Among the principal uses of powder metallurgy are the following:

Refractory metals. The manufacture of parts of high melting temperature or refractory metals such as tungsten, thorium, thorium oxide, etc. can be carried out only by powder metallurgy because of the extreme brittleness of these materials.

Cemented carbides. Tungsten carbide, titanium carbide, and tantalum carbide particles are mixed with a binder such as cobalt, pressed to shape, and then sintered at a temperature above the melting point of the matrix metal.

Porous bearings. Copper, tin, graphite, or other metals are used for bearings. After sintering, the bearings are sized and impregnated with oil by a vacuum treatment.

Porous metal filters. Bronze, carefully graded powder, is molded with no pressure and sintered to produce a very porous metal useful for noise damping, filtering of gases, and metering.

Motor brushes. Copper is added to graphite to provide adequate brush strength; tin or lead may also be used to improve wear resistance.

Magnets. Small magnets can be produced from iron, aluminum, nickel, and cobalt when properly combined in powder form. Ferrite magnets used in computer memory cores are sintered from powder.

Contact parts. Electric-contact parts that are wear resistant, refractory, and have good electrical conductivity can be produced from a variety of metal powders.

Gears and rotors. Accurate small gears and pump rotors of iron can be produced.

Other uses of powder metallurgy. Rods and slugs of nuclear reactor materials such as uranium, uranium oxide, zirconium, and hafnium are produced by powder metallurgy under carefully controlled atmospheres. The parts are then **clad** with less reactive metal for mechanical and corrosion protection.

Forging Processes

ELEMENTS OF FORGING. In all forging processes metal is heated to a plastic state. It can then be readily formed by pressure or impact into a predetermined shape. Forging processes may be roughly classified as flat die (smith) forging, drop forging, upset forging, press forging, roll forging, and rotary swaging.

Hot-working or forging operations are done at temperatures above the recrystallization temperature of the metal so that annealing and work hardening cancel each other. For steel the temperature must be above the critical range and the work is usually started at temperatures ranging from 2,200 to 2,500°F. For nonferrous metals the temperatures are much lower depending on the alloy composition. The plastic temperature range varies widely for different kinds of metals and alloys.

Forming metals to shape by forging is about the same in the various forging processes. All are commercially proven processes capable of changing raw materials, such as bars and billets, to a predetermined shape. The advantages claimed for these processes include:

1. Decreased porosity of metal.
2. Refinement of coarse grains.
3. Generally improved physical properties.
4. The breaking up and uniform distribution of the impurities in the form of inclusions.
5. Minimum finish allowances necessary.

SMITH FORGING. Flat die or smith forging consists of hammering the heated metal between flat dies in a steam hammer. It is similar to hand forging as practiced by a blacksmith in that no impression dies are used and considerable skill is required to shape the metal. The nature of the process is such that close accuracy is not obtained nor can complicated shapes be made. Forging hammers are made in the single or **open frame** type for light work and the **double housing** type for heavier service. The process is used primarily for short-run jobs and repair work which do not warrant the expense of special tooling. Forgings, ranging from a few to over 200,000 lb. are made by flat die forging.

DROP FORGING. Drop forgings are produced by shaping the hot plastic metal in **closed impression dies.** Impact blows of the ram, carrying one of the dies, strike the plastic metal held on the base in the other die, compelling it to conform to some planned shape. Both dies are held in perfect alignment. To insure proper flow of the metal during these intermittent blows, the operation is divided into a number of steps. Each step changes the form in a gradual fashion controlling the flow of the metal until the final shape is attained. The number of steps required varies according to the size and shape of the part, the forging qualities of the metal, and the tolerances required. For products of large or complicated shapes a preliminary shaping operation may be required, using more than one set of dies.

A finished forging will have a thin projection of excess metal extending around it at the parting line. This excess metal is provided to insure complete filling of the dies and is removed in a separate **trimming press** immediately after the forging operation. Small forgings may be trimmed cold. Care must be taken in the trimming operation not to distort the part. The forging is usually held uniformly by the die in the ram and pushed through the trimming edges. Punching operations may also be done while trimming.

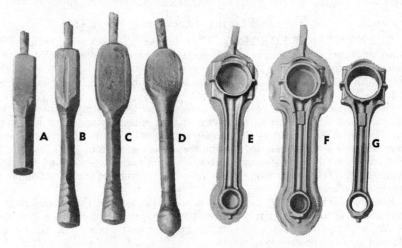

Fig. 8.   Sequence of forging operations for a connecting rod.

Forging Operations. Figure 8 (Forging Industry Association) illustrates a series of operations performed in the forging of a connecting rod. The raw material is a square billet of steel on which a small projection is prepared, as at A. It then goes through a **fullering** operation B to reduce the cross-section and lengthen the forging and the large end is flattened as at C. A second **rolling impression** D gathers the stock to proper proportion and cross section by rotating the forging while hammering. The blank then undergoes a **blocking** operation E which forms the rod into a shape determined by the contours of

the dies. The final forging operation with the surrounding flash metal is shown
at F. The flash metal is removed in a **trimmer die** leaving a completed forging
G ready for subsequent heat treating and machining operations.

   Types of Hammers. Three types of drop-forging hammers are used—the
board hammer, the **air-lift hammer,** and the **steam hammer.** The first two
are gravity-type hammers in which the force delivered by the hammer is determined by the height of the drop whereas in the steam hammer, the hammer is
driven down by steam pressure.

   In a gravity-type hammer the impact pressure is developed by the force of
the falling ram and die as it strikes upon the lower fixed die. The **board drop
hammer,** illustrated in Fig. 9 (Interstate Drop Forge Co.), is one type which

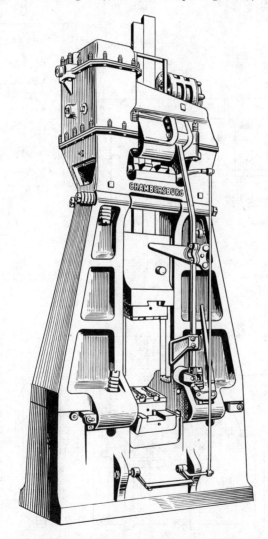

Fig. 9. Gravity-type drop hammer.

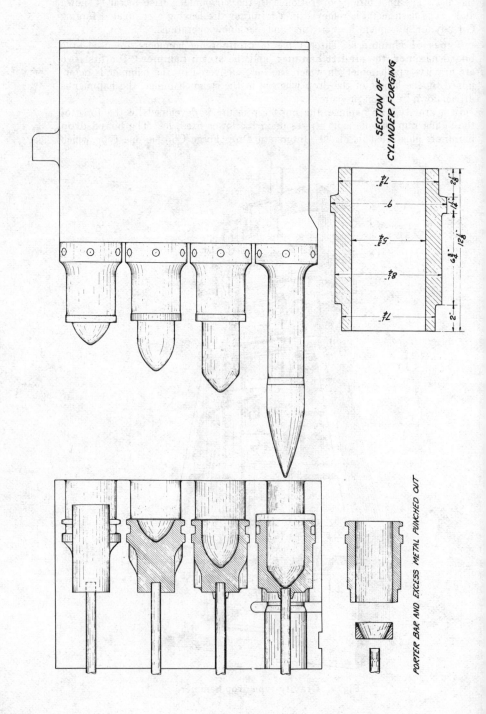

SECTION OF CYLINDER FORGING

PORTER BAR AND EXCESS METAL PUNCHED OUT

has several hardwood boards attached to the hammer for lifting purposes. After the hammer has fallen, rollers engage the boards and lift the hammer a desired amount (ranging up to 5 ft.). When the stroke is reached the rollers spread and the boards are clamped in place. The operator can release the hammer by stepping on a treadle. As long as the treadle is kept depressed, the cycle is repeated and up to 70 blows/min. may be achieved. The force of the blow is entirely dependent upon the height of the drop and the weight of the hammer which seldom exceeds 8,000 lb.

The **air lift hammer** also depends on gravity drop but employs a single-acting compressed air cylinder to lift the weight. For any fixed weight hammer, the force of the blow is determined by the length of the stroke which may be varied readily by the operator. This allows rapid action and the use of heavier hammers ranging from 500 to 10,000 lb. The forging cycle is also adaptable to programming and numerical control.

The **steam hammer** has a much sturdier construction but is similar to the air-lift hammer in that the ram is raised by steam pressure. It is also **driven downward** by steam pressure and can therefore deliver blows of greater force. The force can be controlled precisely by accurate throttling of the steam to give a range of from very light to very heavy blows. Forging hammers range from 1,000 to 50,000 lb. falling weight. **Anvils** are much heavier than **rams** and thus require massive foundations and sturdy frames.

Other types of hammers include **vertical counterblow hammers, impacters, helve hammers,** and **trip hammers.**

Upon completion all forgings are covered with scale and must go through a cleaning operation. This can be done by pickling in acid, shot peening, or tumbling, depending on forging size and composition. If some distortion has occurred in forging, a sizing or straightening operation may be required. Controlled cooling is usually provided for large forgings and, if certain physical properties are necessary, provision is made for subsequent heat treatment.

Drop forging defects which may occur include mis-alignment of the dies, scale inclusions on the surface, and seams in the forgings caused by the metal folding over during the operation, but such defects can be controlled.

UPSET FORGING. Parts produced by the forging machines (upsetters or headers) are made by pressing or squeezing the plastic metal into enclosed dies rather than by the impact force utilized by the drop hammer. Upset forging machines are double-acting and of horizontal construction with the dies at one end. A heated bar of stock is placed between a fixed and a movable die which grips the bar firmly when closed. A portion of the bar projects beyond the die for the upsetting operation by the header rams. The cavity impression on the end of this ram squeezes the plastic metal and forces it to fill the die cavity. For some products the heading operation may be completed in one position, but in most cases the work is progressively placed into different positions in the die. The impressions may be in a punch, gripping die, or both. In most instances these forgings do not require a trimming operation. Modern forging machines are used not only in initial stages of complex forging processes, but also to produce such products as bolts, bearing racer, spindles, and gear blanks.

Progressive piercing, or internal displacement, is the method frequently employed on upset forging machines for producing parts such as artillery shells and radial engine cylinder forgings. A typical sequence of operations is shown in Fig. 10 (Ajax Mfg. Co.). Round blanks of a predetermined length for a single cylinder are first heated to forging temperature. To facilitate handling the blank, a **porter bar** is then pressed into one end. In the following three operations the blank is upset and progressively pierced to a heavy-bottomed cup. In the last operation a tapered-nosed punch expands and stretches the metal into the end of the die, frees the porter bar and punches out the end slug. Large cylinder barrels weighing over 100 lb. can be forged in this manner.

Parts produced by this process vary in weight up to several hundred pounds. The dies are not limited to upsetting; they may also be used for piercing, punching, trimming, or extrusion.

PRESS FORGING. Like the upset forging process, press forging employs a slow squeezing action in deforming the plastic metal as contrasted to the rapid impact blows of the drop hammer. Presses are of the vertical type and may be either mechanically or hydraulically operated. The **mechanical presses,** which are the faster and most commonly used, range in capacities from 500 to 10,000 tons, and provide a fixed stroke. The **hydraulic presses** can be adjusted to various speeds, pressures, and dwell time and have a variable stroke. They are usually slow moving and activated either directly by high-pressure pumps or by hydraulic fluid from accumulators which are pressurized by high-pressure pumps. The larger presses are of the latter or **hydropneumatic** type and can run to as high as 50,000 tons.

Closed impression dies are used for producing press forged products which are finished in one to three steps in most cases. The maximum pressure in these presses is built up at the end of the stroke which forces the metal into shape. Dies are often mounted as separate units, but all of the cavities may be put into a single block. For small forgings, individual die units are more convenient. There is some difference in the design of dies for various metals. Forgings of brass and bronze can be made with less draft than steel; consequently more complicated shapes can be produced. These alloys flow well in the die and are readily extruded. Most press forgings are symmetrical in shape, having surfaces which are quite smooth and provide a closer tolerance than obtained by a drop hammer. Therefore, sizing operations are not needed in the press forging process.

ROLL FORGING. Roll forging machines are primarily adapted to reducing and tapering operations on short lengths of bar stock. The rolls on the machine are not completely circular but have from 50 to 75 percent cut away to permit the stock to enter between the rolls. The circular portion of the rolls is grooved according to the shaping to be done, as shown in Fig. 11. When the rolls are in open position the operator places the heated bar between them, retaining it with tongs. As the rolls rotate, the bar is gripped by the roll grooves and pushed toward the operator. When the rolls open, the bar is pushed back and rolled again or is placed in the next groove for subsequent forming work. By rotating the bar 90° after each roll pass, there is no opportunity for flash to form.

Examples of **roll-forged parts** include axles, crowbars, knife blades, chisels, tapered tubing, and ends of leaf springs. This method is often used to supply single preform shapes for subsequent forging by other forging operations. Parts made in this fashion have a smooth-finished surface and tolerances equal to other forging processes. The metal is hot-worked thoroughly and has good physical properties. Because of the high cost of the rolls this method is used only where there is a large volume of production.

ROTARY SWAGING. Rotary swaging is a means of reducing the ends of bars, tubing, or round stock by repeated hammering from rotating dies as they open and close on the work. The head of the machine, shown in Fig. 12, is

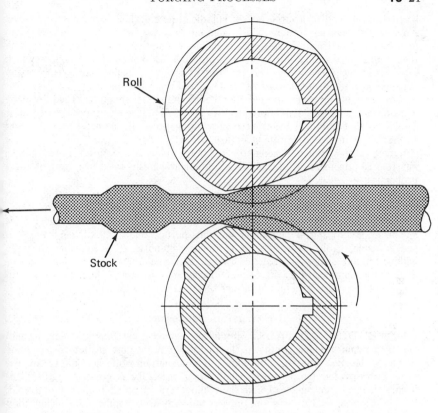

Fig. 11. Schematic diagram of roll forging.

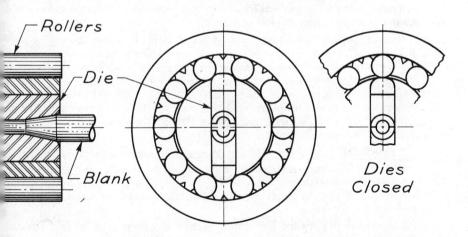

Fig. 12. Operation of dies in a swaging machine.

located at the end of a large hollow spindle through which the work can move. Around the end are rollers which contact the dies when the machine is in operation. Centrifugal force opens the dies to permit entry of the work; they then remain open only until the opposite ends are contacted by rollers which force them together again. The rapidity of this pressing or hammering action depends on the rotational speed of the head and the roller spacing.

Most swaging is done on cold metal although it is sometimes used for hot-working when severe reductions are to be made. The effect on the metal is similar to that obtained by other cold-working processes, namely, hardness and strength are increased, grain structure distorted, and a clean smooth surface obtained. Hot-working is necessary on some operations and for some metals, but this results in an appreciable loss in surface finish, tolerance, and physical properties.

Basic operations done by rotary swaging include reducing, pointing, forming, and attaching. The finished product in all cases is round but the size varies. Examples include pointing bars or valve needles, forming axle ends, gear shift levers, golf club shafts and other tubular parts, and attaching of fittings to ends of steel cables. By using a stationary mandrel, it is possible to forge hollow tubular shapes. Capacities range from small machines having a maximum work diameter of ⅛₆ in. to much larger ones capable of handling stock up to around 6 in. The production rate of these machines is high.

Extrusion Processes

EXTRUDING METALS. Metals are extruded by forcing them through dies in a manner which results in the metal having a cross section of the same shape as the die. Copper, aluminum, and magnesium and their alloys are extruded at elevated temperatures where their plasticity is greatest. Other materials such as zinc and lead and their alloys are extruded both hot and cold. By the use of special presses and erosion-resistant dies, stainless and nickel alloy steel have been extruded at temperatures as high as 2,400°F. Beryllium is fabricated into many structural shapes by extrusion presses. All of these extrusion processes require lubrication and it was impossible to extrude steel until **glass** was found to be a suitable lubricant.

Advantages and Disadvantages of Extrusion. Extruded shapes have replaced rolled, forged, and cast shapes in many instances because extrusion dies are low in cost, machining on the section is kept to a minimum, fine surface finishes are obtained, good dimensional accuracy is inherent, and as compared to rolling mill equipment, costs are relatively low. The principal disadvantage of the process is the amount of material that must be thrown away or remelted as a result of the oxidized metal being pulled through the orifice, particularly near the end of each extrusion cycle.

TYPES OF EXTRUSION. Basic extrusion processes are as follows.

Direct or Forward Extrusion. Figure 13a illustrates the process of forcing a heated solid block of metal, a billet, through a die or orifice by means of hydraulic pressure, the product being a continuous length of metal of uniform cross section.

Inverted or Backward Extrusion. Figure 13b shows a hot or cold process, depending on the material being extruded, in which the die is mounted on a hollow ram which is pushed into the billet instead of the billet being pushed through the die, as in the direct process.

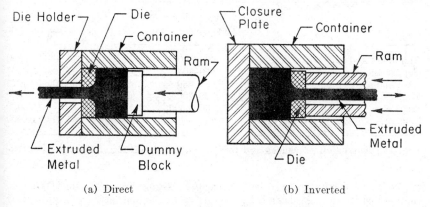

(a) Direct   (b) Inverted

Fig. 13. Direct and inverted extrusion processes.

Impact Extrusion. This method is used primarily in the manufacture of short tubes of soft alloy. Toothpaste, shaving cream, and paint pigment tubes are examples of this process. In this method, thin slugs of metal may be pictured as squirting up around the punch upon impact in the same manner as **flash** around forgings. Fig. 14 illustrates the Hooker process for extruding small

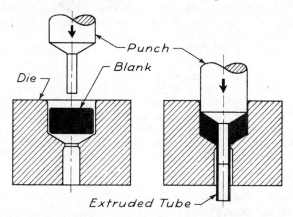

Fig. 14. Method of cold-impact extrusion of metals.

tubes or cartridge cases. In this process, slugs are impact-extruded downward through a die opening and harder metals may be used. Cartridge cases are sometimes made by this process.

EXTRUSION EQUIPMENT. Figure 15 indicates the approximate temperature and pressures used to extrude materials by the **direct** and **inverted** processes. In most cases the addition of an alloying material necessitates a higher extrusion pressure. The speed at which an extrusion is made may vary from 1 to 900 ft. per min., depending upon the metal, the cross section, extrusion temperature, and extrusion pressure.

Hydraulic Presses. Most extrusions are made on hydraulic presses. The presses are either **vertical** or **horizontal** depending on the direction of travel

Material	Extrusion Temperature, °F.	Extrusion Pressures, psi
Aluminum and alloys...............	600–800	15,000–100,000
Copper and alloys.................	500–800	35,000–90,000
Lead and alloys	450	18,000–60,000
Magnesium and alloys.............	450–700	20,000–100,000
Steel.............................	1,700–2,500	120,000
Zinc and alloys..................	400–600	40,000–110,000

Fig. 15. Extrusion pressures and temperatures for direct and inverted extrusion.

of the ram. Vertical presses have capacities ranging from 300 to 1,000 tons, they are easier to align, have a higher rate of production, and require less floor space. Often, however, the need is for floor pits or headroom if long extrusions are to be made.

Horizontal presses are more common. They range from 1,500 to 12,000 tons. The billet bottom cools more rapidly and as a result warping and non-uniform wall thicknesses occur. These presses are used for most commercial bar and tube extrusions.

Dies. The tools and dies used for extruding must be able to withstand high stresses, thermal shock, and oxidation. High alloy steels are commonly used for dies. Flat or square dies with 90 degree half-die angles are typical, but improved conditions result when the angles are 45 degrees. At the smaller angle, die pressures are lower and die life is increased. The exact die configuration depends on the metal being extruded.

The shape of the die determines the outside shape so that the advantage of the extrusion process in the manufacture of long, uniform structural shapes, pipes and tubing, as well as diving boards and model railroad tracks is obvious. However extrusions are also used to produce numerous thin items. For example, aluminum letters and numbers used for signs and displays are produced as extrusions and then cut into thin slices for the individual letter. Similarly intricately shaped ornaments and small hardware can be manufactured by first producing a long extrusion of the basic shape and then slicing the extrusion into smaller parts. Small gears, brackets, and clamps are only a few of the items that can be produced in this fashion.

EXTRUDING PLASTICS. The extrusion process is one that is adapted especially to the molding of uniform sections as in plastics.

Types of Plastics. Although both thermosetting and thermoplastic materials are processed in this manner, the thermosetting materials are not well adapted to this process because they harden very rapidly. Products made by extrusion include garden hose, structural and decorative sections, conduit, safety glass sheeting, bristles, filaments for plastic fabrics, and insulating materials extruded directly on wire and cable.

Continuous Process. Most extrusion molding is done in a **screwtype machine** and is a **continuous process.** Handling equipment for removing the extruded product is essential because of the speed of operation. A diagram of a typical process is shown in Fig. 16. Granular material is fed into the hopper and then forced through the heated cylinder by the screw. As it is pushed through by the screw it is melted to a thick viscous mass and in this condition is forced through the die orifice and formed to shape. As it leaves the machine it is rapidly cooled by air or water, hardening as it rests upon the handling equipment.

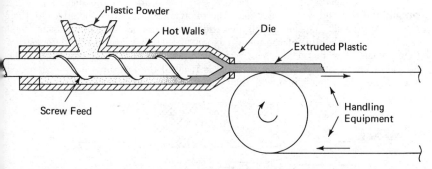

Fig. 16. Continuous extrusion of plastic.

The **design of the screw** is important and varies considerably for different materials. Clearance should be small and the depth of screw shallow to permit uniform heating of the material. The rate of flow through the cylinder is controlled by the pitch of the screw. Cylinders are usually heated with oil although steam or electricity can be used. Temperatures vary with the materials being processed and must be closely controlled.

Metal Stamping and Forming

BASIC OPERATIONS. In the stamping process the size or shape of a metal part is altered by the application of forces which cause plastic flow in the material. These stresses may be tensile, compressive, shearing, bending or any combination. Stamping and forming operations are usually cold-working processes.

STAMPING AND FORMING MACHINES. Forming operations can be carried out on presses, hammers, rolls, and drawbenches. A press is a device containing two platens upon which tools and dies are mounted. Rolls consist of at least two counter-rotating rolls with an adjustable gap such that the metal passing through the roll gap is reduced in thickness. Drawbenches involve pulling the material through dies to effect the reduction.

Types of Presses. There is a wide variety of presses suited to many types of operations. They are usually classified according to the manner in which the power is applied, the purpose of the press, or the design features of the frame. There are many **hand-operated** presses which are satisfactory for jobbing work on thin sheet metal. These machines are mostly for general purpose operations such as bending, shearing, seaming, and simple punching. The majority of production presses are **power-operated** using various mechanisms such as cranks, eccentrics, screws, gears, knuckle-joints, cams, toggle joints, or hydraulic means to effect the necessary pressure. A large variety of operations can be performed by these presses depending on their design and the type of dies used. Classification according to the **types of press operations** performed is as follows:

1. **Shearing**—blanking, trimming, cutting off, punching, perforating, notching, slitting, shaving, and lancing.
2. **Bending**—angle bending, curling, folding, seaming, and straightening.

3. **Drawing**—cupping, forming tubes, embossing, and reducing.

4. **Squeezing**—coining, sizing, flattening, upsetting, riveting, and extruding.

BLANKING AND SHEARING. Blanking and shearing are performed on fast operating crank or eccentric presses. The **blades** of the shear are called **punch** and **die** and a short distance, the **clearance**, separates the edges at their closest approach. As the punch descends upon the metal, the pressure first reaches the elastic limit of the metal. Upon further pressure plastic deformation takes place. The metal is stressed in shear, and fracture starts on both sides. As the ultimate strength of the material is reached the fractures have progressed and, if the clearance is correct, they meet at the center of the sheet. Fig. 17

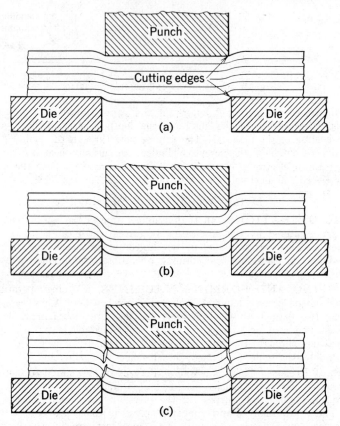

Fig. 17. **Progression of punch through metal showing plastic deformation and fracture.**

from Vidosic (Metal Machining and Forming Technology) shows the final stage of elastic strain (a), plastic deformation (b), and fracture (c). Short strokes and high shearing force are necessary for hard and brittle metals, whereas long strokes and greater clearance must be allowed for soft, ductile metals because of the plastic deformation.

Blanking operations usually apply to fairly large areas where the sheet metal is to be cut to a certain shape. Punching holes in metal, notching metal from edges, or perforating are all similar operations, but the metal removed varies. **Trimming** is the removal of "flash" or excess metal from around the edges of a part and is essentially a blanking operation. **Shaving** is similar to trimming but involves the removal of less metal. **Slitting** is making incomplete cuts in a sheet as with a scissors. If a hole is partially punched and one side bent down as in a louver, it is called **lancing**. These operations may all be done on presses of the same type and differ only in the dies that are used.

BENDING AND FORMING. Bending and forming may be performed on the same type of equipment as that used for shearing—namely, crank, eccentric, and cam operated presses. In processes involving bending, the metal is stressed in both tension and compression at values below the ultimate strength of the metal. As in a press brake, simple bending implies a straight bend across the sheet of metal. Other bending operations such as curling, seaming, and folding are similar but slightly more involved.

In designing a rectangular bar for bending, one must determine how much metal should be allowed for the bend since the outer fibers are elongated and the inner ones are shortened. During the operation the neutral axis of the bar is moved in toward the compression side which throws more of the fibers in tension. The entire thickness is slightly decreased, with the width being increased on the compression side and narrowed on the other. Correct lengths for bends can be determined by empirical formulas but they are influenced considerably by the physical properties of the metal. Metal which has been bent retains some of its original elasticity and there is some elastic recovery after the punch is removed. This is known as **spring back.** The fibers in compression expand slightly and those in tension contract, the combined action resulting in a slight opening up of the bend. Spring back may be corrected by overbending by an amount such that when the pressure is released the part will return to the required shape.

In the forming and cutting of thin sheet aluminum and magnesium into relatively shallow parts, use of the **Guerin process,** utilizing thick pads of rubber for the punch, reduces tooling costs. Rubber proves satisfactory in this process because of its similarity to a fluid when properly restrained. Thick pads of rubber are mounted on the ram of the press and confined in a container which extends just below the pads. Appropriate, low-cost **forming dies** made of steel, plastic, aluminum, or Masonite are placed on the platen. As the ram moves down and the rubber is confined, the force of the ram is exerted evenly in all directions, resulting in the sheet being pressed against the die block, as shown in Fig. 18. This process is limited to the cutting of annealed aluminum up to 0.051 in. thick. For bending and forming the limit is around 0.1875 in. thick. Stainless steel has also been used successfully with this process. Sharp corners cannot be formed by this process since rubber will not flow into the corners of the die. **Urethane** has been used in much the same manner as rubber.

A modification of the Guerin process, called the **Wheelon process,** uses an expanding fluid cell to force the rubber pad against the sheet metal which in turn is formed around the male die.

The **Hydroforming process** (Cincinnati Milacron, Inc.) uses a rubber diaphragm, a pressure chamber, and a punch to form sheet metal parts. Figure 19 shows how a dome shape is produced by Hydroforming. The advantage of this process is that there is very little thinning of the metal due to the high pressure pressing the diaphragm against the punch.

DRAWING. Drawing is similar to bending or forming except that it requires greater plastic flow in the metal. The stresses involved exceed the yield strength of the metal and the resulting plastic deformation allows the metal to conform to the punch. These stresses cannot exceed the ultimate strength without cracks developing. Most drawn parts start with a flat plate of metal. As the punch is forced into the metal severe tensile stresses are induced in the sheet while it is formed about the punch. At the same time the outer edges of the sheet which have not engaged the punch are in compression and undesirable wrinkles tend to form. In most cases this must be counteracted by a blank-holder or **pressure plate** which holds the flat plate firmly in place.

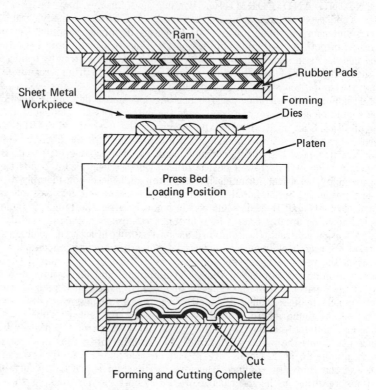

Fig. 18. Forming and cutting sheet metal by the Guerin process.

In simple drawing operations of relatively thick plates, the plate thickness may be sufficient to counteract wrinkling tendencies and the operation may be done in a **single-acting press.** Additional draws may be made on the cup-shaped part, each one elongating it and reducing the wall thickness.

Most drawing, involving the shaping of thin metal sheets, requires the use of **double-acting presses** so that the sheet may be held in place as the drawing progresses. Presses of this type vary considerably in performance but usually two slides are provided, one within the other. One slide moves to the sheet ahead of the other and holds it in place. The motion of the slide is controlled by a toggle or cam mechanism in connection with the crank. Hydraulic presses are also well adapted for drawing because of their relatively slow action, close speed control, and the uniform pressure obtainable during the drawing. Figure 20 (Hydraulic Press Mfg. Co.), shows the sectional diagram of drawing dies mounted on a hydraulic press. In this case the punch and blank-holder ad-

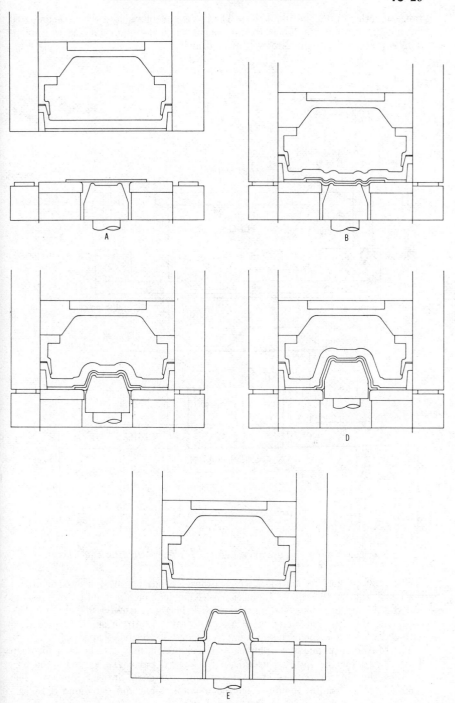

Fig. 19. The Hydroforming process.

vance together until the blank is reached. As the blank and die are contacted the punch continues its downward movement while the blank-holding ring maintains contact with the blank edges during drawing.

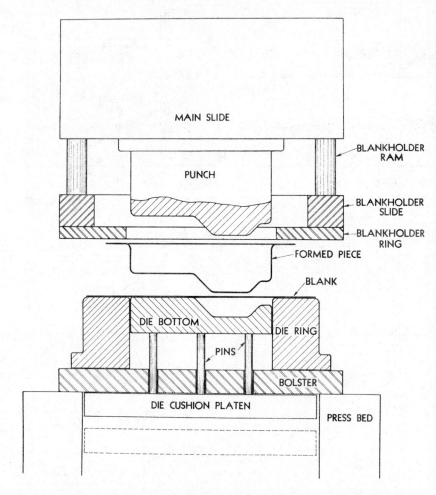

Fig. 20. Punch and die arrangement for deep drawing sheet steel.

The limit to the size and depth of a draw depends on the tensile strength of the blank and its thickness. Reduction is dependent on the hold-down pressure and the die smoothness. The maximum shell height is usually not greater than 65 percent of the shell diameter in a single draw. Multiple steps may be used with or without intermediate annealing to achieve the desired shapes.

The **Marform process**, employing a rubber punch, may also be used for deep drawing and for the forming of irregularly shaped parts. It combines the features of the Guerin process and the conventional deep drawing process. The confined rubber pad is mounted on the movable slide and, surrounded by the blank holding plate, the punch is below. As the slide descends, the rubber pad contacts the blank and clamps it securely against both the top of the punch and the surrounding plate. As the operation progresses, the blank is formed over the punch and at the same time sufficient pressure is exerted over the unformed section to prevent wrinkles. Any tendency for tearing of the metal, in the drawn part, is materially reduced by the pressure exerted against the metal as the operation progresses. During the drawing operation, the downward movement of the blank holding plate is opposed by pressure pins which can be controlled to exert any pressure.

ROLL FORMING. Continuous cold roll-forming is accomplished by a series of mating rolls in line which progressively form strip metal as it is fed continuously through the machine. The number of **roll stations** required on a machine depends upon the intricacy of the part being formed; for a simple angle two pairs may be used, whereas for a tube some six or more may be required. Most of the roll stations are in a horizontal position although vertical rolls may also be required for guiding, straightening, or assisting in the forming operation. As the products emerge from the machine, other operations such as cutting to length, seam welding, or additional bending may be incorporated.

A sequence of forming operations for a window screen section is shown in Fig. 21 (Yoder Co.). The vertical center or **pass line** is first established so that

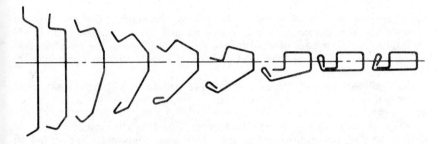

Fig. 21. Sequence of operations in the rolling of a window screen section.

the number of bends on either side is about the same. Forming usually starts at the center and progresses out to the two edges as the sheet moves through successive roll passes. The amount of bending at any one roll station is limited. If the bending is too great it carries back through the sheet and affects the action at the preceding roll station. Corner bends are limited to a radius of the sheet thickness.

In terms of capacity for working mild steel, standard machines form strips up to 0.156 in. thick by 16 in. wide. Special units have been made for much heavier and wider strip steel. The process is rapid since most machines operate at around 100 ft. per min. It is applicable to the forming of products having sections requiring a uniform thickness of material throughout their entire length. Unless production requirements are high, the cost of the machine and tooling cannot be justified.

Bending-roll machines are also used for fabricating smaller lots of cylinders, cones, ovals, and the like. Sheet and plate steel up to a thickness of 1 in. may be used, with power-operated floor-mounted rollers handling the larger sizes and manually operated bench rollers used for the lighter gage sheet metals. In either case the bending is accomplished by passing the metal between horizontally mounted bending rolls. The rolls are usually made up of two pinching rolls and a bending roll. Spacing between the rolls is adjustable to accommodate the various thicknesses of metal. The cylindrical bodies of tanks are often produced in quantity using bending rolls.

COINING AND EMBOSSING. Embossing and coining, illustrated by Begeman and Amstead (Manufacturing Processes) in Fig. 22, are similar in ap-

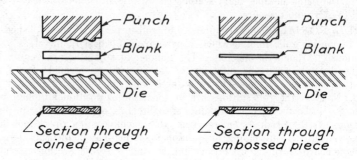

Fig. 22. Coining and embossing.

pearance; yet the action in the metal differs. **Coining** is performed in an enclosed die and the metal flow is restricted in a lateral direction. It is accomplished by an impact or compressive force which causes the metal to flow in the shallow configurations of the blank being coined. Coining operations are restricted to soft metals and alloys.

Embossing is more of a forming or drawing operation. The metal is stretched or formed according to the configuration as in the dies. Both the punch and die have the same configuration so there is little squeezing of the metal to change its thickness as would be the case in coining. The punch is usually relieved so that it touches the metal only at the place being embossed. The action in making deep configurations is similar to that of drawing and there is often some decrease in metal thickness.

SPIN FORMING. Spin forming is a process in which a sheet of metal is shaped by pressing it against a form while it is rotating. **Conventional hand spinning** is performed on a high-speed lathe as illustrated in Fig. 23a from Hammond et al. (Engineering Graphics). The workpieces are forced against hard wood forms called **mandrels** attached to the lathe faceplate. The tools used are usually blunt hand tools supported on a compound rest.

Spin forming may also be performed using hardened steel rollers as the tool. These rollers travel parallel to and at a preset distance from a rotating mandrel so that the contours of the finished part are identical with that of the mandrel. Note that the diameter of the blank is the same as the largest diameter of the finished part (Fig. 23b). The advantage of this **shear spinning** is that a more precise finished product is possible, springback is all but eliminated, the metal is hardened in the process of forming, and the tensile strength is greatly increased.

Spin forming is often a finishing process in which parts initially processed by casting, drawing, stamping, etc., receive their final shaping operation. Spin formed objects must be cylindrically symmetrical (as are turned objects). This process can be used for both non-ferrous metals and some steels. It is often

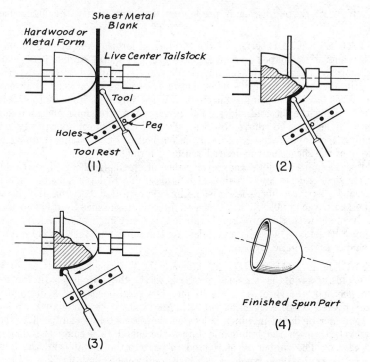

Fig. 23a. Conventional hand spinning operation.

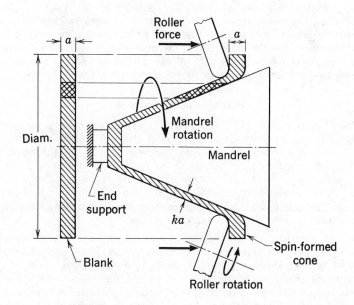

Fig. 23b. Elements of shear spinning.

used in place of deep drawing if production does not justify the high cost of tooling.

METAL STRETCHING. In the forming of large sheets of thin metal involving symmetrical shapes or double-curve bends, a metal stretch press can be used effectively. A single die mounted on a ram is placed between two slides which grip the metal sheet. The die has a movement in a vertical direction and the slides have a horizontal movement. Large forces of from 50 to 150 tons are provided for the die and slides. The process is a stretching one and causes the sheet to be stressed above its yield strength and deform plastically while conforming to the die shape. This is accompanied by a slight thinning and narrowing of the sheet and the desired section is formed.

Adapted to short-run jobs, inexpensive dies of wood, Kirksite (a lead alloy), plastic, or steel can be used. Although it is desirable to have a highly polished lubricated die surface, fiberglass is sometimes substituted between the die and sheet metal. Large double-curved surfaces, difficult to produce by other methods, are easily made with this process. Scrap loss is fairly high as considerable material must be left at the ends and sides for trimming and there is a limitation to the shapes that can be formed.

EXPLOSIVE FORMING. Metals may be deformed into shape by the use of explosives and the pressure or shock waves they produce. In the most common method, the sheet or plate of metal is positioned over a die and the two are covered with a sheet of rubber. An explosive is detonated within a closed container of liquid in which is located the die-metal assembly. The shock wave travels through the liquid and forces the metal to assume the shape of the die cavity. The rubber sheet is used to protect the metal from the liquid.

Both **high explosives** (such as TNT, nitroglycerine, and RDX/TNT) and **low explosives** (such as black powder) are used. The basic difference in the processes involves the pressure-time characteristics of the two types of explosives. High explosives reach very high pressures very quickly and sustain them for only a very brief period. Pressures up to 4 million psi may last for a few microseconds. On the other hand, low explosives, which burn rather than explode, may reach pressures of only 40,000 psi but sustain the pressure for a much longer period thus applying greater total impulse to the workpiece.

Machining Processes

IMPORTANCE OF METAL MACHINING. Machining processes remove metal from a parent body in the form of chips. This is done on a power-operated machine tool equipped to hold work and tool in the proper relationship during the cutting operation; its prime function is to change shape, provide proper surface finish and obtain dimensional accuracy. Many different machines and tools are required because of the great variation in size, shape, and accuracy of products to be machined. Machining can hold closer tolerances than casting, forging, or other similar forming operations.

Types of Cutting Operations. Conventional machining is essentially a finishing operation. Its objective is to bring a part to some predetermined dimension by turning, shaping, threading, drilling, boring, milling, sawing, broaching, or grinding. Only two motions are used to perform these cutting operations: either the tool or the work **reciprocates** (shaping, planing, broaching, etc.) or **rotates** (turning, drilling, milling, etc.). The tool is either fed into the work, as in a planer, or the work moved against the tool, as in a shaper. With these simple relationships available in machine tools, any common type of surface can be formed or generated.

The simplest form of cutting tool is the single-pointed type. (See section on Tools, Jigs, and Fixtures.) When operating in a lathe, the tool is supported rigidly and the work revolves against the edge of the tool. To cut metal efficiently, the cutter tool must be ground with correct angles for the particular metal, and the shearing action requires a keen edge to provide good surface finish.

Machinability. This term expresses the following cutting properties of a material: length of tool life, power required to make a cut, surface finish, and cost of removing a given amount of metal. Standard cutting tests are not infallible, since machinability is influenced by the coolant, cutting speed, feed, and tool angles.

The forces acting upon cutting tools, the materials used, the life of tools, computations for cutting speeds, and economic factors in tooling are discussed in the section on Tools, Jigs, and Fixtures.

TURNING. Turning is normally done by machines which remove material by rotating the work against a cutter. The lathe, an extremely versatile machine tool, can turn work held between centers, in a chuck, on a face plate, or in a collet. Facing, reaming, boring, drilling, threading, knurling, grinding, and a number of other operations can be performed on a lathe equipped with proper attachments. Lathes vary in size, design, method of drive, arrangement of gears, and purpose.

Lathes. The least complicated of all lathes are called **speed lathes.** As the name implies, they operate at high speeds (600 to 3,000 rpm) and are adaptable to such high-speed operations as woodworking, metal spinning, and polishing. Simple hand tools are used, and light cuts are made.

Engine lathes are the general all-purpose lathes of the metal working industry, differing from speed lathes in that they have controls for spindle speeds· and means for feeding the tool. They are more rigid and are equipped with supports and controls for fixed cutting tools. The spindle of these machines may be driven by belts or gears powered by individual motors or line shafts. Engine lathes handle work up to 40 in. in diameter and have beds from 4 to 12 ft. in length.

Bench lathes, as the name suggests, are essentially small engine lathes which are bench-mounted. They are adapted to small work, having a maximum diameter of 9 in. at the face plate, and are often used for precision work.

Toolroom lathes, the most modern of engine lathes, are equipped with all the accessories necessary for toolroom work. These accurate machines have geared heads with a wide range of spindle speeds. Steady rests, quick change gears, feed rods, taper attachments, draw-in collets, and coolant pumps are standard attachments which make them capable of performing precise operations in the fabrication of small tools, test gages, and dies.

Duplicating lathes are machines on which the movement of the tool is controlled by a tracer mechanism which follows a template. The operator need not be skilled, and such products as drive shafts, axles, piston rods, and pump impellers are turned out rapidly and cheaply.

Turret and automatic lathes. The development of turret and automatic lathes has made interchangeable manufacture what it is today. Building the skill of the operator into these machines has made it possible to employ unskilled mechanics for production of identical parts in large quantities. The outstanding characteristic of these lathes is that they permit setting up tools in the proper sequence for performing consecutive operations.

Horizontal turret lathes are designed for use with bar stock or as chucking machines which hold the work piece in chucks or collets. They may be either ram or saddle type. **Ram machines** allow the turret to slide back and forth on the saddle, which is clamped to the lathe bed. **Saddle units** are arranged so that the saddle moves with the turret. Automatic turret lathes utilize a hydraulic system to manipulate the tools systematically and at predetermined feeds.

Vertical turret lathes have the characteristic turret mounted in a vertical position and resemble a horizontal turret unit standing on the headstock end. The spindle is replaced with a table or chuck rotating in the horizontal position. This machine was developed to facilitate the mounting, holding, and machining of heavy parts. **Automatic vertical turret lathes** are units which automatically control the rate and direction of feed, change the spindle speed, and index the turret. Once the cycle of operations is preset, the operator need only load the parts to be machined and unload the finished work.

Automatic vertical multistation lathes are machines in which several spindles operate simultaneously. For example, a six-station machine would have five working spindles and one loading position. The work is held in chucks or fixtures which index under the spindles where separate operations are performed.

Automatic lathes automatically feed the tools to the work and withdraw them after the cycle is complete. Work may be fed to the machine either manually or through a magazine. Lathes of this type can have front, rear, and end tool slides, or, in special cases, they may be equipped with tool slides at angles. Each tool slide has its individual feed and power source.

The **automatic screw machines** are essentially turret lathes designed to use only bar stock. The turret is fed into the work, withdrawn, and indexed to the next position by means of a drum cam located beneath the turret. Stock is clamped in the collet, released at the end of the cycle, and fed against the stock stop automatically by a cam.

Multispindle automatic screw machines use three, four, six, eight, or nine spindles with corresponding cross slides. A tool position is located opposite each spindle so that a number of operations can be performed simultaneously. These machines are not limited to the fabrication of screws, but may be used to produce a wide variety of small parts.

MILLING. Milling machines remove metal by feeding the work against a rotating multi-blade cutter. The milling cutter, having no motion other than rotation, is circular-shaped. As the work is passed under this cutter, each tooth removes a small amount of metal, resulting in a continuous cut completed in one pass. In addition to cutters of this type, which are supported on an arbor, other cutters may be held in the arbor socket. A few of the many milling cutters available include side milling cutters, metal slitting saws, angle milling cutters, form milling cutters, end mill cutters, T-slot cutters, and inserted tooth cutters. As is evident from this large variety, milling machines perform a great many operations and are considered the most versatile of all machine tools.

Milling Machines. The two basic types of milling machines are the **vertical millers** with cutters that rotate about a vertical axis, and the **horizontal machines** with cutters that rotate about a horizontal axis. The work table may be of the fixed bed type, that has only longitudinal movement, or the column and knee type, that has longitudinal, transverse, and vertical motion. Movement of the table may be controlled manually, mechanically, or hydraulically.

Plain milling machines of the column and knee type are usually equipped with power feeds for all table movements, are of the horizontal type, and may be used for slotting, facing, plain, and form cutting. Universal millers (Fig. 24) are plain milling machines with the added feature of a table swivel which

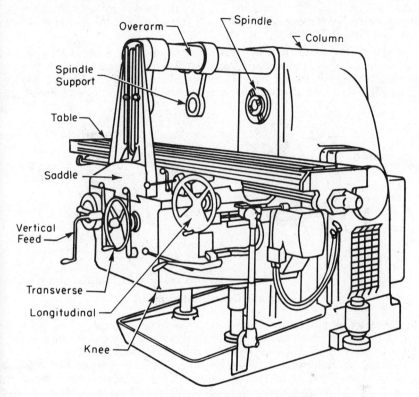

Begeman and Amstead. Manufacturing Processes

Fig. 24. Universal milling machine.

permits cutting of spirals and angular milling. Ram-type universal machines, which can do conventional, horizontal, angular, or vertical milling, have a vertical spindle that moves in and out and can be adjusted to any angle between horizontal and vertical.

Planer-type milling machines resemble a planer and carry the work on a long table having only a longitudinal movement. Machines with multiple spindles and fixed beds are designated as fixed-bed milling machines and may be simplex, duplex, or triplex, depending upon the number of spindles used.

Duplicators, die sinkers, profiling machines, and pantograph machines are among the many special purpose machines available for duplicating and high-production work. They use various types of controls and sensing mechanisms, including templates, cams, and magnetic tapes to guide the cutting tool.

GRINDING. Grinding refers to the abrading or wearing away by friction of a material. In metal cutting, grinding is accomplished by forcing the work against a rotating abrasive wheel. The action of the wheel is similar to that of a milling cutter in that it produces minute chips of metal. Grinding is the only method of cutting extremely hard materials. It has the added advantage of producing smooth finishes to accurate dimensions in a short time, and it requires very little pressure, permitting its use on light work which would otherwise spring away. See the discussion of abrasive cutting tools in the section on Tools, Jigs, and Fixtures.

Grinding Machines. Grinding machines are especially designed for cylindrical, internal, surface, and tool grinding, as well as for such special grinding operations as cutting-off, honing, lapping, and superfinishing. **Cylindrical grinders** (Fig. 25) either hold the work between centers or are of the centerless type.

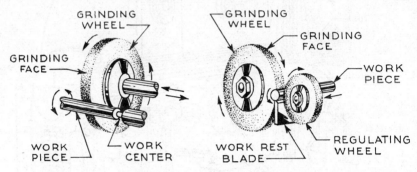

Begeman and Amstead. Manufacturing Processes

Fig. 25. Cylindrical grinding.

In the centerless process, the work is supported by an arrangement of the work rest, a regulating wheel, and the grinding wheel itself. Internal grinding is performed with the work stationary, held in a chuck or by means of rolls.

Rotating or reciprocating tables under vertically or horizontally rotating spindles are utilized in **surface grinding. Tool grinders** are designed to sharpen tools or to grind such tools as drills, milling cutters, or tool bits. Special purpose high-production machines are designed for such jobs as snagging castings, cutting-off, and crankshaft grinding.

Honing, lapping, and **superfinishing** are processes in which surface finish is improved by removal of very minute particles of metal.

Superfinishing is a surface-improving process which removes undesirable fragmentation metal, leaving a base of solid crystalline metal. An abrasive stone, similar to those used in honing, is utilized in finishing round work; oscillated $\frac{1}{8}$–$\frac{1}{4}$ in. over the revolving workpiece at about 425 cycles per min., it produces the required finish.

In superfinishing flat work, a rotating cup-shaped abrasive stone is used, with the work resting on a revolving table. The scrubbing action of these systems results in a mirror-like surface on the part. This refinement is a means of reducing wear and seizure problems in bearings.

SHAPING AND PLANING. Planers and shapers are used to generate plane surfaces with the use of single-pointed tools. The shaper gives the tool a reciprocating motion and moves the work across the path of the tool, thus producing a plane surface. The planer, which can handle large work, imparts reciprocating motion to the work and feeds the tool across it. Planers and shapers may be driven either hydraulically or mechanically.

Shapers. Shapers are classified by the plane in which the tool-carrying ram moves, either vertical or horizontal, and by the tool action, either push-cut or draw-cut. These basic tools are found in all toolrooms, die shops, and small manufacturing plants because they are fast, flexible, and can cut external and internal keyways, spiral grooves, gear racks, dovetails, T-slots, etc. **Universal shapers** are equipped with table swiveling and tilting arrangements for accurate machining of angles. They vary in size from small bench machines with strokes of 7 or 8 in. to heavy duty models with 36-in. strokes.

Planers. These tools can produce plane surfaces which are horizontal, vertical, or at an angle. They are most commonly used for making long straight cuts on thick material. The **double housing planer** is the usual type and is used for routine planer operations. The **open-side planer**, utilizing a single-column tool support, is particularly adapted to handling wide work, while the **pit-type planer** is designed for extremely large work. The latter has a stationary bed, and the tool is moved over the work. **Plate and edge planers** are used in the fabrication of heavy steel plates for pressure vessels and armor plate. They are similar to the pit-type in that the plate is stationary while the carriage, carrying both tool and operator, moves along the work.

DRILLING AND BORING. Producing a hole in an object by forcing a rotating drill against it is known as **drilling**, while **boring** is the operation of enlarging an existing hole. **Reaming** refers to the finishing of a hole to an accurate size by means of a fluted tool called a reamer, which should never remove more than about 0.015 in. of metal. Other operations include **counterboring, centering, countersinking,** and **spot-facing.**

Drilling Machines. Machines for drilling may be classified as portable, sensitive, upright, radial, gang, multispindle, and special-purpose. **Portable drills** are compact units, air or electric motor driven, which are hand held and operated. The **sensitive drill,** often referred to as a drill press, consists of an upright standard, a horizontal table, and a vertical spindle. The drill is manually fed into the work through a wheel or lever control. **Upright drills** are similar to sensitive drills, except that they have power feeds and are capable of handling larger work.

Machines which permit the drill to be swiveled at angles and moved accurately from one position to another are known as **universal radial drills.** These units are used for drilling holes in large castings and forgings which are too heavy to move about under the drill spindle. A series of sensitive drills with individual feeds and a common table is referred to as a **gang drill.** This production tool may be set up so that a hole can be drilled, counterbored, tapped, and spot faced in a series of operations without changing tools.

Multispindle drilling machines were developed for the purpose of drilling several holes simultaneously. They are production machines which, when once set up, drill numerous parts accurately. Multispindle drilling is ordinarily used in the manufacture of interchangeable parts.

Many drilling operations where high production is paramount require the use of special purpose machines. **Automatic-transfer** processing machines have been developed for producing the large number of holes in automobile engine cylinder heads. **Deep-hole drilling** operations, as encountered in the fabrication of rifle barrels, long spindles, and some connecting rods, are done on special machines with long single-fluted drills. The machines automatically retract and re-enter the drills when chips or high resistances are met.

The **jig borer,** designed and constructed for precision work on jigs and fixtures, is similar in appearance to a drill press or a light milling machine and is adaptable to drilling, boring, and end milling.

The **vertical boring mill** is quite similar in appearance to a vertical turret lathe; the horizontal work table revolves while single-point tools execute horizontal facing, vertical turning, or boring. These machines are used in fabrication of large pulleys, flanges, flywheels, and other heavy circular parts. **Horizontal boring machines** drill and enlarge holes by rotating the tool against stationary workpieces. These units are utilized in the same class of work as vertical boring mills.

THREADING. A screw thread is a ridge of uniform section, in the form of a helix, on the surface of a cylinder. Threads may be external or internal, tapered or straight, and of forms such as square, vee, acme, and so forth.

Methods of Thread Making. Several methods are used to produce threads, with such factors as available equipment, workpiece accuracy, cost, etc., determining the best method for a particular case. Vidosic (Metal Machining and Forming Technology) lists the most common methods as follows:

External threads
  Cut or chased on a lathe
  Thread-cutting die
  Rolled between dies
  Special threading machine tools
  Cut on a milling machine
  Grinding
  Cast

Internal threads
  Cut on a lathe
  Cut with a tap
  Cut on a milling machine
  Cast

Taps and dies are discussed in the section on Tools, Jigs, and Fixtures. Although all forms of threads may be cut on the engine lathe, this method is normally used only when small quantities or special forms are required. Cutting dies are used manually or in automatic die heads on turret lathes, the latter method being used when high production is involved. In small lots internal threads are cut manually with taps. Numerous automatic tapping devices including collapsible taps are available for mass-production threading.

SAWING. An important operation in any shop is the sawing of bars and other shapes for subsequent operations. Nearly all machine tools can perform some cutting-off operation, but special machines can do the work better and faster. Metal saws for power machines are made in circular, straight, or continuous shapes.

Reciprocating saws are the most popular for cutting bar stock and similar operations. Simple in construction, they may have mechanical or hydraulic drives and numerous types of feed mechanisms; they vary in size from the simple light duty crank-driven reciprocating hacksaw to the heavy-duty hydraulically driven models. The hacksaw-type blades operate at from 50 to 150 surface ft. per min.

Circular metal saws, similar to metal slitting saws, use solid, inserted teeth, or segmental-type blades; they operate at from 25 to 80 surface ft. per min.

Steel friction disks operating at 18,000 to 25,000 ft. per min. melt their way through I-beams and other ferrous materials. **Abrasive disks,** on the other hand, with surface speeds from 9,000 to 16,000 ft. per min. can be used for almost all materials.

Band-sawing machines are extremely versatile, capable of sawing, filing, polishing, and friction cutting. Available in a variety of sizes, with throats up to 48 in., they have wide usage in contouring of dies, jigs, cams, templates, and other parts which formerly had to be made with other machine tools or, at greater expense, by hand.

BROACHING. Broaching is the operation of removing metal with an elongated tool, whose successive teeth of increasing size, cut a fixed path. Begeman and Amstead (Manufacturing Processes) show a simple broach for sizing a hole (Fig. 26). Each part is completed by one stroke of the broach, the last

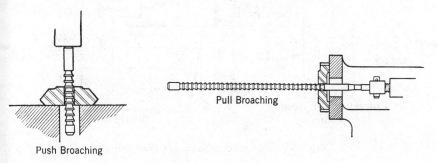

Pull Broaching

Push Broaching

Fig. 26. Broaching a hole.

teeth performing the finishing and sizing operation. Broaches may be pushed or pulled across the work. The process is equally effective whether the work is held stationary or moved across the broach. Broaches are available for surfacing, keying, slotting, cutting splines (including helical splines), and gear cutting (see section on Tools, Jigs, and Fixtures).

The exceptionally high rate of production of broaching machines makes them useful in mass production. Cutting time is a matter of seconds, rapid loading and unloading of fixtures is feasible, and they may be used for either internal or external cuts. Tolerances maintained are suitable for interchangeable manufacture. The limiting factors of broaching machines are the cost of tools, the rigid supports necessary for tools and work, and the fact that the surface to be broached cannot be obstructed.

ELECTROMACHINING. New and exotic materials and intricately shaped products have led to the development of machining and fabricating methods that do not rely on the conventional cutting processes to remove metal. The methods take advantage of long-known electrical phenomena such as electrolysis and spark erosion to do the work. Two so-called "chipless" processes are

Electrodischarge machining (EDM)
Electrochemical machining (ECM)

Pressure or stress on the workpiece is not required and a solid tool does not make contact with the workpiece; however, electrical energy must be supplied during the metal removal process.

Electrodischarge Machining (EDM). The basic elements of an electrodischarge machining system are shown in Fig. 27 from Vidosic (Metal Machining and Forming Technology). The basic operation involves producing a spark between the workpiece and the tool across the gap. Metal removal is effected by the melting of the workpiece at the contact point of the spark. The small globule of metal is then removed by the dielectric. The dielectric also serves to cool the workpiece and tool. The spark is not continuous but occurs in rapid succession at a frequency of up to 10 kHz.

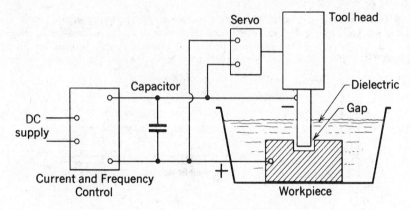

Fig. 27. Basic electrodischarge machining.

According to Lindberg (Materials and Manufacturing Technology) the advantages and disadvantages of EDM are:

Advantages:
1. Any material, regardless of hardness or strength can be machined provided it can conduct electricity.
2. Any shape that can be produced in a tool can be duplicated in the workpiece.
3. Since no mechanical force is required even the most delicate materials can be machined by EDM without distortion.
4. Since no mechanical force is applied by the tool very delicate tool materials and shapes can be used to produce fine details.

Disadvantages:
1. The workpiece and tool must both conduct electricity.
2. EDM is slow; whenever possible the shape or hole is roughed out before using EDM.
3. Because of the intense heat, thermal distortion is a problem.

Electrochemical Machining (ECM). This method differs from EDM in that chemical energy combines with electrical energy to do the "cutting." ECM is very similar to electroplating except that the material is dissolved off the workpiece and carried away by the electrolyte, not deposited on the tool (cathode). The tool must contain a hole so that electrolyte can be pumped under pressure through the gap between workpiece and tool. The entire arrangement of tool and workpiece is contained in an enclosure to prevent splattering the electrolyte all over the machine. Figure 28 from Vidosic (Metal Machining and Forming Technology) shows the elements of an ECM system.

The advantages and disadvantages are similar to those for EDM with the following exceptions.

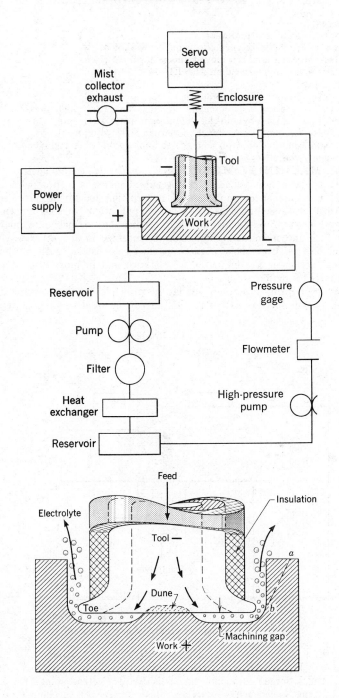

Fig. 28. Basic electrochemical machining system.

Advantages:
1. There is no significant tool wear.
2. There are no adverse thermal effects on the workpiece.
3. The faster the material is removed, the better the surface.
4. Material removal is faster than with EDM.

Disadvantages:
1. The tools are more difficult to design.
2. Because the electrolyte is pumped through the gap at high pressure special fixtures must be used to hold the workpiece in place.
3. ECM basic equipment costs much more than EDM equipment.
4. The most common electrolyte used, sodium chloride, is highly corrosive to equipment, workpiece, fixtures, etc.

OTHER MACHINING METHODS. As technological advances are made in other fields and as the demands for specialized solutions to production problems arise, the number of sophisticated non-cutting machining processes increases. Most of these processes are merely applications of long-known phenomena, and most are restricted in their application to machining materials that are either uneconomical or impossible to machine in any other way. The removal of material from the workpiece in most cases is the result of **thermal** or **chemical** action or a combination of the two and in most cases electrical energy initiates the action.

The **electroshaping** process is similar in principle to electrochemical machining. A system such as shown in Fig. 29 can be used to produce a shape

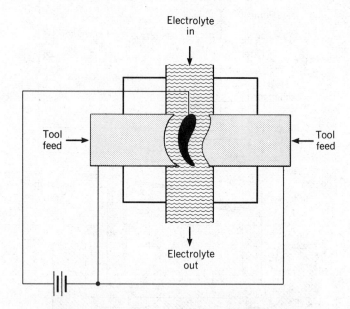

Fig. 29.   Schematic of electroshaping system.

duplicating that of the electrodes. Dimensional tolerances within 0.002 in. can be maintained. For some shapes, such as the turbine blade shown in the figure, the part is rough forged first and then finished by electroshaping. On a production basis this is less expensive than finish forging or machining by conventional methods.

The **electrolytic grinding (EAG)** process (Fig. 30) is an application of Faraday's law of electrolysis. In effect, material is **deplated** from the workpiece. Usually aluminum oxide is employed as the insulating grit but diamonds have been used effectively for grinding tungsten carbide. The grit not only acts as insulation between the metal wheel and the workpiece but also scrubs away particles of decomposed metal from the surface of the workpiece. More than 90 percent of the metal removed from the workpiece is a result of the electrolytic process, with only 10 percent being "scrubbed" away.

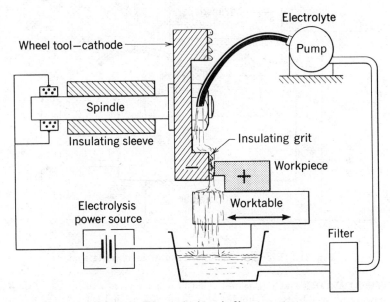

Fig. 30.   Electrolytic grinding system.

Chemical milling removes metal by chemical action. Basically it is an etching process with the chemical reaction carefully controlled. As with other etching processes, precleaning, masking, etching, and stripping are among the operations that must be performed.

By vibrating a tool against a workpiece at ultrasonic frequencies and with amplitudes in the order of grain size, "chips" of material can be dislodged from the workpiece. In **ultrasonic machining** an abrasive material is used between workpiece and tool and it is the direct contact of tool and abrasive against the workpiece that does the actual cutting. This requires that some pressure be applied to the tool itself to force it against the workpiece. Hard brittle materials are very effectively machined in this fashion but with soft materials the abrasive has a tendency to become embedded in the material causing little or no "chipping." Frequencies of about 20 kHz and amplitudes of 0.0003 in. are typical. The elements of an ultrasonic machining system are shown in Fig. 31 from Vidosic (Metal Machining and Forming Technology).

Very small diameter holes may be drilled using the **plasma-jet process.** The drilling is actually accomplished through thermal action produced by a directed, concentrated jet of electrons. This stream of electrons has sufficient energy to melt the material around a small area and vaporize it away. To avoid interference from other molecules surrounding the material, the plasma-jet process is performed in a vacuum chamber. Holes as small as 0.0008 in. have been "drilled" accurately using this method.

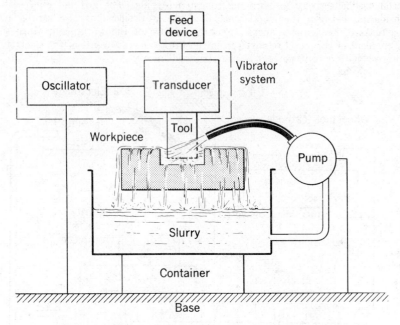

Fig. 31.   Elements of an ultrasonic machining system.

WOODWORKING. Because wood is relatively easy to machine, the basic woodworking processes are simple; like metalworking, hard woods are less workable than soft woods. The requirements of high production and the need for cost reducing techniques lead to the use of multipurpose machines of considerable complexity. Koch (Wood Machining Processes) notes that the plywood, hardboard, fiberboard, and particle-board industries have generated a new class of machines that did not exist prior to 1930. Since 1930 the laminating industry has developed new and important uses of wood, and by so doing has created a demand for new and specialized woodworking machines.

Basic Woodworking Tools. The fundamental woodworking operation is cutting. A woodworking shop making all types of cuts and joints and performing finishing operations might contain the following **basic tools:**

1. Circular or table saw for cross-cutting, ripping, dadoing, etc.
2. Shaper or planer for finishing operations.
3. Jointer for joining.
4. Drill press for drilling, routing, mortising, etc.
5. Wood lathe for turning operations.
6. Sander for final finishing.

A production shop can use these same tools or combinations of these for making such products as flooring in one pass.

The cutting speeds and feeds of woodworking machinery are determined by the hardness and moisture content of wood, direction of the grain, and the cutter angle and edge spacing. Most woodworking cutters revolve at a fixed speed, and the amount of material removed is determined by the feed which in turn determines the finish. Spindle speeds and feed ranges are shown in Fig. 32.

Machine	Approx. Range (Spindle Speed)	Range of Feeds
Circular saws 9″ to 20″ diam. ...	2,000–3,600 rpm	50–300 fpm
Band saws	7,000–9,000 fpm	55–225 fpm
Planers (matchers, molders)....	3,600–7,200 rpm	20–90 fpm
Sanders (drum)	1,200–1,800 rpm	12–36 fpm
Routers	10,000–20,000 rpm and up	Hand and automatic
Tenoners	3,600 rpm	17–60 fpm
Lathe	Up to 4,000 rpm	Hand and automatic
Borers	1,200–3,600 rpm	2–35 strokes/min.
Mortisers	900–3,600 rpm	7–70 strokes/min.

Since most woodworking machines operate at constant spindle speeds and utilize various diameter cutters, spindle speeds are given in lieu of cutting speeds.

Fig. 32. Operational speeds and feeds for various woodworking machines.

Sawing. Wood sawing is accomplished by circular, band, and jig saws. Circular saws are of two general types, the **tilting-arbor** or **table saw** and the **radial** or **cut-off saw.** The variety, or universal, saw utilizes a solid tooth blade which revolves about a tilting arbor located below the surface of the table. The work may be hand fed or power fed into this saw which is, with proper attachments, capable of cutting grooves, moldings, rabbets, tenons and dadoes, as well as miter joints and other angles.

The **radial** or **cut-off** saw is typified by the circular saw blade rotating about an arbor which is located on an overarm above the work table. This saw is particularly adaptable to cross-cut and cut-off work and is also capable of producing most of the cuts made on the table saw. A variation of the radial saw known as the **straight-line ripper** uses a blade, or series of blades on an overarm, together with a chain drive, to power feed the work through the saw.

Band saws, utilizing continuous blades of a number of types, are especially adaptable to cutting curved edges and may be used for resawing and certain ripping operations as bevel ripping. These saws have the advantages of being able to cut short radii and produce irregular-shaped and beveled holes. Production machines of this type are equipped with feeding devices, and some that are equipped with multiple blades have attachments—known as **gang saws**—for duplicating parts. One advantage of the band saw is that it produces a small amount of kerf.

The **jig saw** is used for cutting irregular-shaped objects and producing irregular holes by means of a vertically reciprocating blade similar to a hacksaw blade. The use of this tool is limited to light material, which is usually hand fed because it is a light-weight low-power tool.

Turning and Boring. The basic turning tool of the woodworking industry is the hand lathe, which revolves the workpiece at speeds up to 6,000 rpm. This machine is used for short-run work and pattern-making employing numerous hand tools known as gouges, chisels, and parting tools to produce round parts. For this reason it is known as a **tool-point type** of lathe. As in the turning of metal, the work may be held between centers, on face plates, or in chucks.

Modern mass-production techniques have changed the design of the wood lathe to the point where table legs, bed posts, bowling pins, lamp bases, etc. are produced in a matter of seconds. This change has been accomplished by replacing the chisels and gouges with revolving cutters which are fed into the rotating work piece automatically. The **automatic shaping lathe** and the **copying lathe**, with proper attachments and synchronization, can produce perfect squares, casket corners, square-contoured legs, gun stocks, shoe lasts and other articles that are asymmetrical in longitudinal as well as cross section. The relative machinability of hardwoods in the turning process is given in Fig. 33 (after Davis, in Koch, Wood Machining Processes).

Kind of Wood	Fair-to-Excellent Turnings (percent)	Basis of Comparison†	Kind of Wood	Fair-to-Excellent Turnings (percent)	Basis of Comparison†
Walnut, black	91	1	Yellow-poplar	81	1
Beech	90	1	Birch	80	1
Oak, chestnut	90	1	Maple, bigleaf	80	2
Mahogany	89	1	Ash	79	1
Pecan	89	1	Magnolia	79	1
Alder, red	88	2	Tupelo	79	1
Cherry, black	88	2	Chinkapin	77	2
Madrone	88	2	Hackberry	77	1
Chestnut	87	1	Maple, soft	76	1
Laurel, California	86	2	Blackgum	75	1
Sweetgum	86	1	Cottonwood	70	1
Oak, white	85	1	Basswood	68	1
Sycamore	85	1	Aspen	65	2
Hickory	84	1	Elm, soft	65	1
Oak, red	84	1	Gumbo-limbo	60	2
Maple, hard	82	1	Buckeye	58	1
Oak, tanbark	81	2	Willow	58	1

* Spindle speed 3300 rpm.
† Basis No. 1: Average for three moisture content levels, 6, 12, and 20 percent. Basis No. 2: Tested at 6 percent moisture content only.

Based on Davis. Machining and Related Characteristics
of United States Hardwoods

Fig. 33. Relative machinability of hardwoods in the turning process.

Veneer cutting is a wood-turning operation in which a thin sheet is rotary cut from a log. These thin sheets are layered with the grain crossed for additional strength. Outside sheets of decorative hardwood are made by huge knives slicing the wood into thin sheets which are subsequently bonded to the outside of the laminated pile.

Wood-boring machines are especially adapted to producing, enlarging, or truing holes in wood. This operation can be accomplished on simple drill presses, single spindle borers, or universal spindle, gang-boring machines. **Routing** operations are also performed on these machines. Some typical routing operations are shown in Fig. 34 from Koch.

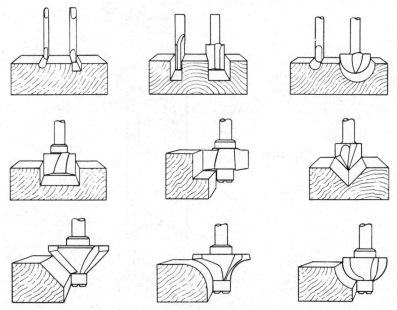

Fig. 34.   Representative cuts machined on the router.

All boring-type machines use revolving drill bits or tools especially designed for wood. The drill or boring tool may be fed manually for short-run jobs or automatically for high production. Power feeding mechanisms may be either hydraulic, pneumatic, or mechanical.

Mortising machines produce square, rectangular, or triangular cavities in wood. They are normally referred to as **hollow-chisel mortisers,** so named because the hole is produced by a hollow chisel inside of which is a revolving drill. The drill produces a round hole, and the square or shaped chisel follows up to finish the hole (Fig. 35). Mortising machines are either single spindle for job shops or multispindle for production work.

The **wood-milling machine** is a special adaptation of the vertical milling machine used in metalworking. Such units are equipped with a vertical rotating spindle located over an adjustable table. Shaping, routing, joining, boring, and drilling tools are attached to this spindle and the work, attached to the table, is fed against the cutter. Wood millers are used in the manufacture of wood patterns where intricate, irregular shapes are frequently encountered.

Joining. The operation of producing joints in wood is known as joining. Joining includes the making of both end joints and edge joints, as well as some edge-surfacing operations. **Tenoning, rabbeting, beveling, tapering, chamfering,** etc. can be done on a jointer. The **jointer** is a machine with a high-speed cutter revolving about a horizontal axis located below the table, as seen in Fig. 36. There are no rollers or pressure feeding mechanisms on the simple jointers; this allows the work to be cut on only one side at a time and is an excellent means of eliminating warp. The jointer, unlike the planer, will produce straight edges but not parallel sides. If parallel surfaces are to be produced, the work should first be jointed, then planed. With proper cutters, shaping operations such as molding and routing can be performed on the jointer.

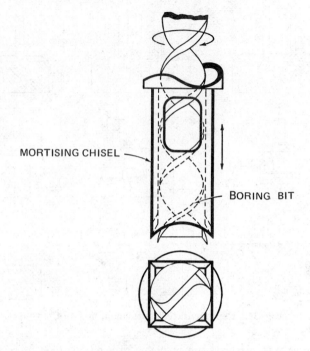

Fig. 35. Hollow mortising chisel and matching boring bit.

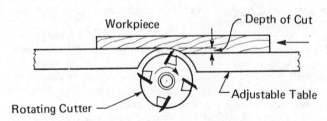

Fig. 36. Joining operation.

Such other joining operations as dovetailing, doweling, and fly-cutting may be accomplished on a shaper, drill press, or other special machines designed for the job.

Special machines designed for **tenoning operations** are either single-end or double-end tenoners. When fully equipped, they take stock cut to length and completely finish either, or both, ends for a number of joints. The production tenoner is equipped with feed chains, pressure beams, cut-off saws, spindle tilting adjustments, cope units, dado attachments, shaping spindles, and a number of other refinements applicable to mass production.

Surfacing and Sanding. Wood-surfacing machines include planers, sanders, and other tools which perform surface-finishing operations. The planer is most commonly used for finishing flat surfaces that must be parallel. The single **surfacer** or **planer**, Fig. 37, has one cutter head which rotates about a hori-

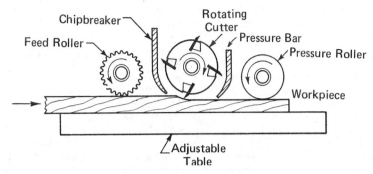

Fig. 37. Planer or single surfacer.

zontal axis over the work, finishing one side of the material and producing work to a specified thickness. **Double surfacers** or **planers,** with cutter heads both above and below the workpiece, finish both sides of the work.

Sanding machines include drum, disk, vertical spindle, and belt sanders, the name being descriptive of the form in which the sanding medium is used. While irregular shapes can be sanded on most of these machines, the belt sander is probably the most versatile in that it can sand flat, curved, or irregular surfaces both on edges and plane surfaces. Belt sanders are used extensively for sanding doors, moldings, table legs, table tops, and numerous other items.

Abrasive tumbling, or barrel finishing, is used to produce a clean well-sanded surface on all areas of the workpiece. The parts best suited to this type of finishing are small with no sharp corners or extensive flat surfaces. Abrasive tumbling is used for small turnings, such as tool and cooking utensil handles and short dowel parts. The process involves placing the parts and an abrasive material, usually dry pumice, in a rotating or vibrating barrel. The tumbling process takes from 2 to 8 hours. Large batches of parts may be finished at one time depending on the size of the barrel and the amount of abrasive used.

Shaping and Molding. Shaping and molding are essentially peripheral milling processes. The purpose of the shaping operation is to cut an edge pattern on the end, side, or periphery of the workpiece. The purpose of the molding operation is to machine long or short pieces of lumber into special forms as shown in Fig. 38. Because of the ease of set up, molding operations are often performed in relatively short runs for any given pattern. In some plants as many as twenty different patterns may be run in a single day.

Fig. 38. Cross sections of typical moldings.

MACHINING PLASTICS. Most plastics can be machined using conventional machine tools. The important factors to consider in machining plastics is their softness and heat sensitivity. Whenever possible compressed air or water-soluble oil coolants should be used during the machining process. Because plastics have greater thermal expansion than metals extra clearances must be provided when working with them.

Plastics can be turned on woodworking or metal-cutting lathes. High speeds are advisable for most plastics with surface speeds as high as 3,000 sfpm recommended for styrenes and acrylics. The cutting tool should have as shallow a top rake angle as possible (0 to −5°) to produce a scraping action rather than a cutting action. The depth of cut and feed should be light. Measures must be taken to reduce the amount of plastic dust produced during the machining process since the dust presents a health hazard. In some cases exhaust systems may be required.

Composite, reinforced, and laminated plastics are very abrasive and require special attention during machining. High-speed tools or carbide bits are recommended.

Automatic Machines and Numerical Control

AUTOMATIC MACHINES. High production levels and labor-saving techniques require that machine tools perform as many operations simultaneously or consecutively as possible, without the necessity of manual handling or setting up for each operation. Automatic, or semi-automatic, operation is characterized by one or more of the following: automatic speed and feed regulation, gaging, workpiece loading, ejecting, push-button operation, programming, and electronic controls. Common automatic machine tools include the lathes, screw machines, and cold-headers.

Positioners. A positioner is defined by Keller (Unit-Load and Package Conveyors) as any mechanical device used to orient the item being conveyed into a machine tool or to change its position while being conveyed; or it may be some form of transfer that alters the position of the item while achieving the transfer (see the discussion of Transfer Machines that follows). Fully automatic devices are available that can upend or rotate parts in any preset position desired. A basic element of any automatic position system is the sensor which determines the orientation of the workpiece as it enters the positioner and reorients it so that it enters the machine tool in the proper position.

Automated positioners are capable of reaching a predetermined position through programmed point-to-point travel or by continuous path control.

Transfer Machines. Vidosic (Metal Machining and Forming Technology) describes a transfer machine as an automatic machine tool consisting of a number of basic machine tool operations linked together automatically to perform various machining and gaging operations in sequence. The distinctive features of this type of machine is that one or more cutting operations may be performed at each machine station and the workpiece moves from one station to another automatically. During the transfer process positioning may also take place so that the orientation of the workpiece does not remain constant as it passes through the transfer machine. Transfer machines may be straight-line or circular. Where space limitations require, straight-line transfer machines may have one or two right-angle bends.

The workpiece in a transfer machine is moved from station to station on a carrier or fixture usually travelling on ways. At each station the workpiece is accurately indexed and held rigidly for the operation to be performed. All movements and operations are done automatically. A single operator may be stationed at a control board covered with indicating lights that inform him of all operations going on in the machine. He alone may be responsible for the operation of the entire machine.

Unless production requirements are very high, as for example in the production of automobile engines, the initial expenses for this type of machine may not be warranted.

NUMERICAL CONTROL OF MACHINE TOOLS. Numerical control machining may be defined as an operation in which servo motors replace human operators in the positioning of the workpiece, and the positioning and operation of the tools. The major disadvantage of automation in production is the requirement of high production rates and standardization of product. Thus most automated machine tools are uneconomical for small runs, custom made products, or job-lots. Numerical control not only makes such production possible but also economically desirable.

Principles of Control. According to Leone (Production Automation and Numerical Control) a numerical control machine-tool system consists of three control phases—tape reading, controlling, and position measuring. **Tape reading** involves converting the information represented by holes in a tape into information usable by the control system. **Position measuring** involves monitoring the positions of the machine tool elements at all times. Transducers attached to the machine elements themselves perform this operation. Figure 39 from Leone shows the basic elements comprising the control function.

Punch Tape Instructions. Machine instructions are usually put on paper or plastic tape. The perforations in the tape represent specific instructions to the machine system through the use of a machine language. Standards have been established and others proposed for the size of the tape and the language and format used so that tapes could be interchanged between different systems. The standard tape is 1 in. wide with eight information channels across its width. Formats for both positioning and contouring are set for most conditions (Electronic Industries Association Standards RS-244, 273, and 274). The tape format for a typical positioning operation is shown in Fig. 40 (Leone, Production Automation and Numerical Control).

Position and Contour Control. Numerical controls fall into two basic categories: positioning systems and contouring systems. In the positioning systems, also called **point-to-point systems,** the workpiece and tool positioning operations are performed in discrete movements that can be described by means of position coordinates. If there is movement in more than one axis, coordination of the movements in the different axes is not necessarily available. In most cases the tool is moved from one point to another for the actual machining operation which is performed only after the positioning operation ceases. Machining operations such as milling may be performed during the actual movement of the tool during the positioning operation. Such operations are most often done in a straight line or in a random path between the two programmed points. Though it has definite limitations in speed and accuracy, a contouring operation may be simulated by partial coordination of the two coordinate positioning axes.

In the true contouring systems, also called **continuous-path** systems, the work and tool movements follow a prescribed path continuously under control. This involves coordination of all axes of movement necessary for the prescribed path. Not only are the physical components of such a system more sophisti-

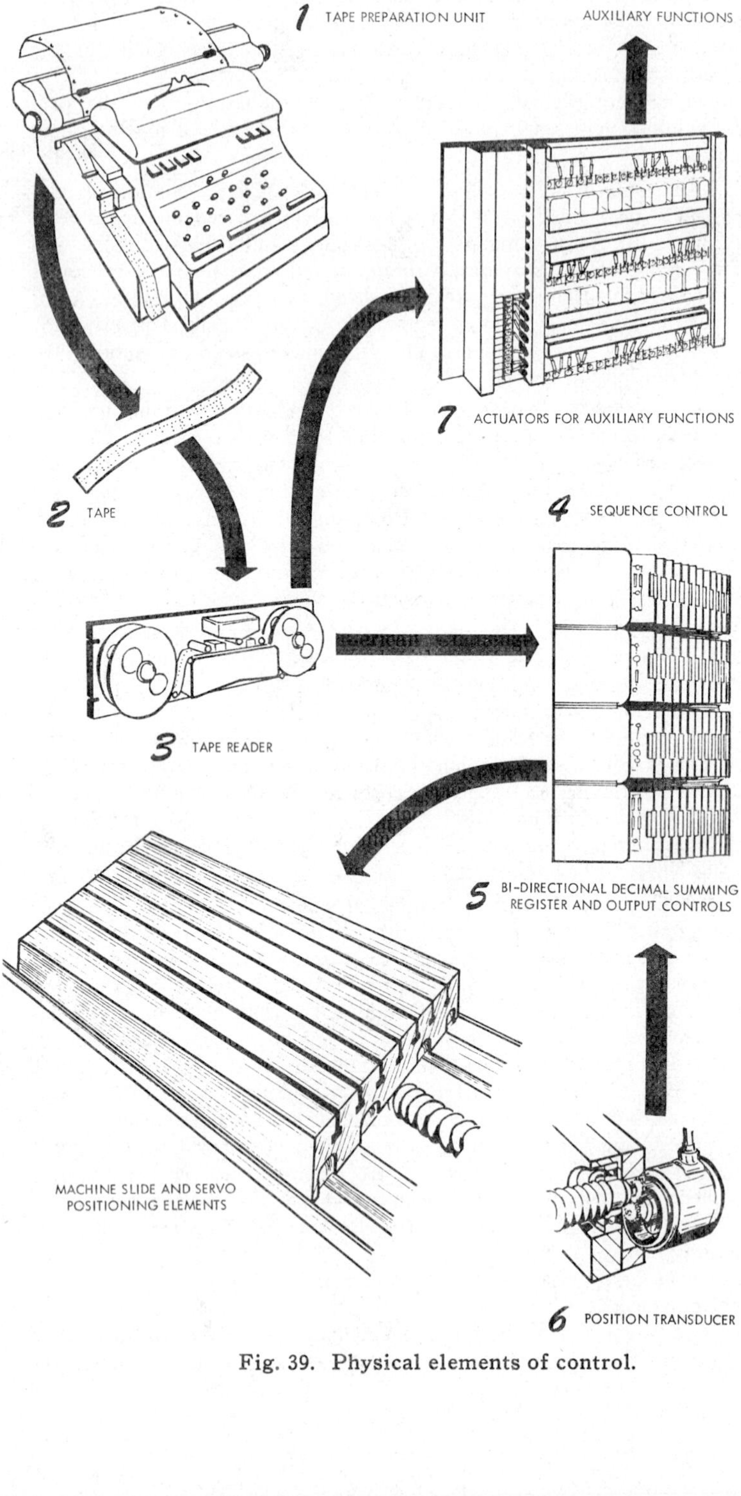

Fig. 39. Physical elements of control.

CODE	MISCELLANEOUS FUNCTIONS
m00	STOP
m08	COOLANT ON
m09	COOLANT OFF
m21	TOOL OFFSET
m23	MISC. AUX. 1
m24	MISC AUX. 2
m25	MISC AUX. 3 (Creep off)
m26	X, Y TABLE OFFSET
m30	END OF TAPE

CODE	PREPARATORY FUNCTIONS
g80	CANCEL CYCLE (Mill or tap off)
g81	RAPID
g82	DWELL
g84	TAP (Spindle reverse)
g86	MILL ON
g88	RETRACT TO LIMIT

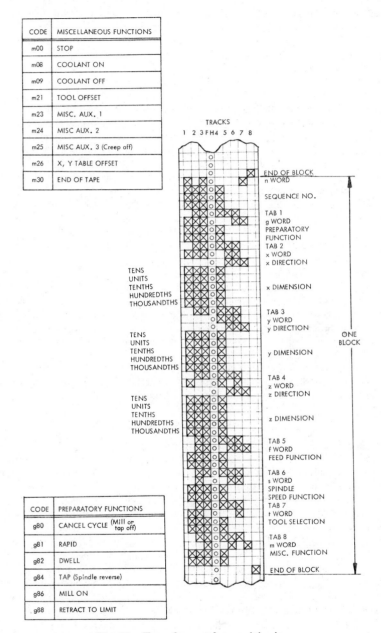

Fig. 40. Tape format for positioning.

cated but the information describing the path is more complex. Several approaches are used to describe the path in contouring work including a breakdown of the curve into small straight lines or easily definable curves such as circles, parabolas, hyperbolas, etc.

Further complexity is introduced by the fact that the prescribed path is delineated by the center of the cutting tool whereas the actual cutting takes place at the cutting edge. Thus the diameter of the cutting tool must be taken into account and the path adjusted to compensate for this dimension.

Accuracy and Precision. Since the accuracy of numerical control machining is not dependent upon the reaction speed of the operator nor his sensitivity to deviation from the norm, extremely high accuracy is possible. The response of properly functioning electronic controls is vastly superior to that of humans; thus the ability to attain highly accurate results is easier and more economical with numerical controls than by manual means.

Although one may question the accuracy of a given numerical control system there is hardly any doubt about the ability of these systems to reproduce or repeat an operation with almost no deviation. This near infallibility of reproduction, or precision, is one of the important advantages of numerically controlled machine tools.

Retrofitting. Modifying an existing machine tool to adapt it to numerical control is called **retrofitting.** Leone (Production Automation and Numerical Control) finds that retrofitted systems rarely approach the performance levels of new integrated systems. This is because machines that are specifically built to be tape controlled often have features that cannot be easily duplicated in retrofit machines without very costly modifications. Also machines that are retrofit are apt to be worn in such places as ways, gears, slides, drives and other points basic to the accuracy of the machine and these are the very elements that must be controlled. Thus the number of machines being retrofit is relatively low when compared to the number of new numerical control installations.

Welding Processes

ELEMENTS OF WELDING. Welding is the fusion or uniting of two pieces of metal by heat or a combination of heat and pressure. This important process is so highly developed and diversified that it is a necessity in almost every metalworking operation. Casting repair, building erection and repair, piping, bridge building, fabrication of pressure vessels, sheetmetal working, and the manufacture of household appliances, automobiles, airframes, and guided missiles all rely on some form of welding.

Classification of Processes. Welding processes may be divided into three major groups: non-pressure fusion processes, pressure processes, and brazing processes. The gas, arc, electron beam, laser, and thermit processes involve fusion or melting of the metal but no pressure. Pressure welding which includes forge, friction, inertia, spot, seam, and butt welding does not involve melting. Brazing and soldering comprise the last category.

Classification of Joints. All welding processes are facilitated by the **parent metal preparations**—grinding, machining, wire brushing, sand blasting, or degreasing. Impurities in the weld tend to decrease the joint strength and cause such imperfections as poor penetration, gas pockets, cracks, slag inclusions, excess oxidation, and poor appearance. Soundness and appearance of any weld are also improved by proper "fit up" and **alignment of the parts** prior to, and during, welding. Choice of the proper process for the particular welding job is of utmost importance. Information on procedure and welder qualification can be found in the Welding Handbook, published by the American Welding Society. Various codes and specifications for pressure vessels and special processes are also available from the "Codes, Standards and Specifications," American Welding Society. The principal types of joints used in most welding processes are shown in Fig. 41 (Lincoln Electric Co.). The forms may vary slightly according to the thickness of the material, but all can be used for either gas or arc welding.

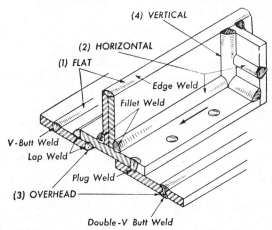

Lincoln Electric Co.

Fig. 41. Basic types of welded joints and their positions.

ARC WELDING. Arc welding is a welding process in which melting is obtained by heat produced from an electric arc between the work and the electrode. Contact is first made between the work and electrode to create an electric circuit, and then, by separating the conductors, an arc is formed. The arc attains a temperature of $9,000°-10,000°F$. causing the adjacent metal to melt almost instantly.

Arc welding equipment produces either alternating or direct current. Most a-c machines are simple transformers with output ratings in the neighborhood of 200–300 amperes at 18–25 volts, 60 Hz. Some special-purpose a-c machines have high-frequency attachments. D-c machines are either motor generator sets, gasoline-engine or electric-motor driven, or a-c rectifiers. These machines are built with capacities up to 600 amperes., having an open circuit voltage of 50 to 95 volts. The machines most widely used are rated at 200 amperes, with an actual output of from 40 to 250 amperes and a closed circuit of 18 to 25 volts. **Rectifier**-type d-c welders, readily available, are widely used. Some models provide both a-c and d-c output at the option of the operator.

Arc Welding with Coated Electrodes. In this operation, molten metal from the electrode is forced onto the base metal by the action of the arc. This characteristic makes it possible to perform overhead welding. There are a number of electrodes generally designated as heavy coated. Specifications for mild steel arc welding electrodes are listed in Fig. 42 (American Welding Society). Fusion is produced by the heating effect from the arc between the coated electrode and the workpiece. Shielding is obtained from the coating, the electrode

MANUFACTURING PROCESSES AND MATERIALS

MILD STEEL ARC-WELDING ELECTRODES

Electrode Classi- fication Number	Type of Coating or Covering	Capable of Producing Satisfactory Welds in Positions Shown*	Type of Current
E45 Series: Minimum Tensile Strength of Deposited Metal in Non-Stress-Relieved Condition 45,000 Psi			
E4510 E4520	Sulcoated or Light coated	F, V, OH, H H-Fillets, F	Not specified, but generally dc, straight polarity (electrode negative).
E60 Series: Minimum Tensile Strength of Deposited Metal in Non-Stress-Relieved Condition 60,000 Psi or Higher			
E6010	High cellulose sodium	F, V, OH, H	For use with dc, reversed polarity (electrode positive) only.
E6011	High cellulose potassium	F, V, OH, H	For use with ac or dc, reversed polarity (electrode positive).
E6012	High titania sodium	F, V, OH, H	For use with dc, straight polarity (electrode negative), or ac.
E6013	High titania potassium	F, V, OH, H	For use with ac or dc, straight polarity (electrode negative).
E6015	Low hydrogen sodium	F, V, OH, H	For use with dc, reversed polarity (electrode positive) only.
E6016	Low hydrogen potassium	F, V, OH, H	For use with ac or dc, reversed polarity (electrode positive).
E6018	Iron powder, low hydrogen	F, V, OH, H	For use with ac or dc, reversed polarity.
E6020	High iron oxide	H-Fillets, F	For use with dc, straight polarity (electrode negative), or ac, for horizontal fillet welds; and dc, either polarity, or ac, for flat-position welding.
E6030	High iron oxide	F	For use with dc, either polarity, or ac.
E70 Series: Minimum Tensile Strength of Deposited Metal in As-Welded Condition 70,000 Psi or Higher			
E7014	Iron powder, titanis	F, V, OH, H	For use with dc, either polarity or ac.
E7015	Low hydrogen sodium	F, V, OH, H	For use with dc, reversed polarity (electrode positive) only.
E7016	Low hydrogen potassium	F, V, OH, H	For use with ac or dc, reversed polarity (electrode positive).
E7028	Iron powder, low hydrogen	F, H-fillets	For use with ac or dc, reversed polarity.

* The abbreviations F, H, V, OH, and H-Fillets indicate welding positions (for electrodes $\frac{3}{16}$ in. and under except in classifications E6015 and E6016, where electrodes $\frac{5}{32}$ in. and under are used):

F—Flat.
H—Horizontal.
H-Fillets—Horizontal fillets.

V—Vertical.
OH—Overhead.

Fig. 42. Electrode classification chart.

being used as filler metal. The heavy coatings stabilize the arc, prevent formation of oxides and nitrides, provide a slag to protect the weld, slow the cooling rate, and perform metallurgical refining operations. This **shielded-metal arc process** is applicable to all metals, providing the proper electrode is used, and is the most widely used of all welding processes.

Impregnated Tape Metal Arc Welding. In this method, an electrode encased in an impregnated tape wrapping is used. Fusion or coalescence is produced in the same manner as in the shielded metal arc process. Because the tape is wrapped around the bare wire just ahead of the arc, an automatic machine is required. Excellent welds are produced by this process.

Atomic Hydrogen Arc Welding. Here, single-phase a-c current is utilized to produce the arc between two tungsten electrodes. Hydrogen is introduced into the arc through electrode clamps or holders. As the hydrogen enters the arc, the molecules are broken up into atoms which recombine into molecules of hydrogen outside the arc. This reaction is accompanied by the liberation of intense heat, attaining a temperature approximating 11,000°F. Fig. 43 shows the electrode holder and the manner in which the hydrogen is supplied to the arc.

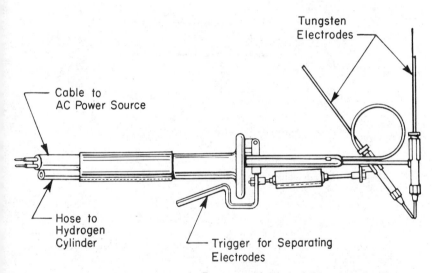

Begeman and Amstead. Manufacturing Processes

Fig. 43. Atomic hydrogen arc-welding electrode holder.

This process, whether used automatically or manually, has three distinct **advantages**: the arc temperature is very high, the tool holder can be moved about without extinguishing the arc, and the work and electrode are shielded by an atmosphere of hydrogen. Operating cost is somewhat more than for other arc welding processes. The atomic hydrogen process produces metallurgically sound welds in materials such as stainless steel, tool steels, heat resisting alloys, and other alloys normally difficult to weld. **Filler metal** can be added as needed in either bare or fluxed form. Resulting welds are clean, smooth, free from scale, and respond well to heat treatment.

Inert-Gas-Shielded Metal Arc Welding. This is a process in which coalescence is produced by heat from an arc between a metal electrode and the work, which is shielded by an atmosphere of either argon or helium. Filler metal may or may not be used. Tungsten electrodes are generally used because their melting point is high and they are not consumed in the inert atmosphere. Addition of filler metal requires use of a separate welding rod. Automatic or manually operated holders which use a consumable wire electrode are available. Inert-gas-shielded metal arc welding is especially adaptable to welding aluminum, magnesium, beryllium, copper, and stainless steels and is suitable for welding almost all metals without the use of fluxes. A tungsten electrode torch used in this process is illustrated in Fig. 44 (Linde Div., Union Carbide Corp.).

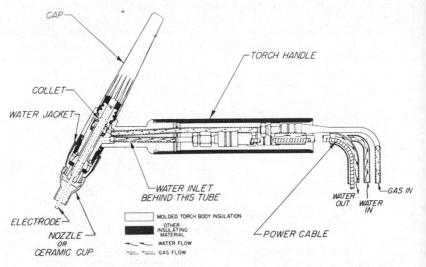

Fig. 44.   Hand-welding torch for inert-gas-shielded arc welding.

Submerged Arc Welding. This technique is so named because the metal arc is shielded by a blanket of **granular fusible flux** during welding. The consumable welding electrode is automatically fed through the welding head and into the vee-groove, as shown in Fig. 45 (Linde Div., Union Carbide Corp.). As molten metal is formed in the joint, some of the granular flux is melted. The latter floats on top of the weld and, upon cooling, solidifies into a protective coating over the weld. High welding currents which permit rapid welding speed and metal transfer can be used. This process is used primarily on flat welds where high production and deep penetration are required. While particularly adaptable to low carbon and alloy steels, the process may also be used on many nonferrous materials.

Shielded-Stud Arc Welding. This method is used to end-weld metal studs to flat surfaces. It is a d-c process utilizing a pistol-shaped gun to hold the stud, which in turn, is shielded by a ceramic ferrule. Heat is generated by the arc developed between stud and plate. After heating, pressure is utilized to effect the weld. Such welding is used in shipbuilding and industrial applications involving metal fasteners.

Unshielded Metal Electrode Arc Welding. In unshielded metal electrode arc welding, coalescence is obtained by heat produced from an arc between a bare electrode and the workpiece.

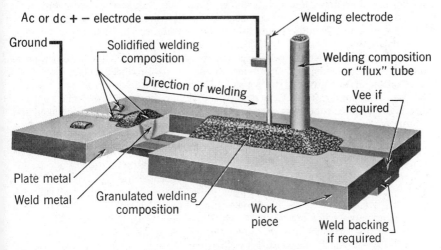

Fig. 45. Submerged arc welding.

Bare metal and contact electrode arc welding are welding processes similar to the shielded metal electrode method but differ in that the electrode is either bare or coated with a current-conducting flux. The contact electrode is covered with powdered metal and some flux. These electrodes maintain proper arc length when held in contact with the work. Contact electrode coatings melt at a lower rate than the core wire; this results in formation of a deep cup in the end of the electrode which prevents it from freezing to the work.

Contact electrodes are used when it is desirable to deposit a great deal of metal rapidly in a horizontal position. These electrodes are not adaptable to overhead or position welding, and they operate at higher currents than equivalent size coated rods.

Carbon Electrode Arc Welding. In carbon electrode arc welding, heat is obtained from an arc generated between a carbon electrode and the workpiece. The weld may be shielded by an inert gas, the combustion of a solid flux fed into the arc, a blanket of flux on the workpiece, or a combination of these. When twin carbon electrodes are used, heating is obtained from the arc produced between the two electrodes. The unit is used manually, much like an oxyacetylene torch. Pressure and filler metal may or may not be used. Steel, galvanized iron, and cast iron are the metals most frequently welded by this process.

GAS WELDING AND CUTTING. Gas welding includes all of the processes in which gases are combined to produce a hot flame. Although the **oxyacetylene process** is most widely used because the flame temperature reaches about 4,200°F., natural gas, hydrogen, propane, and butane can be used in combination with oxygen or air. The fact that gas welding processes may be used for welding, brazing, cutting and machining, and are portable, adds to their universal usage.

Oxyacetylene Gas Welding. Acetylene is a compound of carbon and hydrogen (C_2H_2) obtained by the reaction of calcium carbide and water. This highly combustible gas is stored in cylinders of 300 cu. ft. or less at pressures not exceeding 250 psi in combination with acetone. The maximum safe pressure for usage and storage of acetylene without acetone is 15 psi. Acetylene generators are available, but in the interest of safety and portability, the gas in cylinders is preferred.

Oxygen is a colorless, odorless gas which, although it will not burn, supports combustion and increases flame temperature. It is produced commercially by liquefaction of air, by heating certain oxides or by the electrolysis of water. Oxygen, at around 2,000 psi, is stored in cylinders ranging from small ones to long high-pressure tubes mounted on trailers. Extreme care must be taken not to store oxygen near oil, grease, or other combustible materials.

The **welding torch** is a device for mixing low pressure gases and delivering them to the welding tip where they are burned. The resulting oxyacetylene flame may, with excess oxygen, be oxidizing; with approximately a one-to-one ratio, be neutral; or, with excess acetylene, be carburizing. The neutral flame is used in most applications. Steel pipe, plate, and sheet are fusion welded with a neutral flame using a steel filler rod without flux. High carbon steel requires a carburizing flame, while brass is best welded with an oxidizing flame.

Oxyacteylene welding has the advantage of being inexpensive, portable, and versatile. Particularly adapted to welding of wrought iron and plain carbon steel, it can also be used for many nonferrous metals, flame hardening, and hard-surfacing of materials.

Air-Acetylene Welding. In this gas welding process, oxygen is replaced with air, and the resulting torch is not unlike a Bunsen burner. Therefore the flame temperature is necessarily low and, as a consequence, the uses are limited to low temperature brazing and soldering operations.

Oxyhydrogen Welding. The oxyhydrogen flame, produced with equipment similar to that used in oxyacetylene welding, operates at about 3,700°F. Larger torch tips are employed and control of the gas mixtures is critical in this process. It is used primarily for welding thin sheets and metals with low melting points, as well as for some brazing. A reducing atmosphere results in good quality welds.

Pressure Welding. Pressure gas welding utilizes an oxyacetylene flame to heat the surfaces to be joined in a butt welding operation. As can be seen in Fig. 46, from Begeman and Amstead (Manufacturing Processes), the joint ends

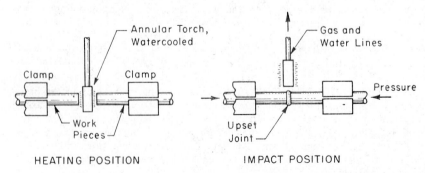

HEATING POSITION   IMPACT POSITION

Fig. 46. Pressure gas welding.

are heated, the torch is removed, and pressure is applied to produce coalescence. A variation of this method starts with the abutting areas together while heat is applied, pressure being added when the proper upsetting temperature is reached. Pressures vary from 1,500 psi to 4,000 psi, depending upon the metal to be joined and the size and type of joint.

Pressure welding is useful primarily in the joining of rods, tubes, rails, and pipes. It is adaptable to high-production and automated operations. It is employed also in the welding of dissimilar metals, such as high speed steel and carbon steel.

Flame Cutting. Flame cutting of steel has developed into an important process in the steel fabrication industry. Simple hand or automatic torches with portable accessory equipment make gas cutting an economical process where accuracy is not paramount.

The tip of a cutting torch is so arranged that a number of preheating oxy-acetylene flames circle the central passage through which oxygen passes. The steel is first preheated to around 1,800°F.; then a jet of oxygen is directed on it. Instantaneously the steel is burned into an iron oxide slag. Hydrogen, natural gas, or propane can also be used for preheating.

Gas cutting is used either manually or automatically to prepare plate edges, shape large or small parts, scarf rounds, or to cut plate, bar, or sheet. This process can be used for automatic cutting with multiple torches. Material up to 30 in. thick can be torch-cut. Cast iron, nonferrous alloys, and high manganese alloys are not readily cut by this process.

Flame machining is the term used to describe the operation of removing metal with a cutting torch without severing the piece.

RESISTANCE WELDING. In electrical resistance welding, the flow of a controlled heavy electric current through a high resistance joint produces local heating sufficient to cause fusion of the joint when pressure is applied. The current used is usually 60 Hz ac supplied at 220/440 volts, which is stepped down in a transformer to a high amperage at 4 to 12 volts for the welding circuit. Direct current, stored energy ac damped oscillation, stored energy d-c surge, or stored energy d-c may also be used. Proper **electronic devices** are necessary in all applications to control the time and amperage of the current. Pressure can be applied by pneumatic, mechanical, or hydraulic means. This pressure must be coordinated with the flow of current.

Spot Welding. Spot welding is a resistance welding process in which two or more sheets of metal are held between metal electrodes. Spot, seam, and projection welding are illustrated in Fig. 47. The **welding cycle**, controlled electronically, starts with a "squeeze time"—the interval during which pressure is applied with no current passing. Next, while the material is under pressure, current flows during "weld time." So that the weld may solidify, pressure is maintained for a period after the current passes; this is known as "hold time." Where repetitive welds are made, the interval between welds is known as "off time."

Spot welding is the simplest form of resistance welding and most other types are derived from it. It is a production method usually employing stationary machines. The electrodes, however, may be attached with proper cables and made portable. A large variety of machines make spot welds in thin sheet (0.001 in.) or in materials up to ½ in. thick. Multiple spot welders have been developed to make several spots at one time.

Nearly all metals can be fabricated by spot welding. Spot welds are widely

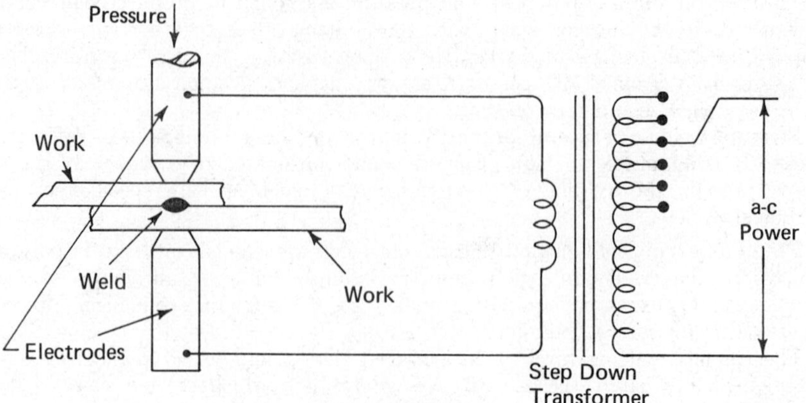

(a) Spot Welding

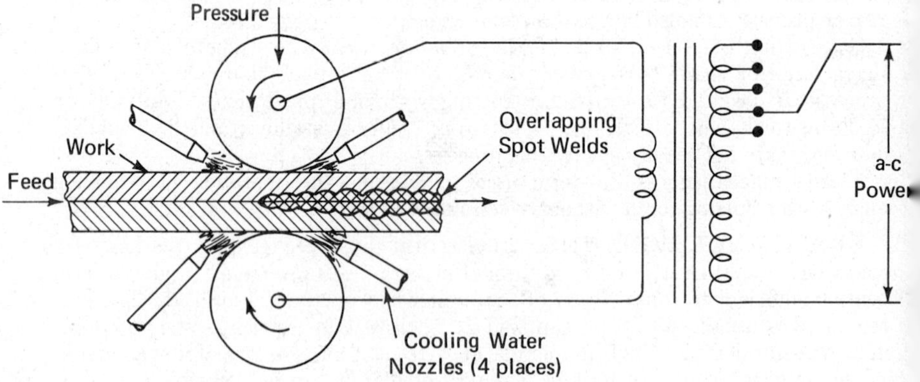

(b) Seam Welding

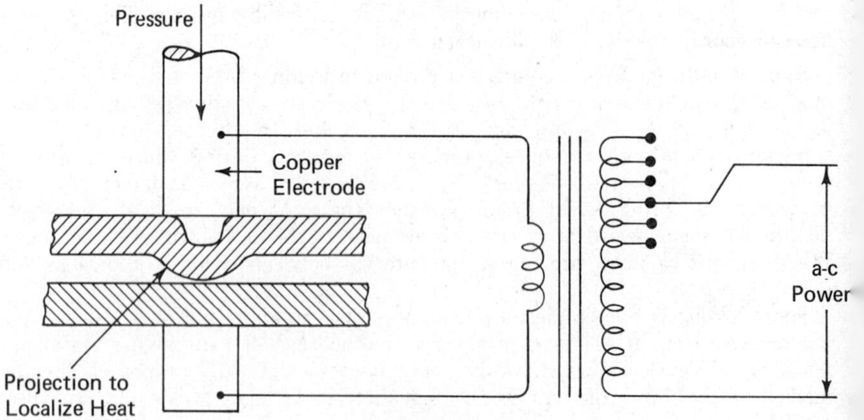

(c) Projection Welding

Fig. 47. Spot, seam, and projection welding.

used in manufacture of automobiles, refrigerators, metal toys, airframes, and numerous other sheet metal applications.

Seam Welding. If the electodes in a spot welder are replaced with mechanized pressure rollers and current is passed, the resulting welds are called either roll spot welds or seam welds. The flow of current is regulated and synchronized with the movement of the material to produce from 5 to 14 spots per inch. The number and size of spots are determined by the type of material and its thickness. The seam weld, shown in Fig. 47, is so named because the welds overlap and produce a gas- and water-tight joint. Seam welding is an extremely rapid process capable of producing continuous welds at speeds of 200 in. per min. and up. Metal containers, automobile mufflers, stove pipes, refrigerator cabinets, and gasoline tanks are fabricated by this method.

Projection Welding. While seam welding is considered as a series of overlapping spot welds, projection welding might be described as multiple spot welds. One sheet is dimpled or embossed while the second sheet is flat. Both sheets are held under pressure between flat electrodes or platens while current is passed to produce coalescence. The **projection spots** are usually of a diameter equal to the thickness of the sheet and project about 60 percent of the sheet thickness. Practically all metals which are spot weldable may be projection welded.

Butt Welding. Resistance welded butt joints are usually made by one of the three processes illustrated in Fig. 48, utilizing clamps to hold the material

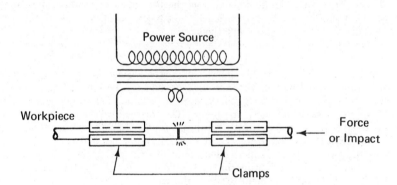

Fig. 48. Elements of butt welding.

while the current passes. Welds are produced with pressure after the plastic temperature has been reached. **Flash welding** is accomplished by bringing the abutting surfaces into light contact, drawing an arc to melt the metal, then applying pressure to forge the joint. **Upset welding** starts with the abutting surfaces in solid contact and depends upon contact resistance to produce heat prior to the application of forging pressure. **Percussion welding** requires that an air gap be present just prior to the discharge of a heavy current which melts the surfaces to be joined. Simultaneously, a percussive or impact blow brings the surfaces together to join them.

These processes are all applicable to welding of rods, pipes, and tubes. Flash welding is the most popular and versatile, while percussion welding is particularly adaptable to welding heat-treated parts and dissimilar metals, since penetration is only about 0.01 in.

BRAZING AND SOLDERING. Brazing and soldering are metal joining processes in which nonferrous alloys are used to join either similar or dissimilar metals. The brazing or soldering alloy must melt at a temperature below the melting point of the base metal. To be effective, both processes require clean metal and joints with proper fit. Brazing differs from soldering in that it uses alloys of copper, silver, or aluminum which melt at temperatures above 800°F. The strength of the joint depends upon the area of overlap into which the filler metal has penetrated by capillary action. Soldering utilizes lead and tin base alloys that melt at temperatures from 300° to 700°F. The parts are soldered using irons, furnaces, solder pots, induction heaters, and hot plates. The solder is melted and flows throughout the solder joint by the heat of the joint itself.

Types of brazing are usually defined by the different methods of heating or applying the braze metal: **torch brazing** uses the oxyacetylene or fuel gases and the typical oxyactylene torch; **carbon arc brazing** uses twin carbon electrodes; **furnace brazing** uses gas or electrically heated furnaces; **induction brazing**, uses the resistance of a part or joint to the passage of electric current; **dip brazing**, uses a molten metal or a chemical dip furnace; and **flow brazing** involves pouring molten metal into the joint.

OTHER WELDING METHODS. There are many other methods of welding which are used because of their special properties or the characteristics of the welded assembly. Some of these are forge, thermit, friction, laser, electron beam, and ultrasonic welding.

Forge Welding. Forge welding was the first form of welding and for many centuries the only one in general use. The process consists of heating the metal in a forge to a plastic state and then uniting it by pressure or hammer blows. It is used with low carbon steels and wrought iron. Butt-welded steel pipe is made by a similar technique.

Thermit Welding. In thermit welding, coalescence is produced by heating with a superheated metal. Pressure may or may not be used. The super-heated metal is obtained by igniting a mixture of fine aluminum powder and iron oxide with a fuse made of magnesium ribbon. The mixture reacts to produce purified iron or steel at a temperature of approximately 4,500°F. with a slag on top. The pieces to be welded are enclosed in a refractory mold usually made by the lost-wax process. As the superheated metal fills the mold, the weld is produced. Thermit welding is used primarily in the field repair of thick sections, such as tracks, which would be difficult to weld otherwise.

Friction Welding. Friction welding depends upon pressure and heat to form a solid-state bond between two pieces of metal. One piece is clamped in a collet or chuck and rotated at very high speeds. The other piece is held stationary. At the start, the two pieces are so positioned that they rub together and friction causes heating and cleans the two surfaces. At a given moment, the two pieces are forced together and the rotation stopped. The welds produced by this process are competitive with other methods such as butt welding, hot pressure welding, and **cold welding.**

Laser Welding. Lasers (light amplification by stimulated emission of radiation) generate energy of high intensity in a concentrated beam visible to the eye as intense light. They have rapid and deep heating and penetrating powers. Lasers are used for precision welding where large heat-affected zones are undesirable and for metals that might be difficult to weld.

Electron Beam Welding. A beam of electrons can be used to produce a very narrow and deep weld. It is expensive because of the large power supplies and the need for a vacuum chamber for the workpiece and electron gun. Because of this vacuum, it is very useful for welding very reactive metals such as titanium.

Utrasonic Welding. Ultrasonic welding is carried out with or without any external heat source. Vibrational energy in the form of ultrasonic vibrations is imparted to the metal to be welded using apparatus similar to electric resistance spot welding equipment. The electrodes transmit vibrations, not current, to the workpiece. Another type of equipment used resembles a soldering iron and provides vibrations at its tip. This is often used for aid in breaking up oxide films which prevent cold welding as in the soft soldering of aluminum. It is a combination of vibratory disruption of the oxide layers and frictional rubbing of the metal together under the ultrasonic action that assist the fusion process.

Assembly Methods

IMPORTANT FACTORS IN ASSEMBLY. Most manufactured products are the result of two or more assembled components. There are many types of assembly methods and devices including machine screws, bolts, self-tapping screws, rivets, glue adhesives, and welding. In order to simplify the selection of the best method of assembly for a product, it is necessary to consider:

1. Frequency of disassembly and reassembly.
2. Loading on fastener and how applied.
3. Types of material being joined.
4. Purpose of the fastener other than for holding power.
5. Environmental conditions associated with the particular assembly.
6. Ease of making a particular assembly.
7. Improved efficiency of a product, such as the benefits derived by reduced use of rivets on airplane wings.
8. The influence of fastener upon final appearance of a product.

SCREWS AND BOLTS. When machine elements must be disassembled and reassembled, the best means of assembly is usually a **screw fastener.** Various types of threads are available for fasteners. Fine threads are used in preference to coarse threads when the part is subjected to vibration, as in automotive and aircraft components. Coarse threads can be produced at a lower cost and assembly is faster. The strength of a screw fastener is based upon the root area of the threads.

Screws are threaded fasteners with heads of various types as illustrated in Fig. 49. Screws are designed to screw into a threaded hole and not through it. At times, however, screws are used like bolts with a nut to secure an assembly. The term "machine screw" applies to small screws and is often used interchangeably with the term "cap screw."

Set screws are threaded members that are screwed through one part and prevent relative motion with a second part due to a locking action. They are secured with a screw driver or a recessed head wrench.

Self-tapping screws are threaded fasteners which are sufficiently harder than the material into which they are screwed to cut their own threads, and thus eliminate a tapping and, in some cases, a drilling operation. They function best with sheet metal, plastics, and wood materials. **Drive screws** are a special type self-tapping screw.

Bolts are threaded fasteners that are secured with nuts during assembly. Although they vary in length and have different types of heads and finish, most bolts are threaded a length of 1½ to 2 diameters. Various types are machine

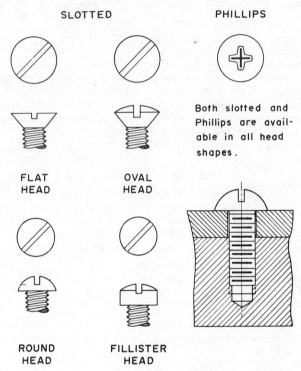

SLOTTED   PHILLIPS

Both slotted and
Phillips are avail-
able in all head
shapes.

FLAT   OVAL
HEAD   HEAD

ROUND   FILLISTER
HEAD   HEAD

Hammond, et al. Engineering Graphics

Fig. 49. Machine screws.

bolts, carriage bolts, stove bolts, stay bolts, eye bolts, and u-bolts (Fig. 50).

Studs are threaded on both ends and are used where through bolts are impractical or undesirable. The assembly is made by driving (screwing) the stud in one part, the second part of the assembly being held to the first by a nut. Studs are especially adapted to securing cylinder heads and removable covers.

Nuts are mating parts to bolts and are classified as full, jam, castellated, slotted, or wing nuts, according to function.

Locknuts, speednuts, or **threadlocks** are used as vibration-proof holding devices for screw threads. There are many devices such as cotterkeys, elastic stop nuts, Palnuts, etc., to prevent vibration and other forces from causing parts to unscrew. The use of deformed threads and fiber inserts in a nut are popular methods of making nuts withstand a tendency to unscrew.

Washers are of two types: (1) flat washers, which provide a seat for a bolt, screw head, or nut; and (2) lockwashers, which serve to retard the tendency of a bolt or screw to loosen.

RIVETING AND STAKING. Both riveting and staking are fastening or assembling methods.

Riveting (Fig. 51) is the process of upsetting the head or point of a pin, rod, or bolt (called a **rivet**) in a punch. Most rivets are made of wrought iron or soft steel, although for certain applications, rivets may be made from copper or

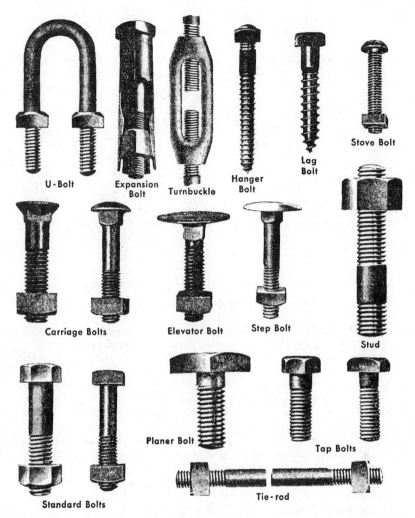

U-Bolt

Expansion
Bolt

Turnbuckle

Hanger
Bolt

Lag
Bolt

Stove Bolt

Carriage Bolts

Elevator Bolt

Step Bolt

Stud

Planer Bolt

Tap Bolts

Standard Bolts

Tie-rod

Fig. 50. Bolts.

aluminum. Larger rivets are set at elevated temperatures, and small rivets are normally set cold. Various representative types of rivet are shown in Fig. 52. These include drive, tubular, and split rivets.

Staking (Fig. 51) is a similar operation in that the metal of one part of an assembly is upset or peened in such a fashion as to make a tight fit with the second part. Staking is used extensively in the assembly of watch and clock components.

STAPLING, TACKING, AND STITCHING. These are methods using various types of wire fasteners for joining materials. Stapling is an operation of joining two or more pieces of material by means of a preformed wire staple which is clinched. The familiar office stapler used for joining pieces of paper illustrates the principle. Wire stapling is also used to join two pieces of sheet

metal or the fastening of wood, cloth, or paper to a metal sheet. The operation
is economical and very rapid up to 400 cycles per min. **Tacking** is similar to
stapling except that the staple is not clinched. **Stitching** is similar to stapling
except that the staples are not preformed but are made, as they are used, from
a spool of wire. Metals over $\frac{1}{8}$ in. thick can be penetrated with wire staples.

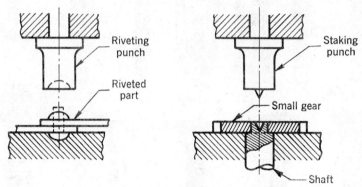

Fig. 51. Riveting and staking operations.

SEAMING AND CURLING. Sheet metals are joined by these operations.
Seaming, as illustrated in Fig. 53, is the process of joining two pieces of sheet
metal by bending them in such a way that a permanent assembly is obtained.
There are many types of seams that can be made on hand and power seaming
presses. The seams may be made pressure tight such as the one shown in Fig.
53 which is used for metal cans. To make the double seam shown, **edge
flanging, curling,** and **flattening** operations must be performed. Automatic
machines are available for this and other seam operations.

SHRINK FITS. A shrink fit is made when two parts are assembled by
means of a severe interference fit, the interference being eliminated during assembly by dimension changes resulting from heating or cooling the components.
Examples are the shrinking of steel rims on cast iron wheels and the shrinking
of liners in large bore guns.

ADHESIVES. Adhesives cover a wide field of materials designed to produce joints to hold materials together.

Glue is a product used principally for bonding woods. There are six types of
glue; animal, liquid, starch, blood albumin, casein, and synthetic resins.

Adhesives are used in the bonding of almost all materials including metals.
The term **adherents** applies to the materials to be bonded together. The aircraft and automotive industries use "structural-type" adhesives extensively as
do many other industries. The joints produced by plastic or adhesive bonding
are strongest in shear, as in the case of riveted or spot-welded construction.
The strength compares favorably with other fabrication methods due to the
stresses being distributed over the entire bonding area. Adhesives are readily
adaptable to mass production, although they do have the disadvantage of not
being an instantaneous process. Most adhesives are applied in the liquid or
plastic state by brush, roller, spray, or spatula; dry adhesives are applied in
stick or powder form. Both dry and liquid can be applied mechanically. In

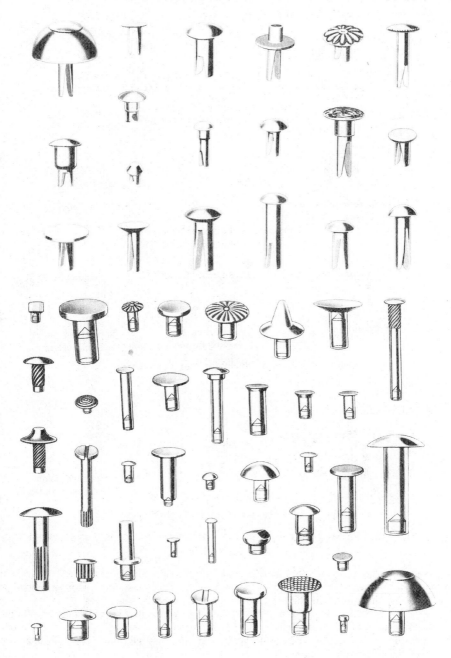

Fig. 52.   Representative rivets used in manufacturing.

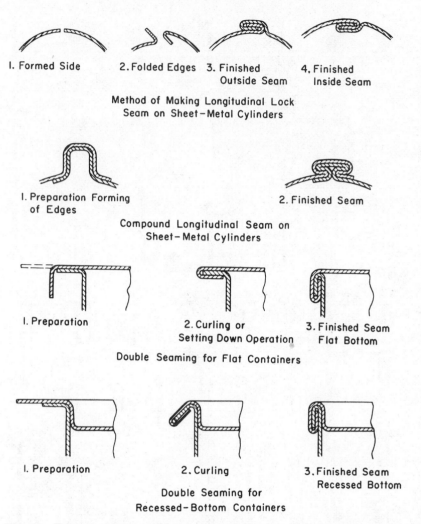

I. Formed Side   2. Folded Edges   3. Finished   4. Finished
Outside Seam   Inside Seam

Method of Making Longitudinal Lock
Seam on Sheet–Metal Cylinders

I. Preparation Forming   2. Finished Seam
of Edges

Compound Longitudinal Seam on
Sheet–Metal Cylinders

I. Preparation   2. Curling or   3. Finished Seam
Setting Down Operation   Flat Bottom

Double Seaming for Flat Containers

I. Preparation   2. Curling   3. Finished Seam
Recessed Bottom

Double Seaming for
Recessed–Bottom Containers

Fig. 53. Typical seams used in the manufacturing of light-gage metal containers.

general, the thinner the adhesive coat, the stronger the joint. After application of the adhesive, the adherents are usually held together until sufficient bond is made; the application of heat is used if necessary to set the adhesive.

The selection of a proper adhesive depends upon the following factors: ability to bond the adherents, resistance to effects of its surroundings, ease of application, expense, appearance of joints, and the possibility of future disassembly of the adherents.

Four general groups of adhesives are used in these applications:

1. **Thermoplastic adhesives.** These adhesives never become permanently hard and can be made ineffective by the application of heat. They have limited use, since their softness, particularly at elevated temperatures, induces joint

creep. Examples of this type adhesive are acrylics, cellulose nitrates, oleo-resins, and polyvinyl alcohols and acetates.
2. **Thermosetting adhesives.** These adhesives become permanently hard when the cure is complete; however, elevated temperatures do cause some loss in strength. They are the most used when the joint is under a stress, and include such plastics as epoxies, phenolics, alkyds, and formaldehydes.
3. **Elastomeric adhesives.** Although similar to thermoplastic adhesives, these adhesives are less sensitive to temperature. They have relatively low strength unless they are combined with thermoplastic or thermosetting materials. Examples are the natural, reclaimed, and synthetic rubbers, and silicones.
4. **Adhesive alloys.** Adhesive alloys are blends made of several basic resins and are designed to exhibit the best characteristics of the components. They usually consist of phenolic blended with such materials as vinyl, neoprene, or polyvinyl butyral.
Several other materials such as the asphalts and shellac are used for general purpose bonding agents as are some of the glues.

Protective Coatings

PURPOSE OF COATING. With very few exceptions, any marketable product must be surface finished. Often the primary purpose of a coating or finish is only to increase the appearance and sales value of the item, but coatings must be used on most materials to give permanent resistance to destructive influences, including wear, weather, corrosive atmospheres, and electrolytic decomposition.

METHODS OF CLEANING. Before a metal product can be coated, it is necessary to prepare the surface properly to improve adhesion and enhance the appearance. Parts are cleaned by different methods depending upon material, size, and surface peculiarities, but there are three basic ways in which most metal products are cleaned: mechanical, chemical, and electrolytic.

Mechanical Cleaning Methods. The three basic mechanical cleaning methods are (1) tumbling or blasting, (2) abrasive cleaning or polishing, and (3) ultrasonic cleaning.

Tumbling consists of rotating the parts to be cleaned or finished in a tumbling barrel. These barrels are of various sizes and shapes and usually employ an additional solid material to increase the cleaning action. Such materials as steel or cast iron shot or "stars," aluminum oxide pellets, and abrasive materials are examples of tumbling "abrasives." Often a carrier or lubricant is used as tumbling progresses. For nonferrous materials, a high-velocity stream of water and sand is played against the parts as they tumble. The amount of material loaded in a tumbler is an important factor, since it regulates the intensity of the tumbling action. Tumbling is usually employed only for cleaning purposes, but it can be employed as a means of providing a surface (not a protective finish) which may or may not be coated in a later operation. For example, mirror-type finishes can be produced on small die cast parts by suitable tumbling action.

Sandblasting and shotblasting are accomplished with a "gun," the principle of which involves passing air through a mixing chamber at high velocity. This produces low pressure in a mixing chamber which draws sand, pellets, or shot into the chamber. They pass out of the muzzle of the gun with air acting as a carrier and are forced against the workpiece. The choice of "abrasives" is dependent upon the metal being blasted and the surface finish required. Often parts which do not have sufficient strength to be tumbled may be blast cleaned. In the case of nonferrous castings, water and oils are sometimes used in the place of air. There are also systems that employ impeller wheels to force the sand or shot against the workpiece. For the most part, blasting is only a cleaning method for metals, but some novel, nonprotective finishes can be produced by the method.

Abrasive belts, wheels, and stones are used extensively in cleaning, polishing, and finishing operations. Coarse abrasive wheels and belts are used for rough cleaning operations, and fine-grained polishing belts and wheels are used to obtain a smoother finish. Mirror-type finishes can be produced by polishing or cleaning with successively finer grit, followed by cloth wheel or belt polishing with a fine abrasive of tripoli or alumina in an oil or grease carrier. Such polishing is called **buffing.** Especially prepared buffing compounds can be used to color some metals a moderate amount during the operation. Buffing offers little in the way of a protective finish, but for corrosion-resistant materials it is usually the final finishing operation.

Ultrasonic cleaning consists of immersing the material to be cleaned in a bath of abrasive or solvent which is agitated by applying ultrasonic vibrations to the tank. The vibrations together with abrasive or solvent action loosen and remove surface contaminants. Ultrasonic cleaning is used extensively in the jewelry industry because of its simplicity, its ability to remove tiny particles trapped in recesses, and its economy.

Chemical Cleaning Methods. For cleaning materials to be coated, chemical methods are the most frequently used. While they do not take off flash or burr, they do remove dirt, grease, most scale and, in a few cases, a very small amount of surface metal. Chemical washing is often employed at elevated temperatures with a mechanical agitation system to improve the cleaning efficiency. Chemical cleaning compounds may be classified as follows:

1. **Petroleum solvents** include gasoline and kerosene and are generally used to remove oils and grease. The action depends upon the material being removed going into solution with the solvent. Stoddard's petroleum solvent is a popular type. Most plants are prevented from using solvents of this type because of the fire hazard.

2. **Chlorinated solvents** include carbon tetrachloride, trichloroethylene, and other chlorinated hydrocarbons. These solvents, slightly more expensive than petroleum solvents, accomplish, in general, about the same results but do not present the fire hazard. The fumes, particularly in vapor degreasing with chlorinated solvents, must be controlled for plant safety, since they are highly toxic.

3. **Alkali cleaners,** composed of approximately 4 to 8 percent of an alkali salt and the remainder water, are the most common cleaning media. Cleaners of this type are usually employed in a tank or with steam. Sodium carbonate (soda ash) is the most popular alkali used. Often a wetting agent is used to increase the efficiency of the cleaner. There is no fire hazard, nor are there toxic fumes generated with this method. Steam cleaning devices used in garages to clean automobile engines and white side-wall tires are examples of this process.

4. **Emulsion cleaners** are oil-soluble soaps mixed with a petroleum or chlorinated solvent, and depend for action upon the ability of the emulsion agent to disperse the foreign materials as the solvent acts.

5. **Pickling** is a type of cleaning which utilizes an acid and water solution. A 5-to-20-percent solution of sulfuric acid is the most common solution used, although hydrochloric (or muriatic) acid often gives better results. Since sulfuric acid is less expensive and its vapors are not excessively corrosive, it is used more frequently. Considerable caution must be used in choosing a pickling solution, since the acid attacks the base metal and may cause em-

brittlement. Pickling is particularly useful in the removal of scale from steel after rolling or heat treatment.

6. **Electro-cleaning** or **electropolishing** is accomplished by immersing the part to be cleaned in a tank containing the proper electrolyte. The part to be cleaned is connected to the positive terminal of a direct current power supply and the tank or another piece of metal serves as the negative electrode. Electropolishing is the reverse of electroplating and metal atoms are ionized and leave the surface of the part being cleaned. This method is very useful for cleaning parts of ultra-high vacuum systems. Again, safety precautions must be taken depending on the electrolytes used.

ORGANIC COATINGS. Of all the coating processes, organic coatings are the most used. The coating is intended to cover the surface of a metal object and prevent the metal from coming into contact with the environment which could be corrosive.

Types. Although the terms paint, enamel, shellac, lacquer, and varnish no longer can be concisely defined as they once could, they are still the basic trade terms used. Many finishes are a mixture of two or more such products. Figure 54 lists various organic coatings together with typical applications and service conditions.

Normal Outdoor Exposure

Oil-base paints	Buildings, vehicles, bridges maintenance
Alkyds	Trim, metal finishes
Amino-resin-modified alkyds	Automotive, aluminum siding
Acrylics	Automotive

Marine Atmosphere

Alkyds, chlorinated rubber, phenolics, vinyls, vinyl-alkyds	Superstructures and shore structures
Urethanes	Marine varnishes

Chemical Fumes

Epoxies, chlorinated rubber vinyls, urethanes	Chemical processing equipment

High Temperatures

Alkyds	Motor blocks (200°F max.)
Amino-resin-modified alkyds	Driers, stove parts (200°F max.)
Epoxies	Motors, piping (250°F max.)
Silicones	Stove parts, roasters (550°F max.)

Fig. 54. Organic coating applications.

Principal Advantages. Advantages of these coatings include ease of application, decorative effect, and protection of the base metal from the elements of corrosion, wear, and weathering. Many special finishes can be produced, such as crinkle and luminous coatings.

Methods of Application. Organic coatings are generally applied by brushing, dipping, or spraying. Many finishes are now sprayed hot to increase the percent of solids deposited, and to reduce the solvent cost and overspray losses. **Electrospraying** involves electrically charging the paint and the workpiece. The paint particles sprayed out are charged and are attracted to the oppositely charged surface of the workpiece. Spraying "around corners" and in blind holes is possible with this method. An extremely uniform coating of paint can be obtained using electrospraying.

Before an organic coating can be applied, the base metal must be clean and, in some cases, special **pre-coatings** or surface treatments must precede the application of the final coating. For steel, red lead and zinc chromate primers are very successful as pre-coatings.

METALLIC COATINGS. Metal coatings are often used in place of organic coatings or plastic coatings. They can be applied in many different ways.

Types. There are basically two types of metallic coatings. The coating metal is either more noble (more resistant to corrosion) than the base metal or it is more active (less resistant to corrosion). The former type of coating is called a **noble** or **passive** coating while the latter type is called a **sacrificial** coating.

Metallic paints are primarily fine particles of the metal suspended in an organic vehicle such as varnish. Examples of metallic paints include the silver (aluminum) and gold (bronze) paints.

Zinc coatings on steel are sacrificial coatings. The zinc corrodes in preference to the underlying steel and if such a **galvanized** coating is scratched, the exposed steel will not rust. These are low-cost coatings that have reasonable appearance and good wearing properties. An improved appearance, known as the **spangle effect**, can be produced by small additions of tin and aluminum. Zinc baths are usually maintained at about 850°F. Rolls, agitators, and metal brooms are used to remove the excess zinc from the product. Continuous and automatic processes are used for sheet and wire coating. Zinc coatings can also be produced by **spraying** molten zinc on steel, by **sheradizing**, which is the tumbling of the product in zinc dust at elevated temperatures, and by **electroplating**.

Tin coatings are applied extensively to sheet steel to be used for food containers. In this application, the tin behaves as a sacrificial coating. In fact, tin-can manufacturers use approximately 90 percent of the tin produced. Although many tin coatings are now applied by electro-tinning, which is a process of immersing parts in an electrolyte and passing a current from the electrodes to the work, the hot-dip method is still used considerably. Tin is applied easily by dipping at temperatures of approximately 600°F without affecting the base metal. In most cases, the tin coating is approximately 0.0001 in. thick as compared to about 0.00003-in. thicknesses in electro-tinned sheets. Porosity is greater, however, in plated tin coatings, and when used for food, a lacquer seal is necessary.

Noble coatings on steel if scratched allow the base metal to corrode and rust. Thus, they behave like paints when scratched.

Hot Dipping. Hot dipping involves immersing the metal part in a container of molten metal. The process is simple and relatively inexpensive. The base material must be chemically and physically clean to allow for proper adhesion of the coating. Sheet steel, chain-link fence material, and outdoor hardware are typical products that are coated by this process (zinc and/or tin alloy coatings).

Electroplating. Electroplating is a process in which a direct current is passed between an anode (+) and a cathode (−), the two being immersed in a tank of metallic salt solution called the **bath**. The metal to be plated is made the cathode, and the material to be deposited is made the anode. When current is applied, the anode transfers some of its mass to the bath and an equal amount of the material is deposited on the cathode which is the part being plated. Although the process is simple in theory, good results are dependent upon careful selection of the bath salts. Concentration, current density, temperature, and metal cleanliness are other important factors. A brighter finish results from a finer crystalline pattern being deposited, a harder finish from higher applied currents, and softer finishes from elevated bath temperatures. The thickness of deposit is controlled by the amount of current flowing; thus, a longer time or a higher current per unit of time will give increased deposits.

Most metals, with the exception of cadmium and zinc, offer protection from corrosion only if they completely cover the base metal. Porosity, except in the case of these two materials, must be closely controlled to keep contaminants from attacking the base metal.

In order to get the best service out of some plated metals, it is necessary to use one or more different materials as **"pre-plates"** before the final plating operation. This produces proper ductility, appearance, and corrosion-resistant qualities. For instance, chromium is usually plated over nickel which has been plated over copper; the thickness of the chromium is only about 0.00002 in.

The fundamental physical characteristics of the principal coatings deposited by electrolytic action are shown in Fig. 55. Precious materials, such as gold, platinum, and silver, are also applied by electroplating.

Oxide Coatings. With a few important exceptions, oxide coatings enhance the appearance of a product but do little to offer resistance to destructive influences.

Anodizing is an oxidation process developed for aluminum. An electrolyte of sulfuric, oxalic, or chromic acid is employed with the part to be anodized as the anode. Since the coating is produced entirely by oxidation and not by plating, the oxide coating is a permanent and integral part of the original base material. Although the coating is hard, it is porous which is an advantage from a decorative standpoint. The oxide coating enables organic coatings and dyes to be successfully applied to the surface of aluminum. Modern aluminum tumblers and pitchers are examples of this process. Magnesium is anodized in a somewhat similar manner.

Clad Materials. Clad materials are metal composites in which the properties of different metals are combined in the composite. A typical example of cladding involves commercial purity aluminum. An inner layer of high-strength aluminum alloy provides the structural properties and outer layers of corrosion resistant pure aluminum provide corrosion protection. Clad products include sheet, rod, and wire forms. Claddings of dissimilar metals offer other advantages besides corrosive resistance provided proper fabrication into the composite form is possible.

Spraying. Metallizing is the process of spraying molten metal upon other materials by means of a metal spray gun which resembles a paint spray gun. Actually, wire or powder is fed into an oxyacetylene flame, which melts the material to be sprayed, and compressed air forces the molten metal against the surface being coated. Since the bond between the coating and the base metal is purely one of mechanical interlocking, the surface must be properly roughened and cleaned. Metallizing is used to create decorative finishes, to build up worn parts, and for adding materials to the parent metal that might better resist destructive influences. Metals sprayed on in this manner have a density of approximately 80 percent as compared to the same metal in the form of a casting. These coatings are lower in strength but are hard.

| Plating Material | Application | Appearance | | Abrasion Resistance |
		Initial	After Exposure	
Cadmium	Rust protection. Improve appearance. Provide soldering surface.	White lustrous	Grayish-white matte	Very poor
Brass	Improve appearance. Provide soldering surface.	Satin yellow to bronze	Black to green	Fair
Chromium	Improve appearance. Abrasion resistance.	Bluish white (matte or mirror finish)	Unchanged	Fair to excellent (depending on thickness and finish)
Copper				
Acid	Base for nickel plating.	Salmon red	Black to green	Fair
Cyanide	Base for nickel plating (but not directly on steel).	Salmon red	Black to green	Fair
	Base for oxidized finish.	Not applicable		Fair
Nickel	Rust protection. Improve appearance. On steel over copper; directly on copper, brass, or zinc.	Yellowish-white (mirror or matte finish)	Dark to brown	Good
Tin	Corrosion resistance on copper and brass.	Frosty white	Grayish	Very poor
Zinc				
Acid	Rust protection.	Bluish satin	Dark gray	Poor
Cyanide	Rust protection.	Matte white	Dirty gray	Poor

Some surfaces can be buffed to lustrous finish after plating.

Fig. 55. Characteristics of electroplated coatings.

PLASTIC COATINGS. Protection against the corrosive nature of liquids and gases can be obtained by bonding plastics or rubber to metal surfaces.

Types. Vinylidene chloride or **Saran** is often used to protect steel. It is expensive and is therefore used mostly in the chemical industry where severe corrosive environments are found. It is a substitute for rubber or neoprene.

Vinyl or **polyethylene** are used in tape form to protect buried metal structures such as pipelines, valves, and auxiliary equipment exposed to the soil.

Tetrafluorethylene or **Teflon** is one of the most stable plastics. It resists most chemicals, including aqua regia, boiling concentrated acids and alkalies, gaseous chlorine, and all organic solvents up to 480°F. Teflon is not strong and it tends to creep readily under stress. It is useful for linings, gaskets, and diaphragms.

Methods of Application. Plastic coatings are applied in various ways. Usually an adhesive is necessary to bond sheets or strips of the plastic to the base metal but the inertness of the plastic often makes such bonding difficult.

Some plastics can be applied in the liquid state or in a solvent and then caused to set. Insulation on electrical wires is normally applied by extrusion processes.

MISCELLANEOUS COATINGS. Other types of coatings used in industrial and laboratory applications to protect metals include **natural oxides, paper** and **fabric** coatings, and special chemical treatment.

Natural oxides produced upon exposure to the atmosphere protect aluminum, steel, copper, and copper alloys (such as bronze) from corrosion. Very thin films of these oxides form on the surface and act to prevent further base metal corrosion. The scratching of this film down to base metal would result in conversion of the exposed base metal to the natural oxide and thus restore the protective coating.

In certain cases these natural oxides can be utilized and produced in greater thicknesses as in **anodized** aluminum.

Paper and fabric coverings of metal parts to prevent corrosion requires that the adhesive and coating together act as a barrier to air, moisture, and bacterial fungus. For temporary protection during storage, impregnated papers and fabrics can be used to seal off the metal. This coating is often combined with greases and other corrosion preventing materials.

Parkerizing is a process for making a thin phosphate coating on steel which acts as a base or primer for enamels and paints. In this process the steel is dipped in a solution of manganese di-hydrogen phosphate for about 45 minutes. **Bluing** is a process of dipping steel or iron in a 600°F. molten bath of nitrate of potash (saltpeter) for from 1 to 15 minutes. There are many salts that can be used to color brass and steel by dipping at elevated temperatures, but most of these have limited application and differing degrees of permanence.

Materials—Characteristics and Applications

IRONS AND STEELS. Commercial irons and steels are essentially alloys of the metallic elements iron and carbon. The amount of carbon, the other alloying elements present in addition to iron and carbon, and the rate of solidification and cooling determine the structure and properties of the resultant metal and whether it is a steel or cast iron.

Types and Characteristics. Castings made directly from the blast furnace melt are known as **pigs** or **pig-iron** and normally weigh 50 to 200 lb. They are seldom used for commercial castings but are available for remelting. The following are the most common types of irons in commercial use.

Gray iron is a cast iron in which the carbon and silicon content are high and the carbon is present in the form of graphite flakes. The cooling rate and chemistry of the metal determine its physical properties. By varying the rate of cooling once solidification has taken place, the strength of the resulting gray iron can be varied. Production of gray iron castings ranks among the top of all metals based on tonnage. The raw materials are low in cost, economical to melt, and relatively few castings are lost to scrap. The vibration damping properties of gray iron make it useful as machine tool beds. Gray cast iron has very good machinability, and there is a lubricating effect due to the graphite flakes. It also has very good wear resistance.

White iron is produced by casting and allowing more rapid solidification than is normally used for gray iron. It contains less carbon and silicon than gray iron but more carbides and consequently is extremely hard. The carbon appears in the form of iron carbide rather than as graphite. It is used mainly for hard surfaces on gray iron castings where the insertion of local chills in the mold produce these regions to resist wear. There are few commercial uses for white iron as such. It is used for railroad car brakes and as an intermediate stage in the production of malleable iron.

Malleable iron is made from white cast iron by annealing it at temperatures of from 1,500 to 1,850°F over a period of several days. In the course of this annealing, the iron carbide breaks up and rosettes of graphite are formed. These castings have good shock resistance and machinability and find wide use in the railroad, pipefitting, and agricultural implement industries.

Nodular iron is made from gray cast iron by ladle additions of magnesium or cerium which allow the carbon to form in the melt into spherical nodules rather than flakes. The metal exhibits physical characteristics between that of steel and gray iron and these properties may be varied by heat treatment. It competes with malleable iron in some applications. Due to its ductility and high strength, nodular iron is an excellent material for many machine parts.

Steel is an alloy of iron and carbon. Other alloying elements are present in small quantities because they are difficult to remove during the steelmaking process or because they are deliberately added to produce certain properties in the metal.

Steels are usually given identification code numbers. These are known as AISI (American Iron and Steel Institute) designations and consist of four digits. The first two digits denote the principal alloying element, and the last two digits represent the carbon content of the steel, expressed in hundredths of a percent of carbon (Fig. 56, from the American Iron and Steel Institute).

1XXX	Plain carbon steels. The first X is the approximate amount of the principal alloying element; for example, 1080 represents a steel containing 0.80% carbon and no other alloying element. 1100 represents a steel containing approximately 1% of manganese.
2XXX	Nickel steels.
3XXX	Nickel plus chromium steels.
40XX	Molybdenum steels.
41XX	Chromium molybdenum steels.
43XX	Nickel chromium molybdenum steels.
46XX 48XX	Nickel and molybdenum steels.
5XXX	Chromium steels.
6XXX	Chromium vanadium steels.
7XXX	Tungsten steels.
8XXX	National emergency steels containing much less expensive alloying elements such as nickel, chromium, and molybdenum.
9XXX	Silicon manganese, manganese, and nickel chromium manganese steels.

Fig. 56. AISI steel designations.

Each steel has a definite use based on the alloy content and the heat treatment which it has received. In general, steels are very strong and work hardened, which permits their use in many types of structural application. The ability to heat treat steels to obtain properties that range from very hard to very soft allows the forming of these materials into useful shapes and the use of steel in many different applications.

ALLOY STEELS. Alloy steels are steels that contain considerable amounts of other elements to impart properties that are particular to this alloy.

High-speed tool steels, hot-work steels, and stainless steels are examples of such alloys.

Types and Applications. **High-speed steels** are primarily alloys of iron and carbon in which tungsten, molybdenum, chromium, and vanadium have been added so as to form carbides of these various elements. It is the presence of these carbides that gives rise to very hard materials with high heat resistance. These high-speed steels are used for drill bits and in tool steel applications where they perform very well at rather high temperatures. Care must be taken that they are not overheated to the point where the carbides tend to dissolve.

Stainless steels fall into three groups, the ferritic, martensitic, and austenitic stainless steels. It is only the latter group which is truly resistant to corrosion. These alloys contain chromium and nickel to various degrees. The ferritic stainless steels normally contain from 12 to 25 percent chromium, the martensitic steels from 12 to 18 percent chromium, and the austenitic steels contain 18 percent chromium and 9 percent nickel or 18 percent chromium and 12 percent nickel. The martensitic steels are used for such applications as knife blades or cutlery because they tend to hold a very good edge. The bulk of all stainless steel belongs to the austenitic group. Kitchen utensils, sinks, and cabinets are made of this type. There is a protective natural oxide coating which forms on the surface of these steels to protect the base metal against corrosion. It is the combination of nickel and chromium which produces this particular passive layer.

NON-FERROUS METALS. The non-ferrous metals consist of all alloys which are not based on iron. **Aluminum, copper** and **copper alloys, nickel alloys, magnesium, titanium,** and many other metals belong to this group.

Aluminum and Aluminum Alloys. The properties that make aluminum and its alloys the most economical for a wide variety of uses are: appearance, ease of fabrication, mechanical properties, corrosion resistance, electrical and thermal conductivity, and low density. Nearly all aluminum alloys contain small additions of an alloying element. These alloying elements may be copper, manganese, silicon, zinc or others.

Commercially pure aluminum is often used in the cold-worked state because of its corrosion resistance imparted by a layer of natural oxide. It is also used as the outer layer of composite materials or clad materials.

Age hardenable alloys are aluminum alloys in which a precipitation reaction or an aging reaction causes an increase in hardness as a function of time during heat treatment after fabrication. This gives rise to aluminum alloys which can be just as strong as many steels.

Aluminum is used in both the wrought condition and in the cast condition. Usually, a particular alloy is good in only one of these two forms.

Copper and Copper Alloys. The major properties of copper and copper alloys are electrical and thermal conductivity, corrosion resistance, machinability, resistance to fatigue, malleability, formability, and strength. One of the most useful properties of copper is its very high electrical conductivity; most electrical cables, wires, and busbars are copper, although aluminum is competitive in many applications. The ease in joining copper and copper alloys by welding, brazing, or soldering techniques gives it the edge in many cases. Copper and its alloys also have very good casting properties and are often used in the cast as well as in the wrought condition.

Commercially pure copper is available in both the tough-pitch form and as oxygen free high conductivity (OFHC) copper. These are used mainly for electrical applications with the OFHC copper being preferred because of its greater electrical conductivity.

Brasses are alloys of copper and zinc containing various amounts of zinc. Among the many brasses commonly used are gilding metal (5% zinc), cartridge brass (30% zinc), and Muntz metal (40% zinc). Each of the brasses has its own particular properties which suit it for particular applications. Gilding metal is often used in coinage, as a base for gold plate, and in costume jewelry. Cartridge brass, as its name implies, is used mainly in military applications. Muntz metal is used for heavy structural sheets as heat exchanger tubing and as decorative panels.

Some brasses contain small amounts of tin and other elements to make them more suitable for naval and marine applications.

Bronzes are alloys of copper and tin which are used generally in either the wrought or cast condition. They contain from 5 to 12 percent tin and varying percentages of other alloying elements. Bronzes have excellent corrosion resistance and good fatigue resistance and are often used in marine applications and for pump rods, springs, and diaphragms.

Beryllium copper is an alloy of copper and beryllium which can be age hardened to very great strengths. It is normally used in the wrought condition and although it is expensive, it is non-sparking and is used to manufacture tools for the petrochemical industry. It also has high fatigue resistance and is used for springs and diaphragms as well as electrical parts.

Nickel Alloys. Nickel alloys are used for their resistance to corrosion or high temperature oxidation.

Monel is an alloy of about 65 percent nickel and 35 percent copper and has high strength and good corrosion resistance. It is used for marine applications and for kitchenware.

Inconel is an alloy of nickel, chromium, and iron which is used because of its good creep resistance and resistance to high temperature oxidation.

Nichrome is an alloy of nickel and chromium used as resistive heating wire in furnaces because of its high electrical resistivity.

Other Important Metals. Magnesium is approximately one fourth as heavy as iron, and has excellent machinability. It is often alloyed with aluminum, manganese, zinc, copper, nickel, iron, and silicon. It is used in both the wrought and cast state, with the former being preferred. It has replaced aluminum in many applications where light weight and high strength are critical. Magnesium stepladders are common and magnesium is widely used in aircraft applications.

Titanium has a density which is less than that of steel, but it has a very high strength and high melting temperature. For this reason, it is used in weapons and aircraft applications. Titanium and titanium alloys have very good corrosion resistance and are used where light weight, corrosion resistance, and high melting temperature are important. Normally materials are in the wrought form.

PLASTICS. Plastic materials are classified as either **thermosetting** or **thermoplastic.** When heated thermoplastic materials soften, become a viscous liquid, and flow under pressure. On cooling a thermoplastic hardens and maintains a stable form. The process of heating and cooling can be repeated many times and each time the material flows when heated. Thermosetting resins or plastics undergo polymer cross-linking on the first heating. This cross-linking of the polymer chains causes the plastic to become rigid and fixed in shape.

Once heated and cross-linked, it cannot be made to flow. Thermosetting materials, therefore, have better long-time dimensional stability than thermoplastics but are more difficult to fabricate.

Silicones are the only commercial family of inorganic plastics. They are thermosetting materials.

Types and Properties. There are many different plastics and based on their properties each has different uses.

ABS represents acrylonitrile, butadiene, and styrene. It is a strong, tough composite thermoplastic consisting of rubber particles in a plastic matrix.

Acrylic is a family of glossy thermoplastics manufactured from acrylates and methacrylates. Polymethyl with acrylate is used widely for automobile tail light lenses. Acrylics can be made tough, strong, and opaque and are used as replacements for ivory piano keys.

Cellophane is regenerated (as opposed to natural) cellulose in the form of film or sheet.

Cellulose acetate is a very stable derivative of cellulose and has high strength. It is used for motion picture film and can be fabricated into sheets or molded to form flashlight cases, knobs, and toys. **Cellulose acetate butyrate** is a thermoplastic and can be produced in all colors, is tough, absorbs little moisture, and exhibits exceptional dimensional stability. It is commonly extruded and is used for pen and pencil barrels, steering wheels, helmets, and gas and water piping. **Cellulose proprionate** is similar to cellulose acetate butyrate and has similar uses. **Ethyl cellulose** is a very tough thermoplastic and is used for football helmets.

Chlorinated polyether is a thermoplastic with exceptionally good corrosion resistance.

Diallyl phthallate is a thermosetting resin used for electrical insulation.

Epoxy is a name of a family of thermosetting materials widely used for casting and encapsulating and as adhesives. The resins, when cured, are low in shrinkage and have good dielectrical, chemical, and physical properties. As adhesives they bond well to both metals and glasses.

Fluorocarbons are thermoplastics based on fluorine and carbon and contain no hydrogen atoms in contrast to most plastics which are mainly carbon and hydrogen. The fluorocarbons such as FEP are chemically inert, have low surface friction, and good dielectrical properties.

Ionomer is a class of thermoplastics that adhere to metals, wood, and glass and provide oil-resistant coatings.

Melamine-formaldehydes are thermosetting resins used in heat-resistant counters and dishware.

Nylon is a polyamide engineering thermoplastic used often in the form of high strength fibers and filaments. In solid form it is used for low friction and wear-resistant bushings, grommets, bearings and rollers.

Parylene represents a family of engineering thermoplastics which are formed into films for high-vacuum and cryogenic applications.

Phenol-formaldehyde is the oldest and most common thermosetting resin. Its low cost and heat resistance makes it useful for automobile distributor caps, cabinets, and other molded objects. It can be produced in a variety of colors and can be cast. It is often called **phenolic.**

Phenory is another engineering thermoplastic which forms clear blow-molded containers. It can retain odors and flavors and is FDA approved for food and drug applications.

Polybutylene is a thermoplastic that has good resistance to cold flow and stress-cracking. It is used for drinking water piping and is approved by the National Sanitation Foundation.

Polycarbonate is a tough thermoplastic used to replace glass in glazing applications where breakage is a problem.

Polyethylene ranks first in volume of production. There are two types: a stiff, strong material called high-density or low-pressure polyethylene; and the other a flexible, low melting type known as low-density or high-pressure polyethylene. Products may be made by molding or extrusion. It is used for trays, fabrics, and packages.

Polyamide is a thermoplastic which is usable at very high temperatures.

Polyphenylene oxide is a thermoplastic with exceptionally low dielectric loss and chemical resistance over a wide temperature range.

Polypropylene is a thermoplastic which is higher melting and more rigid and abrasion resistant than polyethylene. It is low in cost and is often used for interior automobile trim.

Polystyrene is a large volume thermoplastic, second only to polyethylene. It is a glossy, general purpose material easily used in injection molding and extrusion. It is a good substitute for rubber because of its dielectrical properties. It is also used for dishes and boxes. ABS and SAN resins are even better in resistance to heat and impact.

Polysulfone is a thermoplastic used in areas of potential fire hazard, for example aircraft cabins, kitchen range hardware, etc., where its self-extinguishing, low-toxicity and low smoke-generation properties are most critical.

Polyurethane is used in foams, fibers, and coatings.

Polyvinyl acetate is a rubbery polymer used in adhesives, chewing gum, and paints.

Polyvinyl alcohol is soluble in water but leaves a film impervious to oils, fats, and gases.

Polyvinyl butyral is a rubbery thermoplastic which adheres to glass and can absorb large amounts of energy on deformation. It is used as the interlayer in safety glass.

Polyvinyl chloride (PVC) is a large volume thermoplastic. It is highly resistant to many solvents and will not support combustion. It is used in many products in place of rubber.

Polyvinyl fluoride and **polyvinylidene chloride** are film forming thermoplastics which provide good weathering characteristics and chemical resistance.

Silicones are thermosetting materials used in coatings, elastomers, and encapsulation.

Urea formaldehyde is a thermosetting resin with good colorability, hardness, and resistance to scratches. It is used in tableware, light fixtures, and as an adhesive.

Composite Plastics. Composites are plastic materials consisting of two constituent materials of different properties. These two components are mixed together in such a way as to form a two-phase structure. Examples are glass fibers embedded in plastics, ABS resins, and mechanical mixtures of rubber and plastics. The combination of the dispersed and the matrix materials often results in a material having the most desirable properties of both.

WOOD. Wood is an organic substance composed principally of cellulose and lignin. Common native woods are obtained from the broadleaf trees which produce the so-called hardwoods and the cone bearing trees which produce soft woods.

The specific gravity of wood varies from 0.3 to 0.9 and the weight of most structural woods runs from 20 to 40 pounds per cubic foot.

The principal woods used are ash, cedar, Douglas fir, hemlock, hickory, locust, maple, oak, pine, redwood, and spruce. These have various physical properties, but certain generalizations may be made. Wood is strongest in tension parallel to the grain. It is weakest in tension across the grain. For air-dried Douglas fir, the values of these two tensile strengths are about 14,000 and 3,000 psi respectively.

Since the soil, climate (moisture and temperature), and the amount of growing space exert a profound influence on the growth of trees, clear wood of any species will undergo variations in mechanical properties due to changes in the environment and the rate of growth.

Large wood beams and columns always contain some defects such as checks and knots. The effects of these defects must be taken into account when trying to determine the strength of these large timbers and whether they meet specifications or not. The American Society for Testing and Materials has specifications which cover most of these defects.

Developments in the field of adhesives and bonding have brought about a group of wood-base materials utilizing chips, shavings, and other wood scrap. These materials are treated, bonded together, and pressed into sheets and panels sometimes called **hardboard.** By varying the type of wood used and the method of bonding, **reconstituted wood** can be produced with a number of different properties. The sheets and panels are widely used in home building and for decorative panelling.

RUBBER. Although both **natural** and **synthetic** rubber are available, most of the rubber in use today is synthetic. Some of the synthetic rubbers are Thiokol, Neoprene, GR-S, Buna N, butyl, and silicone rubbers.

GR-S is used in automobile tires and is a copolymer of butadiene and styrene. It can be produced in varying degrees of hardness and may be mixed or compounded with natural rubber for tires.

Neoprene is a chloroprene polymer which has a good resistance to oils, heat, and sunlight. It is used for insulation, tubes, shoe soles, and as a binder to hold together the abrasive particles of grinding wheels.

Butyl rubber has many properties similar to natural rubber and is often used for the manufacture of inner tubes and for high temperature applications.

Thiokols are very resistant to oils, gas, and sunlight and are often used in applications similar to Neoprene. Objects molded out of this material are quite resilient.

CERAMICS. Ceramic materials contain atomic arrangements of metallic and non-metallic elements. There are many ceramic materials because there are many possible combinations of metal and non-metal atoms and many different forms of their compounds. Ceramics have properties that are different from polymeric or plastic materials and metallic materials.

Van Vlack (Elements of Materials Science) lists among the more common engineering ceramics, glass, brick, stone, concrete, abrasives, porcelain enamels, refractories, dielectric insulators, and non-metallic magnets.

The largest tonnage of ceramics is in the form of glasses. The next largest is **lime** and **cement** products especially building materials. Whitewares include

porcelain, pottery, and other compositions used for toilet bowls, sinks, and other household goods.

Structural clay products are brick and tile and include sewer and water pipes. **Refractories** are fire clay products and magnesite, chromite, silica, and other oxide bricks.

Abrasives are usually oxides or carbides.

TOOLS, JIGS, AND FIXTURES

CONTENTS

CONTENTS (*Continued*)

TOOLS, JIGS, AND FIXTURES

Nature of Tools

DEFINITIONS. Tools, jigs, and fixtures are used by industry to accurately reproduce items under production conditions. Each of these will be discussed in relation to its classification, design, materials, capabilities, applications, and economics.

Tools as discussed in this section are those directly concerned with material removal. The operating mechanisms frequently referred to as machine tools are discussed in the section on Manufacturing Processes and Materials, while gages and gaging equipment are detailed in the section on Inspection.

Tools are differentiated from jigs and fixtures as follows:

A **tool** is a device or element designed and used for the express purpose of removing material from a workpiece under controlled or stable conditions.

A **jig** is a device for locating, supporting, and holding a workpiece and for accurately guiding the tool to the workpiece.

A **fixture** is a device for supporting, locating, and holding a workpiece.

PURPOSE. Accurate tooling, resulting in **reproducibility** and **interchangeability** of parts or products, is one of the primary purposes of the tooling effort in modern industry. Accurate tooling also reduces the need for highly skilled production workers in many operations. It transfers the needed skill from the worker to the machine and markedly decreases the chances of error. Less skilled workers may be utilized without jeopardizing the quantity and/or quality of the product.

CHOICE OF A CUTTING TOOL. A machining process is no more efficient than its cutting tool. Good cutting tools must include sufficient strength to maintain a sharp cutting edge, sufficient resistance to wearing of the cutting edge, and sufficient hardness to prevent picking up chips. A tool having a favorable balance of these factors is considered to have good cutting ability.

However, if it is to give maximum production with a minimum of trouble and maintenance, it is necessary that the tool be:

1. Made of the proper material for the cutting task.
2. Given the correct hardening heat treatment.
3. Accurately designed and produced.
4. Correctly applied to the workpiece.
5. Used with proper coolants and lubricants.

CLASSIFICATION OF CUTTING TOOLS. Cutting tools vary according to the operation, workpiece, and machining process. Generally, cutting tools fall into two classes: single point and multipoint.

Single-Point Cutting Tools. A single-point cutting tool is defined by St. Clair (Design and Use of Cutting Tools) as one which has one effective cutting edge and removes excess material from the workpiece along this cutting edge. Single-point tools may be classed as follows.

A **ground tool** is one in which the cutting edge is formed by grinding the end of a piece of tool steel stock.

A **forged tool** is one in which the cutting edge is formed by rough forging before hardening and grinding.

A **tipped tool** is one in which the cutting edge is a small tip of high-grade material welded to a shank of lower grade material.

A **bit tool** is one in which a high-grade material of square, rectangular, or other shape is held mechanically in a tool holder.

Single-point tools may also be classed as either **left-cut** or **right-cut** tools. Right-cut tools are those in which the cutting edge is on the right side when viewed from the point end of the tool. Similarly, a left-cut tool has its cutting edge on the left side when viewed from the point end of the tool.

Multipoint Cutting Tools. These are defined by expanding St. Clair's definition of single-point tools: Multipoint cutting tools are those that have more than one effective cutting edge to remove excess material from the workpiece. Since most multipoint tools can be considered combinations of single-point tools, they may be classified in much the same categories as single-point cutting tools.

Tool Materials and Cutting Fluids

TYPES OF MATERIALS. The selection of a cutting tool material is based on the properties of the material to be machined, the finish desired, and cutting temperatures under which it must operate. The diversity of production practices, specifications, and engineering materials has imposed severe demands upon cutting tools. To accommodate these many conditions, a wide range of tool materials has been developed. The most common cutting-tool materials are:

1. Plain carbon steel.
2. High-speed steel.
3. Cast nonferrous alloys.
4. Sintered carbides.
5. Ceramics.
6. Diamonds.

Plain carbon steel, known for a long time and improved about the middle of the eighteenth century, was the only tool material available until the end of the nineteenth century. Each new material that has been developed has tended to supplement, rather than replace the older materials. Progress from the carbon steels through the high-speed steels, cast alloys, carbides, ceramics, and diamonds is associated with the demand for increasing wear resistance, hot hardness (the ability to operate effectively at high cutting speeds), and a high-quality finish of the workpiece surface.

Plain Carbon Steel. The carbon content of plain carbon steels varies from 0.60 to 1.30 percent. Small percentages of silicon and manganese are retained for their beneficial effects, with small amounts of sulfur and phosphorus remaining as impurities. These steels can be readily hardened by quenching in water to a surface hardness of over Rockwell C60.

Because of their wear resistance, steels in the 1.00 to 1.30 percent carbon range are employed in the manufacture of milling cutters, twist drills, and turning and forming tools used in machining wood, magnesium, aluminum, and brass. These materials are relatively free cutting so that the temperature at the cutting edge can be kept comparatively low (below 400° F). This permits the cutting tool to maintain its sharp edge and original hardness.

Carbon tool steels can be sharpened to a very keen edge and are widely used in engraving tools, files, and reamers. These tools are usually operated at relatively low cutting speeds and light cutting duty. In machining free-cutting workpiece materials at low cutting speeds, the life of carbon steel tools is so great that sometimes there is no advantage in using the more costly tool steels.

At temperatures up to 400° F, high-carbon tool steels are harder than high-speed tool steels. At temperatures above 400° F, carbon steels lose their hardness rapidly and do not regain it upon cooling. Since tool temperatures increase with an increase in either cutting speed or feed, faster, heavier machining operations entailing tool temperatures in excess of 400° F, are not compatible with the use of carbon steel tools.

In general, the composition of plain-carbon tool steel is as follows:

Carbon	0.60 to 1.30%
Silicon	0.30% max.
Manganese	0.30% max.
Phosphorus	0.30% max.
Sulfur	0.30% max.
Iron	Balance

High Manganese Steel. These usually contain 10 to 14 percent manganese and are cast into shape. They have superior resistance to abrasion and wear and long tool life. Besides the austenitic characteristics of ductility and toughness, the surface hardness increases under severe service conditions of impact and high loads.

High-Speed Steel. The name "high speed" steel describes a wide variety of tool steels containing alloying elements, principally tungsten, to improve cutting properties (Fig. 1).

Tool steels used at a maximum cutting speed of about 35 sfpm are sometimes called "semi-high-speed" steels. The **major constituents** of these steels are tungsten and manganese. Tool performance is superior to that of plain carbon tools due primarily to an increase in wear resistance.

The most common type of high-speed steel, known as 18–4–1, contains from 0.55 percent to 0.75 percent carbon, 18 percent tungsten, 4 percent chromium, and 1 percent vanadium. Tungsten and chromium form carbides of extreme hardness, molybdenum increases hot hardness, and vanadium imparts wear resistance and smaller grain size to the steel.

The **most significant characteristic** of this steel is the retention of the cutting-edge hardness up to a temperature of 1100°F. An increase in the vanadium content to 2 or 3 percent, with a corresponding increase in carbon, results in a tool material of increased hot hardness and resistance to abrasion better adapted to machining harder materials.

Super-High-Speed Steel. High-carbon, high-vanadium, super-high-speed steels, developed in 1939, now constitute a major share of high-speed tool production. These tools are used where maximum edge strength, toughness, and abrasion resistance are required. They are particularly well adapted to interrupted cutting or high-rake angle machining of the super alloys, refractory metals, and ultra-high strength steels.

According to the Society of Manufacturing Engineers, super-high-speed steels offer 4 to 10 times as much resistance to grinding as the conventional grades and also exhibit higher hot hardness. This excellent wear resistance and hot hardness is due to the extreme hardness of the vanadium carbides in the steels. These vanadium carbides are harder than tungsten carbide or aluminum oxide. Although the proportion of vanadium carbide to total carbide is generally less than 20 percent in conventional high-speed steels, it is well over 50 percent in the

Steel Type	Percent Element Composition											Remarks
	C	Si	Mn	P	S	Co	Cr	V	W	Mo	Fe	
1	0.55–0.75	0.32	0.20	0.011	0.009		4.0	1.00	18.0		Balance	General roughing and finishing work on most types of materials.
2	0.66	0.42	0.32	0.029	0.025		4.0	2.00	14.0		Balance	
3	0.75–0.85	0.32	0.32	0.02	0.014		4.0	2.00–3.25	18.0		Balance	
4	0.70	0.25	0.15	0.02	0.02	5.00–8.00	4.0	2.00	13.0		Balance	Tougher. Used with heavier cutting.
5	0.65–0.80	0.15	0.41	0.021	0.008	3.00–5.00	4.0	1.00	18.5	0.18	Balance	
6	0.70					6.00–9.00	4.0	1.90	18.0	0.75	Balance	Rough cutting cast iron and nonferrous alloys.
7	0.70					10.0–13.0	4.0	2.00	19.0	0.75	Balance	Great resistance to abrasion. Used on gritty material.
8	0.70–0.85						4.0	0.90–1.50	1.25–2.00	8.00–9.50	Balance	Tungsten saving alternate. Same as 18-4-1.
9	0.70–0.90						4.0	1.50–2.25		7.50–9.00	Balance	Tungsten saving alternate. Same as 18-4-1.
10	0.70–0.90					4.0	4.0	1.0	1.5	8.0	Balance	Tungsten saving alternate. Same as 18-4-1.
11	0.75–0.90					4.50–5.50	4.0	1.05–1.35		8.00–9.00	Balance	Tungsten saving alternate. Same as 18-4-1.
12	0.75–0.90						4.0	1.25–2.00	6.00	5.00	Balance	Tungsten saving alternate. Same as 18-4-1.

Fig. 1. High-speed tool steels—composition and properties.

super-high-speed steels. The notable increase in the proportion of vanadium carbide in the hardened steel as compared with the annealed state indicates that this extremely hard carbide is also very stable. The hardness of super-high-speed steel closely approaches sintered carbide and approximates some ceramic materials.

Toughness is a combination of high yield or fracture strength and high shock resistance. Because of the greater ductility inherent in a wrought product, super-high-speed steels exhibit much greater bend strength than either cemented-carbide or ceramic tool materials by factors varying between 2 to 1 and 8 to 1. They also exhibit the greatest rupture strength, not only among tool materials, but among all wrought steels presently known. Thus, marked improvement in wear resistance and hot hardness is obtained with little sacrifice in toughness.

Subzero temperature treatment (deep freezing) and surface treatments such as liquid nitriding, oxidation, chromium plating, sulfidizing, liquid honing or vapor blasting, and superfinishing are frequently given to high-speed steel tools after all hardening and tempering operations have been completed, so as to retain full hardness in the parent material. Increased surface hardness and/or changed frictional characteristics serve to prevent seizing and thus lead to improved performance. It has been demonstrated that tool finishes of 3 microinches or finer can sometimes improve tool life 10 times over that obtained with hand-ground tools.

Cast Nonferrous Alloys. Cast nonferrous cutting-tool alloys contain 28 to 31 percent chromium, 35 to 46 percent cobalt, 16 to 18 percent tungsten, 0 to 5 percent tantalum, 1.8 to 2.3 percent carbon, 0.2 to 0.8 percent boron, 1 to 4.5 percent iron, 0 to 4.5 percent vanadium, and 0 to 0.7 percent molybdenum.

The typical **chemical composition** of these alloys and their trade names are shown in Fig. 2.

Trade Name	Figures in Percent									
	Cr	W	Co	C	B	Mo	V	Fe	Ni	Ta
Stellite J. metal	35	20	40							
Stellite 98-M2	30	18	35	1.8	0.8	0.7	4.5	4.5	3.0	
Stellite 2400	31	18	40	2.3	0.2	0.3	2.4	4		
Rexalloy	32	18	45	2	0.2			1		
Braecast	31	18	44	2	0.8			2		
Excelite	32	18	45	2	0.2			2		
Tantung G	28	16	46	2	0.2			2		5

Fig. 2. **Cast nonferrous cutting tool alloys.**

The hardness of these alloys is inherent, that is, they require no heat-treatment to develop it. They have exceptionally high hot-hardness properties. After they are cast, they can be reheated to temperatures as high as 1700°F with only a slight loss of hardness. They are also highly resistant to abrasive wear. Their high hardness in the as-cast condition make them difficult to machine. It is therefore necessary to grind them to final size and shape. They make excellent inserts for tipped tools. Since their hardness does not change with higher temperatures they can be brazed or welded to the steel tool shank with little danger of decreasing the hardness of the cast bit. They are often cast into different shapes to fit specially shaped tools.

While cast alloy cutting tools are usually recommended for deep roughing operations, they can also be used for any machining operations on all types of steel, iron, copper, brass, bronze, and aluminum alloys. It is generally recommended that the cutting surface speed of cast alloy tools be 25 to 80 percent greater than that used for high-speed steel tools. Machine tools should be free from vibration and well supported and the overhang of the tool should be held to a minimum. Coolants and lubricating cutting compounds are not necessary and are usually used only to obtain a special finish. Cast alloy tools are generally recommended for the middle range of machining, that is, between that of high-speed steel and cemented carbide tools. They have longer life between grinds than the high-speed steels and shorter life than the cemented carbides.

Sintered Carbides. The hardest and most stable carbide known is **tungsten carbide.** To produce a carbide tool, powdered tungsten is heated with lamp-black or other carbonaceous material at a heat where chemical combination takes place and tungsten carbide is formed. This is then ballmilled, usually with cobalt balls and sometimes in a mill lined with cobalt, until a thin layer of cobalt, a highly satisfactory binding material, surrounds each particle of carbide. The powder is then pressed at high pressures to form a blank and baked at a relatively low temperature to increase its strength. The blanks are then ground to slightly larger than final shape, to allow for shrinkage. Finally the shaped blanks are heated to just below the melting point of the cobalt, about 2600°F, where the cobalt partially fuses together, holding the hard carbides in place. All heating operations take place in an atmosphere of hydrogen so that neither the tungsten nor the carbon is oxidized.

Variations of the above practice permit changes in composition and properties of the final product. Use of titanium or tantalum in place of part of the tungsten results in a tool that has superior steel cutting properties. Control of the cobalt content yields either hard and brittle or softer and tougher tools. When large amounts of tantalum are used, nickel rather than cobalt is used as a binder. It is necessary to sinter tantalum tools in a vacuum to prevent a reaction between hydrogen and tantalum. These variations in composition and in powder particle size have permitted the production of sintered-carbide tools with specific properties and account for the large variety of grades produced by the manufacturers.

Figure 3 adapted from Swinehart, ed. (Cutting Tool Material Selection) lists the **eight machining classifications** in the "C" systems together with their specific applications.

The approximate composition of the most common tungsten carbide tool is 88 percent tungsten, 6 percent carbon, and 6 percent cobalt. The carbon is all in the form of tungsten carbide. Variations from this base composition include varying the cobalt from 3 percent to almost 17 percent, substituting varying amounts of tantalum or titanium carbide for tungsten carbide, and substituting nickel for all or part of the cobalt.

These changes affect the hardness, strength, density, and toughness of the finished carbide tool. There are also changes in heat conductivity, coefficients of friction, and fineness of edge obtainable; however, the latter is the result of processing methods rather than composition.

Due to the low tensile strength and high cost, carbides were originally used only as small tips or inserts brazed to a tougher, less costly material. Since carbide costs have been reduced considerably, heavier cross sections are now used economically. For example, in certain types of inserted-blade face-milling cutters,

Tool Properties	Designation	Application	Increasing hardness and wear resistance	Increasing strength and binder content
Abrasion Resistant	C-1	Roughing cuts—cast iron and nonferrous materials		
	C-2	General purpose—cast iron and nonferrous materials		
	C-3	Light finishing—cast iron and nonferrous materials		
	C-4	Precision finishing—cast iron and nonferrous materials		
Crater and Deformation Resistant	C-5	Roughing cuts—steel		
	C-6	General purpose—steel		
	C-7	Finishing cuts—steel		
	C-8	Precision finishing—steel		

Fig. 3. Classification of carbides for machining applications.

the blade is made entirely of carbide instead of an alloy-steel blank with a small, brazed carbide tip. Although this has not increased the time between grinds it does have several advantages over the older practice.

Brazing, even though accomplished with low melting point silver solder, may subject the carbide to detrimental stresses in cooling because of the difference in the rates of contraction and thermal conductivity between the carbide and the supporting material. Such stresses alone may not be large enough to cause fracture, but when normal cutting stresses are superimposed, the resultant stress may be sufficient to cause cracking. Mechanical means of securing carbides in cutting tools can eliminate this possibility. However, brazing is a widely used technique which when properly done is very satisfactory.

The life of carbide tools is adversely affected by deflection and vibration. The frequency of vibration can be increased and the amplitude of vibration reduced by decreasing the unsupported length of the tool tip and also by increasing the thickness of the shank.

Despite their hardness and abrasion resistance, carbides can be ground readily if the proper equipment and techniques are employed. Silicon carbide or diamond wheels (operated at about 5,000 sfpm) must be used since aluminum oxide abrasive particles are not hard enough. An open, porous, vitrified wheel, ensuring a constant breaking action or resharpening of the wheel face, is needed to avoid checking or cracking of carbides through excessive localized heat. Because of rapid wear, silicon carbide wheels should be used only for rough grinding. Wheels should be used dry or completely wet. Partial or incomplete use of a fluid results in unequal heating, which is very detrimental to the carbides. Machine grinding with silicon carbide is not recommended except in a few instances. **Rough grinding** can be done by hand with the aid of adjustable angle tables, fixtures, and templates. To prevent chipping at the cutting edge it is recommended that carbide cutting tools be ground with the abrasive grains moving in a direction from the cutting edge back along the carbide.

Diamond wheels should be used for **finish-grinding** only the carbide tip, since the shank steel will load the diamond wheel. Both off-hand grinding of single-point tools and machine grinding of single or multiple-point tools can be done practically and economically with a diamond wheel. Not more than 0.0004 in. of carbide should be removed at a pass. Lapping with boron carbide or diamond dust is sometimes employed to obtain very smooth surfaces and keen edges for machining aluminum or magnesium alloys, while honing the cutting edge serves to strengthen it rather than make it microscopically sharp.

Electrolytic grinding of carbide tools can reduce diamond-wheel costs considerably. It has been reported that diamond wheels in combination with electrolytic grinding will outlast ten diamond wheels used conventionally. Grinding of carbides by the electric-discharge method is done with a brass or other metallic wheel without an abrasive.

Although carbide tools have been responsible for great increases in production they will probably never replace the various tool steels and cast alloys. Tool steels and cast alloys will continue to fill definite needs in certain machining applications. Carbon tool steels are still in use, having survived competition from the successive introduction of air-hardening high-speed, and super-high-speed steels, cast alloys, carbides, and ceramics. Carbides should be considered as augmenting, rather than supplanting, the older tool materials.

Ceramics. Ceramic tool materials are best described as oxides, ranging from single oxides to complex combinations. Tool inserts or tips of ceramic materials are manufactured by either sintering or hot pressing. Aluminum oxide, constituting up to 99 percent of the total mixture, is the major ingredient of most cutting-tool ceramics, the remainder consisting of small impurities.

The principal elevated-temperature properties of alumina are high hardness, chemical inertness, and resistance to wear. The high hardness and wear resistance of alumina are the main reasons for its use in machining cast iron and hardened steel at high cutting speeds. The high wear resistance also permits machining of long cylindrical surfaces without taper. The inertness of alumina to iron at high temperature prevents welding of the tool to steel or cast iron workpieces and contributes to the production of good surface finish. Ceramic tools are satisfactory for general machining of steel where there are no heavy interrupted cuts and where negative rake angles can be used. Typical feeds are 0.020 ipr for cast iron, 0.017 ipr for heat treated steel, and 0.021 ipr for soft steel. *Depth of cut* usually depends on the amount of stock to be removed; cuts up to half the width of the insert should not shorten the life of a ceramic tool excessively.

Diamonds. A large percentage of the world's diamond output is used in industry as a powder for grinding and polishing. Whole stones are used for wire drawing, core drilling, and truing abrasive wheels. Less than 10 percent of the total diamond production finds its way into such machining operations as boring, turning, and milling. These hard stones are employed in machining precision bearings, pistons, calender rolls, hard rubber, fiber, celluloid, soft and light metals, plastics, porcelain, natural and artificial stone, and in some cases even hardened steel.

Diamonds, because of their perfect crystalline structure, exhibit unusual properties, including very low compressibility, high modulus of elasticity, great resistance to abrasion, great breaking strength, high heat conductivity and melting point, and chemical inertness. As a class, the diamond is the hardest substance known, but there are noticeable variations in properties between and within individual stones. In most industries, **synthetic diamonds** are a suitable substitute for natural diamonds.

Diamonds are principally used to machine materials too difficult for tool steels, to obtain mirror-like surfaces on nonferrous metals, or to finish work to extremely accurate limits. Diamonds are particularly suitable for these applications because, while they are extremely hard and wear resistant, they are also capable of receiving a polished surface comparable with an optical flat and maintaining sharply defined cutting edges. Because of their expense and brittleness, diamond cutting tools are used only for special applications. The use of X-ray diffraction techniques to determine the hardness direction of a diamond is generally considered highly advantageous since it can lead to a notable improvement in the cutting performance. Diamond cutting tools do not smear the workpiece surface crystals in the boundary zone; instead they cut right through the granular structure, producing a smooth surface that is especially desirable in bearings and other close-fitting rotating or sliding parts. Rough diamonds with natural edges can sometimes be utilized in tools, but the shaped stone is generally preferable.

Diamonds are used only as finishing tools where operations typically call for small chip loads, a shallow depth of cut, and high cutting speeds. If successful performance is to be obtained, rigid machines with negligible vibration and bearing play are an absolute necessity. In most machining applications, diamonds possess extremely long tool life; carelessness in handling rather than tool wear often is the cause of tool failure.

Three types of diamonds are used for making tools: crystallized or gemstones; **bortz**, a round or imperfectly crystallized form; and **carbonados** or black diamonds, which are impure aggregates of small diamond crystals (most diamond cutting tools are made from bortz).

CUTTING FLUIDS. Any substance—liquid, gas, or solid—applied to a tool during a cutting operation to facilitate removal of chips is a cutting fluid. Vidosic (Metal Machining and Forming Technology) states that such fluids must accomplish one or more of the following:

1. Reduce friction at chip-tool and work-tool interfaces (lubrication—important at low cutting speeds).
2. Help carry away the heat generated (cooling—important at high speeds).
3. Remove chips and debris to protect finish.
4. Protect the finished surface from corrosion.

Removal of chips from the immediate tool area is quite important. Heat contained by chips is prevented from being transferred to the tool or workpiece, and the finished surface is freed of debris that might otherwise harm it. Furthermore, in operations where free chip flow is restricted as, for instance, in tapping, broaching, and boring, flushing with a cutting fluid may be the only answer.

To be of practical value cutting fluids should possess the following qualities:

1. They should present no fire or accident hazard.
2. They should not emit obnoxious odors or vapors harmful to the operator, workpiece, or surrounding area.
3. They should not cause skin irritation.
4. They should have a long life, free of excessive oxide formation that might clog circulating systems.
5. They should be suitable for a variety of cutting operations and materials to reduce the number of fluids needed by any single consumer.
6. They should be of low viscosity to permit easy separation from impurities and chips collected.
7. They should be transparent, so that an operator can clearly view tool and work—very important where high dimensional accuracy and fine finish are required.
8. They should be easily removed from surfaces of workpiece and machine.

Use of cutting fluids results in the following **benefits**:

1. Power consumption is reduced.
2. Tool life is increased.

3. Surface finish and dimensional precision are improved.

4. Higher cutting speeds and feeds with the same cutting tool are possible.

Types of Cutting Fluids. Numerous requirements are imposed upon cutting fluids by the varied machining situations. Figure 4 lists the cutting fluids recom-

Work Material	Type of Operation	Cutting Fluid
Aluminum alloys	Heavy cuts	Soluble oils
	Light cuts	Kerosene base oils
	Finishing	Mineral oil with friction additives
Copper alloys	Heavy cuts	Straight mineral oil
	Light cuts	Dry
Ferrous	General cutting	Straight soluble oil
	Light cuts	Synthetic coolants
	Threading	Sulphonated oils
High-temperature alloys	General cutting	Soluble oils
Inconels	General turning	Synthetic coolants
Nickel alloys	General cutting	Non-sulphorized mineral oils
Ni-resist irons	General machining	Water base emulsions
	Finishing	Friction additives
Nonferrous metals	General cutting	Straight soluble oils
Plastics	General machining	Soluble oils
Stainless steel	Light cuts	EP-soluble oils
Steel	Heavy cuts	Dry
	Low speeds	EP-soluble oils
Titanium	Turning	Synthetic coolants

Fig. 4. Cutting fluid recommendations.

mended for various work materials and machining operations (Vidosic, Metal Machining and Forming Technology).

Factors in Cutting Tool Design

OBJECTIVES. The principal objectives of good tool design are to maximize cutting efficiency and minimize costs. Cutting efficiency involves obtaining easy chip flow, high productivity, dimensional accuracy, and proper surface finish with the least amount of input power.

SINGLE-POINT TOOL NOMENCLATURE. To understand the importance of **tool form** and correct **cutting** angles it is essential to know the various tool parts and nomenclature.

Parts of the ordinary single-point cutting tool are defined as follows (Fig. 5):

The **shank** is the main body of the tool as distinct from the cutting portion and is the portion on which the point is formed or the tip or bit supported.

The **neck** is an extension of the shank but of a reduced sectional area.

The **face** is the portion of the tool seen by the operator looking down at the tool from above, assuming that the tool is set horizontally. The face is more exactly defined by Woldman and Gibbons (Machinability and Machining of Metals) as the surface across which the turnings or chips travel as they are formed.

The **base** is the surface on which the tool rests as it is positioned in the tool post or toolholder.

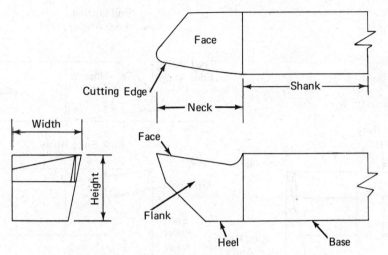

Fig. 5. Single-point cutting tool nomenclature.

The **heel** is the surface which is at the end of the base adjacent to the cutting edge and directly below it when the tool is in a horizontal position. The heel can also be called the lower face.

The **cutting edge** or **lip** is the portion of the face edge along which the chip is separated from the workpiece.

The **flank** is the surface adjacent to the cutting edge and behind it when the tool is held horizontal.

Single-Point Tool Angles. Tool angles are of two general types: angles which depend upon the shape of the tool and angles which do not depend upon the shape of the tool. The former are referred to as **standard angles** and the latter as **working angles**. Standard angles are generally those machined in a shop (Fig. 6).

Standard Angles. The **back-rake angle** is the angle between the face of the tool and a line parallel to the base of the shank in a plane parallel to the center line of the point and at right angles to the base. The angle is positive if the face of the tool slopes downward from the point toward the shank and is negative if the face slopes upward toward the shank. The back-rake angle turns the chip away from the workpiece.

The **side-rake angle** is that angle between the tool face and a line parallel to its base. It is measured in a plane at right angles to the base and at right angles to the center line of the point of the tool. The side-rake angle determines the thickness of the tool behind the cutting edge.

Increasing the rake angle facilitates chip flow and escape and minimizes the size and effect of the built-up edge, cutting temperature, cutting forces, and power consumption. In addition surface finish is generally improved. Excessive rake, however, weakens the tool and may induce tool chatter.

The **end-relief angle** provides clearance between the workpiece and the tool. The angle lies between the surface of the flank immediately below the point and a line drawn perpendicular to the point from the base.

The **side-relief angle** provides clearance between the flank of the tool and the workpiece surface. It is determined in much the same manner as the end

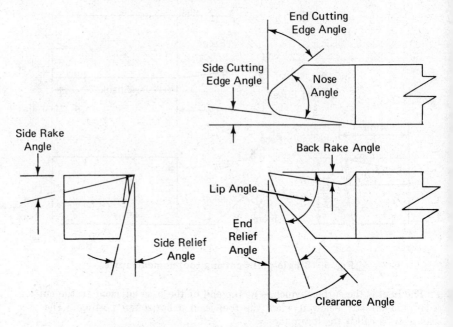

Fig. 6. Single-point cutting tool—standard angles.

relief angle except that it is measured in a plane at right angles to the center line of the point of the tool.

The **nose angle** is often specified as a nose radius. It is the angle between the side-cutting edge and the end-cutting edge. The nose angle removes the fragile corner of the tool, increases tool life, and improves surface finish.

The **side-cutting edge angle** is responsible for turning the chip away from the finished surface.

The **end-cutting edge angle** provides clearance between the tool cutting edge and workpiece.

The **clearance angle** is that between the portion of the flank adjacent to the base and a plane perpendicular to the base. Increasing the clearance angle results in freer cutting action, minimizes tool forces, and decreases cutting temperatures. Improper clearance angles may cause chatter or cause excessive tool wear.

Working Angles. The working angles are those formed between the tool and the workpiece and depend on the **position** of the tool with respect to the workpiece. These angles, as discussed below, have a direct bearing on tool chatter and surface finish as well as tool life.

The **setting angle** is that made by the straight portion of the shank of the tool with the finished surface of the work. An increase in the setting angle increases cutting-edge engagement and cutting pressure as well as power consumption and the tendency to produce chatter. The gain in tool life, however, frequently offsets these disadvantages.

The **true-rake angle,** under actual cutting conditions, is the slope of the tool face toward the base from the active cutting edge in the direction of chip flow. It is a combination of the back-rake and side-rake angles. The setting of the tool, the feed of the workpiece, and the depth of cut may effect variations in the true-rake angle.

The **cutting angle** is that between the face of the tool and a tangent to the machined surface at the point of contact of the tool with the workpiece. It equals 90° minus the true-rake angle.

Chip Breakers. Tool geometry is affected by chip breakers in the sense that they alter the tool face. Chip breakers are provided in single-point tools when a long continuous chip is encountered in machining. This type of chip presents the dual problem of a hazard to the operator and a threat to the surface finish of the workpiece. The chip breaker is intended to break the chip as it is sheared from the parent material. The most frequently used type of chip breaker is one provided by grinding a depression or **gullet** in the face of the tool immediately behind the cutting edge. Typical chip breakers utilized in breaking up long continuous chips are shown in Fig. 7.

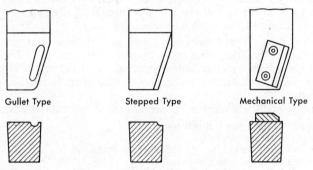

Gullet Type   Stepped Type   Mechanical Type

Fig. 7.   Typical chip breakers.

Tool Holders and Tool Posts. The cutting tool must be held rigidly in position in the correct location and at the proper angle with respect to the work. Sometimes cutting edge and tool holder are one and the same.

The tool base or tool holder may rest on a rocker base. This allows many adjustments and settings of the cutting tool. However, a vertical adjustment only, using shims, is preferred, because mounting is more rigid, and rake and clearance angles cannot be changed (such changes may allow cutting procedures detrimental to the tool). The tool holders are in turn mounted in a tool post. The best cutting conditions are obtained when the tool holder is held short in the tool post and the tool is held short in the tool holder. The farther the cutting tool projects, the greater the leverage and the more the spring of the tool. This causes chattering and other adverse cutting conditions and should be avoided wherever possible.

MULTIPOINT TOOLS. Multipoint cutting tools are basically a series of single-point tools mounted in a holder and operated in such a manner that all cutting edges follow essentially the same path across the workpiece. Multipoint tools are generally discussed in relation to the operations that each performs.

Boring Tools. Boring tools in their simplest form are single-point tools secured by some mechanical means to a boring bar. Borers can be used in either of two ways to remove material: (1) the boring bar and cutter can be rotated and fed into the workpiece or (2) the workpiece can be rotated while the bar remains stationary.

Design of tool angles of the boring tool are extremely important. Chip clearance and disposal are more critical in boring than in most other machining operations. It is essential that the clearance angle be large enough on a boring tool to reduce cutting temperatures, lessen tool wear, and minimize power consumption. The side-cutting edge angle must be varied for each type of boring operation. The end-relief angle must be sufficient to clear the bore surface; hence, this angle must be increased as the bore size is decreased. An excessive end-relief angle is not recommended because the cutting edge will be weakened. Back- and side-rake angles must act together to turn chips toward the center of the bore and away from the finished surface.

Boring tools can become progressively more complex by using several boring cuttings on the same boring bar or head to perform more than one operation simultaneously.

Forged boring cutters are generally used for small hole operations. Larger holes are generally handled by boring cutters that have one or more adjustable tools. The pilot on the end of the bar aids in guiding the tool and in keeping the tool from exceeding the depth of a blind hole.

Drills. A drill is a fluted end cutting tool used to initiate or enlarge a hole in solid material. Standard practice is to rotate the drill as it feeds into the workpiece. On a lathe or other similar machine the drill may be either rotated while being fed into stationary work or held stationary while being fed into rotating work.

Drills are generally classified in one or several of the following ways:

1. The material from which it is made.
2. The method of manufacture.
3. The length, shape and type of helix, flute, or shank.
4. The diameter of the hole it will make.

For example, there are several types of drill shanks—taper and straight being those most generally used. Other examples include low-helix drills (generally having comparatively thin webs), high-helix drills (wide flutes and narrow lands), V-shaped drills, and sub-land drills, all having different types of flutes.

The American Society for Metals (Metals Handbook, Vol. 3. Machining. Lyman, ed.) suggests that the following factors be considered when selecting a drill:

1. Composition and hardness of work material.
2. Rigidity of the tooling setup.
3. Dimensions of the hole to be drilled.
4. Type of machine used to rotate the drill or the workpiece.
5. Whether the drill is used for originating or enlarging holes.
6. Tolerances on the hole to be drilled.
7. Whether related operations, such as countersinking or spot facing, must be performed with the drilling operation.
8. Cost.

Figure 8 depicts twist drill terminology as given in ANSI Standard B94.11-1967 (Twist Drills, Straight Shank and Taper Shank Combined Drills and Countersinks).

The **point angle** of most general purpose drills is 118°. Drills such as the low-helix type have 90° points to provide for maximum chip space. As the hardness of the material increases, the included point angle also increases, sometimes to as much as 145°.

Lip relief angles of approximately 10° to 14° are those most generally used. Smaller diameter drills require greater relief angles. When drilling hard material the lip relief angle is generally decreased to strengthen the cutting lips. High penetration rates in soft materials require increased relief angles to facilitate chip removal.

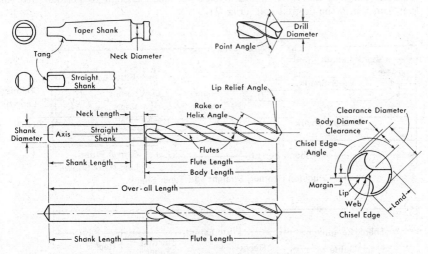

Fig. 8. Twist drill nomenclature.

Web thinning is necessary after sharpening a drill to maintain the original web thickness at the point. Web thickness should be kept between 10 and 12 percent of the drill diameter.

Drills are frequently classed according to and designed for particular applications. Low-helix drills are generally used in soft materials where the reduced number of helical twists facilitates chip removal. These drills are generally designed to break chips into small pieces and are well suited to applications in which a large volume of chips are generated.

High-helix drills are made with narrow lands and wide flutes and are generally applied to those operations in which relatively deep holes are drilled in nonferrous metals.

Screw machine drills are made short in overall length with short flutes. These drills are used where maximum rigidity is required without loss of cutting ability, and especially where tough or hard materials are being drilled.

Straight flute drills are well adapted for drilling brass and other soft materials since they do not grab the workpiece during drilling.

Step drills have two or more diameters, produced by grinding various steps on the diameter of the drill. The steps usually consist of square or angular cutting edges. Step drills are used extensively in applications requiring multiple diameter holes. The simplest form of a step drill is a combination drilling and countersinking tool.

Gun drills have a single V-shaped flute and a single cutting face sharpened so as to form two cutting angles. During drilling these angles produce two chips broken into short lengths for easy expulsion from the workpiece. The body of the drill is generally a steel tube through which a cutting fluid passes under pressure to cool and lubricate the cutting area and flush out chips. Gun drills are generally used when extremely accurate holes are required.

Reamers. Reamers are used for enlarging and finishing holes to size. They are designed principally for operations where only a moderate amount of stock is to be removed from the hole walls. The cutting ends of these tools have relieved chamfers which extend to a diameter small enough to permit entry of the reamer into the desired hole (Fig. 9).

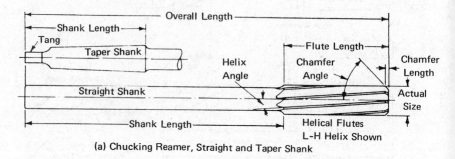

(a) Chucking Reamer, Straight and Taper Shank

(b) Hand Reamer, Pilot and Guide

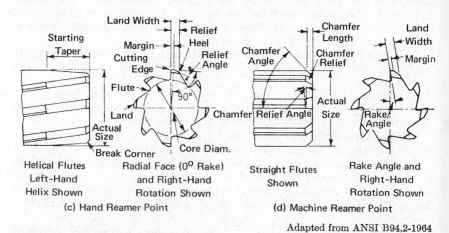

(c) Hand Reamer Point

(d) Machine Reamer Point

Adapted from ANSI B94.2-1964

Fig. 9. Reamer terminology.

The **angle of chamfer** is the part of the reamer that does the initial cutting. It is ground on every flute uniformly so that there is clearance behind each chamfered cutting edge. Finish cutting is done by the flute edges.

The **body** is made up of several flutes and lands (the land being the part between each flute). A margin runs along the top of each land, with a relief in the back of the margin to facilitate chip flow.

The **shank** is the driving end of the reamer and fits into the chuck or holder of the machine in which it is to be used.

The two basic types of reamers are hand reamers and chucking or machine reamers.

Care should be taken never to utilize a hand reamer in a machine operation, and vice versa. Differences in cutting angles between the two will seriously affect tool life and surface finish when used in the wrong application. Other classifications of reamers include:

Solid reamers of one piece carbide construction.

Carbide-tipped reamers having carbide cutting edges brazed or bonded onto the end of a solid body.

Inserted-blade reamers that employ mechanically held replaceable carbide blades of solid or tipped construction. The blades are often adjustable.

Expansion reamers that can be expanded slightly to compensate for wear or resharpening.

Adjustable reamers having carbide blades that can be adjusted individually away from or toward the reamer axis.

Chucking or **machine reamers** consist of two types: rose and fluted. **Rose reamers** have teeth that are beveled off on the end and relieved. Cutting is done only on the end. The lands are nearly as wide as the grooves, but are not relieved. It does not cut a particularly smooth hole, but it is used to bring a hole to within a few thousandths of its final size. The hole may then be finished with a hand reamer. **Fluted reamers** have more teeth for a given diameter than a rose reamer, with narrower lands that are relieved along their entire length. The front ends of the teeth are beveled or rounded and then relieved. It is a valuable finishing reamer in cases where extreme accuracy is not essential.

Shell reamers are generally used only when the tool is held stationary and the workpiece rotates. It is usually used in finishing operations.

Planer and Shaper Tools. Planers and shapers generally use single-point cutting tools modified to fit the particular machining operation.

Carbide-tipped tools with removable bits are generally preferred in planing operations. Since planer toolheads are bulky and awkward to handle, replaceable carbide inserts for these tools are made to be removed, replaced, or adjusted without the need to move the heavy tool shank or holder. Several planer tools may be used simultaneously to complete the desired job. Planing tools are usually classed as either roughing or finishing tools.

Back-rake angles for **roughing tools** range from 0° to 15° according to the American Society for Metals. The more difficult it is to machine the workpiece metal the more negative the back-rake angle should be made. This applies to the side-rake angle. The harder it is to machine the workpiece, the smaller the nose angle should be made. A cutting angle of about 25° to 35° is preferred. Anything smaller will cause excessive shock; anything greater will produce long chips. Side clearance angles are kept small to strengthen the tool.

Finishing tools are generally broad nosed; feeds should be less than the width of the cutting tool.

Shaping tools are much the same in design as planing tools, except that they are smaller in size. There is no rocker in the tool posts of the shaper, hence the tool cannot be adjusted for clearance; the proper clearance angles must be ground on the tool. A side clearance angle of 4° is maximum, generally 2° or 3°

is sufficient. Too much front clearance will quickly dull the cutting edge; 4° is recommended by A.S.M. A small nose radius is also essential in shaper tools.

Broaches. Broaches are linear-travel tools that remove material by being pushed or pulled across an external or internal surface. They are much like planers and shapers in their operation, except that all material is removed on a single stroke. The broach can remain fixed as the workpiece moves against it (like a planer) or it can move against a fixed workpiece (like a shaper). Broaches are quite expensive and their use is usually confined to high production operations.

Broaches are generally classed into two types according to their application: surface broaches and internal broaches. Both are available in many different sizes and shapes. Since the broach makes only one pass these shapes are quite intricate. Broaches have teeth (cutting edges) that taper and gradually increase in height or diameter until the finished dimension is obtained with the last few sets of teeth (Fig. 10). Most broaches contain chip breakers on the initial teeth.

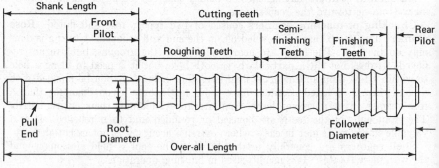

Essential features and nomenclature of broaches as illustrated by an
internal pull broach for cutting round holes

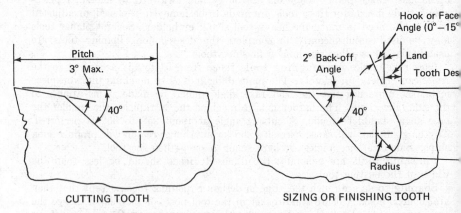

Fig. 10. Broach cutting angles.

Donaldson and LeCain (Tool Design) have suggested a general formula for pitch of broach teeth which is much used:

$$P = 0.35\sqrt{L}$$

where P = pitch
        L = length of surface to be broached

Another rule of thumb says the pitch should be made as long as possible. Other factors to be considered in determining pitch include the shape of the broach, its length, the chip thickness per tooth desired, and the type of material being machined.

All angles are very similar to those in other cutting tools except the back-off angles which are generally quite small. This is done to avoid excess loss of size in resharpening.

Milling Cutters. Milling cutters are revolving multipoint cutting tools. They are generally classed into three basic types: peripheral, face, and end mills.

Peripheral mills are so named because cutting is done mainly by teeth on the periphery of the mill. They are used primarily for finishing workpiece surfaces.

Face mills have teeth on the flat end surface of the tool. Cutting on face (or side) mills is done at right angles to the axis of rotation of the tool.

End mills have cutting edges both on the face and on the periphery; hence, they may be used in place of either face mills or peripheral mills. End mills are either shank type or arbor mounted.

The American Society of Metals (Metals Handbook, Vol. 3. Machining. Lyman, ed.), discusses milling cutter nomenclature in the following list of terms as they apply to plain milling cutters (Fig. 11):

1. **Outside diameter** is the diameter of a circle passing around the peripheral cutting edges and is used to calculate surface speed from spindle speed.
2. **Root diameter** is the diameter of a circle tangent to the roots of the teeth fillets.
3. **The tooth** is the part of the cutter starting at the body and ending with the peripheral cutting edge. Inserted teeth are called **blades.**
4. **Tooth face** is the surface of the tooth between the fillet and the cutting edge, where the chip slides during cutting.
5. **Land** is the area behind the cutting edge on the tooth that is relieved to avoid interference.
6. **Flute** is the space for chip flow between the teeth.
7. **Fillet** is the radius at the bottom of the flute to promote chip flow and curling.

The following terms refer to tooth design:

1. **Peripheral cutting edge** is the edge aligned principally in the direction of the cutter axis. In peripheral milling it is the edge that removes metal.
2. **Face cutting edge** is the metal-removing edge on a face mill that travels in a plane perpendicular to the axis. It is the edge that sweeps the milled surface in normal face milling.
3. **Tooth angle** is the angle included between the face and the land of the cutter tooth. This angle should be as large as possible to provide maximum tooth strength and better dissipation of heat.
4. **Radial rake angle** (shown as positive in Fig. 11) is the angle between the tooth face and a radial line passing through the cutting edge in a plane perpendicular to the cutter axis.
5. **Clearance angle** is the angle included between the land on the back or flank of the milling cutter tooth, and the tangent to the periphery of the cutter at

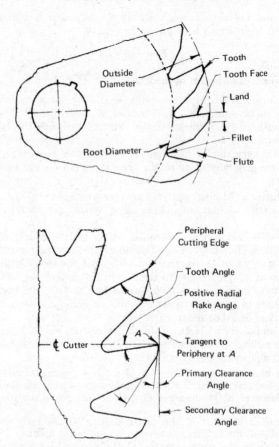

Fig. 11. Milling cutter tooth design nomenclature.

the cutting edge. Clearance angles are always positive and are usually divided into primary and secondary angles.

Hobbing Cutters. Hobbing is a special milling operation of cutting gear teeth. According to the A.S.M., hobbing is a generating process in which both the cutting tool and the workpiece revolve in a constant relation as the hobbing cutter is being fed across the face width of the gear blank (Fig. 12).

Tapping and Threading Tools. Taps are tools used to cut internal threads in a workpiece; threading dies (chasers) cut external threads on cylindrical workpieces. Although both perform threading operations, their geometry is usually quite different and distinct.

The following nomenclature is applicable to taps and dies (Figs. 13 and 14).

Angle of thread is the angle included between the sides of the thread measured in an axial plane.

Axis of tap is the longitudinal centerline of the tap.

Base of thread is the bottom section of a thread; the greatest section between the two adjacent roots.

The body is the threaded and fluted portion of tap.

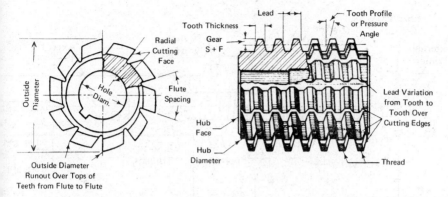

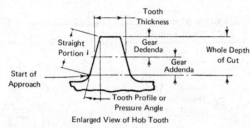

Fig. 12. Hob nomenclature.

Chamfer is the tapered outside diameter at the front end of the threaded section.

Crest is the top surface joining the two sides of the thread.

Cutting face, also called the **front flank,** is the front part of the threaded section of the land.

Depth of thread is the distance between the top of crest and the base or root of thread measured perpendicular to the axis of tap.

External (male) center is the cone-shaped end of a small tap.

Flute is the groove that makes possible the cutting faces of the threads and facilitates chip passage and lubrication.

Heel, also called the **rear flank,** is the back part of the threaded section of the land.

Helix is the curve formed on any cylinder, especially a right circular cylinder, by a straight line wrapped round the cylinder; for example an ordinary screw thread.

Helix angle is the angle made by the helix of the thread at the pitch diameter with a plane perpendicular to the axis of the tap.

Internal (female) center is a small drilled and countersunk hole at the end of the tap.

Land is the threaded web between flutes.

Major diameter is the diameter of the tap measured from top-of-crest to top-of-crest when viewed in profile.

Minor diameter is the diameter of the tap measured from base-of-thread to base-of-thread.

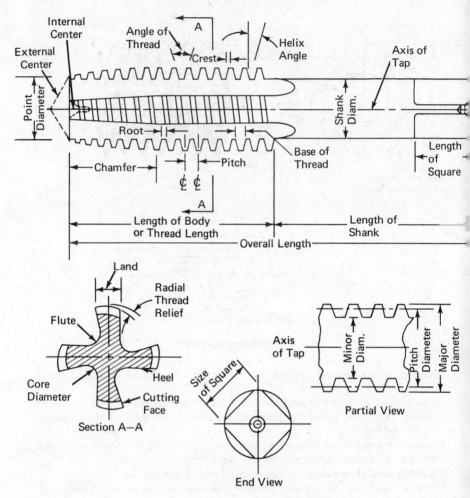

Fig. 13. Tap nomenclature.

Pitch diameter is the diameter of an imaginary cylinder passing through the thread profiles such that the groove width at that point is equal to one-half the thread width.

Point diameter is the outside diameter at the front end of the chamfered portion.

Root is the bottom surface joining the sides of two adjacent threads.

Shank is that portion of the tap behind the threaded and fluted section.

Square is the square end of the top shank.

Abrasive Cutting Tools. An abrasive cuts much like single- and multipoint cutting tools, with some important exceptions. Each grain of an abrasive in contact with the work acts as a tiny cutting tool. Single grain studies and examination of **swarf** left from grinding operations reveal that the chips cut

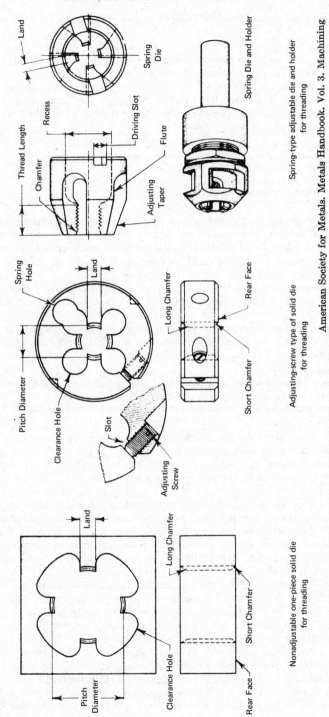

Fig. 14. Threading die nomenclature.

American Society for Metals. Metals Handbook. Vol. 3. Machining from the workpiece are miniature replicas of the chips formed by an ordinary cutting tool.

Grinding is a process which employs either bonded or loose abrasives to generate new surfaces. The individual grains, bonded into a single body, present a large number of randomly arranged cutting edges.

Grinding wheels are identified by their composition and properties. As specified by ANSI Standard B74.13-1970 Markings for Identifying Grinding Wheels and Other Bonded Abrasives, the characteristics of a wheel appear in the wheel identification code:

1. Abrasive type (A for aluminum oxide, S for silicon carbide, etc.)
2. Grain (size of abrasive particle—mesh numbers)
3. Grade (A for very soft to Z for very hard)
4. Structure numbers (grain spacing—dense, open, etc.)
5. Bond (V for vitrified, S for silicate, etc.)
6. Manufacturers' private numbers.

Because silicon carbide has low resistance to attrition when cutting steel, and its grains dull quickly, aluminum oxide is preferred as an abrasive for grinding wheels. For hard materials, such as hardened steel, small grains are preferred because more grains can be brought to bear on the material, there being the possibility of only small bites, due to the hardness.

Since hardened steel would tend to dull an abrasive rapidly, a soft grinding wheel that would readily release grains when they become dull would be preferable.

Dense spacing, for the reason of bringing as many cutting edges to bear as is possible on relatively impregnable materials (such as hardened steel) is desirable.

The three common grinding media are:

(1) **Grinding wheels.**
(2) **Coated abrasives** consisting of a thin layer of abrasive grains bonded to a flexible backing made of paper, cloth, or plastic material.
(3) **Loose grains** of fine mesh used with a lubricant, such as water, kerosene, or oil.

The individual abrasive grains of a grinding wheel, as those of a coated abrasive, are in contact with the work only for a very short period of time. The heat generated during the course of cutting by each grain of a grinding wheel tends to be transmitted to the bond of the wheel and, unless coolants are used, the build-up of heat is quite rapid. In the case of a coated abrasive, the heat generated during the cut is similarly transmitted to the bond. However, by the time any point of the belt or other form, has made a complete revolution and returned to the work area, the heat has been dissipated to the surrounding air, and the coated abrasive surface is therefore cooler than that of a grinding wheel.

The **principal grinding wheel bonds,** each best suited to a particular application, are vitrified, silicate, resinold, rubber, and shellac bonds.

The **vitrified bond** is clay fired and melted in a kiln to achieve a glass-like consistency. It is not affected by water, oils, acids (an important fact to know when coolants or cleaning compounds are used in the grinding), or common temperature conditions. For these reasons, most bonds are vitrified. A **silicate bond** is essentially water glass (sodium silicate) hardened by baking. It is more friable than a vitrified bond and gives a cooler cut. **Rubber bonds** are composed of fairly hard vulcanized rubber. Wheels made with this bond type are strong and dense and can be made very thin. The **resinold bond** is a synthetic, organic, or plastic compound. Being strong, fairly flexible, and cool cutting, it can be run at high speeds. A **shellac bond** produces high finishes on such products as cam shafts and mill rolls and cuts smoothly on hardened steels and thin sections.

Truing is necessary when the effect of wear produces concave or convex surfaces on the periphery of the wheel. Industrial diamonds are used for truing worn wheels. The diamond (or many small diamonds) is set on the end of a bar and applied only at angles of 3° to 15° in the horizontal plane and 30° in the vertical. Small (0.001 inch) cuts are made to avoid overheating the stone and the bar is turned slightly from time to time to present a new face to the work. Most grinding machines have a built-in truing device.

Dressing a wheel refers to the removal of metal particles embedded in the spaces between the abrasive grains or the fracturing and tearing away of grains dulled by abrasive action. In both cases the effect is to "sharpen" the wheel, that is, improve its cutting action. There are many types of wheel dressers; most are inexpensive, simple, and hand or machine operated. Essentially, all they do is turn with the wheel, inclined at a small angle, effecting a wiping or shearing across the wheel face. Crush dressing uses a wheel with a desired grain pattern cut into its face. The wheel is applied to the abrasive wheel and the crush wheel's high points crush the abrasive grains it comes in contact with, thus leaving a duplicate of the crush wheel face on the grinding wheel.

Tool Classification and Identification

SYSTEMS FOR IDENTIFYING TOOLS. There are three types of systems commonly used to identify tools:

1. **Numeric system** which uses straight number sequences or special number classifications.
2. **Alphabetic system** which uses straight letter sequences or special combinations (called **mnemonics**) suggestive of the tool's name or application.
3. **Combination numeric and alphabetic system.**

There are many variations of each system. It is good practice to use the same system throughout the plant for identifying jigs, fixtures, gages, and instruments.

NUMERIC SYSTEMS. There are two basic numeric systems used for tool identification: straight numeric and classified. In some cases the choice of a system depends on the type of identification system used by other departments, such as accounting and purchasing.

Straight Numeric System. In this system all special tools are numbered consecutively. An **index** is kept of these numbers containing information as to the nature of the tool and the operation for which it was designed. The usual drills, taps, and reamers are not given numbers but are specified by name. A serious **disadvantage** of this system is that it is rarely practical to store tools according to their numbers; often two consecutive numbers represent tools of a very different nature and size.

In some plants tool numbers are placed on working drawings. The worker has only to list the numbers of the tools he needs and present the list to the toolroom. Since the tools are often stored according to type, the crib attendant usually must refer to the index to determine the nature of the tools. After a period of time, however, the attendant will become familiar with the more commonly used items and reference to the index will be less frequent. In such systems it is absolutely essential that all working drawings be kept as up to date as possible.

Classified Numeric System. In this system each general class of tools is assigned a number, usually a single digit. Every tool in a particular class has,

---

Key to General Classes of Tools

1. Cutting tools	All tools that work by cutting off material, except chisels and blanking tools
2. Measuring devices	All gages and instruments of precision
3. Jigs and fixtures	All tools used for duplicating work such as jigs, fixtures, etc.
4. Impact tools	All tools that work by impact, including chisels, hammers, etc.
5. Wrenches	All tools that work by causing rotation
6. Holding tools	Clamps of all kinds, mandrels, nuts, dogs, etc.
7. Fire tools	All tools that are used for melting, heating, welding, etc.
8. Transportation tools	All tools that are used in moving materials
9. Miscellaneous	All tools not otherwise classified

Fig. 15. An example of a general tool classification system.

as its first number, this key digit (Fig. 15). Each general class may then be subdivided indefinitely. Typically the key number is followed by a decimal point and the numbers signifying the subdivisions are written as decimals, resembling in some respects the Dewey decimal system used to classify books. An illustration of subdivisions under the cutting tool classification is shown in Fig. 16.

1.	Cutting tools
1.1	Cutting tools; drills
1.11	Cutting tools; drills, twist
1.112	Cutting tools; drills, twist; taper shank
1.1123	Cutting tools; drills, twist; taper shank; high speed
1.2	Cutting tools; taps;
1.21	Cutting tools; taps; U. S. standard;
1.212	Cutting tools; taps; U. S. standard; machine shank
1.2123	Cutting tools; taps; U. S. standard; machine shank; high speed

---

Fig. 16. An example of subdivisions under the general tool classification.

Simply by adding digits to the key number, it is possible to subdivide the classification to as small a detail as desired. The most obvious shortcoming of this system is that no category can contain more than ten variables. Unfortunately this is not enough to take care of all the classes of tools, in some categories. For example, cutting tools (1.) would normally include drills, taps, reamers, counterbores, countersinks, milling cutters, hack saws, threading dies, lathe tools, broaches, hobs, and many more. Another shortcoming of this system is the difficulty of remembering the identifying numbers especially when they contain more than six digits. Errors in transcribing numbers on record sheets and drawings increase as the number of digits in any number increases.

LETTER SYSTEMS. The two basic types of letter systems are: straight letter and mnemonic. Although straight letter systems are used, letters are not as natural or logical to arrange in consecutive order as are numbers. The strength of a straight letter system is the fact that for each main classification twenty-six variables are possible, almost three times as much as is possible with a numeric system. Experience has shown however that letter combinations are more difficult to remember than number combinations of comparable length unless some particular order or association is suggested by the letters.

Mnemonic Systems. These use combinations of letters in such a way as to suggest the tool being described. Thus a memory (mnemonic) factor is introduced in the classification itself. To classify a tool in a mnemonic system, each general tool classification is first assigned a letter identifying the class: for example, A for abrading tools; B for blanking tools; M, measuring devices, and so on. If more than one classification begins with the same letter use a letter denoting some distinctive sound or standard letter symbol closely associated with the tool class. Next, make up a key sheet with subdivisions for each class. To do this write the letter and the tool class it represents at the top of a sheet of paper. On this sheet list all major subdivisions of the general classification, assigning each subdivision a letter based on the initial letter, or some other identifiable letter of the tool name. The subdivision sheet will resemble Fig. 17.

M—Measuring Devices
All gages and instruments of precision

M		M	
	A—		N—Indicators
	B—Bevels		P—Protractors
	C—		Q—Squares
	D—Dividers		R—Rules
	E—End measuring rods		S—Scale weighing
	F—Reference disks		T—Timing device
	G—Gages		U—Plumb bobs
	H—Pressure gages		V—Verniers
	J—		W—
	K—Electrical and electronic		X—Heat measuring
	L—Levels		Y—
	M—Meters		Z—Miscellaneous

Fig. 17. First stage of mnemonic classification of measuring devices.

Further subdivisions can then be made by using similar sheets to record each main sub-category. Letters O and I are omitted because they are likely to be confused with figures.

As an example, suppose a 5–5½-in., external, adjustable, limit gage of the caliper type is to be classified. On the group head sheet M, Fig. 17, under G appears the subdivision "Gages." A gage head sheet is made out and on this sheet, under C, place "Caliper gages." Next the head sheet for caliper gages (MGC) is made out. Adding the letter X for "External gages" gives MGCX. On another sheet list the different types of external caliper gages. Letter L is used for limit gages. On the MGCXL sheet under J enter adjustable gages. This classifies the particular gage in question to its smallest variable. The complete symbol for a 5–5½-in., external, adjustable, limit gage is MGCXLJ 5–5½ in. The steps are indicated in the accompanying breakdown.

MG   A—
        B—
        *C—Caliper
        D—Depth

MGC   A—
      B—
      C—Combined external
      D—
      *X—External

MGCX A—
      B—Bar type
      C—Type
      D—
      *L—Limit

MGCXL A—
      B—
      C—
      D—Double-end
      *J—Adjustable

In a like manner any tool may be classified. Letter Z of each head sheet is left for special tools of that division which cannot be otherwise classified.

TOOL MARKING METHODS. The tool identification symbol should be permanently and clearly marked in a conspicuous place on the tool. Cutting tools designated by size should have tool size added to symbol. Special tools for a particular manufactured part are often stamped with that part number. If the tool is soft, or has a soft spot, the symbol should be stamped with a die. If the tool is hard or delicate, so that it might be injured by stamping, the symbol should be etched. **Electrical, mechanical,** or **chemical etching** devices will pay for themselves in a short time where much tool marking is done. Size of characters should be large enough to be easily read. One-eighth to one-quarter inch high is satisfactory. Small characters cause loss of time in identifying the tool in the hands of the operator, or when replacing in storage. Characters on cutting tools, such as drills, reamers, counterbores, and the like, should be such that they can be read in any light in an ordinary shop. Figures stamped on by manufacturers are sometimes difficult to distinguish at the machine requiring the tool to be carried to a strong light before the symbol can be read. To accommodate large characters it is usually necessary to grind flats upon shanks of round tools. These flats serve the double purpose of permitting the use of satisfactory characters and also of indicating instantly the location of the tool size. Both features conserve time where it is necessary to verify size of tools prior to use. Very small tools may be kept in paper or plastic envelopes on which their number is printed.

Large plants engaged in mass production prefer not to stamp part numbers on small tools because the tools may be used for a number of different parts and confusion would result.

Tool Life and Metal Cutting

METHODS OF SPECIFYING TOOL LIFE. Tool life is generally defined as the **time between tool resharpenings** or **tool replacements**. Armarego and Brown (Machining of Metals) state "A tool that no longer performs the desired function is said to have failed and hence reached the end of its useful life. At such an endpoint the tool is not necessarily unable to cut the workpiece but is merely unsatisfactory for the purpose required. The tool may therefore be resharpened and used again, it may be used on less restrictive machining operations, or it may be disposed of." The most common methods of specifying tool life are as follows:

1. Elapsed time of machine operation. The tool may actually be cutting only during a portion of the machine cycle.
2. Actual cutting time, or the time the tool is in contact with the workpiece. This is the most common method of defining tool life.
3. Volume of material removed.
4. Number of pieces produced. This is common in shop parlance, but finds very little application in tool life testing.
5. "Taylor" speed, or equivalent cutting speed. This refers to the cutting speed obtained under a given set of conditions for a given cutting time, usually 60 minutes.
6. Relative cutting speed. This is a relative tool life based on a standard such as developed in item 5 above. It is used to rate both tools and workpiece materials.

DEFINING TOOL FAILURE. Tool failure is defined by characteristics of either the tool or the workpiece:

1. **Flank wear** or **clearance face wear** on the cutting tool (experienced with either roughing or finishing cuts). This is a common method of expressing tool failure.
2. **Crater wear** or **rake face wear** on the cutting tool (usually experienced with roughing operations). This is caused by chips impinging on the face behind the cutting edge. The result is a hole or crater in the tool face.
3. **Total tool destruction.** This is the point at which the tool ceases to cut or the wear rate tends toward infinity. It is sometimes a criterion for evaluating roughing performance when using high-speed steel tools.
4. **Tool point fracture.** This occurs when the cutting force is too large or the tool is applied too suddenly. The more brittle the tool material, the more likely such failure will occur. Intermittent cutting, high frequency vibration, or chatter can accelerate this type of fracture. Negative rake angles have been used largely to provide stronger tool points for the more brittle cutting tool materials.
5. **Chipping of the cutting tool.** In this case, fine pieces of the tool material break off along the cutting edge.
6. **Formation of cracks,** usually at right angles and next to the cutting edge, on the face or flank of the tool. This often results from the excessive heat generated by carbide tools operating at excessive speeds.
7. **Temperature failures at the cutting edge.** As high temperatures develop, the tool becomes too soft, loses its strength, stops functioning properly, and may shear off a part of the cutting edge. Such failure occurs quite rapidly at very high cutting speeds.
8. **Degradation of surface finish on the workpiece** below some specified limit. This criterion is difficult to employ because surface finish often does not vary uniformly with tool wear. This is usually a combination of excessive crater wear, flank wear, or chipping.
9. **Inability to maintain workpiece dimensions** as cutting progresses due to increased tool wear.
10. **Decrease in feed rate,** evidenced in fixed force operations such as band sawing, air-feed drilling, and various belt grinding and snagging operations.
11. **Increase in power requirements** due to tool dulling.

The first two are manifestations of wear normally encountered. The rest represent failures due to faulty use and/or faulty tools.

Figure 18 shows how **progressive tool wear** occurs. Part (a) illustrates the tool at the beginning of a cut with a sharp tool. Note the start of a built-up edge. Part (b) shows the tool, after some length of time, rounded at the edge and with a crater beginning to form. The machined surface indicates the wear through a poorer surface finish. Part (c) shows serious flank and crater wear, the workpiece finish is very poor, and the tool is about to fail.

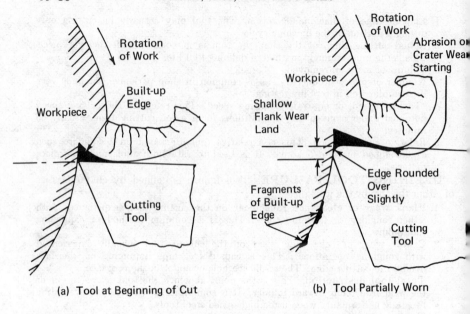

(a) Tool at Beginning of Cut

(b) Tool Partially Worn

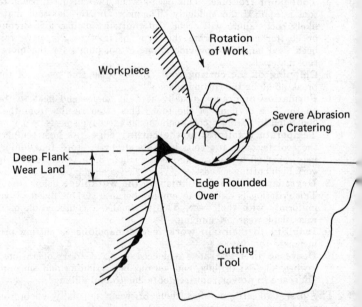

(c) Tool Badly Worn and Approaching Failure

Fig. 18. Progressive wear in a cutting tool.

Normal tool wear generally may be caused by:

1. The **gouging action** of the hard particles (chips) in the metal being cut. The rate of wear depends on the number, size, and the relative hardness of the particles and the tool.
2. The formation and subsequent breaking of **instantaneous welds** between the chip and the tool.

Welds formed during the machining operation may be either pressure welds or temperature welds.

At low cutting speeds, the surface temperature is low and the welds are all of the pressure type, but as the speed is increased, some temperature welds occur, and finally all the welds are of the temperature type and the built-up edge practically disappears. The tendency to form welds of either type decreases with increased cutting speed because of the time required for plastic flow to occur even on an atomic scale.

Rake face wear or crater formation may be encountered with high-speed steel, cast nonferrous metal or sintered-carbide tools. A cup develops just behind the cutting edge (Fig. 18b and c) which increases in size and gradually approaches the cutting edge from behind until the edge breaks off resulting in sudden tool failure. The wear is approximately proportional to the pressure of the tool, except at the tip of the tool which may be protected by the built-up edge formation.

Clearance wear is due to abrasion on the flank below the cutting edge and it is customary for it to appear when turning abrasive materials of low ductility (such as cast iron) with carbon steel, high-speed steel, cast nonferrous metal, or sintered-carbide tools; also when turning steel with carbon-steel tools. Clearance wear increases uniformly with the duration of operation. It is greatly affected by variation in the rigidity of workpiece and tool.

TYPES OF CHIP FORMATION. Chips are commonly considered as being of three types: discontinuous, continuous, and continuous with build-up. The three basic types are found in all kinds of machining operations, the type of chip formed on any particular occasion depending on work material, cutting speed, tool shape, depth of cut, lubricant, and so on.

The **discontinuous type** or segmental chip is formed when metal is removed intermittently in separate pieces. It is often encountered with brittle materials, such as cast iron. Generally, there is some deformation before shearing occurs but in its extreme form the metal hardly changes its shape at all. The resultant surface finish is poor, the power requirements comparatively high, and the tool wear excessive.

The **continuous type** chip is removed in a continuous ribbon. It is typical of ductile materials, particularly when cutting at high speeds and is the most efficient type of cutting, both for minimum power required and for superior finish. It is therefore preferred despite the difficulties of handling (or breaking) the long ribbons of metal that are cut from the workpiece. Chips breakers on the cutting tools aid in breaking up the chip for easier disposal.

The **continuous chip with build-up** is similar to the pure continuous chip, except that a hard metal builds up on the face of the tool immediately behind the cutting edge and acts as a very rough cutting tool, usually causing a deterioration in the surface finish of the workpiece.

The built-up edge is caused by temperature welding due to friction between the chip and the tool. The built-up material consists of workpiece particles that have been severely deformed and hardened. When the tool is actually cutting, however, the built-up edge becomes so hot that it softens and an extremely rapid cycle of building up and breaking away occurs.

GENERAL CUTTING OPERATIONS. Material is removed from a workpiece by either a roughing cut or a finishing cut. When a workpiece is roughed out, it is fairly near the shape and size required, but enough metal has been left on the surface to finish to the exact size.

Workpiece material such as steel bars, forgings, castings, etc., should be obtained in a size and shape convenient for machining. Ideally, the workpiece should be completed with one roughing and one finishing cut. When only a comparatively small amount of material is to be removed, a finishing cut may be all that is required.

The greater part of the excess material should be removed with a reasonably heavy **roughing cut.** The machinist's task is to remove excess stock as fast as he can without leaving a surface too torn and rough and without warping the workpiece.

The purpose of the **finishing cut** is to make the work smooth and accurate. It is a much finer cut involving the removal of comparatively little metal; a sharp tool and a high degree of accuracy in measurement is generally required. Whether roughing or finishing, the machine must be set for the given job. Consideration must be given to the size and shape of the workpiece, the material, the kind of tool used, and the nature of the cut to be made.

In operations where the workpiece is stationary, **cutting speed** refers to the number of feet that a point on the circumference of the rotating tool travels in one minute. The **feed** is the distance that the work advances into the rotating tool (or vice versa) in one minute, usually measured in inches per minute. The term "cut" refers to the **depth of cut.**

Suppose a cylinder of machine steel 2 inches in diameter is put in a lathe and a cut made reducing the diameter to $1\frac{7}{8}$ inches. Regardless of the speed or feed, the depth of the cut is 1/16 inch.

Cutting Speed and Feed Calculations. Cutting speed has previously been defined as the rate at which a point on the circumference of the workpiece or the cutting edge of the tool travels. When a multiple point cutter is turning and the workpiece is fed into the cutter, such as in milling, the circumference is the extreme path of the tool. The diameter of a workpiece or cutting tool is usually specified, however, not the circumference.

Since cutting speed is expressed in feet per minute (fpm) and most diameters are expressed in inches, it is necessary to convert to feet and multiply by π to obtain the actual speed.

$$\text{Cutting speed (fpm)} = \frac{\pi \times \text{Diameter (inches)} \times \text{Spindle speed (RPM)}}{12}$$

$$v = \frac{\pi \, dN}{12} \text{ feet per minute}$$

Since $\pi/12$ is close to 1/4, a good approximation of the cutting speed could be found from

$$v = \frac{dN}{4} \text{ feet per minute}$$

The **feed rate** (F) in inches per minute is the product of the feed in inches per revolution (f) multiplied by the number of revolutions per minute:

$$F = f \times N \text{ inches per minute}$$

Optimum Machining Practices. At slow cutting speeds, tool forces are markedly higher and tools with a low tensile strength, such as ceramics, are likely to chip or crack. At higher cutting speeds or with harder workpieces, the feed and the ratio of feed to depth of cut becomes more important. Deeper cuts at lighter feed are better than lighter cuts at higher feeds. The feed rate for each job should be considered separately for each production process and must often be established by trial.

Depth of cut usually depends on the amount of stock to be removed. This too must be determined either by established machine standards or by trial.

Tests by A. B. Albrecht of the Monarch Machine Tool Company provided the basis for the cutting relationships shown in Fig. 19.

Cook (Manufacturing Analysis) conducted experiments which indicated the following:

    a. Tool forces are only slightly affected as the cutting speed is increased.
    b. Horsepower consumed during cutting increases as the cutting speed is increased.
    c. Tool forces increase as the depth of cut is increased.
    d. Horsepower consumed increases as the rate of feed is increased.
    e. Tool forces increase as the tool temperature increases.
    f. Tool forces increase as the rate of feed is increased.

TOOL-LIFE TESTS. Since many metalworking companies interested in tool economy carry on tests to determine tool life, the American National Standards Institute has developed standards of tool-life tests for evaluating the machinability of single-point cutting tools, cutting fluids, or materials cut (Life Tests for Single-point Tools of Sintered Carbide—ANSI B94.38–1956 (R1971) and Life Tests of Single-point Tools Made of Materials Other Than Sintered Carbides—ANSI B94.34–1946.

The American Society of Mechanical Engineers has published the following formula connecting tool life, cutting speed, and other variables (ASME Manual on Cutting of Metals):

$$V = \frac{C}{t^{n_1} L^{n_2} t M^{n_3}}$$

where
    V = cutting speed, feet per minute (fpm)
    t = chip thickness, inches
    L = length of tool engaged in cutting, inches
    M = tool life
    C = constant
    n_1, n_2, and n_3 = experimental indices (see Fig. 20)

Although nomographs can be developed for finding depth of cut, feed, tool material, work material, cutting speed, and tool life, the results are very approximate. Further, it is difficult to make allowance for the effects of different cutting fluids, tool shapes, and metallurgical structures, all of which may exert a considerable effect.

The two general methods for obtaining machinability ratings based on tool life are:

    1. To obtain tool life when cutting under standardized conditions at a constant cutting speed, as in turning cylindrical test bars.
    2. To obtain tool life when cutting at a uniformly increasing or decreasing speed, as in facing on a taper.

The formula for expressing the relation between cutting speed and tool life between grinds for a given tool, material, feed, and depth of cut (method 1, above) is as follows.

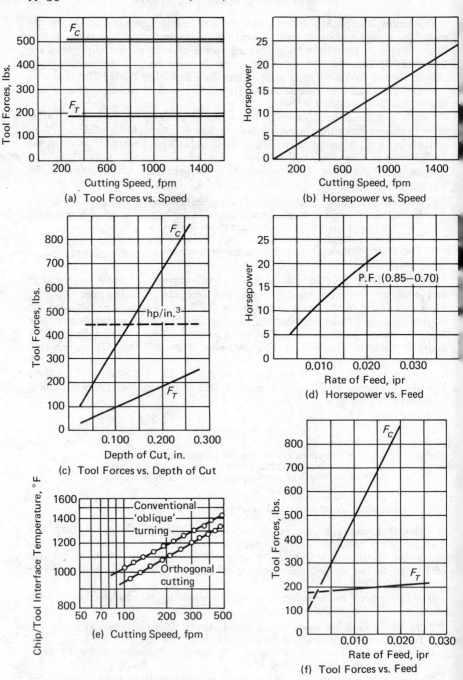

Fig. 19. Factors influencing tool forces, feeds, cutting speeds, and power requirements.

Tool Material	Work Material	n_1	n_2	n_3	
				$t > 0.015$ in.	$t \leq 0.015$ in.
Carbon steel	Steel	0.67	7.2	0.20	0.075
	Cast iron	0.43	7.2	0.075	0.075
High-speed steel	Steel	0.67	2.3	0.125	0.10
	Cast iron	0.43	2.3	0.10	0.10
Tungsten carbide	Steel	0.67	1.0	0.16	0.13
	Cast iron	0.43	1.0	0.13	0.13

Fig. 20.   Constants for A.S.M.E. tool life formula.

$$VT^n = C$$

where V = cutting speed (sfpm)
T = tool life or duration of cut between grinds (minutes)
C = a constant, depending on conditions, representing the cutting speed for a tool life of 1 minute
n = the slope of the tool-life straight line on log-log paper

The Society of Manufacturing Engineers (Fundamentals of Tool Design, Wilson, ed.) suggests that the tool life obtained at a given cutting speed is influenced by the dimensions of the cut. The general empirical relation between the cutting speed for a chosen tool life (say, 60 minutes) and the feed and depth of cut is recognized to be ordinarily of the following form:

$$V = \frac{C}{d^x f^y}$$

where V = equivalent cutting speed (cutting speed for a given tool life, surface feet per minute)
C = a constant whose value depends on other machine variables and work material variables
f = feed per revolution (or feed per tooth)
d = depth of cut, inches
x and y = constants whose values depend on the workpiece material

In practice, average values for the exponents are, for machining of steel, $x = 0.14$ and $y = 0.42$; for machining of cast iron, $x = 0.10$ and $y = 0.30$. Thus, the equation above becomes

For steel: $V = \dfrac{C}{d^{0.14} f^{0.42}}$ feet per minute

For cast iron: $V = \dfrac{C}{d^{0.10} f^{0.30}}$ feet per minute

According to the S.M.E., this equation emphasizes two very important general facts about the machining of metals:

1. As feed or depth is increased, the cutting speed must be decreased to keep the tool life constant.

2. However, when this is done, the amount of metal removed by the tool during the same given life is considerably increased. This is especially true of an increase in the depth of cut, because of the very low value of the exponent of d.

The general rule is: the combination of deep cuts and high rates of feed with a low cutting speed allows a large amount of metal to be removed during a given tool life.

Deep cuts and moderate feeds are desirable for long tool life for a given amount of metal removal. As might be expected, the length of cutting edge in action has less effect on tool life than the undeformed chip thickness or feed.

Equations on page 35 can be used for direct calculation of suitable cutting speeds for machining with various feeds and depth of cut. Values for C are given in Fig. 21 and values for $d^{0.14}$, $f^{0.42}$, and $d^{0.10}$, $f^{0.30}$, in Fig. 22. Cutting

	Machining Steel with Carbide Tools			Machining Cast Iron with Carbide Tools, Avg.	
Brinell	High Grade	Medium Grade	Older Types	Nodular Iron	Flake Graphite
100	161	122	65	—	18.3
125	120	91	48	—	15.4
150	95	72	38	—	12.2
175	81	61	32	17	9.6
200	67	50	27	9.6	7.7
225	56	43	23	6.9	6.1
250	50	37	20	5.4	4.6
275	43	33	17	5.0	3.5
300	39	29	16	5.0	

Kronenberg. Tool Engineer, Vol. 8

Fig. 21. **Values of cutting-speed constant** C **(60-minute tool life).**

speeds, V, can thus be arrived at by simple multiplication and division based on the previous equations.

Example 1: Carbon steel of 200 Brinell hardness is to be machined using a depth of cut of 0.250 in., a feed of 0.025 ipr, and a high-grade carbide tool. The desired tool life is 60 minutes. What cutting speed should be used?

Solution: From Fig. 21, $C = 67$. From Fig. 22, $d^{0.14} = 0.824$ and $f^{0.42} = 0.213$. Therefore

$$V = \frac{67}{0.824 \times 0.213} = 381 \text{ sfpm}$$

The size of the cut must be varied (light, intermediate, and heavy) to give specific tests under commercial conditions attending the use of the tool. Feed is represented in inches per revolution of work or cutter. Suggested cuts are:

1. 0.005–0.010-in. depth of cut and 0.002-in. feed.
2. 0.100-in. depth of cut by 0.0125-in. feed.
3. 1/8-in. depth of cut by 0.020-in. feed.
4. 3/16-in. depth of cut with 0.005-in. or 0.010-in. feed.
5. Maximum of 1/4-in. depth of cut with 0.03-in. to 0.05-in. feed.

If the edge of the tool is chamfered to give an appreciable approach angle, the chip thickness can be decreased for the same feed and depth of cut, and hence for the same rate of metal removal. Alternatively, the feed and hence the removal rate can be increased for the same depth of cut and chip thickness. A greater length of cutting edge comes into action when an appreciable approach angle is employed and hence more power may be required, even for the same rate of metal removal.

Machining Steel				Machining Cast Iron			
Depth of Cut d, in.	$d^{0.14}$	Feed f, ipr	$f^{0.42}$	Depth of Cut d, in.	$d^{0.10}$	Feed f, ipr	$f^{0.30}$
0.010	0.525	0.001	0.055	0.010	0.630	0.001	0.125
0.020	0.575	0.002	0.072	0.020	0.675	0.002	0.155
0.040	0.635	0.004	0.100	0.040	0.725	0.004	0.190
0.060	0.675	0.006	0.125	0.060	0.755	0.006	0.216
0.080	0.700	0.008	0.130	0.080	0.775	0.008	0.235
0.100	0.725	0.010	0.145	0.100	0.795	0.010	0.250
0.140	0.760	0.014	0.165	0.140	0.821	0.014	0.277
0.180	0.785	0.018	0.184	0.180	0.843	0.018	0.330
0.200	0.800	0.020	0.195	0.200	0.850	0.020	0.308
0.250	0.824	0.025	0.213	0.250	0.870	0.025	0.330
0.300	0.845	0.030	0.230	0.300	0.886	0.030	0.346
0.350	0.862	0.035	0.242	0.350	0.900	0.035	0.365
0.400	0.878	0.040	0.256	0.400	0.916	0.040	0.380
0.450	0.894	0.045	0.270	0.450	0.923	0.045	0.394
0.500	0.905	0.050	0.284	0.500	0.933	0.050	0.405
0.750	0.960	0.075	0.335	0.750	0.971	0.075	0.460
1.000	1.000	0.100	0.380	1.000	1.000	0.100	0.500

Kronenberg. Tool Engineer, Vol. 8

Fig. 22. Numerical values for $d^{0.14}$ and $f^{0.42}$ (steel) and for $d^{0.10}$ and $f^{0.30}$ (cast iron).

ECONOMIC FACTORS IN TOOLING. Manufacturing is composed of many functions, each contributing either directly or indirectly to the cost of production. An analysis of the tooling costs listed below would reveal those factors which can be varied to decrease costs.

1. **Idle and loading costs** reflecting noncutting time of normal production. In job shops such time may consume 40 to 80 percent of the total cycle time. Automatic loading and unloading can reduce this time and serve to keep the operator at a high level of productivity.

2. **Cutting costs** reflecting actual tool cutting times. On highly repetitive machining operations cutting time represents a major portion of the total cycle time. Costs may be decreased by increasing the cutting speed.

3. **Tool changing costs** reflect the time it takes to remove, replace, and reset the tool and put the machine back into production. Regrinding or sharpening time is included if the operator does his own grinding.

4. **Tool grinding costs** include depreciation of the original tool cost as well as the cost of labor and grinding equipment.

COST RELATIONSHIPS. The four cost factors listed above are related as shown in Fig. 23. The terms shown in the diagram expressing the cost relationships are defined as follows.

$K_1 =$ Direct labor rate $+$ overhead rate (in dollars per minute). Includes depreciation, power, insurance, and maintenance.

$K_2 =$ Tool cost per grind. Includes tool depreciation and regrinding cost (in dollars per tool). This is dependent upon the tool.

$L =$ Length of cut (in inches).

$D =$ Diameter of cut (in inches).

$V =$ Cutting speed (in surface feet per minute).

$f =$ Feed per revolution (in inches).

$Vt^n = C$, cutting-speed–tool-life relation.

$T =$ Tool life (in minutes).

$t =$ Tool-changing time (in minutes).

$n =$ Slope of cutting speed vs. tool-life curve plotted on log-log paper.

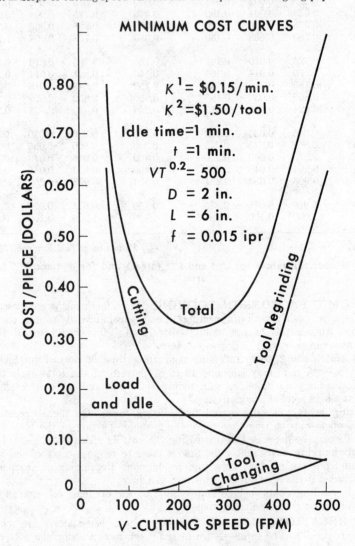

Fig. 23. Production costs.

Cost Formulas.

1. Idle cost per piece = $K_1 \times$ idle time per piece (includes loading, unloading, tool approach, etc.).

2. Cutting cost per piece = $K_1 \times$ cutting time per piece = $K_1 \dfrac{\pi DL}{12fV}$

3. Tool-changing cost per piece = $K_1 \times$ tool failures per piece $\times$ tool-changing time, or $K_1 \dfrac{\pi DLV^{(1/n-1)}}{12fC^{1/n}} \times t$

4. Tool-regrinding cost = $K_2 \times$ tool failures per piece = $K_2 \dfrac{\pi DLV^{(1/n-1)}}{12fC^{1/n}}$

5. Total cost per piece = $K_1 \times$ idle time $+ K_1 \dfrac{\pi DL}{12fV} + K_1 \dfrac{\pi DLV^{(1/n-1)}}{12fC^{1/n}} \times t + K_2 \dfrac{\pi DLV^{(1/n-1)}}{12fC^{1/n}}$

6. Cutting speed for minimum cost per piece =

$$V = \frac{C}{\left[\left(\dfrac{1}{n} - 1\right)\left(\dfrac{K_1 t + K_2}{K_1}\right)\right]^n}$$

or

$$V = C \left(\frac{n}{1-n}\right)^n \left(\frac{\text{Cost of labor and overhead per minute}}{\text{Cost of changing and regrinding a tool}}\right)^n$$

7. Tool life for minimum cost per piece =

$$T = \left(\frac{1}{n} - 1\right)\left(\frac{K_1 t + K_2}{K_1}\right)$$

8. If the original cost and regrinding cost of tools are negligible, or of less importance than the maximum production rate, then K_2 will be negligible, and

$$\text{Tool life} = \left(\frac{1}{n} - 1\right) t = T_{\text{max. prod.}}$$

and

$$\text{Cutting speed} = V = \frac{C}{\left[\left(\dfrac{1}{n} - 1\right) t\right]^n} = \frac{C}{T^n_{\text{max. prod.}}}$$

Jigs and Fixtures

FUNCTIONS. Mass production methods demand a fast, easy method of positioning work for accurate, reproducible drilling, milling, and boring.

A **jig** is a device which guides the cutting tool, and holds the workpiece.

A **fixture** is a holding device which supports the workpiece in a fixed orientation with respect to the tool.

Jigs and fixtures may be large (airplane fuselages are built on "picture frame" fixtures) or very small (as in watchmaking). Their use is limited only by job requirements and the imagination of the designer.

Proper jig and fixture design may be summarized by the following list of "should-be" characteristics: dimensionally accurate, rugged, adaptable, economical, salvageable, attractive to the user, accessible, simple, safe, foolproof, portable, and easy to construct.

LOCATING WORKPIECES. Each jig or fixture is composed of a number of elements of which **clamps** and **locators** are the most important to consider in connection with locating the workpiece.

Degrees of Freedom. Locating elements place the workpiece in essentially the same position cycle after cycle. In this sense, the locator provides a reference point from which all sizing or spacing may be accomplished. A free body may have three degrees of freedom of translation and three degrees of freedom of rotation. It is the purpose of locators to restrict these six degrees of freedom in order to give points of reference.

A workpiece would lose three degrees of freedom if placed and maintained on the locators lettered (A) shown in Fig. 24.

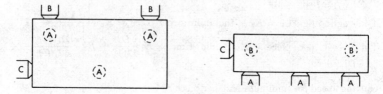

Fig. 24.   Placement of locators.

If the workpiece is then brought in contact with locators (B), two more degrees of freedom are restricted. The final degree of freedom is lost when the workpiece is brought in contact with locator (C). In the example shown in Fig. 24, the **reference surfaces** could be the surface resting on the (A) locators, the surface resting on the (B) locators, or the surface resting on the (C) locator. Thus, the placement of locators, as shown in Fig. 24, insures reproducible location of the workpiece each time the surfaces are brought in contact with the locators.

Influence of Workpiece Surface. The form of the locator selected is influenced to a large degree by the condition of the reference surface. **Finished surfaces** can be located on a plane rather than suspended on points. However, even if a high finish and true surface is available as a reference surface, there is danger of misalignment from minute chip particles or dirt lodging between adjacent planes. **Rough surfaces** are given as few points of contact as deemed necessary for stability of the part. Three points form a plane and the introduction of additional points may offer greater stability, but also introduce the risk of unevenness with the plane formed by three points. Identification of reference surfaces must be made prior to the selection of locators, since the condition of the surface is so influential in choosing locators.

TYPES OF LOCATORS. The type of locator selected for application is largely a function of the shape of the workpiece and more particularly the shape of the reference surface. Fig. 25 shows various types of edge and bottom locators commonly employed.

Edge locators must provide recognition for the problem of burrs on the workpiece and the presence of chips in the jig or fixture as the result of previous cutting operations. Misalignment resulting in spoilage can be the outgrowth of the jig or fixture designer's failing to provide this needed clearance for edge and bottom locators.

Pin locators are widely used where holes are used as reference points. In addition to the allowance both for burrs on the workpiece and for the presence of chips or dirt, consideration must be given to placing the workpiece over the locating pin.

When the reference surface is the outside diameter of a cylindrical part, the **V-locator** provides the most desirable means of locating. The 90° angle between faces on the V gives the best cylindrical surface locator. Care must be exercised when using the V-locator to insure that the direction of the force introduced by the tool bisects the included angle of the V-locator. This practice insures positive seating. For instance, the only center of the lateral surface of a cylindrical piece located by a V-locator is the one bisecting the included angle.

On occasion it is necessary to center a piece of other than cylindrical shape. In this situation a **centralizer** or **radial locator** is used to position the workpiece in some fixed relationship to the work.

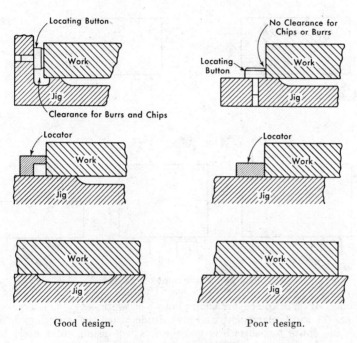

Good design.   Poor design.

Fig. 25. Application of edge and bottom locators.

CLAMPS AND CLAMPING. The purpose of clamping is twofold. First, it must hold the workpiece firmly against the locators provided, and second, it must resist all forces introduced by the operation. **Placement of the clamp** is extremely important in the design of jigs or fixtures. Care must be taken to apply the clamp in an area that is directly opposed by the locators. A poorly placed clamp can cause warping of the workpiece or introduce a turning moment that would tend to unseat the workpiece. Examples of desirable clamp locations are shown in Fig. 26.

The shaded area shown in Fig. 26(a) offers the correct site for applications of a vertical clamp. If the clamping force is applied outside this shaded area it would result in either part deformation or a tilting of part away from the base locators. The forces F_1 shown on Fig. 26(b) and (c) would introduce a turning moment, whereas if the clamping forces F_2 were introduced at the points designated, the part would be seated firmly on the locators. If a single clamp is used to secure the workpiece, its line of action must lie within the extreme points of contact, such as is shown at F_3 in Fig. 26(d). Fig. 26(e) again demonstrates the need for the application of the clamping force F_4 in such a manner so as to insure contact at the locator points. The use of a **single clamp,** such as shown in

Fig. 26(d), is not only the most economical, since there is but one clamp to tighten, but also the best method for insuring positive contact on all locators. When employing a clamping system similar to those shown in Fig. 26(b) and (c), the tightening of one side clamp may lift the workpiece from the locators on the adjacent side. If the first side clamp is tightened securely the second side clamp may never seat the workpiece against the opposing locators.

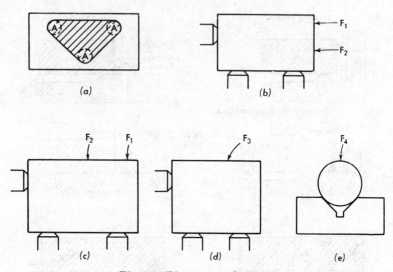

Fig. 26. Placement of clamps.

Influence of Workpiece on Clamping. The workpiece, to a large extent, determines the **form, placement,** and **magnitude** of the clamp employed. If, for instance, the entire top surface is machined, a side clamp must exert both a horizontal and vertical force in order to assure a firm holding action. Finished surfaces that must remain free of marks or scratches require a form of clamp that will not mar the finish. Workpieces of a fragile nature limit the magnitude of the clamping force to that which would not deform or warp the part. Rough surfaces permit the use of either **hold-down pins** that dig into the work or **serrated-faced clamps** that give more positive holding action. It is with these considerations for the workpiece that the jig and fixture designer must approach the problem of clamping or holding.

JIG AND FIXTURE DESIGN CONSIDERATIONS. In addition to consideration for locators and clamps employed in the design of jigs and fixtures, there is need for design consideration of a number of other factors for most efficient functioning.

Chip Clearance. Space must be provided for proper chip clearance in the design of jigs or fixtures. Chips are a constant threat to proper alignment of the workpiece housed in a jig or fixture. Chips also present a realistic threat to desired surface finish in a machining operation. The thorough designer must make adequate provision for the presence of chips when planning a jig or fixture. Chip clearance should be provided between the bottom of the bushing and the top of the workpiece, as shown in Fig. 27. This chip clearance at the top of the workpiece keeps the chips from packing up in the drill flutes and marring the surface of the hole drilled. As a rule of thumb, the upper chip clearance is equal to the drill diameter used. The space provided between the base of the workpiece and the jig by the presence of the base locators, as shown in Fig. 27, reduces the hazard of misalignment of the part as the result of a chip remaining in the jig after cleaning.

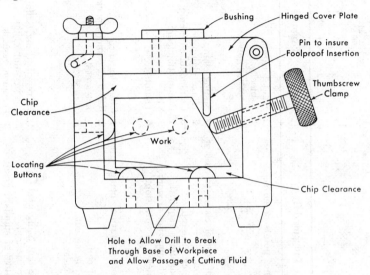

Fig. 27. Example of drill jig design.

Foolproof Operation. The pin shown on the underside of the jig hinged cover plate in Fig. 27 is an example of a device which is used to insure foolproof insertion of a workpiece in either a jig or fixture. Should the part be inserted in any other manner than that shown, it would be impossible to close the hinged cover plate. Thus, by warning the operator of improper referencing of the workpiece, costly mistakes in part loading can be avoided. Devices for foolproof workpiece insertion should always be considered in jig and fixture design.

Mechanical Forces on Workpieces. The forces introduced by the drilling setup shown in Fig. 27 would be twofold. There would be an **axial force** downward that would be transmitted through the part to the base locators which would seat the part firmly on the locators. A **torque** introduced by the cutting action of the drill would tend to rotate the part. This force would be transmitted through the part to the side locators since the workpiece is held firmly on the locators by the thumbscrew clamp. To insure positive clamping, an additional clamp would be utilized on the front side of the jig shown in Fig. 27. This clamp is not shown in order not to obscure the workpiece in the jig.

DESIGN CHECKLIST. A checklist for evaluating jig or fixture design was prepared by Tilles (American Machinist, vol. 95). This checklist (Fig. 28) provides for the orderly analysis of jig and fixture design and avoids the omission of pertinent details.

ECONOMICS OF JIGS AND FIXTURES. Roe (Mechanical Engineering, vol. 63) developed a group of simple formulas to be used in determining the economic results from the use of jigs and fixtures. He states that the eco-

Factor	Definition	Influenced By	Example
Location	Establishing desired relationship between work piece and jig or fixture.	1. Reference points or surfaces.	1. Sight location. 2. V-blocks. 3. Locating pins.
Clamping	Use of a mechanical device to maintain the relationship established above during the operation being performed.	1. Ratio of cost of clamp to cost of operation. 2. Frequency of setup. 3. Clamping force required. 4. Size and condition of work. 5. Convenience of operation.	1. Sliding clamp. 2. Hinge clamp. 3. Cam-actuated clamp. 4. Pneumatic clamp. 5. Vise. 6. Thumbscrew.
Chip control	Minimizing the adverse effects of chip and burr formation on the functioning of the jig or fixture.	1. Type of operation being performed. 2. Form of chip. 3. Material being cut. 4. Use of coolant.	1. Chip clearance. 2. Relief at corners, pockets, or contact areas. 3. Chip slides.
Positioning	Establishing desired relationship between jig or fixture and the cutting oil.	1. Machine tool being used. 2. Tolerance desired. 3. Availability of standard positioning devices. 4. Use of tool guides.	1. Universal tilting table. 2. Box or tumble jigs. 3. Trunnion mounts. 4. Index fixtures. 5. Bushings.

Standardization	Use of standardized components wherever possible.	1. Sizes commercially available. 2. Amount of tooling manufactured. 3. Type of part needed.	1. Bushings. 2. Hand knobs. 3. Locating pins. 4. Crank levers.
Ease of operation	Facility with which the operator is able to use the completed jig or fixture. Minimum requirements of time, effort, dexterity.	1. Positioning. 2. Clamping device selected. 3. Size and shape. 4. Location of manual components.	1. Quick-acting clamps or pneumatic devices where possible. 2. All levers within easy reach. 3. Quick location.
Safety	Absence of hazards in use of jigs or fixtures.	1. Clamping device selected. 2. Chip control. 3. Machine tool being used.	1. No protruding sharp edges. 2. Hands not required to operate jig or fixture while close to cutting tool. 3. Ample clearance for knuckles and fingers.
Cost	Initial investment required to design and build the jig or fixture.	1. Number of parts to be made. 2. Size and shape of part. 3. Tolerances desired. 4. Materials specified.	1. All components as simple as possible. 2. Standard parts used wherever possible. 3. Replaceable parts where subject to wear.

Fig. 28. Checklist of factors in jig and fixture design.

nomic problem involved in any situation centers around one or more of the following questions:

1. How many pieces must be run to pay for a fixture of given estimated cost to show a given estimated saving in direct labor cost per piece? For instance, how low a run will justify a fixture costing $400 to save 3 cents on direct labor cost of each piece?
2. How much may a fixture cost which will show a given estimated unit saving in direct labor cost on a given number of pieces? For instance, how much can be put into a fixture to "break even" on a run of 10,000 pieces, if the fixture can save 3 cents on direct labor cost of each piece?
3. How long will it take a proposed fixture, under given conditions, to pay for itself, carrying its fixed charges while so doing? For instance, how long will it take a fixture costing $400 to pay for itself if it saves 3 cents on direct labor cost per unit, production being at a given rate?

Questions 1, 2, and 3 assume that savings just balance the expense. There is another practical question:

4. What profit will be earned by a fixture of a given cost, for an estimated unit saving in direct labor cost and given output? For instance, what will be the profit on a $200 fixture if it will save in direct labor cost 3 cents a piece on 10,000 pieces?

These questions involve something more than simple arithmetic for an answer. While the credit items for the fixture depend mainly on the number of pieces machined, the debit items involve time and number of setups required, i.e., whether pieces are run off continuously or in a number of runs.

Formulas. Roe (Mechanical Engineering, Vol. 63) developed his formulas for jigs and fixtures from general equations for the efficiency of equipment which were formulated by the Materials Handling Division of the American Society of Mechanical Engineers. The original equations provide a basis for the economic analysis of industrial equipment, and for the determination of probable profit on any proposed installation for a given situation and cost of performance.

Roe modified the equations to make them applicable to tool equipment. He originally presented nine formulas which have since been reduced to four by leaving out factors for power cost, unamortized value of equipment displaced (less its scrap value), and savings through increased production, which are small values.

To be of practical value formulas should be as simple as possible and yet reflect essential conditions, and should be easily applied. The final equations meet these conditions for most fixtures. They take into account number of pieces manufactured, saving in unit labor cost, overhead on labor saved, the cost and frequency of setups, interest on investment, taxes, insurance, upkeep, and depreciation.

Roe's symbolization follows:

$$N = \text{Number of pieces manufactured per year}$$

(Note that N is the number of pieces manufactured *in a year*, not per run, except for the case of a single run of less than one year's duration.)

Debit Factors:

> A = Yearly percentage allowance for interest on investment. Either original cost or depreciated value of investment may be used.
> B = Yearly percentage allowance for such fixed charges as insurance, taxes, etc.
> C = Yearly percentage allowance for upkeep.

$1/H$ = Yearly percentage allowance for depreciation and obsolescence on the basis of uniform depreciation, where H is the number of years required for amortization of investment out of earnings.

I = Estimated cost of the equipment or fixture, i.e., cost installed and ready to run, including drafting and tool room time, material and tool room overhead, in dollars.

Y = Yearly cost of setups, in dollars. This value should include expense for taking down the apparatus and putting machine into normal condition.

Credit Factors:

S = Yearly total saving in direct cost of labor in dollars
  = Ns

where s = Savings in unit labor cost.

T = Yearly total savings in labor overhead, in dollars
  = St

where t = Percentage of overhead on the labor saved.

V = Yearly gross operating profit, in excess of fixed charges, in dollars.

To "break even," the yearly operating savings equal total fixed charges per year.

$$N = \frac{I\left(A + B + C + \dfrac{1}{H}\right) + Y}{s(1 + t)}$$

$$I = \frac{Ns(1 + t) - Y}{A + B + C + \dfrac{1}{H}}$$

$$V = (Ns)(1 + t) - Y - I\left(A + B + C + \frac{1}{H}\right)$$

$$H = \frac{I}{Ns(1 + t) - Y - I(A + B + C)}$$

Items A, B, and C, once settled upon, need change little. If the plant has the practice of requiring new equipment to pay for itself in a definite time H, say 2 years, depreciation $1/H$ may be added to the other carrying charges, making a single percentage factor for the term $(A + B + C + 1/H)$ which can be used until management deems that changed conditions require modification.

It is recommended that in authorizing expenditures for all fixtures and tools above some established minimum cost, an **estimate** be made of

1. Cost of the fixture.
2. Output of the fixture.
3. Profit or saving from it.

When it is put into operation, the actual results should be checked with these estimates.

Such a procedure will give a check on quality of tool designing. If tool costs are overrunning estimates and output and savings are falling short, the facts will be shown. If tool work is good, the management will know it, and have means for measuring the profit obtained.

The Tool Crib

LOCATION. There are three basic approaches to general tool storage:

1. **Central tool crib** system has a large control crib in which all factory tools are concentrated and from which they are issued.
2. **Subtool-crib** system provides a central, or head, tool crib used more in the capacity of a storeroom than as an issuing tool crib. Each department, under

this system, has a tool crib which carries only tools used by that department. When these tool cribs need additional tools, they requisition them from the main tool storeroom.

3. Flexible plan whereby a **movable tool crib** is provided with tools for the entire plant so that, upon signal or upon receipt of tool issue slips, it can go to the department needing tools and make deliveries from its shelves.

Tool storage determination is based on:

1. **Size and layout of factory.** Cost of tool distribution from a single central tool crib may be excessive in large factories.
2. **Allocation of space to departments.** One section of a factory is sometimes given over to assembly and testing, while other sections are used for fabrication of parts. Relatively few tools are needed in the former, while the latter may require the use of every kind of tool that the factory owns. Obviously, the cribs should be placed nearest to the departments that use the most tools.
3. **Nature of product.** Some products, such as those produced by the aerospace industry, lend themselves to complete fabrication, assembly, and testing in a single area. Many companies have one or more articles of this kind. In cases where a department functions independently of the rest of the factory, it may be advisable to have a tool crib solely for its use.
4. **Methods of operation.** Products manufactured determine methods of operation to follow. In factories where process methods are used, such as textile mills or cement plants, tool requirements are distinctly different from those of a highly developed machine shop. Where pre-set tooling is used with numerically controlled machines movable tool cribs have proven most advantageous.
5. **Skill of operators employed.** Highly-skilled operators take pride in good tools and are generally careful in their use. Machine attendants, however, are not always so careful, and provisions must be made accordingly.

In general, the location of a tool crib is governed by facilitation in receiving and checking out tools. Tool delivery, however, should not conflict with general transportation activities in the factory. Neither are tool cribs always put where they would be most convenient for the issuing of tools, especially where they would be located in the middle of a prime manufacturing area and thus interfere with the flow of work and with plant personnel.

Men selected for responsible operation of a tool crib system should be chosen with regard to a knowledge of, and familiarity with, tools, appreciation of responsibilities of the work, and suitable personal characteristics. Preferably they should be men with a good understanding of machine processes. A classification for such personnel is: foremen, tool inspectors, clerk, group attendants, laborers, and possibly apprentices.

LAYOUT OF TOOL CRIB. The layout of a tool crib depends upon the area allocated to it and the shape and relative dimensions of this area. Layouts are sometimes expedients, but often can be definitely planned to give the greatest efficiency in tool storing, issuing, receiving back, inspecting, reconditioning, replacing on shelves, and record-keeping.

Determining Space Requirements. General principles which aid in the proper allocation of space for each class of tools are:

1. The tools should not, as a general rule, be stored over 6 ft. high. The top sections of bins should be left empty in the beginning to take care of additional tools of each class which are sure to accumulate as time goes on.
2. Tool sizes are given in dimension, number, weight, etc. For example, twist drills are made in size increments of 1/64 in. Therefore, a rack for drills should

have a space for each 1/64 in. from smallest to largest size carried, provided 1/64-in. increments are used throughout the entire range.

3. Aisle space between bins should be ample for free passage of men and machines through them. Thirty inches for bin aisles and 36 in. for main aisles have been found suitable minimums.

It is evident that the space required varies with the number of tools to be handled, as well as the variety of each class of tool. Space assigned should be limited to that which is necessary. At the same time it should be ample to avoid the tendency to overcrowd the crib and thus hamper its effectiveness.

CHARGE-OUT SYSTEMS. It is important to keep records of the issues of tools because of the following factors:

1. Value of tools, which makes it imperative to prevent their loss and secure their return.
2. Need for their use on other jobs to keep production going, thus necessitating prompt replacement in the tool crib, or location in the shop if they have to be recovered for immediate transfer.
3. Keeping tools continually in good condition, by getting them back for prompt inspection and maintenance.
4. Replacing tools that are lost, broken, or damaged, by placing purchase or manufacturing requisitions at once, especially when replacement requires a long lead time.
5. Fixing responsibility for the possession and care of the tools to prevent carelessness on the part of workmen.
6. Maintaining a history of use to guide in future purchasing (as to vendor) and in determining tool life, replacement time cycles, and quantities to carry on hand.

There are a number of methods under which tools are charged out, each having features making it particularly useful under certain circumstances. These methods are:

1. Single-check system.
2. Double-check system.
3. Triplicate tool-slip systems.
4. Charge plate method.

Single-Check System. Under the single-check system the worker securing a tool from the tool crib, either in person or by delivery service, gives for it a brass check with his number stamped on it. This check is hung on a hook by the place where the tool is kept, or it can be hung on a board indexing the tools. If the tool is wanted, the toolkeeper knows who took it, but not when or for what job, or whether the check is on the right or wrong hook, or if some other worker used a fellow-worker's check, and there is no record of how many tools the worker may have withdrawn. This plan is used in only the simplest cases.

Double-Check System. The double-check system is similar to the single-check system and has the same characteristics and faults. Its only other feature is that the second check is hung on a hook under the worker's number and there it can be seen how many (but not which) tools he has secured.

Triplicate Tool-Slip System: Spring-Clip Boards. One variety of the triplicate tool-slip system makes use of spring-clip holders as record compartments, the tool slips being filed under the clips, as stores requisition slips are sometimes handled. When an employee calls or sends for a tool, he fills out a tool order (Fig. 29) in triplicate with one writing, using inexpensive carbon-backed forms —original white, duplicate yellow, triplicate pink. The operator writes in his clock number, date, and tool drawn, and signs his name. The bin number is entered by the crib attendant, and the tool is given to the workman with the yellow copy which he keeps in a slip-holder on his machine or bench. The white copy acknowledges receipt of the tool and is retained in the employee's compartment under the clip bearing his number. The pink copy is filed in the tool compartment under the clip for that particular tool, in a cabinet that may be closed and locked. Employee's records usually are filed on wall boards giving visibility to the slips of 60 men on each side, so that those accumulating too many tools are identified at a glance. During a rush, the attendant hands out tools and yellow slips rapidly but delays filing slips until the peak is over. This plan speeds up service and permits men to return to work.

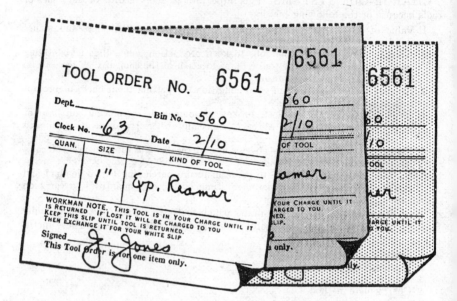

Fig. 29. Typical tool order made out in triplicate.

The tool crib attendant, under this method, can easily see what tools have been issued to each operator, and the exact location of every tool checked out. When the employee returns a tool he also returns his yellow copy of the tool order. If the tool is in good condition, the tool crib attendant removes the corresponding white slip, matching the serial number which appears on all three copies of each set, and the employee immediately destroys them. His responsibility for that particular tool is ended. Then, or later, the attendant removes the pink copy from the tool record compartment, thus indicating that the tool has been returned, but not destroying the record of the transaction. Pink copies are filed to provide valuable data for purposes described below. The tools need not be replaced in the bins at once, but may be held for further inspection or repair. Thus the accumulation of defective tools in the bins, which may occur under the single-check system, is avoided.

A broken or defective tool is accepted by the tool crib attendant only when accompanied by an **explanatory report** from the foreman of the employee returning it. The report may be made on the employee's yellow copy of the tool order or preferably on a report identifying the cause of the damage. If carelessness or inefficiency of the employee was responsible, a copy may be filed at the back of his compartment. Such breakage may be reduced by warning the small percentage of offending employees against continued carelessness, instructing them in tool use, or transferring them to other work. Occasionally the reports provide the basis for raising the working efficiency of entire departments.

When a defective tool is returned for repair, the report may serve as a **repair order**, indicating the department to which costs are to be charged. When a tool has to be scrapped, the report may provide the accounting department with the basis for charging it against the department responsible and crediting it to the tool investment account. If it must be replaced by a new tool, an explanatory report may be attached to the requisition, justifying the issuance of the new tool from stock.

Valuable data concerning tool economics, tool design, setups, etc., are gained from analysis of these reports. If tools are found to be inferior, defective, or unsuited to the work, reports may be filed in the tool compartments behind the pink slips. Evidence thus accumulated may indicate to the production and purchasing departments that certain brands or types of tools last longer than others, or function better under specific conditions.

A **tabbed tool inventory card**, placed at the back of each tool compartment, with the entire top visible, shows the name and description of the tool, its location, maximum and minimum quantities or reordering points, and purchase price or appraised or depreciated value, and provides columns for receipts, scrapping, and remaining balances, with dates. The quantity of each tool on hand is the number in the bin at any time plus the number in use as shown by the pink slips. Control of tool requisitions or purchases and inventory are thus readily carried on.

Tool activity, or the frequency with which tools are issued, is secured by accumulating pink copies of tool orders behind the inventory cards when tools are returned. Once a month the slips for each kind of tool are counted, the number is recorded on the back of the inventory card, and the slips are destroyed. After a few months, this information provides a factual basis for removing obsolete tools and reducing the inventory of slow-moving tools. Inventory reduction in average cribs frequently varies from 35 to 50 percent.

When tools are issued together as a kit, the transaction can be handled on one tool order. The combination is given the equivalent of a tool number, such as Kit or Setup No. 25. The individual tools in the kit are charged against that number by the attendant to maintain the accuracy of the tool records. The employee is given a standard list of tools in the setup for which he is responsible. When an employee requisitions tools from several cribs, his use of tools may be controlled through some one crib to which he is assigned, or through all of them. The former plan is advantageous if the employee leaves and is checked out, since the records of all tool withdrawals are in one place.

Three separate records are kept: a record card with the employee's name, filed by employee number, on which tool loss and breakage are recorded; the tool inventory control card, on which ordering, receiving, and balance-on-hand information is kept; and the parts cards, on which standard and special tools needed to make the part are listed. The tool inventory control card should always show a balance equal to the quantity of tools loaned, as recorded on slips, plus the number still in the crib.

In operation, the employee fills out a **tool loan order** in duplicate. The original is filed with his employee name card, the duplicate with the tool inventory control. Upon return of the tool he receives the original. The duplicate is pulled for later recording on a tool usage report.

The employee is responsible for all tools for which signed originals are found in his record pocket. If he should break a tool he shows it to his foreman and receives a broken or **damaged tool report** made out in duplicate. Both copies are turned in with the tool, posted to the employee's name card, and deducted from the tool inventory control. At the end of the month a **broken-tool distribution report** by departments and shifts is prepared for the management. Thus any inefficient departments and careless employees will be shown up by the records.

The part name record is kept in the tool crib as a listing of the standard and special tools, jigs, and fixtures required for making certain parts. When a part is to be made the planning department merely notifies the tool crib and all necessary tools are gathered so that production can proceed without delay.

Charge Plate Method. To save repeated filling out of standard information on tool issue slips, a charge plate system can be used to print the employee's name, number, department number and date much in the manner of the standard credit card systems. Plates for each tool carried are also made, from which bin labels and cards for a tool index system are printed. When tools are to be issued, the corresponding plates for employee and tool are taken out and run off on **two-section tool slips,** each section showing the full information from these plates. The employee section of the card has with it a perforated section for the employee to sign, acknowledging receipt of the tool. The employee portion of the card is filed in the employee index file behind his name or number, and the tool portion in the tool index behind the tool number, symbol, or name. Upon return of the tool the employee is given his receipt stub, the remainder of the card serving as a memorandum to remove the corresponding card from the tool index file, thus clearing the record. The used cards may be summarized to indicate tool activity as in other systems.

Another plan used provides plates for parts to be manufactured, giving data identifying the part, data on material for it, routing, and number of operations. These plates are used to print data on manufacturing orders, material requisitions, identification tags, labor tickets, tool notices, etc. The tool notices for the respective operations on a part are filed in job envelopes and go with the envelopes, through scheduling and production control, to the successive operating departments, in each of which the work dispatcher (or foreman) removes the tool notice—if tools are needed—for the operations to be performed there and sends them to the tool crib with work order card on which the tools required are listed and detailed. The tools are then charged out and supplied and, after the operation is done, are returned to the crib with a filled-out finished work order for identification. The tools are replaced in storage, the charges and notices are canceled, and the work order is forwarded to the production control department for recording.

Tool Inventories

DETERMINING TOOL QUANTITY REQUIREMENTS. Figures showing the number of individual tools in each group to be kept in the tool crib should be worked out from actual experience as to how many of each are generally used. **Standard tool lists** give valuable aid in this work. After these figures are determined, a study should be made of the most economical ordering units either by purchase or from manufacture within the factory: for example, on ⅝-in. machine taps, the best manufacturing quantity is about 100 and the most economical purchase unit is one gross or above. However, this quantity is too great to keep in the tool crib, because if the crib is allowed too many tools of any specific item, the tendency is to give out new tools instead of using the old ones until they are worn out. Therefore, any excess of tools should be kept in a separate central tool crib or in the regular storeroom and disbursed to the tool crib as needed to keep its stock to the specified quantity.

CHECKING THE INVENTORY. A complete inventory of every tool and appliance entrusted to the tool crib should be maintained. This inventory should be perpetual in character, additions to the tool-crib stock being promptly entered on it, and lost, broken, or discarded tools removed from it. The number of tools on inventory should at all times correspond with the sum of the number of tools and employees' checks, representing tools, in the tool crib. The inventory should be checked on a regular program by counting tools in the successive sections of the tool crib, in sequence as opportunity offers, the entire tool crib to be covered at intervals of from three to six months.

The methods for keeping a perpetual tool inventory do not differ in principle from those that apply to perpetual inventory of stores (see section on Materials Management).

Metrication

USE OF METRIC UNITS. Sizes and standards of tools, fasteners, materials and the like employed in American industry are based on the **English system** of inches and pounds. Within this system both fractional and decimal dimensions are in common use. Complete conversion to the metric system means not only conversion of inch units to equivalent metric units, but conversion of standard inch-size tools, fasteners, and materials to **standard metric-sizes.** It is this problem that will absorb the attention of tool engineers and machine designers for many years to come.

Metrication has been underway on a voluntary basis in many companies during recent years, especially where international trade is involved. Under the momentum of a government directed program it is expected that American industry will join the other major industrial nations in the general use of the SI (metric) units.

CAPITAL INVESTMENT ANALYSIS

CONTENTS

CAPITAL INVESTMENT ANALYSIS

Capital Budgeting

IMPORTANCE OF INVESTMENT DECISIONS. The future success of a company depends on the investment decisions it makes today. The importance of this is underscored by the fact that most important investment decisions are approved by a company's chief executive officer or board of directors. It is at this level that the total needs of the company can be brought into focus and the proper strategies developed to cope successfully with the future. Investments of lesser magnitudes are often delegated to lower levels of management. Although the impact on the company differs, the focus of thinking is the same. The decisions as to whether the commitments of resources, time, or money, are worthwhile in terms of the benefits to be expected are the essence of the capital budgeting concept.

CRITERION FOR DECISION MAKING. The primary motivation for making capital investments in machinery and equipment is to maximize profits. A good capital budgeting program is able to measure and rank individual proposals according to their investment worth. Such a ranking, or **cutoff point** within it, will guide acceptance or rejection of the individual projects. In order to perform such an analysis the expected benefits from investments must be measured in dollars. Benefits which cannot be measured in dollars should also be considered and certainly will influence the decision to accept or reject the proposal. But, as stated by Oakford (Capital Budgeting), "All other things being equal, the alternative selected will be that which promises the maximum long-run monetary benefit to the owners of the enterprise for which the decision is made."

Return on Investment. Several yardsticks are used to measure investment worth. One of the most common of these is the return on investment. Return on investment is also implicit in such measures as present value and the M.A.P.I. urgency rating. Putnam, Barlow, and Stilian (Unified Operations Management) state "The basic yardstick that a company should use in planning for the future is the incremental return on investment; in other words, the additional profit that can be anticipated for the additional investment." Measures such as "degree of necessity" or payback period are either too subjective or inaccurate to reflect the true productivity of capital.

Minimum Acceptable Return. After investment projects have been measured and ranked in terms of return on investment, a decision should be reached as to the minimum return acceptable to the company. This cutoff point is related to the cost and supply of available money for investments. The manner in which this minimum acceptable return is calculated is a controversial subject among financial experts. Various assumptions about the cost and supply of money lead to differing results. The calculations typically cover the range of 3 to 15 percent after taxes.

Many practical businessmen and successful companies tend toward the higher cutoff rates for investment but avoid being unrealistic by setting the minimum rate so high as to preclude attractive investments. It must be remembered that many investments, such as a company cafeteria, will yield no return. The income producing investments thus have to support the non-income producing investments. Furthermore, too many low-yield investments tend to diminish the financial attractiveness of the company.

NEED FOR CONSISTENCY. Even a cursory review of the literature on capital budgeting will indicate a proliferation of methods used to evaluate investment projects. Most of the methods in current use are discussed later in this section.

It could be argued that the method of analysis is not important and that any one of several acceptable methods could be used indiscriminately. If the executives making the decision are intimately familiar with the investment proposals, this argument is undoubtedly correct. However, in large or decentralized business organizations, it is almost impossible for the top executives to maintain an intimate knowledge of the details of investment decisions which they must ultimately approve or disapprove. These executives must rely upon the evaluations and recommendations of their subordinates. Bierman and Smidt (The Capital Budgeting Decision) point out that "In order to make reasonable choices in weighing alternative investments, it is increasingly necessary that various proposals be evaluated as nearly as possible on some uniform, comparable basis. In such circumstances, although the measure of economic worth of an investment should never be the sole factor considered in making a final decision, it may play an increasingly important part in the majority of the investments under consideration by the firm."

Factors Influencing Equipment Decisions

MANAGEMENT CONSIDERATIONS. There are numerous factors which influence equipment decisions. The relative importance of these factors will vary among companies and will vary with time within a company. These factors are related to:

1. Manufacturing strategy
2. Engineering and technical evaluation
3. Economic evaluation
4. Risk evaluation, and
5. Sociological considerations

Manufacturing strategy influences a company's decision in the trade off between general purpose equipment, specially built equipment, and automation.

Engineering and technical evaluations are brought to bear during engineering for new manufacturing, consideration of machine replacement factors, and consideration of the technical factors in equipment selection. These evaluations have a direct influence on the economic evaluation by identifying many of the relevant cost and risk factors.

An **economic evaluation** is essential since the return on the investment is usually the primary motivation for acquiring equipment. Methods of economic analysis are discussed on subsequent pages. The **risk** associated with the investment is analyzed to assess the desirability of making the investment. The **sociological** factors involve such things as the environmental and legal aspects, employee displacement, and the like.

MACHINE REPLACEMENT FACTORS. Technological advances require a continuous review, renewal, and transformation of a company's productive facilities. All machines and capital goods equipment are subject to wear and tear, which results from use. **Obsolescence** may occur any time, even a few months after the installation of a new piece of equipment. Existing equipment must be kept on the defensive, compelled always to justify its tenure against the challenge of new and more modern machines. If any piece of existing equipment fails to meet this challenge, it must be displaced, regardless of its age or condition, and regardless of whether it is physically worn out. A **re-equipment policy** that fails to give full recognition to obsolescence is bound to lead to poor manufacturing processes and inefficiency. It is not enough to examine one's equipment to see if it is still in usable condition or not. The criterion must be whether it is sufficiently modern to compete with the newest and most productive equipment of its kind now in use.

Every company naturally wants to keep its productive facilities completely modern. However, equipment investment is too high to warrant the discarding of existing facilities every time there is an improvement. Therefore, the **installed capacity** of industry necessarily lags, on the average, far behind the best that is currently available. The problem is not to eliminate the lag but to hold it to the lowest limits that are economically justifiable. Thus the primary task of re-equipment policy is a continuous and timely transformation of installed facilities.

Replacement Plans. Machines may be replaced according to a definite program or only when some production problem arises. Problems may involve the amount and quality of work being obtained from existing equipment or the adequacy of existing equipment to produce a new product. In any case, it is necessary to develop some plan of investigation. Initially the plan will establish a **checklist of points** on which to evaluate the present and proposed replacement machines from the standpoint of technical suitability and cost-saving features.

The **points requiring attention** fall into two classes: the physical and technical operating characteristics of the machines; and the cost and financial factors. The equipment engineer will determine: first, the operating characteristics and limitations of the existing machines, and second, what substitute equipment might be installed and what manufacturing advantages would probably result. He may find no reasons for making a change.

Checklist for Replacement Analysis. Usually some technical advantages are discovered, especially since machine and equipment manufacturers are continually improving their existing products and bringing out improved models or new basic machinery. Frequently, these advantages can be measured in terms of costs. While a cost accountant might make this cost study, the points are so interconnected with operating factors that it is better for an equipment engineer to make the investigation. Occasionally desired operating advantages are obtained even though no immediate substantial cost saving can be demonstrated.

The following questions should be considered in connection with the cost study:

1. Is the present equipment worn out?
2. Is it obsolete?
3. Is it inadequate from the standpoint of:
   a. Range or size of work
   b. Speed of operation
   c. Accuracy or degree of fineness of work
   d. Strength or rigidity for heavier operations
   e. Rate of output
   f. Power
4. Has it been made unsuitable by other changes in equipment in the plant, as, for example, the setting up of a new product line or installation of other machines working to closer tolerances?

5. Can its operations be more readily done if combined with other operations on an automatic machine?
6. Does it lack the controls, special attachments, and safety features of newer kinds of equipment?
7. Will a new machine do not only the present work but also other kinds of work which the present machine cannot handle?
8. Will a new machine replace hand operations or bench work?
9. Will a new machine have special advantages from the standpoint of:
   a. Ease of setup
   b. Convenience of operation
   c. Safety, such as guards, stop buttons, etc.
   d. Reliability in performance

TECHNICAL FACTORS IN EQUIPMENT SELECTION. There are certain technical factors to be surveyed prior to selecting equipment which are pertinent to whether the equipment is for new manufacturing or for replacement of existing equipment. The survey should encompass the following:

1. Equipment **design and performance.**
   a. Product design (shape, size, and proportion)
   b. Cutting tool design and application
   c. Type of motion and its control
2. Equipment **construction and operation.**
   a. Function of various components
   b. Parts or units that grasp and rotate the workpiece
   c. Parts that hold and move the tool
   d. Units that control the speed and direction of motion
   e. Selection and application of standard tools
   f. Working characteristics of the material
   g. Type of operations to be performed
3. Equipment **operating procedures.**
   a. Number of setups for standard operations
   b. Number of setups requiring use of attachments for special operations
   c. Methods of performing multiple operations in sequence
4. Factors influencing **efficient machine production.**
   a. Feed-to-speed ratios
   b. Microstructures of materials being machined
   c. Effect of cutting compounds on production, tool life, and finish
   d. Power for optimum productivity
5. The **physical condition** of equipment:
   a. Bearings, gearing, clutches, attachments, and accessories
   b. Lubrication system
   c. Hydraulic system
   d. Controls
   e. Rigidity for high speed operations
   f. Number of times machine has been reconditioned
   g. Nature of repairs rendered
   h. Time involved in manipulating controls

RISK EVALUATION. Any proposed investment project is, by necessity, dealing with assumptions about the future. This always involves uncertainty and risk. Risk is present even in the most routine equipment replacement projects.

It is the element of uncertainty which spells the difference between an attractive economic evaluation on paper and the real world of profit and loss. There are a number of means of dealing with uncertainty on a quantitative basis. Hertz (Harvard Business Review, vol. 42) in summarizing the methods takes the position that most are insufficient in one way or another. The best method is the one which gives a clear picture of the relative risk and probable odds of coming out ahead or behind.

Reducing the error in estimates by making **more accurate forecasts** is desirable, however it does not truly deal with the question of uncertainty. An estimate is still an estimate and the future is still unknown. **Empirical adjustments** to estimates based on historical comparisons of estimated-versus-actual results may be made to improve the accuracy of investments. These adjustments should be made for **bias**, not uncertainty. There is a danger that indiscriminate use of empirical adjustments may hide good investments.

Many companies will insist on a higher rate of return on high risk investments. Thus the return is proportional to the risk the company takes. The expected high return provides some incentive for taking the risk. While this is a perfectly sound position to take, the approach does not quantitatively identify the risks involved or the probability of achieving the expected return.

Three-Level Estimates. The development of **three-level estimates** is a relatively simple means to quantify the risk involved in an investment. Pessimistic, expected, and optimistic estimates are made for each of the factors of the analysis (for example, cost reduction, production level, etc.) in order to give the range of the possible return as well as the expected return on investment.

The most sophisticated approach goes far beyond the three-level method by developing the **likelihood of occurrence** for the entire range of possible returns from the investment. An estimate of the range of values for each of the pertinent factors of the analysis is made. Within that range an estimate is made of likelihood of occurrence of each value. All the values of all the factors are then combined to produce the **probability distribution** of the entire range of the possible return on investment.

Identifying Costs and Savings

ECONOMIC INVESTIGATION. It is important to estimate all costs and savings in order to properly evaluate the economic desirability of an investment. Checklists (Fig. 1) are helpful as a guide but there is no substitute for a **perceptive investigation** at all levels of management to insure all economic factors have been considered. The failure to consider all the costs can lead to the rejection of a desirable investment or the acceptance of an undesirable one.

CASH FLOW ANALYSIS OF INVESTMENT. A correct estimate of the earnings of a project must be based on the simple principle that the earnings are measured by the total added earnings or savings from making the investment as opposed to not making it. Joel Dean (Harvard Business Review, vol. 32) points out:

The proper bench mark for computing earnings on a project is the best alternative way of doing the job; comparison therewith will indicate the source and amount of the added earnings. Project costs should be unaffected by allocations of existing overheads but should cover all the changes in total overhead (and other) costs that are forecasted to result from the investment, but nothing else—nothing that will be the same regardless of whether the proposal is accepted or rejected.

The value of a proposed investment depends on its future earnings. Hence, the earnings estimate should be based on the best available projections of future volume, wage rates, and price levels. Earnings should be estimated over the economic life of the proposed facilities. Because project earnings vary in time shape, and because this will affect the rate of return, the earnings estimates should reflect the variations in the time trend of earnings.

Modern accounting orientation is characterized by cash flow, with increased attention to cash budgets and direct costing. Consideration of cash flows for

1. *Net Investment*
   A. Add
      1. Installed cost of project—capitalized
      2. Installed cost of project—expensed
      3. Working capital required for project
   B. Subtract
      1. Disposal value of assets to be returned by project
      2. Investment avoided by project
      3. Rebuilding or reconditioning costs avoided by project
   C. Also consider
      1. Salvage value at end of useful life of project
      2. Termination of working capital at end of useful life
   D. Do not consider
      1. Original cost of equipment to be replaced
      2. Book value of equipment to be replaced (except for effect on income taxes)

2. *Net Return*
   A. Net increase in revenue due to change in
      1. Volume of output
      2. Price of product
   B. Decreases due to net change in
      1. Direct labor
      2. Indirect labor
      3. Fringe benefits
      4. Maintenance
      5. Tooling
      6. Scrap and rework
      7. Downtime
      8. Floor space
      9. Power
      10. Taxes and insurance
      11. Supplies
      12. Subcontracting
      13. Inventory
      14. Federal income tax
   C. Do not consider
      1. Arbitrary burden or overhead charges

Fig. 1. Checklist of costs and savings.

investment analysis fits in well with these concepts. Depreciation of capital assets does not represent a cash transaction to the company. It is merely a book transaction made by the accountants to the company's financial statements. Since it represents a non-cash expense it is not directly considered in the analysis of investment projects. The same is true of the **unamortized costs** (that is, remaining book value) of equipment which may be retired. Removing the unamortized cost from the company books is merely a bookkeeping transaction. If, however, the old equipment is sold, this represents a very real cash transaction and is directly considered in the analysis.

Even though depreciation and unamortized costs are non-cash items and thus do not directly enter into the cash flow analysis, they both have a very direct effect on the income taxes the company pays. Taxes are a very real cash transaction. These non-cash transactions thus have an indirect effect on cash flows through income taxes. This will be covered later in this section.

Advantage of Cash Flow Method. One of the main advantages of the cash flow procedure is that it avoids certain difficult problems underlying the measurement of corporate income which necessarily accompanies the accrual method of accounting. Bierman and Smidt (The Capital Budgeting Decision) list the following problems:

1. In what time period should revenue be recognized?
2. What expenses should be treated as investments and therefore capitalized and depreciated over several time periods?
3. What method of depreciation should be used in measuring income as reported to management and stockholders (as distinct from income measurement for tax purposes)?

4. Should LIFO (last-in, first-out), FIFO (first-in, first-out) or some other methods be used to measure inventory flow?
5. What costs are inventorial? Should fixed, variable, direct, indirect, out-of-pocket, unavoidable, administrative, or selling costs be included in evaluating inventory?

The determination of the cash flows of the project's earnings is only half the problem. The proper amount for the investment must be determined.

Dean (Harvard Business Review, vol. 32) suggests that the investment on which the rate of return is to be based include the added outlay due to the adoption of the project as opposed to rejecting it and adopting an alternative requiring a lower investment. To this he would add any additional investment in working capital, auxiliary facilities, and research and promotional expense. If the project involves the transfer of existing facilities, the opportunity cost of the facility should be added to the amount of the investment.

Even though the quantitative analysis of a project's investment worth is based solely on cash flows, corporate earnings cannot be neglected in the final decision-making process.

Identifying Applicable Cash Flows. Bierman and Smidt (The Capital Budgeting Decision) give the following example of some of the problems that may arise due to failure to identify the proper cash flows.

The research and development section of a large chemical manufacturing firm submitted technical data on a new product to one of the firm's operating divisions. After investigation of the product by the engineering, production, and sales staffs, the operating division management decided that the product should be added to their line. Since existing facilities were not adequate for the production of the new product, a capital appropriation request for the new plant and equipment was submitted for review and approval to the firm's executive committee. On review by the executive committee the following deficiencies were uncovered: (1) the appropriation request did not include an estimate of the working capital requirements that would be required to operate the new plant and market the resulting product; (2) one of the raw materials required in the new process would be purchased from another operating division of the company, and the increased output of that division would have required additional plant and equipment expenditures by the supplier division; (3) the new product was partially competitive with one of the company's existing products, and the decline in the profit potential from this existing product had not been taken into consideration; (4) distribution of the new product would require acquisition of additional storage facilities, since demand for the product was seasonal, but efficient production would require a steady rate of production. The proposal was returned to the operating division for further study. After additional investigation it was determined that the company could most effectively utilize the new product by licensing other manufacturers to produce and market it.

INITIAL INVESTMENT. The entire cost of the initial investment should be considered. The **investment base** includes all of the cash outlays which would have to be spent if the investment is approved rather than rejected in favor of some other alternative. One of the primary concerns during the decision to approve or disapprove an investment is how much it will cost to achieve the results promised. Whether the investment dollars are earmarked capital or expense is irrelevent as these are the dollars in total which must be spent. Tax benefits which result from expensing certain items rather than capitalizing them should be reflected in the savings analysis.

All too often a narrow view of an investment is taken. Frequently one expenditure leads to another. An investment which appears to be attractive is made and then it is realized that **additional expenditures** are necessary to support the investment.

A company had a series of manual labor operations which could be modernized. It was believed that the return on investment would be attractive. On an operation-by-operation basis the company analyzed the investment required to replace the manual labor operations and estimated the savings from the new equipment. Each equipment investment could support itself with a good rate of return. After the equipment was installed it was learned that a new electric substation and bus ducts would have to be built because of the increased electrical load. This investment yielded no return.

The above company should have considered the modernization program in total by looking at the total projected savings from the program compared to the total investment including the substation and bus ducts.

In considering the outlays required for plant and equipment it is possible to lose sight of the fact that **working capital** may be needed to operate an investment project. Working capital is as much a part of an initial investment as is the hardware required. As such it must share in earning a return on investment. As previously noted the total additional investment required by the approval of a project must be considered. Working capital for plant expansion or new product projects has the effect of increasing the initial investment required. For most equipment replacement projects, however, the change in working capital is usually negligible.

The additional investment in auxiliary activities should not be neglected. **Research** and **promotional expenses** required to start a new product or expand plant production are also a part of the initial investment.

FACTORS INFLUENCING NET RETURN. The net return from an investment project includes all the added earnings, savings, expense reductions, or expense avoidance that may be expected from making the investment as opposed to not making it. The net return from a project will replace the initial investment outlay and provide an additional amount that can be considered the return on the investment.

The savings analysis should be comprehensive and realistic. If an old machine is to be replaced with a new one to reduce maintenance costs it should not be assumed that the new one will run indefinitely without maintenance. The future maintenance cost of the new machine should be estimated. If a new machine requires less skilled operators but more highly skilled maintenance men than previously, it should be allowed for in the savings estimate. Developing the projected annual costs for all expected conditions is a difficult, demanding, and tedious task but critical to the proper evaluation of the project.

Overhead Costs. Project costs are unaffected by **allocations of overhead costs.** If certain overhead costs are expected to actually change due to the investment decision, this should be factored into the analysis. Overhead costs which are not expected to change except on the accountant's cost allocation ledger are not pertinent to the investment decision. If, for example, a department is replacing a piece of equipment, the department foreman's salary does not enter into the analysis. If, however, the project involves adding an entire new department, the foreman's salary is relevant to the decision.

Effect on Existing Products. The analysis should not overlook some of the less obvious effects from the operation of the proposed project. If the project involves a new product which is partially competitive with an existing product, then the decline in the profit potential of the existing product should be considered. If the project involves making parts which are currently supplied by another division of the company, the impact of the reduced production volume should be analyzed. If a proposed investment will serve as a partial substitute for any other investments to which a company is already committed or which it is considering, its economic impact should be investigated.

Effect of Increased Revenues. This is often overlooked during the analysis of a project's net return. For example, one of the steps in a company's fabrication process involved a highly proprietary method of welding. The company, therefore, never subcontracted the operation. The engineering department developed a technically improved welding technique which would also reduce the welding time and labor cost 15%. The labor savings would amount to $12,000 per year. To use the new technique, however, involved modifying their existing 10 welding machines at a cost of $20,000 each. When the company was awarded a sizable contract for their product which would insure full utilization of the welding equipment for several years the engineering manager felt the time was right to request capital funds to change over to the new process. Even though the new welding technique was technically better, the president of the company was not enthused about a $200,000 investment which promised to save $12,000 per year for a technological lifetime estimated to be ten years. He did, however, direct the engineering department to analyze all production operations to determine what would be required to increase production 10%. After an extensive analysis it was found that most equipment had at least 10% more capacity available. Only a few machining operations, which could be expanded for an investment of $30,000, and the welding operation, limited the production increase. Furthermore it was determined that the company's overhead costs would remain virtually unaffected by the expansion. Further analysis indicated the increased volume would generate $1,000,000 per year in additional revenues. Costs would go up $400,000 per year for labor, materials, and some overhead operations. The president then directed the engineering manager to proceed on a priority basis to implement the $230,000 welding modification and equipment expansion program to increase production with the anticipation of receiving a $612,000 per year return on his investment.

The above case is an example of the **break-even** effect in business. Above the break-even point costs do not increase as rapidly as revenues. When bottleneck equipment or volume expansion programs are under consideration, the effect of increased revenues should always be examined.

ESTABLISHING ALTERNATIVES. No matter how carefully the costs and savings of an investment proposal is analyzed, if the right alternative is not considered, the right decision cannot be made. Management must always search for alternatives to insure that all possibilities are explored during the decision-making process.

Computing the relative cost and savings of an investment involves an explicit comparison of two alternatives. The size of the added cost and savings depends on the alternative that is used as the basis of comparison. Bierman and Smidt (The Capital Budgeting Decision) stress that almost any investment can be made to seem worthwhile if it is compared with a sufficiently bad alternative. There may be a significant saving to be realized from scrapping obsolete equipment and replacing it with modern, automated equipment. It may, however, be better to subcontract the part or operation. The savings from new equipment relative to the alternative of subcontracting may not justify the purchase of he equipment. On a larger scale it may not be desirable to invest in reducing the manufacturing cost of a marginal product if the savings as compared to the alternative of dropping the product do not justify the investment.

A decision model in **matrix format** representing the various alternatives and their associated outcomes and in which each element may be given in symbolic terms has been suggested by Morris (Analysis of Management Decisions).

INFLUENCE OF INCOME TAX AND DEPRECIATION ON CASH FLOWS. It is sometimes felt that the adjustment of investment analysis for corporate income taxes is academic. The argument is that the underlying value of a project is obscured (rather than revealed) by allowing for tax effects, and that the ranking of investments will be the same whether or not they are deflated for taxes. Joel Dean (Harvard Business Review, vol. 32) points out the two fallacies of this argument.

(a) In order to apply tenable acceptance standards such as the company's outside cost of capital, it is necessary to measure rate of return after taxes, rather than before taxes.

(b) The impact of taxes differs depending on the time shape of the project, and the after-tax ranking of proposals will differ significantly from their before-tax ranking if taxes are correctly taken into account in computing rate of return. For example, the tax effects of accelerated amortization can convert a border-line project into a highly profitable investment opportunity.

There are two non-cash expenses which affect income taxes and thus cash flows. One is **depreciation** which varies depending on the depreciation method chosen. A second is the **book value** or the **unamortized cost** of a capital asset to be retired. The income tax liability is computed by applying the income tax rate to the additional taxable income where depreciation of the investment and the write off of unamortized costs are allowable deductions for tax purposes.

From the above it can be seen that the higher the depreciation taken for income tax purposes, the lower the income tax will be and thus the greater the after-tax cash proceeds. Cash proceeds for a period are increased by the allowable depreciation times the tax rate. This is often referred to as the "depreciation tax shield." The write off of unamortized costs also generates a "tax shield."

To illustrate this consider the following example: Revenues = $8,000; other costs = $4,000; depreciation for tax purposes = $1,000; unamortized costs written off = $1,000;

Taxable income = $8,000 − 4,000 − 1,000 − 1,000 = $2,000
Income tax (52% tax rate) = $2,000(0.52) = $1,040
After-tax cash proceeds = $8,000 − 4,000 − 1,040 = $2,960

or

= 0.48($8,000 − 4,000) + 0.52($1,000) + 0.52($1,000)
= $1,920 + $520 + $520 = $2,960

Note that the depreciation tax shield = 0.52($1,000) = $520, and the savings from write off of unamortized costs = 0.52($1,000) = $520.

The after-tax cash proceeds are calculated for each year of the project's life in the same manner using the appropriate depreciation value for tax purposes for each year. The depreciation differs each year if one of the accelerated depreciation methods is used.

SEPARATION OF ECONOMIC EVALUATION FROM FINANCING ARRANGEMENTS. One of the most common difficulties in the evaluation of investment projects is the confusion which arises when an investment proposal is presented accompanied by a specific financing plan. Satchell (Financial Executive, vol. 34) points out that it is a fallacy to include the cash flows associated with the financing costs and capital repayment schedules in computing the rate of return of an investment opportunity. "If the financing costs and capital repayment schedule are interwoven with the cash flows promised by a project, the result is a rate-of-return computation that does not indicate either (1) the true economics of the project, or (2) the desirability of the financing arrangement." Frequently this misleading approach to project evaluation is used and advocated with the justification that there are no alternatives. Only the rarest kind of investment project would provide absolutely no financing alternatives or variations.

Investment projects are commonly presented with a **financing plan** in connection with leasing arrangements. Plant or equipment may be acquired by making a small initial payment with the balance paid as **lease payments.** By considering the initial payment as the investment and subsequent payments as a cost of the project, the project analysis may show a very attractive return on investment. The true case may be that the project, analyzed independently of the lease arrangement, yields a low return on investment. Furthermore the cost of money obtained by the lease may be more expensive than alternate sources of money.

Consider an investment opportunity which called for a $10,000 investment in equipment which would return $730 per year after taxes for ten years plus $500 per year depreciation tax shield for a total after-tax cash flow of $1,230 per year:

Year	Cash Flow	4% Present Value Factor	Discounted Cash Flow
0	−$10,000	1.000	−$10,000
1–10	+ 1,230/yr.	8.111	+ 10,000
			0

Using the internal rate of return method, it can be seen that the return on investment is only 4%, clearly an unattractive proposition. Consider the same proposal except that it is financed with a lease arrangement. The terms of the lease call for an initial payment of $2,000 plus annual payments of $1,088 at the end of each year for 10 years. Since lease payments are a tax deductible expense, the after-tax cash flow is $1,000 for the initial payment and $730 less $544 or $186 per year for 10 years:

Year	Cash Flow	13% Present Value Factor	Discounted Cash Flow
0	− $1,000	1.000	−$1,000
1–10	+ 186/yr.	5.433	+ 1,010
			$ 10

The investment now appears to be more attractive, yielding more than a 13% return. The inherent economics of the investment has not changed, it is merely obscured by commingling the financing arrangement with the project economics.

It should also be noted that cash disbursed for interest is normally excluded from the cash flow computation used in analyzing investments since the interest factor is taken into consideration by discounting for the **time value of money.** To also include the cash disbursement for interest would result in **double counting.**

SUNK COSTS. The book value of economically obsolete equipment has acted as one of the greatest obstacles to sound replacement analyses. Some managements dictate the principle that any investment for replacement purposes must not only be profitable in its own right but must also recoup the unamortized value of its predecessor. The theory behind this thinking involves the question of the disposition of the unamortized value, or sunk cost. This line of reasoning, however, is completely erroneous.

In any cost analysis, the cost factors to be considered are those that occur in the present and future. Any costs that have occurred in the past are merely of historical interest and are not included in the analysis. Grant and Ireson (Principles of Engineering Economy) state: "From the viewpoint of an economy study, a past cost should be thought of as a **sunk cost**, irrelevant in the study except as its magnitude may somehow influence future receipts or disbursements or other future matters."

Methods of Economic Analysis

METHODS AND TERMINOLOGY. One of the difficulties encountered in the study of machinery and equipment economics is the proliferation of methods of analysis and the terminology used to describe the various methods.

The major methods of analysis are present value, internal rate of return, dual rate discounted cash flow, MAPI investment formula, minimum annual cost, capitalized cost, payback, and various book value return-on-investment methods. The term **discounted cash-flow** method is used to describe the general approach to investment analysis used in the present-value and internal-rate-of-return methods.

The internal rate of return, present value, and MAPI formula are the methods most widely used by progressive companies. Gerstenfeld and Carroll (Faculty Working Paper, Boston University College of Business Administration) conducted a survey of the capital investment analytical methods used by 271 manufacturing companies. Over 60 percent of the companies used the internal-rate-of-return method, either alone or in combination with other methods. The MAPI formula was used alone or in combination with other methods by 9 percent of the companies. An industry analysis shows that over 30 percent of the companies in the machinery industry used the MAPI formula. This is to be expected since the formula was designed by the Machinery and Allied Products Institute. The present value method was not found to be widely used in spite of the support it receives from theorists. It was used by less than 8 percent of the companies.

The payback method, a very poor measure of an investment's worth, was the sole method of evaluation used by 22 percent of the companies surveyed. This method is used more by small companies than by large companies. The book-value-return-on-investment methods, which are relatively poor measures of the investment, do not appear to be in very wide use.

The dual-rate, discounted cash-flow method is mathematically complex but will probably gain increasing acceptance as the next level of sophistication in analytical techniques. The minimum-annual-cost and capitalized-cost methods are rather limited approaches of questionable value.

Present Value. The present value method, or present worth as it is often referred to, discounts the expected capital outlays at a rate considered satisfactory to the company in order to find the present value of the outlays. The present value of the expected cash earnings is found in the same way. If the present value of the earnings is greater than the present value of the outlays, the project should be accepted. This method of analysis yields unambiguous, mathematically correct answers for all problems, deals with cash flows, and accounts for the time value of money in the future distribution of cash flows. It is a relatively easy method to use, especially when dealing with complex projects with unusual cash flows. The present-value method does require that management specify in advance a **discount rate** to be used by the analyst for all capital equipment evaluations. There is little agreement between authorities however, as to how a company should determine the discount rate or even if the rate should be fairly high or low. The selection of the proper discount rate is discussed at length on subsequent pages.

Internal Rate of Return. This is also referred to as the return on investment, yield of investment, interest rate of return, investors method, present value return on investment, discounted cash flow, profitability index, and marginal efficiency of capital. The approach expresses each project's estimated value as a single overall annual rate of return on investment.

In simple terms, it is the annual income provided by the investment (after allowing for the recovery of the initial capital) divided by the initial investment. The internal-rate-of-return method is an **analytical technique** for determining this annual rate of return while explicitly taking into account the time value of money and unequal cash flows during the expected life of the investment. Mathematically, the rate of return is equal to that discount rate which causes the present value of the expected cash earnings of the project to exactly equal the project's initial cash outlay. If the rate of return on a project is greater than a return rate considered satisfactory to the company, the project should be accepted. While the theory may seem complex, the analytical method, as outlined on subsequent pages, is straightforward.

One **advantage of this method** is that many businessmen find it easier to think in terms of rate of return rather than in terms of present value. Another advantage is that the procedure can be used by an analyst without management establishing in advance a precise discounting rate. Once the rate of return of the project has been calculated it can be decided at the appropriate management level whether or not this rate is satisfactory. The internal-rate-of-return method does have **theoretical shortcomings.** The method will slightly overstate the return on very attractive investments. For certain unusual cases it may yield an ambiguous solution of two rates of return. Furthermore, as rates of return are usually calculated in practice they are not strictly comparable from one alternative to another due to the difference in the life of various investment projects and the opportunities for reinvesting funds returned by projects with the shorter lives. These handicaps are not sufficient to keep this method from being satisfactorily used by many well-managed companies. Solomon (The Management of Corporate Capital) concludes that "the rate of return is a useful concept that enables us to express the profitability of an investment proposal as a single explicit value. This value automatically adjusts for differences in the time pattern of expected cash outflows and inflows. It is also independent of the absolute size of the project. Thus it provides a useful standard by which all types of projects—large and small, long-run and short-run—can be ranked against each other in relative terms and also against the company's cost of capital, in order to judge their absolute worth."

Dual Rate. This is a discounted cash-flow method which deals explicitly with the shortcomings noted for the present-value and internal-rate-of-return methods. A project's estimated value is expressed in familiar terms of rate of return. A discounting rate need not be established in advance. The method will not overstate the return on very attractive investments and will always yield mathematically correct answers.

The dual-rate method divides the annual cash flow from an investment project into two components. One part is for a **sinking fund** to recover the investment; the balance of the cash flow after deducting for the sinking fund is for income and is used to determine the project's rate of return. If the project's rate of return is greater than a rate considered satisfactory to the company, the project should be accepted. Hunt (Financial Analysis in Capital Budgeting) states that "methods which do not separate sinking fund from income fall into errors which arise from the use of one rate, which implies the judgment that the (sinking fund and minimum acceptance) rates shall be the same. The rate computed by the discounted cash flow method assumes that each project may find an equally attractive sinking fund; the value computed by the net-present-value method assumes that only the criterion rate will be available for the sinking fund."

MAPI Investment Formula. This method considers the advantage of making an investment in replacement equipment today compared to the alternative of waiting until later. The procedure ranks investment projects by their after-tax return. The MAPI method is primarily directed toward **machinery replacement decisions,** although it can be adapted for general-purpose use. The formula utilizes a mathematical model which factors in technological obsolescence and depreciation of current equipment.

Minimum Annual Cost. This method is based on the premise that the least annual cost for a given production rate is the most profitable. Therefore, the choice between alternatives is based strictly on the minimum annual operating cost which includes a capital recovery factor at a stipulated minimum interest rate. Grant and Ireson (Principles of Engineering Economy) consider the method equivalent to the present-value method but easier to understand.

Capitalized Cost. In this method alternatives are analyzed by comparing the sums of the capitalized annual cost and the in-place cost. The capitalized cost is calculated by dividing the annual cost by the acceptable financial return. In the use of this technique care must be taken to avoid misinterpreting the figures used. The figures merely measure the **magnitude of difference** in alternatives and should not be construed as anything else. The method is used most frequently in public projects rather than in industrial analysis.

Payback Method. The criterion in this method is the number of years it will take to recover a project's initial investment from the funds generated by the project. Since it is presumably desirable to recover funds invested in a project as soon as possible, a short payback period is desirable. Bierman and Smidt (The Capital Budgeting Decision) consider this to be a very poor measure of a project's desirability as it does not take into account a project's earning life. Proceeds from the investment received after the payback period are completely ignored. This method is not recommended except, possibly, as a crude screening device prior to more accurate calculations. Terborgh (Business Investment Management) believes its use as a decision-making criterion probably hinders the profitability of American industry.

Other Methods. There are numerous methods of analysis, usually called **return on investment** or **accounting return on investment,** which utilize definitions of income and capital recovery in the same manner as an accountant dealing with financial statements. The rate of return is usually the ratio of the project's income divided by its initial investment although there are many variations of this, each one with some justification of its validity. None of the variations succeed in bringing the timing of the cash proceeds into the analysis.

PRESENT - VALUE AND INTERNAL - RATE - OF - RETURN METHODS. The present-value and internal-rate-of-return methods have received considerable attention as management tools for proposed investment projects. Henrici (Harvard Business Review, vol. 48) notes that the increasing use of such terms as discounted cash flow, ranking of investments, and cutoff rates reflects general acceptance of quantitative, mathematical methods in the decision-making process.

Present value and internal rate of return are both discounted cash flow methods. The term "discounted cash flow" implies a consideration of the **time value of money** by discounting a project's future proceeds and the consideration of the project's **cash flows** rather than its income as defined by accountants. It should be noted that **depreciation** as a concept of recovering the wasting assets of a project is not utilized in the discounted cash-flow method. Recovery of the total investment is achieved through a hypothetical sinking fund compounded at the discount rate and is developed automatically through the process of discounting the project's cash flows. These methods are computed on the basis of cash flows after taxes as these after-tax dollars are what determine the ultimate benefit to the company.

Example of Methods. A simple example will serve to illustrate the present-value and internal-rate-of-return methods. Assume there is an investment project with the following after-tax cash flows:

Year	Cash Flow
0	−$12,000
1	+ 10,000
2	+ 5,000

A **minus sign** is used to indicate a net outflow or expenditure of cash and a **plus sign** to indicate a net cash inflow or cash gain. The $12,000 investment is made at the beginning of year 1, a $10,000 cash gain is expected at the end of year 1, and a $5,000 gain is expected at the end of year 2.

A table of present values (Fig. 2) is used to compute the present value of this investment at a 10 percent discount rate. From this table it is seen that

$1.00 received 1 year from now discounted at 10 percent is worth $0.909
$1.00 received 2 years from now discounted at 10 percent is worth $0.826

(All numbers are rounded off to three decimal places.)
The present value of the investment, assuming a 10% discount rate, is:

Cash Flow (1)	Present Value Factor (2)	Present Value (3) = (1) × (2)
−$12,000	1.000	−$12,000
+ 10,000	0.909	+ 9,090
+ 5,000	0.826	+ 4,130
	Net present value	+$ 1,220

n	8%	8½%	9%	10%	11%
1	0.925 926	0.921 659	0.917 431	0.909 091	0.900 901
2	0.857 339	0.849 455	0.841 680	0.826 446	0.811 622
3	0.793 832	0.782 908	0.772 183	0.751 315	0.731 191
4	0.735 030	0.721 574	0.708 425	0.683 013	0.658 731
5	0.680 583	0.665 045	0.649 931	0.620 921	0.593 451
6	0.630 170	0.612 945	0.596 267	0.564 474	0.534 641
7	0.583 490	0.564 926	0.547 034	0.513 158	0.481 658
8	0.540 269	0.520 669	0.501 866	0.466 507	0.433 926
9	0.500 249	0.479 880	0.460 428	0.424 098	0.390 925
10	0.463 193	0.442 285	0.422 411	0.385 543	0.352 184
11	0.428 883	0.407 636	0.387 533	0.350 494	0.317 283
12	0.397 114	0.375 702	0.355 535	0.318 631	0.285 841
13	0.367 698	0.346 269	0.326 179	0.289 664	0.257 514
14	0.340 461	0.319 142	0.299 246	0.263 331	0.231 995
15	0.315 242	0.294 140	0.274 538	0.239 392	0.209 004
16	0.291 890	0.271 097	0.251 870	0.217 629	0.188 292
17	0.270 269	0.249 859	0.231 073	0.197 845	0.169 633
18	0.250 249	0.230 285	0.211 994	0.179 859	0.152 822
19	0.231 712	0.212 244	0.194 490	0.163 508	0.137 678
20	0.214 548	0.195 616	0.178 431	0.148 644	0.124 034
21	0.198 656	0.180 292	0.163 698	0.135 131	0.111 742
22	0.183 941	0.166 167	0.150 182	0.122 846	0.100 669
23	0.170 315	0.153 150	0.137 781	0.111 678	0.090 693
24	0.157 699	0.141 152	0.126 405	0.101 526	0.081 705
25	0.146 018	0.130 094	0.115 968	0.092 296	0.073 608
26	0.135 202	0.119 902	0.106 393	0.083 905	0.066 314
27	0.125 187	0.110 509	0.097 608	0.076 278	0.059 742
28	0.115 914	0.101 851	0.089 548	0.069 343	0.053 822
29	0.107 328	0.093 872	0.082 155	0.063 039	0.048 488
30	0.099 377	0.086 518	0.075 371	0.057 309	0.043 683
31	0.092 016	0.079 740	0.069 148	0.052 099	0.039 354
32	0.085 200	0.073 493	0.063 438	0.047 362	0.035 454
33	0.078 889	0.067 736	0.058 200	0.043 057	0.031 940
34	0.073 045	0.062 429	0.053 395	0.039 143	0.028 775
35	0.067 635	0.057 539	0.048 986	0.035 584	0.025 924
40	0.046 031	0.038 266	0.031 838	0.022 095	0.015 384
45	0.031 328	0.025 448	0.020 692	0.013 719	0.009 130
50	0.021 321	0.016 924	0.013 449	0.008 519	0.005 418
55	0.014 511	0.011 255	0.008 741	0.005 289	0.003 215
60	0.009 876	0.007 485	0.005 681	0.003 284	0.001 908
65	0.006 721	0.004 978	0.003 692	0.002 039	0.001 132
70	0.004 574	0.003 311	0.002 400	0.001 266	0.000 672
80	0.002 119	0.001 464	0.001 014	0.000 488	0.000 237
90	0.000 981	0.000 648	0.000 428	0.000 188	0.000 083
100	0.000 455	0.000 286	0.000 181	0.000 073	0.000 029

Fig. 2. Sample table

12%	13%	14%	15%	20%	n
0.892 857	0.884 956	0.877 193	0.869 565	0.833 333	1
0.797 194	0.783 147	0.769 468	0.756 144	0.694 444	2
0.711 780	0.693 050	0.674 972	0.657 516	0.578 704	3
0.635 518	0.613 319	0.592 080	0.571 753	0.482 253	4
0.567 427	0.542 760	0.519 369	0.497 177	0.401 878	5
0.506 631	0.480 319	0.455 587	0.432 328	0.334 898	6
0.452 349	0.425 061	0.399 637	0.375 937	0.279 082	7
0.403 883	0.376 160	0.350 559	0.326 902	0.232 568	8
0.360 610	0.332 885	0.307 508	0.284 262	0.193 807	9
0.321 973	0.294 588	0.269 744	0.247 185	0.161 506	10
0.287 476	0.260 698	0.236 617	0.214 943	0.134 588	11
0.256 675	0.230 706	0.207 559	0.186 907	0.112 157	12
0.229 174	0.204 165	0.182 069	0.162 528	0.093 464	13
0.204 620	0.180 677	0.159 710	0.141 329	0.077 887	14
0.182 696	0.159 891	0.140 096	0.122 894	0.064 905	15
0.163 122	0.141 496	0.122 892	0.106 865	0.054 088	16
0.145 644	0.125 218	0.107 800	0.092 926	0.045 073	17
0.130 040	0.110 812	0.094 561	0.080 805	0.037 561	18
0.116 107	0.098 064	0.082 948	0.070 265	0.031 301	19
0.103 667	0.086 782	0.072 762	0.061 100	0.026 084	20
0.092 560	0.076 798	0.063 826	0.053 131	0.021 737	21
0.082 643	0.067 963	0.055 988	0.046 201	0.018 114	22
0.073 788	0.060 144	0.049 112	0.040 174	0.015 095	23
0.065 882	0.053 225	0.043 081	0.034 934	0.012 579	24
0.058 823	0.047 102	0.037 790	0.030 378	0.010 483	25
0.052 521	0.041 683	0.033 149	0.026 415	0.008 735	26
0.046 894	0.036 888	0.029 078	0.022 970	0.007 280	27
0.041 869	0.032 644	0.025 507	0.019 974	0.006 066	28
0.037 383	0.028 889	0.022 375	0.017 369	0.005 055	29
0.033 378	0.025 565	0.019 627	0.015 103	0.004 213	30
0.029 802	0.022 624	0.017 217	0.013 133	0.003 511	31
0.026 609	0.020 021	0.015 102	0.011 420	0.002 926	32
0.023 758	0.017 718	0.013 248	0.009 931	0.002 438	33
0.021 212	0.015 680	0.011 621	0.008 635	0.002 032	34
0.018 940	0.013 876	0.010 194	0.007 509	0.001 693	35
0.010 747	0.007 531	0.005 294	0.003 733	0.000 680	40
0.006 098	0.004 088	0.002 750	0.001 856	0.000 273	45
0.003 460	0.002 219	0.001 428	0.000 923	0.000 110	50
0.001 963	0.001 204	0.000 742	0.000 459	0.000 044	55
0.001 114	0.000 654	0.000 385	0.000 228	0.000 018	60

of present value of 1.

The net present value is positive indicating the project is expected to return more than the 10 percent discount rate established as a criterion. Thus the project should be accepted.

To compute the internal rate of return of the investment it is necessary to find the rate of discount that causes the sum of the present values of the cash flows to be equal to zero. As a first trial assume the rate is 10 percent. In the preceding it was found that the present value using 10 percent is a positive $1,220. To decrease the present value of the cash flows, increase the rate of discount.

As a second trial try 20 percent as the rate of discount:

Cash Flow	Present Value Factor	Present Value
−$12,000	1.000	−$12,000
+ 10,000	0.833	+ 8,330
+ 5,000	0.694	+ 3,470
	Net present value	−$ 200

The net present value is negative, indicating that the 20 percent rate of discount is too large. Thus try a lower rate for the next estimate. Try 15 percent:

Cash Flow	Present Value Factor	Present Value
−$12,000	1.000	−$12,000
+ 10,000	0.870	+ 8,700
+ 5,000	0.756	+ 3,780
	Net present value	+$ 480

The new present value is now positive; therefore the discount rate falls somewhere between 15 percent and 20 percent. The rate which would make the net present value equal to zero can be approximated by interpolation. The internal rate of return of this investment is found to be approximately 18 percent. The project should be accepted if the criterion rate of acceptance is 10 percent.

Figure 3 gives the present value of a uniform series of $1.00 payments received at the end of each year. Consider a project with an estimated savings of $10,000 a year for 20 years. From Fig. 3 the present value of $1.00 per year for 20 years discounted at 10 percent is $8.514 (rounded to three decimal places). Thus the present value of the $10,000 annual savings for 20 years at a 10 percent discount rate is $85,140 ($10,000 × 8.514).

Time Value of Money. Future receipts should always be discounted. The purpose of discounting the cash flows expected from an investment is to determine whether the investment yields more cash than alternative uses of the same amount of money.

The present value of $1.00 payable in two years is the amount of money necessary to invest today at compound interest in order to have $1.00 in two years. This will depend upon the rate of interest at which the money will grow and the frequency at which it will be compounded. Assume, for example, a 4 percent interest compounded annually. One dollar invested today at 4 percent compounded annually would grow to $1.04 ($1.00 × 1.04) at the end of one year and $1.0816 ($1.04 × 1.04) in two years. Since $1.00 will grow to $1.0816 in two years, $0.925 ($1.00 ÷ 1.0816) will grow to $1.00 in two years. Thus the present value of $1.00 payable in two years at 4 percent interest compounded annually is $0.925.

Present Value of Tax Shields. The Internal Revenue Code of 1954 provides a choice of depreciating a new asset by using straight-line, sum-of-the-years' digits, or double straight-line declining balance depreciation. The choice of a depreciation method will affect the profitability of the investment. To select the best depreciation method the following equation for computing the after-tax cash proceeds may be used. It divides the after-tax cash proceeds into two parts—the first part is independent of the depreciation method, and the second part depends only on the depreciation method.

$$\text{After-tax cash proceeds} = (1 - t)(r - e) + (td)$$

where t = tax rate
  r = revenues
  e = expenses other than depreciation
  d = depreciation

The net present value of after-tax cash proceeds for each depreciation method is the sum of the present value of the first term plus the present value of the second term. The different depreciation methods may be tried to determine the one which gives the highest present value.

Bierman and Smidt (The Capital Budgeting Decision) have developed tables to aid in selecting the optimum depreciation method. They further point out that the cost of money, high tax rates, and long-lived assets, combined with accelerated depreciation for tax purposes, can lead to situations where the presence of a salvage value in a proposed investment is undesirable. To illustrate, consider a 10 percent discount rate, a 50 percent tax rate, a life of 20 years for the asset, and a tax depreciation method that allows a company to write off a depreciable asset in five years. In this case $100 of depreciable assets may be worth more than $100 of terminal value (salvage value). The present value of $100 of terminal value due in 20 years is $14.90 ($100 × 0.149).

If there were no salvage value there would be $100 of additional depreciable assets which would reduce taxes a total of $50, or $10 per year. The present value of $10 per year for five years (at 10%) is $37.91 ($10 × 3.791).

It can be seen that the tax deduction is worth more than the salvage value with the facts as given, and such facts are reasonable and close to reality. Thus the ideal situation from the point of view of the investor would be to write off the investment for tax purposes as if it had no salvage and then wait and see if any salvage will develop. The taxpayer is going to be better off with a conservative estimate of salvage. In addition, other things being equal, an expenditure that can be charged off as an expense now for tax purposes is more desirable than one that must be depreciated for tax purposes over a number of years. Under the present tax code, it may be more desirable to increase income through research rather than through increasing plant and equipment.

Comparison of Methods. Authorities differ on whether the present-value method or the internal-rate-of-return method is superior for **capital equipment evaluation.** The internal-rate-of-return method can, in certain instances, lead to an ambiguous mathematical solution. This occurs when a project produces savings in the early years followed by an additional investment (for example, a

n	8%	8½%	9%	10%	11%
1	0.925 925	0.921 659	0.917 431	0.909 091	0.900 901
2	1.783 265	1.771 114	1.759 111	1.735 537	1.712 523
3	2.577 097	2.554 022	2.531 295	2.486 852	2.443 715
4	3.312 127	3.275 597	3.239 720	3.169 865	3.102 446
5	3.992 710	3.940 642	3.889 651	3.790 787	3.695 897
6	4.622 880	4.553 587	4.485 919	4.355 261	4.230 538
7	5.206 370	5.118 514	5.032 953	4.868 419	4.712 196
8	5.746 639	5.639 183	5.534 819	5.334 926	5.146 123
9	6.246 888	6.119 063	5.995 247	5.759 024	5.537 048
10	6.710 081	6.561 348	6.417 658	6.144 567	5.889 232
11	7.138 964	6.968 984	6.805 191	6.495 061	6.206 515
12	7.536 078	7.344 686	7.160 725	6.813 692	6.492 356
13	7.903 776	7.690 955	7.486 904	7.103 356	6.749 870
14	8.244 237	8.010 097	7.786 150	7.366 687	6.981 865
15	8.559 479	8.304 237	8.060 688	7.606 080	7.190 870
16	8.851 369	8.575 333	8.312 558	7.823 709	7.379 162
17	9.121 638	8.825 192	8.543 631	8.021 553	7.548 794
18	9.371 887	9.055 476	8.755 625	8.201 412	7.701 617
19	9.603 599	9.267 720	8.950 115	8.364 920	7.839 294
20	9.818 147	9.463 337	9.128 546	8.513 564	7.963 328
21	10.016 803	9.643 628	9.292 244	8.648 694	8.075 070
22	10.200 744	9.809 796	9.442 425	8.771 540	8.175 739
23	10.371 059	9.962 945	9.580 207	8.883 218	8.266 432
24	10.528 758	10.104 097	9.706 612	8.984 744	8.348 137
25	10.674 776	10.234 191	9.822 580	9.077 040	8.421 745
26	10.809 978	10.354 093	9.928 972	9.160 945	8.488 058
27	10.935 165	10.464 602	10.026 580	9.237 223	8.547 800
28	11.051 078	10.566 453	10.116 128	9.306 567	8.601 622
29	11.158 406	10.660 326	10.198 283	9.369 606	8.650 110
30	11.257 783	10.746 844	10.273 654	9.426 914	8.693 793
31	11.349 799	10.826 584	10.342 802	9.479 013	8.733 146
32	11.434 999	10.900 078	10.406 240	9.526 376	8.768 600
33	11.513 888	10.967 813	10.464 441	9.569 432	8.800 541
34	11.586 934	11.030 243	10.517 835	9.608 575	8.829 316
35	11.654 568	11.087 781	10.566 821	9.644 159	8.855 240
40	11.924 613	11.314 520	10.757 360	9.779 051	8.951 051
45	12.108 402	11.465 312	10.881 197	9.862 808	9.007 910
50	12.233 485	11.565 595	10.961 683	9.914 814	9.041 653
55	12.318 614	11.632 288	11.013 993	9.947 106	9.061 678
60	12.376 552	11.676 642	11.047 991	9.967 157	9.073 562
65	12.415 983	11.706 140	11.070 087	9.979 607	9.080 644
70	12.442 820	11.725 757	11.084 449	9.987 338	9.084 800
80	12.473 514	11.747 479	11.099 849	9.995 118	9.088 758
90	12.487 732	11.757 087	11.106 354	9.998 118	9.090 151
100	12.494 318	11.761 336	11.109 102	9.999 274	9.090 626

Fig. 3. Sample table of

12%	13%	14%	15%	20%	n
0.892 857	0.884 956	0.877 193	0.869 565	0.833 333	1
1.690 051	1.668 102	1.646 661	1.625 709	1.527 778	2
2.401 831	2.361 153	2.321 632	2.283 225	2.106 481	3
3.037 349	2.974 471	2.913 712	2.854 978	2.588 735	4
3.604 776	3.517 231	3.433 081	3.352 155	2.990 612	5
4.111 407	3.997 550	3.888 668	3.784 483	3.325 510	6
4.563 757	4.422 610	4.288 305	4.160 420	3.604 592	7
4.967 640	4.798 770	4.638 864	4.487 322	3.837 160	8
5.328 250	5.131 655	4.946 372	4.771 584	4.030 967	9
5.650 223	5.426 243	5.216 116	5.018 769	4.192 472	10
5.937 699	5.686 941	5.452 733	5.233 712	4.327 060	11
6.194 374	5.917 647	5.660 292	5.420 619	4.439 217	12
6.423 548	6.121 812	5.842 362	5.583 147	4.532 681	13
6.628 168	6.302 488	6.002 072	5.724 476	4.610 567	14
6.810 864	6.462 379	6.142 168	5.847 370	4.675 473	15
6.973 986	6.603 875	6.265 060	5.954 235	4.729 561	16
7.119 630	6.729 093	6.372 859	6.047 161	4.774 634	17
7.249 670	6.839 905	6.467 420	6.127 966	4.812 195	18
7.365 777	6.937 969	6.550 369	6.198 231	4.843 496	19
7.469 444	7.024 752	6.623 131	6.259 331	4.869 580	20
7.562 003	7.101 550	6.686 957	6.312 462	4.891 316	21
7.644 646	7.169 513	6.742 944	6.358 663	4.909 430	22
7.718 434	7.229 658	6.792 056	6.398 837	4.924 525	23
7.784 316	7.282 883	6.835 137	6.433 771	4.937 104	24
7.843 139	7.329 985	6.872 927	6.464 149	4.947 587	25
7.895 660	7.371 668	6.906 077	6.490 564	4.956 323	26
7.942 554	7.408 556	6.935 155	6.513 534	4.963 602	27
7.984 423	7.441 200	6.960 662	6.533 508	4.969 668	28
8.021 806	7.470 088	6.983 037	6.550 877	4.974 724	29
8.055 184	7.495 653	7.002 664	6.565 980	4.978 936	30
8.084 986	7.518 277	7.019 881	6.579 113	4.982 447	31
8.111 594	7.538 299	7.034 983	6.590 533	4.985 372	32
8.135 352	7.556 016	7.048 231	6.600 463	4.987 810	33
8.156 564	7.571 696	7.059 852	6.609 099	4.898 842	34
8.175 504	7.585 572	7.070 045	6.616 607	4.991 535	35
8.243 777	7.634 376	7.105 041	6.641 778	4.996 598	40
8.282 516	7.660 864	7.123 217	6.654 293	4.998 633	45
8.304 498	7.675 242	7.132 656	6.660 515	4.999 451	50
8.316 972	7.683 045	7.137 559	6.663 608	4.999 779	55
8.324 049	7.687 280	7.140 106	6.665 146	4.999 911	60

present value of annuity of 1.

major overhaul) in the later years. The technique of setting the positive discounted flow equal to the negative discounted flow does not give a meaningful measure of its **profitability.** There are techniques designed to yield the correct rate of return in all cases but Solomon (Management of Corporate Capital) suggests it is far simpler to analyze the investment using the present-value method. There is also a problem associated with the internal-rate-of-return method tending to overstate the rate of return on very attractive investments. This is due to the implicit assumption that funds generated by a proposal can be reinvested in the business at the same rate of return as the proposal itself.

The internal-rate-of-return method can be used in addition to the present-value method in order to dramatize the desirability of an investment. Even though a project discounted at 10 percent may show a positive present value it is very useful to know if the yield is 60 percent or 10.1 percent. It gives a good measure of the "elbow room" or "urgency" of the project. The present-value method does not do this. A positive present value of $500 can result from a $1,000 investment or a $1,000,000 investment.

The internal-rate-of-return method although not theoretically pure, does provide the practical businessman with a significant advantage. It provides a **consistent method of analysis** throughout the organization which is completely independent of top management's current view of the proper cutoff rate of return. Authorities on the subject differ widely on how to compute the proper cutoff rate.

The concept of a rate of return is usually better understood than present value. Rate of return is a familiar term associated with everyday activities—stock portfolios, savings accounts, and automobile financing. The economic analyst concerned about the relatively minor mathematical problems associated with the internal-rate-of-return method may well consider the dual-rate method.

SELECTION OF PROPER CUTOFF RETURN OR DISCOUNT RATE. The selection of a minimum acceptable return on investment or a discount rate establishes a **criterion for decision making.** If the present-value method is used, a rate must be selected to discount future cash flows. The decision rule is then to accept all projects which show a positive present value. If the internal-rate-of-return-on-investment method is used, the projects which show a yield greater than the cutoff rate should be accepted.

The proper way to select this cutoff return is not obvious and is a matter of some controversy. The main points of view are summarized below.

One viewpoint holds that any profit is beneficial to a company. If a company can borrow money at 3 percent (after taxes) and make investments at 4 percent (after taxes), the company experiences a **net beneficial gain.** Therefore the after-tax interest rate on borrowed funds should be the cutoff rate of return on investments.

Others contend that this is not realistic since a company's total financing consists of both equity and debt financing. Raising capital through borrowing must eventually be followed by an equity issue to maintain a proper balance. The cost of equity financing from the current stockholders' point of view is the current earnings per share divided by the market price of the stock plus a factor for anticipated earnings growth. From this point of view the **cost of equity financing** is considerably more expensive than debt financing. The proper cost of capital is the average cost of both weighed by the proportion of each in the capital structure of the company. Therefore the minimum acceptable rate of return on investments is the weighted average cost of capital. Any return less than this will not be sufficient to attract capital funds to the firm.

Other authorities take the position that the average cost of capital is too low for the minimum return on investment. They suggest the use of an **opportunity rate** which is the cost of equity capital. They point out that a company has the option of investing in its own stock. Therefore, if internal investment projects do not yield as great a return as their own stock, the project should not be accepted and surplus funds should be used to buy stock either in their own company or in other similar risk companies. Thus the cost of equity capital becomes the minimum acceptable return on investment. The notion of the proper cutoff rate often varies in time as business and economic conditions change.

SIMPLIFIED RETURN-ON-INVESTMENT PROCEDURES. In most cases the manual calculation of the after-tax internal rate of return is a tedious trial-and-error procedure. **Forms and graphs** have been developed by many companies in order to reduce the time and effort required. An example of a manual procedure is outlined below.

Manual Procedure. Figure 4 may be used to summarize the data for manual calculations of the return on investment. Provision is made for showing the cash flow for each individual year.

Column 5 shows the net cash flow for each year in the life of the investment. The net cash flow each year will be either negative (cash outflow, e.g. investment in equipment) or positive (cash inflows, e.g. savings). Through a trial-and-error process using the table of present values (Fig. 2) the discount rate which causes the algebraic sum of the present values of cash flows in column 5 to equal zero can be determined. This discount rate is equal to the project's internal rate of return on investment.

DUAL-RATE, DISCOUNTED CASH-FLOW METHOD. Annual return on investment (ROI) by definition is the net income per year (E) divided by the initial investment (I). Net income is the balance left after the direct outlays have been paid and the capital brought up to the same point of value as it was at the beginning. Thus annual net income equals the annual after-tax cash flow (CF) minus a factor to recover the initial capital (A). Thus,

$$\text{ROI} = \frac{E}{I} = \frac{CF - A}{I}$$

Conceptually the annual capital recovery factor is put into a sinking fund which will equal the initial investment by the end of the life of the investment. In conventional accounting systems the sinking fund amount is the annual depreciation. No compounding of the sinking fund is allowed. The internal-rate-of-return method provides for a sinking fund compounded at the internal rate of return of the project. Provision for the sinking fund and compounding is done automatically when using the present-value tables. This process overstates the return of very attractive investments. A firm fortunate enough to have a project which yields 50 percent return on the investment is not likely to be able to reinvest the proceeds from the project in a sinking fund compounded at 50 percent per year. This, however, is the assumption built into the use of the present value tables. In contrast the dual-rate method considers the proceeds from an investment are placed into a sinking fund which is compounded at a rate selected to be equal to the company's average rate of return on invested capital.

The two rates used in this **analytical approach** are a rate for compounding the hypothetical sinking fund and a rate to evaluate the desirability of the investment. The sinking fund rate is selected to be equal to the ratio of the firm's after-tax cash flow to gross invested capital. This is the company's aver-

Capital Cost Analysis — Manual Procedures

Calculation of the % Return on Investment   Project No. _____

Year	Total Investment (1)	Total Savings (2)	Tax Credits (3)	Tax Payments (4)	After Tax Cash Flow (5)
0					
1					
2					
3					
4					
5					
6					
7					
8					
9					
10					
11					
12					
13					
14					
15					
16					
17					
18					
19					
20					
Total					

INSTRUCTIONS:

Column (1) = C + D + W
Column (2) = A + H + J + K + N
Column (3) = (T × D) + (T × C*) + (I × C) − (I × K)
Column (4) = (T × N) + (T × J) + (T × K*) +
              .25 (A − B) + .25 (H − C**)
Column (5) = (2) + (3) − (1) − (4)

T   = Tax Rate
I   = Investment Credit Rate
C*,K* = Annual Depreciation Charge
C** = Balance of C* in Year (F), Column (3)

SCHEDULE # 1 CODE

A = CASH SALVAGE VALUE
B = REMAINING BOOK VALUE
C = INSTALLED COST − CAPITALIZED
D = INSTALLED COST − EXPENSED
H = FUTURE SALVAGE VALUE
J = INVESTMENT AVOIDED − EXPENSE
K = INVESTMENT AVOIDED − CAPITAL
N = ANNUAL SAVINGS
W = WORKING CAPITAL

Fig. 4. Summary sheet for calculating return on investment.

age rate of return on invested capital. The second rate, independently chosen, is the minimum acceptable return on investment desired by the company.

A project is analyzed to determine the net investment and the after-tax cash income for each year of its life. The after-tax cash flow is divided into two components. One component is an equal annual annuity to a sinking fund to recover the investment. The sinking fund is compounded annually at the sinking fund rate and is established to be equal to the **net investment** by the end of the investment's life. The balance of the annual after-tax cash flow, after deducting the sinking fund annuity, is the amount remaining for income. The project's rate of return on investment is the annual income divided by the net investment. If the project's rate of return is greater than the minimum acceptable rate of return, the project should be accepted.

To illustrate the method assume a firm's average rate of return on invested capital is 6 percent. Consider an investment of $10,000 which promises an after-tax cash flow of $4,000 per year for 5 years. An annuity of $1,774 per year for 5 years deposited into a sinking fund compounded at 6 percent will recover the $10,000 investment at the end of 5 years (see Fig. 5). Thus the project's return on investment is 22.3 percent $\left(\dfrac{\$4,000 - 1,774}{\$10,000}\right)$. In this simplified example the method is very easy. It tends to become mathematically more complex when the annual cash flows are not equal or the investment is not a single payment at the beginning of the first year. The use of a computer is recommended for the calculations.

Procedure for Analysis. The full procedure is described by Hunt (Financial Analysis in Capital Budgeting) as follows:

1. For each time period in the life of a project, schedule the net after-tax inflows and outflows of cash.
2. If the investment is not a single cash payment at the beginning of the first year of the project, establish the equivalent of a single cash payment by using the sinking fund rate to develop the present value of all negative inflows. Inflows are measured from the amount of the annual sinking fund found in the next step. Some iteration between steps 2 and 3 may be necessary.
3. Determine the annual sinking fund amount to restore the initial investment at the sinking fund rate. Subtract the annual sinking fund amount from the net cash flows.
4. All remaining flows, if not zero, should be positive, measured from the sinking fund amount. If these flows are not equal each year they must be spread into the equivalent of equal annual flows.
5. Divide the equal annual cash flow for income by the initial investment. The result is the project's rate of return on investment.

Example of the Procedure. Consider an investment with the following after-tax cash flows:

Year	Cash Flow
0	−$10,000
1	5,000
2	5,000
3	− 1,000
4	500
5	5,000

Sinking fund rate = 6%
Minimum acceptable ROI = 10%

n	1%	2%	3%	4%	5%
1	1.000 000	1.000 000	1.000 000	1.000 000	1.000 000
2	0.497 512	0.495 050	0.492 611	0.490 196	0.487 805
3	0.330 022	0.326 755	0.323 530	0.320 349	0.317 209
4	0.246 281	0.242 624	0.239 027	0.235 490	0.232 012
5	0.196 040	0.192 158	0.188 355	0.184 627	0.180 975
6	0.162 548	0.158 526	0.154 598	0.150 762	0.147 017
7	0.138 628	0.134 512	0.130 506	0.126 610	0.122 820
8	0.120 690	0.116 510	0.112 456	0.108 528	0.104 722
9	0.106 740	0.102 515	0.098 434	0.094 493	0.090 690
10	0.095 582	0.091 327	0.087 231	0.083 291	0.079 505
11	0.086 454	0.082 178	0.078 077	0.074 149	0.070 389
12	0.078 849	0.074 560	0.070 462	0.066 552	0.062 825
13	0.072 415	0.068 118	0.064 030	0.060 144	0.056 456
14	0.066 901	0.062 602	0.058 526	0.054 669	0.051 024
15	0.062 124	0.057 825	0.053 767	0.049 941	0.046 342
16	0.057 945	0.053 650	0.049 611	0.045 820	0.042 270
17	0.054 258	0.049 970	0.045 953	0.042 199	0.038 699
18	0.050 982	0.046 702	0.042 709	0.038 993	0.035 546
19	0.048 052	0.043 782	0.039 814	0.036 139	0.032 745
20	0.045 415	0.041 157	0.037 216	0.033 582	0.030 243
21	0.043 031	0.038 785	0.034 872	0.031 280	0.027 996
22	0.040 864	0.036 631	0.032 747	0.029 199	0.025 971
23	0.038 886	0.034 668	0.030 814	0.027 309	0.024 137
24	0.037 073	0.032 871	0.029 047	0.025 587	0.022 471
25	0.035 407	0.031 220	0.027 428	0.024 012	0.020 952
26	0.033 869	0.029 699	0.025 938	0.022 567	0.019 564
27	0.032 446	0.028 293	0.024 564	0.021 239	0.018 292
28	0.031 124	0.026 990	0.023 293	0.020 013	0.017 123
29	0.029 895	0.025 778	0.022 115	0.018 880	0.016 046
30	0.028 748	0.024 650	0.021 019	0.017 830	0.015 051
31	0.027 676	0.023 596	0.019 999	0.016 855	0.014 132
32	0.026 671	0.022 611	0.019 047	0.015 949	0.013 280
33	0.025 727	0.021 687	0.018 156	0.015 104	0.012 490
34	0.024 840	0.020 819	0.017 322	0.014 315	0.011 755
35	0.024 004	0.020 002	0.016 539	0.013 577	0.011 072
40	0.020 456	0.016 556	0.013 262	0.010 523	0.008 278
45	0.017 705	0.013 910	0.010 785	0.008 262	0.006 262
50	0.015 513	0.011 823	0.008 865	0.006 550	0.004 777
55	0.013 726	0.010 143	0.007 349	0.005 231	0.003 667
60	0.012 244	0.008 768	0.006 133	0.004 202	0.002 828
65	0.010 997	0.007 626	0.005 146	0.003 390	0.002 189
70	0.009 933	0.006 668	0.004 337	0.002 745	0.001 699
80	0.008 219	0.005 161	0.003 112	0.001 814	0.001 030
90	0.006 903	0.004 046	0.002 256	0.001 208	0.000 627
100	0.005 866	0.003 203	0.001 647	0.000 808	0.000 383

Fig. 5. Sample table of sinking fund factor

5½%	6%	6½%	7%	7½%	n
1.000 000	1.000 000	1.000 000	1.000 000	1.000 000	1
0.486 618	0.485 437	0.484 262	0.483 092	0.481 928	2
0.315 654	0.314 110	0.312 576	0.311 052	0.309 538	3
0.230 294	0.228 591	0.226 903	0.225 228	0.223 568	4
0.179 176	0.177 396	0.175 635	0.173 891	0.172 165	5
0.145 179	0.143 363	0.141 568	0.139 796	0.138 045	6
0.120 964	0.119 135	0.117 331	0.115 553	0.113 800	7
0.102 864	0.101 036	0.099 237	0.097 468	0.095 727	8
0.088 839	0.087 022	0.085 238	0.083 486	0.081 767	9
0.077 668	0.075 868	0.074 105	0.072 378	0.070 686	10
0.068 571	0.066 793	0.065 055	0.063 357	0.061 697	11
0.061 029	0.059 277	0.057 568	0.055 902	0.054 278	12
0.054 684	0.052 960	0.051 283	0.049 651	0.048 064	13
0.049 279	0.047 585	0.045 940	0.044 345	0.042 797	14
0.044 626	0.042 963	0.041 353	0.039 795	0.038 287	15
0.040 583	0.038 952	0.037 378	0.035 858	0.034 391	16
0.037 042	0.035 445	0.033 906	0.032 425	0.031 000	17
0.033 920	0.032 357	0.030 855	0.029 413	0.028 029	18
0.031 150	0.029 621	0.028 156	0.026 753	0.025 411	19
0.028 679	0.027 185	0.025 756	0.024 393	0.023 092	20
0.026 465	0.025 005	0.023 613	0.022 289	0.021 029	21
0.024 471	0.023 046	0.021 691	0.020 406	0.019 187	22
0.022 670	0.021 278	0.019 961	0.018 714	0.017 535	23
0.021 036	0.019 679	0.018 398	0.017 189	0.016 050	24
0.019 549	0.018 227	0.016 981	0.015 811	0.014 711	25
0.018 193	0.016 904	0.015 695	0.014 561	0.013 500	26
0.016 952	0.015 697	0.014 523	0.013 426	0.012 402	27
0.015 814	0.014 593	0.013 453	0.012 392	0.011 405	28
0.014 769	0.013 580	0.012 474	0.011 449	0.010 498	29
0.013 805	0.012 649	0.011 577	0.010 586	0.009 671	30
0.012 917	0.011 792	0.010 754	0.009 797	0.008 916	31
0.012 095	0.011 002	0.009 997	0.009 073	0.008 226	32
0.011 335	0.010 273	0.009 299	0.008 408	0.007 594	33
0.010 630	0.009 598	0.008 656	0.007 797	0.007 015	34
0.009 975	0.008 974	0.008 062	0.007 234	0.006 483	35
0.007 320	0.006 462	0.005 694	0.005 009	0.004 400	40
0.005 431	0.004 700	0.004 060	0.003 500	0.003 011	45
0.004 061	0.003 444	0.002 914	0.002 460	0.002 072	50
0.003 055	0.002 537	0.002 101	0.001 736	0.001 432	55
0.002 307	0.001 876	0.001 520	0.001 229	0.000 991	60
0.001 748	0.001 391	0.001 103	0.000 872	0.000 688	65
0.001 328	0.001 033	0.000 801	0.000 620	0.000 478	70
0.000 769	0.000 573	0.000 424	0.000 314	0.000 231	80
0.000 448	0.000 318	0.000 225	0.000 159	0.000 112	90
0.000 261	0.000 177	0.000 120	0.000 081	0.000 054	100

(annuity whose compound amount is 1).

	Project Facts		First Trial			Second Trial				
Year	Cash Flow	P.V. Factor @ 6%	Income	Capital	P.V. of Capital	Sinking Fund	Income	Capital	P.V. of Capital	Year
0	−$10,000	1.000	$0	$10,000	$10,000	$0	$0	$10,000	$10,000	0
1	+5,000		5,000	0	0	1,923	3,077	0	0	1
2	+5,000		5,000	0	0	1,923	3,077	0	0	2
3	−1,000	0.840	0	1,000	840	1,923	0	2,923	2,455	3
4	+500	0.792	500	0	0	1,923	0	1,423	1,127	4
5	+5,000		5,000	0	0	1,923	3,077	0	0	5
					$10,840				$13,582	

Present Value of Capital Investment: $10,840

Sinking Fund To Recover Capital: $10,840 × 0.1774 = $1,923 (Second Trial Sinking Fund)

$13,582 × 0.1774 = $2,409 (Third Trial Sinking Fund)

	Project Facts		Ninth Trial			
Year	Cash Flow	P.V. Factor @ 6%	Sinking Fund	Income	Capital	P.V. of Capital
0	−$10,000	1.000	$0	$0	$10,000	$10,000
1	+5,000		2,608	2,392	0	0
2	+5,000		2,608	2,392	0	0
3	−1,000	0.840	2,608	0	3,608	3,031
4	+500	0.792	2,608	0	2,108	1,670
5	+5,000		2,608	2,392	0	0
						$14,701

(Third to eighth trials not shown)

Present Value of Capital Investment: $14,701

Sinking Fund To Recover Capital: $14,701 × 0.1774 = $2,608

Since $2,608 = $2,608, iteration stops.

Smoothing the Irregular Income

Income	P.V. Factor @ 10%	P.V. of Income	Year
$0			0
2,392	0.909	$2,174	1
2,392	0.826	1,976	2
0			3
0			4
2,392	0.621	1,485	5
		$5,635	

Present Value of Income: $5,635

P.V. Factor for Five-Year Uniform Series @ 10%: 3.791

Equivalent Equal Annual Income: $1,486 (5,635 ÷ 3.791)

Fig. 6 Example of dual rate calculation

The mathematics for this project is shown in Fig. 6. The return on investment for this project is 14.86 percent ($1,486 ÷ $10,000). It should be accepted since it exceeds the minimum acceptable return on investment.

MAPI INVESTMENT FORMULA. This method of analysis was developed by George Terborgh of the Machinery and Allied Products Institute. The "new" MAPI formula outlined by Terborgh (Business Investment Management) is an improved and refined version of his original formula (Dynamic Equipment Policy).

A projection of the future economic benefits of an investment project based on an analysis of the benefits during the first year is built into the MAPI formula. Terborgh considers that the main use of the investment formula is in the appraisal of minor investment projects, "particularly those—the most numerous category in most organizations—that concern the reduction of costs and the improvement of methods in existing operations." He goes on to elaborate that "if a major project does not lend itself to quantitative estimates, an investment formula cannot be applied. If, on the other hand, it justifies the time required to develop the necessary estimates **for both present and future,** these estimates naturally supersede formula projections. A formula finds its principal usefulness where the present, or initial, benefits of a project can be estimated, but where it is impractical, because of the time required or for other reasons, to develop the numerous future estimates required for a nonformula solution of the project. As noted, this condition occurs most frequently with minor projects."

MAPI Worksheet. The MAPI worksheet in Fig. 7 shows the summary of the analysis to determine the required investment, the annual operating advantage of the investment project, and the computation of the return on investment. The analysis is made between operating with the new equipment as compared to operating with the existing equipment for one more year or a longer period of comparison.

MAPI Charts. Terborgh (Business Investment Management) lists the four factors that must be specified to use the MAPI charts:

1. The comparison period
2. The estimated service life
3. The estimated terminal salvage value, if any (as a percentage of the asset's cost)
4. The depreciation method used for tax purposes

The comparison period is determined by the life of the shorter-lived alternative. This is usually the alternative of keeping the existing equipment one more year. The following four MAPI charts are all for a **one-year comparison** and the following income tax depreciation methods: sum-of-digits (Fig. 8, page 32), double-declining balance (Fig. 9, page 33), and straight-line (Fig. 10, page 34). Expensing is shown in Fig. 11, on page 35.

The **estimated service life** of the project is the period the asset in question is expected to be kept, not its full physical life. The estimate of service life should not be arbitrarily shortened to allow for obsolescence. Full allowance for deterioration and obsolescence is built into the MAPI formula and does not need to be made a second time by abridging the life estimate. The **terminal salvage value** must be estimated if it is at all significant. Since this is difficult to do accurately so far in advance, all that can be done is to make an informed guess. The MAPI charts assume an income tax rate of 50 percent during the life of the investment.

PROJECT NO._____

MAPI SUMMARY FORM
(AVERAGING SHORTCUT)

PROJECT_____

ALTERNATIVE_____

COMPARISON PERIOD (YEARS)   (P)_____

ASSUMED OPERATING RATE OF PROJECT (HOURS PER YEAR)
                                                             _____

I. OPERATING ADVANTAGE
(NEXT-YEAR FOR A 1-YEAR COMPARISON PERIOD,* ANNUAL AVERAGES FOR LONGER PERIODS)

A. EFFECT OF PROJECT ON REVENUE

		INCREASE	DECREASE
1	FROM CHANGE IN QUALITY OF PRODUCTS	$	$
2	FROM CHANGE IN VOLUME OF OUTPUT		
3	TOTAL	$ X	$ Y

B. EFFECT ON OPERATING COSTS

4	DIRECT LABOR	$	$
5	INDIRECT LABOR		
6	FRINGE BENEFITS		
7	MAINTENANCE		
8	TOOLING		
9	MATERIALS AND SUPPLIES		
10	INSPECTION		
11	ASSEMBLY		
12	SCRAP AND REWORK		
13	DOWN TIME		
14	POWER		
15	FLOOR SPACE		
16	PROPERTY TAXES AND INSURANCE		
17	SUBCONTRACTING		
18	INVENTORY		
19	SAFETY		
20	FLEXIBILITY		
21	OTHER		
22	TOTAL	$ Y	$ X

C. COMBINED EFFECT

23 NET INCREASE IN REVENUE (3X−3Y) $_____
24 NET DECREASE IN OPERATING COSTS (22X−22Y) $_____
25 ANNUAL OPERATING ADVANTAGE (23+24) $_____

* Next year means the first year of project operation. For projects with a significant break-in period, use performance after break-in.

Fig. 7. MAPI summary form.

Depreciation

NATURE OF DEPRECIATION. Any simple definition of depreciation is subject to shades of opinion and can become sophisticated and controversial. This is largely due to a shift in emphasis from the older, traditional engineering appraisal of **useful life** to the financial accounting viewpoint which is shaped

II. INVESTMENT AND RETURN

A. INITIAL INVESTMENT

26 INSTALLED COST OF PROJECT $ _____
      MINUS INITIAL TAX BENEFIT OF $ _____   (Net Cost) $ _____
27 INVESTMENT IN ALTERNATIVE
      CAPITAL ADDITIONS MINUS INITIAL TAX BENEFIT $ _____
      PLUS: DISPOSAL VALUE OF ASSETS RETIRED
              BY PROJECT * $ _____ $ _____
28 INITIAL NET INVESTMENT (26—27) $ _____

B. TERMINAL INVESTMENT

29 RETENTION VALUE OF PROJECT AT END OF COMPARISON PERIOD
      (ESTIMATE FOR ASSETS, IF ANY, THAT CANNOT BE DEPRECIATED OR EXPENSED. FOR OTHERS, ESTIMATE
      OR USE MAPI CHARTS.)

Item or Group	Installed Cost, Minus Initial Tax Benefit (Net Cost) A	Service Life (Years) B	Disposal Value, End of Life (Percent of Net Cost) C	MAPI Chart Number D	Chart Percentage E	Retention Value $\left(\dfrac{A \times E}{100}\right)$ F
$						$

      ESTIMATED FROM CHARTS (TOTAL OF COL. F) $ _____
          PLUS: OTHERWISE ESTIMATED $ _____ $ _____
30 DISPOSAL VALUE OF ALTERNATIVE AT END OF PERIOD * $ _____
31 TERMINAL NET INVESTMENT (29—30) $ _____

C. RETURN

32 AVERAGE NET CAPITAL CONSUMPTION $\left(\dfrac{28-31}{P}\right)$ $ _____

33 AVERAGE NET INVESTMENT $\left(\dfrac{28+31}{2}\right)$ $ _____

34 BEFORE-TAX RETURN $\left(\dfrac{25-32}{33} \times 100\right)$ % _____
35 INCREASE IN DEPRECIATION AND INTEREST DEDUCTIONS $ _____
36 TAXABLE OPERATING ADVANTAGE (25—35) $ _____
37 INCREASE IN INCOME TAX (36×TAX RATE) $ _____
38 AFTER-TAX OPERATING ADVANTAGE (25—37) $ _____
39 AVAILABLE FOR RETURN ON INVESTMENT (38—32) $ _____
40 AFTER-TAX RETURN $\left(\dfrac{39}{33} \times 100\right)$ % _____

* After terminal tax adjustments.

Fig. 7. Concluded.

by income tax considerations as well as implications of financial reporting to the public. As Massé (Optimal Investment Decisions) points out: "Obsolescence anticipates wear and tear. Our cities are prisoners of their narrow streets, built to the scale of pedestrians and hansom cabs. In the same way, investment, after having done immense service to mankind, may turn into a hampering restriction, a pretext for clinging to outworn strictures. Every investment is, to a certain extent, a fossilizing process."

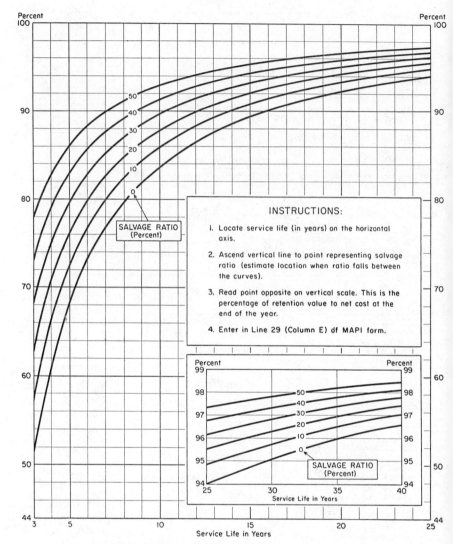

Fig. 8. MAPI chart, sum-of-digits tax depreciation.

Depreciation accounting as described by the American Institute of Certified Public Accountants (Accounting Terminology Bulletin No. 1) is:

. . . a system of accounting which aims to distribute the cost or other basic value of tangible capital assets, less salvage (if any), over the estimated useful life of the unit (which may be a group of assets) in a systematic and rational manner.

Coughlan and Strand (Depreciation—Accounting, Taxes, and Business Decisions) observe:

A machine's service may be terminated by casualty or economics and thus is to a degree controlled by chance, location, general business trends, and cycles. The service

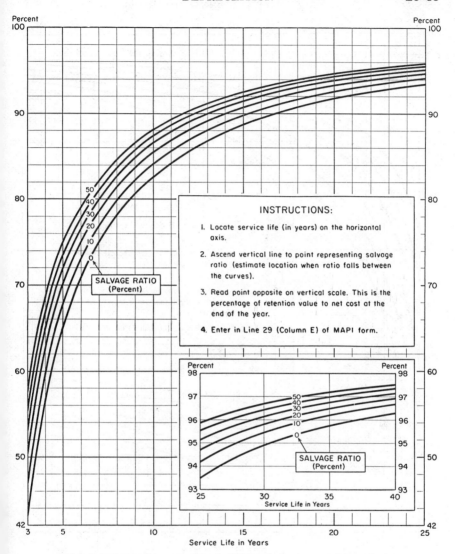

Fig. 9. MAPI chart, double-declining-balance tax depreciation.

life of any single unit of plant, such as a truck or a power saw, when first purchased and put into service, is manifestly impossible to predict with absolute accuracy. Yet since many thousands of trucks and power saws have been in service in the past, the average life of all newly installed trucks and power saws may be predicted with a good level of accuracy.

Bonbright (Valuation of Property) states that depreciation involves decrease in value, amortized cost, impaired serviceability, or the difference between the value of an old asset and a hypothetical new one taken as a standard of comparison. He claims that the many other meanings given to depreciation are merely variations of the above four basic concepts.

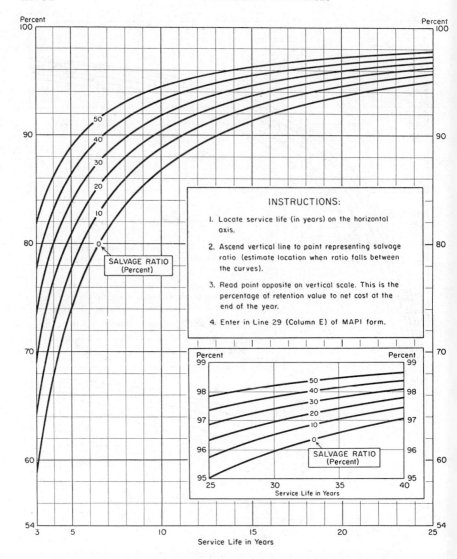

Fig. 10. MAPI chart, straight-line tax depreciation.

COMPUTING ANNUAL DEPRECIATION ALLOWANCES. The methods for computing the annual depreciation allowances shown below are accepted accounting practices and approved by the federal tax regulations. Many companies use one method of depreciation for tax purposes and a different method for financial reporting.

Straight-Line. The most common form of determining depreciation is the straight-line method. The annual deduction for depreciation in this method is obtained by dividing the cost or other basis of the asset (less any estimated salvage value) by the number of years estimated as the useful life of the property, or

$$D = \frac{I - S}{N}$$

where D = annual depreciation allowance
       I = in-place cost of equipment
       S = estimated salvage value at the end of its service life
       N = estimated years of service life

A depreciation rate may be obtained by dividing 100 percent by the estimated years of service life. Thus an asset with an estimated life of 20 years would have a 5 percent (or 0.05) straight-line rate.

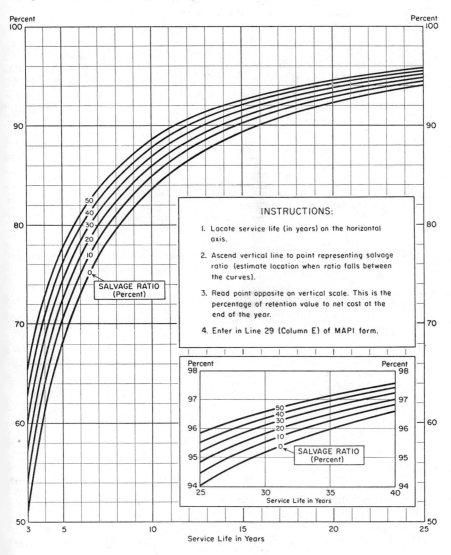

Fig. 11. MAPI chart, expensing.

Donaldson and Pfahl (Corporate Finance) warn that if a company requires the sum of depreciation and maintenance (or repair) charges to remain uniform during the entire life of the asset, the straight-line method should not be used. This is because maintenance charges on a piece of machinery are usually low when the machine is new and increase sharply as the machine gets older. In addition, this method does not reflect usage of the equipment. Thus when the machine is used intensively, the depreciation is understated; when the machine receives light use, the depreciation is overstated.

Sum-of-the-Years'-Digits. This is sometimes called the sum-of-digits or the years'-digits method. This method, considered the most liberal allowed under federal tax regulations, provides a more rapid recovery of the cost or other basis over the first half of the asset's life than other methods of depreciation. The largest deduction occurs in the first year of the estimated life. According to Coughlan and Strand (Depreciation—Accounting, Taxes, and Business Decisions) the amount of depreciation is found using a fraction whose **numerator** is the number of years remaining in the life of the asset (at the beginning of the year for which the depreciation is being calculated), and whose **denominator** is the sum of all the years comprising the asset's life. The **formula for this method** is

$$D = \frac{(R)(I - S)}{1 + 2 + \cdots + N} = \frac{2R(I - S)}{N(N + 1)}$$

where R = years remaining in estimated life (calculated at the beginning of the year for which the depreciation is being determined)
I = in-place cost of equipment
S = estimated salvage value at the end of its service life
N = estimated years of service life

As an example, the yearly depreciation of an asset costing $2,000 with a salvage value of $500 and an estimated life of 5 years will be as follows:

Sum-of-the-years'-digits = $1 + 2 + 3 + 4 + 5 = 15$

Year	Cost	Cost Less Salvage	Years of Estimated Life Remaining	Depreciation Rate	Depreciation
1	$2,000	$1,500	5	5/15	$ 500
2			4	4/15	400
3			3	3/15	300
4			2	2/15	200
5			1	1/15	100
				15/15	$1,500

Declining-Balance. This is also known as the double-declining-balance method or the DDB method. In this method a fixed percentage is applied to the undepreciated cost or other basis over the life of the asset. The maximum percentage permitted by law is twice the rate calculated by the straight-line method. As an example, consider an asset that cost $2,000. The estimated life is 5 years. Under straight-line depreciation the depreciation rate would be

$$\frac{100}{5} = 20\%, \quad \text{or} \quad 0.20$$

Thus using the declining-balance method a fixed percentage equal to $2 \times 0.20 = 0.40$ would be permitted on the undepreciated balance of the asset each year.

Year	Cost	Depreciation Rate	Depreciation	Undepreciated Amount (Cost less Depreciation)
1	$2,000	0.40	$ 800	$1,200
2		0.40	480	720
3		0.40	288	432
4		0.40	173	259
5		0.40	104	155
			$1,845	

Donaldson and Pfahl (Corporate Finance) cite as advantages of this method the fact that heavy depreciation charges are made in the early years of an asset when resale value tends to drop sharply; during the latter years when maintenance charges can be expected to be high, depreciation charges are low. Thus near uniform charges can be expected over the years for the total expenses of the machine.

One disadvantage of this method is that it is more difficult than the others to calculate. Note also that the asset can never be fully depreciated, i.e., a salvage value must be implied or known although it is not used in the calculation. The greatest advantage of this method is gained when the salvage value of the asset is relatively high or when it is to be disposed of prior to the end of its service life.

Sinking Fund. This method can be defined as the uniform annual charge that must be established so that the depreciated amount plus its accumulated compound interest (predetermined) will equal the capital to be recovered at some fixed future time. The predetermined interest to be assigned is not related in any way to the return on investment, but rather is an estimate of the market value of money. This method is such that as the interest rate approaches zero, the depreciation rate approaches a straight line. The disadvantage of this method is the fact that depreciation charges are very light in the early years and grow increasingly heavier. This method, therefore, is contrary to accepted practices in machinery and equipment depreciation.

The sinking fund method has ceased being an important factor in depreciation accounting.

GUIDELINES FOR DEPRECIATION. Guidelines have been established by the Internal Revenue Service (Revenue Procedure 62-21) for use in determining the service life of most types of structures, machinery, and equipment used in industry. Figure 12 gives some examples of service lives for depreciable assets used in manufacturing.

GUIDELINES FOR MANUFACTURING

In general, a single guideline class is specified for each manufacturing industry. This single guideline class includes all depreciable property that is not covered by another guideline class. Thus a single industry guideline class includes production machinery and equipment; power plant machinery and equipment; special equipment; and special-purpose structures. While more than one guideline class is specified for a particular industry, each guideline class covers that portion of the total depreciable property appropriate to the class.

Years

Aerospace Industry. Includes the manufacture of aircraft, spacecraft, rockets, missiles, and component parts. 8

Apparel and Fabricated Textile Products. Includes the manufacture of apparel, fur garments, and fabricated textile products except knitwear, knit products, and rubber and leather apparel. 9

Chemicals and Allied Products. Includes the manufacture of basic chemicals such as acids, alkalis, salts, and organic and inorganic chemicals; chemical products to be used in further manufacture, such as synthetic fibers and plastics materials, and finished chemical products such as pharmaceuticals, cosmetics, soaps, fertilizers, paints and varnishes, explosives, and compressed and liquefied gases. 11

Excludes the manufacture of finished rubber and plastics products.

Electrical Equipment.
  (a) Electrical Equipment. Includes the manufacture of electric household appliances, electronic equipment, batteries, ignition systems, and machinery used in the generation and utilization of electrical energy. . 12
  (b) Electronic Equipment. Includes the manufacture of electronic communication, detection, guidance, control, radiation, computation, test, and navigation equipment and components thereof. 8

Excludes manufacturers engaged only in the purchase and assembly of components. These manufacturers are included under guideline class (a).

Fabricated Metal Products. Includes the manufacture of fabricated metal products such as cans, tinware, hardware, metal structural products, stampings, and a variety of metal and wire products. 12

Glass and Glass Products. Includes the manufacture of flat, blown, or pressed glass products, such as plate, safety and window glass, glass containers, glassware, and fiberglass. 14

Excludes the manufacture of lenses.

Knitwear and Knit Products. Includes manufactured knitwear, knit products. 9

Lumber, Wood Products, and Furniture. Includes the manufacture of lumber, plywood, veneers, furniture, flooring, and other wood products. 10

Excludes logging and sawmilling and the manufacture of pulp and paper.

Machinery Except Electrical Machinery, Metalworking Machinery, and Transportation Equipment. Includes the manufacture of machinery such as engines and turbines; farm machinery; construction and mining machinery; food products machinery; textile machinery; woodworking machinery; paper industries machinery; compressors; pumps; ball and roller bearings; blowers; industrial patterns; process furnaces and ovens; office machines; and service industry machines and equipment. 12

Fig. 12. Guidelines for estimating service lives of depreciable assets in manufacturing.

Years

Metalworking Machinery. Includes the manufacture of metal cutting and forming machines and associated jigs, dies, fixtures, and accessories.   12

Paper and Allied Products.
  (a) Pulp and paper. Includes the manufacture of pulp from wood, rags, and other fibers and the manufacture of paper and paperboard from pulp.   16

Excludes paper finishing and conversion into cartons, bags, envelopes, and similar products.

  (b) Paper Finishing and Converting. Includes paper finishing and conversion into cartons, bags, envelopes, and similar products.   12

Petroleum and Natural Gas.
  (a) Drilling, Geophysical and Field Services. Includes the drilling of oil and gas wells on a contract, fee, or other basis and the provision of geophysical and other exploration services. Includes oil and gas field services, such as chemically treating, plugging, and abandoning wells and cementing or perforating well casings.   6

Excludes integrated petroleum and natural gas producers which perform these services for their own account.

  (b) Exploration, Drilling and Production. Includes the exploration, drilling, maintenance, and production activities of petroleum and natural gas producers. Also includes gathering pipelines and related storage facilities of such producers.   14

Excludes gathering pipelines and related storage facilities of pipeline companies.

  (c) Petroleum Refining. Includes the distillation, fractionation, and catalytic cracking of crude petroleum into gasoline and its other components.   16

  (d) Marketing. Includes the marketing of petroleum and petroleum products. Also includes related storage facilities and complete service stations.   16

Excludes petroleum and natural gas trunk pipelines and related storage facilities. Excludes natural gas distribution facilities.

Plastics Products. Includes the manufacture of processed, fabricated, and finished plastics products.   11

Excludes the manufacture of basic plastics materials.

Printing and Publishing. Includes printing, publishing, lithographing, and printing services such as bookbinding, typesetting, photoengraving, and electrotyping.   11

Rubber Products. Includes the manufacture of finished rubber products and the recapping, retreading, and rebuilding of tires   14

Other Manufacturing. Includes the manufacture of products not covered by other guideline classes in this group, such as the manufacture of fountain pens and jewelry. Also includes production of motion picture and television films, waste reduction plants, ginning of cotton, and the manufacture of musical instruments, including organs, pianos, and violins.   12

Excludes property used in the manufacture of products for which this guideline is clearly inappropriate. The depreciable life of such property shall be determined according to the particular facts and circumstances.

Fig. 12. Concluded.

Equipment Replacement Program

IMPORTANCE OF PROGRAM. The economic evaluation of investment opportunities is of limited value if it is not utilized as part of an equipment replacement program. Its use in connection with a replacement program brings order out of the chaos of replacement. The replacement analysis, as part of an established policy, singles out which equipment to replace first; therefore, the capital investment is always yielding the maximum return. A well-functioning replacement program is an effective tool to insure that investment opportunities are not missed.

In spite of this, R. G. Murdick, reporting on the results of a survey regarding the equipment replacement policies of manufacturing companies (Automation, vol. 12) found that 56 percent of the respondents have no special replacement policy. Even among those firms included in the top 500 U.S. manufacturers, 37 percent have no special policy and 10 percent more have only an unwritten policy. These results suggest that the majority of manufacturing firms do not attach the degree of importance to this subject which it deserves.

ESTABLISHING A REPLACEMENT PROGRAM. A successful replacement program will generally consist of the following elements:

1. The assignment of responsibility for initiating replacement proposals, for analysis of proposals, and for the final replacement decision.
2. Classification of replacements by purpose.
3. Adequate record keeping.
4. Coordination with all production functions.
5. Post completion audits.

ASSIGNMENTS OF RESPONSIBILITY. In the purchase of equipment, whether for replacements or new installations, many production functions might be concerned: manufacturing, production control, methods engineering, standardization, quality control, plant engineering, purchasing, finance, safety engineering, and sometimes product engineering. Their responsibilities from the various standpoints, are listed in Fig. 13. Selection, therefore, should undergo the scrutiny of each of these departments at some stage of the procedure so that all the factors relating to the acquisition and use of new or replacement machines receive proper consideration.

Where no special organization is set up to handle equipment problems, as may be the case in small plants, initiation of the study and recommendations usually come from the shops—the production control manager, or some executive or member of the firm supervising manufacturing operations. The problems are faced as they arise. There are seldom any planned methods for dealing with questions of equipment in the smaller organizations.

According to Murdick's survey (Automation, vol. 12) most recommendations for machinery replacement (80 percent) originate with manufacturing executives and supervisors below the officer level down through the foreman level. About 35 percent originate with the middle management group including plant managers, plant superintendents, chief engineers, and equivalent titles. About 20 percent originate with the foremen, and other direct supervisors. Replacement decisions, on the other hand, are made at the officer level in 70 percent of the firms reporting. Most of the remainder are made either by a general manager, division manager, or high-level manufacturing manager. Responsibility for **establishing and implementing** replacement policy may be assigned to:

1. The methods engineering or tool engineering group.
2. A standardization or equipment committee.
3. A screening procedure.

Function	Nature of Responsibility in the Selection
Product Engineering	Suitability of new equipment for processing parts of the size, shape, etc., which may be designed, and materials which are to be worked on the equipment. Degree of accuracy obtainable on new machines.
Production Control	Suitability, adequacy, and capacity of equipment from standpoint of getting out production. Changes which may be necessary in methods, routings, times, scheduling, etc. Effect on workers—training, wage-rate changes, etc. Relation of new equipment to present equipment from the standpoint of alternate routings, carrying of overloads, etc.
Manufacturing	Adequacy and capacity of new equipment for the work. Relation to present equipment. Convenience, reliability, and safety of operation. Changes which may be necessary in methods. Required training of, and supervision over, workers.
Methods Engineering	Changes in processing which the equipment may introduce. Tooling required for new equipment—jigs, fixtures, dies, small tools, etc.
Standardization	Standardization of materials, parts, etc., which may be made possible by new equipment. Standardization of routine paper-work procedures, or changes in existing standards, which may be brought about. Standardization of processing methods which may be introduced.
Inspection and Quality Control	Nature, amount, and kind of inspection required on work from the new machines. New inspection equipment (electronic gages, etc.) which the installation may make possible. Degree of accuracy obtainable on the equipment.
Plant Engineering	Requirements as to moving in and installing the new equipment (foundations, floor loads, doorway and elevator sizes, etc.). Power requirements. Maintenance and repairs (location for convenience in repairing, stocks of repair parts necessary, etc.).
Purchasing	Securing authorization for the purchase, with proper approval signature. Securing proper specifications for the new equipment. Negotiation of an adequate and legal purchase contract, including any necessary guarantees of performance, provisions for acceptance tests, etc.
Finance	Keeping equipment purchases within the budget, or seeing that proper authorization is secured for buying any special items. Checking the effect of the installation on cost methods and cost data. Amount of commitment and methods for meeting it. Time when payment must be made to obtain discount.

Fig. 13. Responsibilities of various production functions in selection of equipment.

However, even though a responsible group has been created for carrying out the replacement program, key personnel from all divisions must participate if the program is to be effective.

METHODS ENGINEERING PLAN. An effective way to handle problems of equipment engineering is through either methods engineering or a tool engineering department. Either, or both, if they exist concurrently, may report to the production department head or the plant manager, or they may form part of the general industrial engineering organization and report to a chief engineer. The responsibility setup to adopt is that which will produce the best results in the particular company. Eventually, in most cases, recommendations must meet with the approval of the plant manager.

Under the methods engineering plan, which is often set up to report to the head of production, all factors concerning the effective selection and application of equipment receive proper attention in a department well organized to handle such duties. The work can be carried on as a full-time job of competent engineering specialists who will assume full responsibility for the results as far as their studies and recommendations are concerned. These men should be encouraged and aided in keeping abreast of the latest technical developments in their field.

Equipment Reports and Records. Details of the methods of implementing an equipment plan, studies made, records kept, forms used, recommendation procedures, and other activities vary with the plant and industry, the plans of the management, and the type of problems faced. In any case, **adequate reports** should be made, used, indexed, and filed with a statement as to action taken, in order that studies will not be repeated and that information on the equipment is readily available to all departments which may need it. The equipment records can also be kept by methods engineering, and in that case all reports on machine inspection, condition, maintenance, alteration, moving, etc., should likewise clear through it for extraction of necessary notations.

Familiarity with Plant Conditions. An important advantage of placing the replacement function under methods engineering is that it is closely associated with all **time and motion study** work and **job standards.** Hence, it is thoroughly familiar with equipment in action under operating conditions. It is also concerned with the tooling of jobs—including the design, making, and use of all jigs, fixtures, appliances, and small tools required and the extent to which such devices may be profitably utilized. Its surveys cover the methods of processing and the layout of equipment. It is familiar with machine capacities, ranges of work, routing, operations performed, and conditions surrounding the process. Its equipment studies and recommendations, therefore, will fit in with plant operation to the most effective degree, and it will be definitely interested in following up each machine replacement or installation to see that the equipment is used as intended and gives both the operating and cost-saving results planned.

Equipment salesmen should be interviewed in this department. Even at times when no equipment changes are contemplated, it is advisable to keep in touch with equipment manufacturers through their salesmen to get news of machine developments and the way equipment is being efficiently utilized in other plants. The **purchasing department's** relation to the buying of equipment is usually only commercial, since it is not often possible to have an expert machinery buyer in the purchasing department. At any rate, he would have to be in intimate touch with the whole plant at all times. Usually the technical questions regarding the purchase of equipment are left to methods engineering and the shops, and, if the purchase is within the budget and receives the approval of a financial officer, the order will be put through under the standard purchasing procedure.

COMMITTEE PLAN OF SELECTING EQUIPMENT. When a company is of such size that the equipment problem is of major importance, executives must divide their time among many activities of an operating nature, and usually no one man can adequately investigate and decide on the selection of new machinery. At the same time, it may not be considered necessary to set up a department for the purpose. The advisability of integrating the equipment needs of the entire plant is of great enough importance to warrant group action. The solution, therefore, may be to adopt a committee plan of handling the problem. In this way the requirements of all departments may be met with due consideration to the demands of the company as a whole. This will also insure that all factors concerning any particular installation will be considered.

Standardization Committee. Larger and more extensively organized companies sometimes have a standardization committee, usually reporting to the plant manager, to which equipment matters may be referred. While such committees are concerned largely with standardization of materials and processes, these questions often involve equipment problems, and therefore these studies can be conveniently referred to the committee. **Equipment standardization** is highly desirable from the standpoint of uniformity of equipment, interchangeability of machines in layouts, reduction of stocks or machine repair parts, development of the most efficient use of equipment, taking and using of time studies, training of operators, efficiency of labor, routing of work, preparation of operation and route sheets, machine loading, and dispatching of work. However, as discussed above, many jobs can be more profitably handled with specially built equipment, and the decision to adopt standard machinery or to utilize specially built machinery must be reached by the committee in the light of all of the conditions.

These factors have a significant bearing on the kinds of materials used and the methods of processing, both of which obviously must be standardized as part of any plan of mass or straight-line production.

Equipment Committee. A still further development in larger companies is the establishment of an equipment committee, which concerns itself mainly with investigations of possible new or replacement machines for improved processing, and to which all suggestions of this nature are referred for study. This committee in most cases reports to the factory manager, and its recommendations are usually accepted unless financial reasons prevent their adoption.

Advantages and Disadvantages of Committee Plan. In all these committee plans, the membership represents most if not all departments concerned with the acquisition or use of the equipment. To this extent, the final selection meets the requirements and falls within the limitations of all divisions—operating and financial—having an interest in the problem. A frequent disadvantage, however, is that no members of the committee are specialists in equipment. Many of them are well informed about the technical and operating factors of their industry and plant, but from the standpoint of product rather than efficiency of the machinery and devices used for processing. The work of committees, moreover, is regarded as a side issue apart from regular operation of the plant. The committees meet only when necessary, the members finish their work as quickly as possible because of other obligations, and carry on any outside studies largely in a perfunctory and often inadequate and superficial manner.

SCREENING PROCEDURE PLAN. The methods for assignment of replacement responsibility explained above are not in as general use as the screening plan. The reason is that the screening procedure seems to be the natural occurrence in the development of a company. A company which used this method had the following **normal replacement pattern:**

1. The **foreman** recommended a replacement when he felt it was justifiable.
2. The **plant superintendent** and **foreman** jointly decided on a specific replacement.
3. The **plant manager's** approval was the superintendent's authority to originate a purchase order.
4. The **vice-president in charge of manufacturing** then was the final authority unless the nature was such that the top management would be involved.

However, this procedure was not fixed because replacement could be initiated by (1) the production department, (2) any of the engineering departments, (3) the methods or time study department, and (4) the vice-president in charge of manufacturing.

Capital equipment expenditures at National Gypsum Company are controlled by the following procedure:

1. Capital budgets are prepared annually by each operating division taking into consideration:
   a. Equipment replacement
   b. Capacity needs for future years
   c. Improving production methods
   d. Relating production capacity with short and longer term forecasts
2. Total capital budget is reviewed and approved by the Board of Directors.
3. After approval, the capital budget becomes the control medium for submission of individual appropriation requests prior to activation. Submission follows the sequence:
   a. Plants initiating the request
   b. Preparation of engineering details
   c. Marketing and sales forecasts
   d. Financial staff developing economics
4. Review of capital requests and approval for appropriating funds follows:
   a. Request for appropriating of budgeted funds under $25,000 are reviewed and approved by divisional management.
   b. Request for appropriating of budgeted funds of $25,000 and over are reviewed by divisional management and the corporate staff with approval by Chief Executive Officer.
   c. Request for appropriating of unbudgeted funds of $250,000 and over are reviewed by divisional and corporate management with final approval by the Board of Directors.

COORDINATION WITH ALL PRODUCTION FUNCTIONS. With the large number of functions likely to become involved with machine replacements, it is important that the individual or department charged with the primary responsibility provide the necessary coordination between them. As described in the previous articles, each of these functions or departments will have its own responsibilities toward machine replacements. Many of these responsibilities will conflict with one another thus requiring that someone with ability and authority coordinate the needs of all in the best interest of the project.

CLASSIFICATION OF REPLACEMENTS. An important part of any replacement program is the classification of capital expenditures. This classification will vary from one company to another, depending upon the nature of the product, the competitive situation, the amount of money available or obtainable for replacements and new equipment and other factors pertaining to the industry. Norton Company (Iron Age, vol. 198) has ten categories of capital spending:

Corporate Image	Customer Service
Safety	Reliability
Working Conditions	Maintenance
Quality	Cost Reduction
Development	Increased Revenue

The Ford Motor Company (Machinery, vol. 72) has four broad categories:

Replacement of worn out, inefficient, or obsolete tools
Process and product improvement
New model programs
Planned capacity increases

The purpose of these classifications is to know just what benefits will be reaped from capital outlays.

EQUIPMENT OPERATION RECORDS. Adequate record keeping is a prerequisite to a successful program. However, according to Murdick's survey, less than 60 percent of the firms surveyed maintained historical records on either all machines or those in excess of a stated value. The remainder kept no records at all or kept records only on selected types of equipment. Thus it would appear that a first step in establishing a replacement program for many companies would be to institute a detailed record keeping system for capital equipment.

As a minimum, these records should contain a detailed record of **repair and maintenance expenditures** over the life of the machine. It is also desirable in a great many instances to include records of the duration and frequency of breakdowns and any chronic failures. When the time comes to determine the need or to justify a replacement, further records concerning production rates, setup costs, tooling costs, rejections, rework, material handling, etc. will prove invaluable.

TYPICAL REPLACEMENT PROGRAM. Huber (Machinery, vol. 72) describes a successful replacement program operated by the Bellefontaine plant of the Rockwell Manufacturing Company. In this plant, the **industrial engineering department** has been charged with the responsibility of keeping abreast of improved manufacturing techniques and capabilities of new equipment as well as replacing and adding production machines as they are needed.

Comprehensive records are maintained for each piece of production equipment. As soon as an increase is noted in the scrap rate or maintenance cost of a piece of equipment, consideration is given to overhauling or replacing it.

Before any new machine is purchased to replace existing equipment or expand production capacity, an extensive analysis is made to justify the expenditure. This analysis will include an estimated utilization of the new equipment, its useful life, and its cost. Based on a study of all the parts to be processed on the new equipment, computations are made of annual savings due to:

Less reworking	Reduced maintenance
Faster setups	Reduced inspection
Improved material handling	

If the new machine is to replace an existing one, the cost and savings are also compared to the cost of overhauling the old machine and its estimated useful life after overhaul.

In order to be approved, a replacement proposal must meet the minimum corporate return on investment objective of 15 per cent of total assets before taxes.

A post completion audit is conducted from six months to one year after the installation of the equipment to assure that the estimated savings are being realized.

POST-COMPLETION AUDITS. The books should not be closed on any capital expenditure until management has ascertained whether or not the objectives of the expenditure have actually been realized. This may be accomplished by means of the post-completion audit which is little more than a review of a capital project for the purpose of comparing actual and estimated results. Mock (Management Accounting, vol. 49) outlines **three primary purposes** for a managerial comparison of actual and predicted performance of completed projects:

1. To improve future decisions by knowledge gained from past projects.
2. To ensure realistic estimates.
3. To improve operating management performance.

Kemp (Management Accounting, vol. 47) states that the post-completion audit is performed for two reasons:

1. To help improve the project evaluation process.
2. Where appropriate, to indicate remedial action.

Responsibility for preparation of these audits is generally placed with an accounting group responsible either to the controller or to management. Since one of management's objectives of the audit is to ensure realistic estimates, it is important to have someone, independent of the estimators, check the accuracy of the forecasts. However, where the results of a project are difficult to verify due to its complexity, it may be desirable to use a post-audit team composed of both the people involved in the original justification study and one or more outsiders.

Management must also decide which projects to audit. In making this decision, the costs of the audit relative to the expected benefit should be considered. Mock mentions several factors which affect the possible benefits:

1. The importance of the project
2. The extent that the project represents new ideas
3. The similarity of the project to possible future capital expenditures
4. The quality of the estimates made in the pre-investment evaluation of the project

A more comprehensive approach to selecting projects for audit is to audit all large projects and a sampling of those of lesser importance.

SYSTEMATIC REPLACEMENT ANALYSIS. There is no substitute for systematic analysis before deciding on replacement policies and practices. Partially valid figures can often lead to erroneous conclusions. Every effort must be made to include all valid factors in the analyses.

Much work has been done to provide management with adequate tools to guide their decisions. These tools and techniques have been described earlier in this section. No one program can be prescribed to suit all situations in all companies. It is important that each firm, after analyzing its own situation and needs, develop a replacement program utilizing those tools and techniques which result in the best performance.

PLANT MAINTENANCE

CONTENTS

CONTENTS (*Continued*)

PLANT MAINTENANCE

Scope of Maintenance

OBJECTIVES OF PLANT MAINTENANCE. The purpose of maintenance is to retain the plant including buildings, grounds, equipment, and production machinery in the as-built condition as nearly as possible. It is the task of the maintenance force to plan and schedule work so as to anticipate and prevent interruptions in operations and perform this function at the least cost. The work assigned to the maintenance department usually includes removal and installation of equipment and planning and scheduling of work. If new construction and alterations to building and equipment are added to maintenance duties, separate records should be kept since such work is normally considered capital expenditure rather than maintenance.

FUNCTIONS OF MAINTENANCE DEPARTMENT. The size of the plant, in most cases, determines what functions are allocated to the maintenance department. In small plants, the function is likely to be more inclusive than in large plants where decentralization is necessary. Fig. 1 lists the maintenance functions in a typical plant. In all cases the work includes inspections, adjustments, repairs, replacement, and operation of the shops performing the various jobs.

IMPORTANCE OF PLANNED MAINTENANCE. The importance of adequate planning in the maintenance function cannot be overemphasized. If maintenance is not adequately planned, the cost of maintenance becomes abnormally high. Only a systematic approach to maintenance can reduce this cost.

Factors Influencing Maintenance. The modern practice of organizing maintenance work to prevent interruption of operations is the outgrowth of several factors.

1. Increased mechanization which decreased direct labor cost per output unit has required at least a portion of the gain be spent on maintenance of equipment. It has been found uneconomical to retain large maintenance staffs for emergencies that planning and systemized inspections can avoid. The newer techniques of **operations research,** especially queuing theory, and the application of statistical methods can aid in the optimization of the maintenance force and materially reduce maintenance cost.
2. Close management of production, with minimum stocks between operations or direct flow between machines, has made interruptions to production costly. In process industries the problem of material-between-machines manifests itself in the size of tanks or accumulator drums between operations. Even where workers are paid on an output basis, regulations often prescribe payment for waiting time when workers are delayed on the job through no fault of their own. Lost profits from production stoppage frequently exceed the cost of idle labor.
3. Failure to deliver on time, with the possible loss of future business, may result from interruptions to operations.
4. Preventive maintenance or correction of defective conditions, not only decreases the cost of repairs but also maintains the quality and capacity of

machinery. New techniques, such as non-destructive testing, greatly aid this effort.

5. Utility and service expenses for steam, electricity, gas, water, and the like are reduced by a continuous maintenance program.
6. Craftsmen who specialize in specific areas increase the reliability of work done and lower overall cost.
7. Adequate planning of maintenance operations will insure that needed spare parts and materials are on hand in stores.

PLANT ENGINEERING AND MAINTENANCE. A term commonly used to designate the maintenance function in its broad sense is **plant engineering.** Similarly, the individual in charge of plant engineering is likely to be called the **plant engineer.** A plant engineer's range of work encompasses the functions shown in Fig. 1. In the modern plant, it is necessary to continually introduce new and improved serviec equipment, new and improved operating methods, and additional machinery to perform the ever increasing task of maintenance. A knowledge of the new techniques and equipment in the maintenance field is necessary if the maintenance function is to keep step with progress in other production functions. Although selection of manufacturing equipment is of primary concern to production engineering, it is necessary that the plant engineer be consulted on service requirements, power and utility requirements, and other specifications that affect the maintenance function. Since the plant engineer's duties require a wide technical knowledge, he must not only be a good executive, but must also be the type of individual who keeps up to date technically. Although large multi-plant operations often centralize production engineering it is usually impossible to maintain a parallel organization with maintenance. Good organization dictates that a responsible individual with decision-making powers be present at each site to make the many and varied maintenance decisions necessary to the continuing operation of the plant.

Equipment Design and Selection. Both the frequency and the amount of maintenance required by a machine hinge to a large extent on the basic design of the equipment. Ruggedness of construction, balance of moving parts, and accessibility for repairs are all stressed in the theory of design, but differing ideas of machine builders on first cost and operating efficiency complicate the choice of equipment; new alloys, synthetics, closer specifications of materials and tolerances, and improved elements of construction bring about improved designs and lengthened periods of operation. The plant engineer must keep posted on developments and sources of supply, since selection of equipment is the first step in maintenance. Sales engineers are valuable sources of information; the plant engineer should arrange with the purchasing manager to be notified when visiting salesmen have useful information bearing on the maintenance of equipment.

Maintenance Organization and Functions

ORGANIZING MAINTENANCE OPERATIONS. Carson (Southern Engineering, vol. 84) lists the four organizational approaches to industrial maintenance as follows:

1. Centralization by trade.
2. Maintenance coverage by area.
3. A combination of central and area administration.
4. Use of one man skilled in two or more trades; the craftsman concept; for example, a combination millwright and welder, or a combination electrician and instrument mechanic.

Basic Maintenance Organization. Regardless of the type of maintenance organization used, it is common practice to have a single person responsible for

1. **Building Construction and Maintenance**
   a. Masonry—foundations, walls, permanent partitions, plastering, tiling
   b. Steel work—columns, beams, stairways, windows, fire-escapes
   c. Floors—concrete, plank, wood block, mastic, steel plate, or gratings
   d. Service mains—water, gas, steam, compressed air, oil, piping for solutions used in production
   e. Heating, ventilating, air conditioning—piping, ducts, radiators
   f. Carpentry and wood construction
   g. Painting
   h. Plumbing
   i. Roofing and tinning. cleaning
   j. General building upkeep—hardware, glazing
   k. Minor construction. (In large plants, sometimes major construction is carried on.)
   l. Inspection of construction done by outside contractors

2. **Mechanical Equipment Maintenance**
   a. Steam power equipment
   b. Steam-heating, ventilating, and air-conditioning equipment
   c. Millwright work—shafting, pulleys, drives, equipment installation, moving, set-up, alignment, removal
   d. Compressed air equipment
   e. Heat-treating and furnace equipment
   f. Machine and operating equipment installation and repairs
   g. Lubrication of machinery and equipment
   h. Sheet-metal and welding work for maintenance or special construction
   i. Materials handling equipment
   j. Storeroom equipment setup
   k. In some cases, factory layout—at any rate carrying out relayout plans
   l. Mechanical meters, gages, recording devices and instruments

3. **Electrical Equipment Maintenance**
   a. Electrical power plant equipment, transformers, etc.
   b. Wiring, conduits, switch boxes, cutouts, power outlets
   c. Electric lighting
   d. Motors—rewinding, commutator repairs, etc.
   e. Alarm, signaling, call, and communication systems

3. **Electrical Equipment Maintenance (Cont'd)**
   f. Private telephone systems
   g. Electrical equipment, tools, furnaces, etc.
   h. Lightning protection devices
   i. Electronic control
   j. Electric meters, and instruments, gages, recording devices
   k. Battery charging

4. **Plant Safety, Fire and Theft Protection, and Other Services**
   a. Safety guards and all safety installations
   b. Floor marking in plant, road marking, and walkways outdoors
   c. Warning signs
   d. Railroad and roadway crossing protection
   e. Watchmen's service—fire, burglary, etc.
   f. Fire-fighting equipment—yard hydrants, sprinkler systems, trucks, ladders, axes and other tools, hose, lanterns, pails, extinguishers
   g. Janitor services and general cleanliness
   h. General plant housekeeping and clean-up

5. **Yard and Ground Maintenance**
   a. Railroad tracks, switches, trestles, etc.
   b. Roadways, walkways, paving, concreting
   c. Tunnels and conduits
   d. Sheds and other yard structures
   e. Outdoor crane structures
   f. Poles for cables, wiring, etc.
   g. Fences
   h. Outdoor signs
   i. Rigging
   j. General yard layout
   k. Outdoor storage areas and facilities
   l. Landscaping and gardening—lawns, trees, shrubbery
   m. Parking facilities
   n. Yard drainage
   o. General yard cleanliness and good housekeeping
   p. Collection and disposal of refuse, rubbish, ashes. etc.
   q. Snow removal, road sanding

In all cases this work includes maintenance inspections, and adjustments, repairs, replacements, and the operation of shops for various kinds of work.

Fig. 1. Typical plant maintenance functions.

the operations of the entire department. The title of the person will vary depending upon the company or industry but such titles as Maintenance Manager, Chief Engineer, and Plant Engineer, are among those in common use.

The basic alignment of top supervision necessary to effectively control the maintenance function is shown in Fig. 2. Such an alignment is independent of the departmental approach and can be used just as well with centralized, area, or combination organizations.

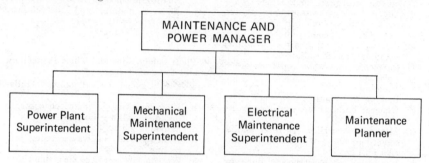

Fig. 2. Basic alignment of maintenance supervision.

The organization of the department below the top supervisory level will reflect the type of maintenance coverage selected by the company.

Each of these organizations will offer its own peculiar advantages as well as limitations.

Before affixing the structure of the maintenance organization, management must carefully weigh the known advantages and disadvantages of each type organization.

The basic people needed in a typical plant would seem to be the Manager, who will be charged with the total responsibility. Then, a Power Plant Superintendent, a Mechanical and Electrical Maintenance Superintendent and the Maintenance Planner, or Controller, as the case may be, will be required. These people will have much to do in setting up their departments, plus assisting with the setting up of the organization so that their time will not be wasted. In fact, this will prove to be the period when duties will be more demanding than at any other time, since none of these people will be in a position to delegate responsibility.

Centralization by Trade. Centralized maintenance simply means that the various maintenance skills are segregated, and the supervisor responsible for each skill is usually a **master craftsman** of that trade. Such an organization may well assume the structure depicted in Fig. 3. Some of the advantages of central maintenance coverage are as follows:

1. A direct responsibility is affixed to a qualified individual for the performance of each craft.
2. The flexibility of stopping the job of least importance, millwide, when manning an emergency breakdown is easily afforded.
3. Since the supervisor is a master craftsman of the trade, he is competent to instruct in method, and capable of determining very accurately the efficiency of his people in performing work.
4. When hiring new personnel, the central trade supervisor is in a better position to select the best qualified people.
5. Being responsible for a highly skilled craft is a constant challenge to the supervisor to stay abreast of modern discoveries in his line of work.

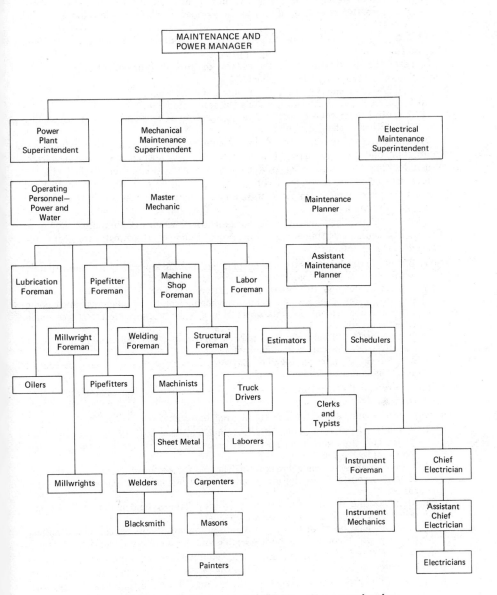

Fig. 3. Centralization by trade maintenance organization.

6. The knowledge and attitude of the crew is affected to the point that a spirit of competition is developed among the workers which results in a highly skilled department.

This type organization does pose some problems and it is necessary that these be carefully weighed before affixing approval.

Some of the disadvantages in the central maintenance organization are:

1. There is a tendency toward overstaffing.
2. It tends to promote specialization among the tradesmen.
3. It fosters a clannish attitude among the tradesmen.
4. A central maintenance organization requires much more accuracy in scheduling than either of the other three.

Maintenance Coverage by Area. A maintenance organization which has more than one craft reporting to one supervisor, is termed **area maintenance**. When looking at maintenance organization can totally divorce itself from the influence of this type coverage. Noting Fig. 3, it will be seen that this centralized organization shows more than one craft responsible to one supervisor, a semblance of area maintenance.

Like the centralized maintenance organization, area maintenance, shown in Fig. 4, has both advantages and limitations. Some of the most prominent advantages are as follows:

1. All major trades necessary to keeping an area in operation are immediately available to the area maintenance supervisor.
2. It aids in keeping the emergency experience in line, in that when operation declares a job to be an emergency, the manpower to do the work must come from a schedules job in that area. Hence, each work request receives much closer scrutiny.
3. It promotes a closer relationship between the various tradesmen, and consequently trade lines are not so finely drawn.
4. Men working so close together day in and day out soon begin to second guess one another.

These are the more prominent advantages afforded by the area maintenance organization.

Now, note some of the limitations of area maintenance.

1. It tends to promote laxity in scheduling, i.e., the foreman has access to all trades and therefore if he misses an assist, he can always pull the man and send him to the job.
2. To cover an emergency breakdown, the manpower must come from the job of least importance in the area, and not millwide as with the central organization.
3. Men working so close together day by day have a tendency to tell one another their outside problems, seek advice, etc.
4. Close association of men could promote antagonism, and hence, ill will among the crew which could cause both schedule and production interruptions.

Area and Central Combination. The combination area and central maintenance organization, seen in Fig. 5, is the most widely used in all types of industry. This organization structure affords more latitude than the others, with the advantages of both area and central. Some of the outstanding advantages of a combination area and centralized organization are as follows:

1. The prime major trades are available to the area maintenance supervisor.
2. The scheduling must be more accurate since the area supervisor is depending on help from outside his own bailiwick.
3. Like the area coverage, operating supervision is reluctant to classify a job as an emergency since the manpower must come from other scheduled jobs in their own area.

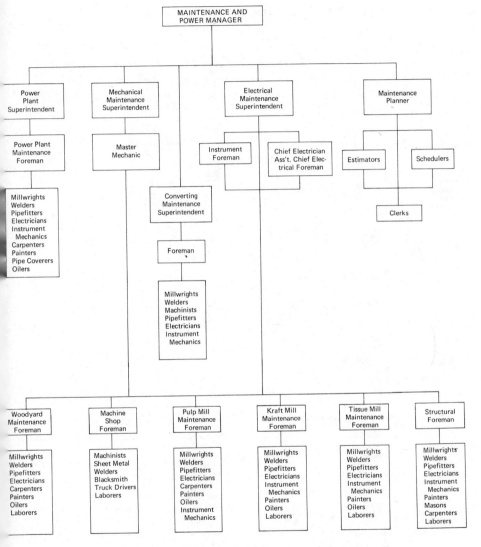

Fig. 4. Organization based on area maintenance.

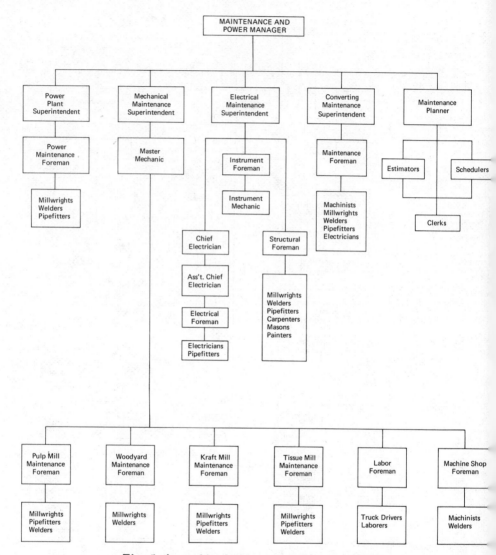

Fig. 5. A combined area and trade organization.

4. It promotes the same close relationships between tradesmen, as described under area maintenance, which affords the same advantages and limitations.

Although this organizational structure is apparently the most effective and likewise the most economical and efficient, it, like the others, does pose some problems. These include the following:

1. The area supervisor does not have all trades necessary to maintain, his area.
2. Again, in covering the emergency breakdown, it must be covered with manpower from the job of least importance area wide as opposed to mill wide.
3. The objections voiced under area maintenance concerning the close relationship of the people would also apply under area-central combination, only to a lesser degree.
4. The crew must have unusually skilled supervisors, or the journeyman mechanic will be dictating methods and times in respect to complicated repairs.

Once a policy has been established and the methods become entrenched, changes are costly.

Multi-Craft Tradesmen. This type maintenance structure is a partial resurrection of bygone days. The modern version has not as yet been accepted by industry as a whole and is still in what might be termed the research stage.

From all appearances, this type organization appears to have merit well worth consideration. Its basic organizational chart should take either the form of straight area maintenance or the combination area and central, depending on how many crafts it is safe to combine, the number of people required in the non-integrating crafts, and the availability of the superskilled persons required to man a plant.

First, it would be well to determine the crafts which would lend themselves to the effort with minimum training. It would seem that the following trades are closely enough related to comply with the needs of most industries:

1. Electricians, electronic technicians, and instrument mechanics.
2. Machinists, millwrights, welders, blacksmith, and sheet metal mechanics.
3. Carpenters, masons, and painters.
4. Pipefitters, pipe coverers, and welders.
5. Auto mechanics, diesel mechanics, and oilers.
6. Laborers, and truck drivers.

Using the above alignment, we see a possibility of combining eighteen skills with the use of six combination tradesmen. Considering this maintenance structure on the organization chart on a straight area basis, we see, in Fig. 6, a totally different situation.

It seems as though this type maintenance organization would afford some interesting advantages, some of which follow:

1. More flexibility in scheduling, i.e., in scheduling a pipefitter to repair a steam leak, it would not be necessary to send a welder.
2. When workload demanded, there would be less conflict when calling on one combination man to help another, i.e., the electrician is a good conduit man, therefore he would require little training in assisting the pipefitter, or vice versa.
3. Such an organization should be capable of operating with a minimum of eighty percent of the normal crew under the organizational structures.
4. When manning the emergency breakdown, less interruptions would occur in the regular schedule.
5. As a consequence, it would be possible to stay more closely to the planned schedule.

This type organization would pose its problems, some much like the area structure, others new and different. They include the following.

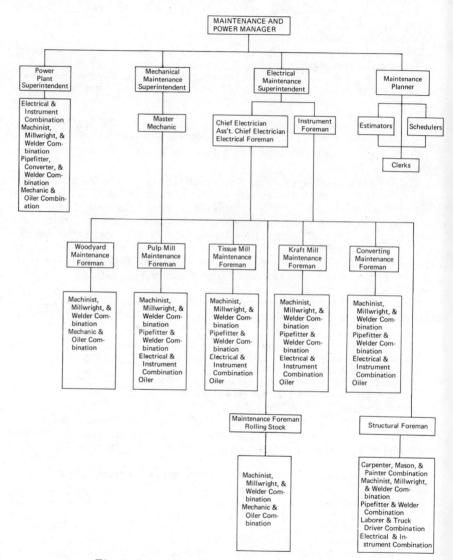

Fig. 6. A multi-craft tradesmen's organization.

1. This type tradesman would be hard to come by or to keep.
2. Obviously, he would come at a higher price.
3. The turnover would probably be higher since these type workers are in demand by management as supervisors.
4. People with the qualifications necessary to hold down the superskilled jobs are usually ambitious and also have a tendency to insubordination.

It must be agreed that regardless of the type organization, the main goal is effectiveness. And, for the most part, the effectiveness of an organization has to do with how well the people are sold on the company and its policies.

SIZE OF MAINTENANCE FORCE. A primary requisite for adequate maintenance is sufficient men—but not an excessive number—of each craft to meet the demands under peak loads. It is unwise to budget too closely, but in the interest of plant discipline as well as economy, excessive personnel should be avoided. Where peak loads can be foreseen but the nature of the jobs prevents spreading work over a suitably long period, **temporary shifting** of men from production to maintenance may remove the need for carrying a large maintenance crew. Where maintenance work has not been set up on an organization basis, past history is an unreliable guide; frequently, unnecessarily large crews have been retained. On the other hand, an aging plant may require an increase in the crew to avoid the danger of undermaintenance.

A satisfactory solution of the problem requires a **man-hour rating of the work and an annual program of jobs,** as discussed later under planning of maintenance activities. Many jobs recur only at long intervals; hence the accumulation of adequate data may take a year or more. As an initial guide in setting up the maintenance force, a general over-all ratio to determine the proper number of workers may help, but such a ratio must be used with caution.

In budgeting, it is important to specify what personnel belong to maintenance. Automated equipment requires that **roving mechanics** be assigned to keep machinery in adjustment for output within specified tolerances. Such personnel customarily report to operating management and are classed as production personnel, not maintenance.

Ratio of Maintenance Force to Total Production Force. As equipment units become larger, heavier, and more intricate, the proportion of maintenance workers increases, but the claim is made that, in a given type of manufacturing, approximately equal ratios are found in well-managed plants. Where the ratio is heavier than normal for the particular type of industry, the figures suggest there may be faulty organization, inadequate supervision or inspection, lack of a maintenance order system and records, or inadequate budgetary control.

A breakdown by **crafts** is useful in adjusting the maintenance force to the annual level of activity. Comparisons between plants, even when the product is similar, should be made with caution.

Where the pattern is fairly constant, managements prescribe varying degrees of **cleanliness** and **plant protection.** A food concern welcoming many visitors may have a modernistic building in a landscaped setting, with the advertising value justifying somewhat large expenditures for groundsmen, floor-polishers, painters, and other workers, far beyond any reasonable necessities of manufacturing. **Location** has a bearing. Northern plants with yards, sidings, and roadways have heavier snow removal expense. Plants in open areas have added yard expenses for the parking and protection of employees' cars.

Mechanical crafts items are so peculiar to each plant that general ratios must be used warily.

The general conclusion is that where the plant engineer has reliable data from closely comparable plants with which he is familiar, such existing satisfactory **maintenance personnel ratios** can be of assistance in establishing the proper size of maintenance crews.

Job Classifications. Most plants have craft classifications which include

Boilermakers	Welders	General laborers
Burners	Carpenters	Machinists
Lead burners	Painters	Masons
Tinners	Electricians	Concrete workers
Insulators	Meter and instrument men	Pipefitters
Plumbers	Riggers	Motorized equipment operators

Specialized industries may include many other classifications.

There is a trend toward creating a more comprehensive classification of craftsmen skilled in more than one skill. For instance, it is obvious that a burner-welder-pipefitter is a more effective maintenance man than a worker skilled in only one of the trades. Where union contracts are in force craft boundaries are often a matter of intensive labor-management negotiations. However where restrictive covenants do not prevent this grouping of classifications, there is a definite advantage in taking such action even if it involves training or retraining. Idle time of workers is greatly reduced and the flexibility achieved by such a move greatly facilitates maintenance scheduling. It is a definite advantage to the worker in that he becomes proficient in a broader range of skills.

Maintenance Expense Ratios. The ratio of maintenance expense (labor and material) to the value of plant equipment is a good indicator of the adequacy of maintenance. One advantage to using expense ratios is that, in budgetary control and expense allocation, the same base can be used as for the calculation of depreciation. As a means of comparison between plants, the investment ratio is open to the same objections as the plant-force ratio. Size and age of units, intensity of use, and design of equipment may justify widely differing ratios on two machines with identical functions. Yet, in practice, similar, well-managed plants often show closely similar amounts in the annual percentage of investment spent on maintenance. Where a satisfactory maintenance expense ratio can be established it can be useful in adjusting craft personnel to changes in the amount of equipment to be maintained, and in allowing for maintenance expense in estimating the operating cost of new acquisitions.

ORIGINS OF WORK ORDERS. Maintenance work orders originate from two points. The first, and most desirable origin, is from the maintenance force itself. This is usually called **preventive maintenance** and is set up on a regular schedule so as to cause the least disruption of production. The second point of origin is the operating department that requests maintenance because of the failure or breakdown of equipment. Work orders from the latter source usually require emergency maintenance. It is usually the most expensive type of work because adequate planning was not possible nor was it possible to assure that all tools and equipment for the job were on hand. One of the purposes of maintenance planners is to minimize this type of operation.

To avoid conflicting claims and promises, all work should be cleared and scheduled by a single authority, normally the plant engineer or his representative. Requests from all departments should clear through this central agency although exceptions may be made. Where a maintenance man has a standing service assignment on a group of machines or an area of the plant minor repairs or adjustments may be made without clearing through the central agency. However, the areas of "minor repairs or adjustments" must be clearly defined and understood to prevent individual interpretation by the maintenance personnel. Even when maintenance men are assigned to specific areas or machines, accurate records should be kept of their jobs so that data will be available for future forecasting and scheduling purposes.

ASSIGNMENT OF MAINTENANCE WORK. Efficient assignment of maintenance work is dependent upon accurate time standards. In turn these standards are dependent to a large extent upon past work records.

No maintenance work should be done without a written order and without some standard being assigned to the job. The planner should assign jobs to each craftsman or group of craftsmen. The record of each job done is based on the maintenance control function's knowledge of how long that job should take. In emergency situations the job order may be issued immediately after the emergency condition has been corrected. Where verbal instructions are given, every attempt should be made to document them as promptly as possible. This is particularly true where it is necessary to reassign maintenance people temporarily to another area until the emergency has been eliminated. Often this type of data is helpful in planning the procedure to follow should the emergency reoccur.

When maintenance jobs must be performed while machines are stopped work should be done, if possible, before or after regular working hours. It is often necessary to arrange for a part of the maintenance crew to work other than the regular shift. In this way, work can be done while all or a portion of the plant is shut down. This plan requires careful planning and scheduling of maintenance work and acceptable arrangements with the men doing this work, but establishes equality of working hours without overloading the work force or requiring overtime except for emergencies. Special care should be taken to insure that supervision for the off-schedule craftsman is of the same quality and quantity as is provided during the regular shift.

Manpower Control. Carson (Southern Engineering, vol. 83) recommends that the maintenance controller or department head should possess the following qualifications:

1. On-the-job experience in all phases of maintenance and construction work.
2. Full knowledge of each maintenance craft.
3. Ability to estimate manpower and material requirements for any type of repair or construction job.
4. Experience in designing forms and establishing policy and procedure for the operation of the department.
5. Experience in scheduling and coordinating the work of maintenance people.
6. A working knowledge of accounting, storeroom, and purchasing procedures.

The maintenance crew in the highly automated industry comprises a large part of the total number of employees. With the present trend toward increased automation, an even greater increase is inevitable, unless controls are installed and rigidly adhered to.

Careful studies of the maintenance crew in many industries have revealed less than fifty percent efficiency. The maintenance cost of the finished product is greater than ten percent.

Due to a scarcity of qualified personnel, some industries have systematized with people of limited experience, in an effort to control these high costs.

The installation of an effective maintenance control system takes time. There are many factors to be considered such as:

1. Size of the plant.
2. Quantity, condition, and location of the equipment.

3. Size and locations of the raw material storage areas.
4. Plan peak production capacity.
5. Number of hours and shifts the equipment operates.
6. Normal expected life of the plant buildings, and equipment.
7. Size, structure, and training of the maintenance crew.
8. Contractual agreements between management and labor.
9. Budgeted expenditure allowed for maintenance.

The speediest results will be realized if the control of manpower is made the first objective. This control should begin with the understanding that as the overall program takes form, changes will be made. Manpower control should start on a temporary basis through a daily planning meeting. The maintenance and operating department supervisors should meet with the maintenance controller and his scheduler to align the work.

Maintenance Time Report. Maintenance department hourly employees need instruction in the proper method of reporting daily manhours.

The form used should be designed for automatic data processing equipment, and should provide for reporting time by individual pieces of equipment or job order.

Job Estimate. Job estimates should include both labor hours and supplies. Work order cost limits should be established for securing management approvals before the work is done. Properly policed, this part of the control system will keep top management advised of anticipated changes in process and major expenditures for repairs and alterations.

The Written Job Order. The written job order is the source document for all maintenance labor and supply charges. The information for time reports, supplies, purchased services, and any other expense is taken from this source. Therefore, it is necessary that all information be correct and legible for accurate costing and reporting. The work requested should be as descriptive and accurate as the writer can make it. The estimating, the trades needed, and the preparations are based on this statement.

The form should provide space for noting repairs made other than those covered in the original request. This may serve to explain any unfavorable variance from the original estimate.

There are many variations of written job order procedures. However, one thing should hold true in all cases: The department superintendent who will be charged with the cost should be aware of what that cost should be before the job is started. He should also have a copy of every active job order for work pending in his department. The department superintendent is a valuable advisor to the maintenance scheduler.

Planning Maintenance Schedules. From the standpoint of minimizing maintenance costs perhaps the most important activity is that of adequate planning of maintenance schedules. Maintenance schedules usually fall into two classes, the long range schedule and the daily adjustment of the long range schedule. These schedules must consider the necessity of performing **emergency jobs** which come up with varying degrees of frequency depending on the type of industry involved. The major characteristic of each of the schedules is the necessity of making frequent changes and adjustments to meet the current situation. Schedule-of-work boards or Gantt charts are probably still the greatest aid used today in planning maintenance operations. In larger plants, electronic data processing equipment is beginning to be used to solve the information retrieval problem.

Routine inspection, except for longer jobs on heavy equipment and chemical plant processing equipment, perhaps may more conveniently be handled through some visual records system from which job orders, if necessary, could be written or the time schedule developed and put on the daily work program sheet. From this work sheet, the total load on the maintenance department can be built up to the labor hours available and the men finally assigned. Maintenance jobs are often short and usually isolated in contrast to production jobs. Scheduling, therefore, is more immediate and must be more flexible than in production. Nevertheless the jobs must be carefully scheduled to make the best use of the craftsmen's time and must be performed on schedule, or at once, in emergency, to avoid interruptions to production or plant shutdowns.

Mechanisms for Scheduling Work. To control maintenance work on jobs other than short routine operations which would take too much time and be too costly to post on work boards, some kind of board such as is used to schedule production is useful. Departments or equipment requiring maintenance work can be posted down the left-hand side of the board. Across the top of the board a date scale (Saturdays and Sundays included) can be entered. If the board is of the pocket, groove insert, spring clip, or a similar type, work order forms, or an identification card or slip, can be put on the board opposite the department or equipment to be worked on and under the date when the job is to be done. If the work will require several days, a series of copies of the order or identification cards can be posted under the respective dates.

Obviously, the job order number, name or number of equipment, and class and kind of work to be done would be shown on the job order or identification slip. As jobs are done, the slips can be removed, any required data entered, and the slips filed for posting to permanent records. By having the department or equipment names on cards (unless there is always work at such places), the listing down the left side of the system can be kept flexible and the list kept short. The date entries also may be on strips or cards so that the board may be reposted when the last few days currently showing are approached.

An alternative plan is to use a **Gantt chart**, which is essentially the above method recorded on paper.

Scheduling of Inspection Work. For regular inspection and adjustment or merely minor repair of building structures, building services, fire protection and safety devices, and manufacturing equipment, the systems just discussed are not necessary. The **visible index systems** of the hinged, center panel post, vertical card, or insert types, are well adapted to such short-cycle, repetitive jobs. The advantages of these more flexible systems are the absence of expensive posting time and "pencil work," ease of making quick changes, high degree of flexibility, ease of locating data, and graphic control features.

A form of equipment index used for the scheduling, control, and recording of inspection work in maintenance is illustrated in Fig. 7. This system is used in the Carborundum Company and consists of visible index cards for each piece of equipment. The area near the top of the card is ruled off into 52 spaces for weeks of the year, and down the left of the area are spaces to enter the years. The squares are for checking the weeks when the work was done. If the inspection card is marked O.K., a check mark is entered in the proper square. If any repairs are made, a number is put in the square for the week, and in the next lower section of the card, beside the corresponding number, details of this work are given. Along the lower section of the card, visible below overlapping cards above in the file, the number and name of the equipment, its location, and the inspection interval are inserted, thus indexing the card. At the bottom edge are the numbers 1 to 52. Signals moved along in the transparent card holder are

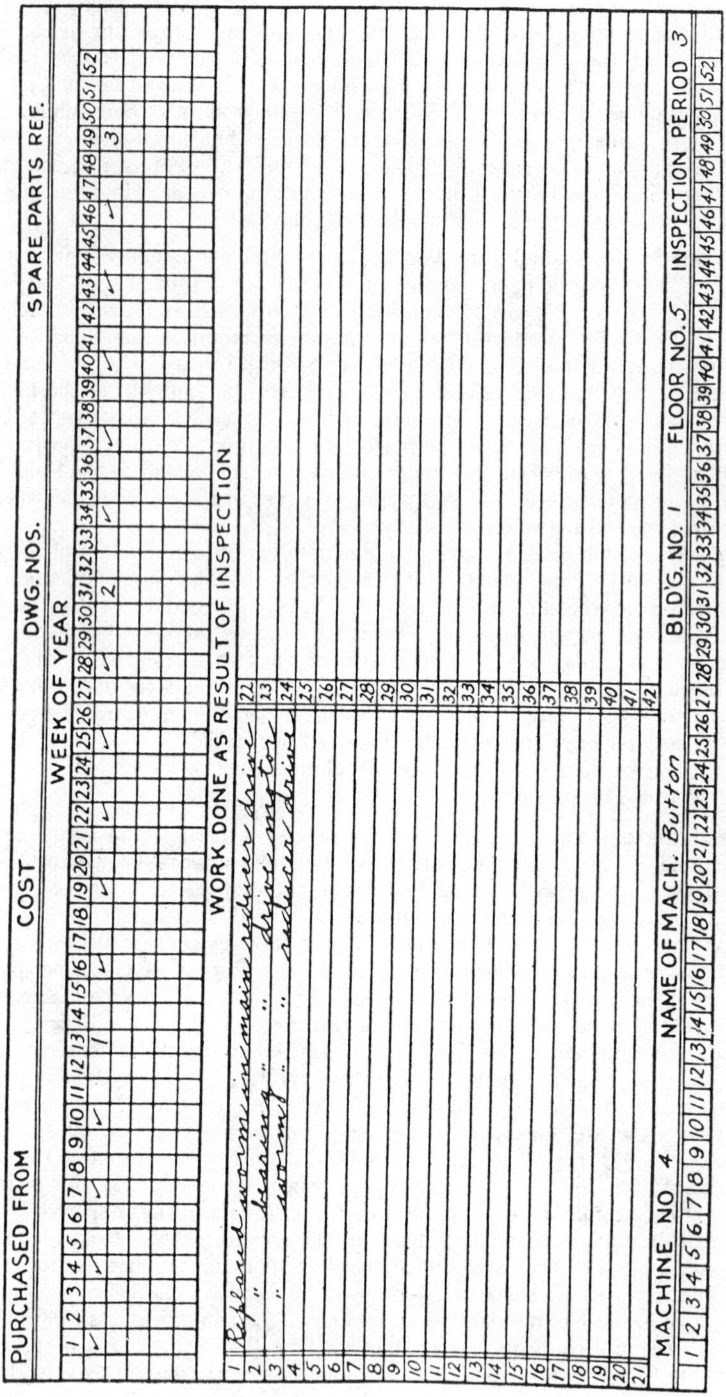
PURCHASED FROM COST DWG. NOS. SPARE PARTS REF.

WEEK OF YEAR

1 2 3 4 5 6 7 8 9 10 11 12 13 14 15 16 17 18 19 20 21 22 23 24 25 26 27 28 29 30 31 32 33 34 35 36 37 38 39 40 41 42 43 44 45 46 47 48 49 50 51 52

✓ ✓ ✓ ✓ 1 ✓ ✓ ✓ ✓ ✓ 2 ✓ ✓ ✓ ✓ ✓ 3

WORK DONE AS RESULT OF INSPECTION

1 Replaced worm in main reducer drive

2 " bearing " " drive motor

3 " worm " " reducer drive

4 5 6 7 8 9 10 11 12 13 14 15 16 17 18 19 20 21

22 23 24 25 26 27 28 29 30 31 32 33 34 35 36 37 38 39 40 41 42

MACHINE NO 4 NAME OF MACH. Button BLD'G. NO. 1 FLOOR NO. 5 INSPECTION PERIOD 3

1 2 3 4 5 6 7 8 9 10 11 12 13 14 15 16 17 18 19 20 21 22 23 24 25 26 27 28 29 30 31 32 33 34 35 36 37 38 39 40 41 42 43 44 45 46 47 48 49 50 51 52

Fig. 7. Equipment index of an inspection system.

set at the week of the year when the next inspection is to occur, regular inspection being indicated by a green signal and special inspection by a red signal.

A brief description of the equipment, and data on spare parts, supplies, special items, and the kind of inspection necessary may also be indicated on the card. Additional records or information may be entered on the back. It is preferable to make out a separate card of this kind for each distinct kind of inspection work on each piece of equipment, such as mechanical inspection, electrical inspection, lubrication, or oil change. While the entries may be made on the same card and distinguished by colored signals and checks or entries, such a plan is likely to be confusing.

Charts such as that shown in Fig. 8 are sometimes used for scheduling inspection. When they are used as schedule boards, the listing of equipment may be in permanent entries, and replaceable sections may be provided for the actual scheduling by dates. These sections may be removed when the time is past, and replaced by others with future dates. Preparation of an entire new chart is therefore unnecessary.

DAILY PLANNING AND SCHEDULING. In its simplest form, daily planning for a small maintenance force is best done by the plant engineer or master mechanic. He will maintain a list of his men grouped by craft leaders and assistants, check on work progress during the day, and, toward the end of the day, have the next day's assignments lined up. A **bulletin board** is often used to show assignments and location of men. But however simple the system, it is vital to maintain some means to ensure regular attention to important equipment. Coordination of maintenance work is improved by planning each day for all assignments to be made on the following day. Objection is sometimes made that emergencies will make such a plan unworkable. Actually, frequent emergencies indicate either undermaintenance or poor control. Some emergencies will always occur, but they can be reduced to between 10 and 20 percent of the total jobs. The daily work plan shows the location and jobs of each crew and makes it easier to get emergency help with the least possible disruption. Cooperation of department heads decreases last minute requests.

The usual practice is that **forenoons are used by supervisors to acquaint themselves with the status of all work** in progress and current operating conditions. Any maintenance requests from operating departments for work the next day should be on the dispatcher's desk by noon; any received after a deadline, which may be 2 P.M., are considered emergency orders, with a count kept of all emergency orders and their origin. Meanwhile, the **dispatcher** will have checked for routine work due and the status of various crafts as to work ahead. If there is a shortage or excess of immediate jobs for any craft, he will notify the master mechanic of the situation for his consideration while on the tour of inspection. Any conferences on the immediate program with the plant engineer, department heads, or chief executive may be arranged around the lunch hour. A regular **daily planning conference** should be an inviolable routine, with 2:30 or 3 P.M. a suitable hour. Its purpose is to make complete plans for the following day. The dispatcher outlines the job program and priorities, and releases only jobs on which all items are available. The master mechanic assigns men best suited to the nature of the work. Assignments can be aided by using the list of craft leaders with their regular helpers, followed by lists of men used for general work, marking assigned jobs against men to prevent duplicating or missing assignments. A full day's work should be assigned to each man. Job orders may be prepared ahead, but issued at this time, any time allowances shown being checked by the master mechanic. When the work is well organized, assignments and issue of

Week Ending

DEPT., BLDG., EQUIPMENT	CYCLE	JAN 5	12	19	26	FEB 2	9	16	23	MAR 2	9	16	23	30	APR 6	13	20	27	MAY 4	11	18	25	JUNE 1	8	15	22	29	JULY 6	13	20	27	AUG 3	10	17	24	31
Bldg. 47																																				
Elevators	4 wk	●				●				●								●				●														
Office Ltg fixtures	16 wk			●																●																
Shop Ltg fixtures	16 wk			●																●																
20-T crane	4 wk		●				●				●				●				●				●													
RG3-pumps (12)	8 wk			●								●								●																
Grounds																																				
Fence (paint, repair)	6 mo													●																						
Sidewalks	6 mo															●																				
Class I roads	3 mo							●													●														●	
Class II roads	6 mo																●																			
Pond	2 times Spring & Fall																		●																	

Fig. 8. Inspection schedule.

orders for a crew of 50 men can be done in less than an hour. In very large plants the procedure may be modified, the dispatcher having separate sessions in turn with the millwright foreman, electrical foreman, and yard foreman.

Job tickets for the next day may be distributed to craft leaders before quitting time as they come in to the tool crib to turn in tools. Hence, job orders should be at the tool crib not later than fifteen minutes before quitting time. This method saves time and avoids the confusion of organizing crews each morning. It is also useful to notify the crib attendant about tools and materials needed, with requisitions accompanying job orders, so that he may have the tools and materials ready for the jobs. Any minor adjustments needed in the morning may be readily handled by always having some postponable work in the daily schedule.

Procedures. Where work takes maintenance crews over scattered buildings and floors, the master mechanic will need a daily **reference sheet.** Operating foremen and the superintendent may wish notices of equipment under repair. To meet such needs, after work assignments are made, a list of jobs slated for the next day may be made up. This list may be made up on a mimeographed blank giving the names of craft leaders, the jobs on which they have been assigned, and where these jobs are located—equipment numbers, building floors, and bay numbers. Copies can be made for the plant engineer, superintendent, tool crib attendant, and dispatcher. This plan also aids in locating men when wanted.

Job tickets will be returned after they are checked by foreman or inspector, and will show the time taken. They will require reconcilement with clock time cards, a check which is usually made by the cost clerk or payroll clerk. Originals may be used for cost and payroll records. The clerk can pick up the corresponding batch of duplicates kept by the dispatcher and mark on them the time actually used (omitting start and stop times) and labor charges, then return the batch of duplicates to the dispatcher for use in maintenance records. A **summarized daily maintenance payroll** may be prepared by the cost clerk and sent to the plant engineer for signature and forwarding to the works manager. This payroll sheet serves as a constant check on the number of maintenance personnel. Job order numbers will identify the work being done on jobs for which appropriations have been made, so that any such expenditures can be shown separately to account for any temporary increase over normal.

Reports on Construction. The customary practice on new construction is to engage outside contractors, but the plant engineer may be responsible for seeing that the plans are carried out, and may exercise the function of the architect's supervision and check progress against schedule. **Plant alterations** may have to be carried out in stages to minimize interruptions to production. Such work is difficult to contract, and if the plant has the nucleus of a construction crew, it may be preferable to have the plant engineer handle the whole job. Where the engineer is responsible for construction cost, the representative practice is for him to report periodically on the percentage of the job physically completed and the actual labor cost to date against estimate, with a report on material cost to date indicating any substantial materials charged but not used. If the work is planned and scheduled and materials are allotted, a check of progress against due dates may provide all the control needed.

Filing Completed Job Orders. Records of work done are often required for future reference and play an important part in control. Many jobs will be completed on one ticket. Where several tickets are issued, a simple suspense file may hold the tickets until the last one comes in, showing that the job has been checked off as complete. The tickets may then be stapled together. Before filing them, entries should be made on the equipment record (described in succeeding paragraphs) showing the date, total labor cost, total material cost, and other important notations. A suggested method of filing job and material tickets is by equipment number, because a search for information will most frequently relate to a particular piece of equipment. Where information on craft activity is wanted, the best source will be a file of the daily work programs.

DEVELOPMENT OF THE LONG-TERM PROGRAM. The principal function of long-term planning is to provide a basis for a stabilized maintenance force and to arrange major jobs so that peak loads do not develop into emergencies.

Where standards or good experience data are available, the most desirable procedure is to assemble a schedule of equipment and buildings and compile a **man-hour summary of normal maintenance work** required by each craft, separating craftsman hours and helper hours in each. In addition, experience and plant policy should permit an estimate of installation work, alteration, and removal work. Total estimated man-hours divided by expected normal annual work-hours per man will give the minimum average force. This number of workers is then translated into a minimum skeleton crew which, because of practical minimums in certain crafts, will probably be somewhat over the figure estimated for the force as a whole.

The next step is to consider **emergency work and definitely known peak loads.** The first item envisages work imposed on the maintenance crew by storms, heavy snowfall, freezing of lines, power failures, and the like. Even when all reasonable precautions have been taken and disasters provided against, extra patrolling, repair, and inspection must be expected. The second item of known peak loads consists of such work as overhauling of major pieces of equipment, outside painting and roof work done in good weather, plant shutdown for changes in production setup, and, in a large plant, considerable yard maintenance on trackage, water, and other service lines, etc.

REDUCING PEAK LOADS. Usually it will be found that by planning ahead, **peak loads can be considerably reduced,** both by preparation and by improved organization of the job itself. If the most advantageous arrangement still imposes too great a load to complete in the desired time, the plant operating personnel should be considered as a possible source from which to secure the added manpower. Where a shutdown will be of long duration, this source should be drawn upon in any case to reduce the shutdown time. Consideration should also be given to the use of **contract labor** for maintenance. This alternative offers real relief to plants with highly fluctuating requirements. Contract maintenance is discussed later in this section.

Other methods of insuring an adequate force, without having idle men or excessive overtime work, are:

1. Adding maintenance department facilities to carry on repair work in the plant instead of sending out to service shops. A pipefitter may repair valves, salvage fittings, and make nipples. An electrician may inspect and recondition motors.
2. Carrying maintenance craftsmen on the payroll in an operating capacity, with an understanding that they may be called upon to do maintenance work when required.
3. Listing men who can be borrowed or engaged on call, either from friendly contractors or service shops, from among former employees, or through application lists, etc.

It is important to develop an **annual program** to insure the continuous and useful employment of maintenance personnel. Otherwise temporary layoffs and capricious reassignments will lower morale and often result in the permanent loss to the company of highly skilled craftsmen. At the same time, permitting employees to loaf or keeping men busy through artificial "make work" projects is bad for discipline and an inefficient and wasteful use of manpower.

ECONOMIC MAINTENANCE CYCLES. Scheduling of heavy maintenance and inspection may be done with the aid of a Gantt-type chart. Past records and experience can be used to determine the most economical cycle to use for major overhauls. In some cases governmental regulations stipulate what these cycles should be. Usage often is the basis for maintenance cycles. Automobile and truck mileage, tonnage of mill output, units delivered from production lines, hours of operation, are all common units employed in scheduling such maintenance as lubrication, disassembly and inspection, replacement of parts and the like. However, where the rate of usage is fairly constant it has been found advantageous to put inspections on a calendar date basis. This tends to produce an even maintenance work load.

To establish a **machine maintenance cycle** it is necessary to determine the point of maximum economy as well as the point at which hazards develop. Such points can be determined only by a study of the machine performance with respect to quantity and quality of output; frequency of adjustments and minor repairs; and the probable time serious failures may be expected. Detailed study of the cycle, causes of failure, and cost of major overhauls may lead to the use of improved construction elements, changes in operating practice, more specific inspection practice, and—by increasing the frequency of minor replacements—lengthening of the major cycle.

For example, in an oil refinery, the on-stream time may be limited by pending still-tube failure through erosion. Substitution of alloy steel tubes at critical points may extend tube life, and heat-exchanger fouling may become the limit. A better design of heat exchanger may further extend on-stream time till diminished heat transfer becomes the limiting factor. At this point maintenance involves only the cleaning of the still tubes.

From the above discussion it is clear that a maintenance program is not static, but should be in continual process of improvement. An annual review is a useful guide in directing efforts to steps promising the best results.

OPERATING COMMITTEES. In the case of many maintenance problems, the joint concern of engineering and production departments in good upkeep leads some companies to appoint a standing committee whose duty it is to consider desired improvements and make recommendations. Such committees are useful in fostering mutual understanding and cooperation. They function most effectively when called upon to discuss definite problems, preferably with a tentative but not final proposal ready to be discussed. The usual members will be the works manager, plant engineer, superintendent, and production control supervisors, with foremen asked to attend when concerned with the proposal under consideration.

MAINTENANCE IN CONTINUOUS-PROCESS INDUSTRIES. Special considerations enter into the planning and management of maintenance in plants which necessarily are in continuous 24-hour operation, such as blast furnaces, coke ovens, cement plants, and oil refineries, where operation may continue for months without interruption.

A.M. 0 1 2 3 4 5 6 7 8

Noon | 5 | 0 | 5 | Noon | 8 | 5

No. 1 Still — 1 — 2 — 3

No. 2 Still — 4 — 5 — 6

Transfer Lines — 7 — 8 9

Separator — 10 — 11 — 12 — 13

Low Pressure Tower — 14 — 15 — 16 — 17

High Pressure Tower — 18 — 19 — 20 21

Exchangers Sched. A — 22 — 23 — 24 — 25

Exchangers Sched. B — 26 27 — 28 29 — 30 31 — 32 33

Pumps Schedule C — 34 — 35 36 — 37 38 — 39

Compressor — 40 — 41 — 42

SHIFT MEN

Shift Foremen — Shutting down → — General Supervision All Inspection Work and Inspection Records

Still Foremen — Shutting down → — General Supervision All Turbining and Equipment Cleaning

1st Helper — 2 — 16 — 41 — 8-12-20

2nd Helper

Preparing Equipment for Maintenance Men

1-42 — Regular Scheduled Maintenance Operations

1. Much of the maintenance work can be done only while the plant is shut down.
2. Shutdown usually puts an expensive sequence of equipment out of revenue-producing activity.
3. The plant often has tight commitments on material deliveries and on shipments, so emergency shutdowns are commercially objectionable.
4. Emergency shutdowns are a great strain on the personnel and lead to unsafe methods of operation; besides, they are more expensive than planned shutdowns.
5. Efficient disposition of the maintenance force is difficult at best. Emergency conditions may demand complete disregard of cost.

These reasons make it especially desirable to plan the entire maintenance program on a **long-term basis.** Every piece of equipment should be scheduled for inspection and overhaul within safe limits of uninterrupted service. Standard methods for the overhaul should be developed, together with a careful estimate of man-hours required for the complete handling of the repair work, from take over by the maintenance crew to release again for operation. Total man-hours for annual maintenance work should be determined, classified by crafts, and separated into work during shutdowns and work while running. Commercial schedules will indicate the times during the year in which the plant can be shut down.

Determining Size of Crew. From man-hours required, and calendar hours available, the requisite size of crew can be decided upon. An effort must be made to reconcile the size of crew that can work while the plant is running with the size of crew needed to complete the work during shutdown hours. Alteration and construction work itself may be partially divided into running and shutdown work. Where the indicated "steady crew" is insufficient to complete the shutdown work in the available time the situation may be handled as a peak load problem.

Where such a peak load exists, special attention should be given to work methods, organization, tools, and conveniences. Expensive tools and rigging may be justified less by direct labor saved than by making it feasible to work with a smaller but stabilized year-round crew. Extra care in preparation and in inspection of materials will prevent delays. Proper instructions will ensure correct performance.

MAINTENANCE WORK DURING SHUTDOWNS. Besides the management aids on planning suggested for ordinary maintenance work, special **shutdown charts** should be made for major jobs requiring several hundred man-hours and involving several pieces of equipment with various crews. A skeleton outline of this kind of chart is shown in Fig. 9. The upper section of chart shows equipment availability and when worked on, the lower section shows disposition of crew. Such charts require considerable work but lead to fruitful results. They show up spots needing attention and reveal possibilities of reducing shutdown time, especially if discussed with master mechanic, foremen, and operating men. It has been found possible to plan so that most men are kept on their own shift during shutdown.

Planning of the shutdown should always include provision for necessary inspections of pumps, turbines, pipelines and the like serving the shut-down unit, and other equipment which can be opened for inspection only at such time. These pieces of equipment are usually not inspected as frequently as the main operating unit. Usually the best method to follow is to distribute such auxiliary inspections over the total number of annual shutdowns and make a proportionate number of auxiliary inspections at each shutdown. This practice not only makes for approximately constant off-production time, but reveals any abnormal condition more promptly than if one extra-long shutdown is made annually. Whatever such off-production time may be, once it is determined as a plant necessity, the company policy should respect it as essential to maintenance of assets and ability to produce on schedule. Infringement by pressure from the sales or the purchasing departments should not be permitted.

Many chemical plants and petroleum refineries make use of network analysis techniques, such as CPM and PERT, in order to effectively schedule maintenance time for processing units. The aim is to properly sequence the utilization of the various crafts and keep the duration of the shutdown to a minimum. In addition the interrelationships shown graphically by the network diagram serve as a psychological impetus for craftsmen to complete their job as soon as possible.

Reassignment of Manpower. When a particular portion of a plant is being shut down for maintenance, large parts of the maintenance force are usually assigned to this effort. This often means that maintenance in the remainder of the plant is being deferred. Naturally an unforeseen condition may arise which could require withdrawal of some maintenance forces from the scheduled shutdown area. Therefore a **contingency plan** should exist for the reassignment of craftsmen so as to disrupt the scheduled shutdown the least. By planning for these alternatives far in advance, logical decisions can be made and possible costly errors made under pressure can be avoided.

Coordinating Production and Maintenance Operations. Previously it was suggested that operating committees be established and given the responsibility of ascertaining when equipment should come out of service for maintenance. It is important that maintenance not be made subordinate to the production function. Frequently problems arise where the next level of hierarchy in the organization structure must make a decision. In these cases, maintenance and production must each present their case as equals. These problems are usually of an economic nature involving the possible loss of production and the risk of prolonging maintenance. It is often necessary to document these expense items in order to determine the alternative with the least cost to the plant. In cases where the maintenance cycle is very long, that is, where plans are made far in advance, the compromise between production and maintenance causes no problems. Most problems leading to conflicts arise when the cycle is short.

Experience has taught production that delaying necessary maintenance can result in abnormally long downtimes; therefore, production is usually sympathetic to the demands of the maintenance organization. The **quantitative techniques of operations research** offer maintenance useful tools with which to predict when equipment should come out of service. The field of reliability and maintainability offers the maintenance organization further aid in scheduling maintenance operations and integrating problems of mutual interest with the production function.

Establishing a Maintenance Program

RESPONSIBILITY FOR ASSIGNING WORK. In the establishment of a maintenance program it is necessary to delineate responsibility for assigning work. In small plants the plant engineer usually performs this duty but as the responsibility grows it is necessary to have one or more planners who perform this work on a daily basis. These individuals must be thoroughly familiar with the working of the plant because often priorities must be assigned to jobs and changes must be made on a short notice. Occasionally the job grows to be big enough so that use can be made of electronic data processing equipment. Many plants still use traditional graphical control methods such as Gantt charts, bar charts and the like.

The Plant Engineer. Since the plant engineer is typically responsible for maintenance in small plants he usually assumes the task of assigning work to the maintenance personnel. In the medium- or large-size plant there is usually an individual or group of individuals whose duty it is to make work assignments. This responsibility may seem fairly routine except that priorities are subject to quick change as emergencies arise. It then is the responsibility of the planners with the backing of the plant engineer to change work schedules to cope with the emergency situations.

Group Supervisors. If maintenance craftsmen are assigned to group supervisors the supervisors become responsible for making work assignments and decentralized maintenance organization is possible. The advantage of this situation lies in the fact that the group supervisor is closer to the jobs to be done and he is better able to assign individuals to jobs which have first priority at the moment. As a result, greater flexibility to change is achieved. Efficiency is increased by not having to go through centralized schedulers. One disadvantage of this plan is that some areas may be in great need of craftsmen while in other areas craftsmen may be idle or engaged in low priority jobs. While the plant engineer is generally responsible for plant condition and maintenance procedure, it is often advisable to relieve him and his master mechanic of duties pertaining to the daily routine of work assignments. Such duties may be delegated to an assistant, who will function as job order clerk or maintenance dispatcher. He will be responsible for all maintenance records of work done and schedules of work to be done, for checking material requirements, and for finding out whether these materials are on hand in stores. If any necessary materials are not carried, he must put through purchase requisitions for them. He will actually issue orders and requisitions, securing approvals as specified, and report to the plant engineer as to the relation between personnel available and work ahead.

PLANNING MAINTENANCE WORK. In small plants where the maintenance load justifies only a small staff of craftsmen, the master mechanic will usually require only a regular routine maintenance schedule. He will receive requests for nonroutine work from foremen and ordinarily manage his crew by itinerant supervision during the day and review in the late afternoon. New assignments would be distributed at quitting time as men punch out at the time clock.

However in a modern active plant especially where reconstruction and relayout is going on continually, it becomes extravagant not to plan carefully. To promote economy, maintenance manpower and projects must be completely coordinated.

To facilitate orderly procedures, appropriate routines are needed, as follows:

1. **Approval** of the plant engineer may be required when estimated cost of labor and material exceeds a given maximum, usually about $50. The plant engineer may set a detailed policy covering classes of expenditure he desires to follow up personally, or equipment slated for early replacement. On major alterations or acquisitions, approval of expenditures and appropriation of funds by the superintendent or other plant executive may be specified. The limit for jobs done without executive approval commonly ranges from expenditures of $250 to $500, or about 5 percent of monthly budget.

2. An **equipment record** is vital, to show the maintenance history of each important piece of equipment.

3. A **schedule of work ahead** is required. This schedule will be made up of three classes of work: (a) jobs which can be definitely planned well ahead, such as inspections and routine jobs, (b) jobs which may vary with con-

ditions but nevertheless must be fitted in at approximate times, and (c) jobs which must be done as emergencies arise. To fit in the latter jobs, flexibility in the master schedule is necessary.

4. A **file** in which work orders and data on forthcoming jobs to be done can be kept according to the date on which such jobs must be planned and scheduled.
5. A **daily work program,** with regular work, and emergency jobs for which orders or requests have just been received.
6. A **daily work force report** showing the disposition of maintenance men according to the kind and location of jobs to which they have been assigned for the day.
7. An **estimate of man-hours** for jobs to be done. Much help can be gained by keeping a file of completed jobs for reference. On work with repetitive elements, standards, as explained later, can be gradually established.
8. A system of **long-term planning,** or annual planning, can be adopted to build up a stabilized maintenance force and a well-balanced maintenance program.

Routine Inspection and Maintenance. One objective of **inspection** is to determine by observation the need for maintenance or repair; on the other hand, the objective of **maintenance** is to correct by physical action the defects uncovered by inspection. It follows then that an active preventive maintenance program requires an active inspection program. Under such a plan emergency maintenance is held to a minimum. In many plants, particularly chemical and petrochemical, the equipment inspection group is entirely separate from the maintenance organization. Typically, the inspection group has the following responsibility:

1. Determine the safe operating conditions of equipment such as piping systems, pressure vessels, heat exchanges, etc. Inspect such items as safety valves and other protective equipment. Determine the life of equipment from physical and nondestructive testing procedures.
2. Formulate procedures to safeguard individuals from accidents, and equipment from improper operation.
3. Inspect incoming material. This includes spare parts and replacements as well as incoming raw material.
4. Establish the useful life of systems and equipment so that long range maintenance schedules can be formulated.
5. Recommend when systems or operating conditions should be altered or designed.
6. Conduct tests to determine the weak points in machine systems and recommend ways of improving or eliminating these points.

Standby Crews and Emergency Work. In plants where downtime is very expensive standby maintenance crews are often used. This situation is particularly true in process industries which operate 24 hours a day and where the failure of any component would disrupt the plant to an appreciable extent. In general, standby crews are used where the ratio of the cost of downtime to the cost of maintaining standby crews is high.

Standby crews are expected to perform only **emergency maintenance** that occurs during their shifts. At other times these craftsmen may be assigned routine work that needs to be done within their own craft. For example, meter and instrument mechanics may be employed reworking control valves while serving on standby maintenance duty.

Operations research techniques, such as queueing theory, are used to determine the number of craftsmen needed to handle emergency work during off shifts.

Master Plan for Maintenance. A master plan for the maintenance function should indicate the allocation of resources necessary for operating over a specified period of time. In the hard goods plant the master plan is usually made up for 12 months in the future. In the process industry where the time between ideas and the production of products is a long cycle, the maintenance master plan might be as long as 5 or 10 years in the future. The master plan may take any of the following forms, or combinations of a number of them:

1. A chart similar to that shown in Fig. 9 is the usual form for master plans in the process industry.
2. Tabulation of requirements, in days, for each craft and area of the plant is another method of indicating future maintenance department requirements. The input to this tabulation can be gotten from the chart referred to in Item 1 above. The tabulation would show that for the period of say June to August of a certain year, X man-hours of machinist labor would be necessary, Y man-hours of pipe fitter labor, etc. From this tabulation maintenance management can ascertain whether the available labor resources are sufficient to meet anticipated demands.
3. A master plan could use regression analysis, to formulate a mathematical model with maintenance labor as the dependent variable. Maintenance management would then have a formula with which to predict the number of men needed within each craft for some time in the future.
4. A financial budget for the maintenance department is, in effect, a master plan since maintenance management must use these resources to perform its job. The master plan is then the allocation of funds from this budget for equipment, material, and labor.

All of the above plans have one thing in common in that they must be continually updated to reflect changes in production as well as changes in the maintenance organization itself.

DAILY SCHEDULING PROCEDURE. An essential part of maintenance control is an efficient daily scheduling procedure. The variables which might affect this scheduling procedure are priority of jobs, need for emergency repairs, and availability of craftsmen and equipment. The efficiency with which the scheduling procedure is carried out is reflected in the percentage of time that the craftsmen actually devote to productive work and the number of "crises" which may or may not arise from time to time. In simple terms the all important goal of maintenance should be: get the necessary work done on time!

No matter how far in the future master schedules are projected, the final test of the system is the day-to-day allocation of work to craftsmen. Experience in this type of work and knowledge of the particular plants operation, or **historical data,** are necessary for the efficient scheduling of any maintenance group. Queueing theory and the Monte Carlo method are modern approaches that can be used effectively to quantify the problem of maintenance scheduling and planning.

Work Orders. If the maintenance operation is viewed as a system, then input information for the system can originate from two points: the schedule as developed by the maintenance department (which includes all preventive maintenance) and the work order. The work order could include emergency repairs as well as those repairs which the production personnel requests.

FORM OF WORK ORDERS. Maintenance work orders may be either of a specific type involving one project or of a blanket type for repeated and more routine maintenance jobs. A typical maintenance work order for specific work is shown in Fig. 10 (Planning, Controlling, and Accounting for Maintenance, NAA Accounting Practice Report No. 2). Such forms provide space for a brief description of the work requested, the person requesting the work, an estimate of the time it will require, an estimate of the cost, the location and number of the equipment to be worked on, the date wanted, date completed, order number, approval space, and similar data which varies from company to company.

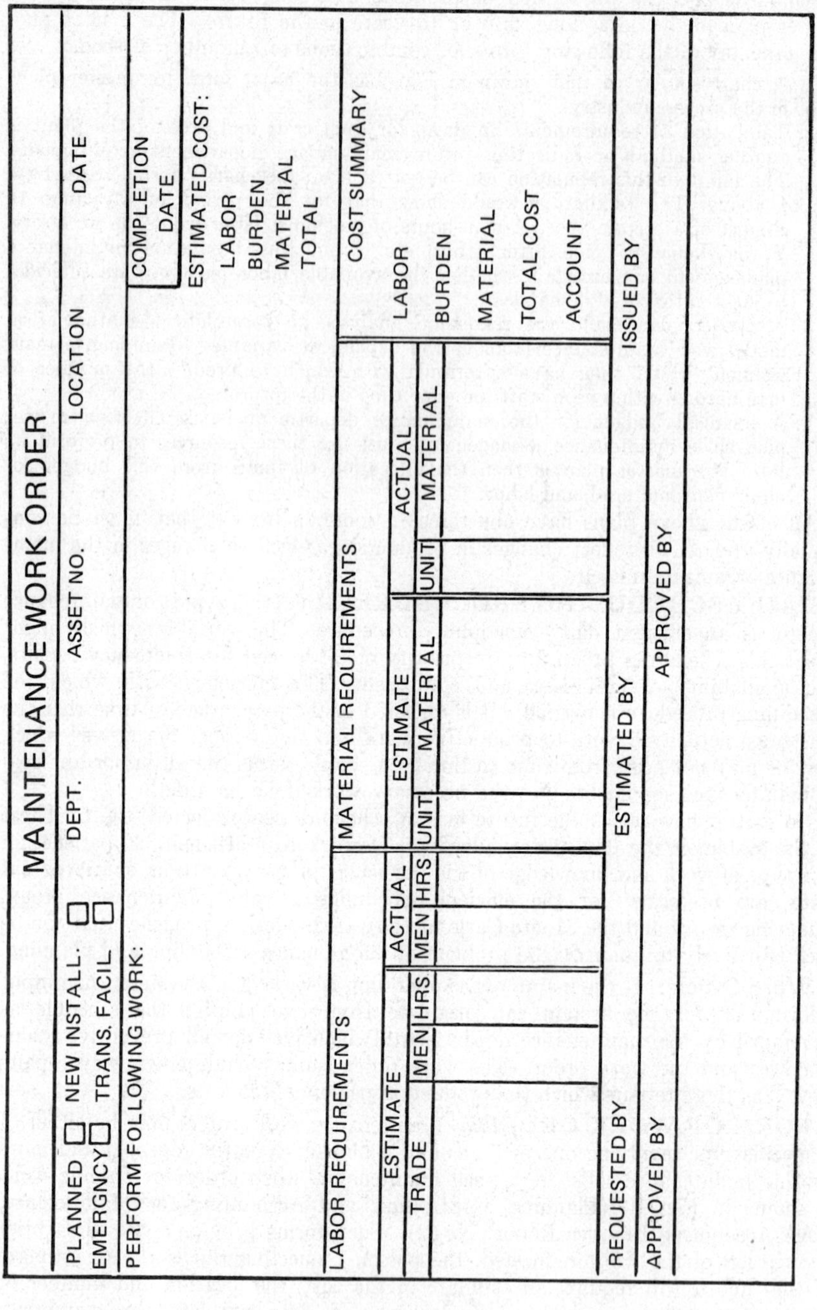

Fig. 10. Maintenance work order.

Not only does the work order initiate work requests, but, after being given a priority rating, it is used for appropriate scheduling of maintenance work, serves as a form on which project costs can be collected, and provides a running record of the status of maintenance in process. When the job is completed, it becomes the basic record from which maintenance costs on each item of equipment can be developed and supplies the financial data needed for formal accounting, planning, and control reports.

A maintenance job order routine for major, minor, and emergency jobs is shown in Fig. 11.

Figure 12 shows a form designed for use where work often requires several men in the crew. It is a combination work order and time ticket. The form also shows suggested items to be checked before issuing the order:

1. Is equipment available and safe for maintenance men?
2. Is needed material on hand in the storeroom?
3. What tools are to be used?
4. Are sketches or drawings to be given out with the job?
5. Is there any deadline for completion?

Forms sometimes have the order on the face and time entries on the back. This practice is not recommended. Shop copies are often folded to keep the face clean, and the back becomes soiled. Moreover, subsequent reference, checking, etc., is simpler with data all on the face. Two copies usually are made, a white carbon being kept at the order desk and a buff original going out on the job. If the organization is large, with many orders, and triplicates are needed, the form can be made to fit the billing-type register.

In most cases, **written instructions on job orders** will and should be quite brief. More elaborate instructions will usually be covered by a standard practice procedure or a memorandum to the master mechanic. The space provided on Fig. 12 has proved ample for all regular work. Note that equipment and location designations avoid the need for explaining further on the order where the job is. Material requisitions are made to match. They are attached to the job order and further diminish the need for explanations on the job order. Materials carried in stores should be coded for convenience of identification and withdrawal.

MATERIAL REQUISITIONS. A material requisition form should be made out as a receipt for all withdrawals of items from stores. In most cases, the exact requirements for a maintenance job can be predetermined, but provision must be made for items for which need is disclosed only after dismantling the equipment. For this reason the master mechanic or inspector should be provided with a requisition book.

The blanks should have spaces for entry of account number and equipment number to be charged with the material. Requisitions are often numbered serially. Two copies usually are sufficient, the original being issued to the worker with the job order and the carbon being retained for office use. Authority to issue materials requisitions for repairs and renewals will ordinarily be vested in the master mechanic or inspector. On major alteration and installation work, the plant engineer will authorize the preparation of a bill of materials from which the requisitions will be written. Where heavy incoming materials are delivered directly to the job without physically passing through stores, records nevertheless must be kept of their arrival and use. Often one simple requisition showing vendor, vendor's invoice number, and the value will suffice. Such a step insures proper entries in maintenance cost records and stores records. Fig. 11 indicates graphically the representative procedure.

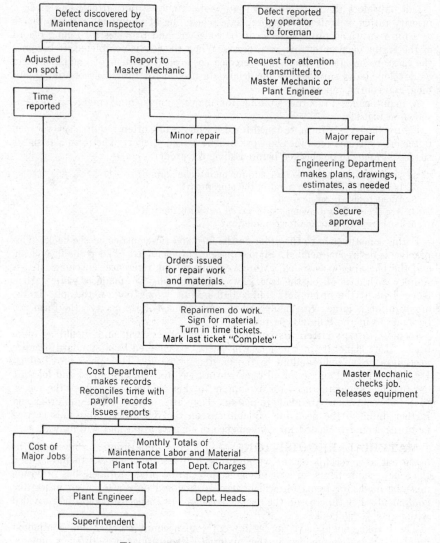

Fig. 11. Maintenance order routine.

USE OF ELECTRONIC DATA PROCESSING EQUIPMENT IN MAINTENANCE. Electronic data processing (EDP) is useful in plant engineering in storing data which can be retrieved quickly and inexpensively. From the maintenance standpoint, the following operations can be performed better by electronic data processing equipment than by the traditional manual procedures:

1. Recording machinery downtime.
2. Recording machinery running time.
3. Processing and calculating equipment cost.
4. Scheduling maintenance inspections and repair work.

5. Reporting work backlogs.
6. Balancing work loads of maintenance crews.
7. Spare parts inventory control.
8. Scheduling by critical path methods and PERT.

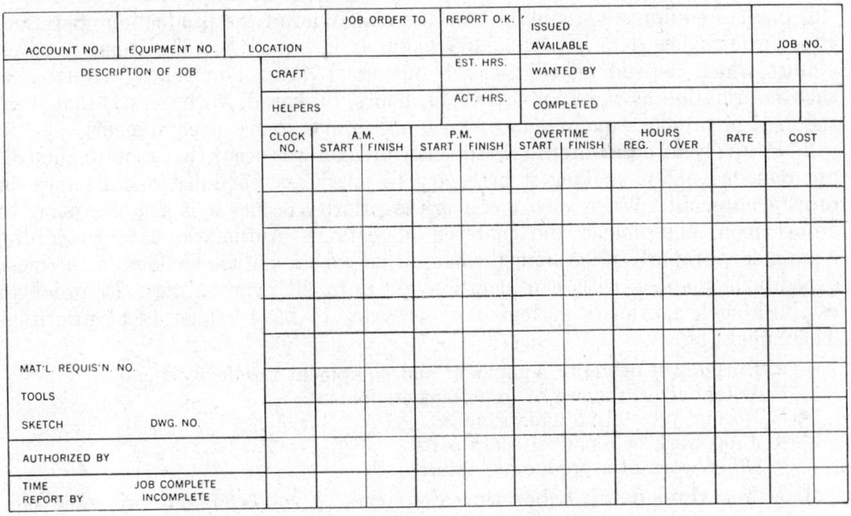

ACCOUNT NO.	EQUIPMENT NO.	LOCATION	JOB ORDER TO	REPORT O.K.	ISSUED AVAILABLE		JOB NO.
DESCRIPTION OF JOB		CRAFT		EST. HRS.	WANTED BY		
		HELPERS		ACT. HRS.	COMPLETED		

	CLOCK NO.	A.M. START	FINISH	P.M. START	FINISH	OVERTIME START	FINISH	HOURS REG.	OVER	RATE	COST
MAT'L. REQUIS'N. NO.											
TOOLS											
SKETCH DWG. NO.											
AUTHORIZED BY											
TIME REPORT BY JOB COMPLETE INCOMPLETE											

Fig. 12. Combined job order and time ticket.

The EDP equipment required for the above operations is often existing in the plant and being used for other purposes, such as technical computations and payroll operations. The added work load required by the maintenance department normally need not overburden the computer time schedule. The computer programmer must either work very closely with experienced maintenance people or must himself be entirely familiar with maintenance operations. If facilities are not available in the plant then the programs can be run on leased equipment or by purchasing time from computer utilities.

Note that the maintenance systems designed for the computer are very similar to the non-EDP systems. The principal advantage comes from the instant recall of data possible with EDP equipment. The plant engineer can make much more accurate decisions when information is immediately available to him on an up-to-date basis. The aim of the system is to gather together all basic information necessary for analysis and decision-making in the maintenance program.

As with most automatic data processing a code number must be assigned to describe the information parameters needed in the system. Each item of equipment must be given an identification as does each department, type of repair etc. Other code numbers are needed to identify such operations as steam cleaning, greasing, and lubrication, in addition to normal maintenance, emergency maintenance, and preventive maintenance.

Typical maintenance operations suitable for programming on data processing equipment are:

1. **Scheduling of maintenance operations.** Plants with large amounts of equipment on varying maintenance schedules can obtain printouts at any desired time for future maintenance schedules. Preventive maintenance operations make up the bulk of the items which are programmed, but sufficient forces must

be held in reserve for emergency maintenance operations which are certain to occur. The information necessary for such a scheduling system includes the projected date of the next regularly scheduled preventive maintenance operation, experience on previous runs of the same equipment, availability of spare parts for the necessary maintenance, and most important, the effect that downtime of this piece of equipment would have on the remainder of the production operation. Past performance data is particularly valuable in large scale operations research studies which depend on considerable historical data. Particularly valuable is such information as the number of man hours, broken down by craft, that were required in previous maintenance efforts on similar items of equipment.

2. **Priority of equipment.** Frequently breakdowns occur that require immediate remedial action making it necessary to postpone scheduled maintenance on other equipment. When such a shifting of priority occurs it is also necessary to adjust men, equipment, and material allocations. Automatic data processing equipment, properly programmed can see that such a shift is handled in the most expeditious manner with consideration given to all eventualities. In order to establish such a priority system it is necessary to have categories of priorities. These might be:

    a. Breakdown of major equipment and equipment unsafe to operate.
    b. Potential breakdown of important equipment.
    c. Normal preventive maintenance.
    d. Shop work or repair of spare parts.
    e. Other work of a productive nature.

3. **Labor time data.** Labor time data can be collected and processed efficiently by making use of a punched card. When a work order is received a card is issued, coded for the particular piece of equipment requiring maintenance and the craft required for the job. This card may also carry other information from the work order including the nature of the defect or the cause of the breakdown. The assigned craftsman fills in the card after completing the job, describing the exact nature of the service rendered, the parts required and the actual time spent on the job. The card is so arranged as to make encoding of the craftsman's information simple. The card is returned to the EDP key punch center where the new data is fed into the system, processed and correlated with existing data, and immediately made available for use in future scheduling.

4. **Backlog.** Electronic data processing can be used to maintain a continuous log not only of jobs to be done but also of the requirements for man hours (by craft), equipment, materials, parts, and so forth. The system must have the ability to compare estimates of performance with actual performance on the job and compare the difference, if any, between the two. If, for example, the actual man hours spent is less than the estimated man hours then the difference must be subtracted from the man-hours backlog; conversely if the actual time was more than the estimated time, the difference in man hours must be added to the backlog.

EDP Surveillance of Performance. There are a number of ways which maintenance can be gauged using data from electronic data processing equipment. Normally performance is measured in some sort of a ratio such as estimated time versus actual time. Performance can also be measured in terms of job progress reports where percentage of completion is estimated periodically and compared with that which was originally estimated for the particular time. The value of the EDP equipment in these instances is the immediate retrieval aspect and the ability to calculate the ratios automatically and frequently, so that, immediate corrective action can be taken.

EDP Surveillance of Equipment. EDP equipment can maintain surveillance of production and maintenance equipment much in the same manner that it does for labor craft man hours. In addition the EDP equipment can maintain records which are very important in any preventive maintenance program or any operations research type studies which require large amounts of historical data. The type of data usually includes the run time of equipment, downtime, reason for downtime, and cost of repair.

When **line item budgets** are used and maintenance cost for specific pieces of equipment is spelled out, then the EDP equipment can maintain a running schedule and budget and the printout can show remaining funds or overruns with respect to each piece of equipment.

Newbrough (Effective Maintenance Management) notes that, "with the use of EDP, it is a very simple matter to issue monthly summary reports of the Three Tops Tens, those ten machines which accumulate the greatest repair cost for the month, those ten which have the greatest amount of downtime for the month, and those ten which are down most frequently during the month. Concentration on the correction of deficiencies in these machines will rapidly improve both production and maintenance operations."

EDP Analysis. EDP analysis in the maintenance area has a very broad application. Almost any reporting and surveillance task can be automated.

The first step in such a systems analysis approach is the justification of existing systems. Once this has been done the systems should then be computerized. Care must be exercised so that useless data and reports are not generated.

Critical path scheduling and PERT have important applications in any maintenance scheduling operation. Except for the simplest systems, however, analysis is impractical without EDP equipment. The frequent and immediate printout aspects of EDP make possible real time decisions when those decisions can be of the most value.

Newbrough notes that "EDP reports can assist the plant engineer in improving operation to an extent that far outweighs the cost of the reports. Such reports can substantiate the need for the disposal of old worn-out production equipment and acquisition of new equipment, for additional maintenance shop areas and equipment, and for other changes that will increase the effectiveness of the maintenance effort."

PLANT ALTERATIONS. Many companies make alterations to their existing plant or process coincidentally with maintenance. In these cases, it becomes necessary for the plant maintenance and plant construction groups to coordinate their efforts. Frequently the difference is merely a bookkeeping one because the same crafts, and indeed the same individuals, may be working on maintenance and construction work at the same time. In separating maintenance from construction the general rule of thumb is that any work which increases the capacity of the plant or causes its products to be better is classified as a capital expenditure, or construction, effort. Maintenance is considered as work necessary to continue the efficient operation of the plant.

When alterations are scheduled at the same time as maintenance operations, both must receive the same meticulous preparation. Frequently, construction forces do not feel the same pressure to complete their work as do maintenance forces. However this attitude must not interfere with the return of the production unit to operation. Equipment should not be removed from service or process lines shut down until all material necessary for the maintenance and construction efforts are on hand. Although this will sometimes delay the start

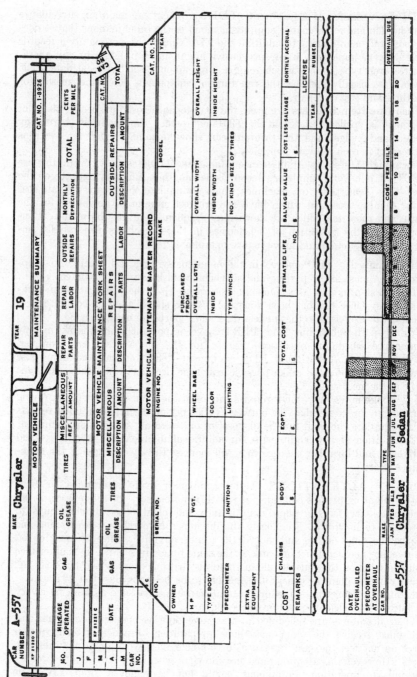

Fig. 13. Motor vehicle maintenance master record, worksheet, and summary.

of the work, too much reliance on vendor's promised delivery dates often results in even greater losses of time.

It is the practice in many plants to perform maintenance work without the benefit of engineering drawings when such jobs involve rerunning sections of pipe, altering building structural members, recircuiting electrical systems, and the like. However, when maintenance work makes drawings of the existing machinery or plant obsolete, it is necessary to update the **record drawings.** Construction work is usually characterized by the existence of a set of drawings to indicate to the field forces exactly how the work is to be done. A source of conflict may be introduced in the assignment of crafts on projects where some construction work is being done along with the usual maintenance efforts. If there is a maintenance supervisor and a construction supervisor, the responsibility of the craftsmen to either one or the other should be made very clear.

Record Keeping

CLASSIFICATION OF MAINTENANCE OPERATIONS. Maintenance Operations are usually classified into three components: production equipment, physical plant, and service facilities. Production equipment includes all machinery and tools necessary for the processing and fabrication of the product. One of the best methods for keeping production equipment records is the **unit log.** In the unit log, each piece of production equipment, whether it is a machine tool or a chemical processing unit, has its own daily record in which all information pertinent to the operation of the equipment is entered. The log includes the time that a piece of equipment is shut down, the reason for the shutdown, and the action necessary to restore it to production. Although the log is maintained by operating personnel, maintenance people should either obtain copies of the information periodically or maintain a dual set of records emphasizing the maintenance information only. The log also includes such information as the actual production output of the equipment so that the effects of overloading and underloading on maintenance can be ascertained.

Physical Plant Maintenance. It is often impossible to separate the maintenance of the physical plant from the maintenance of production equipment. In most cases physical plant maintenance is the major portion of the job of the maintenance supervisor and the forces assigned to him. Because of the nature of the work, maintenance involving electrical services, plumbing, roofing, masonry, carpentry, and other trades is best done by outside contractors.

A discussion of separate aspects of physical plant maintenance will be found in this section under Maintenance Methods.

From the record-keeping standpoint, definite files should be kept describing services done to the physical plant, the name of the contractor, the period of any guarantees, the cost of the work, the dates when the work was started and completed, and any other information which could be helpful to planning future physical plant maintenance.

Service Facilities Maintenance. A record of maintenance of motor trucks, kept by many fleet owners, is typified by the Kardex system illustrated in Fig. 13. Three forms comprise the record: the master, the worksheet, and the summary.

The master record provides space for entering information describing the vehicle. Over it rides the worksheet, on which expenses to the vehicle are posted as they occur. At the end of each month the columns on the worksheet are totaled and recapped to the summary record in the upper pocket.

On the visible margin a Graph-A-Matic signal at the right indicates the cost per mile of operation. By watching this signal in comparison to those on other records for the same class of vehicle, a decision can be made as to whether the vehicle should be sold or overhauled. If an overhaul is decided upon, the date is entered in pencil in the space at the extreme right. At the left, a ¼-in. signal indicates the month when tire inspections, etc., are due.

Recorders like those used on farm tractors are not favored in road truck operation—the control figure is mileage-recorded on an odometer, and this must be kept functioning. Mileage at which oil change is due should be marked on the truck. Truck and trailer manufacturers offer, through appointed service centers, routine check-up service with notification to the owner of any major maintenance required.

The setting up of an **equipment record system** calls for good judgment. Individual records on small inexpensive units of equipment are seldom necessary. They may be handled by means of a general departmental equipment record and account.

Records for special services, such as maintenance of belting, are kept on special forms and often require the use of special job order forms on which data may be entered by the dispatcher or the maintenance man at the time the work is performed.

Gydesen (Factory, vol. 115) reports the use of two simple cards, an instruction card and a lubrication record, in one plant lubrication program. The instruction card, permanently attached to each machine, is made up by process engineers in the production planning department on the basis of manufacturers' recommendations. It identifies the machine by name and number and carries a color code and oil specification number, showing what oils are needed and listing every lubrication point and frequency. The back of the card indicates oil-change periods for both hydraulic and lubricating reservoirs. The color code is also painted on all lubrication containers and equipment. The quarterly lubrication record card provides separate spaces for punching the dates on which both lubrication and inspection take place. The lubrication foreman spot-checks the cards, covering his whole zone every quarter.

Need for Record Keeping. Prior to a discussion of these three components of maintenance it would be helpful to emphasize the importance of accurate and complete records on maintenance operations. Modern concepts dictate that preventive maintenance programs and other maintenance operations which can be analyzed through the use of operations research techniques be made on the basis of historical or existing data. Much of maintenance analysis is concerned with questions: when will the equipment fail? If we were reasonably sure that the equipment would not fail, we would not be concerned with the design of a maintenance program. Since little or no equipment falls into the nonfailure category, it becomes necessary for us to attempt to predict when the equipment will require maintenance in the future. This effort usually entails a study of the past performance of the equipment for which a graphical presentation, such as a frequency distribution, is usually helpful. These types of analyses cannot be attempted without accurate historical data. Frequently, analyses are attempted only to realize that the existing data is insufficient or that data does not meet the needs of the system. In this case, it is first necessary to design a data collecting system before analysis is attempted. Typically the time needed for gathering the data exceeds the analysis time itself.

All of these problems dictate that a formalized procedure be designed to accumulate and store maintenance data so that it can be of some use. Rather than merely designing forms and taking information and filing forms, it is better to give some kind of systematic thought to evolving a procedure where this information will be readily available. It goes without saying that accumulating information for which there is no use is a costly process. There does not appear to be any uniform format for collecting and retaining this data which can be applied either within a single industry or among similar plants.

Equipment Records. There are two main kinds of equipment records. One is kept for the purpose of recording data on the equipment itself—name, number or symbol, date of purchase, cost installed, maker and model, location in plant, changes or additions, current condition, perhaps a calculation or estimate of current value, and record of final disposal. Such a record forms the basis of efficient maintenance work and should be kept in the plant engineer's office, where current entries can be made when the equipment is altered, repaired, moved, or disposed of. The record forms a **history of the machine** and a basis for certain production planning and reliability and operations research studies.

Another use of such a record is for **accounting purposes.** In making financial reports and in preparing tax returns, the amount and value of equipment in the plant must be ascertained or estimated closely, and an equipment record system is therefore imperative. Since depreciation enters into the calculation of equipment value and the determination of tax payments, an equipment record provides for the correct and systematic entry of data which alter the value of machines, such as additions to or removal from the machines of attachments, etc., and deterioration due to normal use, contingencies, or gradual obsolescence. The accounting department may therefore wish to have the same, or a similar, record at the plant engineer's office. In other cases, a single record may be kept in the one department or the other, for the use of both departments. However, it is better to keep it in the plant engineer's office, especially when detailed entries of maintenance work are made, as they should be. In this case the accounting department needs should be provided for on the records, and it is likewise necessary that care be exercised to see that data for this purpose are kept posted.

A second kind of equipment record is that used for the purpose of **entering all details of maintenance and repair work** on equipment. In this same connection a signaling system can be provided on the records to handle regular or periodic inspection and adjustment work for preventive maintenance on equipment. This record calls for data of a kind different from the equipment history card discussed above, which is suitable only for general maintenance control and for accounting purposes. Either or both of the records may be kept on cards or in a loose-leaf record book.

A typical equipment record card is shown in Fig. 14 and is described (Planning, Controlling, and Accounting for Maintenance, NAA Accounting Practice Report No. 2) as follows:

This is a visible-type record and is indexed by equipment numbers, subindexed by location or operating center. It will be noted that this record shows complete information and specifications of the unit. It also shows date of acquisition, original cost, service requirements or rating, and, if it is a part of a composite equipment group, the other units are indicated. This is necessary to allow group inspections under the preventive maintenance program. The card provides for recording all maintenance costs and, when those costs recur too often and in a total which is out of proportion to the unit's cost, age, or service, it signals the need for special study or replacement case survey. The cards also provide for location changes so that the unit's history will follow the original card.

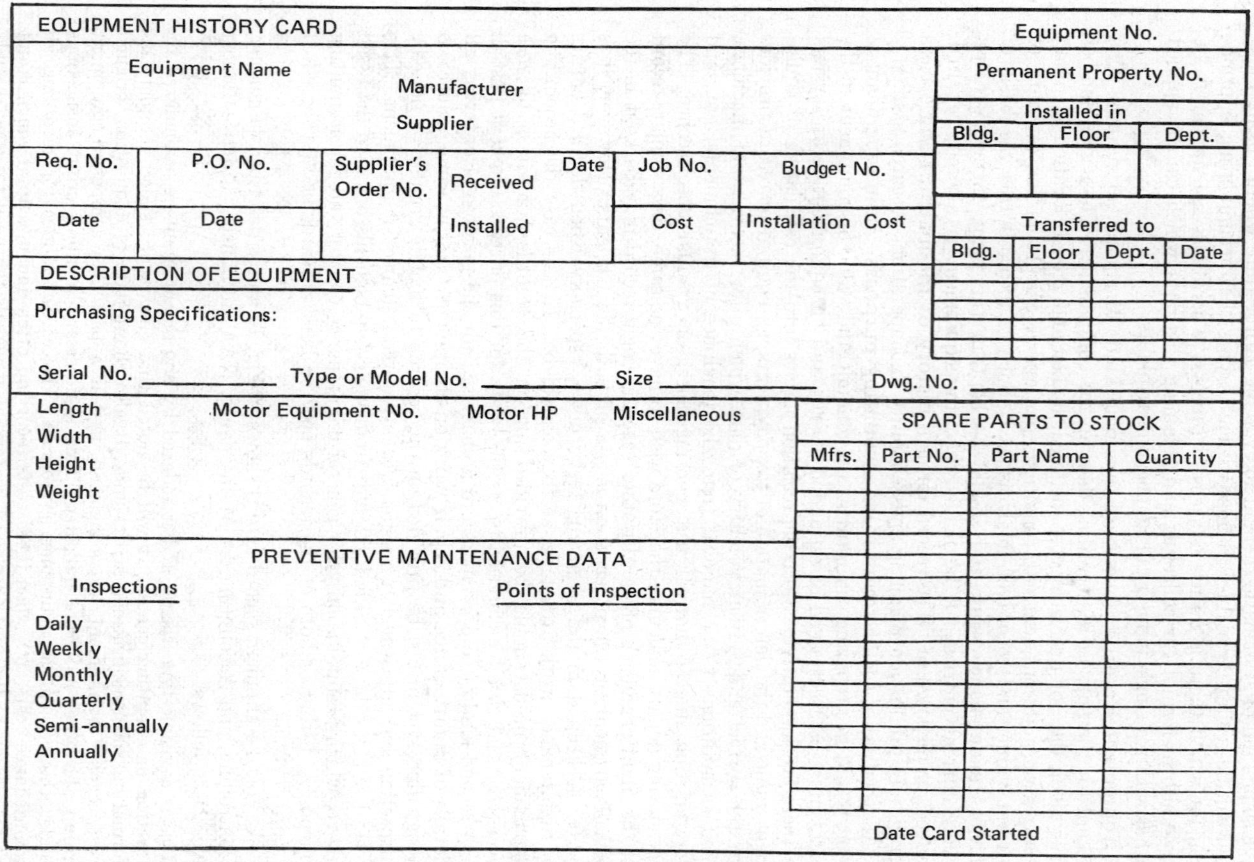

EQUIPMENT HISTORY CARD

Equipment No.

Equipment Name

Manufacturer

Supplier

Permanent Property No.

Installed in

Bldg.	Floor	Dept.

Transferred to

Bldg.	Floor	Dept.	Date

Req. No.	P.O. No.	Supplier's Order No.	Received	Date	Job No.	Budget No.
Date	Date		Installed		Cost	Installation Cost

DESCRIPTION OF EQUIPMENT

Purchasing Specifications:

Serial No. _____ Type or Model No. _____ Size _____ Dwg. No. _____

Length _____ Motor Equipment No. _____ Motor HP _____ Miscellaneous

Width

Height

Weight

SPARE PARTS TO STOCK

Mfrs.	Part No.	Part Name	Quantity

PREVENTIVE MAINTENANCE DATA

Points of Inspection

Inspections

Daily

Weekly

Monthly

Quarterly

Semi-annually

Annually

Date Card Started

Fig. 14. Equipment record card.

The back of the card carries the **repair and maintenance history,** with columns for date, hours, labor hours, labor cost, material cost, total cost, description of work done, and the numbers of parts used.

One company uses a record of the same general nature as the one in Fig. 14, but with different arrangement of entries and a few more details. A feature of this latter record is the listing—along the two visible edges of the foldover form —of numbers from 1 to 50 on one edge, and 51 to 99 on the other edge. Plain and punched hole signals over this double scale show the department where the equipment is currently located, the plant number being printed on the signal itself. Thus, a machine may be in Plant 10 (number of signal) and in Department 33 of that plant. A **current equipment inventory** by plants or departments can thus be quickly taken. The inventory-type or historical record, useful in part for certain accounting data, is the simplest form to keep.

COMBINED EQUIPMENT DATA AND MAINTENANCE COST RECORDS. Fig. 15 shows a type of card on which general information may be entered for quick reference. The design of a card for universal use throughout a plant is difficult and such a card should have a blank space for special entries. Unusual cases may be handled on the 5 x 8-in. card shown in Fig. 15 by entering information on a blank 4 x 6-in. card and stapling it to the regular card, of which only the top and right-hand side will then be used.

A suggested **method of supervising maintenance cost** is to use two files. The plant engineer or master mechanic will have the equipment card file. The maintenance dispatcher will have a job order file, in which each piece of equipment or

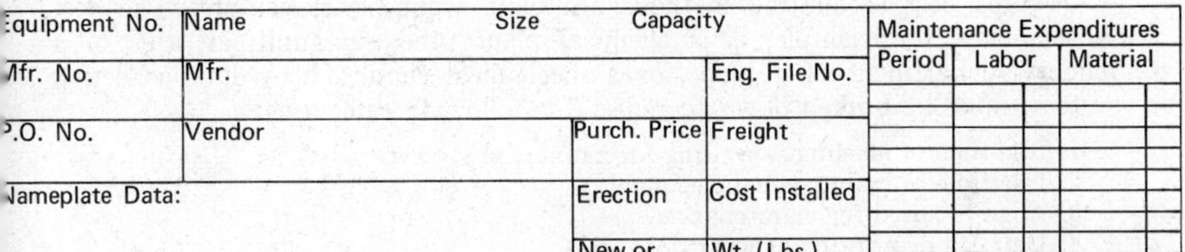

Equipment No.	Name		Size	Capacity	Maintenance Expenditures		
					Period	Labor	Material
Mfr. No.	Mfr.			Eng. File No.			
P.O. No.	Vendor		Purch. Price	Freight			
Nameplate Data:			Erection	Cost Installed			
			New or Used	Wt. (Lbs.)			
			Height	Floor Space			
Date New Installed			Location in Plant				

Fig. 15. Plant engineer's equipment record.

plant unit has a tab card, as shown in Fig. 16. As jobs are completed, entries are made on this card and job tickets are filed behind it. Expenditures may be totaled annually, quarterly, or at shorter intervals, and the total transferred to the equipment card file, at which time the plant engineer may review the record.

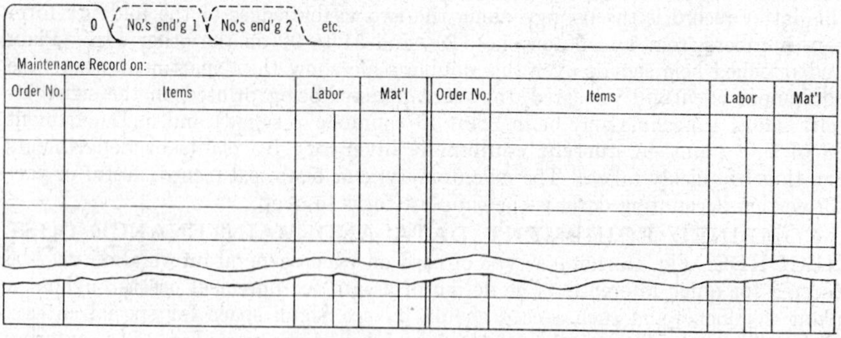

0 No.'s end'g 1 No.'s end'g 2 etc.								
Maintenance Record on:								
Order No.	Items	Labor	Mat'l	Order No.	Items	Labor	Mat'l

Fig. 16. Equipment record of maintenance expenditures.

USE OF COST ACCOUNTING DATA. The establishment of standards, distribution analysis, or operations research techniques are all facilitated if some prior thought is given to the detail with which data are taken and costs records maintained.

Accepted cost accounting methods are built around cost centers. The cost centers may, for example, be a chemical plant processing unit, an area of a refinery, department of a plant, or a mechanical shop. These data collecting systems must be broken down to reflect the following categories:

1. Idle time of machinery waiting for maintenance crew.
2. Idle time of maintenance personnel.
3. Time required for maintenance.
4. Detailed description of work done.
5. Parts required.
6. Labor, by crafts, required for each activity.
7. Costs incurred because productive unit is down.
8. Reason for productive unit becoming idle (not a cost item).

The accounting code system used to record maintenance costs must be reviewed to enable the above mentioned cataloging of expenditures. Once the code system has been revised, extreme care must be taken over a considerable length of time. This care should take the form of assuring that all personnel using the system understand and make use of the new list of charges. Perhaps the most difficult aspect of any cost data collecting system is that of assuring that the charges actually represent what was spent.

If the amount of data is large and if data processing equipment is available, punched cards offer a real solution as a means of getting out of the data "jungle." Sophisticated programs can be written which yield the statistical parameters that are useful in operations research techniques. Rapid recall processes facilitate frequent surveillance of productivity, preventive maintenance programs, work simplification and methods improvement efforts.

Caution must be exercised in requesting such detailed information as is described here. If use is actually made of it, well and good. But collecting data is often a costly process. Not only is it costly in dollars and cents, but costly in that the accuracy of the figures really desired is suspect. If an individual is conscious of the use to which data are put, he will give more care to reporting them.

Maintenance Work Measurement and Improvement

MAINTENANCE WORK MEASUREMENT. Work measurement, in the industrial engineering context, takes many forms but not all are particularly applicable in the plant maintenance field. Most measurement methods make extensive use of historical time data accumulated over a long period of time. Unless some prior thought had been given to the ultimate use of the data, it is often found that the information is in a form that is not readily usable. Usually the breakdown by category is too broad and not detailed sufficiently for analysis. When this situation exists, it is necessary to redesign the data collecting processes and embark upon a new program to amass the required information in a suitable form.

For example, assume that a plant had a large number of pumps of similar size and type that are maintained periodically. The record of maintenance times is broken down as follows: blanking off the piping, removing the casing, internal maintenance, replacing the casing, and repiping the pump. It will be recognized that all operations but internal maintenance are more or less the same for all pumps, therefore, experience with relatively few pumps will yield standards good enough for estimating purposes. Further analysis would be necessary to determine the range of times for internal maintenance depending upon the type of service in which the pumps are used.

Work Sampling. One of the most widely used work measurement techniques in plant maintenance is work sampling. It is used to determine both craft assignments and labor effectiveness. The technique, described in detail in the section on Work Measurement and Time Study, involves many random observations of the craftsmen performing their jobs.

The observer records the activity of the men at the time the observation is made. Typical activity citations are: productive work, waiting for instructions, idle. After many observations the results are tabulated and a percentage of total activities is calculated for each activity. If the observations have been representative, then a good breakdown of the craftsman's time is available.

Work sampling is best used on jobs which are nonrepetitive in nature. It is comparatively inexpensive to perform whereas continuous or stop-watch studies usually require considerable amounts of time and effort.

A simple form of work sampling can also aid in ascertaining how supervisory people, particularly foremen, spend their time. The results are then used to realign the schedule and assignments in order to make the best use of the individual's time.

JOB ANALYSIS AND STANDARDS. In planning work for the maintenance men, it is essential to know what constitutes a fair assignment. In shops with unplanned maintenance, especially where repairmen have a wide range of work, job assignments without adequate instruction or supervision invariably lead to excessive cost of work. Time studies of jobs hitherto unsupervised frequently lead to the development of new work methods or improved tools which alone produce decided improvement. When, in addition, a **standard time** is set up for performance, still greater savings can be attained. The measurement of maintenance work in a large plant with a wide range of equipment necessarily requires many months of sustained effort, but can be very rewarding.

MAINTENANCE FORCE PRODUCTIVITY. Where production workers are paid on some form of incentive basis, the same reasoning which dictated this step, and also the desirability of equal treatment of all workers, leads to the consideration of incentives for maintenance workers.

The objection is often made that maintenance work is not sufficiently standardized, but experience shows that, over an annual period, the largest part of such work is repetitive and therefore lends itself to method study and time study. The standards may not be so precise as in continuous production, but can be set up with a workable degree of accuracy. As to special jobs, in any case they should be studied in advance and methods and times should be outlined in connection with the planning of the work. Such estimates can be used as temporary standards.

This represents a fair summary of competent opinion. Although there is wide disparity of opinion on the advisability of incentives, there is general agreement that maintenance work can be measured in terms of standard labor hours and with a margin of error of less than 10 percent and in some cases to an average plant accuracy better than 5 percent—which is closer than many production work standards. It is generally agreed that measurement and reporting will of themselves, properly administered, raise productivity to a high level, and it is frequently contended that the further gain from introducing incentives is not sufficient to repay the added administrative expense, plus accuracy demands and possible friction when bonus earnings are unsatisfactory.

Mann (Journal of Industrial Engineering, vol. 17) gives a number of quantitative indices that reflect maintenance productivity. A good overall indicator of productivity is the ratio

$$\frac{\text{Total Cost of Labor Man-hours Required for Maintenance}}{\text{Dollars Investment}}$$

This index is satisfactory where the investment in the plant or facility is fluctuating. If the investment is increasing and the maintenance expenditure remains stable, then productivity is increasing. However, it is reasonable to expect maintenance costs to rise as facilities age. Taking this into consideration it is possible for the investment to remain static, maintenance costs to rise, and productivity to rise also.

If the investment does not fluctuate, then the ratio

$$\frac{\text{Maintenance Cost}}{\text{Machine, Area, or Unit}}$$

might be a satisfactory index. The age variable would have to be considered as in the previous ratio.

If a plant has a number of key activities which account for a significant percentage of maintenance costs, it may be well to create an index. If the activities do not vary from year to year, then the differential costs could represent a change in productivity. In this case, maintenance costs could be material and labor or merely labor alone.

ESTABLISHING PROPER WORK METHODS. Standard procedure does not always involve setting of time standards. Establishing proper work methods, instructions as to items to observe, and fixing of responsibility may be covered by standard practice instructions.

Instructions should describe in detail how the work is to be performed and the points needing special attention. Hence the instructions are readily enforced and contribute much to reduce maintenance trouble and cost.

Where maintenance studies indicate plant deficiencies, they should be made the subject of a report to management. For example, improvements were made on the periodic cleaning of a large steam boiler, but maintenance reports on the steam system revealed poor blowdown practice, with priming and boiler compound carried over into the steam system. The appropriate cure was complete elimination of the job by treatment of water before feeding to the boiler.

Training Programs. Mann (Journal of Industrial Engineering, vol. 17) indicates that maintenance training programs manifest themselves in two ways: a craftsman training program and a program for training maintenance management. The **craft training program**, referred to as the CTP, is usually undertaken for the following reasons:

1. Shortage of trained labor pool.
2. Rapidly advancing technology.
3. Conversion of plant to different units of production.
4. Realization by management that increased skills mean reduced maintenance costs.

The **maintenance management program** is intended to train supervisory personnel in scheduling, assigning and utilizing manpower and equipment, minimizing maintenance costs, and the like.

It can be seen that the needs of each plant within each industry could vary greatly.

Shortages of trained craftsmen result in the creation of the CTP by the plant needing the men. Federal, state, and local authorities have also assumed some responsibility in this field. Entering students range from completely inexperienced to trained craftsmen. Usually the latter group must only be oriented to the methods and materials of the plant.

The CTP is also important because of rapidly advancing technology. In many industries the materials and methods of production are caught in the spiral of rapid change. The fields of metal spraying, laser machining, numerical and direct digital control are but a few examples of new ways to do old jobs. The maintenance forces must be introduced to this technology as soon as possible with the least disruption of normal work schedules.

A craft training program is needed when production units are changed. Such units can be a chemical processing plant, specialized machine tools, or the like. In the case of new processing units, pilot plant operations often give a clue to problems likely to be encountered. Many manufacturers work closely with customers when new types of machine tools are installed. Factory training programs, when available, are extremely helpful.

Preventive Maintenance

NATURE AND OBJECTIVES. Preventive maintenance is defined by Reece (Plant Engineering, vol. 16) as periodic inspection or checking of facilities to uncover conditions that may lead to production breakdowns or harmful depreciation, and the correction of these conditions while they are still in the minor stage. The preventive maintenance program should be started off modestly setting short term, reasonably attainable goals. A degree of confidence is established. Quantitative evaluations can be made by management. In order to set up an effective PM plant the following data should be obtained:

1. Length of time unit has been in operation.
2. Length of time needed to repair unit.
3. Detailed description of work done on unit; time required to perform each phase of job.

4. Total elapsed time from unit failure to beginning of repair work.
5. Total elapsed time waiting for delivery of parts, tools and equipment, or specialized labor during which no repair work is done on unit.

A suitable unit of time, usually hours, is selected for the above data.

Once a sufficient amount of data has been accumulated an analysis is made to discover trends or determine standards of performance.

Since the main point of a preventive maintenance program is the creation of effective schedules, item 3 must yield valid data on which to base standards. If there is doubt that the work was performed with the best motion pattern, a comprehensive motion study of the process can be done. Naturally maintenance schedules should have the agreement and cooperation of operating personnel.

The optimum point at which equipment should be removed from service can be determined from the well known minimization curve. This curve is a plot of the average cost of downtime versus the duration of operation. Figure 17 shows a typical minimization curve.

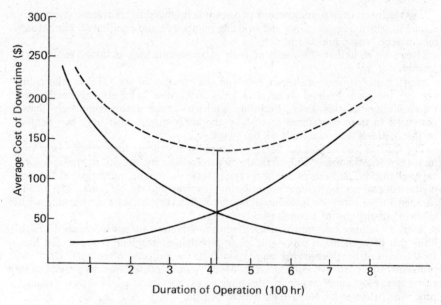

Fig. 17. An example of a minimization curve.

A detailed list of activities to be performed on any piece of equipment that might need servicing should exist. This list must be continuously updated as the equipment approaches its scheduled maintenance date. Material should be on hand; modifications or additions that are planned should be engineered. In this way the most effective maintenance can be obtained with the minimum downtime. This concept is well known and practiced in the process industries.

DESIGN OF MACHINERY AND EQUIPMENT. A major source of maintenance work is the failure of machines, equipment, or structures to withstand wear, abuse, or attrition. Although good design may add to the first cost this is offset by savings in maintenance, reduced labor costs, and savings in the cost and storage of replacement parts. Modern materials now permit low friction coatings to be applied permanently to surfaces previously requiring regular careful lubrication. Casting processes have been developed to allow one piece fabrication of intricate shapes of great strength yet light weight.

Good machinery and equipment design aims at minimizing deterioration. But a successful design also considers the service man. If lubrication is not automatic, service fittings should be easily reached. Parts should be replaceable with a minimum amount of disassembly. The plant engineer approving new equipment must be able to anticipate possible maintenance problems under his own plant's operating conditions.

The ideal of **preventive maintenance** is to make renewals before the failure of equipment and remedy minor defects before they lead to major repairs. The major sources of physical deterioration are:

1. Impact.
2. Vibration.
3. Corrosion.
4. Wear.
5. Abuse.

Impact may be essential (e.g. operation of a stamping press) in which case design should provide for shock-absorbing elements to isolate the machine.

Unless required for operation of equipment (e.g. sorters, vibrating conveyors, shakers) **vibration** can be almost entirely eliminated by design. It may be possible to put mass in stationary parts and lightness and strength in moving parts. The machine may be dynamically balanced. Where necessary separate foundations are provided. Proper cam design can often smooth out motions to diminish jar. Bearings with very close tolerances diminish wear caused by impact and vibration.

Corrosion may be prevented or delayed by using more resistant material, by applying a protective coating, by correcting a corrosive environment, or by introducing a sacrifice element. Among the metals, stainless steel, red and yellow brass, and arsenical copper are often suitable for preventing corrosion. The various chrome-nickel iron alloys have differing properties for varying applications. Smoke stacks are glass-lined to prolong their life. Where high first cost prohibits the use of special material, protection against corrosion may be achieved by suitable paint, lacquer, asphalt, waterproofing, or bituminous coating; steel parts may be protected by galvanizing, parkerizing, sherardizing, or bonderizing. Iron pipe subject to electrolytic corrosion may be protected by a thin coating of concrete. Where corrosion is due to abnormal environment, appropriate protection may lie in attention to the conditions. Thus, de-aeration of boiler feed water will help prevent oxidation. The **sacrifice element** technique ranges from passing corrosive water over finely divided iron with subsequent filtration, to more elaborate systems of electrolytic anodes.

Wear may be controlled through the proper design of velocity and area factors, the careful selection of materials, the use of lubricants, and the like. For example, rubber tubing may have greater resistance to abrasion of sharp particles than hardened steel. Similarly, many plastics withstand erosion better under some conditions than metal (e.g. bearings, gears, conveyor troughs, etc.).

Abuse is a problem of supervision. Constant training in the proper use of equipment is one method by which damaging treatment of apparatus can be reduced. But stern measures against violators of known rules may be necessary in cases of repeated negligence or willful destruction. Abuse manifests itself in any or all of the sources of physical deterioration noted above.

INSPECTION FOR MAINTENANCE. Successful preventive maintenance depends in large degree upon an adequate inspection program. The ideal of preventive maintenance is to prevent defects from developing, correct minor defects before they require major repairs, and make replacements before the failure of equipment. Maintenance inspection is the means of translating the ideal into practice. A good inspection program will include plant and equipment, detect defects, and report when replacements must be made. For inspection guidance, equipment may be classified as follows:

CLASS A. Outage results in widespread and costly interruptions to production, high costs of emergency services, and high mechanical expenses.
CLASS B. Outage primarily results in high mechanical expenses.
CLASS C. Outage affects low-cost machinery with no direct tie-in to general production schedule.

Preventive maintenance of Class A and Class B equipment concentrates on avoiding unscheduled shutdown. Regular inspection reports and records condition. Wear-tolerance points are established. Sufficient scheduled downtime is provided for both inspection and minor repairs and major overhauls. On Class C equipment, ordinary good maintenance procedure is used and equipment is allowed to operate until its condition requires shutdown for major work.

In large plants there may be a **maintenance inspection foreman,** reporting directly to the plant engineer. In smaller plants the plant engineer may lay out inspection schedules and assign inspection jobs to the various crafts. Items to be considered in organizing inspection are:

1. Detailed instruction on what is to be inspected, measurements required, and tolerance or service limits.
2. Frequency of inspections and coordination with other maintenance and production operations.
3. Assignment of inspection work to appropriate men.
4. Provision for inspection records and a follow-up system.

Instructions and Procedures. Both general and specific kinds of instructions may be used. The general program will fix plant policy on the various kinds of inspections to be made; what defects to look for in bearings, gears, motors, control panels, and other elements of construction; when and how to report renewal requirements; other inspection rules developed from plant experience.

Where there is a wide range of equipment, the above procedures may be formulated into a book of inspection rules, with alternate kinds of inspection of similar equipment, each under a separate rule number. Reference to rule number will simplify the issuance of specific directions for work required on a given inspection job.

Timing Inspections. The objective of timing is to space inspections as far apart as possible to reduce cost and yet stay within safe limits of time during which major defects ordinarily develop. Many inspections can be dovetailed into maintenance work, as when equipment is opened for cleaning or taken apart for adjustment and repair. Periods of accessibility and convenience are often a factor. Thus, general heating equipment may have a thorough check in early spring to list all work to be done in summer while it will be out of use.

Some companies distinguish between **visual or external inspection** and **testing or checking.** Thus, a main steam line might be visually inspected weekly for absence of leaks and outward tightness of insulation, and annually checked with instruments to determine insulation efficiency. Motors may be inspected monthly for cleanliness, commutator condition, and normal heat rise, but more thoroughly checked with ammeter and megohmeter every 3 to 6 months, depending on severity of service.

Initial frequency of inspection will be determined by judgment and general experience with the kind of equipment in use. Inspection and maintenance records will show when the frequency should be changed.

Periods for Inspection of Buildings and Equipment. In general, **time-tables** like the following (Carborundum Company) will help to determine the frequency of inspections:

Buildings or groups of buildings should be listed separately in the file, and inspection periods set for intervals of from 6 months to 1 year, depending upon climate, age, foundations, and equipment housed. Inspection of building proper should cover in detail foundations, walls, columns, girders, building joints, etc.

Roofs should be listed separately in file, and inspection periods should be set from 6 months to 1 year, depending on climate, age, and construction.

Floors should be carried in separate groupings in file by buildings or groups of buildings and inspected, depending upon their use, in periods ranging from 3 months to 2 years.

Paint should be checked at stated intervals, 6 months or more, taking into consideration protection, light reflecting capacity, and cleanliness.

Electrical power transmission equipment should be inspected in periods of 3 months or less to insure dependability.

Power-control equipment should be listed separately or in groups or territories and inspected according to use, in periods ranging from 4 to 12 weeks.

Heating equipment and low-pressure steam lines should be covered by thorough inspections every month.

High-pressure steam equipment should receive attention in semi-monthly or weekly periods.

Protection equipment, such as sprinkler lines, fire apparatus, and accessory equipment, should be checked over thoroughly in intervals of from 3 to 6 months. Some portions, however, require inspection more frequently, as daily or weekly.

Elevators require at least monthly inspections, covering both mechanical and electrical equipment.

Material handling equipment, such as hoists (air and electric), must be covered by monthly inspections. Process material handling equipment should be inspected in semi-monthly periods, depending upon usage.

Transmission equipment, such as line shafting, should be covered by monthly inspections, with alignment checked at least every 3 months. Other heavily loaded transmission equipment should be inspected weekly.

Equipment such as machine frames, rolls, foundations, and bases should be thoroughly inspected for flaws and checked for alignment in periods ranging from 2 to 6 months.

Oil-well bearing or any oil-reservoir equipment should have oil removed, equipment flushed with kerosene, and clean oil put in in periods ranging from 3 months to a year, depending upon service. Practice of replenishing oil supply as it becomes low is unsound.

Drinking water systems should be inspected every day; likewise **toilets** and **washbowls,** the latter sometimes twice a day.

A typical schedule of routine inspection is shown in Fig. 18.

Panel controls should be inspected and cleaned every 3 weeks, according to one manufacturer with over 5,000 motors on magnetic controls. Inspection includes seeing that boxes are tight and that any open knockout holes are plugged. Panels are placed at maximum convenient height to minimize fouling by oil haze. **Electronic controls** require weekly attention.

Maintenance Equipment and Supplies

TIME- AND LABOR-SAVING DEVICES. The following labor-saving devices are applicable to the more common kinds of work found in a plant.

Continuously

All construction
Piping (air, water, hydraulic, gas, oil, steam, etc.)
Electrical (wiring, cables, conduits, panels, etc.)
Roofs
Fire exits and doors
Toilets, sinks, and drinking fountains
Fixed ladders and hatches
General safety

Weekly

General production areas
Pressure tanks
Ventilation systems
(including fans, exhausters, blowers)

Monthly

Material handling (cranes, jibs, conveyors, chain blocks, hoists, special handling devices, etc.)
Elevators and dumbwaiters
Switchboards and motor control centers
High-voltage power cables
Power transformers
Ovens and furnaces

Monthly (Cont'd)

Kitchen equipment
Maintenance department area including tools and equipment

Quarterly

Floors (condition and loading)
Utility tunnels
Sewers
Fuel tanks
Lighting fixtures

Semiannually

Outdoor substation, power lines, poles, etc.
Buildings (including fire escapes)
Fences
Bridges
Walks and roadways
Stacks
Manholes

Note: Inspection procedures involving such items as fire-fighting apparatus and systems, security systems, specialized processes, etc., are covered by other departments.

Fig. 18. Typical schedule of routine maintenance inspection.

Painting is customarily done with suitable paint-spraying equipment, using extensions for high parts of walls to avoid scaffolding. On outside walls, a running track is often put in during construction, from which to suspend painters' and cleaners' scaffolds. Power-driven wire brushes are used for preparing surfaces.

Cleaning work can often be lightened by use of suitable hot sprays or of dipping tanks with cleaning compounds. Power scrubbing machines are used on large floor areas. Industrial-type vacuum cleaners and sweeping machines are essential for removing dust, grime, and general litter. Good housekeeping practices simplify cleaning work. Ultrasonic cleaning can be used for lighting fixtures and other equipment.

Roof and tower work requires hoisting of materials and tools. The permanent installation of simple davits will save rigging. A portable power-winch reduces the manpower needed for hoisting and the waiting time of the crew aloft.

Portable power tools, such as chipping hammers, riveters, drills, and wrenches find wide application. The plant system of compressed air and electric lines should provide ample tapping facilities to avoid the need for extreme lengths of hose and cable.

Opening and closing of equipment, such as cylinder heads, manholes, headers, and turbines, may be expedited by the use of power wrenches. Pneumatic impact wrenches will handle the heaviest bolting work in a fraction of the time of handwork and with fewer men in the bolting crew.

Welding usually requires both gas- and arc-welding portable units and a good supply of screens. A supply of different kinds of welding rod should be kept; building up worn parts is a great time and cost saver.

Moving or lifting heavy equipment in an area unserviced by a crane is facilitated by a rail or ring anchored to the building structure above the equipment, to which a chain hoist can be attached, or by the use of a portable crane. Heavy machines can be placed on special dollies or rollers for moving.

Alignment of machines, shafting, etc., calls for levels, transits, and indicators adapted to needs. Tolerance and feeler gages should be used to measure space, and shims should be provided to control allowable play.

Belt repair requires a special room or crib arranged for belt storage, a repair bench and tools, clamps and scales according to kinds and amount of belting in use.

Kinds of ladders used are worth special study to minimize the work of moving them and to provide safe working conditions for one-man jobs. Rolling stepladders are available which are self-locking when stepped on. A modification of the lift truck can travel to the job and swing an elevated platform in a wide radius to reach overhead equipment.

Replacing fluorescent or metallic vapor lamps singly is costly. Modern practice calls for complete relamping and fixture cleaning of a large area at one time, followed by only minor replacements of exceptionally damaged lamps. Cleaning is generally done by a two-man crew with special washing equipment.

Electric trucks are parked daily at charging stations.

Emergency repairs for reducing downtime of equipment such as furnaces or kilns may call for blowers, wooden sandals, inhalators, and special clothing to permit work under otherwise intolerable conditions.

The safety committee should be encouraged to keep posted on current practices and should be informed of all contemplated improvements or changes so that safe practices and safety guards or other precautions may be properly planned and instituted before operation under the new methods is started.

STORAGE AND ISSUE OF TOOLS AND EQUIPMENT. The regulations governing maintenance tools must be as meticulously observed as those for tools and equipment used in production, because a large inventory is often involved. Usually the maintenance department has its own **tool crib** and issues and maintains its own tool equipment under a system similar to that for production tool cribs. The tool crib must be kept locked, and only the regular attendants, or possibly the responsible head of the off-shift maintenance crew, should have keys and be allowed to enter. Others must call at the window for withdrawals of equipment, or the equipment, like heavy or bulky issues of materials, may be delivered to the job.

A tool record is necessary to keep track of all implements used in maintenance. While the men will have regular tool kits, whenever they need special tools for a job they should fill out a requisition or deposit tool checks for the item. The tool is charged out by either a double check method or the McCaskey Register method, whereby the tool is charged to the workman under his name or number and also is recorded as out at its regular storage place or under its class designation (and perhaps its number) in a file system. These charges are canceled when the tool is returned in proper condition. If all right, it is replaced in its location. If dulled, it is sharpened and replaced. When worn or damaged, it is repaired under a repair order or tag, if possible, or if beyond repair is scrapped and a replacement secured if necessary. Tools damaged through a worker's inexcusable carelessness often are charged against him. New tools or equipment should be inspected and approved before being placed in the tool crib.

STORAGE AND ISSUE OF MATERIALS AND SUPPLIES. The usual arrangement is for the maintenance department to have its own storeroom, although in a small plant the production storeroom may handle maintenance stores as well. Separation is better in medium-sized and large plants. The items are thus available in a central place and at all hours for withdrawal by the maintenance man.

Storeskeeping procedures are the same as in the case of production items, and may be supervised, in general, by the chief storeskeeper. Actual direction and handling of the work, however, will be taken care of in the maintenance department.

Stores Record System. The stores record system should follow the same general principles, stores grouping methods, and postings as prevail for regular stores. Usually a simple record of receipts, withdrawals, and balances of each item, and each size of item where varieties are kept, is sufficient. An order point should be established for each item regularly carried, and when the quantity falls to this point, a purchase requisition should be filed with the purchasing department for replenishment. For items carried in the regular storeroom, however, replenishments can be made from regular stores with greater convenience and without affecting the operation of the central stores control. A requisition on the regular storeroom in each case will suffice, and the transfer will be recorded on the records of both departments.

All incoming materials and supplies should be regularly inspected in a place set aside for the purpose, and reports of quantity, kind, and condition should be made out.

Items of materials and supplies which are not stocked must be obtained on purchase requisitions placed with the purchasing department, just as in the case of production activities. Those left over from miscellaneous jobs are carried on unclassified stores lists.

Materials for maintenance work should be issued only upon duly authorized requisitions posted to the records and then used for cost purposes. Inventory control should be as strict as for regular production stores.

The Maintenance Storeroom. The maintenance storeroom is under the control of a storeskeeper and, to avoid keeping men waiting, an attendant must always be on hand during working hours. At starting and quitting time, a relief man is needed to help, since the heaviest withdrawals and returns often occur then. Also, the storeskeeper needs a substitute during lunch hour. A junior clerk is often put on the stores records and may help with the issuing. Sometimes a maintenance helper may be picked for a tour of duty in stores and to learn storeskeeping.

In the storeroom orderly arrangement is essential. Layout should follow the rules for good storing. Items carried in quantity should have separate bins. Those stored in limited amounts can be separated by bin dividers, or merely by placing them in boxes side by side on the shelves. The most frequently used items should be near the issue window, less used items farther back. Heavy or bulky items belong in lower bins or on the floor. Bar stock can be kept on racks in line with the issue window or a small sliding door, for convenience in delivery to the workmen.

The storeroom should be kept locked at all times, no one should be admitted without due authorization, and pass keys should be issued only to the regular storeskeepers and perhaps to the responsible head of the maintenance work carried on during shutdown hours, so that essential needs may be served should an emergency arise.

Maintenance Methods

RECOMMENDED PRACTICES. While among different manufacturing plants there is less standardization of maintenance methods than of production methods, nevertheless certain practices are in wide use. It is advisable, therefore, to study the practices of representative companies doing the same kinds of work under much the same circumstances before planning maintenance operations. The plant engineer should also see that his organization has access to the latest maintenance techniques as described in articles in industrial magazines, in manufacturers' pamphlets and instruction sheets. An exchange of ideas with other plants is recommended because the information given and received is not of a competitive nature. Good maintenance is a means of eliminating waste, an undertaking in which all plants can well afford to cooperate.

The discussion that follows is limited to the essential factors in the maintenance of buildings and equipment. The information is grouped by classes of maintenance work.

BUILDINGS. Maintenance of buildings calls for a variety of work and is best considered by parts. **Outside walls** should be inspected for cracks and openings around windows, and for disintegration of mortar joints. Necessary repairs may be effected by using cement mortar for pointing open joints, and possibly a mastic calking compound about windows. **Outside painting** of steel work, wood, sheet metal, steel sash, etc., is usually required every 2 or 3 years. **Windows** should be washed at least twice a year, once in the fall when broken panes should be repaired, defective putty replaced, and any necessary painting done.

A roofing manufacturer or contractor is best qualified to repair the usual **tar and gravel and built-up roof.** He usually possesses a thorough knowledge of all types of roofs and roofing problems and has the necessary equipment to do an efficient job. If the roof is protected by a surety bond, the roofer responsible should in all cases be called. Summer heat, frost action of winter months, severe storms, vibration, atmospheric gases, or moisture may cause the roof to leak, flashing to deteriorate, and joints of gutters to open up. The **roof drainage system** requires protection against clogging with debris; screens, gratings, and traps need periodic cleaning.

Foundations and footings need to be checked for settlement and imperviousness to water. The condition and safety of the building above is dependent upon them. **Floors** are of many kinds, each calling for different care. Concrete floors are repaired by patching, wood floors by replacement, and mastic floors by filling cracks or applying heat to close cracks. Prevention of accidents and facilitation of industrial truck traffic are aided by floor maintenance.

ELEVATORS. Various kinds of elevator maintenance are offered by manufacturers. The simplest covers an ordinary examination and report to the owner on the condition of apparatus. Another type of service provides examination at regular intervals, including lubrication of apparatus and adjustment of parts. Still another service provides these same features and in addition the replacing, without additional charge, of small items, such as carbon and copper contacts, springs, washers, etc. A complete service covers regular examination, cleaning, lubricating, adjusting, furnishing of all parts required, making repairs, and including new cables which may be needed during the life of the contract. This service is at a fixed cost, and assures continuous operation and safety. Emergency service is available day or night. The American National Standard Safety Code for Elevators (ANSI A17.-1965 and annual supplements) and Practice for the Inspection of Elevators (ANSI A17-1960; A17.2a-1965; A17.2b-1967) give complete information covering construction, inspection, maintenance, and operation.

Where a large number of elevators is to be maintained, it has been found feasible to select trainees from among the elevator operators and teach them the various elements of maintenance of motors, hoisting gear, signal wiring, safety devices, and cable inspection and replacement.

HEATING AND VENTILATING EQUIPMENT. Periodic inspections should disclose any operating deficiencies in piping systems, radiators, valves, and traps. Minor and necessary repairs may be made as needed, but others may be postponed until the summer months, when the entire system can be given attention and made ready for the next heating season. Parts of ventilating equipment, such as fans, motors, pumps, etc., may need daily attention; other parts require inspection and adjustment less frequently. Humidifiers in use should be checked daily.

AIR-CONDITIONING SYSTEMS. Special problems of maintenance are generally centered in the cooling towers and spray rooms. This equipment is particularly subject to corrosion; all metal work should be rust-proofed with extreme care and precautions taken against flaking off of paint. Inspection is not always easy, and maintenance should be completed ahead of peak season with equipment in condition to run through the hot season without dismantling. Where make-up water is added, or where water picks up dust in the washer, adequate bleeding and flushing of recirculated water should be provided to ensure low concentration of solids and flush out sludge.

Shutdown for the winter and exposure to low temperatures are causes of deterioration in water towers.

Duct-work, when properly installed without irregular protrusions to catch dust, tends to be self-cleaning, with any dirt and dust picked up in the return caught by the filters. Filter inspection and renewals should be put on a schedule determined by individual experience according to season. Contact filters will remove most dust and dirt. When requirements are exceptional, electrostatic equipment may be required. This is expensive to install but not to operate.

SCHEDULED MAINTENANCE OF LIGHTING EQUIPMENT. Regularly scheduled maintenance of lighting equipment brings about economies in the use of materials, longer life of fixtures, and better illumination. Dirt on lamps and reflecting surfaces reduces illumination 30 to 50 percent, and in many cases alters light distribution.

SANITARY FACILITIES. Maintenance of **toilets and plumbing fixtures** generally consists of proper janitor service and attention during the day. Matrons on duty in women's rest rooms will report to the proper maintenance department any fixtures out of order, and a posted bulletin in all toilets should contain the phone number of the department to be notified. Periodic inspection will be made by the maintenance department. A cleaning department will thoroughly clean all fixtures every night and check the mechanical functioning of all faucets and equipment. **Floors of toilet rooms** should be scrubbed at these times. Locker rooms should be cleaned daily and scrubbed weekly, or oftener, as required. **Supplies** of towels, soap, and toilet paper should be replenished each day, containers being provided for soiled towels. Attractiveness and upkeep of sanitary facilities influences employees to treat them properly, minimizes repairs, and promotes sanitary habits and conduct.

Where employees bring lunches, insistence on deposit of all scraps and paper in closed **metal containers,** prompt clean-up of the lunchroom, and incineration of trash will prevent vermin nuisance.

Drinking Water Supply. Fountains should be of impervious, vitreous material, with jets of nonoxidizing material. A daily or semi-daily cleaning should be thorough, not only of the fixtures but also of the area immediately adjacent. **Vending machines** to dispense beverages and packaged items are in wide use, commonly installed by a vending machine operator who services the machines. The service operator is trained to know and remedy commonly occurring defects and in case of major trouble reports the condition for the district supervisor to send a special mechanic or replacement unit. Part of the machines' revenue sometimes reverts to shop recreation or welfare committee, and the machines thus retain the good will of employees.

FIRE PROTECTION EQUIPMENT. Fire extinguishers, most of which depend upon their effectiveness in the generation of CO_2 gas, should be recharged immediately after use, and at least once a year. Any damaged or frozen extinguishers should be reconditioned by the maker. Unlined linen **hose** deteriorates when left wet. It should be hung vertically after use or test, to drain and dry out. **Fire buckets** must be kept filled. Addition of calcium chloride or salt will protect the water from freezing. **Sprinkler systems** must be protected from freezing, and heads kept free from corrosion, paint, or dust deposits. **Water supply** should be checked as to quantity and proper working of pumps and equipment. **Dry-pipe systems** must be checked for leakage of valves which may cause ice to form in pipes and prevent their functioning when needed.

Yard hydrants may freeze due to leakage or poor drainage. They may be thawed out by using steam or hot water with unslaked lime added. Hydrants should be flushed and oiled about twice a year. **Fire doors and fire windows** should be inspected to see that they work freely and smoothly and that all automatic devices are in order. Materials and equipment must not be piled against them.

PRODUCTION EQUIPMENT. Each plant develops its own policy on maintenance of productive equipment. The common task of maintenance is the installation of machinery, aligning and securing it to a foundation, connecting the power supply to it, and finally testing it. Where the vendor sends a demonstrator to get the new machine into operation, the master mechanic should check over the instruction book and the machine, discuss the maintenance procedure with the demonstrator, and file a report on items to inspect or service and the intervals at which such work should be done.

Other common duties include:

1. **Checking equipment** mounting, alignment, bearings, vibration, clearances, functioning of control devices, and safety.
2. **Replacing** worn bearings, bushings, packing, gaskets.
3. Replacing or **repairing** worn parts; regrinding knives, rollers, wear-plates, etc.
4. **Inspecting** general conditions and reporting on work to be done and cost estimates as required.

The need for and extent of a machine shop for repairs will depend on the relationship between quantity of work and cost of tools, availability of outside service shops, the nature of work, the occurrence of emergencies, and the number of mechanics on the permanent maintenance staff.

POWER AND HEATING PLANTS. The power plant crew may do some maintenance work, but good practice has the stationary engineer report to the plant engineer on all equipment to ensure a complete program of maintenance for the power plant. The use of chemicals in **water treatment** may cause foaming and carryover of solids; appropriate inspections of turbines, traps, heating coils, etc., should be scheduled and any accumulations removed; kerosene injection while running, or as a wash when dismantled, will soften deposits on turbine blades and aid in cleaning. Extreme care should be used in **cleaning boiler tubes** with pneumatic turbines; scaling occurs much faster on a turbined tube than on new ones, but careful cleaning will increase tube life. A check on the maintenance costs of the entire steam plant may indicate the advisability of installing a water-softening plant.

COMPRESSED-AIR EQUIPMENT. The compressed-air system is peculiarly susceptible to abuse and overloading and is affected by the humidity of intake air. Receivers and lines should have **traps** to discharge accumulated moisture. Where drain pots are used, they should be blown out daily or oftener. **Outside air lines** may freeze in cold weather if moisture is not removed from compressed air prior to circulation. Expansion of air in actuating air tools results in a sharp temperature drop which may cause freezing if the air is not dry. Besides periodic overhauling of the compressor, a frequent check on demand is advisable. Installation of an **auxiliary receiver** near points of surge demand may obviate the need for enlarging the line or compressor capacity.

PIPELINES. Continuous inspection of pipelines will keep maintenance work at a minimum. New lines should be closely watched, especially after temperature changes, to check the functioning of expansion joints, swings, and bends. Lines buried in the ground should be checked for electrolysis by earth removal to expose short sample sections. The fifth year after installation should usually be soon enough for the first inspection, but local experience dictates the frequency of testing. **Exposed steam lines** should be insulated to the degree dictated by possible economy from reduced heat radiation and insulation kept in weatherproof condition.

YARDS, GROUNDS, AND WALKS. These require continued attention to keep in safe, neat condition. Walks, runways, outdoor platforms, and stairs require the same treatment as floors. Motor-driven sweepers and snow removers are required for maintaining large parking areas, roadways, docking facilities, etc.

HYDRAULIC SYSTEM MAINTENANCE. Some principal points to watch in hydraulic systems are:

Overheating Oil. Plugged cooler or lack of cooling water, defective pump, internal leakage, air leakage into suction, excessive discharge pressure, wrong or insufficient oil, or plain overload.

Loss of Pressure. Valve trouble, leaks, worn or defective pump, or wrong oil.

Erratic Action. Dirt in system, leaks, or wrong oil. It is necessary to use correct oil and determine proper inspection and oil-change routine.

LUBRICATION TECHNIQUES. It is good practice to make the plant engineer responsible for all lubrication throughout the plant, with authority to specify lubricants and frequency of attention, all lubrication work being done by his maintenance men. The objective should be beyond mere avoidance of trouble. Engineering attention to the mechanisms of lubrication, power transmission losses, cost of oils and greases, cleanliness, etc., has often yielded remarkable reductions of overhead expense. Consideration should be given to:

1. Type of lubricating mechanism.
2. Selection of lubricants.
3. Lubrication schedules.
4. Control of lubricant consumption.

Types of Lubricants. Selection of lubricants is often a highly technical matter, but the following general rules are useful:

Hand oiling. If bearing conditions are such that oil is wasted, grease cups are indicated. A slight excess of grease prevents dust and grit from entering the bearing. If waste is not abnormal, heavy-bodied oil may be used, preferably compounded oil —mineral oil with about 5 percent to 15 percent fixed oil, such as acidless tallow oil, added.

Drop-feed. If specific pressure in bearing is high, a compounded oil is preferred. Straight mineral oil is satisfactory for moderate or low pressure.

Ring oiling. Since there is constant flooding, and oil is agitated, no compounded oil should be used, as the fixed oil content would cause gumming. Straight mineral oil is to be used.

Splash oiling. Oil should be light in body.

Circulation oiling. Since oil is constantly circulating, it must be adapted to temperatures reached and to withstand exposure to oxidation. It should be suitable for filtering and re-use.

In general, machinery builders' and lubrication equipment makers' recommendations should be followed. To avoid the nuisance and waste of stocking several oils of almost identical properties, a **survey of oil requirements should be made.** Suitable selection of oils to carry in stores can then be made, with steps between specifications small enough so that requirements for any case can be fairly closely matched by one of the oils stocked. Oil stores can then be managed on a regular maximum-minimum stores and purchase system, and ordered in economical quantities.

One company reports reduction by this method to 4 grades of oil stocked, instead of 14 grades accumulated by unsupervised specifications; through this study the use of high-priced oil was also reduced.

Exceptionally severe service may require a combination of a special lubricating system and a special lubricant.

Lubrication Schedules. The amount of attention given to lubrication is governed by the type of oiling device used and by severity of service. In ring, bath, and splash methods of oiling, the governing factor is oil deterioration. In six months or less, most oils form a jellylike sludge or a sediment. The prescribed treatment is to drain the oil completely, flush out the bearing with a solvent, and refill with fresh oil. Hand oiling may be required daily on high speed work, weekly on low-speed shafting, etc. Automatic systems, whether gravity or force feed, should have at least daily inspection of reservoir levels.

It is often practical to assign lubrication operations to an electrician or a millwright helper as part of his inspection routine. Assigning lubrication to maintenance results in efficient care of equipment especially where machinery is divided into territories for maintenance purposes.

Methods of Lubricating. Hand oiling can provide lubrication without excessive oil waste if used with spring-cap oil cups, oil of the proper viscosity, and oil cans provided with devices to control dispensing. Open oil holes should not be used. If hand oiling is done carefully, greasy floors around equipment, unsightly machines, and product contamination can be avoided. An automatic oiler that feeds excess lubricant continuously but recovers, filters, and recirculates it, adding make-up oil as required, is highly effective and can reduce oil consumption to one half of that used in simpler wick fed or bottle oilers. Where severe service requires a constant stream of lubricant delivered by continuous pumping, the system should be protected against failure of either the lubricant supply or the pumping system. Automatic alarms, auxiliary standby pumps, and emergency power supplies may be required in certain cases. Where fully automatic recirculating systems for individual machines are not economical, a centralized feeder system may be used. A tubing and metering network delivers lubricant to each machine and in the proper quantity. This reduces service time and more important, even fittings and bearings inaccessible for hand oiling can be properly lubricated.

Storing and Dispensing Oil Supplies. Control of oil consumption begins with proper storage and distributing facilities. The stores department should arrange all **oil drums in a central storage room** and equip them with suitable tapping facilities to avoid either contamination or waste. The prevailing practice in shops is to provide the oiler with a pushcart of lock-up type, in which to carry suitable oil cans plainly marked with the type of oil contained, and kerosene for flushing bearings and cleaning tools. A **fire extinguisher** may be carried in clips outside the cart, available at all times. Oil may be drawn from the storeroom on requisitions. The method of charging to the operating departments should be based on the territory assigned. That is, **consumption should be measured** only for large groups as a whole, the charges being divided pro rata according to the number and size of units in each department.

Where the amount of oil consumed warrants, standards may be set for the **oil budget** of the oilers. A distinction should be made between devices such as bath and ring oiling, where renewal is made on a time basis, and automatic or bottle oiling, where consumption is proportionate to activity. In the latter case, plant man-hours may furnish the best gage of expected oil consumption.

Good Housekeeping

ADVANTAGES OF GOOD HOUSEKEEPING. Among the important advantages of good housekeeping are:

1. Production rate increased because of the orderly, businesslike condition of departments, removal of obstacles to production, etc.
2. Production control made easier. Materials and parts do not get lost or mixed. Speed of removal of work and less banking of rough or processed materials are corollaries of good order. It is easier and quicker to check operations and get data for records.
3. Inspection work takes on a high character. Quality control of work follows order and cleanliness control of conditions.
4. Materials and parts conserved and salvaged. All unused materials or parts, spoiled work, scrap, etc., are removed to proper places.
5. Time saved. Search for tools, work, etc., eliminated. Workers have more room to operate freely. No time lost in clean-ups to get space in which to work.
6. Floor areas are cleared for production instead of being littered with rubbish or crowded with unnecessary banks of work.
7. Maintenance and repair work facilitated. Repairmen can get at machines, do not have to clean them of dirt and grease, have room in which to do the work.
8. Safety protection made more certain. Elimination of crowded quarters makes machine operation safer. Clear, clean floors cut down stumbling and tripping, and slipping on greasy or oily spots. Clear traffic aisles reduce collisions of trucks, running into workers, knocking over piled materials, etc.
9. Fire protection improved. Fire hazards and spontaneous combustion are removed. Areas are cleared for quick exit, and for room to get at and fight any fires. Carelessness with matches is avoided.
10. Cleaning costs reduced. Janitors can do their work faster and better. It is cheaper to keep dirt down than to remove long-time accumulations.
11. Morale is heightened. Workers used to decent conditions at home become more interested in the plant when cleanliness and order are enforced.

There are no disadvantages in cleanliness and orderliness. No arguments can be advanced against a program of this kind which cannot be disproved by any executive or production engineer who has put methods of plant good housekeeping into operation. Neither merit nor profit attaches to being dirty or disorderly, and no excuse exists for the toleration of such conditions. If they exist, they are a direct reflection on the character of the plant management.

Responsibility for Good Housekeeping. While department foremen are supposed to instruct and supervise their workers in keeping departments and workplaces orderly, neat, and clean, there are many factors which require more attention and service than can be given in the normal course of shop operation. Some equipment and certain facilities are not in direct personal use by individual workers and therefore no one is directly responsible for their care.

In some cases—toilets and washrooms, drinking fountains, safety equipment, etc.—periodic inspection and minor attention may be already in effect. There are likewise scheduled times for cleaning of floors, washing of windows, cleaning and check-up of lighting systems, in many plants. Some of the work is done by janitors who may report to an operating supervisor in the regular line organization. But the plant engineering department, by the very nature of its duties, is called upon to maintain so many of the plant facilities that much of the actual housekeeping work comes under its regular operations. If there is a possibility that the plant engineer can take over the whole responsibility and handle it more successfully than any other available executive, the work should be made a part of his regular assignment.

INSPECTIONS AND REPORTS. Inspections should be periodic, announced in advance, and followed by published reports or announcements, to keep the laggards in line. Once a year a special campaign should be run prior to one of the inspection periods. Memorandums to the foremen and supervisors will keep them informed of matters that have escaped their attention. **Detailed plans** must be prepared for correcting bad conditions and for regularly maintaining cleanliness and order in spots that give particular trouble. As far as possible all the work done should be done in and by the departments where the wrong conditions exist. Otherwise the responsibility will be passed on to the maintenance department, although the latter's job may be one of upkeep and repairs rather than order and cleanliness in manufacturing operations.

Crisp, printed slogans such as:

A PLACE FOR EVERYTHING AND EVERYTHING IN ITS PLACE.

A CLEAN AND ORDERLY PLANT IS A SAFE PLANT;
A SAFE PLANT KEEPS ORDERLY AND CLEAN.

may sound trite but they are surprisingly effective in fostering awareness on the part of the employees. Other slogans can be easily developed (perhaps with awards for the best suggestions).

CHECKLIST FOR MAINTAINING A WELL-KEPT PLANT. The points upon which continued orderliness and neatness in the plant are maintained may be summarized by the following checklist. Modifications may be made to fit the list to the needs of any particular plant.

Walls, Windows, Ceilings.

1. Walls should not be used for storage of materials, such as pipe, small fittings, wire, cord or string, wiping rags, etc.
2. Unnecessary bulletin boards, production boards, work order or time ticket clips, charts, pictures, etc., should be taken down. Notices and information should go on properly placed bulletin boards, not on the wall itself.

3. In places where dusty or dirty operations are done, the walls should be cleaned several times a year. Vacuum cleaning is preferable to brushing. Where practical, walls should be washed and hosed clean.

4. Shop as well as office walls in many cases should be painted when cleaning no longer removes the dirt, or when illumination and work visibility can thus be improved.

5. Storage of items along the ceiling is unsightly and dangerous. Ceilings, like walls, should be cleaned, and painted when badly soiled. Better work illumination results.

Aisles, Exits, Stairways.

1. Traffic lines painted along aisles will keep them clear of stored materials and prevent accidents by marking off work areas.

2. Exits and surroundings should be kept clean of all storage so that traffic of trucks and workers will not be impeded.

3. Doors get dirty from contact with trucks, clothing, and hands. They should be washed occasionally and frequently touched up, at least, with paint.

4. Stairways must be kept clear of all materials, equipment, rubbish and dirt. They need daily cleaning. If corners are painted white, they will be kept free from papers, dirt, cigarette butts, chewing gum, and other unsightliness.

Floors.

1. Oil, waste material, paint, dirt, and other accumulations under machines can be caught in pans or on sheet metal, scrap fabric, or paper which can be regularly removed and replaced. Oil-soaked floors cause accidents and are fire risks.

2. Painted lines will mark off machine and work areas to prevent cluttering and congesting them with materials.

Toilets, Washrooms, Locker Rooms, Showers.

1. Daily cleanings and rubbish removal are imperative to good housekeeping. Floors should be washed. Frequent painting is of definite aid in good sanitation.

2. Rubbish cans, receptacles for used paper towels, neat racks for wet towels and cloth, and other facilities aid in cleanliness.

3. Daily use of inoffensive disinfectants keeps such rooms germ- and vermin-free.

4. Good light and ventilation add to the cleanliness.

5. Soap dispensers are cleaner and more economical than soap cakes.

Drinking Fountains, Beverage Dispensers.

1. Drinking fountains should be washed daily.

2. Signs should be put up, if necessary, to discourage use of the fountains as depositories for chewing gum.

3. Dispensers of bottled or bulk beverages should be replenished regularly and kept clean. If the beverage is in bulk form, the unit should be washed off with hot water periodically.

4. For bottled items, receptacles should be provided for empties and caps. Otherwise they will be thrown around the floor.

Manufacturing Equipment.

1. Dirt and oil accumulations should be removed from machines daily, in most cases, sometimes several times a day, if the work must be kept clean.

2. Painting is necessary on many kinds of equipment for protection and good illumination.

3. Chips from metalworking, cuttings from fabrics, and other collections should be dumped directly into containers, if possible, rather than brushed off to the floor.

4. Tools, gages, etc., should not be left on machines when work is completed. Finished or unfinished parts should be kept in trays, racks or bins not stored on machines.

5. If the machines have highly burnished or plated parts, these should be cleaned occasionally with suitable polishes.

6. Work tables on textile and other machines should be cleaned and waxed periodically with proper polishes.

General Equipment.

1. Tables needed for manufacturing operations should be kept free from collections of finished or unfinished parts, scrap, tools, rubbish, etc. They also require periodic cleaning and polishing.

2. Racks for work should not have clothing, rags, equipment, or odds and ends hung on them.

3. Work benches often accumulate all kinds of tools and accessories used in the department, spare parts, bolts and nuts, bits of material, etc., as well as rags, papers, old records, bottles, and all sorts of rubbish. This area of good housekeeping is the place where it is often hardest to bring about improvements, for the employee obviously must have some place to keep certain personal belongings. Examples of order and cleanliness in benches, and pointed propaganda, will cure much of the trouble and release considerable accumulations of materials and even tools and instruments.

4. Cabinets, like benches, collect all sorts of items which do not belong in them. Signs on the cabinets marked, "This cabinet is for .... only. Please keep it in order," and occasional inspections get rid of the junk and reserve the space for proper use.

Safety Installations.

Besides regular inspection for service and to see whether they are in use, safety installations need attention for their condition and cleanliness. Guards draped with rags, and other cluttering and uncleanliness of safety devices discourage their use.

Fire Protection.

1. Fire-fighting equipment should be kept clear of all materials, work in process, boxes and packing cases, unused machinery, or any other obstacles to immediate access. Hand extinguishers, fire pails, axes, fire hose, emergency lights, etc., should be kept neat and in order, immediately available in proper places, as well as undergoing periodic inspection for condition.

2. All such items should be cleaned and kept polished or painted. City fire departments find that preparedness is increased by the well-known meticulous care used to keep equipment clean, bright, polished, and painted.

Unused Manufacturing Equipment, Tools, Etc.

1. All equipment which is obsolete or worn out should at least be removed from the manufacturing floor, and stored in a central place, marked with an identification tag stating its name, number, make, department used in, date and reason for removal, intended disposal, and perhaps its estimated value. The removal record should be noted on the equipment card, so that the machine can be replaced in use if needed later. If the equipment is of no further use, it should be sold or broken up for scrap. Such a plan enables departments to be kept in better order and to gain more floor area for rearrangement of processes. Also, the departments can be more easily kept clean. Equipment should never be stored within the department, in passageways, under stairs, or in odd spots where there happens to be vacant space.

2. Broken, worn, and obsolete tools should not be allowed to collect in or around workplaces, where they are in the way and indicate lack of order. They should be removed for repair or scrapping.

Salvage Items.

1. Paper which can be re-used should be collected in regular containers provided for the purpose. Scrap paper, if possible kept sorted at point of origin, can be put into other containers and collected.

2. Boxes, crates, cartons, and other packing containers which can be re-used should be promptly collected and removed to storage in a reclaiming department. The rubbish trucks can make the collections and deliveries.

3. Used rags and waste may be reclaimed by washing and should be removed promptly for this purpose because they constitute fire hazards.

4. Discarded materials and spoiled work should be taken over by the salvage department daily to keep operating departments free from disorder, and also to control and account for the causes and costs of these losses. Spoiled work should have a release from the foreman or inspector.

Scrap Removal.

Regular scrap items, such as chips, borings, or other waste materials from operations can be collected, if possible, in boxes or pans right at points of origin, or can be periodically cleaned up and taken to a department receptacle. The different items should be kept separated, particularly different kinds of metals. Where such items as borings can be used in the company foundry, strict discipline on use of proper containers to send scrap to the foundry properly segregated has produced considerable savings.

Rubbish and Garbage Removal.

Floor sweepings consisting only of dirt, miscellaneous rubbish such as small quantities of sawdust, rags too soiled to reclaim, odds and ends of twisted or broken nails or other small items, and soiled paper and cartons should be collected in waste cans instead of accumulating under machines or benches or in some corner. If employees eat their lunches in the departments or buy candy, sandwiches, etc., from traveling vendors, the refuse should at once be deposited in cans to prevent attracting vermin. These cans should be emptied into traveling trash bins twice daily. The cans should be kept clean and in good condition. Periodically they should be washed and repainted.

Storage Areas.

1. In operating departments, white lines painted on floors will keep stored materials in proper places at machines.

2. Marking off floor areas in storerooms for items which must be kept on the floor, likewise will keep these items within proper boundaries and be conducive to orderly piling and regular contour of storage piles.

3. Proper storage layout, with like items together, and regular stacking on shelves will facilitate locating and issuing stores, making physical checks, adding new stock, etc.

4. Orderly storage, with everything neatly in its place, will avoid loss of materials, reordering items already on order, ordering when stocks are already high, etc.

Plant Yard.

1. If for no other reason than the resulting advertising value and public good will, plant yards visible from railroads, highways, and nearby manufacturing or residential areas should be kept neat, orderly, and free from trash, rubbish, and dumps. Even stored bulk materials can be kept trimmed, with the contents all within allotted areas. Yard storage areas, therefore, should be given as much attention as inside areas.

2. Tracks, walks, and roadways must be kept free from obstacles, rubbish, and growths of weeds and coarse grass. The limits should be neatly marked. Traffic and clearance lines also are often effective and are good safety precautions.

3. Outsides of buildings should be kept repaired, painted, if necessary, and clean. The effects of such attention and yard cleanliness are reflected in the higher production and better morale of workers. Company orderliness adds prestige to employees in the plant.

General Cleanliness and Order.

1. Aside from isolated attention to separate factors, there are general matters of order and cleanliness which require attention. In offices, accumulations of unnecessary or unused papers and other items should be thrown out, or stored in proper cabinets or files. Scraps should be thrown into waste baskets, not on the floor.

2. In shop and headquarters offices, mats are sometimes needed to remove dirt from the shoes of men who come in from the yard, foundries, etc.

3. Where smoking is permitted, ashtrays or sandboxes should be provided for butts.

4. All bulletin boards should be kept in good trim. When notices get old, they should be removed. It is better to have separate boards for permanent notices, or to put such items into employee manuals. Current notices on other boards will then be read. Posting should be neatly prepared and squared with the edges of the board. Prompt removal of past events keeps the board orderly and up to date.

INDUSTRIAL SAFETY

CONTENTS

CONTENTS (*Continued*)

SECTION 22

INDUSTRIAL SAFETY

Importance of Safety Organization

ECONOMIC IMPACT OF ACCIDENTS. Occupational accidents in 1969, given in data issued by the National Safety Council (Accident Facts), resulted in a loss of production time equivalent to 245,000,000 man-days. Of this total, 45,000,000 man-days were lost by those actually injured, and 200,000,000 man-days were lost by fellow workers in helping the injured, in repairing damage caused by the accidents, and in temporary loss of efficiency by those witnessing serious accidents. Not including the value of property damaged in non-injury accidents other than fires and the indirect losses of all fires, the total cost for work accidents in 1969 approximated $9,000,000,000. This expense to industry amounted to approximately $110 per worker.

The loss to industry, from accidents, may be seen more dramatically when fatality figures are considered. These are summarized in Fig. 1 (National Safety Council Accident Facts), for the year 1969.

Kind of Accident	Total Number of Workers Killed (Approximate)
All accidents	56,800
Occupational	14,200
At home	7,500
In motor vehicles	27,000
Public (except motor vehicles)	8,100

Fig. 1. One year's fatal accidents to workers, on the job and away from work, in the United States.

Although millions of workers are able to return to a job after an accidental injury, their efficiency is impaired temporarily (sometimes permanently). Tens of millions of workers lose time to get the necessary first-aid treatment for minor preventable injuries. These and other direct losses continue to drain away industry's productive effort, particularly under a continuous-flow or closely coordinated production system. An accident may slow the production of one plant, and the delay of its product may prevent a customer plant from operating at full speed, and so on, until a single accident may have widespread effects.

The importance of safety should require no more forceful demonstration than is given by the above figures. However, they picture only a part of the economic effect which accidents have on industry. Manufacturers are threatened economically by the rising number of law suits and greater indemnities per claim for alleged negligence in equipment designs. Robb and Philo (Lawyers Desk Reference) state "Hundreds of thousands of workers are injured in industrial accidents each year because of negligence or breach of warranty on the part of

22·1

the manufacturers. . . . Manufacturers' recognition of their duty to research, design, guard, and warn will only be realized when hundreds of thousands of suits are filed on behalf of crippled workers and the imagination and creativeness of the plaintiff's trial bar is brought to bear on the problems of raising the standards of negligent industries . . ."

The relationship between business economics and safety effectiveness may be stronger than is evident now. Grimaldi (Journal of the American Society of Safety Engineers, vol. 10) points out that studies indicate "a low accident rate, like efficient production, is an implicit consequence of managerial control." Where operating costs were comparatively high for the business studied, the accident rate also tended to be abnormally high and vice versa.

MANAGEMENT RESPONSIBILITY. The fifty states, the territories, and the Federal Government have placed **legal responsibility** on employers for the safe operation of their establishments. Unlike the old employer's liability laws which required the determination of responsibility for a worker's injury, workmen's compensation laws in the United States and most other nations now hold the employer liable, as a matter of social policy. It is now universally accepted that medical care and compensation within the limits of the law of the state having jurisdiction are due workers injured in the course of their employment. This expense is viewed as a cost of doing business and the need for controlling the impact of such expense is considered by the legislators as a motivation for employers to maintain effective work safety activities.

W. J. McClung of Bethlehem Steel observed (National Safety Council Transactions), "Accident prevention is a major task of management. The creation of a safety-minded, earnest, competent organization is our economic and social duty." Grimaldi notes (Journal of the American Society of Safety Engineers, vol. 10), that in those plants where management's safety responsibility is well understood, the plant manager is the focus of the plant safety program. He regards the field of accident prevention as a source of rewarding human contact devoted to keeping employees free from injuries, stopping losses, and earning new profits. Such a manager opposes the traditional viewpoint that safety is a routine function with only a modest relationship to business effectiveness. He is aware that injuries are usually the result of a failure to control the inherent risks in his business. To him the need for profitable production is related to the necessity for maintaining output while limiting preventable accidental loss. An effective control effort must therefore embrace all aspects of the business' loss potential.

The quality of a manufacturing plant's safety effort is strongly affected by the production manager's attitude toward safety. Simonds (in Simonds and Grimaldi, Safety Management) states that "it should be agreed that a process will not be considered safe simply because it is possible for an employee to carry it on without injury by exercising continuous diligence, being every instant alert and safety minded. Workplaces and processes must be engineered for safety first. . . . There still will be plenty of need for everything that can be done to motivate, teach and control employees so that there will not be unsafe acts." The basic **safety program elements** are stated simply by H. W. Heinrich (Industrial Accident Prevention—A Scientific Approach) to be:

1. Creating and maintaining an interest in safety.
2. Fact finding through periodic inspection of plant and equipment.
3. Finding remedies for unsafe practices and mechanical hazards.

FEDERAL SAFETY–HEALTH LAW. The Occupational Safety and Health Act of 1970 brings under Federal coverage the vast majority of workers in the nation. It authorizes the Federal Government to set and enforce safety and health standards for all places of employment affecting interstate commerce and to enforce the standards with criminal and civil penalties for violations.

Standard Setting. The Secretary of Labor is required to promulgate existing **national consensus standards** and **established Federal standards,** unless he determines that those standards will not improve safety and health of employees.

At any time after the effective date of the Act, the Secretary may promulgate new standards, or modify or revoke any existing standards. He may also promulgate emergency temporary standards if he determines that employees are exposed to grave dangers from toxic or physically harmful agents or new hazards. Temporary variances from standards are authorized to give an employer sufficient time to come into compliance. Variances may be granted without time limits if the Secretary finds that an employer is using safety measures which are as safe as those required in a standard. Adversely affected parties may seek judicial review of standards promulgated by the Secretary.

Enforcement. Following an inspection, the Secretary or his representative is authorized to issue a citation to an employer for violation of the Act. Employers will have 15 days to contest the citation, or any proposed penalty, by appealing to the Occupational Safety and Health Review Commission. The Commission is authorized to affirm, modify, or vacate the Secretary's action, or to direct other appropriate relief. Orders of the Commission are subject to review by U. S. Courts of Appeals.

In the event an inspection discloses an imminent safety or health danger, the Secretary is authorized to seek an injunction in a U. S. District Court to restrain the violation. The Court may prohibit employees from entering the premises where the imminent danger was found. The Act authorizes civil penalties for violations and criminal penalties for willful violations resulting in death.

State–Federal Relations. States which desire to set and enforce their own occupational safety and health standards may submit plans for State administration to the Secretary of Labor for approval.

Responsibility of the Department of Health, Education and Welfare. The Act authorizes the Secretary of HEW, in consultation with the Secretary of Labor and other agencies, to conduct research, experiments, and demonstrations relating to occupational safety and health, to develop criteria for the establishment of occupational safety and health standards, to publish data on occupational illnesses, and to conduct inspections necessary to carry out these responsibilities.

TYPES OF SAFETY ORGANIZATIONS. Many persons influence safety directly. The chief executive, production managers, maintenance personnel, product and process research and development engineers, the comptroller, personnel or industrial relations employees, the plant physician, the foreman, as well as the individual production employee can affect safety qualitatively. Factors that govern plant safety effectiveness include the choice of equipment, procedures and materials, budgets, the selection of personnel, the kind of leadership provided, and the work habits of the individual.

When a safety specialist is employed (full or part-time) to direct the plant safety program his function usually is to assist line managers to control the safety of their operations. Ordinarily he does not assume any management prerogative, but develops information and materials which enable managers to optimize their authority in the pursuit of safety. A safety organization may be set up in several ways to suit the size and need of the plant. T. O. Armstrong, in Industrial Safety (Blake, ed.), discusses three basic **organization types**:

1. Safety work carried on wholly through the **line organization.** Usually found in small organizations where the safety director's work is assigned to a line executive as an additional duty. The chief advantage is that it places responsibility for safety directly on each operating head as part of the production function. Its weakness lies in the fact that supervisory personnel may be occupied with other pressing production problems. Care must be taken to provide those responsible for safety with adequate additional time and facilities.

2. Safety work directed by a full-time **safety director** reporting to a major executive. This type of organization is used mostly by large companies, and the safety director may have one or more assistants or subordinates.

3. Safety work carried on primarily by **committees** set up for the purpose. This type is usually found in companies too small to justify the use of a full-time safety director; it may show the weaknesses inherent in committee-type organization.

The National Safety Council (Accident Prevention Manual) shows (in Fig. 2) the safety program activities and usual organizational relationships where safety work is carried on as a staff function through the line organization with the safety supervisor.

Safety Director and Safety Department. The safety man, whether called safety engineer, safety specialist, safety inspector, or safety director, is the representative of management in accident prevention activities. The chief operating executive in effect directs the plant safety effort. His attitude towards safety, the degree to which he implements the accident prevention effort, his follow up of the activities of his subordinates, and his insistence on competent information from the appointed safety director are principally responsible for the plant's safety effectiveness. Other managers may be energetic or casual in their pursuit of safety depending on the chief executive's attitude towards its achievement.

The safety director's actual **position in the plant organization** varies with the general organization of the individual plant. A survey by the National Safety Council (Accident Prevention Manual) shows that, of the safety directors in American industry, 44.8 percent report directly to a member of top management, 19.8 percent report to the factory, production, or works manager, and 30.4 percent report to the industrial relations manager. The remaining five percent report to the managers of various departments (labor, insurance, security, medical, etc.). An earlier survey by the American Society of Safety Engineers (Accident Prevention Manual) reports that the frequency rate and severity rate are appreciably lower in plants where the safety director reports to top management.

It is generally accepted that practically every phase of an employee's life has a bearing on accident prevention. For this reason the safety director is often placed under the jurisdiction of the industrial relations department, and also because this department usually includes the medical and other functions with which the safety director must be in contact. In any event, the safety director must occupy a position of such rank that the channel to all departments is available to him.

While **final responsibility** for safety rests entirely with the "line," and foremen are recognized as the key men in safety, some plants have developed a completely organized department whose members spend all of their time on safety. This method is considered more effective in some cases since it provides a focal point for implementation of all accident prevention activities. Authority for such important matters as shutting down dangerous jobs, designating the type of safety device that must be used, and, at least, parallel authority on many items of methods, personnel, training, and compensation policy that affect safety can be delegated to this department by top management. Administration of any necessary discipline must be left to foremen and supervisors, but that part of the workers' activities bearing on safety is the province of the safety man. He must be continuously in the factory and thoroughly familiar with the hazards present and the methods of correcting them. It is obvious that with such an authoritarian safety department there is overlapping with line authority. Therefore from the standpoint of sound management, a department of this type is not advised in most cases.

A safety department should have a reasonably **central location,** near the dispensary if possible, and with sufficient space available to provide for group meetings. Facilities must be available for storing and displaying samples of safety equipment, posters, safe-practice references, equipment catalogs, and for the necessary records, files, and charts of the department.

Safety Committees. A safety committee or group of plant committees is common to many plant safety organizations. They are subject to the usual problems of committee operation. (See section on Plant Organization.) There is also some question about their basic value in the safety program. Harvey (Journal of the American Society of Safety Engineers, vol. 9) found that committees "tended to detract from individual managers' responsibility for the safety of their departments" and that "employees felt remote from the activities of the committee." A substitute measure, safety observers, was suggested to secure more widespread employee participation in plant safety.

Where safety committees are established they require executive initiative, control, and support to be effective. No hard-pressed manager may assume that his accident prevention problems are solved by activating a committee and assigning responsibility for safety to it. The carefully established committee, however, can promote interest and participation in the safety effort, and through group study and recommendation it can strengthen the judgment of the executive. Important at all times, but particularly at the outset, is a **written statement** of:

1. **Scope,** or mission, of the committee activity.
2. **Extent** of committee authority, including budgeted funds, if any.
3. **Procedure** as to time and frequency of meetings, attendance requirements, agenda, records to be kept, and reports to be submitted.

Committee size should be kept small for effective work but large enough to include the breadth of knowledge and experience involved in the plant's operation. In the larger plants these two objectives often are obtained by forming one committee on the executive level and one or more committees at the departmental or shop level. Coordination of the committee activities, encouragement of their effectiveness, and provision of assistance in their studies are leading functions of the safety director. Figure 2 is a model of the customary safety program organization. It indicates the relationships between safety committees, management, the safety supervisor and others important to the safety effort.

Safety Observers. As a means of obtaining greater employee participation than is possible with a committee, safety observer programs have been proposed. Harvey (Journal of the American Society of Safety Engineers, vol. 9) reports that observers are volunteers requested by supervisors to serve a four-month term

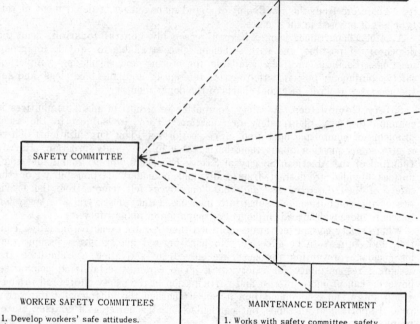

Fig. 2. Safety program activities

SAFETY SUPERVISOR

1. Serves in a staff capacity without line authority.
2. Coordinates safety activities.
3. Keeps and analyzes accident records.
4. Conducts educational activities for supervisors at all levels.
5. Conducts activities for stimulating and maintaining interest of employees.
6. Develops employee safety education programs.
7. Serves on safety committee, usually as secretary.
8. Supervises and appraises accident investigations.
9. Plans and directs a regular program of safety inspections.
10. Checks for compliance with applicable safety laws and codes.
11. Issues regular reports showing safety performance and accident trends.

FOREMEN

1. Inspect for compliance with safe work practices and safety rules.
2. Train men to work safely.
3. Responsible for safety of their crews.
4. Responsible for a safe workplace, good housekeeping, proper light and ventilation, safe piling; also, enforce wearing of protective clothing and equipment.
5. Responsible for obtaining prompt first aid to injured.
6. Report and investigate all accidents and correct causes.
7. Serve on safety committee.
8. Hold crew safety meetings.
9. Discuss safety with individual employees.

EMPLOYEE

1. Works in accordance with accepted safe practices.
2. Reports unsafe conditions and practices.
3. Observes safety rules and regulations.
4. Serves on safety committees.
5. Makes safety suggestions.
6. Does not undertake jobs he does not understand.

and organization relationships.

inspecting their departments routinely each week for hazards. A **checklist of hazards** is provided the observer as a reminder of things to look for. Observed hazards are recorded on the check sheet and given the supervisor who then notes on the sheet the action taken. DeReamer (Modern Safety Practices) states that the observer program promotes a better working relationship between management and employees, but he cautions that care must be taken to avoid the assumption that hazards can be detected and corrected simply with safety observer teams.

Accident Prevention Program

BASIC STEPS IN A SAFETY PROGRAM. Many plans, some of them far reaching, have been formulated for the prevention of accidents. Regardless of the size of the company and the kind of work in which it is engaged, most useful plans are based on **fundamental principles** which start with simple steps undertaken usually in the following order:

1. **Obtain cooperation of plant manager.** The manager's desire for safety achievement must be clearly visible through his action to achieve it.
2. **Obtain cooperation of superintendent.** The superintendent must make safety an integral part of the operating organization.
3. **Appoint safety director.** One man must be designated to develop and codify needed plant safety information, methods and procedures, and counsel on their implementation.
4. **Analyze accident records.** After his appointment, the safety director should analyze the accident reports for the past year or two to learn, if possible, the how, who, where, when, and why of each accident.
5. **Hold meeting of operating executives.** All foremen, superintendents, and operating heads should then be summoned to a general meeting presided over by the manager or general superintendent.
6. **Make inspection of operations.** Following this meeting each foreman should make a complete inspection of his department.
7. **Start mechanical safeguarding.** The safeguarding program should then be developed and carried out, making sure that the most serious conditions are corrected first.
8. **Make general announcement.** Then, and not until then, should the workers be acquainted with the accident prevention plan.
9. **Organize educational work.** Formulate a program to maintain interest and supply information on safety to management, foremen, and workers.
10. **Consider engineering revision.** Consider methods for improving machinery, equipment, and processes to eliminate hazards and increase production efficiency.

DEVELOPING THE PLAN. Safety planning starts with the plant manager. He must make safety a necessary part of the production processes, operations, and procedures. Every foreman and worker should know unequivocally from the manager's position that the company intends to have a safe plant.

Management must evidence its sincerity through its **actions.** To the worker mechanical guards, good lighting, an orderly plant, insistence on compliance with established safe work practices are visible examples of management's concern for safety. If it seems that management is willing to defer implementing the obvious safety requirements, employees often feel they are privileged to do likewise.

Superintendents and foremen must be persuaded by the plant manager to strive for safety, otherwise the shop employees are not likely to deliver the desired safety performance. Safety achievement, like the fulfilling of any business objective, requires the firm, dedicated and wise **leadership** that typifies successful management in any situation.

Planning should include the **sound management practice** of establishing long range achievement objectives, and short term goals designed to bring the objectives toward fulfillment. Supervisory personnel, from plant manager down to foreman, should be followed up to determine progress towards achievement. The safety director function is the source of the information required to measure progress. It should also develop and codify the materials needed to implement the plant's safety program.

Role of the Safety Director. One man must be responsible for the development and codification of safety information, counseling managers on safety implementation methods, and developing progress measurement data in every plant, regardless of its size or the type of safety organization used. It may be advisable or necessary for the manager himself to carry this responsibility, or he may delegate it to an assistant whose duties and qualifications will determine whether he should be known as safety engineer, safety director, safety inspector, or some equally significant title. (For the sake of uniformity, he is referred to here as the safety director.)

The following are **general functions** performed by the safety director whether he is a line executive or head of a separate department:

1. Responsibility for formulating and administering the safety program.
2. Advise on all safety matters for the guidance of management and all departments.
3. Report to management periodically on the safety program (monthly, weekly, daily).
4. Maintain accident records.
5. Supervise or advise on safety training.
6. Coordinate with medical department on proper placement of employees.
7. Personal plant inspection.
8. Keep informed on the latest safety information.
9. Check for compliance with laws and insurance company recommendations.
10. Consult with government agencies and insurance companies on safety problems.

In addition to a knowledge of safety, the successful safety director must have nearly every **personal qualification** that is to be found in successful men in all walks of life. He should have vision, initiative, persistence, judgment, diplomacy, leadership, and, above all, sympathy. If he is given the title "engineer" he should have had an engineering education or its equivalent in actual experience. Such **professional qualification** is unquestionably required in large organizations or where difficult technical problems arise.

An important feature of accident prevention work is that the safety director should know the men he is working for and with, because much of his success will depend upon the manner of his contact and his dealing with the men. The accident prevention work which is most constructive and most lasting is often accomplished by **getting other men to do the work.** Sometimes this participation is brought about by suggestion, perhaps by a direct request or as a personal favor, and again by an order from the manager or superintendent. Whatever the means, the result will be that the man who does the work is interested to a greater degree than he would be otherwise, and feels a personal responsibility for his share of the work. If handled tactfully, this technique will work with the plant engineering department and the plant executives, as well as with the workers.

Analysis of Accident Records. It is assumed in this discussion that no safety work has been attempted in the company other than the safeguarding required by an insurance company or the state factory inspector. In accepting his position, the safety director will have satisfied himself that the management is sincere in its desire to prevent accidents and that it proposes to follow every reasonable and practicable plan for securing the cooperation of its employees in safety work. The first step which he should take after his appointment is to start an analysis of the company's accident reports for the past two or more years. While making these analyses of accident records, the safety director should take advantage of every opportunity to establish personal and close relations with the superintendent, foremen, and other plant executives. Doing so may prevent unpleasant misunderstandings later on.

In many instances it is found that records have been kept with insufficient information to **identify causes.** So far as possible, the new safety director should search out this information by inquiry and interview. A knowledge of causes is essential to accident prevention. Using the procedures suggested in this section will result in a systematic treatment.

Plant Inspections. When an industrial plant embarks on a new or intensified safety program, a thorough plant inspection is desirable. The **environment** of the work force must be known; the entire operation must be considered, and indirect production as well as direct production facilities and methods must be observed and evaluated. Management may assist the safety director in this inspection by utilizing the services of fire and casualty insurance company inspectors, state industrial commission inspectors, or industrial consultants (firms or individuals).

A consulting inspector will submit an **independent report.** The safety director, however, should accompany the consultant, if employed, and the foreman of each department on the inspection. The foreman should make the inspection and file the **departmental report.** The safety director's duty is simply to help by suggesting ideas to the foreman, by making sure that the most important things are not overlooked, and by encouraging the foreman to correct any of the conditions within his control.

The safety director should not be overly concerned if the foremen fail to note all the unsafe conditions that should be corrected. If none of the serious conditions is overlooked, many of the lesser items can be "caught" at later dates.

Provision should also be made for **systematic reinspection** of all plant departments and the plant as a whole. The safety director, the line supervision, maintenance men, and department safety committees all should be brought into the activity of safety inspection on a systematic basis. This subject is more fully discussed elsewhere in this section.

Elimination of Hazards. After all inspection reports are turned in, the safety director should help the superintendent determine which **safety recommendations** made by the foremen and consultant should be carried out and in what order. Many recommendations should be referred back to the foremen from whom they originated, with orders to "go ahead." Others may have to be referred to the master mechanic or to some other person or department for necessary action. The recommendations which seem impracticable, or on which favorable action cannot be taken at once, should be discussed either at another meeting of all foremen or with the individual foremen by whom they were submitted.

Carrying out this part of the program satisfactorily will not only eliminate the majority of the accident hazards which are within the control of the management, but it will also impress upon the minds of the workers the fact that the company is sincere in promoting safety and willing to do its full part toward that end.

Communication with Employees. Safety develops through the teamwork of management, plant engineers, supervisors, and workers. To effect this desirable teamwork, management must use all the best techniques known in the communication skills. Only after the company has made a definite start in its correction of unsafe conditions should an effort be made to secure the full cooperation of the work force. The worker needs visible evidence that management is not merely emphatic, but truly sincere, in its desire for safety.

The first step is to acquaint the supervisory force and then the workers with the fact that the company is starting an **organized effort to prevent accidents.** In doing so, management should emphasize its position: that it will do everything in its power to make dangerous conditions safe; that it expects workers to do everything in their power to perform their work safely; and that it intends to do its best to inform workers how to do their work safely and under no circumstances require or permit work to be done in an unsafe manner or with unsafe conditions.

These facts may be communicated to the workers through personal letters from the management, through announcements posted on the bulletin boards, through the plant publication, at departmental meetings, or at a general mass meeting. Such a step is necessary to give publicity to the plan and to arouse enthusiasm for carrying it out. Without the cooperation of the workers, the plan will fail.

Training Programs. In all but a few plants the actual training of the worker in performance of his job is done by the foreman or some more experienced worker. Although many trades require apprenticeships, the above is still true and the training in safe working methods received by the new employee may be slighted, or non-existent, depending upon the individual foreman or leader. DeReamer (Journal of the American Society of Safety Engineers, vol. 11) says it is not enough to tell a supervisor he is responsible for safety. They must be taught to recognize hazards and to take appropriate action. The **nature and scope** of this responsibility must also be taught. Supervisor safety training must cover these basic accident prevention elements:

1. Development of safe working conditions.
2. Personalized safety training to create safe work habits.
3. Enforcement of safety rules.

It is pointed out that institutional or mass educational methods are deficient as a means of instructing the worker. He must be taught on an individualized basis what to do and how to do it if he is to be fully impressed with the necessity of complying. Pressures are strong and constant for training to achieve satisfactory quantity and quality of work. The objective of a safety training program is to improve the ability of foremen and leaders in the training of new workers in safe performance in all aspects of the job.

There are many facets of **supervisory safety training.** A training director, if present in the organization, should be responsible for the overall program, but should coordinate closely with the safety director for training in safety. The safety director must make certain that the foremen learn the principles of effective **job-instruction** and their application.

Through careful **safety analysis of jobs,** plans can be developed whereby workers can be instructed to do their jobs in the safe way. This requires:

1. Getting the worker's attention to safety.
2. Telling the worker in the language of the "shop" what the hazards are.

3. Demonstrating the proper, safe procedure.
4. Observing the worker on the job.
5. Correcting his procedure until safety habits are formed.
6. Following up to be certain that familiarity does not breed laxity.

Besides the foreman's verbal and manual training, other methods may be used. On production cards, tally sheets, or time cards for jobs, a short, concise **safety message** may be included. For example, on a press job where there is a definite hazard of operation, there may be written or printed on the card, "Job must not run unless guard is in place," or, "Use safety pliers on this job," or, "This job must not turn over, it is rated for start and stop." On portable tools such as grinders, either air or electric, **name plates** should be fastened to the machine indicating such information as, "Do not use wheel larger than 4 in. in diameter," or, "Goggles must be worn while using this tool." Again, where automatic-push-button control devices are used for series operation, a sign should read, "All buttons must be open while working this job unless otherwise authorized by your foreman." A collection of safety information placed directly in front of the worker at the time he is to use the device is more effective than a book of rules, a previous warning on the bulletin board, or verbal instructions by the foreman.

The promotion of such safe practices throughout the entire organization develops safe thinking for future accident prevention. The use of **red danger tags** on machines shut down for the correction of some accident hazard has proved valuable in many instances. A tag reading "Danger" or "Do Not Use," as in Fig. 3, on a machine arouses inquisitiveness among the employees,

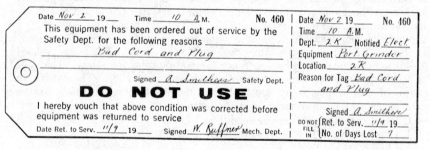

Fig. 3. **Warning tag (red) attached to equipment ordered out of service.**

and even the foremen from other departments, who usually stop to look at the tag to discover why the machine is not in operation. Its use also has a much-desired effect on the production departments, which urge the repair division to get the machine back into service. Since the machine has originally been shut down by the safety department, the advertising feature is obvious.

The safety director provides the specialized safety knowledge concerning materials, machines, and methods that enables the foreman to be sure his instruction is "safe." In some companies the safety director sends to each foreman, at regular intervals, material he can use in periodic ten-minute **safety talks** with his group. This material may be an accident case (within the company or outside) which has illustrative points for discussion. Other materials may be special topics on safety issues the director considers timely as well as items which enable appropriate conversations about safety with individual workers. Simonds (in Simonds and Grimaldi, Safety Management) says "Casual short conversations in which the foreman asks about a worker's family, his vacation, a hobby or how a machine is working are good means of . . . (passing on) a reminder about certain safe practices . . ."

ESTABLISHING AND ENFORCING SAFETY RULES. To set up a standard body of safety rules for plant practice is obviously difficult considering the variety of industrial operations and procedures. The plant can prepare, with some ease however, **standard operating procedures** based on those established by companies engaged in similar work. The valuable exchange of safety ideas and contacts made through local and national organizations, together with the data made available by casualty insurance companies and state labor departments, will be found to be a continued source of vital and up-to-date information.

Safety rules cannot bring about a reduction in accidents unless provision is made for **enforcement.** Enforcement, however, is not entirely a matter of discipline. One of the first steps toward rule observance is that all supervisors must become familiar with the rules and follow them continually. Their good example can do much to influence the workers. The best weapon of enforcement is patience and perseverance rather than threats, insults, and discharge slips. Some workers, however, will resist even the best efforts of the foremen and safety advisers, and stern warnings should be given for such deliberate violations. If, after a short layoff for a willful violation of an important rule, the worker persists in such disobedience of safety rules, a long layoff or discharge is in most cases required. When labor groups or unions represent employees, these groups should be consulted and an agreement made upon methods of enforcement to be used when the occasion arises. Continual offenders always become involved in accidents if their tendencies are not corrected.

KEEPING THE SAFETY PROGRAM ACTIVE. There are many ways in which safety may be kept actively before employees. Posters, bulletin boards, banners, etc., are familiar and may have value, but their use must be carefully controlled if they are to be effective. Action posters describing current events, together with a safety message in the same area of display may create interest in both, but they must be changed frequently or they will be passed by unnoticed. Pictures of well-known workers who have accident-free records or devices which have saved employees from injury, should be displayed on well-lighted and conspicuous bulletin boards.

Congratulatory messages from management for excellent safety records never go unnoticed by workmen. Personal letters mailed to new employees stressing the value of safety are often an important aid to training. These letters should preferably originate in the executive office, so that workers will be made to feel that the "big boss" is vitally interested in the safety and welfare of his employees.

Plant publications and house organs can perform an important service in selling safety to their readers. Photographs of well-known "buddies" with their remarks about safety, particularly if they have been personally involved in accidents, attract much attention. Publication, through this medium, of records of progress in safety in various departments, together with suitable remarks from company officials, brings about a spirit of cooperation. Since most of these publications find their way into the homes, the messages may be read and re-emphasized many times.

Contests and Awards. A familiar method of stimulating optimal performance is to create **competitive situations.** Generally these occur in the form of contests although measured performance with the prospect of positive recognition for success and negative recognition for failure, possesses competitive elements which are equivalent and often superior in effect to a contest.

Many varieties of contests are employed by companies to stimulate interest in safety. The contests may be among companies, such as those conducted by the National Safety Council within each industry. They may be among departments within the company. Or there may be individual competition among employees for prizes, awarded according to the duration of their accident-free records. Frequently companies try to interest employees in seeing how long the entire company (or division or department) can go without a lost-time accident. The record, posted prominently and kept current daily, can be effective in communicating management's concern for safety and in arousing employee interest.

One of the **dangers of contests** is that of ill feeling and letdown when a serious accident finally occurs. After some months of accident-free work, if a serious accident takes place, the need for starting over again may be discouraging. Minimizing the letdown requires pre-planned action that will pick up interest promptly. A new emphasis generally is indicated. There are other dangers too. Often the contest is too heavily relied on to generate safety awareness. In such cases it is forgotten that the most powerful persuader of employee interest is his supervisor. Frequent expressions of the importance of safety (to the man as well as the company) and insistence on compliance with established safety procedures generally are most productive. However, contests may have significant justification in circumstances where there is little opportunity for close supervision of employee performance. Trucking operations are an example in point.

Awards for outstanding performance differ among companies. In some cases **grand prizes** of substantial value (watches, vacations, etc.), are given. Generally, however, the prizes given employee groups have only nominal value (cigarette lighters, barbecue tools, etc). Often a sincere letter written personally by the company's chief executive recognizing an earnest safety effort is as rewarding as a token prize.

Exhibits. An effective way of obtaining the interest and cooperation of employees is to make a well-prepared safety exhibit. Tools, goggles, safety shoes, etc., which have actually been damaged in service as protective devices are educational and stimulating items for exhibition. Smashed safety pliers, broken goggles, damaged safety shoes are always helpful in cases where those who must be "shown" are concerned. Each article definitely becomes an object lesson to employees, many of whom may have rebelled when requested to use such protective devices. While it is not good psychologically to be gruesome, effective results have been obtained by such expedients as, in one case, exhibiting a handful of hair torn from a girl's head by a machine accident.

Safety Instruction and Conferences. There are courses in safety engineering, and for training industrial supervisors in safety, conducted by organizations such as the National Safety Council's Industrial Safety Training Institute, which are well worth the time and attention of foremen, leaders, and safety inspectors. With such basic training, a well-defined program can be set up to fit the various individual needs. When a well-regulated program has been designed to fit in with the safety department's or safety director's plans, a high degree of cooperation with all departments should result.

More and more, management will find that **general meetings**, involving plant matters, should include the safety director. Committees considering new methods, the use of new materials, production setups, purchase of new machinery, and other problems should realize that nearly every phase of their proposed plans has items which require the consideration of safety. Factory layout departments, considering the setting up of new manufacturing facilities, or the relayout of existing floor areas, benefit from discussing the hazards in the storage and handling of manufacturing materials and of work in process. Plant vehicles, materials on the floor in traffic aisles, dangerous turns in aisles, and many other details need review when designing a safe and productive layout. **Die and tool design** may bring about many hazards which do not always show up on the drawing board, and a safety director who is continually looking for such hazards can offer suggestions concerning safe clearance areas in dies, the use of dial feeds, chutes, and automatic feeds which not only produce a safer tool to work with but also permit in most instances an increase in production. A slightly higher first cost which reduces costly accidents and increases production is certainly money well spent.

Discussion of hazard problems, between the foreman and his crew, or the presenting of pertinent general safety knowledge to the work force as a body, are common **informational meetings.** Weekly ten-minute "tail gate sessions," during which the foreman reviews a new job and its safety requirements, analyzes with his people the significance for them of an accident that has occurred, or discusses, periodically, critical operating hazards and their control are standard in many companies. Occasional **group meetings** involving larger segments of the work force also are familiar in industry. Usually these are brief, scheduled at convenient times and often employing visual presentation methods. Employees are likely to find films, for example, more interesting at times than a talk. It is desirable however for the arranger of the meeting to provide authoritative direction so that the audience will be led to identify local applications of the ideas presented in the film.

Eliminating Safety Hazards

SAFETY STANDARDS FOR MECHANICAL SYSTEMS. Any safety department, safety director or other executive specifically responsible for safety should obtain, have available, and become familiar with the published **safety standards** relating to the types of equipment present in the plant. Among the several agencies publishing valuable safety materials are:

1. American National Standards Institute (ANSI).
2. American Society of Safety Engineers (ASSE).
3. National Safety Council.
4. American Insurance Association.
5. National Fire Protection Association (NFPA).
6. Bureau of Mines, U.S. Department of the Interior.
7. National Bureau of Standards, U.S. Department of Commerce.
8. U.S. Department of Labor.

Only in comparatively recent years have machine-tool builders included adequate provisions for safety in their design. Proper guarding before that time, for the most part, was left to the customer, since lack of uniformity in state regulations often meant special guards to meet the respective requirements. The user, however, in every case must carefully check each item, from the mechanical transmission of power to the machine up to the point of operation, on through each step in the process to the delivery of a finished or partly finished part. **Mechanical power transmission apparatus** includes all shafting, belting, pulleys, gears, starting and stopping devices, and other moving parts of such machinery. Reliable starting and stopping equipment must be provided for the safe and efficient operation of power transmission apparatus.

Mechanical Equipment. Many accidents have been caused by belts. The National Safety Council (Accident Prevention Manual) recommends that all V-belts, round belts, and rope drives be enclosed and that all belts with metal lacings or fastenings be guarded. In addition, any belt traveling in excess of 250 ft. per min. and less than 7 ft. from the floor level should be guarded to a minimum distance of 7 ft. above the working area. These recommendations must of course be considered in the light of many other factors, such as load, strength, tension, and the proximity of workmen to the belt. In general, it is best to enclose all belts where possible. The construction of the **belt guard** must be carefully considered, using expanded metal, perforated or solid sheet metal, or wire mesh on a frame of angle iron secured to the floor or the machine. The construction of the guards should be such that they can be easily removed by maintenance men only. Access doors or hinged sections should be provided for making adjustments to the moving parts, and, where possible, lubricating points should be provided with an extension through the guard so that it need not be removed or opened for oiling during operation of the machine.

Mechanical guarding of **rotating parts** such as shaft couplings, collars, keys, setscrews, pulleys, gears and sprockets, and chains will do much to prevent injury to workers. The same care should be used in making guards for such moving elements as is taken for belts and any other moving parts. **Standard practices** should be set up calling for the use of revolving collars of the cylindrical type, without projecting screws and bolts, and setscrews set flush with or countersunk beneath the surface of the metal part in which they are inserted. Pulleys, gears, sprockets, and chains which are 7 ft. or less from the floor or working platform, or are exposed to contact, should be guarded, and the guards should provide complete safety in case chains, etc., should break while in motion. The more intricate guarding of the machine tool itself, or of processing equipment such as emery wheels, belt sanders, power presses, conveyors, portable power tools, should be considered separately and more intensively, since it usually becomes the hazard nearest to the operator.

Equipment which depends on **high pressure** for power must necessarily receive careful study when designed, and piping, cylinders, and other devices should have an adequate factor of safety to protect the operators and any others near such equipment. Even small units, where processing operations require hole punchers, riveters, jig clamps, etc., should have guards over the operating cylinders and other vital parts, since a small blowout will emit a fine stream of liquid at such a high pressure as to do considerable damage if it strikes the human body.

Point-of-Operation Hazards. Many means exist for the control of the hazardous point of operation on such familiar machines as the power press, power brake, power saws, etc. The choice of any single device depends on the production process, installation and operating costs, maintenance requirements, and other factors. The National Safety Council (Accident Prevention Manual) as well as other sources are available to assist in making a practical decision. In general the ideal means of protecting operators from point-of-operation dangers is to design the setup so that operation points are inaccessible to the operator.

Grinding, Buffing, and Polishing. Special instructions or regulations covering the use and guarding of grinding wheels have been formulated by and are available from grinding-wheel manufacturers, state labor departments, the National Safety Council, and other sources. In general, all surrounding conditions should be carefully studied to make the use of such equipment safe. Operators should always wear suitable **goggles** or **shields** to protect their eyes. Dust generated by dry abrasives is a health hazard and should be removed at the point of origin by an efficient exhaust system.

Grinding wheels should be handled carefully to avoid damage. Manufacturers of grinding wheels ordinarily are pleased to provide **inspection recommendations** which enable the detection of damaged wheels, thereby avoiding the hazard of "exploding" wheels. Whenever a wheel breaks, a careful check should be made to determine the cause of the break. Inspection should also be made to make sure that the hood guard has not been damaged or the spindle and flanges sprung out of true or out of balance.

Work should not be forced against a cold wheel, but applied gradually, giving the wheel an opportunity to warm and thereby minimizing the danger of breakage. This precaution applies to starting work in the morning in cold rooms and to new wheels which have been stored in a cold place. Great care should be taken to avoid striking the wheel a side blow, as when grinding castings suspended on chain blocks. Grinding on the flat sides of straight wheels is often hazardous and should not be allowed when the sides of the wheel are appreciably worn or when any considerable or sudden pressure is brought to bear against the sides.

Only competent men should mount, inspect, and care for grinding wheels and, preferably, specific men should be assigned to do all this work.

Hand Tools. Statistics show that hand tools cause a large portion of the injuries in industry. Mishandling hand tools, neglecting to keep them in proper condition, and leaving them in dangerous places are frequent causes of accidents. It is of prime importance, in the effort to prevent accidents, cut tool costs, and maintain a high rate of production, that only the best materials be used for making hand tools. Chisels, punches, drifts, etc., made from poor stock soon become dull, and their heads mushroomed and cracked. Hence, only tools made from the most suitable grades of tool steel should be employed: not necessarily the special expensive brands of alloy steels, but proper grades for the various purposes. Probably the greatest contribution to the reduction of accidents from hand tools comes from **proper maintenance.** Tools with cracked handles, mushroomed heads, improper tempering, etc., should be promptly removed from service and repaired, as a start in eliminating accidents from such sources.

Woodworking Machinery. The variety of operations performed on woodworking machines, the high speed and sharpness of the cutting tools, and the comparatively light weight of the wood being worked on combine to produce high accident rates on such equipment. The range of uses of certain woodworking machines makes satisfactory guarding particularly difficult and increases the temptation of workers to operate without guards. The natural course, particularly where the volume of work is not large, is to save in first cost of equipment by performing many different operations on one machine. Actually, frequent changes in machine setup often are so costly in wasted time and in poor work that true economy usually lies in **providing enough machines to limit the operations on each to closely similar work.** It is possible, for instance, to avoid grooving (dadoing) on circular saws or to avoid using two tools that are operated by the same control, as an auger and a grinder on the same spindle. An absolute fundamental for safety in all such combinations is to arrange the drive so that only one tool can be operated at a time.

The high speed at which woodworking machines are operated often causes excessive vibration unless the machines are properly designed and well made, and unless bearings are properly maintained and tools correctly set.

It is important to enclose completely all belts, pulleys, clutches, gears, sprockets, spindles, and reciprocating parts and to provide practical (and where possible, automatic) **safeguards** for the point of operation of all woodworking machines. No class of machines, with the possible exception of power presses, presents such difficult safeguarding problems. There are many protective devices on the market for use on various woodworking machines, each of which, however, has certain limitations. Careful study should be given to all operations performed on every machine, and a type of guard selected or made that will be practical and effective for each. It is best to secure the cooperation of the operator and his foreman, for not only does a good operator "know his machine," but, once he and his supervisor put their minds to it, they are likely to contribute just the ideas needed. Homemade guards, if well designed and well constructed, are satisfactory, but, if the chief objective is to get something cheap, they are likely to prove unduly expensive in the long run through failure to prevent accidents. A guard that gives a false sense of security may actually be worse than no guard at all.

Machines can best be guarded by the manufacturer. In the purchase of new machinery, specifications should call for guards on all driving parts, as well as for the point of operation. Guards should perform specific functions or meet definite requirements.

ELECTRICAL EQUIPMENT AND SYSTEMS. Generators, power lines, wiring, transformers, etc., should be constructed and installed with the highest standard of safety. All manufacturers of such apparatus recognize this important factor, and the installation and maintenance of such equipment should follow authoritative engineering practice and established regulations and codes, such as the **National Electrical Code.** The **smaller tools and electrical devices** (portable drills, grinders, welding transformers, heating devices, etc.), because of the large amount of handling they receive, require constant checking by competent electricians or maintenance technicians. Electric shocks of as low as 110 volts, caused by faulty wiring, careless handling, incomplete repair, etc., may cause serious or fatal accidents. Because of the increasing use of portable electrical tools, workers must be taught to take proper precautions with low voltage as well as high voltage. The replacement of **attachment plugs** on small tools and receptacles throughout the plant in accordance with the Standard for Dimensions of Caps, Plugs, and Receptacles (ANSI C73 series) will not only provide for the automatic grounding of electrical devices, but will also prevent the accidental connection of a tool to improper voltage.

Grounding should be provided on all exposed metallic noncurrent carrying parts of electrical equipment which carry 150 volts to ground and preferably in 110-volt circuits as well. The total resistance of the grounding wire and its connection with the ground should not exceed 3 ohms for water pipe connections, 25 ohms for buried or driven grounds. Ground wires sometimes are not covered with insulation so that their continuity can be checked readily, visually. Good practice also calls for ground wires to be run in conduit as part of the normal distribution and branch circuit wiring.

CHEMICAL HAZARDS. While uncontrolled physical hazards generally cause sudden injury, chemical hazards may affect employee health gradually over a long period of time and result in **severe permanent injury.** Sudden injuries may also result from fire, explosion, and violent chemical reactions. In general, chemicals can cause unsafe working conditions through contamination of the atmosphere or through direct corrosive and poisonous action on the body.

Handling and Storing Chemicals. Caustics and highly volatile solvents are used in many industries. Because their use is frequently incidental to other operations, many employers and workers give little consideration to the dangers involved. Acids, alkalis, and many industrial solvents in concentrated form, and sometimes even in milder forms, may cause injury in four different ways:

1. Burning, from direct contact with skin or eyes, or indirectly through clothing.
2. Fume poisoning or suffocation. Some chemical fumes are poisonous if inhaled. Others when in sufficient concentration in the atmosphere will fail to sustain life.
3. Poisoning when taken internally.
4. Fires or explosions resulting from their improper handling and storage.

Many chemicals also give rise to dermatitis, especially if the employee is allergic to the materials.

Proper protective clothing and devices often are essential when handling chemicals. Rubber and neoprene gloves, boots, wooden clogs, rubber aprons, tight-fitting goggles preferably with rubber face pads, face shields, and acid-proof hoods, as well as specific safe-handling equipment are common protective means. High-grade respirators, of a proper type suited for the kind of hazard to which the worker is exposed, must be available for protection against fumes.

Good housekeeping is essential wherever chemicals are stored or handled. Chemical burns caused by workers tripping or slipping while handling containers can often be avoided if floors are kept in good repair and free from grease and oil. Good lighting is also an aid in the prevention of accidents. A rigid system of orderliness in the plant and in the operations of the workers is essential to reduce the frequency of injuries. Labels, warning tags, and clear instructions can prevent improper handling and storage of dangerous materials. A very high standard of safety instruction is required in all chemical plants and in plants where large amounts of chemicals are used.

Controlling Atmospheric Hazards. Simonds and Grimaldi (Safety Management) outline methods for eliminating or reducing the dangers associated with atmospheric contaminants.

1. Substitute less hazardous compounds.
2. Revise process or operation.
3. Segregate hazardous processes.
4. Enclose hazardous processes.
5. Provide local exhaust systems (see also "Exhaust Systems" in this section).
6. Upgrade general ventilation.
7. Design, alter, and maintain buildings and equipment to achieve better control.
8. Use special methods, such as "wetting" for dust control.
9. Require personal protective equipment.
10. Educate employees to use established safe working methods.

In many cases the reduction of atmospheric hazards is connected with the particular problem of fire and explosion prevention, discussed elsewhere in this section.

MATERIAL HANDLING. Handling materials causes a large percentage of the compensable accidents in industry. Many companies have found it advantageous to study the work of handling materials as thoroughly as time studies are taken on other operations.

Manual Handling. In **handling materials by hand,** sometimes a slight change from the usual method of grasping a piece, carrying it, and setting it down will bring about greater safety as well as greater efficiency. Many workers are injured because they do not know the safe method of lifting. Back strains and hernia are likely to develop if workers bend at the waist when leaning over to pick up a heavy or oddly shaped object. Hand protection, such as hand leathers and gloves that will resist rough usage, will often reduce injuries from rough or sharp materials. A large variety of "homemade" devices or specially designed tools has helped to improve handling methods as well as to eliminate accidents. Tote boxes, hand trucks to carry gas cylinders, adjustable die trucks for die setters, hand trucks for handling sacks, barrels, boxes, etc., and many other devices are supplied primarily to assist workers to accomplish their tasks safely and without strain and fatigue.

Cranes and Other Hoisting Apparatus. Too many hazards are involved to be fully discussed here, but precautions to observe on the operating floor below the cranes and hoists merit attention, such as avoiding overhanging loads, and observing the warnings of the crane operator and the floor men, who should be trained in safe practices. Complete periodic inspection of cranes and hoists should be rigidly enforced. Special safety meetings should be held for crane men and truck and tractor operators.

Equipment purchased from a reputable maker, including devices such as block and fall, chain hoists, air hoists, jib cranes, portable floor cranes, crabs and winches is usually designed with safety in mind. All these devices are as safe as the operator who handles them and men with rigging and handling experience should always operate such equipment. These employees should be taught the **safe limits** of the apparatus, because there is a general tendency to overload such equipment on account of its flexibility in difficult places.

Conveyors. Serious hazards develop with conveyors unless proper precautions are observed. No adjustments or repairs of any kind, including oiling, should be made while a conveyor is in motion, unless the oiler does not come within dangerous proximity to moving parts. Many accidents have been caused by starting conveyors without first giving warning to men who might be making repairs or adjustment or oiling the machinery. A **lock** or **warning sign** should be provided for repair men to block the control until everyone is clear. Control devices should be installed at frequent intervals in all power-driven conveyors for stopping the conveyor in case of an accident or other emergency. A safe-footing rubber matting or other anti-slip floor surface should be provided at the loading and discharging stations, and any material which might be spilled from conveyors should be removed immediately. The speed should be such that a worker will have ample time to place material in position without losing his balance, and material being moved should not project over the side of the conveyor or be likely to fall. Where conveyors travel over work areas, protective barriers strong enough to catch and hold falling objects should be installed. Many types of conveyors are used: gravity, chute, roller, belt, chain, etc., all of which must be carefully studied when installed to provide the proper safety for workers and for those passing by. Bridges should be provided where it is necessary to cross over conveyors, and overhead conveyors should have inclines to take them up over aisles or passageways to clear any traffic beneath. Guard rails, toe-boards at floor openings, and complete housing of the power drives are likewise necessary. Since the conveyor is a moving device and usually in the open, near workers, every precaution must be taken to insure its safe operation.

Motorized Vehicles. Electric or gasoline-driven trucks and tractors, because they traverse or cross aisles, passageways, roadways, elevator approaches, etc., at many points in a plant, present particular hazards. Floors must be kept in good condition and aisles must be properly marked off and kept free of materials to prevent injury to truck drivers and to employees working beside the aisles. Operators of trucks and tractors must be taught not to exceed a stated **safe speed limit.** Entrances to tunnels, crossovers, sharp turns and corners should be clearly marked with warning signs, and warning gongs or horns should be sounded by the operator at such places. The truck itself, of course, should be kept in good mechanical condition, brakes checked, switching devices, etc., maintained in proper repair, and a suitable substantial guard provided to protect the operator's legs.

In-plant trucks must be considered a special accident problem even though they present hazards similar to motor vehicles generally. For example, lift trucks require protective considerations when operating on shipping docks. Many fatal accidents have occurred when the fork-lift truck fell between the dock and the transport van it was entering, when the van moved. Setting of the van's brakes and chocking its wheels, as well as providing a dock plate, bridging the space (although only a couple of inches) between the dock and the van for the lift truck to cross upon, are customary precautions. A canopy-type guard to protect the fork-lift operator from sliding or falling loads when this lift is elevated, is generally an essential requirement.

EXHAUST SYSTEMS. Dusts, gases, vapors, and fumes generated by industrial equipment and processes constitute special classes of hazards. In addition, high temperature, high humidity, or a combination of these contribute, through worker distraction, to accidents and loss of production. Control of the workers' atmosphere for health, safety, and comfort is accomplished by exhaust systems and complementary ventilation with clean uncontaminated air. The safety director and plant engineer should be familiar with any city, state, or federal laws applying to **air-borne contaminants,** either as released by processes in the plant or exhausted from the plant into the community. Simple exhausting of contaminated air from the interior of the plant is often inadequate:

1. It must not be harmful or hazardous to other areas of the plant or to neighboring property holders.
2. It must be exhausted in a manner to prevent its being redrawn into the plant by natural or forced ventilation.

In general it is desirable to eliminate air pollution at its sources.

Mechanical exhaust systems should conform with applicable laws and regulations as well as authoritative engineering practice. **Local exhaust systems** (a combination of hoods, ducts, filter, and centrifugal fan, for example) placed at the point of operation in a process usually are preferable to general exhaust systems. They control the contaminant more practically in most cases.

Hoods. Hoods are intended to control the concentration of contaminants in the work room air by drawing off, at their point of origin, the objectionable dusts, fumes, vapors, etc. The effectiveness of a hood depends on the proximity of the hood to the contamination source (without interfering with the worker) and **capture velocity** of the air flow at the point of contamination. The air flow into the hood must be sufficiently high to prevent the contaminant from dispersing into the work room and the flow should draw clean air across the breathing zone of the worker. Figure 4 contains recommended capture velocities published by the American Conference of Governmental Industrial Hygienists (Industrial Ventilation, A Manual of Recommended Practice).

The capture velocity is primarily a function of the mass flow of air, not the velocity in the duct or hood. For example, at a distance of twice the diameter from the suction end of a round unflanged duct the air velocity drops to less than 2 percent of duct velocity. Providing a **flange** at the suction end of a duct will increase the effectiveness of the duct through better air-flow control. Additional **baffles** may be placed so that there is no possibility of the worker affecting the desired air flow.

Conditions of Dispersion of Contaminant	Capture Velocity (ft. per min.)	Examples
Released with practically no velocity or into relatively still air	50–100	Evaporation from tanks, degreasing, etc.
Released at low velocity or into moderately still air	100–200	Spray booths, intermittent container filling, low-speed conveyor transfers, welding, plating, pickling
Active generation into zone of rapid air movement	200–500	Spray painting in shallow booths, barrel filling, conveyor loading, crushers.
Released with high initial velocity into zone of very rapid air movement	500–2,000	Grinding, abrasive blasting, tumbling

Fig. 4. Range of capture velocities recommended for dusts, fumes, smokes, mists, gases, and vapors released at various types of operations.

Hood installation requires consideration of the work position in relation to the air flow. The hood should be so located that there is no chance of the contaminant entering the worker's breathing zone. If a worker is required to bend over open tanks containing volatile and toxic materials, he should be provided with a suitable filtering respirator.

Construction of Ducts. The **air velocity** in all parts of a duct must be maintained above certain minimum values to prevent the contaminant from settling out. Generally, speeds up to 2,000 ft. per min. are recommended for fumes; 4,000 ft. per min. for average industrial dusts; 5,000 ft. per min. and up for heavy dusts. The ducting should not have sharp changes in direction, shape, or cross-section area.

The **cross-section area** of any main duct should equal the sum of all branch areas. Some local codes require an excess of 20 percent in the main duct for possible future additions. A circular cross section is preferred with narrow, rectangular shapes least desirable. All laps should be made in the direction of air flow. Materials of construction include riveted or welded sheets of plain or galvanized steel, aluminum, stainless steel, copper, plastics or plastic- and rubber-lined metals. The choice of materials requires consideration of strength, first cost, operating costs due to maintenance, and replacement cost entailed by corrosive conditions.

Fire Safety. Wherever ducts pass through fire walls an automatic closing damper should be provided. If the system is to carry flammable gases or explosive dusts, the entire system should be electrically grounded. If there are any gaps in the system, such as at joints, the parts should be connected with a **grounding strap** preferably made of copper. Under certain conditions it may be advisable to install an inert-gas or dry-powder fire-control system concurrently in the ducting. Also, if one system is to handle more than one type of contaminant, it must be made certain that the mixing of these contaminants in the main duct will not lead to a dangerous condition. It is advisable to submit detail plans of any duct system to the fire insurance company for appraisal.

Clean Air Supply. An air supply containing vapors, fumes, and dusts that is exhausted must be replaced by fresh air. Fresh air, sometimes filtered, heated, or cooled, can be supplied either to the general work area or in the immediate vicinity of a contamination source. General area ventilation may disturb the collecting ability of hoods if there are several in the same area.

The selection of an **air cleaner** depends on the nature of the contaminant, its concentration, and the total volume of air handled. Heavy dusts generally are separated in centrifugal or cyclone separators; and lighter, fine-grain dusts in cloth or fibrous filters or electrostatic precipitators. Fumes and smokes behave as extremely fine dust particles and are controlled by precipitators, wet collectors, and cloth or fibrous filters. Gases and vapors may be removed by absorption, adsorption, combustion, or condensation.

Two general types of fans are available: the **axial flow,** which will move large volumes of air against low static pressures; and the **centrifugal flow,** which can be designed to move large volumes against medium or high static pressures. The fan and motor should be selected only after calculation of the total duty, cubic feet per minute, and static pressure of the exhaust system and any filtration equipment.

WELDING. The nature of the safety measures required for welding depends on the kind of welding equipment used. **Gas welding** includes the particular hazards of handling and using materials capable of explosion. **Electric arc** and **resistance welding** have the particular hazards of electric shock and radiation, harmful not only to the eyes but also to the unprotected body. All types of welding have the hazards attendant to high temperatures: severe burns, fume inhalation, and property damage by fire.

Gas Welding and Cutting. These processes include the use of oxygen and acetylene, oxygen and hydrogen, and other combinations of oxygen with some suitable fuel gas, which present hazards (ANSI Standard for Safety in Welding and Cutting, Z49.1-1967). Acetylene with oxygen gives a much higher temperature than the other combustible gases, and for this reason it is generally employed for gas welding. Many hazards are present in the manufacture and use of acetylene, and extreme care must be exercised at all times. The use of homemade acetylene generators has caused many accidents. Generating equipment should be procured from reliable makers who furnish instructions for safe operation. These instructions should be posted where they will be seen and followed.

Most of the gas used for welding and cutting is purchased in **cylinders.** These cylinders should be manufactured and filled in accordance with I.C.C. specifications and regulations and should be properly marked. They should never be handled with a magnet, and a suitable cradle should be provided if they are moved by crane. Careful handling at all times is essential. Knocks, falls, and rough handling are likely to damage the cylinder, valves, or fuse plugs and cause leakage. Cylinders always should be secured from falling. Since acetylene gas at pressures above 15 lb. per sq. in. is dangerous, it is absorbed in acetone in the pressure cylinders. Acetylene cylinders, therefore, should be set on end for several hours before use to give free vapor opportunity to collect at the top. Oxygen cylinders should have the same careful handling, and no grease or oil should be used on any of the fittings. Oil and grease in the presence of oxygen under pressure may ignite violently and cause serious accidents. All built-in lines to be used for oxygen must therefore be degreased carefully before use.

Regulators, gages, hose and hose connections, and torches should be kept in perfect condition at all times and rigid inspection enforced for their careful handling. Suitable clothing and, particularly, adequate eye protection are essential for welders, and the proper shade of lenses for goggles and hoods will afford ample protection from the light rays. Such equipment should be of high quality, even if the cost is higher. Aprons, shoes, gloves, etc., are required for complete protection. It is important that the operators of such equipment be fully qualified by adequate training and be given all safety precautions to avoid accidents from this type of welding.

Arc Welding. This is a fusion welding process in which the welding heat is obtained from an electric arc formed between an electrode and the base metal, the heat of the arc being approximately 6,000°F. The welding voltage in most cases is low, but it can become a serious hazard, particularly in locations which are damp or wet. The usual precautions should be taken against coming in contact with live conductors by the use of insulated equipment and personal protective equipment. Where welding current is used for arc welding, and the operator is required to work in a metal-enclosed space, **protective relays** should be installed on the circuit so that the operator will not be exposed to an open-circuit voltage of more than 50 volts.

The fact that low voltages are employed should not cause negligence on the part of the operator. Fatal injuries have occurred when persons contacted the frames of welding machines energized by short circuits. Frames of all portable electric welding machines operated from electric power circuits should be effectively grounded.

The electric arc produces **high intensity ultraviolet and infrared rays** which have a harmful effect on the eyes and skin under continued and repeated exposure. The effect of exposure of the skin to the direct rays of the electric arc is very similar to sunburn. It may be very uncomfortable and even painful but causes no permanent injury. Ultraviolet rays do not usually cause permanent injury to the eyes unless by continued and repeated exposure, but temporary effects may be quite painful. Even short exposures have caused painful results and disability. Infrared rays are the heat rays of the spectrum. They do not cause permanent injury to the eyes except from excessive exposure.

It is necessary, therefore, to provide full protection at all times in the presence of the arc, both while the operator is engaged in actual welding and while he is observing welding operations. The operator and his assistant must use **protective hand shields** or **helmets** which protect the skin of the face and neck, and which are also equipped with suitable filter glass (shades 6 to 14) that will provide adequate eye protection. In the selection of **filter glass** it is necessary to depend on laboratory tests, as the transmission of ultraviolet and infrared radiation cannot be determined by visual inspection. Depth of color does not necessarily indicate removal of the invisible radiation which may be injurious to the eyes. Reliable dealers are able to supply filter glasses which have been shown by tests to conform to requirements of the ANSI Practice for Occupational and Educational Eye and Face Protection (Z87.1-1968).

During arc welding operations, certain **gases, fumes, and dusts** are evolved by the heat of the arc, depending on the type of welding rods used, the base metal being welded, and whether or not the base metal is coated with such material as oil, tar, salt, paint, lead, zinc. Some of the gases may include oxides of nitrogen, ozone, carbon dioxide, carbon monoxide, sulfur dioxide, and phosgene. Some of the metallic and mineral substances that may be found in the fumes and dust include iron, zinc, lead, copper, manganese, selenium, silica, arsenic, titanium, and fluorine. Poisoning due to the presence of some of these substances in the fumes has been reported although evidence based on actual cases is rare.

One of the principal health hazards presented by electric welding is **lead poisoning.** If painted or lead-coated materials are cut or welded, the lead volatilizes and may be breathed, causing lead poisoning. Also "zinc chills" may result from breathing fumes when welding zinc, zinc alloys or galvanized metal.

It is difficult to obtain definite data concerning the effects of various gases, fumes, and dusts generated in electric welding operations; consequently, it is usually necessary to make sure that employees do not breathe them. Where welding is carried on outdoors, or in large, well-ventilated shops, and nontoxic materials are involved, experience shows that welding operators suffer no harmful effects. But in other cases the hazard usually is minimized by providing **efficient ventilation and local exhaust systems.** Where harmful concentrations of gases, fumes, and dusts are generated, it is preferable to provide local exhaust systems to remove such substances at their point of origin, particularly where welding is done in confined areas such as small rooms, welding booths, tanks, boilers. In many cases where welding operations are permanently located, the entire booth may be ventilated by an exhaust system such as is used for spray coating booths, or an adequate exhaust pipe may be provided, connected to a central duct system. Portable exhaust systems are also available for this purpose.

Where, because of the intermittent nature of the work or for other good reasons, it is impossible for gases, fumes, and dusts to be kept below their toxic limits by means of general ventilation or by local exhaust systems, welding operators should be required to wear **special respiratory protective equipment** approved for such purposes by the U.S. Bureau of Mines. Supplied-air respirators, such as air-line respirators, hose masks with or without blowers, or self-contained oxygen-breathing apparatus, are recommended for use in confined areas and other locations where high concentrations of toxic substances are encountered.

The present safe practice is to **ventilate all welding operations** which are carried on in relatively small enclosed or restricted space, as in tanks, boilers, pressure vessels, compartments and holds of vessels, because of the possibility of an accumulation of toxic and explosive gases, and the possibility of an oxygen deficiency. Where there is any question, tests should be made to determine the presence of toxic and explosive gases, and, if such gases are found in harmful concentrations, the area should be thoroughly cleaned and ventilated and again tested before permitting welders to enter. Artificial ventilation may be necessary. Safety lamps and other instruments are available for tests for oxygen deficiency.

Resistance Welding. This is a metal-fabricating process in which the fusion temperature is generated at the joint by the resistance to the flow of electric current. When the welding temperature has been reached, the electric circuit is opened and mechanical pressure is applied to complete the weld. The three fundamental factors of resistance welding, therefore, are current, time, and pressure, each of which must be accurately controlled.

The principal **hazards in the operation of resistance welding** equipment include lack of point of operation guards, flying hot metallic particles, handling materials, unauthorized adjustments and repairs, and possible electrical shock. The hazards involved vary greatly with the type of equipment being used, and the kind of work being performed. A careful job analysis should be made of the operations on each welding machine to determine the safeguards and personal protective equipment that will be most appropriate for each job.

On many kinds of resistance welders, particularly automatic and semi-automatic equipment, serious **point-of-operation hazards** exist similar to point-of-operation hazards on punch and forming presses. The possibility of finger amputation requires the installation of guards or devices that will enclose the point of operation or otherwise make it impossible for the operator to reach into the danger zone. In most cases welding machine operators should use some form of face and eye protection to guard against flying hot metallic particles. Goggles with clear lenses and side shields provide effective protection for most resistance welding; however, goggles with filter lenses (shades 1 to 5) may be necessary on some special operations. Face shields to protect the face and neck from hot sparks are desirable and should be of fire-resistant material.

The hazard of **flying sparks** can also be eliminated by installing a shield guard at the point of operation. Such guards should preferably be of transparent material such as safety glass or cellulose acetate. Where such guards are not used, it is good practice to erect some type of shield to prevent injury to other employees who may be passing the machine.

On many operations, leather or canvas gloves and aprons are desirable for preventing burns from hot sparks and avoiding cuts and scratches. In addition, some operators wear leather sleeves primarily to avoid burning their clothing and also to guard against small skin burns. Woolen outer clothing is preferable to cotton as protection against burns.

PLANT HOUSEKEEPING FOR SAFETY. The millwright and maintenance-of-equipment department should keep all machinery in first-class operating condition, but many times the safe condition of equipment is forgotten or given low priority. The safety director (and management) should bring forcefully to the attention of the maintenance supervisor the necessity of keeping all equipment in a safe condition. The matter cannot be left entirely to their care however. A periodic check by a group of their own personnel or by plant safety committee members or the safety department itself is necessary. A well-formulated **plant housekeeping plan** should accomplish three objectives: (1) elimination of accident and fire hazards; (2) conservation of space, time, material, and effort; and (3) improvement of employee morale. (See section on Plant Maintenance.)

Statistics show that a high proportion of industrial accidents are directly traceable to falls, falling objects, and mishandling of materials, and that such accidents are often a direct result of disorder in the plant. Therefore, the problem of industrial housekeeping is a major one, and in the interests of safety, morale, and efficiency it requires careful consideration.

Good housekeeping has often been summarized by the phrase "A place for everything and everything in its place." If management fails to provide the "place," the employee finds adherence to this principle impossible. Providing the "place" should be carried out in its broadest sense. The start of a good program is an analysis of physical plant facilities and a determination of the adequacy of existing equipment, such as shelves, bins, storage rooms, work places.

Building Maintenance. Stairways should be kept clear of all materials and should be properly illuminated. Accidents resulting from tripping and falling are more likely to occur in these locations than on level surfaces. To comply with fire regulations, as well as good safety procedure, handrails of suitable height should be provided.

Aisles and passageways should be kept clear at all times for the safety of pedestrian traffic and trucks. Materials should not be permitted to project into aisles, and the latter should be clearly marked off, either by painted lines, inlaid tile, or other method.

Floors should be kept free from holes, uneven boards, and obstructions, especially where the floors form parts of aisles or walking places. Materials used in floors should be considered from the safety standpoint in the erection of new buildings or in repair of old structures. Small objects, such as scraps of metal, nails, tools, should not be allowed to lie on the floors or in passageways. Oil, grease, chips, and other sources of hazards which could be the cause of slipping or falling should be promptly removed.

All unnecessary hangings and trappings on walls should be removed. Windows should be in good working order, panes clean, and those cracked or broken replaced. Ceilings should be inspected for loose plaster, and skylights kept clean and in good order.

All workbenches, aisleways, and stairways should be suitably lighted and free from shadows. Night lights should be provided throughout departments so that watchmen will not be exposed to tripping and falling accidents. Exit lights should be placed at all emergency doors and exits. The location of fire-fighting apparatus should also be suitably illuminated.

Waste cans with self-closing covers should be provided to hold cotton waste in each machine department. Two cans should be at each location, one for clean, and the other for soiled, waste.

Proper operation and maintenance of **elevator equipment** are essential factors in a good housekeeping program. Storage of any kind of material in elevators should be prohibited, and floors should be kept clear of debris. Elevators which do not stop level with factory floors create a tripping hazard. Elevators should be equipped with interlocking devices which will prevent them from leaving landings while gates are open. Serious accidents have resulted from employees opening hoistway doors, expecting to find elevators at floor level, and falling down the shaft as a result.

Tool Cribs and Storerooms. Small hand tools and implements should not be permitted to lie about where they may be the cause of slipping or tripping accidents. Special tool houses or rooms for picks, shovels, trowels, and similar implements for excavation and construction work may be provided. Some companies provide tool drawers and shelves for both outside and inside jobs. When these storage places are located conveniently, employees will use them, and the temptation to leave equipment around at workplaces where it would cause accidents, or be lost, is greatly reduced.

In the storage of bulky objects, such as rods, pipe, lumber, many companies provide specially designed racks or guards to prevent the material from shifting and falling or rolling into aisles or places where men may be working. Small-sized material is usually put on shelves or in bins.

SAFETY INSPECTIONS. Safety inspections are one of the principal means of locating accident causes. They assist in determining what safeguarding is necessary to eliminate or otherwise remove hazards before accidents and personal injuries occur. Prompt safeguarding of hazards is one of the best methods for management to demonstrate to employees its interest and sincerity in accident prevention work. Inspections, however, should not be limited to unsafe physical conditions but should include unsafe practices. One company recommends that for each inspection made for unsafe conditions three should be made for unsafe practices.

It should be remembered that safety inspections are not made primarily to find out how many things are wrong, but rather to determine, to the extent possible, whether safety is being optimized. It is often assumed that unsafe conditions and practices, not reported by the inspectors, have little significance or they would have been identified. The plant management would be well advised, therefore, to maintain a questioning posture toward the degree of safety effectiveness which the inspection appears to indicate, particularly when it seems that there are no, or only minor, safety problems. The attitude of the inspection team should be one of helpfulness in correcting conditions so as to bring the plant up to accepted and approved standards and make a safer and more helpful place in which to work. One where the working environment is such that operations can be conducted economically, efficiently, and safely.

Regular Inspections. It often is best to plan unannounced periodic inspections (for example, one per month), so that safety conditions are maintained at a high standard throughout the plant. Surveys should include all **means of egress** from the building. All exits, fire towers, fire escapes, halls, fire alarm systems, emergency lighting systems, and places seldom used should be thoroughly inspected to determine their adequacy and readiness for an emergency.

Another kind of periodic inspection is that **required by state and local laws.** These include the inspection at regular intervals of elevators, boilers, unfired pressure vessels, and other special hazards. Such equipment, however, is not usually inspected by plant employees but by outside inspectors, perhaps from casualty insurance companies, because of the special training necessary to qualify for this type of work. Usually a specific schedule of inspections is prescribed by law.

Chains, cables, ropes, and other equipment subject to severe strain in handling heavy materials should be inspected at regular intervals, and a **careful record** kept of each inspection. Some state regulations require such inspections and records. This type of equipment should be stenciled or otherwise marked for ease of identification. Some companies require that all portable electric tools and extension cords be sent to the electrical department, say, between the first and tenth of each month. The electrical department inspects the tools, makes necessary repairs, and attaches a colored tag to the tool or cord showing the month the equipment was last inspected. A different colored tag is used for each month. Any tool or cord found without the proper tag is sent to the electrical department at once.

Other types of equipment, such as **cranes, hoists, presses, ladders, and power trucks,** require periodic inspection. Any equipment used in the field also requires frequent and periodic inspections. Such inspections should be ordered by the proper plant executives, and the safety director should prepare a working schedule so that the correct intervals of inspection can be maintained.

Along with scheduled inspections, a careful survey should be made as to the **adequacy and safety of equipment in the plant.** Recommendations should be made for replacement of defective and obsolete equipment, as well as for the purchase of any additional equipment that may be necessary. Such recommendations should be followed up until the corrections are completed. As new processes and products are added to the manufacturing system, inspections may show that new accident or fire hazards have been introduced which require individual treatment; for example, special fire extinguishing devices.

One of the common kinds of inspections is "spot inspections," made at **intermittent intervals** as the need arises, including unannounced inspections of particular departments, pieces of equipment, or small work areas. Such inspections made by the safety department tend to keep the supervisory staff alert to find and correct unsafe conditions before they are spotted by the safety inspector.

Inspection of Specific Hazards. The need for intermittent inspections is frequently vindicated by accident tabulations and analysis. Should the analysis show an unusual number of accidents for a particular department or location, or an increase in certain kinds of injuries, special inspections should be made to determine the reasons for the increase and what must be done to remove the hazards.

In preparing for an inspection, it is advisable to **analyze all accidents for the past several years** so that special attention can be given those conditions and those locations known to be accident producers. Where accurate accident statistics are kept, such data are usually available in monthly and annual reports. Wherever an accident has occurred, it may take place again unless the unsafe condition is corrected. Experience gained in correcting a hazard at one location will be helpful in safeguarding similar conditions in other locations. Inspections should not be confined to those places where serious injuries have occurred. Even no-injury accidents and near-accidents often point to causes of possible future injuries.

Wherever there is a **suspected health hazard,** a special inspection should be made to determine the extent of the hazard and what precaution or mechanical safeguarding is necessary to provide and maintain safe conditions. These inspections usually require air sampling for the presence of toxic fumes, gases, and dust, testing of materials for toxic properties, or the testing of ventilation and exhaust systems for efficiency of operation.

Safety Inspectors. The number of safety inspectors in any plant depends a great deal on the size of the plant and the kind of industry. Large plants with well-organized accident-prevention programs usually employ a staff of full-time inspectors who work directly under the safety supervisor. Large plants may also have a number of specially designated employees who spend part of their time on inspections. Also, there are usually employee inspection committees which assist in this kind of work.

Plants too small to employ a full-time safety director and assistant inspectors depend on inspections made by maintenance men and supervisors. Frequently an employee carries out the duties of a safety director on a part-time basis and makes periodic inspections. Many plants depend entirely on **inspection service** supplied by casualty insurance inspectors and also state factory inspectors. More frequent inspections, however, are usually necessary than are provided by these agencies.

SAFETY REQUIREMENTS—NEW EQUIPMENT. Cooperation between the safety director and the purchasing agent is important. The purchasing agent is not concerned with the educational and enforcement activities of safety, but with its engineering activities. It is his duty to purchase the various items of machinery, tools, equipment, and materials used in the establishment, and it is his responsibility—at least in part, and often to a considerable degree—to see that safety receives adequate attention in all purchases.

For this purpose he should be familiar with the workplaces and processes as well as the hazards of plant departments. He will want to know where and why accidents are happening and whether or not machinery, tools, or materials are at fault. He will not undertake to purchase any article until he has a thorough knowledge of its strength and work efficiency, and whether it can be used by the workers with the highest possible degree of safety.

Unsuspected Hazards in Purchased Items. Often it is surprising to find that many items have a more important bearing upon safety than would be at first suspected. Particular attention should be given to the purchase of all personal protective equipment, all equipment provided for the movement of suspended loads or for the movement and storage of materials, all miscellaneous substances and fluids used for processing which might constitute or increase a fire or health hazard, and similar items. But investigation also will show that **unsuspected hazards** may lie in very ordinary items, such as the commonest kinds of hand tools, tool racks, cleaning rags, the types of paint to be applied to shop walls and machinery, reflectors, and even bill files. Characteristics such as maximum load strength, long life without deterioration, sharp, rough, or pointed surfaces or edges, the need for frequent adjustment, ease of maintenance, effect of fatigue upon the employees, and hazards to the workers' health are among the many factors requiring attention.

The following are a few **examples of hazards** attending purchased items that were thought to be safe. Because a small hammer had been improperly annealed, a man's eye was lost when a piece of metal from its head flew 20 ft. and struck him as he sat before his own well-guarded machine. Goggles supplied to one group of workers were found to have such imperfections in the lenses that they caused eye strain and headache, which led to fatigue and accidents. The toes of a laborer were crushed when the safety shoe he was wearing had an inferior cap and collapsed under a weight that should have been supported easily by a well-made shoe.

Safeguards on New Equipment. When an order for equipment is about to be placed, the purchasing agent who cooperates with the safety director will not consider any machine that has been only partly guarded by the manufacturer and that therefore will have to be fitted with makeshift safeguards after it has been installed. He will also be particularly careful to see that any purchased machine complies fully with the safety regulations of the state in which it is to be operated, for safety requirements vary widely in different states. The safety man will assist in every way to make tests on new equipment, and, in fact, periodic checks on regular equipment, to assure the highest degree of safety by making certain that the best equipment available is being utilized. Many times it will be discovered that **special safety equipment** such as safety pliers, tongs, tweezers, stands for holding portable tools, spark curtains, are not readily available on the market, and much time may be lost by having them made specially by outside firms. By close contact between safety department and mechanical departments, many such devices can be made in the plant in a few hours, thus removing the hazard and in many cases maintaining production which otherwise would be at a standstill. All safety equipment and, in fact, all machinery to which safety equipment has been attached at the suggestion of the safety director should be carefully tested by, or in the presence of, the safety director and the mechanical division before it is released to the production department. Where portable tools (pliers, tweezers, etc.) are provided, they should be systematically followed up by the safety man to check on their performance from the standpoint of safety.

PERSONAL PROTECTIVE EQUIPMENT. The primary approach in any competent safety effort is the maintenance of the physical environment so that accidents cannot occur. But it sometimes is necessary, for economic or other expeditious reasons, to **safeguard workers** from possible hazards by providing them with certain personal protective equipment. In a machine shop, for example, it would be desirable to remove all sources of flying particles that endanger the eyes. However, this usually is too expensive to accomplish. Protective eye wear therefore is required. Since personal protective equipment must be relied on to some degree, there often is a temptation to consider it first as a safety measure, rather than to investigate the removal of the hazard at its source.

A wide variety of personal protective devices, designed for specific applications is available. Discriminating care should be taken therefore to assure that such purchased equipment will have the required quality and applicability.

Administrative practice generally is to give the worker (gratis) personal protective equipment that he is required to use as a condition of employment. Recommended equipment is generally purchased by the worker himself often at an employee store with payment made through payroll deductions. Sometimes a partial rebate of the cost is made by the employer.

Accident Records and Investigations

INJURY RECORDS AND REPORTS. Successful accident prevention by an employer requires a good system of recording employee injuries. No modern executive would expect profits without adequate records of production, costs, and sales. Accident records serve a similar purpose for safety.

An injury report should be submitted to management at least once each month. Certain items are a "must" on such reports, other items may be included from time to time, or special emphasis may be given to a section of the report where a weakness in accident prevention work is noted. The **essential report items** are:

1. **Frequency rate (number of disabling injuries per million man-hours worked).** The use of the term "disabling injuries" is preferred to "lost-time" injuries. A disabling injury is defined as "a work injury which results in death, permanent total disability, permanent partial disability, or temporary total disability" (ANSI Method of Recording and Measuring Work Injury Experience. Z16.1-1967).
2. **Severity rate (number of man-days lost per 1,000,000 man-hours worked).** The number of man-days lost is to be computed in accordance with the same standard (ANSI Z16.1).
3. **Total time lost** or charged for major injuries.
4. **Costs of accidents**; direct cost and indirect cost.

Since such a report is the basis for management evaluation, **comparisons** should be presented wherever practicable. It is possible to show trends by comparing current performance with (a) last month; (b) same month of previous year; (c) a six- or twelve-month total; (d) performance of similar industries as reported by National Safety Council (Accident Facts) or by U.S. Bureau of Labor Statistics. Further comparisons can be made between similar departments or plants of the company. The National Safety Council's **monthly summary of injuries form** (Fig. 5) summarizes the necessary information for ease of comparison.

NATIONAL SYSTEM FOR RECORDING AND REPORTING OCCUPATIONAL INJURIES AND ILLNESSES. The record-keeping re-

Monthly Summary

Company _____ Plant_____

Period	No. of Non-Dis-abling Injuries	No. of Man-Hours Worked	Avg. No. of Em-ployees	NUMBER OF DISABLING INJURIES								FREQUENCY RATE	
				Deaths & Perm. Total		Perm. Partial		Tempo-rary Total		Total		This Period	Same Period Last Year
				Curr.	Adj.	Curr.	Adj.	Curr.	Adj.				
Jan.													
Feb.													
Cum.													
Mar.													
Cum.													
Apr.													
Cum.													
May													
Cum.													
June													
Cum.													
July													
Cum.													
Aug.													
Cum.													
Sept.													
Cum.													
Oct.													
Cum.													
Nov.													
Cum.													
Dec.													
YEAR													

Fig. 5. Form for monthly

quirements of the Occupational Safety and Health Act of 1970 resulted in a new method of defining and classifying work injuries. Changes in ANSI Z16.1 were under study. It is likely that the reportable injury classification will be widened to include cases which do not lose time, but where a doctor is visited more than once.

NONDISABLING ACCIDENTS. Records of non-disabling accidents, even if only first aid is required, may also be maintained. A report showing trends involving this type of accident is often of value. The small company or small department may not have sufficient "experience" to indicate its true safety performance if measured by disabling accidents alone. The type of accident, the agency, and the cause of "disabling" accidents often correlate with the type, agency, and causes of those which are nondisabling. Where possible, the monthly and annual reports should summarize and compare the frequency of nondisabling accidents between departments and plants both currently and for prior periods. Agency, accident type, and cause also should be compared.

of Injuries, 19——

_____ Department _____

TIME CHARGES						SEVERITY RATE		COST	
Deaths & Perm. Total		Permanent Partial		Temporary Total		Total	This Period	Same Period Last Year	This Period	Same Period Last Year
Curr.	Adj.	Curr.	Adj.	Curr.	Adj.					

summary of injuries.

Finally, when a department or plant has achieved a good safety record, it becomes desirable to report the number of **accident-free days** and the number of **man-hours since the last disabling accident.** E. I. du Pont de Nemours & Co., Inc., which has set many safety records, places considerable emphasis on preventing off-the-job injuries involving loss of working time. Records are kept, reports are submitted, and competition helps keep workers safety-conscious both on and off the plant site.

ACCIDENT INVESTIGATION. Accident investigation is of major importance. Its purpose should be to develop better means for carrying on the accident prevention program. Otherwise, as fast as one hazard is detected and removed, another may develop and eventually result in an accident of even greater proportions.

In most organizations an investigation of some kind is made of each accident resulting in death or injury to an employee. However, accidents which might have caused death or personal injury, but which by a stroke of luck did not harm anyone, often are unreported, or when they are reported are rarely investigated.

Members of the National Safety Council, who make a practice of investigating all accidents, claim there is no more justification for assuming that a noninjury accident will not hurt anyone if it happens again than there is in drawing the same conclusion regarding an accident involving personal injury or death.

Purposes of Investigation. The principal purposes of an accident investigation are:

1. To ascertain the cause or causes so that measures may be taken to prevent similar accidents. These measures may include mechanical improvements, better supervision, instruction of workmen, and sometimes discipline of the person found guilty.
2. To secure publicity among the workmen and their supervisors for the particular hazard, and for accident prevention in general by directing attention to the accident, its causes and results.
3. To ascertain facts bearing on legal liability. Investigations for this purpose only, however, will not always suffice for future accident prevention purposes. But an investigation for preventive purposes may disclose facts that are important in determining liability. In this discussion the investigation is considered from the standpoint of safety, not liability.

Investigation Procedures. Depending on the importance of the accident and other conditions, the investigation may be made by one or more of the following persons or groups:

1. The foreman.
2. The safety engineer or inspector.
3. The workmen's safety committee.
4. The general safety committee.
5. A court of inquiry, board of inquiry, or jury.
6. In accidents involving special features it is often advantageous to call in an engineer from the insurance company or appropriate government agency to assist.

Each investigation should be made as soon after the accident as possible. A delay of even a few hours may permit important evidence to be destroyed or removed, intentionally or unintentionally. The results of the inquiry should also be made known quickly, as their publicity value in the safety education of workmen and supervisors is greatly increased by promptness.

Fairness is an absolute essential. The value of the investigation is largely destroyed if there is any suspicion that its purpose or result is to "whitewash" anyone or to "pass the buck." A "verdict" which places the blame on the workman, especially on the man who was injured, is likely to be scoffed at unless the personnel of the committee or jury arriving at the decision includes a generous proportion of fellow workmen having good standing among their associates. Perhaps even more important is the attitude of the safety department or other company representatives in making the inquiry. No one should be assigned to this work unless he has earned a reputation for fairness and is tactful in gathering the evidence. No browbeating of witnesses, either in private inquiry or in public, should be tolerated.

An accident causing death or some serious injury should obviously be investigated, but the near-accident that might have caused death or serious injury is equally important from the safety standpoint. Any **epidemic of accidental injuries** demands immediate special study. A particle of emery in the eye, or a scratch from handling sheet metal, may be a very simple case. The immediate cause is obvious, and the loss of time may not exceed a few minutes, but, if cases of this or any other kind occur frequently in the plant or in any one department, an investigation should be made to determine the underlying causes.

Sequence of Investigations. The investigation procedure usually covers such questions as: What happened? Why did it happen? How can a similar occurrence be prevented?

Four important steps customarily are taken:

1. The safety office makes an immediate preliminary investigation at the scene of the accident to get all the facts.
2. Later, as an intermediate step, the job is analyzed carefully.
3. A formal investigation is made by a committee composed of the manager or his assistant, an employee, an observer, and the safety specialist whose duty it is to assemble all the facts and place responsibility.
4. Recommendations are later made by the safety office to prevent recurrence of the accident.

The **preliminary investigation** proceeds as follows:

1. The first-aid records are consulted to determine what happened, what the employee was doing and where he was working.
2. The safety specialist goes to the scene of the accident, questions all the workers in the area, takes pictures of all conditions.
3. The injured man is interviewed at once, if his condition permits, to get his story before he has a chance to change it.

Later the **intermediate investigation** is made:

1. The injured man is again questioned to detect any change in his story.
2. A detailed study of the work methods of men on similar jobs and of similar equipment is made.
3. A study of the experience of other companies in similar cases is also made.
4. An investigation is made of the safety appliances in use.
5. The safety records of the injured man and his foreman are checked.

Finally the **formal investigation** is convened:

1. A meeting is held in the Plant Manager's office to establish in the mind of the employee that management is interested.
2. The investigating group is composed of the general superintendent employee representative, and a foreman in the same line of work, chosen by the safety man. The safety man assumes the chairmanship, questions the witnesses, and then presents the evidence.

A **follow-up** by the safety office seeks to prevent similar accidents:

1. Reports of the investigation are sent to the heads of departments.
2. Suggestions to correct similar conditions in other departments are submitted.
3. The report, in general, is publicized.
4. Related operations to detect similar hazards that may result in an accident are studied.
5. The accident is used as a subject for discussion in foremen's meetings.
6. A special bulletin is published each month listing all accidents.
7. A tickler system is used to follow up recommendations and see that they have been put into effect.
8. An exchange of accident experiences is carried on with similar industries.

ANALYSIS OF ACCIDENTS. The data necessary for analysis should include as a minimum:

1. Date of accident.
2. Name of injured.
3. Specific occupation of injured.
4. Nature of injury.
5. Details of accident.
6. Identity of hazard.
7. Preventive action taken.

Standard **injury report forms** well suited to accident and analysis are exemplified in Fig. 6 and are available from the Workmen's Compensation insurance carrier or the various state industrial departments.

W.C.B. CASE NO.	CARRIER'S CASE NO. AND	CODE NO.	DATE OF ACCIDENT

(ENTER CASE NUMBERS, IF KNOWN, IN ABOVE SPACES) (Include Zip Code in All Addresses)

	NAME	ADDRESS
1. EMPLOYER		
2. INSURANCE CARRIER		
3. INJURED PERSON		EMPLOYEE'S S. S. ACCT. NO.

(First Name)   (Middle Initial)   (Last Name)   (Home Address)

EMPLOYER

4. Nature of business: (State principal products manufactured or sold or services rendered)

5. Place where accident occurred:

ACCIDENT

6. Date of accident: _____, 19__, Day of Week _____ Hour of Day _____ A.M. _____ P.M.

7. (a) Date disability began: _____, 19__, _____ Hour of Day _____ A.M. _____ P.M.

   (b) Was injured paid in full for this day? _____

8. Name of foreman

9. When did you or foreman first know of injury?

10. Names and addresses of witnesses:

INJURED PERSON

11. (a) Marital status: _____ (b) Sex _____

12. Age: _____ 13. Did you have on file employment certificate or permit? _____

14. Occupation: (a) Job title for which employed: _____

           (b) Occupation when injured: _____

15. (a) How long employed by you? _____ (b) Piece or time worker? _____

   (c) Hours per day: _____ (d) Days per week: _____

16. Earnings in your employ: (a) Rate per: Hour $ _____ Day $ _____ Week $ _____ Month $ _____

   (b) Total earnings paid during year prior to date of accident: (include bonuses paid, value of board, lodging, etc.) $ _____ Average per week: $ _____

   (c) Bonuses or premiums paid and included in item 16 (b) above: $ _____ (d) Estimated value of board, lodging, or other advantages in addition to wages: (included in item 16 (b) above) $ _____

   (e) Calendar weeks in past 52 in same kind of work as at time of injury:

17. State nature of injury and part or parts of body affected: (as "Injury to Chest, etc.)

NATURE OF INJURY OR OCCUPATIONAL DISEASE

18. Did you provide medical care? _____ If so, when? _____
19. Name and address of physician: _____
20. Name and address of hospital: _____
21. Probable length of disability: _____
22. (a) Has employee returned to work? _____ (b) If so, give date: _____
    (c) At what occupation? _____ (d) At what weekly wage? $ _____
    NOTE: **Form C-11 must be filed each time there is any change in the employment status as reported in item 22 above.**

FATAL CASES

23. Has injured died? _____ (a) If so, give date of death: _____
    (b) Name and address of nearest relative: _____
24. (a) What was employee doing when accident occurred? (Describe briefly as "loading truck," "operating press," "shoveling dirt," "painting with spray gun," "walking downstairs," etc.) _____
    (b) Where did accident occur? (Specify whether in street, factory yard, on loading platform, in factory, etc.)

CAUSE OF ACCIDENT OR OCCUPATIONAL DISEASE

25. How was accident or occupational disease sustained? (Describe fully, stating whether injured person slipped, fell, was struck, etc. and what factors led up to or contributed to accident. Use additional sheets if necessary.)
26. (a) What specific machine, tool, appliance, gas, liquid, or other substance or object was most closely connected with this accident or occupational disease? _____
    (b) If mechanical apparatus or vehicle, what part of it? (State if gears, pulley, motor, etc.)
27. Were mechanical guards or other safeguards (such as goggles) provided? _____ (a) Were they in use at time of accident? _____ (b) Was machine, tool, or object defective? _____ If so, in what way? _____

FIRM NAME: _____

SIGNED BY: _____

Official Title

C-2

DATE OF THIS REPORT: _____

Fig. 6. Standard form for employer's first report of injury, suitable for accident analysis.

Report of ______________

Address ______________

ACCIDENT ANALYSIS CHART

(LIST ALL INJURIES)

Period ______________ to ______________

TABULATION OF DISABILITIES

1. Temporary total disab. .... ______
2. Permanent partial disab. .... ______
3. Deaths and perm. total disab. . . ______
4. TOTAL DISAB. (1, 2, & 3) ... ______
5. Man-hours worked .... ______
6. Freq. rate $\frac{\text{(line 4} \times 1{,}000{,}000)}{\text{(line 5)}}$ ______
7. First-aid cases .... ______

ACC. No.	DATE OF ACC.	NAME OF INJURED	OCC. OR DEPT.	INJURY: Nature and part of body	INJURY: Class (1,2,3,7)	INJURY: Days Lost	DESCRIPTION OF ACCIDENT (use more than one line if needed) Give exact details—employee actions, equipment and part of equipment involved, contributing conditions. Avoid terms "carelessness" and "inattention."	CORRECTIVE ACTION TAKEN

Fig. 7. Accident analysis chart.

Classifying Data. There has always been confusion in the terminology used in reporting and analyzing accidents. Listed under "causes" are slips and falls, burns, slivers, punch presses, and other miscellaneous designations. These terms are used without regard for their correct meaning. Such misuse of words greatly reduces the value of accident records and causes confusion in prevention work. Detailed information on the plan and forms recommended for the purpose of collecting and analyzing accidental injuries to industrial employees should be in possession of the safety director.

The American National Standard Method of Recording Basic Facts Relating to Nature and Occurrence of Work Injuries (ANSI Z16.2-1962 [R1969]) provides a means for uniformly describing and **coding the factors in accidents.** If, for example, a man working on a woodworking saw had his hand caught in the blade because he ignored the rule of using a "pusher," the accident would be described as follows:

1. Agency part (substance most closely related to accident)—Saw, circular, cut-off—Code 00427.
2. Unsafe mechanical condition—None—Code YY.
3. Type of accident—Caught in—Code 2.
4. Unsafe act—Using hands instead of tool—Code 32.
5. Unsafe personal factor—Willful disregard of instructions—Code 00.

The use of this standard will organize the recording of data for its most effective application. Experience has shown that collection and tabulation of accident information in this manner are helpful in aiding management to determine causes, types of accidents, and hazardous locations for proper corrective measures.

The National Safety Council's Accident Analysis Chart (Fig. 7) has been designed for the **smaller company having fewer total accidents.** All injuries, including those requiring only first aid, are recorded on this form.

Fire Prevention

FIRE AND PANIC. The control of possible fire losses and prevention of injuries in industry as a result of fires can be expressed in **four fundamental principles:**

1. Arrange and maintain the physical condition of the plant to prevent or minimize the possibility of fires.
2. Provide for the prompt detection and extinguishing of fires. All but a very few fires begin in a small, easily extinguished manner.
3. Provide means for confining any fire to as small an area as possible.
4. Provide and maintain personnel exit facilities that will be safe either in fire or panic.

The safety director and plant engineer should become familiar with the provisions of local fire laws and building codes and enlist the service and advice of the fire insurance carrier. Other agencies able to render valuable assistance are the National Fire Protection Association, the Engineering Department of the American Insurance Association, and the Engineering Division of Associated Factory Mutual Insurance Companies.

DANGER OF FAMILIARITY WITH ORDINARY HAZARDS. The common or ordinary hazards are the presence of **combustible solids, flammable liquids,** and **electrical equipment,** which are present in everyone's daily life. In small quantities and in normal use their presence and behavior are accepted by most people. It is when flammables and a source of ignition occur simultaneously that a fire starts. When the quantity of ignitables is large or unusual temperature conditions prevail, as they often do in industry, the fire hazard increases. For adequate fire prevention it is essential that the process of fire itself be understood along with the more common fire hazards.

Fire depends on oxygen (usually from air), fuel, and heat. Without all three elements fire will not occur. Heat is essential not only to ignite the fuel but also to maintain its temperature above kindling. If any one of the three basic elements is removed, a fire will die out. In ordinary circumstances, the three are kept separated or under control; in time, their continuing normal behavior encourages an unwarranted disregard of their fundamental hazards. A combination of congested materials, a careless match, a hot spark, or overheated electrical equipment—and suddenly a fire has started which, unless quickly discovered and extinguished, may rage out of control.

CLASSIFICATION OF HAZARDS. The special hazards of particular industries are beyond the scope of this section. Many of these special hazards are the subject of standards established by industrial associations or by fire protection associations. The eight **most common fire hazards** in industry, generally, are:

1. Smoking (and matches).
2. Poor housekeeping.
3. Defective heating equipment.
4. Defective or inadequate electric equipment and wiring.
5. Open flames.
6. Spontaneous ignition.
7. Lightning.
8. Explosive atmospheres.

Smoking. Smoking, together with the unsafe disposal of matches, pipe embers, cigar and cigarette butts, is the major cause of all fires in the United States. It may be necessary to exclude all matches, automatic lighters, and smoking equipment from particularly hazardous areas, even an entire plant. In other operations, it is wise to make an **analysis of the plant** to determine those areas where smoking and the use of matches would be particularly hazardous and those in which smoking might be permitted. Smoking is such a widespread habit that its absolute prohibition requires continual reminder and rigorous enforcement. In instances where large areas must be designated as **NO SMOKING**, thought should be given to the possibility of providing and maintaining a small area or room where it may be permitted with safety.

Housekeeping. Proper storage of materials and removal of waste materials are important in prevention of fires. Large quantities of combustible materials should be stored with consideration for limiting the spread of a possible fire. Small quantities should be stored in **covered metal bins** or cans. Rubbish of all kinds should be removed daily, if possible, and while awaiting disposal should be confined in covered metal bins or some fire-safe sprinklered structure. Combustible waste and rubbish should be burned in a furnace or incinerator or in such manner as to prevent the spread of fire by flying sparks. Provision also should be made for the elimination or prompt removal of **drippings of flammable oils, greases, and fluids.** Exhaust ducts should be kept clean of any accumulation of combustible dusts or condensed flammable vapors.

Heating and Electrical Equipment. Overheated surfaces and defective heating equipment have caused many serious fires. Shafting and overheated bearings, friction between belts and wood structures, all have contributed to fire losses. All **high-temperature equipment** should be insulated or isolated from combustible material. Long-term heating of wood or fiberboard structures gradually lowers their ignition temperature to the danger point. It should be remembered that stove and furnace flues can ignite combustible structures by overheating alone and may leak explosive gases into the plant if in bad repair.

Electrical equipment should be installed and maintained according to local codes or to the National Electrical Code (National Fire Protection Association). Common hazards of electric equipment in bad condition are:

1. Short-circuit ignition of combustible materials.
2. Overload or high resistance heating of defective wiring in contact with combustible material.
3. Electric motors, inadequately protected against "stall" conditions, consequently developing temperatures sufficient to ignite their insulation.
4. Sparking of electrical contacts and motors in combustible atmospheres.

In addition to these, the hazards of poor housekeeping and overheated surfaces often are present in the use of electric heating equipment.

Open Flames. The use of open flames, welding or burning torches, blowtorches, and the like should be preceded by careful preparation of the area in which they are to be used. Combustibles or flammable materials should be removed or protected by metal shields or by asbestos, glass fabric, or flameproofed canvas blankets. When such precautions are not possible a **fire guard** should be present with appropriate fire extinguishers at hand before open flames are used. (In welding operations, sparks have caused substantial numbers of fires when they have lodged undetected in a combustible location, such as the spaces between floorboards. Welding areas, after work, require close scrutiny to be certain no ignition sources remain).

Spontaneous Ignition. Spontaneous ignition of combustible materials results from a process of slow oxidation with insufficient ventilation to remove the resulting heat. When the ignition temperature of the material is reached, fire results. This process is accelerated by the presence of moisture, decomposing organic materials, chemical reactions, and any external heating. Control involves: better housekeeping, storage of large quantities in a well-ventilated, cool, dry area, and keeping small quantities in covered metal containers.

Lightning. Fire losses due to lightning in the United States amount to many tens of millions of dollars annually. There is greater hazard in certain areas of the country where thunderstorms are frequent, but no area probably is entirely free of the hazard. Protection of property depends on the provision of an adequate **conductor path** from above the property to the electrical ground. Lightning rods, collectors, down connectors, and grounding method must everywhere create a path of less electrical resistance to the "stroke" than that through any other part of the property. Adequate grounding is sometimes difficult, particularly in shallow soils containing little moisture. According to the American National Standard Lightning Protection Code (ANSI C5.1-1969), steel-frame or reinforced-concrete structures can provide adequate protection if overhead collectors are bonded to the metal frame or reinforcing, if the metal-to-metal contact is continuous, and if an adequate grounding connection is made.

Explosive Atmospheres. A flammable dust, gas or vapor—present in air in proportions which enable an ignition to accelerate throughout the entire volume—is an explosive atmosphere. The Factory Mutual Association (Handbook of Industrial Loss Prevention) lists a large number of flammable materials and their **explosive limits** in percent by volume in air.

Preventive measures usually begin with limiting a flammable dust, gas, or vapor to a concentration in air below its explosive limit. This ordinarily requires extensive use of local exhaust systems and, in the case of dusts, frequent cleaning to minimize the hazard. Where possible, operations using or producing flammable dusts, gases, or vapors should be performed in enclosed equipment designed and exhausted according to applicable codes and authoritative practice. Buildings with high explosion hazards, such as plastics plants, are provided generally with extensive exhaust equipment. They are often constructed so that explosive pressures will push out windows or wall sections in a predetermined manner thus limiting the overall blast effect.

Ignition prevention generally consists of appropriate ventilation-exhaust systems, as well as the elimination of ignition sources. Non-ferrous materials are used for truck wheels, bucket conveyors, hand tools, etc. to prevent the striking of sparks. Electrical equipment is designed, installed, and used according to the electrical code provisions for the flammable hazard classification involved. Precautions must be taken against overloading electrical equipment, smoking, excessive friction in mechanical equipment, static electricity, sparking, and the like. Normal sources of high temperatures, such as grinding and welding operations, open flames, and other heating processes should be avoided or conducted in carefully shielded areas.

Inasmuch as a flammable gas or vapor dispersed in air will form an explosive mixture only over a definite range of concentrations characteristic of the material, **combustible gas and vapor detectors** are useful means of determining the safeness of space suspected of containing a combustible gas or vapor mixture. Such detection instruments determine the concentration of individual gases or vapors in terms of the **lower explosive limit.** Concentrations above the upper explosive limit must be regarded as extremely hazardous since dilution of the mixture with air will bring it into the explosive range. Both portable and fixed detectors are available. The latter can be equipped with alarms and graphic recorders. The more complex detection systems can control auxiliary ventilation equipment as well. Permanently installed detectors have the advantage of continuous monitoring.

FIRE-SAFETY ORGANIZATION. One person in the plant organization should be responsible for the fire safety of plant and personnel. In the small plant, this person may be the plant manager or one of his first-line assistants. The responsibility may be combined with that for accident prevention or it may be a specific activity assigned to the master mechanic or plant engineer. In larger plants, a full-time position, reporting directly to the manager, plant engineer, or other responsible manager is recommended.

Fire Brigade. The organization of a fire brigade, full-time, stand-by, or a combination, depends on: the size of the plant, whether compact or dispersed; the type of production; and the availability and amount of help which can be expected from a public fire department. The **plant fire chief** should have his authority clearly defined. He should have access to management to advise on fire safety and should be consulted prior to major equipment changes or new construction. The fire chief should be consulted in the selection, location, and installation of all fire-warning and fire control equipment. These arrangements are essential to his proper organization and training of a fire brigade.

Members of the fire brigade should be selected and trained with the following in mind: all must be able to take training, be able-bodied, and be available for instant fire-duty. Shift operation of the plant requires trained men available on each shift. The **brigade training program** must be purposeful and regular. The National Fire Protection Association (Industrial Fire Brigade Training Manual) advises that the course should be prepared to meet the needs of the plant: i.e., its special hazards and its equipment for fire control. Drills and classroom instruction should be given on company time if at all possible, and continuing purposeful training should be given each member at least twice a month. Much help in training can be secured from outside agencies such as state fire schools, state fire marshals, and the plant's fire insurance carrier.

Fire Drills. Fire drills should be part of the training both of fire brigade members and of the other plant personnel. It is not essential that all drills of the fire brigade be accompanied by an **evacuation drill** of plant personnel, but such drills usually are advised. Drills by the fire brigade should be thorough and should closely approximate actual fire conditions. The men should become familiar with the interior arrangement of all buildings, basements, attics, and storage rooms and the location of stairways, exits, and the like. They should have actual practice with laying hose, making hydrant connections, and carrying and moving hose lines under pressure. All members of the brigade should understand the application of various types of extinguishers on different types of fires and have practice in their use. At the conclusion of a drill, all equipment should be returned to its designated location in readiness for use or stand-by equipment should be substituted.

Management should anticipate the possibility of a fire emergency and evacuation of personnel. All areas of the plant should be studied to determine a **safe exit route for all personnel.** All operations and processes that could constitute hazards if unattended should be located, studied, and provided for by shut-down plans or procedures. In larger plants, room captains, floor chiefs, and searchers should be appointed. The evacuation plan for the plant as a whole and for each building should be published and explained to all employees. Then, and only then, should a simulated fire drill and evacuation be attempted.

FIRE PROTECTION EQUIPMENT. The selection and placement of fire-fighting equipment should be undertaken jointly with the local fire department and the fire insurance company. The two reasons for this procedure are:

1. Technical factors which depend on the particular industry are involved in the selection and placement of equipment.
2. Insurance rates are based not only on the hazard but also on the protection available.

Therefore specialized advice should be sought in the selection and placement of all fire-fighting equipment. Undoubtedly, the most recommended fire protection method is the provision of a fixed system of **automatic sprinklers** with an adequate and reliable source of water. This recommendation, however, does not exclude the provision of numerous other facilities needed for special hazards. **Portable fire extinguishers** of appropriate types and sizes should be readily available in all plant areas, as should hose reels and fire hydrants connected to the water mains.

Detectors and Alarms. The primary requirement in all fire extinguishment and control, however, is early detection of the fire. "Control and extinguish while still small" should be the primary aim. Detection by personnel should be followed immediately by notice to the plant fire chief by telephone or by some manually actuated **alarm system.** The fire chief may then take steps to dispatch the fire brigade and/or call the municipal fire department. In addition to detection by personnel, detection can be made **automatic** by several types of devices and systems approved by the Underwriters Laboratories, and Factory Mutual Laboratories. Most detectors are **thermostatic** and are actuated by temperature rise, rate of rise, or fixed temperature. The thermostatic devices may be located in any part of the plant and may be used in conjunction with automatic sprinkler systems, sounding an alarm shortly before or at the time the sprinklers go into action. For certain applications, **smoke detectors** are available. The detecting devices should be so located as to be most sensitive to possible fires and yet free of false-alarm possibilities or damage by plant operations. Since most fire alarm and detection systems depend on electrical power it is important that they be connected to a continuous power supply to insure uninterrupted service during an emergency. These power supplies may take the form of independent battery systems or standby emergency systems to which the fire alarm system is automatically connected when the normal power supply fails.

Sprinklers. Fixed sprinkler systems may be of four types: the wet-pipe, the dry-pipe, the pre-action (normally dry-pipe), and the deluge systems. Where freezing conditions are not present, the **wet-pipe system** may have considerable advantage in quickness of operation and possibly in first cost. Damage to valuable merchandise by leaks sometimes dictates the use of another system. In the **dry-pipe system,** air pressure is maintained in the piping system. A sudden drop in pressure occurs when a fusible link in a sprinkler head operates, the drop in pressure actuates the valve to the water header allowing water to flow throughout the system, to the open sprinkler, and thence to the fire. A disadvantage is the delay between action of the sprinkler head and arrival of water through the system. The **pre-action system** normally gives an earlier warning, floods the system, and delays opening the sprinkler head until a further temperature rise occurs. This may be desirable if personnel are present and able to control a small fire with portable extinguishers. A **deluge system** is also "dry-pipe" but with open sprinklers. The purpose is to wet down at once an entire area in which a fire may originate. A quick-opening valve between the system and the water supply may be manually operated but usually is controlled by a thermostatic fire detector.

In addition to water systems, there are a number of **chemical and inert-gas fixed and automatic systems** for use with special hazards. These systems are of particular value in areas where flammable fluids and chemicals preclude the use of water. It should be remembered that all automatic systems must be in a condition of readiness to give the protection intended. Regular inspection, testing, and maintenance are absolutely essential.

Fire mains and **hydrants** with standard 2½-in. fire hose connections should be installed outside of the plant buildings and in the plant yard. Generally, such hydrants should be located so that two hose lines, neither over 250 ft., can reach every part of the building interior. If at all possible, the yard mains should be separate from the main supplying the standpipe and sprinkler system of the plant. Inside the plant, provision should be made for 1-in. or 1½-in. hose lines connected to the standpipe or sprinkler mains. All hose reels, reel carts, and "ever-ready" reels should carry hose of the same size and type with uniform standard couplings. All hoses, outdoor or indoor, should be carefully drained, dried, and replaced after every use. Even if not used, all hoses should be inspected and tested annually under full pressure.

Fire Extinguishers. To control and extinguish the small fire, there is available a variety of portable and semiportable extinguishers. Each of the three types of fires requires somewhat different extinguishers:

1. **Type A** fires, combustible materials; water, water sprays, soda-acid, and foam extinguishers.

2. **Type B** fires, flammable liquids, greases, etc., where a blanketing effect is needed; foam, carbon dioxide, vaporizing liquid, and dry chemical extinguishers.
3. **Type C** fires, flame in "live" electrical equipment; carbon dioxide, vaporizing liquid, and dry chemical extinguishers.

Access to the extinguishers should be possible at all times, and the location of all units should be vividly marked. The extinguishers should be distributed around the plant in areas according to the type of hazard that exists. The presence in the area of flammable liquids or electrical equipment indicates that units for type B or type C fires only should be provided. Each extinguisher should be **marked** for the type of fire on which it is to be used and **periodically tested** to insure that it is fully charged and in good working order. All workers should be instructed in the proper use of extinguishers and the hazards of using them improperly.

BUILDING FIRE SAFETY INTO A PLANT. Building fire safety into a plant is never an easy task. A great deal of technical knowledge must be applied to the arrangement of the plant, its facilities, and its type of construction. Consideration must be given to the fire hazards of production processes and production materials; the possibilities of segregating high-hazard conditions to protect them more completely or limit damage in event of fire. **Fire-wall partitions** should be used to limit the area extent of a possible fire. Automatic devices should close any openings through such walls, doorways, conveyor-ways, or ventilating ducts in the event of fire on either side of the wall. It should be recognized that plant building construction may be a very important factor in achieving overall fire safety, but under some circumstances the building may be no more than a furnace if large quantities of combustible or flammable materials are present with inadequate protection or extinguishing means.

In discussing what many consider the most disastrous industrial fire to date where six men lost their lives and the company suffered a direct loss of $55,000,-000, the National Fire Protection Association Quarterly (1953) describes the fire as follows and gives an analysis of the interrelated causes for the fire and the heavy loss resulting:

The Livonia, Michigan, General Motors fire was started by sparks from an oxy-acetylene cutting torch operated by a contractor's welding crew. Fire started in an overhead conveyor drip-pan, 10 ft. 8 in. above the floor. The pan was sheet-metal 2 ft. wide and approximately 120 ft. long with a 2-inch lip on either side. It extended beneath a long monorail conveyor which dipped down sufficiently at a dip-tank to immerse metal parts in a 97.7 degree F. flash point rust-inhibiting liquid. The drip pan contained a layer of flammable liquid drippings, probably less than ¼ inch deep.

The fire, when discovered, was immediately attacked with two carbon-dioxide extinguishers from a ladder belonging to the welding crew. The fire was about to be extinguished when the extinguishers were emptied and flames spread the length of the drip pan. Other carbon-dioxide and chemical extinguishers were brought within a few minutes. A 1½-inch hose line also was put in use. However, due to the location of the fire overhead about all that could be accomplished was to prevent fire from extending into the dip tank. The oily condensate on the steel roof members in a nearby heat treatment area ignited, adding heat to the roof deck.

Soon hot tar and asphalt were flowing through cracks between strips in the heat-warped roof deck and igniting. The fire then spread laterally behind the increasing area of melted tar that oozed through the roof. Fires broke out on machinery, in flammable liquid containers and on the wood floor.

The Michigan State Fire Marshall believes that there was a delay of 15 to 20 minutes in notifying the Livonia Fire Department. A spokesman for General Motors fixes this elapsed time as six minutes.

The causes are presented approximately in descending order of importance:

1. An undivided fire area of 1,502,500 square feet (34.5 acres) in which absence of fire walls and roof vents denied access for fire fighting and prevented localization of heat and smoke.
2. Inadequate sprinkler protection (only 20 percent of the total area protected— no sprinklers where the fire started).
3. Incompletely engineered process. Fire protection for the dip tank did not protect the drip pan. Due to the process oily deposits had a tendency to accumulate and increase the fire hazard of the several heat treating areas, yet this condition does not appear to have received engineering attention.
4. Unprotected steel construction, in particular the thin steel (roof) deck that did not offer sufficient insulation between banking heat and the built-up roof covering to prevent asphalt from melting and dripping through joints of the heat-warped deck. Steel trusses collapsed in a matter of minutes.
5. Use of oxy-acetylene torch under unsafe conditions.
6. Lack of an effective private fire brigade.
7. Delayed fire department notification.

Disaster Control

ORGANIZATION. The purposes of a disaster control organization in an industrial plant are:

1. In advance of an actual emergency, to anticipate the problems of the emergency. To recommend and assist in providing facilities and personnel trained suitably to meet the emergency. To establish a liaison with the community disaster control organization.
2. In time of emergency, to safeguard the lives of the workers, and to limit the damage to the buildings and machinery of the plant.

Simonds and Grimaldi (in Safety Management) state that "regardless of whether industry is confronted with enemy action, or whether disaster may result from such causes as flood, fire or major accidents, much the same basic control measures for possible emergency conditions will apply."

It is essential that competent individuals be assigned the responsibility of accomplishing the objectives of disaster control. A framework of authority and responsibility for various phases of the program should be established at the outset. A **disaster control officer** should be responsible for their activities throughout the plant. He should be responsible for disaster control administration and training of the organization and the coordination of the plant disaster control activities with those of the community. As far as possible, existing service departments and functional personnel should be utilized for related responsibilities during emergency situations.

Most plants have plant protection, fire-brigade, first-aid, maintenance, and custodial sections. The personnel of these and similar sections form the nucleus of the **emergency organization.** They continue to serve in normal capacities until or unless emergencies arise. Well-planned training permits them to cope with disaster emergencies without extensive adjustments.

In general, the executive of the company who is responsible for developing the disaster control plan and for training personnel should also be responsible for operation of the organization in time of emergency. **Key assistants** should be assigned to him and, since many factories operate more than one shift, **deputies** should be appointed for each to serve during absence or disability of their superior. The several facilities and functions to be provided for in a medium- or large-size factory are:

1. Safety, shelter, and evacuation.
2. Communications, warnings, alarms.
3. Fire fighting.
4. Rescue and damage control.
5. Health and medical (first aid) facilities.
6. Detection and decontamination (atomic, biological, and chemical measures).
7. Essential plant utilities and services.
8. Other functions indicated by special conditions.

CASUALTY CARE. After an emergency, rescue and first-aid services must be immediately available. Most factories maintain medical dispensaries or first-aid stations. These may form the nucleus of an **emergency casualty station,** but provision should be made in advance for an alternate or additional station in a well-protected area. All such stations should be equipped with at least the minimum first-aid supplies. Provision should be made for the prompt removal and transportation of seriously injured employees to a hospital. Coordination with the community hospital and ambulance services should be arranged for in advance.

DAMAGE CONTROL AND REPAIR OF FACILITIES. One of the major causes of damage to the industrial plant resulting from a disaster is the spread of fire. Since effects of the disaster may have disrupted the established water lines and power sources, quick action with portable equipment is essential. At the same time **equipment repair crews** should exert every effort to make emergency repairs of water lines and pumping equipment. The fire brigade and plant engineering forces should be instructed in their responsibilities for mutual action in such an emergency.

Adequate facilities for normal fire fighting probably would be inadequate in time of a major disaster. Each plant might have to rely entirely on its own forces and equipment without help from the local city department. For this reason, each industrial plant may find it necessary to provide its own large-volume water tank or reservoir and one or more heavy-duty pumping engines.

In order to effect repairs to damaged plant facilities as rapidly as possible; a supply of spare parts is essential. The variety and extent of inventory to be maintained will vary with each factory, but may be approximated by a careful study conducted by the plant engineer and maintenance forces. The organization for actual **repair work** should be an expansion of the existing maintenance and erection forces. Many production workers have the skills required for such work and with experienced supervision rapidly fit into a repair force. Round-the-clock operation should be possible.

CIVIL DEFENSE. In discussing civil defense, Simonds and Grimaldi (Safety Management) say that "One complication that seems to weaken the motivation for engaging in a civil defense program is the feeling of hopelessness, shared by many, that the next war will destroy everything on earth. A cool appraisal of all the facts, however, indicate this would not be so. In fact, a sound basic disaster control plan with certain added considerations for protection against expected radio-active fallout (if nuclear weapons are used) appears to be quite practical."

Plants in certain essential industries may require civil defense measures beyond the disaster control programs outlined above. In those cases plant officials should consult with local and national civil defense authorities concerning such activities as evacuation of personnel, dispersal of plants and facilities, shelter areas, fallout shelters, contamination control, and the like.

ACKNOWLEDGMENTS

In the preparation of the PRODUCTION HANDBOOK the vast literature of industrial production and related fields has been thoroughly researched. With full appreciation for the value and significance to the field of the contributions made by the authors and publishers of these works, we wish to give special acknowledgment to the following sources cited in the Third Edition of the PRODUCTION HANDBOOK.

Acknowledgment is also gratefully made to professional, governmental, and industrial organizations for use of material from their journals, bulletins, reports, transactions, and research papers.

THE RONALD PRESS COMPANY
Publishers

ABRAMOWITZ, I. *Production Management—Concepts and Analysis for Operation and Control.* New York: The Ronald Press Co., 1967.

ABRUZZI, A. *Work, Workers, and Work Measurement.* New York: Columbia University Press, 1956.

Accident Facts. Chicago: National Safety Council, 1970.

Accident Prevention Manual. 5th ed. Chicago: National Safety Council, 1964.

Advanced Management

AIIE Transactions

ALJIAN, G. W. (ed.). *Purchasing Handbook.* 2d ed. New York: McGraw-Hill Book Co., 1966.

AMERICAN CONFERENCE OF GOVERNMENTAL INDUSTRIAL HYGIENISTS. *Industrial Ventilation.* A Manual of Recommended Practice. 8th ed. Lansing, Mich.: Committee on Industrial Ventilation, 1964.

AMERICAN INSTITUTE OF CERTIFIED PUBLIC ACCOUNTANTS. *Accounting Research and Terminology Bulletins.* Final edition. 1961.

AMERICAN INSTITUTE OF INDUSTRIAL ENGINEERS. *14th Annual Conference Proceedings.* 1963.

American Machinist

AMERICAN PRODUCTION AND INVENTORY CONTROL SOCIETY. *APICS Dictionary of Production and Inventory Control Terms.* 2d ed. Chicago: American Production and Inventory Control Society, 1966.

AMERICAN SOCIETY OF MECHANICAL ENGINEERS. *ASME Manual of Cutting of Metals.* 2d ed. New York: American Society of Mechanical Engineers, 1952.

AMERICAN SOCIETY FOR METALS. Lyman, T. (ed.). *Metals Handbook.* Vol. 3. Machining. 8th ed. Cleveland, O.: American Society for Metals, 1967.

AMRINE, H. T., RITCHEY, J. A., and HULLEY, O. S. *Manufacturing Organization and Management.* 2d ed. Englewood Cliffs, N.J.: Prentice-Hall, Inc., 1966.

ANSOFF, H. I. *Corporate Strategy.* New York: McGraw-Hill Book Co., 1965.

APICS Quarterly Bulletin

APPLE, J. M. *Lesson Guide Outline on Material Handling Education.* Pittsburgh: Material Handling Institute, 1970.

APPLE, J. M. *Plant Layout and Materials Handling.* 2d ed. New York: The Ronald Press Co., 1963.

ARMAREGO, E. J. A., and BROWN, R. H. *The Machining of Metals.* Englewood Cliffs, N.J.: Prentice-Hall, Inc., 1969.

ASME Transactions

Automation

AWAD, E. M. *Business Data Processing.* Englewood Cliffs, N.J.: Prentice-Hall, Inc., 1965.

AYRES, J. A. (ed.). *Decontamination of Nuclear Reactors and Equipment.* New York: The Ronald Press Co., 1970.

A·1

BACHMAN, P. W. *Research for Profit.* New York: The Ronald Press Co., 1969.

BARNARD, C. I. *The Functions of the Executive.* Cambridge, Mass.: Harvard University Press, 1938.

BARNES, R. M. *Motion and Time Study, Design and Measurement of Work.* 6th ed. New York: John Wiley & Sons, Inc., 1968.

BAUMEISTER, T. (ed.). *Standard Handbook for Mechanical Engineers.* 7th ed. New York: McGraw-Hill Book Co., 1967.

BEGEMAN, M. L., and AMSTEAD, B. H. *Manufacturing Processes.* 6th ed. New York: John Wiley & Sons, Inc., 1969.

BIERMAN, H., and SMIDT, S. *The Capital Budgeting Decision. Economic Analysis and Financing of Investment Projects.* 3d ed. New York: The Macmillan Co., 1971.

BIRN, S. A., CROSSAN, R. M., and EASTWOOD, R. W. *Measurement and Control of Office Costs.* New York: McGraw-Hill Book Co., 1961.

BLAKE, R. P. (ed.). *Industrial Safety.* 3d ed. Englewood Cliffs, N.J.: Prentice-Hall, Inc., 1963.

BOLZ, H. A., and HAGEMANN, G. E. (eds.). *Materials Handling Handbook.* New York: The Ronald Press Co., 1958.

BONBRIGHT, J. C. *Valuation of Property.* New York: McGraw-Hill Book Co., 1937.

BOWMAN, E. H., and FETTER, R. B. *Analysis for Production and Operations Management.* 3d ed. Homewood, Ill.: Richard D. Irwin, Inc., 1967.

BRIGGS, A. J. *Warehouse Operations: Planning and Management.* New York: John Wiley & Sons, Inc., 1960.

BRIGHT, J. R. *Automation and Management.* Cambridge, Mass.: Harvard Business School, 1958.

BRIGHT, J. R. (ed.). *Technological Forecasting for Industry and Government: Methods and Applications.* Englewood Cliffs, N.J.: Prentice-Hall, Inc., 1968.

BROUHA, L. *Physiology in Industry.* New York: Pergamon Press, 1960.

BROWN, R. G. *Decision Rules for Inventory Management.* New York: Holt, Rinehart & Winston, Inc., 1967.

BROWN, R. G. *Smoothing, Forecasting and Prediction of Discrete Time Series.* Englewood Cliffs, N.J.: Prentice-Hall, Inc., 1963.

BROWN, R. G. *Statistical Forecasting for Inventory Control.* New York: McGraw-Hill Book Co., 1959.

BUFFA, E. S. *Models for Production and Operations Management.* New York: John Wiley & Sons, Inc., 1963.

Business Management

CASHIN, J. A., and OWENS, G. C. *Auditing.* 2d ed. New York: The Ronald Press Co., 1963.

CHURCHMAN, C. W., ACKOFF, R. L., and ARNOFF, E. L. *Introduction to Operations Research.* New York: John Wiley & Sons, Inc., 1957.

COCHRAN, W. G. *Sampling Techniques.* 2d ed. New York: John Wiley & Sons, Inc., 1963.

CONANT, J. B. *On Understanding Science.* New Haven: Yale University Press, 1947.

COOK, N. H. *Manufacturing Analysis.* Reading, Mass.: Addison-Wesley Publishing Co., Inc., 1966.

COUGHLAN, J. D., and STRAND, W. K. *Depreciation—Accounting, Taxes, and Business Decisions.* New York: The Ronald Press Co., 1969.

CROSBY, P. B. *Cutting the Cost of Quality.* Boston: Industrial Education Institute, 1967.

DALE, E. *Planning and Developing the Company Organization Structure.* New York: American Management Association, 1965.

DAVIS, E. M. *Machining and Related Characteristics of United States Hardwoods.* Tech. Bull. No. 1267. Madison, Wis.: U. S. Dept. of Agriculture, Forest Products Lab., 1962.

DEARBORN, D. C., et al. (eds.). *Spending for Industrial Research.* Cambridge, Mass.: Harvard University Press, 1951.

DeREAMER, R. *Modern Safety Practices.* New York: John Wiley & Sons, Inc., 1958.

Distribution Worldwide

DODGE, H. F., and ROMIG, H. G. *Sampling Inspection Tables: Single and Double Sampling.* 2d ed. New York: John Wiley & Sons, Inc., 1959.

DONALDSON, C., and LeCAIN, G. H. *Tool Design.* 2d ed. New York: McGraw-Hill Book Co., 1957.

DONALDSON, E. F., and PFAHL, J. K. *Corporate Finance.* 3d ed. New York: The Ronald Press Co., 1969.

DOYLE, L. E. *Manufacturing Processes and Materials for Engineers.* Englewood Cliffs, N.J.: Prentice-Hall, Inc., 1961.

DRAPER, N. R., and SMITH, H. *Applied Regression Analysis.* New York: John Wiley & Sons, Inc., 1966.

DRIEBEEK, N. J. *Applied Linear Programming.* Reading, Mass.: Addison-Wesley Publishing Co., Inc., 1969.

DUNCAN, A. J. *Quality Control and Industrial Statistics.* 3d ed. Homewood, Ill.: Richard D. Irwin, Inc., 1965.

EILON, S. *Elements of Production Planning and Control.* New York: The Macmillan Co., 1962.

EISENHART, C., HASTAY, M. W., and WALLIS, W. A. (eds.). *Selected Techniques of Statistical Analysis for Scientific and Industrial Research and Production and Management Engineering.* New York: McGraw-Hill Book Co., 1947.

Engineering Journal

ENGINEERS JOINT COUNCIL—ENGINEERING MANAGEMENT COMMISSION. *Professional Income of Engineers.* 1970.

EVEN, A. D. *Engineering Data Processing System Design.* New York: Van Nostrand Reinhold Co., 1960.

Factory

FACTORY MUTUAL ASSOCIATION. *Handbook of Industrial Loss Prevention.* New York: McGraw-Hill Book Co., 1959.

FAYOL, H. *General and Industrial Management.* London: Sir Isaac Pitman & Sons, Ltd., 1949.

FEIGENBAUM, A. V. *Total Quality Control—Engineering and Management.* New York: McGraw-Hill Book Co., 1961.

FELLER, W. *Introduction to Probability Theory and Its Applications.* Vol. 1. 3d ed. New York: John Wiley & Sons, Inc., 1968.

Financial Executive

FLOYD, W. E., and WELFORD, A. T. (eds.). *Symposium on Fatigue.* London: H. K. Lewis & Co., Ltd., 1953.

FORRESTER, J. *Industrial Dynamics.* Cambridge, Mass.: M. I. T. Press, 1961.

FRY, T. C. *Probability and Its Engineering Uses.* 2d ed. New York: Van Nostrand Reinhold Co., 1965.

GENERAL SERVICES ADMINISTRATION. *Warehouse Operations Handbook.* Washington, D.C.: Government Printing Office, 1953.

GENTILE, E. C., JR. (ed.). *Data Communication in Business.* New York: American Telephone and Telegraph Co., Inc., 1965.

GEORGE, C. S. *Management in Industry.* Englewood Cliffs, N.J.: Prentice-Hall, Inc., 1964.

GERSTENFELD, A., and CARROLL, D. Faculty Working Paper. Boston, Mass.: Boston University College of Business Administration, 1968.

GILBRETH, F. B., and GILBRETH, L. M. *Applied Motion Study. A Collection of Papers on the Efficient Method to Industrial Preparedness.* New York: The Macmillan Co., 1917.

GILBRETH, F. B., and GILBRETH, L. M. *Primer of Scientific Management.* New York: Van Nostrand Reinhold Co., 1912.

GOMBERG, W. *A Trade Union Analysis of Time Study.* 2d ed. Englewood Cliffs, N.J.: Prentice-Hall, Inc., 1955.

GRAICUNAS, V. A. Relationships in Organization in *Papers on the Science of Administration.* Ed. by L. Gulick and L. Urwick. New York: New York Institute of Public Administration, 1937.

GRANT, E. L. *Statistical Quality Control.* 3d ed. New York: McGraw-Hill Book Co., 1964.

GRANT, E. L., and IRESON, W. G. *Principles of Engineering Economy.* 5th ed. New York: The Ronald Press Co., 1970.

GREENE, J. H. *Production Control: Systems and Decisions.* Homewood, Ill.: Richard D. Irwin, Inc., 1965.

HADLEY, G., and WHITIN, T. M. *Analysis of Inventory Systems.* Englewood Cliffs, N.J.: Prentice-Hall, Inc., 1963.

HALPIN, J. F. *Zero-Defects—A New Dimension in Quality Assurance.* New York: McGraw-Hill Book Co., 1966.

HAMMOND, R. H., BUCK, C. P., ROGERS, W. B., WALSH, G. W., JR., and ACKERT, H. P. *Engineering Graphics—Design, Analysis, Communication.* 2d ed. New York: The Ronald Press Co., 1971.

HANSSMANN, F. *Operations Research in Production and Inventory Control.* New York: John Wiley & Sons, Inc., 1962.

HARE, V. C., JR. *Systems Analysis: A Diagnostic Approach.* New York: Harcourt Brace Jovanovich, Inc., 1967.

HARRIS, C. M. *Handbook of Noise Control.* New York: McGraw-Hill Book Co., 1957.

Harvard Business Review

HARVEY, A. *The Systems Approach in Distribution Engineering.* Proceedings of the AIIE 18th Annual Conference, Toronto, 1967.

Heating, Piping and Air Conditioning

HEINRICH, H. W. *Industrial Accident Prevention—A Scientific Approach.* 4th ed. New York: McGraw-Hill Book Co., 1959.

HERZBERG, F. *Work and the Nature of Man.* New York: World Publishing Co., 1966.

HERZBERG, F., MAUSNER, B., and SNYDERMAN, B. *The Motivation to Work.* 2d ed. New York: John Wiley & Sons, Inc., 1959.

HERZFELD, C. M. (ed.). *Temperature: Its Measurement and Control in Science and Industry.* Vol. 3. New York: Van Nostrand Reinhold Co., 1963.

HESKETT, J. L., IVIE, R. M., and GLASKOWSKY, N. A., JR. *Business Logistics—Management of Physical Supply and Distribution.* New York: The Ronald Press Co., 1964.

HOLDEN, P., FISH, L. S., and SMITH, H. L. *Top Management Organization and Control.* New York: McGraw-Hill Book Co., 1951.

HOLT, C. C., *et al. Planning Production, Inventories, and Work Force.* Englewood Cliffs, N.J.: Prentice-Hall, Inc., 1960.

HOROWITZ, J. *Critical Path Scheduling—Management Control Through CPM and PERT.* New York: The Ronald Press Co., 1967.

HUNT, P. *Financial Analysis in Capital Budgeting.* The George H. Leatherbee Lectures. Cambridge, Mass.: Harvard Business School, 1964.

ILLUMINATING ENGINEERING SOCIETY. *Lighting Handbook.* 4th ed. New York: Illuminating Engineering Society, 1966.

Industrial Engineering

Industrial Fire Brigade Training Manual. Boston: National Fire Protection Association, 1970.

Industrial Management Society Bulletin

Industrial Quality Control

Industry and Power

Instruments and Control Systems

Iron Age

JENKINS, C. H. *Modern Warehouse Management.* New York: McGraw-Hill Book Co., 1968.

JONES, B. M. *Centrifugal Molding and Casting.* Technical Paper SP66–67. Dearborn, Mich.: Society of Manufacturing Engineers, 1966.

Journal of the American Society of Safety Engineers

Journal of Industrial Engineering

Journal of the Royal Statistical Society

JURAN, J. M. (ed.). *Quality Control Handbook.* 2d ed. New York: McGraw-Hill Book Co., 1962.

KELLER, H. C. *Unit-Load and Package Conveyors—Application and Design.* New York: The Ronald Press Co., 1967.

KISH, J. L., JR. *Business Forms—Design and Control.* New York: The Ronald Press Co., 1971.

KISH, J. L., JR., and MORRIS, J. *Microfilm in Business.* New York: The Ronald Press Co., 1966.

KOCH, P. *Wood Machining Processes.* New York: The Ronald Press Co., 1964.

KOEPKE, C. A. *Plant Production Control.* 3d ed. New York: John Wiley & Sons, Inc., 1961.

KRAUS, M. N. *Pneumatic Conveying of Bulk Materials.* New York: The Ronald Press Co., 1968.

KRICK, E. V. *Methods Engineering.* New York: John Wiley & Sons, Inc., 1962.

KURILOFF, A. H. *Modern Aspects of Manufacturing Management.* Ed. by I. R. Vernon. Dearborn, Mich.: Society of Manufacturing Engineers, 1970.

LEONE, W. C. *Production Automation and Numerical Control.* New York: The Ronald Press Co., 1967.

LINDBERG, R. A. *Materials and Manufacturing Technology.* Boston: Allyn & Bacon, Inc., 1968.

LINDGREN, B. W. *Statistical Theory.* 2d ed. New York: The Macmillan Co., 1968.

LITTERER, J. A. *The Analysis of Organizations.* New York: John Wiley & Sons, Inc., 1965.

LOWRY, S. M., MAYNARD, H. B., and STEGEMERTEN, G. J. *Time and Motion Study and Formulas for Wage Incentives.* 3d ed. New York: McGraw-Hill Book Co., 1940.

LUCKIESH, M. *Light, Vision, and Seeing.* New York: Van Nostrand Reinhold Co., 1944.

MACE, M. L., and MONTGOMERY, G. G., JR. *Management Problems of Corporate Acquisitions.* Cambridge, Mass.: Graduate School of Business Administration, Harvard University, 1962.

Machinery

MACNIECE, E. H. *Production Forecasting, Planning, and Control.* New York: John Wiley & Sons, Inc., 1961.

MAGEE, J. F., and BOODMAN, D. M. *Production Planning and Inventory Control.* 2d ed. New York: McGraw-Hill Book Co., 1967.

MALLICK, R. W., and GAUDREAU, A. T. *Plant Layout: Planning and Practice.* New York: John Wiley & Sons, Inc., 1951.

Management Accounting

Management of New Products. Management Research Department Report. Cleveland, O.: Booz, Allen, and Hamilton, 1960.

Management Science

Manufacturing Engineering and Management (formerly *Tool and Manufacturing Engineer*)

MARIEN, R. *Forms Design, Ideas for Management.* Cleveland, O.: Systems and Procedures Association, 1960.

MASLOW, A. H. *Motivation and Personality.* New York: Harper & Row, 1954.

MASSÉ, P. *Optimal Investment Decisions.* Englewood Cliffs, N.J.: Prentice-Hall, Inc., 1962.

Material Handling Engineering

MATERIAL HANDLING INSTITUTE. *Booklet No. 3.*

MAYNARD, H. B., *et al. Practical Control of Office Costs.* Greenwich, Conn.: Management Publishing Corp., 1960.

MAYNARD, H. B., STEGEMERTEN, G. J., and SCHWAB, J. L. *Methods-Time Measurement.* New York: McGraw-Hill Book Co., 1948.

McCORMICK, C. P. *The Power of the People.* New York: Harper & Row, 1949.

McCORMICK, E. J. *Human Factors Engineering.* 2d ed. New York: McGraw-Hill Book Co., 1964.

McGREGOR, D. *The Human Side of Enterprise.* New York: McGraw-Hill Book Co., 1960.

McKINSEY & Co., INC. *A Study of Western Electric's Performance.* New York: American Telephone and Telegraph Co., 1969.

Mechanical Engineering

Metalworking News

MILES, L. D. *Techniques of Value Analysis and Engineering.* New York: McGraw-Hill Book Co., 1961.

MILLER, I., and FREUND, J. E. *Probability and Statistics for Engineers.* Englewood Cliffs, N.J.: Prentice-Hall, Inc., 1965.

MODER, J. J., and PHILLIPS, C. R. *Project Management with CPM and PERT.* New York: Van Nostrand Reinhold Co., 1964.

Modern Industry

Modern Materials Handling

MOOD, A. M., and GRAYBILL, F. A. *Introduction to the Theory of Statistics.* 2d ed. New York: McGraw-Hill Book Co., 1963.

MOONEY, J. D. *The Principles of Organization.* Rev. ed. New York: Harper & Row, 1947.

MOONEY, J. D., and REILEY, A. C. *Onward Industry.* New York: Harper & Row, 1931.

MOORE, F. G. *Manufacturing Management.* 4th ed. Homewood, Ill.: Richard D. Irwin, Inc., 1969.

MOORE, J. M. *Plant Layout and Design.* New York: The Macmillan Co., 1962.

MORRIS, W. T. *The Analysis of Management Decisions.* Rev. ed. Homewood, Ill.: Richard D. Irwin, Inc., 1964.

MORROW, R. L. *Motion Economy and Work Measurement.* New York: The Ronald Press Co., 1957.

MUNDEL, M. E. *Motion and Time Study: Principles and Practice.* 4th ed. Englewood Cliffs, N.J.: Prentice-Hall, Inc., 1970.

MURRELL, K. F. H. *Human Performance in Industry.* New York: Van Nostrand Reinhold Co., 1965.

MUTHER, R. *Systematic Layout Planning.* Boston: Industrial Education Institute, 1961.

N. A. A. ACCOUNTING PRACTICE REPORT. No. 2. *Planning, Controlling, and Accounting for Maintenance.* New York: National Association of Accountants, 1956.

N. A. A. Bulletin (formerly *N.A.C.A. Bulletin*)

NADER, R. *Unsafe At Any Speed. The Designed-in Dangers of the American Automobile.* New York: Grossman Publishers, 1965.

NADLER, G. *Motion and Time Study.* New York: McGraw-Hill Book Co., 1955.

NADLER, G. *Work Design—A Systems Concept.* Rev. ed. Homewood, Ill.: Richard D. Irwin, Inc., 1970.

NATIONAL ASSOCIATION OF PURCHASING MANAGEMENT. *Guide to Purchasing.* New York: National Association of Purchasing Management, 1965.

National Electrical Code. Boston: National Fire Protection Association, 1971.

National Fire Protection Association Quarterly

National Safety Council Transactions

NEWBROUGH, E. T. *Effective Maintenance Management.* New York: McGraw-Hill Book Co., 1967.

NIEBEL, B. W. *Motion and Time Study.* 4th ed. Homewood, Ill.: Richard D. Irwin, Inc., 1967.

OAKFORD, R. V. *Capital Budgeting.* New York: The Ronald Press Co., 1970.

Office

Operation and Flow Process Charts. Standard 101. New York: American Society of Mechanical Engineers, 1949.

Operations Research with Special Reference to Non-Military Application. Committee on Operations Research. Washington, D.C.: National Research Council—National Academy of Sciences, 1951.

PATTON, A. *Men, Money, and Motivation.* New York: McGraw-Hill Book Co., 1961.

PECK, L. G., and HAZELWOOD, R. N. *Finite Queueing Tables.* New York: John Wiley & Sons, Inc., 1958.

PERT/Cost Guide. Washington, D.C.: National Aeronautics and Space Administration, 1962.

Physiological Reviews

Plant Engineering

Plant Layout Manual. Manville, N.J.: Johns-Manville Corp., 1959.

PLOSSL, G. W., and WIGHT, O. W. *Production and Inventory Control: Principles and Techniques.* Englewood Cliffs, N.J.: Prentice-Hall, Inc., 1967.

PRITSKER, A. A. B. *GERT: Graphical Evaluation and Review Technique.* Santa Monica, Calif.: The RAND Corp., 1966.

Proceedings of the APICS Annual Conference. Washington, D.C.: American Production and Inventory Control Society, 1964, 1965, 1966.

Proceedings of the Institute of Radio Engineers

Product Engineering

Production

Production Engineer

Purchasing

PUTNAM, A. O., BARLOW, E. R., and STILIAN, G. N. *Unified Operations Management.* New York: McGraw-Hill Book Co., 1963.

Quarterly Journal of Experimental Physiology

QUICK, J. H., DUNCAN, J. H., and MALCOLM, J. A., JR. *Work-Factor Time Standards.* New York: McGraw-Hill Book Co., 1962.

RAMLOW, D. E., and WALL, E. H. *Production Planning and Control.* Englewood Cliffs, N.J.: Prentice-Hall, Inc., 1967.

REED, R., JR. *Plant Layout: Factors, Principles, and Techniques.* Homewood, Ill.: Richard D. Irwin, Inc., 1961.

RICHARDS, M. D., and GREENLAW, P. S. *Management Decision Making.* Homewood, Ill.: Richard D. Irwin, Inc., 1966.

RICHMOND, S. B. *Operations Research for Management Decisions.* New York: The Ronald Press Co., 1968.

RICHMOND, S. B. *Statistical Analysis.* 2d ed. New York: The Ronald Press Co., 1964.

ROBB, D. A., and PHILO, H. M. *Lawyer's Desk Reference.* Rochester, N.Y.: The Lawyer's Cooperative Publishing Co., 1965.

ROBERTS, E. B. *The Dynamics of Research and Development.* New York: Harper & Row, 1964.

ROBICHAUD, B. *Understanding Modern Business Data Processing.* New York: McGraw-Hill Book Co., 1966.

ROETHLISBERGER, F. J., and DICKSON, W. J. *Management and the Worker.* Cambridge, Mass.: Harvard University Press, 1939.

RYAN, T. A. *Work and Effort.* New York: The Ronald Press Co., 1947.

SAATY, T. L. *Elements of Queueing Theory with Applications.* New York: McGraw-Hill Book Co., 1961.

ST. CLAIR, L. J. *Design and Use of Cutting Tools.* New York: McGraw-Hill Book Co., 1952.

S.A.M. Journal

SCHEELE, E. D., WESTERMAN, W. L., and WIMMERT, R. J. *Principles and Design of Production Control Systems.* Englewood Cliffs, N.J.: Prentice-Hall, Inc., 1962.

SCHNEIDER, P. *ABC's of ADP.* National Association of Wholesalers Lecture Series. September–October, 1966.

SCIENTIFIC AMERICAN. *Information.* San Francisco: W. H. Freeman & Co., 1966.

SERVAN-SCHRIEBER, J. *The American Challenge.* New York: Atheneum, 1968.

SHAW, A. G. *The Purpose and Practice of Motion Study.* London: Harlequin Press Co., Ltd., 1952.

SHERMAN, H. *It All Depends. A Pragmatic Approach to Organization.* University, Ala.: University of Alabama Press, 1966.

SHUMARD, F. W. *Primer of Time Study.* New York: McGraw-Hill Book Co., 1940.

SIMONDS, R. H., and GRIMALDI, J. V. *Safety Management.* Rev. ed. Homewood, Ill.: Richard D. Irwin, Inc., 1963.

SITLER, R. *How to Plan a Stockroom Layout.* Materials Handling Manual. No. 1. Laughner, V. H. (ed.). Boston: Boston Publishing Co., 1955.

SOCIETY FOR THE ADVANCEMENT OF MANAGEMENT. *Rating of Time Studies.* New York: Society for the Advancement of Management, 1952.

SOCIETY OF MANUFACTURING ENGINEERS. Wilson, F. W. (ed.). *Fundamentals of Tool Design.* Englewood Cliffs, N.J.: Prentice-Hall, Inc., 1962.

SOCIETY OF MANUFACTURING ENGINEERS. *Value Engineering in Manufacturing.* Englewood Cliffs, N.J.: Prentice-Hall, Inc., 1967.

SOLOMON, E. *The Management of Corporate Capital.* New York: The Free Press, 1959.

Southern Engineer

SPRINGER, C. H., HERLIHY, R. E., and BEGGS, R. I. *Advanced Methods and Models.* Vol. 2. Homewood, Ill.: Richard D. Irwin, Inc., 1965.

P. STANLEY ASSOCIATES. *Control Chart Techniques.* Chicago: P. Stanley Associates, Inc., 1967.

STEINER, G. A., and RYAN, W. G. *Industrial Project Management.* New York: The Macmillan Co., 1968.

STOCKTON, R. S. *Basic Inventory Systems: Concepts and Analysis.* Boston: Allyn & Bacon, Inc., 1965.

STUDY ON RESEARCH AND DEVELOPMENT STAFF STABILITY. *Motivating Factors in Engineering Employment.* New York: Deutsch, Shea, & Evans, Inc., 1958.

STUMPFF, R. *Getting Your Money's Worth from EDP in Inventory Control.* Proceedings of the 1966 National Technical Conference. Washington, D.C.: American Production and Inventory Control Society, 1967.

SWINEHART, H. J. (ed.). *Cutting Tool Material Selection.* Dearborn, Mich.: Society of Manufacturing Engineers, 1968.

TAYLOR, F. W. *Shop Management.* New York: Harper & Row, 1911.

Tech Engineering News

TERBORGH, G. *Business Investment Management.* 2d ed. Washington, D.C.: Machinery and Allied Products Institute, 1967.

TERBORGH, G. *Dynamic Equipment Policy.* New York: McGraw-Hill Book Co., 1949.

THUESEN, H. G. *The Development of a Memo-activity Camera.* NSF Grant 17674. Stillwater: Oklahoma State University, 1962.

THUESEN, H. G., and FABRYCKY, W. J. *Engineering Economy.* 3d ed. Englewood Cliffs, N.J.: Prentice-Hall, Inc., 1964.

TIMMS, H. L. *The Production Function in Business.* Rev. ed. Homewood, Ill.: Richard D. Irwin, Inc., 1966.

TOAN, A. B., JR. *Using Information to Manage.* New York: The Ronald Press Co., 1968.

TOTTEN, D. L. *Applications of Quantitative Techniques in Facilities Planning.* M.S. Thesis. Atlanta: Georgia Institute of Technology, 1967.

Transportation and Distribution Management

U.S. DEPARTMENT OF LABOR. *Collective Bargaining Provisions.* Bulletin 908-3.

VAN VLACK, L. H. *Elements of Materials Science.* 2d ed. Reading, Mass.: Addison-Wesley Publishing Co., Inc., 1964.

VIDOSIC, J. P. *Elements of Design Engineering.* New York: The Ronald Press Co., 1969.

VIDOSIC, J. P. *Metal Machining and Forming Technology.* New York: The Ronald Press Co., 1964.

VOICH, D., and WREN, D. *Principles of Management.* New York: The Ronald Press Co., 1968.

VON NEUMANN, J., and MORGENSTERN, O. *The Theory of Games and Economic Behavior.* Rev. ed. Princeton, N.J.: Princeton University Press, 1955.

WESTING, J. H., and FINE, I. V. *Industrial Purchasing.* 2d ed. New York: John Wiley & Sons, Inc., 1961.

WIENER, N. W. *Cybernetics—Control and Communication in the Animal and the Machine.* New York: John Wiley & Sons, Inc., 1948.

WILLIAMS, C. (ed.). *Proceedings of the 8th Annual Conference on the Administration of Research.* New York: New York University Press, 1955.

WITHINGTON, F. G. *The Use of Computers in Business Organizations.* Reading, Mass.: Addison-Wesley Publishing Co., Inc., 1966.

WOLDMAN, N., and GIBBONS, R. *Machinability and Machining of Metals.* New York: McGraw-Hill Book Co., 1951.

WOODSON, W. E., and CONOVER, D. W. *Human Engineering Guide for Equipment Designers.* 2d ed. rev. Berkeley: University of California Press, 1965.

Work Study

Organizations and Governmental Agencies

ACME VISIBLE RECORDS, INC.
ADDRESSOGRAPH-MULTIGRAPH CORP.
AEROSOL RESEARCH CO.
AJAX MANUFACTURING CO.
AMERICAN CAST IRON PIPE CO.
AMERICAN CHAIN AND CABLE CO.
AMERICAN FOUNDRYMEN'S SOCIETY
AMERICAN INSURANCE ASSOCIATION
AMERICAN IRON AND STEEL INSTITUTE (AISI)
AMERICAN MANAGEMENT ASSOCIATION (AMA)
AMERICAN NATIONAL STANDARDS INSTITUTE (ANSI)
AMERICAN PRODUCTION AND INVENTORY CONTROL SOCIETY (APICS)
AMERICAN SOCIETY OF MECHANICAL ENGINEERS (ASME)
AMERICAN SOCIETY FOR QUALITY CONTROL (ASQC)
AMERICAN SOCIETY OF SAFETY ENGINEERS (ASSE)
AMERICAN SOCIETY FOR TESTING AND MATERIALS (ASTM)
AMERICAN TELEPHONE AND TELEGRAPH CO., INC.
AMERICAN WELDING SOCIETY (AWS)
ASSOCIATED FACTORY MUTUAL INSURANCE COMPANIES
ASSOCIATION OF AMERICAN RAILROADS
BAUSCH & LOMB INC.
BELL & HOWELL CO.
THE BENDIX CORP.
BOICE GAGES INC.
BUREAU OF MINES, U.S. DEPARTMENT OF THE INTERIOR
CHARLES BRUNING CO., INC.
CARBORUNDUM CO.
CINCINNATI MILACRON INC.
CONVEYOR EQUIPMENT MANUFACTURERS ASSOCIATION
CURTISS-WRIGHT CORP.
A. B. DICK CO.
DIEBOLD, INC.
E. I. DUPONT DE NEMOURS & CO., INC.
EATON CORP.
ELECTRONIC INDUSTRIES ASSOCIATION (EIA)
THE EMERSON CONSULTANTS, INC.
ENGINEERS JOINT COUNCIL (EJC)
FORGING INDUSTRY ASSOCIATION
GENERAL AMERICAN TRANSPORTATION CORP.
GENERAL ELECTRIC CO.
GENERAL MOTORS CORP.

Honeywell Inc.
H. P. Hood & Sons
Illuminating Engineering Society (IES)
International Business Machines Corp.
International Ladies Garment Workers' Union (ILGWU)
Interstate Commerce Commission
Interstate Drop Forge Co.
Jeffrey Manufacturing Co.
Johns-Manville Products Corp.
Lake Placid Work Simplification Conferences
Liberty Mutual Insurance Co.
Logan Co.
Machinery and Allied Products Institute (MAPI)
Magna Visual Inc.
Material Handling Institute (MHI)
Methods-Time-Measurement Association
Monarch Machine Tool Co.
Monroe Division of Litton Business Systems, Inc.
Moore Business Forms Inc.
National Aeronautics and Space Administration
National Bureau of Standards (NBS)
National Fire Protection Association (NFPA)
National Gypsum Co.
National Research Council—National Academy of Sciences
National Safety Council
National Wooden Pallet and Container Association
New York State Workmen's Compensation Board
Otis Elevator Co.
Pyrofilm Corp.
Rex Chainbelt Inc.
RTE Corp.
Scherr-Tumico Inc.
Society for Advancement of Management (SAM)
Society of Manufacturing Engineers (SME)
Sperry Rand Corp.
The Standard Register Co.
Edwin B. Stimpson Company, Inc.
Underwriters Laboratories
Union Carbide Corp.
U.S. Department of Commerce
U.S. Department of Defense
U.S. Department of Labor
Visual Control Associates
Westinghouse Electric Corp.
Wheeldex Inc.
Wofac Co.
Xerox Corp.

INDEX

(Boldface numbers, followed by a dot, refer to sections; lightface numbers following are the pages of the section.)

1

LIST OF SECTIONS

Alphabetical